KIRK-OTHMER ENCYCLOPEDIA OF

CHEMICAL TECHNOLOGY
Fifth Edition

VOLUME 6

KIRK-OTHMER ENCYCLOPEDIA OF CHEMICAL TECHNOLOGY, FIFTH EDITION
EDITORIAL STAFF

KIRK-OTHMER ENCYCLOPEDIA OF

CHEMICAL TECHNOLOGY

Fifth Edition

VOLUME 6

Kirk-Othmer Encyclopedia of Chemical Technology
is available Online in full color and with additional content at
http://www3.interscience.wiley.com/cgi-bin/mrwhome/104554789/HOME.

⟨W⟩WILEY-INTERSCIENCE

A John Wiley & Sons, Inc., Publication

Library of Congress Cataloging-in-Publication Data:

Kirk-Othmer encyclopedia of chemical technology. – 5th ed.
     p. cm.
Editor-in-chief, Arza Seidel.
"A Wiley-Interscience publication."
Includes index.
  ISBN 0-471-48494-6 (set) – ISBN 0-471-48517-9 (v. 6)
  1. Chemistry, Technical–Encyclopedias. I. Title: Encyclopedia of chemical technology. II. Kroschwitz, Jacqueline I.
  TP9.K54 2004
  660′.03–dc22   2003021960

CONTENTS

CONTRIBUTORS

C. D. Azzara, *The Pennsylvania State University, Hershey Foods Corporation, Hershey, PA,* Chocolate and Cocoa

David A. Berry, *U.S. Department of Energy, Morgantown, WV,* Coal Gasification

Mary C. Blackburn, *Chemical Market Associates, Inc., Houston, TX,* Chlorine

Tilak Bommaraju, *Process Technology Opt, Inc., Grand Island, NY,* Chlorine

Karl Booksh, *Arizona State University, Tempe, AZ,* Chemometrics

J. Bouzas, *The Pennsylvania State University, Hershey Foods Corporation, Hershey, PA,* Chocolate and Cocoa

Dady B. Dadyburjor, *West Virginia University, Morgantown, WV,* Coal Liquefaction

Glen Dammann, *Krupp Uhde GmbH, Dortmund, Germany,* Chlorine

Budd L. Duncan, *Olin Corporation, Norwalk, CT,* Chloric Acid and Chlorates

Cecil Dybowski, *University of Delaware, Newark, DE,* Chromatography

Robert Guglielmetti, *Université de la Méditerrannée, Marseille, France,* Chromogenic Materials

David S. Hage, *University of Nebraska, Lincoln, NE,* Chromatography, Affinity

J. C. Hickman, *The Dow Chemical Company, Freeport, TX,* Chloroethylenes and Chloroethanes

Michael T. Holbrook, *Dow Chemical USA, Midland, MI,* Chloroform

James Hower, *University of Kentucky, Lexington, KY,* Coal

Kevin L. Houghton, *ICI Chemicals and Polymers Ltd., Aberdeen, United Kingdom,* Chlorinated Paraffins

Mary Kaiser, *E.I. du Pont de Nemours, Wilmington, DE,* Chromatography

Pravin Khandare, *Occidental Chemical Corporation, Dallas, TX,* Chlorotoluenes, Ring

Charles B. Kreutzberger, *PPG Industries, Inc., Monroeville, PA,* Chloroformates and Carbonates

Ramesh Krishnamurti, *Occidental Chemical Corporation, Dallas, TX,* Chlorobenzenes

Bruce F. Lipin, *US Geological Survey, Reston, VA,* Chromium and Chromium Alloys

Zhenyu Liu, *Institute of Coal Chemistry, Shanxi, China,* Coal Liquefaction

Gary W. Loar, *McGean-Rohco, Inc., Cleveland, OH,* Chromium Compounds

Rebeca Lopez-Garcia, *Tate & Lyle, London, United Kingdom,* Citric Acid

Benno Lüke, *Krupp Uhde GmbH, Dortmund, Germany,* Chlorine

George Lunn, *Center for Drug Evaluation and Research, Food and Drug Administration, Rockville, MD,* Chromatography, Liquid

Kennric A. Marshall, *The Dow Chemical Company, Freeport, TX,* Chlorocarbons and Chlorohydrocarbons, Survey

Sudhir K. Mendiratta, *Olin Corporation, Norwalk, CT,* Chloric Acid and Chlorates

James A. Mertens, *The Dow Chemical Company, Freeport, TX,* Chloroethylenes and Chloroethanes

Haydn H. Murray, *Indiana University, Bloomington, IN,* Clays, Survey; Clays, Uses

Thomas F. O'Brien, *Consultant, Media, PA,* Chlorine

Billie J. Page, *McGean-Rohco, Inc., Cleveland, OH,* Chromium Compounds

John F. Papp, *US Geological Survey, Reston, VA,* Chromium and Chromium Alloys

André Samat, *Université de la Méditerrannée, Marseille, France,* Chromogenic Materials

Karl W. Seper, *Occidental Chemical Corporation, Grand Island, NY,* Chlorotoluenes, Benzyl Chloride, Benzal Chloride and Benzotrichloride

Lawrence J. Shadle, *U.S. Department of Energy, Morgantown, VA,* Coal Gasification

Gregory C. Slack, *Clarkson University, Potsdam, NY,* Chromatography, Gas

Gayle Snedecor, *The Dow Chemical Company, Freeport, TX,* Chloroethylenes and Chloroethanes

Nicholas H. Snow, *Seton Hall University, South Orange, NJ,* Chromatography, Gas

Ron Spohn, *Occidental Chemical Corporation, Dallas, TX,* Chlorotoluenes, Ring

Apryll M. Stalcup, *University of Cincinnati, Cincinnati, OH,* Chiral Separations

Madhava Syamlal, *Fluent Inc., Lebanon, NH,* Coal Gasification

Hugo O. Villar, *Triad Therapeutics, Inc., San Diego, CA,* Chemoinformatics

Frank Vogt, *Arizona State University, Tempe, AZ,* Chemometrics

B. L. Zoumas, *The Pennsylvania State University, Hershey Foods Corporation, Hershey, PA,* Chocolate and Cocoa

CONVERSION FACTORS, ABBREVIATIONS, AND UNIT SYMBOLS

SI Units (Adopted 1960)

The International System of Units (abbreviated SI), is implemented throughout the world. This measurement system is a modernized version of the MKSA (meter, kilogram, second, ampere) system, and its details are published and controlled by an international treaty organization (The International Bureau of Weights and Measures) (1).

SI units are divided into three classes:

BASE UNITS

length	meter[†] (m)
mass	kilogram (kg)
time	second (s)
electric current	ampere (A)
thermodynamic temperature[‡]	kelvin (K)
amount of substance	mole (mol)
luminous intensity	candela (cd)

SUPPLEMENTARY UNITS

plane angle	radian (rad)
solid angle	steradian (sr)

DERIVED UNITS AND OTHER ACCEPTABLE UNITS

These units are formed by combining base units, suplementary units, and other derived units (2–4). Those derived units having special names and symbols are marked with an asterisk in the list below.

[†] The spellings "metre" and "litre" are preferred by ASTM; however, "-er" is used in the *Encyclopedia*.

[‡] Wide use is made of Celsius temperature (t) defined by

$$t = T - T_0$$

where T is the thermodynamic temperature, expressed in kelvin, and $T_0 = 273.15$ K by definition. A temperature interval may be expressed in degrees Celsius as well as in kelvin.

Quantity	Unit	Symbol	Acceptable equivalent
*absorbed dose	gray	Gy	J/Kg
acceleration	meter per second squared	m/s^2	
*activity (of a radionuclide)	becquerel	Bq	1/s
area	square kilometer	km^2	
	square hectometer	hm^2	ha (hectare)
	square meter	m^2	
concentration (of amount of substance)	mole per cubic meter	mol/m^3	
current density	ampere per square meter	A/m^2	
density, mass density	kilogram per cubic meter	kg/m^3	g/L; mg/cm^3
dipole moment (quantity)	coulomb meter	$C \cdot m$	
*dose equivalent	sievert	Sv	J/kg
*electric capacitance	farad	F	C/V
*electric charge, quantity of electricity	coulomb	C	$A \cdot s$
electric charge density	coulomb per cubic meter	C/m^3	
*electric conductance	siemens	S	A/V
electric field strength	volt per meter	V/m	
electric flux density	coulomb per square meter	C/m^2	
*electric potential, potential difference, electromotive force	volt	V	W/A
*electric resistance	ohm	Ω	V/A
*energy, work, quantity of heat	megajoule	MJ	
	kilojoule	kJ	
	joule	J	$N \cdot m$
	electronvolt[†]	eV[†]	
	kilowatt-hour[†]	$kW \cdot h$[†]	
energy density	joule per cubic meter	J/m^3	
*force	kilonewton	kN	
	newton	N	$kg \cdot m/s^2$

[†]This non-SI unit is recognized by the CIPM as having to be retained because of practical importance or use in specialized fields (1).

Quantity	Unit	Symbol	Acceptable equivalent
*frequency	megahertz	MHz	
	hertz	Hz	1/s
heat capacity, entropy	joule per kelvin	J/K	
heat capacity (specific), specific entropy	joule per kilogram kelvin	$\text{J/(kg} \cdot \text{K)}$	
heat-transfer coefficient	watt per square meter kelvin	$\text{W/(m}^2 \cdot \text{K)}$	
*illuminance	lux	lx	lm/m^2
*inductance	henry	H	Wb/A
linear density	kilogram per meter	kg/m	
luminance	candela per square meter	cd/m^2	
*luminous flux	lumen	lm	$\text{cd} \cdot \text{sr}$
magnetic field strength	ampere per meter	A/m	
*magnetic flux	weber	Wb	$\text{V} \cdot \text{s}$
*magnetic flux density	tesla	T	Wb/m^2
molar energy	joule per mole	J/mol	
molar entropy, molar heat capacity	joule per mole kelvin	$\text{J/(mol} \cdot \text{K)}$	
moment of force, torque	newton meter	$\text{N} \cdot \text{m}$	
momentum	kilogram meter per second	$\text{kg} \cdot \text{m/s}$	
permeability	henry per meter	H/m	
permittivity	farad per meter	F/m	
*power, heat flow rate, radiant flux	kilowatt	kW	
	watt	W	J/s
power density, heat flux density, irradiance	watt per square meter	W/m^2	
*pressure, stress	megapascal	MPa	
	kilopascal	kPa	
	pascal	Pa	N/m^2
sound level	decibel	dB	
specific energy	joule per kilogram	J/kg	
specific volume	cubic meter per kilogram	$\text{m}^3\text{/kg}$	
surface tension	newton per meter	N/m	
thermal conductivity	watt per meter kelvin	$\text{W/(m} \cdot \text{K)}$	
velocity	meter per second	m/s	
	kilometer per hour	km/h	
viscosity, dynamic	pascal second	$\text{Pa} \cdot \text{s}$	
	millipascal second	$\text{mPa} \cdot \text{s}$	
viscosity, kinematic	square meter per second	$\text{m}^2\text{/s}$	
	square millimeter per second	$\text{mm}^2\text{/s}$	

Quantity	Unit	Symbol	Acceptable equivalent
volume	cubic meter	m^3	
	cubic diameter	dm^3	L (liter) (5)
	cubic centimeter	cm^3	mL
wave number	1 per meter	m^{-1}	
	1 per centimeter	cm^{-1}	

In addition, there are 16 prefixes used to indicate order of magnitude, as follows

Multiplication factor	Prefix	symbol	Note
10^{18}	exa	E	
10^{15}	peta	P	
10^{12}	tera	T	
10^9	giga	G	
10^6	mega	M	
10^3	kilo	k	
10^2	hecto	h[a]	[a]Although hecto, deka, deci, and
10	deka	da[a]	centi are SI prefixes, their use
10^{-1}	deci	d[a]	should be avoided except for SI
10^{-2}	centi	c[a]	unit-multiples for area and
10^{-3}	milli	m	volume and nontechnical use of
10^{-6}	micro	μ	centimeter, as for body and
10^{-9}	nano	n	clothing measurement.
10^{-12}	pico	p	
10^{-15}	femto	f	
10^{-18}	atto	a	

For a complete description of SI and its use the reader is referred to ASTM E380 (4) and the article UNITS AND CONVERSION FACTORS which appears in Vol. 24.

A representative list of conversion factors from non-SI to SI units is presented herewith. Factors are given to four significant figures. Exact relationships are followed by a dagger. A more complete list is given in the latest editions of ASTM E380 (4) and ANSI Z210.1 (6).

Conversion Factors to SI Units

To convert from	To	Multiply by
acre	square meter (m^2)	4.047×10^3
angstrom	meter (m)	$1.0 \times 10^{-10\dagger}$
are	square meter (m^2)	$1.0 \times 10^{2\dagger}$
astronomical unit	meter (m)	1.496×10^{11}

†Exact.

To convert from	To	Multiply by
atmosphere, standard	pascal (Pa)	1.013×10^5
bar	pascal (Pa)	$1.0 \times 10^{5\dagger}$
barn	square meter (m^2)	$1.0 \times 10^{-28\dagger}$
barrel (42 U.S. liquid gallons)	cubic meter (m^3)	0.1590
Bohr magneton (μ_B)	J/T	9.274×10^{-24}
Btu (International Table)	joule (J)	1.055×10^3
Btu (mean)	joule (J)	1.056×10^3
Btu (thermochemical)	joule (J)	1.054×10^3
bushel	cubic meter(m^3)	3.524×10^{-2}
calorie (International Table)	joule (J)	4.187
calorie (mean)	joule (J)	4.190
calorie (thermochemical)	joule (J)	$4.184^\dagger$
centipoise	pascal second (Pa·s)	$1.0 \times 10^{-3\dagger}$
centistokes	square millimeter per second (mm^2/s)	$1.0^\dagger$
cfm (cubic foot per minute)	cubic meter per second (m^{3}s)	4.72×10^{-4}
cubic inch	cubic meter (m^3)	1.639×10^{-5}
cubic foot	cubic meter (m^3)	2.832×10^{-2}
cubic yard	cubic meter (m^3)	0.7646
curie	becquerel (Bq)	$3.70 \times 10^{10\dagger}$
debye	coulomb meter (C·m)	3.336×10^{-30}
degree (angle)	radian (rad)	1.745×10^{-2}
denier (international)	kilogram per meter (kg/m)	1.111×10^{-7}
	tex‡	0.1111
dram (apothecaries')	kilogram (kg)	3.888×10^{-3}
dram (avoirdupois)	kilogram (kg)	1.772×10^{-3}
dram (U.S. fluid)	cubic meter (m^3)	3.697×10^{-6}
dyne	newton (N)	$1.0 \times 10^{-5\dagger}$
dyne/cm	newton per meter (N/m)	$1.0 \times 10^{-3\dagger}$
electronvolt	joule (J)	1.602×10^{-19}
erg	joule (J)	$1.0 \times 10^{-7\dagger}$
fathom	meter (m)	1.829
fluid ounce (U.S.)	cubic meter (m^3)	2.957×10^{-5}
foot	meter (m)	$0.3048^\dagger$
footcandle	lux (lx)	10.76
furlong	meter (m)	2.012×10^{-2}
gal	meter per second squared (m/s^2)	$1.0 \times 10^{-2\dagger}$
gallon (U.S. dry)	cubic meter (m^3)	4.405×10^{-3}
gallon (U.S. liquid)	cubic meter (m^3)	3.785×10^{-3}
gallon per minute (gpm)	cubic meter per second (m^3/s)	6.309×10^{-5}
	cubic meter per hour (m^3/h)	0.2271

†Exact.
‡See footnote on p. ix.

To convert from	To	Multiply by
gauss	tesla (T)	1.0×10^{-4}
gilbert	ampere (A)	0.7958
gill (U.S.)	cubic meter (m³)	1.183×10^{-4}
grade	radian	1.571×10^{-2}
grain	kilogram (kg)	6.480×10^{-5}
gram force per denier	newton per tex (N/tex)	8.826×10^{-2}
hectare	square meter (m²)	$1.0 \times 10^{4\dagger}$
horsepower (550 ft·lbf/s)	watt (W)	7.457×10^{2}
horsepower (boiler)	watt (W)	9.810×10^{3}
horsepower (electric)	watt (W)	$7.46 \times 10^{2\dagger}$
hundredweight (long)	kilogram (kg)	50.80
hundredweight (short)	kilogram (kg)	45.36
inch	meter (m)	$2.54 \times 10^{-2\dagger}$
inch of mercury (32°F)	pascal (Pa)	3.386×10^{3}
inch of water (39.2°F)	pascal (Pa)	2.491×10^{2}
kilogram-force	newton (N)	9.807
kilowatt hour	megajoule (MJ)	$3.6^{\dagger}$
kip	newton (N)	4.448×10^{3}
knot (international)	meter per second (m/S)	0.5144
lambert	candela per square meter (cd/m³)	3.183×10^{3}
league (British nautical)	meter (m)	5.559×10^{3}
league (statute)	meter (m)	4.828×10^{3}
light year	meter (m)	9.461×10^{15}
liter (for fluids only)	cubic meter (m³)	$1.0 \times 10^{-3\dagger}$
maxwell	weber (Wb)	$1.0 \times 10^{-8\dagger}$
micron	meter (m)	$1.0 \times 10^{-6\dagger}$
mil	meter (m)	$2.54 \times 10^{-5\dagger}$
mile (statue)	meter (m)	1.609×10^{3}
mile (U.S. nautical)	meter (m)	$1.852 \times 10^{3\dagger}$
mile per hour	meter per second (m/s)	0.4470
millibar	pascal (Pa)	1.0×10^{2}
millimeter of mercury (0°C)	pascal (Pa)	$1.333 \times 10^{2\dagger}$
minute (angular)	radian	2.909×10^{-4}
myriagram	kilogram (Kg)	10
myriameter	kilometer (Km)	10
oersted	ampere per meter (A/m)	79.58
ounce (avoirdupois)	kilogram (kg)	2.835×10^{-2}
ounce (troy)	kilogram (kg)	3.110×10^{-2}
ounce (U.S. fluid)	cubic meter (m³)	2.957×10^{-5}
ounce-force	newton (N)	0.2780
peck (U.S.)	cubic meter (m³)	8.810×10^{-3}
pennyweight	kilogram (kg)	1.555×10^{-3}
pint (U.S. dry)	cubic meter (m³)	5.506×10^{-4}

†Exact.

To convert from	To	Multiply by
pint (U.S. liquid)	cubic meter (m^3)	4.732×10^{-4}
poise (absolute viscosity)	pascal second (Pa·s)	$0.10^{\dagger}$
pound (avoirdupois)	kilogram (kg)	0.4536
pound (troy)	kilogram (kg)	0.3732
poundal	newton (N)	0.1383
pound-force	newton (N)	4.448
pound force per square inch (psi)	pascal (Pa)	6.895×10^{3}
quart (U.S. dry)	cubic meter (m^3)	1.101×10^{-3}
quart (U.S. liquid)	cubic meter (m^3)	9.464×10^{-4}
quintal	kilogram (kg)	$1.0 \times 10^{-2\dagger}$
rad	gray (Gy)	$1.0 \times 10^{-2\dagger}$
rod	meter (m)	5.029
roentgen	coulomb per kilogram (C/kg)	2.58×10^{-4}
second (angle)	radian (rad)	$4.848 \times 10^{-6\dagger}$
section	square meter (m^2)	2.590×10^{6}
slug	kilogram (kg)	14.59
spherical candle power	lumen (lm)	12.57
square inch	square meter (m^2)	6.452×10^{-4}
square foot	square meter (m^2)	9.290×10^{-2}
square mile	square meter (m^2)	2.590×10^{6}
square yard	square meter (m^2)	0.8361
stere	cubic meter (m^3)	$1.0^{\dagger}$
stokes (kinematic viscosity)	square meter per second (m^2/s)	$1.0 \times 10^{-4\dagger}$
tex	kilogram per meter (kg/m)	$1.0 \times 10^{-6\dagger}$
ton (long, 2240 pounds)	kilogram (kg)	1.016×10^{3}
ton (metric) (tonne)	kilogram (kg)	$1.0 \times 10^{3\dagger}$
ton (short, 2000 pounds)	kilogram (kg)	9.072×10^{2}
torr	pascal (Pa)	1.333×10^{2}
unit pole	weber (Wb)	1.257×10^{-7}
yard	meter (m)	$0.9144^{\dagger}$

†Exact.

Abbreviations and Unit Symbols

Following is a list of common abbreviations and unit symbnols used in the Encyclopedia. In general they agree with those listed in *American National Standard Abbreviations for Use on Drawings and in Text* (*ANSI Y1.1*) (6) and *American National Standard Letter Symbols for Units in Science and Technology* (*ANSI Y10*) (6). Also included is a list of acronyms for a number of private and government organizations as well as common industrial solvents, polymers, and other chemicals.

Rules for Writing Unit Symbols (4):

1. Unit symbols are printed in upright letters (roman) regardless of the type style used in the surrounding text.
2. Unit symbols are unaltered in the plural.
3. Unit symbols are not followed by a period except when used at the end of a sentence.
4. Letter unit symbols are generally printed lower-case (for example, cd for candela) unless the unit name has been derived from a proper name, in which case the first letter of the symbol is capitalized (W, Pa). Prefixes and unit symbols retain their prescribed form regardless of the surrounding typography.
5. In the complete expression for a quantity, a space should be left between the numerical value and the unit symbol. For example, write 2.37 lm, *not* 2.37 lm, and 35 mm, *not* 35 mm. When the quantity is used in an adjectival sense, a hyphen is often used, for example, 35-mm film. *Exception:* No space is left between the numerical value and the symbols of degree, minute, and second of plane angle, degree Celsius, and the percent sign.
6. No space is used between the prefix and unit symbol (for example, kg).
7. Symbols, not abbreviations, should be used for units. For example, use "A," not "amp," for ampere.
8. When multiplying unit symbols, use a raised dot:

$$N \cdot m \text{ for newton meter}$$

In the case of $W \cdot h$, the dot may be omitted, thus:

$$Wh$$

An exception to this practice is made for computer printouts, automatic typewriter work, etc, where the raised dot is not possible, and a dot on the line may be used.

9. When dividing unit symbols, use one of the following forms:

$$m/s \quad or \quad m \cdot s^{-1} \quad or \quad \frac{m}{s}$$

In no case should more than one slash be used in the same expression unless parentheses are inserted to avoid ambiguity. For example, write:

$$J/(mol \cdot K) \quad or \quad J \cdot mol^{-1} \cdot K^{-1} \quad or \quad (J/mol)/K$$

but *not*

$$J/mol/K$$

10. Do not mix symbols and unit names in the same expression. Write:

joules per kilogram   *or*   J/kg   *or* $J \cdot kg^{-1}$

but *not*

joules/kilogram   *nor*   Joules/kg   *nor* $Joules \cdot kg^{-1}$

ABBREVIATIONS AND UNITS

A	ampere	AOAC	Association of Official Analytical Chemists
A	anion (eg, HA)		
A	mass number	AOCS	American Oil Chemists' Society
a	atto (prefix for 10^{-18})		
AATCC	American Association of Textile Chemists and Colorists	APHA	American Public Health Association
		API	American Petroleum Institute
ABS	acrylonitrile–butadiene–styrene	aq	aqueous
abs	absolute	Ar	aryl
ac	alternating current, *n*.	*ar-*	aromatic
a-c	alternating current, *adj*.	*as-*	Asymmetric(al)
ac-	alicyclic	ASHRAE	American Society of Heating, Refrigerating, and Air Conditioning Engineers
acac	acetylacetonate		
ACGIH	American Conference of Governmental Industrial Hygienists		
		ASM	American Society for Metals
ACS	American Chemical Society	ASME	American Society of Mechanical Engineers
AGA	American Gas Association		
Ah	ampere hour	ASTM	American Society for Testing and Materials
AIChE	American Institute of Chemical Engineers	at no.	atomic number
AIME	American Institute of Mining, metallurgical, and Petroleum Engineers	at wt	atomic weight
		av(g)	average
		AWS	American Welding Society
		b	bonding orbital
AIP	American Institute of Physics	bbl	barrel
		bcc	body-centered cubic
AISI	American Iron and Steel Institute	BCT	body-centered tetragonal
		Bé	Baumé
alc	alcohol(ic)	BET	Brunauer-Emmett-Teller (adsorption equation)
Alk	alkyl		
alk	alkaline (not alkali)	bid	twice daily
amt	amount	Boc	*t*-butyloxycarbonyl
amu	atomic mass unit	BOD	biochemical (biological) oxygen demand
ANSI	American National Standards Institute		
		bp	boiling point
AO	atomic orbital	Bq	becquerel

C	coulomb
°C	degree Celsius
C-	denoting attachment to carbon
c	centi (prefix for 10^{-2})
c	critical
ca	circa (Approximately)
cd	candela; current density; circular dichroism
CFR	Code of Federal Regulations
cgs	centimeter-gram-second
CI	Color Index
cis-	isomer in which substituted groups are on some side of double bond between C atoms
cl	carload
cm	centimeter
cmil	circular mil
cmpd	compound
CNS	central nervous system
CoA	coenzyme A
COD	chemical oxygen demand
coml	commerical(ly)
cp	chemically pure
cph	close-packed hexagonal
CPSC	Consumer Product Safety Commission
cryst	crystalline
cub	cubic
D	debye
D-	denoting configurational relationship
d	differential operator
d	day; deci (prefix for 10^{-1})
d	density
d-	*dextro-*, dextrorotatory
da	deka (prefix for 10^{-1})
dB	decibel
dc	direct current, *n.*
d-c	direct current, *adj.*
dec	decompose
detd	determined
detn	determination
Di	didymium, a mixture of all lanthanons
dia	diameter

dil	dilute
DIN	Deutsche Industrie Normen
dl-; DL-	racemic
DMA	dimethylacetamide
DMF	dimethylformamide
DMG	dimethyl glyoxime
DMSO	dimethyl sulfoxide
DOD	Department of Defense
DOE	Department of Energy
DOT	Department of Transportation
DP	degree of polymerization
dp	dew point
DPH	diamond pyramid hardness
dstl(d)	distill(ed)
dta	differential thermal analysis
(*E*)-	entgegen; opposed
ϵ	dielectric constant (unitless number)
e	electron
ECU	electrochemical unit
ed.	edited, edition, editor
ED	effective dose
EDTA	ethylenediaminetetra-acetic acid
emf	electromotive force
emu	electromagnetic unit
en	ethylene diamine
eng	engineering
EPA	Environmental Protection Agency
epr	electron paramagnetic resonance
eq.	equation
esca	electron spectroscopy for chemical analysis
esp	especially
esr	electron-spin resonance
est(d)	estimate(d)
estn	estimation
esu	electrostatic unit
exp	experiment, experimental
ext(d)	extract(ed)
F	farad (capacitance)
F	fraday (96,487 C)

f	femto (prefix for 10^{-15})	hyd	hydrated, hydrous
FAO	Food and Agriculture Organization (United Nations)	hyg	hygroscopic
		Hz	hertz
		^{i}(eg, Pr^i)	iso (eg, isopropyl)
fcc	face-centered cubic	i-	inactive (eg, i-methionine)
FDA	Food and Drug Administration	IACS	international Annealed Copper Standard
FEA	Federal Energy Administration	ibp	initial boiling point
		IC	integrated circuit
FHSA	Federal Hazardous Substances Act	ICC	Interstate Commerce Commission
fob	free on board	ICT	International Critical Table
fp	freezing point		
FPC	Federal Power Commission	ID	inside diameter; infective dose
FRB	Federal Reserve Board		
frz	freezing	ip	intraperitoneal
G	giga (prefix for 10^9)	IPS	iron pipe size
G	gravitational constant $= 6.67 \times 10^{11} \text{N} \cdot \text{m}^2/\text{kg}^2$	ir	infrared
		IRLG	Interagency Regulatory Liaison Group
g	gram	ISO	International Organization Standardization
(g)	gas, only as in $H_2O(g)$		
g	gravitatonal acceleration		
gc	gas chromatography	ITS-90	International Temperature Scale (NIST)
gem-	geminal		
glc	gas–liquid chromatography		
g-mol wt; gmw	gram-molecular weight	IU	International Unit
		IUPAC	International Union of Pure and Applied Chemistry
GNP	gross national product		
gpc	gel-permeation chromatography	IV	iodine value
GRAS	Generally Recognized as Safe	iv	intravenous
		J	joule
grd	ground	K	kelvin
Gy	gray	k	kilo (prefix for 10^3)
H	henry	kg	kilogram
h	hour; hecto (prefix for 10^2)	L	denoting configurational relationship
ha	hectare		
HB	Brinell hardness number	L	liter (for fluids only) (5)
Hb	hemoglobin	l-	$levo$-, levorotatory
hcp	hexagonal close-packed	(l)	liquid, only as in NH_3(l)
hex	hexagonal	LC$_{50}$	conc lethal to 50% of the animals tested
HK	Knoop hardness number		
hplc	high performance liquid chromatography	LCAO	linear combnination of atomic orbitals
HRC	Rockwell hardness (C scale)	lc	liquid chromatography
		LCD	liquid crystal display
HV	Vickers hardness number	lcl	less than carload lots

LD_{50}	dose lethal to 50% of the animals tested	N	newton (force)
LED	light-emitting diode	N	normal (concentration); neutron number
liq	liquid	N-	denoting attachment to nitrogen
lm	lumen		
ln	logarithm (natural)	n (as n_D^{20})	index of refraction (for 20°C and sodium light)
LNG	liquefied natural gas		
log	logarithm (common)		
LOI	limiting oxygen index	n (as Bu^n),	normal (straight-chain structure)
LPG	liquefied petroleum gas	n-	
ltl	less than truckload lots	n	neutron
lx	lux	n	nano (prefix for 10^9)
M	mega (prefix for 10^6); metal (as in MA)	na	not available
		NAS	National Academy of Sciences
M	molar; actual mass		
$\overline{M}_w$	weight-average mol wt	NASA	National Aeronautics and Space Administration
$\overline{M}_n$	number-average mol wt		
m	meter; milli (prefix for 10^{-3})	nat	natural
		ndt	nondestructive testing
m	molal	neg	negative
m-	meta	NF	*National Formulary*
max	maximum	NIH	National Institutes of Health
MCA	Chemical Manufacturers' Association (was Manufacturing Chemists Association)	NIOSH	National Institute of Occupational Safety and Health
MEK	methyl ethyl ketone	NIST	National Institute of Standards and Technology (formerly National Bureau of Standards)
meq	milliequivalent		
mfd	manufactured		
mfg	manufacturing		
mfr	manufacturer		
MIBC	methyl isobutyl carbinol	nmr	nuclear magnetic resonance
MIBK	methyl isobutyl ketone		
MIC	minimum inhibiting concentration	NND	New and Nonofficial Drugs (AMA)
min	minute; minimum	no.	number
mL	milliliter	NOI-(BN)	not otherwise indexed (by name)
MLD	minimum lethal dose		
MO	molecular orbital	NOS	not otherwise specified
mo	month	nqr	nuclear quadruple resonance
mol	mole		
mol wt	molecular weight	NRC	Nuclear Regulatory Commission; National Research Council
mp	melting point		
MR	molar refraction		
ms	mass spectrometry	NRI	New Ring Index
MSDS	material safety data sheet	NSF	National Science Foundation
mxt	mixture		
μ	micro (prefix for 10^{-6})	NTA	nitrilotriacetic acid

NTP	normal temperature and pressure (25°C and 101.3 kPa or 1 atm)	pwd	powder
		py	pyridine
		qv	quod vide (which see)
NTSB	National Transportation Safety Board	R	univalent hydrocarbon radical
O-	denoting attachment to oxygen	(*R*)-	rectus (clockwise configuration)
o-	ortho	*r*	precision of data
OD	outside diameter	rad	radian; radius
OPEC	Organization of Petroleum Exporting Countries	RCRA	Resource Conservation and Recovery Act
o-phen	*o*-phenanthridine	rds	rate-determining step
OSHA	Occupational Safety and Health Administration	ref.	reference
		rf	radio frequency, *n*.
owf	on weight of fiber	r-f	radio frequency, *adj*.
Ω	ohm	rh	relative humidity
P	peta (prefix for 10^{15})	RI	Ring Index
p	pico (prefix for 10^{-12}	rms	root-mean square
p-	para	rpm	rotations per minute
p	proton	rps	revolutions per second
p.	page	RT	room temperature
Pa	Pascal (pressure)	RTECS	Registry of Toxic Effects of Chemical Substances
PEL	personal exposure limit based on an 8-h exposure	*s*(eg, Bu*s*); *sec-*	secondary (eg, secondary butyl)
pd	potential difference	S	siemens
pH	negative logarithm of the effective hydrogen ion concentration	(*S*)-	sinister (counterclockwise configuration)
		S-	denoting attachment to sulfur
phr	parts per hundred of resin (rubber)	*s-*	symmetric(al)
p-i-n	positive-intrinsic-negative	S	second
pmr	proton magnetic resonance	(s)	solid, only as in H_2O(s)
p-n	positive-negative	SAE	Society of Automotive Engineers
po	per os (oral)		
POP	polyoxypropylene	SAN	styrene-acrylonitrile
pos	positive	sat(d)	saturate(d)
pp.	pages	satn	saturation
ppb	parts per billion (10^9)	SBS	styrene–butadiene–styrene
ppm	parts per milion (10^6)	sc	subcutaneous
ppmv	parts per million by volume	SCF	self-consistent field; standard cubic feet
ppmwt	parts per million by weight		
PPO	poly(phenyl oxide)	Sch	Schultz number
ppt(d)	precipitate(d)	sem	scanning electron microscope(y)
pptn	precipitation		
Pr (no.)	foreign prototype (number)	SFs	Saybolt Furol seconds
pt	point; part	sl sol	slightly soluble
PVC	poly(vinyl chloride)	sol	soluble

soln	solution	*trans-*	isomer in which
soly	solubility		substituted groups are
sp	specific; species		on opposite sides of
sp gr	specific gravity		double bond between
sr	steradian		C atoms
std	standard	TSCA	Toxic Substances Control
STP	standard temperature and		Act
	pressure (0°C and	TWA	time-weighted average
	101.3 kPa)	Twad	Twaddell
sub	sublime(s)	UL	Underwriters' Laboratory
SUs	Saybolt Universal seconds	USDA	United States Department
syn	synthetic		of Agriculture
^{t} (eg, Bu^t),	tertiary (eg, tertiary	USP	*United States*
t-, tert-	butyl)		*Pharmacopeia*
T	tera (prefix for 10^{12}); tesla	uv	ultraviolet
	(magnetic flux density)	V	volt (emf)
t	metric to (tonne)	var	variable
t	temperature	*vic-*	vicinal
TAPPI	Technical Association of	vol	volume (not volatile)
	the Pulp and Paper	vs	versus
	Industry	v sol	very soluble
TCC	Tagliabue closed cup	W	watt
tex	tex (linear density)	Wb	weber
T_g	glass-transition	Wh	watt hour
	temperature	WHO	World Health Organization
tga	thermogravimetric		(United Nations)
	analysis	wk	week
THF	tetrahydrofuran	yr	year
tlc	thin layer chromatography	(*Z*)-	zusammen; together;
TLV	threshold limit value		atomic number

Non-SI (Unacceptable and Obsolete) Units		Use
Å	angstrom	nm
at	atmosphere, technical	Pa
atm	atmosphere, standard	Pa
b	barn	cm^2
bar†	bar	Pa
bbl	barrel	m^3
bhp	brake horsepower	W
Btu	British thermal unit	J
bu	bushel	m^3; L
cal	calorie	J
cfm	cubic foot per minute	m^3/s
Ci	curie	Bq
cSt	centistokes	mm^2/s
c/s	cycle per second	Hz
cu	cubic	exponential form

†Do not use bar (10^5 Pa) or millibar (10^2 Pa) because they are not SI units, and are accepted internationally only in special fields because of existing usage.

Non-SI (Unacceptable and Obsolete) Units		Use
D	debye	$C \cdot m$
den	denier	tex
dr	dram	kg
dyn	dyne	N
dyn/cm	dyne per centimeter	mN/m
erg	erg	J
eu	entropy unit	J/K
°F	degree Fahrenheit	°C; K
fc	footcandle	lx
fl	footlambert	lx
fl oz	fluid ounce	m^3; L
ft	foot	m
ft · lbf	foot pound-force	J
gf den	gram-force per denier	N/tex
G	gauss	T
Gal	gal	m/s^2
gal	gallon	m^3; L
Gb	gilbert	A
gpm	gallon per minute	(m^3/s); (m^3/h)
gr	grain	kg
hp	horsepower	W
ihp	indicated horsepower	W
in.	inch	m
in. Hg	inch of mercury	Pa
in. H_2O	inch of water	Pa
in.-lbf	inch pound-force	J
kcal	kilo-calorie	J
kgf	kilogram-force	N
kilo	for kilogram	kg
L	lambert	lx
lb	pound	kg
lbf	pound-force	N
mho	mho	S
mi	mile	m
MM	million	M
mm Hg	millimeter of mercury	Pa
mμ	millimicron	nm
mph	miles per hour	km/h
μ	micron	μm
Oe	oersted	A/m
oz	ounce	kg
ozf	ounce-force	N
η	poise	$Pa \cdot s$
P	poise	$Pa \cdot s$
ph	phot	lx
psi	pounds-force per square inch	Pa
psia	pounds-force per square inch absolute	Pa
psig	pounds-force per square inch gage	Pa
qt	quart	m^3; L
°R	degree Rankine	K
rd	rad	Gy
sb	stilb	lx
SCF	standard cubic foot	m^3
sq	square	exponential form
thm	therm	J
yd	yard	m

BIBLIOGRAPHY

1. The International Bureau of Weights and Measures, BIPM (Parc Saint-Cloud, France) is described in Ref. 4. This bureau operates under the exclusive supervision of the International Committee for Weights and Measures (CIPM).
2. *Metric Editorial Guide (ANMC-78-1)*, latest ed., American National Metric Council, 900 Mix Avenue, Suite 1 Hamden CT 06514-5106, 1981.
3. *SI Units and Recommendations for the Use of Their Multiples and of Certain Other Units (ISO 1000-1992)*, American National Standards Institute, 25 W 43rd St., New York, 10036, 1992.
4. Based on IEEE/ASTM-SI-10 *Standard for use of the International System of Units (SI): The Modern Metric System* (Replaces ASTM380 and ANSI/IEEE Std 268-1992), ASTM International, West Conshohocken, PA., 2002. See also www.astm.org
5. *Fed. Reg.*, Dec. 10, 1976 (41 FR 36414).
6. For ANSI address, see Ref. 3. See also www.ansi.org

C

Continued

CHEMOINFORMATICS

1. What is Chemoinformatics?

Different definitions of chemoinformatics (1) have been given but, within the context of this article we will view it broadly as the management, analysis, and dissemination of data related to chemical compounds. Chemoinformatics results from the application of methods in information technologies to problems in chemistry.

During the last decade, chemoinformatics has become one of the essential tools in the early stages of pharmacological and agrochemical discovery. The reason for its importance is rooted in the emergence of high throughput screening and high throughput chemical synthesis as the dominant technologies for the discovery of starting points for chemical optimization (2). The use of robotics for screening and large chemical libraries has resulted in extremely large volumes of data that require informatics management.

Initially, chemical collections, commonly referred to as libraries, comprised small numbers of distinct chemicals (3). Nowadays, much larger compound collections are routine throughout the chemistry-based industries. The need to evaluate ever expanding libraries requires the development of tools for storage of the information generated, data analysis, the identification of trends in the data, and their eventual correlation to the structural and physicochemical properties of the compounds. In the near future, the challenges in this area will be compounded by the integration of developments in genomics, proteomics, and bioinformatics (4). Chemoinformatics work is and will continue to be multidisciplinary, because it acts at the interface between chemistry and informatics, as well as the multiple disciplines that use it.

Some avenues of research in chemoinformatics evolve from observations made as its tools are applied. An example is provided by diversity analysis. In the past, it was observed that if the compounds evaluated in high throughput screening showed a high degree of structural similarity, the result would be sparsely successful, or have a limited numbers of related hits (5). Consequently, the design of the libraries for screening based on chemical diversity ideas became a crucial step for lead discovery. Methods that provide objective measures of the dissimilarity among compounds to be acquired or synthesized (6,7) are part of the chemoinformatics realm, which was developed to avoid the repeated evaluation of the same chemical classes (8). Because of the wide range of subjects, all of these aspects of chemoinformatics will be discussed only briefly.

The main purpose of chemoinformatics is to provide tools for the efficient management of information, a critical step in any decision making process. Chemoinformatics transforms data into information and subsequently into knowledge, thus greatly facilitating all aspects of chemical research. This field is continually expanding and the number of applications and tools available is very large. Therefore, only some of the many algorithm approaches will be described here, with a particular emphasis on analysis of chemical information.

2. Chemical Information Storage

Perhaps the most important task in the creation of a chemical database is the definition of the fields to be stored. The type and scope of the information to be stored should be pondered carefully at the onset of a project. The database design requires particular attention, because errors or lack of foresight when creating it are painful to correct as the systems are deployed throughout the organization. Depending on the type of information, the project may need to be restarted. Yet, such foresight is challenging because chemical databases used in research are continuously evolving together with the data collected. Even at the earlier stages of the project, input from the end-users is a requisite for the design of any database.

The distinctive feature of a chemical database is that it allows the storage and retrieval of structural information as well as textual or numerical information on a chemical. All datatypes, including chemical, numerical, and textual information, can be combined when querying a chemical database. Simple queries may include combinations of datatypes such as 'Display all compounds with an imine functionality that cost less than a given amount and are currently in inventory', or 'Display all benzimidazoles that have been made between 1971 and 1983 that are still available', or 'Display all thiazoles that show no cytotoxicity at 10 μm concentrations'. Without the chemical structure component, the same searches could not be done within the framework of textual or numerical datatypes alone.

Representation of chemical structures in a computer searchable form requires the adoption of special formats. While multiple formats have been used over the years, the dominant chemoinformatics software provides a relationship between structure and either tables or lines that are intrinsically computer searchable. The two dominant file formats for structural representation are the SMILE strings and the MOL files, discussed below.

2.1. Line Notations. Alternatives to the valence representation of the molecular structure in the form of lines or strings have been pursued for decades, even before the use of computers in chemical information storage and management. Earlier attempts included the well-known Wiswesser line notation, or Bielstein's ROSDAL (9). Currently, SMILES (Simplified Molecular Input Line Systems) is one of the most popular notations in this class. The SMILES notation was developed by Weininger and co-workers and is commonly associated with Daylight software (10–12).

In a SMILES string, each atom is identified by its element symbol, as well as additional information that is placed into brackets, including chirality and net charge. Single bonds are not made explicit, double bonds are indicated as "=", while the triple bonds are shown as "#". Aromatic bonds are represented by ":", bond alternancy or more commonly the aromaticity of a ring, is indicated by using lower case letters for the atoms in aromatic rings. For salts, the smiles string of the ion and the counterion are connected by a dot. In all cases, hydrogens are not made explicit, unless required to establish isomerism. Examples of these representations are shown in Figure 1.

The representation of rings requires two steps (12). First, one bond per ring is broken in such a way that an acyclic structure results. There is always a way to break one bond per ring in a structure so that the result is an acyclic molecule. Second, broken bonds are numbered and the string for the resulting acyclic structure is written. In the resulting string, the numbers assigned to each broken bond are placed next to the adjacent atoms. Examples are also shown in Figure 1.

Geometric isomerism is indicated by the use of both the forward and backslash. Before and after a double bond, if the same type of slash is used to show bonds attached to the double bond, then the arrangement of the centers is *trans*, while if opposite, the atoms are *cis*. Optical isomerism is indicated by the use of "@" or "@ @"; if the order of the atoms attached to the chiral center are ordered anticlockwise, or clockwise, respectively.

One of the limitations of the SMILE strings is that there could be more than 1 equivalent string for the same molecule, because they depend on the internal numbering system of the structure used. One given structure will not always yield the same representation, but it will depend on the algorithm used. A canonical representation is one where rules are defined to the extent that only one string is the correct representation for any chemical structure. "Unique SMILES" (USMILES) are such representations (12).

2.2. Table Representation. The most common alternative to the string or line notation is the use of connectivity or connection tables. File formats that incorporate connection tables have been described in detail in the literature, but are commonly associated with the software developed by MDL Inc. An example of a file that incorporates a connection table is shown in Figure 2 (13,14).

MOL (molecule) and SD (structure data) files contain connectivity tables. Both types of files have a "counts' line", after comments lines that specify the total number of atoms in the file, the number of bonds, atom lists, and information on chirality. In addition, since the format of the files has been mildly modified with time, the version of the file is included. The line is followed by an atom

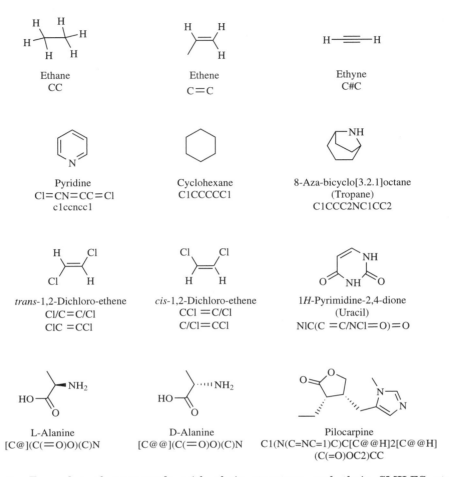

Fig. 1. Examples of compounds with their structure and their SMILES string underneath.

block that contains the atom symbol, charge stereochemistry, attached hydrogens for each atom, and a set of Cartesian coordinates for each atom. In two-dimensional (2D) representations, these coordinates can be used to plot a flat molecular structure. However, the coordinate fields can be used to store a three-dimensional (3D) representation of the molecule, in cases where a spatial arrangement of atoms is available. The ability to store (3D) information in some cases could be an advantage of table representations. The bond block specifies the atoms connected, the bond type, and any stereochemistry or topology associated with the bond. Atom list blocks and a structural text descriptor are also part of the file, though not always explicitly.

The actual connection table follows the atom block, where each bond is represented by each atom and the bond order. Molecular properties including net charge, radical character, isotope, etc, are stored in subsequent lines that

```
Commentdate

 17 17  0  0  0  0  0  0  0  0  0 V2000
       6.3565    2.4220    0.0000 C   0  0  0  0  0  0  0  0  0  0  0  0
       5.6664    3.6329    0.0000 N   0  0  0  0  0  0  0  0  0  0  0  0
       4.2653    3.6329    0.0000 C   0  0  0  0  0  0  0  0  0  0  0  0
       5.6540    1.2110    0.0000 C   0  0  0  0  0  0  0  0  0  0  0  0
       7.7700    2.4220    0.0000 C   0  0  0  0  0  0  0  0  0  0  0  0
       3.5626    2.4220    0.0000 O   0  0  0  0  0  0  0  0  0  0  0  0
       3.4924    5.2324    0.0000 C   0  0  0  0  0  0  0  0  0  0  0  0
       2.0665    5.2324    0.0000 N   0  0  0  0  0  0  0  0  0  0  0  0
       7.7411    0.0000    0.0000 C   0  0  0  0  0  0  0  0  0  0  0  0
       8.4768    1.2275    0.0000 C   0  0  0  0  0  0  0  0  0  0  0  0
       6.3441    0.0000    0.0000 C   0  0  0  0  0  0  0  0  0  0  0  0
       8.4933    3.6619    0.0000 C   0  0  0  0  0  0  0  0  0  0  0  0
       4.3273    1.2110    0.0000 C   0  0  0  0  0  0  0  0  0  0  0  0
       1.4011    6.4599    0.0000 C   0  0  0  0  0  0  0  0  0  0  0  0
       1.3556    4.0504    0.0000 C   0  0  0  0  0  0  0  0  0  0  0  0
       0.0000    6.4599    0.0000 C   0  0  0  0  0  0  0  0  0  0  0  0
       2.0334    2.8435    0.0000 C   0  0  0  0  0  0  0  0  0  0  0  0
  2  1  1  0  0  0  0
  3  2  1  0  0  0  0
  4  1  2  0  0  0  0
  5  1  1  0  0  0  0
  6  3  2  0  0  0  0
  7  3  1  0  0  0  0
  8  7  1  0  0  0  0
  9 11  2  0  0  0  0
 10  5  2  0  0  0  0
 11  4  1  0  0  0  0
 12  5  1  0  0  0  0
 13  4  1  0  0  0  0
 14  8  1  0  0  0  0
 15  8  1  0  0  0  0
 16 14  1  0  0  0  0
 17 15  1  0  0  0  0
  9 10  1  0  0  0  0
M  END
>    <MOL_ID> (2)
346

>    <generic_name> (2)
LIDOCAINE [U;INN]

>    <cas> (2)
137-58-6

>    <source> (2)
Astra, Sweden

$$$$
```

Fig. 2. Example of SD File. The first lines are identical to a MOL file.

start with an M and a word that indicates the property contained in that line. The MOL terminates in a 'MEND' line. While MOL and SD files are identical at this point, SD files also allow the storage of other properties associated with the molecule, as well as multiple molecules in a single file.

Molecular properties provided by the user that are to be stored in an SD file are indicated by a '>' sign followed by the property name in between brackets ('< >') as shown in Figure 2. The information for each molecule is separated by a blank line and "$$$$".

Other file formats centered on the connectivity tables are available. Reaction data (RD) files (14), are similar to the basic SD file but are able to contain structural data for the reactants and products of a reaction, as opposed to individual molecules.

3. Chemical Information Retrieval: Data Searching

The use of computer readable formats to store structural information is the key to generating software that will be capable of searching such data. The search and display of textual and numerical data can be done with Boolean operators (AND, OR, LESS THAN, GREATER OR EQUAL TO, etc). However, the unique feature of a chemical database is in the handling of structural information, and that is where we will focus our discussion.

Different types of searches on structural information can be carried out (15,16). Two-dimensional information is searched differently from 3D. Two-dimensional searches can be done with the purpose of (*1*) identifying an exact chemical structure; (*2*) identifying a molecule or molecules that contain a given structural feature, commonly referred to as a substructure search; and (*3*) to search for molecules that look like those in another used as a query, which is described as a similarity search. Three-dimensional searches can be carried out to identify molecules that have predetermined pharmacophoric features in a correct spatial arrangement, with or without explicit knowledge of the 3D structure of the target.

3.1. Two-Dimensional   Searches. *Structural   and   Substructural Searches.*   The problem of identifying two identical chemical structures could be relatively straightforward, if a canonical representation of a molecule is used (12). However, table representations are not canonical, as atoms could be ordered in a certain sequence in the database, and entered in a different way in the query. Such comparison would require going through all possible permutations of the atoms, which is computationally prohibitive even for relatively small molecules. Algorithms that are more efficient have to be implemented to search in real time, and are the centerpiece of the searching for structural information.

Significant gains can be achieved by the use of filters to limit the number of structures to be explicitly compared. In that way, the expensive exhaustive searches would only be needed on smaller numbers of compounds instead of the entire database. A series of computed properties can be stored as the compounds are loaded into a database and can then be used to establish a similarity or identity to a query fragment or structure.

Structural keys, such as MOLSKEYS, are properties evaluated as compounds are loaded into the database (16) that could be used to reduce the number of pairwise comparisons. Structural keys are binary strings (set of zeros and ones) that indicate whether a given characteristic is present in the molecule. For example, if the molecule is charged, a predetermined bit in the string will be set, a different one will indicate the presence of aromatic amines, while yet another bit could be set by the presence of a carbonyl group. In the end, a vector is created that shows the presence or absence of predefined features. The resulting strings can be compared efficiently by multiple algorithms. For example, look-up tables are created that list all compounds having a particular bit set. Compounds present in all the lists that depend on the query can subsequently be pairwise compared. While not unique for a compound, the comparison of fingerprints greatly reduces the number of structures that have to be compared pairwise. Software, such as Chemfinder (17) or MDL's ISIS relies on the use of structural keys.

Alternatives to the structural keys are molecular fingerprints (12). Contrary to the structural keys, in the case of the fingerprints, there is no preassigned meaning to each bit. Fingerprints are also bit strings but are deprived of a direct meaning as found with structural keys. The process of generating fingerprints is initiated by an exhaustive enumeration of all linear patterns in a molecule, from a list of atoms to paths up to a determined length, which is typically seven bonds. The number of conceivable paths could be extremely large, which makes the assignment of each path a position in a predetermined bit string prohibitive. Instead, each pattern serves as a seed to a pseudonumber generation, the output of which is a set of bits. In more technical terms, the pattern is hashed. The set of bits is then composed in a series of Boolean operations that result in the actual fingerprint. Fingerprints can be handled like structural keys with Boolean operations for structure related searches (12).

The search for a substructure is another common problem in chemoinformatics. The substructure search is the process of finding particular fragments or patterns in a molecule. For table notations, the problem is similar to the identification of complete structures because they also use structural keys, but in this case, only the list of features present (bits set to one) need to be analyzed. The software retrieves compounds from the look-up tables that are associated with all of the features. The fingerprints can be used to match part of the structure as well. A connectivity table can be generated for the fragment and can verify if it is contained in other objects in the database.

Line notations have the additional challenge that the queries are only a part of a structure. SMILES strings are designed for complete structures. Consequently, to construct a query (molecular fragment) a different notation is needed. In the case of SMILES strings, similarly built SMARTS provide the query language (12).

Similarity Searches.   Often times, searches are not for specific compounds or compounds containing a given molecular fragment exactly. Searches could be for compounds that are similar to others or are variants of others in different ways. Such a search requires that the similarities between two molecules be quantified, which is not a simple endeavor. For example, molecules could be similar biologically, have a similar arrangement of key functional groups in space, or be similar in physicochemical characteristics. For database searches, similarity is understood to be structural. In this section, we will be limited to the 2D similarities. Later in this article we will consider other issues on similarity search.

As already described, a compound can be represented by a collection of qualitative properties that describe general aspects of the structure, in the form of a structural key, such as MDLs MOLSKEYS. The similarity between two molecules can be reduced to measuring the similarity between the two binary strings that represent each molecule. Mathematical methods exist that permit such evaluation. Perhaps the most commonly used parameter to measure similarity between binary representations is the Tanimoto Coefficient (18,19). The Tanimoto coefficient is the ratio of bits set (ie, equal to 1) in both molecules to the total number of bits set in either structure. Figure 3, should help us to understand this definition. An alternative to the use of the Tanimoto coefficient is the XOR (exclusive OR) operator, which is simply a count of the number of positions at which the bitstrings for both molecules differ. This operator is also known as city block or Hamming distance, and can be shown to be identical to the square of the Euclidean distance between the two binary strings.

A generalization of the Tanimoto coefficient is the Tversky similarity (20). The coefficient is defined as

$$(Q \text{ AND } M)/(\alpha\, Q \text{ AND } (\text{NOT } M) + \beta\, M \text{ AND } (\text{NOT } Q) + (Q \text{ AND } M))$$

If α and β are set equal to 1, it reduces to the Tanimoto coefficient. If $\alpha > \beta$ means that the features of the query are weighted more heavily and this is commonly referred to as a "superstructure-likeness" search. In the opposite case, where $\alpha < \beta$ produces a "substructure likeness" search, in that case the completely embedded structures have a higher similarity. Super- and substructure likeness searches are part of the major software that runs chemical databases.

3.2. Three-Dimensional Searches: Virtual Screening. The understanding of small molecule protein interactions is key to pharmaceutical and modern agrochemical research because nearly all compounds interact with proteins to elicit their biological activity. Methods to predict those interactions have become of paramount importance. Virtual screening is the process that permits the selection of the compounds that are most likely to interact with a potential target, from a much larger set that is either available or computationally created. Three-dimensional searches are a key component of the virtual screening process (21) because the interaction of a small molecule with a protein depends on the spatial arrangement of their functional groups. Three-dimensional searches can be done based exclusively on the structure of the ligands, without explicit knowledge of the protein target, or they could be done based on the structure of the protein target or direct drug design (22,23). The former is commonly denoted as indirect drug design (24,25), because the process requires the development of hypotheses about the preferences of the protein target of unknown structure based on the types of ligands it binds and those that it does not bind.

The spatial arrangement of functional groups in small molecules is critical in either case: direct or indirect design. The storage of molecular structure becomes important to carry out searches. The creation and maintenance of databases of 3D structures of compounds is essential for chemoinformatics work.

	P1	P2	P3	P4	P5	P6	P7
A	1	2	3	4	5	6	7
B	1	2	1	4	5	6	3
C	2	1	7	7	7	6	5
D	7	1	1	1	2	3	4
E	2	1	3	4	5	6	7
F	7	6	5	4	3	2	1
G	6	7	4	5	2	3	1
H	7	7	6	6	1	1	1
J	5	5	5	6	2	2	2
K	3	4	2	2	4	4	6
L	4	3	2	2	6	6	4
M	1	3	2	5	4	7	6
N	5	1	3	4	6	7	7
O	5	7	6	7	7	1	2
P	7	4	4	3	3	3	2
Q	6	6	6	5	5	5	1

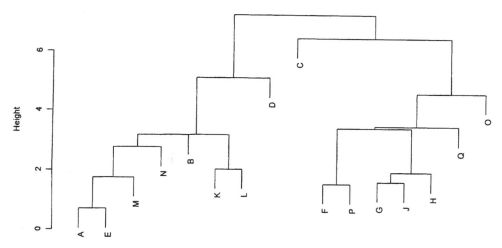

Fig. 3. Examples of bitstrings for a molecule M and a query Q. Tanimoto coefficient is 7/10 or 70% similarity between Q and M, while XOR is 10.

3.3. Construction of Three-Dimensional Chemical Databases.

Structural information about the small molecule is necessary to build a 3D chemical database. The information can be obtained from experimental sources, such as the Cambridge Crystallographic Database. The Cambridge Structural Database (26) (CSD) contains crystal structure information for >230,000 organic and metallorganic compounds. All of these crystal structures have been analyzed using X-ray or neutron diffraction techniques. Alternatively, computational techniques can be used to convert 2D representations of a molecule (a SMILES string or a MOL or SD file) to a 3D structure.

The conversion programs apply a knowledge base to construct the 3D structure. The pioneer program for the rapid conversion of 2D to 3D structures is CONCORD (27), which combines rules with energy estimation procedures in an attempt to produce the lowest energy 3D conformation for each structure. The procedure uses a fragment-based approach, where different portions of each molecule are constructed separately, and pieced together to form a complete molecule. Tables are used for acyclic bond lengths and angles; torsions are assigned to generate optimum interactions between atoms that are not directly linked. For rings, bond lengths and angles are calculated using preassigned rules, and assignment of gross conformations for each ring provides a framework for the torsions. A strain minimization function removes clearly problematic areas in the resulting structure. The structures can be further optimized using molecular mechanics with other programs.

CORINA (28), like CONCORD, is also a rule- and data-based algorithm, but with a superior ring handling technique, as well as organometallics. CORINA has been reported to have a higher conversion rate than CONCORD and other similar software for this purpose. A different approach to the problem is provided by Converter (21). It uses a distance geometry approach, coupled with upper and lower interatomic distance bounds, together with topological rules to generate the 3D structure. The procedure is somewhat slower than the purely rule-based algorithm, and may be less suitable for the conversion of large datasets or if structural information has to be generated as it will be used. That will be the case for software based on-line notations that do not store spatial information but generate it as required.

Conformational Flexibility, Conformational Searches, and Flexible Searches. Conformational flexibility of the molecules presents an additional problem for the 3D searches in chemical databases (25). If only one structure is stored, the searches will be incomplete, because conformations other than the one stored may satisfy the search criteria. A solution to the problem is to store multiple conformations for every compound.

The explicit storage of conformations requires a plan for sampling the small molecule conformational space. The problem of conformational searching in computational chemistry has attracted significant attention for many years and has been the subject of multiple, excellent reviews (29). The approaches devised have relied, for the most part, on the use of molecular mechanics to sample the conformational space of a molecule, by means that include genetic algorithms, stochastic searches, and distance geometry, among others. Since storing all conformations of a molecule is impractical, which conformations should be retained is also a significant issue for all but the most rigid compounds. Many of the conformations generated could be redundant, or essentially identical to others already stored.

One approach commonly used to decide which conformers should be retained has been the generation of extensive conformational libraries, followed by their clustering into related families. The clustering is done using the root-mean-square (rms) deviation in the Cartesian coordinates of the different conformers. Typically, one representative from each cluster is selected.

A more efficient alternative for sampling, which alleviates the time that is required to carry out a full conformational search, is poling (30). This method avoids oversampling of regions of conformational space that have been explored. Still, only individual conformations are stored or scrutinized. Poling represents a significant improvement in terms of the time and amount of conformational space that is sampled compared to other procedures. However, both poling and clustering store only representatives of a much larger set. The set of representative conformers selected may not include the exact spatial arrangements in a query, even though they may be accessible to the molecules under scrutiny in conformations not stored.

With different variants, torsional (21,31) fitting is a technique to deal with the issue of missed conformations that may satisfy a 3D query. Different implementations of the technique, such as the directed tweak algorithm, involve the optimization of rotatable bonds so that they would meet the elements of the query if possible. For a large chemical database, torsional fitting is a process that is computationally demanding, as it involves an optimization step. Filters are applied to limit the number of molecules that must undergo the torsional search. Filters applied depend on the software implementation, or on the protocol defined by the user. Two-dimensional screens ensure that the molecule contains the correct functional groups and are advantageous as a preliminary step. Determination of upper and lower bounds for the distances accessible to the critical atom pairs are common strategies to reduce the amount of sampling required. Inclusion of torsional flexibility increases the number of hits that are retrieved in the search procedure, but with added computational cost.

Building Queries: Pharmacophore Generation and Validation. A query needs to be properly defined before carrying out a search. The type of query and the search strategy will depend on the knowledge that is available on the structure of the target. Indirect drug design techniques can be used to identify pharmacophores when no knowledge about the protein structure is available. Pharmacophores are the collection of relevant groups in the small molecule that can be responsible for the observed biological response. Beyond drug or agrochemical discovery, structure–property relationships, ie, in material science research, can be utilized.

Pharmacophore patterns could serve as queries to identify more molecules that satisfy them, and possibly the pharmacological activity that the pharmacophore summarizes. Their spatial arrangement is called a pharmacophoric pattern, whereas the position of complementary groups on the protein would be designated as a protein or pharmacophore map. A variety of techniques, from simple SAR to computation of quantum mechanical properties, can be used to define a pharmacophoric pattern (24,25). Once the pattern is found, the search of 3D databases can be undertaken, using software such as ISIS (32), UNITY (27), APEX-3D (25), or ALADDIN (12).

Excellent reviews in the area of pharmacophore design and validation exist (33,34). For the most part, automated pharmacophore generation techniques are in use, such as DISCO (35), Catalyst (25), MolMod (36), and FlexS (37) to name but a few. Pharmacophoric points are identified, but in addition, other features can be part of a pharmacophore, such as receptor points or excluded volumes.

The process of identifying a pharmacophore requires consistent biological information determined under identical conditions. The set should contain active and inactive compounds to serve as controls. If the compounds are flexible, a conformational study of each of them is also required. After the conformational libraries are available for each active compound, conformations are sought that maximize the similarities among the compounds of the same biological activity. Those features found common to the active compounds should be absent in the inactive compounds that are structurally related, which greatly reduces the total number of possibilities. From the study, one or a number of models could result that can be used for database searching. In the initial stage, the properties used to define pharmacophores are distances between functional groups within the molecules, which can be grossly classified as hydrophobic centers, hydrogen-bond donors, hydrogen-bond acceptors, charged centers, etc, and are referred to as pharmacophore elements. When similar spatial arrangements of pharmacophore elements are found in any of the low energy conformations of the active compounds that are absent in the inactive compounds, the structures of the active compounds can be overlaid providing a pharmacophore map. The conformation that provides such common spatial arrangement of the pharmacophore elements in each molecule is its bioactive conformation.

In some instances, the identification of pharmacophore points on the molecules themselves is not possible. While the functional groups in a series of different molecules may not be occupying the same relative portion within the molecule, it may be possible to orient them in such a way that they point to a putative external point in a similar manner. These points, external to the molecule, could also be part of a pharmacophore, and are commonly referred as receptor points. These points reflect that the pharmacophore elements are those that by indirect evidence appear to be interacting with the protein target.

Additional information about the active compounds could be incorporated as part of the pharmacophore, and can be used during the search in chemical databases. The most straightforward is to look for patterns in computed partition coefficients that may discriminate the active from the inactive compounds. In addition, other shape or electronic properties can serve to differentiate active from inactive compounds, and can be incorporated during the searches.

After a pharmacophore has been established, it is possible to carry out statistical studies to determine the relative importance of the different properties in the pharmacophores. Analysis such as CoMFA (38) or CoMSIA (39) or Hopfinger's molecular shape analysis (40) do provide information that could be used to rank order the hits that come out of the database. The hits can also be ranked based on their goodness of fit to the pharmacophore features (25).

Most of the work carried out to define pharmacophores has been done using small sets of compounds, in part, due to the need to ensure the homogeneity of the biological data imposed a constraint. Most automated pharmacophore determination software assumes that the compounds under study are binding or activating the target using a similar mechanism and site of action. The advent of high throughput screening as a dominant force in drug discovery has meant that much larger datasets are now available for analysis, but those sets are not necessarily homogeneous. The methods that were used to develop pharmaco-phores are not suitable to address the issues posed by large heterogeneous datasets. One solution has been the selection of homogeneous subsets for which detailed pharmacological information is possible. Methods that take into account the heterogeneous nature of the data as well as larger sample sets have been developed, using new means to carry out the analysis (41,42).

Docking and Target Structure-Based Database Searches. Whenever the structure of the target protein is known, it can be used in the process of searching chemical databases. The search is done by attempting to fit the small molecule into the known protein structure, commonly referred to as dock-ing. In the first step, docking aims to predict how a small molecule can interact with a macromolecular target, and subsequently attempts to score how well the small molecule complements the binding site. The ultimate goal is to predict if and how a given small molecule can favorably interact with a protein.

A large number of different algorithms have been proposed for the auto-mated docking of molecules (43,44), including DOCK (45,46), AutoDOCK (47,48), FLExX (27,49), and GOLD (26,50). While originally, the methodologies for docking were rigid-body matches, the most common implementations opti-mize the small molecule in the cavity of the site, and the way in which that is achieved constitutes one of the most significant differences among the different algorithms. A variety of optimization techniques including shape matching to genetic algorithms, evolutionary programming, or simulated annealing are used to that end.

While the above methods use complete molecular structures to carry out the searches, fragment-based methods are also common (51). Whole molecule meth-ods are based on molecules that are part of the 3D chemical database, but fragment-based methods build new molecular structures in the site from substructures. Fragment-based methods can place a seed molecule in a cavity, and attach other groups in a stepwise manner, building up the desired structure. Another possibility is to place key functional groups complementing the features of the protein, and attempt to connect them into a single structure. LUDI is the most common among such build up procedures (52). The program connects fragments that dock into specific sites of a receptor, such as hydrogen-bond donating or accepting or hydrophobic residues. Fragments come from predefined libraries that can easily be customized. Once the fragments are positioned, they may be linked together using linear groups. If a seed fragment is placed in the binding site, the program can be used to add functionality that complements the site. The program may even perform a preliminary evaluation of synthetic acces-sibility of the linkage required, one of the major problems of drug design. Several other methods have been described.

Once the small molecule has been docked, the goodness of its fit should be determined. The prediction of binding affinity, or at least a correct rank order, is currently one of the most challenging problems in ligand design (53). The pur-pose is to prioritize the hits obtained from a computer program from a 3D search of structures in a database. This is another area where the different methods diverge, as several scoring methods are described (54,55). Some scoring methods such as free energy perturbation provide relative scores that are quite accurate but very computationally demanding, and therefore ill suited for virtual screening where hundreds of thousands or millions of compounds are typically computationally screened. Most scoring is done using force field calculations or empirical free energy scoring functions. Another possibility is the use of consensus scoring, where different scoring techniques are computed simultaneously, and a weighted average is taken as the parameter to rank order the goodness of binding (44).

For the most part, only limited efforts have been placed in dealing with the flexibility of the protein or the binding site. Some attempts have been made to take snapshots during a molecular dynamics trajectory, and to use the different structures, or averaged structures, for docking. The treatment of scoring functions and protein flexibility are major shortcomings at this time, which are attracting active research.

Examples of Applications and the Success of Virtual screening. High throughput screening methodologies in the pharmaceutical industry became so dominant due to a perceived lack of success in the area of structure-based drug design (56). Since then, the use of more sophisticated techniques facilitated their success, particularly when it was closely coupled to structural information (49,57,58).

Human immunodeficiency virus HIV protease and neuramidase inhibitors were derived from the use of computational tools and structural information. Various other enzyme inhibitors were also successfully designed by using a combination of structure-based and computer-aided drug design as well. Pharmacophore-based approaches resulted in the design of metalloprotease, tyrosine kinase inhibitors, and integrin receptor antagonists.

Nowadays, all the tools of computational chemistry, molecular design, and chemoinformatics are integrated into the process of designing new products. Soon it will be difficult to identify compounds largely derived based on those techniques, as they will be entwined with other discovery technologies. The acceptance of the tools of chemoinformatics and virtual screening is pervasive, and most discovery projects use them to the extent that is required.

3.4. Diversity Searches. The need to carry out diversity searches emerged in the pharmaceutical and agrochemical industry because of the advances in automation in biological screening and high throughput chemical synthesis. Those technologies posed a new set of questions to be asked from a chemical database. Initially, chemical collections for screening were obtained from internal libraries in the pharmaceutical industry. Those libraries had been created over the years by medicinal chemistry efforts, and consisted of large numbers of analogues on a few chemical families. Lack of success in the approach (5) was attributed to the relatively small size of the libraries studied, and to the lack of variety prosessed by those chemical collections. The need to detect redundancies in a chemical library led to the concept of diversity analysis.

With few exceptions (59), diversity analysis of chemical collections evolved from the methodologies of structure-based drug design. Consequently, diversity analysis has been heavily dependent on the computation of physicochemical or structural properties (8,60). Chemical diversity is mostly associated with methods that allow the determination of how well libraries could represent portions of chemical property space. The scattering or clumping of representatives in a library can be surrogate indicators of the probability that that library would provide multiple, singular, or no hits for a set of targets. The challenges associated with the determination of chemical diversity are quite varied. The most significant issues are the selection of properties and the algorithms that are to be used to determine and select diverse compound sets.

Properties for Diversity Analysis. Properties in use for diversity analysis include the use of computed physical properties for the compounds. A compound can be described by a collection of global properties, such as the octanol–water partition coefficient, its pK_a parameters related to molecular size and shape, counts of hydrogen-bond donor and acceptor centers, among others (61). These properties can be combined with topological descriptors, such as molecular connectivity indexes that encapsulate information about the 2D structure of the molecule, its structural complexity, and some simple measure of its electronic character (62,63). From them, shape and flexibility parameters can be generated.

While the structural keys were originally designed to make searches of chemical databases more efficient (8,61), they also play a significant role in the analysis of chemical diversity, and chemical similarity. Structural keys are discrete valued descriptors contrary to the global physicochemical properties, and therefore require different tools for their analysis. Some of those methods have been implemented in the Catalyst software (64), etc.

Three-dimensional properties can also be used and provide another view of the diversity of a compound collection. The structures necessary can be generated by the same methods indicated above for the creation of 3D databases. Three-dimensional properties for diversity analysis include the BCUT parameters (65). These parameters involve three types of matrices, where the diagonal elements are based on the atomic charges, polarizability, and hydrogen-bond donor or acceptor capabilities for each atom, respectively. The off-diagonal elements are based on 2D or 3D information, including functions of interatomic distance, overlaps, computed bond-orders, etc. All these parameters can be computed using semiempirical molecular orbital programs. The lowest and highest eigenvalues that result from the diagonalization of these three matrices are considered to reflect most aspects of the molecular structure. Methods must be developed for rationally deciding which BCUT values (eigenvalues) would be best for representing the chemical diversity of a given population of compounds. The analysis is part of the DiversitySolutions (27) software, whose efficiency has been reported in the literature. However, even at the 2D level, BCUT parameters can be satisfactory for diversity analysis.

Counts on the possible spatial arrangement of chemical groups (66,67) are alternatively used to determine the pharmacophores accessible to a molecule. Pharmacophoric centers commonly associated with intermolecular interactions are typically included, such as hydrophobic centers, charged centers and hydrogen-bond donor and acceptor centers. Once the machine recognizes those groups, the distances between each of the centers are recorded. Distances among the pharmacophoric centers assume continuous values, but when variations in distance are small, they may be considered equivalent. For this reason, binning of pharmacophoric patterns is used by the major commercial software for this approach. In this case, the pharmacophore-based representation of a molecule is still a binary string that indicates the presence or absence of a certain combination of pharmacophoric centers at a certain range of distances.

Other 3D descriptors are based on the Comparative Molecular Field Analysis (CoMFA), which is particularly useful for series of related compounds (8). The scores from docking compounds against a set of random proteins have also been utilized to check the diversity of compounds (68). The set of scores determined for each protein constitutes a descriptor. The approach was inspired in a previous attempt to utilize experimental binding data for diversity assessment (59,69).

Algorithms for Diversity Analysis (70).   Regardless of the set of properties considered, molecules are represented by vectors either of continuous values, or as bitstrings in the case of structural keys or pharmacophore analysis. Hence, each molecule in the set is represented by a vector, or as a point in a high-dimensional space, or a "chemical space". The similarity between two molecules can be measured calculating a distance between the two points that represent each molecule. If the properties are binary, then the distance can be computed using the Tanimoto coefficient or the XOR distance, if the properties are continuous, an Euclidean distance can be used.

The distances generated for a set of molecules can be used as a similarity matrix, when the distance between every pair of molecules is determined. A similarity matrix can then be used to select a subset of compounds that are as diverse as possible for the set of properties under consideration. However, the relation between compounds will critically depend on the chemical representation adopted because compounds that appear to be different for a set of properties can be very much alike under a different set of descriptors (71). The selection of a property space is therefore an important issue when analyzing for diversity and requires careful consideration. The selection of properties is mostly based on the ability of the set chosen to segregate compounds of different pharmacological profile (60).

Clustering techniques provide the means to group compounds in sets that have similar properties (72). There are different types of clustering algorithms, but one of the most commonly employed in many arenas and also in chemoinformatics work, is the hierarchical agglomerative. Step after step the method clusters successively more distant compounds. In the first step, the two most similar compounds are grouped together forming a cluster. The next set of closely related points or clusters are linked together, and the procedure continues until all points are part of a single cluster. The representation of the clustering process is a dendrogram (a classification tree) that goes from the individual points or singletons to all compounds in a single cluster, a representation that is common in other disciplines. An example is shown in Figure 4. The number of clusters is

	Bitscreen	**Bits Set**
Query	1,0,0,0,1,1,0,0,0,0,1,0,1,0,0,1,0,0,0,1,1,0,0	8
Molecule	1,0,1,0,0,1,0,0,0,0,1,0,1,0,0,1,0,0,0,1,1,1,0	9
Query AND Molecule	1,0,0,0,0,1,0,0,0,0,1,0,1,0,0,1,0,0,0,1,1,0,0	7
Query OR Molecule	1,0,1,0,1,1,0,0,0,0,1,0,1,0,0,1,0,0,0,1,1,1,0	10

Fig. 4. Dendrogram: A table of properties (P1 to P7) for a series of compounds (A–Q). When a cutoff for similarity (height) of 2 is, eg, selected, compounds A, E, and M belong to a cluster, while G, J, H form a different one. Some compounds have clusters of their own, denoted as singletons. Compounds B, C, and D are examples of that category.

determined by the degree of similarity that is considered significant. Once a set of clusters is defined, a compound can be selected from each cluster, resulting in a diversified set.

A clustering strategy that has been widely used in chemoinformatics is the Jarvis Patrick algorithm (73). In simple terms, the algorithm creates a list of its nearest neighbors for each point. Two points belong to the same cluster if they are in each other's list of nearest neighbors, and they share a number of common neighbors. Multiple technical reasons make it a preferred choice for the problem of compound selection. Beyond their commonly reported speed and efficiency, it can automatically deal with nonspherical clusters efficiently. Hierarchical methods can also deal with nonspherical clusters, but prior knowledge about the distribution is required.

Nonclustering methods are also used for diversity selection. Cell-based methods are part of the DiverseSolutions software (27), where the space defined by the chemical properties selected is partitioned into cells. The occupancy of each cell is determined based on the properties of the compounds (74,75). The advantage of cell-based techniques is that they provide a uniform sampling and the areas of property space that are not represented in the library can readily be identified, providing a simple representation of the completeness of the chemical library. D-optimal design is also a technique that was applied to this problem (61). However, its use has been less significant because of its tendency to select compounds unequally from the entire chemical space, and show a bias for points at the edge distribution.

Software such as C2-Diversity (Accelrys, San Diego) provides a broad assortment of properties and a variety of methods for selecting such diversity (64). Pipeline Pilot offers an alternative of innovative software architecture that computes processing, analysis, and mining of large volumes of data through a user-defined computational protocol (76).

Contrary to similarity comparisons, where success is measurable by the number of compounds that share the desired profile possible from among those chosen using the metric, the goodness of diversity algorithms are harder to characterize. If the goal is simply to remove redundancy from a chemical library, even the simplest of algorithms can fit the requirement. If, on the other hand, the goal is to increase the hit rate for a library of related targets, or for any random set of targets, the best strategy to utilize could be different. The overlap and similarity of the software currently available for library design can be a major challenge, since most different packages are fragmented. The fragmentation is the result of two trends. On one hand, there is the acquisition of software from academic sources or by formal mergers and acquisitions that the major chemoinformatics software vendors have undergone. On the other hand, there is the commercial tendency to fragment the software, to customize to specific needs, which in practice results in redundancy as frequently more than one package is required.

Design of Chemical Libraries. Few, if any, hits from massive and truly random screening libraries could be evolved into starting points for product development. Additional constrains have been imposed to the value ranges of the properties being considered, which are consistent with their intended use, effectively limiting the chemical space taken into account. Lists of exclusions are common when the compounds are intended for screening where reactive groups are removed from the searches. The same types of properties have also been used to tailor a library for drug discovery, in an attempt to look for drugs in an area of chemical space that is relevant to medicinal chemistry (77,78). For the most part, these efforts have been focused around the structural character-istics of drugs (79–81) and the design of libraries for screening has now centered on the use of compounds similar to drugs, or "drug-like" molecules.

The nature of the constraints to be imposed on chemical space was derived by the statistical analysis of the properties of marketed drugs or of compounds that have undergone human testing. The types of chemical functionalities found in them, and their physicochemical characteristics, serve to define the acceptable range of properties that are characteristic of drugs. The argument has been that the properties of compounds that have been in late stage trials reflect what is biologically compatible. The study of successful compounds has been part of the attempt to predict their absorption, metabolism, distribution, excretion (ADME), as well as their toxicology. Prediction of those properties is currently receiving significant attention in the chemoinformatics field since it represents a major bottleneck in drug discovery and development processes.

Within this realm, the "rule of 5" (82) has gained acceptance (83). These rule was developed by a simple analysis of databases of compounds that had undergone clinical trials. It was concluded that poor permeation or absorption were more common when: there were >5 hydrogen bond donors; >10 hydrogen-bond acceptors; the calculated Log P was >5, and the molecular weight was >500. The cutoffs for each of four parameters are multiples of five. Thus its name. The rule does not cover compounds that are actively transported. This simple chemometric exercise has affected the design of ligands for target pro-teins. However, the approach is not without its criticisms (84,85).

Statistical rules to predict solubility, oral availability, and permeability are part of the repertoire of chemoinformatics tools [86–88]. Those properties can be quickly computed with packages such as QikProp (89), or derived based on global molecular properties, such as those provided by the ACD Labs package (90) and that are straightforward to implement. Many other relations are commonly used to predict blood-brain barrier permeation, cell permeability, or to estimate stabi-lity and pharmacokinetic parameters.

Important efforts have been made in the area of using experimental infor-mation to predict oral absorption. The iDEA (In Vitro Determination for the Estimation of ADME) simulation system is a computational model developed to predict human oral drug absorption based on its solubility and permeability, which is empirically determined (91).

The determination of metabolism is an extremely complex issue, where the different isozymes of cytochrome P450 play an all-encompassing role. Prediction of the sites of metabolism (regioselectivity) for this enzyme can be done by eval-uating the electronic tendencies for oxidation of all the potential sites within the substrate molecule (92–95). However, this is only a preliminary approach, and complex simulations are still required to carry out metabolism prediction accu-rately. Biological processes are quite complex and cannot be simulated entirely in-silico. The real value in chemoinformatics resides in the derivation of simple rules. Whenever those rules are possible, they result in biases in compound selec-tion toward candidate compounds that are more likely to succeed.

3.5. Toxicology Prediction. The prediction of the toxicological character of a compound is of foremost importance throughout the chemical industry, and extends from the agrochemical and pharmaceutical industry to the environmental and food chemistry. A large number of approaches have been adopted and are employed to study the compound toxicity problem (96).

One of the most popular packages for that purposes is TOPKAT (64,97) which is a self-contained computational toxicology package that uses 2D descriptors and statistical models to generate reliable toxicological profiles of organic chemicals, one at a time.

An alternative to the statistical analysis of properties are knowledge-based systems that are computer programs to organize relevant experimental data to help a user make decisions about concrete issues. They require the use of a systematic database of information from which rules are derived, which allows the prediction of the property to be scrutinized. HazardExpert predicts different toxicity effects of compounds such as carcinogenic, mutagenic, teratogenic, membrane irritation, and neurotoxic effects (93). The knowledge base was developed based on the list of toxic fragments reported by more than 20 lead experts. This software also predicts bioaccumulation as well as bioavailability based upon predicted physicochemical values. It is a rule-based system using known toxic fragments collected from in vivo experimentation. DEREK is also a rule-based approach that can make predictions about a large set of toxicological properties including carcinogenicity, irritancy, lachrymation, neurotoxicity and thyroid toxicity, teratogenicity, respiratory and skin sensitization, and mutagenicity (98).

MULTICASE and CASE programs (99,100) can automatically identify molecular substructures that have a high probability of being relevant or responsible for the observed biological activity of a learning set comprised of a mix of active and inactive molecules of diverse composition. New, untested molecules can then be submitted to the program, and an expert prediction of the potential activity of the new molecule is obtained.

4. Chemical Databases

Chemical information itself is abundant, and the Chemical Abstract Service (CAS) has maintained the most comprehensive resource, in this field (101). Sci-Finder provides a desktop research tool that allows the exploration of research topics, with little training, containing information on >33 million substances. The STN service provides specialized information on >200 different subjects.

CrossFire Beilstein (32) has extensive information on bioactivity and physical properties that makes it particularly useful when undertaking biological research. The database also provides information on the ecological fate of compounds.

Other databases are also worth noting because they contain information that is more specific. There are vast numbers of commercial databases with different focus. One of the most common is the ACD (Available Chemicals Directory) (32), which contains information on price and availability on >300,000 compounds. This information includes not only a 2D representation of the molecule, but also 3D models that make it useable for pharmacophore searches or for docking purposes. Many chemical vendors also provide catalogs of rare chemicals in 2D format, but that can be readily converted to a 3D database using the methods described previously.

Chemical databases also deal with reactivity information. As described above, reaction databases have special characteristics that arise from the need to handle reactants and products. Multiple databases exist for this purpose. SpresiReact database is available from InfoChem GmbH and contains 2.5 million different reactions (12,32). It contains 1.8 million individual molecules that appear as components of the reactions, journal references, yields, and reaction conditions information. RefLib (32) is another broad collection of novel organic synthetic methodologies that covers functional group transformations, metal-mediated chemistry, and asymmetric syntheses, as well as reactions from Theilheimer's *Synthetic Methods of Organic Chemistry*. An electronic version of the entire series of *Organic Syntheses*, ORGSYN, contains general synthetic methods and proven compound preparations. Similarly, Derwent's *Journal of Synthetic Methods*, has been condensed in the RX-JSM database. *Methods in Organic Synthesis* (MOS) is a selective current awareness database derived from a bulletin of the same name, published by the Royal Society of Chemistry. This database focuses on important new methods in organic synthesis and comprises >3300 reactions per year, dating back to 1991. BioCatalysis is a selective, thematic database that focuses on chemical synthesis using biocatalysts, including pure enzymes, whole cells, catalytic antibodies, and enzyme analogues. The Failed Reactions database (63) is a unique compilation of reactions with unexpected results, which may involve an unexpected product, an immediate further reaction, or simply no reaction.

Combinatorial chemistry has played a significant role in the making of compounds for material sciences as well as for the agrochemical and pharmaceutical industries. The SPORE (32) and Solid-Phase Synthesis (63) databases include data particular to solid-phase organic synthesis, such as information on polymeric materials, linkers, solid supports, and protecting groups. The Protecting Groups database (32) provides information on methods for protection, deprotection with the ability to search generically, based on functional groups, protected groups, tolerated groups, and reaction conditions. Bunin's book *The Combinatorial Index* has also been put into electronic form. Other databases on chemical reactivity also exist

Apart from chemistry resources, there are a large number of content databases with information specific about different areas (102,103) all linked by their chemical structure, which include material sciences, agrochemical, physicochemical, and biological activity. These databases are provided by a variety of solution providers, offering different products, which may complicate the search for an appropriate one.

5. Data Analysis and Presentation

Data mining is crucial when large amounts of data are generated and is a trend observed in many industries, including the pharmaceutical industry (104). The idea is to integrate a number of visualization, statistical analysis tools through graphical user interfaces.

One of the most popular packages for this purpose is Spotfire (105), where chemical structure data can be combined with data from different sources, in order to provide insights into the property of interest, biological or physical. These software tools allow the users to manipulate variables and large quantities of data, and integrate them with simple statistical tools such as graphs, decision trees, and scatter plots. The software allows the user to merge data from diverse sources into a single screen with the ability to visualize trends. The software is in use in the pharmaceutical, but also in the energy, specialty chemical, and semiconductors industries.

DIVA (63) is a package with a similar purpose that was developed specifically for the pharmaceutical industry. DIVA allows users to retrieve and work with chemical structures, assay results, and other chemical and biological data in one convenient spreadsheet. Powerful easy-to-use tools for data integration, visualization, analysis, and reporting save time and allow researchers to get more value from their data.

6. Economics of Chemoinformatics

The economic impact of chemoinformatics is two sided, as it is an industry that produces software, but it also greatly affects the productivity of all chemistry research and development. On one hand, during the year 2000, the overall market for chemoinformatics, bioinformatics, and simulation software was ~1.3 billion. The number is in circulation and is based on assuming a spending of ~7% of the R&D budget of $15 billion in informatics services for the composite of pharmaceutical, specialty chemical, and agrochemical markets. About 90% of that amount is spent in-house, giving an estimate of ~$150 million for third parties. The numbers are poised for significant growth on a yearly basis.

However, the market is remarkably fragmented. The reasons for the fragmentation involve the nature of the business, where technological innovation is key and there are low barriers to establish new ventures. New players with an interesting application can create a niche from which they can grow. Thus, the established vendors face competition from nonprofit organizations, in-house solutions, and other technology providers, such as IBM, SGI, or Agilent.

No publicly traded company can be labeled truly chemoinformatics pure play. In many cases, the company has other business associations or forms part of a major conglomerate that makes the analysis more complex. Two companies that have a strong chemoinformatics component, Tripos and Pharmacopeia, also have associated molecular modeling software and chemistry research. Another equally important player in the area is MDL, Inc., a subsidiary of the major publishing conglomerate Elsevier.

Pharmacopeia's software revenue for the 2001 third quarter rose 20% compared to its 2000 third quarter of $21.5 million, which included the effects of acquisitions. For the 9-month year-to date period, Tripos recorded $32.5 million in revenues compared to $16.2 million in 2000, an increase of 101%. However, this number is an aggregate of financial transactions and other nonsoftware business. The results for both companies reflect the growth rate of the industry, which is picking up pace. As of the end of the third quarter 2001, the market capitalization for Tripos was 127.7 million versus 334.4 million for Pharmacopeia.

Two revenue models exist in the industry. On one hand, there are the companies that simply sell software and services into the drug discovery market, and on the other hand, there are companies with research collaborations with major pharmaceutical companies, with upside potential if royalties are retained. For software providers, the preferred model is that of software licensing and maintenance fees. In some cases, and due to the steep licensing fees, yearly license payments have been adopted. However, this limited the perceived value of those companies and imposed restrictive market caps. In an effort to improve their valuation, companies moved to provide other services as well. Through mergers and acquisitions during the late 1990s, Pharmacopeia, a chemistry services provider for the pharmaceutical industry acquired MSI, Inc., while Tripos, a software provider, acquired Receptor Research a small chemistry services provider based in the United Kingdom.

Consolidation is not new to the industry. Accelrys, the software division of Pharmacopeia, is the result of a number of mergers and acquisitions, the most recent being that of the Oxford Molecular Group, based in the United Kingdom in 2000 for $22 million. The acquisition of Trega by Lion Biosciences has resulted in a different model where there is forward integration and where a company in the area of genomics software acquired a provider of tools and content.

The developments in genomics and proteomics are likely to produce a new wave of software tools that more closely integrate the tools of bioinformatics and chemoinformatics. Structural proteomics is also giving rise to a new crop of companies that aim at structural chemistry and drug discovery, many with in-silico components. This is an arena where the landscape is rapidly changing, with a yearly growth of up to 40%.

From a different angle, it is different to gauge the importance and productivity gains due to chemoinformatics. However, robotics and automation, as well as multiparametric analysis, would not be possible without the intensive use of computers in chemistry.

BIBLIOGRAPHY

1. F. K. Brown, *Annu. Rep. Med. Chem.* **33**, 375, (1998).
2. J. Drews, *Science* **289**, 1960 (2000).
3. L. M. Kauvar and E. Laborde *Curr. Op. Drug Disc. Divers* **1**, 66 (1998).
4. J. Bajorath, *Drug Discov. Today* **6**, 989 (2001).
5. R. Lahana, *Drug Discov. Today* **4**, 447 (1999).
6. L. A. Thompson and J. A. Ellman, *Chem. Rev.* **96**, 555 (1996).
7. J. C. Hogan, Jr., *Nature Biotech.* **15**, 328 (1996).
8. Y. C. Martin, R. D. Brown, and M. G. Bures, in E. M. Gordon and J. F. Kerwin, Jr., eds, *Combinatorial Chemistry and Molecular Diversity in Drug Discovery*, Wiley-Liss, 1998. p. 369.
9. J. M. Barnard, in: P. von R. Schleyer, N. L. Allinger, T. Clark, J. Gasteiger, P. A. Kollman, H. F. Schaefer, and P. R. Shreiner, eds, *Encyclopedia of Computational Chemistry*, Wiley, Chichester, U.K. Vol 4, (1998), p. 2818.

10. D. Weininger, *J. Chem. Inf. Comput. Sci.* **28**, 31 (1998).
11. C. A. James, D. Weininger, and J. Delany, Daylight Theory Manual, Daylight Chemical Information Systems, Inc. Santa Fe, N. Mex.
12. http://www.daylight.com.
13. A. Dalby, J. G. Nourse, W. D. Hounshell, A. K. I. Gushurst, D. L. Grier, B. A. Leland, and J. Laufer, *J. Chem. Inf. Chem. Sci.* **32**, 244 (1992).
14. http://www.mdli.com/downloads/ctfile/ctfile_subs.html.
15. G. M. Downs and P. Willett, *Rev. Comput. Chem.* **7**, 1 (1996).
16. ISIS/Base 2.1.4; MDL Information Systems, Inc., San Leandro, Calif.
17. http://www.cambridgesoft.com/products.
18. P. Willett and V. A. Winterman, *Quant. Struct-Act. Relat.* **5**, 18 (1986).
19. S. L. Dixon and R. T. Koehler, *J. Med. Chem.* **42**, 2887 (1999).
20. P. H. A. Sneath and R. R. Sokal, *Numerical Taxonomy*, W. H. Freeman & Co. San Francisco, Calif. 1973.
21. A. C. Good and J. S. Mason, *Rev. Comput. Chem.* **7**, 67 (1996).
22. P. J. Gane and P. M. Dean, *Curr. Opin. Struct. Biol.* **10**, 401 (2000).
23. D. A. Gschwend, A. C. Good, and I. D. Kuntz, *J. Mol. Recog.* **2**, 175 (1996).
24. G. H. Loew, H. O. Villar, and I. Alkorta. *Pharm. Res.* **10**, 475 (1993).
25. Y. Kurogi and O. F. Guner, *Curr. Med. Chem.* **8**, 1035 (2001).
26. http://www.ccdc.cam.ac.uk.
27. http://www.tripos.com/software.
28. J. Sadowski and J. Gasteiger, *Chem. Rev.* **93**, 2567 (1993).
29. http://cmm.info.nih.gov/modeling/guide_documents/conformation_document.html.
30. A. Smellie, S. L. Teig, and P. Towbin, *J. Comp. Chem.* **16**, 171 (1995).
31. T. Hurst, *J. Chem. Inf. Comput. Sci.* **34**, 190 (1994).
32. http://www.mdli.com.
33. A. K. Ghose and J. J. Wendolowski, *Persp. Drug Discov. Design* **9**, 253 (1998).
34. G. Klebe, in H. Kubinyi, ed, *3D QSAR in drug design: theory methods and applications*, Escom, Leiden, 1993, p. 173.
35. Y. C. Martin, M. G. Bures, E. A. Dandur, J. De Lazier, I. Lico, and P. A. Pavlik, *J. Comput. Aided Drug Des.* **7**, 83 (1993).
36. D. H. Harris and G. H. Loew, *Bioorg. Med. Chem.* **8**, 2527 (2000).
37. C. Lemmen, T. Lengauer, and G. Klebe, *J. Med. Chem.* **41**, 4502 (1998).
38. R. D. Cramer, D. E. Patterson, and J. D. Bunce, *J. Am. Chem. Soc.* **110**, 5959 (1988).
39. G. Klebe and U. Abraham, *J. Comput. Aided Drug Des.* **13**, 1 (1999).
40. A. J. Hopfinger, S. Wang, S. Tokarshi, B. Jin, M. Albuquerque, P. Madhav, and C. Duraiswami, *J. Am. Chem. Soc.,* **119**, 10509 (1997).
41. A. J. Hopfinger and J. S. Duca, *Curr. Op. Biotechnol.* **11**, 97 (2000).
42. X. Chen, A. Rusinko III, A. Tropsha, and S. S. Young, *J. Chem. Inf. Comput. Sci.* **39**, 887 (1997).
43. R. E. Babine and S. L. Bender, *Chem. Rev.* **97**, 1359 (1997).
44. C. Bissantz, G. Folkers and D. Rognan, *J. Med. Chem.* **42**, 4759 (2000).
45. I. D. Kuntz, J. M. Blaney, S. J. Oatley, R. Langridge and T. E. Ferrin, *J.Mol. Biol.* **161**, 269 (1982).
46. http://www.cmpharm.ucsf.edu/kuntz/dock.html.
47. D. S. Goodsell and A. J. Olson, *Proteins* **8**, 195 (1990).
48. http://www.scripps.edu/pub/olson-web/doc/autodock.
49. M. Rarey, B. Kramer, T. Lengauer, and G. Klebe, *J. Mol. Biol.* **261**, 470 (1996).
50. G. Jones, P. Wilett, R. C. Glen, A. R. Leach, and R. Taylor, *J. Mol. Biol.* **267**, 727 (1999).
51. H. J. Bohm, *Persp. Drug Discov. Design* **3**, 21 (1995).
52. G. Schneider, O. Clement-Chomiene, L. Hilfinger, P. Schneider, S. Kirsch, H. J. Bohm, and W. Neidhart, *Angew. Chem. Int. Ed. Engl.* **39**, 4130 (2000).

53. I. Muegge and M. Rarey, *Rev. Comput. Chem.* **17**, 1 (2001).
54. T. Hansson, J. Marelius, and J. Aqvist, *J. Comput. Aided Mol. Des.* **12**, 27 (1998).
55. M. Schapira, M. Trotov, and R. Abagyan, *J. Mol. Recog.* **12**, 177 (1999).
56. R. M. Snider, *Science* **251**, 435 (1991).
57. H. Kubinyi, *J. Recept. Signal Transduct. Res.* **19**, 15 (1999).
58. H. Kubinyi, *Curr. Op. Drug Disc. Dev.* **1**, 4 (1998).
59. S. L. Dixon and H. O. Villar, *J. Chem. Inf. Comput. Sci.* **38**, 1192 (1998).
60. R. D. Brown and Y. C. Martin *J. Chem. Inf. Comput. Sci.*, **37**, 1 (1997).
61. E. J. Martin, J. M. Blaney, M. A. Siani, D. C. Spellmeyer, A. K. Wong, and W. H. Moos, *J. Med. Chem.* **38**, 1431 (1995).
62. L. H. Hall and L. B. Kier, *Rev. Comput. Chem.*, **2**, 367 (1991).
63. R. A. Lewis, J. S. Mansonand and I. M. Mc Lay, *J. Chem. Inf. Comput. Sci.* **37**, 599 (1997).
64. http://www.accelrys.com.
65. R. S. Pearlman and K. M. Smith, *J. Chem. Inf. Comput. Sci.* **39**, 28 (1999).
66. H. Matter and T. Potter, *J. Chem. Inf. Comput. Sci.* **39**, 1211 (1999).
67. J. S. Manson, I. Morize, P. R. Menard, D. L. Cheney, C. Hulme, and R. F. Labaudiniere, *J. Med. Chem.* **42**, 3251 (1999).
68. H. Briem and I. D. Kuntz, *J. Med. Chem.* **39**, 3401 (1996).
69. J. N. Weinstein, T. G. Myers, P. M. O'Connor, S. H. Friend, A. J. Fornace Jr., K. W. Kohn, T. Fojo, S. E. Bates, L. V. Rubinstein, N. L. Anderson, J. K. Buolamwini, W. W. van Osdol, A. P. Monks, D. A. Scudiero, E. A. Sausville, D. W. Zaharevitz, B. Bunow, V. N. Viswanadhan, G. S. Johnson, R. E. Wittes, and K. D. Paull, *Science* **275**, 343 (1997).
70. D. C. Spellmeyer and P. D. J. Grootenhius, *Annu. Rep. Med. Chem.* **34**, 287 (1999).
71. H. O. Villar and R. T. Koehler, *Mol. Div.* **5**, 13 (2000).
72. P. Wilett, *Pers. Drug Disc. Design* **7–8**, 1 (1997).
73. R. A. Jarvis, E. A. Patrick, *IEEE Trans. Comput.*, *C-22*, 1025 (1973).
74. D. Schnur, *J. Chem. Inf. Comp. Sci.* **39**, 36 (1999).
75. P. R. Menard, J. S. Mason, I. Morize, and S. Bauerschmidt, *J. Chem. Inf. Comp. Sci.* **38**, 1204 (1998).
76. http://www.scitegic.com.
77. E. J. Martin and R. E. Critchlow, *J. Comb. Chem.* **1**, 32 (1999).
78. A. K. Ghose, A. K. Viswanadhan, and J. J. Wendolowski, *J. Comb. Chem.* **1**, 55 (1999).
79. W. P. Ajay Walters, and M. A. Murcko, *J. Med. Chem.* **41**, 3314 (1998).
80. J. Sadowski and H. A. Kubinyi, *J. Med. Chem.* **41**, 3325 (1998).
81. R. T. Koehler, S. L. Dixon, and H. O. Villar, *J. Med. Chem.* **42**, 4695 (1999).
82. C. A. Lipinski, F. Lombardo, B. W. Dominy, and P. J. Feeny, *Adv. Drug Delivery Res.*, **23**, 3 (1997).
83. R. A. Lipper, *Modern Drug Disc.* **2**, 55 (1999).
84. T. I. Oprea and J. Gottfries, *J. Mol. Graph. Model.* **17**, 26 (1999).
85. T. I. Oprea, *J. Comput-Aided Mol. Design* **14**, 251 (2000).
86. F. Yoshida and J. G. Topliss, *J. Med. Chem.* **43**, 2575 (2000).
87. P. Sternberg, K. Luthman, and P. Artursson, *J. Control. Release* **65**, 231 (2000).
88. D. E. Clark, *Comb. Chem. High Throughput Screen.* **4**, 477 (2001).
89. http://www.schroedinger.com.
90. http://www.acdlabs.com.
91. http://www.lionbioscience.com/solutions/idea.
92. http://www.camitro.com.
93. http://www.compudrug.com.
94. G. H. Loew and D. L. Harris, *Chem. Rev.* **100**, 407 (2000).
95. P. R. Chaturvedi, C. J. Decker, and A. Odinecs, *Curr. Op. Chem. Biol.* **5**, 452 (2001).

96. D. F. V. Lewis, **3**, 173 (1993).
97. K. Enslein, V. K. Gombar, and B. J. Black, *Mutat. Res.* **305**, 47 (1994).
98. http://lhasa.harvard.edu.
99. G. Klopman, *Quant. Struct.-Act. Rel.* **11**, 176 (1992).
100. http://www.multicase.com.
101. http://www.cas.org.
102. http://www.scivision.com.
103. http://www.pharmacopeia.com/corp/IR/inv_day_2001/Accelrys.pdf.
104. R. Wedin, *Mod. Drug Disc.* **2**, 39 (1999).
105. http://www.spotfire.com.

HUGO O. VILLAR
Triad Therapeutics

CHEMOMETRICS

1. Introduction

Chemometrics (1–3) or more general multivariate regression methods (4,5) are applied in many research fields from social science to measurement techniques. There are two competing and equivalent nomenclature systems encountered in the chemometrics literature. The first, derived from the statistical literature, describes instrumentation and data in terms of "ways" that an analysis is performed. Here a "way" is constituted by each independent and nontrivial factor that is manipulated with the data collection system. Multiway techniques (the section Multiway Analysis) have been investigated and applied to hyphenated measurement techniques (6). For example, with excitation–emission matrix fluorescence spectra, three-way data are formed by manipulating the excitation-way, emission-way, and the sample-way. Implicit in this definition is a fully blocked experimental design where the collected data forms a cube with no missing values. Equivalently, a second nomenclature is derived from the mathematical literature where data are often referred to in terms of "orders". In tensor notation (7) a scalar is a zero-order tensor, a vector is first order, a matrix is second order, a cube is third order, etc. Hence, the collection of excitation–emission matrix fluorescence data would form a third-order tensor. However, it should be mentioned that the "way" and "order" based nomenclature are not directly interchangeable. By convention, "order" notation is based on the structure of the data collected from each sample. Analysis of collected excitation-emission fluorescence, forming a second-order tensor of data per sample, is referred to as second-order analysis compared to three-way analysis. In this work the "way" based notation will be adopted.

Although there is a vast area of chemometric applications, analytical chemistry has been chosen to exemplify the principles. Especially optical measurement techniques are usually multivariate and are appropriate for a descriptive discussion of chemometrics. The first question that arises when introducing chemometrics is What is chemometrics? Simply put, chemometrics is the application of mathematical and statistical methods to the analysis of chemical data. However, it should be stressed that chemometrics is more than a subdiscipline of mathematics or statistics. The key to artfully practicing chemometrics is to extend the limitations of classical mathematics and statistics by understanding, and relying upon, the constraints that chemistry places on possible solutions to a statistically posed question. As Wold noted the impact of chemometrics is in problem solving not data analysis: chemometricians "must remain chemists and adapt statistics to chemistry instead of vice versa (8)." Along these lines Booksh and Kowalski (9) and Brown (10) define chemometrics more as an information science that can be applied to many physical science disciplines. Chemometrics is a truly interdisciplinary science that does draw from mathematics, statistics, and information science; however, the tools from these disciplines cannot be directly applied without sound knowledge and understanding of the chemical system in question. Many statistical tools are useless to chemometricians because the underlying assumptions are violated in the chemical system. Concurrently, a chemometrician could develop very useful 'statistical tools' that cannot be generalized beyond the chemical system in question. Furthermore, the distinction between "statistical significance" and "practical significance" cannot be reliably made without an understanding of both statistics and chemistry. A broad overview of chemometric techniques—without going into depth—has been published recently (11).

From a chemometric standpoint, data and instrumentation can be classified based on the dimensionality of the data set obtained. Instrumentation can be designed to generate a single datum of information per sample analyzed, an ordered vector of data per sample analyzed, or a linked matrix of data per sample analyzed. In general, the higher the dimensionality or number of ways of the data set, the more powerful the instrument. And, consequently, more powerful data analysis methods can be applied to higher directional data set. The different ways of data are presented in Figure 1. A more complete discussion on the interrelationship between data structure and analyzability can be found in references 9 and 12.

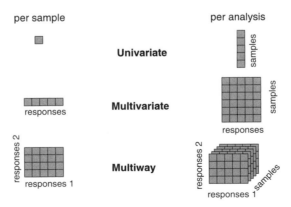

Fig. 1.   Matrix representation of data structure for three classes of data.

The most basic type of data is univariate data. Examples of univariate are data collected from a pH meter or single-channel photometer. Univariate data is the lowest dimensionality of data—a univariate instrument returns a zero-dimensional (zero-order) data tensor. Consequently, a collection of data from a univariate sensor forms a vector and is said to be one-way data. That is it varies in only one way: sample-to-sample.

The majority of literature in chemometrics addresses analysis of multivariate data. Examples of multivariate data include chromatograms or spectra. Analysis of a single sample with a multivariate instrument yields a one direction (first-order) data tensor or vector. A collection of samples forms a two directional matrix and is said to be two-way data because it varies from sample to sample and wavelength to wavelength.

Multiway data are formed, eg, by hyphenated instrumentation such as gas chromatography–mass spectrometry (gc–ms) and excitation–emission matrix spectrometers. Analysis of a single sample yields a two-dimensional (2D)(second-order) tensor (matrix) of data. The key to having true multiway data is that one instrument (or order) must modulate the other instrument (or order). For example ultraviolet–visible–infrared (uv–vis–ir) is not multiway data because the uv–vis and ir spectra of a molecule do not modulate each other. A collection of sensor readings during the progression of a batch process form multiway data (sensors by time). There is no upper limit on the number of "ways" that could go into a data set. Conceivably, one could employ an online high performence liquid chrometography (HPLC)–uv–vis spectrometer to monitor a series of batch processes. A four-way data tensor results: wavelength by chromatographic retention time by time in the batch by batch.

2. Linear Regression Analysis

2.1. Notation and Fundamental Mathematical Tools. *Orthogonal matrices* have the following property: $\mathbf{U}^{-1}{}_{(K \times K)} = \mathbf{U}^{\mathrm{T}}$ or $\mathbf{U} \cdot \mathbf{U}^{\mathrm{T}} = \mathbf{1}_{(K \times K)}$. For rectangular $\mathbf{U}_{(N \times K)}$ matrices consisting of orthonormal rows or columns, a similar property holds:

x	Scalar—upper case italics represent fixed values, ie, J samples; lower case italics represent variables, ie, the jth sample.
$\mathbf{x}$	Column vector.
$\|\mathbf{x}\|_2$	2-Norm of a vector, ie, it's Euclidian length.
$\mathbf{X}_{(N \times K)}$	Matrix with N rows and K columns.
$\mathbf{x}^{\mathrm{T}}, \mathbf{X}^{\mathrm{T}}$	Transposed vector (row vector), transposed matrix.
$\mathbf{X}^{-1}, \mathbf{X}^{+}$	Inverse matrix (if existent), pseudoinverse or Moore-Penrose pseudoinverse (4,13) for pseudoinverting rank-deficient or rectangular matrices (see below).
$\mathbf{X}$	Third-order tensors—a character with a subscript (or subscripts) is assumed to be the appropriate elements from a higher dimensional data matrix. For example, $\mathbf{X}_k$ is the kth slice of the tensor $\mathbf{X}$.
$\hat{x}$	Estimate of the true value $\mathbf{x}$.

$$\mathbf{U} \cdot \mathbf{U}^{\mathrm{T}} = \mathbf{1}_{(N \times N)} \quad \text{and} \quad \mathbf{U}^{\mathrm{T}} \cdot \mathbf{U} = \mathbf{1}_{(K \times K)} \tag{1}$$

A *singular value decomposition* (SVD) (13,14) of a matrix

$$\mathbf{X}_{(K \times N)} = \mathbf{U}_{(K \times K)} \cdot \mathbf{S}_{(K \times K)} \mathbf{P}^{\mathbf{T}}_{(K \times N)} \tag{2}$$

factorizes an arbitrary matrix $\mathbf{X}$ into two orthogonal matrices $\mathbf{U}$ and $\mathbf{P}$ as well as a diagonal matrix $\mathbf{S} = \mathrm{diag}(s_1 \ldots s_K)$. Usually, the singular values s_k are order decreasingly; the columns of $\mathbf{U}$ and the rows of $\mathbf{P}^{\mathbf{T}}$ are ordered accordingly. If $\mathbf{X}$ is rank-deficient, ie, the rank is $R < \min(N,K)$, only R singular values are unequal to zero $s_R < \min(N,K) \neq 0$. Hence, the matrices on the right-hand side of (eq. 2) can be a downsized without loss of information to

$$\mathbf{X}_{(K \times N)} = \mathbf{U}_{(K \times R)} \cdot \mathbf{S}_{(R \times R)} \cdot \mathbf{P}^{\mathbf{T}}_{(R \times N)} \tag{3}$$

A *matrix inversion* based on SVD is known to be numerically very stable (14). A full-rank square matrix is inverted utilizing the properties of orthogonal matrices:

$$\mathbf{X}^{-1}_{(K \times K)} = \left[\mathbf{U}_{(K \times K)} \cdot \mathbf{S}_{(K \times K)} \cdot \mathbf{P}^{\mathbf{T}}_{(K \times K)} \right]^{-1} = \mathbf{P} \cdot \mathbf{S}^{-1} \cdot \mathbf{U}^{\mathbf{T}} \tag{4}$$

For rectangular or singular $\mathbf{X}$ the *Moore-Penrose pseudoinverse* (4,13) has been defined. It combines the ideas of equations (3 and 4): A SVD of X (eq. 2) is calculated followed by downsizing of the three matrices to R, the number of nonzero singular values (eq. 3). This enables the inversion of S, then equation 4 is formally applied

$$\mathbf{X}^{+}_{(N \times K)} = \left[\mathbf{U}_{(K \times R)} \cdot \mathbf{S}_{(R \times R)} \cdot \mathbf{P}^{\mathbf{T}}_{(R \times K)} \right]^{-1} = \mathbf{P} \cdot \mathbf{S}^{-1} \cdot \mathbf{U}^{\mathbf{T}} \tag{5}$$

2.2. Univariate Regression. Two different types of variables are used in regression analysis: predictor or x variables and response or y variables (4). The x variables can be observed but not controlled; values of the y variables are determined by the values of the corresponding x variables. The simplest system is a linear univariate system, which relates the predictor variable x via a proportionality constant a directly to the response variable y, the target figure:

$$y = a \cdot x \tag{6}$$

In spectroscopy, eg, the predictor variable would be the absorption A of a sample at a certain wavelength, the concentration c of a certain chemical in the sample plays the role of the response variable. According to Beer's law, the absorption

$$A = (L \cdot e) \cdot c \tag{7}$$

is directly proportional to the concentration. In this case, the proportionality constant is the inverted product of absorption path length L and the chemical's molar extinction coefficient e at the considered wavelength. Since A is the measur and acquired to determine c the following equation is the analogue to eq. 6:

$$c = \frac{1}{L \cdot e} \cdot A$$

In order to incorporate a constant offset, eg, a constant background absorption A_0, the model (eq. 6) is extended to

$$y = a_0 + a_1 \cdot x \qquad (8)$$

or, respectively,

$$c = A_0 + \frac{1}{L \cdot e} \cdot A \qquad (9)$$

A model is *linear* in the regression sense, if it is linear in the model parameters a_0 and a_1. A polynomial model

$$y = a_0 + a_1 \cdot x + a_2 \cdot x^2 + \cdots + a_q \cdot x^q \qquad (10)$$

is also linear in this meaning; $y = a_0 \cdot \exp(-a_1 \cdot x)$, eg, is not linear.

The model parameters a_0 and a_1 (eq. 8) are usually not known and must be determined experimentally by means of a calibration. One would prepare, eg, two samples with known concentrations $c_1{}^{\text{cal}}$ and $c_2{}^{\text{cal}}$ of the target analyte and would measure the absorbances $A_1{}^{\text{cal}}$ and $A_2{}^{\text{cal}}$ of both calibration samples at the chosen wavelength:

$$y_1^{\text{cal}} = a_0 + a_1 \cdot x_1^{\text{cal}} \quad \text{or} \quad c_1^{\text{cal}} = A_0 + \frac{1}{L \cdot e} \cdot A_1^{\text{cal}}$$

$$\qquad (11)$$

$$y_2^{\text{cal}} = a_0 + a_1 \cdot x_2^{\text{cal}} \quad \text{or} \quad c_2^{\text{cal}} = A_0 + \frac{1}{L \cdot e} \cdot A_2^{\text{cal}}$$

These two equations set up an equation system with two unknowns $a_0 = A_0$ and $a_1 = 1/L \cdot e$, which can be solved. Now, unknown samples can be analyzed by measuring their absorption $x_{\text{meas}} = A_{\text{meas}}$. a_0 and a_1 are used then in equations 8 and 9, respectively, for determining concentration $y_{\text{meas}} = c_{\text{meas}}$. Unfortunately, there are always measurement errors ε disturbing the true model (eq. 8). Instead of the undisturbed model (eq. 8) one has to deal with:

$$y = a_0 + a_1 \cdot x + \epsilon \qquad (12)$$

Since every measurement is affected by a different and unpredictable error ε including $K > 2$ calibration samples

$$(x_1^{\text{cal}}, y_1^{\text{cal}}), \ldots, (x_K^{\text{cal}}, y_K^{\text{cal}}) \qquad (13)$$

does not solve this problem—there are always more unknowns than equations. Hence, a_0 and a_1 cannot be derived from the correct model (eq. 8) by solving a set of calibration equations (eq. 11).

To overcome this problem and to get a workable solution, one has to accept errors in the model parameters and estimate them by means of a least-squares fit. The estimated parameters are denoted by $\hat{a}_0$ and $\hat{a}_1$. As will be discussed below, $\hat{a}_0$ and $\hat{a}_1$ will be determined from error affected calibration data in such a way, that they are a good compromise explaining the calibration set (eq. 13) as accurate as possible. The conditions for good estimates will be discussed in more detail in the section Statistical Background of Regression Analysis. The idea behind linear least-squares regression is to determine estimators $\hat{a}_0$ und $\hat{a}_1$ from a calibration set (eq. 13) such that the sum of squared errors S is minimized:

$$S = \sum_{i=1}^{K} \epsilon_i^2 = \sum_{i=1}^{K} (y_i^{cal} - [\hat{a}_0 + \hat{a}_1 \cdot x_i^{cal}])^2 \qquad (14)$$

In order to derive the minimum of $S(\hat{a}_0, \hat{a}_1)$, partial derivatives of equation 14 with respect to $\hat{a}_0$ und $\hat{a}_1$ are calculated and set to zero. In theory, this results in an extremum, which could also be a maximum, however, $\hat{a}_0$ und $\hat{a}_1$ can be chosen so out of the way that S can reach basically any value. For all practical purposes a minimum of S is obtained.

$$\frac{\partial S}{\partial \hat{a}_0} = 2 \cdot \sum_{i=1}^{K} (-1)(y_i^{cal} - [\hat{a}_0 + \hat{a}_1 \cdot x_i^{cal}]) = 0$$

$$\frac{\partial S}{\partial \hat{a}_1} = 2 \cdot \sum_{i=1}^{K} (-x_i^{cal}) \cdot (y_i^{cal} - [\hat{a}_0 + \hat{a}_1 \cdot x_i^{cal}]) = 0 \qquad (15)$$

Die addends containing the response variable y are transferred to the right side of the equation system (eq. 15). Also the matrix notation is used from here on

$$\begin{pmatrix} \sum_{i=1}^{K} 1 & \sum_{i=1}^{K} x_i^{cal} \\ \sum_{i=1}^{K} x_i^{cal} & \sum_{i=1}^{K} x_i^{cal} \cdot x_i^{cal} \end{pmatrix} \cdot \begin{pmatrix} \hat{a}_0 \\ \hat{a}_1 \end{pmatrix} = \begin{pmatrix} \sum_{i=1}^{K} 1 \cdot y_i^{cal} \\ \sum_{i=1}^{K} x_i^{cal} \cdot y_i^{cal} \end{pmatrix} \qquad (16)$$

Equation system 16 is solved resulting is the estimates $\hat{a}_0$ and $\hat{a}_1$ of the true model parameters a_0 and a_1. Now, $\hat{a}_0$ and $\hat{a}_1$ are used in equation 8 instead of a_0 and a_1 for evaluating unknown samples [see paragraph after equation 11]:

$$y_{meas} = \hat{a}_0 + \hat{a}_1 \cdot x_{meas} \qquad (17)$$

This procedure can easily be extended to handle polynomial models (eq. 10).

   2.3. Figures of Merit for Univariate Chemical Analysis. One essential task in chemometrics is comparing quantitatively the performance of multiple types of chemical sensors or of multiple options for calibration models. In order to objectively perform these comparisons, it is useful to have quantifiable criteria on which discussions are based. Examples of such quantifiable criteria include speed, cost, reliability, precision, sensitivity, selectivity, and detection limit of analysis (15). While speed, cost, and reliability weigh heavily in pragmatic decisions of which instrumental technique to employ for a particular application, these figures of merit are not intrinsic to a given instrumental method. For example, the cost and speed of analysis largely depends on the number of samples to be analyzed; with large quantities of samples economic and time savings can be achieved through bulk purchasing and automation. However, the final four figures of merit are intrinsic to the application of an instrumental method and are directly related to the instrumental response for a particular set of analytes.

The selectivity is the fraction of the instrumental signal that is unique to the analyte. Assuming the instrument is "zeroed" to remove any baseline,

$$\text{SEL} = \frac{r_a}{r} \tag{18}$$

where r_a is the instrumental signal of just the analyte and r is the instrumental signal of the sample. The SEL is a value between 0 and 1 with SEL $= 1$ being a fully selective sensor. For univariate calibration, an instrument must be fully selective. Otherwise, a bias will be imbedded in the prediction of future samples. There is no way, based only on statistical analysis of collected data, to determine the contribution or existence of nonselective interferents in any given 'unknown' sample.

The sensitivity is the change in instrumental response r with respect to changes in analyte concentration

$$\text{SEN} = \frac{\partial r}{\partial c}$$

For univariate linear calibration, this is the slope of the calibration curve, ie, the constant a_1 in equation 8. The precision of a method is best expressed in the signal to noise ratio (S/N),

$$\text{S/N} = \frac{r_a}{e}$$

where e is a measure of the reproducibility of replicated measurements. In many cases, the measurement reproducibility is not concentration dependant, eg, if thermal noise limited analyses. In this case S/N will not vary with analyte concentration.

The limit of detection (LOD) is defined by the International Union of Pure and Applied Chemists and the American Chemical Society to be the smallest amount of a chemical that can be reasonably detected by a given analytical method (16). Of the many ways to calculate the LOD, determine $r_a = y_{meas} = c_{meas}$ for given a signal, x_{meas} (eq. 17), equal to three standard deviations of replicated instrumental blanks is the most straightforward. Assuming the calibration model is linear, univariate, and free of instrumental offset [$a_0 = 0$ (eq. 8)], the detection limit can be expressed as

$$\text{LOD} = \frac{3 \cdot e}{a_1} = \frac{3 \cdot e}{\text{SEN}} \tag{19}$$

where a_1 (eq. 8) is the slope of the calibration (16).

2.4. Multivariate Linear Regression (MLR). *Calibration.* If the response variable y is linearly dependent on several predictor variables the univariate approach (see section Univariate Regression) has to be extended to the multivariate method named multivariate least-squares regression or multilinear regression (4,17). In the following, only two predictor variables $x_{(1)}$ and $x_{(2)}$ are used to keep the discussion concise, but the concept is easily extended to more x variables. Equation 12 is replaced in a multivariate case by

$$y = a_0 + a_1 \cdot x_{(1)} + a_2 \cdot x_{(2)} + \epsilon \tag{20}$$

In case of two predictor variables, a fit plane is determined by the calibration—in higher dimensional applications a hyperplane is obtained.

As an example, the concentration of a chemical might be additionally temperature dependent, ie, an experimenter has to measure the absorbance $A = x_{(1)}$ at a certain wavelength and the temperature $T = x_{(2)}$ of a sample. From these two reading he/she can calculate the concentration $c = y$. For this purpose, three model parameters have to be estimated: $a_0 = A_0$ the background absorption, $a_1 = 1/L \cdot e$ [see discussion after equation 11], and a temperature coefficient $a_2 = 1/\tau$. For this estimation $K \geq 3$ calibration samples $(x^{\mathrm{cal}}_{(1)1}, x^{\mathrm{cal}}_{(2)1}, y^{\mathrm{cal}}_1), \ldots,$ $(x^{\mathrm{cal}}_{(1)K}, x^{\mathrm{cal}}_{(2)K}, y^{\mathrm{cal}}_K)$ have to be provided. They are obtained in this example from measuring the absorption A and the temperature T of samples with known chemical concentration.

To estimate the three model parameters in

$$y^{\mathrm{cal}} = \hat{a}_0 + \hat{a}_1 \cdot x^{\mathrm{cal}}_{(1)} + \hat{a}_2 \cdot x^{\mathrm{cal}}_{(2)} \tag{21}$$

the sum of squared errors [cf. equation 14]

$$S = \sum_{i=1}^{K} \epsilon_i^2 = \sum_{i=1}^{K} \left(y^{\mathrm{cal}}_i - \left[\hat{a}_0 + \hat{a}_1 \cdot x^{\mathrm{cal}}_{(1)i} + \hat{a}_2 \cdot x^{\mathrm{cal}}_{(2)i} \right] \right)^2$$

is minimized. For this purpose the three partial derivatives are calculated and set to zero equivalent to the univariate case (see section Univariate Regression). This results in the following equation system:

$$\begin{pmatrix} \sum_{i=1}^{K} 1 & \sum_{i=1}^{K} x^{\mathrm{cal}}_{(1)i} & \sum_{i=1}^{K} x^{\mathrm{cal}}_{(2)i} \\ \sum_{i=1}^{K} x^{\mathrm{cal}}_{(1)i} & \sum_{i=1}^{K} x^{\mathrm{cal}}_{(1)i} \cdot x^{\mathrm{cal}}_{(1)i} & \sum_{i=1}^{K} x^{\mathrm{cal}}_{(1)i} \cdot x^{\mathrm{cal}}_{(2)i} \\ \sum_{i=1}^{K} x^{\mathrm{cal}}_{(2)i} & \sum_{i=1}^{K} x^{\mathrm{cal}}_{(2)i} \cdot x^{\mathrm{cal}}_{(1)i} & \sum_{i=1}^{K} x^{\mathrm{cal}}_{(2)i}(2)^{\mathrm{cal}}_i \end{pmatrix} \cdot \begin{pmatrix} \hat{a}_0 \\ \hat{a}_1 \\ \hat{a}_2 \end{pmatrix} = \begin{pmatrix} \sum_{i=1}^{K} 1 \cdot y^{\mathrm{cal}}_i \\ \sum_{i=1}^{K} x^{\mathrm{cal}}_{(1)i} \cdot y^{\mathrm{cal}}_i \\ \sum_{i=1}^{K} x^{\mathrm{cal}}_{(2)i} \cdot y^{\mathrm{cal}}_i \end{pmatrix}$$

$$\tag{22}$$

This equation can be written in a more compact notation allowing easier operations on the data in the remainder. Therefore, the following vectors and matrices are defined

$$\mathbf{y}^{\mathrm{cal}} = (y^{\mathrm{cal}}_1 \cdots y^{\mathrm{cal}}_K)^{\mathrm{T}} \qquad \mathbf{1} = (1 \cdots 1)^{\mathrm{T}} \qquad \mathbf{a} = (a_0\ a_1\ a_2)^{\mathrm{T}}$$

$$\mathbf{x}^{\mathrm{cal}}_{(1)} = \left(x^{\mathrm{cal}}_{(1)t} \cdots x^{\mathrm{cal}}_{(1)K} \right)^{\mathrm{T}} \qquad \mathbf{x}^{\mathrm{cal}}_{(2)} = \left(x^{\mathrm{cal}}_{(2)1} \cdots x^{\mathrm{cal}}_{(2)K} \right)^{\mathrm{T}} \qquad \mathbf{X}^{\mathrm{cal}}_{(K \times 3)} = \left[\mathbf{1}\ \mathbf{x}^{\mathrm{cal}}_{(1)}\ \mathbf{x}^{\mathrm{cal}}_{(2)} \right]$$

$$\tag{23}$$

By using these definitions, equations 20 and 22 can be rewritten in matrix notation

$$\mathbf{y}^{\text{cal}} = \mathbf{X}^{\text{cal}} \cdot \mathbf{a} + \epsilon \tag{24}$$

$$\mathbf{X}^{\text{T cal}} \cdot \mathbf{X}^{\text{cal}} \cdot \hat{\mathbf{a}} = \mathbf{X}^{\text{T cal}} \cdot \mathbf{y}^{\text{cal}} \tag{25}$$

The estimate of the fit parameters $\hat{\mathbf{a}}$ can finally be determined by multiplying the inverse of the covariance matrix from the left:

$$\hat{\mathbf{a}} = (\mathbf{X}^{\text{T cal}} \cdot \mathbf{X}^{\text{cal}})^{-1} \cdot \mathbf{X}^{\text{T cal}} \cdot \mathbf{y}^{\text{cal}} \tag{26}$$

Equation 25 is known as the normal equation of a least-square problem. If the covariance matrix is singular, equation 26 has to employ the Moore-Penrose pseudoinverse [see the section Notation and Fundamental Mathematical Tools, equation 5 and the following Supplementary Topics section]:

$$\hat{\mathbf{a}} = \mathbf{X}^{\text{+ cal}} \cdot \mathbf{y}^{\text{cal}} \tag{27}$$

Prediction. A response variable $y_{\text{meas}} = c_{\text{meas}}$ of an unknown sample, a concentration value, eg, can be predicted from an unknown data set $\mathbf{x}_{\text{meas}} = (1 \; x_{(1)} = A \; x_{(2)} = T)^{\text{T}}_{\text{meas}}$ comprising in the given example measured values for the absorbance A and temperature T by

$$y_{\text{meas}} = \hat{\mathbf{a}}^{\text{T}} \cdot \mathbf{x}_{\text{meas}} \tag{28}$$

An offset free model $a_0 = 0$ (eq. 20) can always be obtained by mean centering (see also section Data Pretreatment—Mean Centering and Scaling).

For this purpose, the first row of equation 22 is rewritten

$$k \cdot \hat{a}_0 + k \cdot \hat{a}_1 \cdot \bar{x}^{\text{cal}}_{(1)} + k \cdot \hat{a}_2 \cdot \bar{x}^{\text{cal}}_{(2)} = k \cdot \bar{y}^{\text{cal}}$$

The bar on top of the variables indicates mean values. Dividing this equation by k and solving for $\hat{a}_0$ results in

$$\hat{a}_0 = \bar{y}^{\text{cal}} - \hat{a}_1 \cdot \bar{x}^{\text{cal}}_{(1)} - \hat{a}_2 \cdot \bar{x}^{\text{cal}}_{(2)}$$

This equation for $\hat{a}_0$ is used in equation 21

$$y^{\text{cal}} - \bar{y}^{\text{cal}} = \hat{a}_1 \cdot \left(x^{\text{cal}}_{(1)} - \bar{x}^{\text{cal}}_{(1)} \right) + \hat{a}_2 \cdot \left(x^{\text{cal}}_{(2)} - \bar{x}^{\text{cal}}_{(2)} \right)$$

This mean centered model is reduced by one parameter, and hence one degree of freedom. Now, the whole multivariate least-squares procedure described above is performed on predictor and response variable subtracted by their mean values. If mean centering is applied, equation 28 must also incorporate mean centering

$$y_{\text{meas}} = \hat{\mathbf{a}}^{\text{T}} \cdot (\mathbf{x}_{\text{meas}} - \bar{\mathbf{x}}^{\text{cal}}) + \bar{y}^{\text{cal}}$$

For deriving equation 24 only two predictor variables and a bias were assumed. If N predictor variables have to be included into the calibration model, $\mathbf{X}^{\mathrm{cal}}$ must be augmented by additional columns containing the appropriate calibration values, eg, absorbances at several wavelength positions.

If M response variables, concentrations of several chemicals, eg, have to be determined, $\mathbf{y}^{\mathrm{cal}}$ and $\mathbf{a}$ (eq. 24) have to be augmented by one column per response variable:

$$
\begin{aligned}
\lfloor \mathbf{y}_1^{\mathrm{cal}} \cdots \mathbf{y}_M^{\mathrm{cal}} \rfloor &= \mathbf{X}^{\mathrm{cal}} \cdot [\mathbf{a}_1 \cdots \mathbf{a}_M] + \epsilon \\
\mathbf{Y}_{(K \times M)}^{\mathrm{cal}} &= \mathbf{X}_{(K \times N)}^{\mathrm{cal}} \cdot \mathbf{A}_{(N \times M)} + \epsilon
\end{aligned}
\tag{29}
$$

Instead of estimating a model vector $\mathbf{a}$ by means of equation 26, a model matrix $\mathbf{A}$ is estimated by

$$
\hat{\mathbf{A}} = (\mathbf{X}^{\mathrm{cal}^{\mathrm{T}}} \cdot \mathbf{X}^{\mathrm{cal}})^{-1} \cdot \mathbf{X}^{\mathrm{cal}^{\mathrm{T}}} \cdot \mathbf{Y}^{\mathrm{cal}}
\tag{30}
$$

Unknown predictor variables or measurement vectors $\mathbf{x}^{\mathrm{meas}}$ are evaluated then by

$$
\begin{pmatrix} y_1^{\mathrm{meas}} \\ \vdots \\ y_M^{\mathrm{meas}} \end{pmatrix} = \hat{\mathbf{A}}^{\mathrm{T}} \cdot \mathbf{x}^{\mathrm{meas}}
\tag{31}
$$

Supplementary Topics

1. The concept of pseudoinverses (the section Notation and Fundamental Mathematical Tools) is closely related to MLR applications (eq. 27). For the following short discussion, $\mathbf{X}^{\mathrm{cal}}$ is assumed to have rank $R < \min(K, N)$. Hence, the covariance matrix in equation 26 cannot be inverted. In equation 32, the "inversion" of a rectangular matrix $\mathbf{P}$, which consists of orthonormal columns, is performed in the sense of equation 1. The reader has to keep in mind, that the following is done in a descriptive way without being mathematically thorough.

$$
\begin{aligned}
(\mathbf{X}^{\mathrm{T\,cal}} \cdot \mathbf{X}^{\mathrm{cal}})^{-1} \cdot \mathbf{X}^{\mathrm{T\,cal}} &= \left(\mathbf{P} \cdot \mathbf{S} \cdot \underbrace{\mathbf{U}^{\mathrm{T}} \cdot \mathbf{U}}_{=1} \cdot \mathbf{S} \cdot \mathbf{P}^{\mathrm{T}} \right)^{-1} \mathbf{P} \cdot \mathbf{S} \cdot \mathbf{U}^{\mathrm{T}} \\
&= \mathbf{P} \cdot \mathbf{S}^{-2} \cdot \mathbf{P}^{\mathrm{T}} \cdot \mathbf{P} \cdot \mathbf{S} \cdot \mathbf{U}^{\mathrm{T}} \\
&= \mathbf{P} \cdot \mathbf{S}^{-1} \cdot \mathbf{U}^{\mathrm{T}} \\
&= \mathbf{X}^{+\,\mathrm{cal}}
\end{aligned}
\tag{32}
$$

2. The computation of the pseudoinverse $\mathbf{X}^{+\mathrm{cal}}$ in equation 27 involves a SVD of $\mathbf{X}^{\mathrm{cal}}$ (eq. 5) (section Notation and Fundamental Mathematical Tools), which can be computed in a very reliable way. However, the SVD algorithm takes a lot of computation power: The number of executed floating point

operations (flops), ie, additions and multiplications, of a widely used (14) SVD algorithm is given in Ref. 13 to be

$$\text{flops}\left\{\text{SVD}(\mathbf{X}^{\text{cal}}_{(K\times N)})\right\} = 14\cdot K\cdot N^2 + 8\cdot N^3 \tag{33}$$

According to equation 33, it is evident, that it takes fewer flops to decompose the transposed matrix $\mathbf{X}^{\text{cal}^{\text{T}}}$ than $\mathbf{X}^{\text{cal}}$, whenever $N > K$ since

$$\text{flops}\left\{\text{SVD}(\mathbf{X}^{\text{cal T}}_{(N\times K)})\right\} = 14\cdot N\cdot K^2 + 8\cdot K^3 < 14\cdot K\cdot N^2 + 8\cdot N^3$$
$$= \text{flops}\left\{\text{SVD}(\mathbf{X}^{\text{cal}}_{(K\times N)})\right\} \tag{34}$$

After transposing $\mathbf{X}^{\text{cal}^{\text{T}}}$, one obtains the same matrices from the SVD but in reversed order and transposed. Rearranging and transposing them to get the correct pseudoinverse $\mathbf{X}^{+\text{cal}}$ is in almost all cases much faster than decomposing the original matrix $\mathbf{X}^{\text{cal}}$.

3. Another important property of $\hat{\mathbf{a}}$ is that it is determined in such a way that the vectors $\mathbf{X}\cdot\hat{\mathbf{a}}$ and ε, ie, the residuum, are orthogonal to each other

$$(\mathbf{X}\cdot\hat{\mathbf{a}})^{\text{T}}\cdot\epsilon = (\mathbf{X}\cdot\hat{\mathbf{a}})^{\text{T}}\cdot(\mathbf{y} - \mathbf{X}\cdot\hat{\mathbf{a}})$$
$$= \hat{\mathbf{a}}^{\text{T}}\cdot\mathbf{X}^{\text{T}}\cdot(\mathbf{y} - \mathbf{X}\cdot(\mathbf{X}^{\text{T}}\mathbf{X})^{-1}\cdot\mathbf{X}^{\text{T}}\cdot\mathbf{y})$$
$$= \hat{\mathbf{a}}^{\text{T}}\cdot\left(\underbrace{\mathbf{X}^{\text{T}} - \mathbf{X}^{\text{T}}\cdot\mathbf{X}\cdot(\mathbf{X}^{\text{T}}\mathbf{X})^{-1}\cdot\mathbf{X}^{\text{T}}}_{=0}\right)\cdot\mathbf{y}$$
$$= 0$$

This general fact will be of importance for PLS (the section Partial Least Squares).

2.5. Data Pretreatment—Mean Centering and Scaling. The success of multivariate and multiway data analysis often depends on the application of data pretreatment to remove, scale, or standardize the sources of observed variance. The methods described in this section are applicable to univariate, multivariate, and multiway data analysis strategies. Like all tools, the use and power of each preprocessing methods should be understood before it is applied. Pretreating data, if done properly, can bring out desired information. Likewise, pretreating data, if done improperly, can obscure any desired information embedded in the data.

"Mean centering" and "variance scaling" (18) are often performed on multivariate data without much thought to the consequences of these actions. Mean centering removes the average, or mean, response of a given variable or sample. This translates the variance of the data set to be centered about the ordinate axis. Variance scaling normalizes each variable, or sample, such that the data's variance becomes unity. This places the data on a unit sphere. When mean centering and variance scaling are both applied to a collection of data, the data is said to be 'auto scaled.' Auto scaling places the data on a unit sphere centered about the origin of the multivariate space of the data.

There are specific instances when mean centering and variance scaling should and should not be applied to a data set. In general, mean centering aids in interpretation of factor analysis (FA) models and construction of calibrations. By removing the mean of the data set, often one less factor is required for analysis. An exception may occur when the data is collected under 'closure' (19,20). Closure exists when the sum of the variables or concentrations is constrained to equal a preset value. The most common type of closure is seen in mixture analysis when the sum of percent composition of all detectable species is constrained to equal 100%. Other examples may occur when improper experimental designs are employed. When closure exists, mean centering will not always eliminate a factor. In these instances, the errors introduced by estimating the mean of the data set are not offset by the gains associated with a simpler model.

Mean centering is applied by subtracting the mean spectrum of the data set from every spectrum in the data set. For a data set $\mathbf{R}_{(I \times J)}$ of I samples, each of J predictor variables like discrete digitized wavelengths, the mean centered jth wavelength of the ith sample is defined by

$$^{mc}R_{i,j} = R_{i,j} - \left(\sum_{j=1}^{J} R_{i,j} / J \right) \tag{35}$$

In a multivariate sense, this preprocessing method translates the collection of data to the origin of the multivariate space, where analysis will be performed. The practical consequence of mean centering data is often a more simple and interpretable regression model. In effect, mean centering removes the need for an intercept from the regression model. Consequently, since fewer terms in the regression model may need to be estimated, estimated analyte concentrations may be more precise following mean centering the data. It should be noted that mean centering does not always yield the most precise calibration model. Each calibration method should be tested on mean centered and nonmean centered data.

The effect of mean centering is demonstrated in Figure 2. Figure 2**a** presents raw NIR spectra of the 40 cornflour samples, while Figure 2**b** presents the mean-centered spectra. Although the spectra do not appear to be visually interpretable, none of the variance within the data set has been altered. The major effect of mean centering is removing the broad sloping background from the data collection. The effect of mean centering on principal component based models is shown in the cartoons of Figure 2**c** and **d**. The data cloud in the upper right corner of Figure 2**c** is translated to the origin of the J dimensional space. The arrows of Figure 2**c** and **d** present the direction of greatest variance from the origin. For the nonmean centered data, the direction of greatest variance is the mean of the spectra. With mean centered data, the direction of greatest variance is now the direction of greatest variance *within* the data set. Consequently, more of the information content of a data set can usually be described with a simpler model if the data is mean centered.

When a data set is variance scaled all variables, or samples, are given equal weight in determining the factors of the model. This may be beneficial when variables with small variance have greater predictive variance than variables with larger variance. A prime example is seen in fusing data measurements with

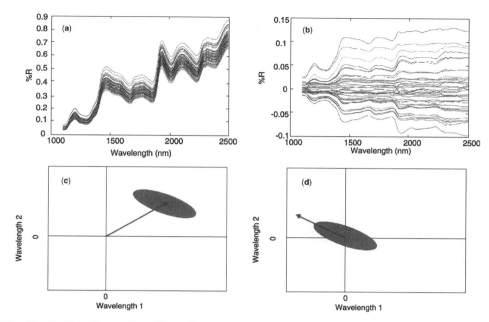

Fig. 2. In this figure the effect of mean centering the data set can be seen. (**a**) The raw data for 40 NIR spectra from corn flour samples. (**b**) The same spectra after mean centering. (**c**) Pictorial representation of a data cloud in two-dimensional space without mean centering. The arrow represents the first PC. (**d**) Pictorial representation of the same data cloud after mean centering; note that the data cloud becomes centered on the ordinate axis. The arrow represents the first PC.

drastically different scales (ie, physical measurements like temperature and pressure with spectroscopic data). However, in most spectroscopic, chromatographic, or electrochemical analyses, the measurement is chosen such to be most sensitive to the analyte of interest. Here, it would not be favorable to give equal weight to background noise in uninformative measurements as is given to measurements with maximum analyte sensitivity.

Variance scaling is applied to the jth wavelength of every spectrum by division of the standard deviation of the jth wavelength over all spectra in the calibration set. Thus, by variance scaling, the impact each variable has in determining the parameters of the calibration model is equalized. Variance scaling is best employed when the variance of a particular wavelength has no correlation to the useful information content of that particular wavelength. Variance scaled data gives equal weight to all wavelengths, regardless of whether they represent a vibrational overtone, scattering, or just baseline noise. Consequently, variance scaling is seldom beneficial for spectroscopic calibration. However, in instances where the analytically useful signal is very weak compared to other signals, variable scaling can be essential.

The effect of variance scaling and auto scaling are shown in Figure 3 for the cornflour spectra. Variance scaling of the corn flour spectra have little superficial effect seen in Figure 3a. However, a close inspection would show that the spread of data at each wavelength is much more uniform across the spectra. This is more

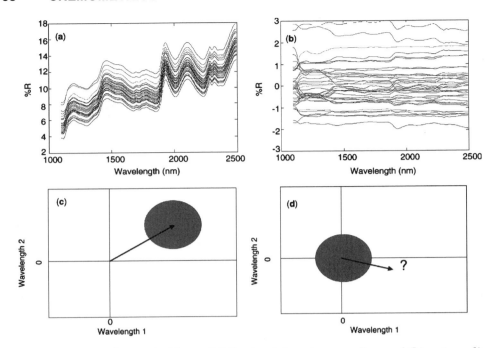

Fig. 3. The same data as in Figure 2 following (**a**) variance scaling and (**b**) auto scaling. Pictorial representations of a data cloud following (**c**) variance scaling and (**d**) auto scaling. The arrow represents the first PC. For variance scaling the data is transformed to lie on a sphere with the same variance in each direction. For auto scaled data, the sphere is translated to the origin of the coordinate axes.

readily seen in Figure 3**b**, where variance scaling is applied to the mean-centered spectra. Thus auto scaling the data. Figure 3**c** shows that variance scaling transforms the oblong data cloud into a unit circle. The direction of greatest variance from the origin is still the mean spectrum. Auto scaling translates the unit circle to the origin of the data space. The direction of greatest variance is now determined by the internal variance of the data with each wavelength having equal weight (regardless of the original magnitude of internal variance).

A third type of scaling often employed is scaling each variable or sample to unit area. This scaling is successfully applied to samples when matrix or sampling effects alter the measurement efficiency of a method. Examples include sample-to-sample variance due to sample thickness in reflectance spectroscopy and effective pathlength in other optical methods. Unit area normalization obscures the absolute concentrations of analytes but preserves the relative concentration of constituents between and among samples. Therefore, absolute calibration cannot be performed unless the calibration set is constrained by closure once the data is normalized.

2.6. Statistical Background of Regression Analysis. As was discussed in the section Univariate Regression, the true model parameters cannot be determined since the experimental calibration set is affected by measurement errors. Instead of the correct model parameters estimates have to be determined and used for prediction of unknown data sets. The question discussed in this section is: How reliable are these estimates? It has been proven (5), that the expectation values of the least squares estimates are unbiased, ie, $E[\hat{a}_i] = a_i$, if the three Gauss-Markov conditions are fulfilled.

1. ie, the expectation value of all n measurement errors $E[\varepsilon_n] = 0 \; \forall n$ (eq. 12) is zero. By means of this condition, it is guaranteed that the assumed fit model is appropriate for the measured data. This condition prevents, eg, that a parabola is fitted to a cubic relationship between prediction and response variables.

2. $E[\varepsilon^2_n]$ is equal $\forall n$ measurement points ie, the error of the measurement data is independent from the values of predictor variable. This type of error is called homoscedastic. If errors are heteroscedastic, the least-squares fit would be more influenced by large predictor variable values than by small.

3. The errors in different measurements of the predictor variable(s) are uncorrelated $E[\varepsilon_n \cdot \varepsilon_m] = 0 \; \forall n \neq m$, this means in the spectroscopic example that the measurement errors of the absorption at different wavelength positions are uncorrelated.

3. Bilinear Chemometric Methods

The reported successes of multivariate chemical analysis are based on three facts. (1) Most, if not all chemical processes are multivariate in nature. Consequently, to be able to effectively perform in a multivariate world, multivariate data must be collected and analyzed. (2) Even if only a single piece of information is needed from a chemical system it is very difficult to design a sensor that is fully selective to that property of interest. Therefore, to circumvent the lack of fully selective sensors, arrays of partially selective sensors can be constructed that rely on multivariate analysis methods to extract the information of interest. (3) There are inherent advantages associated with the redundancy of data when there are many more variables measured per sample than samples collected.

3.1. Classical Least Squares (CLS) versus Inverse Least Squares (ILS). This discussion on CLS versus ILS approaches is based on reference 21. Again a spectroscopic application was used in this discussion since physical meaningful objects help to understand the methods better. The difference between both techniques lies in the approach, which will be explained by means of the Beer's law (eq. 7). The absorbance spectra are written in a matrix $\mathbf{A}_{cal} = \mathbf{X}^{cal}$ (eq. 23), the concentrations in a matrix $\mathbf{C}_{cal} = \mathbf{Y}^{cal}$ (eq. 29). Replacing both items back to predictor and response variables enables transforming this discussion to applications other than spectroscopy.

The physics oriented CLS approach considers the measured spectra as products of molar extinction coefficients $\mathbf{K}$ (unit spectra) and concentrations $\mathbf{C}_{cal}$. The spectral errors are contained in $\mathbf{E}_A$:

$$\mathbf{A}_{cal} = \mathbf{C}_{cal} \cdot \mathbf{K} + \mathbf{E}_A \tag{36}$$

The CLS calibration step estimates $\mathbf{K}$ by means of a multivariate least-squares procedure equivalent the one presented in the section Multivariate Linear Regression

$$\hat{\mathbf{K}} = (\mathbf{C}_{cal}^{T}\mathbf{C}_{cal})^{-1} \cdot \mathbf{C}_{cal}^{T} \cdot \mathbf{A}_{cal} \tag{37}$$

and evaluates unknown spectra by

$$\hat{\mathbf{c}}_{meas} = (\hat{\mathbf{K}} \cdot \hat{\mathbf{K}}^{T})^{-1} \cdot \hat{\mathbf{K}} \cdot \mathbf{a}_{meas} \tag{38}$$

If $\mathbf{C}_{cal}$ or $\hat{\mathbf{K}}$ are singular the corresponding pseudoinverse $\mathbf{C}^{+}_{cal}$ or $\hat{\mathbf{K}}^{+}$ (eq. 5) (the section Notation and Fundamental Mathematical Tools) has to be used.

ILS uses a less intuitive calibration routine, which follows the introduction into multivariate least-squares fits as given in the section Multivariate Linear Regression:

$$\mathbf{C}_{cal} = \mathbf{A}_{cal} \cdot \mathbf{P}_{(N \times M)} + \mathbf{E}_{C}$$

In this approach, $\mathbf{P}$ contains calibration coefficients, which relate the spectral intensities to concentration of chemicals. The parameter $\mathbf{E}_c$ contains random concentration errors. This regression matrix $\mathbf{P}$ is purely a mathematical construct and has no physical meaning. A calibration step estimates

$$\hat{\mathbf{P}} = (\mathbf{A}_{cal}^{T} \cdot \mathbf{A}_{cal})^{-1} \cdot \mathbf{A}_{cal}^{T} \cdot \mathbf{C}_{cal} \tag{39}$$

which can then be used for predicting unknown samples:

$$\hat{\mathbf{c}}_{meas} = \hat{\mathbf{P}}^{T} \cdot \mathbf{a}_{meas}$$

Both methods have advantages and drawbacks: CLS minimizes spectral errors—ILS minimizes concentration errors. Usually, the spectroscopic data contain more noise then the calibration concentrations, which can be determined by very precise reference methods. Hence, the CLS calibration (eq. 37) is the more appropriate one compared to the ILS calibration (eq. 39) since CLS calibration is based on the precisely known model. The ILS method, however, uses the less precise calibration based on noisier spectral data. Nonetheless, ILS is supposed to be the superior approach for practical applications since it only needs calibration concentrations of the analytes of interest. In order to make CLS a good predictor calibration concentrations of all analytes must be included that are expected during the measurement process. This restriction is severe, especially in process monitoring where usually a huge number of absorbers are involved. For such applications it is unfeasible or even impossible to determine calibration concentrations of all of them. This is emphasized by means of Figure 4: Second derivative uv spectra obtained from gaseous samples containing different concentrations of NH_3, NO, and SO_2 (22) have been analyzed by CLS and PCR—a ILS based approach (see the section Principal Component Analysis and Principal Component Regression). For demonstration purposes of how CLS fails, only NH_3 and NO had been calibrated although SO_2 was contained in the

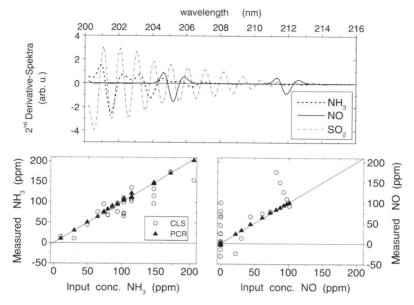

Fig. 4. Comparison of CLS (see section Classical Least Squares) versus PCR (see section Principal Component Analysis (PCA) and Principal Component Regression (PCR)) applied to second derivative spectroscopy in case of incomplete calibration information. Calibration concentrations of gaseous NH_3 and NO have been used—but not of SO_2 (incomplete information), which has also been contained in some calibration samples, though.

calibration samples, too. It is obvious that the CLS calibration cannot handle the SO_2 features strongly overlapping the features of the calibrated analytes. PCR on the other side, is able to calibrate the SO_2 features implicitly and determines correct concentration results. In many applications, eg, spectroscopic estimation of octane number or Reed vapor pressure, no "spectrum" of the extrinsic property would exist. Hence, ILS methods have to been applied.

3.2. Principal Component Analysis (PCA) and Principal Component Regression (PCR).

Factor analysis (FA) is employed to aid in visualization of sample (time) dependent trends and measurement (sensor) dependent trends in a multidimensional data space. In general, factor analysis does not give a physically meaningful model—only correlations among samples and measurements are determined. However, FA methods have been modified to apply constraints and assumptions based on previous knowledge of the chemical system being analyzed. These modified FA methods are useful for determining the underlying instrumental and/or sample (time) profiles of the chemical constituents of a process. Perhaps the most commonly applied method of FA is principal component analysis–regression (PCA–PCR) (1–3,23,24)—an ILS based approach. The PCA method only extracts the principal component (PC) or loading vectors by means of which unknown measurement data will be represented. It takes a second step to relate such an abstract data representation to chemical properties, concentrations, eg. Both steps together are PCR and will be discussed in the following.

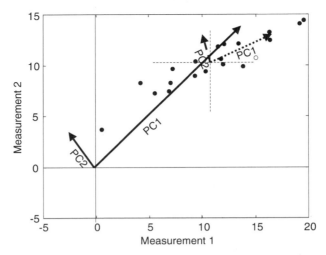

Fig. 5. Graphic representation of principal component analysis. For the data set (red dots) principal components (blue lines) are defined to start at the origin of the coordinate axis system. The first PC describes the main source of variance in the data. For uncentered data, the first PC generally points from the origin of the coordinate axis through the center of the data set; the second PC describes the direction of greatest variance orthogonal to the first PC. For mean centered data, the center of the data set is translated to the origin of the coordinate system; the first PC then describes the direction of greatest variance.

The goal of PCA is to identify the major sources of correlated variance in a collection of data. Once these sources of variance have been identified, they can be exploited to aid in the visualization of the major trends throughout the data collection. The data collection can be reduced from a complicated multidimensional representation to a more easily visualized two- or three dimensional space that describes the majority of the variance (information) in the data collection.

The conceptual idea behind PCA is presented in Figure 5. The largest direction of variance in the data collection is the first PC. The second PC is defined to describe the maximum amount of variance in the data collection while constrained to be orthogonal to the first PC. Consequently, each additional PC is also defined to maximize variance described while constrained to be orthogonal to all preceding PCs. Note that the PCs are defined as vectors originating at the origin of the coordinate space. Therefore, the PCs are dependent on the average value of the data collection; translating the data cloud to a different point in the coordinate space changes the direction of the PCs. For this reason, the data collection is often translated to be centered about the origin of the coordinate space (see the section Data Pretreatment—Mean Centering and Scaling). However, the location of the data collection does not affect the ability of PCA to model the data variance. Only the ease of interpreting the model is affected.

There is a difference between factor analysis and calibration methods: Factor analysis extracts underlying factors or model by means of which the analyzed data can be described—calibration, however, extracts such factors and relates them to chemical or physical properties. A calibration enables a prediction of chemical or physical properties of unknown future samples, factor analysis analyzes the presented (calibration) data only.

Calibration. This discussion of PCR makes wide use of the multivariate least-squares fit concepts (the section Multivariate Linear Regression) and the same notation is used. To exemplify the discussion for spectroscopy, the corresponding spectroscopic items are mentioned in parenthesis. During the calibration process, K calibration samples (spectra) are acquired. The values of N different predictor variables $x_{(1)} \ldots x_{(N)}$ (absorption at different wavelength positions) are measured for each of these K calibration samples. These N predictor variable values are concluded in N calibration vectors $\mathbf{x}_{(1)}^{cal} \ldots \mathbf{x}_{(N)}^{cal}$ (eq. 23) comprising K values each, one for every calibration sample. These calibration vectors define a calibration matrix $\mathbf{X}_{(K \times N+1)}^{cal} = [1 \ \mathbf{x}_{(1)}^{cal} \ldots \mathbf{x}_{(N)}^{cal}]$ (eq. 23). If mean centering was applied, the first column of ones is not needed, ie, $\mathbf{X}_{(K \times N)}^{cal} = [\mathbf{x}_{(1)}^{cal} \ldots \mathbf{x}_{(N)}^{cal}]$.

Calculation of PCs can be accomplished by a singular value decomposition of $\mathbf{X}^{cal}$ (SVD, eq. 2—see the section Notation and Fundamental Mathematical Tools and Fig. 6):

$$\mathbf{X}_{(K \times N)}^{cal} = \mathbf{U}_{(K \times K)} \cdot \mathbf{S}_{(K \times K)} \cdot \mathbf{P}_{(K \times N)}^{T} = \mathbf{T}_{(K \times K)} \cdot \mathbf{P}_{(K \times N)}^{T} \qquad (40)$$

The K orthonormal PCs $\mathbf{p}_k$ each consisting of N loading values are contained in the rows of $\mathbf{P}^{T} = [\mathbf{p}_1 \ldots \mathbf{p}_K]^{T}$. The corresponding scores of the calibration spectra are hold in the columns of $\mathbf{T} = \mathbf{U} \cdot \mathbf{S}$. A scores vector, ie, a column of $\mathbf{T}$, contains K weight factors determining how strong which PC contribute to the corresponding calibration sample (calibration spectrum). Often $\mathbf{P}^{T}$ and $\mathbf{S}$ are multiplied, what is not done here in order to retain orthonormal PCs enabling less costly computations in the following. Calculating the PCs is a PCA, relating the PCs to chemical information extends PCA to PCR.

Due to noise contained in the calibration data $\mathbf{X}^{cal}$, the "chemical" rank of $\mathbf{X}^{cal}$ is usually equal to $\min(K,N)$ although there are only $R < \min(K,N)$ chemical meaningful PCs. The number of linear independent influences on the calibration samples determines the number of chemical meaningful PCs.

Especially for process monitoring applications when only incomplete information about the calibration samples is available, the decision about the true

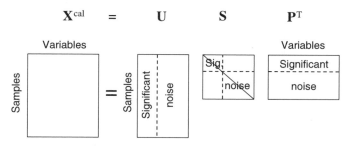

Fig. 6. With PCA the singular value decomposition is often employed to decompose a data matrix $\mathbf{X}^{cal}$ into three submatrices $\mathbf{U}$, $\mathbf{S}$, and $\mathbf{P}^{T}$. These three matrices can be partitioned into columns that describe systemic, often chemical, sources of variance and columns that describe only random measurement related noise.

"chemical" dimension R of the calibration model is rather difficult. If all K PCs would be included, the PCA calibration model would usually be overfitted and hence the evaluation results would be downgraded (25). The parameter R is usually determined by methods presented in the section Model Selection. After finding R, $\mathbf{P}_{(R \times N)}^{\mathrm{T}}$ and $\mathbf{T}_{(K \times R)}$ are downsized to the number of significant PCs without changing the notation in the following. Evaluating an unknown data set $\mathbf{x}_{\mathrm{meas}}$ is a two-step process: In the first step, $\mathbf{x}_{\mathrm{meas}}$ is projected onto the PCs in order to determine its scores vector $\mathbf{t}_{\mathrm{meas}}$. The second step relates these scores to chemical information (concentration of the calibrated analytes). The scores itself have no chemical meaning, however, the wanted pieces of chemical information are linear combination of the scores. This mapping matrix $\mathbf{B}$ from scores to chemical information has to be extracted from the calibration set, too

$$\mathbf{Y}_{(K \times M)}^{\mathrm{cal}} = \mathbf{T}_{(K \times R)} \cdot \mathbf{B}_{(R \times M)} \Rightarrow \hat{\mathbf{B}} = (\mathbf{T}^{\mathrm{T}} \mathbf{T})^{-1} \cdot \mathbf{T}^{\mathrm{T}} \cdot \mathbf{Y}^{\mathrm{cal}} \qquad (41)$$

By comparing equation 41 to eqs. 29 and 30 reveals that $\hat{\mathbf{B}}$ plays the role of $\hat{\mathbf{A}}$ and $\hat{\mathbf{T}}$ the one of $\mathbf{X}^{\mathrm{cal}}$ in this application of multivariate least squares fit. Now the calibration is finalized.

From a different standpoint, these PCs span a R-dimensional subvector space of the N-dimensional vector space of real numbers R^N. The scores are the coordinates of the N-dimensional predictor variable vectors. All features contained in the future unknown predictor vectors will be found in this subvector space. However, the PCs have no physical or chemical meaning—they are linear combinations of all physical or chemical properties present in the calibration samples. The same is true for the score vectors.

Prediction of Unknown Samples.   Based on a PCR calibration (see section Calibration) unknown data sets $\mathbf{x}_{\mathrm{meas}}$ can be evaluated by

$$
\begin{aligned}
\mathbf{x}_{\mathrm{meas}} &= \mathbf{P} \cdot \mathbf{t}_{\mathrm{meas}} + \epsilon \\
\hat{\mathbf{t}}_{\mathrm{meas}} &= \underbrace{(\mathbf{P}^{\mathrm{T}} \mathbf{P})^{-1}}_{=1} \cdot \mathbf{P}^{\mathrm{T}} \cdot \mathbf{x}_{\mathrm{meas}} = \mathbf{P}^{\mathrm{T}} \cdot \mathbf{x}_{\mathrm{meas}} \\
\mathbf{y}_{\mathrm{meas}} &= \mathbf{B}^{\mathrm{T}} \cdot \hat{\mathbf{t}}_{\mathrm{meas}}
\end{aligned}
\qquad (42)
$$

The MLR analogon to the last line of equation 42 is equation 31.

The power of the PCA–PCR approach lies herein: One needs only to know calibration information on the wanted response variables $\mathbf{Y}^{\mathrm{cal}}$ (eq. 29) even if there are plenty of other unknown influences affecting the values of the predictor variables—the algorithm determines by itself an appropriate calibration model, ie, the PCs $\mathbf{P}$ (eq. 40) and a transform matrix $\mathbf{B}$ (eq. 41). However, all influences occurring during the evaluation of unknown data sets $\mathbf{x}_{\mathrm{meas}}$ must also be present during calibration for being implicitly calibrated. In contrast to this, conventional multivariate least squares regression (the section Multivariate Linear Regression) needs the user to include all influences explicitly into an appropriate calibration model $\mathbf{X}^{\mathrm{cal}}$.

An Experimental Example.   Second derivative uv spectroscopy has been used in this example of how PCR can implicitly calibrate imperfect measurement

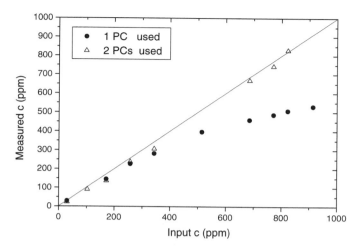

Fig. 7. Concentration errors (solid dots) determined with a one PC model obtained from single compound samples—improved precision (hollow triangles) obtained with a two PC model obtained from the same data set.

data. It was shown in (22) how a linear relation between concentrations and derivative spectra is obtained. As an example gaseous ammonia samples in the concentration range 0–1000 ppm had been prepared and analyzed by means of a PCR. The first calibration set contained two samples (0 and 103 ppm) only which is in theory sufficient for such simple applications. Just one significant PC was found. However, it was found that Beer's law (eq. 7) is not valid over the concentration range aimed at since the measured concentrations are falling short over 300 ppm (Fig. 7).

As is demonstrated by means of Figure 8, this is not a problem of PCR but of the employed measurement technique. Dividing the measured spectra by the concentration values of the sample derives extinction or unit spectra. In Figure 8, three experimentally determined extinction spectra are shown obtained from low, medium, and high concentration samples. In absence of systematic measurement errors, all three would be the same. However, with increasing concentration the peak height is not increasing linearly, furthermore the shape of the extinction spectra is smeared out and the peaks height ratio to each other changes.

In a nutshell, there is more then one influence, ie, concentration, defining the measurement data. Hence, a calibration model including just one predictor variable vector or PC is not appropriate. The CLS (the section Classical Least-Squares versus Inverse Least Squares) would not be able to lift this problem since this algorithm is not flexible enough—it allows only for as many dimensions of the calibration model as analytes have been calibrated. Not so PCR: After increasing the calibration set to three calibration spectra (30, 514, 914 ppm), two significant PCs were found (Fig. 9).

Including both significant PC resulted in clearly improved concentrations (Fig. 7). This second PC enabled an adjustment of the systematic measurement errors. The nonlinear cooperation of the two factors is demonstrated by means of

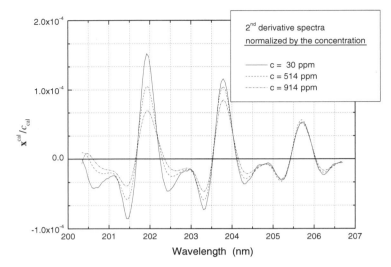

Fig. 8. Comparing three normalized second derivative spectra (extinction spectra) of gaseous NH_3 diluted in N_2 (22), ie, derivative spectra divided by the concentration of the samples. If Beer's law had been applicable, all normalized spectra would be equivalent.

Figure 10, which shows the concentration dependency of the corresponding scores. The scores of the first PC increase with increasing concentration but this increase is slowed with increasing concentration. The second loading is also increasing for concentrations below ~500 ppm due to enhancing the NH_3 peak near 202 nm. Above, its importance is decreasing because of the altering spectrum shape, which is getting more similar to the first factor. At ~800 ppm the second loading is zero, ie, for this concentration the first factor is proportional to the NH_3 derivative spectrum. Above this threshold, the second loading is

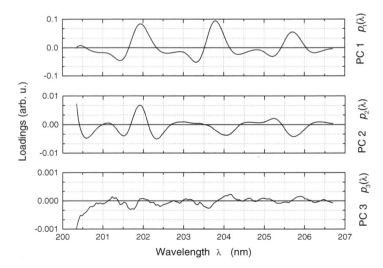

Fig. 9. Two significant PCs and one irrelevant PC obtained from the calibration samples comprising one analyte and systematic measurement errors, ie, nonlinearities (Fig. 8).

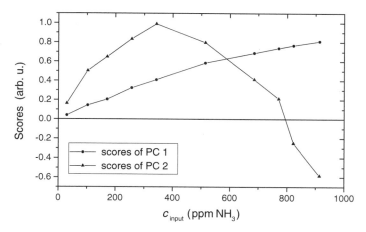

Fig. 10. Nonlinear relationship of the scores of PC 1 and PC 2 (Fig. 9) with increasing concentration.

getting negative, ie, compared to the other two peaks the 202 nm peak is suppressed further by the second factor.

3.3. Partial Least Squares (PLS). Partial least-squares regression (PLS) (21,26–28) has been employed since the early 1980s and is closely related to PCR and MLR. In fact, PLS can be viewed as a compromise midway between PCR (the section Principal Component Analysis and Principal Component Regression) and MLR (29) (the sections Multivariate Linear Regression and Classical Least-Squares versus Inverse Least Squares and 3.1). In determining the decomposition of $\mathbf{X}^{cal}$ and consequently removing unwanted random variance, PCR is not influenced by knowledge of the calibration set's response variables $\mathbf{y}^{cal}$ and $\mathbf{Y}^{cal}$, respectively. Only the variance in $\mathbf{X}^{cal}$ is employed to determine the loading vectors. Conversely, MLR does not factor $\mathbf{X}^{cal}$ prior to regression; all variance correlated to response variables is employed for estimation. PLS determines each loading vector to simultaneously optimize variance described in $\mathbf{X}^{cal}$ and correlation with $\mathbf{y}^{cal}$. The PLS loading vectors are rotations of the PCA PCs for a slightly different optimization criterion. In fact, numerous algorithms exist that are optimized for various sizes of $\mathbf{X}^{cal}$ (30,31).

PLS has two distinct advantages compared to PCR. First, PLS generally provides a more parsimonious model than PCR. The PCR calculates factors in decreasing order of $\mathbf{X}^{cal}$-variance described. Consequently, the first factors calculated, that have the least imbedded errors, are not necessarily most useful for calibration. On the other hand, the first few PLS factors are generally most correlated to concentration. As a result, PLS achieves comparable calibration accuracy with fewer loading vectors in the calibration model. This further results in improved calibration precision because the first factors are less prone to imbedded errors than are lower variance factors.

Second, the PLS algorithm is often faster to implement and optimize for a given application than is the PCR algorithm. The PLS calculates the factors one at a time. Hence, only the loading vectors needed for calibration are determined. The PCR, employing the singular value decomposition, calculates all possible loading vectors for $\mathbf{X}^{cal}$ prior to regression. For large data sets that require relatively few factors for calibration, PLS can be significantly faster than PCR.

PLS extracts iteratively as much variance form $\mathbf{X}^{\text{cal}}$ as it can correlate to the response variable values $\mathbf{y}^{\text{cal}}$. The explained information is subtracted from $\mathbf{X}^{\text{cal}}$ and $\mathbf{y}^{\text{cal}}$—the residuals enter the next iteration step then. Since every iteration operates on the residual of the previous step, the extracted loading vectors are mutually orthogonal to each other (the section Supplementary Topics). The PLS determines a calibration model for every response variable independently—if there are several response variables to be predicted, the algorithm presented in the following has to be run the several times using the $\mathbf{y}^{\text{cal}}$ vectors one after the other.

Calibration.   The iterative PLS calibration algorithm is presented first followed by a step-by-step discussion of it.

The following discussion explains how the hth iteration works:

1. Initialization of the algorithm with the mean centered (see section Data Pretreatment—Mean Centering and Scaling) original data $\mathbf{X}^{\text{cal}}$ and $\mathbf{y}^{\text{cal}}$.

	Model	Least-squares estimate
1. initialization:	$\mathbf{E}_0 = \mathbf{X}_{(K \times N)}{}^{\text{cal}}$ and $\mathbf{e}_0 = \mathbf{y}$ (mean centered, see section Data Pretreatment—Mean Centering and Scaling) $h = 1$ iteration counter	
2. determine a weight loading vector $\mathbf{w}_h$ (new predictor variable vector):	$\mathbf{E}_{h-1} = \mathbf{e}_{h-1} \cdot \mathbf{w}_h^{\mathrm{T}} + \epsilon_{\mathrm{E}}$	$\hat{\mathbf{w}}_h = \mathbf{E}_{h-1}^{\mathrm{T}} \cdot \mathbf{e}_{h-1} \cdot (\mathbf{e}_{h-1}^{\mathrm{T}} \cdot \mathbf{e}_{h-1})^{-1}$
3. normalize $\hat{\mathbf{w}}_h$:	$\hat{\mathbf{w}} = \frac{\hat{\mathbf{w}}}{\|\hat{\mathbf{w}}\|_2}$	
4. determine the corresponding scores vector $\mathbf{t}_h$:	$\mathbf{E}_{h-1} = \mathbf{t}_h \cdot \hat{\mathbf{w}}_h^{\mathrm{T}}$	$\hat{\mathbf{t}}_h = \mathbf{E}_{h-1} \cdot \hat{\mathbf{w}}_h \underbrace{(\hat{\mathbf{w}}_h^{\mathrm{T}} \cdot \hat{\mathbf{w}}_h)^{-1}}_{=1}$ $\hat{\mathbf{t}}_h = \mathbf{E}_{h-1}^{\mathrm{T}} \cdot \hat{\mathbf{w}}_h$
5. relate scores to residues of the response variable $\mathbf{e}_{h-1}$:	$\mathbf{e}_{h-1} = \hat{\mathbf{t}}_h \cdot v_h + \epsilon_e$	$\hat{v}_h = (\hat{\mathbf{t}}_h^{\mathrm{T}} \cdot \hat{\mathbf{t}}_h)^{-1} \cdot \hat{\mathbf{t}}_h^{\mathrm{T}} \cdot \mathbf{e}_{h-1}$
6. determine a loading vector $\mathbf{b}_h$	$\mathbf{E}_{h-1} = \hat{\mathbf{t}}_h \cdot \mathbf{b}_h^{\mathrm{T}} + \epsilon_{\mathrm{E}}$	$\hat{\mathbf{b}}_h = (\hat{\mathbf{t}}_h^{\mathrm{T}} \cdot \hat{\mathbf{t}}_h)^{-1} \cdot \hat{\mathbf{t}}_h^{\mathrm{T}} \cdot \mathbf{E}_{h-1}$
7. determine new residues	$\mathbf{E}_h = \mathbf{E}_{h-1} - \hat{\mathbf{t}}_h \cdot \hat{\mathbf{b}}_h^{\mathrm{T}}$ $\mathbf{e}_h$ $= \mathbf{e}_{h-1} - \hat{v}_h \cdot \hat{\mathbf{t}}_h$	
8. if $\|\mathbf{e}_h\|_2 < \min$, then $R = h$ else $h \to h+1$	END go to step 2	

2. Determine a loading vector $\hat{\mathbf{w}}_h$, which is used to estimate the scores $\hat{\mathbf{t}}_h$. This is done in such a way that as much variance of $\mathbf{X}^{\text{cal}}$ (for $h = 1$) or the remaining residuals $\mathbf{E}_{h-1}$ (for $h > 1$) is extracted as can be explained by the response variable $\mathbf{y}^{\text{cal}}$ (for $h = 1$) or the remaining residuals $\mathbf{e}_{h-1}$ (for $h > 1$). For $h = 1$, PLS is a CLS [(the section Classical Least Squares versus Inverse Least Squares), especially equation 37 for $M = 1$ with $X^{\text{cal}} = A^{\text{cal}}$

and $\mathbf{y}^{cal} = \mathbf{C}^{cal}{}_{(K \times M = 1)}$] and there is actually a physical meaning for $\hat{w}_h$: PLS estimates the pure component—in spectroscopic applications this would be the pure component or more precisely molar extinction spectrum, ie, $\mathbf{e}(\lambda)$ (eq. 7). This function of wavelength $\mathbf{e}(\lambda)$ should not be confused with the residual vector $\mathbf{e}_h$ used in the PLS algorithm above. This estimation, however, will be poor if more than one analyte is contained in the calibration samples. The following iterations extract correction terms.

3. Normalize $\hat{w}_h$ to get an orthonormal basis, this simplifies step 4 and gives every vector the same weight during evaluation.

4. Now the scores $\mathbf{t}_h$ for $\hat{w}_h$ are determined describing how strong $\hat{w}_h$ are present in $\mathbf{E}_{h-1}$. $\hat{w}_1$ estimates the pure component spectrum in the aforementioned spectroscopic application; $\mathbf{t}_1$ is a first order approximation of the calibration concentrations.

5. This step the equivalent to equation 41 in PCR: Relate the scores to chemical meaningful items, concentrations, eg—or more precisely the hth contribution to the concentrations.

6. Now the loading vectors $\hat{w}_h$ have to be "updated" to $\mathbf{b}_h$ incorporating how much of the response variables (concentrations) have actually been explained by the scores. The parameter $\hat{w}_h$ could not be used in step 4 as final loading vector since the scores $\hat{t}_h$ had to be estimated first. The difference between $\mathbf{t}_h$ and $\hat{t}_h$ makes this step and the definition of $\mathbf{b}_h$ necessary.

7. Subtract the extracted and explained information from the residues of the previous step.

8. The algorithm is aborted, if, eg, all information on the calibration's response variable is explained. Otherwise a new iteration is started.

Prediction of Unknown Samples. The prediction algorithm for unknown samples is initialized with the mean-centered predictor and response variables. Then the sample is projected onto the $\hat{w}_h$ loading vectors in order to estimate the scores value $t_h{}^{meas}$. By means of the constant $\hat{v}_h$ relating scores and response variable a back transformation is done from the scores representation to physical meaningful objects, concentrations for instance. Step by step the wanted response variable y is updated to the final value. The prediction algorithm has the same number of iterations as the calibration.

---

1. initialization
$$e_0^{meas} = \mathbf{x}_{meas} - \bar{\mathbf{x}}^{cal} \quad \text{and} \quad y_{meas} = \bar{y}^{cal}$$
$$h = 1 \quad \text{iteration counter}$$

2. project the residues of the unknown predictor variable vector onto the new predictor variable vector:
$$t_h^{meas} = \hat{\mathbf{w}}_h^{T} \cdot e_{h-1}^{meas}$$

3. update the estimate of the response variable:
$$y_h = y_{h-1} + \hat{v}_h \cdot t_h^{meas}$$

4. determine a new residual of the predictor variable vector:
$$e_h = e_{h-1} - \hat{\mathbf{b}}_h \cdot t_h^{meas}$$

5. $h \rightarrow h + 1$
6. if $h \leq R$, go to step 2 else END

3.4. Model Selection. As stated in the sections Principal Component Analysis and Principal Component Regression and Partial Least Squares, the most difficult part of PCR and PLS is to determine the dimension R of the calibration model. The singular values s_K (eq. 2) arranged in decreasing order could be used at least for PCR. As was stated in the section Notation and Fundamental Mathematical Tools, there are R singular values unequal to zero. However, due to the noise no singular value is exactly zero except if mean centering was applied since a degree of freedom was lost. In that case the last one is zero. In most applications the singular values drop to very small values belonging to non-significant PCs with a more or less pronounced step. Plotting these singular values can help to get a first guess at least. But this works only in limited cases and often a more sophisticated method is needed.

Usually cross-validation (1) is applied that excludes one of the calibration samples for determining the number R of significant PCs. This is done by including iteratively more PCs and estimating the values of the response variables. Since the true value(s) y^{cal} or $\mathbf{y}^{cal}$ are known for this excluded calibration samples, the estimated value(s) $\hat{y}^{cal}$ or $\hat{\mathbf{y}}^{cal}$ can be compared to the true values. The procedure of excluding a calibration sample and estimate its response variable(s) is done for all calibration samples successively. The number of PCs achieving the closest estimates is chosen for future evaluation.

If a sufficient quantity of calibration samples is available, the best method for selecting and validating a model is to divide the calibration set into three subsets. One set is employed to construct all of the models to be considered. The second set is employed to choose the best model in terms of accuracy and precision. The third set is employed to estimate the performance of the chosen model on future data. There are three statistics often employed for comparing the performances of multivariate calibration models: root-mean-squared error of calibration (RMSEC), root-mean-squared error of cross-validation (RMSECV), and root-mean-squared error of prediction (RMSEP). All three method are based on the calculated root mean squared error (RMSE)

$$\text{RMSE} = \left(\sum_{k=1}^{K} (y_k^{cal} - \hat{y}_k^{cal})^2 / K \right)^{1/2} \tag{43}$$

where RMSEC, RMSECV, and RMSEP differ in the determination of $\hat{y}^{cal}$. The best estimate of future performance of a calibration model is the RMSEP. Estimates $\hat{y}^{cal}$ in the RMSEP are determined by applying the calibration model to a subset of data that was not employed in determining the model parameters. The RMSEP may be calculated for a 'validation set' in order to determine the optimal number of factors in a model or to a 'test set' in order to test the performance of the optimal model on future data. If an external subset of data is not available to optimize the calibration model, the RMSEP can be estimated by the RMSECV. The concentration estimates of equation 43 are determined in cross-validation. RMSEC is a measure of how well the calibration model fits the calibration set. This is potentially the least informative of the three statistics. The RMSEC is an extremely optimistic estimation of the model performance. In the limit, if every factor were included in the calibration model, the RMSEC would be zero.

Hence, RMSEC is always decreasing with number of factors. As more factors are included in the calibration model, the model begins to fit the random errors imbedded in the spectra and concentrations. Therefore, the RMSEC will always decrease as more factors are added. However, new samples not included in the calibration set will have a different realization of random errors. Therefore, the calibration model will not fit these errors to the same degree as the errors in the calibration set. When extra factors that mostly describe random errors are included in the calibration model, these factors will introduce the errors in future samples and the RMSECV and RMSEP may increase. Therefore, RMSECV is a better estimate of future performance of model prediction than is RMSEC. This so-called overfitting is well described for PCR in reference 25—similar facts apply to PLS.

The performances of the three statistics are evident in Figure 11a–d presenting the RMSEC, RMSECV, and RMSEP versus number of factors for PLS calibration of moisture, oil, protein, and starch, respectively. All spectra were preprocessed by MSC and mean centered. The optimal number of factors may be estimated by statistical tests applied to the RMSE, choosing the first minima in the plot, or choosing the global minimum in the plot (3).

Unfortunately, it is often difficult to obtain reliable calibration samples, which are hence too valuable for testing the calibration model only. Furthermore, the dimension of a calibration model defined cross-validation is fixed and cannot be adjust to certain data. Hence, for spectroscopic applications a fine-tuning

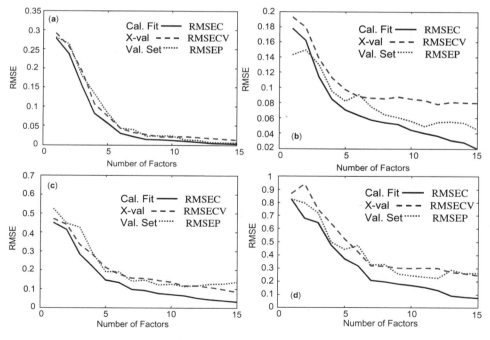

Fig. 11. Plots of RMSEC, RMSECV and RMSEP for prediction of (**a**) moisture, (**b**) oil, (**c**) protein, and (**d**) starch from NIR spectra of cornflour samples.

approach was proposed (32) adjusting the number of used PCs to every measurement spectrum individually. This method can be extended to other multivariate data sets, too. By means of this method no valuable calibration data must be excluded for testing purposes and the calibration model is flexible. In general, there are three types of PCs: Primary PCs are the most important ones modeling the major and true spectroscopic features in a spectrum. The secondary PCs are needed for correcting imperfect measurements like drifts (33) or washed out features (22). Tertiary PCs are due to noise and should not be included into the calibration model to prevent overfitting (25). Starting with one PC the number of PCs is increased stepwise by one. This defines a reduced model. The variance of the residual spectrum obtained at every step of this iteration is F-tested against the variance of the residual spectrum obtained with full model including all PCs. The number of PCs is increased until the F-test cannot find significant differences between the restricted and full model. At that point both methods have the same predictability. By means of this the algorithm reduces overfitting and still extracts all relevant information. By means of synthetic data it was proven that the algorithm selects the correct number of PCs, if noise level is reasonable and if sufficient calibration samples are provided. This fine-tuning of the calibration model could also be applied advantageously to different experimental spectra sets.

3.5. Target Factor Analysis. As mentioned in the section Calibration, the principal components (PCs) derived from PCA do not necessarily describe single, physically meaningful, effects. That is, while a set of data may consist of the NIR spectra of hydrocarbon mixtures, the PCs of the data set are not constrained to be NIR spectra of the constituent hydrocarbons. However, the multivariate space defined by the principal components is the same as the multivariate space defined by the pure (true) spectra of the chemical constituents of the data set plus any other forms of systematic variance. The difference is the basis (see last paragraph in the section Calibration) used for representation: The PCs are rotated versions of the pure component spectra. Target factor analysis (TFA) is a method of testing whether the spectrum of a hypothesized chemical constituent, as defined by an assumed or recorded spectrum, lies in the PC space of the model. If the hypothesized constituent does lie in the PC space, the associated spectrum x_{meas} can be expressed as a linear combination of the PCs [rows of $\mathbf{P}^{T}{}_{(R \times N)}$ (eq. 40)]:

$$x_{meas} = \mathbf{P} \cdot \mathbf{t} \tag{44}$$

The coefficients of the linear combination are the scores $\mathbf{t}$ of $\mathbf{x}_{meas}$. $\mathbf{t}$ can be calculated by regressing, i.e. projecting, the target spectrum, $\mathbf{x}_{meas}$, onto the orthonormal PCs $\mathbf{P}^{T}$:

$$\hat{\mathbf{t}} = (\mathbf{P}^{T} \cdot \mathbf{P})^{-1} \cdot \mathbf{P}^{T} \cdot x_{meas} = \mathbf{P}^{T} \cdot x_{meas}$$
$$\hat{x}_{meas} = \mathbf{P} \cdot \mathbf{P}^{T} \cdot x_{meas} \tag{45}$$

Whether $\mathbf{x}_{meas}$ lies in the vector space spanned by the PCs is tested by comparing $\mathbf{x}_{meas}$ with $\hat{x}_{meas}$. If $\mathbf{x}_{meas}$ and $\hat{x}_{meas}$ are determined to be sufficiently similar by a statistical or empirical test (3), $\mathbf{x}_{meas}$ is an element of the PCs vector space. If $\mathbf{x}_{meas}$ is not a member of this vector space, the regression [first line in equation 45] estimating $\mathbf{t}$ determines a wrong vector $\hat{\mathbf{t}}$. In return, $\hat{\mathbf{x}}_{meas}$ will be significantly different from $\mathbf{x}_{meas}$. Analogous equations can be constructed that project sample targets onto the scores $\mathbf{T}$ (eq. 40) of $\mathbf{X}^{cal}$. Recently, methods had been developed which extracts the parts of $\mathbf{x}_{meas}$ causing it not being an element of the PC vector space (34,35). Such algorithms can be applied to analyze unknown spectra qualitatively for detecting and correcting of uncalibrated interferences.

3.6. Locally Weighted Regression.   The global linear models calculated by PCR or PLS are not always the best strategy for calibration. Global models span the variance of all the samples in the calibration set. If the data are nonlinear, then the linear PCR and PLS methods do not efficiently model the data. This happens for instance, if a linear Beer's law (eq. 7) type relationship between predictor variables (spectra) and response variables (predicted chemical properties) does not hold. One option is to use nonlinear calibration methods employing a global, nonlinear model (the section Nonlinear Methods). The second option is to employ linear calibration methods on small subsets of the data. The locally weighted regression (LWR) philosophy assumes that the data can be efficiently modeled over a short span with linear methods (36–40). The first step in LWR is to determine the Q calibration samples that are most similar with the unknown sample to be analyzed. Similarity can be defined by distance between samples in the spectral space (38), by projections into the principal component space (39), and by employing estimates of the property of interest (40). Once the Q nearest standards are determined, either PLS or PCR is used to calculate the calibration model. The LWR has the advantage of often employing a much simpler and more accurate model for estimation of a particular sample. However, there are three disadvantages associated with LWR. First, two parameters must be optimized for LWR, number of local samples and number of factors, compared to only the latter one for PLS and PCR. Second, a new calibration model must be determined for every new sample analyzed. Third, LWR often requires more samples than PCR or PLS in order to build meaningful, local calibration models.

3.7. Nonlinear Methods.   There are numerous nonlinear, multivariate calibration methods described in the chemometric literature. These methods can be divided into two classes. Alternating Conditional Expectations (ACE) (41,42) and Projection Pursuit (PP) (43) seek to transform the nonlinear data such that a linear calibration model is appropriate. Similarly, Global Linearizing Transformations (GLT) is employed to optimally linearize data prior to factor analysis by PCA (44,45). On the other hand, nonlinear-PLS (NPLS) (46,47). Multivariate Adaptive Regression Splines (MARS) (48,49) and Artificial Neural Networks (ANN) (50) determine nonlinear global models that span the entire range of samples. While impossible to provide sufficient detail for each method in this section, some general comments regarding the application of these nonlinear methods are warranted.

Specific nonlinear methods have been compared and contrasted over a wide variety of linear and nonlinear calibration applications (51–53). No single method has demonstrated systematic superiority to the other methods. The safe conclusion is that calibration method superiority is application dependent (54,55) When the underlying type of nonlinearity implicit in the calibration method matches the latent nonlinearity in the data, the method will optimally model the data. This assertion has been supported by the improvement in calibration performance when theoretical instrument response functions replace the sigmoid transfer function in ANN calibration (56).

Nonlinear methods are much more prone to "over-fitting" the calibration model than linear approaches. Overfitting occurs when the calibration begins to employ random variance (instrumental errors) for determining calibration parameters. The flexibility of the nonlinear models and the relatively large number of parameters that need to be estimated are the primary cause for this phenomenon. Consequently, the more complicated the model, the more prone the method is to overfitting (ie, ANN vs. PCR). A decision tree based on Occums Razor has been proposed to aid chemists in choosing among the nonlinear methods) (54) Linear and nonlinear calibrations were linked in a hierarchical web. The hierarchy is based on nested models and degrees of freedom required to calculate the model. Simple, linear models are at the top of the hierarchy; complex, nonlinear methods are at the bottom. It is recommended that to guard against overfitting and spurious modeling of the data, the method nearest the top of the hierarchy that provides sufficient calibration reliability for the application be employed. That is, use the simplest model that works.

3.8. Multivariate Curve Resolution (MCR).

Where TFA (the section Target Factor Analysis) allows the analysis to test for the presence of a hypothesized constituent, TFA is limited in the ability to estimate the spectral profile of any constituents in the data set. This is due to the fact that TFA requires that the spectral profile of the target is available for target testing. If the profile is unavailable, TFA cannot be performed. On the other hand, multivariate curve resolution (MCR) methods allow for the estimation for both the hypothesized and unknown constituents in the data matrix, ie, spectral or chromatographic profile of the separated constituent as well as concentration profiles. Usually MCR techniques are applied in spectroscopy and chromatography. The rotational ambiguity of the decomposition in (eq. 40) is circumvented by making assumptions regarding the nature of the true constituent spectral profiles and sample profiles. These assumptions are translated into constraints applied to the iterative factorization of $\mathbf{X}^{cal}$. Once additional constraints are applied to the factors of $\mathbf{X}^{cal}$, the factors are not true principal components. These factors are properly described as intrinsic factors, but not PCs.

Numerous constraints have been applied to the iterative factorization of $\mathbf{X}^{cal}$ in order to enhance the probability that the determined factors will be physically meaningful. Perhaps, the most common constraint is nonnegativity of estimated spectral and sample profiles (57–68). This constraint is based on the common sense notion that the factorization of $\mathbf{A} = \mathbf{X}^{cal}$ (eq. 36) (see section Classical Least Squares versus Inverse Least Squares) should lead to positive estimates of extinction coefficients or unit spectra (rows of $\mathbf{K}$) and concentrations $\mathbf{C}_{cal}$. In neither case would the true profile likely contain negative values. Another common spectral constraint employs assumptions regarding the content of $\mathbf{X}^{cal}$. If the spectral profile of one or more of the assumed chemical constituents is known, the factorization of $\mathbf{A} = \mathbf{X}^{cal}$ can be constrained to contain the assumed spectral profiles in the solution. It is also possible to employ assumptions regarding the interrelationship among the samples.

For resolving overlapping chromatographic peaks, Gaussian or unimodal elution profiles are assumed for the rows of $\mathbf{X}^{cal}$ (60,62,63) ie, there is for sure only one maximum in the chromatogram. Concurrently, the presence of samples that contain only one compound may be successfully postulated for chromatographic or kinetic data (62,64). This is referred to as the "uniqueness" constraint. If the concentration of one or more compound is known in any of the particular sample, the resolved profiles can be constrained to reflect this information. For kinetic data, the sample profiles can be constrained to fit a class of differential equations that reflect the postulated reaction pathway (65,66). The validity of the assumed reaction can be tested based of the ability of the data to fit this model.

Of course, application of other constraints and the combination of multiple constraints are possible. The constraints resulting from these assumptions are particularly powerful when well-ordered data, such as kinetics or chromatographic data, are analyzed. Constraining does not ensure that physically meaningful profiles will be determined. In general, application of constraints only reduces the range of feasible solutions where, ideally, the true profiles will lie within this range. The more constraints properly applied to the decomposition, the tighter the estimated range of profiles will resemble the true profiles. However, if a constraint is improperly imposed the estimated profiles will yield erroneous profiles, for instance if nonnegativity when in fact the profile should have negative values.

Practically, iterative MCR methods are capable of resolving spectra and concentrations from complicated, multianalyte mixtures without a priori information aside from constrains. Iterative MCR methods employ the bilinear factorization model

$$\mathbf{A}_{(K \times N)} = \mathbf{C}_{(K \times M)} \cdot \mathbf{P}^{\mathrm{T}}_{(M \times N)} + \mathbf{E}_{(K \times N)} \tag{46}$$

where the K mixture spectra or chromatograms are written in the rows of $\mathbf{A}$. The rows of C contain the analyte concentrations for the samples, rows of $\mathbf{P}^{\mathrm{T}}$ hold the spectral or chromatographic profiles of the pure single analytes at unit concentration, $\mathbf{E}$ is the residuals matrix.

$\mathbf{C}$ and $\mathbf{P}^{\mathrm{T}}$ are estimated by an alternating least squares (ALS) (67,68) algorithm: This algorithm starts with an estimate of either $\mathbf{C}$ or $\mathbf{P}^{\mathrm{T}}$. Assuming $\hat{\mathbf{C}}$ is employed for initialization, estimates of the spectral profiles are calculated based on MLR (the section Multivariate Linear Regression). Either a constrained least-squares fit is employed or the constraints are imposed after $\hat{\mathbf{P}}^{\mathrm{T}}$ is calculated directly by least squares:

$$\hat{\mathbf{P}}^{\mathrm{T}} = (\hat{\mathbf{C}}^{\mathrm{T}} \cdot \hat{\mathbf{C}})^{-1} \cdot \hat{\mathbf{C}}^{\mathrm{T}} \cdot \mathbf{A} \tag{47}$$

or utilizing the pseudoinverse (see section Notation and Fundamental Mathematical Tools and Supplementary Topics) when necessary

$$\hat{\mathbf{P}}^{\mathrm{T}} = \hat{\mathbf{C}}^{+} \cdot \mathbf{A}$$

If, eg, nonnegativity is applied, all negative entries of $\hat{\mathbf{P}}^{\mathrm{T}}$ are set to be zero. Once the constrained estimate of $\hat{\mathbf{P}}^{\mathrm{T}}$ is calculated, $\hat{\mathbf{P}}^{\mathrm{T}}$ is employed to update the estimate of $\hat{\mathbf{C}}^T$. As with calculating $\hat{\mathbf{P}}^T$, constraints can be imposed during the calculation of $\hat{\mathbf{C}}^T$ by a constrained least squares method, or after the estimation of $\hat{\mathbf{C}}$ by ordinary least squares

$$\hat{\mathbf{C}} = \mathbf{A} \cdot \hat{\mathbf{P}} \cdot (\hat{\mathbf{P}}^T \cdot \hat{\mathbf{P}})^{-1} \quad \text{or} \quad \hat{\mathbf{C}} = \mathbf{A} \cdot \hat{\mathbf{P}}^+ \tag{48}$$

The method iterates by alternating calculating constrained updates of $\hat{\mathbf{P}}^T$ (eq. 47) and $\hat{\mathbf{C}}$ (eq. 48) back and forth until further refinement does not significantly change the model.

3.9. Outlier Detection. Two important statistics for identifying outliers in the calibration set containing K samples are the "sample leverage"and the "studentized residuals". A plot of leverage versus studentized residuals makes a powerful tool for identifying outliers and assigning probable cause. The sample leverage is a measure of the influence, or weight, each sample has in determining the parameters of the calibration model. Samples near the center of the calibration set (average samples) will have a relatively low leverage compared to samples at the extreme edges of the experimental design and outliers. The sample leverage is determined by

$$h_k = 1/K + \mathbf{u}_k^T \cdot \mathbf{u}_k \tag{49}$$

where $\mathbf{u}_k$ is the row of associated matrix $\mathbf{U}$ (eq. 40) with the R significant principal components for the kth sample. Consequently, the sample leverage ranges from 0 for a sample in the center of an infinitely large calibration set to 1 for an extreme sample in a small data set.

The studentized residual is an indication of how well the calibration model estimates the analyte property in each sample. The studentized residual is similar to the Student's t-statistic; the estimation error of each sample is converted to a distance in standard deviations away from zero. An additional term is often added to the calculation to correct for the weight each sample has in determining the calibration model. The studentized residual is increased for samples with a large leverage; this is known as the studentized leverage corrected residuals. The studentized leverage corrected residuals are calculated by

$$t_k = \frac{|c_k - \hat{c}_k|}{\sigma \sqrt{1 - h_k}} \tag{50}$$

where

$$\sigma = \sqrt{\frac{\sum_{k=1}^{K} (c_k - \hat{c}_k)^2}{K - R - 1}} \tag{51}$$

with R being the number of PCs in the calibration model.

The plot of studentized leverage corrected residuals versus sample leverage provides insights into the quality of each calibration sample (Fig. 12). Samples with low leverages and low studentized residuals are typical samples in the

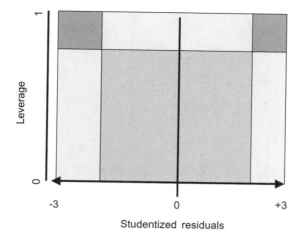

Fig. 12. Leverage-studentized residuals plots can be used to determine suspect data points.

calibration set. Data in the green region are generally "good" data and there is little statistically valid reason to remove any of these data. Data in the far-left and -right yellow regions have large fit errors and small leverages. These points are suspect in that they may have concentration errors or be mislabeled. Data in the upper yellow region are suspect due to spectral anomalies, but might have a high leverage just because they have an extreme concentration. Data in the red regions are most likely to be bad and should probably be removed.

4. Multiway Analysis

4.1. Introduction. Multiway analysis became popular in the late 1970s in the psychometric literature. Psychologists employed the multiway models primarily for factor analysis in order to determine intrinsic factors in large, complex data sets. However, as chemical instrumentation advanced with automated data collection, chemists began to acquire large, multiway data sets. In 1980, Hirschfeld listed 66 instruments capable of generating multiway data (69). Geladi cataloged the manners in which multiway data can be collected in chemical applications (12). Since a different notation has been defined in literature for multiway analysis compared to bilinear chemometrics the standard multivariate notation will be used in this section.

There are six classes of three-way data and four of these classes can be appropriately modeled with the basic trilinear, or PARAFAC (PARAllel FACtor analysis) (70–72) model. The PARAFAC decomposes the data cube (Figs. 1, 13) into N sets of triads, $\hat{x}$, $\hat{y}$, and $\hat{z}$ (see right part of Fig. 14). The elements of a trilinear $I \times J \times K$ data cube $\mathbf{R}$ can be presented as

$$R_{i,j,k} = \sum_{n=1}^{N} X_{i,n} \cdot Y_{j,n} \cdot Z_{k,n} + E_{i,j,k} \qquad (52)$$

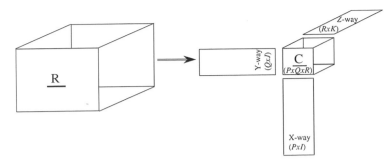

Fig. 13. The Tucker3 model (the section Tucker3 Models) is a generalization of the PAR-AFAC model (the section Multiway Curve Resolution—PARAFAC/CANDECOMP). The Tucker3 model decomposes the data cube into three sets of spectral and concentration profiles, like the PARAFAC model. However, the Tucker3 model additionally employs a core cube C that governs the mixing between the three spectral and concentration profiles. If the core matrix is all zeros except for having ones along the superdiagonal, the Tucker3 model reduces to the PARAFAC model.

Here, N refers to the rank of the model, ie, number of factors employed by the model. The parameter N must be determined (73) by the user before the algorithm is started. The residual data cube E contains the errors, which cannot be modeled.

To give an example: PARAFAC was used in combination with excitation–emission matrix (EEM) spectroscopy (74–76). The EEM spectroscopy uses a broadband light source usually in the uv–vis range to excite naturally fluorescent analytes, eg, aqueous solution of pesticides and polycyclic aromatic hydrocarbons. By means of two perpendicular spectrographs the excitation and emission spectra are projected in a perpendicular fashion onto a focal plane array (FPA). In other words, emission spectra measured after excitation at different wavelength are measured by the rows the FPA. The 2D $I \times J$ EEM spectra

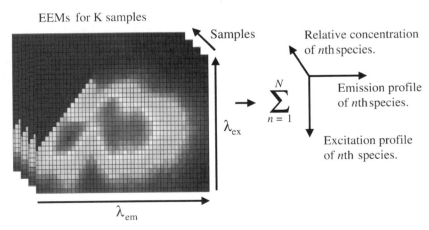

Fig. 14. Pictorial representation of collection of data as it can be factored into pure spectral and concentration profiles by PARAFAC.

obtained from K different samples are stacked to form a three-way data set $\mathbf{R}$ (Fig. 14). The PARAFAC is applied then in order to extract the excitation spectra $\mathbf{X}$, the emission spectra $\mathbf{Y}$, and the concentration profiles $\mathbf{Z}$ of the analytes from $\mathbf{R}$. Since the results are unique to a scaling factor (see discussion below), the excitation and emission spectra of the analytes are normalized to area one; the concentrations or multiplied then with the inverse of the spectra scaling factors. Another example would be obtaining a data cube from a LC-uv/vis device. Each $\hat{x}_n$ would correspond to one of the true N chromatographic profiles, each $\hat{y}_n$ to one of the true spectroscopic profiles, and each $\hat{z}_n$ to the relative concentrations in the K samples.

In general, the number and form of factors are not constrained to be representative of any physical reality. With two-way factor analysis, PCA, this is often referred to as the rotational ambiguity of the factors; there is a continuum of factors that satisfy the PCA model and equivalently describe the data. This is different for three-way analysis. If the following four conditions are given, the factors $\hat{x}$, $\hat{y}$, and $\hat{z}$ of a chemical component are accurate and unique estimates of the true underling factors $\mathbf{x}$, $\mathbf{y}$, and $\mathbf{z}$ except for a scaling constant:

1. The true underlying factor in each of the three modes is independent from the state of the other two modes.
2. The true underlying factor in any of the three modes cannot be expressed by linear combinations of the true underlying factors of other components in the same mode.
3. Linear additivity of instrumental responses among the species present is given.
4. The proper number of factors N is chosen for the model.

4.2. Multiway Curve Resolution—PARAFAC/CANDECOMP.

PARAFAC is originally based on the work of Kroonenberg (77) and as CANDECOMP (canonical decomposition) on the work of Harshman (78). In either case, the two base algorithms are practically identical. The PARAFAC uses an alternating least squares (ALS) based algorithm for multivariate curve resolution (the section Multivariate Curve Resolution) applied to three-way data sets.

The PARAFAC/CANDECOMP algorithm (79) stores iteratively improved estimates for the X-way, Y-way, and Z-way information in matrices $\mathbf{X}_{(I \times N)}$, $\mathbf{Y}_{(J \times N)}$, and $\mathbf{Z}_{(K \times N)}$. Before the algorithm is presented, six additional matrices have to be defined. Three of which ($\mathbf{R}^A$, $\mathbf{R}^B$, $\mathbf{R}^C$) contain elements of $\mathbf{R}$ and remain unaltered—$\mathbf{A}$, $\mathbf{B}$, and $\mathbf{C}$ will be updated in each iteration step. To keep this discussion concise, these definitions are formally introduced beforehand:

- $\mathbf{R}^C_{(I \cdot J \times K)}$ is a matrix constructed by unfolding the K slices of $\mathbf{R}$ in the XY-plane containing the elements

$$R^C_{(j-1)I+i,k} = R_{i,j,k}$$

- $\mathbf{C}_{(I \cdot J \times N)}$ is formed from the N columns of $\hat{\mathbf{X}}$ and $\hat{\mathbf{Y}}$ with elements

$$C_{(j-1)I+i,n} = \hat{X}_{i,n} \cdot \hat{Y}_{j,n} \tag{53}$$

- $\mathbf{R}^{B}{}_{(I \cdot K \times J)}$ is a matrix constructed by unfolding the J slices of $\mathbf{R}$ in the XZ plane containing the elements

$$R^{B}_{(k-1)I+i,\,j} = R_{i,j,k}$$

- $\mathbf{B}_{(I \cdot K \times N)}$ is formed from the N columns of $\hat{\mathbf{X}}$ and $\hat{\mathbf{Y}}$ with elements

$$B_{(k-1)I+i,\,n} = \hat{X}_{i,n} \cdot \hat{Y}_{j,n} \tag{54}$$

- $\mathbf{R}^{A}{}_{(J \cdot K \times I)}$ is a matrix constructed by unfolding the I slices of $\mathbf{R}$ in the YZ plane containing the elements

$$R^{A}_{(k-1)J+j,\,i} = R_{i,j,k}$$

- $\mathbf{A}_{(J \cdot K \times N)}$ is formed from the N columns of $\hat{\mathbf{Y}}$ and $\hat{\mathbf{Z}}$ with elements

$$A_{(k-1)J+j,\,n} = \hat{Y}_{j,n} \cdot \hat{Z}_{k,n} \tag{55}$$

Step 0: Initial guess of the $\mathbf{X}$ and $\mathbf{Y}$ starting profiles—this can be random numbers (80), or an eigenproblem based algorithm like the Direct Trilinear Decomposition (DTLD) (81), or a priori information about the samples. From these two matrices the $\mathbf{Z}_{(K \times N)}$ profiles will be calculated in Step 1 of the algorithm.

Step 1: Employing equation 53, updated estimates of the Z-way data are determined by solving:

$$\mathbf{R}_{C} = \mathbf{C} \cdot \mathbf{Z}^{T}$$

This is done by a multivariate least-squares fit (eq. 30) (the section Multivariate Linear Regression) $\hat{\mathbf{Z}}^{T} = (\mathbf{C}^{T} \cdot \mathbf{C})^{-1} \cdot \mathbf{C}^{T} \cdot \mathbf{R}_{C}$ or by using the pseudoinverse (eq. 5) (the section Notation and Fundamental Mathematical Tools):

$$\hat{\mathbf{Z}}^{T} = \mathbf{C}^{+} \cdot \mathbf{R}_{C} \tag{56}$$

In the remainder only the pseudoinverse will be mentioned.

Step 2: Employing equation 54 updated estimates of the Y-way data are determined by solving:

$$\mathbf{R}_{B} = \mathbf{B} \cdot \mathbf{Y}^{T}$$
$$\hat{\mathbf{Y}}^{T} = \mathbf{B}^{+} \cdot \mathbf{R}_{B} \tag{57}$$

Step 3: Employing equation 55 updated estimates of the X-way data are determined by solving:

$$\mathbf{R}_{A} = \mathbf{A} \cdot \mathbf{X}^{T}$$
$$\hat{\mathbf{X}}^{T} = \mathbf{A}^{+} \cdot \mathbf{R}_{A} \tag{58}$$

Step 4: Update $\mathbf{A}$, $\mathbf{B}$, and $\mathbf{C}$ by using the new estimates $\hat{\mathbf{X}}$, $\hat{\mathbf{Y}}$, and $\hat{\mathbf{Z}}$ in equations 53–55, respectively.

Step 5+: The algorithm proceeds iteratively, cycling through equations 56–58, until the convergence criterion is satisfied.

Two more topics remain to be discussed: The initialization of **X** and **Y** (Step 0) as well as the stopping criterion of the iteration (Step 5).

The PARAFAC algorithm is sensitive to the starting guess of the solution for $\hat{X}$ and $\hat{Y}$. This results from PARAFAC often becoming trapped in local minima and, hence, not converging to the global optimum least squares solution. Furthermore, the PARAFAC algorithm can become delayed in "swamps" far from the optimum solution (82). Although this markedly increases the analysis time, when employing a random starting value, multiple initial guesses should be considered. The solution for each starting value will be different; however, if all or most of the solutions are similar, it is safe to assume that PARAFAC has converged to near the global optimal solution. The convergence time for PARAFAC can be improved by initializing the algorithm with guesses near the optimal solution. These guesses can come from DTLD or reference spectra of species either known or highly suspected to be in the data set. Care should be employed when utilizing the DTLD solutions since DTLD often yields significant imaginary components in predicting X- and Y-way factors. The problems caused by initializing PARAFAC with imaginary components can be circumvented by employing the real components of $\hat{X}$ and $\hat{Y}$ from DTLD or the absolute values of $\hat{X}$ and $\hat{Y}$ from DTLD.

Two popular convergence criteria for the PARAFAC algorithm are based on changes in the residuals (unmodeled data) between successive iterations and changes in the predicted profiles between successive iterations. In the first case, the algorithm is terminated when the root average of the squared residuals between successive iterations agree to within an absolute or relative tolerance, say 10^{-6}. While such fit based stopping criteria are conceptually easy to visualize, a faster method for determining convergence relies on the correlation between the predicted X-, Y-, and Z-way profiles between successive iterations. When the product of the cosines between successive iterations in the X-, Y-, and Z-modes approach arbitrarily close to 1, say within 10^{-6}, the algorithm is terminated. The cosine in the X way is determined by unfolding the $I \times N$ matrices $\hat{X}_{old}$ and $\hat{X}_{new}$ into a column vectors $\hat{x}_{old}$ and $\hat{x}_{new}$. The $\cos \theta_X$ is defined as

$$\cos \theta_X = \frac{x_{old}x_{new}}{\sqrt{(x_{old}x_{old})(x_{new}x_{new})}} \tag{59}$$

The other two terms, $\cos \theta_Y$ and $\cos \theta_Z$ are defined equivalently. Convergence in all three modes is implied, if

$$\cos \theta_X \cdot \cos \theta_Y \cdot \cos \theta_Z > 1 - 10^{-6}$$

since at least $\cos \theta > 1 - 10^{-6}$ is obtained for all X, Y, and Z ways.

Mitchell and Burdick sight, besides speed, an additional benefit to correlation based convergence (82). In cases when two factors are highly correlated in one ore more of the three ways, ALS methods may become mired in "swamps" where the fit of the model changes slightly but the correlation between the predicted X, Y, and Z ways change significantly between successive iterations. After many iterations the ALS algorithm will then emerge from the "swamp" and the residuals and estimated profiles will then both rapidly approach the optimum. Hence, correlation based convergence is more resistant to inflection points in the error response surface when optimizing the model.

4.3. Tucker3 Models. The generalization of the PARAFAC model is the Tucker3 model (83,84). The PARAFAC model is intrinsically linear model and straightforward application thus assumes linear interactions and behavior of the samples. While many of the systems of interest to chemists contain nonlinearities that violate the assumptions of the models, the PARAFAC model forms an excellent starting point from which many subsidiary methods are constructed to incorporate nonlinear behavior into calibration models constructed from three-way data collected with hyphenated methods. The trilinear model is actually a specific case of the Tucker3 model. The Tucker3 model is best understood by viewing a graphical representation such as in Fig. 13. A data cube **R** is decomposed into three sets of factors, $\hat{x}$, $\hat{y}$, and $\hat{z}\hat{z}$, as with PARAFAC. However, the Tucker3 model differs from the PARAFAC model in two key ways. The number of factors N in each way of the Tucker3 model is not constrained to be equal. Also, the Tucker3 model employs a small core cube, **C**, that governs the interactions among the factors. A non zero element at the pth, qth, rth position of the core **C** dictates an interaction between the pth factor in the X way, the qth factor in the Y way, and the rth factor in the Z way. This permits modeling of two or more factors that might have, eg, the same chromatographic profile but different spectral and concentration profiles (85,86). If there are the same number of factors in each way, and **C** is constrained to only have nonzero elements on the super diagonal, then the Tucker3 model is equivalent to the PARAFAC model.

One alternating least-squares algorithm for estimating the parameters of the Tucker3 model is Tuckals (for TUCK Alternating Least Squares). This iterative Tuckals algorithms proceeds similarly to the PARAFAC/CANDECOMP algorithm except instead of cycling through three sets of parameters, four sets of parameters must be successively updated, $\hat{X}$, $\hat{Y}$, $\hat{Z}$, and **C**. However, where with PARAFAC just the number of factors in the model N needs to be preassumed, with Tucker3 the three dimensions of the core array P, Q, and R, need to be assumed.

4.4. Solution Constraints. ALS algorithms are more flexible than rank annihilation based algorithms (87) since constraints (cf. the section Multivariate Curve Resolution) can be placed onto the solutions derived from ALS methods. The ALS algorithms implicitly constrain the estimated profiles to lie in the real space. Rank annihilation methods may fit factors with imaginary components to the data. Ideally, constraints are not needed for ALS to achieve accurate, meaningful concentration and spectral profile estimates. However, the presence of slight nonlinear interactions among the true underlying factors, of highly correlated factors, or of low signal to noise in the data will often result in profile estimates that are visually unsatisfying and large quantitative errors are derived from the model. These effects can often be minimized by employing constraints to the solutions that are based on a priori knowledge or assumptions of the data structure, eg, prior knowledge of sample concentrations or spectral profile characteristics.

Perhaps the most common constraint consciously placed on the PARAFAC or Tucker3 models is nonnegativity. When one of the modes represents concentrations, chromatographic profiles, or in many cases spectra, constraining the solutions to yield only nonnegative profile estimates often improves the quantitative and qualitative accuracy of the models. Care should be taken when applying nonnegativity constraints since some spectral effects, such as absorbance and quenching in fluorescence, can be manifested, detected, and modeled as negative profiles. Nonnegative estimates of the three-way profiles can be obtained by replacing the least squares update of any given profile with the nonnegative least squares (NNLS) solution that is well defined in the mathematics literature (88). The method described in Ref. 88 is readily available as a Matlab function. The downside of this method is that it is numerically intensive compared to computing the regular least-squares solution for each update.

A second constraint often applied in three-way calibration of chromatographic data is unimodality. This constraint exploits the knowledge that chromatographic profiles have exactly one maximum. Unlike NNLS, there is now method to calculate the true unimodal least-squares update during each iteration. Instead a search algorithm must be implemented that finds the maximum of each profile and assures that from that maximum all values are monotonically decreasing.

The third common constraint is based on a priori knowledge of the three-way profiles. In this case, the known relative concentrations of the standards or the known spectral profiles of one or more components can be fixed as part of the solution. In the Tucker3 model, it is common to restrict some of the potential interactions between factors when they are known not to exist. Care must be employed when applying fixed values to the solutions as the scaling of the factors must still be taken into account.

5. Selected Topics

5.1. Background Spectrum Correction.
Background correction methods are often employed in spectroscopic applications to remove broad features from the data set. These features hinder calibration as a large source of variance compared to the analyte or as a seemingly random source of variance that consumes many factors in the model. Examples include fluorescence background in Raman spectroscopy and scattering backgrounds in near-ir reflectance spectroscopy.

Simple efforts at background correction include derivatives, polynomial curve fitting, and Fourier Transform (FT) filtering (89). Derivatives remove the portion of a background that can be modeled by a low order polynomial. Taking the first derivative of a spectrum removes the baseline offset. The second derivative removes the linear approximation of the background (and the analyte signal). However, in spite of digital filters for simultaneously smoothing the data while calculating the derivatives (90), the S/N rapidly declines with each derivatization. Polynomial curve fitting is useful when there are regions of the spectra that contain only background variance. These regions must be distributed across the entire spectrum such that the background can be modeled. The FT filtering removes both low and high frequency variance across the spectrum. It is assumed that the lowest frequency signal is the background and the highest frequency signal is random instrumental errors. Problems may occur with FT filtering due to poorly chosen apodization functions applied to the signal or insufficient ability to distinguish between the signal and the background. This will lead to distortion of the analyte signal.

Multiplicative scatter correction (MSC) was developed to reduce the effect of scattered light on diffuse reflection and transmission NIR spectra (91,92). This method has also shown utility as a means of removing varying background spectra with nonscattering origins. Consequently, MSC sometimes appears as multiplicative signal correction. The basic application of MSC is presented here. However, a more advanced version of MSC exists that assumes a unique scattering model for different regions of the spectra (93).

Scattering theory states that scattering should have a multiplicative effect on reflection (and transmittance) spectra. That is, the observed spectra will contain a broad, changing background from differential scattering at each wavelength. In Fig. 15**a** it is apparent that the largest source of variance within the NIR reflectance spectra of the 40 cornflour samples is derived from scattering. Assuming a multiplicative model for the scattering, the scattering profile in a spectrum can be deduced from a plot of the spectrum of a standard scatterer versus a given spectrum at each wavelength. And ideal 'standard' would have no NIR absorbance (or transmittance) features; however, the mean spectrum from a collection of similar samples will suffice. Fig. 15**b** presents the plot of the intensity of each wavelength for the mean of 40 calibration spectra versus two of the individual calibration spectra. Note that one is scattering more than the average spectrum and one is scattering less than the average spectrum. The plot for each of these two samples lies about a line with a little variation around the line. The difference between each sample and the best fit line through each sample in Figure 15**b** can be interpreted as the chemical signal and the best fit line gives the spectrum of the scattering in the sample. Consequently, the scattering is determined by regressing each spectrum onto the mean spectrum, where the scattering at the jth wavelength of a sample can be modeled by

$$x_j = a + b\bar{x}_j + \epsilon_j \tag{60}$$

with a and b being constant for all J wavelengths in the sample. The scatter corrected data is determined by the scaled deviations about the regression

$$x_{j,\mathrm{MSC}} = (x_{j,raw} - a)/b \tag{61}$$

The corrected spectra for the 40 calibration NIR cornflour samples are shown in Figure 15c. For correction of future samples, the mean of the calibration set may be employed as the scatter standard. Figure 15d shows the corrected spectra of 20 cornflour spectra that were not included in the calibration set. Evident from these figures is that the spectral features are not distorted by

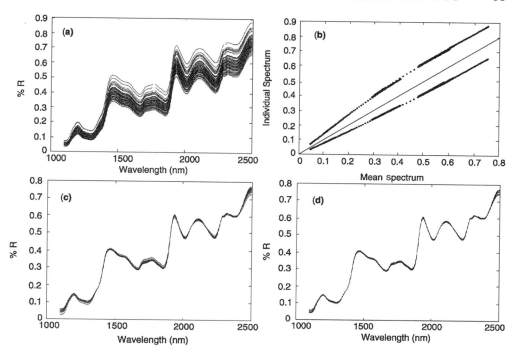

Fig. 15. Demonstration of multiplicative scatter correction. **(a)** The major source of variance for the 40 NIR cornflour spectra is due to scattering affecting the spectral baseline. **(b)** Relationship between two spectra and the mean of the 40 spectra. **(c)** MSC applied to calibration data: corrected spectra are found by the residual of the 40 spectra after regression against the mean of the 40 spectra. **(d)** MSC applied to future data: corrected spectra are found by the residual of the 20 future spectra after regression against the mean of the 40 calibration spectra.

MSC contrasted to scatter correction by calculating the second derivative of each spectrum.

A different approach explicitly including drifts into the calibration was proposed to artificially extend the set of PCs with the so-called pseudo PCs (94). Advantage is taken from the fact that background drifts are usually very broad compared to the more localized absorption features. These pseudo-PCs have been defined to be polynomials up to a user selectable order, however, other linear independent functions could be used, too. It was shown that this combined set of PCs and pseudo PCs is able to determine considerably improved concentration results from highly drift affected spectra compared to a conventional PCR. An example utilizing uv derivative spectroscopy of aqueous samples (95) is given in Figure 16. Several weeks after performing the calibration the zero-point concentrations of three aromatics hydrocarbons have been monitored >13 h. Since considerable drifts occurred due to an instable uv light source, the concentration errors without drift correction equal 10% of the measurement range. Most of these concentration errors could be removed based on pseudo-PCs.

An alternative (33) to pseudo-PCs utilizes a similar idea: Polynomials are fitted to the regular PCs and subtracted from them. In this case drift effects,

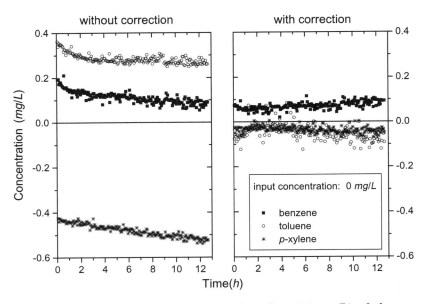

Fig. 16. Comparing the concentration zero points (input 0 mg/L) of three aromatic hydrocarbons dissolved in water (measurement range 0–5 mg/L) obtained from uv derivative spectroscopy (22) without and with drift corrections by means of pseudoprincipal components (94).

which can be modeled by polynomials, are orthogonal to the PCs. This is due to the fact the "corrected" PCs, ie, original PCs minus fit polynomials, are the residuals of these fits. Hence, polynomial like drifts up to the considered order are orthogonal to the corrected PCs (cf. section 3 in Supplementary Topics, chapter Multivariate Linear Regression) and cannot influence the concentration results. The pseudo-PCs method extracts additional information, ie, an estimate of the drift spectrum. The approach fitting polynomials to the PCs is computational less expensive since it is done just once during the calibration. This can be advantageous is computation resources are limited.

5.2. Instrument Standardization. One practical concern with multivariate calibration and prediction is the transport and stability of the calibration models. Ideally, a calibration model can be constructed in the laboratory on a bench-top instrument, then the model can be applied to many similar instruments in the field. Also, once a model is successfully transferred to the field, it will be robust to changes in instrumental sensitivity and alignment. Of course, the goal of a universal transferable and robust instrument–model has not been achieved. Seemingly identical spectrometers have slight wavelength resolution, and sensitivity differences that can prohibit reliable distribution of the calibration model among numerous instruments. Also, time-dependent instrumental drift eventually can render the calibration model obsolete for whichever instrument the model was constructed.

Individual calibration of each instrument is not an acceptable solution to the problem of model distribution. Calibration may be an expensive, time-consuming task when many calibration samples are needed, the calibration samples are not readily transportable, or the instrument is not easily accessible in the process stream. Concurrently, it is also unacceptable to repeat an entire calibration procedure whenever there are minor changes in the instrumental character.

Instrumental standardization (96–99) strives to solve the problems derived from instrumental differences when constructing one calibration model for multiple instruments. The instrumental standardization philosophy is to construct the best model possible on one instrument then to build a second model that will transform the spectra from other instruments to appear as if they were recorded on the first instrument. Usually, this transfer function can be reliably calculated with less effort.

One standardization method popular in the literature is Piecewise Direct Standardization (PDS) (99–102). With PDS, a set of transfer samples is analyzed on both the original instrument and the instrument to which the calibration model will be transferred. It is best if the transfer samples are a subset of the calibration set; however, other surrogate samples may be employed. A separate transfer function is determined for each wavelength in the spectra by least squares regression using neighboring wavelengths as the independent variables. That is, a local subset of variables measured on the second instrument is employed to build a model that predicts what each measurement would have been if it were measured with the first instrument. This method accounts for shifts and intensity changes over a small spectral window. The drawback of PDS is that success of the standardization is dependent on choice of the transfer samples. The transfer samples must be identical when measured on each instrument and the set of samples must span the space of all encountered spectral changes between the two instruments. Therefore, the choice and number of transfer samples must be optimized by the analyst.

A more useful method of standardization would not require transfer samples to be analyzed. There have been two approaches to this problem. When it can be safely assumed that the only spectral shifts (ie, wavelength or retention time) occur a PCA based method of standardization may be employed (103,104). The spectral (or time) indexes are shifted such that the projection of each sample into the PC space defined by the original instrument is optimized. A more general method based loosely on MSC has also demonstrated success when there are relatively minor performance differences between the original and second instruments (105,106). Here a local selection of wavelengths from each spectrum is regressed against the mean spectrum to build a transfer function. Consequently, the spectra from the second instrument are not transformed to look like the spectra from the first instrument. Instead, spectral responses from both instruments are transformed to lie in a common multidimensional space.

5.3. Optical Computation. Most spectrometer concepts include moving parts like interferometers or scanning gratings. Such moving parts, however, limit the ruggedness of a field analyzer and the time resolution of the concentration runs. The strong point of such spectrometers combined with chemometric software packages is their versatility. For many applications this is not needed, though. In process analytics, eg, a measurement device is usually applied to one very specific task not needing versatility at all. Mechanical stability and good time resolution is of greater importance. In order to overcome both mentioned drawbacks it was proposed to design so-called multivariate optical elements (MOE) (107–111). MOE are specially designed interference filter in a beam split-ter arrangement. The light is emitted from the source, transmitted through the sample and split by the MOE into a transmission part and a reflection part. The idea is to design the transmission spectrum of the interference filter such that it is an imprint of a PC onto a transmission offset. This offset is necessary to enable positive and negative features of the PC. The transmission of light through such an interference filter followed by generating a signal in the detector element resembles the projection of a measured spectrum onto a PC. The transmission through the filter replaces the multiplication of loadings with measurement points of a spectrum; the detector integrating over all wavelengths replaces the summation part of calculating a scalar product. What is left to do is subtract-ing the transmission offset mentioned above from the results. For this purpose, the interference filter had been placed in a 45° arrangement. Then the transmis-sion and the reflection spectrum can be measured by means of two detectors arranged in perpendicular lines of sight. Calculating the difference signal of both detectors cancels the transmission offset.

5.4. Artificial Neural Networks Combined with Variable Selection.
The measurement technique surface plasmon resonance (spr) (112) is sensitive for analyzing refractive indexes of liquids or vapors. Since the matrix, water, eg, and a dissolved analyte have different refractive indices the refractive index of a sample is concentrations dependent. A change of the samples' refrac-tive index is measured by a highly nonlinear wavelength shift of the plasmon absorbance. However, since only one property of a sample, ie, the refractive index, is measured, binary or ternary mixtures cannot be investigated without experimental adjustments. A polymer coating of the spr sensor head was pro-posed resulting in different, time-dependent enrichment or desorption processes depending on the molecule size. That means different analytes cause a time- and analyte-dependent change of the spr spectra. This idea was applied in references 113 and 114 to measure binary samples of two chlorofluorocarbons and ternary mixtures of alcohols, respectively. Time series of spr spectra monitoring different sorption–desorption behaviors of the analytes were evaluated then by means of a neural network. Inputs into the neural net are the wavelength shifts measured at preselected points of times (variables). Usually, one wants a high time resolution, ie, a large number of variables, in order to capture fast and similar responses and not to lose information. However, there are several disadvantages of using a lot of information like hiding meaningful variables by irrelevant vari-ables or overfitting. Furthermore, danger to change the correlation is increased with the number of variables and many variables mean increased computation time for the neural net training.

Full-connected neural networks employing a large number of variables are prone to overfitting (113). Hence, so-called growing neural nets were applied in Refs. 113 and 114 resulting in sparse, nonuniform structures optimized to a spe-cific problem. The growing of a feedforward back-propagation network is started with one not having hidden neurons or connections. Then one neuron is added at a time, which is connected to one output neuron and two other neurons such that the error decrease regarding the training data is maximized. However, the out-come of this procedure is still dependent on the way the calibration data are split into training and monitoring data. To overcome this ambiguity two strategies have been proposed: (*1*) To grow neural networks on a rather large number of different training–monitoring sets in parallel. Ranking the variables considering their importance follows this. The number of net growings in which it was selected determines the importance of a variable. The final network is grown in a second step by iteratively adding variables to it in order of decreasing importance until the addition of a variable does not increase the predictability of this final network anymore. (*2*) A certain training–monitoring set is defined and a small number of nets are grown with different initial weights. The best of these is chosen to be the initial topology for the second, different training–monitoring set. Again a number of networks with different initial weights are grown and the best one is selected. This is continued until the topology of the best network does not change anymore. It was found that the procedure (*2*) resulted in better generalizations. Application of such grown networks to binary mixtures resulted in convincing concentration prediction.

As an alternative for variable selection, a genetic algorithm (115–118) has been used in Ref. 114 for selecting the optimum subset for neural networks based on the procedure (*1*) discussed above. A genetic algorithm is applied to a rather large number of different training–monitoring sets in parallel resulting in a set of neural nets. Again, the variables are ranked in decreasing importance and added in the second step one after the other to a final network until the predictability is not improved anymore. The prediction of concentrations could be considerably improved using five selected variables compared to using all 50.

BIBLIOGRAPHY

"Chemometrics" in *ECT* 4th ed., pp. 837–869, by Deborah Illman, University of Washington; "Chemometrics" in *ECT* (online), posting date: December 4, 2000, by Deborah Illman, University of Washington.

1. H. Martens and T. Næs, *Multivariate Calibration*, 2nd ed., John Wiley & Sons, Inc., New York, 1991.
2. I. T. Jolliffe, *Principal Component Analysis*, 2nd ed., Springer-Verlag: New York, 2002.
3. E. Malinowski, *Factor Analysis in Chemistry*, 3rd ed., John Wiley & Sons, Inc., New York, 2002.
4. N. R. Draper and H. Smith, *Applied Regression Analysis*, 3rd ed., J. Wiley & Sons, Inc., New York, 1998.
5. A. Sen and M. Srivastava, *Regression Analysis—Theory, Methods, Applications*, Springer Verlag, New York, 1990.
6. *Anal. Bio. Chem.* **373**(6), (2002). Special review issue.
7. D. Burdick, *Chemom. Intell. Lab. Syst.* **28**, 229 (1995).
8. S. Wold, *Chemom. Intell. Lab. Syst.* **30**, 109 (1995).
9. K. S. Booksh and B. R. Kowalski, *Anal. Chem.* **66**, 782A (1994).
10. S. D. Brown, *Chemom. Intell. Lab. Syst.* **30**, 49 (1995).
11. P. Hopke, *Anal. Chim. Acta* **500**(1–2), 365 (2003).
12. P. Geladi, *Chemom. Intell. Lab. Syst.* **7**, 11 (1989).

13. G. H. Golub and C. F. Van Loan, *Matrix Computations*, 3rd ed., Johns Hopkins University Press, Baltimore, 1996.

14. W. H. Press, B. P. Flannery, S. A. Teukolsky, and W. T. Vettering, *Numerical Recipes in C++: The Art of Scientific Computing*, 2nd ed., Cambridge University Press, Cambridge, 2002.

15. J. K. Taylor, *ChemTech.* **16**, 763 (1986).

16. G. L. Long and J. D. Winefordner, *Anal. Chem.* **55**, 712A (1983).

17. D. L. Massart, B. G. M. Vandegonste, L. M. C. Buydens, S. DeJong, P. J. Lewi, and J. Smeyers, *Handbook of Chemometrics and Qualimetrics*, Elsevier, Amsterdam, The Netherlands, 1997.

18. R. Bro and A. Smilde, *J. Chemom.* **17**, 16 (2003).

19. R. J. Pell, M. B. Seasholtz, and B. R. Kowalski, *J. Chemom.* **6**, 52 (1992).

20. M. B. Seasholtz, B. R. Kowalski, *J. Chemom.* **6**, 103 (1992).

21. D. M. Haaland and E. V. Thomas, *Anal. Chem.* **60**, 1193 (1988).

22. F. Vogt, U. Klocke, K. Rebstock, G. Schmidtke, V. Wander, and M. Tacke, *Appl. Spec.* **53**, 1352 (1999).

23. W. Egan, W. Brewer, and S. Morgan, *Appl. Spectrosc.* **53**, 218 (1999).

24. F. Vogt, M. Karlowatz, M. Jakusch, and B. Mizaikoff, *Analyst* **128**, 397 (2003).

25. J. Mandel, *Am. Stat.* **36**, 15 (1982).

26. P. Geladi and B. R. Kowalski, *Anal. Chim. Acta* **185**, 1 (1986).

27. A. Lorber, L. E. Wangen, and B. R. Kowalski, *J. Chemom.* **1**, 19 (1987).

28. R. Marbach and H. M. Heise, *TRAC* **11**, 270 (1992).

29. M. Stone and R. J. Brooks, *J. R. Stat. Soc. B.* **52**, 337 (1990).

30. S. de Jong, *Chemom. Intell. Lab. Syst.* **18**, 251 (1993).

31. R. Manne, *Chemom. Intell. Lab. Syst.* **2**, 187 (1987).

32. F. Vogt and B. Mizaikoff, *J. Chemom.* **17**, 346 (2003).

33. F. Vogt, H. Steiner, and B. Mizaikoff, *Appl. Spec.* 2003, submitted for publication.

34. F. Vogt and B. Mizaikoff, *J. Chemom.* **17**, 225 (2003).

35. F. Vogt and B. Mizaikoff, *Anal. Chem.* **75**, 3050 (2003).

36. T. Naes and T. Isaksson, *Appl. Spectrosc.* **46**, 34 (1992).

37. K. S. Johnston, S. S. Lee, and K. S. Booksh, *Anal. Chem.* **69**, 1844 (1997).

38. W. S. Cleveland and S. J. Devlin, *J. Am. Stat. Assoc.* **83**, 596 (1988).

39. T. Naes, T. Isaksson, and B. R. Kowalski, *Anal. Chem.* **62**, 664 (1990).

40. Z. Wang, T. Isaksson, and B. R. Kowalski, *Anal. Chem.* **66**, 249 (1994).

41. L. Brieman and J. H. Friedman, *J. Am. Stat. Assoc.* **80**, 580 (1985).

42. I. E. Frank and S. Lanteri, *Chemom. Intel. Lab. Syst.* **3**, 301 (1988).

43. J. H. Friedman and W. Steutzle, *J. Am. Stat. Assoc.* **76**, 817 (1981).

44. M. M. C. Ferreira, W. C. Ferreira, and B. R. Kowalski, *J. Chemom.* **10**, 11 (1996).

45. S. Winsberg and J. O. Ramsey, *Psychometrika* **48**, 575 (1984).

46. I. E. Frank, *Chemom. Intel. Lab. Syst.* **8**, 109 (1990).

47. S. Wold, *Chemom. Intel. Lab. Syst.* **14**, 71 (1992).

48. J. H. Friedman, *Ann. Stat.* **19**, 199 (1991).

49. S. Sekulic and B. K. Kowalski, *J. Chemom.* **6**, 199 (1992).

50. J. Freeman and D. Skapura, *Neural Networks—Algorithms, Applications and Programming Techniques*, Addison-Wesley Publishing Company, New York, 1991.

51. S. Sekulic, M. B. Seasholtz, Z. Wang, B. R. Kowalski, S. E. Lee, and B. R. Holt, *Anal. Chem.* **65**, 835A (1993).

52. I. E. Frank, *Chemom. Intel. Lab. Sys.* **27**, 1 (1995).

53. K. S. Booksh and B. R. Kowalski, *Anal. Chim. Acta* **348**, 1 (1997).

54. M. Gerritsen, J. A. van Leeuwen, B. G. M. Vandeginste, L. Buydens, and G. Kateman, *Chemom. Intel. Lab. Sys.* **15**, 171 (1992).

55. M. B. Seasholtz and B. R. Kowalski, *Anal. Chim. Acta* **277**, 165 (1993).

56. Z. Wang, J. N. Hwang, and B. R. Kowalski, *Anal. Chem.* **67**, 1497 (1995).
57. W. H. Lawton and E. A. Sylvestre, *Technometrics* **13**, 617 (1971).
58. H. Gampp, M. Maeder, C. J. Meyer, and A. D. Zuberbuhler, *Talanta* **32**, 1133 (1985).
59. Y-Z Liang, R. Manne, and O. M. Kvalheim, *Chemom. Intell. Lab. Syst.* **14**, 155 (1992).
60. W. Windig, *Chemom. Intell. Lab. Syst.* **16**, 1 (1992).
61. R. Bro and S. De Jong, *J. Chemom.* **11**, 393 (1997).
62. R. Tauler, I. Marques, and E. Casassas, *J. Chemom.* **12**, 55 (1998).
63. W. C. Bell, K. S. Booksh, and M. L. Myrick, *Anal. Chem.* **70**, 332 (1998).
64. S. P. Gurden, R. G. Bereton, and J. A. Groves, *Chemom. Intell. Lab. Syst.* **23**, 123 (1994).
65. E. A. Sylvestre, W. H. Lawton, M. S. Maggio, *Technometrics* **16**, 353 (1974).
66. R. I. Shrager, *Chemom. Intell. Lab. Syst.* **1**, 59 (1986).
67. J. Saurina, S. Hernandez-Cassou, R. Tauler, and A. Izquierdo-Ridorsa, *J. Chemom.* **12**, 183 (1998).
68. P. Gemperline and E. Cash, *Anal. Chem.* **75**, 4236 (2003).
69. T. Hirschfeld, *Anal. Chem.* **52**, 297A (1980).
70. R. Bro, *Chemom. Intell. Lab. Syst.* **38**, 149 (1997).
71. C. Andersen and R. Bro, *J. Chemom.* **17**, 200 (2003).
72. N. Faber, R. Bro, and P. Hopke, *Chem. Intell. Lab. Syst.* **65**, 119 (2003).
73. R. Bro and H. Kiers, *J. Chemom.* **17**, 274 (2003).
74. A. Muroski, K. Booksh, and M. Myrick, *Anal. Chem.* **68**, 3534 (1996).
75. R. JiJi, G. Cooper, and K. Booksh, *Anal. Chim. Acta* **397**, 61 (1999).
76. R. JiJi, G. Andersson, and K. Booksh, *J. Chemom.* **14**, 171 (2000).
77. P. M. Kroonenberg, *Three-mode Principal Component Analyses. Theory and Applications*, DSWO Press, Leiden, 1983.
78. R. A. Harshman, UCLA Working Paper on Phonetics, Vol. 16, 1970, pp. 1–84.
79. C. Andersson and R. Bro, *Chemo. Intell. Lab. Syst.* **52**, 1 (2000).
80. R. A. Harshman and M. E. Lundy, "The PARAFAC model for Three-Way Factor Analysis and Multidimensional Scaling," in H. G. Law and co-workers, eds., *Research Methods for Multimode Data Analysis*, Praeger, New York, 1984.
81. E. Sanchez and B. R. Kowalski, *J. Chemom.* **4**, 29 (1990).
82. B. C. Mitchell and D. S. Burdick, *J. Chemom.* **6**, 155 (1992).
83. V. Pravdova, F. Estienne, B. Walczak, and D. L. Massart, *Chemom. Intell. Lab. Syst.* **59**, 75 (2001).
84. J. Ten Berge and A. Smilde, *J. Chemom.* **16**, 609 (2002).
85. A. K. Smilde, R. Tauler, J. M. Henshaw, L. W. Burgess, and B. R. Kowalski, *Anal. Chem.* **66**, 3345 (1994).
86. A. K. Smilde, Y. Wang, and B. R. Kowalski, *J. Chemom.* **8**, 21 (1994).
87. C.-H. Ho, G. D. Christian, and E. R. Davidson, *Anal. Chem.* **50**, 1108 (1978).
88. C. L. Lawson and R. J. Hanson, *Solving Least Squares Problems*, Prentice-Hall, Englewood, Cliffs, New York, 1974.
89. Q. Ding, G. W. Small, and M. A. Arnold, *Appl. Spectros.* **53**, 402 (1999).
90. A. Savitzky and M. Golay, *Anal. Chem.* **36**, 1627 (1964).
91. P. Geladi, D. MacDougall, and H. Martens, *Appl. Spectrosc.* **39**, 491 (1985).
92. C. E. Miller, S. A. Svendsen, and T. Naes, *Appl. Spectrosc.* **47**, 346 (1993).
93. T. Isaksson and B. R. Kowalski, *Appl. Spectros.* **47**, 702 (1993).
94. F. Vogt, K. Rebstock, and M. Tacke, *Chemom. Intell. Lab. Syst.* **50**, 175 (2000).
95. F. Vogt, M. Tacke, M. Jakusch, and B. Mizaikoff, *Anal. Chim. Acta* **422**, 187 (2000), Erratum: *Anal. Chim. Acta* **431**, 167 (2001).
96. Y. Wang, M. J. Lysaght, and B. R. Kowalski, *Anal. Chem.* **64**, 562 (1992).
97. C. S. Chen, C. W. Brown, and S. C. Lo, *Appl. Spectros.* **51**, 744 (1997).
98. J. Lin, *Appl. Spectrosc.* **52**, 1591 (1998).

99. P. J. Gemperline, J. H. Cho, P. K. Aldridge, and S. S. Sekulic, *Anal. Chem.* **68**, 2913 (1996).

100. Y. Wang, M. J. Lysaght, and B. R. Kowalski, *Anal. Chem.* **64**, 562 (1992).

101. C. S. Chen, C. W. Brown, and S. C. Lo, *Appl. Spectros.* **51**, 744 (1997).

102. J. Lin, *Appl. Spectrosc.* **52**, 1591 (1998).

103. K. S. Booksh, C. M. Stellman, W. C. Bell, and M. L. Myrick, *Appl. Spectros.* **50**, 139 (1996).

104. B. J. Prazen, C. E. Bruckner, R. E. Synovec, and B. R. Kowalski, *J. Microcol. Sep.* **11**, 97 (1998).

105. T. B. Blank, S. T. Sum, S. D. Brown, and S. L. Monfre, *Anal. Chem.* **68**, 2987 (1996).

106. S. T. Sum and S. D. Brown, *Appl. Spectrosc.* **52**, 869 (1998).

107. M. Nelson, J. Aust, J. Dobrowolski, P. Verly, and M. Myrick, *Anal. Chem.* **70**, 73 (1998).

108. O. Soyemi, D. Eastwood, L. Zhang, H. Li, J. Karunamuni, P. Gemperline, R. Synowicki, and M. Myrick, *Anal. Chem.* **73**, 1069 (2001).

109. M. Myrick, O. Soyemi, H. Li, L. Zhang, and D. Eastwood, *Fresenius J. Anal. Chem.* **369**, 351 (2001).

110. M. Myrick, O. Soyemi, J. Karunamuni, D. Eastwood, H. Li, L. Zhang, A. Greer, and P. Gemperline, *Vibrat. Spec.* **28**, 73 (2002).

111. O. Soyemi, F. Haibach, G. Frederick, P. Gemperline, and M. Myrick, *Appl. Spec.* **56**, 477 (2002).

112. K. Johnston, S. Yee, and K. Booksh, *Anal. Chem.* **69**, 1844 (1997).

113. F. Dieterle, S. Busche, and G. Gauglitz, *Anal. Chim Acta* **490**, 71 (2003).

114. F. Dieterle, B. Kieser, and G. Gauglitz, *Chem Intel. Lab. Syst.* **65**, 67 (2003).

115. C. Lucasius and G. Katerman, *Chem Intel. Lab. Syst.* **19**, 1 (1993).

116. C. Lucasius and G. Katerman, *Chem Intel. Lab. Syst.* **25**, 99 (1994).

117. B. Smith and P. Gemperline, *Anal. Chim. Acta* **423**, 167 (2000).

118. R. Leardi, *J. Chemom.* **15**, 559 (2001).

FRANK VOGT
KARL BOOKSH
Arizona State University

CHIRAL SEPARATIONS

1. Introduction

Chiral separations are concerned with separating molecules that can exist as nonsuperimposable mirror images. Examples of these types of molecules, called *enantiomers* or *optical isomers*, are illustrated in Figure 1. Although chirality is often associated with compounds containing a tetrahedral carbon with four different substituents, other atoms, such as phosphorus or sulfur, may also be chiral. In addition, molecules containing a center of asymmetry, such as hexahelicene, tetrasubstituted adamantanes, and substituted allenes or molecules with hindered rotation, such as some 2,2′ disubstituted binaphthyls, may also be

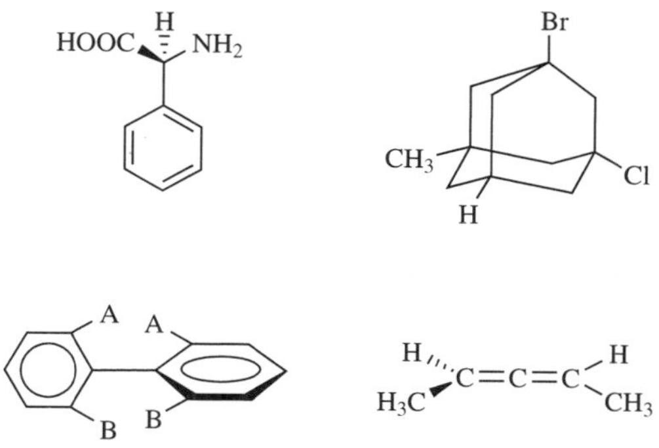

Fig. 1. Examples of chiral molecules.

chiral. Compounds exhibiting a center of asymmetry are called *atropisomers*. An extensive review of stereochemistry may be found under PHARMACEUTICALS, CHIRAL.

Although scientists have known since the time of Louis Pasteur (1) that optical isomers can behave differently in a chiral environment (eg, in the presence of polarized light), it has only been since about 1980 that there has been a growing awareness of the implications arising from the fact that many drugs are chiral and that living systems constitute chiral environments. Hence, the optical isomers of chiral drugs may exhibit different bioactivities and/or biotoxicities.

In the case of enantiomerically pure chiral drugs, the possibility of racemization or inversion either *in vivo* or during storage cannot be ruled out. Ibuprofen is an example of a chiral drug which undergoes rapid inversion *in vivo* (2). In addition, there are several examples of achiral (or *prochiral*) drugs being biotransformed into chiral entities. In some cases, the enantiomeric ratios produced by laboratory animals may differ from that produced in humans. For example, cimetidine (**1**), used to treat peptic ulcers and marketed as Tagamet, is achiral. However, cimetidine sulfoxide (**2**), one of its major metabolites, is chiral by virtue of oxidation of the sulfur atom to a sulfoxide (the lone pair of electrons on the sulfur constitutes the fourth group). In humans, the (+) enantiomer predominates (2.4:1) but in rats, although (+)-cimetidine sulfoxide is produced in excess, the enantiomeric ratio approaches racemic (1.3:1). This raises the question of the suitability of rats as appropriate test models for this particular drug (3).

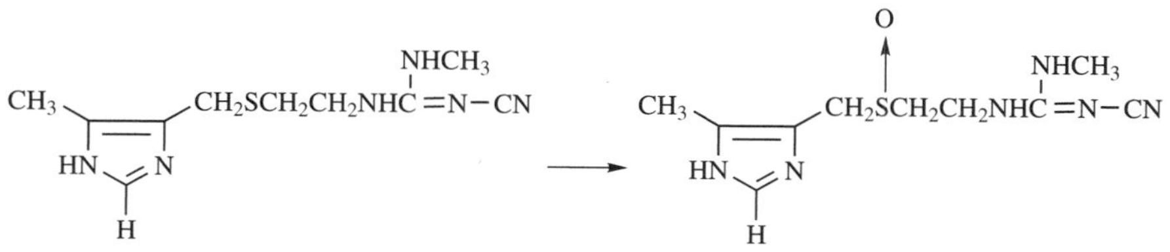

For those drugs that are administered as the racemate, each enantiomer needs to be monitored separately yet simultaneously, since metabolism, excretion or clearance may be radically different for the two enantiomers. Further complicating drug profiles for chiral drugs is that often the pharmacodynamics and pharmacokinetics of the racemic drug is not just the sum of the profiles of the individual enantiomers.

Although it might seem that administration of enantiomerically pure substances would always be preferred, the diuretic indacrinone (3), is an example of a drug for which one enantiomer mediates the harmful effects of the other enantiomer (4). (+)-Indacrinone, the diuretically active enantiomer or *eutomer*, causes uric acid retention. Fortunately, the other enantiomer (*distomer*) causes uric acid elimination. Thus, administration of a mixture of the two enantiomers, although not necessarily racemic, may have therapeutic value.

(3)

Although a great deal of the work currently being done in chiral separations is related to pharmaceuticals, the agricultural and the food and beverage industries are affected as well. For instance, several chiral pesticides are used commercially. It is possible that the enantiomers may differ in their persistence in the environment and their effectiveness against specific pests. For example, the neurotoxic action of the pesticide, ethyl-4-nitrophenyl phenylphosphono thionate (EPN), resides almost entirely in the S enantiomer while the desired insecticidal activity resides entirely in the R enantiomer (5). This raises the question of whether the pesticide may be safer and more effective if applied as an enantiomerically pure formulation. In the food and beverage industry, many of the constituents that confer flavor or aroma in foods and beverages are chiral. For instance, the configuration of the 4-alkyl-substituted γ-lactones responsible for much of the flavor in fruits is almost exclusively R (6). Often, the two enantiomers have very different aromas or flavors. The presence of any of the "unnatural" enantiomer may confer an "off-flavor" to the substance and may be indicative of racemization under adverse storage conditions, adulteration, or formulation from nonnatural sources.

The growing awareness of the implications of chirality to the pharmaceutical industry has spurred tremendous effort toward stereoselective synthetic strategies and the development of new chiral catalysts. However, the enantiomeric purity of these substances or their chiral precursors needs to be determined. Also, there are many chiral compounds for which no stereospecific synthetic pathways have been devised. Thus, there is a tremendous need not only for analytical scale (<5 − 10 mg), but bulk-scale chiral separations as well. Whether analyzing drugs or synthetic precursors for enantiomeric purity, monitoring biological or environmental samples for chiral discrimination or trying to enantioresolve kilogram quantities of a racemic drug, there are a variety of reasons for performing chiral separations. The purpose of the separation dictates, to some extent, the method employed.

Traditionally, chiral separations have been considered among the most difficult of all separations. Conventional separation techniques, such as distillation, liquid–liquid extraction, or even some forms of chromatography, are usually based on differences in analyte solubilities or vapor pressures. However, in an achiral environment, enantiomers or optical isomers have identical physical and chemical properties. The general approach, then, is to create a "chiral environment" to achieve the desired chiral separation and requires chiral analyte–chiral selector interactions with more specificity than is obtainable with conventional techniques.

A variety of strategies have been devised to obtain chiral separations. Although the focus of this article is on chromatographically based chiral separations, other methods include crystallization and stereospecific enzymatic-catalyzed synthesis or degradation. In crystallization methods, racemic chiral ions are typically resolved by the addition of an optically pure counterion, thus forming diastereomeric complexes.

Enzymatically based methods depend on the stereospecificity of an enzyme-catalyzed reaction, such as lipase-catalyzed esterification, to degrade enantioselectively the unwanted enantiomer or to produce the desired enantiomer. Because only one enantiomer undergoes the reaction, the subsequent separation is reduced to separating two different species. For example, in the case of enzyme-catalyzed esterification, the originally difficult enantiomeric separation is reduced to the separation of the ester of one optical isomer from the alcohol or acid of the other optical isomer of the original starting material, and may be accomplished using a variety of conventional separation methodologies (7). One disadvantage of enzymatically based methods is that only one enantiomer is obtained and there is usually no analogous method for producing the opposite enantiomer.

An alternative method of creating a chiral environmental is to derivatize a chiral analyte with an optically pure reagent, thus, producing diastereomers. The resultant diastereomers, containing more than one chiral center, have slightly different melting and boiling points and can often be separated using conventional methods. A number of chiral derivatizing agents, as well as the types of compounds for which they are useful, have been developed and are listed in Table 1. Limitations of this approach include lack of suitable functionality in the analyte that can be derivatized with an appropriate enantiomerically pure derivatizing agent, unavailability of a suitable derivatizing agent of sufficiently high or at least known optical purity, difficulty of removing the derivatizing group after the desired separation has been accomplished, enantiodiscrimination during derivatization, potential racemization either during derivatization or removal or the chiral derivatizing group (which is not always possible), and the additional validation required to confirm that the enantiomeric ratio of the final product corresponds to the original enantiomeric ratio.

2. Use of Chiral Additives

Another method for creating a chiral environment is to add an optically pure chiral selector to a bulk liquid phase. Historically, many of the chiral selectors

Table 1. **Analyte Functional Groups and Chiral Derivatizing Reagents**

Analyte functional group	Derivatizing agent	Product	Examples of derivatizing agents
carboxylic acid (acid or base catalyzed)	alcohol amine	ester amide	(−)-menthol 1-phenylethylamine 1-(1-naphthyl)ethylamine
amine (1°)	aldehyde	isoindole	o-phthaldialdehyde–2-mercaptoethanol
amine (1° and 2°)	anhydrides	amide	γ-butyloxycarbonyl-L-leucine anhydride O,O-dibenzoyltartaric anhydride
	acyl halides	amide	(R)-(−)-methylmandelic acid chloride α-methoxy-α-trifluoromethylphenyl- acetyl chloride
	isocyanates	urea	α-methylbenzyl isocyanate 1-(1-naphthyl)ethyl isocyanate
	isothiocyanate	thiourea	2,3,4,6-tetra-O-acetyl-β-D-glucopyrano- syl isothiocyanate α-methylbenzyl isothiocyanate
(1°, 2°; can N-dealkylate 3°)	chlorofor- mates	carbamate	(−)-menthyl chloroformate (+)-1-(9-fluorenyl)ethylchloroformate
alcohols	acyl halides	ester	(−)-menthoxy acid chloride (S)-O-propionylmandelyl chloride
	anhydrides	ester	(S,S)-tartaric anhydride
	chloroformate	carbonate	(−)-menthyl chloroformate
	isocyanate	carbamate	α-methylbenzyl isocyanate

currently available as chiral stationary phases for high performance liquid chromatography originated as chiral mobile-phase additives, particularly in thin-layer chromatography (tlc). Chiral additives have several advantages over chiral stationary phases and continue to be the predominant mode for chiral separations by tlc (8) and capillary electrophoresis (ce) (9). First of all, the chiral selector added to a bulk liquid phase can be readily changed. The use of chiral additives allows chiral separations to be done using less expensive, conventional stationary phases. A wider variety of chiral selectors are available to be used as chiral additives than are available as chiral stationary phases, thus, providing the analyst with considerable flexibility. Finally, the use of chiral additives may provide valuable insight into the chromatographic conditions and/or likelihood of success with a potential chiral stationary-phase chiral selector. This is particularly important for the development of new chiral stationary phases because of the difficulty and cost involved.

Chiral additives, however, do pose some unique problems. Many chiral agents are expensive or are not commercially available, and therefore, must be synthesized. The presence of the chiral additive in the bulk liquid phase may also interfere with detection or recovery of the analytes. Finally, the presence of enantiomeric impurity in the chiral additive may add analytical complications (10).

2.1. Thin-Layer Chromatography. Thin-layer chromatography (tlc) offers several advantages for chiral separations and in the development of new chiral stationary phases. Besides being inexpensive, tlc can be used to screen mobile-phase conditions rapidly (ie, organic modifier content, pH, etc), chiral selectors, and analytes. Several different analytes may be run simultaneously on the same plate. Usually, no preequilibration of the mobile phase and stationary phase is required. In addition, only small amounts of mobile phase, and therefore, chiral mobile-phase additive, are required. Another significant advantage is that the analyte can always be unambiguously found on the tlc plate.

Two mechanisms for chiral separations using chiral mobile-phase additives, analogous to models developed for ion-pair chromatography, have been proposed to explain the chiral selectivity obtained using chiral mobile-phase additives. In one model, the chiral mobile-phase additive and the analyte enantiomers form "diastereomeric complexes" in solution. As noted previously, diastereomers may have slightly different physical properties such as mobile phase solubilities or slightly different affinities for the stationary phase. Thus, the chiral separation can be achieved with conventional columns.

An alternative model has been proposed in which the chiral mobile-phase additive is thought to modify the conventional, achiral stationary phase *in situ*, thus, dynamically generating a chiral stationary phase. In this case, the enantioseparation is governed by the differences in the association between the enantiomers and the chiral selector in the stationary phase.

Several different types of chiral additives have been used including (1R)-(−)-ammonium-10-camphorsulfonic acid (11), cyclodextrins (12,13), proteins, and various amino acid derivatives such as N-benzoxycarbonyl-glycyl-L-proline as well as macrocyclic antibiotics (14). Chiral counterions such as (1R)-(-)-ammonium-10-camphorsulfonic acid and N-benzoxycarbonyl-glycyl-L-proline have been used under normal phase conditions (eg, ca 2.5 mM with 1 mM triethylamine in methylene chloride on a diol tlc plate), promoting ion-pair associations (15). In contrast, the cyclodextrins (16,17), proteins, and amino acid derivatives have been used exclusively under aqueous mobile phase conditions. In the case of the cyclodextrins, the limited aqueous solubility of β-cyclodextrin ($\sim$0.17 M at room temperature), the most commonly used cyclodextrin, can be enhanced by using a saturated urea solution. In addition, it is recommended that 0.6 M NaCl can be used to stabilize the binder by which the stationary phase is attached to the glass support (18).

Chiral separation validation in tlc may be accomplished by recovering the individual analyte spots from the plate and subjecting them to some type of chiroptical spectroscopy such as circular dichroism or optical rotary dispersion. Alternatively, the plates may be analyzed using a scanning densitometer. Scanning densitometers irradiate the surface of the plate at a specified wavelength (in the ultraviolet or visible regions) and can measure the intensity of the reflected beam. A trace of the reflected beam vs distance has the general appearance of a chromatogram and an example is shown in Figure 2 (19). The relative peak heights or areas of the two enantiomers obtained at two or more different wavelengths should remain constant because the extinction coefficients of the enantiomers are identical at every wavelength.

2.2. Capillary Electrophoresis. Capillary electrophoresis (ce) or capillary zone electrophoresis (cze), a relatively recent addition to the arsenal of analytical techniques (20,21), has also been demonstrated as a powerful chiral separation method. Its high resolution capability and lower sample loading relative to hplc makes it ideal for the separation of minute amounts of components in complex biological mixtures (22,23).

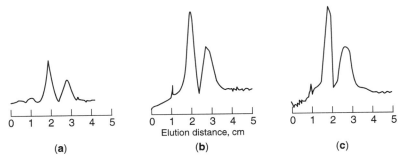

Fig. 2. Tlc densitometer scans showing the resolution of isoproterenol on a hptlc silica-gel plate obtained using a mobile phase consisting of 6.8 mM (1R)-(-)-ammonium-10-camphorsulfonic acid in 75:25 (v/v) methylene chloride:methanol. (**a**) 254 nm, (**b**) 275 m, (**c**) 300 nm.

In a ce experiment, a thin capillary is filled with a run buffer and a voltage is applied across the capillary. Although a complete treatment of the fundamental principles of ce is beyond the scope of this article, it can be said that the underlying impetus for separations in ce is, in general, derived from the fact that charged species migrate in response to an applied electric field proportionately to their charge and inversely proportionately to their size. Thus, given equivalent charges, lighter analytes have higher electrophoretic mobilities than heavier analytes and, given equivalent sizes, more highly charged species have higher mobilities than lesser charged or neutral species. In fact, neutral species have no intrinsic electrophoretic mobility. Species having opposite charges have electrophoretic mobilities in opposing directions.

Chiral separations by ce have been performed almost exclusively using chiral additives to the run buffer. The advantages of this approach are identical to the advantages mentioned previously with regard to using chiral mobile-phase additives in tlc. Many of the chiral selectors used successfully as mobile-phase additives in tlc and as immobilized ligands in hplc have been used successfully in ce including proteins (24), native (25) and functionalized cyclodextrins (26,27), various carbohydrates (28,29), assorted functionalized amino acids (30), chiral-ion pairing agents (31), and macrocyclic antibiotics (32). Other ce chiral selectors which have not been used as immobilized chiral selectors in hplc include bile salts (33), chiral surfactants (34), and dextran sulfate (35).

Although chiral ce is most commonly performed using aqueous buffers, there has been some work using organic solvents such as methanol, formamide, N-methylformamide or N,N-dimethylformamide with chiral additives such as quinine (36) or cyclodextrins (37,38). Nonaqueous ce requires that the background electrolyte be prepared using organic acids (eg, citric acid or acetic acid) and organic bases (eg, tetraalkylammonium halides or tris(hydroxy-methyl)-aminomethane).

Theoretical models (39,40), as expressed in equation 1, where μ represents the mobility

$$\mu_1 - \mu_2 = \frac{\left[\mu_{1,c} - \mu_{1f}\right]\left[K_1 - K_2\right][CA]}{\left[1 + K_1[CA]\right]\left[1 + K_2[CA]\right]} \tag{1}$$

of the analyte in the free and complexed states, K represents the binding constants of enantiomers 1 and 2 and [CA] is the molar concentration of the additive, reveal that, in general, two conditions must be met to achieve chiral separations by ce. First, there must be differences in the binding constants of the two enantiomers with the chiral selector. Second, because the intrinsic electrophoretic mobilities of the enantiomers are identical in the free state, there must be a significant difference in the mobilities of the analyte in the complexed and free state. Chiral selectors have generally been shown to be the most effective when the intrinsic electrophoretic mobility of the additive is in the opposite direction of the intrinsic electrophoretic mobility of the analyte.

Chiral separations by ce is a rapidly growing field and offers the analyst tremendous flexibility with regard to chiral selector choice. In addition, because the additive is typically in the run buffer, there is virtually no column preequilibration. However, ce instruments tend to cost more than most chromatographic systems, sample capacity is much smaller for ce than for an analogous hplc method, sample recovery is not a trivial problem in ce and run-to-run reproducibility for ce tends to be much worse than for most chromatographic methods. Nevertheless, the flexibility, minimal sample and/or chiral selector required and the extremely high resolving power of ce ensure that this technique will continue to play an important role in chiral separations in the future.

3. Chiral Stationary Phases

Most chiral chromatographic separations are accomplished using chromatographic stationary phases that incorporate a chiral selector. The chiral separation mechanisms are generally thought to involve the formation of transient diastereomeric complexes between the enantiomers and the stationary phase chiral ligand. Differences in the stabilities of these complexes account for the differences in the retention observed for the two enantiomers. Often, the use of a chiral stationary phase allows for the direct separation of the enantiomers without the need for derivatization. One advantage offered by the use of chiral stationary phases is that the chiral selector need not be enantiomerically pure, only enriched. In addition, for chiral stationary phases having a well understood chiral recognition mechanism, assignment of configuration (eg, R or S) may be possible even in the absence of optically pure standards. However, chiral stationary phases have some limitations. The specificity required for chiral discrimination limits the broad applicability of most chiral stationary phases; thus there is no "universal" chiral stationary phase. The cost of most chiral columns are typically much higher ($\sim 3\times$) than for conventional columns. In contrast to conventional chromatographic columns, chiral stationary phases are generally not as robust, require more careful handling than conventional columns and usually, once column performance has begun to deteriorate, cannot be returned to their original performance levels. In many cases, chromatographic column choice or mobile-phase optimization for chiral stationary phases is not as straightforward as with conventional stationary phases. In conventional chromatography, there is usually a well-behaved relationship between retention and mobile phase composition or column temperature. However, in many of the chiral stationary phases, the stationary phase present a multitude of different types of sites with not necessarily equivalent populations for interaction with the analytes. In the case of some liquid chromatographic stationary phases, the different types of sites may result in normal phase type behavior under very nonpolar mobile phase conditions and reversed-phase type behavior under highly polar mobile phase conditions. The multiplicity of types and numbers of sites also confounds thermodynamic considerations (41). Often, there is a narrow window of mobile-phase conditions under which enantioselectivity is observed and these conditions may be unique for a particular chiral analyte. Thus, for many of the chiral stationary phases, adequate chiral recognition models, used to guide selection of the appropriate column for a given separation, have yet to be developed. Column selection, therefore, is often reduced to identifying structurally similar analytes for which chiral resolution methods have been reported in the scientific literature or chromatographic supply catalogues and adapting a reported method for the chiral pair to be resolved.

An additional complication, sometimes arising with the use of chiral stationary phases, may occur when the analytes either exist as *conformers* or can undergo inversion during the chromatographic analysis. Figure 3 illustrates a typical chromatogram obtained for oxazepam, one of the chiral benzodiazepines that can undergo ring opening and inversion at the chiral center (42). As can be seen from Figure 3, the peaks appear to have a plateau between them and are sometimes referred to as "Batman peaks". This effect can sometimes be suppressed by lowering the column temperature. Although the appearance of Batman peaks is not unique to chiral separations, the specificity of chiral analyte–chiral selector interactions may increase the frequency of their occurrence.

3.1. Thin-Layer Chromatography.
Chiral stationary phases have been used less extensively in tlc as in high performance liquid chromatography (hplc). This may, in large part, be due to lack of availability. The cost of many chiral

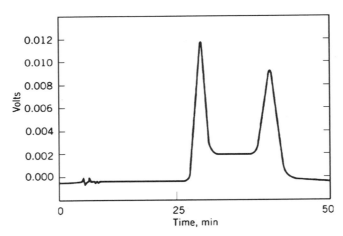

Fig. 3. The chiral separation obtained for oxazepam on a sulfated cyclodextrin hplc column (4.6 mm ID × 25 cm) using a 10% acetonitrile/buffer (25 mM ammonium acetate, pH 7).

selectors, as well as the accessibility and success of chiral additives, may have inhibited widespread commercialization. Usually, nondestructive visualization of the sample spots in tlc is accomplished using iodine vapor, uv or fluorescence. However, the presence of the chiral selector in the stationary phase can mask the analyte and interfere with detection (43).

Chiral stationary phases in tlc have been primarily limited to phases based on normal or microcrystalline cellulose (44,45), triacetylcellulose sorbents or silica-based sorbents that have been chemically modified (46) or physically coated to incorporate chiral selectors such as amino acids (47,48) or macrocyclic antibiotics (49) into the stationary phase.

Of the silica-based materials, only the ligand-exchange phases are commercially available (Chiralplate, tlc plates are available through Alltech Associates, Inc.) Supelco, Inc., the Aldrich Chemical Company, and Bodman Industries are all based on ligand exchange. Typically in the case of the ligand-exchange type tlc plates, the ligand-exchange selector is comprised of an amino acid residue to which a long hydrocarbon chain has been attached (eg, (2S,4R,2'RS)4-hydroxy-1-(2-hydroxydodecyl)proline) (50). The hydrocarbon chain of the functionalized amino acid is either chemically bonded to the substrate or intercalates in between the chains of a reversed phase-stationary phase thus immobilizing the chiral selector. The bidentate amino acid chiral selector is thought to reside close to the surface of the stationary phase and participates as a ligand in the formation of a bi-ligand complex with a divalent metal ion (eg, Cu^{2+}) and the chiral bidentate analyte (Fig. 4). Analytes enantioresolvable using ligand exchange are usually restricted to 1,2-diols, α-amino acids, α-amino alcohols, and α-hydroxyacids (51,52). Again, differences in the stabilities of the diastereomeric complexes thus formed give rise to the chiral separation.

3.2. High Performance Liquid Chromatography. Although chiral mobile phase additives have been used in high performance liquid chromatography (hplc), the large amounts of solvent, thus chiral mobile phase additive, required to pre-equilibrate the stationary phase renders this approach much less attractive than for tlc and is not discussed here.

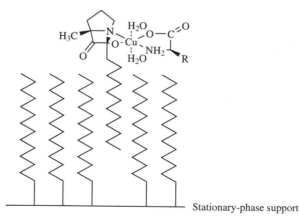

Fig. 4. A ligand-exchange chiral selector complexed with a chiral analyte.

Table 2. **Classes of Hplc Chiral Stationary Phases**

Column chiral selector	Typical mobile phase conditions	Typical analyte features required
pirkle	nonpolar organic; 2-propanol–hexane	π-acid or π-basic moieties for charge transfer complex; hydrogen-bonding or dipole stacking capability near chiral center
protein	phosphate buffers	aromatic near chiral center; organic acids or bases; cationic drugs
cyclodextrin	aqueous buffers; polar organic	good "fit" between chiral cavity or chiral mouth of cyclodextrin and hydro-phobic moiety; hydrogen-bonding capability near chiral center
ligand exchange	aqueous buffers	α-hydroxy or α-amino acids near chiral center; can do nonaromatic
chiral crown ether	0.01 N perchloric acid	primary amines near chiral center; can do nonaromatic
macrocyclic antibiotics	aqueous buffers, nonpolar and polar organic	amines, amides, acids, esters; aromatic; hydrophobic moiety
cellulosic and amylosic	nonpolar organic	aromatic

The last decade has seen the commercialization of a large number of different types of chiral stationary phases including the cyclodextrin phases (53), the chirobiotic phases (54), the π–π interaction phases (55,56), the protein phases (57–61), as well as the cellulosic and amylosic phases (62,63) and chiral crown ether phases (64,65). Currently, there are over 50 different chiral columns that are commercially available for hplc. Table 2 briefly summarizes the types of columns available as well as typical applications and mobile-phase conditions. Each of these chiral stationary phases are very successful at separating large numbers of enantiomers, which in many cases, are unresolvable using any of the other chiral stationary phases. Unfortunately, despite the large number and variety of chiral stationary phases currently available, there remains a large number of enantiomeric compounds that are unresolvable by any of the existing chiral stationary phases. In addition, incomplete understanding of the chiral recognition mechanisms of many of these chiral stationary phases limits the realization of the full potential of the existing chiral stationary phases and hampers development of new chiral stationary phases.

4. Ligand-Exchange Phases

Among the earliest reports of chiral separations by liquid chromatography were based on work done by Davankov using ligand exchange (66). These types of columns are available from Phenomenex, J. T. Baker, and Regis Technologies, Inc. As noted previously in the discussion regarding ligand exchange in tlc, chiral separations by ligand exchange in hplc is accomplished using bidentate amino acid ligands, immobilized on a chromatographic substrate, and a divalent metal cation which participates in the formation of a diastereomeric complex with a bidentate chiral analyte and the ligand. Although almost any amino acid can form the basis for the chiral selector, proline and hydroxyproline exhibit the most widespread utility. Also, although other metals can be used, copper(II) is usually the metal of choice and is added to the aqueous buffer mobile phase.

The dependence of chiral recognition on the formation of the diastereomeric complex imposes constraints on the proximity of the metal binding sites, usually either an hydroxy or an amine α to a carboxylic acid, in the analyte. Principal advantages of this technique include the ability to assign configuration in the absence of standards, enantioresolve nonaromatic analytes, use aqueous mobile phases, acquire a stationary phase with the opposite enantioselectivity, and predict the likelihood of successful chiral resolution for a given analyte based on a well-understood chiral recognition mechanism.

5. Pirkle Phases

The first commercially available chiral column for liquid chromatography was introduced in 1980. This was the first generation of the "Pirkle phases", named after their originator, and was based on N-(3,5-dinitrobenzoyl)phenylgly-cine which was immobilized on a silica support (67). Of all of the commercially available chiral stationary phases for liquid chromatography, the chiral recognition mechanism for the "Pirkle" phases are among the best understood. Chiral recognition on Pirkle phases is thought to depend upon complimentary interactions between the analyte and the selector. These interactions may be $\pi-\pi$, steric, hydrogen-bonding, or dipole–dipole interactions and contribute to the overall stability of the diastereomeric association complexes that form between the individual enantiomers and the chiral selector in the stationary phase. The $\pi-\pi$ interactions arise through the association of aromatic systems with complementary electron withdrawing (eg, nitro) and electron donating (eg, alkyl) substituents. The electron-deficient aromatic system is often referred to as π-*acidic*; the electron-rich system is usually referred to as π-*basic*. Three unique interactions emanating from the chiral centers of the analyte and their chiral ligand in the stationary phase, seem to be required for successful chiral recognition. A model invoking three unique points of interaction is sometimes referred to as the *3-point interaction model* first proposed by Dalgliesh (68). To promote analyte–selector interactions, functional groups are often introduced into the analyte through achiral derivatization. For example, amines may be derivatized with 3,5-dinitrobenzoyl chloride to introduce a π-acid aromatic group to promote diastereomeric complexation with a π-basic (R)-N-(2-naphthyl)-alanine chiral selector in the stationary phase. Derivatization often has the additional benefit of enhancing solute solubility.

Nonpolar organic mobile phases, such as hexane with ethanol or 2-propanol as typical polar modifiers, are most commonly used with these types of phases. Under these conditions, retention seems to follow normal phase-type behavior (eg, increased mobile phase polarity produces decreased retention). The normal mobile-phase components only weakly interact with the stationary phase and are easily displaced by the chiral analytes thereby promoting enantiospecific

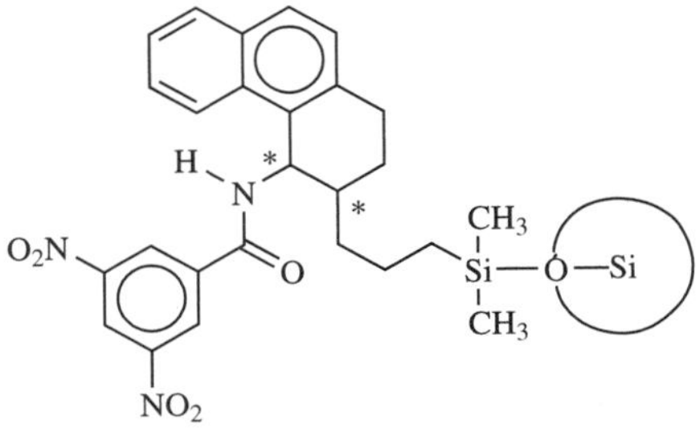

Fig. 5.   The structure of the chiral selector in the Whelk-O-1 chiral stationary phase.

interactions. Some of the Pirkle-types of phases have also been used, to a lesser extent, in the reversed phase mode.

Reciprocity, an important concept introduced by Pirkle (69,70), exploited the notion that analytes that were well resolved using a particular chiral selector would likely be good candidates for chiral selectors to enantioresolve analytes similar to the original chiral selector. For instance, the first generation Pirkle phase incorporating N-(3,5-dinitrobenzoyl)phenylglycine was very successful at enantioresolving compounds containing naphthyl moieties near the stereogenic center. This insight spawned a second generation of Pirkle phases based on N-(2-naphthyl)-α-amino acids (71). These phases were very successful at enantioresolving analytes containing a 3,5-dinitrobenzoyl group, such as 3,5-dinitrophenyl carbamates, and ureas of chiral alcohols and amines (72). These columns are available through a variety of sources including Phenomenex, Regis Technologies, Inc., J. T. Baker, Inc., and Supelco, Inc.

The structure of the Whelk-O-1 phase, the most recent addition to this type of chiral stationary phase, is illustrated in Figure 5. This selector has a wedge-like chiral surface with one edge offering the π-basic tetrahydrophenanthrene ring system; the other edge is comprised of a 3,5-dinitrobenzoyl π-acidic moiety. The amide linkage between the two ring systems presents dipole stacking and hydrogen-bonding interaction sites. The presence of both π-acid and π-base features, as well as the inherent rigidity of the chiral selector, confers greater versatility than any of the previous Pirkle-type phases, imposing fewer constraints on both analyte structural features required for successful enantioresolution and mobile phase conditions. Indeed, this chiral stationary phase has demonstrated considerable chiral selectivity for naproxen, warfarin, and its _p_-chloro analogue under nonaqueous reversed-phase conditions (73) and reversed-phase conditions (74,75). An additional advantageous feature of this phase is its availability with either the (R,R) or (S,S) configuration, thus, permitting the enantiomeric elution order to be readily changed. The small size of the chiral selector also promotes fairly high bonded ligand densities in the stationary phase, which coupled with the high enantioselectivities often achieved with these phases, facilitates their use for preparative-scale separations (76).

6. Cyclodextrin Phases

Cyclodextrins are macrocyclic compounds comprised of D-glucose bonded through 1,4-α-linkages and produced enzymatically from starch. The greek letter which

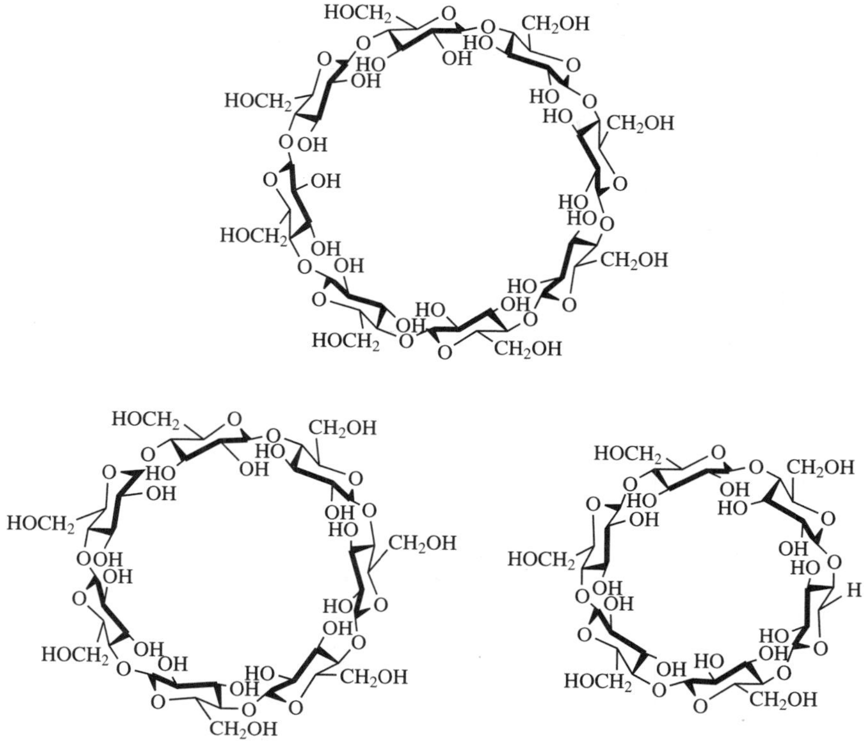

Fig. 6. The structure of the three most common cyclodextrins.

proceeds the name indicates the number of glucose units incorporated in the CD (eg, $\alpha = 6$, $\beta = 7$, $\gamma = 8$, etc). Cyclodextrins are toroidal shaped molecules with a relatively hydrophobic internal cavity (Fig. 6). The exterior is relatively hydrophilic because of the presence of the primary and secondary hydroxyls. The primary C-6 hydroxyls are free to rotate and can partially block the CD cavity from one end. The mouth of the opposite end of the CD cavity is encircled by the C-2 and C-3 secondary hydroxyls. The restricted conformational freedom and orientation of these secondary hydroxyls is thought to be responsible for the chiral recognition inherent in these molecules (77).

Among the most successful of the liquid chromatographic reversed-phase chiral stationary phases have been the cyclodextrin-based phases, introduced by Armstrong (78,79) and commercially available through Advanced Separation Technologies, Inc. or Alltech Associates. The most commonly used cyclodextrin in hplc is the β-cyclodextrin. In the bonded phases, the cyclodextrins are thought to be tethered to the silica substrate through one or two spacer ligands (Fig. 7). The mechanism thought to be responsible for the chiral selectivity observed with these phases is based on the formation of an inclusion complex between the hydrophobic moiety of the chiral analyte and the hydrophobic interior of the cyclodextrin cavity (Fig. 8). Preferential complexation between one optical isomer and the cyclodextrin through stereospecific interactions with the secondary hydroxyls which line the mouth of the cyclodextrin cavity results in the

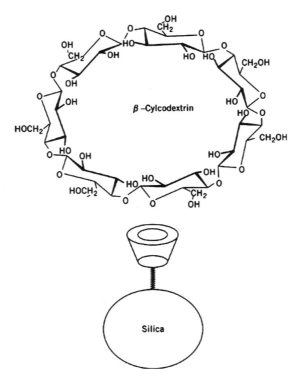

Fig. 7. A tethered cyclodextrin and the structure of β-cyclodextrin, the most common cyclodextrin used as a bonded ligand in liquid chromatography.

enantiomeric separation. Unlike the Pirkle-type phases, enantiospecific interactions between the analyte and the cyclodextrin are not the result of a single, well-defined association, but more of a statistical averaging of all the potential interactions with each interaction weighted by its energy or strength of interaction (80).

Vast amounts of empirical data suggest that chiral recognition on cyclodextrin phases in the reversed phase mode require the presence of an aromatic moiety that can fit into the cyclodextrin cavity, that there be hydrogen bonding

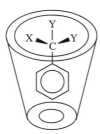

Fig. 8. A hydrophobic inclusion complex between a chiral analyte and a cyclodextrin.

groups in the molecule, and that the hydrophobic and hydrogen-bonding moieties should be in close proximity to the stereogenic center. Chiral recognition seems to be enhanced if the stereogenic center is positioned between two π-systems or incorporated in a ring.

Most of the chiral separations reported to date using the native cyclodextrin-based phases have been accomplished in the reversed-phase mode using aqueous buffers containing small amounts of organic modifiers. However, polar organic mobile phases have gained in popularity recently because of their ease of removal from the sample and reduced tendency to accelerate column degradation relative to the hydroorganic mobile phases (81). In these cases, because the more nonpolar component of the mobile phase is thought to occupy the cyclodextrin cavity, the analyte is thought to sit atop the mouth of the cyclodextrin much like a "lid".

Limitations with the chiral selectivity of the native cyclodextrins fostered the development of various functionalized cyclodextrin-based chiral stationary phases, including acetylated (82,83), sulfated (84), 2-hydroxypropyl (85), 3,5-dimethylphenylcarbamoylated (86) and 1-naphthylethylcaarbamoylated (87) cyclodextrin. Each of the glucose residues contribute three hydroxyl groups to which a substituent may be appended; thus, each cyclodextrin contributes multiple sites for derivatization. Typical degrees of substitution per β-cyclodextrin (with 21 hydroxyls) range from three to ten. Hence, there are many residual hydroxyls on each cyclodextrin.

The substituents of these functionalized cyclodextrins seem to play a variety of roles in enhancing chiral recognition. In some cases, the substituent may only serve to enlarge the chiral cavity or may provide alternative interaction sites. For instance, in the case of the naphthylethylcaarbamoylated cyclodextrin, the naphthyl ring provides a π-basic site and the carbamate linkage provides additional hydrogen bonding and dipole interaction sites not available with the native cyclodextrin. On the sulfated cyclodextrin phase, the sulfate group presents the potential for ion-pair formation unavailable with the native cyclodextrin. The introduction of 2-hydroxypropyl- and 1-naphthylethylcarbamoyl substituents incorporates additional stereogenic centers onto the cyclodextrin. In some cases, the configuration of the substituent dominated the enantiomeric elution order. However, in other cases, the enantiomeric elution order was independent of the configuration of the substituent. In addition, in some cases, the chiral selectivity of the cyclodextrin seemed to synergistically augment the chiral selectivity of one configuration of the substituent while antagonizing the chiral selectivity of the oppositely configured substituent. A particularly attractive feature of these functionalized cyclodextrins is that many of them exhibit enantioselectivity under hydroorganic reversed phase, as well as normal phase and polar organic mobile-phase conditions, and that each set of conditions can provide chiral separations for analytes which are not resolved under any of the other type of mobile-phase conditions. Further, the chromatographic mode (eg, reversed phase to normal phase) can be readily changed with no deleterious impact on chiral recognition as long as routine care is taken to avoid problems with solvent immiscibility. The naphthylethyl-carbamoylated cyclodextrin phase was considered to be one of the first "multimodal" chiral stationary phases (88).

7. Cellulosic and Amylosic Phases

Cellulose and amylose are comprised of the same glucose subunits as the cyclo-dextrins. In the case of cellulose, the glucose units are attached through 1,4-β-linkages resulting in a linear polymer. In the case of amylose, the 1,4-α-linkages, as are found in the cyclodextrins, are thought to confer helicity to the polymeric chain.

As mentioned previously, cellulosic phases as well as amylosic phases have also been used extensively for enantiomeric separations more recently (89,90). Most of the work in this area has been with various derivatives of the native car-bohydrate. The enantioresolving abilities of the derivatized cellulosic and amylo-sic phases are reported to be very dependent on the types of substituents on the aromatic moieties that are appended onto the native carbohydrate (91). Table 3 lists some of the cellulosic and amylosic derivatives that have been used. These columns are available through Chiral Technologies, Inc. and J. T. Baker, Inc.

With the exception of the microcrystalline cellulose I triacetate (92) and tri-benzoate materials, which are sufficiently robust to be used directly as packing material, most of the commercially available cellulosic and amylosic phases are comprised of mixtures of exhaustively derivatized polymers which are coated onto large pore γ-aminopropyl silica. These coated polymeric phases exhibit admirable enantioselectivities, but as is the case with all commercially available chiral stationary phases, they also have some potential disadvantages. The large polymer size requires the use of fragile, large pore silica. The fact that the chiral selector for these phases is coated onto the silica sometimes restricts the types of mobile phases that can be used. In addition, the secondary structure of the poly-mer, which seems to be important in the chiral recognition mechanism, may be altered irreversibly by storing the columns in polar solvents leading to disastrous consequences for chiral separations. The polymeric nature of the chiral selectors for these phases and the importance of the secondary structure also hamper the development of models for the chiral recognition mechanism for these phases (93). Despite these factors, the cellulosic and amylosic phases have enjoyed tre-mendous success at enantioresolving structurally diverse compounds (94,95) and have some of the highest capacities of all the chiral commercially available chiral stationary phases, thus, rendering them among the most suitable for preparative chromatography (96).

Table 3. **Carbohydrate Derivatives Used as Hplc Chiral Stationary Phases**

Cellulosic	Amylosic
triacetate	
tribenzoate	
tribenzylether	
tricinnamate	
triphenylcarbamate	triphenylcarbamate
tris-3,5-dichlorophenylcarbamate	
tris-3,5-dimethylphenylcarbamate	tris-3,5-dimethylphenylcarbamate
tris-1-phenylethylcarbamate	tris-1-phenylethylcarbamate

The chiral recognition sites on these polymeric carbohydrate phases are thought to be channels or grooves in the polymer matrix and that analytes are included into these channels. Evidence for inclusion is provided by the enhanced chiral recognition observed for many analytes as the steric bulk of the alcohol mobile phase modifier increases. Chiral recognition seems to require the presence of an aromatic ring, for $\pi-\pi$ interactions, and polar sites of unsaturation or hydrogen bonding functionalities. As in the case of the Pirkle-type phases, these chiral stationary phases are usually used in the normal phase mode and mobile phases typically consist of hexane and 2-propanol although there have been some reports of these phases being used in the reversed phase mode (97).

8. Protein-Based Phases

Proteins, amino acids bonded through peptide linkages to form macromolecular biopolymers, used as chiral stationary phases for hplc include bovine and human serum albumin, α_1-acid glycoprotein, ovomucoid, avidin, and cellobiohydrolase. The bovine serum albumin column is marketed under the name Resolvosil and can be obtained from Phenomenex. The human serum albumin column can be obtained from Alltech Associates, Advanced Separation Technologies, Inc., and J. T. Baker. The α_1-acid glycoprotein and cellobiohydrolase can be obtained from Advanced Separation Technologies, Inc. or J. T. Baker, Inc.

In most cases, the protein is immobilized onto γ-aminopropyl silica and covalently attached using a cross-linking reagent such as N,N'-carbonyldiimidazole. The tertiary structure or three dimensional organization of proteins are thought to be important for their activity and chiral recognition. Therefore, mobile phase conditions that cause protein "denaturation" or loss of tertiary structure must be avoided.

Typically, the mobile phases used with the protein-based chiral stationary phases consist of aqueous phosphate buffers (98). Often small amounts of organic modifiers, such as methanol, ethanol, propanol, or acetonitrile, are added to reduce hydrophobic interactions with the analyte and to improve enantioselectivity. In some cases, dramatic changes in chiral recognition occur when small amounts of organic modifiers, such as N,N-dimethyloctylamine or octanoic acid are added to the mobile phase. It is thought that these additives may be playing an active role in enhancing chiral recognition through absorption of the organic modifier onto the protein which induces conformational changes in the overall tertiary structure of the protein. In these cases, the *allosteric* modifier-mediated conformational changes in the protein are thought to enhance chiral recognition by a variety of mechanisms including changes in the accessibility of various stereospecific sites on the protein or obstruction of nonstereospecific sites (99).

As in the case of the cyclodextrin and amylosic and cellulosic phases, the chiral recognition mechanism for these protein-based phases is not well understood. In some cases, it is thought that analytes may form inclusion complexes with hydrophobic pockets within the biopolymeric matrix. These hydrophobic interactions may couple with hydrogen bonding, electrostatic interactions, and $\pi-\pi$ or dipole stacking to individual amino acid residues, thus, contributing to stereospecific orientational constraints within the hydrophobic pockets.

Optimization of chromatographic conditions and selection of analytes that can be successfully resolved on these phases is usually done empirically. In addition, the large molecular weight of these biopolymers dictates that the amount of chiral selector that can be immobilized on the column packing material is very small. Although the protein is large, relative to the analytes, the actual region of the protein that affects the chiral separation may be very small. Thus, the capacity (amount of material resolvable during a single chromatographic run) of these columns is generally fairly small ($<\sim$0.1 mg) and the columns are easily overloaded.

An interesting application of the protein-based phases is various protein binding and displacement experiments which can be done fairly routinely (100). For instance, the chiral selectivity of chiral stationary phases derived from the serum albumin, one of the most abundant blood proteins which functions as a transport protein, from different animal species including rabbit, rat and human has been compared (101). This work suggests that differences in the enantioselectivity, toward a particular drug, of a column derived from human serum albumin and a column derived from some other animal serum albumin might be indicative that a particular species might not be a good animal model during drug development, thus, obviating the need for animal testing.

Chiral separations on protein-based phases may also provide useful information on drug interactions. For instance, the effect of the individual enantiomers of warfarin on the enantioselectivity of human serum albumin toward benzodiazepinones has been studied using a human serum albumin column with warfarin as a mobile phase additive (102).

9. Chirobiotic Phases

The chirobiotic chiral stationary phases (103,104) are based on macrocyclic antibiotics such as vancomycin (**4**) and teicoplanin (**5**).

(**4**)

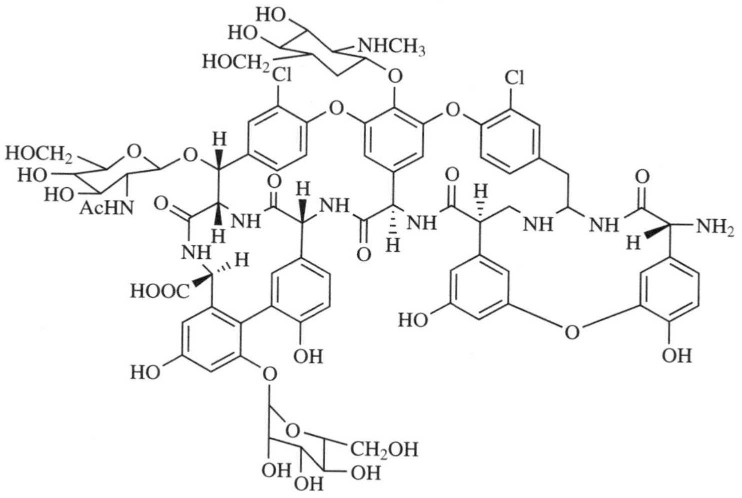

(5)

These chiral selectors, originally used as chiral additives in capillary zone electrophoresis, incorporate aromatic and carbohydrate, as well as peptide and ionizable moieties. The presence of aromatic groups, allowing for $\pi-\pi$ interactions, and the macrocyclic rings, offering potential inclusion complexation, give these phases some of the advantages of the protein-based phases (eg, peptide and hydrogen bonding sites) and the carbohydrate-based phases but with greater sample capacity and greater mobile phase flexibility. Indeed, these phases seem to be truly "multimodal" in that they have demonstrated chiral selectivity in the normal, polar organic, and reversed-phase modes. In the normal and polar organic phase modes, $\pi-\pi$ interactions, and dipole stacking are thought to play a predominant role in chiral selector–analyte interactions. In the reversed-phase mode, hydrogen bonding, inclusion complexation and, for charged analytes, electrostatic interactions are thought to dominate the interactions. In addition, the use of such well-defined chiral selectors facilitate method development and optimization. These columns are commercially available through Advanced Separation Technologies, Inc. and Alltech Associates.

10. Chiral Crown Ether Phases

Chiral crown ethers based on 18-crown-6 (Fig. 9) can form inclusion complexes with ammonium ions and protonated primary amines. Immobilization of these

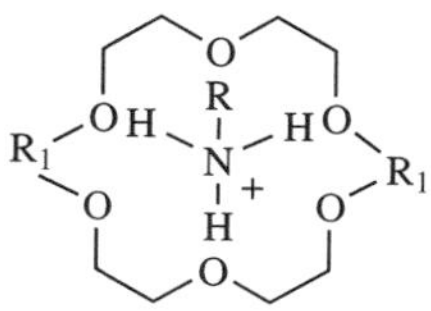

Fig. 9. An inclusion complex formed between a protonated primary amine and a chiral crown ether.

chiral crown ethers on a chromatographic support provides a chiral stationary phase which can resolve most primary amino acids, amines, and amino alcohols. However, the stereogenic center must be in fairly close proximity to the primary amine for successful chiral separation (105,106). Significantly, the chiral crown ether phase is unique in that it is one of the few liquid chromatographic chiral stationary phases that does not require the presence of an aromatic ring to achieve chiral separations. Although chiral recognition seems to be enhanced for analytes containing either bulky substituents or aromatic groups near the stereogenic center, only the presence of the primary amine is mandatory.

Mobile phases used with this stationary phase are typically 0.01 N perchloric acid with small amounts of methanol or acetonitrile. One significant advantage of these phases is that both configurations of the chiral stationary phase are commercially available and can be obtained from J. T. Baker Inc. and Chiral Technologies, Inc. (Crownpak CR).

11. Chiral Synthetic Polymer Phases

Chiral synthetic polymer phases can be classified into three types. In one type, a polymer matrix is formed in the presence of an optically pure compound to molecularly *imprint* the polymer matrix (Fig. 10) (107,108). Subsequent to the polymerization, the chiral template is removed, leaving the polymer matrix with chiral cavities. The degree of cross-linking in the polymer matrix and degree of association between the template molecule and the monomer, is governed by the type and concentration of the monomer, the concentration of the template, the solvent and temperature or pressure under which polymerization takes place. All play a role in the chiral selectivities achieved with these phases. The selectivities achieved with these phases are generally excellent, thus, facilitating semi-preparative separations. However, the applicability of these chiral stationary phases are generally limited to the analyte upon which the phase is based and a limited number of analogues. In addition, these types of phases generally exhibit poor efficiency in large part because the polymeric matrix contributes

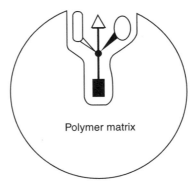

Fig. 10. The relationship between a chiral template molecule and the polymeric matrix formed in the presence of the template molecule.

to nonsterespecific binding. Advantages of this approach include the ability to prepare reciprocal phases and the predictability of the enantiomeric elution order.

Another type of synthetic polymer-based chiral stationary phase is formed when chiral catalyst are used to initiate the polymerization. In the case of poly (methyl methacrylate) polymers, introduced by Okamoto, the chirality of the polymer arises from the helicity of the polymer and not from any inherent chirality of the individual monomeric subunits (109). Columns of this type (eg, Chiralpak OT) are available from Chiral Technologies, Inc., or J. T. Baker Inc.

A third type of synthetic polymer-based chiral stationary phase, developed by Blaschke (110), is produced when a chiral selector is either incorporated within the polymer network (111) or attached as pendant groups onto the polymer matrix. Both are analogous to methods used to produce polymeric chiral stationary phases for gc. The polymers can be either coated onto a silica substrate, comonomers bearing silane functional groups may be added for subsequent reaction with the silica, or the silica may be chemically modified to incorporate monomer-bearing silanes. More recently, L-valine-3,5-dimethylanilide has been bonded to a poly(glycidylmethacrylate-co-ethylenedimethacrylate polymer which formed the underlying substrate (112). Chemical bonding of the polymer to the substrate eases the mobile phase restrictions imposed on the coated chiral polymer stationary phases.

In general, the synthetic polymeric phases seem to have polarities analogous to diol-type phases and a wide range of mobile phase conditions have been used including hexane, various alcohols, acetonitrile, tetrahydrofuran, dichloromethane and their mixtures, as well as aqueous buffers.

12. Chiral Separation Validation for Hplc

Chiral separations present special problems for validation. Typically, in the absence of spectroscopic confirmation (eg, mass spectral or infrared data), conventional separations are validated by analyzing "pure" samples under identical chromatographic conditions. Often, two or more chromatographic stationary phases, which are known to interact with the analyte through different retention mechanisms, are used. If the pure sample and the unknown have identical retention times under each set of conditions, the identity of the unknown is assumed to be the same as the pure sample. However, often the chiral separation that is obtained with one type of column may not be achievable with any other type of chiral stationary phase. In addition, "pure" enantiomers are generally not available.

Most commonly, uv or uv–vis spectroscopy is used as the basis for detection in hplc. When using a chiral stationary phase, confirmation of a chiral separation may be obtained by either monitoring the column effluent at more than one wavelength or by running the sample more than once. The same mobile-phase conditions are used, but monitoring is done at different wavelengths. Because enantiomers have identical spectra in an achiral environment, the ratio of the peaks for the two enantiomers should be independent of wavelength.

Although not absolute proof of a chiral separation, this approach does provide strong supporting evidence.

As in tlc, another method to validate a chiral separation is to collect the individual peaks and subject them to some type of optical spectroscopy, such as, circular dichroism or optical rotary dispersion. Enantiomers have mirror image spectra (eg, the negative maxima for one enantiomer corresponds to the positive maxima for the other enantiomer). One problem with this approach is that the analytes are diluted in the mobile phase. Thus, the sample must be injected several times. The individual peaks must be collected and subsequently concentrated to obtain adequate concentrations for spectral analysis.

Alternatively, a chiroptical spectroscopy can be used as the basis for detection on-line using commercially available optical rotary dispersion or circular dichroism-based detectors. Optical rotary dispersion instruments are analogous to refractive index-based detectors for conventional chromatography in that they are universal, do not require the presence of a chromophore in the analyte and have the least sensitivity of the optical detectors. Circular dichroic detection, although more sensitive than optical rotary dispersion-based detection, requires not only the presence of a uv chromophore in the analyte, but that the chromophore be not too distant from the asymmetric center of the analyte. Figure 11**a** illustrates a simulated chromatogram for an enantiomeric separation obtained using a conventional absorption detector. Figure 11**b** illustrates a simulated chromatogram for an enantiomeric separation using circular dichroic detection. Both types of chiroptical detectors produce positive and negative peaks for the two enantiomers. However, neither chiroptical detector can distinguish a fair separation (Fig. 11**d**) from a poor separation (Fig. 11**f**) in which there is considerable overlap of the two peaks. This is because the signals generated by the two enantiomers have opposite signs, and thus, any overlap causes cancellation of signal. Further, peak overlap results in nonlinear detector response vs concentration. Therefore, some other detection method must be used in conjunction with either of these types of detection. Nevertheless, as can be seen from Figure 11**f**, chiroptical detection can be advantageous if there is considerable overlap of the two peaks. In this case, chiroptical detection may reveal that the leading and tailing edges of the peak are enantiomerically enriched which may not be apparent from the chromatogram obtained with nonchiroptical detection (Fig. 11**e**).

Another method for validating chiral separations by lc is to couple the chromatographic system to a mass spectrometer. In mass spectrometry, high energy ions are used to bombard molecules exiting the column. The impact of the high energy ions causes the molecules to "fragment" into various ions which are then sent to a "mass discriminator". The ion fragments are detected and a fragmentation pattern or mass spectrum is reconstructed. Enantiomers have identical fragmentation patterns. Hence, identical fragmentation patterns for two peaks in the chromatogram confirms a chiral separation.

13. Chiral Stationary Phases for Gas Chromatography

Although chiral stationary phases for gas chromatography (gc) were introduced before liquid chromatographic chiral stationary phases, development of gc chiral

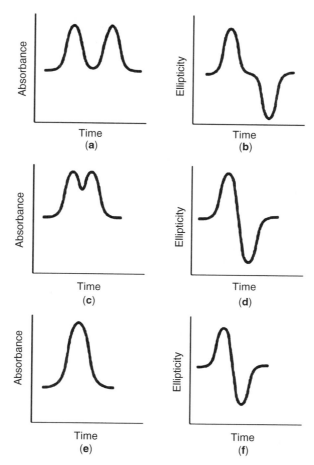

Fig. 11. Simulated chromatograms of chiral separations obtained using nonchiroptical detection (**a, c, e**) and chiroptical detection (**b, d, f**) illustrating the effect of peak overlap on the resultant chromatogram.

stationary phases lagged behind for a variety of reasons. First of all, analysis by gc requires that the analyte be volatile and thermally stable. This condition often requires that the analyte be derivatized with an achiral reagent prior to chromatographic analysis to enhance sample volatility. In some cases, derivatization may actually enhance detector response (eg, trifluoroacetylation amplifies electron capture detection) or chiral interactions. However, it should be noted that the presence of more than one type of functionality (eg, amine and alcohol) in the analyte with differing reactivities toward the derivatizing agent may add additional complications. Typical achiral derivatizing reagents, as well as the appropriate functionality required in the analyte, are listed in Table 4.

The use of gas chromatography for chiral separations was also hampered because the high column temperatures typically used in gc tend to accelerate racemization of the stationary phase, thus, decreasing column longevity. The high column temperatures typically used in gc also tend to accelerate racemization of the analyte. In addition, the differences in the stabilities of the

Table 4. Analyte Functional Groups and Typical Achiral Derivatizing Reagents

Type of derivatizing agent	Analyte functional group	Examples of derivatizing agents
alkyl silyl	alcohols thiols carboxylic acid amines	N-trimethylsilylimidazole N,O-bis(trimethylsilyl)- trifluroracetamide
acyl, haloacyl or anhydride	alcohols amines amides oximes thiols ketones	acetic acid heptafluorobutyryl chloride trifluoroacetyl chloride
alcohol	carboxylic acids	methanol
alkyl halides	carboxylic acids	methyl bromide
diazoalkyl	carboxylic acids sulfonic acids phenols	diazomethane
isocyanate	alcohols amines hydroxy acids	isopropylisocyanate
phosgene	β-amino alcohols β-amino thiols diols N-methylamino acids	
alkyl hydroxylamine	ketones	methylhydroxylamine

diastereomeric complexes formed between the enantiomers and the stationary phase tends to be overcome by the high column temperatures. Finally, preparative scale separations are generally harder to implement in gc than in hplc. However, the inability of most liquid chromatographic methods to resolve chirally small nonaromatic compounds that are frequently used as chiral synthetic building blocks, as well as improvements in gc column technology, has led to renewed interest in chiral stationary phases for gc.

Gc chiral stationary phases can be broadly classified into three categories: diamide, cyclodextrin, and metal complex.

13.1. Diamide Chiral Separations. The first chiral stationary phase for gas chromatography was reported by Gil-Av and co-workers in 1966 (113) and was based on N-trifluoroacetyl (N-TFA) L-isoleucine lauryl ester coated on an inert packing material. It was used to resolve the trifluoroacetylated derivatives of amino acids. Related chiral selectors used by other workers included n-dodecanoyl-L-valine-t-butylamide and n-dodecanoyl-(S)-α-(1-naphthyl)-ethylamide. The presence of the long alkyl groups lowered chiral selector volatility thus reducing but not entirely eliminating column bleed and improving column longevity.

The first commercially available chiral column was the Chiralsil-val (Fig. 12), which was introduced in 1976 (114) for the separation of amino acid-type compounds by gas chromatography. It is based on a polysiloxane polymer containing chiral side chains incorporating L-valine-t-butylamide. The polysiloxane back-

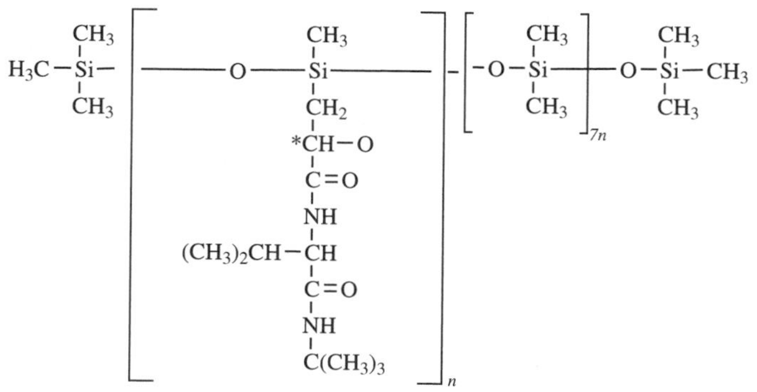

Fig. 12. Structure of Chiralsil-val.

bone improved the thermal stability of these chiral stationary phases relative to the original coated columns and extended the operating temperatures up to 220°C. The column is effective for the separation of perfluoroacylated and esterified amino acids, amino alcohols, and some chiral sulfoxides. Another polysiloxane-based chiral stationary phase incorporating L-valine-(R)-α-phenylethylamide appended onto hydrolyzed XE-60 was found to be particularly successful at resolving perfluoroacetylated amino alcohol derivatives (115). Through judicious choice of derivatizing agent, chiral separations were obtained for a wider range of compounds, including amino alcohols, α-hydroxy acids, diols and ketones, than had previously been obtainable using these types of stationary phases (116).

The chiral recognition mechanism for these types of phases was attributed primarily to hydrogen bonding and dipole–dipole interactions between the analyte and the chiral selector in the stationary phase. It was postulated that chiral recognition involved the formation of transient five- and seven-membered association complexes between the analyte and the chiral selector (117).

On each of these amino acid-based chiral stationary phases, the configuration of the most retained enantiomer corresponds to the configuration of the chiral selector in the stationary phase (eg, L-amino acids were retained longer on the N-trifluoroacetyl (N-TFA) L-isoleucine lauryl ester column). Thus, configuration of the analytes can be assigned even if no optically pure material is available as long as optically pure standard materials are available for structurally related compounds. Another advantage is that stationary phases incorporating the chiral selector with either configuration may be readily prepared or are commercially available. Thus, the elution order may readily be reversed by using a column containing the chiral selector with the opposite configuration, thus, providing another tool for chiral separation validation. Also, quantitation of a very small peak in the presence of a very large peak is generally easier if the smaller peak elutes first.

13.2. Metal Complex. Complexation gas chromatography was first introduced by V. Schurig in 1980 (118) and employs transition metals (eg, nickel, cobalt, manganese or rhodium) complexed with chiral terpenoid ketoenolate ligands such as 3-trifluoroacetyl-1R-camphorate (**6**), 1R-3-pentafluoro-benzoyl-camphorate or 3-heptafluorobutanoyl-(1R,2S)-pinanone-4-ate. In most cases, the chiral selector is dissolved in a polymer matrix coated on the interior walls of a capillary column. This class of chiral columns is particularly adept at enantioresolving some olefins and oxygen-containing compounds such as ketones, ethers, alcohols, spiroacetals, oxiranes, and esters. Many of these compounds lack suitable functionality for derivatization with chiral reagents, and thus, are not amenable to diastereomer formation. Unfortunately, as is the case with many of the chiral stationary phases, the chiral recognition mechanism is not sufficiently refined to allow for prediction of analyte absolute configuration on the basis of retention times except for a very limited number of cases. Nevertheless, these columns allow for the direct chiral separation of compounds that are important synthetic precursors and may be difficult to separate by any other method.

(**6**)

13.3. Cyclodextrins.   As indicated previously, the native cyclodextrins, which are thermally stable, have been used extensively in liquid chromatographic chiral separations, but their utility in gc applications was hampered because their highly crystallinity and insolubility in most organic solvents made them difficult to formulate into a gc stationary phase. However, some functionalized cyclodextrins form viscous oils suitable for gc stationary-phase coatings and have been used either neat or diluted in a polysiloxane polymer as chiral stationary phases for gc (119). Some of the derivatized cyclodextrins which have been adapted to gc phases are 3-O-acetyl-2,6-di-O-pentyl, 3-O-butyryl-2,6-di-O-pentyl, 2,6-di-O-methyl-3-O-trifluoroacetyl, 2,6-dipentyl, 2-O-methyl-3,6-di-O-pentyl, permethyl, permethylhydroxypropyl, perpentyl, and propionyl. Several of these are available commercially. For instance, Advanced Separation Technologies, Inc., Alltech Associates, J. & W. Scientific, and Supelco, Inc., all carry cyclodextrin-based chiral gc columns. Although these derivatized cyclodextrins are often coated, neat, onto the capillary column inner walls, some work has been done to tether the cyclodextrins to a polysiloxane backbone to enhance the thermal stability of the resultant phase (120). Some of the separations obtained with these materials are quite remarkable and include compounds such as halogenated alkanes (Fig. 13) (121), alcohols, alkenes, bicyclic compounds, and simple alkanes.

Although the chiral recognition mechanism of these cyclodextrin-based phases is not entirely understood, thermodynamic and column capacity studies indicate that the analytes may interact with the functionalized cyclodextrins by either associating with the outside or mouth of the cyclodextrin, or by forming a more traditional inclusion complex with the cyclodextrin (122). As in the case of

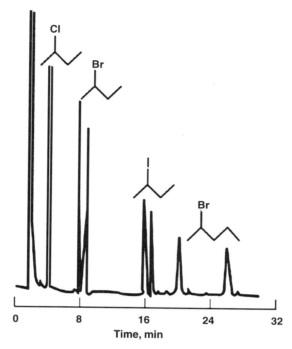

Fig. 13. Enantiomeric separations of monohalohydrocarbons on a 2,6-*O*-dipentyl-3-*O*-trifluoroacetyl-γ-cyclodextrin coated capillary column (10 m, 0.25 mm ID). Column temperature, 30°C; nitrogen carrier gas, 20.7 kPa (3 psi).

the metal-complex chiral stationary phase, configuration assignment is generally not possible in the absence of pure chiral standards.

14. Chiral Separation Validation for Gas Chromatography

The special problems for validation presented by chiral separations can be even more burdensome for gc because most methods of detection (eg, flame ionization detection or electron capture detection) in gc destroy the sample. Even when non-destructive detection (eg, thermal conductivity) is used, individual peak collection is generally more difficult than in lc or tlc. Thus, off-line chiroptical analysis is not usually an option. Fortunately, gc can be readily coupled to a mass spectrometer and is routinely used to validate a chiral separation.

15. Conclusions

The field of chiral separations has grown explosively into a well-developed specialty within separation science in the last two decades. Considerable effort in the field has thus far been directed toward solving analytical separation problems and in the design and development of new chiral separation methods. However, in the future, chiral separations may have the potential to provide an invaluable tool for probing intermolecular interactions, increasing our understanding of the chemistry behind some biological processes and disease states (123,124). Also the efficacy of drug therapy can be enhanced by eliminating or minimizing untoward side effects (125) produced by distomeric ballast and allowing for better control in the application of chiral or prochiral entities to biological or environmental systems.

BIBLIOGRAPHY

1. L. Pasteur, *Comptes Rendus de l'Academie des Sciences* **26**, 535 (1848).
2. W. J. Wechter, D. G. Loughhead, R. J. Reischer, G. J. Van Giessen, and D. G. Kaiser, *Biochem. Biophys. Res. Comm.* **61**, 833 (1974).
3. R. A. Kuzel, S. K. Bhasin, H. G. Oldham, L. A. Damani, J. Murphy, P. Camilleri, and A. J. Hutt, *Chirality* **6**, 607 (1994).
4. S. A. Tobert, *Clin. Pharmacol. Ther.* **29**, 344 (1981).
5. H. Ohkawa, *Bull. Environ. Contamin. Toxicol.* **18**, 534 (1977).
6. H. G. Schmarr, A. Mosandl, and K. Grob, *Chromatographia* **29**, 125 (1990).
7. T. Aleibi-Kolbah, G. Félix, and I. W. Wainer, *Chromatographia* **35**, 264 (1993).
8. D. W. Armstrong, J. R. Faulkner, Jr., and S. M. Han, *J. Chromatogr.* **452**, 323 (1988).
9. T. J. Ward, *Anal. Chem.* **66**, 632A (1994).
10. C. Pettersson, A. Karlsson, and C. Gioeli, *J. Chromatogr.* **407**, 217 (1987).
11. J. D. Duncan, D. W. Armstrong, and A. M. Stalcup, *J. Liq. Chromatogr.* **13**, 1091 (1990).
12. D. W. Armstrong, F.-Y. He, and S. M. Han, *J. Chromatogr.* **448**, 345 (1988).
13. A. D. Cooper, T. M. Jeffries, and R. M. Gaskell, *Anal. Proc.* **29**, 258 (1992).
14. D. W. Armstrong and Y. Zhou, *J. Liq. Chromatogr.* **17**, 1695 (1994).
15. C. Pettersson and G. Schill, *J. Liq. Chromatogr.* **9**, 269 (1986).
16. M.-B. Huang, H.-K. Li, G.-L. Li, C.-T. Yan, and L.-P. Wang, *J. Chromatogr. A* **742**, 289 (1996).
17. D. W. Armstrong, J. R. Faulkner, Jr., and S. M. Han, *J. Chromatogr.* **452**, 323 (1988).
18. D. W. Armstrong, F.-Y. He, and S. M. Han, *J. Chromatogr.* **448**, 345 (1988).
19. J. D. Duncan, D. W. Armstrong, and A. M. Stalcup, *J. Liq. Chromatogr.* **13**, 1091 (1990).
20. W. G. Kuhr, *Anal. Chem.* **62**, 403R (1990).
21. B. L. Karger, *Amer. Lab.*, 23 (Oct. 1993).
22. J. W. Jorgenson and K. D. Lukacs, *Anal. Chem.* **53**, 1298 (1981).
23. A. S. Cohen, A. Paulus, and B. L. Karger, *Chromatographia* **24**, 15 (1987).
24. P. Sun, N. Wu, G. Barker, and R. A. Hartwick, *J. Chromatogr.* **648**, 475 (1993).
25. S. Terabe, K. Otsuka, and H. Nishi, *J. Chromatogr.* **666**, 295 (1994).
26. Y. Y. Rawjee and G. Vigh, *Anal. Chem.* **66**, 619 (1994).
27. A. M. Stalcup and K. H. Gahm, *Anal. Chem.* **68**, 1360 (1996).
28. A. M. Stalcup and N. M. Agyei, *Anal. Chem.* **66**, 3054 (1994).
29. H. Soini, M. Stefansson, M.-L. Riekkola, and M. V. Novotny, *Anal. Chem.* **66**, 3477 (1994).
30. P. Gozel, E. Gassman, H. Michelson, and R. N. Zare, *Anal. Chem.* **59**, 44 (1987).
31. A. M. Stalcup and K. H. Gahm, *J. Microcolumn Sep.* **8**, 145 (1996).
32. D. W. Armstrong, K. Rundlett, and G. L. Reid, *Anal. Chem.* **66**, 1690 (1994).
33. T. O. Cole, M. J. Sepaniak, and W. L. Hinze, *J. High Resolut. Chromatogr. Chromatogr. Commun.* **13**, 570 (1990).

34. Y. Mechref and Z. El Rassi, *Chirality* **8**, 515 (1996).
35. N. M. Agyei, K. H. Gahm, and A. M. Stalcup, *Anal. Chim. Acta* **307**, 185 (1995).
36. A. M. Stalcup and K. H. Gahm, *J. Microcol. Sep.* **8**, 145 (1996).
37. R. S. Sahota and M. G. Khaledi, *Anal. Chem.* **66**, 1141 (1994).
38. F. Wang and M. G. Khaledi, *Anal. Chem.* **68**, 3460 (1996).
39. S. A. C. Wren and R. C. Rowe, *J. Chromatogr.* **603**, 235 (1992).
40. A. Guttman, A. Paulus, A. S. Cohen, N. Grinberg, and B. L. Karger, *J. Chromatogr.* **448**, 41 (1988).
41. S. Jönsson, A. Schön, R. Isaksson, C. Pettersson, and G. Petterson, *Chirality* **5**, 505 (1992).
42. A. M. Stalcup, S. Gratz, and Y. Jin, unpublished results.
43. L. Witherow, T. D. Spurway, R. J. Ruane, I. D. Wilson, and K. Longdon, *J. Chromatogr.* **553**, 497–501 (1991).
44. H. T. K. Xuan and M. Lederer, *J. Chromatogr.* **635**, 346 (1993).
45. H. T. K. Xuan and M. Lederer, *J. Chromatogr.* **645**, 185 (1993).
46. C. A. Brunner and I. W. Wainer, *J. Chromatogr.* **472**, 277 (1989).
47. R. Bhushan and V. Parshad, *J. Chromatogr.* **721**, 369 (1996).
48. R. Bhushan and I. Ali, *Chromatographia* **35**, 679 (1993).
49. R. Bhushan and V. Parshad, *J. Chromatogr.* **736**, 235 (1996).
50. U. A. Th. Brinkman and D. Kamminga, *J. Chromatogr.* **330**, 375 (1985).
51. V. Mathur, N. Kanoongo, R. Mathur, C. K. Narang, and N. K. Mathur, *J. Chromatogr.* **685**, 360 (1994).
52. M. Remelli, R. Piazza, and F. Pulidori, *Chromatographia* **32**, 278 (1991).
53. D. W. Armstrong and W. DeMond, *J. Chromatogr. Sci.* **22**, 411 (1984).
54. D. W. Armstrong, Y. Tang, S. Chen, Y. Zhou, C. Bagwill, and J. R. Chen, *Anal. Chem.* **66**, 1473 (1994).
55. W. H. Pirkle, J. M. Finn, B. C. Hamper, J. L. Schreiner, and J. R. Pribish, *Am. Chem. Soc. Symp. Ser. No. 185*, Chapt. 18, 1982.
56. W. H. Pirkle and P. G. Murray, *J. Liq. Chromatogr.* **13**, 2123 (1990).
57. J. Hermansson and M. Eriksson, *J. Liq. Chromatogr.* **9**, 621 (1986).
58. G. Schill, I. W. Wainer, and S. A. Barkin, *J. Liq. Chromatogr.* **9**, 641 (1986).
59. S. Allenmark, *J. Liq. Chromatogr.* **9**, 425 (1986).
60. M. Okamoto and H. Nakazawa, *J. Chromatogr.* **508**, 217 (1990).
61. T. Miwa, H. Kuoda, S. Sakashita, N. Asakawa, and Y. Miyake, *J. Chromatogr.* **511**, 89 (1990).
62. R. Isaksson, P. Erlandsson, L. Hansson, A. Holmberg, and S. Berner, *J. Chromatogr.* **498**, 257 (1990).
63. Y. Okamoto, R. Aburatani, K. Hatano, and K. Hatada, *J. Liq. Chromatogr.* **11**, 2147 (1988).
64. T. Shinbo, T. Yamaguchi, K. Nishimura, and M. Sugiura, *J. Chromatogr.* **405**, 145 (1987).
65. M. Hilton and D. W. Armstrong, *J. Liq. Chromatogr.* **14**, 9 (1991).
66. V. A. Davankov, A. A. Kurganov, and A. S. Bochov, *Adv. Chromatogr.* **22**, 71 (1983).
67. W. H. Pirkle, J. M. Finn, J. L. Schreiner, and B. C. Hamper, *J. Am. Chem. Soc.* **103**, 3964 (1981).
68. C. E. Dalgliesh, *J. Chem. Soc.*, 3940 (1952).
69. W. H. Pirkle, D. W. House, and J. M. Finn, *J. Chromatogr.* **192**, 143 (1980).
70. W. H. Pirkle and T. C. Pochapsky, *J. Am. Chem. Soc.* **108**, 352 (1986).
71. W. H. Pirkle, T. C. Pochapsky, G. S. Mahler, D. E. Corey, D. S. Reno, and D. M. Alessi, *J. Org. Chem.* **51**, 4991 (1986).
72. W. H. Pirkle, G. Mahler, and M. H. Hyun, *J. Liq. Chromatogr.* **9**, 443 (1986).
73. W. H. Pirkle and C. J. Welch, *Tetrahedron Assym.* **5**, 777 (1994).

74. C. J. Welch, T. Szczerba, and S. R. Perrin, *J. Chromatogr. A* **758**, 93 (1997).
75. Regis Technologies Application Guide, pp. 30, 31.
76. C. J. Welch and S. R. Perrin, *J. Chromatogr.* **690**, 218 (1995).
77. T. J. Ward and D. W. Armstrong, "Cyclodextrin Stationary Phases" in M. Zief and L. J. Crane, eds., *Chromatographic Chiral Separations*, Marcel Dekker, Inc., New York, 1988, p. 131.
78. S. M. Han and D. W. Armstrong, in A. M. Krstulovic, ed., *Chiral Separations by HPLC*, John Wiley & Sons, Inc., New York, 1989, 208–287.
79. D. W. Armstrong, T. J. Ward, R. D. Armstrong, and T. E. Beesley, *Science* **232**, 1132 (1986).
80. R. E. Boehm, D. E. Martire, and D. W. Armstrong, *Anal. Chem.* **60**, 522 (1988).
81. D. W. Armstrong, S. Chen, C. Chang, and S. Chang, *J. Liq. Chromatogr.* **15**, 545 (1992).
82. A. M. Stalcup, J. R. Faulkner, Y. Tang, D. W. Armstrong, L. W. Levy, E. Regalado, *Biomed. Chromatogr.* **5**, 3 (1991).
83. P. Camilleri, C. A. Reid, and D. T. Manallack, *Chromatographia* **38**, 771 (1994).
84. A. M. Stalcup and K. H. Gahm, *Anal. Chem.* **68**, 1369 (1996).
85. A. M. Stalcup, S. Chang, D. W. Armstrong, and J. Pitha, *J. Chromatogr.* **513**, 181 (1990).
86. D. W. Armstrong, A. M. Stalcup, M. L. Hilton, J. D. Duncan, J. R. Faulkner, and S. C. Chang, *Anal. Chem.* **62**, 1610 (1990).
87. A. M. Stalcup, S. C. Chang, and D. W. Armstrong, *J. Chromatogr.* **540**, 113 (1991).
88. D. W. Armstrong, M. Hilton, and L. Coffin, *LC-GC* **9**, 647 (1992).
89. H. Hopf, W. Grahn, D. G. Barrett, A. Gerdes, J. Hilmer, J. Hucker, Y. Okamoto, and Y. Kaida, *Chem. Ber.* **123**, 841 (1990).
90. Y. Okamoto, Y. Kaida, R. Aburatani, and K. Hatada, *J. Chromatogr.* **477**, 367 (1989).
91. Y. Okamoto, K. Hatano, R. Aburatani, and K. Hatada, *Chem. Lett.*, 715 (1989).
92. J. M. Jansen, S. Copinga, G. Gruppen, R. Isaksson, D. T. Witte, and C. J. Grol, *Chirality* **6**, 596 (1994).
93. T. Shibata, I. Okamoto, and K. Ishii, *J. Liq. Chromatogr.* **9**, 313 (1986).
94. D. T. Witte, F. J. Bruggeman, J. P. Franke, S. Copinga, J. M. Jansen, and R. A. De Zeeuw, *Chirality* **5**, 545 (1993).
95. Y. Okamoto, T. Ohashi, Y. Kaida, and E. Yashima, *Chirality* **5**, 616 (1993).
96. J. Wagner, H.-J. Hamann, W. Döpke, A. Kunath, and E. Höft, *Chirality* **7**, 243 (1995).
97. M. Tanaka, H. Yamazaki, and H. Hakusui, *Chirality* **7**, 612 (1995).
98. J. Iredale, A.-F. Aubry, and I. W. Wainer, *Chromatographia* **31**, 329 (1991).
99. T. A. G. Noctor, I. W. Wainer, and D. S. Hage, *J. Chromatogr.* **577**, 305 (1995).
100. D. S. Hage, T. A. G. Noctor, and I. W. Wainer, *J. Chromatogr. A* **693**, 23 (1995).
101. G. Massolini, A.-F. Aubry, A. McGann, and I. Wainer, *Biochem. Pharm.* **46**, 1285 (1993).
102. E. Domenici, C. Bertucci, P. Salvadori, and I. W. Wainer, *J. Pharm. Sci.* **80**, 164 (1991).
103. D. W. Armstrong, Y. Tang, and S. Chen, *Anal. Chem.* **66**, 473 (1994).
104. D. W. Armstrong, Y. Liu, and K. H. Ekborgott, *Chirality* **7**, 474 (1995).
105. T. Shinbo, T. Yamaguchi, K. Nishimura, and M. Sugiura, *J. Chromatogr.* **405**, 145 (1987).
106. M. Hilton and D. W. Armstrong, *J. Liq. Chromatogr.* **14**, 9 (1991).
107. B. Sellergren, M. Lepistö, and K. Mosbach, *J. Am. Chem. Soc.* **110**, 5853 (1988).
108. L. Fischer, R. Müller, and B. Ekberg, *J. Am. Chem. Soc.* **113**, 9358 (1991).
109. Y. Okamoto, K. Suzuki, K. Ohta, K. Hatada, and H. Yuki, *J. Am. Chem. Soc.* **101**, 4763 (1979).

110. G. Blaschke, W. Bröker, and W. Fraenkel, *Angew. Chem.* **98**, 808 (1986).

111. S. G. Allenmark, S. Andersson, P. Möller, and D. Sanchez, *Chirality* **7**, 248 (1995).

112. Y. Liu, *Anal. Chem.* **69**, 61 (1997).

113. E. Gil-Av, B. Feibush, and R. Charles-Sigler, *Tetrahedron Lett.* **10**, 1009 (1966).

114. H. Frank, G. J. Nicholson, and E. Bayer, *J. Chromatogr. Sci.* **15**, 174 (1974).

115. W. A. König, I. Benecke, and S. Sievers, *J. Chromatogr.* **217**, 71 (1981).

116. W. A. König and E. Steinbach, and K Ernst, *J. Chromatogr.* **301**, 129 (1984).

117. B. Feibush, A. Balan, B. Altman, and Gil-Av, *J. Chem. Soc. Perkin II*, 1230 (1979).

118. V. Schurig, *Chromatographia* **13**, 263 (1980).

119. H. P. Nowotny, D. Schmalzing, D. Wistuba and V. Schurig, *J. High Resolut. Chromatogr. Chromatogr. Commun.* **12**, 383 (1989).

120. D. W. Armstrong, Y. Tang, T. Ward, and M. Nichols, *Anal. Chem.* **65**, 1114 (1993).

121. W.-Y. Li, H. L. Jin, and D. W. Armstrong, *J. Chromatogr.* **509**, 303 (1990).

122. A. Berthod, W. Li, and D. W. Armstrong, *Anal. Chem.* **64**, 873 (1992).

123. D. W. Armstrong, J. Zukowski, N. Ercal, and M. Gasper, *J. Pharm. Biomed. Anal.* **11**, 881 (1993).

124. J. Roboz, E. Nieves, and J. F. Roland, *J. Chromatogr.* **500**, 413 (1990).

125. I. W. Wainer, J. Ducharme, C. P. Granvil, H. Parenteau, and S. Abdullah, *J. Chromatogr. A* **694**, 169 (1995).

GENERAL REFERENCES

C. F. Poole and S. K. Poole, *Chromatography Today*, Elsevier Science Publishers B.V., Amsterdam, The Netherlands, 1991.

I. W. Wainer, *Drug Stereochemistry: Analytical Methods and Pharmacology*, Marcel Dekker, Inc., New York, 1993.

W. A. König, *The Practice of Enantiomer Separation by Capillary Gas Chromatography*, Hüthig Verlag, Heidelberg, Germany, 1987.

P. Schreier, A. Bernreuther, and M. Huffer, *Analysis of Chiral Organic Molecules*, Walter de Gruyter & Co., Berlin, Germany, 1995.

G. Subramanian, *A Practical Approach to Chiral Separations by Liquid Chromatography*, VCH, Weinheim, Germany, 1994.

APRYLL M. STALCUP
University of Cincinnati

CHLORIC ACID AND CHLORATES

1. Chloric Acid

Chlorates are salts of chloric acid [7790-93-4], $HClO_3$.

1.1. Physical Properties. Aqueous chloric acid is a clear, colorless solution stable when cold up to ca 40 wt% (1). Upon heating, chlorine [7782-50-5], Cl_2, and chlorine dioxide [10049-04-4], ClO_2, may evolve. Concentration of chloric

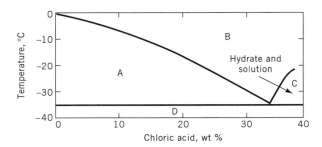

Fig. 1. Solubility of chloric acid in water where A represents ice and chloric acid solution; B, solution; C, chloric acid hydrate and solution; and D, a eutectic of ice and the hydrate (2).

acid by evaporation may be carried to >40% under reduced pressure. Decomposition at concentrations in excess of 40% is accompanied by evolution of chlorine and oxygen [7782-44-7] and the formation of perchloric acid [7601-90-3], $HOClO_4$, in proportions approximating those shown in equation 1.

$$8\,HClO_3 \longrightarrow 4\,HClO_4 + 2\,H_2O + 3\,O_2 + 2\,Cl_2 \tag{1}$$

Impurities such as chloride ion or other reducing agents generate chlorine dioxide when the chloric acid solution is heated. Transition-metal ions do not affect the stability of pure chloric acid at room temperature. Thirty-five percent solutions of $HClO_3$ have been shown to be stable for 20 days at room temperature containing up to 1000 ppm Ni^{2+}, 800 ppm Zn^{2+}, 700 ppm Fe^{3+}, or 600 ppm Cr^{3+} (2). The solubility of chloric acid in water is shown in Figure 1.

Chloric acid, a strong acid, has $pK_a = -2.7$ (3). It is a strong oxidizing agent, $E_0 = 1.175$ V with ClO_2 as the reduction product (4). The heat of formation is -99.2 kJ/mol (-23.7 kcal/mol) and the Gibbs free energy of formation is -3.3 kJ/mol (-0.79 kcal/mol) for both chloric acid and the chlorate ion (4).

1.2. Chemical Properties.   Chloric acid is a strong acid and an oxidizing agent. It reacts with metal oxides or hydroxides to form chlorate salts, and it is readily reduced to form chlorine dioxide.

$$ClO_3^- + 2\,H^+ + e^- \longrightarrow ClO_2 + H_2O \tag{2}$$

Titanium, Hastelloy (grades C22 and C276), and 316 stainless steel all exhibit corrosion rates of less than 0.08 mm/yr at room temperature in 35 wt% chloric acid solutions (2).

1.3. Manufacture.   Chloric acid is the precursor for generation of chlorine dioxide for pulp bleaching and other applications (see BLEACHING AGENTS), and is formed *in situ* by reaction of sodium chlorate [7775-09-9], $NaClO_3$, and a strong acid, eg,

$$H_2SO_4 + 2\,NaClO_3 \longrightarrow Na_2SO_4 + 2\,HClO_3 \tag{3}$$

Any chloride present in the solution is oxidized to chlorine by the chloric acid.

$$4\,H^+ + 2\,Cl^- + 2\,ClO_3^- \longrightarrow 2\,ClO_2 + Cl_2 + 2\,H_2O \tag{4}$$

In order to eliminate or reduce the sodium sulfate by-product of reaction (3), activity has focused on the direct manufacture of chloric acid.

Emerging technologies for the commercial manufacture of chloric acid fall into three categories: (1) generation of high purity chloric acid by thermal decomposition of pure solutions of hypochlorous acid [7790-92-3], HClO (5).

$$5\,HOCl \longrightarrow HClO_3 + 2\,Cl_2 + 2\,H_2O \tag{5}$$

The chlorine generated as a by-product is recovered. Stable solutions of up to 40% by weight chloric acid are generated by evaporative concentration. The chloric acid obtained is free of metal cations and chloride and sulfate anions; (2) generation of chloric acid by passing a solution of sodium chlorate through a cation ion-exchange resin (6,7). Electrochemical methods for producing chloric acid from sodium chlorate have also been developed (see ELECTROCHEMICAL PROCESSING). These include electrodialysis (qv) methods employing discrete anion and cation membrane separated compartments or bipolar membranes (8,9) (see MEMBRANE TECHNOLOGY). An alternative electrochemical process employs perfluorinated cation-exchange membrane bounded cell compartments to produce high concentration chloric acid–sodium chlorate mixtures. The mixture is evaporated under vacuum to crystallize out sodium chlorate and produce a low sodium content chloric acid (10). All of the electrochemical routes to chloric acid produce sodium hydroxide, oxygen, and hydrogen as coproducts. The resulting $HClO_3$ solution contains some dissolved sodium chlorate as well as the impurities that were present in the initial sodium chlorate solution; (3) and hypochlorous acid can be oxidatively electrolyzed to chloric acid (11,12). Chloric acid prepared by oxidizing HOCl is both metal- and chloride-ion free. It can be reduced to chlorine dioxide without the formation of solid by-products or chlorine. This reduction can be conducted electrochemically (7,13,14) or chemically.

1.4. Shipment.   Solutions of greater than 10 wt% chloric acid may be shipped using the label, "oxidizing substance, liquid, corrosive, n.o.s.," and using identification number UN3098, packing group II.

1.5. Health and Safety Factors.   Chloric acid is a strong oxidizing agent and concentrated solutions ignite organic matter on contact. The acid must be stored apart from reducing agents and organic materials. Concentrated solutions are corrosive to the skin (1). It is a strong irritant by ingestion and inhalation (21).

1.6. Uses.   Chloric acid is formed *in situ* by reaction of sodium chlorate and a strong acid during chlorine dioxide production. Stoichiometric amounts of sodium salts are also formed as a by-product. The use of chlorine dioxide for pulp (qv) bleaching applications is growing and disposal of the by-product solids is a primary environmental concern. Use of chloric acid to generate chlorine dioxide can eliminate this problem. A process for bleaching pulp which employs chloric acid as the oxidizing agent, in the absence of a transition metal catalyst, has been reported (12).

Chloric acid also has found limited applications as a catalyst for the polymerization of acrylonitrile (qv) [107-13-1], C_3H_3N, and in the oxidation of cyclohexanone [108-94-1], $C_6H_{10}O$ (22) (see CYCLOHEXANOL AND CYCLOHEXANONE).

2. Sodium and Potassium Chlorate

2.1. Physical Properties.

The physical properties of sodium chlorate [7775-09-9] and potassium chlorate [3811-04-9], $KClO_3$, are summarized in Table 1 (23). The solubilities of these chlorates in water are given in Figure 2 (24−26).

Table 1. **Physical Properties of Sodium and Potassium Chlorates**

Properties	$NaClO_3$	$KClO_3$
molecular weight	106.44	122.55
crystal system	cubic	monoclinic
mp, °C	248−260	356−368
dec pt, °C	265	400
density, g/mL	2.487^a	2.338^b
n_D^{20}	1.515	1.440
affinity towards water	hygroscopic	nonhygroscopic
enthalpy of fusion, ΔH_{fus}, kJ/mol^c	21.3	
molar heat capacity, J/(mol · K)^c	$100^{b,d}$	99.8^b
standard enthalpy of formation, kJ/mol^c		
crystals	−365.8	−391
ideal soln of unit activity	−344.1	
standard entropy J/(mol · K)^c crystals	123.4	143
ideal solution of unit activity	22.3	
enthalpy of dissolution,^{e} kJ/mol^c	21.6	40.9

^{a}At 25°C.
^{b}At 20°C.
^{c}To convert J to cal, divide by 4.184.
^{d}From 298 to 533 K.
^{e}1 mol of chlorate per 200 mol H_2O at 25°C.

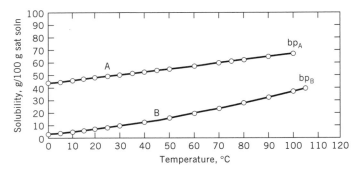

Fig. 2. Aqueous solubility of A, sodium chlorate and B, potassium chlorate where bp_A represents 122°C, the boiling point of a saturated solution of sodium chlorate is 122°C; bp_B represents the boiling point, 104°C, of potassium chlorate solution.

Table 2. **Electrical Conductivity of Aqueous Sodium Chlorate Solutions (ohm · m^{-1})**

Concentration, g/L	Temperature		
	20°C	40°C	60°C
100	6.2	8.9	11.8
200	10.4	14.9	19.7
300	13.4	18.9	25.0
400	15.0	21.5	28.5
500	15.7	22.7	30.3
600	15.5	23.1	30.8
750		21.7	29.7

The electrical conductivity of a pure aqueous sodium chlorate solution is given in Table 2. Additional data are given (29). Table 3 summarizes the solubility data for two aqueous chlorate–chloride systems (30–32).

2.2. Chemical Properties. On thermal decomposition, both sodium and potassium chlorate salts produce the corresponding perchlorate, salt, and oxygen (34). Mixtures of potassium chlorate and metal oxide catalysts, especially manganese dioxide [1313-13-9], MnO_2, are employed as a laboratory source of oxygen. The evolution of oxygen starts at ~70°C and becomes rapid at 100°C, below the fusion point (35). The molten chlorates are powerful oxidizing agents. Mixtures of chlorates and organic materials have been employed as explosives. However, because of extreme shock sensitivity and unpredictability, such mixtures are not classed as permissible explosives in the United States. Chlorates also form flammable and explosive mixtures with phosphorus, ammonium compounds, some metal compounds, and some metal salts (36). Chlorates in neutral and alkaline solutions at room temperature do not show oxidizing properties. Concentrated acidic solutions of chlorates are strong oxidants as a result of chloric acid formation and may also liberate chlorine dioxide gas.

2.3. Manufacture. Most chlorate is manufactured by the electrolysis of sodium chloride solution in electrochemical cells without diaphragms. Potassium chloride can be electrolyzed for the direct production of potassium chlorate

Table 3. **Mass Ratio of Crystalline Chloride and Chlorate Salts in Equilibrium with an Aqueous Solution**

Temperature, °C	Solution system, kg/kg of water^a			
	NaCl	NaClO$_3$	KCl	KClO$_3$
−9.8	0.270	0.360	0.2466	0.0056
10	0.249	0.499	0.3123	0.0144
30	0.2125	0.706	0.3703	0.0321
50	0.1785	0.958	0.4226	0.0635
70	0.1495	1.238	0.4651	0.1162
100	0.1245	1.85	0.518	0.2588

^{a}Densities of the NaClO$_3$–NaCl and KClO$_3$–KCl aqueous solutions are given in Ref. 33.

(37,38), but because sodium chlorate is so much more soluble (see Fig. 2), the production of the sodium salt is generally preferred. Potassium chlorate may be obtained from the sodium chlorate by a metathesis reaction with potassium chloride (39).

The sodium chlorate manufacturing process can be divided into six steps: (1) brine treatment; (2) electrolysis; (3) crystallization and salt recovery; (4) chromium removal; (5) hydrogen purification and collection; and (6) electrical distribution. These steps are outlined in Figure 3.

The production of sodium chlorate is very energy intensive requiring between 4950–6050 kW·h of electricity per metric ton of sodium chlorate produced (40). More than 95% of the energy is used in the electrolysis step. Increases in energy cost have resulted in use of highly efficient noble metal coated titanium anodes and elimination of the less efficient graphite anodes (41). The by-product hydrogen generated from the cell is also recovered for its fuel value. Advances in electrical bus connection design have also been incorporated to reduce the cell-to-cell voltage drop. The use of noble metal anodes requires that hardness, ie, Ca^{2+} and Mg^{2+} ions, and metals be removed from the sodium chloride brine, hence the brine treatment technology that was originally developed for chlor-alkali industry has now become an integral part of sodium chlorate manufacture (see ALKALI AND CHLORINE PRODUCTS). Sodium chlorate manufacturing technology now incorporates a low chloride–chlorate solution manufacture coupled with a chromium removal system, or the use of a crystallizer to produce crystal chlorate as the final project.

Electrolysis. The overall chemical reaction is

$$NaCl + 3\,H_2O \longrightarrow NaClO_3 + 3\,H_2 \tag{6}$$

The reaction in equation 6 requires six Faradays to produce one mole of chlorate. The reaction is endothermic, $\Delta H = 224$ kcal/mol (53.5 kcal/mol) of chlorate or

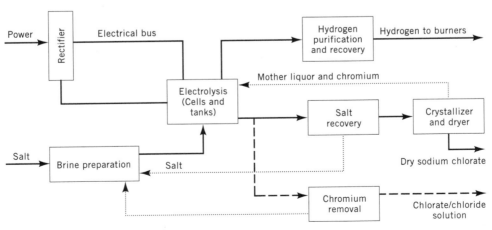

Fig. 3. Schematic of the key steps in a sodium chlorate plant where (···) represents recycle streams and (− − −), process for solution product. Most plants produce crystalline product.

2.43 kW · h/kg. In practice, it takes about 5 kW · h of energy to produce a kilogram of sodium chlorate. The remaining energy is lost to electrolyte solution resistance and heat.

In the sodium chlorate cell, free chlorine is formed at the anode:

$$2\,Cl^- \longrightarrow Cl_2 + 2\,e^- \tag{7}$$

The E_0 at 25°C is 1.36 V vs a normal hydrogen electrode (NHE). The chlorine reacts in the boundary layer and hydrolyzes to form HOCl and HCl

$$Cl_2 + H_2O \rightleftharpoons HOCl + HCl \tag{8}$$

which reacts in the bulk electrolyte to form chlorate (42). Hypochlorous acid, HClO, or hypochlorite ions can be converted to chlorate by two separate simultaneous reactions: by the decomposition of hypochlorite ion and free hypochlorous acid (eq. 9) (43) and by the electrochemical formation of chlorate by the anodic oxidation of hypochlorite (eq. 10).

$$2\,HClO + ClO^- \longrightarrow ClO_3^- + 2\,Cl^- + 2\,H^+ \tag{9}$$

$$6\,ClO^- + 3\,H_2O \longrightarrow ClO_3^- + 4\,Cl^- + 1.5\,O_2 + 6\,H^+ + 6\,e^- \tag{10}$$

The reaction described in equation 9 occurs at 100% current efficiency; current efficiency of that in equation 10 is only 66.7% (44,45).

The most favorable conditions for equation 9 are temperature from 60–75°C and pH 5.8–7.0. The optimum pH depends on temperature. This reaction is quite slow and takes place in the bulk electrolyte rather than at or near the anode surface (46–48). Usually 2–5 g/L of sodium dichromate is added to the electrolysis solution. The dichromate forms a protective Cr_2O_3 film or diaphragm on the cathode surface, creating an adverse potential gradient that prevents the reduction of OCl^-; to Cl^-; ion (46). Dichromate also serves as a buffering agent, which tends to stabilize the pH of the solution (47,48). Chromate also suppresses corrosion of steel cathodes and inhibits O_2 evolution at the anode (49–53).

There are other parallel electrochemical reactions that can occur at the electrodes within the cell, lowering the overall efficiency for ClO^-;$_3$ formation. Oxygen evolution accounts for about 1–3% loss in the current efficiency on noble metal-based electrodes in the pH range 5.5–6.5.

$$2\,H_2O \longrightarrow O_2 + 4\,H^+ + 4\,e^- \tag{11}$$

Where E_0 at 25°C is 1.23 V vs NHE at pH = 0, and 0.876 V vs NHE at pH = 6. Oxidation of chlorates to perchlorate can also occur if the cell voltage increases above 6.5 V, or if the chloride concentration is depleted below about 80 g/L.

$$ClO_3^- + H_2O \longrightarrow ClO_4^- + 2\,H^+ + 2\,e^- \tag{12}$$

In addition to the electrochemical reactions, there are some undesirable nonelectrolytic reactions that produce chlorine and oxygen gases, thus lowering the current efficiency (18). Oxygen generated from decomposition of hypochlorite species in bulk accounts for about half of the oxygen involved. The decomposition of the hypochlorite species is catalyzed by low levels of ionic metal impurities, specifically Ni^{2+}, Co^{2+}, Ir^{3+}, and Ir^{4+}. The current efficiency in the chlorate cell can be monitored by analyzing the exit gas stream and can be calculated by using the following formula:

$$\text{chlorate efficiency} = (100 - 3\,W_O - 2\,W_{Cl})/(100 - W_O - W_{Cl}) \qquad (13)$$

where W_O = vol % of oxygen in exit gas and W_{Cl} = vol % of chlorine in the exit gas. The approximate energy consumption P in $(kW \cdot h)/t$ of a chlorate cell is a function of current efficiency CE expressed as a fraction and cell voltage V in volts

$$\frac{1509V}{CE} \qquad (14)$$

Current efficiency depends on operating characteristics, eg, pH, temperature, and cell design, and is generally in the 90–98% range. The cell voltage is a function of electrode characteristics and electrolyte conductivity and can be expressed as

$V =$ (thermodynamic decomposition voltage of the anode and the cathode)

+ (anode overvoltage) + (cathode overvoltage)

+ (ohmic drop between the anode and cathode resulting from
the electrolyte/gas mixture)

+ (ohmic drop in the electrical connections and hardware)

Typical energy requirement and operating condition are summarized in Table 4.

Table 4. **Electrical Energy Requirement for a Chlorate Cell Using Steel Cathodes and Pt–Ir Anodes**[a]

current density, kA/m^2	2–3
current efficiency, %	94
cell voltage components, V	
thermodynamic decomposition	1.71
anode overvoltage	0.05
cathode overvoltage	0.94
ohmic drops[b]	0.80
average cell voltage, V	3–3.50
electrical energy requirement, $kW \cdot h/t$[c]	5700
operating conditions	
temperature, °C	80
solution composition, g/L	
NaCl	150
NaOCl	3–5
$NaClO_3$	450
$Na_2Cr_2O_7$	2–5

[a]Ref. 54.
[b]The gap is from 3 to 5 mm.
[c]Per metric ton of chlorate formed.

Electrolyzer System. The basic criteria for electrolyzer system design is to minimize capital and operating costs. There are about a dozen electrolyzer system configurations being used for sodium chlorate manufacture. These combinations range from high capital/low operating cost to low capital/high operating cost (55–92). Electrolyzer systems have four basic components: an electrolysis zone, a reaction zone, a cooling zone, and a circulation zone.

In the electrolysis zone, the electrochemical reactions take place. Two basic electrode configurations are used: (*1*) monopolar cells where the same cell voltage is applied to all anode/cathode combinations; and (*2*) bipolar cells where the same current passes through all electrodes (Fig. 4). To minimize the anodic oxidation of OCl^-;, the solution must be quickly moved out of this zone to a reaction zone. Because the reaction to convert OCl^- to ClO^{3-} (eq. 9) is slow, a relatively large volume reaction zone is required to carry out the reaction. Moreover, because the energy supplied to the cell is about twice the energy required to carry out the reaction, a cooling zone is required. Then, a circulation zone and circulation mechanism are provided.

Many combinations of these component zones have been designed. Some are shown in Figure 4. The combination range from all zones in one vessel to vessels for each zone.

Brine Preparation. Rock salt and solar salt (see CHEMICALS FROM BRINE) can be used for preparing sodium chloride solution for electrolysis. These salts contain Ca, Mg, and other impurities that must be removed prior to electrolysis. Otherwise these impurities are deposited on electrodes and increase the energy requirements. The raw brine can be treated by addition of sodium carbonate and hydroxide to reduce calcium and magnesium levels to below 10 ppm. If further reduction in hardness is required, an ion-exchange resin can be used. A typical brine specification for the Huron chlorate cell design is given in Table 5.

Hydrogen Purification and Recovery. Because the operation of the modern chlorate cell is quite efficient, ie, CE $= 90 - 98\%$, the hydrogen generated from the cell is quite pure and can be recovered for its fuel value. The hydrogen typically contains 2–3% chlorine and less than 2% oxygen by volume. If the cells are operated at lower than a normal pH of from 5.8–6.4, Cl_2 concentration in the hydrogen can be higher. The chlor-alkali industry has developed an extensive hydrogen recovery technology, which has been transferred to chlorate technology. Typically the hydrogen recovery system includes a caustic scrubber to remove chlorine and a compressor or blower to send hydrogen gas to a fuel burner or boiler (93).

Chlorate Recovery and Salt Removal. Prior to chlorate recovery, the residual hypochlorite in the electrolyzer liquor is destroyed by adding a reducing agent such as formate or urea (93,94). The liquor that contains sodium chlorate, chloride, sulfate, and dichromate is filtered to remove any insoluble particles. The liquor is concentrated by evaporation and sodium chloride is precipitated, filtered, and recycled back to the cell. The liquor is then cooled to yield sodium chlorate crystals, which are separated by centrifugation. The chromium containing centrate is returned to the evaporator or cell feed. Since the feed brine contains sodium sulfate, the sodium sulfate can be removed from the system by purging a small amount of filtrate from the crystallizer (95–98). Alternatively, the sodium sulfate from the filtrate can be crystallized out by the method

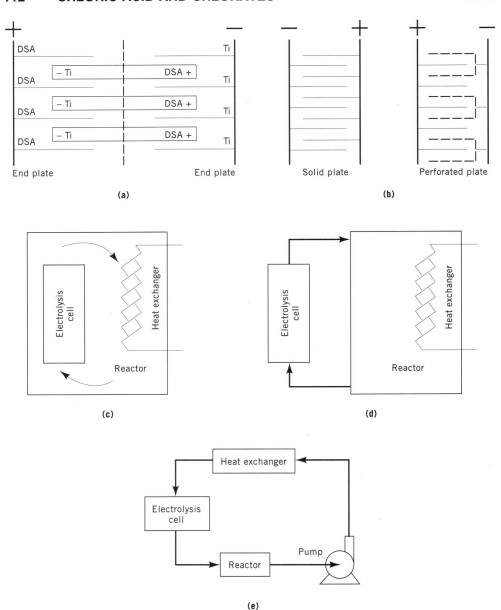

Fig. 4. Sodium chlorate cell designs (**a**) the horizontal bipolar cells used by Huron having narrow gap vertical plates or horizontal mesh where (− − −) represents an insulating partition, DSA is dimensionally stabilized anode, and (**b**) the parallel plate monopolar cells and electrolyzer configurations used by Chemetics, Krebs, etc where (− − −) represents a perforated plate; (**c**) the single vessel system used by DeNora, Huron, and OCC; (**d**) the double vessel system used by Krebs; and (**e**) the three vessel system used by Pennwalt, Ugine-Kuhlmann, and Kemanord. Materials of construction given in Table 5.

Table 5. **Brine Specification for Huron Sodium Chlorate Cell**[a]

Component	Maximum concentration, ppm
Mn	0.01
Ni	0.01
Fe	1.0
Al	1.0
Ba	0.1
Zn	0.01
F	1.0
cyanide	0.1
total organics	3.0
hardness as Ca	4.0[b]
Na_2SO_4	25.0[c]
NaCl	200 to 315[c,d]

[a]Temperature is a maximum of 75°C.
[b]For mild steel cathodes maximum hardness is 2 ppm. Somebrines contain up to 20 ppm.
[c]Units are g/L.
[d]Value varies depending on application.

described in reference 99. Recent advances in chlorine dioxide generation technology require that the sodium chloride content be less than 5% of the solution of sodium chlorate concentration. Sodium chlorate recovered by crystallization meets this requirement.

Chromium Removal System. Chlorate manufacturers must remove chromium from the chlorate solution as a result of environmental regulations. During crystallization of sodium chlorate, essentially all of the sodium dichromate is recycled back to the electrolyzer. Alternatively, hexavalent chromium, Cr^{6+}, can be reduced and coprecipitated in an agitated reactor using a choice of reducing agents, eg, sodium sulfide, sulfite, thiosulfate, hydrosulfite, hydrazine, etc. The product is chromium(III) oxide [1333-82-0], Cr_2O_3 (100–108). Ion exchange and solvent extraction techniques have also been utilized for recycling chromium (109–111). The resulting precipitate is easily filtered using ceramic or Teflon filters without the use of precoats so that the filtered chromium oxide can be returned directly to the chlorate process without further treatment. Once in the chlorate cell, the chromium reoxidizes immediately to hexavalent sodium dichromate. Essentially no chromium ever leaves the system (112). Chromium levels of less than 1 ppm are achieved in the final chlorate product.

2.4. Economic Aspects. North America, Western Europe, and Japan represented 94% of the total sodium chlorate consumption in 2002. The remaining 6% was distributed among South America, other Asia, and the rest of the world.

World operating capacity was 2.8×10^6 t in 2002. North America and Canada represent 69% of this total. Western Europe follows at 25%, South America at 5%, Japan and other Asia at 1% (113).

Table 6 lists U.S. and Canadian producers of sodium chlorate and their capacities (114).

Table 6. **Sodium Chlorate Producers and 1998 Capacities,** $\times 10^3$ t[a]

Producer	Location	Capacity[b]
United States		
CXY	Taft, La.	122
Eka Nobel	Columbus, Miss.	199
Eka Nobel	Moses Lake, Wash.	57
Elf Atochem	Portland, Ore.	53
Georgia Gulf	Plaquemine, La.	24
Huron Tech	Clairborne, Ala.	36
Huron Tech	Augusta, Ga	131
Kerr McGee	Hamilton, Miss.	130
Sterling Pulp Chemicals	Valdosta, Ga.	100
Western Electrochemical	Cedar City, Utah	6
Total		*858*
Canada		
Albchem	Bruderheim, Alberta	75
B.C. Chemicals	Prince George, B.C.	71
CXY	Amherstburg, Ont.	48
CXY	Beauhamois, Que.	44
CXY	Brandon, Man.	103
CXY	Bruderheim, Alberta.	68
CXY	Nanaimo, B.C.	18
Eka Nobel Canada	Magog Que.	150
Eka Nobel Canada	Valleyfield, Que.	113
PCI Chemicals	Delhousie, N.B.	22
St. Anne Chemical	Nackawic, N.B.	10
Sterling Pulp Chemicals	Buckingham, Que.	120
Sterling Pulp Chemicals	Grand Prairie, Alberta	50
Sterling Pulp Chemicals	North Vancouver, B.C.	92
Sterling Pulp Chemicals	Saskatoon, Sask.	50
Sterling Pulp Chemicals	Thunder Bay, Ont.	50
Total		*1084*
Total, North America		*1942*

[a]Ref. 114.
[b]Both solution and crystal material. Commercial production is by electrolysis of a sodium chloride solution.

An estimated 1.87×10^6 t was used in North America in 2002 to generate chlorine dioxide for bleaching of chemical pulp. Of this total, 68% was used in the U.S. and 32% was used in Canada. This represented 99% of demand for sodium chlorate.

EPA cluster rules are in favor of elemental-free-chlorine bleaching agents for pulp rather than total chlorine-free bleaching agents. Chlorine dioxide is generated from sodium chlorate and is a substitute for chlorine gas in bleaching applications. Full implementation of the cluster rules was to be finalized in 2001. The chlorate market is expected to level off and track a modest increase in demand after the full implementation. North American consumption is expected to grow at a rate of 0.9% in 2001–2006. Canada's growth rate is expected at 0.8% (113).

Western Europe uses less sodium chlorate for pulp bleaching and consumption in Japan is small. Japanese paper producers are expected to use elemental-chlorine-free bleaching agents by 2005 despite Japan's weak economy.

2.5. Product Specification. Sodium chlorate can be shipped either as solid crystals or preblended chlorate–chloride solution. A typical specification for technical-grade sodium chlorate is $NaClO_3$, 99.5 wt% min; NaCl, 0.12 wt% max; moisture, 0.20 wt% max; and 5 ppm chromium.

The crystalline sodium chlorate is usually dried in rotary driers to less than 0.2 wt% moisture content and is loaded into shipping containers or stored in moisture-free bins or silos prior to packaging. For conventional chlorine dioxide generators, sodium chlorate is shipped as a solution containing: ca 200 g/L (15 wt%, 3.4 M) sodium chloride; ca 350 g/L (26 wt%, 3.3 M) sodium chlorate; and 130 ppm chromium. Alternatively, for newer chlorine dioxide generators, 600 g/L sodium chlorate; 30 g/L sodium chloride; and less than 30 ppm chromium is used.

2.6. Analytical Methods. Chlorate ion concentration is determined by reaction with a reducing agent. Ferrous sulfate is preferred for quality control (115), but other reagents, such as arsenious acid, stannous chloride, andpotassium iodide, have also been used (116). When ferrous sulfate is used, a measured excess of the reagent is added to a strong hydrochloric acid solution of the chlorate for reduction, after which the excess ferrous sulfate is titrated with an oxidant, usually potassium permanganate or potassium dichromate.

2.7. Health and Safety Facters. Sodium chlorate is harmful if swallowed, inhaled, or absorbed through the skin. Symptoms of inhalation include burning sensation, coughing, wheezing, and laryngitis. Symptoms of ingestion include nausea, vomiting, abdominal pain, cyanasis, and diarrhea (117–119). Acute oral toxicity in laboratory animals for different species are in 1200–1800 mg/kg range. Lethal doses for children are 2 g and for adults are from 15 to 30 g (200–400 mg/kg). Permissible exposure limit for total dust is 15 mg/m^3 and for respirator protection, 5 mg/m^3.

Sodium chlorate and potassium chlorate are human poisons by unspecified routes. They are moderately toxic by ingestion and intraperitoneal routes. They both damage red blood cells when ingested (21).

Chlorates are strong oxidizing agents. Dry materials, such as cloth, leather, or paper, contaminated with chlorate may be ignited easily by heat or friction. Extreme care must be taken to ensure that chlorates do not come in contact with heat, organic materials, phosphorus, ammonium compounds, sulfur compounds, oils, greases or waxes, powdered metals, paint, metal salts (especially copper), and solvents. Chlorates should be stored separately from all flammable materials in a cool, dry, fireproof building.

Flammable resistant clothing such as Nomex should be worn when working with chlorates. Clothing splashed with chlorate solution should be removed before it dries. Shoes and gloves should be rubberized. Leather (qv) should not be worn. Goggles, face shields, and dust respirators should be worn when necessary to protect against dust, splashing, or spillage. Workers should bathe before leaving the working area.

Sodium chlorate does not burn if exposed to fire, but it decomposes to give off oxygen. Consequently only water is effective in the event of a spill or fire.

Table 7. **Hazard Ratings for Sodium Chlorate**

Situation	Nonfire[a]	Fire[a]
health (blue)	1	0
flammability	0	0
reactivity	2	2
other	oxy	oxy

[a]4 = extreme, 3 = high, 2 = moderate, 1 = slight, and 0 = insignificant.

Water cools and dilutes the sodium chlorate. Carbon dioxide, Halon, dry chemical or dry powder types of fire extinguishers are ineffective. EPA classifies sodium chlorate waste as D001 (ignitable waste). The Clean Water Act lists sodium chlorate as a hazardous substance, which if discharged into or upon water, may require immediate response to mitigate danger to public health and welfare. Hazard ratings for sodium chlorate according to the National Fire Protection Association (NFPA) are shown in Table 7.

2.8. Uses. The primary (99%) use of sodium chlorate is in the production of chlorine dioxide for bleaching in the pulp (qv) and paper industry (qv).

Chemical wood (qv) pulp bleached with chlorine dioxide has superior brightness over pulps bleached using other reagents (see BLEACHING AGENTS). The strength of the cellulose (qv) fiber is not degraded; thus a whiter and stronger paper is obtained using chlorine dioxide. When chlorine dioxide is used in place of chlorine for bleaching pulp, the adsorbable organic halides (AOX), which include dioxins, are reduced by as much as 90%. However, chlorine dioxide cannot be shipped and is therefore generated by the pulp producers at the bleaching plant.

The second most important use of sodium chlorate was as an intermediate in the production of other chlorates and of perchlorates.

The agricultural use of sodium chlorate is as a herbicide, as a defoliant for cotton (qv). Magnesium chlorate is used as a desiccant for soybeans to remove the leaves prior to mechanical picking (see DESICCANTS).

Sodium chlorate is used in uranium mixing. This usage was expected to decline sharply. Minor uses of sodium chlorate include the preparation of certain dyes and the processing of textiles (qv) and furs.

Potassium chlorate is used mainly in the manufacture of matches (qv) and pharmaceutical preparation. In pyrotechnics, chlorate salts may be mixed with certain organic compounds such as lactose to give a relatively cool flame, so that certain dyes may be incorporated in the mixture to give colored flares.

3. Other Chlorates

Barium chlorate monohydrate [10294-38-9], $Ba(ClO_3)_2$:H_2O, has colorless monoclinic crystals; mp or loss of water at 120°C; sp gr, 3.18; n_D^{20}, 1.562; is prepared by the reaction of barium chloride [10361-37-2], $BaCl_2$, and sodium chlorate in solution. Barium chlorate precipitates on cooling and is purified by recrystallizing. It is used in pyrotechnics.

Lithium chlorate [13453-71-9], $LiClO_3$, has rhombic needles; mp 124–129°C; decomposes on heating to 270°C. It is one of the most soluble salts known and it is very hygroscopic. $LiClO_3$ is prepared by adding lithium chloride [7447-41-8] to sodium chlorate solution. Sodium chloride precipitates, the liquor is concentrated, and the lithium chlorate is filtered and dried. It has limited use in pyrotechnics.

BIBLIOGRAPHY

"Chlorates" under "Chlorine Compounds, Inorganic" in *ECT* 1st ed., Vol. 3, pp. 707–716, by H. L. Robson, Mathieson Chemical Corp.; "Chloric Acid and Chlorates" under "Chlorine Oxygen Acids and Salts" in *ECT* 2nd ed., Vol. 5, pp. 50–61, by T. W. Clapper and W. A. Gale, American Potash & Chemical Corp.; in *ECT* 3rd ed., Vol. 5, pp. 633–645, by T. W. Clapper, Kerr-McGee Corp.; in *ECT* 4th ed., Vol. 5, pp. 998–1016, by Sudhir K. Mendiratta and Budd L. Duncan, Olin Corporation; "Chlorine Oxygen Acids and Salts, Chloric Acid and Chlorates" in *ECT* (online), posting date: December 4, 2000, by Sudhir K. Mendiratta, Budd L. Duncan, Olin Corporation.

CITED PUBLICATIONS

1. O. Kaemmerer, *Pogg. Ann.* **138**, 399 (1869).
2. B. L. Duncan, Internal Olin communication, 1989.
3. J. A. Dean, ed., *Langes Handbook of Chemistry*, 12th ed., McGraw-Hill Book Co., New York, pp. 5–14.
4. A. J. Bard and co-workers, *Standard Potentials in Aqueous Solutions*, Marcel Dekker, Inc., New York, p. 75.
5. Int. Pat. 91-03421 (Mar. 21, 1991), B. L. Duncan, G. W. Geren, and D. R. Leonard (to Olin. Corp.).
6. U.S. Pat. 3,810,969 (May 14, 1974), A. A. Schlumberger (to Hooker Chemical Corp.).
7. U.S. Pat. 4,798,715 (Jan. 17, 1989), K. L. Hardee and A. R. Sacco (to Eltech Systems Corp.).
8. Intl. Pat. 90-10733 (Sept. 20, 1990), R. M. Berry and R. D. Mortimer (to Pulp and Paper Research Institute of Canada).
9. U.S. Pat. 4,915,927 (Apr. 10, 1990), M. Lipsztajn (to Tenneco Canada, Inc.).
10. U.S. Pat. 5,084,189 (Jan. 28, 1992), J. J. Kaczur, D. W. Cawlfield, K. E. Woodard, Jr., and B. L. Duncan (to Olin Corp.).
11. U.S. Pat. 5,064,514 (Nov. 12, 1992), D. W. Cawlfield, H. J. Loftis, R. L. Dotson, K. E. Woodard, Jr., and S. K. Mendiratta (to Olin Corp.).
12. U.S. Pat. 5,108,560 (Apr. 28, 1992), D. W. Cawlfield, R. L. Dotson, S. K. Mendiratta, B. L. Duncan, and K. E. Woodard, Jr., (to Olin Corp.).
13. U.S. Pat. 5,089,095 (Feb. 18, 1992), D. W. Cawlfield and coworkers (to Olin Corp.).
14. U.S. Pat. 4,767,510 (Aug. 30, 1988), M. Lipstajn (to Tenneco Canada Inc.).
15. U.S. Pat. 2,751,374 (June 19, 1956), A. Creswell (to American Cyanamid Co.).
16. U.S. Pat. 2,983,718 (May 19, 1961), M. Wishman and W. R. Kokay (to American Cyanamid Co.).
17. U.S. Pat. 3,021,301 (Feb. 13, 1961), T. J. Suen and A. M. Schiller (to American Cyanamid Co.).
18. U.S. Pat. 3,208,962 (Sept. 28, 1965), M. Taniyama (to Toho Rayon).

19. Brit. Pat. 1,023,901 (Mar. 30, 1966), (to Japan Exlan Co. Ltd.).
20. A. Borsari and co-workers, *Chem. Ind. (London)*, 524–525, 1979; A. Borsari and co-workers, *J. Mol. Catal.*, 13–20, (1980).
21. R. J. Lewis, Sr., ed., *Sax's Dangerous Properties of Industrial Materials*, 10th ed., John Wiley & Sons, Inc., New York, 2000.
22. U.S. Pat. 5,330,620 (July 19, 1994), R. M. Berry, M. Paleologou, and N. Liebergott (to Pulp and Paper Research Insitute of Canada).
23. W. Gerhartz, ed., *Ullmann's Encyclopedia of Industrial Chemistry*, 5th ed., Vol. A6, VCH Verlagsgesellscaft mbH, D-6940, Weiheim, Germany, 1985, pp. 483–503.
24. W. F. Linke, *Solubilities of Inorganic and Metal-Organic Compounds*, 4th ed., Vol. 2, American Chemical Society, Washington, D.C., 1965, pp. 171, 1015.
25. H. C. Bell, *J. Chem. Soc.* **123**, 2712 (1923).
26. A. V. Babaeva, *Zh. Obshch. Khim.* **59**, 491 (1936).
27. N. D. Nies and R. W. Hulbert, *J. Chem. Eng. Data* **14**, 14 (1969).
28. J. E. Ricci and N. S. Yanick, *J. Am. Chem. Soc.* **59**, 491 (1939).
29. N. V. S. Knibbs and H. Palfreeman, *Trans. Faraday Soc.* **16**, 402 (1920).
30. R. A. Crawford, W. B. Darlington, and L. B. Kliener, *J. Electrochem. Soc.* **117**, 279–282 (1970).
31. A. Nallet and R. A. Paris, *Bull. Soc. Chim. Fr.*, 488–494 (1956).
32. T. S. Oey and D. E. Koopman, *J. Phys. Chem.* **62**, 755–756 (1958).
33. H. Vogt, *Chem. Age India* **26**, 540–544 (1975).
34. J. C. Schmacher, *ACS Manographics* **146**, 77 (1960).
35. H. M. McLaughlin and F. F. Brown, *J. Am. Chem. Soc.* **50**, 782 (1928).
36. S. Morishima and co-workers, *Nippon Kagaku, Kaisui* **9**, 1172–1176 (1991).
37. Fr. Pat. 2630426 (Oct. 27, 1989), D. Marais and J. Collants (to French Demands).
38. F. Wang and W. Wang, *Faming Zhuamli Shenqing Gongkai* (1987).
39. Fr. Pat. 2,630,426 (Apr. 22, 1988), D. Marais and J. Collantes (to Krebs & Cie).
40. B. V. Tilak, E. M. Spore, and J. C. Hanson, paper presented at the *Electrolytic Technology Committee Ad-Hoc Meeting*, U.S. Dept. of Energy and Argonne National Laboratory, Washington, D.C., Mar. 12–13, 1979.
41. U.S. Pat. 3,940,323 (Feb. 24, 1976), H. E. Cook, Jr. (to Hooker Chemicals and Plastics Corp.).
42. M. M. Jaksic, *J. Electro. Chem. Soc.* **121**, 70 (1974).
43. F. Foerster and E. Muller, *Z. Electrochem.* **46**, 23 (1903).
44. M. M. Jaksic, *Electrochem. Acta* **26**, 1127 (1976).
45. R. Bauer, *Chem Ing. Tech.* **34**, 376 (1962).
46. Taniguchi and T. Sekine, *Denki Kagaku* **43**, 715 (1975).
47. M. J. Jaksic, A. R. Despic, and B. Z. Nikolic, *Sov. Electrochem* **8**, 1533 (1972).
48. D. M. Brasher and A. D. Mercer, *Trans. Faraday. Soc.* **61**, 803 (1965).
49. H. Kerst, *Corrosion* **16**, 523 (1960).
50. H. Oranowska and Z. Szklarska-Smizlowska, *Zasch. Metal* **8**, 523 (1972).
51. S. Kowamura, N. Tanaka, and M. Nagayama, *Boshoku Gijutsu* **22**, 500 (1973).
52. I. E. Veselovskaya, E. M. Kuchinskii, and L. V. Morochko, *J. Appl. Chem. USSR* **37**, 76 (1964).
53. H. Vogt, *J. Electrochem. Soc.* **128**, (1981).
54. M. Hazzaa and F. A. Abd El Aleem, *Bull. Electrochem.* **6**, 74–78 (1990).
55. Can. Pat. 904,792 (July 11, 1972), R. M. O. Maunsell (to Electric Reduction Co. of Canada, Ltd.).
56. U.S. Pat. 3,809,629 (May 7, 1974), G. Messner, V. De Noro, and O. De Nora (to Impianti Electtrochimici).
57. Ger. Pat. 2,432,416 (Jan. 30, 1975), T. F. O'Brien and J. R. Hodges (to Penn-Olin Chemical Co.).

58. U.S. Pat. 3,878,072 (Apr. 15, 1975), E. H. Cook, Jr. and A. T. Emery (to Hooker Chemicals and Plastics Corp.).

59. U.S. Pat. 4,046,653 (Sept. 6, 1977), O. De Nora, V. De Nora, and P. M. Spaziante (to Impianti Elettrochimici SpA).

60. U.S. Pat. 4,060,475 (Nov. 29, 1977), D. Fournier and H. Bouregeois (to Rhone-Poulenc Industries (SA).

61. U.S. Pat. 4,087,344 (May 2, 1978), H. V. Casson and R. E. Loftfield (to Huron Chemicals Ltd.).

62. U.S. Pat. 4,159,929 (July 3, 1979), M. P. Grotheer (to Hooker Chemicals and Plastics Corp.).

63. F. Hine, *Kagaku (Kyoto)* **34**(2), 161–162 (1979).

64. U.S. Pat. 4,194,953 (Mar. 25, 1980), D. G. Hatherly (to Erco Industries Ltd.).

65. Ger. Pat. 3,001,191 (July 17, 1980), R. Charvin and J. L. Pignan (to Produits Chimiques Ugine Kuhlmann).

66. B. Jackson, *Sodium Chlorate System for Direct Feed to Chlorine Dioxide Generators and Chlorine Removal; Recycle Process*, Huron Technical Corp. commercial brochure, Feb. 13, 1979.

67. J. E. Coleman, *AIChE Symp. Ser.* **77**(204), 244–263 (1981).

68. Jpn. Pat. 56,123,387 (Sept. 28, 1981), (to Asahi Chemical Industry Co., Ltd.).

69. Jpn. Pat. 57,026,184 (Feb. 12, 1982), (to Chemetics International Ltd.).

70. Jpn. Pat. 57,073,190 (May 7, 1982), (to Chemetics International Ltd.).

71. Eur. Pat. 82,103,139 (Apr. 14, 1982), T. Akazawa, K. Suzuki, and T. Haga (to Hodogaya Chemical Co., Ltd.).

72. U.S. Pat. 4,414,088 (Nov. 8, 1983), J. B. Ford (to Erco Industries Ltd.).

73. U.S. Pat. 4,405,418 (Sept. 20, 1983), R. Takemura (to Ashai Chemical Industry Co., Ltd.).

74. C. G. Rader, E. M. Spore, and J. C. DeLong, *Pap. Int. Chlorine Symp. 3rd Meeting*, 305–314 (1982).

75. U.S. Pat. 4,434,033 (Feb. 28, 1984), J. J. Kaczur, S. A. Iacoviello, and E. G. Miller (to Olin Corp.).

76. U.S. Pat. 4,461,692 (July 24, 1984), C. W. Raetzsch, Jr. (to PPG Industries, Inc.).

77. U.S. Pat. 4,470,888 (Sept. 11, 1984), G. A. Wheaton (to Pennwalt Corp.).

78. W. A. McNeil, *Proc. Electrochem. Soc.* **84**(11) 341–354 (1984).

79. Jpn. Kokai Tokkyo Koho 60,046,384 (Mar. 13, 1985), (to Hodogaya Chemical Co. Ltd.).

80. P. Kohl and K. Lohrberg, *DECHEMA-Monogr.* **98**, 379–387 (1985).

81. Fr. Pat. 2,594,107 (Aug. 14, 1987), C. Welander, T. C. K. Ohlin, and R. P. Jarvis (to Kema Nord Blekkemi AB).

82. U.S. Pat. 4,702,805 (Oct. 27, 1987), J. E. Burkell and I. H. Warren (to CIL Inc.).

83. E. P. Drozdetskaya and V. I. SkripchenkoIzv, *Tekh. Nauki* **a3**, 111–112 (1986).

84. V. Rengarajan, R. Palanisamy, M. Sadagopalan, and K. C. Narasimham, *Bull. Electrochem.* **6**(1), 68–70 (1990).

85. SE Pat. 460,484 (Oct. 16, 1989), A. Ullman and M. Norell (to Eka Nobel AB).

86. Eur. Pat. 430,830 (June 5, 1991), J. C. Millet (to Atochem SA).

87. *Sodium Chlorate Electrolysis*, commercial brochure, Krebskosmo.

88. *The Pennwalt Electrochemical Sodium Chlorate Process*, commercial brochure, Pennwalt, Philadelphia, Pa., 1975.

89. U. Kuhlmann, "Manufacture of Sodium Chlorate by Electrolysis," commercial brochure, *PCUK Process*, 1978.

90. J. E. Coleman, *AiChE Symp. Ser 204* **77**, (1981).

91. D. M. Novak, B. V. Tilak, and B. E. Conway in *Modern Aspects of Electrochemistry*, Vol. 14, Plenum Press, New York, 1982.

92. U.S. Pat. 4,098,671 (1978), G. O. Westerlund.
93. Eur. Pat. 384,860 (Aug. 29, 1990), C. M. Pralus, J. C. M. Chassagne (to Atochem SA).
94. Eur. Pat. 266129 (May 4, 1988), M. Lipsztajn (to Tenneco Canada, Inc.).
95. H. Takeuchi, K. Takahashi, K. Tomita, and M. Imanaka, *J. Chem. Eng. Jpn.* **12**(3), 209–214 (1979).
96. B. Simon, *J. Cryst. Growth* **63**(1), 225–258 (1983).
97. P. S. Chen, P. J. Shlichta, W. R. Wilcox, and R. A. Lefever, *J. Cryst. Growth* **47**(1), 43–60 (1979).
98. B. Simon and J. Pantaloni, *PCH. PhysicoChem. Hydrodyn.* **5**(1), 19–27 (1984).
99. U.S. Pat. 4,702,805 (Oct. 27, 1987), J. E. Burkell and I. H. Warren (to CIL Inc.)
100. U.S. Pat. 4,481,087 (Nov. 6, 1984), P. M. DiGiacomo (to USA).
101. U.S. Pat. 4,804,528 (Feb. 14, 1989), I. H. Warren (to CIL Inc.).
102. U.S. Pat. 4,699,701 (Oct. 24, 1987), M. Lipsztajn (to Tenneco Canada, Inc.).
103. Can. Pat. 1,163,419 (Mar. 13, 1984), T. H. Dexter (to Occidental Chemical Corp.).
104. U.S. Pat. 4,268,486 (May 19, 1981), M. G. Noack and S. A. Manke (to Olin Corp.).
105. U.S. Pat. 4,259,297 (Mar. 31, 1981), J. J. Kaczur and S. K. Mendiratta (to Olin Corp.).
106. Jpn. Pat. 53,066,896 (June 14, 1978), M. Watanabe, T. Nomura and S. Nishimura (to Solex Research Corp. of Japan).
107. U.S. Pat. 4,086,150 (Apr. 25, 1978), B. Kindl and J. G. Atkinson (to Huron Chemicals, Ltd.).
108. Ger. Pat. 2,302,723 (July 26, 1973), H. D. Partridge and J. M. Hildyard (to Hooker Chemical Corp.).
109. Ger. Pat. 2,419,690 (Nov. 14, 1974), T. F. O'Brien (to Penn-Olin Chemical Co.).
110. Ger. Pat. 2,839,894 (Mar. 22, 1979), J. G. Grier and J. R. H. (to Pennwalt Corp.).
111. Fr. Pat. 2,502,136 (Sept. 24, 1982), D. Beutier, C. Palvadeau, G. Pasquier, and M. Dietrich (to Krebs et Cie).
112. U.S. Pat. 3,980,751 (Sept. 14, 1976), F. R. Foulkes (to Huron Chemicals, Ltd.).
113. "Sodium Chlorate," *Chemical Economics Handbook*, Stanford Research Institute, Menlo Park, Calif., Dec. 2002.
114. "Sodium Chlorate, Chemical Profile," *Chemical Market Reporter*, Feb. 22, 1999.
115. A. J. Boyle, V. V. Hughey, and C. C. Casto, *Ind. Eng. Chem. Anal. Ed.* **16**, 370 (1944).
116. I. M. Kolthoff and R. Belcher, *Volumetric Analyses*, 2nd ed., Vol. 3, Interscience Publishers, New York, 1957.
117. *Material Safety Data Sheet*, Eka Nobel Inc., Marietta, Ga., Sept. 1991.
118. *Material Safety Data Sheet* Sigma-Aldrich Corp., Milwaukee, Wi., Apr. 1992.
119. *Cah. Notes* **124**, 1–4 (Dec. 1986).

SUDHIR K. MENDIRATTA
BUDD L. DUNCAN
Olin Corporation

CHLORINATED PARAFFINS

1. Introduction

Chlorinated paraffins with the general molecular formula $C_xH_{(2x-y+2)}Cl_y$ have been manufactured on a commercial basis for over 50 years. The early products were based on paraffin wax feedstocks and were used as fire retardants and plasticizers in surface coatings and textile treatments and as extreme pressure–antiwear additives in lubricants. The development of chlorinated paraffins into new and emerging technologies was constrained principally because of the limitations of grades based on paraffin wax and the lack of suitable alternative feedstocks to meet the demands of the new potential markets.

In the early 1960s the petroleum industry employing molecular sieve technology made available a low cost and plentiful supply of normal paraffin fractions of very high purity. This enabled chlorinated paraffin manufacturers to exploit new applications with a range of products specifically designed to meet the technical and commercial requirements.

The principal feedstocks used today are the normal paraffin fractions C10–C13, C12–C14, C14–C17, and C18–C20 together with paraffin wax fractions of C24–C30, precise compositions may vary depending on petroleum oil source. Chlorination extent generally varies from 30 to 70% by weight. The choice of paraffinic feedstock and chlorine content is dependent on the application.

The availability of alpha olefins has enabled some manufacturers to offer a range of chlorinated alpha olefins alongside their existing range of chlorinated paraffins. Chlorinated alpha olefins are virtually indistinguishable from chlorinated paraffins but do offer the manufacturer a single-carbon number paraffinic feedstock and even greater flexibility in the product range.

2. Chemical and Physical Properties

By virtue of the nature of the paraffinic feedstocks readily available, commercial chlorinated paraffins are mixtures rather than single substances. The degree of chlorination is a matter of judgment by the manufacturers on the basis of their perception of market requirements; as a result, chlorine contents may vary from one manufacturer to another. However, customers purchasing requirements often demand equivalent products from different suppliers and hence similar products are widely available.

The physical and chemical properties of chlorinated paraffins are determined by the carbon chain length of the paraffin and the chlorine content. This is most readily seen with respect to viscosity (Fig. 1) and volatility (Fig. 2); increasing carbon chain length and increasing chlorine content lead to an increase in viscosity but a reduction in volatility.

Chlorinated paraffins vary in their physical form from free-flowing mobile liquids to highly viscous glassy materials. Chlorination of paraffin wax (C24–C30) to 70% chlorine and above yields the only solid grades. Physical properties of some commercially available chlorinated paraffins are listed in Table 1.

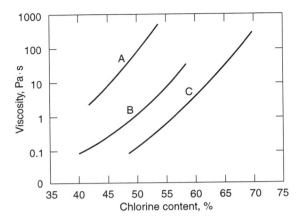

Fig. 1.   Viscosity of chlorinated paraffins at 25°C. Paraffin feedstock: A, wax; B, C14–C17; C, C10–C13. To convert Pa·s to P, multiply by 10.

A key property associated with chlorinated paraffins, particularly the high chlorine grades, is nonflammability, which has led to their use as fire-retardant additives and plasticizers in a wide range of polymeric materials. The fire-retardant properties are considerably enhanced by the inclusion of antimony trioxide.

Chlorinated paraffins are relatively inert and exhibit excellent resistance to chemical attack and are hydrolytically stable. They are soluble in chlorinated solvents, aromatic hydrocarbons, esters, ketones, and ethers but only moderately soluble in aliphatic hydrocarbons and virtually insoluble in water and lower alcohols.

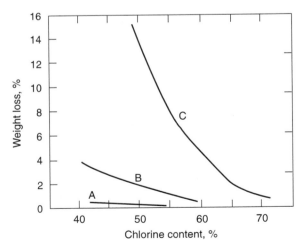

Fig. 2.   Volatility of chlorinated paraffins at 180°C after four hours. Paraffin feedstock: A, wax; B, C14–C17; C, C10–C13.

Table 1. **Physical Properties of Selected Commercial Chlorinated Paraffins**

Paraffin carbon chain length	Nominal chlorine contents, %w/w	Color hazen (APHA)	Viscosity,[a] mPa·s (= cP)	Density,[a] g/mL	Thermal stability,[b] %w/w HCl	Volatility,[c] %w/w	Refractive index
C10–C13	50	100	80	1.19	0.15	16.0	1.493
	56	100	800	1.30	0.15	7.0	1.508
	60	135	3500	1.36	0.15	4.4	1.516
	63	125	11,000	1.41	0.15	3.2	1.522
	65	150	30,000	1.44	0.20	2.5	1.525
	70	200	800[d]	1.50	0.20	0.5	1.537
C14–C17	40	80	70	1.10	0.2	4.2	1.488
	45	80	200	1.16	0.2	2.8	1.498
	52	100	1600	1.25	0.2	1.4	1.508
	58	150	40,000	1.36	0.2	0.7	1.522
C18–C20	47	150	1700	1.21	0.2	0.8	1.506
	50	250	18,000	1.27	0.2	0.7	1.512
Wax	42	250	2500	1.16	0.2	0.4	1.506
>20	48	300	28,000	1.26	0.2	0.3	1.516
	70	100[e]	f	1.63	0.2		

[a]At 25°C unless otherwise noted.
[b]Measured in a standard test for four hours at 175°C.
[c]Measured in a standard test for four hours at 180°C.
[d]At 150°C.
[e]10 g in 100 mL toluene solvent.
[f]Solid, softening point = 95 − 100°C.

Although considered to possess good thermal stability, chlorinated paraffins, if held at high temperatures for prolonged periods, first darken in color and then release detectable quantities of hydrochloric acid. Manufacturers often quote a thermal stability index which is a measure of the quantity of hydrochloric acid released expressed as a percentage by weight after heating the product for four hours at 175°C. Degradation of chlorinated paraffins can also be accelerated at elevated temperatures in the presence of iron, zinc, and dehydrochlorination catalysts. The thermal stability of chlorinated paraffins can be improved by the inclusion of epoxidized compounds which typically are epoxy esters, antioxidants of the hindered phenol-type, and metal soaps.

3. Manufacture

Chlorinated paraffins are manufactured by passing pure chlorine gas into a liquid paraffin at a temperature between 80 and 100°C depending on the chain length of the paraffin feedstock. At these temperatures chlorination occurs exothermically and cooling is necessary to maintain the temperature at around 100°C. Catalysts are not usually necessary to initiate chlorination, but some manufacturers may assist the process with ultraviolet light. Failure to control the reactive exotherm during chlorination may lead to a colored and unstable product. The reaction is terminated by stopping the flow of chlorine when the desired degree of chlorination has been achieved. This is estimated by density, viscosity, or refractive index measurements. The reactor is then purged with air or nitrogen to remove excess chlorine and hydrochloric acid gas. Small quantities of a storage stabilizer, typically epoxidized vegetable oil, may be mixed in at this stage or later in a blending vessel.

In general terms, for each ton of chlorinated paraffin produced, approximately one-half ton of hydrochloric acid is generated. Thus materials of construction must be resistant to acid attack. Reactor vessels were traditionally lined with lead or ceramics but glass-lined mild steel is now preferred. Ancillary equipment such as stirrers, pumps, valves, and pipelines should be of corrosion/acid-resistant material. Good housekeeping is vital as minute traces of metal chlorides entering the process can cause dehydrochlorination leading to discoloration of the chlorinated paraffin. A typical system employed in commercial production is shown in Figure 3.

In order to operate an economically viable chlorinated paraffin business, it is essential to have a profitable outlet for the surplus hydrochloric acid, either through direct sales into the market, or preferably via an oxychlorination unit in an integrated vinyl chloride/chlorinated solvent unit, while still maintaining the option of direct sales.

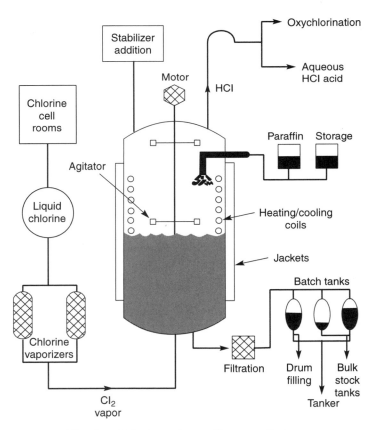

Fig. 3. Chlorinated paraffin manufacture.

4. Shipment and Storage

Liquid chlorinated paraffins are shipped in drums usually lacquer-lined mild steel or polyethylene and in road or rail barrels. Where appropriate larger quantities can be shipped by sea either in deck tanks of conventional cargo ships or in chemical parcel tankers for larger consignments.

The high viscosity of a number of grades generally precludes consideration of bulk supplies unless special transport, heating, pumping, and storage can be made.

Road and rail barrels are usually constructed of stainless steel or lacquer-lined mild steel and may require some provision for heating during cold weather. This is best achieved by submerged coils circulating hot water at about 40°C or low pressure steam, but care must be taken to ensure the surface temperature of the coils does not rise excessively otherwise discoloration of the product may occur.

The main points to be considered when designing a bulk storage installation are (1) the viscosity of all grades of chlorinated paraffin varies sharply with a change in temperature; (2) chlorinated paraffins should not be exposed to temperatures in excess of 40°C for prolonged periods of time; (3) chlorinated paraffin stability can be affected by contact with zinc and iron, therefore tanks of mild steel should generally be lined and galvanized steel avoided. Stainless steel, lacquer, or glass-lined mild steel tanks are recommended. The preferred linings are of the heat-cured phenol—formaldehyde type; and (4) chlorinated paraffins swell certain types of rubber and therefore rubber joints should be avoided. Polytetra-fluorethylene joints are recommended.

As for storage tanks, stainless steel and lacquer-lined mild steel are suitable materials of construction for pipe lines. For pumps, valves, etc, various alloys are suitable, including phosphor bronze, gun metal, Monel, stainless steel, and certain nickel steel alloys. Alloys with high proportions of zinc and tin together with copper and aluminum are not recommended.

5. Economic Aspects

U.S. Producers of chlorinated paraffins are listed in Table 2.

Demand for Chlorinated paraffins in 2000 was 42.9×10^6 kg (97×10^6 lb), in 2001 demand was 42.5×10^6 kg (96×10^6 lb). Expected demand in 2005 is 42.9×10^6 kg (97×10^6 lb). Demand equals production plus imports (negligible).

Table 2. **U.S. Producers of Chlorinated Paraffins and Their Capacities**[a]

Producer	Capacity, $\times 10^6$ kg ($\times 10^6$ lb)
Dover Chemical, Dover, Ohio	40.8 (90)
Keil Chemical, Hammond, Ind.	23.7 (50)
Total	*63.5 (140)*

[a]Ref. 1.

Exports are estimated to be about $4.9–7.3 \times 10^6$ kg ($11–16 \times 10^6$ lb) in recent years (1).

Growth is expected at 0% since applications are mature. Environmental pressure exists against chlorinated materials although no future regulations directly related to chlorinated paraffins is anticipated (1).

The European Union was to adopt restrictions on the use of short chain chlorinated paraffins for metal working and leather finishing. The United States Environmental Protection Agency completed its investigation and concluded there was no need to restrict short-chain chlorinated paraffins in any of its uses (1).

Prices have remained over the period 1996–2001 at \$0.32/kg (\$0.70/lb), 50% chlorine, list drum T.1. del.

In the United States approximately 50% of the chlorinated paraffins consumed domestically are used in metal-working lubricants. Approximately 20% are consumed as plastic additives, mainly fire retardants, and similarly 12% in rubber. The remainder as plasticizers in paint (9%), caulks, adhesives, and sealants at 6%, miscellaneous uses account for 3% (1).

6. Health and Safety Factors

A substantial body of information on the toxicological and environmental effects of chlorinated paraffins has been compiled over the past 20 years, and research is still continuing in both areas.

The acute toxicity of chlorinated paraffins has been tested in a range of animals and was found to be very low (2). A comprehensive study (3) demonstrated that the toxicity of chlorinated paraffins was related to carbon chain length and to a lesser degree chlorine content. The shorter chain-length chlorinated paraffins were more toxic than the longer chain chlorinated paraffins.

Subchronic studies in mice, rats, and rabbits determined the liver as the primary organ for attack by chlorinated paraffins. Mutagenicity, and reproductive and teratology studies revealed no abnormal effects. However, the National Toxicology Program (NTP) in the United States concluded that there was sufficient evidence of carcinogenicity from lifetime studies in rats and mice with one chlorinated paraffin having chain length of 12 carbon atoms and chlorine content 58% by weight. It is listed in the *Fifth Annual Report on Carcinogens*. Parallel studies on a long-chain chlorinated paraffin of average chain length (23 carbon atoms and 42% chlorine) showed no statistical increase in tumors. The NTP studies were reviewed by the International Agency for Research on Cancer (IARC) who concluded that the short-chain chlorinated paraffin was a possible human carcinogen (Cat II B). More recently, an extensive series of experiments (4) have been conducted to study further the biochemistry of the carcinogenic effects of the short-chain chlorinated paraffin C12 and 58% chlorine in rats and mice. The results support an earlier hypothesis (2) that the mechanism responsible for the occurrence of tumors in the liver of rats and mice is of a nongenotoxic nature and is associated with liver growth. This work shows that the effect is accompanied by peroxisome proliferation, and these effects are unlikely to occur in humans.

Because of the nature of some applications in which chlorinated paraffins are used, skin contact is inevitable and therefore an important potential route into the body. Skin absorption studies (5) have shown that chlorinated paraffins are very poorly absorbed through the skin and should not cause significant systemic concentrations.

7. Environmental Aspects

In general, chlorinated paraffins biodegrade; the rate is determined by chlorine content and carbon chain length. Microorganisms previously acclimatized to specific chlorinated paraffins show a greater ability to degrade the compounds than nonacclimatized organisms. Mammals and fish have been shown to metabolize chlorinated paraffins (6).

The acute toxicity of chlorinated paraffins to mammals, birds, and fish is very low (6), but over longer periods of exposure certain chlorinated paraffins have proved to be toxic to some aquatic species. However, the very low water solubility of chlorinated paraffins has made studies on aquatic species complicated. Laboratory experiments in which the chlorinated paraffins had been artificially solubilized showed only the short-chain grades to be toxic at low concentration; other longer-chain grades showed no adverse effects on the majority of aquatic species tested. The degree of solubilization achieved in the laboratory is unlikely ever to be experienced in the environment and is of doubtful environmental relevance (7).

In the United States further information and advice is readily available from the Chlorinated Paraffin Industry Association (CPIA) based in Washington D.C.

8. Applications

Chlorinated paraffins are versatile materials and are used in widely differing applications. As cost-effective plasticizers, they are employed in plastics particularly PVC, rubbers, surface coatings, adhesives, and sealants. Where required they impart the additional features of fire retardance, and chemical and water resistance. In conjunction with antimony trioxide, they constitute one of the most cost-effective fire-retardant systems for polymeric materials, textiles, surface coatings, and paper products. Chlorinated paraffins are also employed as components in fat liquors used in the leather industry, as extreme pressure additives in metal-working lubricants, and as solvents in carbonless copying paper.

8.1. Plasticized PVC. Chlorinated paraffins are employed as secondary plasticizers with fire-retardant properties in PVC and can be used as partial replacements for primary plasticizers (qv) such as phthalates (8) and phosphate esters (9).

By selection of those chlorinated paraffins specifically developed for the PVC industry to match the properties of primary plasticizers, reductions in costs can be achieved without significant change in properties. However, certain aspects can be improved by the inclusion of chlorinated paraffin such as flame resistance, chemical and water resistance, low temperature performance, and the viscosity aging stability in plastisols.

8.2. Metal-Working Lubricants.   A range of chlorinated paraffins are used as components of straight and emulsifiable metal-working lubricants as well as gear oils for industrial and automotive applications. In heavy-duty industrial gears, hypoid gears, metal cutting, and allied operations where high pressures and rubbing action are encountered, hydrodynamic lubrication cannot be maintained. In order to maintain lubrication under such conditions, extreme pressure (EP) or antiwear additives must be added to the lubricant. Such additives contain one or more of the elements of chlorine, sulfur, or phosphorus. Chlorinated paraffins are cost-effective extreme pressure additives and are either used alone or in combination with additives containing sulfur and/or phosphorus according to the application. High chlorine content lubricants are used for severe metal-forming operations such as deep drawing and stamping. This area is the principal outlet for chlorinated paraffins in the United States, accounting for approximately 50% of total sales.

The selection of chlorinated paraffin and the level of additives to a lubricating oil depends on the type of application and the severity of the operation. An approximate guide for the formulation of straight-cutting oils for metal-working is as follows:

Fluid type	Additives
heavy-duty broaching	60% chlorine + 10% fatty acid
very heavy-duty cutting	30% chlorine + 1% sulfur + 10% fatty acid
medium to heavy-duty	5 − 10% chlorine + 1 − 0.5% sulfur + 1 − 5% fatty oil
light to medium-duty	2 − 5% chlorine + 0.2% sulfur + 1 − 5% fatty oil

Metal-forming operations such as deep drawing, stamping, wire drawing, etc, are extremely severe and require large amounts of chlorinated paraffins; often the high chlorine containing grades are preferred. In some applications a mid-range chlorine content grade may be used neat. After deep drawing and stamping of mild steel, components are frequently left unprotected in storage. This can result in corrosion problems. For these applications chlorinated paraffins containing corrosion inhibitors and special stabilizers are available.

8.3. Paints.   Chlorinated paraffins are used as plasticizers for paints based on many types of resins, particularly chlorinated rubber and vinyl copolymers. Chlorinated rubber-based paints are employed in aggressive marine and industrial environments and vinyl copolymer principally for the protection of exterior masonry. The excellent chemical resistance of chlorinated paraffins and their ability to withstand prolonged contact with water makes them ideally suited as plasticizers for the most demanding applications.

8.4. Adhesives and Sealants.   Various grades of chlorinated paraffins are used as nonvolatile inert fire-retardant plasticizers and modifying resins in adhesives and sealants (10). They find wide application in polysulfide, polyurethane, acrylic, and butyl sealants for use in building and construction. The low volatility high chlorine types are also employed in sealants for double- and triple-glazed windows.

8.5. Flame-Retardant Applications. The flame resistance of polyolefins, unsaturated polyester, rubber, and many other synthetic materials can be improved by the inclusion of chlorinated paraffins. The solid 70% chlorine product is the preferred choice in most polymeric systems, but the liquid grades are widely used in rubbers, polyurethane, and textile treatments.

Chlorinated paraffins and modified types are used as solvents in carbonless copying paper production based on the encapsulation of a solution of reactive dyes. Chlorinated paraffins fullfill the technical requirements for a solvent including excellent solvency for the dyes; they do not react with the dyes nor encapsulation material, are immiscible with water, and have low volatility and low odor.

8.6. Fat Liquors for Leather. The addition of a chlorinated paraffin to a sulfated or sulfonated oil offers alternatives to natural oils as fat liquors for leather.

BIBLIOGRAPHY

"Chlorinated Paraffins" under "Chlorine Compounds, Organic" in *ECT* 1st ed., Vol. 3, pp. 781–786, by H. M. Roberts, Imperial Chemical Industries Ltd.; "Chlorinated Paraffins" under "Chlorocarbons and Chlorohydrocarbons" in *ECT* 2nd ed., Vol. 5, pp. 231–240, by D. W. F. Hardie, Imperial Chemical Industries Ltd.; in *ECT* 3rd ed., Vol. 5, pp. 786–791, by B. A. Schenker, Diamond Shamrock Corp.; in *ECT* 4th ed., Vol. 6, pp. 78–87, by Kelvin L. Houghton, ICI Chemicals and Polymers Ltd.; "Chlorinated Paraffins" in *ECT* (online), posting date: December 4, 2000, by Kelvin L. Houghton, ICI Chemicals and Polymers Ltd.

CITED PUBLICATIONS

1. "Chloroparaffins, Chemical Profile", *Chemical Market Reporter*, Aug. 26, 2002.
2. R. D. N. Birtley and co-workers, *Toxicol. Appl. Pharmacol.* **54**, 514 (1980).
3. D. M. Serrone and co-workers, *Fd. Chem. Toxic.* **25**(7), 553–562 (1987).
4. C. R. Elcombe and co-workers, *Mutagenesis* **5**(5), 515–518 (1990).
5. R. C. Scott, *Arch. Toxicol.*(63), 425–426 (1989).
6. J. R. Madely and R. D. N. Birtley, *Environ. Sci. Technol.* **14**, 1215 (1980).
7. I. Campbell and G. McConnell, *Environ. Sci. Technol.* **14**, 1209 (1980).
8. H. J. Caesar "Chlorinated Paraffins as Secondary Plasticizers in PVC," *Chem. Ind.* (Aug. 1978).
9. H. J. Caesar and P. J. Davis "Flame Retardant Vinyl Compounds," *33rd Annual Technical Conference*, Atlanta, Ga., May 6, 1975.
10. K. L. Houghton and M. E. Moss "Chlorinated Paraffins as Plasticizers in Polymer Sealant Systems," *ASC Supplier Short Course*, Nashville, Tenn., May 14–17, 1990.

Kelvin L. Houghton
ICI Chemicals and Polymers Ltd.

CHLORINE

1. Introduction

Chlorine and caustic soda, manufactured by the electrolysis of aqueous sodium chloride (or brine), are among the top ten commodity chemicals, in terms of capacity, in the United States. When chlorine is produced electrolytically, caustic soda is also produced in a ratio of 1.1 units of caustic soda per unit of chlorine. The combination of 1 ton of chlorine and 1.1 tons of caustic soda generated in this process is referred to as an electrochemical unit (ECU). Since one molecule cannot be created in this process without the other, profitability, costs, and margins are evaluated on an ECU basis in the chlor-alkali industry (1–7).

There are, however, chlorine processes that do not produce caustic soda, such as the electrolysis of hydrogen chloride. A small amount of chlorine is produced in conjunction with molten metal, such as sodium and magnesium, and by the electrolysis of potassium chloride solutions to generate potassium hydroxide (KOH).

Almost all new grassroots capacity will utilize membrane technology. During the next 5–10 years, some mercury cell plants and diaphragm cell plants will be converted to membrane or shut down because of poor production economics or environmental issues. Figure 1 represents the various processes that produce chlorine.

There are some processes that produce only caustic soda and no chlorine, such as the chemical caustic soda process that uses soda ash as the raw material. This process accounts for 1–2% of the total world capacity of caustic soda. Total caustic soda capacity in 2000 was ~56 million dry metric tons. Since there are sources of chlorine that do not produce caustic soda, the production ratio of caustic soda to chlorine in any given year in some countries, such as the United States, is not the theoretical ratio of 1.1. The total nameplate capacity for soda ash in 2000 was ~43 million metric tons and capacity for potassium hydroxide was 1.4 million metric tons (5,6).

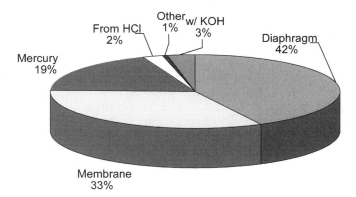

Fig. 1. Technologies for manufacturing chlorine in 2000. Total capacity $= 53 \times 10^6$ t (5,6).

2. Physical Properties

Table 1 presents characteristic properties of chlorine; further details are available (8–11). References 8 and 10, in particular, contain extensive tabulations of data and graphs of some of the more frequently used properties. Here, the data are described by the following equations: The vapor pressure P (in kPa) can be calculated in the temperature range 172–417 K from the equation 1.

$$\ln P = A - B/T - C \ln T + DT + E(F - T) \ln(F - T)/FT \tag{1}$$

where $A = 62.402508$, $B = 4343.5240$, $C = 7.8661534$, $D = 0.010666308$, $E = 95.248723$, and $F = 424.90$. The latent heat of vaporization in the range 0 to $-80°C$, which covers nearly all liquefaction operations is, in kilojoules per kilogram (kJ/kg):

$$\lambda = 268.5140 - 0.63197t - 0.00191245t^2 - 7.7778E{-}06t^3 \tag{2}$$

Table 1. **Physical Constants of Chlorine**

Property	Value
CAS Registry number	[7782-50-5]
atomic number	17
atomic weight (carbon-12 scale)	35.453
stable isotope abundance, atom %	
^{35}Cl	75.53
^{37}Cl	24.47
electronic configuration in ground state	[Ne]$3s^2 3p^5$
melting point, °C	-100.98
boiling point at 101.3 kPa(°C)	-33.97
gas density relative to air	2.48
critical density, kg/m^3	573
critical pressure, kPa	7977
critical volume, m^3/kg	0.001745
critical temperature, °C	143.75
critical compressibility	0.284777
gas density, kg/m^3 at 0°C and 101.3 kPa	3.213
gas viscosity, mPa.s at 20°C	0.0134
liquid viscosity, mPa.s at 20°C	0.346
gas thermal conductivity at 20°C, W/m K	0.00866
liquid thermal conductivity at 20°C, W/m K	0.120
latent heat of vaporization, kJ/kg	287.75
latent heat of fusion, kJ/kg	90.33
heat of dissociation, kJ/mol	2.3944
heat of hydration of Cl$^-$, kJ/mol	405.7
standard electrode potential, V	1.359
electron affinity, eV	3.77
ionization energies, eV	13.01, 23.80, 39.9, 53.3, 67.8, 96.6, and 114.2
specific heat at constant pressure	0.481
specific heat at constant volume	0.357
specific magnetic susceptibility at 20°C, m^3/kg	-7.4×10^{-9}
electrical conductivity of liquid at-70°C, $(\Omega\,cm)^{-1}$	10^{-16}
dielectric constant at 0°C (wavelengths > 10 m)	1.97

The following equations correlate the density, viscosity, and thermal conductivity of saturated liquid (10):

$$\rho = 573 + 1060.6083Y - 160.418Y^2 + 837.08192Y^3 - 247.20716Y^4 \quad (3)$$

$$\ln \mu_l = -2.64947 + 484.686/(T + 12.2345) \quad (4)$$

$$\kappa = 0.2458 - 3.094\text{E}{-}04T - 4.053\text{E}{-}07T^2 \quad (5)$$

where ρ = density, kg/m^3 $Y = (1 - T_r)^{1/3}$

T_r = reduced temperature
μ = viscosity, mPa s (or centipoise)
T = temperature, K
κ = thermal conductivity, W/(m K)
t = temperature, °C

and the transport properties of the gas (at 1atm):

$$C_p = 0.605770219 - 4.6076698\text{E}{-}05T - 41.8722507/T + 2408.76803/T^2 \quad (6)$$

$$\ln \mu_g = -2.40317 - 1117.9/(T + 292.683) \quad (7)$$

$$\ln \kappa = -2.7230 - 1004.4/(T + 202.68) \quad (8)$$

where C_p = heat capacity, kJ/kg K

Chlorine is somewhat soluble in water and in salt solutions, its solubility decreasing with increasing temperature or salt strength. It is partially hydrolyzed in aqueous solution according to equation 9.

$$Cl_2 + H_2O \rightleftharpoons HOCl + H^+ + Cl^- \quad (9)$$

This hydrolysis increases the solubility by reducing the partial pressure corresponding to a given amount of dissolved chlorine. Taking the equilibrium of reaction 9 into account, chlorine's solubility in water at a given temperature can be expressed as a function of its partial pressure by

$$S = HP + (KHP)^{1/3} \quad (10)$$

where
H = Henry's law solubility constant for molecular chlorine
K = equilibrium constant for reaction 9.

Chlorine forms a hydrate, containing 7–8 molecules of water per molecule of chlorine below 9.6°C at atmospheric pressure. These crystals are greenish-yellow in color. Chlorine hydrate forms greenish-yellow crystals below 9.6°C at atmospheric pressure. It does not exist above the quadruple point temperature of 28.7°C. It has a density of ~1.23 and therefore floats on liquid chlorine when deposited in a liquefaction system. Table 2 gives data for the solubility of chlorine in solvents other than water.

Table 2. **Solubility of Chlorine in Various Solvents**[a]

Solvent	Temperature, °C	Solubility
sulfuryl chloride	0	12.0
disulfur chloride	0	58.5
phosphoryl chloride	0	19.0
silicon tetrachloride	0	15.6
titanium tetrachloride	0	11.5
dimethylformamide	0	123[b]
acetic acid (99.84%)	0	11.6[b]
benzene	10	24.7
chloroform	10	20.0
carbon tetrachloride	20	17[c]
hexachlorobutadiene	20	22[c]

[a] Solubility in wt % unless otherwise noted.
[b] g/100 mL.
[c] mol%.

3. Chemical Properties

Martin and Longpre (9) discuss the chemical properties of chlorine. Chlorine usually exhibits a valence -1 in compounds, but it also exists in the formal positive valence states of $+1$ (NaClO [7681-52-9]), $+3$ (NaClO$_2$ [7758-19-2]), $+5$ (NaClO$_3$ [7775-09-9]), and $+7$ (NaClO$_4$ [7601-89-0]).

Molecular chlorine is a strong oxidant and a chlorinating agent, adding to double bonds in aliphatic compounds or undergoing substitution reactions with both aliphatics and aromatics. Tables 3 and 4 present significant industrial reaction products of chlorine and hydrochloric acid (12). Chlorine is very reactive under specific conditions but is not flammable or explosive. Reactions with most elements are facile, but those with nitrogen, oxygen, and carbon are indirect. Chlorine reacts with ammonia to form the explosive compound NCl$_3$. It does not react with hydrogen at normal temperatures in the absence of light. However, at temperatures 250°C, or in the presence of sunlight or artificial light of ~470-nm wavelength, H$_2$ and Cl$_2$ combine explosively to form HCl. Explosive limits of mixtures of pure gases are ~8% H$_2$ and ~12% Cl$_2$ (v/v) (see Fig. 2). These limits depend on temperature and pressure, and they can be altered by adding inert gases such as nitrogen and carbon dioxide. Most plants control the process to keep the hydrogen concentration below 4%.

Dry chlorine reacts combustively with most metals at sufficiently elevated temperatures. Aluminum, arsenic, gold, mercury, selenium, tellurium, and tin react with dry chlorine in gaseous or liquid form at ordinary temperatures; carbon steel ignites at ~250°C, depending on physical shape; and titanium reacts violently with dry chlorine. Wet chlorine is very reactive and very corrosive because of the formation of hydrochloric and hypochlorous acids (see eq. 9). Metals stable to wet chlorine include platinum, silver, tantalum, and titanium. The apparent anomaly in the behavior of titanium, which is stable in wet chlorine but explosive with dry chlorine, is due to the formation of a protective layer of

Table 3. **End Uses of Chlorine**[a]

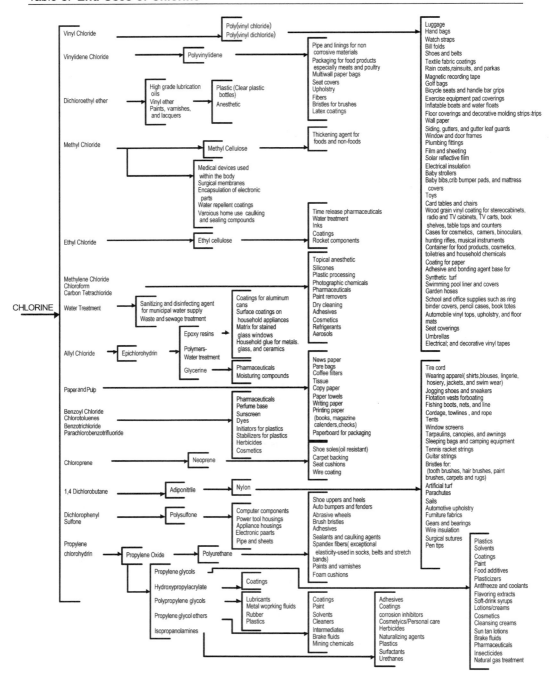

Table 3 (Continued)

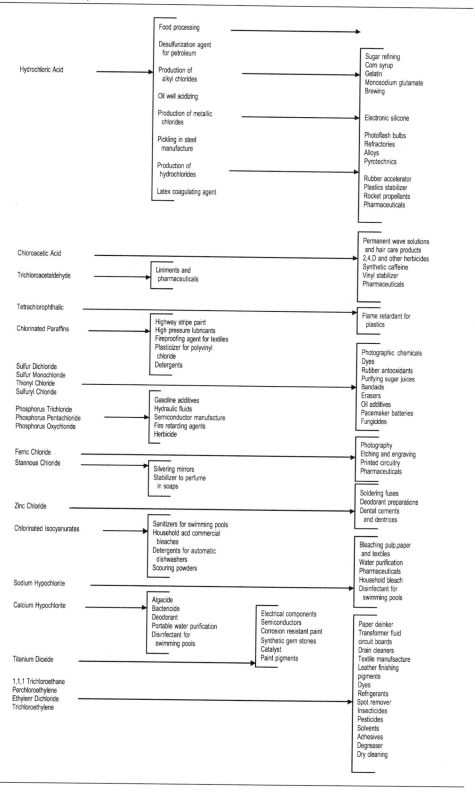

Hydrochloric Acid

Food processing

Desulfurization agent for petroleum

Production of alkyl chlorides

Oil well acidizing

Production of metallic chlorides

Pickling in steel manufacture

Production of hydrochlorides

Latex coagulating agent

Sugar refining
Corn syrup
Gelatin
Monosodium glutamate
Brewing

Electronic silicone

Photoflash bulbs
Refractories
Alloys
Pyrotechnics

Rubber accelerator
Plastics stabilizer
Rocket propellants
Pharmaceuticals

Chloroacetic Acid

Trichloroacetaldehyde

Liniments and pharmaceuticals

Permanent wave solutions and hair care products
2,4,D and other herbicides
Synthetic caffeine
Vinyl stabilizer
Pharmaceuticals

Tetrachlorophthalic

Chlorinated Paraffins

Highway stripe paint
High pressure lubricants
Fireproofing agent for textiles
Plasticizer for polyvinyl chloride
Detergents

Flame retardant for plastics

Sulfur Dichloride
Sulfur Monochloride
Thionyl Chloride
Sulfuryl Chloride

Phosphorus Trichloride
Phosphorus Pentachloride
Phosphorus Oxychloride

Gasoline additives
Hydraulic fluids
Semiconductor manufacture
Fire retarding agents
Herbicide

Photographic chemicals
Dyes
Rubber antooxidants
Purifying sugar juices
Bandaids
Erasers
Oil additives
Pacemaker batteries
Fungicides

Ferric Chloride
Stannous Chloride

Silvering mirrors
Stabilizer to perfume in soaps

Photography
Etching and engraving
Printed circuitry
Pharmaceuticals

Zinc Chloride

Soldering fuses
Deodorant preparations
Dental cements and dentrices

Chlorinated Isocyanurates

Sanitizers for swimming pools
Household acd commercial bleaches
Detergents for automatic dishwashers
Scouring powders

Bleaching pulp,paper and textiles
Water purification
Pharmaceuticals
Household bleach
Disinfectant for swimming pools

Sodium Hypochlorite

Calcium Hypochlorite

Algacide
Bactericide
Deodorant
Portable water purification
Disinfectant for swimming pools

Electrical components
Semiconductors
Corrosion resistant paint
Synthetic gem stones
Catalyst
Paint pigments

Titanium Dioxide

1,1,1 Trichloroethane
Perchloroethylene
Ethylenr Dichloride
Trichloroethylene

Paper deinker
Transformer fluid
circuit boards
Drain cleaners
Textile manufsacture
Leather finishing
pigments
Dyes
Refrigerants
Spot remover
Insecticides
Pesticides
Solvents
Adhesives
Degreaser
Dry cleaning

Table 4. **Operating Characteristics of ELTECH Diaphragm Cells**

Parameter	MDC 29		MDC 55		H2A		H4	
operating range, kA	40	80	75	150	50	80	90	150
chlorine capacity, t/day[a]	1.21	2.43	2.26	4.55	1.51	2.42	2.71	4.54
caustic capacity, t/day	1.36	2.74	2.55	5.14	1.70	2.73	3.06	5.13
hydrogen capacity, m^3/day	399	798	748	1497	498	798	897	1496
current density, kA/m^2	1.38	2.76	1.37	2.74	1.38	2.21	1.40	2.33
cell voltage, V (includes intercell bus)	2.97	3.60	2.97	3.59	2.97	3.35	2.98	3.40
energy consumption, dc kWh/t of Cl$_2$	2363	2847	2363	2839	2363	2655	2371	2694
modified diaphragm life, days	425	200	410	200	410	300	425	375
anode life, year	10–15	8–10	10–15	8–10	10–15	8–10	10–15	8–10
cathode life, year	10–15	5–8	10–15	5–8	10–15	5–8	10–15	5–8
distance between cells[b], m	1.6		2.13		2.32		3.05	

[a] To convert to short ton, multiply by 1.1.
[b] distance from center to center and side by side positioning with bus connected.

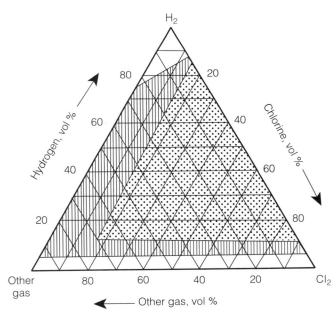

Fig. 2.   Explosive limits of Cl$_2$–H$_2$–other gas mixtures where the area with vertical lines represents the explosive region in the presence of residual gas from chlorine liquefaction (O$_2$, N$_2$, CO$_2$) and the area with black dots refers to the presence of inert gases (N$_2$, CO$_2$).

TiO_2 when the partial pressure of water vapor is sufficiently high. Tantalum is the most stable metal to both dry and wet forms.

Chlorine reacts with alkali and alkaline earth metal hydroxides to form bleaching agents such as NaOCl:

$$Cl_2 + 2\,NaOH \rightarrow NaCl + NaOCl + H_2O \tag{11}$$

Reaction of hypochlorite with ammonia produces hydrazine:

$$2\,NH_3 + NaOCl \rightarrow N_2H_4 + NaCl + H_2O \tag{12}$$

TiO_2 reacts with chlorine in the presence of carbon in the manufacture of $TiCl_4$, an intermediate in the production of titanium metal and pure TiO_2 pigment:

$$TiO_2 + 2Cl_2 + C(\text{or } 2C) \rightarrow TiCl_4 + CO_2(\text{or } 2CO) \tag{13}$$

$SiCl_4$ is produced by a similar reaction with SiO_2.

Chlorine reacts with saturated hydrocarbons and certain of their derivatives by substitution of hydrogen to form chlorinated hydrocarbons and HCl. Thus, methanol and methane chlorinate to form methyl chloride, which can continue to react with chlorine to form methylene chloride, chloroform, and carbon tetrachloride. Reaction of chlorine with unsaturated hydrocarbons results in saturation of the double or triple bond. This is a very important reaction in the production of ethylene dichloride, an intermediate in the manufacture of vinyl chloride and the largest consumer of chlorine:

$$CH_2{=}CH_2 + Cl_2 \rightarrow CH_2Cl{-}CH_2Cl \tag{14}$$

Organic substitution chlorinations utilize only one-half of the chlorine they consume; the other one-half is converted to HCl. While addition chlorinations consume the whole chlorine molecule, the resulting product may be dehydrochlorinated, again releasing one-half the chlorine value as HCl. This is the basis of manufacture of vinyl chloride monomer:

$$CH_2Cl{-}CH_2Cl \rightarrow CH_2{=}CHCl + HCl \tag{15}$$

The amount of HCl that results from these reactions would far exceed the market demand. This situation is relieved by the use of oxychlorination, in which oxygen is mixed with the feedstock and has the effect of converting the HCl byproduct back to chlorine. Oxychlorination is the favored process for ethylene dichloride:

$$CH_2{=}CH_2 + 2\,HCl + \tfrac{1}{2}O_2 \rightarrow CH_2Cl{-}CH_2Cl + H_2O \tag{16}$$

This allows the use of by-product HCl in the process. Alternatively, vinyl chloride manufacture can be by balanced oxychlorination, using the proper amount of fresh chlorine along with the HCl produced in reaction 15.

Direct chlorination of vinyl chloride generates 1,1,2-trichloroethane [79-00-5], from which vinylidene chloride is produced. Hydrochlorination of vinylidene chloride produces 1,1,1-trichloroethane [71-55-6]. This was a commercially important solvent whose use is being phased out under the terms of the Montreal Protocol. Tri- and tetrachloroethylene are manufactured by chlorination, hydrochlorination, or oxychlorination reactions involving ethylene. Aromatic solvents or pesticides such as mono-, di-, and hexachlorobenzene are produced by reaction of chlorine with benzene. Monochlorobenzene is an intermediate in the manufacture of phenol, aniline, and dyes (see CHLOROCARBONS AND CHLOROHYDROCARBONS, TOXIC AROMATICS).

4. Manufacture and Processing

4.1. Historical. In 1774, Swedish apothecary Carl W. Scheele first generated and collected chlorine by reacting manganese dioxide and hydrochloric acid. Scheele also discovered chlorine's bleaching action, after placing some leaves and flowers in a bottle containing the gas. Textile manufacturers in France produced the first commercial liquid chlorine bleach (13), by bubbling chlorine through a potash solution, in 1789.

Salt was first electrochemically decomposed by Cruickshank in 1800, and in 1808 Davy confirmed chlorine to be an element. In the 1830s, Michael Faraday, Davy's laboratory assistant, produced definitive work on both the electrolytic generation of chlorine and its ease of liquefaction. In 1851, Watt obtained the first English patent for an electrolytic chlorine production cell (14).

Through the 1880s and 1890s, producers in England, Canada, and the United States refined chlorine technology: Siemens developed electric generators; the Griesheim Company (Germany) invented the first practical diaphragm cell; Castner produced the first commercially viable mercury cell; and German producers found that water-free chlorine could be safely shipped in ordinary iron or steel pressure vessels. Thus by the early 1900s, chlorine was produced in mercury and diaphragm electrolytic cells and routinely shipped in liquid form. In 1913, Altoona, Pennsylvania became the first of many cities to treat sewage with liquid chlorine. Throughout the early and mid-1900s a wide variety of other chlorine uses were discovered. Presently, chlorine ranks in the top 10 volume chemicals produced in the United States.

Early demand for chlorine centered on textile bleaching, and chlorine generated through the electrolytic decomposition of salt sufficed. Sodium hydroxide was produced by the lime-soda reaction, using sodium carbonate readily available from the Solvay process. Increased demand for chlorine for polyvinyl chloride (PVC) manufacture led to the production of chlorine and sodium hydroxide as coproducts. Solution mining of salt and the availability of asbestos resulted in the dominance of the diaphragm process in North America, whereas solid salt and mercury availability led to the dominance of the mercury process in Europe. Japan imported its salt in solid form and, until the development of the diaphragm and then the membrane process, also favored the mercury cell for production.

4.2. Anodes. Graphite was exclusively used as the anode for chlorine production for >60 years even though it exhibited high chlorine overpotential and dimensional instability caused by the electrochemical oxidation of carbon to CO_2, which led to an increased electrolyte ohmic drop, and hence to high energy consumption during use. In the late 1960s, H.B. Beer's (15) invention of noble metal oxide coatings on titanium substrates as anodes revolutionized the chlor-alkali industry. The most widely used anodes were ruthenium oxide and titanium oxidez-coated titanium, which operated at low chlorine overpotential with excellent dimensional stability. Escalating power costs, triggered by the oil crisis in the mid-1970s, accelerated the transition of the chlor-alkali industry from graphite to metal anodes. Presently, $RuO_2 + TiO_2$ coated titanium anodes are exclusively used as the anodes. These anodes are supplied in the United States only by deNora, Electrode Products Inc. and ELTECH Systems Corporation under the trade name DSA, for dimensionally stable anodes.

The DSA electrode is a titanium substrate coated with titanium and ruthenium oxide mixture containing >30 mol% precious metal oxide, the precious metal loading ranging from 5 to 20 g/m^2. These coatings are formed on titanium mesh substrates by thermal decomposition. Ruthenium and titanium salts dissolved in organic solvent (eg, butanol), are applied to the Ti mesh and fired at 500°C in air to obtain the mixed oxides.

Both RuO_2/TiO_2 based coatings are polycrystalline and structurally complex. The oxides exist in the catalytic layer as $Ti_{(1-n)}Ru_nO_2$, where $n < 1$. The coatings exhibit a "mud-cracked" surface area of 200–500 cm^2 per apparent cm^2, the exact value being a function of the heat treatment and the coating composition. Two coatings that predominate in the industry are the original Beer coating containing a Ti/Ru mole ratio of 2:1 and a three-component coating with a mole ratio of 3Ru/2Sn/11Ti. This coating has a $RuO_2 + SnO_2$ loading of 1.6 mg/cm^2. It generates 22–25% less oxygen than RuO_2/TiO_2 coatings.

The Ru—Ti and Ru—Sn—Ti based compositions were prone to degradation via Ru losses in membrane cell operations, thereby limiting their life to 2–3 years. This problem was overcome with Ru–Ir–Ti coatings with total precious metal loading of 10–12 g/m^2 and Ru/Ir ratio of 1:1 based on mol% or wt%. These coatings are claimed to last >5 years in membrane cell operations (16,17). References (18) and (19) give more details on the various patented compositions and methods of preparation.

DSA anodes exhibit long life, very low operating voltage, and high chlorine efficiency. They are tolerant to a wide range of operating conditions, although performance degradation can occur when the percentage of oxygen generated is high. Harmful conditions include operation at feed brine pH values of >11 and at feed brine NaCl concentrations of <280 g/L. Exposure to caustic or fluorides results in the dissolution of RuO_2 as ruthenate ion or as ruthenium fluoride, and titanium as soluble titanium fluoride. Deposition of foreign matter such as MnO_2 or $BaSO_4$ can result in blockage of active sites on the surface and lead to the failure of the catalytic coatings, as revealed by the increased anode potential during service.

The mechanism of coating failure appears to depend on the type of cell in which the anode is operated. Life in diaphragm cells is at least 12 years; in mercury cells, it is considerably shorter, ~3–4 years. The unavoidable occurrence of minor short circuits, through contact with the mercury cathode, causes gradual physical wear of the anode coating. The limiting factor in membrane and diaphragm cells, in the absence of impurity-related effects, is the loss of RuO_2 by dissolution that accompanies high oxygen production and cell shutdowns

(20,21). A reduction in precious metal loading to <2 g/m^2 or a reduction in ruthenium content to <20 mol% leads to increased anode potentials. When the RuO$_2$ is lost from the surface layers, it is possible to reactivate the coating (22,23). When the total loading becomes too low, it is necessary to recoat the titanium substrate. Another explanation for the enhanced anode potentials is the formation (and build up) of insulating TiO$_2$ layer(s) between the coating and the substrate, which lacks direct evidence.

The anode suppliers recommend a feed brine concentration of 300 g/L and pH < 12, along with the following maximum limits of impurities (noted in parentheses), in ionic form: Hg (40 ppb); Mn (0.01 ppm); heavy metals (0.3 ppm); total organic content (1 ppm); F$^-$ (1 ppm); Ba (0.4 ppm).

4.3. Electrolytic Cell Operating Characteristics. Diaphragm cell technology is the source for generating the largest volume of chlorine, followed by the membrane cell process, and then the mercury process. However, the membrane process will dominate in the future, reflecting its ecological and economical benefits associated with it versus the other technologies. Over the past 10 years, neither new diaphragm nor mercury plants have been built. The electrolytic designs that are currently licensed or used are discussed have with a special emphasis on the membrane cells.

Depending on operating parameters (ie, current density, cell voltage, and current efficiency), the electrical energy consumption in membrane chlor-alkali electrolysis is between 1950 and 2300 kWh/t of chlorine. The amount of energy required to concentrate caustic soda from 32 to 50 wt% varies between 700 and 800 kWh/t (basis: medium pressure steam) of chlorine for a double-effect evaporator. Energy savings may justify investment in a third effect.

The electrical energy consumption in diaphragm chlor-alkali electrolysis is $\sim$10–20% higher than in membrane technology. In addition, diaphragm cells operate at a low caustic soda concentration of $\sim$11wt%, and hence the energy requirements to achieve the commercial concentration of 50% wt of caustic soda are much higher compared to the membrane process. The overall plant efficiency, defined as the ratio of the heat of the reaction to the sum of the endothermic processes and the electrical energy in to the cell, with diaphragm chlor-alkali cells is $\sim$23%, while the corresponding value for membrane cells is $\sim$35–40%.

4.4. Cell Technologies. As stated earlier, there are three primary electrolytic technologies based on mercury, diaphragm and membrane cells, that are used to produce chlorine and caustic. The currently available membrane cell technologies, with a brief description of the mercury and diaphragm cell technologies, are presented below.

Mercury Cells. Although no mercury cell plants were built over the past 15–20 years, $\sim$20% of the world's chlorine (and <6% of the chlorine in the United States) is produced using the mercury cell technology of Uhde, DeNora, Olin, Solvay, Krebs, and others. The mercury cell process (Fig. 3) consists of two electrochemical cells: the electrolyzer and the amalgam decomposer.

The mercury cell has a steel bottom with rubber coated steel sides, as well as end boxes for the brine and mercury feed and exit streams with a rubber or rubber-lined steel covers. Adjustable metal anodes hang from the top and mercury flows on the inclined bottom. The current flows from the steel bottom to the flowing mercury. Sodium chloride brine of 25.5% salt strength, fed from

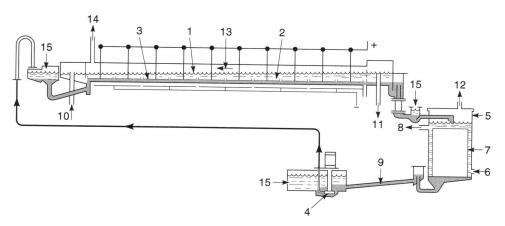

Fig. 3. Mercury cathode electrolyzer and decomposer : (*1*) brine level; (*2*) metal anodes; (*3*) mercury cathode flowing along base plate; (*4*) mercury pump; (*5*) vertical decomposer; (*6*) water feed to the decomposer; (*7*) graphite packing promoting the decomposition of the amalgam; (*8*) caustic liquor outlet; (*9*) denuded mercury; (*10*) brine feed; (*11*) brine exit; (*12*) hydrogen exit from the decomposer; (*13*) chlorine gas space; (*14*) chlorine exit; (*15*) wash water. Courtesy of Chlorine Institute Inc.

the inlet end box, is electrolyzed at the anode to produce chlorine gas, which leaves from the top of the cell. The sodium ions get discharged at the cathode to form sodium amalgam containing 0.25–0.4% Na, which flows out of the outlet end box. The sodium amalgam is subsequently reacted with water in the decomposer, packed with graphite particles, to form 50% NaOH and H_2. The unreacted brine out of the exit end box is resaturated and recycled back to the cells along with the denuded mercury from the decomposer.

Mercury cells operate efficiently because the hydrogen overpotential is high on mercury, favoring the sodium amalgam formation over the hydrogen evolution reaction. However, the brine should be free of impurities such as vanadium or chromium that will lower the hydrogen overvoltage, allowing hydrogen to form in dangerous quantities.

Developments in mercury cell operations over the past 10–15 years have been aimed toward lowering the energy consumption of the cells and minimizing the mercury losses to the air, water, or the products of electrolysis and the decomposer. Most plants operating with the mercury cells have complied with and sometimes surpassed the standards set by the local government and the federal agencies. The Environmental Protection Agency's (EPAs) National Emission Standards for Hazardous Air Pollutants for mercury currently call for a maximum mercury level of 1000 g in air at point source and 1300 g/day as fugitives. These standards are expected to be revised in the year 2002. References (24) and (25) give additional details related to the mercury cell technologies.

Diaphragm Cells. The first chlorine cell incorporating a percolating asbestos diaphragm was designed by E. A. LeSueur in the 1890s. In this cell, brine flowed continuously into the anolyte and subsequently through a diaphragm into the catholyte. The diaphragm separates the chlorine liberated at the anode from the sodium hydroxide and hydrogen produced at the cathode.

Failure to separate the chlorine and sodium hydroxide leads to the production of [sodium hypochlorite [7681-52-9], NaOCl, which undergoes further reaction to sodium chlorate [7775-09-9], $NaClO_3$. The commercial process to produce sodium chlorate is, in fact, by electrolysis of brine in a cell without a separator (see CHLORINE OXYGEN ACIDS, SALTS).

The early cells incorporated a horizontal asbestos sheet as the diaphragm. During the 1920s, this type of cell was the most widely used in the world and a few are still in operation. Subsequently, three basic types of diaphragm cells have been developed: rectangular vertical electrode monopolar cells, cylindrical vertical electrode monopolar cells, and vertical electrode filter press bipolar cells.

Asbestos Diaphragms.   The earliest diaphragms were made of asbestos paper sheets. Asbestos was selected because of its chemical and physical stability and because it is a relatively inexpensive and abundant raw material. The vacuum-deposited asbestos diaphragm, developed in the 1920s, was the diaphragm of choice until 1971, when it was supplanted by the Modified Diaphragm (trademark of ELTECH Systems, Inc.) (26). In its most common form, the Modified Diaphragm contains fibrous polytetrafluoroethylene (PTFE) and a minimum of 75% asbestos. The polymer, following fusion, stabilizes the asbestos, lowers cell voltage, and allows the use of the expandable DSA anode (27), which further lowers the cell voltage (28). The Modified Diaphragm in various formulations is the most common diaphragm in use today.

The toxicological problems associated with asbestos have been widely publicized and an EPA ban was overturned in 1991. Asbestos bans affecting the chlor-alkali industry exist in France, Argentina, and Scandinavia and indirectly in Brazil.

Non-Asbestos Diaphragms.   Numerous patents relating to non-asbestos diaphragms have been issued, and three have demonstrated commercial success sufficient to warrant the conversion of one or more cell rooms.

Polyramix fiber is a zirconia-PTFE deposited separator (29) developed by ELTECH Systems in use in five cell rooms commercially. Three of these cell rooms have been converted to Polyramix diaphragms, while two others are being converted. Polyramix separators offer longer diaphragm life, lower energy consumption, and can be reclaimed, or cleaned, with inhibited HCl to restore performance.

PPG Industries has also developed a non-asbestos diaphragm referred to as the Tephram diaphragm (30). This technology uses vacuum depositing to produce a base diaphragm composed of PTFE fluoropolymer materials and inorganic particulate materials. This diaphragm has also exhibited greatly extended life in comparison to asbestos and equal or better voltage and current efficiency characteristics. PPG uses the Tephram diaphragm in a significant portion of its Chlor/Alkali operations.

Chlor-Alp has developed a synthetic two layer separator consisting of an activated nickel deposited sublayer with a synthetic separator deposited on top; their two cell rooms in France are being converted to that technology (31).

Electrolyzers.   Various designs of bipolar and monopolar diaphragm electrolyzers were developed and used in commercial operations (32). These include the bipolar cells of Dow and the Glanor cells of PPG, and the monopolar cells of the Eltech/Uhde HU-type, which have not been licensed for use during the

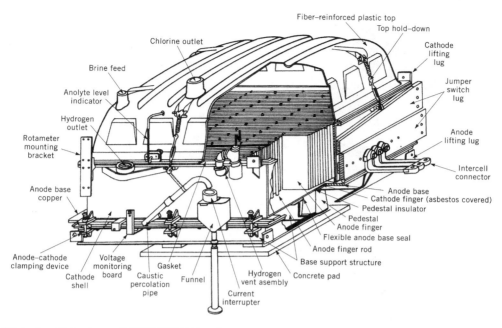

Fig. 4. Cut view of ELTECH H-4 diaphragm cell operating at a nominal rating of 150 kA.

past 15 years. The sole supplier of diaphragm cell technology, at present, is ELTECH Systems, Inc., located in Chardon, Ohio.

ELTECH Monopolar Electrolyzers. ELTECH supplies monopolar diaphragm electrolyzers of two designs: the ELTECH H-type (33,34) shown in Figure 4 and the MDC-type (34,35) in Figure 5.

The H-series of monopolar cells incorporate DSA anodes and operate at high current densities. The H-series employs cathode tubes having both ends open and extending across the cell, similar to the MDC cells. ELTECH has introduced a series of improvements to the design and performance of its monopolar diaphragm cells. The MDC series cathodes have been redesigned to replace the tube sheets and horizontal strap supports with copper corrugations that extend and are welded to the current carrying side plate to which the intercell connectors are bolted. Similar construction, which also features explosion bonded side plates, is offered for the H-series cathodes. The cathodes are also designed so that they can be stress relieved, which, along with the copper internals, extends the life of the cathode by at least 25%. These new cathodes are called ATC for advanced technology cathode. The conventional MDC cathode construction featuring tubes welded through a pair of tube sheets is being phased out.

New anode designs offer lower resistance, and can be operated at zero gap with Polyramix diaphragms (29,30). The low voltage anode (LVA) features a solid conductor bar and 1-mm thick expanders to reduce structure drop. The energy saving anode ESA features expanded mesh with micromesh continuously welded to the substrate, in addition to the LVA improvements, and is designed to operate at zero gap with the Polyramix diaphragm.

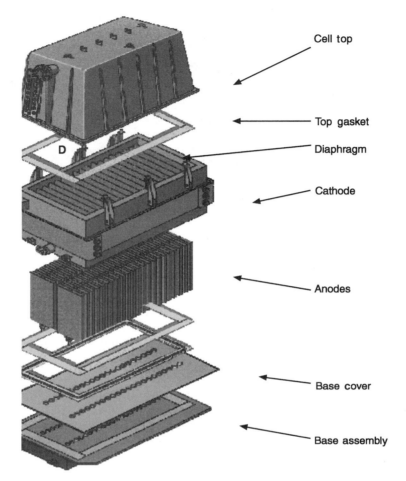

Cell top

Top gasket

Diaphragm

Cathode

Anodes

Base cover

Base assembly

Fig. 5. Exploded view of an ELTECH MDC-55 cell.

The most effective improvements, however, involve rebuilding the cell to increase the electrode areas by reducing the cathode tube thickness and the spacing between the electrodes. Cell areas have been increased from 14–24%, depending on design.

The life of a Polyramix (PMX) diaphragm is several times that of a polymer modified asbestos (PMA) diaphragm. Furthermore, *in situ* diaphragm reclamation procedures have been developed that permit the life of the PMX diaphragm to be extended to match the life of the cathode internals. Titanium base covers (TIBAC), are replacing rubber covers to protect the base. Improved gaskets have been developed to extend the life of the modern diaphragm cell assembly to 5 years or more. Advanced diaphragm cell technology (ADCT), incorporating these improved cell components is claimed to offer electrical energy reduction of 10–15% over conventional diaphragm cell technology using expandable anodes and polymer modified asbestos diaphragms.

Table 5. **Characteristic Features of Current Membrane Technologies**

Supplier	Effective membrane area (m^2)	Max. number of cells per electrolyzer	Circuit load (kA)	Current density (kA/m^2)	Specific energy consumption (kWh/t NaOH, 100%)
Bipolar electrolyzers					
Asahi Kasei, ML 32 or ML 60	2.72 or 5.05	150	5.4–30.3	2–6	2100 (at 4 kA/m^2)
CEC, BiTAC-800	3.28	80	5–19.65	1.5–6	2150 (at 5 kA/m^2)
ELTECH, ExL[B]	1.5	60	2.25–10.5	1.5–7	2100 (at 5 kA/m^2)
Krupp Uhde/Uhde Nora, BL-2.7	2.7	160	5.4–16.3	2–6	2130 (at 5 kA/m^2)
Ineos Chlor Ltd	2.89	160	5.79	6	2190 (AT 6 kA/m^2)
Monopolar electrolyzers					
CEC, DCM 416 × 2	3.03	32	194	4	2250 (at 4 kA/m^2
ELTECH, ExL[Ma]	1.5	30	15–180	1.5–6	2150 (at 5 kA/m^2)
Ineos, FM-21-SP or FM-1500	2 × 0.21	60 or 120	30–480	1.5–4	2140 (at 4 kA/m^2)

[a] Based on the ELTECH MGC-Electrolyzer.

Over 20,000 t/day of chlorine capacity has been licensed with MDC cells. Table 5 lists the operating characteristics of the four most common ELTECH diaphragm cells.

Ion-Exchange Membrane Cells. As stated earlier, the membrane chlor-alkali electrolysis is the state-of-the-art process and is the choice of cell technology for grass root plants and for expansion/conversion projects. The conversion of a mercury or a diaphragm plant to membrane cell operations often is an interesting alternative to building a new plant, as investment costs can be reduced by integrating the membrane electrolyzers into the existing infrastructure.

The key component of a membrane cell is the ion exchange membrane, which should be chemically stable, withstanding the extremely aggressive conditions in an electrolysis cell (ie, chlorine in the anolyte and concentrated caustic in the catholyte) and offering excellent ion-exchange selectivity to meet the electrochemical requirements. The first membrane of such a type was developed by DuPont in the early 1970s (37). Membrane-cell technology was first developed, demonstrated, and commercialized in Japan. In 1978, the first bilayer membrane exhibiting low electrical resistance and high current efficiency was introduced.

In the year 2000, ~33% of the world chlorine production was met by membrane cell technology, and this percentage will continue to grow.

Principles. In a membrane cell, a cation-exchange membrane separates the anolyte and the catholyte, as shown in Figure 6. Ultrapure brine (containing <20 ppb of Ca^{2+} and Mg^{2+} ions combined) with a salt content of ~310 g/L NaCl is fed into the anolyte compartment, where chlorine gas is generated at the anode. The sodium ions, together with the associated water molecules, migrate through the membrane into the catholyte. The membrane effectively prevents the migration of hydroxyl ions into the anolyte. Unlike the separators used in

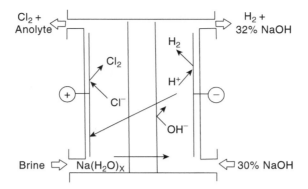

Fig. 6. Principle of the membrane cell.

the diaphragm-cell process, the cation-exchange membrane mostly prevents the migration of chloride ions into the catholyte. The depleted brine from the anode compartment is resaturated with salt to maintain the required salt concentration and fed back to the cells. On the catholyte side, water is electrolyzed at the cathode to form hydrogen and hydroxyl ions, which combine with the sodium ions transported from the anolyte to form caustic soda. Present-day membranes perform optimally at a caustic concentration of $\sim$32–35%. The product caustic concentration is adjusted by adding demineralized water to the recirculating caustic.

Membranes. The membrane is the most critical component of the membrane-cell technology. It determines the current efficiency, cell voltage, and hence the energy consumption for the production of chlorine and caustic. An ideal ion-exchange membrane should exhibit:

- High selectivity for the transport of sodium or potassium ions,
- Negligible transport of chloride, hypochlorite, and chlorate ions,
- Zero back-migration of hydroxide ions,
- Low electrical resistance,
- Good mechanical strength and properties with long term stability (38).

An ion-exchange membrane exhibiting these characteristics was first developed by DuPont, based on perfluorosulfonate polymers (39–46). These membranes achieved high efficiency and low voltage only at low caustic concentrations ($\sim$10–12%). At high caustic strengths, there was significant back-migration of caustic into the anode compartment, which resulted in low caustic and chlorine current efficiencies (47,48). Next, Asahi Glass Company of Japan developed a cation exchange membrane, based on perfluorocarboxylic acid polymer. This showed high caustic efficiencies at high caustic concentrations but exhibited high electrical resistance due to the low water content in the membrane (49). The membrane manufacturers then combined the best features of the perfluorosulfonic acid and perfluorocarboxylic acid polymers (Fig. 7) to develop composite membranes with sulfonic acid layers facing the anolyte side and carboxylic acid layers facing the catholyte side, with an intermediate reinfor-

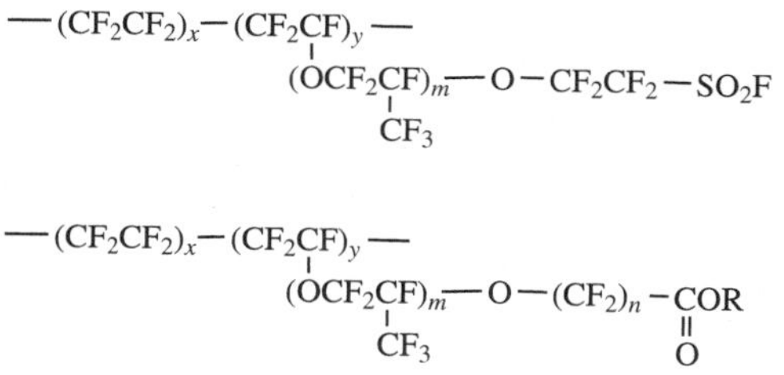

$$—(CF_2CF_2)_x—(CF_2CF)_y—$$
$$(OCF_2CF)_m—O—CF_2CF_2—SO_2F$$
$$CF_3$$

$$—(CF_2CF_2)_x—(CF_2CF)_y—$$
$$(OCF_2CF)_m—O—(CF_2)_n—COR$$
$$CF_3 \qquad \qquad O$$

$m = 0–1 \qquad n = 1–5 \qquad R = Alkyl$

Fig. 7. Structural formulas for perfluorosufonate (top) and perfluorocarboxylate groups (bottom) (57).

cing fabric between them (Fig. 8). It is these composite membranes that are currently used in the membrane cells to achieve high current efficiency and low cell voltage at current densities >5 kA/m². These composite membranes are prepared by lamination of the perfluorocarboxylate and perfluorosulfonate films, by the chemical conversion of the perfluorosulfonic acid to realize a carboxylate layer thickness of 5–10 μm, or by coextension of the two different polymer films, which will provide high efficiency with low voltage penalty.

An important characteristic of the persulfonate membranes is their ability to achieve the desired transport of sodium ions while hindering the migration of the hydroxyl ions (38). Several theoretical descriptions have been proposed to understand the unique transport character of ions and water molecules across these polymeric membranes. Of these the most popular one is the cluster-network model proposed by Gierke based on experimental evidence supporting the model. According to this description, the ions and the sorbed solutions are in clusters as shown in Figure 9. In a 1200-equivalent weight polymer, the clusters are of 3–5 nm in diameter, each containing ~70 ion-exchange sites and 1000 water molecules (50,51).

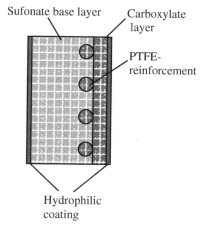

Fig. 8. Structure of a high performance chlor-alkali membrane.

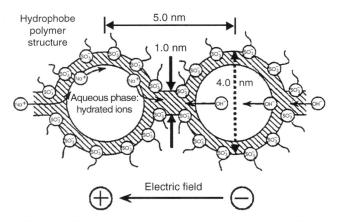

Fig. 9. Principle of the cluster-network model (51).

The counterion, the fixed sites and the swelling water phase are separated from the perfluorocarbon matrix into spherical domains connected by short narrow channels. The fixed sites are embedded in the water phase near the water–fluorocarbon interface. The diameter of these channels is 1 nm. High caustic efficiency, according to this model, is a result of the repulsive electrostatic interaction between the hydroxyl ion and the fixed charges on the surface of the clusters. Hence, the migration of the hydroxyl ion is difficult as it has to overcome a large electrostatic barrier in the channel. This results in high caustic efficiencies. Sodium ion, on the other hand, moves easily from one site to the other because of the favorable potential gradient and the small distances between sites.

A phenomenon of importance during electrolysis is the transport of water by electroosmotic mass transfer, driven by the electrical field between anode and cathode. While Na^+ ions migrate through the membrane in a hydrated state ($\sim$3–5 molecules of water per ion), the OH^- ions cannot pass through the membrane in the opposite direction. As a result, there is a net transport of water to the catholyte. The amount of water carried over into the catholyte is inversely related to the salt concentration in the anolyte.

In the design of a membrane electrolyzer with standard membranes, minimization of the voltage drop across the electrolyte is accomplished by bringing the electrodes close together. However, when the gap is very small, the voltage increases because of the entrapment of gas bubbles between the electrodes and the hydrophobic membrane. This effect is avoided in high performance membranes by coating both sides of the membrane with a thin layer of a porous inorganic material to enhance the membrane's ability to release the gaseous products from its surface. These improved membranes have allowed the development of modern electrolyzers with zero or narrow gap between the membrane and the cathode (Fig. 10). References (14,52,53) address these features in more detail.

Catalytic Cathodes.   The first membrane electrolyzers introduced in the late 1970s were built with low carbon steel cathodes, a material that has been successfully used in diaphragm cells. However, since the early 1980s, the focus

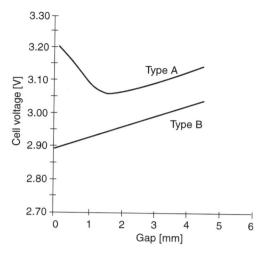

Fig. 10. Comparison conventional (type A) versus high performance (type B) membrane.

of the cathode material has been on nickel because of its stability in concentrated caustic solutions.

Depending on the electrolyzer load, the hydrogen overpotential with carbon steel cathodes is in the range of 300–400 mV. Energy savings, by reducing the overpotential as much as 200–280 mV, are realizable, in principle, by using nickel cathodes with a catalytic coating. The various approaches include using materials that provide enhanced surface area and better electrocatalytic properties than steel. Composites generally chosen for coating are based on nickel or noble metals. They are deposited on the cathode by thermal, plasma, or electrolytic routes using a second component such as aluminum or zinc, which is leached out in NaOH solutions to give a high, electrochemically active surface. Many compositions are mentioned in the literature (25,54–56). However, the coatings that are commercially employed in membrane cells include nickel–sulfur (57), nickel–aluminum (58), nickel–nickel oxide mixtures, and nickel coatings containing the platinum group metals (59). Although catalytic cathode technology is practiced in membrane cells, commercialization is still awaited in diaphragm cells. The technical problems confronting the use of catalytic cathodes in diaphragm cells include selection of a coating technique for the complex cathode assembly that would not adversely influence the structural tolerances involved in the fabrication of cathodes and developing shutdown procedures that would eliminate hypochlorite as quickly as possible to preserve the catalytic activity of the coatings. Such problems do not exist in membrane cells because of the anion rejection properties of the membrane and effective anolyte flushing procedures during shutdowns.

Electrolyzers. Electrolyzers are classified as monopolar or bipolar, depending on the manner in which an electrical connection is made between the electrolyzer elements. In the monopolar type, all anode and all cathode elements are arranged in parallel (see Fig. 11). Such an electrolyzer will operate at a high amperage and low voltage. While the amperage is set by the number of

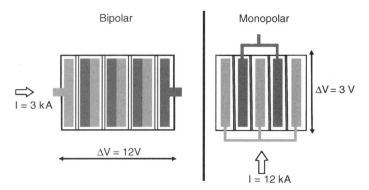

Fig. 11.   Bipolar and monopolar arrangements.

elements in an electrolyzer, the total voltage depends on how many electrolyzers are in an electric circuit. In a bipolar configuration, the cathode of one cell is connected to the anode of the next cell, and thus the cells are configured in series. This scheme of cell assembly results in operation of the cells at a low amperage and high voltage. The bipolar arrangement is advantageous for realizing low voltage drop between the cells. However, problems associated with current leakage and corrosion arise in bipolar operations, since the feed and discharge streams of electrolytes to and from the cells, having different electrical potentials, are hydraulically connected via common manifolds and collectors. These problems can be avoided by properly designing the electrolyzer (eg, by limiting the number of elements per electrolyzer) to limit the overall stack voltage to a safe value. The monopolar design, on the other hand, suffers from the voltage losses occurring in the interelectrolyzer connectors. This inevitable drawback can be minimized by a conservative design. While the bipolar systems allow shutdown for maintenance of a single electrolyzer unit, independently from the rest of the plant, monopolar electrolyzers have to be designed in such a way that an individual electrolyzer can be short circuited, enabling maintenance and membrane replacement without shutting down the entire circuit. It is also possible to combine these two configurations in a hybrid electrolyzer arrangement. This has been used in conversion projects where the electrolyzers had to be configured properly to integrate with the existing rectifiers and their performance.

Commercial Electrolyzers.   All membrane electrolyzers have common general design features like vertical membrane position, stacked elements and usage of similar materials of construction. Nevertheless, there are quite substantial differences in the cell design.

One of general design differences is the manner of achieving effective sealing of the electrolyzer. The most common principle is the filter press arrangement, where tightness is achieved by pressing together all elements of an electrolyzer from both ends. The relatively high sealing-forces are produced by means of hydraulic devices or tie rods. A different approach is the single element design, developed and applied by Uhde, where each element is individually sealed, by a flanged connection between the cathode and the anode semishell. The preassembled elements are stacked to form an electrolyzer, but only moderate forces are applied to ensure electrical contact.

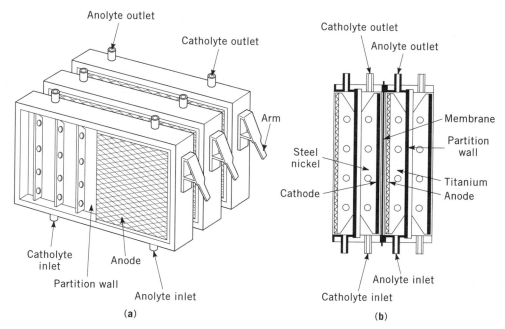

Fig. 12. Schematic of Asahi Kasei Acilyzer-ML bipolar electrolyzer. (**a**) View of cell units; (**b**) structure of cell.

As the membrane development allows increasing current densities, electrolyzer internals have to meet the related effects. Essential design targets are the minimization of structural voltage losses, homogenization of electrolyte concentration and temperature, as well as measures to counter problems related to the increased gas evolution.

The actual electrolyzer designs of the suppliers are discussed below.

Asahi Kasei Industry's Acilyzer-*ML Bipolar electrolyzers.* Asahi Kasei's electrolyzer (Acilyzer-ML), shown in Figure 12, is of the bipolar type, composed of a series of cell frames (49). Each cell consists of a pair of anode and cathode compartments facing each other, having an ion-exchange membrane in between. The anode and cathode compartments are separated by an explosion-bonded titanium-steel or titanium-nickel plate, and vertical support ribs are welded to each side of these partition walls, to which the anode and cathode mesh are spot welded in turn. Each compartment has an electrolyte inlet at the bottom and gas–liquid outlet at the top. These inlets and outlets are connected to the supply and collection headers by PTFE hoses. Both anolyte and catholyte are recirculated back to the electrolyzer from collection tanks. Deionized water is added to the catholyte collection tank to control the caustic concentration, and ultrapurified brine is added to the anolyte collection tank to control the NaCl concentration. A portion of the catholyte is drawn off and is sent to storage or an evaporator for further concentration. A portion of the depleted anolyte is drawn off and is returned to the salt dissolution and primary brine purification system. Supporting arms are attached to both sides of the cell frame, and these frames are hung on the side bars of a hydraulic press.

The electrolyzers of Asahi Kasei are offered in two sizes, each available with either natural or forced circulation. The model ML32, with an effective membrane area of 2.72 m^2, has an annual production capacity of ~15,000 tons of NaOH (100% basis), while the bigger version, ML60, has an effective area of 5.05 m^2 and an annual production capacity of 30,000 tons of NaOH (100% basis). As of 1997 a total production capacity of ~1,832,000 tons of NaOH per year has been in operation or under construction with Asahi Kasei technology.

Chlorine Engineers CME Monopolar Electrolyzers. Chlorine Engineers Corporation (CEC) (a subsidiary company of Mitsui and Company), produces the filter press type, monopolar membrane electrolyzer (60) shown in Figure 13. Uniform electrical current travels into each anode element through titanium-clad, copper-cored conductor rods and current distributors. The current distributor in the electrolyzer serves an additional role as a downcomer, helping the electrolyte self-circulate within the cell to maintain uniformly distributed concentrations as well as efficient gas release. The internal circulation is intended to eliminate the necessity for an external forced recirculation system.

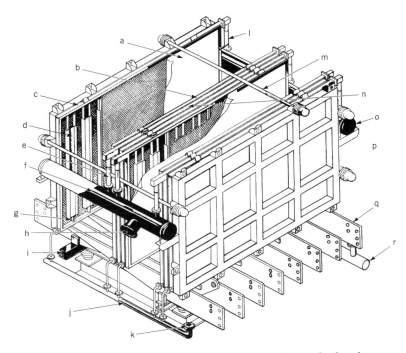

Fig. 13. CME monopolar electrolyzer: (**a**) membrane; (**b**) cathode element; (**c**) half-cathode element; (**d**) current distributor; (**e**) Teflon tube; (**f**) Cl$_2$ + depleted brine manifold; (**g**) conductor rod; (**h**) Cl$_2$ + depleted brine outlet nozzle; (**i**) base frame; (**j**) recycled NaOH manifold; (**k**) recycled NaOH inlet nozzle; (**l**) gasket (the gasket/element ratio is quite small); (**m**) tie rod; (**n**) anode element; (**o**) H$_2$ + NaOH manifold; (**p**) end plate; (**q**) under cell bus bar (simplifies piping around the electrolyzers); (**r**) feed brine manifold.

The gasket thickness sets the electrode spacing, and either finite or zero-gap configuration can be accomplished. The anode frame is titanium and the cathode frame is stainless steel. The CME elements are thicker than competing elements for a lower electrolyte gas void fraction. This feature minimizes the drop in the liquid level during shutdowns. CEC offers electroplated activated cathodes. Gas and liquid exit the cell in the stratified overflow mode, as the liquid level is maintained in the upper cell frame. Semitransparent Teflon tubes are used to monitor the operation visually. The electrical bus bars are installed underneath and at a right angle to the cell elements, requiring no equalizer between electrolyzers. The bus bar can be used as a short circuiting element by changing the connections. As of 1997, the installed capacity with these cells was 2,500,000 mtpy of chlorine.

Chlorine Engineers MBC electrolyzer. Chlorine Engineers retrofitted ELTECH's MDC monopolar diaphragm cell electrolyzers to convert them into membrane cells (61,62). In retrofitting, CEC installs the membrane in the form of a bag that encloses the anodes (membrane bag cell). In the MBC-29 shown in Figure 14, one bag encloses two anodes. The current conductor bar of the anode passes through a hole in the bottom of the membrane bag for connection to the base. The open end of the bag, facing upward, is fixed to the partition plate by a sealing plug.

CEC BiTAC-800 Electrolyzer. The Chemical Engineers Corporation (CEC) adapted a bipolar electrolyzer first developed by Tosoh Corporation. The electrolyzer is of the filter press type and has an effective membrane area of ~3.3 m^2. Up to 80 element frames can be assembled together by spring-loaded tie rods. The electrode frames are made of a special titanium alloy on the anode side, and nickel on the cathode side. The wave-like structure of the

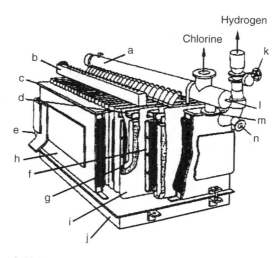

Fig. 14. Cut view of Chlorine Engineers membrane bag cell: (**a**) manifold; (**b**) frame; (**c**) partition plate; (**d**) sealing plug; (**e**) recirculated NaOH inlet; (**f**) cathode; (**g**) anode; (**h**) cathode pan; (**i**) membrane bag; (**j**) base; (**k**) butterfly valve; (**l**) feed brine; (**m**) depleted brine; (**n**) caustic-outlet.

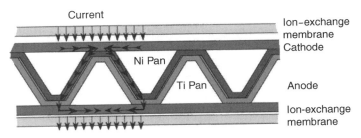

Fig. 15. Cross-section view of a BiTAC element.

cathode and anode pan serves the combined functions of partition plates and conductor ribs, as shown in Figure 15.

The current is led to the anode mainly through nickel to avoid the ohmic losses from the poor electrical conductivity of titanium. The feed and exit streams leave the cells by overflowing into the collecting pipes located at each side of the electrolyzer via external flexible hoses. The membrane cathode gap can be adjusted from zero to a finite value, to meet the requirements of various commercial membranes.

Ineos FM21-SP Monopolar Electrolyzer.   Ineos' FM21-SP monopolar electrolyzer (see Fig. 16) incorporates stamped electrodes. The anode assembly is composed of 2-mm thick titanium panel between compression molded joints of a special cross-linked EPDM elastomer which is a terpolymer of ethylene, propylene, and diene. The cathode assembly consists of a 2-mm thick nickel panel between compression molded joints, also of EPDM.

The anodes and cathodes are assembled between two end plates, up to 60 anodes in the FM21-SP and up to 90 anodes in the larger FM-1500. A key feature of both designs is the elimination of any external piping to the individual cell

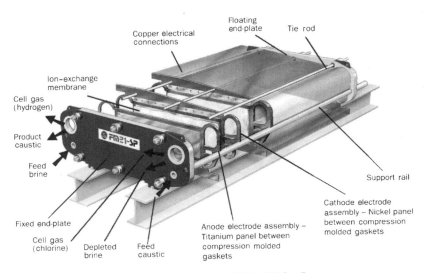

Fig. 16.   Electrolyzer FM21-SP by Ineos.

compartments by the use of a simple but effective internal manifold arrangement. As shown in Figure 16, the anode and cathode panels are designed to form feed and discharge channels when assembled.

The electrolyzer has coated titanium anodes while the cathodes are made of pure nickel, either plain or coated with activated coatings. Both electrodes are pressed from integral sheets of pure metal, and this makes recoating of the electrodes simple and cost effective. Recoated structures can be sent to site prior to electrolyzer refurbishment from a pool of electrodes kept in stock by the electrode supplier for all customers. The removed, worn out electrodes are recoated without interfering with plant operations and are added to the pool of electrodes.

The effective electrode area of one monopolar cell is 0.21 m^2, which makes this electrolyzer very compact (63). The individual, lightweight electrodes are easily handled without the need for lifting devices, and this allows the electrolyzer to be rebuilt or refurbished in a short time.

ELTECH ExL Electrolyzers. ELTECH offers three electrolyzers called ExL^M, ExL^B, and ExL^{DP}. The monopolar ExL^M is the modified version of the earlier MGC electrolyzer (membrane gap cell) (64). Improvements have been made to the anode and cathode design and fabrication techniques, as well as to the gasket and manifold materials. This has resulted in a uniform current distribution across the active membrane area and increased internal circulation. One of the main features of this cell is the double-gasket design. The cathode O-ring is located closer to the liquid than the anode O-ring, which is not in permanent contact with the chlorinated brine, thereby serving as a well protected back up seal. The general cell design is of the filter press type, where the elements are pressed together by tie rods with copper distributors. The intercell connections are from the sides, and the support for the electrolyzer is the copper current redistribution bus between each electrolyzer. The feed streams and products enter and leave the element via the attached manifold devices. Up to 30 membranes with an effective area of ~1.5 m^2 each, can be put together in one electrolyzer providing a production capacity of 9.3 tons of NaOH per day (100% basis) The maximum current density is 7 kA/m^2.

The model ExL^B (Fig. 17) is the bipolar version of the ExL^M. The bipolar arrangement is realized by pressing the nickel cathode pan onto the nickel plated back of the anode pan. Up to 60 elements can be combined in one electrolyzer, to realize a production capacity of 29 tons of NaOH per day (100% basis).

The ExL^{DP} electrolyzer (dense pak) represents a hybrid cell arrangement, combining multiple ExL^B type electrolyzer blocks, each consisting of 2–10 elements, in one electrolyzer unit. The blocks are electrically separated from each other using standard monopolar components.

Uhde Bipolar Electrolyzers, Single Element Design. The characteristic feature of the Uhde membrane electrolyzer is its single element design (65). Each element mainly consists of an anode and a cathode semishell separated by an ion-exchange membrane. Unlike the filter press design concept used in other electrolyzers, each element is individually sealed by means of a separately bolted flange with gasket. This enables a long term storage of fully assembled elements in working order. The elements are suspended on to a steel rack and are pressed together to establish a good electrical contact. High forces are not required to achieve effective sealing in the single-element concept. Up to 160 single elements

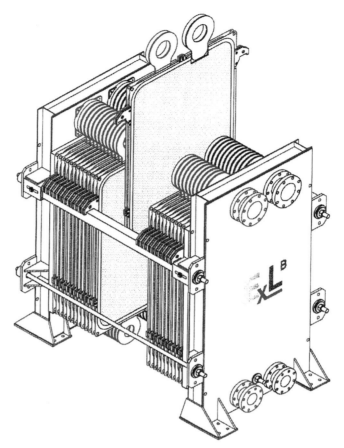

Fig. 17.   Electrolyzer ExL^{B} by ELTECH.

can be combined in one electrolyzer as shown in Figure 18. Currently, single elements of the third generation, as shown in Figure 19, are used. The standard electrolyzer configuration is the bipolar arrangement, although monopolar or hybrid arrangements can be made, as needed by the customer.

The current passes from the back of the cathode wall from one element to the back anode wall of the subsequent element by a series of contact strips. Voltage losses are reduced by means of a laser welded, direct connection between the outer contact strips and, via vertical inner-current conducting plates, the electrodes. Both brine and caustic enter the element via flexible hoses leading to horizontal inner distribution pipes. These provide uniform feed concentration profiles inside each compartment. Internal circulation is achieved and enhanced by two baffle plates located in the anode compartment. The upper, inclined baffle plate provides a constant exposure of brine to the membrane, thereby avoiding the gas-phase blistering of the membrane. In addition, the vertical baffle plate contributes to the uniformity of both, temperature and concentration profile in the compartment.

Arrangement
of Single Elements in a Cell Rack

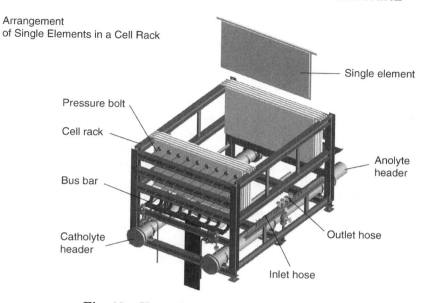

Fig. 18.   Krupp Uhde BM-2.7 electrolyzer.

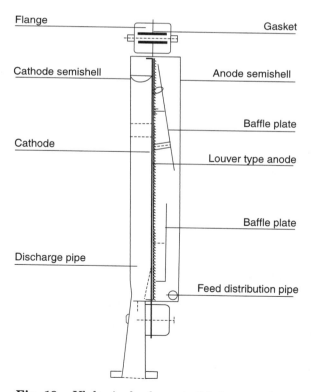

Fig. 19.   Uhde single element, third generation.

The product gases leave the element together with the electrolyte downward via vertical discharge pipes and exit to the collectors via flexible PTFE hoses. All inlet and outlet connections are located at the bottom of the electrolyzer, releasing the space above the cell for maintenance purposes. The mesh-type anode serves as a support for the membrane while a defined gap is maintained between the louver type cathode and the membrane by spacing devices.

The typical electrode active area is 2.7 m^2 (1.8 m^2 also available) and the annual electrolyzer production capacity can be up to 28,000 tons of NaOH (100% basis, with electrolyzers having 160 elements at 6.0 kA/m^2). In 2002 Uhde had 70 plants with a total annual capacity of ~4,800,000 tons of NaOH (100% basis) commissioned or under construction. About one-third of this capacity has come from conversion projects from mercury or diaphragm plants to membrane technology (65). A summary of the current membrane cell technologies is provided in Table 5.

Membrane Cells with Oxygen or Air Depolarized Cathodes.   Substituting an oxygen reduction reaction for the hydrogen evolution reaction at the cathode in chlor-alkali electrolysis will reduce the total cell voltage by ~1 V (theoretically 1.23 V), thereby realizing a substantial saving of electrical energy (66). Figure 20 depicts a schematic of a membrane cell operating with an air-depolarized cathode, referred to as a "Gas Diffusion Electrode" (GDE).

The anode reaction is the same as in a conventional chlor-alkali cell, where the chloride ions are discharged to form chlorine gas product and the sodium ions migrate to the cathode compartment through the ion-exchange membrane. At the cathode, oxygen is reduced to the OH$^-$ ions, which combine with the Na$^+$ ions to form sodium hydroxide. The oxygen consumed in this reaction enters the air cathode compartment either as water-saturated pure oxygen gas or as

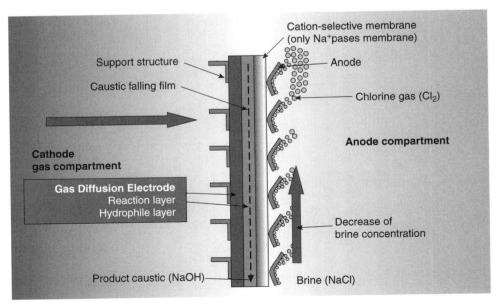

Fig. 20.   Schematic cross-section through a GDE cell, falling film type (67).

air and gets reduced at the porous air cathode. The advantages of using the oxygen reduction reaction include avoidance of costly downstream treatment of hydrogen and absence of the gas void fraction in the catholyte, resulting in a reduced ohmic drop in the cell. However, this scheme requires a high performance air scrubbing system to remove all carbon dioxide from the air in order to protect the air cathode from the accumulation of sodium carbonate and a premature failure of the cells. As the anode side of the process does not differ from the conventional chlor-alkali cell, those components can be used without any modifications.

The cathode reaction for this system is

$$2\,H_2O + O_2 + 4\,e^- \rightarrow 4\,OH^-$$

giving a standard potential 1.23 V lower than that of the decomposition voltage of water to hydroxide ion and hydrogen. Pilot electrolyzers have operated at <2.5 V at a current density of 6 kA/m^2 (67).

The cathode side has several special requirements, the central one being the adjustment of the local differential pressure between the caustic and the oxygen compartment on the other side of the electrode. Due to the porous nature of the GDE, a pressure balance across the electrode has to be established in order to avoid the flow of the fluid from one side to the other. As shown in Figure 21, there is a certain tolerance range of the differential pressure, where the electrode works properly.

The local differential pressure is a function of height, because of the different densities of the fluids, and can be made small using two different approaches. The first one involves splitting the cathode compartment into several horizontal compartments, called gas pockets in which the height of each subcompartment limits the hydrostatic pressure of the caustic to a tolerable value. The lean caustic flows through the pockets successively, by overflowing from one pocket to the next one below. This gas-pocket principle (see Fig. 22) is patented and now

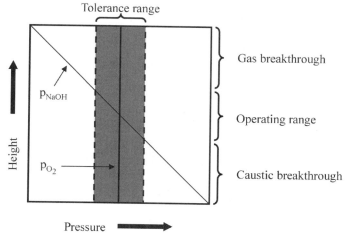

Fig. 21. Differential pressure over a porous GDE-type electrode (51).

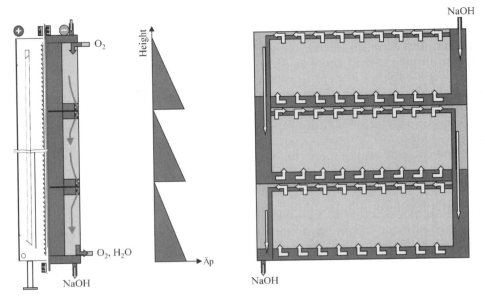

Fig. 22.   Gas-pocket type GDE element.

being tested by the Bayer AG group (68). A second approach to the problem is the falling-film principle (see Fig. 23). The development of this type was initiated by the Hoechst group in the 1980s (69–72), but is presently continued by Uhde. The idea here is to decrease the hydrostatic pressure of the caustic successively by establishing a falling film of caustic between the electrode and the membrane. This is realized by means of a layer of hydrophilic material that is fixed between the anode and the cathode. This design ensures a constant gap between the GDE and the membrane itself. Because of the electroosmotic water transport from the anolyte to the catholyte, the caustic flow increases from the top to the bottom of the cell. A high flow in the hydrophilic layer will lead to a flooding of the GDE, and hence a breakthrough of caustic into the oxygen compartment, caused by the increased differential pressure. This flooding can be avoided by a proper design of the hydrophilic layer. The falling-film technology shows some inherent advantages. Unlike the gas-pocket principle, the falling-film technology does not need an extensive gasket system throughout the surface of the GDE, thereby allowing a simplified design. Furthermore, the GDE can be run close to atmospheric pressure, so it will always operate in the optimum operating pressure range. Both processes show comparable operating data. However, they are still at an experimental level. Nevertheless, the initial results with the GDE system are promising toward achieving energy savings of ∼30% compared to the conventional membrane process.

Figure 24 gives an overview of the evolution of the specific energy consumption of membrane cells. The remarkable improvement in performance reflects the progress made in membrane development and electrolyzer design. This has reduced operating costs by reducing voltage losses and optimizing current efficiencies. Furthermore, the operating current density has been doubled from

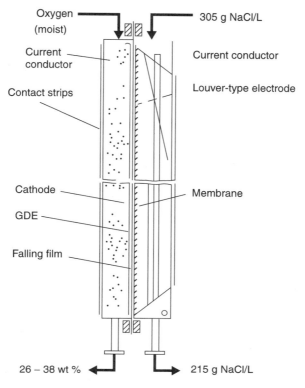

Fig. 23. Falling-film type GDE element (65).

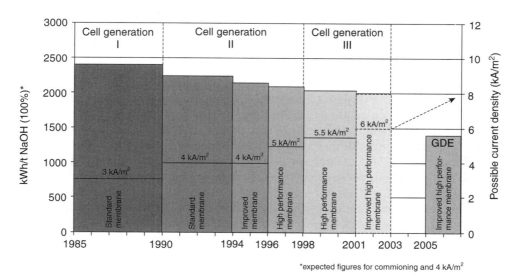

*expected figures for commioning and 4 kA/m²

Fig. 24. Evolution of energy consumption and current density for Uhde membrane technology (67).

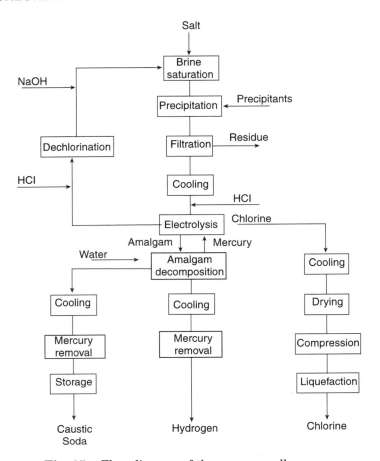

Fig. 25.   Flow diagram of the mercury cell process.

3 to 6 kA/m^2 over the past 20 years, reducing the capital costs associated with the electrolyzers.

4.5. Chlorine Plant Auxiliaries.   The auxiliary systems for each of the three electrolytic chlor-alkali processes, depicted by block flow diagrams (Figs. 25–27). Although they differ somewhat in operation, the processes of brine purification and chlorine recovery are common to each.

Brine Preparation.   There are several different major sources of the fundamental raw material, sodium chloride. It can be recovered from underground rock salt deposits. Methods of recovery include conventional mechanical mining, which has much in common with the mining of coal, and solution mining, in which water or weak brine is forced into a salt deposit to dissolve the material and carry it back to the surface. The former typically costs about twice as much as the latter (73). Salt is also obtained by evaporation of saline waters. Ocean water, for example, contains ~3.5% NaCl. Inland sources may be more concentrated and may contain other minerals of greater value. The Dead Sea and the Great Salt Lake are examples. The salt obtained from these waters is

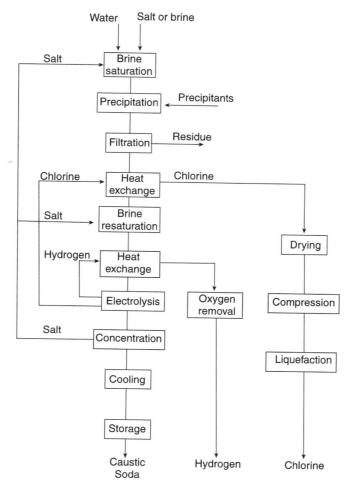

Fig. 26. Flow diagram of the diaphragm cell process.

usually referred to as solar salt, since solar energy provides the driving force for evaporation. There are also surface or near-surface deposits of salt and wild brines near the surface of the earth, but these are relatively minor sources.

However obtained, sodium chloride will have a characteristic set of impurities. The major impurity in nearly all salts is some form of calcium sulfate. While compositions from different sources vary widely, a few generalizations can be made. Solar salt, at least after the common operation of washing, is usually purer than rock salt. It is also more susceptible to caking and mechanical degradation. While rock salt contains more calcium sulfate, it contains in proportion less magnesium. The higher ratio of calcium to magnesium improves performance in the brine treatment process.

Many chlorine plants operate with salt that has already undergone some purification. Most solar salts, eg, are washed to remove occluded liquor and surface impurities. Another example is vacuum purified salt, which is recrystallized

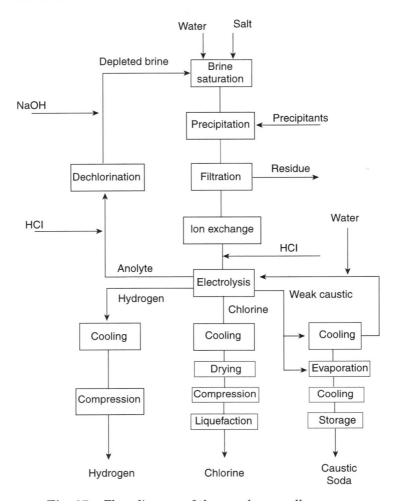

Fig. 27.   Flow diagram of the membrane cell process.

from brine after most of the impurities have been removed by chemical treatment. Table 6 shows typical analyses of the three types of salt discussed above.

The salt may be delivered to a plant as the solid or already dissolved to give a raw brine. Solution-mined material is an example of the latter. As the flow diagrams and the process descriptions that follow will show, only the diaphragm-cell process can operate with a brine feed without special design. When solid salt is

Table 6. **Comparison of Purity of Various Types of Salt**[a]

Component	Rock salt	Washed solar salt	Vacuum salt
sodium chloride	93–99	99	99.95
sulfate	0.2–1	0.2	0.04
calcium	0.05–0.4	0.04	0.0012
magnesium	0.01–0.1	0.01	0.0001

the raw material, a dissolving operation becomes necessary. The partially depleted brine from mercury or membrane cells is recirculated and resaturated with the fresh salt. Handling and dissolving are not technically complex operations, but their scale can be large. A 1000-tons/day chlorine plant requires ~1700 tons/day of salt. Much of the detail in saturator design has to do with the behavior of the solid particles as they continuously become smaller and the removal and handling of undissolved impurities (74).

Brine Purification. Impurities in the brine can affect the performance of any type of cell. The most common cationic impurities, and those present in the greatest concentrations, are the hardness elements, calcium and magnesium. In diaphragm and membrane cells, these can precipitate as the hydroxides when exposed to a sufficiently high pH, which occurs within the diaphragm or membrane as the ions approach the catholyte side. Membranes are especially sensitive to brine impurities, and the membrane-cell process requires a higher degree of brine quality than does either of the other technologies. Where the hardness specification for the brine fed to a diaphragm cell may be ~5 ppm, an efficient membrane cell will require a level of no greater than 20 ppb. Elements whose oxides can form insoluble complexes, particularly in the presence of calcium, also must be controlled. These include silicon and aluminum. Iodine, too, can form the highly insoluble sodium paraperiodate, $Na_3H_2IO_6$ or interact with barium. Table 7 compares the brine specifications for membrane and diaphragm cells. Membrane-cell specifications in particular vary with service conditions and supplier's requirements. Some specifications change with operating current density, and some are interrelated (eg, some cell suppliers allow higher strontium content if the silica level is quite low).

With a different operative mechanism in mercury cells, the same impurities do not cause permanent damage to the chlorine/caustic separator. Instead, traces of metals such as vanadium, molybdenum, chromium, iron, titanium, and

Table 7. Typical Specifications for Feed Brine to Electrolyzers

Component	Membrane process[a]	Diaphragm process[a]
sodium chloride	290–310 g/l	320 g/L[b]
calcium and magnesium	20 ppb	5 ppm
sodium sulfate	7 g/L	5 g/L
silica	5 ppm	0.5 ppm
aluminum	0.1 ppm	0.5 ppm
iron	1 ppm	0.3 ppm
mercury	0.5 ppm	1 ppm
heavy metals	0.2 ppm	50 ppb
fluoride	0.5 ppm	1 ppm
iodine	0.2 ppm	
strontium	0.5 ppm	c
barium	0.5 ppm	c
total organic carbon	1 ppm	1 ppm
pH	2–11	2.5–3.5

[a] All ppm and ppb values represent maximum concentrations allowed.
[b] Minimum.
[c] Included with calcium and Magnesium.

tungsten must be avoided. These metals lower the hydrogen overvoltage at the mercury cathode and permit increased formation of hydrogen. This lowers the current efficiency and creates a hazard in the processing of chlorine.

Impurities that are soluble in the acidic anolyte can become insoluble once they have entered the membrane. In addition to the induced flow of cation impurities, neutral and anion impurities can enter by diffusion and by the considerable water flow from anolyte to catholyte. Internal precipitation of these impurities disrupts the structure of the membrane. This precipitation can increase the voltage drop through the membrane. If it occurs in the cathode-side layer, it will reduce the selectivity of the membrane and therefore the current efficiency of the cell. Even if the impurity is subsequently washed out, the void left behind still contributes to reduced current efficiency. There also are synergistic effects, such as the precipitation of complex compounds, usually involving calcium and magnesium in combination with silica and alumina.

The most important anionic impurity, because of its widespread distribution in salt, is sulfate. Its control becomes increasingly more important as cell technology advances. Refined salts can be purchased at a premium to reduce the magnitude of the problem; sulfate can be preferentially rejected in the dissolving process; or it can be removed from the brine by precipitation, ion exchange, or nanofiltration. There is a growing literature on the subject, much of it available in the General References.

Some brines also contain ammonium ions or organic nitrogen that can be converted to the explosive nitrogen trichloride. Ammonium ions in the brine are removed by treatment with chlorine or hypochlorite. Depending on its severity, this treatment produces NH_2Cl, which can be removed as a gas. Brine purification represents a significant part of chlor-alkali production cost, and this is especially so in the membrane process, with its stringent requirements. Moreover, part or all of the depleted brine from mercury and membrane cells must be dechlorinated before recycle to the cells, at further expense. Further discussion of this operation is noted below.

Brines are treated with Na_2CO_3 to precipitate calcium carbonate and with NaOH to precipitate metal hydroxides, notably that of magnesium. The brine may be heated before treatment to shorten the reaction time and improve the subsequent precipitation and settling of the solids. Calcium carbonate tends to form a dense, coherent precipitate, while magnesium hydroxide forms a floc that is light and fragile. The $Mg(OH)_2$ precipitates can be degraded by agitation and tend to settle slowly. When the two metals are precipitated together, hybrid particles form, and the calcium helps the magnesium to settle more quickly. When the calcium to magnesium ratio is low, the calcium carbonate precipitate is not able to assist settling of the magnesium hydroxide floc. In such cases, flocculants are sometimes added to the brine. These typically are polyelectrolytes that are able to bind together the light, fluffy $Mg(OH)_2$.

After carbonate and caustic treatment, the precipitates settle in a clarifier, where most of the solids are removed as a mud. This mud can be processed further to increase its solid content and to recover most of the occluded chloride. The overflow from the clarifier contains a few tens of ppm of solids, which are removed by filtration. The first step of filtration usually is in beds packed with sand or, particularly in membrane-cell plants, non-siliceous materials such as anthracite or garnet. Most modern plants use pressure filtration, and flow may be up or down through the beds. Precoated polishing filters follow this primary filtration. Leaf and candle filters are common. In membrane-cell plants, it is even more important to avoid siliceous materials in this step, because those in the form of filter aids are amorphous and are more easily soluble. Cellulosic materials are the usual substitute. Membrane-based filters with controlled pore sizes are able to function in this application without filter-aid. Use of sharp back-pulse of brine after deposit of small quantity of solids keeps the filter surface clean and functional.

After filtration, brine usually contains <5 ppm hardness and is suitable for the mercury and diaphragm processes. The membrane process, however, requires brine with <20 ppb hardness, which requires another purification step, based on ion exchange. The resins used have the standard styrene–divinylbenzene backbone, but special exchange groups are necessary. The process is one of softening by replacing undesirable multivalent cations in the brine with sodium or potassium. Because of the extremely high concentration of the latter ions in the brine, standard resins do not have the selectivity required for efficient removal of the other ions. The types used in this application therefore have either iminodiacetate or aminophosphonic pendant groups (See Reference 72). The alkaline brine from the filtration plant passes through beds packed with such a resin. The process is cyclic, with bed regeneration by HCl, to strip off the metal ions, and NaOH, to restore the resin to its active form.

The brine fed to the electrolyzers in all processes is usually acidified with HCl. This neutralizes residual hydroxide and carbonate in the brine and prevents their reaction with chlorine formed at the anodes. The formation of by-products such as oxygen and chlorate is reduced in this way. Excess acid also can react with hydroxyl ions that leak back from the catholyte to the anolyte in the cells. This again reduces oxygen formation and provides a purer gas, and so it serves as a technique for producing chlorine gas suitable for direct use in its primary outlet, the synthesis of ethylene dichloride. Over-acidification in a membrane-cell plant, however, can reverse the hydrolysis of the carboxylate groups that provide high ion selectivity in the membranes. The resulting formation of the nonconductive carboxylic acid form renders the membranes inactive. The process of acidification must be closely monitored to prevent this adverse effect.

As mentioned above, the depleted brine issuing from membrane or mercury cells must be recycled. Dissolved chlorine must first be removed. Most of the chlorine present is in the hydrolyzed form of hypochlorite:

$$Cl_2 + H_2O \leftrightarrow H^+ + Cl^- + HOCl \qquad (17)$$

Addition of HCl to the brine reverses the reaction, and the molecular chlorine formed is much easier to remove, usually by application of a vacuum. There is a certain residual concentration after this dechlorination process, and in a membrane-cell plant it must be removed to protect the ion-exchange resin from oxidation. This is done chemically by addition of a reducing agent, which typically is hydrogen peroxide or a reacted form of SO_2. The product of the reaction is the innocuous chloride ion. When sulfite is the reducing agent, the reaction is

$$SO_3^{2-} + OCl^- \rightarrow SO_4^{2-} + Cl^- \qquad (18)$$

In a mercury-cell plant, on the other hand, complete dechlorination is undesirable, because a certain amount of free chlorine in the circulating brine solubilizes mercury and prevents its loss from the system along with the brine treatment sludge. The chemical dechlorination step therefore is omitted.

Chlorine Processing.   The first step in the chlorine-handling process is the cooling of the gas, which is saturated with water at cell operating temperature. Most cells operate very close to atmospheric pressure, although membrane cells have the capability to run under positive pressures, up to several hundred kilopascals. Direct- and indirect-contact cooling is used. Direct cooling was more common in the past, when metals resistant to wet chlorine gas were not economically available. In this process, the gas from the cells is contacted with a countercurrent stream of brine or cooling water in a column. Indirect exchangers are single-pass devices with titanium tubes and carbon steel shells. Titanium plate exchangers are used in some smaller plants.

Cooling the chlorine incidentally removes most of the water by condensation. Lower temperatures promote the removal of water and reduce the consumption of drying acid in the next step of the process. Particularly in large plants, the use of two coolers in series is common, with the second operating on chilled water. There are, however, basic limitations on the degree of cooling that can be safely applied. At atmospheric pressure, eg, a solid hydrate of chlorine forms at temperatures below 9.6°C. The group of gas hydrates that includes chlorine forms unit cells of forty-six water molecules; the composition of the hydrate is $Cl_2 \cdot nH_2O$, where the value of n usually is taken to be between 7 and 8. Cotton and Wilkinson (75) estimate $n = 7.2$ or 7.3. This hydrate deposits in equipment, disrupting operation and creating a hazard. In titanium equipment, another limitation is the loss of the protective oxide film that prevents reaction between chlorine and the metal when the partial pressure of water is too low. Work sponsored by Euro Chlor (76,77) suggests that the temperature of the gas at atmospheric pressure should not be taken below ~13°C. This result is consistent with other work (78) that indicates that a concentration of 1.0–1.5% water vapor is sufficient to maintain the film.

The water condensed in the cooling process is saturated with chlorine. Chlorine can be removed from this water by acid-assisted vacuum stripping. In a mercury- or membrane-cell plant, the water can be combined with the depleted brine and the two treated together. Diaphragm-cell plants, with no depleted brine recycle, require addition of a dedicated stripper to handle the condensate. The process is basically the same, but the lack of salt in solution means that the chloride ion concentration is lower. As reaction 17 will make clear, more acid must then be added to reverse the hydrolysis reaction and reduce the solubility of chlorine.

After the gas is cooled, the next step in the process is its drying. Concentrated sulfuric acid (93–98%) removes the remaining water, producing a spent acid. Depending on further use or the method of disposal, the strength of this acid may range from 50 to 80% H_2SO_4. Drying takes place in columns that usually provide two to four theoretical stages. Outlet moisture specifications usually are <50 ppm (v/v). Both plate and packed columns are used; some columns are hybrid, with plates mounted above a packed section. Because of the very low net throughput of acid, packed columns require circulation of acid over the beds. The circulating acid can be cooled to remove the heat of absorption of water. Multiple beds may be installed in a single column, or separate single-bed columns may be used. The flows of gas and acid through the columns are countercurrent.

Packed columns in large plants have been built of brick-lined steel, with glass fiber membranes between the bricks and the wall. In smaller units, FRP (fiber-reinforced plastic) is a common material, and this is often reinforced with PVC. In the last stage of drying, where the acid concentration is essentially that of the feed material, unlined carbon steel frequently is used. Ceramic and plastic packing are common, and support and hold-down plates may be ceramic or coated steel.

The dry chlorine gas issuing from the drying system and the wet gas issuing from the cells both contain fine particles in the submicron to ten-micron range. Particles of such small size by definition constitute a mist. In the case of dry gas, the particles are droplets of concentrated sulfuric acid. In the case of wet gas, the particles are those of solutions of salt and sometimes caustic soda. In both cases, the mist can deposit in downstream equipment. Normal practice is to remove these particles before allowing the gas to travel forward from the operation in question. Conventional inertial impact mist eliminators are not highly effective against mists. The standard apparatus therefore contains candles wrapped with a porous fiber. This fiber is tightly constructed and forces the gas through a narrow, tortuous path. The mechanism of collection of the finest particles depends on their Brownian motion. The vessels containing the demisting elements are carbon steel in the case of dry chlorine and FRP or lined steel in the case of wet chlorine (79). The elements are glass in stainless steel mesh enclosures (dry chlorine) or glass or polyester in FRP or PVC enclosures (wet chlorine).

After drying, chlorine usually is compressed to a higher pressure. This compression is done to allow liquefaction of the gas or to deliver it to another process (eg, production of ethylene dichloride). Compressors may be centrifugal, liquid-ring, or reciprocating. The centrifugal compressor is the standard in large plants. Several stages of compression usually are necessary to reach the desired pressure. Because of the combustibility of chlorine in contact with the ferrous metals used in compression systems, temperatures must be strictly limited. Failure to do this has been the cause of a number of fires. Temperatures are limited by mechanical design and by coolers placed before compression and between stages of the compressor.

Centrifugal compressors have a number of advantages in this service, including relatively compact size, high mechanical efficiency, smooth operation, and outstanding reliability. The last of these allows them to be used in many plants without installed spare capacity. Disadvantages include the mechanical demands of their high rotational speed, the complex lubrication and sealing systems required, and the fact that performance depends on molecular weight of the gas. Centrifugal machines also are subject to surge, and hence surge control is an important part of system design.

Liquid-ring compressors in chlorine service are sealed with concentrated sulfuric acid. They are limited in size and have relatively low mechanical efficiencies. Because the rotating acid ring absorbs some of the heat of compression, they are able to operate at higher compression ratios than the other types of compressor without developing dangerous temperatures. Frequently, then, single-stage machines can deliver chlorine directly to the liquefaction process. Before the advent of the centrifugal compressor, a number of fairly large plants depended on batteries of liquid-ring machines. These have gradually converted their operation to the centrifugal type, and the liquid-ring type today is found primarily in smaller plants.

Reciprocating compressors also are used less frequently as primary compressors than they were in the past. Their disadvantages include their close clearances and the pulsating outlet flow of gas. Their main applications now are as interstage boosters in liquefaction systems and as tail/vent gas compressors.

About one-half of the chlorine produced is used as the dry gas and transported directly to the consuming process. The rest is liquefied, stored, and transferred or shipped. Liquefaction usually involves the indirect contact of the compressed gas with a boiling refrigerant. Process conditions are chosen to optimize the total energy consumption of compression of the process gas and compression of the refrigerant (80). Because of the presence of noncondensible gases in the chlorine, complete liquefaction is not possible. These noncondensibles include oxygen formed in the cells, which is the major source of current inefficiency, and carbon dioxide, which results from the decomposition of residual carbonate introduced in the brine purification plant. Air also enters the gas from compressor seals, equipment and pipeline purging, venting of returned shipping containers, and leakage into those parts of the process that are under vacuum.

Any hydrogen present in the chlorine gas as produced is concentrated in the residual gas from liquefaction. Hydrogen is explosive in mixtures with chlorine (see Section 3). The lower explosion limit is very nearly the same as in the case of hydrogen in air, and furthermore the minimum ignition energy is lower in chlorine mixtures. Some plants use liquefiers designed to reduce the probability of explosion and also to contain the energy if an explosion occurs. More often, the process includes dilution of the gas with dry air to keep the hydrogen concentration out of the explosive range. This dilution adds to the noncondensible load and makes liquefaction more difficult. One of the advantages of membrane cells is the lower hydrogen content of the chlorine produced. This reduces the amount of diluent required, or even eliminates the need altogether.

The residual gas from liquefaction is referred to as "tail gas." This can be combined with chlorine-containing gas from plant evacuation systems and with that from returned shipping containers. Methods of recovery of the chlorine value include the manufacture of certain by-products and the absorption of the chlorine in a solvent from which it later is stripped (81). The latter process relied on carbon tetrachloride, whose use is being phased out under the Montreal Protocol. Several recovery units have accordingly been shut down, and enhanced liquefaction under more severe conditions is becoming more common. This reduces the chlorine content in the tail gas to a point at which it can be utilized as one of the by-products listed in the next paragraph or simply destroyed by absorption in alkaline solution.

Manufactured by-products include hydrogen chloride, formed by controlled combustion of chlorine with a slight excess of hydrogen; various forms of bleach, formed by absorption of chlorine in an alkaline material such as caustic soda or lime; and ferric chloride, formed by the reaction of chlorine or HCl with metallic iron and the further oxidation by chlorine of ferrous chloride to the ferric state.

Hydrogen Processing. The hydrogen produced in all electrolytic chlor-alkali processes is relatively pure (>99.9%) and requires only cooling to remove water along with entrained salt and caustic. The heat is often recovered into the brine system. The hydrogen is compressed using water-sealed liquid-ring pumps, Roots-type blowers or reciprocating compressors. It is desirable, although not always possible, that the entire system from the electrolyzers to the compressor suction is under positive pressure to avoid contamination with air, which could form an explosive mixture. Some uses of hydrogen require additional removal of traces of oxygen. This is accomplished by combination of the oxygen with some of the hydrogen over a platinum catalyst. The hydrogen produced in mercury cells is also contaminated with mercury, which must be removed before using the gas. Cooling removes most of the water in the gas, along with a preponderance of the mercury. Final purification is by scrubbing with an oxidizing solution in which the metal is converted to the more soluble mercuric form, or by adsorption on activated carbon. The carbon usually contains an additive such as sulfur to improve the retention of mercury.

The hydrogen can be used for organic hydrogenation, catalytic reductions, and ammonia or methanol synthesis. It can also be used to provide a reducing atmosphere in some applications or burned with chlorine to produce high quality HCl. The HCl is produced in a high temperature burner and then usually cooled and absorbed in chemically resistant graphite equipment to produce 30–38% acid. Some of this acid goes for sale, and some is used within the chlor-alkali plant to acidify cell-feed brine, to regenerate ion-exchange resins, or to assist in the dechlorination of depleted brine and the water condensed from chlorine gas.

Much of the electrical energy consumed in a cell is associated with the production of hydrogen, and so the gas represents a significant amount of energy. Collection of hydrogen for combustion to provide thermal energy, such as in the production of steam, is a very common practice. Developing technologies may offer more effective means of recovery of the energy in the gas. One such technology is the fuel cell, in which hydrogen and oxygen both are consumed in an electrochemical cell to produce water and energy as direct-current electricity.

Sodium Hydroxide Processing. Coproduct sodium hydroxide is usually sold and shipped as a 50% solution. A 73% solution and the anhydrous material are also marketed. The solutions produced by the various types of cell differ markedly, and the purity of the final product and the energy required to concentrate the cell product to 50% are major differentiators among the three technologies. Diaphragm cells produce a relatively weak liquor of ~10–12% NaOH. Impurities that enter with the treated brine, such as sodium sulfate, and impurities generated in the cells, such as sodium chlorate, accompany the caustic. The unconverted NaCl, in concentrations of 13–16%, also is in the liquor. Evaporation of this mixture toward 50% NaOH causes it to become supersaturated in NaCl, and most of the salt drops out of solution. This salt must be removed, usually by centrifuging, and returned to the brine plant for reuse. Cooling of the solution then rejects more NaCl, which is removed by filtration or centrifuging.

The solubility of NaCl in the cooled solution still is 1% or more, and this salt content is the distinguishing characteristic of diaphragm-cell 50% caustic soda.

The capital and operating costs of the evaporation plant are major factors in diaphragm-cell economics. The relatively low quality of the product makes it unsuited for certain applications, and the products of the other types of cell often are sold at a premium as "rayon grade" or "membrane-cell" grade. Because of the energy consumed in evaporation and the resulting higher heat/power ratio, most large diaphragm-cell plants have associated cogeneration facilities.

Mercury-cell caustic from the decomposers is already 50%, and only filtration to remove mercury droplets is required. The purity of the solution is quite high, and no evaporation is required. Membrane cells produce NaOH at different strengths, depending on the type of membrane used. Plants that sell NaOH usually produce a solution in the 30–40% range. Evaporation is much less expensive than in diaphragm-cell plants, with much lower evaporative loads and no need for salt recovery and handling.

Nickel is the standard material of construction for NaOH evaporators. Multiple-effect units are a universal choice in all but the smallest plants. Membrane-cell plants use double- or triple-effect evaporators. With the high evaporative load associated with the diaphragm-cell process, quadruple-effect systems often are justified. When liquor temperatures in the last effect are sufficiently low, stainless steel is sometimes used in place of nickel.

4.6. Other Chlorine Production Processes.
Electrolytic production of chlorine and NaOH from NaCl accounts for a preponderance of the chlorine produced. Other commercial processes are in operation, and several other processes exist that are no longer practiced commercially.

Chlorine from Potassium Hydroxide Manufacture. Mercury and membrane cells can produce KOH instead of NaOH if KCl is used as the feedstock. Chlorine and hydrogen again are coproducts. The combined nameplate capacity for KOH (caustic potash) in 2001 is about 1.7 MM t/year. Yearly production in 2001 was about 1.4 MM tons. The projected growth rate of 4–5% over the next 5 years (82) will strain the capacity by the end of that interval. The steps involved in the process are identical to those described above, but there are some different characteristics. Potassium chloride is more soluble than NaCl, but the concentrations used around the cells are about the same. Control of the feed brine concentration is therefore more of an issue. Mercury cells are generally more sensitive to trace metals in the brine (83). In membrane cells, much less water is transported through the membranes with potassium ions. The resulting change in the water balance requires more flow on both sides of the cell in order to maintain a given set of operating concentrations.

Chlorine from HCl. Reflecting the low cost, bulk chemical status of the two products, HCl, which otherwise is often made from chlorine, can sometimes be economically converted back to chlorine. There are two fundamentally different approaches, electrolysis and oxidation.

Electrolysis of HCl. Electrolytic decomposition of aqueous HCl to generate chlorine and hydrogen follows the overall reaction:

$$2\,HCl \rightarrow H_2 + Cl_2 \tag{19}$$

There are a number of these operations around the world, but only one in the United States, operated by Bayer at Baytown, Texas. Installed capacity is >1000 tons/day of chlorine. The typical electrolyzer as provided by Uhde employs graphite electrodes and a PVC or PVC/PVDF diaphragm. The arrangement is bipolar, with one side of an electrode serving as anode and the other as cathode. The assembly resembles a filter press. The effective electrode area is 2.5 m^2, and the current density is 4–5 kA/m^2. The lower figure results in a cell potential of ~1.9 V, corresponding to 1400–1500 kWh/ton of chlorine. The voltage is considerably lower than that in a brine electrolyzer, and oxygen generation is not a problem. This results in much slower deterioration of the electrodes than was experienced in older chlor-alkali cells with graphite anodes.

HCl is the electrolyte on both sides of the cell. The anolyte contains dissolved chlorine. Some of this diffuses through the diaphragm and is reduced at the cathode. This results in a loss of 2–2.5% in current efficiency. A pressure differential between anolyte and catholyte prevents backflow of hydrogen.

The acid strength in a cell is ~17%, with feed at 21–22%. Depleted acid must be reconcentrated for recycle. Most plants are interested in recovery of the chlorine value of HCl because it is a by-product of a chlorination or phosgenation process. Accordingly, the HCl is available in a gas stream. The weak acid then serves as an absorbent, producing 28–30% acid. This is blended with circulating electrolyte to maintain the desired cell feed concentration. Absorption in boiling HCl helps to dissipate the heat of absorption and reject any organics present in the gas.

Chlorine processing is much the same as in the standard electrolytic process. The water content is lower than it is in chlorine generated from NaCl or KCl brine. This fact reduces the demand on the cooling process. Two-stage cooling with chilled water on the second stage still is practiced, and this has the additional advantage of removing nearly all the vaporized HCl from the gas. The hydrogen gas from the cells is saturated with water vapor and HCl, and it also contains some chlorine. First, it is cooled by direct contact with dilute acid. This keeps the bulk of the HCl within the process. A second tower scrubs the gas with a caustic solution to remove residual HCl and chlorine. Another caustic tower can serve as a safety measure in case of breakthrough from the second tower.

These cells can also be fitted with ion-exchange membranes (84). The operating voltage is ~300 mV lower. The lower permeability of membranes also improves the product purity to ~99.5% for both chlorine and hydrogen. The lack of most of the contaminants that are responsible for membrane deterioration in chlor-alkali electrolysis means that membrane life should be longer in this process.

Another development not yet in commercial practice is the electrolysis of anhydrous HCl to form dry chlorine (85). This depends on a cation-exchange membrane to transfer protons from the anode to the cathode side, where water is added to maintain the water content of the membrane and to carry away any HCl that penetrates the membrane.

Other electrolytic processes use metal chlorides to catalyze the process. None is commercial today, but the metals used in the past include nickel, copper, and manganese. The typical process is cyclic, with electrolysis of the chloride producing chlorine and the metal:

$$MCl_2 \rightarrow M + Cl_2 \tag{20}$$

The metal then reacts with HCl to regenerate the chloride:

$$M + 2\,HCl \rightarrow MCl_2 + H_2 \tag{21}$$

The overall reaction is the same as equation 19.

   Chemical Oxidation of HCl.   Chlorine can also be produced from HCl by the following equilibrium reaction:

$$4\,HCl + O_2 \leftrightarrow Cl_2 + 2\,H_2O \tag{22}$$

Air or oxygen can be used as the oxidizing agent. This reaction is the basis for the Deacon process, which was the first continuous catalytic process carried out on large scale. It was the major source of chlorine before development of the electrolytic route.

   The fundamental problem with the Deacon process is the inability to achieve very high degrees of conversion of the HCl. Reactor gas is a high temperature mixture of HCl, water, chlorine, and oxygen and so is very corrosive. When the oxidant is air, there are also large quantities of nitrogen, which make separations more difficult. At the reaction temperatures used, there is also some volatilization of the metal catalyst.

   Two other processes used commercially in recent years but no longer in operation are the Shell process (86) and the Kellogg process (87,88). The Shell process carries out reaction 22 in the presence of cupric and other chlorides on a silicate carrier. It operated at 60–70% conversion in the range 430–475°C. The Kellogg process used ~1% nitrosylsulfuric acid [7782-78-7] catalyst and a dissimilar material containing a clay desiccant with a reversible water content of ~0.5 wt% and a crystalline structure stable to at least 760°C. It absorbs water as it forms, shifting the equilibrium of equation 22 to the right. The basic reaction was carried out in a fluidized bed in which the solids ran countercurrent to the gaseous reactants at a temperature of 400–500°C and pressures of 300–1200 kPa (3–12 atm). The process fluids are extremely corrosive, and tantalum-lined or plated equipment and piping were used.

   Another variation on HCl oxidation is Mitsui Toatsu's MT Chlor process. This is essentially a Deacon reaction carried out in a fluidized bed, using a chromium oxide catalyst.

   More recently, a new "carrier catalyst" process has been studied in which the basic Deacon reaction is carried out in stages in fluidized beds (89,90). The active catalyst is copper chloride on an inert support. Two-stage operation improves the achievable conversion of HCl to chlorine. The catalyst, in different states of oxidation, is transferred back and forth between the beds. The first stage is the "chlorination" reactor in which HCl reacts with a partly oxidized form of the catalyst to produce copper chloride and water.

$$4\,HCl + 2\,CuO \rightarrow 2\,CuCl_2 + 2\,H_2O \tag{23}$$

In the second (oxidation) stage, the chlorinated catalyst, by reaction with oxygen, reverts to its oxidized form and releases chlorine gas:

$$2\,CuCl_2 + O_2 \rightarrow 2\,CuO + 2\,Cl_2 \qquad (24)$$

The sum of the two reactions is the same as equation 22 with the catalyst in effect "carrying" chlorine from one stage to the other. Running the two stages at different temperatures, in particular with a lower temperature in the chlorination reactor, allows conversions higher than typical Deacon process equilibrium to be obtained. This process has been operated on a small scale in a pilot plant in Spain. A new plant was just announced, using a process developed by Sumitomo.

Chlorine from the Magnesium Process. Magnesium is produced by the fused salt electrolysis of $MgCl_2$ (see MAGNESIUM AND MAGNESIUM ALLOYS). The largest magnesium plant in the United States, that of Dow Chemical at Freeport, Texas, shut-down in 1998, used calcium–magnesium carbonate as a raw material, and the chlorine was recycled within the process. The Rowley, Utah, plant of AMAX Magnesium Corporation produces the metal directly from purified $MgCl_2$ recovered from the Great Salt Lake.

Chlorine from the Titanium Process. Electrolysis of magnesium chloride is a step in the production of titanium. Titanium metal is produced by the reaction of titanium tetrachloride with magnesium (see TITANIUM AND TITANIUM ALLOYS). The magnesium chloride formed in this reaction is electrolyzed, as in the preparation of magnesium. Usually, the chlorine is recycled in the production of $TiCl_4$ from the titanium dioxide ore rutile [1317-80-2]. A variation is the purchase of $TiCl_4$ [7550-45-0] and the sale of the chlorine generated by $MgCl_2$ electrolysis.

4.7. Energy Requirements. The minimum amount of electrical energy required for producing 1 ECU (electrochemical unit), constituting 1 mol of chlorine and 2 mol of caustic is 5.75×10^6 Btu or 1686 kWh for 1 ton of chlorine and 1.1 tons of caustic soda. The corresponding value for 1 short ton of chlorine and 1.1 tons of sodium hydroxide is 5.23×10^6 Btu or 1534 kWh. The total energy consumed in the year 1999 by the chlor-alkali industry is $\sim 48 \times 10^9$ kWh to produce 14.14 MM short tons of chlorine, representing 1.2% of the total annual production of 3800×10^9 kWh (91–93).

Electrolytic Decomposition of Sodium Chloride. Electrolysis of aqueous solutions of sodium chloride [7647-14-5], NaCl (or KCl), commonly called brine, simultaneously generates chlorine and caustic soda (or potash), following the overall chemical reactions:

$$2\,NaCl + 2\,H_2O \xrightarrow{\text{electrical energy}} 2\,NaOH + H_2 + Cl_2 \qquad (25)$$

$$2\,KCl + 2\,H_2O \xrightarrow{\text{electrical energy}} 2\,KOH + H_2 + Cl_2 \qquad (26)$$

Reactions 25 and 26 have a positive free energy change of 422.2 kJ (or 100.95 kcal)/ mol of chlorine at 25^0C and 417.8 kJ (or 99.90 kcal), respectively, at $95°$C. Therefore, energy has to be provided in the form of dc electricity to force the reaction to proceed in the forward direction. The amount of electrical energy required depends on the amount of the product needed and the electrolytic cell parameters, current density and cell voltage—the latter being dictated by the nature of the anode and cathode material, the separator, the interelectrode gap and the cell design.

Production of chlorine and caustic from brine is accomplished in three types of electrolytic cells: the diaphragm, the membrane, and the mercury cell. The distinguishing characteristic of these cells is the manner in which the electrolytic products are prevented from mixing with each other.

The primary electrochemical reactions during the electrolysis of brine are the discharge of the chloride ion at the anode to form chlorine,

$$2\,Cl^- \rightarrow Cl_2 + 2\,e^- \tag{27}$$

and the generation of hydrogen [1333-74-0], and hydroxide ions, OH^-, at the cathode.

$$2\,H_2O + 2\,e^- \rightarrow H_2 + 2\,OH^- \tag{28}$$

Chlorine is generated at the anode in all the three types of electrolytic cells. The cathodic reaction in diaphragm and membrane cells is the electrolysis of water to generate hydrogen as described by equation 28, whereas the cathodic process in mercury cells is the discharge of sodium ion, Na^+, to form sodium amalgam containing 0.2–0.3% sodium.

$$Na^+ + Hg + e^- \rightarrow Na(Hg) \tag{29}$$

This amalgam is subsequently reacted with water in denuders or decomposers to generate hydrogen and caustic.

$$2\,Na(Hg) + 2\,H_2O \rightarrow 2\,NaOH + 2\,(Hg) + H_2 \tag{30}$$

Separation of the anode and cathode products in diaphragm cells is realized using asbestos [1332-21-4], polymer-modified asbestos composite, or non-asbestos material deposited on a foraminous cathode. In membrane cells, the separator is a cation-exchange membrane. Mercury cells require no diaphragm or membrane, the mercury itself acting as a separator.

The catholyte from diaphragm cells typically contains 9–12% caustic soda and 14–16% salt. This cell liquor is concentrated to 50% NaOH in a series of evaporation steps involving three or four stages. Membrane cells produce 30–35% NaOH that is evaporated in two or three stages to produce 50% caustic soda. A 50% caustic soda containing very little salt is made directly in the mercury cell process by reacting the sodium amalgam with water in the decomposers.

Energy Consumption and its Components. Faraday's law states that 96,487 C (1 C=1 amp s) are required, for a single electron process, to produce 1 g equiv weight of the electrochemical reaction product. This relationship determines the minimum coulombic requirements for chlorine and caustic production, in units of kilo ampere hours per metric ton of Cl_2 or NaOH.

$$\text{For } Cl_2 \quad \frac{96,487 \times 1000}{60 \times 60 \times 35.45} = 756.2\,\text{kA h/t}$$

$$\text{For NaOH} \quad \frac{96,487 \times 1000}{60 \times 60 \times 40.00} = 670.3\,\text{kA h/t}$$

$$\text{For KOH} \quad \frac{96,487 \times 1000}{60 \times 60 \times 56.09} = 477.8\,\text{kA h/t}$$

The current efficiency of an electrolytic process (η_{current}) is the ratio of the amount of material produced to the theoretically expected quantity.

Energy consumption, expressed in terms of kilowatt hours per ton of Cl_2 (E_{Cl_2}) or of NaOH (E_{NaOH}), is the popular terminology in the chlor-alkali industry to describe and evaluate cell performance. Calculation of this value requires information related to the operating cell voltage, $V_{\text{cell}} = V_{\text{Total}}/N$, where $N = $ number of cell elements, the current efficiency, η_{current}, and the efficiency of the rectifier used to convert ac power to dc power, $\eta_{\text{rectifier}}$. The energy consumption for producing a metric ton of Cl_2, in units of ac kwh per ton of product is :

$$E_{Cl_2} = \frac{756.2\,V_{\text{cell}}}{\eta_{\text{current}}\eta_{\text{rectifier}}}$$

and that for a ton of NaOH is

$$E_{\text{NaOH}} = \frac{670.3\,V_{\text{cell}}}{\eta_{\text{current}}\eta_{\text{rectifier}}}$$

For KOH, the corresponding value is

$$E_{\text{KOH}} = \frac{477.8\,V_{\text{cell}}}{\eta_{\text{current}}\eta_{\text{rectifier}}}$$

The minimum energy required to produce chlorine, hydrogen and caustic from salt is the same (ie, 1686.32 kWh/ton of chlorine) for all three cell technologies. However, the actual energy consumed is much higher than the minimum. The energy consumed in the mercury-cell process is the greatest because the combined voltages required by reactions 29 and 30 are higher those encountered in the diaphragm- or membrane-cell process.

Electrolysis of brine is endothermic. The overall heat of the reaction is 446.68 kJ or (106.76 kcal)/mol of chlorine and hence the thermoneutral voltage, ie, the voltage at which heat is neither required by the system nor lost by the system to the surroundings, would therefore be 2.31 V. In practice, however, chlor-alkali cells operate in the range of 3.0–3.5 V, at an average chlorine efficiency (CE) of 95%, resulting in heat generation (Q) to the extent of 3960 kJ/kg (1710 Btu/lb) of Cl_2 for a voltage of 3.5 V as:

$$Q = \left[\left(\frac{100}{\text{CE}}\right)(46.05\,\text{V})\right] - \Delta H$$

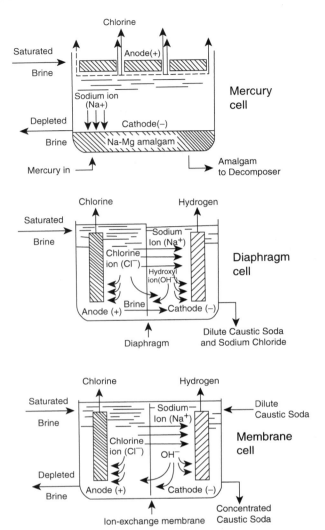

Fig. 28.   Chlorine electrolysis cells.

Heat produced in these cells operating at voltages >2.31 V is generally removed by water evaporation and radiation losses. Figure 28 illustrates the basic principles of the three cell processes and Table 8 summarizes the differences in the cell technologies and their performance.

The values of energy consumption presented in Table 9 are not optimal for the given technology, since the actual value depends on the current density, cell voltage, and current efficiency, which are affected by the process variables. Similarly, the energy for evaporation varies with the type of the evaporator system used. In the case of membrane cells, the energy requirements would be 15–25% lower than those for diaphragm cells—the major savings resulting from caustic evaporation. Thus, the energy needed to concentrate 33% NaOH to

Table 8. **Comparison of Diaphragm, Membrane, and Mercury Cells**

Component	Mercury cell	Diaphragm cell	Membrane cell
anode	$RuO_2 + TiO_2$ coating on Ti substrate	$RuO_2 + TiO_2 + SnO_2$ on Ti substrate	$RuO_2 + IrO_2 + TiO_2$ coating on Ti substrate
cathode	mercury on steel	steel (or steel coated with activated nickel)	nickel coated with high area nickel based or noble metal based coatings
separator	none	asbestos, polymer-modified asbestos, or nonasbestos diaphragm	ion-exchange membrane
cathode product	sodium amalgam	10–12% NaOH + 15–17% NaCl + 0.04–0.05% $NaClO_3$, and H_2	30–33% NaOH + <0.01% NaOH and H_2
decomposer product	50% NaOH and H_2	none	none
evaporator product	none	50% NaOH with ~1.1% salt and 0.1–0.2% $NaClO_3$	50% NaOH with ~0.01% salt
steam consumption	none	1500–2300 kg/t NaOH	450–550 kg/t NaOH
cell voltage, V	4–5	3–4	2.8–3.3
current density, kA/m^2 energy consumption (kWh/ton of Cl_2)	7–10	0.5–3	2–6
electricity for electrolysis	3200–3600	2800–3000	1950–2220[b]
steam for [a] caustic evaporation	0	600–800	200–350[c]
total	*3200–3600*	*3400–3800*	*2150–2570*

[a] 1 ton of steam = 400 kWh.
[b] Load: 3–6 kA/m^2.
[c] MP steam(10 bar a, 200°C) double effect evaporator.

Table 9. **Thermodynamic Decomposition Voltage of Chlor-Alkali Cells**

	Diaphragm/Membrane cell (at 90°C)	Mercury cell (at 80°C)
anode reaction	$2Cl^- \rightarrow Cl_2 + 2e^-$	$2Cl^- \rightarrow Cl_2 + 2e^-$
anode potential, E^0, V	1.36	1.36
cathode reaction	$2H_2O + 2e^- \rightarrow H_2 + 2OH^-$	$Na^+ + e^- \rightarrow Na \, (amalgam)$
cathode potential, E^0, V	− 0.89	− 1.80
overall reaction	$2H_2O + 2Cl^- \rightarrow Cl_2 + 2OH^- + H_2$	$2H_2O + 2Cl^{Hg} \, Cl_2 + 2Na(Hg)$
cell potential, E^0, V	2.25	3.15

50% caustic can vary from 720 kWh/ton of caustic to 314 kWh/ton, depending on whether a single effect or a triple effect is used for caustic concentration.

There are several inefficiencies arising from parasitic reactions occurring at the electrodes and in the bulk, which are described below. The two parasitic reactions offsetting anode efficiency are (1) cogeneration of oxygen [7782-44-7], from the anodic discharge of water,

$$2\,H_2O \rightarrow O_2 + 4\,H^+ + 4\,e^- \tag{31}$$

and (2) electrochemical oxidation of hypochlorite ion, OCl^-, to chlorate, ClO_3^-:

$$6\,OCl^- + 3\,H_2O \rightarrow 2\,ClO_3^- + 4\,Cl^- + 6\,H^+ + 3/2\,O_2 + 6\,e^- \tag{32}$$

The amount of oxygen generated from these reactions depends on the nature of the anode material and the pH of the medium. The current efficiency for oxygen is generally 1–3% when using commercial metal anodes.

At the cathode, water molecules are discharged to form gaseous hydrogen and hydroxide ions, OH^-. Some of the caustic generated in the cathode compartment back-migrates to the anode compartment and reacts with dissolved chlorine, $Cl_{2(aq)}$, to form chlorate as follows:

$$Cl_{2(aq)} + OH^- \rightarrow HOCl + Cl^- \tag{33}$$

$$HOCl + OH^- \leftrightarrow H_2O + OCl^- \tag{34}$$

$$2\,HOCl + OCl^- \rightarrow ClO_3^- + 2\,H^+ + 2\,Cl^- \tag{35}$$

There are two reactions that influence the cathodic efficiency, viz, the reduction of OCl^- and ClO_3^-

$$OCl^- + H_2O + 2\,e^- \rightarrow Cl^- + 2\,OH^- \tag{36}$$

$$ClO_3^- + 3\,H_2O + 6\,e^- \rightarrow Cl^- + 6\,OH^- \tag{37}$$

Although these reactions are thermodynamically possible, they are not kinetically significant under normal operating conditions. Hence, the cathodic efficiency is usually high (>95%) in diaphragm and membrane cells. In mercury cells, the cathodic inefficiency arises from the discharge of H_2 at the cathode as a result of the impurities in the brine. Reactions contributing to anodic inefficiency in mercury cells are the same as in diaphragm or membrane cells.

Current Efficiency.   Current efficiency for caustic production in mercury, diaphragm and membrane cells can be determined by collecting a known amount of caustic over a given time, and from a knowledge of the number of coulombs of electricity passed during that period. The ratio of the amount of caustic collected to that expected from Faraday's law provides the magnitude of the current efficiency. An alternative method involves analysis of the gases evolved during electrolysis and the anolyte composition.

The chlorine and caustic current efficiency expressions (94,95), based on rigorous material balance calculations, for all three cell technologies are shown in equations 38–43. The symbol eta denotes the current efficiency for the product in the subscript and the technology in the superscript, Hg for mercury cells, d for diaphragm cells, and m for membrane cells. Superscripts f and d refer to the feed and depleted streams respectively.

Membrane Cells

$$\eta_{Cl_2}^m = \frac{1 - (2F/I)p(XCl_2^d - Cl_2^f)}{C - 2YCl_2^d} \tag{38}$$

$$\eta_{NaOH}^m = \frac{1 - (F/I)p\left[2(1-C)(LCl_2^d - Cl_2^f) + C(LOH^d - OH^f)\right]}{C - M\left[2(1-C)Cl_2^d + COH^d\right]} \tag{39}$$

where $F =$ Faraday; $I =$ current (amps); $p =$ feed brine flow rate(L/sc):

$$C = 1 + 2(\%O_2 + \%Cl_2); \quad X = Cl^f/Cl^d; \quad Y = 1/Cl^d :$$

$$Cl_2^f = Cl_{Cl_2}^f + C_{HOCl}^f + C_{NaOCl}^f + 3C_{NaClO_3}^f$$

$$Cl_2^d = Cl_{Cl_2}^d + C_{HOCl}^d + C_{NaOCl}^d + 3C_{NaClO_3}^d$$

$$Cl^f = C_{NaCl}^f + 2C_{Cl_2}^f + C_{HOCl}^f + C_{NaOCl}^f + C_{NaClO_3}^f + C_{HCl}^f$$

$$Cl^d = C_{NaCl}^d + 2C_{Cl_2}^d + C_{HOCl}^d + C_{NaOCl}^d + C_{NaClO_3}^d + C_{HCl}^d$$

$$OH^f = C_{HOCl}^f + 2C_{NaOCl}^f + 6C_{NaClO_3}^f + 2C_{Na_2CO_3}^f + C_{NaHCO_3}^f$$
$$+ C_{NaOH}^f - C_{HCl}^f - C_{NaHSO_4}^f$$

$$OH^d = C_{HOCl}^d + 2C_{NaOCl}^d$$

The C values in the above equation refer to concentrations of the species in the subscript, the superscript representing the feed (f) or depleted (d) stream.

Diaphragm Cells.

$$\eta_{Cl_2}^d = \frac{\eta_{H_2}^d + \frac{F}{I}p\left\{6C_{NaClO_3}^f + C \times G\right\}}{C + 6(C_{NaClO_3}^d/C_{NaOH}^d)} \tag{40}$$

$$\eta_{NaOH}^d = \frac{\eta_{H_2}^d + \frac{6F}{I}p\left\{C_{NaClO_3}^f - \left(\frac{C_{NaClO_3}^d}{C_{NaOH}^d}\right)G\right\}}{C + 6(C_{NaClO_3}^d/C_{NaOH}^d)} \tag{41}$$

where

$$G = C_{NaOH}^f + 2C_{Na_2CO_3}^f + C_{NaHCO_3}^f - C_{HCl}^f - C_{NaHSO_4}^f$$

$$\eta_{H_2}^d = 1 - 2(C_{av.Cl_2}^a/C_{NaOH}^d)$$

$$C_{av.Cl_2}^a = \text{concentration of available chlorine in the anolyte}$$

Mercury Cells

$$\eta_{Cl_2}^{Hg} = \cfrac{1}{C - \left(\cfrac{6\Delta C_{NaClO_3} + 2\Delta C_{Av.Cl_2}}{\Delta C_{Cl}}\right)} \tag{42}$$

$$\eta_{NaOH}^{Hg} = \left\{\cfrac{\Delta C_{NaCl} + \Delta C_{NaClO_3} + \Delta C_{NaOCl}}{\Delta C_{Cl}}\right\} \eta_{Cl_2}^{Hg} \tag{43}$$

where,

$$\Delta C_{NaClO_3} = C_{NaClO_3}^{f} - C_{NaCl}^{d}$$

$$\Delta C_{av.Cl_2} = C_{av.Cl_2}^{f} - C_{av.Cl_2}^{d}$$

$$\Delta C_{NaCl} = C_{NaCl}^{f} - C_{NaClO_3}^{d}$$

$$\Delta C_{NaOCl} = C_{NaOCl}^{f} - C_{NaOCl}^{d}$$

$$\Delta C_{Cl} = \left(C_{NaCl}^{f} - C_{NaClO_3}^{f} + C_{Cl_2}^{f} + C_{av.Cl_2}^{f} + C_{HCl}^{f}\right)$$

$$- \left(C_{NaCl}^{d} - C_{NaClO_3}^{d} + C_{Cl_2}^{d} + C_{av.Cl_2}^{d} + C_{HCl}^{d}\right)$$

$C_{av.Cl2}$ = concentration of available chlorine that includes dissolved chlorine, HOCl and NaOCl in the given stream.

Equation 43 assumes that all the sodium amalgam is decomposed to form caustic in the decomposer. Otherwise, appropriate corrections are needed to account for the incomplete decomposition of the sodium amalgam.

The disparity in the chlorine and caustic efficiency values are a result of the alkalinity or acidity in the feed brine as described by equation 44

$$\eta_{NaOH} - \eta_{Cl_2} = \frac{F}{I}\Delta\{J_{NaOH} + 2J_{Na_2CO_3} + J_{NaHCO_3} - J_{HOCl} - J_{HCl} - 2J_{Cl_2}\} \tag{44}$$

where J values represent the difference in the mass flow rates of the species in the subscript between the feed and depleted brine streams.

Equation 40 is the proper material balance expression for calculating the chlorine efficiency of diaphragm cells. However, many approximate versions are practiced (91), the one closest to the material balance expression (eq. 40) being the "six equation":

$$\eta_{Cl_2} = \left[\frac{100}{1 + 2(\%O_2 + \%Cl_2) + 6C_{NaClO_3}(\text{cell liquor})/C_{NaOH}}\right] \tag{45}$$

Current efficiency values based on the "six equation" are higher by $\sim 1.0\%$ than those from equation 40.

Cell Voltage and Its Components: The cell voltage, E, is composed of various terms as

$$E = E^0 + |\text{Overvoltages(anodic and cathodic)}| + \text{ohmic drops} \tag{46}$$

The ohmic drops in the above equation are associated with the electrolyte, the separator, and the hardware.

The minimum voltage required for electrolysis for a given set of cell conditions, such as a temperature of 95°C, is the sum of the cathodic and anodic reversible potentials and is known as the thermodynamic decomposition voltage, $E°$. The parameter $E°$ is related to the standard free energy change $\Delta G°$ for the overall chemical reaction 25 or 26 as

$$\Delta G^0 = -nFE^0$$

where n refers to the number of electrons involved in the primary electrode reaction and F is the Faraday constant, expressed in ampere-hours equiv. The $E°$ values for all three types of cells are presented in Table 10. The $+0.924$ V difference in $E°$ values between diaphragm or membrane and mercury cells arises from the reaction:

$$2\,Na + 2\,H_2O \rightarrow 2\,NaOH + H_2$$

which takes place outside the electrolytic cells. Although, in principle, this voltage is recoverable, this concept has not yet been commercially demonstrated. The $E°$ value for diaphragm or membrane cells at 95°C is 2.23 V for a caustic concentration of 3.5 M. However, the electrolytic cells operate 3.0–3.2 V at current density of 2–3 kA/m^2, not at 2.23 V, because in order to achieve acceptable rates, an additional driving force is required to overcome cell resistances and electrode overvoltages.

Table 10. Components of Chlor-Alkali Cell Voltages

	Diaphragm cell[a]	Membrane cell[b]	Mercury cell[c]
Thermodynamic decomposition voltage			
anode	1.32	1.32	1.32
cathode	0.93	0.93	1.83
Overvoltage			
anode	0.03	0.05	0.1
cathode	0.28	0.10	0.4
Ohmic drops			
solution	0.12	0.20[d]	0.15
diaphragm/membrane	0.38	0.44	
anode and contact to base	0.11		
base	0.06	0.03[e]	0.2
Cathode	0.09		
Cell voltage	3.32	3.07	4.0

[a] Voltages given are for Oxy Tech H4 cell operating at 2.3 kA/m^2 (150 kA and 95°C).
[b] Voltages given are for third generation Uhde's single cell element operating at 5.0 kA/m^2 (90°C).
[c] Voltages given are for De Nora 24M2 system operating at 10 kA/m^2 (270 kA and 80°C).
[d] With 1.2-mm gap.
[e] Total structural IR drop.

Overvoltage.   Overvoltage (η_{ac}) arises from kinetics of the electrode reaction on a given substrate. The magnitude of this value is generally expressed in the form of the Tafel equation:

$$\eta_{ac} = k\log(i/i_0) \tag{49}$$

where k is the slope of the η_{ac} vs $\log i$ curve, i is the applied current density, and i_0 is the exchange current density of the reaction. The quantity i_0 is a measure of the rate of a given reaction, eg, 1 mA/cm^2 for the Cl$_2$ evolution on dimensionally stable electrodes (DSA). Overvoltage can be lowered by increasing the electrochemically active surface area, which reduces the magnitude of i, or by using materials exhibiting high exchange current density.

Ohmic Drops.   Voltage drops across the electrolyte between the electrodes, separator and the hardware constitute a significant portion of the overall cell voltage. The ohmic R drop across the hardware, also called as IR drop, can be calculated from Ohm's law and the relationship:

$$R = \rho l/A \tag{50}$$

where R is the resistance (in Ω) of the conductor of length l with a specific resistance of ρ and cross-sectional area A.

The ohmic drop across the electrolyte and the separator can also be calculated from Ohm's law using a modified expression for the resistance. When gas bubbles evolve at the electrodes, they are dispersed in the electrolyte and the resulting conductivity characteristics of the medium will be different from those of a pure electrolyte. According to Rousar (96), the resistance of the gas-electrolyte mixture R_{mix}, is related to the pure electrolyte, R_{sol}:

$$R_{mix} = R_{sol} (1 + 1.5\epsilon) \tag{51}$$

where ϵ is the gas void fraction, defined as the ratio of the volume of the gas to the volume of the gas plus the volume of the electrolyte. The *IR* drop in brine solution is generally $\sim$30–40 mV/mm at 95°C and a current density of 2.32 kA/m^2. Similarly, for calculating the IR drop across the separator, the separator thickness term has to be modified because the distance between the two faces of a separator such as the asbestos diaphragm is not equal to its thickness. The liquid path is tortuous and the area is limited by finite porosity. Thus, the *IR* drop across the separator would be

$$IR_{sep} = xil\rho \tag{52}$$

where x reflects the tortuosity-to-porosity ratio. Typical values of tortuosity for asbestos diaphragms range from 2.2 to 2.8; porosity varies from 0.7–0.8.

The conductivity of the composite membranes constituting the perfluorsulfonic and perfluorocarboxylic layers is best determined experimentally, although some approximate theoretical descriptions are available (50, 97–99). Its value in the chlor-alkali cell operating environment is in the range of 1.5–2 Ω -cm^2. Note, however, that the membrane manufacturers are constantly striving to develop new compositions to lower the resistance and at the same time achieve better performance and life.

The components of the diaphragm-, membrane-, and mercury-cell voltages presented in Table 10 show that the major components of the cell voltage are the $E°$ term and the ohmic drops.

Direct-Current Electric Power. The operation of a chlor-alkali plant is dependent on the availability of large quantities of dc electric power, which is usually obtained from a high voltage source of alternating current. The lower voltage required for an electrolyzer circuit is produced by step-down transformers. Silicon thyristor rectifiers convert the ac to dc for the electrolysis. A set of rectifiers can supply up to 450 kA. Although these devices can operate at 400 V/device, a peak ac voltage of 1500 V, corresponding to a dc output of 1200 V is not exceeded for safety reasons. Rectifier efficiency ($\eta_{rectifier}$) is generally ~97.5–98.5%. The harmonics fed back to the supply system as a result of thyristor operation are kept within limit by providing suitable harmonic filters.

Older generation high current rectifier systems based on diodes and transformers with on-load tap changers are now being replaced by new generation rectifier systems using thyristors as the devices. A thyristor rectifier system has following advantages over the diode type:

(1) Rapid current regulation due to control by solid-state devices as compared to diode type where the control is by operation of saturable reactors, which are slower in their response.
(2) Smooth control of current and voltage.
(3) Better overall system efficiency.
(4) Compact size and layout.

The thyristor rectifier system, however, generates greater harmonic currents and voltages, and operates at relatively lower power factor. These harmonics can be filtered by providing suitable tuned capacitor banks which also improve the operating power factor of the system.

The unit cost of the dc supply decreases with increasing voltage and amperage (Fig. 29). An electrolytic plant is therefore most economical when as many

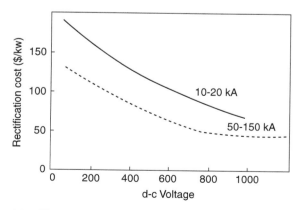

Fig. 29. Variation of rectification costs with voltage (100).

high amperage cells as possible are connected in series. The costs (100) in Figure 29 are for the rectification equipment with optimum medium distribution voltage level delivering ac power to the rectifier-transformer primary. Advances in transformer and rectifier design allow large rectifiers to be connected directly to voltages up to 138 kV. Such a configuration increases the rectification cost by 10–15% while saving plant space, and the cost and energy losses associated with the step-down transformers. These costs were typical for 2001. Various design and performance options result in a variation of ± 15%.

5. Materials of Construction

The choice of construction material for handling chlorine depends on equipment design and process conditions. Dry chlorine, with <50 ppm (w/w) of water, can be handled safely below 120°C in iron, steel, stainless steels, Monel, nickel, copper, brass, bronze, and lead. Silicones, titanium, and materials with high surface areas (eg, steel wool) should be avoided. Titanium ignites spontaneously at ordinary temperatures in dry chlorine; steel reacts at an accelerated rate at temperatures above 120°C, igniting near 250°C (17). The presence of rust or organic substances increases the risk of steel ignition because of the exothermic reactions of these materials with chlorine. Thorough cleaning, degreasing, and drying are essential before commissioning the steel equipment into chlorine service.

With chlorine gas, ordinary carbon steel is safe so long as the temperature is held safely below the ignition temperature. Good practice allows a wide margin to allow for hot spots and to extend the life of the equipment. In the chlorine production process, the only step in which dangerous temperatures normally may occur is compression. Limiting the compression ratio in each stage and cooling the gas before the compressor and between stages keeps the temperature low. In those applications of chlorine that require temperatures above 150°C, materials of construction other than carbon steel must be used.

The atmospheric pressure boiling point of chlorine (−34°C) demands the use of low temperature carbon steels. Because this temperature is possible whenever a higher pressure system is depressurized, most liquid chlorine systems are constructed of these materials. Liquid chlorine usually is stored in vessels made from unalloyed carbon steel or cast steel. Fine grain steel with limited tensile strength is used to facilitate proper welding.

A protective layer of ferric chloride on the metal prevents the reaction of dry chlorine with steel. This layer can be disrupted by excessive temperature (as above), the presence of moisture, or erosion. Accordingly, the velocity of liquid chlorine in piping is limited to 2 m/s. In long-distance transfer lines, a lower limit may be used. Both Euro Chlor (101) and The Chlorine Institute (102) provide specifications for chlorine piping. The latter divides services into six classes, with three different temperature ranges for gas systems and for liquid systems or those systems that may carry liquid. Euro Chlor also has an official procedure for homologation of valves, and various suppliers certify valves to these specifications. Reference 102 directs a reader to specifications for various types of valves.

Wet chlorine gas is handled in fiberglass-reinforced plastics. Special constructions are necessary to achieve optimum performance (103). Rubber-lined steel is suitable for wet chlorine gas up to ~100°C. At low temperature and pressure, PVC, CPVC, and reinforced polyester resins are also used. PTFE, PVDF (polyvinylidene fluoride), and FEP (fluorinated ethylene propylene) are resistant at higher temperatures but suffer from permeability or poorer mechanical properties. Other materials stable in wet chlorine include graphite and glass. Among the metals, titanium is the usual choice for wet-chlorine service. As was the case with steel in dry chlorine, titanium depends on a protective layer (TiO_2) on the metal for its corrosion resistance. Maintenance of this layer requires a minimum water content in the gas (76,78). Tantalum is the most resistive metal over a wide range of conditions. Its limited availability and high price preclude its use in large equipment, but it is quite common in small parts and instrument systems.

Gaskets in both dry gas and liquid chlorine have been made for many years from some form of compressed asbestos. It is an ideal material for the service, but restrictions on the use of asbestos have led to development of alternatives. Both Euro Chlor (104) and The Chlorine Institute (105) have active programs for the evaluation of these alternatives. For wet chlorine gas, rubber or synthetic elastomers are acceptable. PTFE within its serviceable temperature range is resistant to all forms of chlorine. Tantalum, Hastelloy C, PTFE, PVDF, Monel, and nickel are used in thin sections such as diaphragms, membranes, rupture disks, and bellows.

6. Shipment and Storage

There are two basically different approaches to storing chlorine. One is to store it under positive pressure; the other is to refrigerate it and store it essentially at atmospheric pressure. In a producing plant, the former usually is at liquefaction process pressure. The latter is mechanically more complex and requires a compression system for returning vaporized chlorine to the process. It is considered appropriate only for large systems in producers' plants. Pressurized storage tanks require relief and containment systems. It is common to use an empty expansion tank as a receptacle for escaping chlorine. This contains the material released during minor upsets and buffers the flow of gas to a scrubbing system when releases are too large to be completely contained.

The material of construction for atmospheric storage tanks is unalloyed low-temperature carbon steel. These tanks usually are approximately spherical and are contained within a shell. Overcoming differential thermal expansion is an important part of vessel design and fabrication. The annular space contains insulation and is monitored for chlorine leaking from the inner tank. While the temperature of the chlorine under pressurized storage may be above the minimum temperature for conventional steels (−29°C), such low temperatures are encountered whenever pressure is released from a tank. Normal practice therefore is to build these also from low temperature steel. Chlorine Institute Pamphlet 5 summarizes design principles and good practice.

Liquid chlorine has an unusually large coefficient of thermal expansion. The amount stored in a pressurized tank must be limited to allow for this expansion in case the temperature is allowed to rise. A standard practice is to design for a filling density of 1.25 in order to give 5% freeboard at 50°C.

Because stored liquid chlorine frequently represents a plant's hazard with the most serious potential consequences, the trend is to keep storage volumes to a reasonable minimum.

Chlorine loading and unloading are usually by one of two methods, pumping or pressurization with a dry gas. Pumps in large-scale storage systems usually are of the submerged type. This eliminates suction piping and the need for bottom or side connections to the storage tanks. Excess flow valves prevent release of chlorine in case of downstream maloperation or piping failure. Canned pumps are an alternative and are popular in Europe. Pressurization is mechanically simpler but creates the need for disposal of the padding gas, mixed with residual chlorine.

Chlorine is transported in cylinders and ISO (International Standards Organization) containers and by rail and road tankers. It is classified as a nonflammable compressed gas. U.S. Department of Transportation regulations call for a green label. Repackagers of chlorine supply it in small cylinders containing 45.4 or 68 kg. They also, along with some producers, supply ton lots in cylinders. These are pressurized and protected with fusible-plug relief devices. Quantities between 15 and 90 short tons are transported in tankers with covered manholes fitted with special angle valves. In the United States, shipping containers are fitted with special relief devices comprising a diaphragm-protected conventional relief valve mounted above a breaking-pin assembly. Thinking on the matter of relief of transport containers is divided, and in Europe chlorine is shipped without relief devices (106).

7. Economic Aspects

The choice of technology, the associated capital and operating costs for a chlor-alkali plant are strongly dependent on the local energy and transportation costs, as well as environmental constraints. The primary difference in operating costs between membrane, diaphragm, and mercury plants results from variations in electricity and steam consumption for the three processes. Table 9 provides a comparison of the energy consumption involved in the three cell technologies.

Table 11. **Investment for Chlorine Membrane Electrolysis Plant**

Item	Estimated investment (in 1000 US$)
cells	27,200
brine purification	14,000
chlorine processing	16,000
waste gas treatment	2,300
caustic evaporation	6,900
utilities	4,500
rectifiers	10,000
engineering	10,000
total	*90,900*

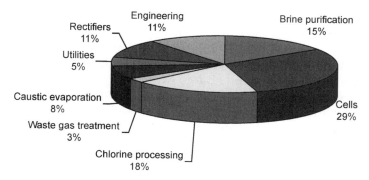

Fig. 30. Distribution of investment for a membrane chlor-alkali plant.

The cost of constructing a plant varies significantly with the actual plant configuration, procurement conditions, and the production capacity. A break-down of the total investment for a grass root membrane electrolysis plant of ~160,000 mtpy of chlorine production capacity is given Table 11 and graphically in Figure 30.

Conversions from mercury to membrane technology require new facilities to protect the membranes from traces of mercury and place restrictions on the use of existing equipment. Diaphragm plants often are more easily adapted to the needs of membrane technology. Those plants that operate on a brine feed will require extensive revisions to achieve a water balance after conversion. See references (25, 107 and 108) for more details related to these capital costs.

7.1. Capacity. In the past, chlorine capacity was located close to its end use markets, which were primarily pulp and paper and vinyls products (ie, ethylene dichloride (EDC) [107-06-2] and vinyl chloride monomer (VCM) [75-01-4]). Other derivative demand was also located in close proximity to the chlorine source. Over time, world trade in chlorine through its derivatives has accelerated. Chlorine facilities have become larger to reduce the overall costs and capture economies of scale, and they are increasingly located in areas with low energy costs. The days of the small chlorine plants servicing a localized market are rapidly coming to an end, except possibly in the developing world.

The regions of North America, West Europe, and Northeast Asia have almost 75% of the world's chlor-alkali capacity (Fig. 31). Of these, only North

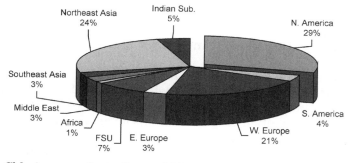

Fig. 31. Chlorine capacity in the world in 2000. Total capacity $= 53 \times 10^6$ t (5,6).

America is and will remain a net exporter of chlorine equivalents. The Middle East possesses only 3% of the world's capacity, but it is a major net exporter of chlorine equivalents in the form of vinyl intermediates such as EDC. The driving force behind the location of new chlor-alkali facilities will be where electricity (energy) is available at a low cost. Large exports of vinyls and competitive electricity costs will reduce the overall costs and capture economies of scale. Although transportation of chlorine derivatives and caustic soda is a consideration, low cost power is the greatest concern.

Growth in chlorine derivatives is the key to understanding the addition of chlorine capacity around the world. Chlorine demand will increase with improving economic conditions, since chlorine is consumed in the manufacture of durable goods such as those used in infrastructure (PVC pipe, windows, and doors, polyurethane insulation). About 35% of total chlorine demand is consumed in vinyls (PVC) through the direct chlorination EDC process. The organic chemicals group, excluding vinyls, accounts for ~20% of the total chlorine demand and includes polycarbonates, toluene diisocyanate (TDI) and methylene (MDI) (for polyurethanes), epichlorohydrin, and propylene oxide. Both the vinyls and organics categories are forecast to require increasing volumes of chlorine, as the global economy improves.

7.2. U.S. Chlorine Market and Growth. The chlor-alkali industry went through a period of capacity rationalization during the mid-1980s. Then, the demand for chlorine derivatives changed from a regional or local market to a more global market. On a world basis, ~4 million metric tons of capacity were shut down during this period. The smaller, higher cost, and less environmentally friendly plants were the ones that were closed. The closure of these smaller plants set the stage for the larger, lower cost plants that have come on line during the 1990s.

Environmental issues concerning chlor-alkali facilities and some of the chlorine derivatives were very prevalent during the 1990s. Mercury cell chlorine plants faced strict limits on mercury emissions and they have, thus far, succeeded in meeting all of the standards placed on them. Mercury cells have been phased out in Japan for environmental reasons and there is an agreed time table for their phase-out in Europe. Several of the chlorine derivatives have encountered their own environmental roadblocks, such as

- Dioxin issues with pulp and paper.
- Banning of some fluorocarbons.
- Carbon tetrachloride being almost eliminated.
- Restricted emissions of chlorinated solvents.
- PVC in bottling and packaging (non durable) markets.

Environmental issues will continue to shape the industry. However, the benefits of chlorine and its derivatives will significantly outweigh its alleged hazards. The chlorine industry will continue to actively promote the use of chlorine, based on its contribution to society and its favorable environmental impact, which is based on "cradle to grave" scientific analysis.

Technology has also played a significant role in the shaping of the chlor-alkali industry. Most of the early plants used mercury cell technology and

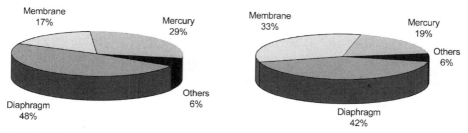

Fig. 32. Chlor-alkali technologies in 1990 and 2000. 1990 Capacity $= 45.0 \times 10^6$ t; 2000 Capacity $= 53.0 \times 10^6$ t. (5,6).

then diaphragm plants took a larger percentage of total world capacity. Today, membrane technology is the technology of choice. This progression was driven by local economic factors and the advancement of technology (Fig. 32). The trend to membrane cells is due to their lower operating costs and to environmental concerns over mercury and asbestos. Figures 33 and 34 show the locations of U.S. chlorine plants in 2000. Table 12 depicts U.S. capacity information for the period 1990–2000.

The data represent elemental chlorine only and do not include chlorine in the form of HCl. More than 95% of total HCl produced in the United States is the result of a chlorine-using process. In general, processes that use chlorine consume about one-half of the chlorine in the process and convert the other one-half to HCl. Hydrogen chloride can be marketed in either the aqueous or the anhydrous form.

Fig. 33. Chlorine producing locations in the United States (5,6). See Fig. 34 for key.

United States
Chlor-alkali producing locations

1 Burkville, Alabama
   GE Plastics (CHL/CAU)
2 Mcintosh, Alabama
   Olin (CHL/CAU)
   Sunbelt(CHL/CAU)
3 Mobile, Alabama
   Oxy (CHL)
4 Muscle Shosis, Alabama
   Oxy (CHL)
5 Delaware City, Delaware
   Oxy (CHL/CAU)
6 Augusta, Georgia
   Olin (CHL/CAU)
7 Rincon, Georgia
   Ft. Howard (CHL/CAU)
8 Mount Vermon, Indiana
   GE Plastics (CHL/CAU)
9 Wichita, Kansas
   Vulcan (CHL/CAU)
10 Calvert City, Kantucky
   Westlake (CHL/CAU)
11 Baton Rouge/ Plaquemine/
   St.Gabriel, Louisiana
   FPC USA (CHL/CAU)
   Dow (CHL/CAU)
   George Gulf (CHL/CAU)
   Pioneer Chlor (CHL/CAU)

12 Convent/Geismar/
   Gramercy/Taft, Louisiana
   Oxy (CHL/CAU)
   Vulcan (CHL/CAU)
   Vulcan C-A (CHL/CAU)
   LaRoche (CHL/CAV)
13 Lake Charles, Louisiana
   PPG (CHL/CAU)
14 Orrington, Maine
   Hotrachem Mfg (CHL/CAU)
15 Vicksburg, Mississippi
   Cedar (CHL)
16 Acme, North Carolina
   Hottrachem Mfg (CHL/CAU)
17 South Kearny, New Jersey
   Kuchne (CHL/CAU)
18 Handerson, Nevada
   Pioneer Chlor (CHL/CAU)
   TIMET (CHL)
19 Niagara Falls, New York
   Du Pont (CHL)
   Olin (CHL/CAU)
   Oxy (CHL/CAU)
20 Ashtabula, Ohio
   Ashta (CHL)
21 Muskogee, Oklahoma
   Ft. Howard (CHL/CAU)
22 Albany, Oregon
   Oregon Metail (CHL)

23 Portland, Oregon
   Atolina (CHL/CAU)
24 Charteston, Tennessea
   Olin (CHL/CAU)
25 Baytown/Deer Park/
   LaPorte, Texas
   Bayer (CHL/CAU)
   Oxy Vinyls (CHL/CAU)
26 Freeport, Texas
   Dow (CHL/CAU)
27 Ingleside, Texas
   Oxy (CHL/CAU)
28 Point Comfort, Texas
   FPC USA (CHL/CAU)
29 Tacoma, Washington
   Pioneer Chlor (CHL/CAU)
30 Green Bay, Wisconsin
   Ft. Howard (CHL/CAU)
31 Port Edwards, Wisconsin
   Vulcan (CHL) (CAU)
32 Natrium, West Virginia
   PPG (CHL/CAU)
33 Granger/Green River, Wyoming
   FMC Wyoming (CAU)

Fig. 34. Key to the numbers in Figure 33.

Aqueous HCl (muriatic acid) is a solution of about two-thirds H_2O and one-third HCl. Anhydrous HCl is used primarily as a source of chlorine for the production of EDC, the precursor of VCM. The VCM production itself accounts for 65% of total production of HCl in the United States, but it is not a source of muriatic acid because all the HCl is recycled through the oxychlorination process to EDC.

Most anhydrous HCl goes into the captive or nonmerchant market. The distinguishing difference between the two forms of HCl is that the anhydrous material moves almost exclusively by pipeline while muriatic acid moves by trucks, tank cars, and barges. Clearly, transporting muriatic acid is expensive because more than one-half of the volume is water. Therefore, the muriatic acid market tends to be regional and there is almost no international trade. The U.S. muriatic acid market is split among the Northwest, Northeast, and Gulf Coast regions. The Northwest and Northeast markets are smaller than that of the Gulf Coast and are supplied largely by acid manufactured by chlor-alkali producers, by direct reaction of hydrogen with chlorine.

The Gulf Coast market is by far the dominant regional market in volume, but it is characterized by significantly more by-product HCl. Large HCl producers on the Gulf Coast include isocyanate and fluorocarbon producers. These industries are significant sources of HCl, and its supply therefore is a function of the demands for their products. Recognizing this is the key to understanding

Table 12. **Chlorine Capacities**[a]

Company United States	Location	Process	Average Annual Capacities, (-000- Metric Tons)											Remarks
			1995	1996	1997	1998	1999	2000	2001	2002	2003	2004	2005	
Ashta	Ashtabula, Ohio	(8)	40	40	40	40	40	40	40	40	40	40	40	from LinChem/KOH
ATOFINA	Portland, Oreg.	(1)						85	45	(170)	(170)	(170)	(170)	
Bayer	Baytown, Tex.	(2)						17	8	(34)	(34)	(34)	(34)	
		(2)					284	310	310	310	310	310	310	Ku Tech
BF Goodrich	Calvert City, Ky.	(4)	62	82	82	82	(82)	40	80	80	80	80	80	KU Tech
		(3)	111	111	83									To Westlake
Cedar	Vicksburg, Miss.	(9)	36	45	45	45	45	45	45	10	(45)	(45)	(45)	from Vertac Chemcial Co.
Dow	Freeport, Tex.	(1)	2204	2264	2431	2431	2431	2431	2431	2431	2431	2431	2431	
		(2)				75	75	450	500	500	500	500	500	Asahi Chem. magnesium
	Plaquemine, La.	(5)	250	250	250	75	(250)	(250)						
		(1)	525	525	525	525	525	525	525	525	525	525	(525)	
		(1)	745	745	745	745	745	745	745	745	745	745	745	Asahi Chem + 100 Planne
		(2)											655	
Du Pont	Corpus Christi, Tex.	(4)												
	Niagara Falls, N.Y.	(5)	77	77	77	77	77	77	77	77	77	77	77	no caustic

Table 12 (*Continued*)

| Company United States | Location | Process | Average Annual Capacities, (-000- Metric Tons) | | | | | | | | | | | Remarks |
			1995	1996	1997	1998	1999	2000	2001	2002	2003	2004	2005	
Elf Atochem	Portland, Oreg.	(1)	170	170	170	170	170	85						built 1947
Fort James	Green Bay, W.S.	(2)	34	34	34	34	34	17						built 1990
	Green Bay, W.S.	(1)			8	(8)								
	Muskogee, Okla.	(2)			5	(5)								
	Rincon, Ga.	(2)			6	(6)								
FPC USA	Baton Rouge, La.	(1)	216	216	216	236	236	236	236	236	236	236	236	Expn Planned; Chl. Engr. T
	Point Comfort, Tex.	(2)	654	654	654	690	798	798	798	798	798	798	798	
Ft. Howard	Green Bay, Wis.	(1)	8	8										built 1968; to Fort James
	Muskogee, Okla.	(2)	5	5										to Fort James
Ft. Howard	Rincon, Ga.	(2)	6	6										built 1990; to Fort James
GE Plastics	Burkville, Ala.	(1)	24	24	24	24	24	24	24	24	24	24	24	built 1987
Georgia Gulf	Mount Vernon, Ind.	(1)	50	50	50	50	50	50	50	50	50	50	50	built 1976
	Plaquemine, La.	(1)	427	427	427	427	427	427	427	427	427	427	427	from Georgia Pacific
Georgia Pacific	Bellingham, Wash.	(3)	82	82	82	82	48	(82)						built 1965 from Brunst P&P
	Brunswick, Ga.	(3)												
Holtrachem Mfg	Acme, N.C.	(2)					8	40	(50)					OxyTech; permanent shutdown

Company	Location		C1	C2	C3	C4	C5	C6	C7	C8	C9	C10	Remarks
Hooker Chemical	Orrington, ME	(3)	48	48	48	48	48	(48)	12				permanent shutdown
		(3)	73	73	73	73	73	50	73	(73)			
	Tacoma, Wash.	(1)											to Oxy
	Taft, La.	(1)											to Oxy some KOH;
Hooker/IMC JV	Niagara Falls, N.Y.	(1)											
Kuehne	S.Kearny, N.J.	(2)					8	*16*	16	16	16	16	ICI Tech.
La Roche	Gramercy, La.	(1)	180	180	180	180	180	*180*	90	(180)	(180)	(180)	from Kaiser
Magnesium	Rowley, Utah.	(5)	18	18	18	18	18	*18*	18	18	18	18	from Renco
Miles	Baytown, Tex.	(4)	20										to Bayer
Niachlor	Niagara Falls, N.Y.	(2)	218										
Olin	Augusta, Ga.	(3)	102	102	102	109	109	*109*	109	109	109	109	built 1965
	Charleston, Tenn.	(3)	236	236	236	245	245	*245*	245	245	245	245	built 1962
	McIntosh, Ala.	(1)	364	364	364	364	364	*364*	364	364	364	364	364
	Niagara Falls, N.Y.	(2)		218	218	218	218	*218*	218	218	218	218	218
Oregon Metallurgical	Albany, Oreg.	(5)	5	5	5	5	5	*5*	5	5	5	5	built 1971
Oxychem	Convent, La.	(1)	336	336	336	336	336	*336*	336	336	336	336	built 1981; From diamond Sh.
Oxychem	Deer Park, Tex.	(1)	248	248	248	248	248	*248*	83				
	Delaware City, Del.	(3)	100	100	100	100	33						built 1958
		(3)	74	74	74	74	74	*74*	74	74	74	74	built 1965; from diamond Sh.

Table 12 (*Continued*)

Company United States	Location	Process	Average Annual Capacities, (-000- Metric Tons)											Remarks
			1995	1996	1997	1998	1999	2000	2001	2002	2003	2004	2005	
		(8)	61	61	61	61	61	61	61	61	61	61	61	from KCl; From diamond Sh.
	Ingleside, Tex.	(1)	455	455	455	513	571	571	571	571	571	571	571	built 1974; From Du Pont
	La Porte, Tex.	(1)	479	500	500	500	167							built 1974; From diamond Sh.
	Mobile, Ala.	(7)	41	41	41	41	41	41	41	41	41	41	41	
	Muscle Shoals, Ala.	(8)	132	132	132	132	132	132	132	132	132	132	132	only KOH after 92; From Di Sham.
	Niagara Falls, N.Y.	(1)	293	305	305	305	305	305	305	305	305	305	305	built 1898; From Hooker/IMC
	Tacoma, Wash.	(1)	105	105	52									From Hooker To Pioneer
		(2)	101	101	50									to Pioneer
	Taft, La.	(1)	473	473	473	473	473	473	473	473	473	473	473	built 1966; From Hooker
Oxy Vinyls LP	Deer Park, Tex.	(2)	115	115	146	177	177	177	177	177	177	177	177	41 on hold
		(1)					165	248	248	(248)	(248)	(248)	(248)	
	La Porte, Tex.	(3)					67	100	100	(100)	(100)	(100)	(100)	
		(1)					333	500	500	500	500	500	500	

196

Pioneer	Henderson, Nev.	(1)	138	138	138	138	138	*138*	138	138	138	138	138	built 1942; From Stauffer
	St. Gabriel, La.	(3)	179	179	179	179	179	*179*	179	179	179	179	179	built 1970; From Stauffer
	Tacoma, Wash.	(1)			53	105	105	*105*	105	17	(105)	(105)	(105)	from Oxy
		(2)			51	101	101	*101*	101	16	(101)	(101)	(101)	from Oxy
PPG	Lake Charles, La.	(1)	1000	1000	1000	1000	1000	*1000*	1000	1000	1000	1000	1000	
PPG	Lake Charles, LA	(3)	180	180	180	180	180	*180*	180	180	180	180	180	
	Natrium, WV	(1)	270	270	270	270	270	*270*	270	270	270	270	270	
Sunbelt	McIntosh, AL	(3)	91	91	91	91	91	*91*	91	91	91	91	91	built 1943
		(2)			52	227	227	*227*	227	227	227	227	227	KU Tech.
TIMET	Henderson, NV	(5)	5	5	5	5	5	*5*	5	5	5	5	5	
Vulcan	Geismar, LA	(1)	248	248	248	248	248	*248*	248	248	248	248	248	
	Port Edwards, WI	(3)	45	45	45	45	45	*37*	35	35	35	35	35	built 1967
		(8)	42	42	42	42	42	*50*	53	53	53	53	53	from KCL + Mercury Cel
	Wichita, KS	(1)	165	165	165	165	165	*165*	165	165	165	165	165	built 1952
		(2)	83	83	83	83	83	*83*	83	83	83	83	83	1st U.S. membrane tech
Vulcan C-A	Geismar, LA	(2)						*98*	195	195	195	195	195	Mitsui jv
Westlake	Calvert City, KY	(2)								155	155	155	155	Asahi Chem Tech
		(3)			28	111	111	*111*	111	(111)				From BFG; DeNora Tech

Table 12 (*Continued*)

Company United States	Location	Process	Average Annual Capacities, (-000- Metric Tons)											Remarks
			1995	1996	1997	1998	1999	2000	2001	2002	2003	2004	2005	
	Lake Charles, LA	(2)	160	160	160	160	27	(160)						+360 delayed indefinitely
Weyerhaeuser	Longview, WA	(1)												built 1957
(1) Diaphragm cell Subtotal			9283	9376	9543	9613	9538	9511	9296	8825	8808	8808	8283	
(2) Membrane cell Subtotal			1216	1216	1299	1530	2013	2552	2633	2695	2679	2679	3334	
(3) Mercury cell Subtotal			1321	1321	1321	1337	1267	1176	1124	913	913	913	913	
(4) From HCl Subtotal			82	82	82	82	82	40	80	80	80	80	80	
(5) With metal production Subtotal			355	355	355	180	105	105	105	105	105	105	105	
(7) From KCl+ Membrane Cell Subtotal			41	41	41	41	41	41	41	41	41	41	41	
(8) From KCl+ Mercury Cell Subtotal			275	275	275	275	275	283	286	286	286	286	286	
(9) Other Subtotal			36	45	45	45	45	45	45	10				
TOTAL	United States		12609	12711	12961	13103	13284	13753	13610	12955	12912	12912	13042	

[a] *Note*: Capacities shown in parentheses refer to units that are not operating; however, the equipment is still present, i.e. mothballed.

198

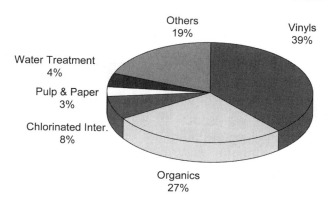

Fig. 35.   Chlorine demand in the United States in 2000. Domestic demand $= 13.2 \times 10^6$ t.
(5,6).

the merchant market. Isocyanate producers have worked to couple capacity and
HCl production with HCl disposal strategies. Fluorocarbon producers have not,
as yet, developed a similar strategy to couple HCl production with its disposal.

Besides its consumption in the vinyls industry, HCl is used in a variety of
industrial applications, ranging from chemical manufacturing to steel production, oil and gas well acidization, and food processing.

Low cost energy and raw materials and well-developed infrastructure make
the United States one of the most attractive locations for chlor-alkali and derivative capacity. The vinyls market represents almost 40% of the domestic chlorine
demand, followed closely by the Others and Organics sectors at 19 and 27%,
respectively (Fig. 35).

Most of the major U.S. vinyls producers are integrated into chlorine or have
access to chlorine through a joint venture partner. Shintech is the notable exception, with chlorine supplied in the form of VCM from Dow. During the past
10 years or so, there has been consolidation in the U.S. vinyls industry, and
the number of PVC producers declined as a result. Many of the remaining producers have strengthened their positions by integrating downstream into compounded and fabricated products. Despite a slowdown in demand growth in
1998–1999 and in late 2000–2001, the United States is, and will continue to
be, a primary supplier of vinyls to the international market. Growth in export
and domestic vinyls demand will increase the chlorine consumption into this
end use by an average of ~4–5% per year through 2005.

In the Organics sector, PO, epichlorohydrin (epi), TDI, MDI, and polycarbonates all will exhibit growth at or above GDP. The total growth for this end use
group is forecast to be 3.8%/year, with PO as the dominant component. Polycarbonates is one of the smaller components but will experience the largest growth
rate. Titanium dioxide is one of the components of the Inorganics market segment, which is included on the graph in the Others category. Chlorine and caustic soda are used in the manufacturing of wood pulp. Because of environmental
pressure, the industry is now moving towards elemental chlorine-free pulp
bleaching, and demand for chlorine from pulp bleaching was expected to be

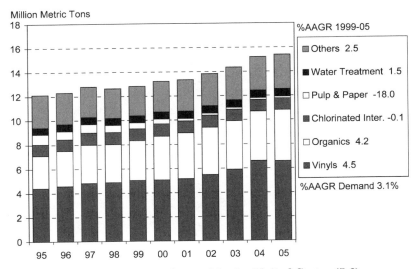

Fig. 36. Chlorine demand in the United States (5,6).

greatly reduced by the end of 2001. Caustic soda, however, will continue to be required in the pulp process.

With U.S. chlorine demand projected to increase faster than chlor-alkali capacity, operating rates will be at a maximum by 2003–2004 (Fig. 36). Additional capacity is expected to be installed to satisfy the anticipated chlorine demand.

The end use pattern for caustic soda (Fig. 37) in the United States closely resembles that for the entire world, where pulp accounts for ~21% of the total

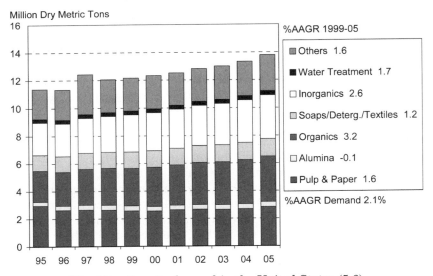

Fig. 37. Caustic demand in the United States (5,6).

United States domestic demand and is forecast to consume a modest 1.6% per year more caustic soda until 2005. Although 68% of the region's pulp capacity is located in the United States, most of the pulp is integrated into paper, in contrast to Canada, where about one-half of the pulp production is sold as market pulp. This means that consumption of caustic soda is impacted by the net trade position of the U.S. paper and paperboard industries. When paper and paperboard imports rise causing domestic paper and paperboard operating rates to decline, pulp producers cannot move the pulp into the market pulp market and pulp operating rates will decline as well.

The alumina sector is a relatively small player in the United States. However, this sector is the destination of more than one-half of the U.S. caustic soda exports. When end use markets of the U.S. caustic soda exports are considered in a total demand analysis, the alumina sector accounts for ~10% of the total U.S. caustic soda production.

The "Others" category includes many miscellaneous consumers, including petroleum refining and caustic soda consumption that can be switched to and from soda ash. There is an estimated 400,000 dry metric tons of caustic soda that could be switched to soda ash for applications such as pulp, sodium silicates, sodium chromates, and soaps/detergents. Caustic soda consumption into these switchable applications has not been anywhere close to the 400,000 dry metric ton level since the early 1990s. Based on Chemical Market Associates, Inc (CMAI) projections of caustic soda and soda ash prices for the next few years, soda ash producers have a larger than usual window of opportunity to secure additional volume from caustic soda. It is surmised that a minimal amount of caustic soda will still be consumed into some switchable applications due to specific circumstances.

Exports from the United States constitute ~12% of the total caustic soda market and will continue to be a key source in supplying caustic soda to the international market. The expected growth in exports of >9%/year will support overall demand growth of 3%/year for caustic soda. The relative net export volume for the United States is illustrated in Figure 38, as the difference between the production and demand bars.

With the start up of new 215,000 dry metric tons per year Vulcan/Mitsui caustic plant during 2000, there is very little additional nameplate capacity scheduled for the future. Formosa has announced a 235,000 dry metric ton membrane unit for Baton Rouge, Louisiana, which is targeted for a 2004 start up. Westlake intends to build a 171,000 dry metric ton membrane unit at Calvert City, Kentucky, replacing its existing 122,000 dry metric ton mercury cell unit.

The domestic requirements for chlorine, to support rising vinyls exports, will require additional chlor-alkali capacity. Higher operating rates and improving margins over the next couple of years will help to justify this additional capacity.

7.3. Chlorine/Caustic Balance. The operating rate of a chlor-alkali unit is determined by the demand for chlorine, which means that the caustic supply is determined by chlorine demand and not caustic demand. In a typical cycle, chlorine demand (and market price) will increase first with improving economic conditions, since chlorine end uses are consumed in long term durable goods such as the construction of infrastructure (PVC pipe, windows, and

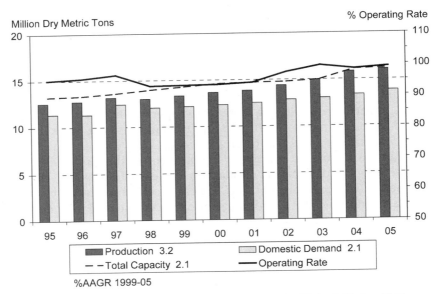

Fig. 38.   Chlorine and caustic demand in the United States (5,6).

doors, polyurethane insulation); ~55% of total chlorine demand is consumed in PVC and polyurethane applications. Caustic soda end uses are more diverse and cause caustic soda demand to react more slowly to improving economic conditions. The resultant impact on caustic soda prices is downward pressure. As chlorine demand rises, prices will rise and more chlor-alkali (chlorine and caustic soda) will be produced. The slower acting caustic soda demand will not be able to consume the additional caustic supply in a timely manner and caustic soda prices will weaken. As caustic soda demand catches up with its supply, the tightening supply/demand balance will provide market support for higher caustic soda prices. The objective of the chlor-alkali producers is to maximize the price of an ECU (electrochemical unit), not just chlorine or just caustic soda prices.

The U.S. production ratio of caustic to chlorine is below the theoretical ratio of 1.1 ton of caustic soda for each one ton of chlorine, because ~4% of the chlorine capacity does not coproduce caustic soda. Most of these units coproduce potassium hydroxide, while some produce a molten metal. The U.S. exports are an important end use for domestically produced caustic soda, and ~12% of the total production is exported.

8. Analytical Methods

Industrial liquid chlorine is routinely analyzed by established procedures (9,13) for moisture, chlorine, other gaseous components, bromine nitrogen trichloride, and mercury. Moisture and nonvolatile matter are determined by evaporation at ambient temperature, followed by gravimetric measurement of the residue (109). Free chlorine levels are estimated quantitatively by thiosulfate titration of the iodine liberated by addition of excess acidified potassium iodide to a gas mixture.

9. Health and Safety Factors

No one should attempt to handle chlorine for any purpose without a thorough understanding of its properties and the hazards involved. Standard safety books and handbooks contain much of this information (110–112). Industry organizations such as The Chlorine Institute and Euro Chlor have active programs to create and maintain a series of pamphlets and manuals that address various issues such as plant operation, piping design and layout, storage system design and practices, transportation, and emergency procedures. Among these, *The Chlorine Manual* (11), which has been approved as an American National Standard, can be recommended as a single source of basic information and listing of important references.

Chlorine is a respiratory irritant and is readily detectable at concentrations <1 ppm in air because of its penetrating odor. Chlorine gas, after several hours of exposure at ∼1 ppm causes mild irritation of the eyes and mucous membranes of the respiratory tract. At high concentrations and in extreme situations, increased difficulty in breathing can result in death through suffocation. The physiological response to various concentrations of chlorine appears in Table 13.

Manufacture of chlorine presents the usual hazards found in the chemical industry and the specific hazard of live electrical equipment at modest to high voltage in the elecrrolysis area. Standard safety gear includes electrically resistant boots and gloves for those working in the electrolysis area and escape respirators for all personnel and visitors in the chlorine production area. Other specialized safety equipment is necessary for various operations, and this is discussed in the appropriate manuals.

A standard feature on chlorine plants is a scrubbing system containing an alkaline solution (usually NaOH). This removes chlorine from vent streams, whether continuously flowing or present only in emergency situations. The low pressure side of the chlorine process usually is protected by water seals or mechanical relief devices, depending on the operating pressure. Any vent

Table 13. **Physiological Response to Chlorine**

Parameter	Parts of chlorine per million parts of air (volume)
least amount required to produce slight symptoms after several hours of exposure	1
least detectable odor	3.5
maximum amount that can be inhaled for 1 h without serious physiological response	4
least amount required for throat irritation	15
least amount required to induce coughing	30
amount causing severe symptoms in 30–60 min	40–60
LD_{50}	
humans, in 30 min	840
rats, in 60 min	290
mice, in 60 min	137
amount expected to affect aquatic life	<0.1

gas from these devices flows to the scrubbing system before discharge to the atmosphere.

10. Uses

Figure 39 depicts the chlorine demand for various end-uses (see also Tables 3 and 4).

Phosgene is made by reacting carbon monoxide with chlorine in the presence of activated carbon. It is the starting material for the manufacture of polycarbonates, TDI, and MDI. Polycarbonate (PC) resins are engineering thermoplastics that are produced by reacting phosgene and bisphenol A (BPA). The PC resins are typically characterized by high impact strength, dimensional stability, transparency, and excellent electrical properties. In addition to general purpose polycarbonate resins, a variety of specialty materials, blends, and alloys are also available. Polycarbonates are widely used, with applications in the automotive industry, glazing, electronics, computers and business machines, and for software, audio and video compact discs (CDs, DVDs).

TDI is a major raw material for producing urethane foam, both flexible and rigid. Flexible urethane foams, by far the largest TDI end use, are used as bedding, furniture cushioning, carpet underlay, and packaging. MDI is used almost exclusively to make polyurethanes. MDI-based rigid urethane foams have the lowest thermal conductivity of any common insulation material. The largest insulation uses for these foams are in construction, refrigerators and freezers, and refrigerated rail cars and trucks. Rigid foams also have excellent buoyancy, and have been used in life-saving gear, swimming pools, sporting goods, and other flotation devices.

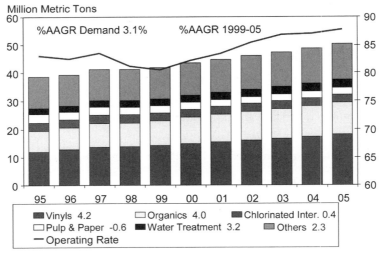

Fig. 39. World chlorine demand (5,6) (% AAGR refers to % average aggregate growth rate).

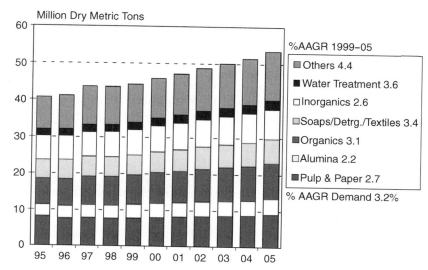

Fig. 40. World caustic demand (5,6) (% AAGR refers to % average aggregate growth rate).

Most epoxy resins are made from BPA and epi [106-89-8]. The chlorination of propylene produces allyl chloride which, when dehydrochlorinated with caustic soda, produces epi. Lime can also be used in this process, but disposal of the calcium chloride by-product creates strong opposition from environmentalist groups. Epoxy resins, used in making protective coatings, laminates, and fiber-reinforced composites, are the largest end use for epi. The second largest end use is synthetic glycerin, produced by the hydrolysis of epi with caustic soda. Production of synthetic glycerin has declined in recent years because of competition from natural material. Glycerin is used in the pharmaceutical, tobacco, cosmetics, and food and beverage industries.

Chlorine and propylene are used in the chlorohydrin process to produce propylene oxide (PO) [75-56-9]. In the chlorohydrin process, propylene reacts with chlorine to make propylene chlorohydrin, which is then dehydrochlorinated with caustic soda to make propylene oxide. An alternative process competing with this route is the direct oxidation of propylene. PO is a highly reactive alkyl epoxide used principally as a chemical intermediate. The largest application for PO is as a raw material for polyether polyols. Polyether polyols are used in polyurethane applications including flexible and rigid foams, elastomers, coatings, adhesives, and reaction injection molding polymers. PO can also be reacted with water to form propylene glycol, a raw material for unsaturated polyester resins. A few applications of unsaturated polyester resins include tubs and showers, gasoline tanks, and boat hulls. Dow is the largest producer of PO using the chlorohydrin process.

Actual declines in the use of chlorine in chlorinated intermediates (chloromethanes and chloroethanes) and pulp and paper are expected because of environmental concerns in these applications.

Chlor-alkali plants are built and operated to satisfy chlorine demand, not caustic demand. Chlorine demand is much more volatile than caustic demand

and reacts more quickly to changes in economic conditions. Over time, it is expected that both chlorine and caustic soda demand will increase by the same average annual rate and both trend with GDP.

Compared to chlorine, the market for caustic soda adapts more slowly to changing economic conditions, because of the diversity of its end uses (Fig. 40). Consumption of caustic soda in pulp mills is the largest single end use at 17%, which is about one-half of the size of the vinyl segment for chlorine consumption.

Alumina is the second largest end use for caustic soda at 8%, and demand for alumina is driven by aluminum metal. About one-third of aluminum is consumed by the transportation sector and ~15% by the construction sector, both of which are dependent on global economic health. The organics category is a fairly large segment and is a collection of many organic chemicals, including PO, epi, TDI and MDI (isocyanates), and polycarbonates.

11. Acknowledgments

The authors would like to thank Manfred Quissek of Q Group Inc. for providing us with the latest information on the transformers and rectifiers, Rick Romine of ELTECH systems for updating us on the developments in the diaphragm cell technologies, and K. Sambamurthy of Krupp-Uhde for encouragement during the course of the manuscript preparation and for providing clarifications on the current developments in the chlor-alkali technologies.

BIBLIOGRAPHY

"Alkali and Chlorine Industries" in *ECT* 1st ed., Vol. 1, pp. 358–430, by Z. G. Deutsch, Deutsch and Loonam; "Alkali and Chlorine Industries" in *ECT* 2nd ed., Vol. 1, pp. 668–758, by Z. G. Deutsch, Consulting Engineer, and C. C. Brumbaugh and F. H. Rockwell, Diamond Alkali Company; "Alkali and Chlorine Products, Chlorine and Sodium Hydroxide" in *ECT* 3rd ed., Vol. 1, pp. 799–865, by J. J. Leddy, I. C. Jones, Jr., B. S. Lowry, F. W. Spillers, R. E. Wing, and C. D. Binger, Dow Chemical U.S.A.; "Chlorine" in *ECT* 1st ed., Vol. 3, pp. 677–681, by D. G. Nicholson, University of Pittsburgh; "Chlorine" in *ECT* 2nd ed., Vol. 5, pp. 1–6, by D. G. Nicholson, East Tennessee State University; "Chlorine" in *ECT* 2nd ed., Supplement, pp. 167–177, by H. B. Hass, Chemical Consultant; "Alkali and Chlorine Products, Chlorine and Sodium Hydroxide" in *ECT* (online), posting date: December 4, 2000, by L. Calvert Curlin, OxyTech Systems, Inc., Tilak V. Bommaraju, and Constance B. Hansson, Occidental Chemical Corporation.

CITED PUBLICATIONS

1. *Chlor-alkali Consulting Service* (includes the monthly Chlor-alkali Market Report), Chemical Market Associates, Inc, Houston, Tex.
2. *World Chlor-alkali Analysis*, 1997, Chemical Market Associates, Inc, Houston, Tex.
3. *World Chlor-alkali Analysis*, 1998, Chemical Market Associates, Inc, Houston, Tex.
4. *World Chlor-alkali Analysis*, 1999, Chemical Market Associates, Inc, Houston, Tex.

5. *World Chlor-alkali Analysis*, 2000, Chemical Market Associates, Inc, Houston, Tex.

6. *World Chlor-alkali Analysis*, 2001, Chemical Market Associates, Inc, Houston, Tex.

7. S. Berthiaume, E. Anderson, and Y. Yoshida, *CEH Marketing Repoprt, Chlorine/Sodium Hydroxide, 733.1000A*, SRI International, Menlo Park, Calif., 2000.

8. R. Kapoor and J. J. Martin, *Thermodynamic Properties of Chlorine*, University of Michigan, Ann Arbor, Mich., 1957.

9. J. J. Martin and D. M. Longpre, *J. Chem. Eng. Data* **29**, 466 (1984).

10. *Properties of Chlorine in SI Units*, 2d ed., The Chlorine Institute, Washington, D.C., 1986.

11. Chlorine Institute, *The Chlorine Manual*, 6th ed., The Chlorine Institute, Washington, D.C., 1990.

12. CRA No, 188.04, *Assessment of the Economic Benefits of Chlor-Alkali Chemicals to the United States and Canadian Economies*, Charles River Associates, Boston, 1993.

13. *Chlorine Handbook*, Diamond Shamrock Chemical Co, Irving, Tex., 1984.

14. D. F. W. Hardie, *Electrolytic Manufacture of Chemicals from Salt*, The Chlorine Institute Inc, New York, 1975.

15. Brit. Pat. 1,147,442 (May 12, 1965), H. B. Beer (to Chemnor Corp.); 1,195,871 (Feb. 10, 1967), H. B. Beer (to ChemnorAG).

16. R. Kotz and S. Stucki, *Electrochim Acta* **31**, 1331 (1986).

17. T. Morimoto, T. Matsubara, and S. Ohashi, *Denki Kagaku* **60**, 649 (1992).

18. S. Trasatti, ed, *Electrodes of Conductive Metallic Oxides*, Parts A and B, Elsevier, Amsterdam, The Netherlands, 1980, 1981.

19. D. M. Novak, B. V. Tilak, and B. E. Conway, in J.O'M. Bockris, B. E. Conway, and R. E. White, eds, *Modern Aspects of Electrochemistry*, Vol. 14, Plenum Press, New York, 1982.

20. T. V. Bommaraju, C.-P. Chen and V. I. Birss, in J. Moorhouse, ed, *Modern Chlor-Alkali Technology*, Vol. 8, Blackwell Science, London, 2001 p. 57.

21. B. V. Tilak, V. I. Birss, J. Wang C.-P. Chen, and S. K. Rangarajan, *J. Electrochem. Soc.* **148**, **D109** (2001).

22. U.S. Pat. 5,948,222 (Sept. 7, 1999), C.-P. Chen and T. V. Bommaraju (to Occidental Chemical Corp.).

23. U.S. Pat. 6,156,185 (Dec. 5, 2000), C.-P. Chen and T. V. Bommaraju (to Occidental Chemical Corp).

24. J. E. Currey and G. G. Pumplin in J. J. MeKetta and W. A. Cunningham, eds., *Encyclopedia of Chemical Engineering and Design*, Vol. 7, Marcel Dekker, Inc, New York, 1978, p. 305.

25. P. Schmittinger, T. Florkiewicz, L. C. Curlin, B. Lüke, R. Scanell, T. Navin, E. Zelfel, and R. Bartsch in *Ullmann's Encyclopedia of Industrial Chemistry*, 6th ed., Wiley-VCH Verlag GmbH, Weinheim, 1999, p. 1.

26. U.S. Pat. 4,410,411 (Oct. 18, 1983), R. W. Fenn, III, E. J. Pless, R. L. Harris, and K. J. O'Leary (to OxyTech Systems, Inc.).

27. U.S. Pat. 3,674, 676 (July 4, 1972), E. J. Fogelman (to OxyTech Systems, Inc.).

28. U.S. Pat. 3,928,166 (Dec. 23, 1975), K. J. O'Leary, C. P. Tomba, and R. W. Fenn, III (to OxyTech Systems, Inc.).

29. T. Florkiewicz, "Diaphragm Cell Improvements", Chlorine Institute meeting, New Orleans, La., March 14, 2001.

30. P. C. Foller, D. W. DuBois, and J. Hutchins, in S. Sealy, ed., "*Modern Chlor-Alkali Technology*", Vol. 7, p.162, Society of Chemical Industry, London, 1998.

31. F. Kuntzburger, D. Horbez, J. G. LeHelloco, and J. M. Perineau, in S. Sealy, ed., *Modern Chlor-Alkali Technology*, Vol. 7, p. 181, Society of Chemical Industry, London, (1998).

32. L. C. Curlin, T. V. Bommaraju, and C. B. Hansson, *Kirk-Othmer Encyclopedia of Chemical Technology*, 4th ed., Vol. 1, John Wiley & Sons, Inc., New York, 1991, p. 938.

33. *Diaphragm Cells*, OxyTech Systems, Inc, Chardon, Ohio, 1988.

34. U.S. Pat. 4,834,859 (May 30, 1989), L. C. Curlin and R. L. Romine (to OxyTech Systems, Inc.).

35. U.S. Pat. 3,591,483 (July 6, 1971), R. E. Loftfield and H. W. Laub (to OxyTech Systems, Inc.).

36. R. L. Romine and R. Matousek, *New and Improved Cell Hardware Designs*, a paper presented at the Electrode Corporation Chlorine/Chlorate Seminar, Sept. 15, 1998.

37. W. G. Grot, *Chem. Ing. Tech.* **44**, (1972) 167.

38. W. G. Grot, *Discovery and Development* of Nafion *Perfluorinated Membranes*, The Castner Medal Lecture, London International Chlorine Symposium, June, 1985.

39. J. L. Hurst, *Implementing Membrane Cell Technology within OxyChem Manufacturing*, International Symposium on Chlor-Alkali Industry, Tokyo, Japan, April, 1986.

40. *Perfluorinated Membranes* for the *Chlor-Alkali Industry*, E. I. du Pont de Nemours & Co., Inc., Wilmington, Del., 1983.

41. D. L. Peet and J. H. Austin, *Nafion Perfluorinated Membranes, Operation in Chlor-Alkali Plants*, Chlorine Institute Plant Managers Seminar, The Chlorine Institute, Tampa, Fla., Feb. 1986.

42. *Nafion 90209, Product Bulletin*, E. I. du Pont de Nemours & Co., Inc., Wilmington, Del., 1988.

43. *Nafion NX-961, Product Bulletin*, E. I. du Pont de Nemours & Co., Inc., Wilmington, Del., 1984.

44. D. L. Peet, *Membrane Durability in Chlor-Alkali Plants*, Electrochemical Society Meeting, Honolulu, Hawaii, Oct. *1987; Proceedings of the Symposium on Electrochemical Engineering in the Chlor-Alkali and Chlorine Industries*, PV. No. 88-2, 1988, pp. 329–336.

45. U.S. Pat. 4,025,405 (1972), R. L. Dotson and K. J. O'Leary (to OxyTech Systems, Inc.).

46. H. Ukihashi, M. Yamabe, and H. Miyake, *Prog. Polym. Sci.* **12**, 229 (1986).

47. *Nafion Perfluorinated Membranes for KOH Production, Nafion Product Bulletin*, E. I. du Pont de Nemours & Co., Inc., Wilmington, Del., 1988.

48. *Nafion Perfluorinated Membranes, Introduction*, E.I. du Pont de Nemours & Co., Inc., Wilmington, Del., 1987.

49. Asahi Chemical Membrane Chlor-Alkali, Asahi Chemical Industry Co., Ltd., Tokyo, Japan, 1987.

50. T. D. Gierke and W. Y. Hsu, *Perfluorinated Ionomer Membranes*, A. Eisenberg and H. L. Yeager, eds. ACS Symposium Series, Washington, D.C., Vol. 180, 1982, p. 283.

51. D. Bergner, M. Hartmann, and R. Staab, *Chem.-Ing.-Tech.* **66**(6) 783 (1994).

52. N. M. Prout and J. S. Moorhouse, *Modern Chlor-Alkali Technology*, Vol. 4, Society of Chemical Industry, Elsevier Applied Science, London, 1990.

53. T. A. Davis, J. D. Genders, and D. Pletcher, *A First Course in Ion Permeable Membranes*, The Electrochemical Consultancy, Hants, U.K., 1997.

54. B. V. Tilak, A. C. R. Murthy, and B. E. Conway, *Proc. Indian Acad. Sci. (Chem. Sci.)* **97**, 359 (1986).

55. D. L. Caldwell, in J. O'M. Bockris, B. E. Conway, E. A. Yeager, and R. E. White, eds., *Comprehensive Treatise of Electrochemistry*, Vol. 2, Plenum Press, New York, 1981, Chapt. 2.

56. F. Hine, B. V. Tilak, and K. Viswanathan in R. E. White, J. O'M. Bockris, and B. E. Conway, eds., *Modern Aspects of Electrochemistry*, Vol. 18, Plenum Press, New York, 1986, Chapt. 5.

57. F. Hine, M. Yasuda, and M. Watanabe, *Denki Kagaku* **47**, 401 (1979).

58. U.S. Pat. 4, 024, 044 (May 17, 1977), J. R. Brannan and I. Malkin (to OxyTech Systems, Inc.).

59. Eur. Pat. Appl. 129, 734 (Jan. 2, 1985), N. R. Beaver, L. E. Alexander, and C. E. Byrd (to the Dow Chemical Co.); Eur. Pat. Appl. 129, 347 (Dec. 27, 1984), J. F. Cairns, D. A. Fenton, and P. A. Izard (to Imperial Chemical Industry PLC).

60. *CME Chlorine Engineers Membrane Electrolyzer*, Chlorine Engineers Corp. Ltd., Tokyo, Japan, 1989.

61. M. Esayian and J. H. Austin, *Membrane Technology for Existing Chlor-Alkali Plants*, presented at The Chlorine Institute, Feb. 1984.

62. *Japan's Chlor-Alkali Producers Save Energy by Retrofiting Diaphragm Cells* (Case History), E. I. du Pont de Nemours & Co., Inc., Wilmington, Del.

63. *FM-21 SP Series Membrane Electrolyzer*, ICI Winnington, Northwich, Cheshire, England, 1989.

64. *MGC Monopolar Membrane Electrolyzer*, OxyTech Systems, Inc., Chardon, Ohio, 1988.

65. *Alkaline Chloride Electrolysis by the Membrane Process*, Krupp Uhde GmbH, Dortmund, Germany, 2001.

66. L. J. Gestaut, T. M. Clere, C. E. Graham, and W. R. Bennett, *Abstract No. 393*; L. J. Gestaut, T. M. Clere, A. J. Niksa, and C. E. Graha, *Abstract No. 124*, Electrochem. Soc. Meeting, Washington, D. C., Oct. 1983.

67. F. Gestermann and A. Ottaviani, *Modern Chlor-Alkali Technol.* **8**, 49 (2001).

68. K. Schneiders; A. Zimmermann; G. Henßen; *Membranelektrolyse—Innovation für die Chlor-Alkali-Industrie*; Forum Thyssen Krupp, 2/2001.

69. DE 19715429 A1, Bayer AG (Erfinder: Gestermann u.a., 15.10.1998, "*Elektrochemische Halbzelle*".

70. EP–Pat. 0150017–1984, EP 0150018–1985; A. G. Hoechst, K.-H. Tetzlaff, D. Schmidt, and J. Russow.

71. K.-H. Tetzlaff and W. Wendel; *Elektrochemische Zellen als Fallfilmapparat*; *Chem.-Ing.- Tech.* 60 (1988) Nr. 7, S. 563.

72. Krupp Uhde GmbH, Dortmund; 2001.

73. M. T. Halbouty, *Salt Domes, Gulf Region, United States and Mexico*, p. 125, Gulf Publishing Co., Houston, Tex., 1967.

74. D. W. Kaufmann, Chapter 14, *Sodium Chloride*, ACS Monograph 145, Reinhold Publishing Co., New York, 1960.

75. E. A. Cotton and G. Wilkinson, *Advanced Inorganic Chemistry—A Comprehensive Text*, 4th ed., John Wiley & Sons, Inc., New York, 1980, pp. 227 and 546.

76. J. A. Walkier, *Modern Chlor-Alkali Technology*, Ed. T. C. Wellington, Elsevier Appl. Science, London, Vol. 5, p. 233, 1992.

77. P. C. Westen, 'How to Use Steel and Titanium Safely,' *Proceedings, Third Euro Technical Chlor Seminar*, Paris, 1993. Reprinted in GEST 93/192.

78. J. S. Grauman and B. Willey, *Chem. Eng.* **105**(8), 106 (1998).

79. T. F. O'Brien and I. F. White, in R. C. Curry, ed., *Modern Chlor-Alkali Technology*, Royal Society of Chemistry, Cambridge, Vol. 6, p. 70, 1995.

80. T. F. O'Brien and I. F. White, S. Sealy, ed., *Modern Chlor-Alkali Technology*, Royal Society of Chemistry, Cambridge, Vol. 7, p. 202, 1998.

81. OxyTech Systems, Inc., *Chlorine Recovery Process*, Chardon, Ohio, (1988).

82. *Personal communication*, R. E. Shamel, Consulting Resources Corp. (2002).

83. K. Haas, *Electrochem. Techn.* **5**, 246 (1967).

84. K. Schneiders and C. Herwig, *Recycle of HCl to Chlorine*, 38th Chlorine Institute Plant Operations Seminar, Houston, Tex., 1995.

85. D. J. Eames and J. Newman, *J. Electrochem. Soc.* **142**, 3619 (1995).

86. C. W. Arnold and K. A. Kolbe, *Chem. Eng. Prog.* **48**, 167 (1971).

87. L. E. Bostwick and C. P. van Dijk, *Hydrogen Chloride to Chlorine: The Kel-Chlor Process*, 20th Chlorine Institute Plant Managers Seminar, New Orleans, La., 1977.

88. The M. W. Kellogg Co., *Hydrocarbon Process.* **60**(11), 143 (1981).

89. H. Y. Pan, R. G. Minet, S. W. Benson, and T. T. Tsotsis, *Ind. Eng. Chem. Res. 1994,* **33**, 2996.

90. U.S. Pat. 4,994,256, Feb. 1991, R. G. Minet, S. W. Benson, and T. T. Tsotsis.

91. M. Doyle and P. Arora, *J. Electrochem. Soc.* **148**, k1-k41 (2001).

92. *Chemical Marketing Reporter*, Sept. 4, 37 (2000); Sept. 11, 41 (2000).

93. http://www.eia.doe.gov/emeu/cabs/usa.html (2001).

94. C.-P. Chen, B. V. Tilak, and J. W. Quigley, *J. Appl. Electrochem.* **25**, 95 (1995).

95. C.-P. Chen and B. V. Tilak, *J. Appl. Electrochem.* **29**, 1300 (1997).

96. I. Rousar, V. Cezner, M. Vender, M. Kroutil, and J. Vachuda, *Chem. Prum.* **17/42**, 466 (1967).

97. W. Y. Hsu, J. R. Barkley, and P. Meakin, *Macromolecules* **13**, 198 (1980).

98. P. C. Rieke and N. Vanderborgh, *J. Membrane Sci.* **32**, 313 (1987).

99. H. Ukihashi, M. Yamabe, and H. Miyake, *Prog. Polym. Sci.* **12**, 229 (1986).

100. M. Quissek, *Personal Communication*, (2001).

101. GEST 79/81, Euro Chlor.

102. Pamphlets 6 and 60, The Chlorine Institute.

103. R. C. Talbot, *FRP Usage in the Chlorine Industry*, Bulletin No. 1704, Ashland Chemical Co., Columbus, Ohio.

104. K. Hannesen "*Materials of Construction for Handling Chlorine,*" Chlorine Safety Seminar, Brussels, 1990.

105. Pamphlet 95, Appendix A, The Chlorine Institute, Washington, D.C., 1997

106. GEST 73/20, Euro Chlor.

107. Y. C. Chen, *Process Economics Program Report*, Stanford Research Institute, Menlo Park, Calif., No. 61, 1970; No. 61A, 1974; No. 61B, 1978; No. 61C, 1982.

108. Abam Engineers, *Final Report on Process Engineering and Economic Evaluations of Diaphragm and Membrane Chloride Cell Technologies*, ANL/OEPM-80-9, Argonne National Laboratories, Argonne, Ill., Dec. 1980.

109. *ASTM Standard Test Method*: E410-86.

110. F. Lees, *Loss Prevention in the Process Industries*, 2 vols., Butterworths, London, 1980.

111. Compressed Gas Association, Inc., *Handbook of Compressed Gases*, 3rd ed., Van Nostrand Reinhold, New York, 1990.

112. R. J. Lewis, Sr., *Sax's Dangerous Properties of Industrial Materials*, 8th ed., Van Nostrand Reinhold, New York, 1992.

GENERAL REFERENCES

D. W. Kaufmann, *Sodium Chloride*, ACS Monograph 145, Reinhold Publishing Co., New York, 1960.

J. S. Sconce, *Chlorine, Its Manufacture, Properties, and Uses*, Reinhold Publishing Corp., New York, 1962.

M. O. Coulter, ed., *Modern Chlor-Alkali Technology*, Ellis Horwood, London, 1980.

C. Jackson, ed., *Modern Chlor-Alkali Technology*, Vol. 2, Ellis Horwood, Chichester, 1983.

K. Wall, ed., *Modern Chlor-Alkali Technology*, Vol. 3, Ellis Horwood, Chichester, 1986.

N. M. Prout and J. S. Moorhouse, eds., *Modern Chlor-Alkali Technology*, Vol. 4, Elsevier Applied Science, London, 1990.

T. C. Wellington, ed., *Modern Chlor-Alkali Technology*, Vol. 5, Elsevier Applied Science, London, 1992.

R. W. Curry, ed., *Modern Chlor-Alkali Technology*, Vol. 6, Royal Society of Chemistry, Cambridge, 1995.

S. Sealey, ed., *Modern Chlor-Alkali Technology*, Vol. 7, Royal Society of Chemistry, Cambridge, 1998.

J. Moorhouse, ed., *Modern Chlor-Alkali Technology*, Vol. 8, Blackwell Science, Oxford, 2001.

W. Grot, in J. I. Kroschwitz, ed., *Encyclopedia of Polymer Science and Engineering*, 2nd ed., Vol. 16, Wiley-Interscience, New York, 1989.

P. Schmittinger, ed., *Chlorine—Principles and Industrial Practice*, Wiley-VCH, Weinheim, 2000.

F. Hine, *Electrochemical Processes and Electrochemical Engineering*, Plenum Press, New York, 1985.

Diaphragm Cells for Chlorine Production, Society of Chemical Industry, London, 1977.

F. Hine, R. E. White, W. B. Darlington, and R. D. Varjian, eds., *Proceedings of the Symposium on Electrochemical Engineering in the Chlor-Alkali and Chlorate Industries*, Proc. Vol. 88-2, The Electrochemical Society Inc., Pennington, N.J., 1988.

F. Hine, B. V. Tilak, J. M. Fenton, and J. D. Lisius, eds., *Proceedings of the Symposium On Performance of Electrodes for Industrial Electrochemical Processes*, Proc. Vol. 89-15, The Electrochemical Society Inc., Pennington, N.J., 1989.

Chlor-Alkali and Chlorate Technology: R. B. Macmullin Memorial Symposium, eds., H. S. Burney, N. Furuya, F. Hine and K.-I. Ota, Proc. Vol. 99-21, The Electrochemical Society Inc., Pennington, N.J., 1999.

TILAK V. BOMMARAJU
Process Technology Optimization, Inc

BENNO LÜKE
Glen Dammann, Krupp Uhde GmbH

THOMAS F. O'BRIEN
Media, PA

MARY C. BLACKBURN
Chemical Market Associates, Inc.

CHLOROBENZENES

1. Introduction

The chlorination of benzene can theoretically produce 12 different chlorobenzenes. With the exception of 1,3-dichlorobenzene, 1,3,5-trichlorobenzene, and 1,2,3,5-tetrachlorobenzene, all of the compounds are produced readily by chlorinating benzene in the presence of a Friedel-Crafts catalyst (see FRIEDEL-CRAFTS

REACTIONS). The usual catalyst is ferric chloride either as such or generated *in situ* by exposing a large surface of iron to the liquid being chlorinated. With the exception of hexachlorobenzene, each compound can be further chlorinated; therefore, the finished product is always a mixture of chlorobenzenes. Refined products are obtained by distillation and crystallization.

Chlorobenzenes were first synthesized around the middle of the nineteenth century; the first direct chlorination of benzene was reported in 1905 (1). Commercial production was begun in 1909 by the former United Alkali Co. in England (2). In 1915, the Hooker Electrochemical Co. at Niagara Falls, New York, brought on stream its first chlorobenzenes plant in the United States with a capacity of about 8200 metric tons per year.

The Dow Chemical Company started production of chlorobenzenes in 1915 (3). Chlorobenzene was the first and remained the dominant commercial product for over 50 years with large quantities being used during World War I to produce the military explosive picric acid [88-89-1].

The Dow Chemical Company in the mid-1920s developed two processes which consumed large quantities of chlorobenzene. In one process, chlorobenzene was hydrolyzed with ammonium hydroxide in the presence of a copper catalyst to produce aniline [62-53-3]. This process was used for more than 30 years. The other process hydrolyzed chlorobenzene with sodium hydroxide under high temperature and pressure conditions (4,5) to product phenol [108-95-2]. The I.G. Farbenwerke in Germany independently developed an equivalent process and plants were built in several European countries after World War II. The ICI plant in England operated until its closing in 1965.

In the 1930s, the Raschig Co. in Germany developed a different chlorobenzene-phenol process in which steam with a calcium phosphate catalyst was used to hydrolyze chlorobenzene to produce phenol (qv) and HCl (6). The recovered HCl reacts with air and benzene over a copper catalyst (Deacon Catalyst) to produce chlorobenzene and water (7,8). In the United States, a similar process was developed by the Bakelite Division of Union Carbide Corp., which operated for many years. The Durez Co. licensed the Raschig process and built a plant in the United States which was later taken over by the Hooker Chemical Corp. who made significant process improvements.

Although Dow's phenol process utilized hydrolysis of the chlorobenzene, a reaction studied extensively (9,10), phenol production from cumene (qv) became the dominant process, and the chlorobenzene hydrolysis processes were discontinued.

With the discontinuation of some herbicides, eg, 2,4,5-trichlorophenol [39399-44-5], based on the higher chlorinated benzenes, and DDT, based on monochlorobenzene, both for ecological reasons, the production of chlorinated benzenes has been reduced to just three with large-volume applications of (mono)chlorobenzene, *o*-dichlorobenzene, and *p*-dichlorobenzene. Monochlorobenzene remains a large-volume product, considerably larger than the other chlorobenzenes, in spite of the reduction demanded by the discontinuation of DDT. But more recent market developments in the 1990's have significantly affected the outlook for MCB and dichlorobenzenes (See the Section titled Economic Aspects).

2. Physical and Chemical Properties

The important physical properties of chlorobenzenes appear in Table 1. Only limited information is available for some chlorobenzenes:

Chlorobenzene	CAS Registry Number	Mol wt	Melting point, °C	Normal boiling point, °C
1,2,3,5-tetrachloro	[634-90-2]	215.9	51	246
pentachloro	[608-93-5]	250.35	85	276

Vapor pressure as a function of the temperature is correlated by the Antoine equation:

$$\log_{10} P(\text{kPa}) = A - B/(T + C) - 0.875097$$
$$(\log_{10} P(\text{mm Hg})) = A - B/(T + C)$$

(1)

where T is the temperature in °C, and A, B, C are the Antoine constants (Table 2).

Nitration of chlorobenzenes, mostly monochlorobenzene in the United States, with nitric acid has wide industrial applications.

$$Cl_n C_6 H_{6-n} + HNO_3 \longrightarrow Cl_n C_6 H_{5-n} NO_2 + H_2O$$

$$n = 1 \text{ to } 5$$

Nitrated monochlorobenzene is used as a building block to produce many other products. There is also some commercial nitration of o-dichlorobenzene in the United States and Western Europe.

3. Manufacture

The production of any chlorinated benzene is a multiple-product operation. Plants for chlorobenzene must produce HCl and some other chlorinated benzenes. Only limited control can be exercised over the product ratios. Chlorinated benzenes can be produced by the vapor-phase chlorination of benzene using air and HCl as chlorinating agents. This was the first stage of the Raschig phenol process (7,8). The energy costs are so high, this process could never have been considered in the past to commercially produce chlorobenzenes as main products. Chlorine and benzene react in the vapor phase at 400 to 500°C to give a different distribution of products (12), but such a process is much more costly than conventional liquid-phase operations.

All of the chlorobenzenes are now produced by chlorination of benzene in the liquid phase. Ferric chloride is the most common catalyst. Although precautions are taken to keep water out of the system, it is possible that the $FeCl_3 \cdot H_2O$ complex catalyst is present in most operations owing to traces of moisture in benzene entering the reactor. This $FeCl_3 \cdot H_2O$ complex is probably the most effective catalyst (13).

Table 1. Physical and Thermodynamic Properties of Chlorobenzenes

	Chlorobenzene	1,2-Dichlorobenzene	1,3-Dichlorobenzene	1,4-Dichlorobenzene	1,2,4-Trichlorobenzene	1,2,3,4-Tetrachlorobenzene	1,2,4,5-Tetrachlorobenzene	Hexachlorobenzene
CAS Registry Number	[108-90-7]	[95-50-1]	[541-73-1]	[106-46-7]	[120-82-1]	[634-66-2]	[95-94-2]	[1118-74-1]
mol wt	112.56	147.005	147.005	147.005	181.45	215.90	215.9	284.80
mp, °C	-45.34	-16.97	-24.76	53.04	17.15	46.0	139.5	228.7
bp at 101.3 kPa[a], °C	131.7	180.4	173.0	174.1	213.8	254.9	248.0	319.3
critical temperature, °C	359.2	417.2	415.3	407.5	453.3	450	489.8	551
critical pressure, kPa[a]	4519	4031	4864	4109	3718	3380	3380	2847
critical density, kg/L	0.3655	0.411	0.458	0.411	0.447	0.40	0.475	0.518
liquid density, kg/L	1.10118	1.3022	1.2828	1.2475	1.44829	1.70	1.833(s)	1.596
viscosity, mPa·s (= cP)	0.756	1.3018	1.0254	1.188	1.008	3.37	1.142	
heat capacity for liquid, J/g[b]	1.339	1.159				1.259		
heat of fusion, J/g[b]	90.33	86.11	85.98	123.8	85.78	64.52	112.2	89.62
heat of vaporization, J/g[b]	331.1	311.0	296.8	297.4	280.0	268.9	221.8	190.8
flash point[c], °C	28	71		67	99		none	
standard heat of formation of liquid[d], J/g[b]	-95.90	-125.23	-145.73	-284.6(s)	-263.1			-460(s)
standard entropy of formation, J/mol/K	197.5 (liq)			175.4 (cryst)				
Thermal conductivity of liquid, W/(m·K)	0.127	0.121		0.105		0.108		
refractive index of liquid, n_D^{25}	1.5219	1.5492	1.54337	1.52849 (55°C)		1.56933		
dielectric constant of liquid	5.621	9.93	5.04	2.41		2.24		
surface tension, mN/m (= dyn/cm)	32.65	36.61	36.20	31.4		38.54		21.6
heat of combustion (25°C), kJ/mol	-3100	-2962	-2955	-2934				
coefficient of expansion of liquid, K^{-1}	95×10^{-5}	110×10^{-5}	111×10^{-5}	116×10^{-5}				
ignition temperature, °C	590	640	>500	>500	>500	>500	>500	

[a] To convert kPa to mm Hg, multiply by 7.5.
[b] To convert J to cal, divide by 4.184.
[c] ASTM method D56-70, closed cup.
[d] Ref. 11.

Table 2. **Antoine Constants**[a] **for Chlorobenzenes**

Chlorobenzene	A	B	C
monochlorobenzene	7.046324	1482.156	224.115
1,2-dichlorobenzene	7.143024	1703.916	219.352
1,3-dichlorobenzene	7.072644	1629.811	215.821
1,4-dichlorobenzene	7.002424	1578.149	208.84
1,2,4-trichlorobenzene	7.136684	1790.267	206.283
1,2,3,4-tetrachlorobenzene	7.159274	1930.023	196.213
1,2,4,5-tetrachlorobenzene	7.284164	2003.495	207.038
hexachlorobenzene	6.66747	1654.17	117.536

[a] See equation 1; 1 kPa = 7.5 mm Hg; log kPa = log mm Hg − 0.875097.

The liquid-phase chlorination of benzene is an ideal example of a set of sequential reactions with varying rates from the single-chlorinated molecule to the completely chlorinated molecule containing six chlorines. Classical papers have modeled the chlorination of benzene through the dichlorobenzenes (14,15). A reactor system may be simulated with the relative rate equations and flow equation. The batch reactor gives the minimum ratio of $(C_6H_{5-n}Cl_{n+1}: (C_6H_{6-n})Cl_n$. This can be approximated by either a plug flow reactor or a multistage stirred reactor. A single-stage, stirred reactor will produce the highest $(C_6H_{5-n})Cl_{n+1}:(C_6H_{6-n})Cl_n$ ratio. If chlorobenzene (mono) is the desired product, control over the dichlorobenzenes to chlorobenzene ratio is effected primarily by controlling the extent of chlorination. The low di:mono ratio is obtained at the expense of energy used in recycling the unreacted benzene.

In the liquid-phase chlorination, 1,3-dichlorobenzene is found only in a small quantity, and 1,3,5-trichlorobenzene and 1,2,3,5-tetrachlorobenzene are undetectable. The ratios of 1,4- to 1,2-dichlorobenzene with various catalysts are shown in Table 3. Iodine plus antimony trichloride is effective in selectively chlorinating 1,2,4-trichlorobenzene to 1,2,4,5-tetrachlorobenzene (22), however, 1,2,4,5-tetrachlorobenzene is of limited commercial significance.

The chlorination reaction is exothermic. The heat liberated is about 1.83 kJ/g Cl₂ (437 cal/g Cl₂). Heat is removed in some cases by circulating the reaction liquid through a suitable cooler (see HEAT EXCHANGE TECHNOLOGY). In other cases, chlorination occurs at the boiling point. The heat of the reaction is removed from

Table 3. **Ratio of 1,4- to 1,2-Dichlorobenzene with Various Catalysts**

Catalyst	Ratio	Reference
FeCl₃	1.4	
several	1.2–1.7	16
FeCl₃, AlCl₃, or SbCl₃ and organic sulfur compounds	2.2–3.3	17
FeS and organic sulfur compounds	2.0–2.65	18
FeCl₃ or SbCl₃ and sulfur	2–4	19
SbCl₃ and sulfur	3–5	20
SnCl₄ and/or TiCl₄ and AlCl₃	2.3	21

the reactor by the vaporizing liquid. The latter procedure has the disadvantage of operating at a higher temperature but has the advantage of allowing a low inventory reactor system, which saves equipment costs, reduces operating hazards, and makes heat recovery possible.

Benzene chlorination reactors are subject to design and operating hazards. Stagnant areas must be avoided in reactor design as they allow chlorination to the tetra- and pentachlorobenzenes. These compounds have low solubility in the liquid and can cause plugging. Another hazard is the equivalent of spontaneous combustion. The temperature can rise locally to a point where the reaction $C_6H_6 + 3\,Cl_2 \longrightarrow 6\,C + 6\,HCl$ can occur, primarily in the vapor phase. The exothermic reaction proceeds out of control and releases large amounts of HCl gas. This phenomenon can also occur when the chlorine concentration builds up in the reactor if the normal chlorination catalyst is inactivated by a cause such as an operational error that allows a sudden input of water.

Because HCl is constantly present in most parts of the equipment, corrosion is always a potential problem. Chlorine and benzene, or any recycled material, must be free of water to trace amounts to prevent corrosion and deactivation of the catalyst. The reactor product contains HCl and iron. In some plants, the product is neutralized with aqueous NaOH before distillation. In others, it is handled in a suitably-designed distillation train, which includes a final residue from which $FeCl_3$ can be removed with the high boiling tars.

Chlorobenzene mixtures behave in distillation as ideal solutions. In a continuous distillation train, heat may be conserved by using the condensers from some units as the reboilers for others thereby, saving process energy.

The dichlorobenzene isomers have very similar vapor pressures making separation by distillation difficult. Crystallization is generally used in combination with distillation to obtain the pure 1,2 and 1,4-dichlorobenzene isomers. The small quantity of 1,3-dichlorobenzene isomer produced is not generally isolated as a pure product. Environmental concerns have led to the use of improved crystallization systems that contain the products with minimal losses to the environment.

HCl is a constant by-product in the manufacture of chlorobenzenes. It is usually recovered by passing the gas stream through a scrubber tower over which a reactor mixture containing chlorination catalyst is circulated. This removes any unreacted chlorine that may have passed through the reactors. The HCl is then passed through one or more scrubbing towers in which high boiling chlorobenzenes are used as the solvent to remove the organic content. The absorbent in the final tower is refrigerated to the lowest possible temperature.

The HCl gas is absorbed in water to produce 30–40% HCl solution. If the HCl must meet a very low organic content specification, a charcoal bed is used ahead of the HCl absorber, or the aqueous HCl solution product is treated with charcoal. Alternatively, the reactor gas can be compressed and passed to a distillation column with anhydrous 100% liquid HCl as the distillate; the organic materials are the bottoms and are recirculated to the process. Any noncondensible gas present in the HCl feed stream is vented from the distillation system and scrubbed with water.

Any plant at times produces unwanted isomers. This requires an incinerator, capable of burning chlorinated hydrocarbons to HCl, H_2O, and CO_2 equipped with an efficient absorber for HCl (see INCINERATORS). An alternative to burning is dechlorination using hydrogen over a suitable catalyst. The ultimate product could be benzene.

$$C_6H_{6-n}Cl_n + n\,H_2 \longrightarrow C_6H_6 + n\,HCl$$

Dechlorination can be done in the vapor phase with palladium, platinum, copper, or nickel catalysts (23–26) or in the liquid phase with palladium catalysts (27). The vapor-phase dechlorination of 1,2,4-trichlorobenzene is reported to give good yields of 1,3-dichlorobenzene (24,26).

Another alternative to burning is rearrangement of the undesired isomers. This technique is practiced extensively in the petroleum industry, for example, in the production of xylene isomers using an HF–BF$_3$ catalyst-extractor system (28) (see BTX PROCESSING). Polychlorinated benzenes are considerably more resistant to rearrangement than are the isomeric hydrocarbon mixtures. Some patents have been issued to cover rearrangements using an aluminum chloride catalyst (29–31). A HF–SbCl$_5$ catalyst system is also reported to be effective in converting dichlorobenzenes to 1,3-dichlorobenzene (32). To date, there have been no reported commercial operations using these technologies.

4. Storage, Shipment, and Handling

Chlorobenzenes are stored in manufacturing plants in liquid form in steel containers. Mono-, 1,2-di- and 1,2,4-trichlorobenzenes are liquids at room temperature and are shipped in bulk in aluminum tank trucks and steel or stainless steel tank cars. In situations where chlorobenzenes are contained in aluminum at elevated temperatures, the product must be clean, ie, nonacidic and the moisture, low. The use of aluminum with a mixture of chlorobenzenes and strong oxidizers should be avoided. Mixtures of orthodichlorobenzene and chlorinated olefins react with aluminum, leading to catastrophic failure of aluminum tanks containing such a mixture. 1,4-Dichlorobenzene is shipped either in molten form in insulated steel tank cars with heater coils, or as flake or granular solid in suitably sealed containers, such as paper bags, fiber packs, or drums. Phenolic linings in all vessels offer protection over a wide range of conditions for all chlorobenzenes as well as the vessels themselves. For drums, the phenolic coating should be modified with epoxy for maximum impact resistance. 1,4-Dichlorobenzene has different labeling classifications depending on its intended use. Regulatory requirements change so the latest regulations should be checked and observed.

Chlorobenzenes are generally considered nonflammable materials with the exception of monochlorobenzene, which has a flash point of 34.5°C and is a flammable solvent based on DOT standards.

Chlorobenzenes are stable compounds and decompose slowly only under excess heating at high temperatures to release some HCl gas and traces of phosgene. It is possible, under certain limited conditions of incomplete combustion or pyrolysis, to form polychlorinated dibenzo-*p*-dioxins (PCDDs) and dibenzofurans (PCDFs) from chlorobenzenes (CHLOROCARBONS AND CHLOROHYDROCARBONS, TOXIC AROMATICS).

5. Health and Safety Factors

In general, all of the chlorobenzenes are less toxic than benzene. Liquid chlorobenzenes produce mild to moderate irritation upon skin contact. Continued contact may cause roughness or a mild burn. Solids cause only mild irritation.
Absorption through the skin is slow. Consequently, with short-time exposure
over a limited area, no significant quantities enter the body.

Contact with eye tissue at normal temperatures causes pain, mild to moderate irritation, and possibly some transient corneal injury. Prompt washing
with large quantities of water is helpful in minimizing the adverse effects of
eye exposure.

The data from some single-dosage oral toxicity tests, expressed as LD_{50}, are
reported in Table 4. The values reported on the order of 1 g/kg or greater indicate
a low acute oral toxicity. In animals, continued ingestion of chlorobenzenes over
a long time can cause kidney and liver damage.

The threshold limit value (TLV), the vapor concentration in ppm by volume,
to which humans may be exposed for an eight-hour working day for many years
without adverse effects, is also reported in Table 4. The saturated vapor concentration of the chlorobenzenes at 20°C listed in Table 4 are well above the TLV
values; therefore well-designed ventilation is required for working areas. A few
kidney and liver damage cases reported may have been caused by repeated exposure to some chlorobenzenes. Fires involving chlorobenzenes liberate HCl and
possibly phosgene. Under certain limited conditions of incomplete combustion
or pyrolysis, it has been reported that chlorobenzenes form 2,3,7,8-tetrachlorodi-
benzo-p-dioxin. When chlorobenzenes are involved in a fire, the proper protective
equipment must be used for personnel involved in fighting the fire.

Table 4. **Toxicity of Chlorinated Benzenes**

Chlorobenzene	Fish[a] toxicity, mg/L[b]	LD_{50} (oral), g/kg	TLV (inhal), ppm[c]	Saturated concentration, ppm by vol at 20°C
monochloro	<3[d,e]	2.9 (rat)	10	11,900
	16	2.8 (rabbit)		
1,2-dichloro	3	2.1 (rat)	25	1,125
		1.8 (rabbit)		
1,4-dichloro	0.7[e,f]	0.5 (rat)	10	1,570
		2.8 (rabbit)		
1,2,3-trichloro	[g]	0.8 (mice)	[g]	[g]
1,2,4-trichloro	3[e]	0.8 (rat)	[g]	[g]
1,2,4,5-tetrachloro	<1	<1 (rat)	[g]	[g]
1,2,3,4-tetrachloro	1.1	[g]	[g]	[g]
pentachloro	0.25[e,f]	1.1 (rat)	[g]	[g]

[a] Fathead unless otherwise noted.
[b] No observed adverse effect at this concentration in H_2O; 72 h static test (33) unless otherwise noted.
[c] Volume per volume of air.
[d] Rainbow trout.
[e] 96 h dynamic test (33).
[f] Bluegill.
[g] No value suggested.

Toxicity to fish is included in the data listed in Table 4. Marine life, particularly fish, may suffer damage from spills in lakes and streams. The chlorobenzenes, because they are denser than water, tend to sink to the bottom and may persist in the area for a long time. However, some data indicate that dissolved 1,2,4-trichlorobenzene can be biodegraded by microorganisms from wastewater treatment plants and also has a tendency to slowly dissipate from water by volatilization (34).

Most recently, the main health emphasis has been on carcinogenic potential. p-Dichlorobenzene was listed by the National Toxicology Program (NTP) as a product that could possible cause cancer. The results were based on mice and rats having large quantities of the chemical forced into their stomachs. The U.S. EPA subsequently reviewed the data and observed that there is no evidence of para-induced cancer. It is believed that the chemical complexes with a protein found only in the male rat, thereby not likely to pose a risk to humans. The male rat protein alpha 2u globulin involvement is linked to the hyaline droplet nephropathy, which is seen in most of the chlorobenzene studies. The Science Advisory Board of the U.S. EPA reaffirmed EPA's position that the data on p-dichlorobenzene demonstrate the phenomenon cannot be used for evaluating human carcinogenic risk from chlorinated benzenes (35). Domestically, the U.S. Consumer Product Safety Commission (CPSC), and internationally, the International Programme of Chemical Safety (IPCS) under the World Health Organization (WHO), in considering the mechanistic data also concluded that p-dichlorobenzene does not produce a human carcinogenic risk.

The listing by the NTP triggers labeling requirements in an organization such as the Occupational Safety and Health Administration (OSHA). However, the EPA classification may trigger different requirements such as no labeling, as in the case of p-dichlorobenzene. States develop their own labeling requirements, therefore the possible use of p-dichlorobenzene and state requirements for proper labeling must be considered.

6. Economic Aspects

Total production of chlorobenzenes in the three principal producing regions of the world amounted to aproximately 400 thousand metric tons in 1988: the United States, 46%, Western Europe, 34%, and Japan, the remainder. Monochlorobenzene accounted for over 50% of the total production of chlorinated benzenes. The largest use of monochlorobenzene worldwide is for the production of nitrochlorobenzene: 41% for the United States' demand, 70% for the Western European demand, and 89% for the Japanese demand in 1988. Currently, an estimated 43% of MCB production is used to make nitrochlorobenzenes which are used to make dye and pigment intermediates, rubber processing chemicals, pesticides, pharmaceuticals, and other organic intermediates. DuPont and Monsanto produced nitrochlorobenzenes until 1993 when DuPont left the business. Since 1993, Monsanto has been the only U.S. producer of these MCB derivatives. Since 1990, U.S. demand for MCB has been in the 200–240 million pound per year range, about the same as it was during the mid-1980s. However, captive usage increased during this time, with a net decrease in merchant market demand. The merchant market for MCB declined from almost 50% of demand in the mid-1980s to 24% in 1993. Additionally, several recent developments have impacted the outlook for MCB in a major way. In 1994, DuPont announced that it will soon cease production of 3,4-dichloroaniline. 4-Aminodiphenylamine, which is traditionally made from MCB, is expected to be made from nitrobenzene instead of MCB by Flexsys, a new Monsanto/Akzo Nobel NV joint venture, starting in 1997. In 1995, Dow Chemical Company announced that from 1996 onward, it would switch its feedstock for diphenyloxide production from MCB to phenol, a move that could reduce demand by 60-65 million pounds. Solvent use of monochlorobenzene in the United States is much greater than in Western Europe and Japan because of its use in herbicide formulations and other agriculture products.

o-Dichlorobenzene is consumed for 3,4-dichloroaniline, the base material for several herbicides, in the United States and Western Europe and is emulsified in Japan for garbage treatment. The greatest market worldwide for p-dichlorobenzene is for deodorant blocks and moth control. A growing use for p-dichlorobenzene is the manufacture of poly(phenylene sulfide) (PPS) resins. The U.S. herbicide market is expected to remain stable with some growth potential in developing countries, but the growth rate for o-dichlorobenzene is not expected to keep up with that of p-dichlorobenzene due to the high growth rate for PPS resins which, according to the resin producers was estimated at over 15% in 1996. In 1995, para demand was expected to be about 80 million pounds, with about 30 million converted into PPS. Bayer AG in Germany is also a large producer of chlorobenzenes but only plays a small role in the merchant market since a large portion of its capacity is consumed internally. Due to these major changes in market outlets for the mono and dichlorobenzenes, keeping isomer balance is expected to pose some challenges to the chlorobenzene producers. The mono/di breakdown varies based on plant design, although some designs allow some flexibility. The split between para and ortho is currently around three parts para to one part ortho. Demand for para is already more than four times that of ortho and is growing relative to mono, so meeting para demand withou creating significant oversupply of mono and ortho is a problem. One potential way out is the so-called para-on-purpose (POP) process. POP is not considered difficult, but is currently too expensive to be viable. Chlorobenzene producers believe POP could be used to meet para demand only if customers were able to meet the high price.

With the exception of use in the manufacture of polymers, markets for chlorobenzenes are mature, and demand is expected to show little if any growth in the next few years.

The chlorobenzene operations in the United States were developed primarily for the manufacture of phenol, aniline, and DDT. However, with the process changes in the production of phenol and aniline, the phase-out of DDT production, and changes in the herbicide and solvent markets, the U.S. production of chlorinated benzenes has shrunk by more than 50% since the total production peaked in 1969. U.S. production of monochlorobenzene peaked in the 1960s and decreased to a low of 101 million kg in 1986 with an 11% and 9% increase, respectively, in 1988 and 1989.

Commercial chlorination of benzene today is carried out as a three-product process (monochlorobenzene and o- and p-dichlorobenzenes). The most economical operation is achieved with a typical product split of about 85% monochlorobenzene and a minimum of 15% dichlorobenzenes. Typically, about two parts of p-dichlorobenzene are formed for each part of o-isomer. It is not economical to eliminate the coproduction of the dichlorobenzenes. To maximize monochlorobenzene production (90% monochlorobenzene and 10% dichlorobenzene), benzene is lightly chlorinated; the density of the reaction mixture is monitored to minimize polychlorobenzene production and the unreacted benzene is recycled.

Producing the chlorobenzenes higher than mono- can pose significant process problems because production must match the market or the unwanted material must be destroyed. Use must be found for the HCl by-product and Cl_2 must be available at a reasonable price.

In 1988, the United States consumption of monochlorobenzene was 120 million kilograms; 42% for the production of nitrochlorobenzenes, 28% for solvent uses, and the remaining 30% for other applications such as diphenyl ether, ortho- and para-phenylphenols, sulfone polymers, and diphenyldichlorosilane, an intermediate for specialty silicones.

The principal use of o-dichlorobenzene is to manufacture 3,4-dichloroaniline (DCA) which is a raw material for several herbicides and for the production of 3,4,4'-trichlorocarbanilide (TCC), a bacteriostat used in deodorant soaps. Some is exported, but the amount is expected to decline as Brazil brings on increased capacity. A modest decline in U.S. consumption between 1989 and 1994 is expected. About 11,400 t were consumed in 1988. The decline in ortho-dichlorobenzene output is due to reduced demand for amide and urea herbicides made from DCA because of loss of market to more selective herbicides in corn and soybeans, the major markets for herbicides. Some of the decline has been offset by small gains as a process solvent, particularly for the manufacture of isocyanates, such as toluene diisocyanate (TDI).

The largest single market and a growing outlet for p-dichlorobenzene in the United States is the production of poly(phenylene sulfide) (PPS) resin. Of 42 million kilograms of p-dichlorobenzene consumed in the United States in 1988, 21% was for PPS. The second largest consumption in the United States of p-dichlorobenzene (16%) is the room deodorant market which is static and likely to remain unchanged. Moth control (11%) is also expected to remain static. However, when the room deodorant and moth control markets in the United States are added together they become the largest consumption, similar to the world market. Exports accounted for about 30% and about 21% remained in inventory.

Prices for the chlorobenzenes fluctuate widely. Some prices fluctuate with the price of benzene. Pricing information is available (36). More recent pricing information for MCB and dichlorobenzenes is also available (37).

Western Europe has a capacity of around 228 thousand metric tons as of January 1989; 76% of that capacity is located in Germany. Most of the capacity is captively consumed in products similar to the United States and also production of other products no longer produced in the United States.

The production of chlorobenzenes in Eastern Europe is concentrated in the former Soviet Union, Poland, and Czechoslovakia. The estimated capacity is 200–250 thousand metric tons; the former Soviet Union has most (230 thousand tons) of this capacity. There is trade between Eastern and Western Europe on mono-chlorobenzene and the dichlorobenzenes, but the net trade balance is probably even at about 20 thousand metric tons. Eastern Europe exported 20 thousand metric tons of monochlorobenzene principally to Germany, France, and the United States.

Japan, as of January 1, 1989, had a total capacity of 28 thousand metric tons of monochlorobenzene and 49 thousand tons of dichlorobenzenes. The Japanese prices have remained fairly constant since 1985. The Japanese consumption of p-dichlorobenzene is 81% for moth control, 11% for PPS resins, and 8% for dye-stuffs. There has been very little export from Japan of chlorobenzenes and imports have been mainly p-dichlorobenzene from the United States, Germany, France, and the United Kingdom.

Brazil has two small producers of chlorobenzenes. One producer has a capacity of 4.8 thousand metric tons. The other producer's facility has a rated capacity of 28 thousand metric tons, which produces mono and ortho for local consumption, and the para may be used in Brazil and possibly exported. A third plant with a 400 metric ton capacity is believed to be on standby.

Canada has no known basic producers of chlorobenzene. There is one company that isolates small quantities of ortho and para from purchased mixed dichlorobenzenes. Some of the isolated product is exported. The primary portion of Canada's chlorobenzenes comes from the United States.

7. Specifications, Analyses, and Quality Control

Trade specifications for the chlorobenzenes are subject to modification by agreement with the customer of each producer's standards. All of the chlorobenzenes show readily separated and identifiable peaks by glc. This method is used exclusively for plant, quality control, and for sales specifications.

Typical analyses in wt% are chlorobenzene: benzene <0.05, dichlorobenzenes <0.1; and 1,4-dichlorobenzene: chlorobenzene and trichlorobenzenes <0.1, 1,2- and 1,3-dichlorobenzene: each <0.5.1,2-Dichlorobenzene is sold as two grades: technical: chlorobenzene <0.05, trichlorobenzenes <1.0, 1,2-dichlorobenzene 80, and other isomers <19.0; and purified, produced by redistilling the technical product in a very efficient still: chlorobenzene <0.05, 1,2,4 trichlorobenzene <0.2; and 1,2-dichlorobenzene 98.0.

8. Uses

8.1. Monochlorobenzene. The largest use of monochlorobenzene in the United States is in the production of nitrochlorobenzenes, both ortho and para, which are separated and used as intermediates for rubber chemicals, antioxidants (qv), dye and pigment intermediates, agriculture products, and pharmaceuticals (Table 5). Since the mid-1980s, there have been substantial exports of both o-nitrochlorobenzene, estimated at 7.7 million kg to Europe and p-nitrochlorobenzene, estimated at 9.5 million kg to the Far East. Solvent use of monochlorobenzene accounted for about 28% of the U.S. consumption. This application

Table 5. **Derivatives of Nitrochlorobenzenes**

Derivative	CAS Registry number	Reactant(s)	Intermediate for
First-step derivatives of p-nitrochlorobenzene			
p-nitrophenol (PNP)	[100-02-7]	caustic hydrolysis	parathion acetyl-*p*-aminophenol, dyes
p-nitroaniline (PNA)	[100-01-6]	ammonia	*p*-phenylenediamine gasoline antioxidants, dyes, rubber chemicals
p-nitrophenetole (PNPt)	[100-29-8]	sodium ethylate	*p*-phenetidine used as ethoxyquin intermediate
4-nitrodiphenylamine (4 NDPA)	[836-30-6]	aniline	*N*-phenyl-*p*-phenylenediamine rubber chemicals
p-chloroaniline (PCA)	[106-47-8]	hydrogen	agriculture chemicals, carbanilidebacteriostats
4,4′-dinitrodiphenyl ether (DNDPO)	[101-63-3]	sodium phenate	oxybisaniline used as polymer intermediate
First-step derivatives of o-nitrochlorobenzene			
o-nitroaniline (ONA)	[88-74-4]	ammonia	fungicide (benomyl), stabilizers, benzotriazole
o-nitrophenol (ONP)	[88-75-5]	caustic hydrolysis	carbofuran and agriculture chemicals
o-nitroanisole (ONAS)	[91-23-6]	sodium methylate	3,3′-dimethoxybenzidine (a pigment intermediate)
3,3′-dichlorobenzidine	[91-94-1]	self	yellow pigments
o-chloroaniline (OCA)	[95-51-2]	hydrogen	agriculture chemicals

involves solvents for herbicide production and the solvent for diphenylmethane diisocyanate manufacture and other chemical intermediates.

Other applications for monochlorobenzene include production of diphenyl-ether, *ortho*- and *para*-phenylphenol, 4,4′-dichlorodiphenylsulfone, which is a primary raw material for the manufacture of polysulfones, diphenyldichlorosilane, which is an intermediate for specialty silicones, Grignard reagents, and in dinitrochlorobenzene and catalyst manufacture.

8.2. *o*-Dichlorobenzene. The principal use of *o*-dichlorobenzene in the United States is the manufacture of 3,4-dichloroaniline [95-76-1], a raw material used in the production of herbicides, although recent developments in the marketplace suggest that this application is likely to see lowered demand. A small amount of 3,4-dichloroaniline is used to produce 3,4,4′-trichlorocarbanilide [101-20-2] (TCC) used as a bacteriostat in deodorant soaps.

8.3. *p*-Dichlorobenzene. *p*-Dichlorobenzene's largest and growing outlet is in the manufacture of poly(phenylene sulfide) resin (PPS). Other applications include room deodorant blocks and moth control, a market which is static and likely to remain unchanged but combined is currently a larger outlet than PPS. Small amounts of *p*-dichlorobenzene are used in the production of 1,2,4-trichlorobenzene, dyes, and insecticide intermediates. Exports have been a principal factor in U.S. production with about 25% exported in 1988.

8.4. *m*-Dichlorobenzene. Isolation of pure *m*-dichlorobenzene [541-73-1], produced at ∼1% in the mixed dichlorobenzenes, is not economical. It is produced by rather exotic chemistry and has established only very specialized uses, believed to be only a few hundred kg per year, because of its high cost and the lack of commercial availability. However, there is potential for *m*-dichlorobenzene in some new experimental agricultural chemicals. In addition to liquid-phase isomerization of the ortho and para isomers in the presence of Lewis acids, such as $AlCl_3$ and $HF-SbCl_5$ (29–32), vapor phase chlorination, either thermal or with UV activation, can give a product with a high proportion of the meta isomer (38–39). There are a number of patents that cover its production (24,26,29–32), but only limited commercial production has been reported to date.

8.5. Other Chlorobenzenes. The market for the higher chlorobenzenes (higher than di) is small in comparison to the combined mono- and dichlorobenzenes. 1,2,4-Trichlorobenzene is used in the manufacture of the herbicide, Banvel. Trichlorobenzenes are used in some pesticides, as a dye carrier, in dielectric fluids, as an organic intermediate and a chemical manufacturing solvent, in lubricants, and as a heat-transfer medium. These are small and decreasing markets.

BIBLIOGRAPHY

"Chlorinated Benzenes" are treated in *ECT* 1st ed. under "Chlorine Compounds, Organic," Vol. 3, "Monochlorobenzene," pp. 812–817, by L. A. Kolker, Kolker Chemical Works, Inc., and N. Poffenberger, The Dow Chemical Company; "*o*-Dichlorobenzene," pp. 817–818, by N. Poffenberger, The Dow Chemical Company; "*p*-Dichlorobenzene," pp. 819–822, by Axel Heilborn, Niagara Alkali Co.; "Chlorinated Benzenes" under "Chlorocarbons and Chlorohydrocarbons" in *ECT* 2nd ed., Vol. 5, pp. 253–267, by D. W. F. Hardie, Imperial Chemical Industries Ltd.; in *ECT* 3rd ed., Vol. 5, pp. 797–808, by Chi-I Kao and N. Poffenberger, Dow Chemical U.S.A.; "Chlorinated Benzenes" in *ECT* 4th ed., vol. 6, pp. 57–100, by James G. Bryant, Standard Chlorine of Delaware, Inc.; "Chlorinated Benzenes" in *ECT* (online), posting date: December 4, 2000, by James G. Bryant, Standard Chlorine of Delaware, Inc.

CITED PUBLICATIONS

1. J. B. Cohen and P. Hartley, *J. Chem. Soc.*, 87, 1360 (1905).
2. D. W. F. Hardie, "A History of the Chemical Industry in Widnes," Imperial Chemical Industries Ltd., 1950, p. 155.
3. M. Campbell and H. Hatton, *Herbert H. Dow: Pioneer in Creative Chemistry*, Appleton-Century-Crosts, Inc., New York, 1951, p. 1114.
4. U.S. Pat. 1,607,618 (Nov. 23, 1926), W. J. Hale and E. C. Britton (to The Dow Chemical Company).
5. W. J. Hale and E. C. Britton, *Ind. Eng. Chem.* 20, 114 (1928).
6. R. M. Crawford, *Chem. Eng. News* 25(1), 235 (1947).
7. Gen. Pat. 539,176 (Nov. 12, 1931), W. Prohl (to F. Raschig GmbH).
8. Gen. Pat. 575,765 (Apr. 13, 1933), W. Prahl and W. Mathes (to F. Raschig GmbH).
9. L. Luttrighaus and D. Ambrose, *Chem. Ber.* 89, 463 (1956).
10. J. D. Roberts and A. T. Bottini, *J. Am. Chem. Soc.* 79, 1458 (1957).
11. D R. Stull, E. F. Westrum, and G. C. Sinke, *The Chemical Thermodynamics of Organic Compounds*, John Wiley & Sons, Inc., New York, 1969.

12. Brit. Pat. 388,818 (Mar. 6, 1933), T. S. Wheeler (to ICI).
13. H. van den Berg and R. M. Westerink, *Ind. Eng. Chem. Fund.* **15**(3), 164 (1976).
14. M. F. Bourion, *Ann. Chim. Paris* **14**(9), 215 (1920).
15. R. B. MacMullin, *Chem. Eng. Prog.* **44**(3), 183 (1948).
16. H. F. Wiegandt and P. R. Lantos, *Industrial Engineering Chemistry* **43**, 2167 (1951).
17. U.S. Pat. 3,226,447 (Dec. 28, 1965), G. H. Bing and R. A. Krieger (to Union Carbide Australia Ltd.).
18. Neth. Pat. 7413614 (Oct. 16, 1974), S. Robota, R. Paolieri, and J. G. McHugh (to Hooker Chemicals).
19. U.S. Pat. 1,946,040 (Feb. 6, 1934), W. C. Stoesser and F. B. Smith (to The Dow Chemical Company).
20. U.S. Pat. 2,976,330 (Mar. 21, 1961), J. Guerin (to Société Anonyme).
21. U.S. Pat. 3,636,171 (Jan. 18, 1972), K. L. Krumel and J. R. Dewald (to The Dow Chemical Company).
22. U.S. Pat. 3,557,227 (Jan. 19, 1971), M. M. Fooladi (to Sanford Chemical Co.).
23. U.S. Pat. 2,826,617 (Mar. 11, 1958), H. E. Redman and P. E. Weimer (to Ethyl Corp.).
24. U.S. Pat. 2,943,114 (June 28, 1960), H. E. Redman and P. E. Weimer (to Ethyl Corp.).
25. U.S. Pat. 2,886,605 (May 12, 1959), H. H. McClure, J. S. Melbert, and L. D. Hoblit (to The Dow Chemical Company).
26. U.S. Pat. 2,866,828 (Dec. 30, 1958), J. A. Crowder and E. E. Gilbert (to Allied Chemical Corp.).
27. U.S. Pat. 2,949,491 (Aug. 16, 1960), J. J. Rucker (to Hooker Chemical Corp.).
28. S. Ariki and A. Ohira, *Chem. Econ. Eng. Rev.* **5**(7), 39 (1973).
29. U.S. Pat. 2,666,085 (Jan. 12, 1954), J. T. Fitzpatrick (to Union Carbide Corp.).
30. U.S. Pat. 2,819,321 (Jan. 7, 1958), B. O. Pray (to Columbia Southern Chemical Corp.).
31. U.S. Pat. 2,920,109 (Jan. 5, 1960), J. W. Angelkorte (to Union Carbide Corp.).
32. Yu G. Erykolov and co-workers, *Zh. Org. Khim* **9**, 348 (1973).
33. *Standard Methods for Examination of Water and Wastewater*, 14th ed., American Public Health Association, Washington, D.C., 1975, p. 800.
34. P. Simmons, D. Branson, and R. Bailey, *"Biodegradability of 1,2,4-Trichlorobenzene"*, paper presented at the *1976 Association of Textile Chemicals and Colorist International Technical Conference*, Chicago, Ill., 1976.
35. J. A. Barter and R. S. Nair, *Review of the Scientific Evidence on the Human Carcinogenic Potential of Para-Dichlorobenzene*, Chlorobenzene Producers Association, Washington, D.C., 1990.
36. W.K. Johnson with A. Leder and Y. Sakuma, *"CEH Product Review"*, Chlorobenzenes Chemical Economics Handbook, SRI International, Menlo Park, Calif., Oct 1989.
37. *"Dichlorobenzene"*, *Chemical Products Synopsis*; A Reporting Service of Mannsville Chemical Products Corp., Adams, NY, Dec 1995.
38. J.W. Engelsma, E.C. Kooyman, and J.R. Van der Bij, *Recueil des Travaux Chimiques des Pays-Bas* **76**, 325 (1957).
39. W. Dorrepaal and R. Louw, *Int. J. Chem. Kinetics* **10**, 249 (1978).

RAMESH KRISHNAMURTI
Occidental Chemical Corporation

CHLOROCARBONS AND CHLOROHYDROCARBONS, SURVEY

1. Introduction

Chlorinated hydrocarbons comprise a family of products widely used throughout the chemical and manufacturing industries. The most common of these products are the chlorinated derivatives of methane, ethane, propane, butane, and benzene. These compounds are used as chemical intermediates and as solvents. Solvent uses include a wide variety of applications, ranging from metal and fabric cleaning operations to reaction media for chemical synthesis. Pressure to reduce emissions to the environment of chlorocarbons, chlorohydrocarbons, and their derivatives has resulted in a decrease in demand for these products in many applications. At the same time, demand for these products has increased in key applications where the environmental impact is not as severe. The result of the changing pattern of end uses for chlorocarbons and chlorohydrocarbons is a transformation of the industry to one focused on the production of chemical intermediates for sustainable products.

2. Physical Properties

Progressive chlorination of a hydrocarbon molecule yields a succession of liquids or solids of increasing density, viscosity, and improved solubility for a large number of inorganic and organic materials. Other physical properties such as flammability, specific heat, dielectric constant, and water solubility decrease with increasing chlorine content. The chlorocarbons and chlorohydrocarbons of greatest industrial importance are listed in Table 1. Physical properties of selected chlorocarbons and chlorohydrocarbons are listed in Table 2.

3. Chemical Properties

Chlorocarbons and chlorohydrocarbons participate is a wide variety of chemical reactions. This versatility makes these products extremely useful in chemical synthesis.

3.1. Substitution Chlorination. The substitution of chlorine atoms for hydrogen atoms in organic molecules is the basis for many important chlorination processes. Substitution chlorination consists of a series of free-radical reactions, which can be further characterized into a series of steps—initiation, propagation, and termination.

The chlorine radicals needed for this chemistry can be generated by thermal, photochemical, or chemical means. The thermal method requires temperatures of at least 250°C to initiate the decomposition of chlorine molecules. Photochemical generation of chlorine atoms is commonly achieved using mercury arc lamps. These lamps generate ultraviolet (uv) light for photochemical processes

Table 1. **Chlorocarbons and Chlorohydrocarbons of Industrial Importance**

IUPAC name	Common name	CAS Registry Number	Chemical formula
chloromethane	methyl chloride	[74-87-3]	CH_3Cl
dichloromethane	methylene chloride	[75-09-2]	CH_2Cl_2
trichloromethane	chloroform	[67-66-3]	$CHCl_3$
tetrachloromethane	carbon tetrachloride	[56-23-5]	CCl_4
chloroethene	vinyl chloride, VCM	[75-01-4]	$CHCl=CH_2$
1,1-dichloroethene	vinylidene chloride	[75-35-4]	$CCl_2=CH_2$
trichloroethene	trichloroethylene	[79-01-6]	$CCl_2=CHCl$
tetrachloroethene	perchloroethylene	[127-18-4]	$CCl_2=CCl_2$
chloroethane	ethyl chloride	[75-00-3]	CH_2ClCH_3
1,2-dichloroethane	ethylene dichloride, EDC	[107-06-02]	CH_2ClCH_2Cl
1,1,1-trichloroethane	methyl chloroform	[71-55-6]	CCl_3CH_3
3-chloro-1-propene	allyl chloride	[107-05-1]	$CH_2=CHCH_2Cl$
2-chlorobutadiene	chloroprene	[689-97-4]	$CH_2=CClCH=CH_2$
chlorobenzene	monochlorobenzene	[108-90-7]	C_6H_5Cl
1,2-dichlorobenzene	o-dichlorobenzene	[95-50-1]	$1,2-C_6H_4Cl_2$
1,4-dichlorobenzene	p-dichlorobenzene	[84348-21-0]	$1,4-C_6H_4Cl_2$

with wavelengths from 300–500 nm. Thermal chlorination is inexpensive and less sensitive to inhibition than the photochemical process. Thermal or photochemical initiation of diatomic chlorine result in the formation of two chlorine radicals, as shown in equation 1.

$$Cl_2 \xrightleftharpoons{hv \text{ or } \Delta} 2Cl\cdot \tag{1}$$

Table 2. **Physical Properties of Selected Chlorocarbons and Chlorohydrocarbons**

Name	Molecular weight	Melting point (°C)	Boiling point (°C)	Density, 25°C (kg/m³)	Refractive index, 20°C
methyl chloride	50.49	− 97.7	−24.09	911 (p > 1 atm)	1.3389
methylene chloride	84.93	− 97.2	40.0	1325	1.4242
chloroform	119.38	− 63.41	61.17	1479	1.4459
carbon tetrachloride	153.82	− 22.62	76.8	1594	1.4601
vinyl chloride	62.50	−153.84	−13.3	911 (p > 1 atm)	1.3700
vinylidene chloride	96.94	−122.56	31.6	1213	1.4249
trichloroethylene	131.39	− 84.7	87.21	1463	1.4773
perchloroethylene	165.83	− 22.3	121.3	1623	1.5059
ethyl chloride	64.50	−138.4	12.3	891	1.3676
ethylene dichloride	98.96	− 35.7	83.5	1256	1.4450
methyl chloroform	133.40	− 30.01	74.09	1338	1.4379
allyl chloride	76.52	−134.5	45.1	939	1.4157
chloroprene	88.54	−130	59.4	956	1.4583
monochlorobenzene	112.56	− 45.31	131.72	1107	1.5241
o-dichlorobenzene	147.00	− 17	180	1306	1.5515
p-dichlorobenzene	147.00	53.09	174	1241	1.5285

Chemical initiation generates organic radicals, usually by decomposition of azo (2) or peroxide compounds (3). The organic radicals react with chlorine to initiate the radical-chain chlorination reaction. The organic radical initation process is demonstrated in equations 2 and 3.

$$[\text{Initiator}] \longrightarrow [\text{Initiator} \cdot] \tag{2}$$

$$\text{Cl}_2 + [\text{Initiator} \cdot] \longrightarrow \text{Cl} \cdot + [\text{Initiator Cl}] \tag{3}$$

Chlorine radicals obtained from the dissociation of chlorine molecules react with organic species to form hydrogen chloride and an organic radical. The organic radical can react with an undissociated chlorine molecule to produce the organic chloride and a new chlorine radical necessary to continue the reaction. This process may repeat many times before the reaction is finally terminated. In the case of the photochlorination of chloroform, this process has been estimated to occur 4800 times before termination (4). Propagation reactions are represented in equations 4 and 5.

$$\text{Cl} \cdot + \text{R} - \overset{|}{\underset{|}{\text{C}}} - \longrightarrow \text{R} - \overset{|}{\underset{|}{\text{C}}} \cdot + \text{HCl} \tag{4}$$

$$\text{R} - \overset{|}{\underset{|}{\text{C}}} \cdot + \text{Cl}_2 \longrightarrow \text{R} - \overset{|}{\underset{|}{\text{C}}} - \text{Cl} + \text{Cl} \cdot \tag{5}$$

Chain terminations occur in a number of ways. One way is the collision between two chlorine atoms, illustrated by the reverse of equation 1. Other representative terminations are given by equations 6 and 7.

$$2\text{R} - \overset{|}{\underset{|}{\text{C}}} \cdot \longrightarrow \text{R} - \overset{|}{\underset{|}{\text{C}}} - \overset{|}{\underset{|}{\text{C}}} - \text{R} \tag{6}$$

$$\text{R} - \overset{|}{\underset{|}{\text{C}}} \cdot + \text{Cl}_2 \longrightarrow \text{R} - \overset{|}{\underset{|}{\text{C}}} - \text{Cl} \tag{7}$$

Oxygen can act as a very strong inhibitor for the chlorine free-radical generation process. The process is inhibited by oxygen to the extent that only a few ppm of oxygen can drastically reduce the reaction rate.

The propagation reactions are highly exothermic and demand careful temperature control by cooling or dilution. Adiabatic reactors with an appropriate diluent are successfully used in industrial processes. Substitution chlorination can take place in either the vapor (5) or the liquid phase (6). In liquid-phase processes, vaporization may make an important contribution to the control of the temperature of the process.

3.2. Addition Chlorination. Chlorination of olefins such as ethylene by the addition of chlorine is a commercially important process and can be carried out either as a catalytic vapor- or liquid-phase process (7). The reaction is influenced by light, the walls of the reactor vessel, and inhibitors such as oxygen. Addition chlorination may proceed by a radical-chain mechanism or an ionic addition mechanism. Ionic addition mechanisms can be maximized and accelerated by the use of a Lewis acid such as ferric chloride, aluminum chloride, antimony pentachloride, or cupric chloride. The chlorination of ethylene with a ferric chloride catalyst is illustrated in equations 8–10 (8).

$$FeCl_3 + Cl_2 \longrightarrow FeCl_4^-Cl^+ \tag{8}$$

$$FeCl_4^-Cl^+ + \underset{\diagup}{\overset{\diagdown}{C}}=C\underset{\diagdown}{\overset{\diagup}{}} \longrightarrow Cl-\overset{|}{\underset{|}{C}}-\overset{|}{\underset{|}{C}}{}^+ \; FeCl_4^- \tag{9}$$

$$Cl-\overset{|}{\underset{|}{C}}-\overset{|}{\underset{|}{C}}{}^+ \; FeCl_4^- \longrightarrow Cl-\overset{|}{\underset{|}{C}}-\overset{|}{\underset{|}{C}}-Cl + FeCl_3 \tag{10}$$

The free-radical addition of chlorine to double bonds, which is strongly inhibited by oxygen, probably proceeds by a radical-chain mechanism due to its reasonably low activation energy barrier (9–13). This reaction is shown in equation 11.

$$Cl\cdot + \underset{\diagup}{\overset{\diagdown}{C}}=C\underset{\diagdown}{\overset{\diagup}{}} \longrightarrow Cl-\overset{|}{\underset{|}{C}}-\overset{|}{\underset{|}{C}}\cdot \tag{11}$$

The radical species formed in this step tends to be unstable, and the reaction is prone to reverse at high temperatures or low chlorine concentration. However, at high chlorine concentration, the propagation reaction may proceed as in equation 12.

$$Cl-\overset{|}{\underset{|}{C}}-\overset{|}{\underset{|}{C}}\cdot + Cl_2 \longrightarrow Cl-\overset{|}{\underset{|}{C}}-\overset{|}{\underset{|}{C}}-Cl + Cl\cdot \tag{12}$$

The addition of chlorine to perchloroethylene is an example of the process described by equations 10 and 11. This process has been demonstrated in the case of photochemical initiation (14).

Chlorination of olefins is an exothermic process. For the case of the chlorination of ethylene to produce EDC, the heat of reaction is 218 kJ/mol.

3.3. Addition of Hydrogen Chloride. The addition of hydrogen chloride to alkenes in the absence of peroxides takes place by an electrophilic addition mechanism. The orientation is in accord with Markovnikov's rule in which the hydrogen atom adds to the side of the double bond that will result in the most stable carbonium ion. The addition occurs in two steps with formation of an intermediate carbonium ion, as shown in equation 13. Addition of the chloride ion (eq. 14) completes the hydrochlorination mechanism.

$$\text{>C=C<} + \text{HCl} \longrightarrow \text{Cl}-\overset{+}{\text{C}}-\text{C}- + \text{Cl}^- \tag{13}$$

$$-\overset{+}{\text{C}}-\text{C}- + \text{Cl}^- \longrightarrow \text{Cl}-\text{C}-\text{C}- \tag{14}$$

Historically, metal chloride catalysts have been used in commercial processes for the hydrochlorination of olefinic derivatives. For example, hydrochlorination of ethylene at temperatures <100°C and in the presence of aluminum chloride yields ethyl chloride. Hydrogen chloride addition reactions have been carried out in the presence of aluminum(III) chloride and nitrobenzene (15,16), and with secondary, tertiary, and quaternary ammonium hydroxides as catalysts (17). Hydrochlorination reactions are most advantageously carried out at ambient temperature with iron(III) chloride as the catalyst and the reaction product as the solvent (18).

The hydrochlorination of olefins is a weakly exothermic reaction with heats of reaction ranging from 4 to 21 kJ/mol. The hydrochlorination of acetylene is more exothermic, at ~184 kJ/mol.

3.4. Elimination of Hydrogen Chloride.
Dehydrochlorination of chlorinated hydrocarbons is useful as a means of producing unsaturated products. Dehydrochlorination can be accomplished by reaction with bases or Lewis acids, catalytic reactions, or by thermal noncatalytic chemistry.

Alkyl halides react with aqueous alkalies such as calcium hydroxide (19,20), sodium hydroxide (21) and magnesium hydroxide (22). In the resulting reaction, HCl is eliminated and an unsaturated bond is introduced into the product. The reaction generally proceeds by a classic E2 elimination mechanism (23–25). In this reaction, the base abstracts a hydrogen atom from a carbon atom. Simulateously, a chloride ion separates from the molecule, leaving a double bond between the carbon atoms (eq. 15).

$$-\overset{\text{Cl}}{\underset{\overset{|}{\text{H}}}{\text{C}}}-\text{C}- \underset{:B}{\longrightarrow} \text{Cl}^- + \text{>C=C<} + \text{H:B} \tag{15}$$

The base elimination of hydrogen chloride takes place in the aqueous phase. The use of phase-transfer catalysts promotes the rate of reaction and the selectivity in many elimination reactions (26). In contrast, the use of a methanol–water mixture results in a reduction of the reaction rate, especially in the presence of an emulsifying agent (27).

In the case of dehydrochlorination of 1,1,2-trichloroethane with a base, the formation of 1,1-dichloroethylene is highly favored over the *cis*- and *trans*-1,2-dichloroethylene isomers. Similarly, the reaction of 3,4-dichloro-1-butene with a base produces the 2-chlorobutadiene isomer rather than the *cis*- and *trans*-1-chlorobutadiene isomers.

Asymmetrically chlorinated ethanes are resistant to dehydrochlorination by bases. Reaction of these compounds with bases proceeds by a substitution reaction mechanism (21,28). For EDC, both the base elimination and substitution reaction mechanisms are reported to occur (29). In the case of 1,1,1-trichloroethane, the primary organic product is sodium acetate (21). Dehydrochlorination of asymmetric chlorinated ethanes is more readily accomplished with a Lewis acid rather than a base. Significant levels of tar formation accompany these reactions. Tar formation is found to be reduced by complexing the Lewis acid catalyst with nitro-aromatic (30) or 1-nitro-alkane (31,32) solvents.

A number of catalysts capable of dehydrochlorinating chlorinated hydrocarbons have been identified. These include alumina, silica, silica-alumina (33), alumina-boria (34), metal sulfates (35,36), and alkali or alkaline earth metal chlorides (37). The mechanism of the dehydrochlorination reaction depends on the ease with which the chlorine atom can be extracted versus the acidity of any abstractable protons, and the relative acidity or basicity of the catalyst (38).

Gas-phase thermally induced dehydrochlorination proceeds by a radical-chain mechanism (39). The reaction is accelerated by radical initiators such as chlorine and retarded or inhibited by olefins and alcohols (40). Addition of small amounts of a chlorinating agent, such as chlorine, promotes radical dehydrochlorination in the gas phase through an abstraction mechanism that results in loss of hydrogen chloride and formation of a double bond. The dehydrochlorination of a chlorinated olefin in the presence of chlorine, as shown in equations 16 and 17, is a typical example.

$$Cl-\overset{|}{\underset{|}{C}}-\overset{|}{\underset{|}{C}}- \; + \; Cl\cdot \; \longrightarrow \; Cl-\overset{|}{\underset{|}{C}}-\overset{|}{\underset{|}{C}}\cdot \; + \; HCl \tag{16}$$

$$Cl-\overset{|}{\underset{|}{C}}-\overset{|}{\underset{|}{C}}\cdot \; \longrightarrow \; {>}C{=}C{<} \; + \; Cl\cdot \tag{17}$$

The dehydrochlorination reaction is endothermic. For example, the dehydrochlorination of EDC to produce vinyl chloride requires 71 kJ/mol.

Subsequent elimination of hydrogen chloride from alkenes can result in the formation of alkynes. These reactions proceed slowly, especially in the case of elimination of vinyl halides.

3.5. Chlorinolysis. Chlorinolysis, also referred to as chlorolysis, is the process of inducing a rupture of carbon–carbon bonds through a combination of saturation or near saturation of the organic reactant with chlorine at elevated temperature. Reaction of C2 (41) and C3 (42) hydrocarbons with excess chlorine at high temperatures can cleave the C–C bonds of the hydrocarbon to give chlorinated derivatives of shorter chain length. Aromatic and aliphatic chlorohydrocarbons containing up to six carbons can be converted to carbon tetrachloride by chloronolysis, though pre-saturation of double bonds in the feedstock with chlorine is usually required in these cases (43).

In the chlorinolysis chemistry environment, substitution chlorination (eqs. 3 and 4), free-radical addition chlorination (eqs. 10 and 11), and free radical hydrogen chloride elimination reactions (eqs. 15 and 16) will occur. These reactions are important in producing species that are unstable enough to result in rupture of a carbon–carbon bond. This generally can occur once a $Cl_2C=Cl–X$ moiety is produced from the starting compound. The carbon–carbon scission is shown in equation 18. This reaction become more facile as the compound becomes more highly chlorinated.

$$Cl-\overset{\cdot}{\underset{\underset{R}{|}}{C}}-\overset{\overset{Cl}{|}}{\underset{\underset{Cl}{|}}{C}}-\overset{\overset{Cl}{|}}{\underset{\underset{Cl}{|}}{C}}-Cl \longrightarrow \overset{Cl}{\underset{Cl}{>}}C=C\overset{Cl}{\underset{Cl}{<}} + \cdot CCl_3 \tag{18}$$

The decomposition reaction in equation (18) has been shown to have a low activation energy (44). Further chlorination of the products of this reaction result in one- and two-carbon chlorinated species. Generally, because these reactions are carried out in a chlorine-rich environment, the products are carbon tetrachloride and perchloroethylene. These two products are in equilibrium and their ratio is determined by the reaction conditions. The interconversion between carbon tetrachloride and perchloroethylene is illustrated in equations 19–22, and is a good example of the types of reactions that occur under chlorinolysis conditions.

$$Cl\cdot + \overset{Cl}{\underset{Cl}{>}}C=C\overset{Cl}{\underset{Cl}{<}} \rightleftharpoons Cl-\overset{\overset{Cl}{|}}{\underset{\underset{Cl}{|}}{C}}-\overset{\overset{Cl}{|}}{\underset{\underset{Cl}{|}}{C}}\cdot \tag{19}$$

$$Cl-\overset{\overset{Cl}{|}}{\underset{\underset{Cl}{|}}{C}}-\overset{\overset{Cl}{|}}{\underset{\underset{Cl}{|}}{C}}\cdot + Cl_2 \rightleftharpoons Cl-\overset{\overset{Cl}{|}}{\underset{\underset{Cl}{|}}{C}}-\overset{\overset{Cl}{|}}{\underset{\underset{Cl}{|}}{C}}-Cl + Cl\cdot \tag{20}$$

$$Cl-\overset{\overset{Cl}{|}}{\underset{\underset{Cl}{|}}{C}}-\overset{\overset{Cl}{|}}{\underset{\underset{Cl}{|}}{C}}-Cl \rightleftharpoons 2\cdot CCl_3 \tag{21}$$

$$\cdot CCl_3 + Cl_2 \rightleftharpoons CCl_4 + Cl\cdot \tag{22}$$

3.6. Oxychlorination. The oxychlorination reaction consists of combining a source of chlorine atoms, such as hydrogen chloride or chlorine, oxygen (air), and a hydrocarbon or chlorohydrocarbon in the presence of a cupric chloride catalyst. This chemistry is of special importance in the manufacture of chlorinated hydrocarbons, since many of these process produce hydrogen chloride as a coproduct. Oxychlorination allows the hydrogen chloride to be captured for conversion to useful products. The oxychlorination process has been reviewed (45).

Oxychlorination chemistry has been proposed to follow an oxidation–reduction type mechansim (46). The reaction steps are illustrated by equations 23–25 for the case of oxychlorination of an olefin.

$$[Cu_2Cl_4] \ + \ \text{\Large >}C=C\text{\Large <} \ \longrightarrow \ [Cu_2Cl_2] \ + \ Cl-\overset{|}{\underset{|}{C}}-\overset{|}{\underset{|}{C}}-Cl \qquad (23)$$

$$[Cu_2Cl_2] \ + \ \tfrac{1}{2}O_2 \ \longrightarrow \ [Cu_2OCl_2] \qquad (24)$$

$$[Cu_2OCl_2] \ + \ 2\,HCl \ \longrightarrow \ [Cu_2Cl_4] \ + \ H_2O \qquad (25)$$

The copper chloride catalyst for oxychlorination is supported for most industrial processes. Generally, gamma-alumina is used for ethylene oxychlorination, but the use of materials such as alumina gels, silica and silicates, diatomaceous earth, and pumice has also been reported (47). The support itself can play a role in the chemistry of an oxychlorination system, and should be considered when developing a catalytic system (48). Also, various additives to the catalyst formulation are known to affect the kinetics and selectivities (49).

Oxychlorination is an exothermic reaction. The heat of reaction for the oxychlorination of ethylene to produce EDC is 238 kJ/mol.

3.7. Chloro Dehydroxylation.
Alcohols react with hydrogen halides to form alkyl halides and water. This chemistry can be carried out in the vapor phase by catalytic means or in the liquid phase by either catalytic or noncatalytic means.

Vapor-phase chlorination of alcohols has been tested over a variety of catalysts. In particular, activated alumina and silica are known to be effective (50). For the conversion of methanol to methyl chloride, the reaction is reported to occur between chemisorbed methanol and gas-phase HCl via an Eley-Rideal mechanism (51). Methanol is proposed to bind to a Lewis acid site on the alumina surface where it undergoes nucleophilic reaction with HCl. Increasing acidity of the catalyst reduces the tendency of the alcohol to undergo dehydration to form ethers (52).

The liquid-phase reaction of primary alcohols and methanol with hydrogen chloride proceeds by a bimolecular nucleophilic substitution mechanism. In the first step, the alcohol is protonated (eq. 26). The products are formed by a bimolecular reaction between the chloride and protonated alcohol (eq. 27).

$$R-OH \ + \ HCl \ \longrightarrow \ ROH_2^+ \ + \ Cl^- \qquad (26)$$

$$Cl^- \ + \ ROH_2^+ \ \longrightarrow \ \left[\overset{\delta-}{Cl}\text{-}\text{-}R\text{-}\text{-}\overset{\delta+}{OH_2}\right] \ \longrightarrow \ RCl \ + \ H_2O \qquad (27)$$

All other alcohols react with hydrogen chloride by a unimolecular nucleophilic substitution mechanism (53). In this case, the protonation of the alcohol in equation 26 is followed by the formation of a carbonium ion, equation 28.

The carbonium ion reacts with the chloride to form the alkyl halide product, equation 29.

$$ROH_2^+ \longleftrightarrow R^+ + H_2O \tag{28}$$

$$R^+ + Cl^- \longrightarrow RCl \tag{29}$$

Metal chloride catalysts are effective for promoting the chlorination of alcohols (54). This reaction is also conducted efficiently without the use of a catalyst (55).

3.8. Hydrolysis of Alkyl Halides. The reverse reaction of chloro-dehydroxylation of alcohols, or hydrolysis to replace a chlorine atom with an alcohol substituent group, can occur under appropriate conditions. This chemistry is used for the conversion of epichlorohydrin, which is derived from allyl chloride, into glycerol. This chemistry is an example of the introduction of chlorine into an organic molecule to allow the easy introduction of a different functionality, in this case an alcohol, at a later step in the synthesis.

3.9. Chlorination of Aromatics. Aromatic compounds may be chlorinated with diatomic chlorine in the presence of a catalyst such as iron, iron(III) chloride, or other Lewis acids (56). The halogenation reaction involves electrophilic displacement of the aromatic hydrogen by halogen. Introduction of a second chlorine atom into the monochloro aromatic structure leads to ortho and para substitution. The presence of a Lewis acid favors polarization of the chlorine molecule, thereby increasing its electrophilic character. Because the polarization does not lead to complete ionization, the reaction should be represented as shown in equation 30.

$$\tag{30}$$

Continuous chlorination of benzene at 30–50°C in the presence of a Lewis acid typically yields 85% monochlorobenzene. Temperatures in the range of 150–190°C favor production of the dichlorobenzene products. The para isomer is produced in a ratio of 2–3 to 1 of the ortho isomer. Other methods of aromatic ring chlorination include use of a mixture of hydrogen chloride and air in the presence of a copper-salt catalyst, or sulfuryl chloride in the presence of aluminum chloride at ambient temperatures. Free-radical chlorination of toluene successively yields benzyl chloride, benzal chloride, and benzotrichloride. Related chlorination agents include sulfuryl chloride, tert-butyl hypochlorite, and N-chlorosuccinimide, which yield benzyl chloride under the influence of light, heat, or radical initiators (see TOLUENES, RING-CHLORINATED; BENZYL CHLORIDE, BENZAL CHLORIDE, AND BENZOTRICHLORIDE).

3.10. Other Chemistry of Chlorocarbons and Chlorohydrocarbons. Chlorocarbons and chlorohydrocarbons undergo a wide variety of reactions that make them suitable intermediates for the production of many useful products.

Fluorination. The chlorine atoms in chlorocarbons can be replaced by fluorine through halogen exchange chemistry. A chlorine containing feedstock is typically reacted with hydrogen fluoride or other suitable inorganic fluoride. The selection of the halogenating agent is determined by the chemical structure of the chlorocarbon and the desired degree of fluorination of the product (57) (see FLUORINATED ALIPHATIC COMPOUNDS).

Silation. Methyl chloride reacts with silicon in the presence of copper and promoters at high temperature to form methylchlorosilanes. Methylchlorosilanes are the building blocks of many silicone products (see SILICON COMPOUNDS, SILICONES).

Reaction with Aluminum. Many chlorinated hydrocarbons react readily with aluminum in the so-called bleeding reaction. A red aluminum chloride-chlorinated hydrocarbon complex is formed. Storage of uninhibited chlorinated solvents in aluminum vessels results in corrosion in a short period of time. For this reason, aluminum should not be used in the construction of storage vessles for chlorinated hydrocarbons. Proprietary organic inhibitors permit commercial use of solvents such as trichloroethylene for cleaning of aluminum.

Oxidation. All chlorinated hydrocarbons are susceptible to pyrolysis at high temperatures. This process liberates hydrogen chloride, water, and carbon dioxide. Thermal oxidation processes have been used as means to destroy chlorinated by-products of chlorohydrocarbon manufacture that cannot otherwise be utilized. The Catalytic oxidation processes are also used for the safe disposal of waste chlorinated hydrocarbons. Processes to convert chlorocarbons to hydrogen chloride, carbon monoxide and hydrogen (syngas) are in development (58) (see EXHAUST CONTROL, INDUSTRIAL; INCINERATION; WASTE TREATMENT, HAZARDOUS WASTE).

Chlorinated hydrocarbons may be partially oxidized. This chemistry is used in the manufacture of dichloroacetyl chloride from vinylidene chloride (59). Partial oxidation can also cause undesired product degradation. Chlorinated alkenes such as trichloroethylene are oxidized in the presence of oxygen when subjected to uv light or heat to yield hydrogen chloride, phosgene, and chlorinated acetyl chlorides (acid derivatives) (60). Vinylidene chloride polymerizes to a solid very quickly if exposed to air without the presence of a proper inhibitor. Formation of a peroxide intermediate is an important step in the initiation of this process.

Saturated aliphatic chlorine derivatives are usually quite stable to oxidation, although 1,1,2-trichloroethane exhibits appreciable oxidation when contrasted with the stable 1,1,1-trichloroethane isomer. The oxidation of chlorinated hydrocarbons usually proceeds via a hydroperoxide mechanism. Alcohols and amines are often added to oxidation-sensitive solvents to minimize this mode of degradation.

Hydrogenation. Chlorocarbons can be hydrogenated in the presence of suitable catalysts in order to replace chlorine atoms with hydrogen atoms. This chemistry can be employed to convert carbon tetrachloride to chloroform or perchloroethylene to trichloroethylene (61).

Polymerization. Olefinic chlorohydrocarbons react to form many useful polymers. The usual mechanism is a free-radical initiated chain reaction. Poly-(vinyl chloride) (PVC), poly(vinylidene chloride), and polychloroprene are all polymers of chlorinated alkenes. The four principal steps of a free-radical chain reaction are initiation, propagation, chain transfer, and termination (see VINYL CHLORIDE POLYMERS; VINYLIDENE CHLORIDE MONOMER AND POLYMERS; POLYCHLOROPRENE).

Reimer-Timann Reaction.   In the Reimer-Tiemann reaction, phenol is reacted with chloroform in the presence of a base to introduce an aldehyde group onto the aromatic ring. Chloroform reacts with the base to form the highly reactive dichlorocarbene radical. Dichlorocarbene reacts with the aromatic ring to produce a substituted benzal chloride as an intermediate. The chlorine atoms on this group are then hydrolyzed to form salicyladehyde (62). The dichlorocarbene radical that is formed from the reaction of chloroform with a base is strongly electrophilic, and hence highly reactive. For this reason, chloroform is considered to be incompatible with strong bases, and contact between the two should be generally avoided.

4. Manufacture

Chlorinated hydrocarbons are produced by a wide variety of chemical processes. Since chlorinated products can be synthesized and reacted by so many different mechanisms, processes have evolved over the years to obtain high yields to desired products. An important aspect of the development of chlorinated hydrocarbon manufacture is the use of available by-products from the manufacture of other chlorinated products as raw materials. These by-products include products of over-chlorination as well as HCl that is coproduced as a part of many processes. This development has great significance in the reduction of the volume of waste products that must be treated. An example of the integration of the use of intermediates and by-products that is possible in the manufacture of chlorinated hydrocarbons is depicted in Figure 1.

4.1. Chlorinated Methanes.   Methyl chloride may be produced from either methanol or methane. The methanol-based route may be conducted in either the vapor or liquid phase. The vapor-phase process typically utilizes a reactor packed with catalyst. The catalyst is alumina or alumina based. Hydrogen chloride is mixed with vaporized methanol and fed to the reactor. The reactor is usually of the shell-and-tube variety, with shell-side cooling to remove the heat of reaction. The reaction is typically carried out at temperatures between 200 and 350°C. Products emerge from the reactor as a vapor and are sent to a quench column. In the quench column, HCl is separated from the methyl chloride product through absorption in coproduct water. The remainder of the process is dedicated to further purification of the product and removal of trace organics from the aqueous acid.

The liquid-phase methyl chloride process is the more widely used process. The liquid-phase process can be designed with multiple reactors to achieve both high methanol and high HCl conversion. In the liquid-phase process, methanol is contacted with HCl at 70–160°C and 200–1100 kPa in a boiling bed reactor (63). The effluent gases pass through a rectifying column, where refluxed methyl chloride drives the water vapor back down and reacts any HCl with by-product DME to form methyl chloride. Typically, 0.5–3% of the methanol is converted to DME.

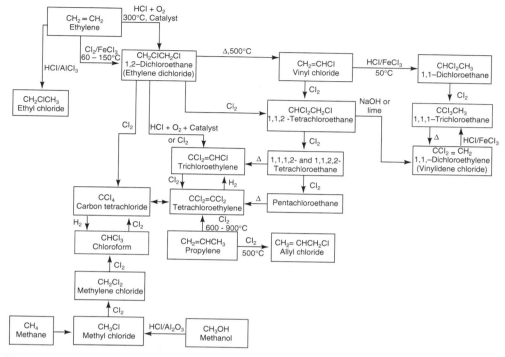

Fig. 1. Example of representative integrated manufacturing process for production of chlorocarbons and chlorohydrocarbons.

Methyl chloride is also produced by the thermal chlorination of methane in the gas phase at a temperature in the range of 490–530°C. Methylene chloride, chloroform, and carbon tetrachloride are formed in this process, along with coproduct HCl (64).

When methyl chloride is produced using the methanol process, methyl chloride is used as a feedstock to a thermal chlorination process to produce methylene chloride, chloroform, and carbon tetrachloride in a process similar to the methane chlorination process. Methyl chloride can also be chlorinated in a free-radical initiated liquid-phase process. Hydrogen chloride is a by-product of the chlorination process. The methanol-based process has the advantage of consuming some or all of the coproduct HCl from the chlorination process, which in effect increases the chlorine utilization of the entire process (see METHYL CHLORIDE; METHYLENE CHLORIDE; CHLOROFORM; CARBON TETRACHLORIDE).

The ratio of chlorine to methane or methyl chloride is varied to control the product distribution from the thermal chlorination process. Typically, the process is operated with excess organic feedstock to ensure complete conversion of the chlorine. Methane (if used as a feedstock), methyl chloride, and also methylene chloride can be recycled to control the temperature of the reactor. In the vapor-phase chlorination process, the temperature is maintained between 350 and 550°C and the residence time is in the range of 3–30 s.

The methane oxychlorination process has also been studied. Production of methylene chloride and chloroform has been the object of some of these studies. However, much of the work in this field has had the goal of activating methane for conversion to higher hydrocarbons by reacting the methane with hydrogen chloride and oxygen to produce to methyl chloride. The methyl chloride is then converted to ethylene or other higher hydrocarbons (65). This technology is not currently practiced on an industrial scale.

4.2. Chlorinated Ethanes and Ethylenes. *Ethyl Chloride.* Ethyl chloride can be produced by the reaction of hydrogen chloride with ethanol, the chlorination of ethane, or the addition of hydrogen chloride to ethylene. The addition of HCl to ethylene is the most commonly practiced commercial process. Aluminum chloride is typically used as the catalyst (67). Ethylene and hydrogen chloride are fed to the process at essentially the stoichiometric ratio. Staged addition of ethylene has been reported to improve overall yield (68). The process is operated at a temperature between 30 and 60°C (see ETHYL CHLORIDE).

EDC and Vinyl Chloride. Vinyl chloride is produced almost exclusively by the thermal dehychlorination of EDC, which is produced from ethylene. Two processes are typically used. EDC may be produced by the direct chlorine addition to ethylene or by the oxychlorination of ethylene with oxygen and HCl.

A typical commercial process for the preparation of 1,2-dichloroethane is the chlorination of ethylene at 60–120°C (68). The direct chlorination process is normally conducted as a liquid-phase reaction, and usually the product EDC is the reaction medium. The addition of chlorine to the ethylene double bond is facilitated through the use of a Lewis acid catalyst such as ferric chloride. The direct chlorination process is characterized by very high yields. The primary by-products are the products of overchlorination of EDC, in particular 1,1,2-trichloroethane. The formation of by-products can be suppressed in a number of ways. Introduction of a small amout of air to the process is one method used to suppress undesired free-radical reactions. Use of co-catalysts has also been reported to be effective. Mixing of the ethylene and chlorine has also been identified as an important factor in maintaining good yield to EDC. The use of multiple reactors has also been employed to optimize yield.

The utility of the oxychlorination process is that it allows coproduct HCl from the manufacture of vinyl chloride and other chlorination processes to be captured for reuse. Most vinyl producers operate so-called "balanced" plants, where the ratio of production of EDC by direct chlorination and by oxychlorination is manipulated to minimize the loss of chlorine from the process as HCl.

The Dow Chemical Company commercialized the first large-scale oxychlorination process for EDC in 1958. Both fixed-tube catalytic reactors and fluidized bed reactors are used commercially for ethylene oxychlorination. Catalyst formulations and carriers may vary, but are generally based on a catalyst of cupric chloride supported on an active carrier such as activated alumina. Typical residence times are between 15 and 22 s, and reactor temperatures range from 230 to 315°C. Yields to 1,2-dichloroethane range from 92 to 97%. By-products include carbon monoxide, carbon dioxide, and minor amounts of chlorinated ethanes and ethylenes.

The ethylene oxychlorination reaction is very exothermic. Heat removal is accomplished by two mechanisms. First, the reactors themselves are cooled by circulating hot oil or steam through the outer jacket of the reactor or through cooling coils in the interior of the reactor. Second, heat removal is augmented by cofeeding an inert diluent to the reactor. Usually this diluent consists of unreacted ethylene along with products of combustion.

Vinyl chloride is produced from EDC by a thermal dehydrochlorination process. Reactions are conducted in cracking furnaces, somewhat similar to the process for hydrocarbon cracking. The gas-phase thermal dehydrochlorination of 1,2-dichloroethane at 350–515°C proceeds by a radical-chain mechanism (39). The reaction is accelerated by radical initiators such as chlorine and retarded or inhibited by olefins and alcohols (40). A typical cracking furnace for the production of vinyl chloride has a residence time of 3–10 s. Per pass conversion of EDC ranges from 40 to 60%. Typical by-products include acetylene and dimerization products of vinyl chloride and acetylene. Overall process yields range from 98 to 99%.

A small amount of vinyl chloride continues to be manufactured by the hydrochlorination of acetylene. The disadvantage of this process is the high cost of acetylene relative to ethylene.

A number of attempts have been made to develop an oxychlorination process to produce vinyl chloride directly from ethane (69,70). None of these proposed processes has been adopted on an industrial scale. The feedstocks of the proposed ethane-to-vinyl chloride processes are ethane, oxygen, and either HCl or chlorine in the presence of a catalyst. Products of the reaction are vinyl chloride and water, along with intermediates such as ethylene, ethyl chloride, and EDC. Undesired by-products include carbon monoxide and carbon dioxide, as well as products of overchlorination. The process has many features of oxychlorination chemistry, though the actual chemistry may be much more complicated (see VINYL CHLORIDE).

1,1,1-Trichloroethane. The production of 1,1,1-trichloroethane has been sharply curtailed because of restrictions imposed by the Montreal Protocol. Chlorination of ethane was at one time used to produce 1,1,1-trichloroethane. The primary commercial process for 1,1,1-trichloroethane production is a two-step process based on vinyl chloride as a feedstock (71). Vinyl chloride is reacted with HCl, typically in the liquid phase in the presence of a Lewis acid catalyst, to produce 1,1-dichloroethane. After purification, the 1,1-dichloroethane is chlorinated, either thermally or by a photochlorination process, to produce the desired product and coproduct HCl. The HCl from the chlorination step is recycled to the hydrochlorination step to keep the overall process in balance. The major by-product of the chlorination step is 1,1,2-trichloroethane. A less commonly used maufacturing route is to hydrochlorinate vinylidene chloride to produce 1,1,1-trichloroethane (see CHLOROETHANES).

Vinylidene Chloride. Vinylidene chloride can be manufactured from 1,1,1- or 1,1,2-trichloroethane. The 1,1,2-trichloroethane route is the more common. If 1,1,1-trichloroethane is used, it is converted to vinylidene chloride by a thermal dehydrochlorination process (71). The dehydrochlorination may be catalyzed or uncatalyzed.

The 1,1,2-trichloroethane feedstock to produce vinylidene chloride may be obtained as a by-product of the manufacture of EDC or 1,1,1-trichloroethane. It may also be produced from EDC by a liquid-phase chlorination process.

The liquid-phase chlorination of EDC to 1,1,2-trichloroethane also results in the coproduction of HCl. Tetrachloroethane isomers are a by-product of the production of 1,1,2-trichloroethane (72).

Once the 1,1,2-trichloroethane has been produced and purified, it is dehydrochlorinated to produce vinylidene chloride. The most common route is to react aqueous sodium hydroxide with the 1,1,2-trichloroethane. This process has the advantage of producing vinylidene chloride at a very high selectivity relative to the production of the *cis*- and *trans*-1,2-dichloroethylene isomers. A small amount of monochloroacetylene is also produced as a by-product of this process. The main disadvantage is that the chlorine that is removed from the feedstock is converted to sodium chloride, which cannot be readily reused.

Vapor-phase thermal dehydrochlorination of 1,1,2-trichloroethane has the advantage of producing the HCl coproduct in anhydrous form so that it is more easily recovered. However, selectivities to vinylidene are typically poor. The by-products of this process are *cis*- and *trans*-1,2-dichloroethylene. These by-products must be rehydrochlorinated and recycled to obtain acceptable overall yields. Attempts have been made to discover a vapor-phase dehydrochlorination catalyst that is highly selective to vinylidene chloride, without commercial success (73). Another route that has received attention is the use of a regenerable base. This technology promises the selectivity of a basic elimination reaction to obtain high yield to vinylidene with the possibility of recovering the HCl from the base in a regeneration step (74) (see Vinylidene chloride monomer and polymers).

Trichloroethylene and Perchloroethylene. Trichloroethylene and perchloroethylene can be produced by many routes. An oxychlorination process has been used quite successfully to coproduce trichloroethylene and perchloroethylene (75). A wide range of chlorinated hydrocarbons may be used as feedstocks in this process. Ethylene, EDC, or the chlorinated by-products of EDC manufacture are the most common feedstocks. Either chlorine or HCl may be used as chlorinating agents in this process. The more heavily chlorinated organic species present in by-products of EDC manufacture can also serve as a chlorine source (76). Recent advances in oxychlorination to produce trichloroethylene and perchloroethylene have demonstrated enhancements in the yield to trichloroethylene (77). The use of by-products of the manufacture of allyl chloride has also been studied for conversion to trichloroethylene and perchloroethylene in an oxychlorination process (78,79).

In a typical operation, 1,2-dichloroethane, hydrogen chloride or chlorine, and oxygen are fed to fluidized bed reactor at 400°C to produce trichloroethylene and perchloroethylene. The catalyst bed consists of cupric chloride and potassium chloride on graphite, diatomaceous earth, or some other suitable carrier. A modified oxychlorination technique known as the Transcat process has been developed by the Lummus Co. (80). The feedstock can be a saturated hydrocarbon or chlorohydrocarbon and the process is suited to the production of C1 and C2 chlorohydrocarbons.

Trichloroethylene is also produced via a thermal chlorination process. The typical feedstock is EDC or the by-products of EDC manufacture. In the thermal chlorination process, trichloroethylene may be produced directly, or the intermediate tetrachlorethane isomers are produced. The tetrachloroethane isomers are subsequently dehydrochlorinated in a separate process to produce trichloroethylene. This procedure is done in a high temperature vapor-phase dehydrochlorination process analogous to the conversion of EDC to vinyl chloride. Introduction of these tetrachloroethane derivatives into a tubular-type furnace at temperatures of 425–455°C gives good yields of trichloroethylene (81). In the cracking of the tetrachloroethane stream, introduction of ferric chloride into the 460°C vapor-phase reaction zone improves the yield of trichloroethylene product.

Liquid-phase cracking processes have also been investigated (82). Hydrogen chloride is a major coproduct of both the thermal chlorination and dehydrochlorination processes for producing trichloroethylene.

Like trichloroethylene, perchloroethylene can be produced in a thermal chlorination process. Historically, the thermal chlorination process for perchloroethylene has been used to coproduce carbon tetrachloride. A well-known commercial process involving this technique is thermal chlorination of propane and chlorinated hydrocarbon feedstocks with chlorine to produce carbon tetrachloride and perchloroethylene with hydrogen chloride as co-product (83). The hydrocarbon feedstock may include hydrocarbons up to C3 and any partially chlorinated derivatives (84). The yields can be varied widely by controlling recycle streams to take advantage of the equilibrium conversion of carbon tetrachloride to perchloroethylene; eg, recycling carbon tetrachloride increases the perchloroethylene yield.

A typical reactor operates at 600–900°C with no catalyst and a residence time of 10–12 s. At this high temperature, chlorinolysis chemistry results in the scission of the C3 intermediates to form the desired products. Yield to carbon tetrachloride and perchloroethylene is ~92–96% based on the chlorine input. Perchloroethylene and carbon tetrachloride can be interconverted through a chemical equilibrium at high temperatures, so that simple manipulation of process conditions can be used to swing production between the two products in order to satisfy market demand. With diminished demand for carbon tetrachloride, producers using the thermal chlorination process have had to find ways to improve selectivities to perchloroethylene relative to carbon tetrachloride.

As with all thermal chlorination processes, the manufacture of perchloroethylene results in the coproduction of anhydrous HCl. A particular advantage of the thermal chlorination process for perchloroethylene is the wide range of feedstocks that can be used. The conditions of this process promote many rearrangment reactions, so that almost any one-, two-, three-, of four-carbon hydrocarbon or chlorohydrocarbon can be used as a feedstock. Advantages include the use of inexpensive raw materials, flexibility of the ratios of carbon tetrachloride and perchloroethylene produced, and waste chlorinated residues can be used as a feedstock to the reactor (84). By-products of this process can include hexachloroethane, hexachlorobutadiene, and hexachlorobenzene (85) (see TRICHLOROETHYLENE; TETRACHLOROETHYLENE; CHLOROCARBONS AND CHLOROHYDROCARBONS SURVEY, TOXIC AROMATICS).

4.3. Chlorinated Propanes and Higher.   *Allyl Chloride.* Allyl chloride is produced from chlorine and propylene in a thermal chlorination process, similar to that used for the production of chlorinated methanes, trichloroethylene, or perchloroethylene. The reaction takes place in the vapor phase under conditions of high temperature, excess propylene, and short residence time to promote selectivity to the dresired product.

Typical reactor conditions are 500°C, a pressure of 69–250 kPa, and a residence time on the order of 1–4 s. The reactor is operated adiabatically. Since the reaction is highly exothermic, excess propylene is fed to act as a diluent and also has the effect of improving selectivity.

HCl is a coproduct of this manufacture of allyl chloride. By-products of the manufacture of allyl chloride include 1-chloropropene, 2-chloropropene, 1,2-dichloropropane, 1,3-dichloropropene, and 2,3-dichloropropene. Additional by-products include more highly chlorinated propanes and propylenes, as well as dimerization products. Some of the by-products of allyl chloride production can be recovered for use as products (86) (see ALLYL CHLORIDE).

Chloroprene. Chloroprene is manufactured by a two-step process. The basic feedstocks are butadiene and chlorine. In the first step, butadiene is chlorinated to produce 3,4-dichloro-1-butene and other isomers. The isomers can be converted to 3,4-dichloro-1-butene in the presence of suitable catalysts in order to enhance the process yield. Historically, the chlorination was a vapor-phase process, but in recent years a liquid-phase chlorination process has also been developed. The liquid-phase process is characterized by higher yields and higher selectivities than the vapor-phase process (87).

The 3,4-dichloro-1-butene intermediate produced in the chlorination process is dehydrohalogenated to form the desired product, 2-chloro-1-butadiene. Aqueous sodium hydroxide is used for the dehydrohalogenation process. This process is very similar to the process for producing vinylidene chloride from 1,1,2-trichloroethane. The reaction can be facilitated by the use of phase transfer catalysts (88) (see CHLOROPRENE).

4.4. Chlorinated Aromatics. Raschig developed the first commercial oxychlorination process to make chlorobenzene in 1928. The chlorobenzene product was then hydrolyzed to phenol. The Durez plant in North Tonawanda, New York, which began production in 1937, used this process. Chlorination of benzene in the presence of a catalyst such as ferric chloride, or oxychlorination with hydrogen chloride, yields primarily monochlorobenzene. Additional chlorination gives the dichloro- isomers and some higher analogues (89) (see CHLORINATED BENZENES).

4.5. Waste Disposal. The desire to improve the overall quality of the environment has resulted in a greater emphasis on waste reduction and minimization in the production and use of chlorocarbons and chlorohydrocarbons. Regulations are in place to control the use and disposal of these substances. Both manufacturers and end-users have been affected by these regulations.

Within the manufacturing environment, waste minimization is typically the first priority, and is achieved through adoption of efficient technologies. Secondarily, wastes and by-products that are generated are reused as feedstocks for other processes whenever possible. For example, trichloroethylene and perchloroethylene can be produced from waste organochlorines from the manufacure of EDC, vinylidene chloride, chlorinated methanes, or allyl chloride. Hydrogen chloride can be recycled for the manufacture of a number of products, such as ethyl chloride, methyl chloride, EDC, trichloroethylene, and perchloroethylene.

Any by-products of manufacture that cannot be utilized must be destroyed in a manner to prevent harm to the environment. These wastes are considred hazardous and are regulated under the federal Resource Conservation and Recovery Act (RCRA). This act regulates the storage, transport, and disposal of hazardous waste. Importantly, all solvent wastes are prohibited from disposal in landfills, including hazardous waste landfills.

Conventional disposal of chlorinated hydrocarbon wastes has been achieved by high temperature incineration. This process produces aqueous HCl, water, and carbon dioxide, and steam is cogenerated. The incineration process has come under scrutiny due to concerns with the production of trace quantities of undesired compounds such as dioxins and furans. As a result, a number of processes for disposal of wastes chlorocarbon wastes have been investigated (see EXHAUST CONTROL, INDUSTRIAL; INCINERATORS; WASTE TREATMENT, HAZARDOUS WASTE).

Use of chlorocarbons and chlorohydrocarbons that results in the generation of waste from processes such as solvent cleaning operations also falls under the RCRA regulation. Small volume users of these products typically work with their distributors for their waste disposal needs. In the most common arrangement, a distributor will return used solvent to a reclaimer. The reclaimer recovers the usable solvent, and then disposes of the remaining material through a commercial incinerator. Each link in the chain of custody of the product from the manufacturer, through the distributor, to the customer and to final disposition must be properly permitted according to applicable federal, state, and local regulations. Hazardous waste haulers must be permitted as well and a manifest system must record the ultimate disposal and be retained by the source.

5. Shipment and Storage

Because of the wide range in properties of the various chlorocarbons and chlorohydrocarbons, persons handling these substances should be trained to recognize and avoid the hazards of the specific compound they are using. For example, methyl chloride is flammable, while the more highly chlorinated compounds are nonflammable liquids. The extreme difference in properties requires very different handling procedures. For example, methyl chloride is typically transported and stored as a liquefied gas under pressure. The following are very general guidelines for the handling of chlorocarbons and chlorohydrocarbons.

Storage tanks should be of minimum carbon steel construction. Aluminum, zinc, and magnesium alloys should generally be avoided. Many chlorinated hydrocarbons react readily with aluminum to form a red aluminum chloride–chlorinated hydrocarbon complex. Aluminum especially should be avoided in the presence of methyl chloride, since explosive methyl aluminum compounds can form. Proprietary organic inhibitors allow commercial use of solvents such as trichloroethylene for cleaning of aluminum. Care should also be taken in the selection of polymeric or elastomeric materials that may come in contact with chlorocarbons. Many rubber compounds are permeable to chlorocarbons.

Storage tanks should be grounded and have provision for adequate pressure relief. Labeling should conform to local requirements to communicate the flammability, reactivity, and toxicity of the material being stored. Storage tanks should also have adequate spill protection. The spill protection system should be designed to prevent cross-contamination of incompatible materials. The vapor space of storage tanks used in chlorocarbon service should be oxygen-free and padded with nitrogen. The tank vent system should be designed to prevent cross-contamination with other chemicals stored at the same facility. Warning signs should also be posted to communicate the dangers of a nitrogen-containing vapor space.

Consideration should be given to the installation of filtration equipment at loading facilities. Also, means to remove contaminants such as water or iron that can contaminate the product during shipment should be available at terminals or other product centers.

Chlorocarbons and chlorohydrocarbons can be shipped by a number of means. The products that are normally liquids are shipped in drums, tank trucks, rail cars, barges, or ocean-going ships.

Commercial use of many chlorinated derivatives imposes stress on the stability of the solvent. Inhibitors classified as antioxidants, acid acceptors, and metal stabilizers are added to minimize these stresses. All the chlorinated derivatives hydrolyze at a slow but finite rate when dissolved in water. Hydrolysis of chlorinated solvents typically liberates hydrogen chloride that can corrode storage containers and commercial metal-cleaning equipment. The liberated hydrogen chloride can be neutralized by an appropriate epoxide to form noncorrosive chlorohydrins.

6. Economic Aspects

World demand for chlorocarbons and chlorohydrocarbons has been strongly influenced by the implementation of improved control technologies in many solvent cleaning applications, and by the shift from chlorofluorocarbons to hydrofluorocarbons in many applications. Products that are generally used as chemical intermediates have tended to demonstrate steady growth.

6.1. Chlorinated Methanes. The consumption data for chlorinated methanes for the United States, Western Europe, and Japan demonstrates the change in emphasis in end uses (90) (Table 3). Since 1985, consumption of methyl chloride has increased dramatically, driven in large part by the growth in the silicones sector. Similarly, chloroform consumption has also grown due to its use as a feedstock for fluoropolymer production. In contrast, methylene chloride usage has declined because of its extensive emissive applications, and carbon tetrachloride consumption has been phased out nearly completely due to regulations introduced by the Montreal Protocol.

Table 3. **Consumption of Chloromethanes, 1000 t**

Product	1987	1990	1993	1997	2000
methyl chloride	565	628	869	1075	1249
methylene chloride	449	392	252	349	328
chloroform	394	450	481	537	554
carbon tetrachloride	873	536	322	41	20
data from (90)					

Table 4. **World Consumption of EDC and Vinyl Chloride, 1000 t**

Product	1993	1994	1995	1996	1997	1998
EDC	30,444	32,277	32,522	35,415	38,014	38,437
vinyl chloride	19,891	21,123	21,702	23,232	24,802	25,206
data from (68, 91)						

Spot pricing for methylene chloride f.o.b. the U.S. Gulf Coast is in the range of $370–460/t. Pricing for chloroform, on the same basis, is $440–530/t.

6.2. Chlorinated Ethanes and Ethylenes. Ethyl chloride manufacture has declined greatly, primarily due to the decrease in the production of tetraethyl lead. U.S. production of ethyl chloride in the year 2000 was estimated at ~36,000 t. About 60% of this volume was exported, much of it for production of tetraethyllead in countries where leaded fuels are still used. This market is expected to decline as these countries transition to the use of unleaded fuels.

EDC is by far the most widely manufactured of the chlorinated hydrocarbons. Virtually all EDC that is produced is used for the production of vinyl chloride. In turn, almost all vinyl chloride that is produced is used for the manufacture of PVC. Vinyl chloride production has grown steadily over the course of the past decade (68) (Table 4). Demand growth for EDC and vinyl chloride is driven by the growth of the PVC business.

Spot pricing for EDC f.o.b. the U.S. Gulf Coast is $340–370/t. Pricing for vinyl chloride, on the same basis, is $500–520/t.

Vinylidene chloride consumption was steady through most of the 1990s (90). However, the recent phase out of HCFC-141b has resulted in a significant drop in the market for vinylidene chloride. Annual U.S. consumption of vinylidene chloride is on the order of 70–80,000 t.

Demand in the United States, Western Europe, and Japan for the products 1,1,1-trichloroethane, trichloroethylene, and perchloroethylene has been driven strongly by legislative initiatives to reduce levels of emissions to the atmosphere (92) (Table 5). Since the mid-1980s, 1,1,1-trichloroethane consumption has dropped dramatically, which is due to the elimination of the use of this product in emissive applications as of January 1, 1996. It is now primarily used only for production of certain hydrochlorofluorocarbons (HCFC-141b and HCFC-142b), laboratory and analytical procedures, and for certain exempt applications. Production of HCFC-141b production is to be phased out by the end of 2002.

Trichloroethylene demand has risen since 1990. The increase in demand for trichloroethylene is primarily due to its use as a feedstock for hydrofluorocarbon

Table 5. **Consumption of 1,1,1-Trichloroethane, Trichloroethylene, and Perchloroethylene, 1000 t**

Product	1987	1990	1993	1996	2000
1,1,1-trichloroethane	577	678	339	199	203
trichloroethylene	276	224	201	249	258
perchloroethylene	610	596	366	275	269
data from (92)					

Table 6. **U.S. Consumption of Chlorinated Benzenes, 1000 t**

Product	1985	1990	1995	1998
chlorobenzene	103	112	103	69
o-dichlorobenzene	17	17	8	9
p-dichlorobenzene	21	25	28	35

data from (89)

production. Growth of this product has been limited by drop in demand as a metal degreaser. Perchloroethylene demand dropped sharply in the early 1990s due to stricter regulation of emissions from the dry cleaning industry. Also, perchloroethylene demand was reduced because it was used as a precursor for many chlorofluorocarbons that were eliminated in the 1990s. Demand has increased slightly in the late 1990s due to the increase in demand for hydrofluorocarbons, some of which can be produced from perchloroethylene (92).

Spot pricing for trichloroethylene f.o.b. the U.S. Gulf Coast is from \$530–620/t. Pricing for perchloroethylene, on the same basis, is \$400–500/t.

6.3. Chlorinated Propanes and Higher. Allyl chloride, like EDC and vinyl chloride, has experienced a relatively constant rate of growth over the past decade. Allyl chloride is used primarily to produce epichlorohydrin. Epichlorhydrin is used in the synthesis of of many chemicals. The growth in consumption of allyl chloride has been driven by increased demand for nearly all end uses. World consumption is on the order of 600–700 thousand metric tons per year.

U.S. chloroprene consumption has decreased slowly over the past decade. Almost all chloroprene that is manufactured is used to produce polychloroprene and other synthetic rubbers. Loss of demand is primarily due to substitution of newer, less expensive polymers (87). Annual U.S. consumption is roughly in the range of 90,000 metric tons per year.

6.4. Chlorinated Aromatics. The United States demand for chlorinated benzenes generally dropped during the 1990s, though the pressures that contributed to this are different than the case of many other chlorocarbons (Table 6). The decline in demand for monochlorobenzene since 1995 was due to a change in the raw material used by Dow Chemical to produce diphenyl ether and phenylphenols. Phenol was substitued for monochlorobenzene. U.S. consumption of o-dichlorobenzene was reduced when DuPont ceased manufacture of 3,4-dichloroaniline. Demand for p-dichlorobenzene has shown steady growth, mainly as a raw material for the production of polyphenylene sulfide resin (89).

Spot pricing for chlorobenzene is ∼\$1200/t, f.o.b.

7. Specifications, Standards, and Quality Control

Chlorocarbons and chlorohydrocarbons are typically produced at very high purities. Typical organic impurities are chlorinated by-products of the manufacturing process that are not completely removed during distillation. These are usually present at levels ranging up to a few hundred parts per million. Water is also commonly present, again at ppm levels.

Metals are generally not present at detectable levels, and generally are a result of contamination during shipping and storage. Most metals, with the exception of iron(III) chloride, are not very soluble in chlorohydrocarbons. Iron(III) chloride can be detected using colorimetric methods. Like metals, nonvolatile matter (NVM) is typically not present in chlorocarbons or chlorohydrocarbons, and its presence is generally a result of contamination.

The presence of acidity in chlorocarbon and chlorohydrocarbon products is generally the result of oxygenated impurities from the manufacturing process. These impurities include compounds such as phosgene and acetyl chloride and its chlorinated dervatives. HCl can also be present in some products. It can be present as a result of incomplete removal during the purification process, or due to degradation of the product.

Alkalinity is used as a specification for chlorocarbon or chlorohydrocarbon products that contain amines as part of the inhibitor package. The presence of alkalinity is an indication that the inhibitor is present and has not been consumed as a result of acidic impurities or product degradation.

8. Health and Safety Factors

The environmental and health effects of chlorinated hydrocarbons have been studied extensively, beginning shortly after these products were developed. The wide use of these compounds as replacements for more flammable solvents in a diverse number of industries raised questions about the potential impact that these products had on humans and the environment. On occasion, the existence of such data has led, in part, to more extensive restriction than with other products.

From an environmental standpoint, issues have been raised about the impact of these compounds both on the atmosphere and in groundwater. Under the Montreal Protocol, emissive uses of 1,1,1-trichloroethane and carbon tetrachloride have already been phased out in the developed world, due to their impact on stratospheric ozone. The ban is nearing in developing countries, with certain countries taking such action before the 2010 date. Further, due to their classification as hazardous air pollutants, regulations, known as Maximum Achievable Control Technology (MACT) Standards, have established controls for emissions from production, processing, and use. For most chlorinated hydrocarbons, ie, perchloroethylene and trichloroethylene, restrictions have also been established to deter groundwater contamination. Such concerns have occurred as a result of their toxicity as well as persistency in the environment. To ensure safety of water supplies, effluent limitation guidelines as well as drinking water standard contaminant levels have been established. Finally, the chlorinated hydrocarbons are also considered hazardous under the U.S. Federal Resource Conservation and Recovery Act (RCRA) as well as similar laws that exist in other countries. Wastes must be stored, transported and disposed of in accordance with applicable RCRA and state requirements.

Challenges continue to be waged about the potential impact that chlorinated hydrocarbon products have on human health. The earliest symptoms associated with these compounds were related to their impact on the central nervous system. Chloroform was initially marketed as an anaesthetic due to this effect. The most significant findings in long-term studies are centered on effects on liver and kidney for trichloroethylene, perchloroethylene, and ethylene dichloride, while methylene chloride effects are mostly associated with the liver and the lungs. For vinyl chloride, the most significant toxic effect is carcinogenicity, specifically angiosarcoma of the liver, which occurs with extremely high exposure to vinyl chloride. As a result of the findings in animal studies, chlorinated hydrocarbons have been the focus of many epidemiology studies. Yet, vinyl chloride is the only compound in this category of chemicals that has been classified as a known human carcinogen by numerous authoritative bodies as a result of the findings of epidemiology studies. These studies have led to the inclusion of many chlorinated hydrocarbons in various assessments of carcinogenicity potential by the International Agency for Research of Cancer (IARC), EPA, MAK Commission, National Toxicology Program, etc. For example, methylene chloride and ethylene dichloride are classified as "2b: possibly carcinogenic"; and trichloroethylene and perchloroethylene as "2a: probably carcinogenic".

To ensure workers health is protected both from potential acute and chronic health hazards, workplace exposure limits have been established in the United States by the Occupational Safety and Health Administration (OSHA) as well as other agencies. Both 8-h time-weighted average (TWA) and short-term exposure limits (STEL) exist for these chlorinated hydrocarbons.

Like all chemicals, chlorinated hydrocarbons do have inherent hazards associated with them. However, the products can be used safely, without concern of significant health risk, if regulatory requirements and manufacturers' directions are followed. As is true for all chemicals, it is nevertheless important to continue to take steps to ensure that workplace exposure be kept as low as practical.

The use of chlorinated hydrocarbons as solvents is gradually declining due to the costs associated with the handling and disposal of these compounds. With improvements made in closed-loop technology systems, both during delivery and use, compliance is achieved more easily albeit with a bigger investment. However, for operations that require the performance characteristics of chlorinated solvents, these systems can support their continued use. The primary use of chlorinated hydrocarbons continues to be as chemical intermediates in the production of a variety of compounds. There is limited overall growth as several of the downstream products are being faced with environmental challenges as well. Fortunately, the EH&S and regulatory hurdles that affect solvent consumption do not have the same ramifications for intermediate applications where there are far fewer issues around emission control.

9. Uses

Because of their widely varying properties, chlorocarbons and chlorohydrocarbons are used in a broad range of applications. Many of these products have excellent solvent properties. For this reason, they continue to be in demand for metal cleaning and vapor degreasing operations and as reaction media for other chemical processes. They are also used as chemical intermediates for a variety of products, from polymers to silicones, fluorocarbons, and other speciality chemicals.

9.1. Chlorinated Methanes. Methyl chloride is used as a raw material for the manufacture of chemical products such as silicones, methyl cellulose, quartenary ammonium compounds, agricultural products, and butyl rubber. Methyl chloride is also the feedstock for the majority of the world's production of methylene chloride and chloroform (see SILICON COMPOUNDS, SILICONES; METHYLENE CHLORIDE; CHLOROFORM).

Methylene chloride is used as both a solvent and as a chemical intermediate. As a solvent, it is used in metal cleaning and degreasing, in adhesive and paint stripper formulations, in the manufacture of certain polymers and films, and as a reaction media for the production of many pharmaceuticals. It is used as an intermediate for the production of hydrofluorocarbon-32 (HFC-32), an ingredient in fluorocarbon blends for refrigerants (see FLUORINATED ALIPHATIC COMPOUNDS; REFRIGERATION).

Chloroform is primarily used in the manufacture of hydrochlorofluorocarbon-22 (HCFC-22). This product is used as a refrigerant, and also as a monomer for fluoropolymer production (see FLOURINE-CONTAINING POLYMERS, POLYTETRAFLUORO-ETHYLENE).

Carbon tetrachloride was mainly used in the production of chlorofluorocarbon-11 (CFC-11) and CFC-12 in the developed world, before the use of these products was phased out. It is still used as a chemical initiator for the production of vinyl chloride from EDC.

9.2. Chlorinated Ethanes and Ethylenes. Ethyl chloride historically was used for the production of tetraethyllead, an anti-knock compound for fuels. Use of tetraethyllead has been eliminated in the United States and many other parts of the world. Ethyl chloride continues to be used for the manufacture of ethyl cellulose. It is also used for the production of alkyl catalysts, as a topical anesthetic, and in aerosol and dye formulations.

More EDC is produced than any other chlorinated hydrocarbon. The majority of EDC that is produced is used to manufacture vinyl chloride. EDC is also used as a raw material to produce other chlorinated ethanes and ethylenes, as well as ethyleneamines. Vinyl chloride is chiefly used as a monomer for the manufacture of PVC. A minor amount of vinyl chloride is used in the manufacture of poly(vinylidene chloride).

Like vinyl chloride, vinylidene chloride is primarily used as a monomer for polymer production. Poly(vinylidene chloride) is a copolymer of vinyl chloride and vinylidene chloride. Vinylidene chloride is also used in the manufacture of HCFCs and certain specialty products and formulations (see FLOURINE-CONTAINING POLYMERS, POLY(VINYLIDENE FLUORIDE)) .

Trichloroethylene and perchloroethylene are both used as vapor degreasers and in the manufacture of hydrofluorocarbon-134a (HFC-134a). Perchloroethylene is an intermediate for many other CFC, HCFC, and HFC products. It is also used extensively as a dry cleaning solvent.

The production of 1,1,1-trichloroethane has been largely phased out. This product was mainly used in vapor degreasing applications, and its use was banned by the Montreal Protocol. It is still used as an intermediate for the production of hydrochlorofluorocarbons HCFC-141b and HCFC-142b. These products are used primarily as blowing agents. The use of HCFC-141b will be phased out at the end of 2002.

9.3. Chlorinated Propanes and Higher. 3-Chloro-1-propylene (allyl chloride) is produced as a chemical intermediate for the manufacture of 1-chloro-2,3-epoxypropane (epichlorohydrin). Epichlorohydrin is in turn used as an intermediate in the production of epoxy resins and glycerol. Epichlorohydrin is also used to produce elastomers and other speciality chemicals. A by-product of allyl chloride manufacture, 1,3-dichloropropene, is used as a soil fumigant for the control of nematodes.

Chloroprene is used to produce neoprene (polychloroprene) and other synthetic rubbers.

9.4. Chlorinated Aromatics. Monochlorobenzene (MCB), *o*-dichlorobenzene (*o*-DCB), and *p*-dichlorobenzene (*p*-DCB) are the major chlorinated aromatic species that are produced on an industrial scale. MCB is used as both a chemical intermediate and as a solvent. As an intermediate, it is used to produce chloronitrobenzene, pesticides, and pharmaceutical products. In solvent applications, MCB is used in the manufacture of isocyanates. Its high solvency allows it to to be used with many different types of resins, adhesives, and coatings.

o-Dichlorobenzene is primarily used for organic synthesis, especially in the production of 3,4-dichloroaniline herbicides. Like MCB, it can be used as a solvent, especially in the production of isocyanates. It is also used in motor oil and paint formulations.

p-Dichlorobenzene is used as a moth repellent, and for the control of mildew and fungi. It is also used for odor control. It is a chemical intermediate for the manufacture of pharmaceuticals and other organic chemicals.

BIBLIOGRAPHY

"Survey" under -Chlorocarbons, "Hydrocarbons" in *ECT* 3rd ed., Vol. 5, pp. 668–676, by W. L. Archer, Dow Chemical, U.S.A.; in *ECT* 4th ed., Vol. 5, pp. 1017–1028, by Daniel J. Reeds, Dow Chemical, U.S.A.; "Chlorocarbons and Chlorohydrocarbons, Survey" in *ECT* (online), posting date: December 4, 2000, by Daniel J. Reeds, Dow Chemical, U.S.A.

CITED PUBLICATIONS

1. Jpn. Pat. 01 308,238 (Dec. 12, 1989), I. Muramoto and N. Sasaki (to Asahi Chemical Industry Co., Ltd., Japan).
2. Jpn. Pat. 59 044,290 B4 (Oct. 29, 1984), Showa (to Tokuyama Soda Co., Ltd., Japan).
3. N. V. Kruglova and R. K. Freidlina, *Dokl. Akad. Nauk SSSR, Ser. Khim.*, (**2**), 388 (1984).
4. R. Pieck and J. C. Jungers, *Bull. Chem. Soc. Belg.* **60**, 357, 1951.
5. G. Chiltz, P. Goldfinger, G. Huybrechts, G. Martins, G. Verbeke, *Chem. Rev.* **63**, 355 (1963).
6. A. I. Semenov, V. V. Lebedev, E. R. Berlin, Yu. A. Treger, A. B. Rabinovich, *Zhur. Prikl. Khimii* **58**(4), 840 (1985).
7. V. N. Antonov, V. I. Rozhkov, and A. A. Zalikin, *Zh. Prikl. Khim.* **60**(6), 1347 (1987).

8. H. P. Rothbaum, I. Ting, and P. W. Robertson, *J. Chem. Soc.*, 980 (1948); D. A. Evans, T. R. Watson, and P. W. Robertson, *J. Chem. Soc.* 1624 (1950).

9. G. Huybrechts, I. Theys, B. Van Mele, *Int. J. Chem. Kin.* **28**, 755–761 (1996).

10. R. S. Timonen, J. J. Russell, D. Gutman, *Int. J. Chem. Kin.* **18**, 1193 (1986).

11. P. C. Beadle, J. H. Knox, *J. Chem. Soc. Faraday Trans. 1* **72**, 1418 (1974).

12. W. B. DeMore and co-workers, JPL Publication 97-4, 100–266 (1997).

13. S. Dusoleil and co-workers, *Trans. Faraday Soc.* **57**, 2197 (1961).

14. J. A. Leermakers and R. G. Dickenson, *J. Am. Chem. Soc.* **54**, 3853, 4648 (1932).

15. U.K. Pat., 1,208,957 (1970), A. Campbell and R. A. Carruthers (to Imperial Chemical Industries).

16. Jpn. Pat. 18,204 (1973), E. Nishihara, K. Suzuki, R. Muraoka, I. Goto, and I. Kaneko (to Asahi Glass Co.).

17. U.S. Pat. 4,049,730 (Sept. 20, 1977), J. K. Baggs (to Exxon Research and Engineering Co.).

18. E. Milchert, W. Goc, J. Stefanska, and W. Pazdzioch, *Ind. Eng. Chem. Res.* **35**, 3290 (1996).

19. I. F. Pimenov and E. V. Sonin, *Zh. Fiz. Khim.* **45**, 2099 (1971).

20. R. Zdenek and J. Vrany, *Chemicky Prumysel* **27**(52), 10 (1977).

21. R. Walraevens, P. Trouillet, and A. Devos, *Int. J. of Chem. Kin.* **6**, 777 (1974).

22. P. W. Sherwood, *Ind. Eng. Chem.* **54**, 29 (1962).

23. J. March, *Advanced Organic Chemistry: Reactions, Mechanism and Structure*, 4th ed., John Wiley and Sons, Inc., New York, 1992, p. 983.

24. R. Bruckner, *Advanced Organic Chemistry:Reaction Mechanisms*, Harcourt Academic Press, New York, 2002, p. 141.

25. F. G. Bordwell, *Acc. Chem. Res.* **5**, 374 (1972).

26. K. A. Kurginyan, *Zhur. Vses. Khim. Ob-va im. D.I Mendeleeva* **31**(2), 44 (1986).

27. J. Svoboda, I. Ondrus, and J. Mazanec, *Petrochemia* **20** (3–4), 120 (1980).

28. G. H. Grant and C. N. Hinshelwood, *J. Chem. Soc.* 258 (1933).

29. K. Okamoto, H. Matsuda, H. Kawasaki, and H. Shingu, *Bull. Chem. Soc. J.* **40**, 1917 (1967).

30. U.S. Pat. 3,304,336 (Feb. 14, 1967), W. A. Callahan (to Detrex Chemical Industries, Inc.).

31. Aust. Pat. 14,649/70 (Nov. 11, 1971), B. H. McConkey and P. R. Wilkinson (to ICI Australia Limited).

32. U.K. Pat. 1,281,278 (July 12, 1972), (to ICI Australia Limited).

33. A. P. Khardin, A. V. Spitsyn, and P. Y. Gokhberg, *Khim. Prom-st. (Moscow)* (4), 208 (1982).

34. I. Mochida and Y. Yoneda, *J. Org. Chem.* **33**(5), 2161 (1968).

35. I. Mochida, A. Kato, and T. Seiyama, *J. Catal.* **18**, 33 (1970).

36. I. Mochida, Y. Anju, H. Yamamoto, A. Kato, and T. Seiyama, *Bull. Chem. Soc. Jpn.* **44**, 3305 (1971).

37. T. Sankarshana and M. B. Rao, *Ind. Chem. Eng.* **31**(1), 38 (1989).

38. I. Mochida, Y. Anju, A. Kato, and T. Seiyama, *J. Org. Chem.* **39**(25), 3785 (1974).

39. D. H. R. Barton, *J. Chem. Soc.* 1624 (1950).

40. N. V. Lashmanova, V. A. Kolesnikov, and R. V. Efremov, *Zh. Prikl. Khim. (Leningrad)* **63**(8), 1899 (1990).

41. F. Gajewski, J. Ogonowski, and J. Rakoczy, *Przem. Chem.* **62**(2), 95 (1983).

42. U.S. Pat. 2,577,388 (Dec. 4, 1951), G. W. Warren (to The Dow Chemical Company).

43. U.S. Pat. 4,211,728, (July 8, 1980), J. G. Guérin (to Produits Chimiques Ugine Kuhlmann).

44. A. Bencsura, V. D. Knyazev, S.-B. Xing, I. R. Slagle, and D. Gutman, *Symp. Int. Combust. Proc.* **24**, 629 (1992).

45. Y. Ito, *J. Soc. Org. Syn. Chem. Jpn.* **23**(1), 33 (1965).
46. E. Cavaterra, *Hydrocarbon Proc.*, Dec. (1988).
47. U.S. Pat. 3,040,109 (June 19, 1962), R. E. Feathers and R. H. Rogerson (to Pittsburg Plate Glass Co.).
48. X. Youchang, Z. Huixin, and R. Wang, *Scientia Sinica* **23**(6), (1980).
49. H. D. Eichhorn, C. Jackh, W. D. Mross, and H. Schuler, *8th Intl. Congress on Catal*, Berlin, 1984.
50. M. A. Thyagarajan, R. Kumar, and N. R. Kuloor, *Ind. Eng. Chem., Proc. Dec. Div.* **5**(3), 209 (1966).
51. E. G. Schlosser, M. Rosseberg, and W. Lendle, *Chem.-Ing.-Tech.* **42**, 1215 (1970).
52. S. Akiyama, T. Hisamoto, T. Takada, and S. Mochizuki, *Catal. Convers. Synth. Gas Alcohols Chem. Proc. Symp.*, 1984.
53. R. Bruckner, *Advanced Organic Chemistry:Reaction Mechanisms*, Harcourt Academic Press, New York, 2002, p. 69.
54. U.S. Pat. 2,091,986 (Sept. 7, 1937), L. C. Holt and H. W. Daudt (to E.I. du Pont de Nemours and Co.).
55. Fr. Pat. 2,270,224, I. O. Larson and J. A. Gentilucci (to E.I. du Pont de Nemours and Co.).
56. J. March, *Advanced Organic Chemistry: Reactions, Mechanism and Structure*, 4th ed., John Wiley and Sons, Inc., New York, 1992, p. 531.
57. J. March, *Advanced Organic Chemistry: Reactions, Mechanism and Structure*, 4th ed., John Wiley and Sons, Inc., New York, 1992, p. 430.
58. U.S. Pat. Appl. 2002/0098133 (July 25, 2002), D. W. Jewell and co-workers (Appl. No. 10/104,703).
59. Y.-C. Yen, J.-T. Huang, and C.-J. Louh, "Pesticides and Intermediatesa," *Process Economics Program Report No. 171*, December 1983.
60. F. W. Breitbarth and S. Rottmayer, *Plasma Chem. Plasma Process* **6**(4), 381 (1986).
61. Jpn. Pat. 05-320076, M. Shinshuke, Y. Masaru, and T. Shin (to A.G. Technology KK).
62. J. March, *Advanced Organic Chemistry: Reactions, Mechanism and Structure*, 4th ed., John Wiley and Sons, Inc., New York, 1992, p. 544.
63. R. H. Schwaar and J. Kwok, *Process Economics Program Review 92-3-3*, April, 1994.
64. U.S. Pat. 3,126,419 (Mar. 24, 1964), W. M. Burks, Jr., (to Stauffer Chemical).
65. U.S. Pat. 6,452,058 B1 (Sept. 17, 2002), A. E. Schweizer, M. E. Jones, and D. A. Hickman (to Dow Global Tecnologies, Inc.).
66. U.S. Pat. 2,469702 (May 10, 1949), C. C. Schwegler and F. M. Tennant (to the Dow Chemical Co.).
67. U.S. Pat. 2,818,448 (Dec. 31, 1957), H. E. O'Connell and J. H. Huguet (to Ethyl Corp.).
68. *Process Economics Program Report 5D – Vinyl Chloride* (March 2000).
69. U.S. Patent 5,087,791 (Feb. 11, 1992), A. J. Magistro (to B.F. Goodrich).
70. World Pat. 9,507,249, I. M. Clegg and R. Hardman (to EVC Technology AG).
71. Jpn. Pat. 56 145,231 A2 (Nov. 11, 1981), Showa (to Asahi-Dow Ltd., Japan).
72. L. A. Wasselle, "Vinylidene Chloride Monomer and Polymers", *Process Economics Program Report No. 105*, June 1976.
73. Russ. Pat. 2,078,071, Yu. V. Belokopygov, N. D. Konovalova, and Yu. A. Serguchev.
74. U.S. Pat. 2,989,570 (June 20, 1961), F. Conrad and M. L. Gould (to Ethyl Corp.).
75. J. F. Knoop and G. R. Neikirk, *Hydrocarbon Processing*, Nov. 1972.
76. U.K. Pat. 1,276,431, W. E. Wimer (to PPG Industries, Inc.).
77. Jpn. Pat. P2001-322954A, S. Handa, M. Hiraoka, K. Yokoyama, and T. Nakamura (to Asahi-Penn Chemical Co., Ltd.).
78. F. F. Muganlinskii and co-workers, *Zhur. Prikladnoi Khim* **49**(6) 1424 (1976).
79. S.U. Pat. 1,817,762 A3 (May 23, 1993), B. B. Mamedov and co-workers.
80. U.S. Pat. 3,949,101 (Apr. 6, 1976), M. C. Sze (to The Lummus Co.).

81. S. Tsuda, *Chem. Eng.* **74** (May 4, 1970).
82. U.S. Pat. 3,691,240 (Sept. 12, 1970) C. E. Kircher, D. R. McAlister, and D. L. Brothers (to Detrex Chemical Industries, Inc.).
83. U.K. Pat. 1,275,040 (Aug. 14, 1968), R. P. Obrecht and co-workers (to Hoechst).
84. Z. Pokorska and co-workers, *Przemysl Chemiczny* **66**(2), 88 (1987).
85. Z. Pokorska, co-workers, *Przemysl Chemiczny* **68**(3), 121 (1988).
86. L. M. Porter and F. F. Rust, *J. Am. Chem. Soc.* **78**, 5571 (1956).
87. S. Berthiaume and M. Yoneyama, "Polychloroprene (Neoprene) Elastomers", *Chem. Econ. Handbook Data Summary*, January, 2002.
88. U.S. Pat. 4,418,232 (Nov. 19, 1983), L. J. Maurin III (to E.I. du Pont de Nemours and Co.).
89. J. G. Lacson, C. Cometta, and M. Yoneyama, "Chlorobenzenes", *Chem. Econ. Handbook Product Review*, December, 1999.
90. E. Linak and G. Toki, "Chlorinated Methanes", *Chem. Econ. Handbook Product Review*, December, 2001.
91. A. M. Jebens and A. Kishi, "Ethylene Dichloride", *Chem. Econ. Handbook Product Review*, January, 2001.
92. E. Linak and G. Toki, "C2 Chlorinated Solvents", *Chem. Econ. Handbook Product Review*, January, 2002.

KENRIC A. MARSHALL
The Dow Chemical Company

CHLOROETHYLENES AND CHLOROETHANES

1. Ethylene Dichloride

1.1. Introduction.

1,2-Dichloroethane [107-06-2], ethylene chloride, ethylene dichloride, CH_2ClCH_2Cl, is a colorless, volatile liquid with a pleasant odor, stable at ordinary temperatures. It is miscible with other chlorinated solvents and soluble in common organic solvents as well as having high solvency for fats, greases, and waxes. It is most commonly used in the production of vinyl chloride monomer.

1.2. Physical and Chemical Properties.

The physical properties of 1,2-dichloroethane are listed in Table 1.

Pyrolysis. Pyrolysis of 1,2-dichloroethane in the temperature range of 340–515°C gives vinyl chloride, hydrogen chloride, and traces of acetylene (3) and 2-chlorobutadiene. Reaction rate is accelerated by chlorine (4), bromine, bromotrichloromethane, carbon tetrachloride (5), and other free-radical generators. Catalytic dehydrochlorination of 1,2-dichloroethane on activated alumina (6), metal carbonate, and sulfate salts (7) has been reported, and lasers have been used to initiate the cracking reaction, although not at a low enough temperature to show economic benefits.

Hydrolysis. Heating 1,2-dichloroethane with excess water at 60°C in a nitrogen atmosphere produces some hydrogen chloride. The rate of evolution is

Table 1. **Properties of 1,2-Dichloroethane**[a]

Property	Value
melting point, °C	−35.3
boiling point, °C	83.7
density at 20°C, g/L	1.2529
n_{D}^{20}	1.4451
viscosity at 20°C mPa·s (= cP)	0.84
surface tension at 20°C, mN/m (= dyn/cm)	31.28
specific heat at 20°C, J/(g·K)[a]	
liquid	1.288
gas	1.066
latent heat of vaporization at 20°C, J/g[b]	323.42
latent heat of fusion, J/g[b]	88.36
critical temperature, °C	290
critical pressure, MPa[c]	5.36
critical density, g/L	0.44
flash point °C	
closed cup	17
open cup	21
explosive limits in air at 25°C, vol %	6.2–15.6
autoignition temperature in air, °C	413
thermal conductivity, liquid at 20°C, W/(m·K)[d]d	0.143
heat of combustion, kJ/g[b]	12.57
heat of formation, kJ/mol[b]	
liquid	157.3
vapor	122.6
dielectric constant	
liquid, 20°C	10.45
vapor, 120°C	1.0048
dipole moment, C·m[e]	5.24×10^{-30}
coefficient of cubical expansion, mL/g, 0–30°C	0.00116
vapor pressure, kPa[f]	
10°C	5.3
20°C	8.5
30°C	13.3
solubility at 20°C, g	
1,2-dichloroethane in 100 g H_2O	0.869
H_2O in 100 g 1,2-dichloroethane	0.160
azeotropes[g], bp, °C	
with 10.5% H_2O	72
with 5% H_2O and 17% ethanol	66.7

[a]See Ref. 11 for additional property data.
[b]To convert J to cal, divide by 4.184.
[c]To convert MPa to atm, multiply by 9.87.
[d]To convert W/(m·K) to Btu·ft)/(h·ft^2·°F), divide by 1.73.
[e]To convert C·m to debyes, multiply by 3×10^{29}.
[f]To convert kPa to mm Hg, multiply by 7.5.
[g]See Ref. 2 for additional binary azeotropes.

dependent on the temperature and volume of the aqueous phase. Hydrolysis at 160–175°C and 1.5 MPa (15 atm) in the presence of an acid catalyst gives ethylene glycol, which is also obtained in the presence of aqueous alkali at 140–250°C and up to 4.0 MPa (40 atm) pressure (8).

Oxidation. Atmospheric oxidation of 1,2-dichloroethane at room or reflux temperatures generates some hydrogen chloride and results in solvent discoloration. A 48-h accelerated oxidation test at reflux temperatures gives only 0.006% hydrogen chloride (9). Addition of 0.1–0.2 wt. % of an amine, eg, diisopropylamine, protects the 1,2-dichloroethane against oxidative breakdown. Photooxidation in the presence of chlorine produces monochloroacetic acid and 1,1,2-trichloroethane (10).

Corrosion. Corrosion of aluminum, iron, and zinc by boiling 1,2-dichloroethane has been studied. Dry and refluxing 1,2-dichloroethane completely consumed a 2024 aluminum coupon in a 7-d study, whereas iron and zinc were barely attacked. Aluminum was attacked less than iron or zinc by refluxing with 1,2-dichloroethane containing 7% water. Corrosion rates in μm/yr (mils penetration per year or mpy) in dry solvent are 0.254 (0.01) for iron and 3.05 (0.12) for zinc. In the wet solvent, the corrosion rate for iron increases to 145 μm/yr (5.7 mpy) and for zinc to 1.2 mm/yr (47 mpy). Corrosion rate for aluminum in the wet solvent is 2.36 mm/yr (92 mpy) as compared to complete dissolution in the dry solvent.

Nucleophilic Substitution. The kinetics of the bimolecular nucleophilic substitution of the chlorine atoms in 1,2-dichloroethane with NaOH, NaOC$_6$H$_5$, (CH$_3$)$_3$N, pyridine, and CH$_3$COONa in aqueous solutions at 100–120°C has been studied (11). The reaction of sodium cyanide with 1,2-dichloroethane in methanol at 50°C to give 3-chloropropionitrile proceeds very slowly. Dimethyl sulfoxide as a solvent for the reaction greatly enhances nucleophilic substitution of the chlorine atom. Further reaction of sodium cyanide at room temperature gives acrylonitrile (qv), CH$_2$=CHCN (12). 1,2-Dichloroethane reacts with toluene in the presence of Friedel-Crafts catalysts such as AlBr$_3$, AlCl$_3$, GaCl$_3$, and ZrCl$_3$ (13). Ammonolysis of 1,2-dichloroethane with 50% aqueous ammonia at 100°C is a primary commercial process for producing ethyleneamines (14).

1.3. Manufacture. 1,2-Dichloroethane is produced by the vapor- (15) or liquid-phase chlorination of ethylene. Most liquid-phase processes use small amounts of ferric chloride as the catalyst. Other catalysts claimed in the patent literature include aluminum chloride, antimony pentachloride, and cupric chloride and an ammonium, alkali, or alkaline-earth tetrachloroferrate (16). The chlorination is carried out at 40–50°C with 5% air or other free-radical inhibitors (17) added to prevent substitution chlorination of the product. Selectivities under these conditions are nearly stoichiometric to the desired product. The exothermic heat of reaction vaporizes the 1,2-dichloroethane product, which is purified by distillation.

$$H_2C\!\!=\!\!CH_2 + Cl_2 \xrightarrow[50-150°C]{FeCl_3} ClH_2C\!-\!CH_2Cl$$

Oxychlorination of ethylene has become the second important process for 1,2-dichloroethane. The process is usually incorporated into an integrated vinyl chloride plant in which hydrogen chloride, recovered from the dehydrochlorination or cracking of 1,2-dichloroethane to vinyl chloride, is recycled to an oxychlorination unit. The hydrogen chloride by-product is used as the chlorine source in the chlorination of ethylene in the presence of oxygen and copper chloride catalyst:

$$2\,H_2C\!=\!CH_2 + 4\,HCl + O_2 \xrightarrow[270°C]{CuCl_2} 2\,ClH_2C\!-\!CH_2Cl + H_2O$$

Reactor designs have included fixed and fluidized beds. A fluidized-bed oxychlorination reactor developed by B. F. Goodrich is claimed to provide very good temperature control (18). A large number of patents deal with the catalyst technology (19–27), which usually includes $CuCl_2$ and minor amounts of by-product inhibitors, such as potassium, sodium, lithium, or magnesium. To reduce oxidation to carbon dioxide and carbon monoxide, plant designs may include two or three reactors in series, with the HCl and oxygen feeds split to the secondary reactor(s) to decrease the C_2H_4:oxygen ratio in each reactor. By-products of this reaction include carbon dioxide, carbon monoxide, ethyl chloride, 1,1,2-trichloroethane, and trichloroacetaldehyde (chloral). The reactor products are usually condensed, unreacted HCl and water are separated from the organics, and the 1,2-dichloroethane is purified by distillation. Unreacted ethylene and oxygen can be recycled to the reactor, or sent to a vent recovery system; however, a purge is needed in the recycle system to prevent inert gas buildup in the recycle stream. Air or pure oxygen can be used; oxygen-based systems lose less ethylene during the inert purge.

1.4. Economic Aspects. A significant portion of ethylene chloride (EDC) is converted to vinyl chloride monomer which is used in the production of poly(vinyl chloride). Worldwide consumption of EDC reached a level of 32×10^6t in 1999. North America is the largest exporter of EDC. The Middle East is also a major exporter. It is cheaper for Asian companies to import EDC rather than make it (28).

Table 2 lists the U.S. producers and their capacities.

Table 2. **U.S. Producers of Ethylene Dichloride and Their Capacities**

Producer	Location	Capacity[a], $\times 10^3$t ($\times 10^6$lb)
Borden	Geismar, La.	535 (1,180)
Dow	Freeport, Tex. (2 plants)	2,041 (4,500)
Dow	Plaquemine, La.	1,043 (2,300)
Formosa Plastics	Baton Rouge, La.	816 (1,800)
Formosa Plastics	Point Comfort, Tex.	1,134 (2,500)
Georgia Gulf	Lake Charles, La.	771 (1,700)
Georgia Gulf	Plaquemine, La.	1,270 (2,800)
Occidental Chemical	Convent, La.	680 (1,500)
Occidental Chemical	Ingleside, Tex.	680 (1,500)
OxyMar	Ingleside, Tex.	1,542 (3,400)
OxyVinyls	Deer Park, Tex.	884 (1,950)
OxyVinyls	La Porte, Tex.	1,814 (4,000)
PHH Monomers	Lake Charles, La.	635 (1,400)
PPG	Lake Charles, La.	726 (1,600)
Vulcan	Geismar, La.	272 (600)
Westlake	Calvert City, Ky.	885 (1,950)
	Total	*15,730 (34,680)*

[a]Ref. 29.

Demand in 2000 was 6.97×10^3t, expected demand in 2004 is 8.13×10^3t. Prices for the period 1995–2000 were a high of $.10/kg and a low of $0.05/kg (29).

Commercial technology using a direct ethylene to vinyl chloride monomer could reduce the demand for EDC. The market was soft in 2001 because of a decline in construction and the slowing of the economy. Growth through 2005 is expected at the rate of 3.5% (29).

1.5. Health and Safety Factors. 1,2-Dichloroethane at high vapor concentrations (above 200 ppm) can cause central nervous system depression and gastrointestinal upset characterized by mental confusion, dizziness, nausea, and vomiting. Liver, kidney, and adrenal injuries may occur at the higher vapor levels. The recommended 1991 AGCIH vapor exposure TWA standard for 1,2-dichloroethane was 10 ppm, with a STEL guideline of 40 ppm. The odor threshold for 1,2-dichloroethane is 50–100 ppm and thus odor does not serve as a good warning against possible overexposure.

1,2-Dichloroethane is one of the more toxic chlorinated solvents by inhalation (30). The highest nontoxic vapor concentrations in chronic exposure studies with various animals range from 100 to 200 ppm (31,32). 1,2-Dichloroethane exhibits a low single-dose oral toxicity in rats; LD_{50} is 680 mg/kg (30). Repeated skin contact should be avoided since the solvent can cause defatting of the skin, severe irritation, and moderate edema. Eye contact may have slight to severe effects.

1.6. Environmental Concerns. Removal of metal chlorides from the bottoms of the liquid-phase ethylene chlorination process has been studied (33). A detailed summary of production methods, emissions, emission controls, costs, and impacts of the control measures has been made (34). Residues from this process can also be recovered by evaporation, decomposition at high temperatures, and distillation (35). A review of the by-products produced in the different manufacturing processes has also been performed (36). Several processes have been developed to limit ethylene losses in the inerts purge from an oxychlorination reactor (37,38).

1.7. Uses. Production of vinyl chloride monomer comprises 94% of EDC use. Three percent goes to the production of ethylene amines; 1% goes to the production of 1,1,1-trichloroethane; 1% goes to the production of vinylidene chloride; 1% goes to miscellaneous uses (includes production of trichloroethylene and perchloroethylene) (29).

Worldwide environmental concerns have caused a reduction in the use of chlorine and chlorinated derivatives. Consumption of chlorinated solvents has been negatively impacted. The use of poly(vinyl chloride) is also under scrutiny (28).

2. Dichloroethylene

2.1. Introduction. 1,1-Dichloroethylene [75-35-4] is more commonly known as vinylidene chloride and is covered in an article in the *Encyclopedia* by that title.

1,2-Dichloroethylene [540-59-0] (1,2-dichloroethene) is also known as acetylene dichloride, dioform, α,β-dichloroethylene, and *sym*-dichloroethylene. It exists as a mixture of two geometric isomers: *trans*-1,2-dichloroethylene [156-60-5] (**1**) and *cis*-1,2-dichloroethylene [156-59-2] (**2**).

The isomeric mixture is a colorless, mobile liquid with a sweet, slightly irritating odor resembling that of chloroform. It decomposes slowly on exposure to light, air, and moisture. The mixture is soluble in most hydrocarbons and only slightly soluble in water. The cis–trans proportions in a crude mixture depend on the production conditions. The isomers have distinct physical and chemical properties and can be separated by fractional distillation.

2.2. Physical and Chemical Properties.   1,2-Dichloroethylene consists of a mixture of the cis and trans isomers, as manufactured. The physical properties of both isomeric forms are listed in Table 3. Binary and ternary azeotrope data for the cis and trans isomers are given in Table 4.

2.3. Manufacture.   1,2-Dichloroethylene can be produced by direct chlorination of acetylene at 40°C. It is often produced as a by-product in the chlorination of chlorinated compounds (40) and recycled as an intermediate for the synthesis of more useful chlorinated ethylenes (41). 1,2-Dichloroethylene can be formed by continuous oxychlorination of ethylene by use of a cupric chloride–potassium chloride catalyst, as the first step in the manufacture of vinyl chloride [75-01-4] (42).

The trans isomer is more reactive than the cis isomer in 1,2-addition reactions (43). The cis and trans isomers also undergo benzyne, C_6H_4, cycloaddition (44). The isomers dimerize to tetrachlorobutene in the presence of organic peroxides. Photolysis of each isomer produces a different excited state (45,46). Oxidation of 1,2-dichloroethylene in the presence of a free-radical initiator or concentrated sulfuric acid produces the corresponding epoxide [60336-63-2], which then rearranges to form chloroacetyl chloride [79-04-9] (47).

The unstabilized grade of 1,2-dichloroethylene hydrolyzes slowly in the presence of water, producing HCl. Although unaffected by weak alkalies, boiling with aqueous NaOH may give rise to an explosive mixture because of monochloroacetylene [593-63-5] formation.

2.4. Storage and Handling.   1,2-Dichloroethylene is usually shipped in 208-L (55 gal) and 112-L (30 gal) steel drums. Because of the corrosive products of decomposition, inhibitors are required for storage. The stabilized grades of the isomers can be used or stored in contact with most common construction materials, such as steel or black iron. Storage stabilized 1,2-dichloroethylene compositions have been reported (48). Contact with copper or its alloys and with hot alkaline solutions should be avoided to preclude possible formation of explosive monochloroacetylene. The isomers do have explosive limits in air (Table 1). However, the liquid, even hot, burns with a very cool flame which

Table 3. **Physical Properties of the Isomeric Forms of 1,2-Dichloroethylene**

Property	Trans	Cis
mol wt	96.95	96.95
mp, °C	−49.44	−81.47
bp, °C	47.7	60.2
density, g/mL	1.2631	1.2917
15°C	1.44903	1.45189
20°C	1.44620	1.44900
viscosity, mPa·s (= cP)		
−50°C	1.005	1.156
−25°C	0.682	0.791
0°C	0.498	0.577
10°C	0.447	0.516
20°C	0.404	0.467
surface tension at 20°C, mN/m (= dyn/cm)	25	28
latent heat of vaporization[a], kJ/kg[b]	297.9	311.7
heat capacity at 20°C, kJ/(kg·K)[b]	1.158	1.176
vapor pressure, kPa[c]		
−20°C	5.3	2.7
−10°C	8.5	5.1
0°C	15.1	8.7
10°C	24.7	14.7
20°C	35.3	24.0
30°C	54.7	33.3
40°C	76.7	46.7
47.7°C	101	66.7
60.25°C		
soly of the isomer in water at 25°C, g/100 g	0.63	0.35
soly of water in the isomer at 25°C, g/100 g	0.55	0.55
steam distillation point at 101 kPa,[c] °C	45.3	53.8
flash point, °C	4	6
explosion limit in air, vol %[d]	5.6–12.8	

[a]At the boiling point.
[b]To convert J to cal, divide by 4.184.
[c]To convert kPa to mm Hg, multiply by 7.5.
[d]A cis-trans mixture (39).

self-extinguishes unless the temperature is well above the flash point. A red label is required for shipping 1,2-dichloroethylene.

2.5. Health and Safety Factors.   1,2-Dichloroethylene is toxic by inhalation and ingestion and can be absorbed by the skin. It has a TLV of 200 ppm (49). The odor does not provide adequate warning of dangerously high vapor

Table 4. **Azeotropes of *trans* and *cis*-1,2-Dichloroethylene Isomers**

Second component	Bp, °C	Binary azeotropes			
		Trans isomer in mixture wt%	Bp of azeotrope, °C	Cis isomer in mixture wt%	Bp of azeotrope, °C
methanol	64.5			87	51.5
ethanol	78.2	94.0	46.5	90.2	57.7
water	100.0	98.1	45.3	96.65	55.3

concentrations. Thorough ventilation is essential whenever the solvent is used for both worker exposure and flammability concerns. Symptoms of exposure include narcosis, dizziness, and drowsiness. Currently no data are available on the chronic effects of exposure to low vapor concentrations over extended periods of time.

1,2-Dichloroethylene appeared frequently in the 1980s literature largely because of its presence at ground water cleanup sites. The continued presence of 1,2-dichloroethylene may be a result of the biotransformation of tetrachloroethylene and trichloroethylene, which are much more common industrial solvents and are likely present because of past disposal practices (50,51).

2.6. Uses. 1,2-Dichloroethylene can be used as a low temperature extraction solvent for organic materials such as dyes, perfumes, lacquers, and thermoplastics (See ref. 52 for example). It is also used as a chemical intermediate in the synthesis of other chlorinated solvents and compounds (40).

Recently several patents have been issued (53–56) describing the use of 1,2-dichloroethylene for use in blends of chlorofluorocarbons for solvent vapor cleaning. One patent describes a method for cleaning acetate-based photographic film with *trans*-dichloroethylene (57). This art is primarily driven by the need to replace part of the chlorofluorocarbons because of the restriction on their production under the Montreal Protocol of 1987.

3. Trichloroethylene

3.1. Introduction. Trichloroethylene [79-01-6], trichloroethene, $CHCl=CCl_2$, commonly called "tri," is a colorless, sweet smelling (chloroformlike odor), volatile liquid and a powerful solvent for a large number of natural and synthetic substances. It is nonflammable under conditions of recommended use. In the absence of stabilizers, it is slowly decomposed (autoxidized) by air. The oxidation products are acidic and corrosive. Stabilizers are added to all commercial grades. Trichloroethylene is moderately toxic and has narcotic properties.

Trichloroethylene was first prepared by Fischer in 1864. In the early 1900s, processes were developed in Austria for the manufacture of tetrachloroethane and trichloroethylene from acetylene. Trichloroethylene manufacture began in Germany in 1920 and in the United States in 1925. Early uses of trichloroethylene were as an extraction solvent for natural fats and oils, such as palm, coconut, and soybean oils. It was later used for decaffeination of coffee, but this use has essentially been replaced by steam processes today. The demand for trichloroethylene was stimulated by the development of the vapor-degreasing process during the 1920s and by the growth of the dry-cleaning industry during the 1930s. By the mid-1950s, perchloroethylene had replaced trichloroethylene in dry-cleaning, and metal cleaning became the principal use for trichloroethylene.

The demand for trichloroethylene grew steadily until 1970. Since that time trichloroethylene has been a less desirable solvent because of restrictions on emissions under air pollution legislation and the passage of the Occupational Safety and Health Act. Whereas previously the principal use of trichloroethylene was for vapor degreasing, 1,1,1-trichloroethane then became the most used solvent for vapor degreasing. The restrictions on production of 1,1,1-trichloroethane [71-55-6] from the 1990 Amendments to the Montreal Protocol on substances that deplete the stratospheric ozone and the U.S. Clean Air Act 1990 Amendments will lead to a phase out of 1,1,1-trichloroethane by the year 2005, which in turn will likely result in a slight resurgence of trichloroethylene in vapor-degreasing applications. The total production, however, will probably stay relatively low because regulations will require equipment designed to assure minimum emissions.

3.2. Physical and Chemical Properties. The physical properties of trichloroethylene are listed in Table 5. Trichloroethylene is immiscible with water but miscible with many organic liquids; it is a versatile solvent. It does not have a flash or fire point. However, it does exhibit a flammable range when high concentrations of vapor are mixed with air and exposed to high energy ignition sources.

The most important reactions of trichloroethylene are atmospheric oxidation and degradation by aluminum chloride. Atmospheric oxidation is catalyzed by free radicals and accelerated with heat and with light, especially ultraviolet. The addition of oxygen leads to intermediates (**3**) and (**4**).

Compound (**3**) decomposes to form dichloroacetyl chloride, which in the presence of water decomposes to dichloroacetic acid and hydrochloric acid (HCl) with consequent increases in the corrosive action of the solvent on metal surfaces. Compound (**4**) decomposes to yield phosgene, carbon monoxide, and hydrogen chloride with an increase in the corrosive action on metal surfaces.

In the presence of aluminum, oxidative degradation or dimerization supply HCl for the formation of aluminum chloride, which catalyzes further dimerization to hexachlorobutene. The latter is decomposed by heat to give more HCl. The result is a self-sustaining pathway to solvent decomposition. Sufficient quantities of aluminum can cause violent decomposition, which can lead to runaway reactions (58,59). Commercial grades of trichloroethylene are stabilized to prevent these reactions in normal storage and use conditions.

Amine-stabilized products, once the predominant grade, are sold today only in limited amounts. Most vapor-degreasing grades contain neutral inhibitor mixtures (60–62) including a free-radical scavenger, such as an amine or pyrrole, to prevent the initial oxidation reaction. Epoxides, such as butylene oxide and epichlorohydrin, are added to scavenge any free HCl and $AlCl_3$. Concern over the toxicity of these epoxides has eliminated the use of epichlorohydrin in the United States during the 1980s and may restrict butylene oxide in the future.

Trichloroethylene is not readily hydrolyzed by water. Under pressure at 150°C, it gives glycolic acid, $CH_2OHCOOH$, with alkaline hydroxides. Strong alkalies dehydrochlorinate trichloroethylene with production of spontaneously

Table 5. **Properties of Trichloroethylene**

Property	Value
molecular weight	131.39
melting point, °C	−86.5
boiling point, °C	87.3
specific gravity, liquid	
20/4°C	1.464
100/4°C	1.322
vapor density at bp, kg/m^3	4.61
n_D	
liquid, 20°C	1.4782
vapor, 0°C	1.001784
viscosity, mPa·s (= cP) liquid	
20°C	0.57
60°C	0.42
vapor at 100°C	0.01246
surface tension at 20°C, mN/m(= dyn/cm)	29.3
heat capacity, J/(kg·K)^a	
liquid at 20°C	938
vapor at 100°C	693
critical temperature, °C	300.2
critical pressure, MPa^b	4.986
thermal conductivity, W/(m·K)	
liquid at 20°C	0.115
vapor at bp	0.00851
coefficient of cubical expansion, liquid at 0−40°C	0.00119
dielectric constant, liquid at 16°C	3.42
dipole moment, C·m^c	3.0×10^{-30}
heat of combustion, kJ/g^a	−6.56
heat of formation, kJ/mol^a	
liquid	−42.3
vapor	−7.78
latent heat of evaporation at bp, kJ/kg^a	238
explosive limits, vol % in air	
25°C	8.0-saturation
100°C	8.0−44.8
vapor pressure^d, kPa^e	
Antoine constants	
A	5.75373
B	1076.67
C	199.991
solubility, g	
H$_2$O in 100 g trichloroethylene	
0°C	0.010
20°C	0.0225
60°C	0.080
trichloroethylene in 100 g H$_2$O	
20°C	0.107
60°C	0.124

[a]To convert J to cal, divide by 4.184.
[b]To convert MPa to atm, divide by 0.101.
[c]To convert C·m to debye, divide by 3.336×10^{-30}.
[d]$\log_{10} P = A - B/(T + C)$, T in °C.
[e]To convert kPa to mm Hg, multiply by 7.5.

explosive and flammable dichloroacetylene. Reaction with sulfuric acid (90%) yields monochloroacetic acid, $CH_2ClCOOH$. Hot nitric acid reacts with trichloroethylene violently, producing complete oxidative decomposition. Under carefully controlled conditions, nitric acid gives trichloronitromethane (chloropicrin) and dinitrochloromethane (63). Dichloroacetylene, C_2Cl_2, can also be formed from trichloroethylene in the presence of epoxides and ionic halides (64).

In the presence of catalysts, trichloroethylene is readily chlorinated to pentachloro- and hexachloroethane. Bromination yields 1,2-dibromo-1,1,2-trichloroethane [13749-38-7]. The analogous iodine derivative has not been reported. Fluorination with hydrogen fluoride in the presence of antimony trifluoride produces 2-chloro-1,1,1-trifluoroethane [75-88-7] (65). Elemental fluorine gives a mixture of chlorofluoro derivatives of ethane, ethylene, and butane.

Liquid trichloroethylene has been polymerized by irradiation with ^{60}Co γ-rays or 20-keV x-rays (66). Trichloroethylene has a chain-transfer constant of <1 when copolymerized with vinyl chloride (67) and is used extensively to control the molecular weight of poly(vinyl chloride) polymer.

A variety of trichloroethylene copolymers have been reported, none with apparent commercial significance. The alternating copolymer with vinyl acetate has been patented as an adhesive (68) and as a flame retardant (69,70). Copolymerization with 1,3-butadiene and its homologues has been reported (71–73). Other comonomers include acrylonitrile (74), isobutyl vinyl ether (75), maleic anhydride (76), and styrene (77).

Terpolymers have been made with vinyl chloride–vinylidene chloride (78) and vinyl acetate–vinyl alcohol (79).

3.3. Manufacture. *From Acetylene.* As late as 1968, 85% of the production capacity in the United States was based on acetylene, but rising acetylene [74-86-2] costs reduced this figure to 8% by 1976 (80), and now most trichloroethylene is made from ethylene [74-85-1], 1,2-dichloroethane [107-06-2], or ethylene dichloride.

The acetylene-based process consists of two steps. First acetylene is chlorinated to 1,1,2,2-tetrachloroethane [79-34-5]. The reaction is exothermic (402 kJ/mol = 96 kcal/mol) but is maintained at 80–90°C by the vaporization of solvent and product. Catalysts include ferric chloride and sometimes phosphorus chloride and antimony chloride (81).

The product is then dehydrohalogenated to trichloroethylene at 96–100°C in aqueous bases such as $Ca(OH)_2$ (82) or by thermal cracking, usually over a catalyst (81) such as barium chloride on activated carbon or silica or aluminum gels at 300–500°C. The yield of trichloroethylene (80) is about 94% based on acetylene. A significant disadvantage of the alkaline process is the loss of chlorine as calcium chloride. In thermal cracking the chlorine can be recovered as hydrochloric acid, an important feedstock in many chemical processes. Since it poisons the catalysts during thermal cracking, all ferric chloride must be removed from the tetrachloroethane feed (81). Tetrachloroethane can also be cracked to trichloroethylene without catalysts at 330–770°C, but considerable amounts of tarry by-products are formed.

Chlorination of Ethylene. Dichloroethane, produced by chlorination of ethylene, can be further chlorinated to trichloroethylene and tetrachloroethylene.

The exothermic reaction is carried out at 280–450°C. Temperature is controlled by a fluidized bed, a molten salt bath, or the addition of an inert material such as perchloroethylene. The residence time in the reactor varies from 2 to 30 seconds, depending on conditions (81). Catalysts include potassium chloride and aluminum chloride (83), Fuller's earth (84), graphite (85), activated carbon (86), and activated charcoal (84).

Maximum conversion to trichloroethylene (75% of dichloroethane feed) is achieved at a chlorine to dichloroethane ratio of 1.7:1. Tetrachloroethylene conversion reaches a maximum (86% conversion of dichloroethane) at a feed ratio of 3.0:1 (81).

Oxychlorination of Ethylene or Dichloroethane. Ethylene or dichloroethane can be chlorinated to a mixture of tetrachoroethylene and trichloroethylene in the presence of oxygen and catalysts. The reaction is carried out in a fluidized-bed reactor at 425°C and 138–207 kPa (20–30 psi). The most common catalysts are mixtures of potassium and cupric chlorides. Conversion to chlorocarbons ranges from 85–90%, with 10–15% lost as carbon monoxide and carbon dioxide (81). Temperature control is critical. Below 425°C, tetrachloroethane becomes the dominant product, 57.3 wt% of crude product at 330°C (87). Above 480°C, excessive burning and decomposition reactions occur. Product ratios can be controlled but less readily than in the chlorination process. Reaction vessels must be constructed of corrosion-resistant alloys.

3.4. Shipping and Storage. Shipment of trichloroethylene is usually by truck or rail car and also in 208-liter (55-gallon) steel drums. Mild steel tanks, if appropriately equipped with vents and vent driers to prevent the accumulation of water, are adequate for storage. Precautions, such as diking, should be taken to provide for adequate spill containment at the storage tank. Seamless black iron pipes are suitable for transfer lines, gasketing should be of Teflon, Viton, or other solvent impermeable material. Centrifugal or positive-displacement pumps made from cast iron, steel, or stainless steel are suitable for use. Aluminum should never be used as a construction material for any halogenated hydrocarbon. Glass containers, amber or green, are suitable for small quantities, such as in a laboratory, but care should be taken for spill containment in the event of breakage.

Containers should bear warning labels against breathing vapors, ingesting the liquid, splashing solvent in eyes or on skin and clothing, and using it near an open flame, or where vapors will come in contact with hot metal surfaces (>176°C).

Precautions in handling any waste products in conformance with federal, state, and local regulations should be included.

Although the flammability hazard is very low, ignition sources should not be present when trichloroethylene is used in highly confined or unventilated areas. Tanks in which flammable concentrations could develop should be grounded to prevent build-up of static electric charges. Under no circumstances should welding or cutting with a torch take place on any storage container or process equipment containing trichloroethylene.

3.5. Economic Aspects. Table 6 lists the U.S. producers of trichloroethylene and their capacities. PPG capacity is flexible and can be swung between TCE and tetrachlorethylene production (88).

Table 6. **U.S. Producers of Trichloroethylene and Their Capacities**[a]

Producer	Location	Capacity, $\times 10^3$ t ($\times 10^6$ lb)
Dow	Freeport, Tex.	59 (130)
PPG	Lake Charles, La.	91 (200)
Total		*150 (330)*

[a] See Ref. 88.

There has been a significant increase in global use of TCE as a feedstock for the production of hydrofluorocarbons, particularly HFC-134a, which is used as a replacement for CFC-12 in refrigerants. This use has climbed steadily over the past few years and will continue to be strong. However as the CFC-12 is largely replaced, growth could drop to 6%/yr.

The use of TCE in metal cleaning is expected to decrease as users have installed new work procedures and equipment to limit fugitive emissions.

Demand for 2001 was 91×10^3 t. Demand for 2005 is expected to be 111×10^3 t. Prices have remained stable at \$0.29/kg (88).

3.6. Specifications and Standards. Commercial grades of trichloroethylene, formulated to meet use requirements, differ in the amount and type of added inhibitor. The grades sold in the United States include a neutrally inhibited vapor-degreasing grade and a technical grade for use in formulations. U.S. Federal Specification O-T-634b lists specifications for a regular and a vapor-degreasing grade.

Apart from added stabilizers, commercial grades of trichloroethylene should not contain more than the following amounts of impurities: water 100 ppm; acidity, ie, HCl, 5 ppm; insoluble residue, 10 ppm. Free chlorine should not be detectable. Test methods have been established by ASTM to determine the following characteristics of trichloroethylene: acid acceptance, acidity or alkalinity, color, corrosivity on metals, nonvolatile-matter content, pH of water extractions, relative evaporation rate, specific gravity, water content, water-soluble halide ion content, and halogen content (89).

The passage of the Resource Conservation and Recovery Act in 1978 and its implementation in 1980 generated an increase in the recycling of trichloroethylene, which, in turn, defined the need for specifications for recycled solvent. Reagents, methods, and kits for detecting trichloroethylene and tetrachloroethylene have been described in a patent (90).

3.7. Health and Safety Factors. Trichloroethylene is acutely toxic, primarily because of its anesthetic effect on the central nervous system. Exposure to high vapor concentrations is likely to cause headache, vertigo, tremors, nausea and vomiting, fatigue, intoxication, unconsciousness, and even death. Because it is widely used, its physiological effects have been extensively studied.

Exposure occurs almost exclusively by vapor inhalation, which is followed by rapid absorption into the bloodstream. At concentrations of 150–186 ppm, 51–70% of the trichloroethylene inhaled is absorbed. Metabolic breakdown occurs by oxidation to chloral hydrate [302-17-0], followed by reduction to trichloroethanol [115-20-8], part of which is further oxidized to trichloroacetic acid [76-03-9] (91,92). Absorbed trichloroethylene that is not metabolized is eventually eliminated through the lungs (93). The OSHA permissible exposure limit (PEL) eight-hour TWA concentration has been set at 50 ppm for eight-hour exposure ACGIH TLV is 50 ppm also (94).

It is estimated that concentrations of 3000 ppm cause unconsciousness in less than 10 minutes (89). Anesthetic effects have been reported at concentrations of 400 ppm after 20-min exposure. Decrease in psychomotor performance at a trichloroethylene concentration of 110 ppm has been reported in one study (100), whereas other studies find no neurotoxic effects at concentrations of 200 ppm (96–99).

Victims of overexposure to trichloroethylene should be removed to fresh air, and medical attention should be obtained immediately. A self-contained positive pressure breathing device should be used wherever high vapor concentrations are expected, eg, when cleaning up spills or when accidental releases occur.

The distinctive odor of trichloroethylene may not necessarily provide adequate warning of exposure, because it quickly desensitizes olfactory responses. Fatalities have occurred when unprotected workers have entered unventilated areas with high vapor concentrations of trichloroethylene or other chlorinated solvents. For a complete description of proper entry to vessels containing any chlorinated solvent, see ASTM D4276-84, Standard Practice for Confined Area Entry (89).

Ingestion of large amounts of trichloroethylene may cause liver damage, kidney malfunction, cardiac arrhythmia, and coma (93); vomiting should not be induced, but medical attention should be obtained immediately.

Protective gloves and aprons should be used to prevent skin contact, which may cause dermatitis (101–103). Eyes should be washed immediately after contact or splashing with trichloroethylene.

The National Cancer Institute reported in 1975 that massive oral doses of trichloroethylene caused liver tumors in mice but not in rats (104). Trichloroethylene was tested again in the 1980s by the National Toxicology Program (NTP) with similar results (105,106). The EPA has classified trichloroethylene as B2, a probable human carcinogen. The International Agency for Research on Cancer (IARC) classifies it as group 3, ie, unclassifiable as to human carcinogenicity. Teratogenicity studies conducted with trichloroethylene by The Dow Chemical Company showed no significant effects on fetal development (107). During the 1980s several epidemiology studies were conducted on worker populations exposed to trichloroethylene (108–110). Each of these studies failed to show a positive link between human exposure in the work place and cancer. Human mutation data has been reported, however (5).

During the 1980s a significant amount of work was done on developing methods for treatment of contaminated groundwater and also on setting standards for trichloroethylene under the Safe Drinking Water Act. The EPA has set a maximum contaminant level goal (MCLG) at 0 based on the animal carcinogenic effects (111). The maximum contaminant level (MCL) is currently set at five micrograms per liter.

3.8. Uses. Approximately 67% of the trichloroethylene produced in the United States is consumed in the vapor degreasing of fabricated metal parts (see METAL SURFACE TREATMENTS); 30% is divided equally between exports and miscellaneous applications and 3% is used as a polymerization modifier (84). A variety of miscellaneous applications include use of trichloroethylene as a component in adhesive and paint-stripping formulations, a low temperature heat-transfer medium, a nonflammable solvent carrier in industrial paint systems, and a solvent base for metal phosphatizing systems. Trichloroethylene is used in the textile industry as a carrier solvent for spotting fluids and as a solvent in waterless preparation dying and finishing operations. Cleaning or drying compositions as a replacement for CFCs has been described (112).

Trichloroethylene was approved for use for many years as an extraction solvent for foods. In late 1977, the Food and Drug Administration (FDA) banned its use as a food additive, directly or indirectly, prohibiting the use in hop extraction, decaffeination of coffee, isolation of spice oleoresins, and other applications. The FDA also banned the use of trichloroethylene in cosmetic and drug products (84).

Trichloroethylene is widely used as a chain-transfer agent in the production of poly(vinyl chloride). There has been a significant increase in the use of trichloroethylene as a feedstock in the production of hydrofluorocarbons, the replacements for chlorofluorocarbons implicated in the depletion of stratospheric ozone. HFC-134a (hydrotetrafluoroethane) is one of these products. It replaces CFC-12 as a refrigerant in the United States. This application has been growing steadily, but will slow as CFC-12 is replaced completely.

4. Tetrachloroethylene

4.1. Introduction. Tetrachloroethylene [127-18-4], perchloroethylene, $CCl_2\!\!=\!\!CCl_2$, is commonly referred to as "perc" and sold under a variety of trade names. It is the most stable of the chlorinated ethylenes and ethanes, having no flash point and requiring only minor amounts of stabilizers.

Tetrachloroethylene was first prepared in 1821 by Faraday by thermal decomposition of hexachloroethane. Tetrachloroethylene is typically produced as a coproduct with either trichloroethylene or carbon tetrachloride from hydrocarbons, partially chlorinated hydrocarbons, and chlorine. Although production of tetrachloroethylene and trichloroethylene from acetylene was once the dominant process, it is now obsolete because of the high cost of acetylene.

4.2. Physical and Chemical Properties. The physical properties of tetrachloroethylene are listed in Table 7. It dissolves a number of inorganic materials including sulfur, iodine, mercuric chloride, and appreciable amounts of aluminum chloride. Tetrachloroethylene dissolves numerous organic acids, including benzoic, phenylacetic, phenylpropionic, and salicylic acid, as well as a variety of other organic substances such as fats, oils, rubber, tars, and resins. It does not dissolve sugar, proteins, glycerol, or casein. It is miscible with chlorinated organic solvents and most other common solvents. Tetrachloroethylene forms approximately sixty binary azeotropic mixtures (113).

Stabilized tetrachloroethylene, as provided commercially, can be used in the presence of air, water, and light, in contact with common materials of construction, at temperatures up to about 140°C. It resists hydrolysis at temperatures up to 150°C (114). However, the unstabilized compound, in the presence of water for prolonged periods, slowly hydrolyzes to yield trichloroacetic acid

Table 7. **Properties of Tetrachloroethylene**

Property	Value
molecular weight	165.83
melting point, °C	−22.7
boiling point at 101 kPa[a], °C	121.2
specific gravity, liquid, at °C	
10/4	1.63120
20/4	1.62260
30/4	1.60640
120/4	1.44865
vapor density at bp at 101 kPa, kg/m^3	5.8
viscosity, mPa·s (= cP)	
liquid, °C	
15	0.932
25	0.839
50	0.657
75	0.534
vapor at 60°C	9900
surface tension, mN/m (= dyn/cm)	
15°C	32.86
30°C	31.27
thermal capacity, kJ/(kg·K)[b]	
liquid at 20°C	0.858
vapor at 100°C	0.611
thermal conductivity, mW/(m·K)	
liquid	126.6
vapor at bp	8.73
heat of combustion	
constant pressure with formation of aq HCl, kJ/mol[b]	679.9
constant volume at 18.7°C, kJ/mol[b]	831.8
latent heat of vaporization at 121.2°C, kJ/mol[b]	34.7
critical temperature, °C	347.1
critical pressure, MPa[c]	9.74
latent heat of fusion, kJ/mol[b]	10.57
heat of formation, kJ/mol[b]	
vapor	−25
liquid	12.5
n_D at 20°C	1.50547
dielectric constant at 1 kHz, 20°C	2.20
electrical conductivity at 20°C, 10^{15} $(\Omega \cdot m)^{-1}$	55.8
coefficient of cubical expansion at 15−90°C, av	0.001079
vapor pressure, kPa[c], at °C	
−20.6	0.1333
13.8	1.333
40.0	5.466
60.0	13.87
80.0	30.13
100.0	58.46
121.2	101.3
solubility at 25°C, mg	
C_2Cl_4 in 100 g H_2O	15
H_2O in 120 g C_2Cl_4	8

[a]To convert kPa to mm Hg, multiply by 7.5.
[b]To convert kJ to kcal, divide by 4.184.
[c]To convert MPa to atm, divide by 0.101.

[76-03-9] and hydrochloric acid. In the absence of catalysts, air, or moisture, tetrachloroethylene is stable to about 500°C. Although it does not have a flash point or form flammable mixtures in air or oxygen, thermal decomposition results in the formation of hydrogen chloride and phosgene [75-44-5] (115).

Under ultraviolet radiation in the presence of air or oxygen, tetrachloroethylene undergoes autoxidation to trichloroacetyl chloride [76-02-8]. This reaction, which accounts for the slow decomposition of tetrachloroethylene under prolonged storage in the presence of light and air or oxygen, is inhibited in commercial products by the addition of amines or phenols as stabilizers. Peroxy compounds (**5**) and (**6**) are intermediates of this autoxidation. Compound (**5**) rear-ranges to form trichloroacetyl chloride and oxygen, whereas compound (**6**) breaks down to form two molecules of phosgene.

$$CCl_2{=}CCl_2 \ + \ O_2 \ \longrightarrow \ \begin{bmatrix} Cl \\ Cl{-}\!\!\diagdown\!\!\triangleright\!\!O^+{-}O^- \\ Cl{-}\!\!\diagup \\ Cl \end{bmatrix} + \begin{bmatrix} Cl \\ Cl{-}\!\!\boxminus\!\!{-}O \\ Cl{-}\!\!{-}O \\ Cl \end{bmatrix}$$

$$\text{(5)} \qquad\qquad \text{(6)}$$

Reaction with hydrogen at 220°C in the presence of reduced nickel catalyst results in total decomposition to hydrogen chloride and carbon. An explosive reaction occurs with butyllithium in petroleum ether solution (116). Tetrachloroethylene also reacts explosively with metallic potassium at its melting point, however it does not react with sodium (117).

Photochlorination of tetrachloroethylene, observed by Faraday, yields hexachloroethane [67-72-1]. Reaction with aluminum bromide at 100°C forms a mixture of bromotrichloroethane and dibromodichloroethane [75-81-0] (118). Reaction with bromine results in an equilibrium mixture of tetrabromoethylene [79-28-7] and tetrachloroethylene. Tetrachloroethylene reacts with a mixture of hydrogen fluoride and chlorine at 225–400°C in the presence of zirconium fluoride catalyst to yield 1,2,2-trichloro-1,1,2-trifluoroethane [76-13-1] (CFC 113) (119).

Tetrachloroethylene reacts with formaldehyde and concentrated sulfuric acid at 80°C to form 2,2-dichloropropanoic acid [75-99-0] (120). Copolymers with styrene, vinyl acetate, methyl acrylate, and acrylonitrile are formed in the presence of dibenzoyl peroxide (121,122).

Tetrachloroethylene is heated at 110–120°C with o-benzenedithiol, in the presence of sodium ethoxide, to form 2,2′-bis-1,3-benzdithiolene (123).

The addition of stabilizers to tetrachloroethylene inhibits corrosion of aluminum, iron, and zinc which otherwise occurs in the presence of water (124). Where water in excess of the solubility limit is present, forming separate layers, hydrolysis and corrosion rates increase. System design and construction materials should consider these effects.

4.3. Manufacture. Many processes have been used to produce tetrachloroethylene. One of the first was chlorination of acetylene (C_2H_2) to form tetrachloroethane, followed by dehydrochlorination to trichloroethylene. If tetrachloroethylene was desired, the trichloroethylene was further chlorinated to pentachloroethane and dehydrochlorinated. This process is no longer used in the United States; Hooker Chemical closed down the last plant in 1978.

In Japan, Toagosei is reported to produce trichloroethylene and tetrachloroethylene by chlorination of ethylene followed by dehydrochlorination. In this process the intermediate tetrachloroethane is either dehydrochlorinated to trichloroethylene or further chlorinated to pentachloroethane [76-01-7] followed by dehydrochlorination to tetrachloroethylene. Partially chlorinated by-products are recycled and by-product HCl is available for other processes.

The following processes are commonly used today.

Chlorination of Ethylene Dichloride. Tetrachloroethylene and trichloroethylene can be produced by the noncatalytic chlorination of ethylene dichloride [107-06-2] (EDC) or other two-carbon (C2) chlorinated hydrocarbons. This process is advantageous when there is a feedstock source of mixed C2 chlorinated hydrocarbons from other processes and an outlet for the by-product HCl stream. Product ratios of tri- and tetrachloroethylene are controlled by adjusting the Cl_2: 2:EDC ratio to the reactor. Partially chlorinated by-products are recycled to the chlorinator. The primary reactions are

$$CH_2ClCH_2Cl + 3\,Cl_2 \longrightarrow Cl_2C{=}CCl_2 + 4\,HCl$$

$$CH_2ClCH_2Cl + 2\,Cl_2 \longrightarrow CHCl{=}CCl_2 + 3\,HCl$$

Chlorination of C1–C3 Hydrocarbons or Partially Chlorinated Derivatives. Tetrachloroethylene and carbon tetrachloride are produced with or without a catalyst at high temperatures (550–700°C) from light hydrocarbon feedstocks or their partially chlorinated derivatives. This is one of the most versatile processes, allowing for a wide range of mixed chlorinated hydrocarbon wastes from other processes to be used as feedstocks. However the phase out of CFC-11 and CFC-12, most of carbon tetrachloride use, has caused producers to maximize perchloroethylene production and minimizing or eliminating production of carbon tetrachloride (125). The large quantities of HCl produced requires integration with other HCl consuming processes. As with the previous process, product distribution is controlled by controlling feedstock ratios, and partially chlorinated by-products are recycled to the chlorinator. As examples, reaction of EDC or propane are shown in the following.

$$3\,CH_2ClCH_2Cl + 11\,Cl_2 \longrightarrow 2\,Cl_2C{=}CCl_2 + 2\,CCl_4 + 12\,HCl$$

$$CH_3CH_2CH_3 + 8\,Cl_2 \longrightarrow Cl_2C{=}CCl_2 + CCl_4 + 8\,HCl$$

Oxychlorination of C2 Chlorinated Hydrocarbons. Tetrachloroethylene and trichloroethylene can be produced by reaction of EDC with chlorine or HCl and oxygen in the presence of a catalyst. When hydrochloric acid is used, additional oxygen is required. Product distribution is varied by controlling reactant ratios. This process is advantageous in that no by-product HCl is produced,

Table 8. U.S. Producers of Tetrachloroethylene and Their Capacities[a]

Producer	Location	Capacity, $\times 10^3$ t ($\times 10^6$ lb)
Dow Chemical	Plaquemine, La.	40.8 (90)
PPG Industries	Lake Charles, La.	90.7 (200)
Vulcan Materials	Geismar, La.	63.5 (140)
Total		*195 (430)*

[a] Ref. 125.

and it can be integrated with other processes as a net HCl consumer. The reactions may be represented as follows:

$$CH_2ClCH_2Cl + Cl_2 + O_2 \longrightarrow Cl_2C\!=\!CCl_2 + 2\,H_2O$$

$$CH_2ClCH_2Cl + 1/2\,Cl_2 + 3/4\,O_2 \longrightarrow CHCl\!=\!CCl_2 + 3/2\,H_2O$$

4.4. Shipping and Storage. Tetrachloroethylene is shipped by barge, tank car, tank truck, and 55-gallon (208-L) steel drums. It may be stored in mild steel tanks that are dry, free of rust, and equipped with a chemical (such as calcium chloride) vent dryer and controlled evaporation vent. Appropriate secondary containment including dikes and sealed surfaces should be provided in accordance with federal and local standards to prevent potential groundwater contamination in the event of a leak. Piping and centrifugal or positive displacement pumps should be constructed of ductile iron or carbon steel with gasket materials made of impregnated cellulose fiber, cork base materials, or Viton resin.

4.5. Economic Aspects. U.S. producers of tetrachloroethylene are listed in Table 8.

Tetrachloroethylene's primary use is as a precursor to manufacture fluorocarbons, particularly HFC-134a, which is used as a replacement for CFC-12 in refrigeration systems. This use has been growing at a rate of 9% in the United States. This rate will probably slow to 4% as the replacement of CFC-12 is complete.

Conversion to more efficient dry-cleaning equipment has resulted in a decline of 65% in the use of tetrachlorethylene as a dry-cleaning agent in the last decade. No alternative has the versatility and low cost of tetrachloroethylene so it will still be used in this application.

Chlorinated solvents are not favored because of their negative impact on the environment and health. Tetrachloroethylene is listed by the EPA as a substitute for 1,1,1-trichlorethane in adhesives, coatings, inks and metal degreasing and as a substitute for CFC-13.

Demand in 2001 was 152×10^3 t. In 2205, demand is expected to be 166×10^3 t. Growth is expected at a rate of 2% through 2005 (125).

Prices for the period 1996–2001 ranged from \$0.15/kg to 0.17/kg (125).

4.6. Specifications. Commercial grades of tetrachloroethylene include a vapor degreasing grade; a dry-cleaning grade; an industrial grade for use in formulations; a high purity, low residue grade; and a grade specifically formulated for use as a transformer fluid. The various grades differ in the amount and type of added stabilizers. U.S. Federal Specification OT-236A covers tetrachloroethylene.

ASTM has established standard test methods to determine acid acceptance, acidity, alkalinity, color, corrosivity to metals, nonvolatile matter content, pH of water extractions, relative evaporation rate, boiling point range, specific gravity, water content, water-soluble halide ions, and halogens (126). Typical commercial grades should not contain more than 50 ppm water, 0.0005 wt% acidity (as HCl), or 0.001 wt % insoluble residue.

4.7. Health and Safety Factors.   Overexposure to tetrachloroethylene by inhalation affects the central nervous system and the liver. Dizziness, headache, confusion, nausea, and eye and mucous tissue irritation occur during prolonged exposure to vapor concentrations of 200 ppm (127). These effects are intensified and include incoordination and drunkenness at concentrations in excess of 600 ppm. At concentrations in excess of 1000 ppm the anesthetic and respiratory depression effects can cause unconsciousness and death. A single, brief exposure to concentrations above 6000 ppm can be immediately dangerous to life. Reversible changes to the liver have been reported following prolonged exposures to concentrations in excess of 200 ppm (128–134). Alcohol consumed before or after exposure may increase adverse effects.

The OSHA permissible exposure limit (PEL) for tetrachloroethylene is 25 ppm (8-h TWA) (135). The American Conference of Governmental Industrial Hygienists (ACGIH) threshold limit value (TLV) is 50 ppm. In addition they recommend a 15 minute, short-term exposure limit (STEL) of 200 ppm (94,136). The odor threshold for tetrachloroethylene ranges from about 5 to 70 ppm (127). Therefore odor alone does not provide adequate warning of potential overexposure in the workplace. Air sampling of the work environment should be performed in order to determine the need for protective equipment. Fatalities have occurred when workers have entered unventilated tanks or equipment containing high vapor concentrations of tetrachloroethylene without utilizing a self-contained breathing apparatus (137). Victims of overexposure should be removed from the area, given artificial respiration or oxygen if necessary, and a physician should be consulted (138).

Repeated exposure of skin to liquid tetrachloroethylene may defat the skin causing dermatitis. When frequent or prolonged contact is likely, gloves of Viton, nitrile rubber, or neoprene should be used, discarding them when they begin to deteriorate. Tetrachloroethylene can cause significant discomfort if splashed in the eyes. Although no serious injury results, it can cause transient, reversible corneal injury. If contact with skin or eyes occurs, follow standard first-aid practices.

Ingestion of small amounts of tetrachloroethylene is not likely to cause permanent injury; however, ingestion of large amounts may result in serious injury or even death. All containers should be properly labeled. If solvent is swallowed, consult a physician immediately. Do not induce vomiting. If solvent is aspirated it is rapidly absorbed through the lungs and may cause systemic effects and chemical pneumonia.

Exposure to tetrachloroethylene as a result of vapor inhalation is followed by absorption into the bloodstream. It is partly excreted unchanged by the lungs

(129,130). Approximately 20% of the absorbed material is subsequently metabolized and eliminated through the kidneys (139–141). Metabolic breakdown occurs by oxidation to trichloroacetic acid and oxalic acid.

Three significant studies have been conducted on the potential carcinogenic effects of tetrachloroethylene in laboratory animals (142–144). Two of these studies showed increases in observed liver and/or kidney tumors at high dosage levels. The third study showed no significant differences between exposed and control groups of animals at inhalation exposure levels up to 600 ppm. Tetrachloroethylene is classified in Group 2B, a "possible human carcinogen" by the International Agency for Research on Cancer (IARC). The National Toxicology Program (NTP) lists tetrachloroethylene as "reasonably anticipated to cause cancer in humans." Pharmacokinetic studies suggest the effects observed in laboratory animals are not directly applicable to humans (145,146). During the early 1990s, the Environmental Protection Agency (EPA), under its *Guidelines for Carcinogenic Risk Assessment*, had not made a final decision on the classification for this chemical (147).

No teratogenic effects were observed in mice and rats exposed to vapor concentrations of 300 ppm. Exposure levels having no effect on the mother are not anticipated to affect the fetus (148).

4.8. Environmental Regulations. Tetrachloroethylene is subject to inventory and release reporting under Title III of the Superfund Amendments and Reauthorization Act of 1986 (SARA). Tetrachloroethylene waste is considered hazardous waste under the Resource Conservation and Recovery Act of 1984 (RCRA). The preferred methods of disposal are through licensed reclaimers or permitted incinerators. The EPA revised the reportable quantity (RQ) for tetrachloroethylene to 100 lbs in 1989. Although tetrachloroethylene does not contribute to smog formation, and the EPA recommended exemption from Volatile Organic Compounds (VOC) regulations in 1983 (149), it continues to be controlled as a VOC. Under the Clean Air Act Amendment of 1990, tetrachloroethylene is considered a hazardous air pollutant. Under this act, the EPA will develop standards to control tetrachloroethylene emissions in dry-cleaning and metal cleaning applications. Under the Safe Drinking Water Act, EPA has established a maximum contaminant level (MCL) of 0.005 mg/L and a goal of 0 mg/L for tetrachloroethylene (150). Packed tower aeration and granular activated carbon are considered the best available technologies for removal of tetrachloroethylene from drinking water.

4.9. Uses. Use of tetrachlorethylene as a feedstock for chlorofluorocarbon production accounts for 65% of demand. Approximately 15% is used in the dry-cleaning industry. Metal cleaning applications account for 10% of consumption and miscellaneous uses account for 10%. Miscellaneous uses include transformer insulating fluid, chemical maskant formulations, and in a process for desulfurizing coal (125).

BIBLIOGRAPHY

"Chlorocarbons and Chlorohydrocarbons, Other Chloroethanes" under "Chlorine Compounds, Organic" in *ECT* 1st ed., Vol. 3, "Ethylidene Chloride, Ethylene Chloride," pp. 760–764, "1,1,1-Trichloroethane," pp. 764–765, "1,1,2-Trichloroethane," pp. 767–771, "Pentachloroethane," pp. 771–773, by J. Searles and H. A. McPhail, E. I. du Pont de Nemours & Co., Inc., "Hexachloroethane," pp. 773–774 by J. Werner, General Aniline & Film Corp., General Aniline Works Division; "Chlorocarbons and Chlorohydrocarbons, Other Chloroethanes," in *ECT* 2nd ed., Vol. 5, pp. 149–170, by D. W. F. Hardie, Imperial Chemical Industries, Ltd.; in *ECT* 3rd ed., Vol. 5, pp. 722–742, by W. I. Archer, Dow Chemical U.S.A; in *ECT* 4th ed., Vol. 6, pp. 11–36, by Gayle Snedecor, The Dow Chemical Company; "Chlorocarbons and Chlorohydrocarbons-Dichloroethylenes" under "Chlorine Compounds, Organic", in *ECT* 1st ed., Vol. 3, pp. 786–787, by J. Werner, General Aniline & Film Corp., Aniline Works Division; "Chlorocarbons and Chlorohydrocarbons-Dichloroethylenes" in *ECT* 2nd ed., Vol. 5, pp. 178–183, by D. W. F. Hardie, Imperial Chemical Industries, Ltd; "1,2-Dichloroethylene" under "Chlorocarbons, -Hydrocarbons" in *ECT* 3rd ed., Vol. 5, pp. 742–745, by V. L. Stevens, The Dow Chemical Company; in *ECT* 4th ed., Vol. 6, pp. 36–40, by James A. Mertens, Dow Chemical U.S.A.; "Chlorine Compounds, Organic (Chlorocarbons and Chlorohydrocarbons-Trichloroethylene)" in *ECT* 1st ed., Vol.3, pp. 788–794, by J. Searles and H. A. McPhail, E. I. du Pont de Nemours & Co., Inc.; "Chlorocarbons and Chlorohydrocarbons (Trichloroethylene)" in *ECT* 2nd ed., Vol. 5, pp. 183–195, by D. W. F. Hardie, Imperial Chemical Industries Ltd; "Chlorocarbons Hydrocarbons (Trichloroethylene)" in *ECT* 3rd ed., Vol. 5, pp. 745–753, by W. C. McNeil, Jr., Dow Chemical U.S.A.; in *ECT* 4th ed., Vol. 5, pp. 40–50, by James A. Mertens, Dow Chemical U.S.A.; "Tetrachloroethylene" under "Chlorine Compounds, Organic" in *ECT* 1st ed., Vol. 3, pp. 794–798, by J. Searles and H. A. McPhail, E. I. du Pont de Nemours & Co., Inc.; "Tetrachloroethylene" under "Chlorocarbons and Chlorohydrocarbons" in *ECT* 2nd ed., Vol. 5, pp. 195–203, by D. W. F. Hardie, Imperial Chemical Industries, Ltd.; in *ECT* 3rd ed., Vol. 5, pp. 754–762, by S. L. Keil, The Dow Chemical Company; in *ECT* 4th ed., Vol. 5, pp. 50– by J. C. Hickman, The Dow Chemical Company; "Chloroethanes" in *ECT* (online), posting date: December 4, 2000, by Gayle Snedecor, The Dow Chemical Company; "Dichloroethylene" in *ECT* (online), posting date: December 4, 2000, by James A. Mertens, Dow Chemical U.S.A.; "Trichloroethylene" in *ECT* (online), posting date: December 4, 2000, by James A. Mertens, Dow Chemical U.S.A.; "Tetrachloroethylene" in *ECT* (online), posting date: December 4, 2000, by J. C. Hickman, The Dow Chemical Company.

1. R. W. Gallant, *Hydrocarbon Process.* **45**(7), 111 (1966).
2. L. E. Horsley, *Azeotropic Data, Advances in Chemistry Series*, No. 6, American Chemical Society, Washington, D.C., 1952; *Azeotropic Data-II*, No. 35, 1962.
3. D. H. R. Barton, *J. Chem. Soc.*, 148 (1949).
4. P. G. Ashmore, J. W. Gardner, and A. J. Owen, *J. Chem. Soc., Faraday Trans. 1* **78**(3), 657–676 (1982).
5. S. Inokawa and co-workers, *Kogyo Kagoku Zasshi* **67**(10), 1540 (1964).
6. S. Kiyonori and A. Shuzo, *Nippan Kagaku Kaishi* (10), 1045 (1974).
7. P. Andreu and co-workers, *Am. Quim.* **65**(11), 931 (1969).
8. U.S. Pat. 2,148,304 (Feb. 21, 1939), J. D. Ruys and H. R. McCombie (to Shell Development Co.).
9. W. L. Archer and E. L. Simpson, *I and EC Prod., Rand D* **16**(2), 158 (June 1977).
10. Y. Chen and X. Jin, *Zheijiang Gongxueyuan Xuebao* (1), 47–53 (1991).
11. K. Okamoto and co-workers, *Bull. Chem. Soc. Jpn.* **40**(8), 1917 (1967).
12. G. E. Ham and J. Stevens, *J. Org. Chem.* **27**, 4638 (1962); U.S. Pats. 3,206,499 (Sept. 14, 1965) and 3,206,499 (Sept. 14, 1965), G. E. Ham (to The Dow Chemical Company).

13. S. Kunichika, S. Oka, and T. Sugiyama, *Bull. Inst. Chem. Res. Kyoto Univ.* **48**(6), 276 (1970).
14. Z. Leszczynski, J. Strzelecki, and D. Zelazko, *Przemyst. Chem.* **44**(6), 330 (1965).
15. E. Lundberg, *Kem. Tidskr.* **96**(10), 34–36, 38 (1984).
16. Ger. Pat. 3,245,366 (June 14, 1984), J. Hundeck, H. Scholz, and H. Hennen (to Hoechst A.-G. Fed. Rep. Ger.).
17. Jpn. Pat. 57,109,727 (July 8, 1982) (to Ryo-Nichi Co., Ltd. Japan).
18. *Hydrocarbon Process. Petrol. Refiner* **44**, 289 (1965).
19. Fr. Pat. 1,577,105 (Aug. 1, 1965), H. Riegel (to Lummus Co.).
20. Fr. Pat. 1,555,518 (Jan. 31, 1969), A. Antonini, P. Joffre, and F. Laine (to Products Chimiques Pechiney Saint-Gobain).
21. Jpn. Pat. 7,133,010 (Sept. 27, 1971), K. Miyauchi, Y. Sato, and S. Okamoto (to Mitsui Toatsu Chemicals Co.).
22. Ger. Pat. 2,106,016 (Sept. 16, 1971), C. H. Cather (to PPG Industries Inc.).
23. Bel. Pat. 900,647 (Mar. 21, 1985) (to BASF A.-G. Fed. Rep. Ger.).
24. U.S. Pat. 4,446,249 (May 1, 1984), J. S. Eden (to Goodrich, B. F., Co. USA).
25. Ger. Pat. 3,607,449 (Sept. 10, 1987), H. D. Eichhorn, W. D. Mross, and H. Schachner (to BASF A.-G. Fed. Rep. Ger.).
26. Ger. Pat. 3,522,473 (Jan. 2, 1987), H. D. Eichhorn, W. D. Mross, and H. Schachner (to BASF A.-G. Fed. Rep. Ger.).
27. Jpn. Pat. 57,136,928 (Aug. 24, 1982) (to Kanegafuchi Chemical Industry Co., Ltd. Japan).
28. "Ethylene Dichloride," *Chemical Economics Handbook*, 651.5000, Stanford Research Institute, Menlo Park, Calif., 2000.
29. "Ethylene Dichloride, Chemical Profile", *Chemical Market Reporter*, March 12, 2001.
30. D. D. Irish, in F. A. Patty, ed., *Industrial Hygiene and Toxicology*, 3rd Revised Ed., John Wiley & Sons, Inc., New York, 1963, pp. 3491–3497.
31. D. D. McCollister and co-workers, *Arch. Ind. Health* (13), 1 (1956).
32. H. C. Spencer and co-workers, *A.M.A. Arch. Ind. Hyg. Occupational Med.* **4**, 482 (1951).
33. U.S. Pat. 4,614,643 (Sept. 30, 1986), E. P. Doane (to Stauffer Chemical Co. USA).
34. J. A. Key, C. W. Stuewe, and R. L. Standifer, *Technical report*, U.S. Environmental Protection Agency, Office of Air Quality Planning and Standards, Washington, D.C., EPA-450/3-80-028c, 363, pp. 1980.
35. G. Scharein, *Hydrocarbon Process., Int. Ed.* **60**(9), 193–194 (1981).
36. J. Schulze and M. Weiser, *Chem. Ind. (Duesseldorf)* **36**(8), 468–474 (1984).
37. Rom. Pat. 89,942 (Aug. 30, 1986), N. Brindas, A. Emanoil, and N. Chiroiu (to Combinatul Chimic, Rimnicu-Vilcea Rom).
38. Belg. Pat. 890,813 (Apr. 1982), B. Gorny and H. Mathais (to Produits Chimiques Ugine Kuhlmann Fr).
39. *NFPA Bulletin*, 325 M, National Fire Protection Association, 1984.
40. M. D. Rosenzweig, *Chem. Eng.* **105**, (Oct. 18, 1971).
41. Jpn. Pat. 7,330,249 (Sept. 18, 1973), H. Takenobu and co-workers (to Central Glass Co., Ltd.).
42. Czarny and co-workers, *Przem. Chem.* **65**(12), 659–661 (1986).
43. G. Berens and co-workers, *J. Am. Chem. Soc.* **97**, 7076 (1975).
44. M. Jones, *Tetrahedron Lett.* (53), 5593 (1968).
45. R. Ausubel, *J. Photochem.* **4**, 2418 (1975).
46. R. Ausubel, *Int. J. Chem. Kinet.* **7**, 739 (1975).
47. U.S. Pat. 3,654,358 (Apr. 4, 1977), J. Gaines (to The Dow Chemical Company).
48. U.S. Pat. 6,153,575 (Nov. 28, 2000), E. M. Gorton and R. D. Olinger (PPG Industries, Ohio).

49. *1990–1991 Threshold Limit Values for Chemical Substances and Physical Agents and Biological Indices*, American Conference of Governmental Industrial Hygienists, Cincinnati, Ohio, 1991, p. 18.
50. J. T. Wilson and B. H. Wilson, *Appl. Environ. Microbiol.*, 242–243 (Jan. 1985).
51. T. M. Vogel and P. L. McCarty, *Appl. Environ. Microbiol.*, 1080–1083 (May 1985).
52. G. Hawley, *The Condensed Chemical Dictionary*, Van Nostrand Reinhold Co., Inc., New York, 1977, p. 279.
53. Jpn. Pat. 02135290 (May 24, 1990), A. Asano and co-workers (to Asahi Glass Co., Ltd.).
54. U.S. Pat. 4961870 (Oct. 9, 1990), J. G. Burt and J. P. Burns (to Allied Signal).
55. U.S. Pat. 4808331 (Feb. 28, 1989), K. D. Cook and co-workers (to E. I. du Pont de Nemours & Co., Inc.).
56. U.S. Pat. 5,654,129 (Aug. 5, 1997), T. L. Taylor.
57. U.S. Pat. 5,478,492 (Dec. 26, 1995), P. Berthelemy, M. Paulus, and R. Pulleman (to Solvay).
58. L. Metz and A. Roedig, *Chem. Ing. Technick* **21**, 191 (1949).
59. W. L. Archer and E. L. Simpson, *Chem. Prof. Polychloroethanes Polychloroalkenes I&EC* Prod. Res. Dev. **167**, 158–162 (June 1977).
60. U.S. Pat. 2,795,623 (June 11, 1957), F. W. Starks (to E. I. du Pont de Nemours & Co., Inc.).
61. U.S. Pat. 2,818,446 (Dec. 31, 1957), F. W. Starks (to E. I. du Pont de Nemours & Co., Inc.).
62. Brit. Pat. 794,700 (May 7, 1958), H. B. Copelin (to E. I. du Pont de Nemours & Co., Inc.).
63. R. B. Burrows and L. Hunter, *J. Chem. Soc.*, 1357 (1932).
64. D. B. Robinson and G. E. Green, *Chem. Ind.*, 214 (Mar. 4, 1972).
65. A. J. Rudge, *The Manufacture and Use of Flourine and its Compounds*, Oxford University Press (for Imperial Chemical Industries Ltd.), Cambridge, Mass., 1962, p. 71.
66. H. L. Cornish, Jr., *U.S. At. Energy Comm.* TID-21388, 1964.
67. J. Pichler and J. Rybicky, *Chem. Prum.* **16**, 559 (1966).
68. Jpn. Pat. 72 45,415 (Nov. 16, 1972), Kimimura, Takayoshi, and S. Wataru (to Hoechst Gosel Co. Ltd.).
69. U.S. Pat. 3,846,508 (Nov. 5, 1974), D. H. Heinert (to The Dow Chemical Company).
70. U.S. Pat. 3,907,872 (Sept. 23, 1975), D. H. Heinert (to The Dow Chemical Company).
71. Ger. Pat. 719,194 (Mar. 26, 1942), H. Kopff and C. Rautenschauch (to I.G. Farbenindustrie, AG).
72. Z. Jedlinski and E. Grzywa, *Polimery* **11**, 560 (1966).
73. Pol. Pat. 53,152 (Feb. 28, 1967), E. Grzywa and Z. Jedlinski (to Zaklady Chemiczne "Oswiecim").
74. S. U. Mullik and M. A. Quddus, *Pak. J. Sci. Ind. Res.* **12**(3), 181 (1970).
75. T. A. DuPlessis and A. C. Thomas, *J. Polym. Sci. Polym. Chem. Ed.* **11**, 2681 (1973).
76. R. A. Siddiqui and M. A. Quddus, *Pak. J. Sci. Ind. Res.* **14**(3), 197 (1971).
77. H. Asai, *Nippon Kagaku Zasshi* **85**, 252 (1964).
78. E. Krotki and J. Mitus, *Polimery* **9**, 155 (1964).
79. Jpn. Pat. 71 01,719 (Jan. 16, 1971), Shimokawa and Wataru (to Hekisto Gosei Co. Ltd.).
80. "C_2 Chlorinated Solvents," in *Chemical Economics Handbook*, Stanford Research Institute, Menlo Park, Calif., 2002.
81. L. M. Elkin, *Process Economics Program*, Chlorinated Solvents, Report No. 48, Stanford Research Institute, Menlo Park, Calif., Feb. 1969.
82. Ger. Pat. 901,774 (Nov. 3, 1940), (to Wacker Chemie, GmbH).

83. U.S. Pat. 2,140,548 (Dec. 30, 1938), J. H. Reilly (to The Dow Chemical Company).
84. Brit. Pat. 673,565 (June 11, 1952), (to Diamond Alkali).
85. U.S. Pat. 2,725,412 (Nov. 29, 1955), F. Conrad (to Ethyl Chemical Co.).
86. Neth. Appl. 6,607,204 (Nov. 28, 1966), F. Sanhaber (to Donau Chemic).
87. Fr. Pat. 1,435,542 (Mar. 7, 1966), A. C. Schulz (to Hooker Chemical).
88. "Trichloroethylene, Chemical Profile", *Chemical Market Reporter*, July 22, 2002.
89. *1990 Annual Book of ASTM Standards*, Section 15, Philadelphia, Pa., 1990.
90. U.S. Pat. 5,994,145 (Nov. 30, 1999), S. W. Stave and co-workers (to Diagnostics, Inc. and EM Industries).
91. B. Soucek and D. Vlachove, *Br. J. Ind. Med.* **17**, 60 (1960).
92. V. Bartonicek, *Br. J. Ind. Med.* **19**, 134 (1962); M. Ogata, Y. Takatsuka, and K. Tomokuni, *Br. J. Ind. Med.* **28**, 386 (1971).
93. D. M. Avaido and co-workers, *Methyl Chloroform and Trichloroethylene in the Environment*, CRC Press, Cleveland, Ohio, 1976.
94. R. J. Lewis, Sr., *Sax's* Properties of Dangerous Materials, 10th. ed., John Wiley & Sons, Inc., New York, 2000.
95. E. O. Longley and R. Jones, *Arch. Environ. Health* **7**, 249 (1963).
96. R. D. Steward and co-workers, *Arch. Environ. Health* **20**, 64 (1970).
97. G. J. Stopps and W. McLaughlin, *Am. Ind. Hyg. Assoc. J.* **29**, 43 (1967).
98. R. J. Vernon and R. K. Ferguson, *Arch. Environ. Health* **18**, 894 (1964).
99. R. K. Ferguson and R. J. Vernon, *Arch. Environ. Health* **29**, 462 (1970).
100. *Fed. Reg.* **54**(12), 2332 (Jan. 19, 1989).
101. K. Kadlec, *Cesk. Dermatol.* **38**, 395 (1963).
102. S. M. Peck, *J. Am. Med. Assoc.* **125**, 190 (1944).
103. J. M. Schirren, *Berufs-Dermatosen* **19**, 240 (1971).
104. *Carcinogenesis Bioassay of Trichloroethylene*, NCI-CG-TR-2, U.S. Dept. of HEW, Washington, D.C., Feb. 1976, p. 197.
105. *Carcinogenesis Bioassay of Trichloroethylene in F344 Rats and B6C3F1 Mice*, NTD 81–84 NIH Publication No. 82–1799, National Toxicology Program (NTP), Research Triangle Park, N.C., 1982.
106. *Toxicological and Carcinogenesis of Trichloroethylene in Four Strains of Rats* (ACI, August, Marchall, Osborne-Mendel), NTP TR 273, NIH Publication No. 88–2529, National Toxicological Program (NTP), Research Triangle Park, N.C., 1988.
107. B. A. Schwetz, B. K. Leong, and P. J. Gehring, *Toxicol. Appl. Pharmacol.* **32**, 84 (1975).
108. S. Shindel and S. Ulrich, *Report of Epidemiologic Study: Warner Electric Brake & Clutch Co.*, South Beloit, Ill., Jan. 1957 to July 1983, Ergotopology Investigative Medicine for Industry, Milwaukee, Wis., Aug. 1984.
109. F. D. Schaumburg, "Banning Trichloroethylene: Responsible Action or Overkill?," *Environ. Sci. Technol.* **24**(1), (1990).
110. L. P. Brown, D. G. Farrar, and C. G. DeRooij, *Health Risk Assessment of Environmental Exposure to Trichloroethylene, Regulatory, Toxicol. Pharmacol.* **11**, 24–41 (1990).
111. *Fed. Reg.* **50** FR 46880 Part III, 46880 (Nov. 13, 1985).
112. U.S. Pat. 6,281,184 (Aug. 28, 2001), P. Michaud and J. J. Martin (to Autofina).
113. L. H. Horsley and co-workers, *Adv. Chem. Ser.* **6**, 32 (1952).
114. W. L. Howard and T. L. Moore, unpublished data, The Dow Chemical Company, 1966.
115. R. P. Marquardt, unpublished data, The Dow Chemical Company, 1964.
116. W. R. H. Hurtley and S. Smiles, *J. Chem. Soc.*, 2269 (1926).
117. L. D. Rampino, *Chem. Eng. News* **36**, 62 (1958).
118. A. Besson, *Compt. Rend.* **118**, 1347 (1894).

119. U.S. Pat. 2,850,543 (Sept. 2, 1958), C. Woolf (to Allied Chemical Corp.).
120. M. J. Prins, *Rec. Trav. Chim.* **51**, 473 (1932).
121. K. W. Doak, *J. Am. Chem. Soc.* **70**, 1525 (1948).
122. F. R. Mayo, F. M. Lewis, and C. Walling, *J. Am. Chem. Soc.* **70**, 1529 (1948).
123. C. S. Marvel, F. D. Hager, and D. D. Coffman, *J. Am. Chem. Soc.* **49**, 2328 (1927).
124. W. Archer and E. Simpson, *IEC Prod. RD* **16**, 158 (June 1977); *ibid.*, 319–325.
125. "Perchloroethylene, Chemical Profiles," *Chemical Market Reporter*, July 29, 2002.
126. *1990 Annual Book of ASTM Standards*, Section 15, Vol. 15.05, ASTM, Easton, Md., 1990.
127. V. K. Rowe and co-workers, *Arch. Ind. Hyg. Occup. Med.* **5**, 556 (1952).
128. R. D. Stewart, *Arch. Environ. Health* **2**, 516 (1961).
129. R. D. Stewart and co-workers, *Arch. Environ. Health* **20**, 224 (1970).
130. R. D. Stewart and co-workers, report number NIOSH-MCOW-ENVM-PCE-74-6, The Medical College of Wisconsin, Milwaukee, Wis., 1974, 172 pp.
131. C. P. Carpenter, *J. Ind. Hyg. Toxic.* **19**, 323 (1937).
132. P. D. Lamson, *Am. J. Hyg.* **9**, 430 (1929).
133. R. Patel, *J. Am. Med. Assoc.* **223**, 1510 (1973).
134. T. C. Tuttle, *Final Report for Contract HSM99-73-35*, Westinghouse Behavioral Services Center, Columbia, Md., p. 124, 1976.
135. *Fed. Reg.* **54**(12), 2332 (Jan. 19, 1989).
136. *1990–1991 Threshold Limit Values for Chemical Substances and Physical Agents, American Conference of Governmental Industrial Hygienists*, 1990.
137. *Standard Practice for Confined Area Entry, D4276-84*, American Society for Testing and Materials, ASTM, Philadelphia, Pa., 1984.
138. *Specialty Chlorinated Solvents Product Stewardship Manual*, 1991 ed, The Dow Chemical Company, Midland, Mich., form 100-6170-90HYC.
139. S. Yllner, *Nature (London)* **191**, 82 (1961).
140. M. Ikeda and co-workers, *Br. J. Ind. Med.* **29**, 328 (1972).
141. M. Ogata and co-workers, *Br. J. Ind. Med.* **28**, 386 (1971).
142. National Cancer Institute, NCI-CG-TR-13, 1977.
143. National Toxicology Program (NTP), TR-311, Research Triangle Park, N.C., 1985.
144. L. W. Rampy, J. F. Quast, B. K. J. Leong, and P. J. Gehring, *Proceedings of the First International Congress on Toxicology*, Academic Press, New York, 1978.
145. D. G. Pegg, J. A. Zempel, W. H. Braun, and P. G. Watanabe, *Toxicol. Appl. Pharmacol.* **51**, 465–474 (1979).
146. A. M. Schumann, J. F. Quast, and P. G. Watanabe, *Toxicol. Appl. Pharmacol.* **55**, 207–219 (1980).
147. *Final Report EPA/600/8-83/0005F*, Environmental Protection Agency, Washington, D.C., 1985.
148. B. A. Schwetz, B. K. J. Leong, and P. G. Gehring, *Toxicol. Appl. Pharmacol.* **32**, 84–96 (1975).
149. *Fed. Reg.* **48**(206), 49,097 (Oct. 24, 1983).
150. *Fed. Reg.* **56**(20), 3536 (Jan. 30, 1991).

GAYLE SNEDECOR
J. C. HICKMAN
The Dow Chemical Company

JAMES A. MERTENS
Dow Chemical USA

CHLOROFORM

1. Introduction

Chloroform [67-66-3] (trichloromethane, methenyl chloride), $CHCl_3$, at normal temperature and pressure is a heavy, clear, volatile liquid with a pleasant, ethereal, nonirritant odor. Although chloroform is nonflammable, its hot vapor in admixture with vaporized alcohol burns with a green tinged flame. Chloroform is miscible with the principal organic solvents and is slightly soluble in water. It is less stable in storage than either methyl or methylene chloride. Chloroform decomposes at ordinary temperatures in sunlight in the absence of air and in the dark in the presence of air. Phosgene is one of the oxidative decomposition products.

Chloroform was discovered in 1831 by Liebig and Soubeirain simultaneously. Liebig obtained chloroform by the action of alkali on chloral, and Soubeirain by reaction of bleaching powder with alcohol or acetone. Guthrie, in the United States, is also alleged to have discovered chloroform in 1831. In 1839, Dumas produced chloroform by heating alkali with trichloroacetic acid. In the following year, Regnault obtained it by chlorinating methyl chloride. Chloroform was first used in medicine as a stimulant, taken internally, and as an inhalant in cases of asthma. In November, 1847, on the suggestion of Waldie, a Liverpool chemist, Simpson used chloroform as a total anesthetic in obstetrics.

Shortly after Simpson's successful use of chloroform in Edinburgh, Fraser began making pure anesthetic chloroform on a small scale in Nova Scotia.

In 1900, the Pennsylvania Salt Manufacturing Co. initiated large-scale production in the United States. The Midland Chemical Co., a subsidiary of The Dow Chemical Company, began to manufacture chloroform by reducing carbon tetrachloride in 1903. Chloroform was one of the first organic chemicals produced on a large scale in the United States.

Chloroform was used chiefly as an anesthetic and in pharmaceutical preparations immediately prior to World War II. However, these uses have been banned. Annual output in both the United States and the United Kingdom was between 900 and 1350 metric tons. During the war, chloroform production in the United States tripled, largely to meet the requirement for penicillin manufacture. Demand for chloroform continued to increase in the postwar period as its technical applications were extended. Consumption continues to increase at a comparatively rapid rate. Chloroform is now used primarily in the manufacture of HCFC-22, monochlorodifluoromethane, a refrigerant, and as a raw material for polytetrafluoroethylene plastics.

2. Physical and Chemical Properties

The physical properties of chloroform are listed in Table 1.

Chloroform dissolves alkaloids, cellulose acetate and benzoate, ethylcellulose, essential oils, fats, gutta-percha, halogens, methyl methacrylate, mineral oils, many resins, rubber, tars, vegetable oils, and a wide range of common organic compounds. A temperature increase occurs when chloroform is mixed

Table 1. **Physical Properties of Chloroform**

Property	Value
mol wt	119.38
refractive index at 20°C	1.4467
autoignition temperature, °C	above 1000
flash point, °C	none
mp, °C	−63.2
$\quad$ 101 MPa^a	−43.4
$\quad$ 507 MPa^a	20.4
$\quad$ 1216 MPa^a	112.6
bp at 101 kPa^b, °C	61.3
sp gr	
$\quad$ 0/4°C	1.52637
$\quad$ 25/4°C	1.48069
$\quad$ 60.9/4°C	1.4081
vapor density at 101 kPa^b, 0°C, kg/m^3	4.36
surface tension, mN/m (= dyn/cm)	
$\quad$ air, 20°C	27.14
$\quad$ air, 60°C	21.73
$\quad$ water, 20°C	45.0
heat capacity at 20°C, kJ/(kg · K)^c	0.979
critical temperature, °C	263.4
critical pressure, MPa^a	5.45
critical density, kg/m^3	500
critical volume, m^3/kg	0.002
thermal conductivity at 20°C, W/(m · K)	0.130
coefficient of cubical expansion	0.001399
dielectric constant, 20°C	4.9
dipole moment, C · m^d	3.84×10^{-30}
heat of combustion, MJ/(kg · mol)^c	373
heat of formation at 25°C, MJ/(kg · mol)^c	
$\quad$ gas	−89.66
$\quad$ liquid	−120.9
latent heat of evaporation at bp, kJ/kg^c	247
solubility of chloroform in water, g/kg H$_2$O	
$\quad$ 0°C	10.62
$\quad$ 20°C	8.22
$\quad$ 30°C	7.76
solubility of water in chloroform, at 22°C, g/kg chloroform	0.806
viscosity, liq, mPa · s (= cP)	
$\quad$ −13°C	0.855
$\quad$ 0°C	0.700
$\quad$ 20°C	0.563
$\quad$ 30°C	0.510

^{a}To convert MPa to atm, multiply by 9.87.
^{b}To convert kPa to mm Hg, multiply by 7.5.
^{c}To convert J to cal, divide by 4.184.
^{d}To convert C · m to debye, divide by 3.336×10^{-30}.

with diethyl ether. Chloroform forms a series of binary azeotropes (1); the azeotrope with water boils at 56.1°C and contains 97.2% chloroform. The ternary azeotrope with ethanol and water boils at 55.5°C and contains 4 mol % alcohol and 3.5 mol% water. At 25°C, chloroform dissolves 3.59 times its volume of carbon dioxide.

Chloroform slowly decomposes on prolonged exposure to sunlight in the presence or absence of air and in the dark in the presence of air. The products of oxidative breakdown include phosgene, hydrogen chloride, chlorine, carbon dioxide, and water. At 290°C, chloroform vapor is not attacked by oxygen. In contact with iron and water hydrogen peroxide is also produced, probably by the following reaction sequence (2):

$$CHCl_3 + O_2 \longrightarrow (Cl_3COOH) \xrightarrow{\text{H}_2\text{O}} Cl_3COH + H_2O_2$$

$$Cl_3COH \longrightarrow COCl_2 + HCl$$

Oxidation with powerful oxidizing agents, eg, chromic acid, results in the formation of phosgene and the liberation of chlorine. Nitrogen dioxide at about 270°C oxidizes chloroform to a mixture of compounds including phosgene, hydrogen chloride, water, and carbon dioxide (3). Ozone forms a blue solution in chloroform and causes rapid decomposition.

Chloroform and water at 0°C form six-sided crystals of a hydrate, $CHCl_3 \cdot 18H_2O$ [67922-19-41], which decompose at 1.6°C. Chloroform does not decompose appreciably when in prolonged contact with water at ordinary temperature and in the absence of air. However, on prolonged heating with water at 225°C, decomposition to formic acid, carbon monoxide, and hydrogen chloride occurs. A similar hydrolysis takes place when chloroform is decomposed at elevated temperature by potassium hydroxide.

$$CHCl_3 + 4\,KOH \longrightarrow HCOOK + 3\,KCl + 2\,H_2O$$

Reaction of chloroform with hydroxide, particularly in the presence of methanol, can result in an explosion.

Chloroform resists thermal decomposition at temperatures up to about 290°C. Pyrolysis of chloroform vapor occurs at temperatures above 450°C, producing tetrachloroethylene, hydrogen chloride, and a number of chlorohydrocarbons in minor amounts (4,5). Pyrolysis in contact with hot pumice is catalyzed by vaporized iodine (1%), resulting in tetrachloroethylene, hexachloroethane, and carbon tetrachloride. Hexachlorobenzene, carbon monoxide, hydrogen chloride, and titanium tetrachloride are formed when chloroform vapor is decomposed by hot titanium oxide. In contact with potassium amalgam or red-hot copper, chloroform reacts to give acetylene.

$$2\,CHCl_3 + 6\,K[Hg] \longrightarrow HC{\equiv}CH + 6\,KCl\,[Hg]$$

Small quantities of ethyl alcohol stabilize chloroform during storage. Various other stabilizers have been proposed, eg, $CH_2{=}CHCH_2CH_2CN$ and methacrylonitrile (6).

Chloroform can be reduced to methane with zinc dust and aqueous alcohol. In the presence of a catalyst or ammonia, the reduction yields methylene chloride as well as methane.

Chloroform reacts readily with halogens or halogenating agents. Chlorination of the irradiated vapor is believed to occur by a free-radical chain reaction (7).

At 225–275°C, bromination of the vapor yields bromochloromethanes: CCl_3Br, CCl_2Br_2, and $CClBr_3$. Chloroform reacts with aluminum bromide to form bromoform, $CHBr_3$. Chloroform cannot be directly fluorinated with elementary flourine; fluoroform, CHF_3, is produced from chloroform by reaction with hydrogen fluoride in the presence of a metallic fluoride catalyst (8). It is also a coproduct of monochlorodifluoromethane from the $HF–CHCl_3$ reaction over antimony chlorofluoride. Iodine gives a characteristic purple solution in chloroform but does not react even at the boiling point. Iodoform, CHI_3, may be produced from chloroform by reaction with ethyl iodide in the presence of aluminum chloride; however, this is not the route normally used for its preparation.

No decomposition occurs when boiling chloroform is in prolonged contact with anhydrous aluminum chloride; a double compound is formed from which unchanged chloroform is liberated by the action of water. With benzene in the presence of aluminum chloride, chloroform reacts to give triphenylmethane, $(C_6H_5)_3CH$, which is also formed from chloroform and phenylmagnesium bromide, C_6H_5MgBr.

In the presence of an alkali metal hydroxide at about 50°C, chloroform condenses with acetone to give 1,1,1-trichloro-2-methyl-2-propanol, [57-15-8] ie, chlorobutanol, chloretone, or acetone–chloroform (9,10). Chlorobutanol is a white crystalline substance with a camphorlike odor; its sedative, anesthetic, and antiseptic properties have given the compound some importance in the pharmaceutical industry.

Chloroform reacts with aniline and other aromatic and aliphatic primary amines in alcoholic alkaline solution to form isonitriles, ie, isocyanides, carbylamines, as shown:

$$CHCl_3 + C_6H_5NH_2 + 3\,KOH \longrightarrow C_6H_5N{\equiv}C + 3\,KCl + 3\,H_2O$$

Phenylisonitrile has a powerful characteristic odor; it is used as a qualitative test (the carbylamine test) for chloroform or primary aromatic amines. Chloroform reacts with phenols in alkaline solution to give hydroxyaromatic aldehydes in the Reimer-Tiemann reaction; eg, phenol gives chiefly p-hydroxybenzaldehyde and some salicylaldehyde (11) (see HYDROXYBENZALDEHYDES).

Chloroform combines with the inner anhydride of salicyclic acid to form a well-defined crystalline double compound (12):

This complex readily liberates chloroform when heated and this reaction has been used to produce very pure chloroform.

Reactivities of several chlorinated solvents, including chloroform, with aluminum, iron, and zinc in both dry and wet systems have been determined, as have chemical reactivities in oxidation reactions and in reactions with amines (11). Unstabilized wet chloroform reacts completely with aluminum and attacks zinc at a rate of >250 µm/yr and iron at <250 µm/yr. The dry, uninhibited solvent attacks aluminum and zinc at a rate of 250 µm/yr and iron at 25 µm/yr.

3. Manufacture

Chloroform can be manufactured from a number of starting materials. Methane, methyl chloride, or methylene chloride can be further chlorinated to chloroform, or carbon tetrachloride can be reduced, ie, hydrodechlorinated, to chloroform. Methane can be oxychlorinated with HCl and oxygen to form a mixture of chlorinated methanes. Many compounds containing either the acetyl (CH_3CO) or $CH_3CH(OH)$ group yield chloroform on reaction with chlorine and alkali or hypochlorite. Methyl chloride chlorination is now the most common commercial method of producing chloroform. Many years ago chloroform was almost exclusively produced from acetone or ethyl alcohol by reaction with chlorine and alkali.

3.1. Methane or Chloromethane Chlorination. This chlorination reaction is generally carried out in the gas phase at 400–500°C making all of the more heavily chlorinated products in substantial yields (see METHYLENE CHLORIDE). It is carried out by mixing the organic feed with gaseous chlorine and heating to initiate the reaction. Once started, this free-radical process rapidly goes to completion with the generation of substantial amounts of heat. This heat is transferred back to heat the feed stream. The reaction is controlled by limiting the amount of one of the feeds, generally the chlorine. The use of two reactors in parallel using different organic feed compositions is known (12,13).

The product selectivity of the chlorination of methane has been adequately described (14) over the whole range of products from methyl chloride to carbon tetrachloride. The presence of partially chlorinated chloromethanes in the reactor feed stream tends to change the product mix toward the more heavily chlorinated products. These partially chlorinated chloromethanes can be gases recycled from the product recovery equipment. Thus there is a fair degree of flexibility in tailoring product selectivity. The most common feed material for this process is methyl chloride. Use of methane as a feed material generally requires the production of the less desirable aqueous HCl and is not widely used industrially. In either case, to maximize chloroform production, product methyl chloride and methylene chloride are usually recycled back to the reactor. Numerous variations have been suggested as ways to increase the selectivity to a desired product. Tokuyama Soda has developed and practices a liquid-phase process based on methyl chloride chlorination (15–17). Light-initiated processes also claim enhanced selectivity (18). A novel process where chlorine atoms, the chain carriers in the free-radical reaction, are generated and transported to a reaction zone has been described (19,20). Fluidized beds have been suggested as a method of handling the large amount of heat generated and affecting selectivity with the proper choice of catalyst (21).

3.2. Oxychlorination of Methane. The oxychlorination of methane with HCl and oxygen has received some attention since the 1970s (22–24), though there are no examples of an industrial process. This can be a coproduct process making all the chloromethanes in significant yields or one that makes primarily methyl chloride. Interest in this route has increased in the past few years because of progress in the methane to light hydrocarbons process.

3.3. Hydrogenation of Carbon Tetrachloride. Carbon tetrachloride can be hydrogenated, ie, hydrodechlorinated, to chloroform over a catalyst (25,26) or thermally (27). Although there are no industrial examples of this process at this time, it will receive more attention as more carbon tetrachloride becomes available as the CFC-11 and -12 markets decline (see CARBON TETRACHLORIDE). Chloroform can be further hydrodechlorinated to methylene chloride (28,29).

A catalyst based on platinum dispersed throughout its support of gamma alumina for the production of chloroform via the hydrodechlorination of carbon tetrachloride has been described (30).

3.4. Reduction of Alcohols or Ketones. The reaction of alcohols and ketones with chlorine and base to give chloroform is well known (31). This was an industrial process for chloroform, but there are no plants currently using this technology. This reaction is possibly an important source of chloroform in the water treating process.

4. Economic Aspects

Table 2 lists the U.S. producers of chloroform and their capacities. Chloroform is coproduced with methylene chloride in direct reaction of methyl chloride with chlorine.

Demand for chloroform was 274×10^6 kg (606×10^6 lb) in 1999. Expected demand for 2003 is 298×10^6 kg (656×10^6 lb) (32). Growth is expected at a rate of 2% through 2003.

Price history for the years 1994 through 1999 was a high of $0.20/kg ($0.44/lb) technical consumers, tanks, dlvd., list and the low was $0.18/kg ($0.395/lb) same basis.

HCFC-22 has grown at a annual rate of 3.2% in recent years as a replacement for chlorofluorocarbons in home and automobile air conditioners and commercial freezers. HCFC is also strong as a feedstock in the preparation of

Table 2. **United States Producers of Chloroform and Their Capacities**[a]

Producer	Capacity, $\times 10^6$ kg ($\times 10^6$ lb)	
Dow, Freeport, Texas	91	(200)
Dow, Plaquemine, La	91	(200)
Vulcan, Geismar, La	73	(160)
Vulcan, Wichita, Kans.	73	(160)
Total	*328*	(720)

[a]Ref. 31.

polytetrafluoroethylene and fluorinated ethylene–propylene. However, use of HCFC as a refrigerant is subject to substitution by other products not subject to legislated restrictions. Beginning in 2010, production will go to feedstock use only and for refrigeration in equipment manufactured prior to 2010. Fluoropolymers will continue to rise because of demand for high performance wire and cable insulation.

5. Specifications and Standards

Technical-grade chloroform generally contains one or more stabilizers, which vary according to specification requirements. The most common is 50 ppm 2-methyl-2-butene [513-35-9]. Other stabilizers are industrial methylated spirit (0.2%), absolute alcohol (0.6–1%), thymol, *t*-butylphenol, or *n*-octylphenol (0.0005–0.01%). A representative technical quality chloroform contains the following amounts of the indicated substances (maximum):

water	50 ppm
acid (as HCl)	10 ppm
methylene chloride	200 ppm
bromochloromethane	300 ppm
carbon tetrachloride	250 ppm
1,2-dichloroethylene	100 ppm
vinylidene chloride	100 ppm
residue (on evaporation at 110°C)	10 ppm
dissolved chlorine	not detectable

6. Analytical Methods

The most widely used method of analysis for chloroform is gas chromatography. A capillary column medium that does a very good job in separating most chlorinated hydrocarbons is methyl silicone or methyl (5% phenyl) silicone. The detector of choice is a flame ionization detector. Typical molar response factors for the chlorinated methanes are methyl chloride, 2.05; methylene chloride, 2.2; chloroform, 2.8; and carbon tetrachloride, 3.1, where methane is defined as having a molar response factor of 2.00. Most two-carbon chlorinated hydrocarbons have a molar response factor of about 1.0 on the same basis.

In the known absence of bromoform, iodoform, chloral, and other halogenated methanes, the formation of phenylisonitrile with aniline provides a simple and fairly sensitive but nonspecific test for the presence of chloroform, the carbylamine test. Phenylisonitrile formation is the identification test given in the *British Pharmacopoeia*. A small quantity of resorcinol and caustic soda solution (10% concentration) added to chloroform results in the appearance of a yellowish red color, fluorescing yellow-green. When 0.5 mL of a 5% thymol solution is boiled with a drop of chloroform and a small quantity of potassium hydroxide solution, a yellow color with a reddish sheen develops; the addition of sulfuric acid causes a change to brilliant violet, which, diluted with water, finally changes to blue (33).

Chloroform may be estimated quantitatively by determining the amount of copper oxide produced when it is warmed with Fehling's solution, which is potassium cupritartrate (34). An alternative procedure consists of heating the chloroform with concentrated alcoholic potassium hydroxide in a sealed tube at 100°C and determining the amount of potassium chloride produced (35).

7. Health and Safety Factors

7.1. Handling.   All persons who have occasion to use or handle chloroform should be thoroughly instructed and adequately supervised in the proper methods of handling the substance to prevent or minimize exposure to the liquid or its vapors and in the proper methods of disposing of this chemical.

Chloroform should be stored in sealed containers in a cool place. Glass containers should be dark green or amber. Bulk storage containers may be constructed of mild or plain steel that is galvanized or suitably lined. Aluminum is not recommended for bulk storage. All bulk storage tanks should be equipped with a vent dryer packed with calcium chloride or other appropriate desiccant to exclude moisture. Alternatively, the tank may utilize a dry inert gas pad with an appropriate pressure vacuum relief valve, which is the recommended procedure, with appropriate disposal of the tank vents. Seamless stainless steel, Teflon, seamless bronze, or seamless steel hose is recommended with asbestos, Teflon, Viton, or Neoprene gaskets (36). Chloroform is transported in drums, truck transports, rail cars, barges, and oceangoing ships.

7.2. Toxicology.   The principal hazard in exposure to chloroform is damage to the liver and kidneys resulting from inhalation or ingestion. Inhalation of high concentrations may result in disturbances of equilibrium or loss of consciousness. Chloroform is mildly irritating to skin and mucous membranes upon contact, and to the alimentary tract upon ingestion. It is believed that medically significant quantities are not absorbed through intact skin.

The toxic effects of chloroform resemble those of carbon tetrachloride. The probable effects of exposure to various atmospheric concentrations of chloroform are summarized in Table 3 (37).

In the past, chloroform was used extensively as a surgical anesthetic, but this use was abandoned because exposure to narcotic concentrations often resulted in sudden death from effects on the heart and circulation or from severe injury to the liver. In addition, chloroform for this and other consumer uses was banned by FDA in 1976 with the discovery that it is carcinogenic in mice (38). When splashed into the eye, chloroform causes local pain and irritation, but serious injury is not expected. Skin contact for single, brief exposures ordinarily causes little or no local irritation.

Repeated or prolonged contact with the skin, especially under clothing, may result in local irritation and inflammation, and at elevated temperatures such as in the presence of an open flame, chloroform decomposes to form by-products, including phosgene, chlorine, and hydrogen chloride, all of which are severe irritants to the respiratory tract.

Table 3. **Effects of Exposure to Chloroform**

Concentration		Response
ppm	mg/L	
205–310	1–1.5	smallest amount that can be detected by smell
390	1.9	endured for 30 min without complaint
1,025	5	definite after effects; fatigue and headache still experienced hours after exposure
1,025	5	dizziness, intracranial pressure, and nausea after 7 min exposure
1,475	7.2	dizziness and salivation after a few minutes exposure
4,100	20	vomiting, sensation of fainting
14,340–16,400	70–80	narcotic limiting concentration

Ingestion of chloroform is followed immediately by a severe burning in the mouth and throat, pain in the chest and abdomen, and vomiting. Loss of consciousness and liver injury may follow depending on the amount swallowed. The tendency of chloroform to produce liver injury is significantly augmented in alcoholics and persons with nutritional deficiencies.

The most serious hazard of repeated exposure to chloroform inhalation is injury to the liver and kidneys. Evidence indicates that in humans, repeated exposure to atmospheric concentrations well below the odor threshold may cause such injury. Industrial experience has shown that daily exposure to concentrations below 100 ppm may result in a variety of nervous system and alimentary tract symptoms, in the absence of demonstrable evidence of injury (39). Injury to the liver is similar to but somewhat less severe than that caused by carbon tetrachloride. Kidney injury is usually associated with but less severe than liver injury.

NIOSH recognizes chloroform as a carcinogen, and recommends a STEL of 2 ppm; ACGIH, TLV TWA is 10 ppm, with an A3 designation. OSHA has a ceiling limit of 50 ppm. WHO has a drinking water guideline of 30 μg/L. IARC classifies chloroform as a group 2B (possibly carcinogenic to humans, and EPA as a B2, probable human carcinogen (40).

Treatment of chloroform poisoning is symptomatic; no specific antidote is known. Adrenalin should not be given to a person suffering from chloroform poisoning.

7.3. Regulations. The NTP database gives the following information. EPA regulates chloroform under the Clean Water Act (CWA), Comprehensive Environmental Response, Compensation, and Liability Act (CERCLA), Federal Insecticide, Fungicide, and Rodenticide Act (FIFRA), Food, Drug, and Cosmetic Act (FD&CA), Resource Conservation and Recovery Act (RCRA), Safe Drinking Water Act (SDWA), and Superfund Amendments and Reauthorization Act (SARA). Chloroform is a toxic pollutant of air and water. EPA has established water-quality criteria for chloroform, effluent guidelines, rules for regulating hazardous spills, general threshold amounts, and requirements for handling and disposal of chloroform wastes. A reportable quantity (RQ) of 10 lb has been established for chloroform under CERCLA and CWA. Chloroform is exempted under FD&CA from tolerances for pesticide chemicals. Chloroform is recognized as an inert ingredient of toxicological concern under FD&CA. A rebuttable presumption against registration of chloroform-containing pesticides has been issued under FIFRA. Chloroform is regulated as a hazardous constituent of waste under RCRA. USEPA requires removal of chloroform from drinking water and establishes a maximum contaminant level (MCL) of 100 mg/L under SDWA. Under EPCRA, EPA identifies chloroform as an extremely hazardous substance and established a threshold planning quantity (TPQ) of 10,000 lb for chloroform. FDA regulates chloroform as an indirect food additive for adhesive components in food packaging materials and as a component of materials that come into contact with food. The use of chloroform in food, drugs (for both humans and animals), and cosmetics for use in cough preparations, liniments, cosmetics, and toothache drops is banned under the Federal Food, Drug, and Cosmetic Act.

8. Uses

About 95% of the chloroform produced goes into the production of HCFC-22 (chlorodifluoromethane [75-45-6]). Of this 95% about 70% is used as a refrigerant and about 30% is used as a starting material in the production of fluoropolymers, such as polytetrafluoroethylene (PTFE). Miscellaneous uses of the remaining 5% of the chloroform production include laboratory reagents and extractive solvents for pharmaceuticals (32).

The miscellaneous uses include extraction and purification of penicillin, alkaloids, vitamins, and flavors, and as an intermediate in the preparation of dyes and pesticides. A biologically active chloroform extract from mangrove plants has been described (41). Use as an antispasmodic, antiarrythmic agent, or anticholinergic agent are possible. Chloroform has also been used as a fumigant and insecticide, in the formulation of cough syrups, toothpastes, liniments, and toothache preparations. These latter uses were banned by the FDA in 1976 (38).

BIBLIOGRAPHY

"Chloroform" in *ECT* 1st ed., Vol. 3, pp. 842–848, by Leonard Stievater, Jr., McKesson & Robbins, Inc., and R. J. Van Nostrand, Brown Company; "Chlorocarbons and Chlorohydrocarbons, Chloroform" in *ECT* 2nd ed., Vol. 5, pp. 119–127, by D. W. F. Hardie, Imperial Chemical Industries, Ltd.; in *ECT* 3rd ed., Vol. 5, pp. 693–703, by D. H. DeShon, Dow Chemical U.S.A.; in *ECT* 4th ed., Vol. 5, pp. 1051–1062 by Michael T. Holbrook, Dow Chemical U.S.A.; "Chloroform" in *ECT* (online), posting date: December 4, 2000, by Michael T. Holbrook, Dow Chemical U.S.A.

CITED PUBLICATIONS

1. I. Mellan, *Source Book of Industrial Solvents*, Reinhold Publishing Corp., New York, 1957, p. 126.

2. R. Neu, *Pharmazie* **3**, 251 (1948).
3. D. V. E. George and J. H. Thomas, *Trans. Faraday Soc.* **58**(470), 262 (1962).
4. G. P. Semeluk and R. B. Bernstein, *J. Am. Chem. Soc.* **76**(14), 373 (1954).
5. W. Ramsay and S. Young, *Jahresber Fort. Chem.* **628** (1886).
6. U.S. Pat. 3,029,298 (Apr. 10, 1962), F. S. Hirsekorn and J. H. Rains (to Frontier Chemical Co.).
7. I. H. Winning, *Trans. Faraday Soc.* **47**, 1084 (1951).
8. A. J. Rudge, *The Manufacture and Use of Fluorine and its Compounds*, Oxford University Press, Inc., New York, 1962, p. 68.
9. C. Willgerodt, *Ber. Deut. Chem. Ges.* **14**, 24, 51 (1881).
10. E. H. Huntress, *Organic Chlorine Compounds*, John Wiley & Sons, Inc., New York, 1948, pp. 285–288.
11. *Ind. Eng. Chem., Prod. Res. Dev.* **26**(2), 158 (1977).
12. Ger. Pat. DD 220022 (Mar. 20, 1985), S. Lippert and co-workers (to VEB Chemiekombinat Bitterfeld).
13. Eur. Pat. 128,818 (Dec. 19, 1984), J. J. Masini and Y. Verot (to Atochem SA).
14. E. T. McBee and co-workers, *Ind. Eng. Chem.* **34**, 296 (1942).
15. S. Akiyama, *Petrotech* **11**(9), 788–790 (1988).
16. T. Hisamoto, *Nikkakyo Geppo* **34**(3 23-3), 11–19 (1981).
17. Jpn. Pat. JP 56 002,922 (Jan. 13, 1981), (to Tokuyama Soda Co., Ltd.).
18. Jpn. Pat. JP 57 142,927 (Sept. 3, 1982), (to Asahi Glass Co., Ltd.).
19. D. Chakravarty and J. S. Dranoff, *AIChE J.* **30**(6 45-4), 986–988 (1984).
20. M. Richard and M. Lenzi, *AIChE J.* **30** (5 74-1), 838–840 (1984).
21. USSR Pat. SU 413,767 (Jan. 15, 1985), G. P. Geid and co-workers.
22. Jpn. Pat. JP 56 013,696 (Mar. 30, 1981), (to Sumitomo Chemical Co., Ltd.).
23. V. N. Rozanov and co-workers, *Khim. Prom-st* **7**, 495–497 (1989); V. N. Rozanov and co-workers, *Khim. Prom-st* **10**, 726–729 (1989).
24. Ger. Pat. DE 2,351,898 (Apr. 25, 1974), M. C. Sze, H. Riegel and H. D. Schindler (to Lummus Co.).
25. U.S. Pat. 3,579,596 (May 18, 1971), C. R. Mullin and C. E. Wymore (to The Dow Chemical Company).
26. L. I. Strunkina and E. M. Brainina, *Izv. Akad. Nauk SSSR, Ser. Khim.* **9**, 2160–2162 (1983).
27. Yu. A. Kolbanovskii, A. S. Chernysheva, and V. S. Shchipachev, *Kinet. Katal.* **29**(5), 1222–1226 (1988).
28. D. A. Dodson and H. F. Rase, *Ind. Eng. Chem., Proc. Res. Dev.* **17**(3), 236–241 (1978).
29. C. J. Noelke and H. F. Rase, *Ind. Eng. Chem., Prod. Res. Dev.* **18**(4), 325–328 (1979).
30. U.S. Pat. Appl. 20020077514 (June 20, 2002), M. Rossi and co-workers.
31. Can. Pat. CA 1,102,355 (June 2, 1981), B. F. Safi and D. Rouleau (to l'École Polytechnique de Montreal).
32. "Chloroform, Chemical Profile," *Chemical Market Reporter*, Oct. 16, 2000.
33. R. Dupouy, *Chem. News* **88**, 37 (1903).
34. E. Baudrimont, *J. Pharm. Chim.* **9**, 410 (1869).
35. L. deSaint-Martin, *Compt. Rend.* **106**, 494 (1888).
36. *Emergency Response and Transportation Equipment Data Sheet*, Dow Chemical U.S.A., Plaquemine, La., Apr. 10, 1990.
37. F. A. Patty, ed., *Industrial Hygiene and Toxicology*, Vol. II, Interscience Publishers, Inc., New York, 1949.
38. *Chem. Week* **18**(14), 17 (Apr. 7, 1976).
39. P. J. R. Challen and co-workers. *Br. J. Ind. Med.* **15**, 43 (1958).

40. J. B. Reid, in E. Bingham, B. Cohrssen, and C. H. Powell, eds., *Patty's Toxicology*, 5th ed., Vol. 5, John Wiley & Sons, Inc., New York, 2001.
41. U.S. Pat. Appl. 20030003171 (Jan. 2, 2003), U. Goswami and N. Fernandes.

MICHAEL T. HOLBROOK
Dow Chemical U.S.A.

CHLOROFORMATES AND CARBONATES

1. Introduction

Discovered by Dumas in 1833 (1), the reaction of phosgene (carbonic dichloride [75-44-5]) with alcohols gives two classes of compounds, carbonic esters and carbonochloridic esters, commonly referred to as carbonates and chloroformates. The carbonic acid esters (carbonates), ROC(O)OR, are the diesters of carbonic acid [463-79-6]. The carbonochloridic esters, also referred to as chloroformates or chlorocarbonates, ClC(O)OR, are esters of hypothetical chloroformic acid [463-73-0], ClCOOH.

The reaction proceeds in stages, first producing a carbonochloridic ester (chloroformate), and then a carbonic acid diester (carbonate). When a different alcohol is used for the second stage, a mixed radical or unsymmetrical carbonate is produced.

$$\underset{\text{Cl}}{\overset{\overset{\displaystyle O}{\|}}{\text{Cl}-\text{C}-\text{Cl}}} + \text{ROH} \xrightarrow{-\text{HCl}} \underset{}{\overset{\overset{\displaystyle O}{\|}}{\text{Cl}-\text{C}-\text{OR}}} \xrightarrow{\text{R'OH}} \underset{}{\overset{\overset{\displaystyle O}{\|}}{\text{R'O}-\text{C}-\text{OR}}} + \text{HCl}$$

An extensive review of the chemistry of chloroformates was published in 1972 (2,3). Over the last 30 years, in excess of 20,000 articles were published on the chemistry and uses of chloroformates and carbonates. Recent interest in chloroformates and carbonates as carbonylating agents has been enhanced due to the increased concern over phosgene transportation and handling. This article briefly reviews important events regarding these materials, especially those related to technology.

2. Chloroformates

In earlier literature, carbonochloridic esters are referred to as chloroformates or chlorocarbonates because of the structural parallel with formic acid [64-18-6], chloroformic acid, and carbonic acid. Before 1972, chloroformates were indexed in *Chemical Abstracts, Eighth Collective Index*, under formic acid, chloroesters; whereas, in the *Ninth Collective Index* (Dec. 1990), they are referred to as carbonochloridic acid esters. Table 1 lists the common names of commercially available carbonochloridates or chloroformates, the CAS Registry Numbers, and the formulas.

Table 1. **Commercial Chloroformates (Carbonochloridates)**

Ester	CAS Registry number	Formula
methyl	[79-22-1]	$ClCOOCH_3$
chloromethyl	[22128-62-7]	$ClCOOCH_2Cl$
ethyl	[541-41-3]	$ClCOOC_2H_5$
1-chloroethyl	[50893-53-3]	$ClCOOCHClCH_3$
2-chloroethyl	[627-11-2]	$ClCOOCH_2CH_2Cl$
2,2,2-trichloroethyl	[17341-93-4]	$ClCOOCH_2CCl_3$
vinyl	[5130-24-5]	$ClCOOCH{=}CH_2$
isopropyl	[108-23-6]	$ClCOOCH(CH_3)_2$
n-propyl	[109-61-5]	$ClCOOCH_2CH_2CH_3$
3-chloropropyl	[628-11-5]	$ClCOOCH_2CH_2CH_2Cl$
allyl	[2937-50-0]	$ClCOOCH_2CH{=}CH_2$
methallyl	[42068-70-2]	$ClCOOCH_2(CH_3)C{=}CH_2$
n-butyl	[592-34-7]	$ClCOOCH_2CH_2CH_2CH_3$
sec-butyl	[17462-58-7]	$ClCOOCH(CH_3)CH_2CH_3$
isobutyl	[543-27-1]	$ClCOOCH_2CH(CH_3)_2$
isoamyl	[628-50-2]	$ClCOOCH_2CH_2CH(CH_3)_2$
n-pentyl	[638-41-5]	$ClCOOCH_2CH_2CH_2CH_2CH_3$
cyclopentyl	[50715-28-1]	$ClCOOC_5H_9$
4-tert-butyl cyclohexyl	[42125-46-2]	$ClCOOC_6H_{10}C(CH_3)_3$
2-ethylhexyl	[24468-13-1]	$ClCOOCH_2CH(C_2H_5)(CH_2)_3CH_3$
2-octyl	[15586-11-5]	$ClCOOCH(CH_3)(CH_2)_5CH_3$
n-decyl	[55488-51-2]	$ClCOOCH_2(CH_2)_8CH_3$
dodecyl	[24460-74-0]	$ClCOOCH_2(CH_2)_{10}CH_3$
myristyl	[56677-60-2]	$ClCOOCH_2(CH_2)_{12}CH_3$
cetyl	[26272-90-2]	$ClCOOCH_2(CH_2)_{14}CH_3$
octadecyl	[51637-93-5]	$ClCOOCH_2(CH_2)_{16}CH_3$
2-methoxyethyl	[628-12-6]	$ClCOOCH_2CH_2OCH_3$
2-phenoxyethyl	[34743-87-8]	$ClCOOCH_2CH_2OC_6H_5$
phenyl	[1885-14-9]	$ClCOOC_6H_5$
p-nitrophenyl	[7693-46-1]	$ClCOOC_6H_4NO_2$
benzyl	[501-53-1]	$ClCOOCH_2C_6H_5$
p-nitrobenzyl	[4457-32-3]	$ClCOOCH_2C_6H_4NO_2$
4-cumylphenyl	[82914-10-4]	$ClCOOC_6H_4C(CH_3)_2C_6H_5$
9-fluorenylmethyl	[28920-43-6]	$ClCOOCH_2C_{13}H_9$
1-naphthyl	[3759-61-3]	$ClCOOC_{10}H_7$
cholesterol	[7144-08-3]	$ClCOOC_{28}H_{45}$
ethylene bis	[124-05-0]	$ClCOOCH_2CH_2OOCCl$
diethylene glycol bis	[106-75-2]	$ClCOOCH_2CH_2OCH_2CH_2OOCCl$
1,6-hexanediol bis	[2916-20-3]	$ClCOO(CH_2)_6OOCCl$
bisphenol A bis	[2024-88-6]	$ClCOOC_6H_4C(CH_3)_2C_6H_4OOCCl$

2.1. Physical Properties. In general, carbonochloridates or chloroformates are clear, colorless liquids with low freezing points and relatively high boiling points ($>100°C$). They are soluble in most organic solvents, but insoluble in water, although they do hydrolyze in water. The lower chloroformates, eg, methyl and ethyl chloroformates, hydrolyze rapidly in water at room temperature, whereas the higher chloroformates, eg, 2-ethylhexyl or aromatic chloroformates, hydrolyze slowly in water at room temperature (4). The physical properties of the most widely used chloroformate esters are given in Table 2 (2).

Table 2. **Physical Properties of Selected Chloroformates**

Chloroformate	Mol wt	Sp gr, d^{20}_4	Refractive index n^{20}_D	Flash point, °C		Viscosity, mPa·s (=cP), 20°C	Bp, °C at		
				TOC[a]	TCC[b]		2.67 kPa[c]	13.3 kPa[c]	101.3 kPa[c]
methyl	94.5	1.250	1.3864	24.4	17.8				71
ethyl	108.53	1.138	1.3950	27.8	18.3				94
isopropyl	122.55	1.078	1.3974	27.8	23.3	0.65	25.3	47	105
n-propyl	122.55	1.091	1.4045	34.4		0.80		57.5	112.4
allyl	120.5	1.1394	1.4223	27.8	31.1	0.71	25	57	
n-butyl	136.58	1.0585	1.4106	52.2	46.0	0.888	44	77.6	
sec-butyl	136.58	1.0493	1.4560	35.6	38.0	0.897	36	69	
isobutyl	136.58	1.0477	1.4079	39.5	34.4	0.88	39	71	
2-ethylhexyl	192.7	0.9914	1.4307		86.0	1.774	98	137	
n-decyl	220.7	0.9732	1.4400	118.3	120.2	3.00	122	159	
phenyl	156.57	1.2475	1.5115	80.0	77.0	1.882	83.5	121	185
benzyl	170.6	1.2166	1.5175	134.0	107.9	2.57	103	123	152
ethylene bis	186.98	1.4704	1.4512		126.0	4.78	108	137	
diethylene glycol bis	231.0	1.388	1.4550	160.0	182.2	8.76	148	180	

[a] Tag open cup.
[b] Tag closed cup.
[c] To convert kPa to mm Hg, multiply by 7.50.

2.2. Chemical Properties. Chloroformates are reactive intermediates that combine acid chloride and ester functions. They undergo many reactions similar to those of acid chlorides; however, the rates are usually slower (5–9). Those containing smaller organic (hydrocarbon) substituents react faster than those containing large organic (hydrocarbon) substituents (4). Reactions of chloroformates and other acid chlorides proceed faster and with better yields when means are employed to remove or capture HCl as it is formed. Classical acid scavengers include alkali hydroxides or tertiary amines, which act as stoichiometric acid acceptors rather than as true catalysts.

Stability. The ester moiety determines thermal stability generally in the following order of decreasing stability: aryl > primary alkyl > secondary alkyl > tertiary alkyl. In terms of mechanistic chemistry the chloroformates that produce stable carbonium ions on thermal decomposition, eg, benzyl, isopropyl, or tertiary butyl, are unstable (10) and can cause increased pressure in closed containers. Thus iron, zinc, and aluminum chlorides, ie, Lewis acids and metal oxides, catalyze decomposition of chloroformates and therefore the chloroformates should be handled in the absence of metals. Chloroformate purification and decolorizing with charcoal is generally avoided due to the tendency to undergo rapid decarboxylation (11), which could lead to a dangerous overpressurization of the reactor vessel. Alkyl chloroformates can be purified easily by distilling in glass vessels, but secondary or benzylic chloroformates have to be distilled under high vacuum in glass vessels. Tertiary chloroformates are too unstable to distill, even under high vacuum. In some situations the instability of chloroformates can be utilized to generate other chemical species. For example, Lewis acid induced decarboxylation of aromatic chloroformates provide access to chloro and fluoro aromatics (12,13).

Chloroformate decomposition can also be initiated by substances that either posses, or can generate (via reaction with the chloroformate), an ionic organic halide. The net product of an unbranched aliphatic radical is an alkyl chloride, while a substituted radical often loses HCl to produce the olefin. The ionic halide degradation is catalytic and the reaction accelerated by heat. Examples of decarboxylation catalysts include tertiary amines such as pyridine or quinoline (14), formamides (15,16), quaternary ammonium or phosphonium halides (17–19), ureas (20), and inorganic oxides (21). In an example, the ionic chloride of a quaternary amine initiates an S_N2 attack on the alpha carbon of the ester to induce decarboxylation and regeneration of the ionic halide. Controlled decarboxylation of chloroformates has been documented as an efficient means to generate alkyl halides (22–24).

$$\text{ClCOOCH}_2\text{CH}_2\text{R} \xrightarrow[\text{amine}]{\text{tertiary}} \text{ClCH}_2\text{CH}_2\text{R} \ + \ \text{CO}_2$$

Reactions with Oxygen Moieties. *Hydroxylic Compounds.* Chloroformates on reaction with water give the parent hydroxy compound, HCl, and CO_2 as well as the symmetrical carbonate formed by the reaction of the hydroxy compound with chloroformate.

$$\underset{\substack{\| \\ O}}{ROCCl} + H_2O \longrightarrow ROH + CO_2 + HCl$$

$$\underset{\substack{\| \\ O}}{ROCCl} + ROH \longrightarrow \underset{\substack{\| \\ O}}{ROCOR} + HCl$$

Alkali Metal Hydroxides.   Addition of base to aqueous chloroformates catalyzes hydrolysis to yield the parent hydroxy compound (25). However, the use of a stoichiometric amount of alkali metal hydroxides can lead to the symmetrical carbonate, especially from aryl chloroformates (26,27).

$$2 \, \underset{\substack{\| \\ O}}{ROCCl} + 4\,NaOH \longrightarrow \underset{\substack{\| \\ O}}{ROCOR} + Na_2CO_3 + 2\,NaCl + 2\,H_2O$$

Aliphatic Alcohols and Thiols.   Aliphatic alcohols on reaction with chloroformates give carbonates and hydrogen chloride. Frequently, the reaction proceeds at room temperature without a catalyst or hydrogen chloride acceptor. However, faster reactions and better yields are obtained in the presence of alkali metals or their hydroxides, or tertiary amines. Reactions of chloroformates with thiols yield monothiolocarbonates (28).

$$\underset{\substack{\| \\ O}}{ROCCl} + R'OH \longrightarrow \underset{\substack{\| \\ O}}{ROCOR'} + HCl$$

$$\underset{\substack{\| \\ O}}{ROCCl} + R'SH \longrightarrow \underset{\substack{\| \\ O}}{ROCSR'} + HCl$$

Heterocylic Alcohols.   Their reactions with chloroformates lead to carbonates. Thus furan- and tetrahydrofuran-derived alcohols give the corresponding carbonates in 75% yield (29). Inorganic bases and tertiary amines as acid acceptors increase the rate and yield in this reaction.

Phenols.   Phenols are unreactive toward chloroformates at room temperature and at elevated temperatures the yields of carbonates are relatively poor (<10%) in the absence of catalysis or quantitative HCl scavengers. Many catalysts have been claimed in the patent literature leading to high yields of carbonates from phenol and chloroformates. Alternate systems include biphasic systems that employ alkali bases and phase transfer catalysts (30). The use of catalyst or an alkali base is even more essential in the reaction of phenols and aryl chloroformates. Among the catalysts claimed are amphoteric metals or their halides (31), magnesium halides (32), activated carbon (33), titanium oxide (34), magnesium or manganese (35), secondary or tertiary amines such as imidazole (36,37), pyridine, quinoline, picoline (38–40), heterocyclic basic compounds (41) and carbonamides, thiocarbonamides, phosphoroamides, and sulfonamides (42).

$$ArOH + \underset{\substack{\| \\ O}}{ROCCl} \xrightarrow[\text{heat}]{\text{catalyst}} \underset{\substack{\| \\ O}}{ArOCOR} + HCl$$

Carboxylic Acids. The reaction product of chloroformates and carboxylic acids is a mixed carboxylic–carbonic anhydride. The intermediate mixed anhydrides are very active acylating agents (43–45), but these agents may be isolated in cold temperatures for producing useful products (46). More often the anhydride is a transient intermediate that leads to the formation of a mixture of ester, carbonate and anhydride. The pathway is strongly dependent upon the anhydride itself and the choice of catalyst (47–50).

$$\underset{\substack{\text{O}\\||}}{\text{ROCCl}} + \underset{\substack{\text{O}\\||}}{\text{R′COH}} \longrightarrow [\underset{\substack{\text{O O}\\||\;||}}{\text{ROCOCR′}}] \xrightarrow{\text{catalyst}} \underset{\substack{\text{O}\\||}}{\text{ROCR′}} + \underset{\substack{\text{O}\\||}}{\text{ROCOR}} + \underset{\substack{\text{O O}\\||\;||}}{\text{R′OCOCR′}} + \text{HCl} + \text{CO}_2$$

Pyrocarbonates or dicarbonates (anhydrides of carbonic acids) have been prepared from the alkali salt of the carbonate as shown (51,52). Modifications include the direct reaction of chloroformate with alkali metal hydroxide in the presence of a phase-transfer catalyst (53–57). Pyrocarbonates are useful as intermediates and protecting groups as well as effective radical scavengers, where they have established utility as polymer stabilizers and as preservatives in beverages such as wine and fruit juices (56,58).

$$\underset{\substack{\text{O}\\||}}{\text{ROCCl}} + \underset{\substack{\text{O}\\||}}{\text{R′OCONa}} \longrightarrow \underset{\substack{\text{O O}\\||\;||}}{\text{ROCOCOR}} + \text{NaCl}$$

Epoxides. Epoxy compounds react with chloroformates to yield β-chloro-substituted carbonates. Ring opening is catalyzed with Lewis acids or sources of chloride anions (49,59,60).

$$\underset{\substack{\text{O}\\||}}{\text{ROCCl}} + \underset{\triangle}{\overset{\text{O}}{}} \xrightarrow{\text{Cl}^-} \underset{\substack{\text{O}\\||}}{\text{ROCOCH}_2\text{CH}_2\text{Cl}}$$

Aldehydes. Aldehydes react with chloroformates in the presence of catalytic pyridine to yield 1-chloro carbonate esters. The reaction is highly dependant upon the stability of the chloroformate to the reaction conditions. The esters are useful intermediates for pharmaceuticals and insecticides (61–64).

$$\underset{\substack{\text{O}\\||}}{\text{ROCCl}} + \underset{\substack{\text{O}\\||}}{\text{HCR′}} \xrightarrow{\text{pyridine}} \underset{\substack{\text{O}\\||}}{\text{ROCOCHR′}}_{\!\!|\;\text{Cl}}$$

Reactions with Nitrogen Compounds. The reaction with ammonia is the classical method for preparing primary carbamates. Excess ammonia is used as an acid acceptor to remove the HCl formed (see CARBAMIC ACID).

$$\underset{\substack{\text{O}\\||}}{\text{ROCCl}} + 2\,\text{NH}_3 \longrightarrow \underset{\substack{\text{O}\\||}}{\text{ROCNH}_2} + \text{NH}_4\text{Cl}$$

Amines. Primary and secondary aliphatic amines also yield carbamates in the presence of excess amine or other acid acceptors such as inorganic bases under conditions analogous to those used in the Schotten-Baumann reaction

(65). Aromatic primary and secondary amines and heterocyclic amines react similarly, although slowly.

$$\underset{\displaystyle ROCCl}{\overset{\displaystyle O}{\|}} + 2\,R'NH_2 \longrightarrow \underset{\displaystyle ROCNHR'}{\overset{\displaystyle O}{\|}} + R'NH_2 \cdot HCl$$

Tertiary amines give crystalline quaternary ammonium compounds (66,67). The acyl ammonium salts provide activation of the carbonyl species and are potential precursors to secondary amines via dealkylation chemistry.

$$\underset{\displaystyle ROCCl}{\overset{\displaystyle O}{\|}} + NR'R''R''' \longrightarrow [\underset{\displaystyle ROCNR'R''R'''}{\overset{\displaystyle O}{\|}}]^+ Cl^-$$

Amino Alcohols. Reaction of chloroformate is much more rapid at the amino group than at the hydroxyl group (5–9). Thus the hydroxy carbamates, which can be cyclized with base to yield 2-oxazolidones, can be selectively prepared (68). Nonionic detergents may be prepared from poly[(ethylene glycol) bis(chloroformates)] and long-chain tertiary amino alcohols (69).

Aminophenols. Reaction of chloroformate with aminophenols (qv) also takes place at the more reactive amino group selectively. Thus *o*-aminophenol [95-55-6] gives benzoxazolone [59-49-4] by cyclization of the intermediate carbamate (70).

Amino Acids. Chloroformates play a most important role for the protection of the amino group of amino acids (qv) during peptide synthesis (71,72). The protective carbamate formed by the reaction of benzyl chloroformate and amino acid (73) can be cleaved by hydrogenolysis to free the amine after the carboxyl group has reacted further. The selectivity of the amino groups toward chloroformates results in amino-protected amino acids with the other reactive groups unprotected (74,75). Methods for the preparation of protected amino acids generally involve a pH stat procedure. These processes have been developed on an industrial scale (76–78). A wide variety of chloroformates have been used that give various carbamates that are stable or cleaved under different conditions.

Acylation. Aryl chloroformates are good acylating agents, reacting with aromatic hydrocarbons under Friedel-Crafts conditions to give the expected aryl esters of the aromatic (Ar) acid (79).

$$\underset{\displaystyle \underset{\displaystyle O}{\|}}{ArOCCl} + Ar'H \xrightarrow{AlCl_3} \underset{\displaystyle \underset{\displaystyle O}{\|}}{ArOCAr'} + HCl$$

However, with aliphatic chloroformates under similar conditions, alkylation takes place (80).

$$\underset{\substack{\parallel \\ }}{ROCCl} \ + \ Ar-H \ \xrightarrow{\text{AlCl}_3} \ R-Ar \ + \ CO_2 \ + \ HCl$$

Chloroformates have also been shown to undergo inter and intramolecular Pd-catalyzed cross-coupling reactions with acetylenes and aryl- and vinylorganotins (81–83).

Dealkylation.   Chloroformates such as vinyl chloroformates (84) are used to dealkylate tertiary amines. Chloroformates are superior to the typical Von Braun reagent, cyanogen bromide, because of increased selectivity producing cleaner products and higher yields. Other chloroformates such as allyl, ethyl, methyl, phenyl, and trichloroethyl have also been used in dealkylation reactions. Although the dealkylation reaction using chloroformates is mostly carried out on tertiary amines, dealkylation of oxygen or sulfur centers, ie, ethers or thioethers, can also be achieved. Commercially available α-Chloroethyl chloroformate [50893-53-3] (ACE-Cl) (85–87) is superior to all previously used chloroformates for the dealkylation reaction.

ACE-Cl has the advantage that the conditions required for ACE removal are much milder, thus expanding the list of functionalities allowed in the amine to be dealkylated. The potential significance of this process in drug congener preparation has been outlined (87–91).

Miscellaneous Reactions. The reaction of chloroformates with hydrogen peroxide or metal peroxides results in the formation of peroxydicarbonates that are used as free-radical initiators of polymerization of vinyl chloride, ethylene, and other unsaturated monomers (92,93).

$$2\ \text{ROCCl} + \text{Na}_2\text{O}_2 \longrightarrow \text{ROC}-\text{OO}-\text{COR} + 2\ \text{NaCl}$$

The reaction of chloroformates with sodium xanthates results in the formation of alkyl xanthogen formates that are useful as flotation agents in extraction of metals from their ores (94).

$$\text{ROCCl} + \text{NaSCOR}' \longrightarrow \text{ROCSCOR}' + 2\ \text{NaCl}$$

Methyleneaziridines undergo nucleophilic ring opening in the presence of alkyl chloroformates to generate enamide products (95).

$$\text{R}'\text{OCCl} + \underset{\text{N}}{\overset{\text{R}}{|}}\!\!\!\bigtriangleup \longrightarrow \text{R}'\text{O}-\text{C}-\text{N}-\!\!\!\begin{smallmatrix}\text{Cl}\\\end{smallmatrix}$$

Alkyl Chloroformates react with HCN in the presence of a tertiary amine, or with cyanide salts under phase-transfer conditions, to give cyanoformate esters. The esters are useful building blocks in pharmaceutical preparations (96–99).

$$\text{ROCCl} + \text{NaCN} \xrightarrow[\text{catalyst}]{\text{Phase transfer}} \text{ROCCN} + \text{NaCl}$$

The reaction of chloroformates with aldehyde and ketone enolates has been shown as a selective route to enol carbonates (100–103).

$$\text{C}_2\text{H}_5\text{OCCl} + \text{O}\!\!=\!\!\bigcirc \xrightarrow[\text{THF, }-78\,^{\circ}\text{C}]{\text{LiTMP}} \text{C}_2\text{H}_5\text{OCO}-\bigcirc$$

2.3. Manufacture. The reaction of phosgene with alcohols or phenols has been thoroughly discussed (2). In general, the availability of a chloroformate is limited only by access to the particular alcohol.

Most alkyl chloroformates, especially those of low molecular weight alcohols, are prepared by the reaction of liquid anhydrous alcohols with molar excess of dry, chlorine-free phosgene at low temperature. Corrosion-resistant reactors, lines, pumps, and valves are required. Materials of construction include glass, porcelain, Hastelloy C, Teflon-lined steel, or chemically impregnated carbon on steel such as Karbate. Temperatures are kept at 0–10°C for the lower alcohols and may rise to 60°C for the higher aliphatic alcohols. Hydrogen chloride is evolved as the reaction proceeds and is then absorbed in a tower after recovering excess phosgene. The reactions are most often run in batch reactors, although some of the high volume chloroformates are produced in cascade-type continuous reactors (104–107) in either cocurrent or countercurrent flow (108,109). The continuous reactors also ensure excess of phosgene at all times since both reactants are added simultaneously, thus minimizing dialkylcarbonate side products. Phenols react with phosgene with difficulty and usually require higher temperature and lead to a fair amount of diaryl carbonate as a side product. Many different catalysts have been used to reduce the reaction temperature and the carbonate side products (110–115).

Unreacted phosgene is removed from the crude chloroformates by vacuum stripping or gas purging. Chloroformates of lower primary alcohols are distillable; however, heavy-metal contamination and charcoal purification should be avoided. As stated earlier, chloroformates generating a stable carbonium ion on decomposition, i.e. secondary or tertiary chloroformates or benzylic chloroformates, are especially unstable in the presence of heavy metals and more specifically Lewis acids and, hence, should be distilled at as low a temperature and high vacuum as possible. The yields of primary chloroformates are usually well above 90%. The secondary chloroformates give yields of 80–90%.

Commercial processes are usually run neat, although solvents such as chloroform, toluene, dioxane, or tetrahydrofuran (THF) are sometimes used to dissolve the starting alcohol or the product chloroformate as may be necessary. Chloroformates of phenols and arylene bisphenols are also made in aqueous base solution, often employing a phase transfer catalyst, pH control, and CH_2Cl_2 as the second phase (116–120). In special cases, chloroformates are prepared solely in cold water (121). In other cases, efficiencies have been improved by utilizing pressure or vacuum in either a solvent and solvent-less process (122,123).

Certain precursor alcohols to chloroformates are synthetically inaccessible. As such, unique synthetic methods have been developed gain access to these materials. For instance, 1-chloroalkyl chloroformates have been synthesized by the reaction of aldehydes with phosgene in the presence of a source of ionic chloride (85,86,124–126). The reaction has been extended to include glyoxylate derived chloroformates (127). Alternatively, chloroalkyl chloroformates can be generated by free radical chlorination of simple alkyl chloroformates, followed by purification (126,128,129).

$$\underset{\substack{\text{O} \\ \|}}{R-C-H} + COCl_2 \xrightarrow{Cl^-} \underset{\substack{\text{O} \quad\;\; \text{O} \\ \|\quad\;\; \|}}{R-CH-O-C-Cl} + Cl^-$$

The synthesis of alkenyl chloroformates has proven to be very challenging both commercially and environmentally. Vinyl chloroformate can be made by the gas-phase pyrolysis of the bis(chloroformate) of ethylene glycol, however, the reaction suffers from low yield due to side reactions (130,131). Direct routes from phosgene and base derived enolates have been entirely unsuccessful. A successful and seemingly viable alternative using mercurials as enolate equivalents is hindered by ecological concerns (132–134).

$$H_2C{=}\overset{R}{\underset{}{C}}{-}O{-}\overset{O}{\underset{}{C}}{-}CH_3 \ + \ HgCl_2/HgO \ \xrightarrow{H_2O} \ ClHgCH_2{-}\overset{O}{\underset{}{C}}{-}R \ + \ Ch_3COOH$$

$$\xrightarrow[\text{nitrobenzene}]{COCl_2} \ H_2C{=}\overset{R}{\underset{}{C}}{-}O{-}\overset{O}{\underset{}{C}}{-}Cl \ + \ HgCl_2$$

In special cases, vinyl chloroformates have been made by treating α-chloro or bromo aldehydes or ketones with zinc dust and phosgene (135). The reaction is limited to aldehydes and ketones that do not contain α-hydrogen.

$$Cl{-}\overset{Cl}{\underset{CH_3}{C}}{-}\overset{O}{\underset{}{C}}{-}H \ + \ COCl_2 \ \xrightarrow{Zn} \ Cl{-}\overset{}{\underset{CH_3}{C}}{=}CH{-}O{-}\overset{O}{\underset{}{C}}{-}Cl$$

Chloroformates have also been generated by the reaction of alcohols with trichloromethyl chloroformate or bis trichloromethyl carbonate (di- and triphosgene, respectively). In the reaction with an alcohol, both reagents serve as chloroformylating agents, and are in effect, a liquid or solid equivalent of phosgene.

A sulfuryl chloride initiated fragmentation of alkyl carbonothioates illustrates a unique indirect entry to synthetically challenging α-substituted acyloxy chloroformates (136), useful in the manufacture of prodrugs (137).

$$\overset{O}{\underset{}{R'C}}O\overset{R}{\underset{}{C}}HO\overset{O}{\underset{}{C}}SR'' \ \xrightarrow[-R^2SCl]{SO_2Cl_2} \ \overset{O}{\underset{}{R'C}}O\overset{R}{\underset{}{C}}HO\overset{O}{\underset{}{C}}Cl$$

2.4. Shipping and Storage. Chloroformates are shipped in nonreturnable 208-L (55-gal) polyethylene drums with carbon steel overpacks or high density polyethylene drums. For bulk shipments, insulated stainless steel tank containers and trucks provide secure protection. Bulk equipment is specially lined for protection from corrosion. Tank truck and rail car quantities are shipped using equipment dedicated for these types of products. Materials such as isopropyl chloroformate, benzyl chloroformate, and sec-butyl chloroformate that require refrigeration are precooled when shipped in bulk containers. Bulk shipments that are pre-cooled must proceed to the destination without layover. Drum shipments of IPCF, BCF, and SBCF must be shipped in refrigerated containers. Many of the chloroformates are only shipped in truckload shipments. The U.S. Department of Transportation (DOT) Hazardous Materials Regulations control the shipments of chloroformates, as described in Table 3.

Chloroformates should be stored in a cool, dry atmosphere, preferably refrigerated, especially for prolonged storage. Drums must be stored out of direct sunlight. Chloroformate transfers to storage tanks or reactors should be made through a closed system, using stainless steel, nickel, glass or Hastelloy pumps, lines, and valves. Contact with iron oxides should be avoided.

Table 3. **Department of Transportation Regulations for Chloroformate Shipment**

Type	DOT hazard class	Subsidiary risk 1	Subsidiary risk 2	Tank truck	Refrigeration
allyl chloroformate	toxic-inhalation hazard	flammable	corrosive	yes[a]	required
benzyl chloroformate	corrosive			yes[a]	required
diethylene glycol bis(chloroformate)	not regulated			yes	needed to pre-serve assay[b]
ethyl chloroformate	toxic-inhalation hazard	flammable	corrosive	[a]	
2-ethylhexyl chloroformate	toxic	corrosive		yes[a]	
hexanediol bischloroformate	not regulated			yes[a]	
isobutyl chloroformate	toxic-inhalation hazard	flammable	corrosive	yes[a]	
isopropyl chloroformate	corrosive	flammable	corrosive	yes[a]	required
methyl chloroformate	toxic-inhalation hazard	flammable	corrosive	yes[a]	
n-propyl chloroformate	toxic-inhalation hazard	flammable	corrosive	yes[a]	
phenyl chloroformate	toxic	corrosive		yes[a]	
sec-butyl chloroformate	toxic-inhalation hazard	flammable	corrosive	yes[a]	required
4-tert-butyl chloroformate	toxic			no	required

[a] Bulk shipments can be made in DOT-approved IMO containers.
[b] Only for shipments through the tropics.

Table 4. **Typical Specifications of Commercial Chloroformates**

Assay	Value
purity, %	98
phosgene, %	<0.1
iron, ppm	<5
acidity as HCl, %	<0.1
alcohol or phenol, %	<2

2.5. Economic Aspects. Most chloroformate production is used captively and production figures are not available. The prices are also not published, but can be obtained by contacting United States and other worldwide producers, such as PPG Industries, Inc., BASF, SNPE, and Hodogaya.

2.6. Specifications and Analysis. Table 4 lists the specifications of commercial chloroformates. The lower boiling chloroformates are analyzed by gas–liquid chromatography. Higher molecular weight chloroformates are first hydrolyzed and then analyzed by titration, using the Volhard method.

2.7. Toxicity. Chloroformates, especially those of low molecular weight, are pungent lachrymators, vesicants, and produce effects similar to those of hydrogen chloride or carboxylic acid chlorides. They can also irritate the skin and mucous membranes, producing severe burns and possible irreversible tissue damage.

Inhalation of vapors of lower chloroformates result in coughing, choking, and respiratory distress, and, with some chloroformates like methyl chloroformate, inhalation can be fatal as a result of the onset of pulmonary edema, which may not appear for several hours after exposure (138). Table 5 gives the acute toxicities of some chloroformates (138–140).

Table 5. **Toxicity Information on Chloroformates: LD_{50} and LC_{50}**

Chloroformate	Oral, mg/kg[a]	Dermal, mg/kg[a]	Inhalation, mg/L[b,c]
methyl	220	>2,500	0.634
ethyl	411	>2,000	<1.62
n-propyl	650	>10,200	1.6
isopropyl	177.8	11,300	1.5
sec-butyl	1,030	>2,025	1.82
isobutyl	2,095	>2,500	1.8
2-ethylhexyl	3,038	>3,038	0.95[d]
allyl	178	1,470	0.3
phenyl	1,581	>3,200	>2.04
benzyl	<5,000	2,065	>2.1
4-*tert*-butyl cyclohexyl	>5,000	>2,000	0.72
ethylene bis	1,100	>2,000	43
diethylene glycol bis	8,13.2	3,400	>2.5

[a] LD_{50} values.
[b] LC_{50} values.
[c] For 1-h exposure (mg/L of air).
[d] For 4-h exposure (mg/L of air).

2.8. Uses. As illustrated in the chemical properties segment, chloroformates are reactive chemical species, and versatile synthetic intermediates. Derivatization of chloroformates with alcohols and amines is commonly practiced, and many industrial uses of the resultant carbamates and carbonates have been described. Chloroformates should be considered as intermediates for syntheses of pesticides, perfumes, drugs, foodstuffs, polymers, dyes, and other chemicals.

Some of these products, eg, carbonates, are used as solvents, plasticizers, or as intermediates for further synthesis. Diethylene glycol bis(chloroformate) [106-75-2] is the starting material for diethylene glycol bis(allyl carbonate) [142-22-3], CR-39, or Nouryset 200 monomer, used in the manufacture of break-resistant optical lenses, which is obtained by the reaction with allyl alcohol [107-18-6] (138). Alternatively, it can be obtained from allyl chloroformate [2937-50-0] and diethylene glycol (139) (see ALLYL MONOMERS AND POLYMERS). Other aliphatic or aromatic bis(chloroformates) are used to make high temperature resistant polycarbonate plastics, the most important example being Lexan or Makrolon (140) (see POLYCARBONATE).

A significant use of chloroformates is for conversion to peroxydicarbonates, which serve as free-radical initiators for the polymerization of vinyl chloride, ethylene, and other unsaturated monomers. The most widely used percarbonate initiators are diisopropyl peroxydicarbonate (IPP), di-2-ethylhexyl peroxydicarbonate (2-EHP), and di-sec-butyl peroxydicarbonate (SBP). The following list includes most of the commercially used percarbonates.

Percarbonate	CAS Registry number
diethyl percarbonate	[14666-78-5]
diisopropyl (IPP) percarbonate	[105-64-6]
di-n-butyl percarbonate	[16215-49-9]
di-sec-butyl (SBP) percarbonate	[19910-65-7]
dicyclohexyl percarbonate	[1561-49-5]
di-4-tert-butylcyclohexyl percarbonate	[26523-73-9]
di-n-hexadecyl percarbonate	[26322-14-5]
di-n-propyl (NPP) percarbonate	[16066-38-9]
di-2-ethylhexyl (2-EHP) percarbonate	[16111-62-9]

Carbamates derived from chloroformates are used to manufacture pharmaceuticals, including tranquilizers (144), antihypotensives, and local anesthetics, pesticides, and insecticides (see CARBAMIC ACID).

Methyl chloroformate is the largest volume chloroformate used in the agricultural industry, primarily in the formation of carbamate functional groups. An important example is the synthesis of the mainstay fungicide Carbendazim (1H-benzimidazole-2-ylcarbamic acid methyl ester [10605-21-7]).

Carbendazim

Another important use of chloroformates is the protection of amino and hydroxyl groups in the synthesis of complex organic compounds such as peptide-based pharmaceuticals (73–77,145–147). The appropriate chloroformates are used in generating alkoxycarbonyl N-protecting groups of amino acids. Common examples of amino acid blocking agents derived from chloroformates include benzyloxycarbonyl (Z or Cbz), 9-fluorenyl-methoxycarbonyl (Fmoc), 4-nitrobenzyloxy carbonyl, and allyloxycarbonyl (Alloc). The industrial significance of chloroformate blocking agents is noted in the synthesis of artificial sweeteners such as aspartame, as well as the synthesis of the antiviral valacyclovir (148).

Carbamate likages of poly(vinyl ether) carbamates, used as detergents additives in gasoline, have been derived from the suitable chloroformate (149–151). Ethyl chloroformate is used in the manufacture of ore flotation agents by reaction with various xanthates (94).

Cholesteric liquid crystal materials, useful as nondestructive indicators, are often derived from cholesterol chloroformate (152–154). Decarboxylation of bis(chloroformates) to alkyl halides is used in the manufacture of the rubber component 1,6-dichlorohexane [2163-00-0] (155–157). Additionally, decarboxylation of alkoxyalkyl chloroformates provides alkyl chloride materials that are useful as surfactant intermediates (158). Bis(chloroformic esters) condense with diamines to give polyurethanes (142) (see URETHANE POLYMERS).

Blowing agents for producing foam rubber, polyethylene, and vinyl chloride are made from chloroformates, hydrazine, and a base. For example, diisopropyl azidoformate [2446-83-5] is made from isopropyl chloroformate (159).

The polymerization of derivatives of vinylic chloroformates has led to a series of interesting monomers (160). However, due to limited supply of the chloroformate raw material, applications are limited to small volume specialty products.

By virtue of their exceptional reactivity, chloroformates are also valued as general purpose derivatizing agents for gas and liquid chromatographic analysis of molecules containing active functionality such as amines and carboxylic acids (161,162).

3. Carbonates

Classically, chloroformates and alcohols or phenols give carbonic diesters. In addition, the higher diesters can be made from the lower ones by alcoholysis or ester interchange by heating the lower diester with a higher alcohol in the presence of acid such as HCl or H_2SO_4, or base such as sodium alcoholate. The driving force for these reactions is the formation of lower alcohols that can be distilled off. Mixed diesters can be prepared by treating a chloroformate with an alcohol or phenol having a different radical.

More recently, preparation of selected carbonic esters by nonphosgene routes, such as the metal catalyzed reaction of CO or NO_x with alcohols, or the catalytic reaction of CO_2 and oxiranes have been preferred. These oxidative carbonylation methods are more economic in many cases and naturally less hazardous than phosgene routes. Carbonates are indexed in *Chemical Abstracts* under carbonic acid, esters. Symmetrical diesters have the prefix di or bis. Unsymme-

Table 6. **Carbonates**

Carbonate	CAS Registry number	Formula
dimethyl	[616-38-6]	$CH_3OCOOCH_3$
diethyl	[105-58-8]	$C_2H_5OCOOC_2H_5$
divinyl	[7570-02-7]	$CH_2{=}CHOCOOCH{=}CH_2$
di-n-propyl	[623-96-1]	$CH_3CH_2CH_2OCOOCH_2CH_2CH_3$
diisopropyl	[6482-34-4]	$(CH_3)_2CHOCOOCH(CH_3)_2$
diallyl	[15022-08-9]	$CH_2{=}CHCH_2OCOOCH_2CH{=}CH_2$
methyl allyl	[35466-83-2]	$CH_3OCOOCH_2CH{=}CH_2$
diisobutyl	[539-92-4]	$(CH_3)_2CHCH_2OCOOCH_2CH(CH_3)_2$
isobutyl propyl	[40882-93-7]	$CH_3CH(CH_3)CH_2OCOOCH_2CH_2CH_3$
di-$tert$-butyl	[34619-03-9]	$(CH_3)_3COCOOC(CH_3)_3$
di-sec-butyl	[623-63-2]	$CH_3CH_2CHCH_3O\text{-}$ $COOCH(CH_3)CH_2CH_3$
di-n-butyl	[542-52-9]	$CH_3CH_2CH_2CH_2O\text{-}$ $COOCH_2CH_2CH_2CH_3$
methyl ethyl	[623-53-0]	$CH_3OCOOC_2H_5$
hexyl methyl	[39511-75-6]	$CH_3CH_2CH_2CH_2CH_2CH_2OCOOCH_3$
pentyl propyl	[40882-94-8]	$CH_3CH_2CH_2CH_2CH_2O\text{-}$ $COOCH_2CH_2CH_3$
1-chloromethyl isopropyl	[79-22-1]	$ClCH_2OCOOCH(CH_3)_2$
1-chloroethyl ethyl	[50893-36-2]	$CH_3CHClOCOOC_2H_5$
1-chloroethyl cyclohexyl	[99464-83-2]	$CH_3CHClOCOOC_6H_{10}$
di-n-octyl	[1680-31-5]	$CH_3(CH_2)_6CH_2OCOOCH_2(CH_2)_6CH_3$
didodecyl	[6627-45-8]	$CH_3(CH_2)_{10}CH_2OCOOCH_2(CH_2)_{10}CH_3$
diphenyl	[102-09-0]	$C_6H_5OCOOC_6H_5$
phenyl allyl	[16308-68-2]	$C_6H_5OCOOCH_2CH{=}CH_2$
vinyl ethyl	[7670-06-1]	$CH_2{=}CHOCOOC_2H_5$
ethyl phenyl	[3878-46-4]	$C_2H_5OCOOC_6H_5$
ethylene	[96-49-1]	$\overset{\frown}{OCOOCH_2CH_2}$
allyl diglycol[a]	[142-22-3]	$O(CH_2CH_2OCOOCH_2CH{=}CH_2)_2$
ditolyl[b]	[41903-18-8]	$CH_3C_6H_4OCOOC_6H_4CH_3$
dibenzyl	[3459-92-5]	$C_6H_5CH_2OCOOCH_2C_6H_5$
di-2-ethylhexyl	[14858-73-2]	$CH_3(CH_2)_3CH(C_2H_5)CH_2O\text{-}$ $COOCH_2CH(C_2H_5)(CH_2)_3CH_3$

[a] Diethylene glycol bis(allylcarbonate).

[b] Diethylene glycol bis(tolylcarbonate).

trical diesters are listed with the two radicals following each other. For example, ethyl phenyl carbonic diester is EtOCOOPh. Table 6 lists commonly used carbonates, their Chemical Abstracts Service Registry Number, and formulas.

3.1. Properties. The physical properties of selected carbonates are given in Table 7. The lower alkyl carbonates are neutral, colorless liquids with a mild sweet odor. Aryl carbonates are normally crystalline compounds with relatively low melting points. Carbonic esters are soluble in polar organic solvents such as alcohols, esters, and ketones, but not soluble in water. An exception is lower molecular weight cyclic carbonates such as ethylene carbonate and propylene carbonate which readily dissolve in water. Several lower aliphatic carbonates form azeotropic mixtures with organic solvents. For instance, dimethyl carbonate

Table 7. **Physical Properties of Selected Carbonates**

Carbonates	Mol wt	Sp gr, d_4^{20}	Refractive index n^t_D	Flash point °C	Viscosity[a] mPa·s (=cP)	Bp, °C[b]
dimethyl	90.08	1.073	1.3697^c	21.7^d 16.7^e	0.664 (20)	90.2
diethyl	118.13	0.975	1.3846^c	46.1^d 32.8^e	0.868 (15)	23.8 (1.33) 69.7 (13.33) 126.8
di-n-propyl	146.18	0.941	1.4022^c	64^e	f	165.5–166.6
diisopropyl	146.18					147.0
diallyl	142.15	0.994	1.4280^c			97 (8.13) 105 (13.33)
di-n-butyl	174.14	0.9244	1.4099^g			166 (97.31)
di-2-ethylhexyl	204.19	0.8974^{20}_{20}	1.4352^g			207.7
diphenyl	214.08	0.8974^{87}_{4}				173 (1.33) 302
diethylene glycol bis(allyl)	274.3	1.143	1.4503	177^h	9 (25)	160 (0.27)
tolyl diglycol	374.4	1.189	1.5229^g			247–248 (0.27)
ethylene	88.06	1.3218^{39}_{4}	1.4158^i			248

^{a} At the temperature noted in parentheses (°C).
^{b} At 101.3 kPa (=1 atm) unless otherwise noted in parentheses in kPa.
^{c} Ref. 30.
^{d} Tap open cup.
^{e} Tag closed up.
^{f} Brookfield no. 1 spindle; rpm (mPa·s) 10(5), 20(6.5), 50(8.0), 100(12.0).
^{g} Ref. 38.
^{h} Cleveland open cup.
^{i} Ref. 42.

and ethylene carbonate form azeotropic mixtures with methanol and ethylene glycol, respectively (163).

3.2. Chemical Properties. The chemistry of carbonates is dominated by a reactivity similar to esters and a tendency to liberate CO_2. Carbonates undergo nucleophilic substitution reactions analogous to chloroformates except in this case, an ^-OR group (rather than chloride) is replaced by a more basic group. Normally these reactions are catalyzed by bases. Carbonates are sometimes preferred over chloroformates because formation of hydrogen chloride as a by-product is avoided, simplifying handling and in some cases eliminating impurities. However, the reactivity of carbonates toward nucleophiles is considerably less than chloroformates. Several recent reviews depict the synthetic utility of carbonates in organic chemistry (164–166).

Reaction with Water. The alkyl carbonate esters, especially the lower ones, hydrolyze very slowly in water when compared to the carbonochloridic esters (chloroformates). Under alkaline conditions, the rates of hydrolysis are similar to those of the corresponding acetic acid esters. The net result is the formation of hydroxy compounds and CO_2.

Reaction with Alcohols and Thiols. The likeness of carbonates to esters is evident in their tendency to undergo transesterification with alcohol and thiols. Transesterification of both cyclic and acyclic carbonates is commonly practiced in industry. The process requires that the equilibrium be shifted in the desired direction. In general, the replacement of one hydroxy with another is matter nucleophilicity, where the more nucleophilic alcohol or thiol replaces the less nucleophilic compound. In cases where the nucleophilicity is similar, the reaction is driven by the removal of the less volatile compound. In this way, the reaction of diaryl carbonate with a low mass alcohol generates dialkyl carbonates and the less nucleophilic phenol. In contrast, low molecular weight dialkyl carbonates can be driven to diaryl carbonates by reaction with phenols and removal of the lower boiling alcohol. The reactions are catalyzed by the usual transesterification catalysts such as sodium metal and Lewis acids (167–168).

$$ROCOR + 2R'OH \underset{\text{Na metal}}{\rightleftarrows} R'OCOR' + 2ROH$$

$$ROCOR + 2R'SH \rightleftarrows R'SCSR' + 2ROH$$

Transesterification has become a convenient method for synthesizing high alkyl, aryl, or alkyl aryl carbonates. Fiber- and film-forming polycarbonates are produced by carbonate exchange involving dialkyl, dicycloalkyl, or diaryl carbonates with alkyl, cycloalkyl, or aryl dihydroxy compounds (169–170).

Reaction with Phenols. Carbonates undergo carbonate interchange with aromatic hydroxy compounds. In cases involving aliphatic carbonates, the reaction is slow and thermodynamically unfavorable. The equilibrium heavily favors the aliphatic ester and high temperatures and very active catalysts are required to drive the reaction.

A model example of phenols reacting with carbonates is evident in the industrial sequence used in the phosgene-free manufacture of bisphenol A polycarbonates. In the initial stage, dimethyl carbonate is reacted with phenol at high temperatures in the presence of Lewis acids or metal complexes to yield diphenyl carbonate (171–174). The process involves complex reaction technology utilizing various reactor zones, azeotropic separation, and recycle loops (175–176). In the next stage of the sequence, purified diphenyl carbonate and bisphenol A [80-05-7] are combined with basic catalysts in a melt reactor. Phenol is liberated from the melt using staged heat and pressure techniques serving to drive the reaction to polymer (see POLYCARBONATE).

$$CH_3O-\overset{O}{\overset{\|}{C}}-OCH_3 + 2PhOH \xrightarrow{Ti(OPh)_4} PhO-\overset{O}{\overset{\|}{C}}-OPh + 2CH_3OH$$

Reaction with Amines and Ammonia.   Carbonates react with aromatic amines, aliphatic amines and ammonia to produce carbamates or ureas (177–178). For example, dimethyl carbonate reacts with ammonia in water to form methyl carbamate useful in coatings applications (179–180). Similar reactions have been used as a choice route to producing carbamate pesticides and bis(urethanes) for polymers (164,181).

$$\underset{\text{ROCOR}}{\overset{\text{O}}{\|}} + 2\,\text{R}'\text{NH}_2 \rightleftharpoons \text{ROH} + \underset{\text{ROCNHR}'}{\overset{\text{O}}{\|}} \overset{\text{R}'\text{NH}_2}{\rightleftharpoons} \underset{\text{R}'\text{NHCNHR}'}{\overset{\text{O}}{\|}} + 2\,\text{ROH}$$

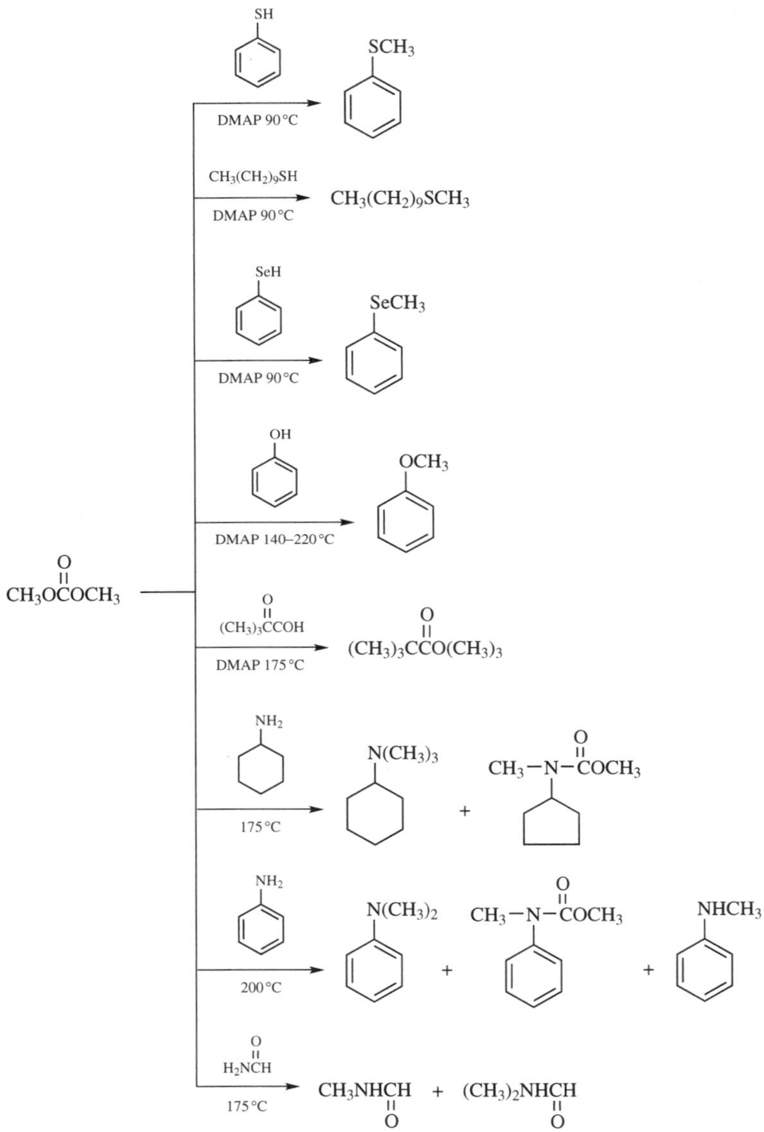

Fig. 1.   Dimethyl carbonate as a methylating agent.

Carbamate esters derived from carbonates undergo thermal elimination of alcohol to yield isocyanates (175,182,183). The method, which primarily employs dimethyl or diphenyl carbonate, is recognized as a phosgene-free alternative to industrially important isocyantes such as methyl isocyante (184,185), TDI, and HMDI (186–189). Compared to current phosgene processes, the carbonate process is generally cost prohibitive, however advantages are found in cases where the isocyanate molecule includes acid sensitive functionality or the processes are set up for alcohol recycle. An example includes isocyantoorganosilanes which are useful in coatings applications (190).

$$(CH_3O)_3Si-R-NH_2 \ + \ CH_3O-\overset{\overset{\displaystyle O}{\|}}{C}-OCH_3 \ \longrightarrow \ (CH_3O)_3Si-R-NH-\overset{\overset{\displaystyle O}{\|}}{C}-OCH_3 \ \xrightarrow[-CH_3OH]{heat}$$

$$(CH_3O)_3Si-R-N=C=O$$

Alkylation. Dialkyl carbonates react with a variety of functional groups to produce the alkylated derivatives. Typical carbonates include dimethyl, diethyl, and dibenzyl carbonates, leading to the respective methylated, ethylated, and benzylated products. Alkylations have been noted on functionality ranging from thioorganics (191), phenols, anilines, amines, oximes (192), carboxylic acids, and silicon dioxide (193) to C alkylation of CH_2 acidic species such as arylacetonitriles and alkyl aryl acetates (164,177–178).

The reaction of dimethyl carbonate as a methylating agent has received a majority of the attention (177,178,194,206). The method offers safety advantages over typical methylating agents, such as hazardous dimethyl sulfate and methyl iodide. Methylation reactions with dimethyl carbonate can be achieved in the liquid phase, usually requiring high temperatures (120–220°C) and further requiring use of catalyst such as 4-dimethylaminopyridine [1122-58-3] (DMAP) or alkali base. Only the sulfur and selenium compounds do not require high temperatures for methylation. Methylations can also be performed by gas–liquid phase-transfer catalysis in the presence of PEG 6000 or in the gas-phase using zeolite catalysts (175,207–208). As seen in Figure 1, the organo sulfur compounds are methylated at the boiling point (90°C) of dimethyl carbonate, whereas methylation (or alkylation with other alkyl groups) of other functional groups requires higher temperatures. This has resulted in the selective methylation of sulfhydryl groups of compounds that contain other substituents that can be alkylated. The other substituents can then be alkylated at elevated temperatures (194). Thus,

Miscellaneous Acylation Reactions. Chloroalkyl alkyl carbonates (described earlier) are more activated acylating agents than their alkyl counterparts.

An extreme example is bis trichloromethyl carbonate or triphosgene, which is often used as a solid source of phosgene (209–214). It is important to note that triphosgene readily decomposes to phosgene at high temperatures, and in the presence of trace metals and small quantities of nucleophilic sources such as chloride ion.

$$Cl_3COCOCCl_3 \xrightarrow{Cl^-} \left[Cl_3COC{-}Cl \right] + COCl_2 \longrightarrow 2\,COCl_2 + Cl^-$$

The less extreme acylating agent, 1-chloroalkyl alkyl carbonate, has been used as an entry to carbamates (213,214).

$$CH_3{-}CH{-}O{-}C{-}O{-}R + HNR'R'' \longrightarrow R'R'N{-}C{-}O{-}R + CH_3CHO + HCl$$

Additionally, the 1-chloroalkyl carbonyl group is used as an acid labile, base stable alcohol protecting group (90,215). The reactivity of the 1-chloroalkyl carbonates also offers entry to fluoroformates that cannot be prepared by halogen exchange of their analogous chloroformate, such as the *tert*-butyl chloroformate (216,217). Thus, *tert*-butyl fluoroformate, which exhibits higher thermal stability than its chloroformate analogue, is used industrially for BOC protection of amino acids and peptides (145,218).

$$Cl_3C{-}CH{-}O{-}C{-}O{-}t{-}C_4H_9 + KF \xrightarrow[\text{in vacuo}]{DMF\,-50\,^{\circ}C} F{-}C{-}O{-}t{-}C_4H_9 + Cl_3C{-}CHO$$

3.3. Manufacture.   The most important and versatile method for producing carbonates is the phosgenation of hydroxy compounds. Manufacture is essentially the same method as chloroformates except that more alcohol is required in addition to longer reaction times and higher temperatures. The products are neutralized, washed, and distilled. The more acidic alcohols are less reactive, and in many cases organic base is included as catalyst. Corrosion-resistant equipment similar to that described for the manufacture of chloroformates is required.

Diaryl carbonates are classically prepared from phosgene via an interfacial process comprising of caustic, triethylamine catalyst and methylene chloride (219,220). The process has reportedly been optimized by elimating solvent and including quaternary amines as phase transfer catalysts (221,222). Over the years, commercially important diphenyl carbonate (DPC) has undergone study to improve production techniques and product quality. A technique that strives for chloride free DPC involves the high temperature phosgenation of phenol with various homogeneous and heterogeneous catalysts. A sample of catalysts includes metal salts (223–226), pyridines (227–229), heterocyclic amines (230–232), phosphorous compounds (233–237), and alumosilicates (238). The phosgenation processes can be carried out continuously in both liquid and gas phases (239–241).

The continuous production of high purity methyl or ethyl carbonate from the alcohol and chloroformates has been patented (242). Chloroformate and alcohol are fed continuously into a Raschig ring-packed column in which a temperature gradient of 72–127°C is maintained between the base and head of the column; HCl is withdrawn at the head, and carbonate (99%) is withdrawn at the base.

Over the past 20 years, the trend is to manufacture carbonates without the use of phosgene. This method has the advantage of avoiding the use of highly toxic phosgene as well as considerably lower cost. The catalytic insertion of CO_2 with oxiranes directly provides the five-membered cyclic carbonate. Oxiranes such as ethylene oxide and propylene oxide undergo insertion at ~150–175°C under pressure with the aid of a quaternary ammonium salt catalyst to yield ethylene carbonate and propylene carbonate respectively (175,243–247). Transesterification with alcohols allows for entry to other acyclic carbonates as well as the production of ethylene or propylene glycol.

Another recent non-phosgene route to carbonates is the oxidative carbonylation of alcohols. The technology involves the catalytic reaction of methanol with carbon monoxide and oxygen to produce dimethyl carbonate. A variety of catalyst systems have been patented (204,247–251). EniChem practices the technology commercially using a copper chloride catalyst and a complex reactor design that involves recycle loops, azeotrope separation, and membrane separation (178,247,251,252). Aromatic carbonates have been made using a similar oxidative carbonylation process, however the technology has yet to be optimized (253–255). Entry to other aliphatic and cyclic carbonates has only been marginally sucessful and continues to be investigated.

$$2\,CH_3OH \;+\; 0.5\,O_2 \;+\; 2\,CuCl \;\longrightarrow\; 2\,Cu(OCH_3)Cl \;+\; H_2O \;\xrightarrow{\;CO\;}$$

$$2\,CuCl \;+\; CH_3O\overset{\overset{\displaystyle O}{\|}}{C}OCH_3$$

Dimethyl carbonate is also manufactured by carbonylation of methylnitrite through a catalytic redox process. The process is practiced commercially in the gas phase by Ube using a palladium supported catalyst system (175,247,256–257).

$$2\,CH_3OH \;+\; 0.5\,O_2 \;+\; 2\,NO \;\longrightarrow\; 2\,CH_3ONO \;+\; H_2O \;\xrightarrow{\;CO\;}$$

$$2\,NO \;+\; CH_3O\overset{\overset{\displaystyle O}{\|}}{C}OCH_3$$

The synthesis of dialkyl carbonates from urea and alcohol with organotin catalyst systems traditionally have been prone to thermal degradation products. However, a recent patent suggest that the problem can be overcome by employing novel organotins in a high boiling polar aprotic solvent such as a polyglycol ether, and distilling away the carbonate as it is formed (258). The operation

Table 8. **DOT Regulations for Carbonate Shipments**

Type	DOT hazard class	Tank truck	Drums
dimethyl carbonate	flammable liquid	yes	yes
diethyl carbonate	flammable liquid	yes	yes
diethylene glycol bis(allyl) carbonate	chemical NOS		yes

claims a continuous system that in theory has the possibility to allow for ammonia recycle back to urea. The net operation of the urea recycle system for dimethyl carbonate would utilize only methanol and carbon dioxide and thus has excellent economic potential.

3.4. Shipping and Storage. Dimethyl and diethyl carbonates are shipped in nonreturnable 208-L (55-gal) polyethylene drums with carbon steel overpack or high density polethylene drums. For bulk shipments, insulated stainless steel tank containers and trucks provide secure protection. Diethylene glycol bis(allyl) carbonate is shipped in drums as above. Diphenyl carbonate is delivered flaked in polyethylene sacks, or by tank car as a melt.

Carbonates are noncorrosive and should be plainly labeled and stored in cool, dry areas away from sources of ignition. The DOT Hazardous Materials Regulations control the shipment of carbonates as described in Table 8.

3.5. Economic Aspects. As in the case of the chloroformates, most of the carbonate production is used captively and production figures are not available. Both DMC and DPC are the largest volume carbonates, used primarily in the synthesis of bisphenol A polycarbonates. The DMC volumes would significantly increase if DMC were to gain interest as an octane enhancing additive in gasoline. Other heavy volume carbonates include DEC, ethylene carbonate, and propylene carbonate. Carbonate pricing in bulk range from $2–5/kg, with higher pricing for specialty esters (259). Major producers of aliphatic and aromatic carbonates include Enichem, Ube, Mitsubishi, Bayer, BASF, and PPG, as well as several sources in China. DPC is produced captively by GE using the Enichem DMC process in conjunction with their polycarbonate operation. Cyclic carbonates are available from Huntsman, Dow, BASF, Degussa, and Equistar.

3.6. Specifications and Analysis. Table 9 lists the specifications of the more important commercial carbonates. Assay is generally determined by gas or liquid chromatography. Water and traces HCl are detected using titration.

3.7. Health and Safety. Unlike chloroformates, diethyl and dimethyl carbonates are only mildly irritating to the eyes, skin, and mucous membranes. Diethylene glycol bis(allyl carbonate) may be irritating to the skin, but it is not classified as a toxic substance; however, it is extremely irritating to the eyes.

Protective clothing, rubber gloves, safety goggles, and adequate ventilation are recommended for all personnel handling high concentrations of carbonates. In case of fire, foam, carbon dioxide, or dry chemical extinguishing agents should be used. However, it is permissible to use a water spray to cool any drums in the vicinity, thus avoiding any spread of the fire.

3.8. Uses. The industrial utility of carbonates are widespread, ranging from pharmaceutical and cosmetic preparations to utility as specialty solvents

Table 9. **Specifications of Commercial Carbonates**

Assay	Dimethyl	Diethyl	Diethylene glycol bis(allyl)
min, %	98	98	94
acidity, max %	0.02	0.02	
water, max %	0.2	0.10	
sp gr, 20°C/4°C	1.070–1.075	0.973–0.977	1.14–1.16
nonvolatile matter, max %	0.01	0.005	1.0 at 150°C[a]
boiling point, °C	90	127	
viscosity, mPa·s (= cP)	0.664/20°C	0.868/28°C	25/25°C
surface tension at 20°C, mNm (=dyn/cm)		26.31	35
mol wt	90.1	118.1	274.3

[a] At 0.7 kPa (=5 mm Hg).

and application in polymers. The major volume carbonates include diphenyl, ethylene, and propylene carbonates as well as the simple aliphatics: diethyl, dimethyl, and dipropyl.

The diethyl ester is used in many organic syntheses, particularly of pharmaceuticals and pharmaceutical intermediates, dyes, and agricultural chemicals. It is also used as a solvent for many synthetic and natural resins, and in vacuum tube cathode-fixing lacquers.

Commercially, dimethyl carbonate (DMC) has emerged as the most important carbonate because of its lower cost from the nonphosgene route. The DMC can be used in the synthesis of important pharmaceuticals, such as the antimicrobial ciprofloxacin (Cipro) where it is used in place of diethyl malonate (260), and in the preparation of guaifenesin, a precursor for the muscle relaxant, methocarbamol (261).

Ciprofloxacin

In many applications DMC is an effective carbonylating agent and is touted as a phosgene replacement (177,251). Huge quantities of bisphenol A polycarbonate are manufactured using a process that involves transesterification of DMC to diphenyl carbonate, followed by ester exchange with bisphenol A to give polycarbonate resin. Temperature resistant polycarbonate resins are used in plastics, the most important example being Lexan or Makrolon (143) (see POLYCARBONATE).

With its high oxygen content, low toxicity, and biodegradability, DMC has been considered as an oxygenate for gasoline. While it has yet to be used in commercial gasoline, hundreds of patents have claimed DMC's usefullness in fuels (247,261–263). Like many carbonates, DMC is an attractive solvent and is noted as a replacement for acetates and halogenated solvents in paint stipping and adhesive applications (175,264–265). Along with other carbonates (ie, methyl ethyl carbonate), DMC is also finding increasing application in the field of rechargable lithium batteries as a nonaqueous electrolyte component (266), as well as a blowing agent in polyurethane foam (267).

Dipropyl carbonate is also an organic intermediate, a specialty solvent, and is used in photoengraving as an assist agent for silicon circuiting. Low molecular weight carbonates are employed to generate −OH terminated aliphatic carbonated by transesterification with lower aliphatic diols (164). The resulting polyalkyl carbonates are useful in thermoplasic urethanes, paints, coatings, medical coatings and fibers, and thermoelastomers (175). Cyclic and aliphatic carbonates in combination with hydrogen peroxide are noted as alternatives to chlorinated solvents in paint and coating removal applications (268).

Diethylene glycol bis(allyl carbonate) polymerizes easily because of its two double bonds and is used for colorless, optically clear castings. Polymerization is catalyzed by the use of diisopropyl peroxydicarbonate [105-64-6] (269,270). Such polymers are used in the preparation of safety glasses, lightweight prescription lenses, glazing cast sheet, and optical cement (see ALLYL MONOMERS AND POLYMERS; POLYCARBONATES).

Specialty carbonates derived from 1-haloalkyl chloroformates are used for the manufacture of pro-drugs that have different properties than the parent drug, yet produce the same physological effect after *in vivo* hydrolysis. Important examples include Bacampicillin, the antibiotic pro-drug of Ampicillin (271), and the reverse transcriptase Viread, used in the treatment of acquired immuno deficiency syndrome (AIDS) (272,273).

Bacampacillin   Viread

Cyclic aliphatic carbonates are also excellent solvents for polymers and resins and are used as additives for hydraulic fluids, cure accelerators of phenol-formaldehyde resins used the plywood and chipboard industry, and cleaning solvents. Cyclic carbonates are also utilized as reactive dilutants for epoxy resins, accelerators in sand molding and polyurethane coatings, thermoset resins (274), metal extraction, and as a gellants in cosmetic applications (164).

BIBLIOGRAPHY

"Carbonic Esters and Chloroformic Esters" in *ECT* 1st ed., Vol. 3, pp. 149–154, by H. L. Fisher, U.S. Industrial Chemicals, Inc.; in *ECT* 2nd ed., Vol. 4, pp. 386–393, by W. M. Tuemmler, FMC Corp.; "Carbonic and Chloroformic Esters" in *ECT* 3rd ed., Vol. 4, pp. 758–771, by E. Abrams, Chemetron Corp.; "Carbonic and Carbonochloridic Esters" in *ECT* 4th ed., Vol. 5, pp. 77–97 by S. B. Damle, PPG Industries, Inc.; "Carbonic and Carbonochloridic Esters" in *ECT* (online), posting date: December 4, 2000, by S. B. Damle, PPG Industries, Inc.

CITED PUBLICATIONS

1. J.-B. Dumas, *C.R. Acad. Sci.* **54**, 225 (1833).
2. M. Matzner, R. P. Kurkjy, and R. J. Cotter, *Chem. Rev.* **64**, 645 (1964).
3. D. N. Kevill, in S. Patai, ed., *The Chemistry of Acyl Halides*, Wiley-Interscience, New York, 1972.
4. S. B. Damle, *Hydrolysis of Chloroformates*, unpublished data.
5. H. K. Hall, Jr., *J. Am. Chem. Soc.* **77**, 5993 (1955).
6. H. K. Hall, Jr., *J. Org. Chem.* **21**, 248 (1956).
7. H. K. Hall, Jr., and P. W. Morgan, *J. Org. Chem.* **21**, 249 (1956).
8. H. K. Hall, Jr., *J. Am. Chem. Soc.* **78**, 2717 (1956).
9. H. K. Hall, Jr., *J. Am. Chem. Soc.* **79**, 5439 (1957).
10. S. B. Damle and J. A. Krogh, *Thermal Stability of Chloroformates*, unpublished data.
11. U.S. Pat. 4,714,785 (Dec. 22, 1987), J. Manner (to PPG Industries, Inc.).
12. N. Lui, A. Marhold, and M. Rock, *J. Org. Chem.* **63**, 2493 (1998).
13. U.S. Pat. 5,874,655 (Feb. 23, 1999), N. Lui and A. Marhold (to Bayer A.-G.).
14. E. S. Lewis, W. C. Herndon, and D. D. Duffey, *J. Am. Chem. Soc.* **83**, 1959 (1961).
15. U.S. Pat. 6,291,731 (Sept. 18, 2001), A. Stamm, J. Henkelmann, and H.-J. Weyer (to BASF A.-G.).
16. Eur. Pat. 645,357 (Oct. 9, 1994), R. Ettl and W. Reuther (to BASF A.-G.).
17. U.S. Pat. 4,814,524 (Mar. 21, 1989), R. Briody and J. Manner (to PPG Industries, Inc.).
18. U.S. Pat. 4,734,535 (Mar. 29, 1988), N. Greif and K. Oppenlaender (to BASF A.-G.).
19. U.S. Pat. 5,196,611 (Mar. 23, 1993), J. Henkelmann *et. al.* (to BASF A.-G.).
20. U.S. Pat. 6,245,954 (June 12, 2001), H.-J. Weyer, A. Stamm, T. Weber, and J. Henkelmann (to BASF A.-G.).
21. Eur. Pat. 625,469 (Nov. 23, 1994) N. Keigo, *et al.* (to Ube Industries, Ltd., Japan).
22. F. Foulon, B. Fixari, D. Piq, and P. Le Perchec, *Tet. Lett.* **38**, 3387, (1997).
23. U.S. Pat. 5,723,704 (Mar. 3, 1988) H. Demail, J.-C. Schweickert, and P. Le Gars (to Societe Nationale des Poudres et Explosifs, Fr.).
24. F. Rigamonti, *Chem. Eng. Sci.* **47**(9–11), 2653 (1992).
25. J. Nemirovsky, *J. Prakt. Chem.* **31**(1), 173 (1885).
26. F. H. Carpenter and D. T. Gish, *J. Am. Chem. Soc.* **74**, 3818 (1952).
27. A. Morel, *Bull. Soc. Chim. (France)* **21**(3), 815 (1899).
28. R. Adams and J. B. Segur, *J. Am. Chem. Soc.* **45**, 785 (1923).
29. J. L. R. Williams, D. D. Reynolds, K. R. Dunham, and J. F. Tinker, *J. Org. Chem.* **24**, 64 (1959).
30. Jpn. Patent 6,271,507 (Sept. 27, 1994) and 7,224,008 (Aug. 22, 1996), O. Takanobu and H. Mizukami and co-workers (to Mitsubishi Gas Chemical Co. Japan).

31. U.S. Pat. 2,362,865 (Nov. 14, 1944), S. Tryon and W. S. Benedict (to General Chemical Co. of New York).
32. U.S. Pat. 3,234,262 (Feb. 8, 1966), R. P. Kurkjy, M. Matzner, and R. J. Cotter (to Union Carbide Corp.).
33. U.S. Pat. 6,175,017 (Jan. 16, 2001), H. Buysch, N. Schon, and G. Jeromin (to Bayer A.-G.).
34. U.S. Pat. 5,527,942 (June 18, 1996), P. Ooms, N. Buysch, and H. Josef (to Bayer A.-G.).
35. U.S. Pat. 3,251,873 (May 17, 1966), R. P. Kurkjy, M. Matzner, and R. J. Cotter (to Union Carbide Corp.).
36. U.S. Pat. 3,211,776 (Oct. 12, 1965), C. W. Stephens (to E. I. du Pont de Nemours & Co., Inc.).
37. Jpn. Patent 9,100,256 (Apr. 15, 1997), Y. Hara, M. Tojima, H. Tsuchisada, and H. Koto (to Mitsubishi Chemical Industries Ltd., Japan).
38. U.S. Pat. 3,170,946 (Feb. 23, 1965), J. R. Kilsheimer and W. H. Hensley (to Union Carbide Corp.).
39. U.S. Pat. 3,275,674 (Sept. 27, 1966), L. Bottenbruch and H. Schnell (to Farbenfabriken Bayer A.-G.).
40. U.S. Pat. 3,211,775 (Oct. 12, 1965), C. W. Stephens and W. Sweeny (to E. I. du Pont de Nemours & Co., Inc.).
41. U.S. Pat. 4,012,406 (Mar. 15, 1977), H. J. Buysch and H. Krimm (to Bayer A.-G.).
42. U.S. Pat. 3,211,774 (Oct. 12, 1965), C. W. Stephens (to E. I. du Pont de Nemours & Co., Inc.).
43. Ger. Pat. 1,133,727 (July 26, 1962), V. Bollert, G. Fritz, and H. Schnell (to Farberfabriken Bayer A.-G.).
44. D. S. Tarbell and N. A. Leister, *J. Org. Chem.* **23**, 1149 (1958).
45. T. B. Windholz, *J. Org. Chem.* **23**, 2044 (1958).
46. T. B. Windholz, *J. Org. Chem.* **25**, 1703 (1960).
47. S. Kim, J. I. Lee, and Y. C. Kim, *J. Org. Chem.* **50**, 560 (1985).
48. G. Barcelo, D. Grenouillat, J.-P. Senet, and G. Sennyey, *Tetrahederon* **46**(6), 1839 (1990).
49. P. Gros, P. Le Perchec, and J.-P. Senet, *Syn. Comm.* **23**(130), 1835 (1993).
50. J.-P. Senet, *The Recent Advance in Phosgene Chemistry* 2, L'Imprimerie GPA à Nanterre, Société Nationale des Poudres et Explosifs, Feb. 1999, p. 137 .
51. G. Sennyey, G. Barcelo, and J.-P. Senet, *Tetrahedron Lett.* **27**(44), 5375 (1986).
52. G. Sennyey, G. Barcelo, and J.-P. Senet, *Tetrahedron Lett.* **28**(47), 5809 (1987).
53. Ger. Pat. 1,418,849 (Mar. 20, 1969), V. Boellert, U. Curtius, G. Fritz, and J. Nentwig (to Bayer A.-G.).
54. D. Plusquellec, F. Roulleau, M. Lefeuvre, and E. Brown, *Tetrahederon* **44**(9), 2471 (1988).
55. U.S. Pat. 5,231,211 (July 27, 1993), R. Tang (to PPG Industries, Inc.).
56. S. B. Damle and R. H. Tang, paper given at *Chemical Specialties U.S.A. 1992*, Cherry Hill, N.J.; "Chemicals and the Environment".
57. U.S. Pat. 5,523,481 (June 4, 1996), M. Pies, H. Fiege, J. Käsbauer, and G. Merz (to Bayer A.-G.).
58. J.-P. Senet, *The Recent Advance in Phosgene Chemistry* 1, L'Imprimerie GPA à Nanterre, Société Nationale des Poudres et Explosifs, Dec. 1997, p. 44.
59. U.S. Pat. 2,518,058 (Aug. 8, 1950), A. Pechukas (to PPG Industries Inc.).
60. Ref. 58, p. 51.
61. Jpn. Pat. 60,252,450 (Dec. 13, 1985), S. Jinbo and co-workers (to Hodogaya Chemical Co., Ltd.).

62. U.S. Pat. 4,652,665 (Mar. 24, 1987), G. Barcelo, J.-P. Senet, and G. Senney (to Societe Nationale des Poudres et Explosifs, Fr.).
63. Eur. Pat. 249,556 (Dec. 16, 1987), J.-P. Senet, G. Sennyey, and G. Wooden (to Societe Nationale des Poudres et Explosifs, Fr.).
64. J. P. Senet, G. Sennyey, and G. Wooden, *Synthesis* **5**, 407 (1988).
65. O. Norman and V. Sonntag, *Chem. Rev.* **52**, 273 (1952).
66. E. Aquino, W. Brittain, and D. Brunelle, *Macromolecules* **25**, 3827 (1992).
67. J. King and G. Bryant, *J. Org. Chem.* **57**, 5136 (1992).
68. E. F. Degering, G. L. Jenkins, and B. E. Sanders, *J. Am. Pharm. Assoc.* **39**, 824 (1950).
69. U.S. Pat. 2,649,473 (Aug. 18, 1953), J. A. Chenicek (to Universal Oil Products).
70. L. C. Raiford and G. O. Inman, *J. Am. Chem. Soc.* **56**, 1586 (1934).
71. E. Wunsch, in *Houben-Weyl*, 4th ed., Vol. XV/1, Georg Thieme Verlag, Stuttgart, Germany, 1974, p. 47.
72. Ref. 50, p. 116.
73. M. Bergmann and L. Zervas, *Ber. Dtsch. Chem. Ges.* **65**, 1192 (1932).
74. U.S. Pat. 4,484,001 (Nov. 20, 1984), J. A. Krogh (to PPG Industries Inc.).
75. U.S. Pat. 4,500,726 (Feb. 19, 1985), J. A. Krogh (to PPG Industries Inc.).
76. S. B. Damle and J. A. Krogh, posters presented at *The Third Chemical Congress of North America*, Toronto, Canada, June 5–10, 1988; *Biol. Chem. Abstr.* **21**, 22, (1988).
77. U.S. Pat. 4,293,706 (Oct. 6, 1981), S. B. Gorman, R. B. Thompson, and E. E. Yonan (to PPG Industries Inc.).
78. U.S. Pat. 3,492,131 (Jan. 27, 1970), J. M. Schlatter (to G. D. Searle & Co.).
79. W. H. Coppock, *J. Org. Chem.* **22**, 325 (1957).
80. S. Yura and T. Ono, *J. Soc. Chem. Ind. (Japan)* **48**, 30 (1945).
81. L. Balas, B. Jousseaume, H. Shin, J.-B. Verlhac, and F. Wallian, *Organometallics* **10**, 366 (1991).
82. R. Adlington, J. Baldwin, A. Gansauer, W. McCoull, and A. Russell, *J. Chem. Soc. Perkin Trans. 1* 1697 (1994).
83. R. Grigg and V. Savic, *Chem. Comm.* 2381 (2000).
84. R. A. Olofson, R. C. Schnur, L. Bunes, and J. J. Pepe, *Tetrahedron Lett.* 1567 (1977).
85. Eur. Pat. 40,153 (Nov. 18, 1981), G. Cagnon, M. Piteau, J.-P. Senet, R. A. Olofson, and J. T. Martz (to Société Nationale des Poudres et Explosifs).
86. J. H. Cooley and E. J. Evain, *Synthesis*, 1 (1987).
87. Ref. 50, p. 141.
88. R. A. Olofson, J. T. Martz, J.-P. Senet, M. Piteau, and T. Malfroot, *J. Org. Chem.* **49**(11), 2081, (1984).
89. U.S. Pats. 4,592,872 and 4,592,874 (June 3, 1986), G. Cagnon, M. Piteau, J.-P. Senet, R. A. Olofson, and J. T. Martz (to Société Nationale des Poudres et Explosifs).
90. R. A. Olofson, *Pure Appl. Chem.* **60**, 1715 (1988).
91. R. A. Olofson and D. E. Abbott, *J. Org. Chem.* **49**, 2795 (1984).
92. U.S. Pat. 2,370,588 (Feb. 27, 1945), F. Strain (to PPG Industries Inc.).
93. F. Strain and co-workers, *J. Am. Chem. Soc.* **72**, 1254 (1950).
94. U.S. Pat. 2,608,572 and 2,608,573 (Aug. 26, 1952), A. H. Fischer (to Minerec Corp.).
95. D. Ennis, J. Ince, S. Rahman, and M. Shipman, *J. Chem. Soc. Perkin 1* **13**, 2047 (2000).
96. M. E. Childs and W. P. Weber, *J. Org. Chem.* **41**, 3486 (1976).
97. Y. Nii, K. Okano, S. Kobayashi, and M. Ohno, *Tet. Lett.* **27**, 2517 (1979).
98. Ref. 58, p. 47.
99. Swiss Pat. 675,875 (Nov. 11, 1990), H. Mettler and F. Previdoli (to Lonza A.-G.).
100. P. F. DeCusati and R. A. Olofson, *Tet. Lett.* **31**(10), 1405 (1990).

101. L. M. Harwood, Y. Houminer, A. Manare, and J. I. Seeman, *Tet. Lett.* **35**(43), 8927 (1994).
102. R. A. Olofson, J. Cuomo, and B. A. Bauman, *J. Org. Chem.* **43**(10) (1978).
103. S. J. Aboulhoda, F. Henin, J. Muzart, and C. Thorey, *Tet. Lett.* **36**(27), 4795 (1995).
104. Jpn. Pat. 56,005,214 (Feb. 4, 1981) (to Mitsubishi Chemical Industries).
105. Ger. Pat. 2,847,484 (Nov. 2, 1980), W. Schulte-Huermann, E. Schellermann, and J. Lahrs (to Bayer A.-G.).
106. U.S. Pat. 3,910,983 (Oct. 7, 1975), K. Merkel, J. Datow, J. Paetsch, H. Hoffmann, and S. Winderl (to BASF A.-G.).
107. U.S. Pat. 4,039,569 (Aug. 2, 1977), F. S. Bell, R. D. Crozier, and L. E. Strow (to Minerec Corporation).
108. Eur. Pat. 75,145 (Mar. 30, 1983), D. Bauer, H. Dohm, W. Schulte-Huermann, and H. Hemmerich (to Bayer A.-G.).
109. Fr. Pat. 1,336,606 (Oct. 20, 1961) (to Imperial Chemical Industries Limited).
110. U.S. Pat. 3,170,946 (Feb. 23, 1965), J. R. Kilsheimer and W. H. Hensley (to Union Carbide Corp.).
111. U.S. Pat. 4,085,129 (Apr. 18, 1978), G. Semler and G. Schaeffer (to Hoechst A.-G.).
112. Fr. Pat. 2,510,989 (Feb. 11, 1983), P. M. Novy (to PPG Industries Inc.).
113. U.S. Pat. 3,211,775 (Oct. 12, 1965), C. W. Stephens and W. Sweeny (to E. I. du Pont de Nemours & Co., Inc.).
114. U.S. Pat. 5,424,473 (June 13, 1995), R. Galvan and M. J. Mullins (to Dow Chemical Company).
115. Eur. Pat. 542,132 (May 19, 1993), H. Koehler, T. Wettling, W. Franzischka, and L. Hupfer (to BASF A.-G.).
116. D. J. Brunelle, D. K. Bonauto, and T. G. Shannon, *Polymer Int.* **37**(3), 179 (1995).
117. U.S. Pat. 6,414,178 (Mar. 21, 2002), J. Silva, D. Dardaris, and T. Fyvie (to General Electric Company).
118. U.S. Pat. 6,268,461 (July 31, 2001), T. Fyvie and J. Silva (to General Electric Company).
119. U.S. Pat. 5,142,008 (Aug. 25, 1992), P. Phelps, E. Boden, and P. Buckley (to General Electric Company).
120. U.S. Pat. 6,392,079 (May 21, 2002), J. Silva, D. Dardaris, and P. Phelps (to General Electric Company).
121. Jpn. Pat. 10,130,205 (May 19, 1998) and 11,076,157 (Oct. 3, 2000) Y. Kashiyama, R. Takei, and S. Handa (to Asahi Penn Chemical K. K., Japan).
122. U.S. Pat. 6,479,690 (Nov. 12, 2002), L. Garel and F. Metz (to Société Nationale des Poudres et Explosifs).
123. U.S. Pat. App. 0082,444 (June 27, 2002), B. Hubert, L. Ferruccio, P. Gauthier, and J.-P. Senet (to Société Nationale des Poudres et Explosifs).
124. Ger. Pat. 3,241,568 (Nov. 10, 1983), R. A. Olofson and J. T. Martz (to Société Nationale des Poudres et Explosifs).
125. U.S. Pat. 5,712,407 (Jan. 14, 1997), C. B. Kreutzberger, S. Eswarakrishnan, and S. B. Damle (to PPG Industries Inc.).
126. Ref. 58, p. 53.
127. M. J. Mulville and J. Gallagher and co-workers, *Synthesis* **3**, 365 (2002).
128. U.S. Pat. 5,298,646 (Mar. 29, 1994), J. A. Manner, F. F. Guzik, and S. B. Damle (to PPG Industries Inc.).
129. Ger. Pat. 3,826,584 (Feb. 16, 1989), F. Mogyorodi and E. Koppany and co-workers (to Eszakmagyarorszagi Vegyimuvek).
130. U.S. Pat. 2,377,085 (May 29, 1945), F. E. Kung (to PPG Industries Inc.).
131. L. H. Lee, *J. Org. Chem.* **30**, 3943 (1965).
132. R. A. Olofson, B. A. Bauman, and D. J. Wancowicz, *J. Org. Chem.* **43**(4), 752, (1978).

133. Ref. 58, p. 81.
134. U.S. Pat. 4,242,280 (Dec. 30, 1980), S. Lecolier, T. Malfroot, M. Piteau, and J.-P. Senet (to Société Nationale des Poudres et Explosifs).
135. M. P. Bowman, R. A. Olofson, J.-P. Senet, and T. Malfroot, *J. Org. Chem.* **55**, 5982 (1990).
136. M. Folkmann and F. J. Lund, *Synthesis* **Dec.** 1159 (1990).
137. J. Alexander, R. Cargill, S. R. Michelson, and H. Schwam, *J. Med. Chem.* **31**, 318 (1988).
138. PPG Industries Inc., unpublished data.
139. N. I. Sax and R. J. Lewis, Sr., eds., *Dangerous Properties of Industrial Materials*, 7th ed., Van Nostrand Reinhold, New York, 1989.
140. L. J. Cralley and L. V. Cralley, eds., *Patty's Industrial Hygiene and Toxicology* Vols. I–III, 3rd ed., John Wiley & Sons, Inc., New York, 1979.
141. U.S. Pats. 2,370,565 and 2,370,566 (Feb. 27, 1945), I. E. Muskat and F. Strain (to PPG Industries Inc.).
142. U.S. Pat. 2,708,617 (May 17, 1955), E. E. Magat and D. R. Strachan (to E. I. du Pont de Nemours & Co., Inc.).
143. D. G. LeGrand and J. T. Bendler, eds., *Handbook or Polycarbonate Science and Technology*, Marcel Dekker, New York, 2000.
144. U.S. Pat. 2,937,119 (May 17, 1960), F. M. Berger and B. Ludwig (to Carter Products).
145. J.-P. Senet, *Chem. Marketing Rep.* **255**(21) 6 (1999).
146. G. Sennyey, *Specialty Chemicals* **Oct.** 364 (1990).
147. J.-P. Senet, *Specialty Chemicals* **Jan.** 12 (1998).
148. U.S. Pat. 4,957,924 (Sept. 18, 1990), L. M. Beauchamp (to Burroughs Wellcome Co.).
149. U.S. Pat. 4,191,537 (Mar. 4, 1980) and U.S. Pat. 4,236,020 (Nov. 25, 1980), R. A. Lewis, L. R. Berkeley, and P. Honnen (to Chevron Research Company).
150. U.S. Pat. 4,197,409 (Apr. 8, 1980), J. E. Lilburn (to Chevron Research Company).
151. U.S. Pat. 4,521,610 (June 4, 1985) and U.S. Pat. 4,695,291 (Sep. 22, 1987), F. Plavac (to Chevron Research Company).
152. H. Kelker and R. Hatz, *Handbook of Liquid Crystals*, Verlag Chemie, 1980.
153. J. I. Ash, *Liquid Crystals for Nondestructive Evaluation*, South-West Research Institute, San Antonio, Tex. NTIAC-78-2 (1978).
154. D. Tsiourvas, T. Felekis, Z. Sideratou, and C. Paleos, *Macromolecules* **35**(16), 6466 (2002).
155. U.S. Pat. 4,587,296 (May 28, 1985), P. G. Moniotte (to Monsanto Europe S. A.).
156. U.S. Pat. 5,442,099 (Aug. 15, 1995), J. Wolpers, K.-H. Nordsiek, J. Monkiewicz, and D. Zerpner (to Huels A.-G.).
157. U.S. Pat. 2001/0025071 A1 (Sep. 27, 2001), T. Fruh, L. Heiliger, and G. E. Muller (to Bayer A.-G.).
158. U.S. Pat. 4,622,431 (Nov. 11, 1986), R. G. Briody and H. C. Stevens (to PPG Industries, Inc.).
159. U.S. Pat. 3,488,342 (Jan. 6, 1970), C. S. Sheppard, H. P. Van Leeuwen, and O. L. Mageli (to Pennwalt Corp.).
160. Ref. 58, p. 92.
161. P. Hesek, *J. Chrom., B.: Biomed. Sci. Appl.* **717**(1 + 2), 57 (1998).
162. R. Bueschges, H. Linde, E. Mutschler, and H. Spahn-Langguth, *J. Chrom. A* **725**(2) 323 (1996).
163. L. H. Horsley, in R. F. Gould, ed., *Azeotrope Data III*, Advances in Chemistry Series 116 ACS, Washington, D.C., 1973.
164. A-A. G. Shaikh and S. Sivaram, *Chem. Rev.* **96**, 951 (1996).
165. J. P. Parrish, R. N. Salvatore, and K. W. Jung, *Tetrahedron* **56**, 8207, (2000).
166. S. Kim and Y. K. Ko, *Heterocycles* **24**(6), 1625 (1986).

167. J. F. Knifton and R. G. Duranleau, *J. Mol. Catal.* **67**, 389 (1991).
168. Ger. Pat. 4,109,236 (Mar. 21, 1991), N. Schon, H.-J. Buysch, E. Leitz, and K.-H. Ott (to Bayer A.-G).
169. U.S. Pat. 3,022,272 (Feb. 20, 1962), H. Schnell and G. Fritz (to Farbenfabriken Bayer A.-G.).
170. U.S. Pat. 5,861,107 (Jan. 19, 1999), H.-J. Buysch, G. Fengler, K.-H. Newman, P. Wagner, M. Melchiors, and W. Hovestadt (Bayer A.-G.).
171. U.S. Pat. 4,609,501 (Sep. 2, 1986), V. Mark (to General Electric Company).
172. Eur. Pat. 780,361 (Dec. 13, 1996), S. Inoki and co-workers (to General Electric Company).
173. U.S. Pat. 4,252,737 (Feb. 24, 1981), H. Krimm, H.-J. Buysch, and H. Rudolph (to Bayer A.-G.).
174. U.S. Pat. 4,410,464 (Oct. 18, 1983), J. E. Hallgren (to General Electric Company).
175. D. Delledonne, F. Rivetti, U. Romano, *Appl. Cat. A: General* **221**, 241, (2001).
176. H.-J. Buysch, *Carbonic Esters, Ullmann's Encyclopedia of Industrial Chemistry*, Wiley-VCH Verlag GmbH, Weinheim, Germany, 2002.
177. Y. Ono, *Pure Appl. Chem.* **68**(2), 376 (1996).
178. P. Tundo and M. Selva, *Acc. Chem. Res.* **35**(9), 706 (2002).
179. U.S. Pat. 5,463,109 (Oct. 31, 1995), K. Nishihira and S. Tanaka (to Ube Industries, Ltd.).
180. U.S. Pat. 6,387,519 (May, 14, 2002), L. G. Anderson and co-workers (to PPG Industries Inc.).
181. U.S. Pat. 5,091,556 (Feb. 25, 1992), C. Calderoni, F. Mizia, F. Rivetti, and U. Romano (to Enichem Synthesis, S.p.A.).
182. U.S. Pat. 5,789,614 (Aug. 4, 1998), Yagii and co-workers (to Daicel Chemical Industries, Ltd.).
183. WO Pat. 9,856,758 (Dec. 17, 1998), R. C. Smith and J. C. Bausor (to Imperial Chemical Industries, UK).
184. U.S. Pat. 4,354,979 (Oct. 19, 1982), V. Schwendemann and D. Mangold (to BASF A.-G.).
185. U.S. Pat. 4,659,845 (Apr. 21, 1987), F. Rivetti, G. Garone, and U. Romano (to Enichem Synthesis, S.p.A.).
186. Eur. Pat. 511,572 (Apr. 21, 1992), H.-J. Buysch *and co-workers* (to Bayer A.-G.).
187. Jpn. Pat. 1,230,550 (Sep. 14, 1989), S. Maomi and co-workers (to Asahi Chem).
188. U.S. Pat. 5,310,849 (May 10, 1994), H.-J. Buysch, K. Konig, A. Klausener, K. Szablikowski, and J. Breckwoldt (to Bayer A.-G.).
189. Y. Wang, X. Zhao, F. Fang, S. Wang, J. Zhang, *J. Chem.Tech. & Biotech.* **76**(8) 857 (2001).
190. U.S. Pat. 5,393,910 (Feb. 28, 1995), J. Y. Mui and M. P. Bowman (to OSi Specialties, Inc.).
191. U.S. Pat. 6,147,257 (Nov. 14, 2000), P. Allegrini, G. Barreca, and E. Rossi (to Zambon Group S.p.A.).
192. U.S. Pat. 5,780,624 (July 14, 1998), H. Wingert and M. Keil (to BASF A.-G.).
193. U.S. Pat. 6,288,257 (Sep. 11, 2001), F. J. Schattenmann and L. N. Lewis (to General Electric Company).
194. S. B. Damle and J. A. Krogh, unpublished work.
195. Y. Tamura, T. Saito, H. Ishibashi, and M. Ikeda, *Synthesis* **641** (1975).
196. U.S. Pat. 3,221,013 (Nov. 30, 1965), D. L. Fields and D. D. Reynolds (to Eastman Kodak Co.).
197. U.S. Pat. 3,590,074 (June 29, 1971), R. Heiss, E. Bocker, B. Homeyer, and I. Hammann (to Farbenfabriken Bayer A.-G.).
198. U.S. Pat. 4,362,670 (Dec. 7, 1982), E. P. Woo (to The Dow Chemical Company).

199. U.S. Pat. 4,192,949 (Mar. 11, 1980), F. Merger, F. Towae, and L. Schroff (to BASF).
200. U.S. Pat. 4,254,276 (Mar. 3, 1981), G. Iori and U. Romano (to Anic, SpA).
201. U.S. Pat. 4,182,276 (Jan. 8, 1980), G. Illuminati, U. Romano, and R. Tesei (to Anic, SpA).
202. U.S. Pat. 4,395,565 (July 26, 1983), U. Romano, G. Fornasari, and S. DiGioacchino (to Anic, SpA).
203. U.S. Pat. 4,326,079 (Apr. 20, 1982), G. Iori and U. Romano (to Anic, SpA).
204. U. Romano, R. Tesei, M. M. Mauri, and P. Rebora, *Ind. Eng. Chem. Prod. Res. Dev.* **19**, 396 (1980).
205. N. Yamazaki, T. Iguchi, and F. Higashi, *J. Polym. Sci.* **17**, 835 (1979).
206. M. Lissel, S. Schmidt, and B. Neumann, *Synthesis*, 382 (1986).
207. M. Selva, A. Bomben, and P. Tundo, *J. Chem. Soc., Perkin Trans. 1* 1041 (1997).
208. Y. Fu, T. Baba, and Y. Ono, *Appl. Catal. A* **166**, 425 (1999).
209. L. Cotarca, P. Delogu, A. Nardelli, and V. Sunjic, *Synthesis* 553 (1996).
210. W.O. Pat. 9,914,159 (Mar. 25, 1999), H. Eckert, B. Gruber, and N. Dirsch (to Dr. Eckert GMBH).
211. L. Cortarca, *Org. Proc. Res. Dev.* **5**(3), 377 (1999).
212. L. Cortarca, S. Mantovani, and P. Delogu, *J. Org. Chem.* **65**, 8224 (2000).
213. G. Barcelo, J.-P. Senet, and G. Sennyey, *J. Org. Chem.* **50**, 3951 (1985).
214. Ref. 58, p. 79.
215. Ref. 50, p. 134.
216. V. A. Dang, R. A. Olofson, P. Wolf, M. Piteau, and J.-P. Senet, *J. Org. Chem.* **55**(6), 1847 (1990).
217. Ref. 58, p. 69.
218. Ref. 50, p. 118.
219. U.S. Pat. 6,420,588 (July 16, 2002), P. J. McKloskey and co-workers (to General Electric Company).
220. U.S. Pat. 5,523,481 (Jun. 4, 1996), M. Pies, H. Fiege, J. Käsbauer, and G. Merz (to Bayer A.-G.).
221. U.S. Pat. 6,348,613 (Feb. 19, 2002), M. Miyamoto and N. Hyoudou (to Mitsubishi Chemical Corp.).
222. U.S. Pat. 6,469,192 (Oct. 22, 2002), T. B. Burnell, P. J. McCloskey, G. Kailasam, and J. A. Cella (to General Electric Company).
223. U.S. Pat. 2,362,865 (1941), S. Tyron and W. S. Benedict (to General Electric Company).
224. U.S. Pat. 3,234,263 (1962), R. R. Kurkjy, M. Matzner, and R. J. Cotter (to Union Carbide Corp.).
225. U.S. Pat. 5,167,946 (1990), M. Mullins, A. T. Chamberlin, and R. Galvan (to Dow Chemical Co.).
226. U.S. Pat. 5,239,105 (Aug. 8, 1993), R. G. Pews and R. G. Bowman (to Dow Chemical Co.).
227. Jpn. Pat. 9,024,278 (Jan. 28, 1995), H. Yoshinori, K. Hideki, and H. Michio (to Mitsubishi Chemical Corp.).
228. Jpn. Pat. 11,005,766 (Jan. 12, 1999), S. Yoshio and co-workers (to Mitsubishi Chemical Corp.).
229. Jpn. Pat. 9,301,931 (Nov. 25, 1997), T. Mitsuhiko, K. Mitsumasa, and T. Hidetaka (to Mitsubishi Chemical Corp.).
230. Jpn. Pat. 9,100,256 (Mar. 15, 1997), H. Yoshinori and co-workers (to Mitsubishi Chemical Corp.).
231. U.S. Pat. 6,348,613 (Feb. 19, 2002), M. Miyamoto and N. Hyoudou (to Mistubishi Chemical Company).

232. U.S. Pat. 4,366,102 (Dec. 28, 1982), G. Rauchschwalbe, K. Mannes, and D. Mayer (to Bayer A.-G.).

233. U.S. Pat. 2,837,555 (June 3, 1958), J. M. Lee (to The Dow Chemical Company).

234. U.S. Pat. 3,251,873 (May 17, 1966), R. P. Kurkjy, M. Matzner, and R. J. Cotter (to Union Carbide Corp.).

235. U.S. Pat. 3,234,262 (Feb. 8, 1966), R. P. Kurkjy, M. Matzner, and R. J. Cotter (to Union Carbide Corp.).

236. U.S. Pat. 3,234,263 (Feb. 8, 1966), R. P. Kurkjy, M. Matzner, and R. J. Cotter (to Union Carbide Corp.).

237. Eur. Pat. 542,117 (Sep. 13, 1995), T. Wettling and J. Henkelmann (to BASF A.-G.).

238. Eur. Pat. 635,476 (Oct. 22, 1997), P. Ooms, N. Schön, and H.-J. Buysch (to Bayer A.-G.).

239. Eur. Pat. 500,786 (Sep. 8, 1995) D. A. Harley, S. S. King, and C. L. Rand (to Dow Chemical Company).

240. U.S. Pat. 5,900,501 (May 4, 1999) P. Ooms, H.-J. Buysch, S. Kühling, and G. Zaby (to Bayer A.-G.).

241. U.S. Pat. 2002/0087022 A1 (Jul. 4, 2002) A. Chrisochoou, S. Kühling, and J. V. Eynde (to Bayer A.-G.).

242. Fr. Pat. 2,163,884 (July 27, 1973) (to Société Nationale des Poudres et Explosifs).

243. Brit. Pat. 2,107,712 (July 10, 1985), J. L. Pounds and R. S. Bartlett (to PPG Industries Inc.).

244. W. J. Peppel, *Ind. Eng. Chem.* **50**, 767 (1958).

245. Eur. Pat. 678,498 (Oct. 7, 1998) C. Menzoza-Frohn and P. Wagner and co-workers (to Bayer A.-G.).

246. Eur. Pat. 499,924 (May 10, 1994) H. J. Buysch and A. Klausener (to Bayer A.-G.).

247. M. Pancheco and C. L. Marshall, *Energy & Fuels* **11**, 2 (1997).

248. Brit. Pat. 2,148,881 (June 5, 1988), S. F. Davison (to BP Chemical Co.).

249. Jpn. Pat. 60,075,447 (Apr. 27, 1985) (to Mitsubishi Gas Chemical Co.).

250. Ger. Pat. 3,045,767 (June 11, 1981), U. Romano, F. Rivetti, and N. DiMuzio (to Anic, SpA).

251. F. Rivetti, U. Romano, and D. Delledonne, *Green Chemistry. Designing Chemistry for the Environment*, ACS Symposium Series no. 626, in P. T. Anatas and T. C. Williamson, Eds., ACS, Washington, D.C., 1996, p. 70.

252. U.S. Pat. 5,686,644 (Nov. 11, 1997) F. Rivetti and U. Romano (to Enichem Synthesis S.p.A.).

253. Ger. Pat. 2,738,488 (Apr. 13, 1978), J. E. Hallgren (to General Electric Co.).

254. Ger. Pat. 2,949,936 (July 3, 1980), J. E. Hallgren (to General Electric Co.).

255. Ger. Pat. 2,738,520 (Apr. 13, 1978), A. J. Chalk (to General Electric Co.).

256. Y. Yamamoto, T. Matsuzaki, K. Ohdan, and Y. Okamoto, *J. Catal.* **161**, 577 (1996).

257. Eur. Pat. 742,198 (Dec. 22, 1999), K. Nishihira (to Ube Industries, Ltd.).

258. U.S. Pat. 5,902,894, J. Y. Ryu (to Catalytic Distillation Technology).

259. *Chem. Mark. Rep.* (May 24, 1999).

260. ES Pat. 2,009,072 (Aug. 16, 1989), I. L. Molina, A. P. Coll, and A. D. Coto (to Union Quimico Farmaceutica S.A.E.).

261. L. A. Shervington and A. Shervington, *Guaifenesin: Analytical Profiles Drug Substances Excipients* **25**, 121 (1998).

262. World Pat. 8,402,339 (June 21, 1984), G. E. Morris (to British Petroleum Co.).

263. U.S. Pat. 4,380,455 (Apr. 19, 1983), H. A. Smith (to British Petroleum Co.).

264. H. Pizzi and A. Stephanou. *J. Appl. Polym. Sci.* **49**, 2157 (1993).

265. U.S. Pat. 4,416,694 (1983), J. Stevenson, J. Machin, and D. L. Dyke (to Foresco International Ltd.).

266. B. Scrosati, *Chim Ind.* **79**, 463 (1997).

267. U.S. Pat. 5,340,845 (Aug. 23, 1994), D. Stefani and F. O. Sam (to Enichem, S.p.A.).
268. U.S. Pat. 2002/0142928 (Oct. 3, 2002), J. R. MacHac, Jr., E. T. Marquis, and S. A. Woodrum (to Huntsman Petrochemical Corporation).
269. F. Strain and co-workers, *J. Am. Chem. Soc.* **72**, 1254 (1950).
270. U.S. Pat. 3,022,281 (Feb. 20, 1962), E. S. Smith (to Goodyear Tire & Rubber Co.).
271. Ref. 58, p. 59.
272. WO Pat. 9,905,150 (Feb. 2, 1999), J. D. Munger, J. C. Rohloff, and L. M. Schultz (to Gilead Sciences, Inc.).
273. WO Pat. 9,804,569 (Feb. 5, 1998), M. N. Arimilli and co-workers (to Gilead Sciences, Inc.).
274. U.S. Pat. 2,370,588 (Feb. 27, 1945), F. Strain (to PPG Industries Inc.).

CHARLES B. KREUTZBERGER
PPG Industries

CHLOROTOLUENES, BENZYL CHLORIDE, BENZAL CHLORIDE, AND BENZOTRICHLORIDE

1. Introduction

The chlorination of toluene in the absence of Lewis Acid catalysts occurs preferentially in the side chain. In the presence of Lewis Acid catalysts, ring chorination is the predominant pathway. Reaction on the side chain is promoted by free-radical initiators such as ultraviolet light, peroxides or Vazo® catalysts. Chlorination of the side chain takes place sequentially and can be controlled to give good yields of the intermediate chlorination products. Small amounts of metal sequestering agents are sometimes used to remove trace amounts of metal ions that promote ring chlorination.

Experimental data taken from the chlorination of toluene in an irradiated continuous stirred tank flow reactor at 111°C result in the product distribution shown in Table 1 (1).

Nearly all of the benzyl chloride [100-44-7], benzal chloride [98-87-3], and benzotrichloride [98-07-7] manufactured is converted to other chemical intermediates or products by reactions involving the chlorine substituents of the side chain. Each of the compounds has a single primary use that consumes the majority portion of the compound produced. Benzyl chloride is utilized in the manufacture of benzyl butyl phthalate, a vinyl resin plasticizer; benzal chloride is hydrolyzed to benzaldehyde; benzotrichloride is converted to benzoyl chloride. Benzyl chloride is also hydrolyzed to benzyl alcohol, which is used in the photographic industry, in perfumes (as esters), and in peptide synthesis by conversion to benzyl chloroformate [501-53-1] (see BENZYL ALCOHOL AND β-PHENETHYL ALCOHOL; CARBONIC AND CARBONOCHLORIDIC ESTERS).

Table 1. **Distributions of Reactor Products[a] from Batch Chlorination of Toluene**

$\dfrac{\text{Mol Cl}_2}{\text{Mol reactant in product}}$	Toluene	Benzyl chloride	Benzal chloride	Benzotrichloride
0.30	0.717	0.271	0.012	
0.51	0.507	0.480	0.013	
0.82	0.250	0.685	0.065	
0.98	0.138	0.744	0.118	
1.19	0.040	0.729	0.231	
1.32	0.030	0.672	0.325	
1.53		0.482	0.503	0.015
1.95		0.105	0.842	0.053
2.18		0.016	0.774	0.210

[a] Mole fractions.

Several related compounds, primarily ring-chlorinated derivatives, are also commercially significant. Compounds such as p-chlorobenzotrichloride, o-chlorobenzotrichloride, 3,4-dichlorobenzotrichloride, 2,4-dichlorobenzylchloride and others can be similarily prepared by the free radical chlorination of the parent ring chlorinated toluene. In the case of p-chlorobenzotrichloride this can be further converted to p-chlorobenzotrifluoride by reaction with HF. p-Chlorobenzotrifluoride is an important intermediate in the manufacture of dinitroaniline herbicides (Trifluralin). Additionally, p-chlorobenzalchloride or o-chlorobenzal chloride can be prepared by side-chain chlorination of p-chlorotoluene and o-chlorotoluene respectively. Para and ortho chlorinated benzal chlorides can in turn be hydrolyzed to p-chlorobenzaldehyde or o-chlorobenzaldehyde which find use as an agricultural intermediate (p-chlorobenzaldehyde) or as an intermediate in preparing optical brightners (o-chlorobenzaldehyde) (2).

2. Physical Properties

Benzyl chloride [(chloromethyl)benzene, α-chlorotoluene], $C_6H_5CH_2Cl$, is a colorless liquid with a very pungent odor. Benzyl chloride vapors are irritating to the eyes and mucous membranes, and it is classified as a powerful lacrimator. The physical properties of pure benzyl chloride are given in Table 2 (3–8). Benzyl chloride is insoluble in cold water, but decomposes slowly in hot water to benzyl alcohol. It is miscible in all proportions at room temperature with most organic solvents. The flash point of benzyl chloride is 67°C (closed cup); 74°C (open cup); autoignition temperature is 585°C; lower flammability limit: 1.1% by volume in air. Its volume coefficient of expansion is 9.72×10^{-4}.

Benzal chloride (MW = 161.03) [(dichloromethyl)benzene, α,α-dichlorotoluene, benzylidene chloride], $C_6H_5CHCl_2$, is a colorless liquid with a pungent, aromatic odor. Benzal chloride is insoluble in water at room temperature but is miscible with most organic solvents.

Benzotrichloride (MW = 195.47) [(trichloromethyl)benzene, α,α,α-trichlorotoluene, phenylchloroform], $C_6H_5CCl_3$, is a colorless, oily liquid with a pungent odor. It is soluble in most organic solvents, but it reacts with water and alcohol.

Table 2. **Physical Properties of Benzyl Chloride, Benzal Chloride, and Benzotrichloride**

Property	Benzyl chloride	Benzal chloride	Benzotrichloride
mol wt	126.58	161.03	195.48
freezing point, °C	−39.2	−16.4	−4.75
boiling point, °C	179.4	205.2	220.6
density, kg/m^3	1113.5$^4_4$	1256$^{14}_{14}$	1374$^{20}_4$
	1104$^{15}_{15}$		
	1100$^{15}_{20}$		
refractive index, n^t_D	1.54124^{15}		
	1.5392^{20}	1.5502^{20}	1.55789^{20}
surface tension,	19.50$^{179.5}$	20.20$^{203.5}$	38.03^{20}
mN/m($=$ dyn/cm)	0.03765^{20}		
dipole moment^a, C·m	6.24×10^{-30}	6.9×10^{-30}	7.24×10^{-30}
diffusion of vapor in air, D_o, cm^2/s	0.066		
vapor density (air $=$ 1)	4.34		6.77
heat of combustion, kJ/mol^b	3708^c	3852^d	3684^d
specific heat at 25°C, (J/kg·K)^b	1444	1377	1206
heat of vaporization, kJ/mol^b	50.1^e	50.4^f	52^g
vapor pressure, °C at kPa^h			
0.13	22.0	35.4	45.8
0.67	47.8	64.0	73.7
1.33	60.8	78.7	87.6
5.33	90.7	112.1	119.8
8.00	100.5	123.6	130.0
13.3	114.2	138.3	144.3
26.7	134.0	160.7	165.6
53.3	155.8	187.0	189.2

^{a} In dilute benzene solution. To convert C·m to debye, divide by 3.336×10^{-30}.
^{b} To convert J to cal, divide by 4.184.
^{c} At constant volume.
^{d} At constant pressure.
^{e} At 25°C.
^{f} At 72°C.
^{g} At 80°C.
^{h} To convert kPa to mm Hg, multiply by 7.50.

For benzotrichloride the flash point is 127°C (Cleveland open cup) and the auto-ignition temperature is 211°C (9).

Binary azeotropic systems are reported for all three derivatives (10). The solubilities of benzyl chloride, benzal chloride, and benzotrichloride in water have been calculated by a method devised for compounds with significant hydrolysis rates (11).

3. Chemical Properties

The reactions of benzyl chloride, benzal chloride, and benzotrichloride may be divided into two classes: (1) reactions taking place on the side chain containing the halogen; and (2) reactions taking place on the aromatic ring.

3.1. Reactions of the Side Chain. Benzyl chloride is hydrolyzed slowly by boiling water and more rapidly at elevated temperature and pressure in the presence of alkalies (12). Reaction with aqueous sodium cyanide, preferably in the presence of a quaternary ammonium chloride, produces phenylacetonitrile [140-29-4] in high yield (13). The presence of a lower molecular-weight alcohol gives faster rates and higher yields. In the presence of suitable catalysts benzyl chloride reacts with carbon monoxide to produce phenylacetic acid [103-82-2] (14–16). Different catalysts in the presence of calcium hydroxide afford double carbonylation to phenylpyruvic acid [156-06-9] (17). Benzyl esters are formed by heating benzyl chloride with the sodium salts of acids, likewise benzyl ethers can be formed by reaction with sodium alkoxides. The ease of ether formation is improved by the use of phase-transfer catalysts (18) (see CATALYSIS, PHASE-TRANSFER).

The benzylation of a wide variety of aliphatic, aromatic, and heterocyclic amines has been reported. Benzyl chloride is converted into a mixture of mono-, di-, and tribenzylamines by reaction with ammonia. N-Phenylbenzyla-mine [103-32-2] results from the reaction of benzyl chloride with aniline. Reaction of benzyl chloride with tertiary amines yields quaternary ammonium salts; with trialkylphosphines, quaternary phosphonium salts; and with sulfides, sulfonium salts are formed.

Benzyl chloride readily forms a Grignard compound by reaction with magnesium in ether with the concomitant formation of substantial coupling product, 1,2-diphenylethane [103-29-7]. Benzyl chloride is oxidized first to benzaldehyde [100-52-7] and then to benzoic acid. Nitric acid oxidizes directly to benzoic acid [65-85-0]. Reaction with ethylene oxide produces the benzyl chlorohydrin ether, $C_6H_5CH_2OCH_2CH_2Cl$ (19). Benzylphosphonic acid [10542-07-1] is formed from the reaction of benzyl chloride and triethyl phosphite followed by hydrolysis (20).

Benzyl chloride reacts with alkali hydrogen sulfides, sulfides, and polysulfides to yield benzenethiol, dibenzyl sulfide, and dibenzyl polysulfide, respectively. With sodium cyanate it forms benzyl isocyanate (21).

Benzyl chloride reacts with benzene in the presence of a Lewis acid catalyst to give diphenylmethane [101-81-5]. It undergoes self-condensation to form polymeric oils and solids (22). With phenol, benzyl chloride produces a mixture of o- and p-benzylphenol.

Benzal chloride is hydrolyzed to benzaldehyde under both acid and alkaline conditions. Typical conditions include reaction with steam in the presence of ferric chloride or a zinc phosphate catalyst (23) and reaction at 100°C with water containing an organic amine (24). Cinnamic acid in low yield is formed by heating benzal chloride and potassium acetate with an amine as catalyst (25).

Benzotrichloride is hydrolyzed to benzoic acid by hot water, concentrated sulfuric acid, or dilute aqueous alkali. Benzoyl chloride [98-88-4] is produced by the reaction of benzotrichloride with an equimolar amount of water or an equivalent of benzoic acid. The reactions are catalyzed by Lewis acids such as ferric chloride and zinc chloride (26). Reaction of benzotrichloride with other organic acids or with anhydrides yields mixtures of benzoyl chloride and the acid chloride derived from the acid or anhydride (27). Benzotrifluoride [98-08-8] is formed by the reaction of benzotrichloride with anhydrous hydrogen fluoride under both liquid- and vapor-phase reaction conditions.

3.2. Aromatic Ring Reactions. In the presence of an iodine catalyst chlorination of benzyl chloride yields a mixture consisting mostly of the *ortho* and *para* compounds. More recently, use of catalytic amounts of zeolites as chlorination catalysts yielded *ortho* and *para* chlorination of benzyl cloride (28). With strong Lewis acid catalysts such as ferric chloride, chlorination is accompanied by self-condensation. Nitration of benzyl chloride with nitric acid in acetic anhydride gives an isomeric mixture containing about 33% *ortho*, 15% *meta*, and 52% *para* isomers (29); with benzal chloride, a mixture containing 23% *ortho*, 34% *meta*, and 43% *para* nitrobenzal chlorides is obtained.

Chlorosulfonation of benzotrichloride with chlorosulfonic acid (30) or with sulfur trioxide (29) gives *m*-chlorosulfonyl benzoyl chloride [4052-92-0] in high yield. Under more driving conditions the 3,5 disulfonic acid adduct can be formed which upon reaction with 2 equivalents of benzotrichloride forms the 3,5 bis-sulfonylchloride adduct of benzoyl chloride [37828-01-6]. This bis-sulfonyl chloride can then be chlorodesulfonated form 3,5-dichlorobenzoyl chloride [2905-62-6], an important agricultural chemical intermediate (32). Nitration with nitronium fluoroborate in sulfolane gives 68% *m*-nitro-benzotrichloride [709-58-0] along with 13% of the *ortho* and 19% of the *para* isomers (33).

Nitrobenzotrichloride is also obtained in high yield with no significant hydrolysis when nitration with a mixture of nitric and sulfuric acids is carried out below 30°C (34). 2,4-Dihydroxybenzophenone [131-56-6], an important component in many uv stabilizers is formed in 90% yield by the uncatalyzed reaction of benzotrichloride with resorcinol in hydroxylic solvents (35) or in benzene containing methanol or ethanol (36). Benzophenone derivatives are formed from a variety of aromatic compounds by reaction with benzotrichloride in aqueous or alcoholic hydrofluoric acid (37).

Benzotrichloride with zinc chloride as catalyst reacts with ethylene glycol to form 2-chloroethyl benzoate [7335-25-3] (38). Perchlorotoluene is formed by chlorination with a solution of sulfur monochloride and aluminum chloride in sulfuryl chloride (39).

4. Manufacture

Benzyl chloride is manufactured by the thermal or photochemical chlorination of toluene at 65–100°C (40). At lower temperatures the amount of ring-chlorinated by-products is increased. The chlorination is usually carried to no more than about 50% toluene conversion in order to minimize the amount of benzal chloride formed (see Table 1). Once toluene is recycled and the overchlorinated products are removed by distillation, overall yield based on toluene is more than 90%. Various materials, including phosphorus pentachloride, have been reported to catalyze the side-chain chlorination. These compounds and others such as amides also reduce ring chlorination by complexing metallic impurities (41).

Liquid-phase batch photochlorination of toluene to benzylchloride reaches a maximum conversion of about 70% after 1.1 moles of chlorine per mole of toluene have been consumed (42). Higher yields of benzyl chloride have been claimed: 80% for low temperature chlorination (43); 80–85% for light-catalyzed chlorination in the vapor phase (44), and 93.6% for continuous chlorination above 125°C in a column packed with glass rings (45).

In commercial practice, chlorination may be carried out either batchwise or continuously. Glass-lined or nickel reactors may be used. Because certain metallic impurities such as iron catalyze ring chlorination and self-condensation, their presence must be avoided. The crude product is purged of dissolved hydrogen chloride, neutralized with alkali, and distilled. Chlorine efficiency is high; muriatic acid made by absorbing the by-product hydrogen chloride in water is usually free of significant amounts of dissolved chlorine.

An 80% yield of benzyl chloride is obtained with sulfuryl chloride as chlorinating agent. Use of NaX-type zeolites are an effective means of directing chlorination to the side chain using sulfuryl chloride as the chlorinating agent (46) Yields of >70% of benzyl chloride are obtained by the zinc chloride-catalyzed chloromethylation of benzene, but formation of bis-chloromethyl ether presents a health hazard for this reaction pathway.

Benzyl chloride undergoes self-condensation relatively easily at high temperatures or in the presence of trace metallic impurities. The risk of decomposition during distillation is reduced by the use of various additives including lactams (47) and amines (48,49). Soluble organic compounds that are reported to function as stabilizers in low concentration include propylene oxide, DMF (50), arylamines (51), and triphenylphosphine (52).

Benzal chloride can be manufactured in 70% yield by chlorination with 2.0–2.2 moles of chlorine per mole of toluene. The benzal chloride is purified by distillation. Benzal chloride is also formed by the reaction of dichlorocarbene (:CCl$_2$) with benzene (53).

Further chlorination at a temperature of 100–140°C with ultraviolet light yields benzotrichloride. The chlorination is normally carried to a benzotrichloride content of greater than 95% with a low benzal chloride content. After purging with inert gas to remove hydrogen chloride, the crude product is utilized directly or purified by distillation. Under batch conditions chlorine efficiency during the latter stages of the chlorination is low. Product quality and chlorine efficiency can be improved by carrying out the chlorination continuously in a multistage system (54). Additives such as phosphorus trichloride are used to complex metallic impurities. Contaminants or reaction conditions that cause darkening and thereby reduce light penetration must be avoided if the chlorination is to be efficient (55). The radiation-initiated chlorination of toluene has also been investigated (56–60).

An understanding of competing reactions in the manufacturing process is important if by-products are to be minimized. Three competing reactions are possible under conditions of the reaction.

4.1. Free-radical Substitution of the Side Chain of Toluene

4.2. Addition to the Aromatic Ring

4.3. Electrophilic Substitution on the Aromatic Ring

An extensive kinetic study of the photochlorination of toluene in a continuous annular reactor has investigated the parameters that effect the product distribution from these reactions (42). Chlorination on the aromatic ring can occur by either addition followed by elimination of HCl or electrophilic aromatic substitution. Both reactions occur at low (40°C) temperature and are promoted by high concentration of chlorine. Electrophilic substitution is catalyzed by traces of metals like iron and aluminum. Formation of ring-chlorinated compounds is markedly increased by lowering the temperature to 40°C and chlorinating in the dark. These products contribute to a high boiling fraction that reduces the yield of side-chain chlorination products.

Free-radical chlorine substitution of the methyl group hydrogens is promoted by elevated temperature (80–130°C), a radical producing light source, and free-radical catalysts like peroxides. Oxygen inhibits the reaction. The ratio of benzyl to benzal to benzotrichloride depends on the ratio of chlorine to toluene in the reaction. From analyses of the product distribution for the free-radical chlorination at 100°C and irradiation with blue light, the relative rates are $k_1/k_2 = 5.9$ and $k_2/k_3 = 5.2$. Blue light (energy maximum at about 425 nm) gives a higher rate of chlorination than ultraviolet (about 370 nm) because it more effectively penetrates a solution containing free chlorine.

5. Handling and Shipment

As is the case during manufacture, contact with those metallic impurities that catalyze Friedel-Crafts condensation reactions must be avoided. The self-condensation reaction is exothermic and the reaction can accelerate producing a rapid buildup of hydrogen chloride pressure in closed systems.

Benzyl chloride is available in both anhydrous and stabilized forms. Both forms can be shipped in glass carboys, nickel and lined-steel drums, and nickel tank trucks and tank cars. Stabilized benzyl chloride can be shipped in unlined and lacquer-lined drums, and tank trucks or cars of construction other than nickel. Glass-lined tanks are the first choice for bulk storage of anhydrous benzyl chloride; lead-lined, nickel, or ceramic tanks can also be used.

Benzyl chloride is classified by DOT as chemicals NO1BN, poisonous, corrosive and a hazardous substance (100 lbs = 45.45 kg). Benzal chloride is classified as poisonous and a hazardous substance (5000 lbs = 2270 kg). Benzotrichloride is classified under DOT regulation as a corrosive liquid NOS and a hazardous substance (10 lbs = 4.5 kg). The Freight Classification Chemical NOI applies. It is shipped in lacquer-lined steel drums and nickel-lined tank trailers. Benzal chloride is handled in a similar fashion.

6. Economic Aspects

Plant capacities for the production of benzyl chloride in the United States is estimated to be 100 million lbs. per year in 1990; in Europe about 108 million lbs per year is produced. Solutia, with plants in Belgium (23,000 t/yr) and Bridgeport, New Jersey (40,000 t/yr), is the world's largest producer. Bayer in West Germany (20,000 t/yr) and Tessenderlo Chemie in Belgium (18,000 t/yr) are also principal producers. Japan does not have significant production of benzyl chloride. Total Western world production in 1990 was approximately 180 million lbs. The list price for benzyl chloride as of June 2000 was $0.69/lb.

Benzotrichloride is produced from total side-chain chlorination of toluene or of residual products from benzyl chloride production. In Western Europe, Bayer has the largest capacity (14,000 t/yr), and there are only two significant producers in the United States: Occidental Chemical in Niagara Falls, New York (20,000 t/yr), and Velsicol Chemical (11,000 t/yr). Total capacity in the Western world is 68,000 t/yr and production of benzotrichloride in 1988 was estimated at 31,500 t. The list price for benzotrichloride is $0.90/lb.

Benzyl chloride and butyl alcohol react with phthalic anhydride in one step to yield benzyl butyl phthalate [85-68-7], a plasticizer made by Monsanto and known by its trade name Santicizer 160.

7. Identification and Analysis

The side-chain chlorine contents of benzyl chloride, benzal chloride, and benzotrichlorides are determined by hydrolysis with methanolic sodium hydroxide followed by titration with silver nitrate. Total chlorine determination, including ring chlorine, is made by standard combustion methods (59). Several procedures for the gas chromatographic analysis of chlorotoluene mixtures have been described (60,61). Proton and ^{13}C nuclear magnetic resonance shifts, characteristic infrared absorption bands, and principal mass spectral peaks have been summarized including sources of reference spectra (62). Procedures for measuring trace benzyl chloride in air (63) and in water (64) have been described.

A gas chromatographic determination of benzotrichloride and related compounds in the work environment, after adsorption on a polymeric adsorbant and desorption with CCl_4 has been reported (65). Trace amounts of benzyl chloride, benzal chloride, and benzotrichloride in environmental samples can be analyzed by Method 8120 of *EPA Manual SW-846* with modifications (66).

8. Health and Safety Factors

Benzyl chloride is a severely irritating liquid and causes damage to the eyes, skin, and respiratory tract including pulmonary edema. Other possible effects of overexposure to benzyl chloride are CNS depression, liver, and heart damage. Table 3 lists some exposure limits.

Benzyl chloride induced a positive mutagenic response in the Ames Assay in strain TA 100 with and without rat liver S-9 metabolic activation. Benzyl

Table 3. **Toxicology of Side-Chain Chlorinated Toluenes**

	Benzyl chloride	Benzal chloride	Benzotrichloride
LD$_{50}$ (rat), mg/kg	1000^{a,b}		6000^{b,c}
LD$_{50}$ (mice), mg/kg		467^{d,e}	
LC$_{50}$ (mice, inhalation 2 h), ppm	80^{f}		
LC$_{50}$ (rat, inhalation), ppm	150^{f}	82^{g}	30^{f}

^{a} Administered subcutaneously in oil.
^{b} Slightly toxic.
^{c} Ref. 63.
^{d} Moderately toxic.
^{e} Ref. 64.
^{f} Ref. 65.
^{g} Ref. 66.

chloride also induced *in vitro* cellular transformation in Syrian hamster embryo cultures and DNA alkylation in several organs of the male mouse following iv administration. In summary, IARC states there is limited evidence that benzyl chloride is carcinogenic in experimental animals; epidemiological data were inadequate to evaluate carcinogenicity to humans (74).

Other toxicological effects that may be associated with exposure to benzyl chloride based on animal studies are skin sensitization and developmental embryo and/or fetal toxicity. A 1988 OSHA regulation has established a national occupational exposure limit for benzyl chloride of 5 mg/m^3 (1 ppm). Concentrations of 160 mg/m^3 (32 ppm) in air cause severe irritation of the eyes and respiratory tract (72). The recommended exposure limit is 1 ppm (15 min). Exposure monitoring protocols are described in OSHA documents.

Vapors of both benzal chloride and benzotrichloride are strongly irritating and lacrimatory. Reported toxicities appear in Table 3. Also, for benzotrichloride, the lowest published lethal dose (frog) is 2150 mg/kg (73), and the toxic dose level (inhalation rats) is 125 ppm/4 h (73).

For all three compounds, biological data relevant to the evaluation of carcinogenic risk to humans are summarized in the World Health Organization International Agency for Research on Cancer monograph which was updated in 1987 (75).

9. Uses

Nearly all uses and applications of benzyl chloride are related to reactions of the active halide substituent. More than two-thirds of benzyl chloride produced is used in the manufacture of benzyl butyl-phthalate, a plasticizer used extensively in vinyl flooring and other flexible poly(vinyl chloride) uses such as food packaging. Other significant uses are the manufacture of benzyl alcohol [100-51-6] and of benzyl chloride-derived quaternary ammonium compounds, each of which consumes more than 10% of the benzyl chloride produced. Smaller volume uses include the manufacture of benzyl cyanide [140-29-4], benzyl esters such as benzyl acetate [140-11-4], butyrate, cinnamate, and salicylate, benzylamine [100-46-9], and benzyldimethylamine [103-83-8], and *p*-benzylphenol [101-53-1].

In the dye industry benzyl chloride is used as an intermediate in the manufacture of triphenylmethane dyes (qv). First-generation derivatives of benzyl chloride are processed further to pharmaceutical, perfume, and flavor products.

Nearly all of the benzal chloride produced is consumed in the manufacture of benzaldehyde. Benzaldehyde (qv) is used in the manufacture of perfume and flavor chemicals, dyes, and pharmaceuticals. The principal part of benzotrichloride production is used in the manufacture of benzoyl chloride (see BENZOIC ACID). Lesser amounts are consumed in the manufacture of benzotrifluoride, as a dyestuff intermediate, and in producing hydroxybenzophenone ultraviolet light stabilizers. Benzotrifluoride is an important intermediate in the manufacture of herbicides, pharmaceuticals, antimicrobial agents, and the lampreycide, 4-nitro-3-(trifluoromethyl)phenol [88-30-2].

Benzyl-derived quaternary ammonium compounds are used widely as cationic surface-active agents and as germicides, fungicides, and sanitizers. Benzyl alcohol is used in a wide spectrum of applications including pharmaceuticals and perfumes, as a solvent, and as a textile dye assistant.

10. Derivatives

10.1. Ring-Substituted Derivatives. The ring-chlorinated derivatives of benzyl chloride, benzal chloride, and benzotrichloride are produced by the direct side-chain chlorination of the corresponding chlorinated toluenes or by one of several indirect routes if the required chlorotoluene is not readily available. Physical constants of the main ring-chlorinated derivatives of benzyl chloride, benzal chloride, and benzotrichloride are given in Table 4.

The 2- and 4-monochloro- and 2,4- and 3,4-dichlorobenzyl chloride, benzal chloride, and benzotrichlorides are manufactured by side-chain chlorination of the appropriate chlorotoluene. p-Chlorobenzotrichloride (1-chloro-4-trichloromethylbenzene) can be prepared by peroxide-catalyzed chlorination of p-toluenesulfonyl chloride or di-p-toluylsulfone (76). 2,4-Dichlorobenzotrichloride (1,3-dichloro-4-trichloromethylbenzene) is obtained by the chlorination of 2-chloro-4-chlorosulfonyltoluene (77).

3,4-Dichlorobenzyl chloride (1,2-dichloro-4-chloromethylbenzene) containing some 2,3-dichlorobenzyl chloride is produced by the chloromethylation of o-dichlorobenzene in oleum solution (78). Chlorination of 2-chloro-6-nitrotoluene at 160–185°C gives a mixture of 2,6-disubstituted benzal chloride and 2,6-dichlorobenzyl chloride (79).

The ring-chlorinated benzyl chlorides are used in the preparation of quaternary ammonium salts and as intermediates for pharmaceuticals and pesticides. p-Chlorobenzyl chloride is an intermediate in the manufacture of the rice herbicide, Saturn ((S-4-chlorobenzyl)-N,N-diethylthiolcarbamate [28249-77-6] (80). The o- and p-chlorobenzal chlorides (1-chloro-2-and 4-dichloromethylbenzenes) are starting materials for the manufacture of o- and p-chlorobenzaldehydes.

The o- and p-monochloro- and 2,4- and 3,4-dichlorobenzotrichlorides are intermediates in the manufacture of the corresponding chlorinated benzoic acids and benzoyl chlorides. Fluorination of the chlorinated benzotrichlorides

Table 4. **Physical Constants of the Main Ring-Chlorinated Derivatives of Benzyl Chloride, Benzal Chloride, and Benzotrichloride**

Benzene derivative	CAS Registry number	Common name	Mp,°C	Bp,°C	n_D^{20}	Density, kg/m³
1-chloro-2-(chloromethyl)	[611-19-8]	o-chlorobenzyl chloride	-17	217	1.5330	1270
1-chloro-3-(chloromethyl)	[620-20-2]	m-chlorobenzyl chloride		215–216[a]		1269.5
1-chloro-4-(chloromethyl)	[104-83-6]	p-chlorobenzyl chloride	31	222	1.5554	
1-chloro-2-(dichloromethyl)	[88-66-4]	o-chlorobenzal chloride		228.5	1.5670[b]	1399
1-chloro-3-(dichloromethyl)	[15145-69-4]	m-chlorobenzal chloride		235–237		
1-chloro-4-(dichloromethyl)	[13940-94-8]	p-chlorobenzal chloride		236[c]		
2,4-dichloro-1-(chloromethyl)	[94-99-5]	2,4-dichlorobenzyl chloride	-2.6	248	1.5761	1407
1,3-dichloro-2-(chloromethyl)	[2014-83-7]	2,6-dichlorobenzyl chloride	39–40	117–119[d]		
1,2-dichloro-4-(chloromethyl)	[102-47-6]	3,4-dichlorobenzyl chloride	37–37.5	241	1.5836	1412
1-chloro-2-(trichloromethyl)	[2136-89-2]	o-chlorobenzotrichloride	29.4	264.3		1519
1-chloro-3-(trichloromethyl)	[2136-81-4]	m-chlorobenzotrichloride		255	1.4461	1495
1-chloro-4-(trichloromethyl)	[5216-25-1]	p-chlorobenzotrichloride		245	1.4463	1495
1,3-dichloro-2-(dichloromethyl)	[81-19-6]	2,6-dichlorobenzal chloride		250		
1,2-dichloro-4-(dichloromethyl)	[56961-84-3]	3,4-dichlorobenzal chloride		257		1518
2,4-dichloro-1-(dichloromethyl)	[134-25-8]	2,4-dichlorobenzal chloride	47–48	155–159[e]		
1,2-dichloro-4-(trichloromethyl)	[13014-24-9]	3,4-dichlorobenzotrichloride	25.8	283.1	1.5886	1591

[a] At 100.4 kPa (753 mm Hg).
[b] At 16°C.
[c] At 100.7 kPa (755 mm Hg).
[d] At 1.87 kPa (14 mm Hg).
[e] At 2.67 kPa (20 mm Hg).

333

produces the chlorinated benzotrifluorides, intermediates in the manufacture of dinitroaniline and diphenyl ether herbicides (81).

2,6-Dichlorobenzal chloride is used in the manufacture of 2,6-dichlorobenzaldehyde and 2,6-dichlorobenzonitrile (82). With the exception of certain products used in the manufacture of herbicides, the volume of individual compounds produced is small, amounting to no more than several hundred tons annually for any individual compound.

10.2. Side-Chain Chlorinated Xylene Derivatives. Only a few of the nine side-chain chlorinated derivatives of each of the xylenes are available from direct chlorination. All three of the monochlorinated compounds, α-chloro-o-xylene [1-(chloromethyl)-2-methylbenzene [552-45-4], α-chloro-m-xylene (1-(chloromethyl)-3-methylbenzene [620-19-9]], and α-chloro-p-xylene [1-(chloromethyl)-4-methylbenzene [104-82-5]] are obtained in high yield from partial chlorination of the xylenes. 1,3-bis(Chloromethyl)benzene [626-16-4] can be isolated in moderate yield from chlorination mixtures (84,84).

The fully side-chain chlorinated products, 1,3-bis(trichloromethyl)benzene [881-99-1] and 1,4-bis(trichloromethyl)benzene [68-36-0], are manufactured by exhaustive chlorination of $meta$ and $para$ xylenes. For the $meta$ compounds, ring chlorination cannot be completely eliminated in the early stages of the reaction. The xylene hexachlorides are intermediates in the manufacture of the xylene hexafluorides and of iso- and terephthaloyl chloride [100-20-9] (see PHTHALIC ACIDS).

1-(Dichloromethyl)-2-(trichloromethyl)benzene [2741-57-3], the end product of exhaustive side-chain chlorination of o-xylene (85) is an intermediate in the manufacture of phthalaldehydic acid [119-67-5].

BIBLIOGRAPHY

"Benzyl Chloride, Benzal Chloride, and Benzotrichloride" under "Chlorine Compounds, Organic" in *ECT* 1st ed., Vol. 3, pp. 822–826 by R. L. Clark and C. P. Neidig, Heyden Chemical Corp.; "Benzyl Chloride, Benzal Chloride, and Benzotrichloride" under "Chlorocarbons and Chlorohydrocarbons" in *ECT* 2nd ed., Vol. 5, pp. 281–289, by H. Sidi, Heyden Newport Chemical Corp.; "Benzyl Chloride, Benzal Chloride, and Benzotrichloride" under "Chlorocarbons, Hydrocarbons (Benzyl)" in *ECT* 3rd ed., Vol. 5, pp. 828–837, by S. Gelfand, Hooker Chemical Corp.; "Benzyl Chloride, Benzal Chloride, and Benzotrichloride" under "Chlorocarbons, -Hydrocarbons (Benzyl Chloride) in *ECT* 4th ed., Vol. 6, pp. 113–126, by Henry C. Lin and Joseph F. Bieron, Occidental Chemical Corporation; "Benzyl Chloride, Benzal Chloride, and Benzotrichloride" in *ECT* (online), posting data: December 4, 2000, by Karl W. Seper, Occidental Chemical Corp.

CITED PUBLICATIONS

1. J. S. Ratcliffe, *Br. Chem. Eng.* **11**, 1535 (1966).
2. U.S. Pat. 5,118,816 (June 2, 1992). V. Kameswaran, et. al. (to American Cyanamid).
3. *Handbook of Chemistry and Physics*, 58th ed., CRC Press Inc., Cleveland, Ohio, 1977–1978, pp. C-522, 523, 527, 528, 738, D-198.

4. *International Critical Tables*, Vol. **5**, McGraw-Hill Book Co., New York, 1929, pp. 62, 111, 169.

5. R. R. Dreisbach, in *Advances in Chemistry Series*, American Chemical Society, Washington, D.C., 1955, 141–143.

6. A. L. McClellan, *Tables of Experimental Dipole Moments*, W. H. Freeman and Co., San Francisco, Calif., 1963, pp. 232, 237, 238, 243.

7. J. Timmermans and Mme. Hennant-Roland, *J. Chim. Phys.* **32**, 501 (1935).

8. D. R. Stull, *Ind. Engr. Chem.* **39**, 525 (1947).

9. Occidental Chemical Corp. MSDS; M7608, Feb. 19, 1991.

10. L. H. Horsley and co-workers, *Azeotropic Data III*, no. 116 in *Advances in Chemistry Series*, American Chemical Society, Washington, D.C., 1973.

11. K. Ohnishi and K. Tanabe, *Bull. Chem. Soc. Jpn.* **44**, 2647 (1971).

12. U.S. Pat. 3,557,222 (Jan. 19, 1971), H. W. Withers and J. L. Rose (to Velsicol Chemical Corp.).

13. Brit. Pat. 1,336,883 (Nov. 14, 1973), H. Coates, R. L. Barker, R. Guest, and A. Kent (to Albright & Wilson, Ltd.).

14. J. K. Stille and P. K. Wong, *J. Org. Chem.* **40**, 532 (1975).

15. Ger. Offen. 2,259,072 (June 20, 1974), M. E. Chahawi and H. Richtzenhain (to Dynamit Nobel AG).

16. Ger. Offen. 2,035,902 (Feb. 4, 1971), M. Foa, L. Cassar, and G. P. Chiusoli (to Montecatini Edison SPA).

17. U.S. Pat. 4,689,431 (Aug. 25, 1987), M. Tanaka and K. Oktsuka (to Nissan Chemical Industries, Ltd.).

18. H. H. Freedman and R. A. DuBois, *Tetrahedron Lett.* **38**, 3251 (1975).

19. Jpn. Kokai 75 62,942 (May 29, 1975), S. Komori.

20. Brit. Pat. 1,366,600 (Sept. 11, 1974), F. J. Harris and H. L. Brown (to Scottish Agric. Ind. Ltd.).

21. Ger. Offen. 2,449,607 (Apr. 30, 1975), Y. Inamoto and co-workers (to Kao Soap Co., Ltd.).

22. H. C. Haas, D. I. Livingston, and M. Saunders, *J. Polym. Sci.* **15**, 503 (1955).

23. U.S. Pat. 3,542,885 (Aug. 18, 1970), A. J. Deinet (to Tenneco Chemicals Inc.).

24. Jpn. Pat. 69 12,132 (June 2, 1969), H. Funamoto (to Kureha Chem. Ind. Co. Ltd.).

25. Jpn. Kokai 73 81,830 (Nov. 30, 1973), K. Shinoda and K. Kobayashi (to Kureha Chem. Ind. Co. Ltd.).

26. Jpn. Kokai 54 019929 (Feb. 15, 1979) (to Nikkei Kako, KK).

27. Jpn. Kokai 61 155350 (July 15, 1986) (to Ihara-Nikkei Kagaku).

28. A. P. Singh, et al. *Catal. Today* **49**, 245 (1999).

29. F. DeSarlo and co-workers, *J. Chem. Soc.*, B719 (1971).

30. U.S. Pat. 3,290,370 (Dec. 12, 1966), E. D. Weil and R. J. Lisanke (to Hooker Chemical Corp.).

31. U.S. Pat. 3,322,822 (May 30, 1967), S. Gelfand (to Hooker Chemical Corp.).

32. U.S. Pat. 3,686,301 (Aug. 22, 1972), S. Lawrence and H. Kirsch (to Rohm & Haas Co.).

33. G. Grynkiewicz and J. H. Ridd, *J. Chem. Soc.*, B716 (1971).

34. U.S. Pat. 3,182,091 (May 4, 1965), O. Scherer, H. Hahn, and N. Munch (to Farb. Hoeschst Akt.).

35. U.S. Pat. 3,769,349 (Oct. 30, 1973), M. Yukutomi, Y. Tanaka, S. Genda, and M. Kitauri (to Kyodo Chemical Co. Ltd.).

36. Ger. Offen. 2,208,197 (Aug. 30, 1973), B. Lachmann and H. J. Rosenkrantz (to Bayer AG).

37. Ger. Offen. 2,451,037 (Apr. 29, 1976), K. Eiglmeier (to Hoechst AG).

38. U.S. Pat. 3,050,549 (Aug. 21, 1962), S. Gelfand (to Hooker Chemical Corp.).

39. M. Ballester, C. Molinet, and J. Castaner, *J. Am. Chem. Soc.* **82**, 4254 (1960).

40. *Faith, Keyes, and Clark's Industrial Chemicals* 4th ed., John Wiley & Sons, Inc., New York, 1975, 145–148.
41. U.S. Pat. 2,695,873 (Nov. 30, 1954), A. J. Loverde (to Hooker Electrochemical Co.).
42. H. G. Haring and H. W. Knol, *Chem. Process. Eng.* **45**, 540, 619, 690 (1964); 46, 38 (1965).
43. G. Benoy and L. DeMayer, *Compt. Rend. 27th Congr. Intern. Chim. Ind.*, Brussels, Belgium, 1954; *Industrie Chim. Belg.* **20**, Spec. No. 160-2 (1955).
44. G. V. Asolkar and P. C. Guha, *J. Indian Chem. Soc.* **23**, 47 (1946).
45. A. Scipioni, *Ann. Chim. (Rome)* **41**, 491 (1951).
46. L. Delaude and P. Laszlo, *J. Org. Chem.* **55**, 5260 (1990).
47. U.S. Pat. 3,715,283 (Feb. 6, 1973), W. Bockmann (to Bayer Akt.).
48. Czech. Pat. 159,100 (June 15, 1975), J. Best and M. Soolek.
49. Brit. Pat. 1,410,474 (Oct. 15, 1975), C. H. G. Hands (to Albright and Wilson Ltd.).
50. Jpn. Kokai 73 05,726 (Jan. 24, 1972), N. Kato and Y. Sato (to Mitsui Toatsu Chemicals Inc.).
51. Jpn. Kokai 73 05,725 (Jan. 24, 1972), N. Kato and Y. Sato (to Mitsui Toatsu Chemicals Inc.).
52. U.S. Pat. 3,535,391 (Oct. 20, 1970), G. D. Kyker (to Velsicol Chemical Co.).
53. Brit. Pat. 1,390,394 (Apr. 9, 1975), A. D. Forbes, R. C. Pitkethly, and J. Wood (to Brit. Petrol. Co. Ltd.).
54. Ger. Offen. 2,152,068 (Apr. 26, 1973), W. Bockmann and R. Hornung; D.T. 2,227,337 (Aug. 28, 1975) (to Bayer AG).
55. Jpn. Kokai 76 08, 223 (Jan. 23, 1976), M. Fuseda and K. Ezaki (to Hodogaya Chemical Co. Ltd.).
56. J. Y. Yang, C. C. Thomas, Jr., and H. T. Cullinan, *Ind. Eng. Chem. Process Res. Develop.* **9**, 214 (1970).
57. H. T. Cullinan, Jr. and co-workers, in Ref. 47, p. 222.
58. B. F. Ives, H. T. Cullinan, Jr., and J. Y. Yang, *Nucl. Technol.* **18**, 29 (1973).
59. W. Kirsten, *Anal. Chem.* **25**, 74 (1953).
60. D. A. Solomons and J. S. Ratcliffe, *J. Chromatog.* **76**, 101 (1973).
61. R. Ramakrishnan and N. Subramanian, *J. Chromatog.* **114**, 247 (1975).
62. J. G. Grasselli and W. M. Richey, eds., *Atlas of Spectral Data and Physical Constants for Organic Compounds*, 2nd ed., Vol. **IV**, CRC Press Inc., Cleveland, Ohio, 1975.
63. B. B. Baker, Jr., *J. Am. Ind. Hyg. Assoc.* **35**, 735 (1974).
64. G. A. Junk and co-workers, *J. Chromatog.* **99**, 745 (1974).
65. H. Matsushita and S. Kanno, *Ind. Health* **17**, 199–206, (1979).
66. V. Lopez-Avila, N. S. Dodhiwala, J. Milones, and W. F. Beckert, *J. Assoc. Off. Anal. Chem.* **72**, 593–602 (1989).
67. N. I. Sax, *Dangerous Properties of Industrial Materials*, 4th ed., Van Nostrand Reinhold Co., New York, 1975.
68. V. V. Stankevich and V. I. Osetrov, *Gigiena i Fisiol. Tr. Proizv. Toksikol., Klinika Prof. Zabolevanii*, 96 (1963).
69. *IARC Monogr. Eval. Carcinog. Risk Chem. Man 11*, 217–223 (1976); *Toxbib.* **77**, 50224 (1977).
70. T. V. Mikhailova, *Gig. Tr. Prof. Zabol* **8**, 14 (1964).
71. *IARC Monogr. Eval. Carcinog. Risk Chem. Man* **29**, 59 (1982).
72. W. F. von Oettingen, *The Halogenated Aliphatic, Olefinic, Cyclic, Aromatic and Aliphatic-Aromatic Hydrocarbons including the Halogenated Insecticides, their Toxicity and Potential Dangers*, DHEW (PHS) Publication No. 414, Washington D.C., U.S. Government Printing Office, 1955, 300–302.
73. H. E. Christensen, ed., *Registry of Toxic Effects of Chemical Substances*, U.S. Dept. of Health, Education, and Welfare, Rockville, Md., 1976.

74. *Code of Federal Regulations* Part 1910. 1000 Occupational and Health Standards – Limits for Air contaminatns, 1998.

75. *IARC Monograph* **29** *Suppl.* 7 (1982).

76. Jpn. Kokai 75 25,534 (Mar. 18, 1975), K. Kobayashi, N. Ishimo, and T. Nobeoka (to Fuso Chemical Co. Ltd.).

77. U.S. Pat. 3,230,268 (Jan. 18, 1966), K. Kobayashi and N. Ishino (to Fuso Chemical Co. Ltd.).

78. Brit. Pat. 951,302 (Mar. 4, 1964), (to Monsanto Canada Ltd.).

79. Ger. Pat. 1,237,552 (Mar. 30, 1967), J. T. Hackmann, J. Yates, T. J. Wilcox, P. T. Haken, and D. A. Wood (to Shell Research Ltd.).

80. U.S. Pat. 3,914,270 (Oct. 21, 1975), K. Makoto, H. Kamata, and K. Masuro (to Kumiai Chem. Ind. Co. Ltd.).

81. F. M. Ashton and A. S. Crafts, *Mode of Action of Herbicides*, John Wiley & Sons, New York, 1973, 10–24, 438–448.

82. U.S. Pat. 3,458,560 (July 29, 1969), R. A. Carboni (to E. I. du Pont de Nemours & Co., Inc.).

83. U.S. Pat. 2,994,653 (Apr. 27, 1959), G. A. Miller (to Diamond Alkali Co.).

84. E. Clippinger, *ACS Petrol. Div. Prep.* **15**(1), B 37 (1970).

85. Ger. Offen. 2,535,969 (Feb. 17, 1977), P. Riegger, H. Richtzenhain, and G. Zoche (to Dyanmit Nobel AG).

Karl W. Seper
Occidental Chemical Corp.

CHLOROTOLUENES, RING

1. Introduction

The ring-chlorinated derivatives of toluene form a group of stable, industrially important compounds. Many chlorotoluene isomers can be prepared by direct chlorination. Other chlorotoluenes are prepared by indirect routes involving the replacement of amino, hydroxyl, chlorosulfonyl, and nitro groups by chlorine and the use of substituents, such as nitro, amino, and sulfonic acid, to orient substitution followed by their removal from the ring.

The first systematic study of the reaction of chlorine with toluene was carried out in 1866 by Beilstein and Geitner. During the next 40 years, many studies were performed to isolate and identify the various chlorination products (1). During the early 1930s, Hooker Electrochemical Co. (Hooker Chemicals & Plastics Corp.) and the Heyden Chemical Corp. (Tenneco) began the manufacture of chlorotoluenes. Hooker Electrochemical Co. was later acquired by Occidental Petroleum Corp. and became the Occidental Chemical Corp. In the mid-1970s, Heyden exited chlorotoluenes production; Occidental exited the buisness in 2000. The current world producers are: Bayer (Germany), Tessenderlo (formerly Enichem in Italy), Ihara, Hodogaya (both in Japan), Zhuzhou, and Danyang (both in China).

Mono- and dichlorotoluenes are used chiefly as chemical intermediates in the manufacture of pesticides, dyestuffs, pharmaceuticals, and peroxides, and as solvents. Total annual production was limited prior to 1960 but has expanded greatly since that time. Chlorinated toluenes are produced in Germany, Japan, Italy and China. Since the number of manufacturers is small and much of the production is utilized captively, statistics covering production quantities are not available. Worldwide annual production of o- and p-chlorotoluene is estimated at several tens of thousands of metric tons. Yearly productions of polychlorotoluene are in the range of 100–1000 tons.

2. Monochlorotoluenes

2.1. Physical Properties.
o-Chlorotoluene [95-49-8] (1-chloro-2-methylbenzene, OCT) is a mobile, colorless liquid with a penetrating odor similar to chlorobenzene. It is miscible in all proportions with many organic liquids such as aliphatic and aromatic hydrocarbons, chlorinated solvents, lower alcohols, ketones, glacial acetic acid, and di-n-butylamine; it is insoluble in water, ethylene and diethylene glycols, and triethanolamine.

p-Chlorotoluene [106-43-4] (1-chloro-4-methylbenzene, PCT) and m-chlorotoluene [108-41-8] (1-chloro-3-methylbenzene, MCT) are mobile, colorless liquids with solvent properties similar to those of the ortho isomer.

Ortho and p-chlorotoluene form binary azeotropes with various organic compounds including alcohols, acids, and esters (2). Oxygen indexes, the minimum percentage of oxygen in an oxygen-nitrogen atmosphere required to sustain combustion after ignition, for the chlorotoluene isomers are ortho 19.2, meta 19.7, and para 19.1 (3). Ortho and p-chlorotoluene form stable ionic complexes with antimony pentachloride (4). They also form complexes with a number of organometallic derivatives, such as those of chromium (5), cobalt (6), iron (7), etc, many of which have synthetic utility. Physical properties of the monochlorotoluene isomers, mol wt 126.59, appear in Table 1 (8–13).

2.2. Chemical Properties.
The monochlorotoluenes are stable to the action of steam, alkalies, amines, and hydrochloric and phosphoric acids at moderate temperatures and pressures. Three classes of reactions, those involving the aromatic ring, the methyl group, and the chlorine substituent, are known for monochlorotoluenes.

Reactions of the Aromatic Ring.   Ring chlorination of o-chlorotoluene yields a mixture of all four possible dichlorotoluenes, the 2,3-, 2,4-, 2,5-, and 2,6-isomers as shown in equation 1 (14).

$$(1)$$

Table 1. **Physical Properties of the Monochlorotoluenes, C_7H_7Cl**

Property	Isomer		
	Ortho	Meta	Para
mp, °C	−35.6	−47.8	7.5
bp, °C	159.2	161.7	162.4
flash point, °C	47	47	49
density[a], kg/m^3			
20°C	1082.5	1072.2	1069.7
25°C	1077.6		1065.1[24.4]
30°C	1072.7		
refractive index[a], n_D^t			
20°C	1.52680	1.5214[19]	1.5211
25°C	1.52221		1.5193[24.4]
surface tension[a], mN/m (= dyn/cm)	334.4[20]		322.4[25]
	323.3[30]		292.2[30]
dielectric constant at 20°C	4.73	5.55	6.20
viscosity (dynamic), mPa·s (= cP)			0.09
dipole moment, C·m[b]	4.80×10^{-30}	5.97×10^{-30}	
heat of vaporization, kJ/mol[c]	43.01	42.18	42.475
vapor density (air = 1)			4.37
vapor pressure, °C at kPa[d]			
0.13	5.4	4.8	5.5
1.3	43.2	43.2	43.8
5.3	72.0	73.0	73.5
13.3	94.7	96.3	96.6
53.3	137.1	139.7	139.8

[a] Superscript indicates temperature.
[b] To convert C·m to debye, divide by 3.336×10^{-30}.
[c] To convert kJ to kcal, divide by 4.184.
[d] To convert kPa to mm Hg, multiply by 7.5.

The principal isomer, 2,5-dichlorotoluene, constitutes up to 60% of the product mixture (15,16). Similarly, nitration of o-chlorotoluene produces a mixture of the four corresponding nitrochlorotoluene isomers. Nitration of p-chlorotoluene gives a mixture of 66% 4-chloro-2-nitrotoluene [89-59-8] and 34% of 4-chloro-3-nitrotoluene [89-60-1], $C_7H_6ClNO_2$, (17). Chlorosulfonation of o-chlorotoluene produces 2-chloro-5-chlorosulfonyltoluene [6291-02-7], (4-chloro-3-methylbenzene-sulfonyl chloride), $C_7H_6Cl_2O_2S$, as the principal product (18). Sulfonation of p-chlorotoluene with 20% oleum gives the 2-sulfonic acid derivative in 68% yield (19). Trifluoromethylation of monochlorotoluenes has been achieved by reaction with carbon tetrachloride and hydrogen fluoride (20). Chloromethylation of o-chlorotoluene gives 2-chloro-4-chloromethyltoluene [2719-40-6] as the sole product. With p-chlorotoluene, a mixture of 4-chloro-2-chloromethyltoluene [34060-72-5] and 4-chloro-3-chloromethyltoluene [34896-68-9], $C_8H_8Cl_2$, is formed in a 63:37 ratio, respectively (21).

Reactions of the Methyl Group. Monochlorotoluenes are widely used to synthesize compounds derived from reactions of the methyl group. Chlorination under free-radical conditions leads successively to the chlorinated benzyl, benzal, and benzotrichloride derivatives (see CHLOROCARBONS AND CHLOROHYDROCARBONS—BENZYL CHLORIDE, BENZAL CHLORIDE, AND BENZOTRICHLORIDE). Oxidation to form chlorinated benzaldehydes and benzoic acids can be performed under both liquid- and vapor-phase conditions (22,23). Catalytic ammoxidation under vapor-phase conditions with oxygen and ammonia produces chlorobenzonitriles (24). Reaction of *p*-chlorotoluene with cyanogen chloride at 650–700°C gives *p*-chlorophenylaceto-nitrile [140-53-4], C_8H_6ClN, as shown in equation 2 (25). Side-chain bromination of *p*-chlorotoluene by bromine catalyzed by lanthanum triacetate is a facile process (26).

$$(2)$$

Halogen Reactions. Hydrolysis of chlorotoluenes to cresols has been effected by aqueous sodium hydroxide. Both displacement and benzyne forma-tion are involved (27,28). *o*-Chlorotoluene reacts with sodium in liquid ammonia to afford a mixture of 67% of *o*-toluidine [95-53-4] and 33% of *m*-toluidine [108-44-1], C_7H_9ClN, as shown in equation 3 (29).

$$(3)$$

With hydrogen sulfide at 500–600°C, monochlorotoluenes form the correspond-ing thiophenol derivatives (30). In the presence of palladium catalysts and car-bon monoxide, monochlorotoluenes undergo carbonylation at 150–300°C and 0.1–20 MPa (1–200 atm) to give carboxylic acids (31). Oxidative coupling of *p*-chlorotoluene to form 4,4′-dimethylbiphenyl can be achieved in the presence of an organonickel catalyst, generated *in situ*, and zinc in dipolar aprotic solvents such as dimethylacetamide (32,33). An example is shown in equation 4.

$$(4)$$

Dehalogenation of monochlorotoluenes can be readily effected with hydrogen and noble metal catalysts (34). Conversion of *p*-chlorotoluene to *p*-cyanotoluene is accomplished by reaction with tetraethylammonium cyanide and zero-valent Group (VIII) metal complexes, such as those of nickel or palladium (35). The reaction proceeds by initial oxidative addition of the aryl halide to the zerovalent metal complex, followed by attack of cyanide ion on the metal and reductive elim-ination of the aryl cyanide. *p*-Methylstyrene is prepared from *p*-chlorotoluene by a vinylation reaction using ethylene as the reagent and a catalyst derived from zinc, a triarylphosphine, and a nickel salt (36).

2.3. Preparation. Monochlorotoluenes have been prepared by chlorinating toluene with a wide variety of chlorinating agents, catalysts, and reaction conditions. The ratio of ortho and para isomers formed can vary over a wide range. Particular attention has been given to studies aimed at increasing the para isomer content owing to its greater commercial significance. The meta isomer can be prepared by either indirect means since only a small amount, <1%, is formed by direct chlorination or by isomerization of the ortho isomer (97). Isolation may then be achieved by selective absorption/desorbtion with various zeolites. (98)

Chlorinations with Elemental Chlorine. Reaction of toluene with chlorine in the presence of certain Lewis acid catalysts including the chlorides of aluminum, tin, titanium, and zirconium give monochlorotoluene mixtures that contain more than 70% of the ortho isomer (37,38). A number of catalyst systems have been developed to enhance the formation of p-chlorotoluene in toluene chlorination. Monochlorotoluenes containing 45–55% of the p-isomer are obtained through the use of certain specific metal sulfides or cocatalyst systems consisting of specific metal salts and sulfur, inorganic sulfides, or divalent sulfur compounds with or without other functional groups (39–45). A growing number of heterogeneous processes that employ zeolite-type catalysts for chlorination of aromatics have been discovered (see CATALYSIS). The majority of these catalysts are synthetic zeolites, specifically the L-type zeolites (46). It is sometimes possible to achieve highly regioselective chlorination of arenes, such as toluene, by the use of certain specific types of zeolites. A catalyst system comprised of TSZ-506, which is a synthetic zeolite, and monochloroacetic acid affords p-chlorotoluene with a selectivity of 75% relative to the ortho isomer at an operating temperature of 70°C (47). One common problem during zeolite-catalyzed chlorination is the structural breakdown of the zeolite lattice because of reaction with hydrogen chloride liberated in the reaction. However, modifications in the synthetic procedure have enabled the preparation of newer types of zeolites that are more resistant to structural deterioration (48).

Noncatalytic ring chlorination of toluene in a variety of solvents has been reported. Isomer distributions vary from approximately 60% ortho in hydroxylic solvents, eg, acetic acid, to 60% para in solvents, eg, nitromethane, acetonitrile, and ethylene dichloride (49,50). Reaction rates are relatively slow and these systems are particularly appropriate for kinetic studies.

Chlorination with Other Reagents. Chlorotoluenes can also be obtained in good yields by the reaction of toluene with stoichiometric proportions of certain Lewis acid chlorides such as iron(III) chloride, as the chlorinating agent (51). Generally, the product mixture contains p-chlorotoluene as the principal component. Several modifications have been proposed to improve product yields (52,53).

Toluene chlorination has also been effected with hydrogen chloride as the chlorinating agent. The reaction is catalyzed by nitric acid under aqueous conditions to give a good conversion and yield on monochlorotoluenes (54). Oxychlorination of toluene with oxygen and hydrogen chloride in the vapor phase over supported copper and palladium catalysts yields chlorotoluene mixtures containing up to 60% of *p*-chlorotoluene along with varying amounts of side-chain chlorinated products (55,56).

Other methods for preparing *p*-chlorotoluene include α-elimination from an organotellurium(IV) halide (57), palladium-catalyzed decarbonylation of 4-methylbenzoyl chloride (58), and desulfonylation of *p*-toluenesulfonyl chloride catalyzed by chlorine (59) or chlorotris(triphenylphosphine)rhodium (60).

Pure monochlorotoluene isomers are prepared by diazotization of the corresponding toluidine isomers followed by reaction with copper(I) chloride (Sandmeyer reaction). This is the preferred method of obtaining *m*-chlorotoluene.

Significant yields of *m*-chlorotoluene have been reported by reacting 3-methyl-phenol with phenylphosphorous pentachloride at 160°C (99). Also, *m*-chlorotoluene as a co-product has been reported by reacting trifluoromethyl benzene in a reagent system of nitryl chloride and titanium tetrachloride (100).

The rate of chlorination of toluene relative to that of benzene is about 345 (61). Usually, chlorination is carried out at temperatures below 70°C with the reaction proceeding at a profitable rate even at 0°C. The reaction is exothermic with ca 139 kJ (33 kcal) of heat produced per mole of monochlorotoluene formed. Chlorine efficiency is high, and toluene conversion to monochlorotoluene can be carried to about 90% with the formation of only a few percent of dichlorotoluenes. In most catalyst systems, decreasing temperatures favor formation of increasing amounts of *p*-chlorotoluene. Concentrations of required catalysts are low, generally on the order of several tenths of a percent or less.

Only trace amounts of side-chain chlorinated products are formed with suitably active catalysts. It is usually desirable to remove reactive chlorides prior to fractionation in order to minimize the risk of equipment corrosion. The separation of *o*- and *p*-chlorotoluenes by fractionation requires a high efficiency, isomer-separation column. The small amount of *m*-chlorotoluene formed in the chlorination cannot be separated by fractionation and remains in the *p*-isomer fraction. The toluene feed should be essentially free of paraffinic impurities that may produce high boiling residues that foul heat-transfer surfaces. Trace water contamination has no effect on product composition. Steel can be used as construction material for catalyst systems containing iron. However, glass-lined equipment is usually preferred and must be used with other catalyst systems.

Both batch and continuous processes are suitable for commercial chlorination. The progress of the chlorination is conveniently followed by specific gravity measurements.

2.4. Handling and Shipment. Monochlorotoluenes are shipped in bulk in steel tank cars and tank trucks. Drum shipments are made using lined or unlined steel drums. Aluminum tanks can be used to store only acid-free material. Under DOT regulations, for transport of over 415 L (110 gal) of monochlorotoluenes, freight classification is combustible liquid NOS, and for truck transport, chemical NOI. The storage vessels are vented to a safe atmosphere and should be protected with suitable diking. Protection against static charge is essential when transferring material. Suitable ventilation should be provided and sources of ignition avoided as the vapor forms flammable mixtures with air.

2.5. Identification and Analysis. A number of analytical methods have been developed for the determination of chlorotoluene mixtures by gas chromatography. These are used for determinations in environments such as air near

Table 2. **Toxicity Parameters for Monochlorotoluenes**

Parameter	Ortho	Para
LD_{50} (rat), mg/kg	2350^a	2100^b
TLV^c, ppm	50	
PEL^d, ppm	50	

[a] Ref. 69.
[b] Ref. 70.
[c] ACGIH, 259 mg/m³, 8 h TWA (71).
[d] OSHA, 250 mg/m³, 8 h TWA.

industry (62) and soil (63). Liquid crystal stationary columns are more effective in separating *m*- and *p*-chlorotoluene than conventional columns (64). Prepacked columns are commercially available. Zeolites have been examined extensively as a means to separate chlorotoluene mixtures (see MOLECULAR SIEVES). For example, a Y-type zeolite containing sodium and copper has been used to separate *m*-chlorotoluene from its isomers by selective absorption (65). The presence of benzylic impurities in chlorotoluenes is determined by standard methods for hydrolyzable chlorine. Proton (66) and carbon-13 chemical shifts, characteristic IR absorption bands, and principal mass spectral peaks are available along with sources of reference spectra (67).

2.6. Health and Safety Factors. Inhalation of high concentrations of monochlorotoluenes will cause symptoms of central nervous system depression. Inhalation studies produced an LC_{50} (rat, 4 h) of 7119 ppm for *o*-chlorotoluene (68). *o*- and *p*-Chlorotoluene are both considered moderately toxic by ingestion (Table 2). A study of the relationship between the electronic structure and toxicity parameters for a series of mono-, di-, and tri-chlorotoluenes has been reviewed (72). A thin-layer chromatographic method has been developed to assess the degree of occupational exposure of workers to chlorotoluenes by determining *p*-chlorohippuric acid [13450-77-6], $C_9H_8ClNO_3$, (*N*-(4-chlorobenzoyl)glycine) in urine samples (73). Health and environmental issues related to chlorinated toluenes are included in the Hazardous Substances Data Bank (101).

A study to isolate and examine the genetic characteristics of bacteria that metabolize chlorotoluenes, such as OCT, PCT, and 2,6-dichlorotoluene, has been reported (74). Two products were isolated from a study of the metabolism of PCT by *Pseudomonas putida*: (+)-*cis*-4-chloro-2,3-dihydroxy-1-methylcyclohex-4,6-diene and 4-chloro-2,3-dihydroxy-1-methylbenzene (75). Enzymatic dehydrogenation of the former compound to the latter was also demonstrated.

3. Higher Chlorotoluenes

3.1. Dichlorotoluenes. There are six possible dichlorotoluene isomers, $C_7H_6Cl_2$, (mol wt 161.03) all of which are known. Physical properties of the dichloro- and other higher chlorotoluenes are given in Table 3.

2,4-Dichlorotoluene (2,4-dichloro-1-methylbenzene) constitutes 80–85% of the dichlorotoluene fraction obtained in the chlorination of PCT with antimony trichloride (76) or zirconium tetrachloride (77) catalysts. It is separated from

Table 3. **Physical Properties of the Higher Chlorotoluenes**

Toluene	CAS Registry number	Mp, °C	Bp, °C	n_{D}^{t}	Density at 20°C, kg/m³
2,3-dichloro	[32768-54-0]	5	208.3	1.5511^{20}	
2,4-dichloro	[95-73-8]	−13.5	201.1	1.5480^{22}	1249.8
2,5-dichloro	[19398-61-9]	5	201.8	1.5449^{20}	1253.5
2,6-dichloro	[118-69-4]		200.6	1.5507^{20}	1268.6
3,4-dichloro	[95-75-0]	−15.3	208.9	1.5471^{20}	1256.4
3,5-dichloro	[25186-47-4]	26	201.2	1.5438^{20}	
2,3,4-trichloro	[7359-72-0]	43−44	244		
2,3,5-trichloro	[56961-86-5]	45−46	229−231		
2,3,6-trichloro	[2077-46-5]	45−46	118^{a}		
2,4,5-trichloro	[6639-30-1]	82.4	$229−230^{b}$		
2,4,6-trichloro	[23749-65-7]	38			
3,4,5-trichloro	[21472-86-6]	45−45.5	$246−247^{c}$		
2,3,4,5-tetrachloro	[1006-32-2]	98.1			
2,3,4,6-tetrachloro	[875-40-1]	92	266−276		
2,3,5,6-tetrachloro	[1006-31-1]	93−94			
pentachloro	[877-11-2]	224.5−225.5	301		

^{a} At 2.4 kPa (18 mm Hg).
^{b} At 95.4 kPa (716 mm Hg).
^{c} At 102.4 kPa (768 mm Hg).

3,4-dichlorotoluene (1,2-dichloro-4-methylbenzene), the principal contaminant, by distillation. Chlorination of OCT with sulfuryl chloride gives mainly 2,4-dichlorotoluene and small amounts of the 2,3 isomer (78). 2,5-Dichlorotoluene (1,3-dichloro-2-methylbenzene) is formed in up to 60% yield in the sulfide-cocatalyzed chlorination of OCT. Purification by recrystallization gives 99% pure product (15,16).

Chlorination of OCT with chlorine at 90°C in the presence of L-type zeolites as catalyst reportedly gives a 56% yield of 2,5-dichlorotoluene (79). Pure 2,5-dichlorotoluene is also available from the Sandmeyer reaction on 2-amino-5-chlorotoluene. 3,4-Dichlorotoluene (1,2-dichloro-4-methylbenzene) is formed in up to 40% yield in the chlorination of PCT catalyzed by metal sulfides or metal halide−sulfur compound cocatalyst systems (80).

2,3-Dichlorotoluene (1,2-dichloro-3-methylbenzene) is present in about 10% concentration in reaction mixtures resulting from chlorination of OCT. It is best prepared by the Sandmeyer reaction on 3-amino-2-chlorotoluene.

2,6-Dichlorotoluene (1,3-dichloro-2-methylbenzene) is prepared from the Sandmeyer reaction on 2-amino-6-chlorotoluene. Other methods include ring chlorination of p-toluenesulfonyl chloride followed by desulfonylation (81), and chlorination and dealkylation of 4-tert-butyltoluene (82) or 3,5-di-tert-butyltoluene (83,84). Isomerization (102) and purification (103) of 2,6-DCT, prepared as a mixture with 2,3-,2,4-, and 2,5-DCT from the FeCl₃ catalyzed chlorination of OCT, can be achieved by the use of zeolites, in a fashion similar to MCT, above.

3.2. Trichlorotoluenes. The chlorination of toluene and o- and p-chlorotoluenes produces a mixture of trichlorotoluenes, (C₇H₅Cl₃, (mol wt 195.48): the 2,3,6-isomer (1,2,4-trichloro-3-methylbenzene) and 2,4,5-trichlorotoluene (1,2,4-trichloro-5-methylbenzene) containing small amounts of 2,3,4-trichlorotoluene

(1,2,3-trichloro-4-methylbenzene) and 2,4,6-trichlorotoluene (1,3,5-trichloro-2-methylbenzene). When toluene is chlorinated in the presence of iron(III) chloride catalyst, a mixture containing nearly equal amounts of 2,4,5- and 2,3,6-trichlorotoluenes is produced (38,85). Chlorination of OCT yields a mixture containing >60% of 2,3,6-trichlorotoluene (86). Reaction of p-toluenesulfonic acid with chlorine and antimony trichloride in chloroform and then sulfuric acid at reflux affords 2,3,6-trichlorotoluene in 89% yield (87). Metal sulfide-catalyzed chlorination of PCT gives trichlorotoluene fractions containing more than 75% of the 2,4,5-isomer (eq. 5) (88). The other chlorotoluenes are available from the Sandmeyer reaction on the corresponding amines. A gas chromatographic study has been conducted to determine the isomer selectivity of stationary phases of different polarity with respect to various chlorotoluenes including the trichlorotoluene isomers (89).

$$(5)$$

Gas-phase ammoxidation of trichlorotoluenes in the presence of catalyst affords the corresponding benzonitrile derivatives (90). In a 28-day feeding study, 2,3,6-trichlorotoluene showed only mild toxicological changes when administered to rats (91).

3.3. Tetra- and Pentachlorotoluenes.
2,3,4,6-Tetrachlorotoluene, $C_7H_4Cl_4$ (mol wt 229.93) (1,2,3,5-tetrachloro-4-methylbenzene), is prepared from the Sandmeyer reaction on 3-amino-2,4,6-trichlorotoluene. 2,3,4,5-Tetrachlorotoluene (1,2,3,4-tetrachloro-5-methylbenzene) is the principal isomer in the further chlorination of 2,4,5-trichlorotoluene. Exhaustive chlorination of p-toluenesulfonyl chloride, followed by hydrolysis to remove the sulfonic acid group yields 2,3,5,6-tetrachlorotoluene (1,2,4,5-tetrachloro-3-methylbenzene) in good yield (92). Pentachlorotoluene (pentachloromethylbenzene), $C_7H_3Cl_5$ (mol wt 264.37), is formed in 90% yield by the ferric chloride-catalyzed chlorination of toluene in carbon tetrachloride or hexachlorobutadiene solution (93). Oxidation of pentachlorotoluene with excess sulfur trioxide, followed by hydrolysis of the intermediate pentachlorobenzyl disulfooxonium hydroxide inner salt produces pentachlorobenzyl alcohol in 91% yield (94). Gas chromatographic separation selectivities of stationary phases of different polarities toward tetrachlorotoluene isomers and pentachlorotoluene have been examined (89).

4. Uses

Chlorotoluenes are used as intermediates in the pesticide, pharmaceutical, peroxide, dye, and other industries. Many side chain-chlorinated derivatives are converted to end products. p-Chlorotoluene is used primarily in the manufacture of p-chlorobenzotrifluoride [98-56-6], a key intermediate in dinitroaniline and diphenyl ether herbicides (95). Other applications include manufacture of *p*-chlorobenzyl chloride, *p*-chlorobenzaldehyde, *p*-chlorobenzoyl chloride, *p*-chlorobenzoic acid, and 2,4- and 3,4-dichlorotoluenes. *p*-Chlorotoluene is an intermediate for a novel class of polyketone polymers (96).

Mono and di-chlorotoluenes have been used as a solvent/desorbent to separate dichlorobenzenes using zeolites (104–107). Separation of 3,5-Dichlorocumene from its isomeric mixture in pure form was achieved by using zeolites and subsequent desorption from the zeolites was effected by using mono or dichlorotouene(108), *o*-Chlorotoluene use as a solvent was identified in the preparation of hydroxynaphthoic acid arylides (109).

Chlorotoluene isomer mixtures, especially those containing a relatively high amount of *o*-chlorotoluene, are widely used as solvents in industry for such purposes as metal-cleaning formulations, railroad industrial cleaners, diesel fuel additives, carbon removal procedures, paint thinners, and agricultural chemicals. Halso 99 and Halso 125 are examples of such solvents. Separation of methanol and methylcarbonate by distillation was effected by *o*-Chlorotoluene as an extractant(110)

2,4-Dichlorotoluene is an intermediate for manufacture of herbicides. It is also used to obtain 2,4-dichlorobenzyl chloride and 2,4-dichlorobenzoyl chloride. 2,4-Dichlorotoluene has been used as a solvent in the preparation of trifluoromethylpyridine derivatives(111). 2,6-Dichlorotoluene is applied as a herbicide and dyestuff intermediate. 2,6-Dichlorotoluene has been used as an entrainer to separate ethanol from isopropanol by extractive distillation(112). 2,3,6-Trichlorotoluene is used as a herbicide intermediate. The other polychlorotoluenes have limited industrial application.

BIBLIOGRAPHY

"Ring-Chlorinated Toluenes" under "Chlorocarbons, Chlorohydrocarbons" in *ECT* 3rd ed., Vol. 5, pp. 819–827, by S. Gelfand, Hooker Chemical & Plastics Corp.; "Ring-Chlorinated Toluenes" in *ECT* 4th ed., Vol. 6, pp. 101–113, by Henry C. Lin and Ramesh Krishnamurti, Occidental Chemical Corporation; "Toluenes, Ring-Chlorinated" in *ECT* (online), posting date: December 4, 2000, by Henry C. Lin and Ramesh Krishnamurti, Occidental Chemical Corporation.

CITED PUBLICATIONS

1. J. B. Cohen and H. D. Dakin, *J. Chem. Soc.* **79**, 1111 (1901).
2. L. H. Horsley and co-workers, *Azeotropic Data III, Advances in Chemistry Series*, No. 116, American Chemical Society, Washington, D.C., 1973, 197–198.
3. G. L. Nelson and J. L. Webb, *J. Fire Flammability* **4**, 325 (1973).
4. R. G. Makitra, Ya. M. Tsikanchuk, and D. K. Tolopko, *J. Gen. Chem. U.S.S.R.* **45**, 1883 (1975).
5. R. S. Bly, K.-K. Tse, and R. K. Bly, *J. Organomet. Chem.* **117**, 35 (1976).
6. V. Galamb, G. Palyi, F. Ungvary, L. Marko, R. Boese, and G. Schmid, *J. Am. Chem. Soc.* **108**, 3344 (1986).
7. A. S. Abd-El-Aziz, C. C. Lee, A. Piorko, and R. G. Sutherland, *Synth. Commun.* **18**, 291 (1988).

8. J. Timmerman, *Physico-Chemical Constants of Pure Organic Compounds*, Elsevier Science Publishing Co., Inc., New York, 1950, 297–298.
9. V. Sedivec and J. Flek, *Handbook of Analysis of Organic Solvents*, John Wiley & Sons, Inc., New York, 1976, 164–168, 398.
10. K. Raznjevic, *Handbook of Thermodynamic Tables and Charts*, McGraw-Hill Book Co., New York, 1976, tables 27-1 and 30-2.
11. R. R. Dreisbach, *Physical Properties of Chemical Compounds I, Advances in Chemistry Series*, No. 15, American Chemical Society, Washington, D.C., 1955, p. 139.
12. A. L. McClellan, *Tables of Experimental Dipole Moments*, W. H. Freeman and Company, San Francisco, Calif., 1963, p. 243.
13. R. M. Stephenson and S. Malanowski, *Handbook of the Thermodynamics of Organic Compounds*, Elsevier Science Publishing Co., Inc., New York, 1987, p. 2270.
14. Eur. Pat. Appl. EP 46,555 (Mar. 3, 1982), G. M. Petruck and R. Wambach (to Bayer A-G).
15. Ger. Offen, 2,523,104 (Nov. 25, 1976), H. Rathjen (to Bayer A-G).
16. U.S. Pat. 4,031,146 (June 21, 1977), E. P. DiBella (to Tenneco Chemicals Inc.).
17. Jpn. Kokai 75 151,828 (Dec. 6, 1975), M. Matsui, T. Kitsukawa, K. Sato, and T. Ogawa (to Mitsubishi Chem. Ind. Co., Ltd.).
18. Ger. Offen. 2,721,429 (Nov. 16, 1978), H. U. Balnk (to Bayer A-G).
19. Y. Muramoto and H. Asakura, *Nippon Kagaku Kaishi* **6**, 1070 (1975).
20. Ger. Offen. 2,837,499 (Mar. 20, 1980), A. Marhold and E. Klauke (to Bayer A-G).
21. E. Kuimova and B. M. Mikhailov, *J. Org. Chem. USSR* **7**, 1485 (1971).
22. Swiss. Pat. CH 645,335 (Sept. 28, 1984), J. Beyrich and W. Regenass (to Ciba-Geigy A-G).
23. B. Chopra and V. Ramakrishnan, *Indian Chem. J. Annu.* **38**, (1972).
24. Jpn. Kokai 81 18,951 (Feb. 23, 1981), K. Sempuku (to Yuki Gosei Kogyo Co., Ltd.).
25. R. A. Grimm and J. E. Menting, *Ind. Eng. Chem. Prod. Res. Div.* **14**, 158 (1975).
26. M. Ouertani, P. Girard, and H. H. Kagan, *Bull. Soc. Chim. Fr.* **9–10**, 327 (1982).
27. A. L. Bottini and J. D. Roberts, *J. Am. Chem. Soc.* **79**, 1458 (1957).
28. M. Zoratti and J. F. Bunnett, *J. Org. Chem.* **45**, 1769 (1980).
29. R. Levine and E. R. Biehl, *J. Org. Chem.* **40**, 1835 (1975).
30. M. G. Voronkov and co-workers, *J. Org. Chem. USSR* **11**, 1118 (1975).
31. Eur. Pat. Appl. EP 283,194 (Sept. 21, 1988), K. Suto, K. Nakasa, M. Kudo, and M. Yamamoto (to Nihon Nohyaku Co., Ltd.).
32. U.S. Pat. 4,263,466 (Apr. 21, 1981), I. Colon, L. M. Maresca, and G. T. Kwiatkowski (to Union Carbide Corp.).
33. R. Vanderessa, J. J. Brunet, and P. Caubere, *J. Organomet. Chem.* **264**, 263 (1984).
34. M. Kraus and V. Bazant, in J. W. Hightower, ed., *Proceedings of the Fifth International Conference on Catalysis*, Palm Beach, Fla., North-Holland Publishing Co., Amsterdam, The Netherlands, 1972.
35. U.S. Pat. 4,499,025 (Feb. 12, 1985), J. B. Davison, R. J. Jasinski, and P. J. Peerce-Landers (to Occidental Chemical Corp.).
36. U.S. Pat. 4,334,081 (June 8, 1982), I. Colon (to Union Carbide Corp.).
37. I.G. Farben Industries, *Reports of the Intermediate Products Commission*, PB-17658, National Technical Information Service, Springfield, Va., 1935–1936, frames 2247–2256.
38. U.S. Pat. 3,000,975 (Sept. 19, 1961), E. P. DiBella (to Heyden Newport Chemical Corp.).
39. Neth. Pat. 6,511,484 (Mar. 3, 1966), (to Hooker Chemicals & Plastics Corp.).
40. U.S. Pats. 4,031,142, 4,031,147 (June 21, 1977), J. C. Graham (to Hooker Chemicals & Plastics Corp.).

41. U.S. Pat. 4,024,198 (May 17, 1977), H. E. Buckholtz and A. C. Bose (to Hooker Chemicals & Plastics Corp.).
42. U.S. Pats. 4,069,263, 4,069,264 (Jan. 17, 1978), H. C. Lin (to Hooker Chemicals & Plastics Corp.).
43. Jpn. Kokai JP 60136576 (July 20, 1985), J. Kiji, H. Konishi, and M. Shimizu (to Ihara Chemical Industry Co., Ltd.).
44. Ger. Offen. DE 3,432,095 (Mar. 6, 1986), H. Wolfram (to Hoechst (A-G)).
45. U.S. Pat. 4,851,596 (July 25, 1989), M. Franz-Josef, F. Helmut, R. Kai, and W. Karlfried (to Bayer A-G).
46. D. W. Breck, *Zeolite Molecular Sieves-Structure, Chemistry, and Use*, John Wiley & Sons, Inc., New York, 1974, p. 257.
47. Eur. Pat. Appl. EP 154,236 (Sept. 11, 1985), T. Suzuki and L. Komatsu (to Ihara Chemical Industry Co., Ltd.).
48. U.S. Pat. 4,794,201 (Dec. 27, 1988), Y. Higuchi and Suzuki (to Ihara Chemical Industry Co., Ltd.).
49. L. M. Stock and A. Himoe, *Tetrahedron Lett.* (13), 9 (1960).
50. L. M. Stock and A. Himoe, *J. Am. Chem. Soc.* **83**, 4605 (1961).
51. P. Kovacic, in G. A. Olah, ed., *Friedel-Crafts & Related Reactions*, Vol. **IV**, Interscience Publishers, Inc., a division of John Wiley & Sons, Inc., New York, 1965, Chapt. XLVIII, 111–127.
52. Jpn. Kokai 74 76,828 (July 24, 1974), (to International Minerals & Chemical Corp.).
53. Ger. Offen. 2,230,369 (Jan. 18, 1973), K. Sawazaki, H. Fujii, and M. Dehura (to Nikkei Kako Co., Ltd. and Sugai Chem. Ind. Ltd.).
54. C. M. Selwitz and V. A. Notaro, *Prepr. Div. Pet. Chem. ACS* **17**(4), E37–46 (1972).
55. Jpn. Kokai 73 81,822 (Nov. 1, 1973), R. Fuse, T. Inoue, and T. Kato (to Ajinomoto Co., Inc.).
56. A. B. Salomonov, P. P. Gertsen, and A. N. Ketov, *Zh. Prikl. Khim.* **43**, 1612 (1970).
57. S. Uemura and S. Fukuzawa, *J. Organomet. Chem.* **268**, 223 (1984).
58. J. W. Verbicky, Jr., B. A. Dellacoletta, and L. Williams, *Tetrahedron Lett.* **23**, 371 (1982).
59. B. Miller, *J. Org. Chem.* **38**, 1243 (1973); U.S. Pat. 3,844,917 (Oct. 29, 1974).
60. J. Blum, *Tetrahedron Lett.* (26), 3041 (1966).
61. P. B. D. DeLaMare and P. W. Robertson, *J. Chem. Soc.*, 279 (1943).
62. T. Bernath, *Gas Waerme Int.* **31**, 338 (1982).
63. D. R. Thielen, P. S. Foreman, A. Davis, and R. Wyeth, *Environ. Sci. Technol.* **21**, 145 (1987).
64. H. Kelker and E. Von Schivizhoffen, in J. C. Giddings and R. A. Kelker, eds., *Advances in Chromatography*, Vol. **6**, Marcel Dekker, Inc., New York, 1968, 247–297.
65. Jpn. Kokai JP 59,176,223(Oct. 5, 1984) (to Toray Industries, Inc.).
66. J. G. Lindberg, G. Y. Sugiyama, and R. L. Mellgren, *J. Magn. Reson.* **17**, 112 (1975).
67. J. G. Grasselli and W. M. Richey, eds., *Atlas of Spectral Data and Physical Constants for Organic Compounds*, 2nd ed., Vol. **IV**, CRC Press Inc., Cleveland, Ohio, 1975, 652–653.
68. Hazleton Laboratories, Project No. 157-147/148; May 10, 1972.
69. Younger Laboratories, Inc., Project No. Y-76-31; Feb. 27, 1976.
70. Springborn Institute, Project No. 3090. Dec. 31, 1980.
71. *Documentation of the Threshold Limit Values for Substances in Workroom Air with Supplements for those Substances Added or Changed Since 1971*, American Conference of Government Industrial Hygienists, 3rd ed., 1971, second printing, 1974, 302–303.

72. I. P. Ulanova, P. N. Dyachkov, and A. I. Khalepo, *Pharmacochem. Libr.* **8** (QSAR Toxicol. Xenobiochem.), 83 (1985).

73. J. Gartzke, D. Burck, P. Schmidt, and G. G. Avilova, *Z. Klin. Med.* **40**, 1701 (1985).

74. P. A. Vandenbergh and R. H. Olsen, *Appl. Environ. Microbiol.* **42**, 737 (1981).

75. D. T. Gibson and co-workers, *Biochemistry* **7**, 3795 (1968).

76. U.S. Pat. 4,006,195 (Feb. 1, 1977), S. Gelfand (to Hooker Chemicals & Plastics Corp.).

77. U.S. Pat. 3,366,698 (Jan. 30, 1968), E. P. DiBella (to Tenneco Chemicals Inc.).

78. T. Tkaczynski, Z. Winiarksi, and W. Markowski, *Przem. Chem.* **58**, 669 (1979).

79. Jpn. Kokai JP 59 206,322 (Nov. 22, 1984), (to Ihara Chemical Industry Co., Ltd.).

80. U.S. Pat. 4,031,145 (June 21, 1977), E. P. DiBella (to Tenneco Chemicals, Inc.).

81. U.S. Pat. 4,721,822 (Jan. 26, 1988), A. Leone-Bay, P. E. Timony, and L. Glaser (to Stauffer Chemical Co.).

82. Brit. Pat. 1,110,030 (Apr. 18, 1968), C. F. Kohll, H. D. Scharf, and R. Van Helden (to Shell Int'l Res. Maat. N. V.).

83. Neth. Appl. 6,907,390 (Nov. 17, 1970), D. A. Was (to Shell Int'l. Res. Maat. N. V.).

84. Jpn. Kokai JP 61 36,234 (Feb. 20, 1986), T. Irie and S. Doi (to Nitto Chemical Industry Co., Ltd.).

85. U.S. Pat. 3,219,688 (Nov. 23, 1965), E. D. Weil and co-workers (to Hooker Chemicals & Plastics Corp.).

86. H. C. Brimelow, L. Jones, and T. P. Metcalfe, *J. Chem. Soc.*, 1208 (1951).

87. F. F. Shcherebina, D. N. Tmenov, T. V. Lysukho, and N. P. Belous, *Zh. Prikl. Khim.* **53**, 2737 (1980).

88. U.S. Pat. 3,692,850 (Sept. 19, 1972), E. P. DiBella (to Tenneco Chemicals, Inc.).

89. V. S. Kozlova and A. N. Korol, *Zh. Anal. Khim.* **34**, 2406 (1979).

90. Jpn. Kokai JP 60 67,454 (Apr. 17, 1985), (to Nippon Kayaku Co., Ltd.).

91. I. Chu, S. Y. Shen, D. C. Villeneuve, V. E. Secours, and V. E. Valli, *J. Environ. Sci. Health, Part B* **B19**, 183 (1984).

92. R. Nishiyama and co-workers, *Yuki Gosei Kagaku Kyokai Shi* **23**, 515, 521 (1965).

93. Jpn. Kokai 70 28,367 (Sept. 16, 1970), M. Ishida (to Kureha Chemical Ind. Co., Ltd.).

94. V. Mark and co-workers, *J. Am. Chem. Soc.* **93**, 3538 (1971).

95. F. M. Ashton and A. S. Crafts, *Mode of Action of Herbicides*, John Wiley & Sons, Inc., New York, 1973, 10–24, 438–448.

96. U.S. Pat. 3,914,298 (Oct. 21, 1975), K. J. Dahl (to Raychem Corp.).

97. A) German Pat. 4314299 (April 30, 1993), Pies, M., Fiege, H., Puppe, L. and Kaeshbauer, J. (to Bayer A.-G). B) German Pat. 3420706 (June 2, 1984), Eichler, K., Arpe, H. J., Baltes, H. and Leupold, E.I. (to Hoechst A.-G). C) Iwayama, K. and Tada, K., *Shokubai* **35**(1), 17–21 (1993). D) Noguchi, Y., Tada, K., Iwayama, K., Kimura, M. and Kanai, T. *Kagaku Kogaku* **57**, 416–418 (1993).

98. A) Eur. Pat. Appl. 107155 (Aug. 7, 1982), Arpe, H.J., Litterer, H. and Mayer, N. (to Hoechst. A.-G). B) Jap. Kokai 31627 (Feb 20, 1982), (to Toray Industries, Inc).

99. E. Bay, D.A. Bak, P.E. Timony and A. Leone-Bay, *J. Org. Chem* **55**, 3415 (1990).

100. G.A. Olah, A.V. Orlinkov, P. Ramaiah, A.B. Oxyzoglou and G.K.S. Prakash, *Russ, Chem. Bl.* **47**, 924 (1998).

101. Available to members at Hazardous Substances Data Bank – http://www.tomescps.com/.

102. Japn Kokai 28363, (Feb. 2. 1999), Iwayama, K. and Kato, H. (to Toray Industries, Inc).

103. A) Japan Kokai 9316015, 9316014, 9316014, 9316013 (Dec 9, 1997), Maeda, M. and Imada, H. (to Toray Industries, Inc).

104. German Pat. 4325484 (February 2, 1995), U. Pentling, H.J. Buysch, L. Puppe, M. Pies and H.I. Paul (to Bayer A.-G.).

105. German Pat. 4218841 (December 16, 1993), U. Pentling, H.J. Buysch, L. Puppe, K. Roehik, R. Grosser and H.I. Paul (to Bayer A.-G.).
106. Japan Pat. 11158093 (June 15, 1999). K. Iwayama and M. Watanabe (to Toray Industries Inc.).
107. Japan Pat. 09188638 (July 22, 1997). M. Suzuki, A. Miyata and M. Ishikawa (to Toray Industries Inc.).
108. Eur. Pat. (July 5, 1989). B. Yamada, M. Kimura and Y. Noguchi (to Toray Industries, Inc.).
109. U.S.S.R. Pat. 1810331 (April 23, 1993), V.D. Bojko, L.Y. Shtejnberg, S.M. Shejn, S.A. Kondratov, B.V. Salov, K.A. Bochenkova and G.N. Sakharova (to Moskovskoe Nauchno-Proizvodstvennoe Ob'edinenie "Niopik").
110. Japn Pat. 06228026 (August 16, 1994). K. Hayashi, T. Mine and M. Mizukami (to Mitsubishi Gas Chemical Co.).
111. Japan Pat. 07304738 (November 21, 1995). N. Tanizawa (to Ihara Chemical Ind Co.)
112. U.S. Pat. 5445716 (August 29, 1995). L. Berg (to Lloyd Berg).

PRAVIN KHANDARE
RON SPOHN
Occidental Chemical Corporation

CHOCOLATE AND COCOA

1. Introduction

The name *Theobroma cacao*, food of the gods, indicating both the legendary origin and the nourishing qualities of chocolate, was bestowed upon the cacao tree by Linnaeus in 1720. All cocoa and chocolate products are derived from the cocoa bean, the seed of the fruit of this tree. Spanish explorers to the New World encountered the use of cocoa beans among the Aztec of Mexico and the Maya of the Yucatan as an ingredient in preparing a chocolate flavored drink much prized by the ruling class. The cultivation of cocoa trees in Mexico was underway for at least 1000 years before the Spanish arrival based on Mayan pottery inscriptions. In the *True History of Chocolate*, credit is given to the even more ancient Olmec civilization in southern Mexico for the first domestication and use of cocoa (1).

The terms cocoa and cacao often are used interchangeably in the literature. Both terms describe various products from harvest through processing. In this article, the term cocoa will be used to describe products in general and the term cacao will be reserved for botanical contexts. Cocoa traders and brokers frequently use the term raw cocoa to distinguish unroasted cocoa beans from finished products; this term is used to report statistics for cocoa bean production and consumption.

2. Standards for Cocoa and Chocolate

In the United States, chocolate and cocoa are standardized by the U.S. Food and Drug Administration under the Federal Food, Drug, and Cosmetic Act. The current definitions and standards resulted from prolonged discussions between the U.S. chocolate industry and the Food and Drug Administration (FDA). These definitions and standards were originally published in the *Federal Register* of December 6, 1944. The current standards for chocolate and cocoa products can be found in the Code of Federal Regulations (CFR), Title 21, Part 163-Cacao Products.

The Food and Agricultural Organization (FAO) and the World Health Organization (WHO) jointly sponsor the Codex Alimentarius Commission, which conducts a program for developing worldwide food standards. The Codex Committee for Cocoa Products and Chocolate has developed standards for chocolate (Codex Standard 87-1981), and cocoa powders and dry cocoa–sugar mixtures (Codex Standard 105-1981). Currently, a proposed draft standard for cocoa and chocolate products is before Codex for review. As a member of the Codex Alimentarius Commission, the United States is obligated to consider all Codex standards for acceptance.

The FDA announced in the *Federal Register* of January 25, 1989 a proposal to amend the U.S. chocolate and cocoa standards of identity. These amendments were in response to a citizen petition submitted by the Chocolate Manufacturers Association (CMA) and to better align U.S. standards with Codex. The new standards published as a final rule in the *Federal Register* of May 21, 1993, allow for the use of nutritive carbohydrate sweeteners, neutralizing agents, and emulsifiers; reduce slightly the minimum milkfat content and eliminate the nonfat milk solids/milkfat ratios in certain cocoa products including milk chocolate; update the language and format of the standards; and provide for optional ingredient labeling requirements. FDA has also received a proposal to establish a new standard of identity for white chocolate. Comments regarding the proposal amendments are under review by FDA, and a final ruling is expected to be issued in the near future.

2.1. White Chocolate. There is no standard of identity published for white chocolate in Title 21, Part 163 of the CFR. FDA published a proposal to establish a standard of identity for white chocolate in the *Federal Register* of March 10, 1997. The proposal was in response to citizen petitions filed by Hershey Foods Corporation (December 1992) and CMA (March 1993) requesting that FDA issue a standard of identity for white chocolate. Current products labeled white chocolate are operating under a temporary marketing permit. The presence of a standard would eliminate confusion over the content of products that are customarily referred to as white chocolate. A standard for white chocolate would also promote regulatory harmonization among nations that have already adopted a standard.

White chocolate has been defined by the European Economic Community (EEC) Directive 75/155/EEC as free of coloring matter and consisting of cocoa butter (not <20%); sucrose (not >55%); milk or solids obtained by partially or totally dehydrated whole milk, skimmed milk, or cream (not <14%); and butter or butter fat (not < 3.5%). In the proposed U.S. standard, white chocolate contains not <20% cocoa butter, not <14% total milk solids, not <3.5% milk fat, and not >55% nutritive carbohydrate sweetener.

3. Cocoa Beans

The cocoa bean is the basic raw ingredient in the manufacture of all cocoa products. The beans are converted to chocolate liquor, the primary ingredient from which all chocolate and cocoa products are made. Figure 1 depicts the conversion of cocoa beans to chocolate liquor, and in turn to the chief chocolate and cocoa products manufactured in the United States, ie, cocoa powder, cocoa butter, and sweet and milk chocolate.

Significant amounts of cocoa beans are produced in ~30 different localities. These areas are confined to latitudes 20° north or south of the equator. Although cocoa trees thrive in this very hot climate, young trees require the shade of larger trees such as banana, coconut, and palm for protection.

New cocoa hybrids and selections have been developed in Malaysia and other countries that produce significantly higher yields in select soil and climate conditions. In addition, high density plantings have demonstrated higher and

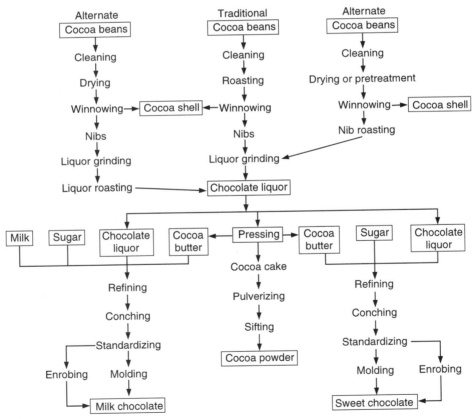

Fig. 1. Flow diagram of chocolate and cocoa production.

earlier yield in Malaysia and the Philippines. Low or no shade cocoa has also proven to increase yields. However, both high density and reduced shade cocoa production requires additional inputs of management and nutrition. Additional inputs to control pests and diseases also may be required.

A cocoa tree produces its first crop in 3–4 years and a full crop after 6–7 years. A full grown tree can reach a height of 12–15 m but is normally trimmed to 5–6 m to permit easy harvest. Because of differences in climate the crop is not confined to one short season but may extend for several months. Indeed some areas have cocoa pods almost all year long with one or two minor peaks. Many areas have peak harvests; for example, in West Africa there is one large main crop (80% or more of total crop) from September to March and a small or medium-sized crop in May. Brazil on the other hand has two crops a year that are almost equal in size.

3.1. Fermentation (Curing).

Prior to shipment from producing countries, most cocoa beans undergo a process known as curing, fermenting, or sweating. These terms are used rather loosely to describe a procedure in which seeds are removed from the pods, fermented, and dried. Some unfermented beans, particularly from Haiti and the Dominican Republic, are also used in the United States.

The age-old process of preparing cocoa beans for market involves specific steps that allegedly promote the activities of certain enzymes. Various methods of fermentation are used to the same end.

Fermentation plays a principal role in flavor development of beans by mechanisms that are not well understood (2). Because freshly harvested cocoa beans are covered with a white pulp rich in sugars, fermentation begins almost immediately upon exposure to air. The sugars are converted to alcohol, and finally to acetic acid, which drains off, freeing the cotyledon from the pulpy mass. The acetic acid and heat formed during fermentation penetrate the skin or shell, killing the germ and initiating chemical changes within the bean that play a significant role in the development of flavor and color. During this initial stage of fermentation, the beans acquire the ability to absorb moisture, which is necessary for many of the chemical reactions that follow.

3.2. Commercial Grades.

Most cocoa beans imported into the United States are one of about a dozen commercial varieties that can be generally classified as Criollo or Forastero. Criollo beans have a light color, a mild, nutty flavor, and an odor somewhat like sour wine. Forastero beans have a strong, somewhat bitter flavor and various degrees of astringency. The Forastero varieties are more abundant and provide the basis for most chocolate and cocoa formulations. Table 1 shows the main varieties of cocoa beans imported into the United States. The varieties are usually named for the country or port of origin.

3.3. Bean Specifications.

Cocoa beans vary widely in quality, necessitating a system of inspection and grading to ensure uniformity. Producing countries have always inspected beans for proper curing and drying as well as for insect and mold damage. Recently, a procedure for grading beans has been established at an international level. This ordinance, reached primarily through the efforts of FAO, has been adopted by Codex as the model ordinance for inspection and grading of beans. It classifies beans into two principal categories according to the fraction of moldy, slaty, flat, germinated, and insect-damaged beans (3).

Table 1. Main Varieties of Cocoa Beans Imported into the United States

Africa	South America	Pacific	West Indies
Ivory Coast	Bahia (Brazil)	Malaysia	Sanchez (Dominican Republic)
Accra (Ghana)	Arriba (Ecuador)	New Guinea	Grenada
Lagos	Venezuelan	Indonesia	Trinidad
Nigeria		Samoa	
Fernando Po			
Sierra Leone			

Cocoa beans are sometimes evaluated in the laboratory to distinguish and characterize flavors. Beans are roasted at a standardized temperature for a specific period of time, shelled, usually by hand, and ground or heated slightly to obtain chocolate liquor. The liquor's taste is evaluated by a panel of experts who characterize and record the particular flavor profile (3).

3.4. Blending.   Most chocolate and cocoa products consist of blends of beans chosen for flavor and color characteristics. Cocoa beans may be blended before or after roasting, or nibs may be blended before grinding. In some cases, finished liquors are blended. Common, or basic beans, are usually African or Brazilian and constitute the bulk of most blends. More expensive flavor beans from Venezuela, Trinidad, Ecuador, etc, are added to impart specific characteristics. The blend is determined by the end use or type of product desired.

3.5. Production.   Worldwide cocoa bean production has increased significantly over the past 10 years from ~2.4 million t in the 1989–1990 crop year to >3.0 million t in 2000. The production share by country has also changed dramatically in the last 10 years. The big gainers were the Ivory Coast and Indonesia. The gains in Indonesia have helped to diversify production and partially shield the market from adverse weather-induced supply shocks. The biggest losers in production share have been Brazil and Malaysia. Table 2 lists production statistics for these countries (4).

3.6. Consumption.   Worldwide cocoa bean consumption has increased over the past 10 years from ~2.3 million t in the 1989–1990 crop year to almost 2.6 million t in 1999–2000 (4). Table 3 gives the annual tonnage of cocoa bean grind in leading countries. Total world cocoa grindings were 3.0 million t in 2000–2001 (4).

Table 2. Production of Raw Cocoa Beans, 10^3t

Region	1989–1990	1999–2000	2000–2001[a]
Ivory Coast	725	1409	1170
Brazil	348	123	152
Ghana	295	437	390
Malaysia	243	45	35
Nigeria	160	165	185
Indonesia	115	410	400
Other	521	472	480
Total	2407	3061	2812

[a] Forecasted data.

Table 3. **Grind of Raw Cocoa Bean, 10^3t**

Main cocoa grinding countries	1998–1999	1999–2000	2000–2001[a]
Netherlands	415	436	450
United States	406	439	438
Ivory Coast	225	235	265
Germany	197	215	227
Brazil	192	202	194
United Kingdom	167	168	162
France	124	142	145
Malaysia	105	115	115
Indonesia	75	85	75
Ghana	65	70	70
Russian Federation	47	60	70
Italy	73	64	65

[a] Estimated.

3.7. Marketing.   Most of the cocoa beans and products imported into the United States are done so by New York and London trade houses. The New York Board of Trade is the parent company for the Sugar, Coffee, and Cocoa Exchange, which provides a mechanism by which both chocolate manufacturers and trade houses can hedge their cocoa bean transactions. Additional information on the functions of the Sugar, Coffee, and Cocoa Exchange can be found on internet websites at http://www.nybot.com or http://www.csce.com. (3) World bean prices have fallen from $1,640 per t in 1997–1998 to $1000 per t in 2001.

4. Chocolate Liquor

Chocolate liquor is the solid or semisolid food prepared by finely grinding the kernel or nib of the cocoa bean. It is also commonly called chocolate, unsweetened chocolate, baking chocolate, or cooking chocolate. In Europe chocolate, liquor is often called chocolate mass or cocoa mass.

4.1. Cleaning.   Cocoa beans are imported in the United States in 70-kg bags. The beans can be processed almost immediately or stored for later use. They are usually fumigated prior to storage.

The first step in the processing of cocoa beans is cleaning. Stones, metals, twigs, twine, and other foreign matter are usually removed by passing beans in a thin layer over a large vibrating screen cleaner. Large objects are retained as the beans fall through a lower screen. The second screen removes sand and dirt that have adhered to the beans. Strategically placed magnets are commonly used to remove small pieces of metal.

4.2. Roasting.   The chocolate flavor familiar to the consumer is primarily developed during roasting, which promotes reactions among the latent flavor precursors in the bean. Good flavor depends on the variety of bean and the curing process used. The bacterial or enzymatic changes that occur during fermentation presumably set the stage for the production of good flavor precursors.

Although flavor precursors in the unroasted cocoa bean have no significant chocolate flavor themselves, they react to form highly flavored compounds. These flavor precursors include various chemical compounds such as proteins, amino acids, reducing sugars, tannins, organic acids, and many unidentified compounds.

The natural moisture of the cocoa bean combined with the heat of roasting cause many chemical reactions other than flavor changes. Some of these reactions remove unpleasant volatile acids and astringent compounds, partially breakdown sugars, modify tannins and other nonvolatile compounds with a reduction in bitterness, and convert proteins to amino acids that react with sugars to form flavor compounds, particularly pyrazines (5). There are ~400–500 compounds that have been identified from volatile and non-volatile fractions of chocolate that contribute to chocolate flavor, including hydrocarbons, alcohols, aldehydes, ketones, esters, amines, oxazoles, and sulfur compounds (6).

Roasting is essentially a cooking process developed by craftsmen who were guided by their senses of smell and taste and their knowledge of how beans of differing degrees of roast behaved in the subsequent processes. The ease and efficiency with which the processes of winnowing, grinding, pressing, and conching can be performed is affected by the degree of roast.

Roasting conditions can be adjusted to produce different flavors. Low, medium, full, and high roasts can be developed by varying time, temperature, and humidity in the roaster. Low roasts produce mild flavors and light color; high roasts produce strong flavors and dark color (7).

Roasters have evolved from the coke-fired rotary drum type to continuous feed roasters. It is traditional to roast cocoa beans with the shell still on. However, other methods of roasting include nib roasting, in which the shell is first removed by a rapid or moist heating step, and liquor roasting (Fig. 1). The newer nib and liquor roasters are designed to subject the cocoa to more uniform heat conditions in addition to minimizing the loss of cocoa butter to the shell. Nib and liquor roasters do not need the high temperatures necessary for whole bean roasting and, therefore, can be considered more energy efficient. Roasting times vary from ~30 to 60 min. Actual temperature of the bean in the roaster is difficult to measure but probably ranges from as low as 70°C to as high as 180°C.

4.3. Winnowing.   Winnowing, often called cracking and fanning, is one of the most important operations in cocoa processing. It is a simple process that involves separating the nib, or kernel, from the inedible shell. Failure to remove shell results in lower quality cocoa and chocolate products, more wear on nib grinding machines, and lower efficiency in all subsequent operations.

Because complete separation of shell and nib is virtually impossible, various countries have established maximum allowable limits of shell by weight; U.S. manufacturers average from 0.05 to 1%. The cocoa shell legal limit is 1.75% in the United States.

The analysis of cocoa shell (8) is given in Table 4. In the United States, shells are often used as mulch or fertilizer for ornamental and edible plants, as animal feed, and as fuel for boilers.

4.4. Grinding.   The final step in chocolate liquor production is the grinding of the kernel or nib of the cocoa bean. The nib is a cellular mass containing

Table 4. **Analyses of Cocoa Shell from Roasted Cocoa Beans**

Component	Shell, %
water	3.8
fat	3.4
ash	
total	8.1
water-soluble	3.5
water-insoluble	4.6
silica, etc	1.1
alkalinity (as K_2O)	2.6
chlorides	0.07
iron (as Fe_2O_3)	0.03
phosphoric acid (as P_2O_5)	0.8
copper	0.004
nitrogen	
total nitrogen	2.8
protein nitrogen	2.1
ammonia nitrogen	0.04
amide nitrogen	0.1
theobromine	1.3
caffeine	0.1
carbohydrates	
glucose	0.1
sucrose	0.0
starch (taka-diastase method)	2.8
pectins	8.0
fiber	18.6
cellulose	13.7
pentosans	7.1
mucilage and gums	9.0
tannins	
tannic acid (Lowenthal's method)	1.3
cocoa-purple and cocoa-brown	2.0
acids	
acetic (free)	0.1
citric	0.7
oxalic	0.32
extracts	
cold water	20.0
alcohol, 85%	10.0

~50–56% cocoa fat (cocoa butter). Grinding liberates the fat locked within the cell wall while producing temperatures as high as 110°C.

Nibs are usually ground while they are still warm from roasting. The original horizontal three-tier stone mills and vertical disk mills have been replaced by modern horizontal disk mills, which have much higher outputs and are capable of grinding nibs to much greater fineness. Two modern machines in particular account for a large percentage of liquor grinding. One uses a pin mill mounted over a roller refiner. The pin mill grinds the nibs to a coarse but fluid liquor. The liquor is delivered to a roll refiner that reduces the particle size to a very fine limit. The second type is a vertical or horizontal ball mill. Coarsely ground nib is fed to the base of a vertical cylinder, which contains small balls in separate compartments. A central spindle causes the balls to rotate at very high speeds, grinding the liquor between them and against the internal wall of the cylinder (9).

5. Cocoa Powder

Cocoa powder (cocoa) is prepared by pulverizing the remaining material after part of the fat (cocoa butter) is removed from chocolate liquor. The U.S. chocolate standards define three types of cocoas based on their fat content. These are breakfast, or high fat cocoa, containing not <22% fat; cocoa, or medium fat cocoa, containing <22% fat but >10%; and low fat cocoa, containing <10% fat.

Cocoa powder production today is an important part of the cocoa and chocolate industry because of increased consumption of chocolate-flavored products. Cocoa powder is the basic flavoring ingredient in most chocolate-flavored cookies, biscuits, syrups, cakes, and ice cream. It is also used extensively in the production of confectionery coatings for candy bars.

5.1. Cocoa Powder Manufacture. When chocolate liquor is exposed to pressures of 34–41 MPa (5000–6000 psig) in a hydraulic press, and part of the fat (cocoa butter) is removed, cocoa cake (compressed powder) is produced. The original pot presses used in cocoa production had a series of pots mounted vertically one above the other. These have been supplanted by horizontal presses that have 4–24 pots mounted in a horizontal frame. The newer presses are capable of complete automation, and by careful selection of pressure, temperature, and time of pressing, cocoa cake of a specified fat content can be produced.

Cocoa powder is produced by grinding cocoa cake. Cocoa cake warm from the press breaks easily into large chunks but is difficult to grind into a fine powder. Cold, dry air removes the heat generated during most grinding operations. Because the finished cocoa powder still contains fat, great care must be taken to prevent the absorption of undesirable odors and flavors.

Commercial cocoa powders are produced for various specific uses and many cocoas are alkali treated, or Dutched, to produce distinctive colors and flavors. The alkali process can involve the treatment of nibs, chocolate liquor, or cocoa with a wide variety of alkalizing agents (10).

Cocoa powders not treated with alkali are known as cocoa or natural cocoa. Natural cocoa has a pH of ~4.8–5.8 depending on the type of cocoa beans used. Alkali processed cocoa ranges in pH from about 6 to as high as 8.5.

6. Cocoa Butter

Cocoa butter is the common name given to the fat obtained by subjecting chocolate liquor to hydraulic pressure. It is the main carrier and suspending medium for cocoa particles in chocolate liquor and for sugar and other ingredients in sweet and milk chocolate.

The FDA has not legally defined cocoa butter, and no standard exists for this product under the U.S. Chocolate Standards. For the purpose of enforcement, the FDA defines cocoa butter as the edible fat obtained from cocoa beans either before or after roasting. Cocoa butter as defined in the *U.S. Pharmacopeia* is the fat obtained from the roasted seed of *Theobroma cacao Linne.*

The Codex Committee on Cocoa Products and Chocolate defines cocoa butter as the fat obtained from cocoa beans with the following characteristics (11):

(1) Free fatty acid content (expressed as oleic acid) not >1.75% m/m.

(2) Unsaponifiable matter not >0.7% m/m, except in the case of press cocoa butter which shall not be >0.35% m/m.

Press cocoa butter is defined as fat obtained by pressure from cocoa nib or chocolate liquor. In the United States, this is often referred to as prime pure cocoa butter. Expeller cocoa butter is defined as the fat prepared by the expeller process. In this process, cocoa butter is obtained directly from whole beans by pressing in a cage press. Expeller butter usually has a stronger flavor and darker color than prime cocoa butter and is filtered with carbon or otherwise treated prior to use. Solvent extracted cocoa butter is cocoa butter obtained from beans, nibs, liquor, cake, or fines by solvent extraction (qv), usually with hexane. Refined cocoa butter is any of the above cocoa butters that has been treated to remove impurities or undesirable odors and flavors.

6.1. Composition and Properties. Cocoa butter is a unique fat with specific melting characteristics. It is a solid at room temperature (20°C), starts to soften ~30°C, and melts completely just below body temperature. Its distinct melting characteristic makes cocoa butter the preferred fat for chocolate products.

Cocoa butter is composed mainly of glycerides of stearic, palmitic, and oleic fatty acids (see FATS AND FATTY OILS). The triglyceride structure of cocoa butter has been determined (12,13) and is as follows:

tri-saturated, 3%;
mono-unsaturated (oleo-distearin), 22%;
oleo-palmitostearin, 57%;
oleo-dipalmitin, 4%;
di-unsaturated (stearo-diolein), 6%;
palmito-diolein, 7%;
tri-unsaturated, tri-olein, 1%.

Although there are actually six crystalline forms of cocoa butter, four basic forms are generally recognized as alpha, beta, beta prime, and gamma. The γ (gamma) form, the least stable, has a melting point of 17°C. It changes rapidly to the α (alpha) form that melts at 21−24°C. At normal room temperature the β′ (beta prime) form changes to the β (beta) form, melting at 27−29°C, and finally, the β form is reached. It is the most stable form with a melting point of 34−35°C (14).

Since cocoa butter is a natural fat, derived from different varieties of cocoa beans, no single set of specifications or chemical characteristics can apply. Codex previously defined the physical and chemical parameters of the various types of cocoa butter (15) (Table 5). However, the Codex Committee recently simplified

Table 5. **Properties and Composition of Cocoa Butter**[a]

Characteristic	Press cocoa butter	Expeller cocoa butter	Refined cocoa butter
refractive index n_D^{40}, °C	1.456–1.458	1.453–1.459	1.453–1.462
melting behavior			
slip point, °C	30–40	30–34	30–34
clear melting point, °C	31–35	31–35	31–35
free fatty acids			
mol% oleic acid	0.5–1.75	0.5–1.75	0–1.75
saponification value			
mg KOH/g fat	192–196	192–196	192–196
iodine value, Wijs	33.8–39.5	35.6–44.6	35.7–41.0
unsaponifiable matter			
petroleum ether % m/m	not > 0.35%	not > 0.40%	not > 0.50%

[a] Contaminants not to exceed 0.5 mg/kg of arsenic, 0.4 mg/kg of copper, 0.5 mg/kg of lead, and 2.0 mg/kg of iron (16).

the standard for cocoa butter and no longer distinguishes between the various types (11).

6.2. Substitutes and Equivalents. In the past 25 years, many fats have been developed to replace part or all of the added cocoa butter in chocolate-flavored products. These fats fall into two basic categories commonly known as cocoa butter substitutes and cocoa butter equivalents. Neither can be used in the United States in standardized chocolate products. However, EU regulations allow for the use of specified vegetable fats in standardized chocolate as long as labeling requirements are met. In conformity with EU regulations only the following vegetable fats may be used in chocolate: illipe, palm oil, sal, shea, kokum gurgi, and mango kernel. However, the regulations allow for an exception—the use of coconut oil in chocolate used for the manufacture of ice cream and similar frozen products.

Cocoa butter substitutes of all types enjoy widespread use in the United States chocolate-flavored products. Cocoa butter equivalents are not widely used because of their higher price and limited supply.

Cocoa butter substitutes do not chemically resemble cocoa butter and are compatible with cocoa butter only within specified limits. Cocoa butter equivalents are chemically similar to cocoa butter and can replace cocoa butter in any proportion without deleterious physical effects (16,17).

Cocoa butter substitutes and equivalents differ greatly with respect to their method of manufacture, source of fats, and functionality; they are produced by several physical and chemical processes (18,19). Cocoa butter substitutes are produced from lauric acid fats such as coconut, palm, and palm kernel oils by fractionation and hydrogenation; from domestic fats such as soy, corn, and cotton seed oils by selective hydrogenation; or from palm kernel stearines by fractionation. Cocoa butter equivalents can be produced from palm kernel oil and other specialty fats such as shea and illipe by fractional crystallization; from glycerol and selected fatty acids by direct chemical synthesis; from edible beef tallow by acetone crystallization; or from domestic fats such as soy and cotton seed by enzymatic interesterification.

7. Sweet and Milk Chocolate

Most chocolate consumed in the United States is consumed in the form of milk chocolate and sweet chocolate. Sweet chocolate is chocolate liquor to which sugar and cocoa butter have been added. Milk chocolate contains these same ingredients and milk or milk solids (Fig. 2).

U.S. definitions and standards for chocolate are quite specific (20). Sweet chocolate must contain at least 15% chocolate liquor by weight and must be sweetened with a nutritive carbohydrate sweetener. Semisweet chocolate or bittersweet chocolate, though often referred to as sweet chocolate, must contain a minimum of 35% chocolate liquor. These products, sweet chocolate and semisweet chocolate, or bittersweet chocolate, are often simply called chocolate or dark chocolate to distinguish them from milk chocolate. Table 6 gives some typical formulations for sweet chocolates (6).

Sweet chocolate can contain milk or milk solids (up to 12% max), nuts, coffee, honey, malt, salt, vanillin, and other spices and flavors as well as a number of specified emulsifiers. Many different kinds of chocolate can be produced by careful selection of bean blends, controlled roasting temperatures, and varying amounts of ingredients and flavors (21).

The most popular chocolate in the United States is milk chocolate. The U.S. Chocolate Standards state that milk chocolate shall contain no <3.39 wt% of milk fat and not <12 wt% of milk solids. Milk chocolate can contain spices, natural

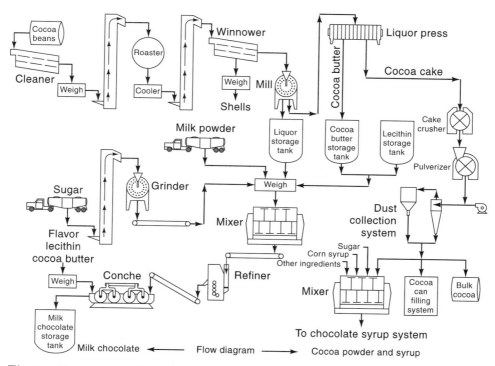

Fig. 2. Process flow diagram for milk chocolate, chocolate syrup, and cocoa powder.

Table 6. **Typical Forumulations for Sweet (Dark) Chocolates**

Ingredient	Formulation, %		
	1	2	3
chocolate liquor	15.0	35.0	70.0
sugar	60.0	50.4	29.9
added cocoa butter	23.8	14.2	
lecithin	0.3	0.3	
vanillin	0.9	0.1	0.1
Total fat	*32.0*	*33.0*	*37.1*

and artificial flavorings, ground whole nut meats, dried malted cereal extract and other seasonings that do not impart a flavor that imitates the flavor of chocolate, milk, or butter. In addition, chocolate liquor content must not be <10% by weight. Some typical formulations of milk chocolate and some compositional values are shown in Table 7 (6).

7.1. Production. The main difference in the production of sweet and milk chocolate is that in the production of milk chocolate, water must be removed from the milk. Many milk chocolate producers in the United States use spray-dried milk powder. Others condense fresh whole milk with sugar, and either dry it, producing milk crumb, or blend it with chocolate liquor and then dry it, producing milk chocolate crumb. These crumbs are mixed with additional chocolate liquor, sugar, and cocoa butter later in the process (22). Milk chocolates made from crumb typically have a more caramelized milk flavor than those made from spray- or drum-dried milk powder.

7.2. Mixing. The first step in chocolate processing is the weighing and mixing of ingredients which is usually a fully automated process carried out in a batch or continuous processing system. In batch processing, all the ingredients for one batch are automatically weighed into a mixer and mixed for a specific period of time. The mixture is conveyed to storage hoppers directly above the refiners. In the continuous method, ingredients are metered into a continuous kneader, which produces a constant supply to the refiners (23). The continuous process requires very accurate metering and rigid quality control procedures for all raw materials.

Table 7. **Typical Formulations for Milk Chocolate**

Ingredient	Formulation, %		
	1	2	3
chocolate liquor	11.0	12.0	12.0
dry whole milk	13.0	15.0	20.0
sugar	54.6	51.0	45.0
added cocoa butter	21.0	21.6	22.6
lecithin	0.3	0.3	0.3
vanillin	0.1	0.1	0.1
Total Fat	*30.9*	*32.6*	*35.0*

7.3. Refining. The next stage in chocolate processing is a fine grinding in which a coarse paste from the mixer is passed between steel rollers and converted to a drier powdery mass. Refining breaks up crystalline sugar, fibrous cocoa matter, and milk solids.

Tremendous advances have been made in the design and efficiency of roll refiners. The methods currently used for casting the rolls have resulted in machines capable of very high output and consistent performance. The efficiency of the newer refiners has also been improved by hydraulic control of the pressure between the rolls and thermostatic control of cooling water to the rolls.

Modern 5-roll refiners with 2-m wide rollers can process up to 2200 kg of paste per hour. The actual output is dependent on the desired particle size.

Particle size is extremely important to the overall quality of sweet and milk chocolate. Hence, the refining process, that controls particle size is critical. Fine chocolates usually have no particles >25 or 30 μm, which is normally accomplished by passing the paste through refiners more than once. However, smooth chocolates can be produced with only a single pass through the refiners if the ingredients are ground prior to mixing. Particle size has a significant effect on both texture and flavor of the finished chocolate.

7.4. Conching. After refining, chocolate is subjected to conching, a step critical to the flavor and texture development of high quality chocolates. Conching is a mixing–kneading process allowing moisture and volatile components to escape while smoothing the chocolate paste. It is one of the less satisfactorily explained parts of the chocolate making process and can embrace a wide range of phenomena, ranging from the relatively simple process of reliquefaction of a newly refined chocolate paste to complex and often controversial processes of flavor development, gloss development, agglomerate reduction, viscosity reduction, and modification of the melting quality.

The name conche derives from the seashell shape of the first really effective conching machine, which consisted of a tank with curved ends and a granite bed on which the chocolate paste from the refiners was slowly pushed back and forth by a granite roller. This longitudinal conche, the development of which is commonly attributed to Rodolph Lindt of Switzerland in 1879, is still used and many experts consider it best for developing subtle flavors.

Several other kinds of conches also are used today. The popular rotary conche can handle chocolate paste in a dry stage direct from the refiners (24). The recently developed continuous conche actually liquifies and conches in several stages and can produce up to 3600 kg of chocolate per hour in a floor area of only 34 m^2.

Conching temperatures range from 55–85°C for sweet chocolate and from 45–55°C for milk chocolate. Higher temperatures are used for milk chocolate if caramel or butterscotch flavors are desired (25).

Conching time varies from a few hours to many days and many chocolates receive no conching. Nonconched chocolate is usually reserved for inexpensive candies, cookies, and ice cream. In most operations, high quality chocolate receives extensive conching for as long as 120 h.

Flavors, emulsifiers, or cocoa butter are often added during conching. The flavoring materials most commonly added in the United States are vanillin, a vanillalike artificial flavor, and natural vanilla (26) (see FLAVORS AND SPICES). Cocoa butter is added to adjust viscosity for subsequent processing.

Several chemical changes occur during conching including a rise in pH and a decline in moisture as volatile acids (acetic) and water are driven off. These chemical changes have a mellowing effect on the chocolate (27).

7.5. Standardizing. In standardizing or finishing, emulsifiers and cocoa butter are added to the chocolate to adjust viscosity to final specifications.

Lecithin (qv), a natural phospholipid possessing both hydrophilic and hydrophobic properties, is the most common emulsifier in the chocolate industry (6). The hydrophilic groups of the lecithin molecules attach themselves to the water, sugar, and cocoa solids present in chocolate. The hydrophobic groups attach themselves to the cocoa butter and other fats such as milk fat. This reduces both the surface tension, between cocoa butter and the other materials present, and the viscosity. Less cocoa butter is then needed to adjust the final viscosity of the chocolate.

The amount of lecithin required falls within a narrow range of $\sim$0.2–0.6% (28). It can have a substantial effect on the amount of cocoa butter used, reducing the final fat content of chocolate by as much as 5%. Because cocoa butter is usually the most costly ingredient in the formulation of chocolate, the savings to a large manufacturer can be substantial.

Lecithin is usually introduced in the standardizing stage but can be added earlier in the process. Some lecithin is often added during mixing or conching. The addition at this point has the added advantage of reducing the energy necessary to pump the product to subsequent operations since the product viscosity is reduced.

Chocolate does not behave as a true liquid owing to the presence of cocoa particles and the viscosity control of chocolate is quite complicated. This non-Newtonian behavior has been described (29). When the square root of the rate of shear is plotted against the square root of shear stress for chocolate, a straight line is produced. With this Casson relationship method (30) two values are obtained, Casson viscosity and Casson yield value, which describe the flow of chocolate. The chocolate industry was slow in adopting the Casson relationship but this method now prevails over the simpler MacMichael viscometer. Instruments such as the Carri-Med Rheometer and the Brookfield and Haake Viscometers are now replacing the MacMichael.

At this stage of manufacture, chocolate may be stored for future use in bulk liquid form if usage is expected to be within 1–2 weeks, or at 43–50°C in a hot water jacketed agitated tank or in solid block form where it can be stored for as long as 6–12 months. Blocks typically weigh between 3 and 30 kg. Storage conditions for block chocolate should be cool and dry, ie, 7–18°C and 40–45% relative humidity. If chocolate has been stored in block form, it can be remelted to temperatures up to 50°C and then processed in the same manner as freshly made liquid chocolate.

At this stage, the chocolate is ready for forming into its final shape after it is tempered. The two most common forms are molding or enrobing.

7.6. Tempering. The state, or physical structure, of the fat base in which sugar, cocoa, and milk solids are suspended is critical to the overall quality and stability of chocolate. Production of a stable fat base is complicated because the cocoa butter in solidified chocolate exists in several polymorphic forms.

Tempering is the process of inducing satisfactory crystal nucleation of the liquid fat in chocolate.

Nucleation tempering of the still molten fat is necessary because the cocoa butter, if left to itself, can solidify in a number of different physical forms, ie, into an unstable form if cooled rapidly, or into an equally unacceptable super stable form if cooled too slowly, as commonly happens when a chocolate turns gray or white after being left in the sun. The coarse white fat crystals that can form in the slowly cooled center of a very thick piece of chocolate are similarly in a super stable form known in the industry as fat bloom.

Control of the polymorphic forms in cocoa butter is further complicated by the presence of other fats such as milk fat. The fat in a chocolate can be likened to the mortar between the bricks in a mason's wall. The solid particles in a well-conched chocolate bed down better than the solids in a coarsely refined and poorly mixed one (31).

A stable crystalline form for chocolate depends primarily on the method used to cool the fat present in the liquid chocolate. To avoid the grainy texture and poor color and appearance of improperly cooled chocolate, the chocolate must be tempered or cooled down so as to form cocoa butter seed crystals (32). This can be accomplished by cooling the warm (44–50°C) liquid chocolate in a water jacketed tank, which has a slowly rotating scraper or mixer. As the chocolate cools, the fat begins to solidify and form seed crystals. Cooling is continued to ~26–29°C, during which time the chocolate becomes more viscous. If not further processed quickly, the chocolate will become too thick to process.

In another method of tempering, solid chocolate shavings are added as seed crystals to liquid chocolate at 32–33°C. This is a particularly good technique for a small confectionery manufacturer who does not produce his own chocolate. However, the shavings are sometimes difficult to disperse and may cause lumps in the finished product (21). Most companies use continuous thin-film heat exchangers for the tempering process.

7.7. Molding.   The liquid tempered chocolate is deposited into a metal or plastic (polycarbonate) mold in the shape of the final product. There are three basic types of molding: solid (or block), shell, and hollow.

Solid chocolate, eg, Hershey's Milk Chocolate Bar, is the most common molding. The chocolate, either milk or dark, is deposited into a mold and the mold passes through a cooling tunnel with a residence time in the tunnel of ~25 min. When the molds emerge from the tunnel, the chocolate is solidified. In addition to solidifying, the chocolate also contracts if it is correctly tempered, thus facilitating the removal of the bar from the mold. The demolded bar is then wrapped and packaged for shipment to the consumer.

Shell molding is a process by which a liquid or soft center is incorporated inside a chocolate shell. Modern equipment codeposits both the center filling and the chocolate shell in one step. The old method, and still the most common, is to form a hollow shell of chocolate in the mold followed by a short cooling tunnel. A filling is then deposited into the shell and sometimes followed by further cooling. A layer of chocolate then is deposited on top of the filling; this layer welds itself to the originally formed shell thus completely encapsulating the filling. The molds are cooled for a third and final time after which the pieces can be removed and wrapped.

Hollow molding as the name implies is a molded product with a hollow center such as Easter eggs, bunny rabbits, and Santa Claus. The molds used in hollow molding are divided in two halves and connected by a hinge. Chocolate is deposited into one-half of the mold. The mold is then closed and rotated so that the liquid chocolate completely coats the inside surface of the mold. After cooling the molds are opened and the piece removed.

7.8. Enrobing. A preformed center such as nougat, fondant, fudge, cookies, etc, is placed on a conveyor belt and passed through a curtain of liquid tempered chocolate. The weight and thickness of the coating adhering to the center is controlled by an air curtain and vibration mechanism located immediately after the chocolate curtain. The now chocolate-coated centers pass into a cooling tunnel with a dwell time between 5 and 10 min. Upon emergence from the cooling tunnel the chocolate-coated pieces are ready for wrapping and packing.

8. Theobromine and Caffeine

Chocolate and cocoa products, like coffee, tea, and cola beverages, contain alkaloids (qv) (1). The predominant alkaloid in cocoa and chocolate products is theobromine [83-67-0], though caffeine [58-08-2] is also present in smaller amounts. Concentrations of both alkaloids vary depending on the origin of the beans. Published values for the theobromine and caffeine content of chocolate vary widely because of natural differences in cocoa beans and differences in analytical methodology. This latter problem was alleviated by the introduction of high pressure liquid chromatography (HPLC), which has greatly improved the accuracy of analyses. HPLC values for theobromine and caffeine in a number of chocolate liquor samples have been published (33) (Table 8). Of the 12 varieties tested, the ratio of theobromine to caffeine varied widely from 2.5:1 for New Guinea liquor to 23.2:1

Table 8. **Variations in Theobromine and Caffeine Content of Various Chocolate Liquors**

Country of origin	Total			
	Theobromine, %	Caffeine %	Alkaloid %	Theobromine/caffeine ratio
New Guinea	0.818	0.329	1.15	2.49:1
New Guinea	0.926	0.330	1.26	2.81:1
Malaysia	1.050	0.252	1.30	4.17:1
Malaysia	1.010	0.228	1.24	4.43:1
Bahia	1.210	0.183	1.39	6.61:1
Main Lagos	1.730	0.159	1.89	10.90:1
Light Lagos	1.230	0.137	1.37	8.98:1
Sanchez	1.570	0.177	1.75	8.87:1
Sanchez (small)	1.250	0.261	1.51	4.79:1
Fernando Po	1.470	0.064	1.53	23.00:1
Tabascan	1.410	0.113	1.52	12.50:1
Trinidad	1.240	0.233	1.47	5.32:1
average	1.240	0.206	1.45	7.91:1
maximum	1.730	0.330	1.89	23.00:1
minimum	0.818	0.064	1.15	2.49:1

Table 9. **Theobromine and Caffeine Content of Finished Chocolate Products**

Product	Theobromine, %	Caffeine, %
baking chocolate	1.386	0.164
chocolate flavored syrup	0.242	0.019
cocoa, 15% fat	2.598	0.247
dark sweet chocolate	0.474	0.076
milk chocolate	0.197	0.022

for that obtained from Fernando Po. Total alkaloid content, however, remained fairly constant, ranging from 1.15 to 1.89%.

The theobromine and caffeine contents of several finished chocolate products as determined by HPLC at Hershey's laboratories are presented in Table 9.

9. Nutritional Properties of Chocolate Products

Chocolate and cocoa products supply proteins, fats, carbohydrates, vitamins, and minerals. The Chocolate Manufacturers Association of the United States (Vienna, Virginia) completed a nutritional analysis from 1973 to 1976 of a wide variety of chocolate and cocoa products representative of those generally consumed in the United States. Complete nutritional data for the various products analyzed are given in Tables 10–14 for analyses conducted by Philip Keeney's laboratory at Pennsylvania State University, and in Table 15 for analyses done at South Dakota State University. Where possible, data on more than one sample of a given variety or type of product are presented.

9.1. Polyphenols.   Chocolate and cocoa have been shown to be a rich source of antioxidant polyphenols. Flavonoids, a major subgroup, have been associated with the development of color and flavor. Flavonoids found in cocoa beans include epicatechin, catechin, and oligomers.

In fact, dark chocolate and cocoa were shown to contain more polyphenols on a dry defatted basis than 23 fruits, vegetables, and beverages (34). Data is growing indicating that polyphenols in the diet can help maintain cardiovascular health. Incorporating chocolate and cocoa in a diet that is rich in other food sources of antioxidants could reduce the risk of cardiovascular disease (35).

10. Economic Aspects

Chocolate consumption on a global basis was ~ $41 billion in 2000. In the United States, Hershey, Mars, and Nestlé control about 70% of the market. For Europe, Nestlé (including Rowntree), Mars, Kraft-Jacob Suchard (Philip Morris), Cadbury, and Ferrero control >70% of the chocolate trade. In Japan, Lotte, Meiji, Fujiya, Morinaga, and Ezaki Glico sell 88% of the chocolate. The reported

Table 10. **Amino Acid Content of Cocoa and Chocolate Products, mg/g**

	Whole beans[c]	Chocolate liquor[a]		Cocoa[b]		Sweet chocolate[d]	Milk chocolate	
		Natural	Dutch	Natural	Dutch		12% MS[e]	20% MS[f]
tryptophan	1.2	1.3				0.6		
threonine	3.5	3.9	3.6	7.7	8.0	1.5	1.8	2.8
isoleucine	3.3	3.8	4.0	7.0	7.4	1.4	2.2	3.5
leucine	5.3	6.0	6.3	11.5	11.3	2.3	3.8	6.1
lysine	4.8	5.1	5.1	8.7	8.3	1.9	2.4	3.9
methionine	0.7	1.1	0.9	2.0	1.7	0.4	0.9	1.4
cystine	1.4	1.1	1.0	2.1	2.1	0.4	0.3	0.4
phenylalanine	4.1	4.9	5.3	9.9	9.7	1.7	2.2	3.6
tyrosine	2.6	3.5	3.6	7.8	8.0	1.2	1.9	3.0
valine	5.1	5.8	6.3	11.1	10.9	2.1	2.7	4.3
arginine	5.0	5.3	5.1	11.3	11.3	1.9	1.2	1.9
histidine	1.6	1.7	1.7	3.4	3.0	0.6	0.7	1.0
alanine	3.8	4.3	4.1	8.7	8.4	1.5	1.4	2.3
aspartic acid	9.2	10.0	9.8	19.1	18.3	3.9	3.5	5.5
glutamic acid	12.8	14.1	14.1	28.0	26.2	5.7	8.5	13.7
glycine	4.0	4.4	4.5	8.3	8.5	1.6	1.0	1.6
proline	3.4	3.7	3.9	7.6	7.5	1.4	3.4	5.5
serine	3.7	4.1	4.0	6.8	8.2	1.5	1.9	3.0
Total AA recovered[g]	*75.5*	*84.1*	*83.3*	*162.0*	*158.8*	*31.6*	*39.8*	*63.4*

[a] Chocolate liquor = 55% fat.
[b] Cocoa = 13% fat.
[c] Whole beans = 48% fat, 5% moisture, 10% shell.
[d] Sweet chocolate = 35% chocolate liquor, 35% total fat.
[e] 12% MS milk chocolate = 12% whole milk solids, 10% liquor, 32% total fat.
[f] 20% MS milk chocolate = 20% whole milk solids, 13% liquor, 33% total fat.
[g] Total AA recovered = sum of individual amino acids (qv).

Table 11. **Composition of Cocoa Beans and Their Products, Whole Weight Basis in** %

	Total solids	Total protein[a]	Cocoa protein[b,c]	Fat	Ash	Total carbohydrates[d]
whole cocoa beans						
Ghana	92.9	10.1	10.1	47.8	2.7	30.3
	94.0	9.8	9.8	51.6	2.6	28.0
	94.5	10.2	10.2	46.4	2.9	33.0
	94.7	10.3	10.3	46.3	3.1	33.0
Bahia	94.0	10.0	10.0	49.3	2.7	30.0
	94.1	10.2	10.2	48.6	2.7	30.6
	95.1	10.2	10.2	48.2	2.7	32.0
	94.9	10.2	10.2	48.4	2.7	31.6
chocolate liquor						
natural	98.4	9.4	9.4	56.2	2.4	28.5
	98.5	10.2	10.2	55.1	2.6	28.6
	98.9	10.1	10.1	57.0	2.4	27.4
Dutch	98.5	9.2	9.2	55.4	3.8	28.5
	98.6	9.2	9.2	55.4	3.8	28.5
	99.2	9.4	9.4	56.0	3.8	28.1
cocoa						
natural	96.3	18.4	18.4	12.8	4.6	56.9
	96.2	18.4	18.4	16.4	4.8	52.9
	97.4	19.8	19.8	12.7	4.5	56.5
Dutch	97.1	17.5	17.5	12.0	8.3	55.9
	97.4	18.3	18.3	14.3	7.4	53.7
sweet chocolate						
	99.6	3.4	3.4	35.1	1.0	59.4
	99.3	3.8	3.8	36.5	1.0	57.3
	99.5	3.6	3.6	35.0	1.0	59.2
milk chocolate						
12%	99.2	4.2	1.0(3.2)	34.7	0.9	59.2
whole milk solids						
	99.5	4.3	1.1(3.1)	30.2	1.0	63.8
	99.6	4.5	1.1(3.4)	32.3	0.9	61.6
	99.5	4.0	1.4(2.6)	29.6	1.0	64.6
20%	98.8	6.6	1.3(5.2)	34.4	1.5	56.1
wholw milk solids						
	99.5	6.5	1.2(5.2)	33.1	1.4	58.3
	99.4	6.8	1.4(5.4)	30.5	1.5	60.4

[a] Total protein = milk protein + cocoa protein.
[b] Cocoa protein = (total nitrogen − milk nitrogen) × 4.7
[c] Milk protein = milk nitrogen × 6.38 appears in parentheses.
[d] Total carbohydrate by difference using cocoa nitrogen × 5.63.

international average for chocolate consumption was 14.2 lb per capita. Per capita chocolate consumption for some leading countries include:

Switzerland	23.11 lb
Germany	22.03 lb
Austria	20.57 lb
Ireland	18.81 lb
Norway	18.74 lb
United States	11.93 lb

Table 12. **Vitamin Content of Various Samples of Cocoa Beans and Chocolate Products,**[a] **Whole Weight Basis, mg/100 g**

Samples	B_1	B_2	Pantothenic acid	Niacin	B_6
whole cocoa beans					
Ghana	0.21	0.16	0.24	0.19	0.22
	0.17	0.18	0.35	1.07	0.21
	0.19	0.18	0.57	0.91	0.18
	0.16	0.15	0.32	0.52	0.01
Bahia	0.14	0.18	0.34	0.46	0.61
	0.17	0.18	0.35	1.13	0.16
	0.13	0.27	0.61	1.00	0.16
	0.16	0.16	0.38	0.81	0.09
chocolate liquor	0.08	0.17	0.20	0.88	0.09
	0.11	0.16	0.27	1.02	0.20
	0.08	0.15	0.17	1.01	0.16
	0.05	0.11	0.15	0.29	0.02
cocoa	0.05	0.19	0.33	1.34	0.17
	0.13	0.23	0.35	1.53	0.17
	0.15	0.22	0.32	1.37	0.24
milk chocolate	0.07	0.10	0.37	0.14	0.02
	0.11	0.24	0.37	0.38	0.02
	0.07	0.16	0.45	0.21	0.07
	0.10	0.25	0.61	0.24	0.08
	0.15	0.33	0.32	1.11	0.20

[a] Vitamin A and C, negligible amounts present.

The leading chocolate companies continue to pursue a global confectionery business strategy with an increase in the early 1990s of confectionery business activity in the Eastern Bloc countries, Russia, China, and South America. Generally as per capita income increases, chocolate consumption increases, and sugar consumption decreases. Consumer demographics, the declining child population, and the increase in consumer awareness of health issues play important roles in the economics of chocolate consumption. Chocolate confectionery busi-

Table 13. **Tocopherols of Chocolate of Cocoa Beans and Chocolate Products, mg/100 g**

	Total tocopherol	Alpha tocopherol
Bahia-Ghana beans	10.3	1.0
liquor, natural	10.9	1.1
liquor, Dutch	10.0	0.8
cocoa butter, natural	19.2	1.2
cocoa butter, Dutch	18.7	1.1
cocoa, natural	2.3	0.2
cocoa, Dutch	2.2	0.2
dark chocolate	6.0	0.7
milk chocolate, 12% milk	5.6	0.7
milk chocolate, 20% milk	6.3	0.7

Table 14. **Fatty Acid Composition of Raw Cocoa Beans and Cocoa Butter**[a]

	Fatty acid[b], mol %					
	14:0	16:0	18:0	18:1	18:2	20:0
cocoa beans						
Ghana	0.16	28.31	34.30	34.68	2.55	
	0.53	30.20	31.88	33.55	3.85	
	0.19	31.72	32.57	32.82	2.70	
	0.23	31.50	32.39	33.06	2.82	
Bahia	0.15	29.29	31.70	35.24	3.62	
	0.12	26.68	32.06	37.90	3.24	
	0.25	33.99	28.80	33.62	3.34	
	0.19	30.91	30.37	35.22	3.31	
natural cocoa butter	0.15	27.08	32.64	35.61	3.63	0.89
	0.19	27.68	32.64	35.03	3.63	0.83
	0.14	28.42	32.55	34.71	3.23	0.95
	0.14	27.29	32.41	35.36	3.70	1.10
Dutch cocoa butter	0.16	27.23	32.69	35.54	3.31	1.07
	0.15	26.63	34.24	34.68	3.52	0.78
	0.15	26.47	33.53	35.45	3.40	1.00

[a] Calculated from peak areas of the gas chromatograms.
[b] Fatty acid is designated by chain length followed by sites of unsaturation.

ness trends include product down-sizing leading to snack size finger foods, increased emphasis on specialty chocolates with concentration on dessert chocolates, chocolate with high liquor content, organic chocolate, chocolate made with cocoa beans from a specific origin, and chocolate brand equity spread into beverages, baked goods, frozen novelties, and even sugar confections.

Table 15. **Mineral Element Content**[a] **of Cocoa and Chocolate Products, mg/100 g**

Product	Ca	Fe	Mg	P[b]	K	Na	Zn	Cu	Mn
raw Accra nibs	59.56	2.50	232.15	385.33	626.70	11.98	3.543	1.930	1.600
raw Bahia nibs	52.73	2.45	229.11	383.33	622.55	13.55	3.423	1.940	2.060
natural cocoa	115.93	11.34	488.51	7716.66	1448.56	20.12	6.306	3.620	3.770
Dutch cocoa	111.41	15.52	475.98	7276.00	2508.58	81.14	6.370	3.610	3.750
chocolate liquor	59.39	5.61	265.23	3996.66	679.61	18.89	3.530	2.050	1.850
12% milk chocolate	106.41	1.23	45.56	159.00	156.64	80.09	0.773	1.020	0.282
20% milk chocolate	174.00	1.40	52.26	207.96	346.33	115.40	1.240	0.126	0.139
dark chocolate	26.33	2.34	93.70	142.90	302.53	18.63	1.500	0.432	0.345

[a] Mean values from duplicate analyses of each of three samples by atomic absorption spectrophotometry.
[b] Total phosphorus—;ash below 550°C (AOAC procedure).

BIBLIOGRAPHY

"Chocolate and Cocoa" in *ECT* 1st ed., Vol. 3, pp. 889–918, by W. Tresper Clarke, Rockwood & Co.; in *ECT* 2nd ed., Vol. 5, pp. 363–402, by B. D. Powell and T. L. Harris, Cadbury Brothers Ltd.; in *ECT* 3rd ed., Vol. 6, pp. 1–19, by B. L. Zoumas, E. J. Finnegan, Hershey Foods Corp.; in *ECT* 4th ed., Vol. 6, pp. 177–199, by B. L. Zoumas and J. F. Smullen, Hershey Foods Corporation; "Chocolate and Cocoa" in *ECT* (online), posting date: December 4, 2000, by B. L. Zoumas and J. F. Smullen, Hershey Foods Corporation.

CITED PUBLICATIONS

1. S. Coe, and M. Coe, *The True History of Chocolate*, Thames and Hudson Inc., New York, 1996. p. 36
2. *Report of the Cocoa Conference, Cocoa, Chocolate and Confectionery* Alliance, London, 1957.
3. New York Board of Trade http://www.nybot.com or http://www.csce.com.
4. International Cocoa Organization (ICCO), Quarterly Bulletin of Cocoa Statistics (2000–2001).
5. G. A. Reineccius P. G. Keeney, and W. Weissberger, *J. Agric. Food Chem.* **20**(2), 202 (1972).
6. J. C. Hoskin and P.S. Dimick, in S.T. Beckett, ed., *Industrial Chocolate Manufacture and Use*, 2nd ed. Blackie New York, 1994, p. 111, chapt. 8.
7. H. R. Riedl, *Confect. Prod.* **40**(5), 193 (1974).
8. A. W. Knapp and A. Churchman, *J. Soc. Chem. Ind. (London)* **56**, 29 (1937).
9. A. Szegvaridi, *Manuf. Confect.* **50**, 34 (1970).
10. H. J. Schemkel, *Manuf. Confect.* **53**(8), 26 (1973).
11. *Report of Codex Committee on Cocoa Products and Chocolate, Codex Alimentarious Commission*, 18th Session, Fribourg, Switzerland, 2000.
12. T. P. Hilditch and W. J. Stainsby, *J. Soc. Chem. Ind.* **55**, 95T (1936).
13. M. L. Meara, *J. Chem. Soc.*, 2154 (1949).
14. S. J. Vaeck, *Manuf. Confect.* **40**(6), 35 (1960).
15. Codex Standards for Cocoa Products and Chocolate, Cocoa Butter Standard 86-1981, *Codex Alimentarius*, Vol. **VII**, 1st ed., Joint FAO/WHO Food Standards Program, 1981.
16. J. Robert Ryberg, *Cereal Sci. Today* **15**(1), 16 (1970).
17. K. Wolf, *Manuf. Confect.* **57**(4), 53 (1977).
18. P. Kalustian, *Candy Snack Industry* **141**(3), 1976.
19. B. O. M. Tonnesmann, *Manuf. Confect.* **57**(5), 38 (1977).
20. Code of Federal Regulations, No. 21, Part 163, *Cacao Products*, Apr. 1, 2001.
21. B. W. Minifie, *Chocolate, Cocoa, and Confectionery: Science and Technology*, AVI, Westport, Conn., 1970.
22. B. Christiansen, *Manuf. Confect.* **56**(5), 69 (1976).
23. H. R. Riedl, *Confect. Prod.* **42**(41), 165 (1976).
24. E. M. Chatt, in Z. J. Kertesz, ed., *Economic Crops*, Vol. 3, Interscience Publishers, Inc., New York, 1953, p. 185.
25. L. R. Cook, *Manuf. Confect.* **56**(5), 75 (1975).
26. H. C. J. Wijnougst, *The Enormous Development in Cocoa and Chocolate Marketing Since 1955*, Mannheim, Germany, 1957, p. 161.
27. J. Kleinert, *Manuf. Confect.* **44**(4), 37 (1964).
28. R. Heiss, *Twenty Years of Confectionery and Chocolate Progress*, AVI, Westport, Conn., 1970, p. 89.

29. E. H. Steiner, *Inter. Choc. Rev.* **13**, 290 (1958).
30. N. Casson, *Br. Soc. Rheo. Bull.* 52, (Sept. 1957).
31. M. G. Reade, UK, personal communication, 1990.
32. W. N. Duck, Ref. 3, p. 22.
33. W. R. Kreiser and R. A. Martin, *J. Assoc. Off. Analy. Chem.* **61**(6), (1978).
34. J.A. Vinson, J. Proch, and L. Zubik, *J. Agric. Food Chem.* **47**(12), 4821 (1999).
35. Y. Wan, J.A. Vinson, T.D. Etherton, J. Proch, S.A. Lazarus, P.M. Kris-Etherton, *Am. J. Clin. Nutr.* **74**, 596 (2001).

B. L. ZOUMAS
C. D. AZZARA
J. BOUZAS
The Pennsylvania State University,
Hershey Foods Corporation

CHROMATOGRAPHY

1. Introduction

Chromatography is a technique used in many areas of science and engineering: petroleum chemistry, environmental studies, foods and flavorings, pharmaceuticals, forensics, and analysis of art objects (see BIOPOLYMERS, ANALYTICAL TECHNIQUES; FINE ART EXAMINATION AND CONSERVATION; FORENSIC CHEMISTRY), for separating and quantifying the constituents of a mixture. Since most chemical processes result in mixtures, separation techniques are essential for a successful characterization of chemical reactions. Most chemical laboratories employ one or more chromatographs for routine chemical analysis (1), and many processes involve the preparative use of chromatography for obtaining pure materials.

The first scientific reports demonstrating chromatographic phenomena appeared in the 1890s. However, the era of analytical chromatography began in 1903 when a paper described the separation and identification of the components of a mixture of structurally similar yellow and green chloroplast pigments in leaf extracts in carbon disulfide passed through a column packed with chalk (2). The technique was seen as potentially valuable for identifying compounds other than by color. In 1906 the term chromatography was coined for these processes, from the combination of two Greek roots "chroma", meaning color, and "graphe", meaning writing (3). Since that time, a wide variety of chromatographic techniques have been developed to provide separation of many mixtures, the components of which differ only subtly from each other, as in the separation of isomers.

Chromatography relies on differential interactions of the components of a mixture with the phases of a chromatographic system to produce separation, e.g. adsorption on the stationary phase. Thus, in addition to providing separation, the study of chromatographic parameters provides a means to determine fundamental quantities describing the interactions between the phases and the components, such as stability constants, vapor pressures, and other thermodynamic data.

The importance of chromatographic processes to science can be gauged in many ways. For example, the 1952 Nobel Prize in chemistry was awarded to A. J. P. Martin and R. L. M. Synge for the development of liquid–liquid chromatography, (4) which led not only to an understanding of the fact that both liquid phases need not move simultaneously to effect separation, but also to development of gas–liquid chromatographic techniques (glc) techniques. A second measure of the importance of chromatography is the number of chromatographic instruments in chemical and biological laboratories (1). By whatever measure, the impact of chromatography on chemistry practice has been immense.

A primary use of chromatography is the analysis of mixtures by passage through a column in which the differential interactions cause the components to pass through at different rates. The measurement of the rate of each material is a means of identification of the material, which is analytical chromatography.

Analysis in the laboratory is only one use of chromatography. On-line chromatographic devices are used as analyzers in many chemical processes in industrial applications. Preparative chromatography, in which the goal is recovery of large amounts of materials after purification by separation, is a major tool for process separations. All forms of chromatography are used for preparative purposes, but liquid chromatography has been of highest value, especially in biological and pharmaceutical applications. In preparative chromatography, the column capacity (the amount of material passed through it) is usually stretched to the limit to allow the greatest quantity of sample to be added to the mobile phase. Since chromatography requires a detection of the material, the detector for preparative chromatography is normally nondestructive.

2. Principles

The principle of chromatographic separation is straightforward. A mixture is allowed to come into contact with two phases, a stationary phase and a mobile phase. The stationary phase is contained in a column or sheet through which the mobile phase moves in a controlled manner, carrying with it any material that may prefer to mix with it. Because of differences in the interactions of the mixture's constituents with stationary and mobile phases (the relative affinity of the constituents), the constituents are swept along with the mobile phase at different rates, so that they arrive at the end of the column at different times. This selective interaction is known as partitioning, and the different components are retained on the column for different times. To determine the retention time of substances on the column, a detector measures either the time required to travel to the end of the column or, as for thin-layer chromatography (tlc), the distance traveled in a fixed time. The detector may be as simple as the human nose or the human eye or as complex as a microsensor. A plot of detector response versus time of travel for a fixed distance is called a chromatogram.

In adsorption chromatography the constituents in the dissolved sample compete with the mobile phase for the active sites on the stationary phase. Retention is determined by how effectively the constituents interact with the active sites. To remove constituents adsorbed on the stationary phase in adsorption chromatrography, the mobile phase's affinity for the stationary phase is increased by changes in its composition, which affects the retention of the analyte.

In preparative chromatography a device may be attached to the end of a column to collect the separated components of a mixture. One must be able to detect the material, so this collection device is in series with the detection device.

The nature of the stationary and mobile phases in a chromatographic experiment determines the efficacy of component separation in a particular mixture. A wide variety of stationary and mobile phases is used. A classification based on the nature of the phases is shown in Figure 1. The stationary phase may be a solid or a liquid supported on a solid. The mobile phase may be a gas, a liquid, or a material such as a supercritical fluid (see SUPERCRITICAL FLUIDS). One names a specific chromatographic technique by naming the mobile and the stationary phase, in that order. Thus, gas–liquid chromatography (glc) uses a gaseous mobile phase in contact with a film of liquid stationary phase.

2.1. Development of the Chromatogram. The term "development" describes the process of performing achromatographic separation. Because the processes that determine retention depend on the nature of the stationary and mobile phases, there are several ways in which separation may be made to occur, and several different ways for development of chromatograms.

Chromatographic techniques may be characterized as frontal, displacement, or elution chromatography. Elution is, by far, the most common, and most people think of elution chromatography upon hearing the word "chromatography".

Frontal chromatography (or frontal analysis) is a technique in which the sample is introduced onto a column continuously. In essence, the sample collected at the end of the column is free of materials that adsorb/absorb on the stationary phase. Once the bed, i.e. the stationary phase, is saturated and no longer removes the adsorbing component, the effluent at the column's end contains

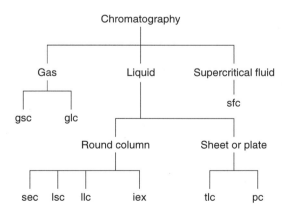

Fig. 1. Classification of chromatographic systems. gsc, gas–solid chromatography; glc, gas–liquid chromatography; sec, size-exclusion chromatography; lsc, liquid–solid chromatography; llc, liquid–liquid chromatography; iec, ion-exchange chromatography; tlc, thin-layer chromatography; pc, paper chromatography; sfc supercritical-fluid chromatography.

these materials. By using an appropriate detector, the condition at which this transition occurs can be determined, as determined by the capacity of the column. Figure 2**a** shows an example of an integral chromatogram from a frontal analysis, in which three adsorbing components gradually saturate the column.

In displacement chromatography a small sample on the column is displaced by a much more strongly held mobile phase. The sample is gradually pushed through the column as the mobile phase advances. As this happens, the components are dispersed into bands that can either be excised to obtain the pure material or displaced from the column. Such techniques are useful for the generation of quantities of pure material. Figure 2**b** shows an integral chromatogram obtained by displacement analysis.

Both frontal and displacement chromatographies suffer a significant disadvantage in that once a column has been used, part of the sample remains on the column, which means it is contaminated for further use. The column must be regenerated before reuse. In elution chromatography all of the sample material is usually removed from the column during the chromatographic process, allowing reuse of the column without regeneration. Most analytical applications of chromatography employ elution methods, in which a small sample is put onto the column, at the column head as a plug or a band. The sample is applied, sometimes by injection, while the mobile phase is moving through the column. Because of the difference in affinities of the sample's components for the stationary phase and the mobile phase, constituents travel through the column at different rates and elute at different times. Figure 2**c** shows a typical differential elution chromatogram.

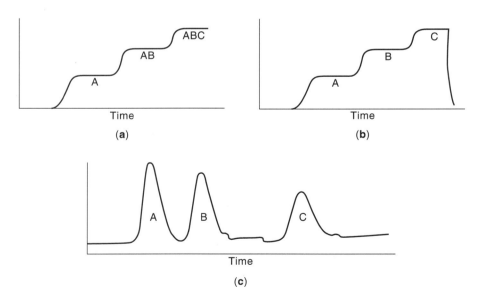

Fig. 2. Chromatograms of a mixture containing three components A, B, and C, where A is less sorbed than B, and B is less sorbed than C; (**a**) frontal analysis; (**b**) displacement analysis; and (**c**) differential elution chromatogram.

3. Gas Chromatography

The most frequently used chromatographic technique is gc, for which instrumentation was first offered commercially in the mid-1950s. Gas chromatographs were the most frequently mentioned analytical instrumentation planned for purchase in laboratory surveys in 1990, and growth in sales has been ~6%/year (1,5).

While solids and liquids are both used as stationary phases in gc, the most commonly used method is glc. Separation in a glc arises from differential partitioning of the sample's components between the stationary liquid phase bound on a porous solid, and the gas phase. In the other variant, gsc, preferential adsorption on the solid or, sometimes, exclusion of materials by size are the means of differentiating between components.

3.1. Packed Columns. The chromatographic column is often described as the heart of the chromatographic system, because it is the single part of the system that must be present to effect a separation. Columns come in a wide variety of sizes and shapes. A schematic drawing of a packed column is shown in Figure 3. They are frequently tubes made of various materials. In these tube is the stationary phase, either coated or bonded chemically to the packing. In general, the larger the diameter of the column, the poorer the separation of the components. Large (> 2-mm internal diameter) columns are often used in preparative chromatography, whereas medium (diameter between 1 and 2 mm) columns are used for analytical chromatography. In glc a packing is coated with a liquid phase, either by chemical bonding or physical coating on porous particles presumed to be inert. In gsc, the packing is usually the stationary phase. The packing is the dominant part of the column, as its interaction with the material determines the retention that underlies the separation process. We discuss this aspect in more detail below.

3.2. Capillary Chromatography. Capillary, or wall-coated open tubular (WCOT), columns are fine tubes having internal diameters in the range of 0.25–0.53 mm. Chromatography with these columns (usually with no packing) gives exceedingly high resolution gc. As the name implies, these columns' walls are often coated with some substance that is, in effect, the stationary phase. Capillary columns are so efficient, they require very little sample, so the system for introduction of a sample onto this sort of column requires the use of a sample splitter, a device to allow a representative aliquot of the sample

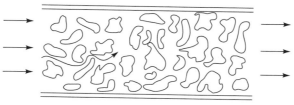

Fig. 3. A packed chromatographic column showing the thin column walls and the irregularly shaped solid support coated with a liquid phase. The arrows indicate the movement of the mobile phase.

onto the column, because the sample capacity of a capillary column is not very great. Wide-bore open-tubular columns (those with column diameters of 0.53 mm) are particularly useful for coupled tandem techniques in which a greater quantity of material is needed for detection: gas chromatography–mass spectrometry (gc–ms), gas chromatography–Fourier transform infrared spectroscopy (gc–ftir), and gas chromatography–nuclear magnetic resonance spectroscopy (gc–nmr), (see ANALYTICAL METHODS, HYPHENATED INSTRUMENTS).

3.3. The Stationary Phase. The stationary phase in an open-tubular column is generally coated or chemically bonded to the wall of the capillary column. In a packed column, it is attached to the packing (or support). For capillary columns, the greater the thickness of the liquid-phase (film) coating, the greater the retarding potential of the column for components attracted to the stationary phase. Chemically bonded stationary phases are usually more resilient than nonbonded phases, tending not to wash out as large amounts of solvent pass through the column, and having much better thermal stability than do nonbonded (coated) phases. In some cases, increased stability of the stationary phase can be achieved by acting on it chemically. The phase can be cross-linked to give more mechanical, chemical, and thermal stability. Cross-linked phases are very stable and can withstand solvent washing to clean the interior of the chromatographic column. Several hundred types of liquid stationary phases are commercially available. These have been used individually or in combination with other liquid phases, inorganic salts, acids, or bases. The selection of stationary phases for a particular application is beyond the scope of this article, however, it is one of the most important chromatographic tasks. Stationary phase selection is discussed at length in books, journal articles, and catalogs from vendors. See *General References* for examples.

3.4. The Support Material. The support is the inert frame onto which the liquid phase is applied. For capillary chromatography, this is the capillary material such as fused silica. The most common support materials for packed-column gc are the diatomaceous earths, the remains of diatoms, single-cell algae. The porous siliceous material has pores ~1 μm in diameter (see DIATOMITE). These materials are typically treated with sodium carbonate to ~900°C. Following this treatment, they are sieved to obtain material of reasonably uniform dimensions. Supports for packed-column chromatography are classified by particle size. Most chromatographic packings are 80/100 mesh (177–149 μm). Supports are washed using acid (HCl) and silanized, i.e. treated with dimethyl-dichlorosilane, DMCS, to reduce the polarity. Silanizing replaces adjacent SiOH groups with nonpolar CH_3 caps. In addition to diatomaceous earths, packed-column supports of carbon, halocarbons, Teflon®, and glass beads are in use by various chromatographers.

3.5. Separation by Molecular Size. The most common solid-phase packing in gsc is a molecular sieve. Molecular sieves are zeolites or carbon sieves that have a regular pore structure and are used almost exclusively for the separation of small molecules such as permanent gases like oxygen, nitrogen, carbon monoxide, argon, and nitric oxide, or low carbon-number hydrocarbons.

Porous organic polymers have been used as gsc packings for separating low molecular weight mixtures containing halogenated or sulfur-containing compounds, water, alcohols, glycols, free fatty acids, ketones, esters (see ESTERS,

ORGANIC), and aldehydes. Such porous polymers usually have a maximum operating temperature lower than many common liquid phases.

Silica, alumina, and other metal oxides and salts have been used as the stationary phase in gsc systems. The applicability of these materials is limited by the difficulty of producing a consistent, resilient, reproducible material.

3.6. Column Tubing.

A packed column is contained in tubing, the composition of which may have a dramatic effect on the separation process, because the sample components may also interact with the walls of the tube. For sensitive compounds such as certain pharmaceuticals, steroids, and pesticides, the standard practice is to use columns packed in glass tubes. The surface of glass is more nearly inert than are the surfaces of metal tubes. Glass columns also have the advantage of being transparent, giving a means to examine the column for degradation, contamination, packing efficiency, and column voids. Glass, however, is more fragile than metal.

Tubes made of metal such as stainless steel, nickel, copper, or aluminum are much more resistant than glass to damage in handling. Stainless steel is often preferred because it is less active than other metals. If corrosive samples are used, however, it is sometimes necessary to contain the column in tubing made of a material such as Teflon®. Columns made of Teflon® generally are difficult to pack. Additionally, connections to the tubing are difficult to make.

For capillary columns, fused silica is the material of choice for the column container. It has virtually no impurities (<1 ppm metal oxides) and tends to be quite inert. In addition, fused silica is relatively easily processed and manufacture of columns from this material is reproducible. Fused-silica columns are externally coated using a protective polyimide layer to improve strength and durability, and to provide a measure of protection against reaction of the silica with water in the environment. The fused-silica column is an inherently straight wire of material, i.e. its resting state is straight, not coiled. To use the material in a chromatographic oven, it must be wound onto a frame that secures it in the coiled configuration. The process of creating a fused-silica column is complex and requires sophisticated, expensive equipment, a high temperature (2000°C) furnace, and a laser-based system for determining the trueness of diameter of the ultimate product.

In addition to fused-silica capillary columns, there are several designs for glass capillary columns, although few are widely used. One system is the so-called porous-layer-open-tubular (PLOT) column, made from glass pulled to a capillary with an internal layer of solid packing. Subsequently this material is removed by etching to produce a column with many pores at the surface of the capillary. Support-coated-open-tubular (SCOT) columns are glass capillary columns having a coating on the column wall. Micropacked columns are glass capillaries packed using small-mesh particles similar to those in liquid–chromatographic columns. There are several different types of capillary column designs.

3.7. The Mobile Phase.

The purpose of the mobile phase, also called the carrier gas, is to transport the sample through the chromatographic column. The selection of carrier gas is often dictated by the type of detector attached to the gas chromatographic system. To achieve the best performance, gases such as nitrogen, carbon dioxide, or argon are used as carriers. If these gases do not permit sufficiently high gas velocities, then a lower molecular weight gas such as helium or hydrogen is used. The purity of the carrier gas is an important consideration because gas passes through the column and impurities could interfere with chromatographic separation or could contaminate the column. Gases used are generally of 99.995% purity or better. Two particularly troubling contaminants are water and air, which can affect the stability of the liquid phase in a packed column. The best compromise for column performance and safety is helium for capillary columns and either helium or nitrogen for packed columns.

3.8. Detectors.   The function of the gc detector is to sense the presence of a constituent of the sample at the outlet of the column. Selectivity is the property that allows the detector to discriminate between constituents. A selective detector responds especially well to compounds of one type, but not to other chemical species. The response is the signal strength generated by a given quantity of material. Sensitivity is a measure of the ability of the detector to register the presence of the component of interest, usually given as the quantity of material that can be detected having a response at twice the noise level of the detector.

By far the most used nondestructive detector is the thermal conductivity detector (TCD). This detector is used for preparative chromatography. Detectors like the TCD are called bulk-property detectors, in that the response is to a property of the overall material flowing through the detector, in this case the thermal conductivity of the stream, which includes the carrier gas (mobile phase) and any material that may be traveling with it. The principle behind a TCD is that a hot body loses heat at a rate that depends on the composition of the material. Most materials have lower thermal conductivities than helium, the typical carrier gas, and most organic materials have similar thermal conductivities, thus the TCD is often used for quantitative analysis of organics with gc. Of course, the thermal conductivities of organic compounds are not exactly the same, so very accurate results require an evaluation of the response factor for each material, essentially an evaluation of the thermal conductivity of each material.

The detector most extensively used in capillary chromatography is the flame-ionization detector (FID). The principle behind its operation is the detection of a current from ions formed when organic materials are burned in a small hydrogen−oxygen flame at the end of the column. Typically a voltage is applied across this region and the small current carried by the ions is detected using a sensitive electrometer. One of the advantages of the FID is that it is, in general, more sensitive than the TCD. In addition, it does not respond to materials such as water, carbon dioxide, carbon monoxide, and most simple sulfur-containing gases. The principal disadvantage of the FID is that it destroys the separated material in the process of detection.

A third detector type is the electron-capture detector (ECD), which is very selective for the detection of highly electronegative compounds in the effluent such as chlorinated hydrocarbons, many pesticides, and polychlorinated biphenyls (see CHLOROHYDROCARBONS). Its principle of action is the interaction of such compounds with electrons emitted from a radioactive source. A detector sensitive to these beta emissions positioned across the stream from a source senses a drop in emissions when the stream contains compounds that capture the emitted electrons. Thus the response to the passage of materials is a loss of signal. The sources generally used in these detectors are nickel-63 or tritium. One disadvantage is the radioactive source, which requires special handling. In addition, the linear response range is not very great and the detector is subject to high background noise, unless care is taken to eliminate column, carrier-gas, or sample contaminants.

The flame-photometric detector (FPD) is selective for organic compounds containing phosphorus and sulfur. In this detector, chemiluminescent species formed in a flame from these materials are detected through a filter by a photomultiplier. The photometric response is linear in concentration for phosphorus, but it is second order in concentration for sulfur.

The alkali flame-ionization detector (AFID), sometimes called a thermionic (TID) or nitrogen–phosphorus detector (NPD), has as its basis the fact that a phosphorus- or nitrogen-containing organic material, when placed in contact with an alkali salt above a flame, forms ions in excess of thermal ionic formation. The ions are detected as a current. Such a detector at the end of a column reports on the elution of these compounds. The mechanism of the process is not clearly understood, but the enhanced current makes this type of detector popular for trace analysis of materials such as phosphorus-containing pesticides.

The ms is a common adjunct to a chromatographic system (see MASS SPECTROMETRY). The combination of a gas chromatograph for component separation and a mass spectrometer (gc–ms) for detection and identification of the separated components is a powerful tool, particularly when the data are collected using an on-line data-handling system. Qualitative information inherent in the separation can be coupled with the identification of structure and relatively straightforward quantification of a mixture's components.

Infrared (ir) spectrometers are used as detectors for gc systems, particularly because the ftir spectrometer allows spectra of the eluting stream to be gathered quickly. The data are valuable alone and as an adjunct to experiments with gc–ms. Gc–ir is a definitive tool for identification of isomers (see INFRARED AND RAMAN SPECTROSCOPY).

Plasma atomic emission spectrometry is also employed as a detection method for gc (see PLASMA TECHNOLOGY). By monitoring selected emission lines, selective detection of materials based on elemental composition can be achieved (see SPECTROSCOPY).

3.9. Theory. Most theoretical models of gas chromatographic processes are based on analogy to processes such as distillation or countercurrent extraction experiments (6). The separation process is viewed as a type of successive partitioning of the components of a mixture between the stationary and mobile phases similar to the partitioning that occurs in distillation columns. In those experiments an important parameter is the number of theoretical plates of which the column may be considered to be composed; the greater the number of theoretical plates, the greater the efficiency of the column for achieving separations of similar components. In gc, the equivalent measure of efficacy is the height equivalent to theoretical plates (HETP), which measures the ultimate ability of the column to separate like components. This quantity depends on many instrumental parameters such as wall or particle diameter, type of carrier gas, flow rate, liquid-phase thickness, etc. The theoretical expression relating HETP to gas velocity is the van Deemter equation:

$$\text{HETP} = A + B/\mu + C\mu$$

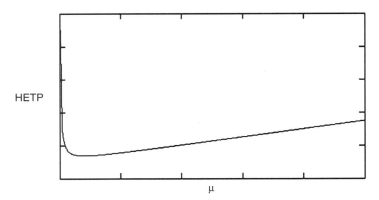

Fig. 4. An example of van Deemter's relation between gas velocity, μ, and HETP number.

where A represents the effects of eddy diffusion, the term containing B represents the effects of molecular diffusion, μ is the linear gas velocity, and C represents the resistance to mass transfer. A plot of HETP versus C gives a hyperbola with a minimum HETP, as shown in Figure 4. This minimum is the optimal gas velocity, μ, where the column operates most efficiently.

3.10. Inlet Systems.   The inlet (or injector) is the means by which the sample is introduced onto the gc column. Sample introduction requires one to create a representative aliquot of the sample at the beginning of the column without degradation or without discrimination among the components of the sample. Most inlets operate on the principle that a sample can be vaporized quickly, assuming it is not already a gaseous material, after being injected from a microliter syringe into a small, heated volume, usually ~50°C hotter than the maximum temperature of the column during the experiment. This vaporized material is quickly swept as a narrow band onto the column by the carrier gas. Once on the column, the components interact with the stationary phase and begin to travel along with the carrier gas at differing rates, depending on the strengths of interaction with the stationary phase.

Sample introduction is critical to the proper operation of a gas chromatograph. To obtain reliable data, a representative sample must reach the column in a narrow band so all components begin the chromatographic process simultaneously. Injection with a syringe is the most widely used technique. However, other means such as injection valves, pyrolyzers, headspace samplers, thermal desorbers, and purge-and-trap samplers are found in various applications.

Inlets for syringe sampling are divided into two categories, depending on the column. For packed columns, all injected material is carried by the mobile phase onto the column, and the inlet is usually an open tube. Sometimes, albeit rarely, the inlet itself may be packed, eg, to assure that the first centimeters of the column do not become contaminated with degradation products or nonvolatile materials that may affect the efficacy of the column.

The second category is inlets for capillary columns. Because of the much smaller capacity of the column, injection requires great care and special techniques. There are four different inlet designs for this purpose: direct, split/split-less, programmed-temperature vaporization, and cool on-column injectors. Direct inlets are generally used with capillary columns of larger diameter and work much as direct injectors for packed columns.

Split/splitless injectors operate in two modes. In the split mode, most of the sample introduced into the inlet goes out a vent that has less resistance to flow than the column. Splitters usually exhaust 90–99% of the volume of the material injected, so that only a small percentage of the injected material enters the column. A primary problem with this injection mode is that the material finding its way onto the column is not always representative of the sample. Discrimination resulting from differential boiling points of the components makes it probable that certain components of the mixture are vented, rather than introduced onto the column. In the splitless mode, the vent is turned off and everything injected goes onto the column. After a short period, the vent is opened and any residual solvent is vented. The splitless mode is found particularly in trace analytical schemes (see TRACE AND RESIDUE ANALYSIS). Splitless sample injection is an art that requires practice to ensure reproducible introduction of sample onto the column.

Programmed-temperature vaporizers are flexible sample-introduction devices offering a variety of modes of operation such as split/splitless, cool-sample introduction, and solvent elimination. Usually the sample is introduced onto a cool injection-port liner so that no sample discrimination occurs. After injection, the temperature is increased to vaporize the sample, and the sample finds its way onto the column.

Cool on-column injection is used for trace analysis. The sample is introduced without vaporization by inserting the needle of the syringe at a place where the column has been previously stripped of liquid phase. The injection temperature must be at or below the boiling point of the solvent carrying the sample. Injection must be rapid and of a very small volume. Cool on-column injection is the most accurate and reproducible injection technique for capillary chromatography, but it is the most difficult to automate.

3.11. Temperature Considerations.

The inlet, detector, and the oven compartment with the column are usually controlled at different temperatures, because each part serves a different function best performed in a specified temperature range. In practice, the maximum oven temperature in the course of an analysis should only be high enough to achieve the desired result in a minimum time. This temperature should be low enough to minimize the probability of column liquid-phase degradation. The injection port's temperature is usually slightly higher than the maximum oven temperature, but low enough to minimize thermal degradation or thermal rearrangement of sample components. Ideally, the thermal energy in the injection port will cause instantaneous vaporization without causing a loss of separation. The detector temperature is usually 10–30°C higher than the injector, but low enough to avoid thermal degradation of the column's liquid phase in that part of the column near the detector.

For materials with a wide boiling range, temperature programming is often used. The initial temperature, if possible, should be near the boiling point of the most volatile component; the final temperature should be near or, if possible, slightly higher than the least volatile component. The heating rate from low to high temperature is usually empirically determined to obtain the most efficient separation in the shortest possible analysis time.

4. Liquid Chromatography

Liquid chromatography (lc) refers to any chromatographic process in which the mobile phase is a liquid. Traditional column chromatography, tlc, pc, ce, cec, and high-performance liquid chromatography (hplc) are members of this class. Modern lc techniques originated in the late 1960s and early 1970s. Developments in hplc were driven by improvements in instrumentation, column packings, and theoretical understanding of the various separation processes involved. For example, use of pressurized mobile phases in place of gravity-driven ones in chromatography greatly shortened the time for separation.

Liquid chromatography is complementary to gc because samples that cannot be easily handled in the gas phase, such as nonvolatile compounds or thermally unstable ones, e.g. many natural products, pharmaceuticals, and biomacromolecules, are separable by partitioning between a liquid mobile phase and a stationary phase, often at ambient temperature. Many separations once done exclusively with gc are nowadays conveniently done with lc.

One advantage of lc is that the composition of the mobile phase, and perhaps of the stationary phase, can be varied during the experiment to provide a means of enhancing separation. There are many more combinations of mobile and stationary phases to effect a separation in lc than one would have in a similar gas chromatographic experiment, where the gaseous mobile phase often serves as little more than a convenient carrier for the components of the sample. In classical column chromatography the usual system consists of a polar adsorbent, or stationary phase, and a nonpolar mobile phase such as a hydrocarbon. In many instances, the polarities of the stationary and mobile phases are reversed for the separation, in which case the technique is known as reversed-phase liquid chromatography.

Paper chromatography (pc) originated in the 1940s and tlc came into use in the 1950s. These techniques are similar in the manner of development of the chromatogram. A chamber is used to isolate the column, which is a piece of filter paper in pc and a glass plate coated with an adsorbent such as silica gel in tlc, from the laboratory environment. The chromatogram is developed as a mobile phase migrates through the column, carrying with it materials from the sample deposited above the initial solvent level. Because of differences in solubility in the mobile phase, the components move at different rates up the column. After this process has proceeded sufficiently, the column is removed from the solvent tank and the mobile phase evaporated. The separated components are visualized elsewhere, e.g. under an ultraviolet (uv) lamp in which various fluorescent bands indicate how fluorescent materials are separated by the movement of the solvent (mobile phase). The distance from the origin to the migrated spot at the end of development is used to calculate a retardation factor (r_f), the ratio of the distance a component travels to the distance the solvent front has traveled. Retardation factors are always ≤ 1 and are used to characterize the partitioning of a component for a particular solvent and stationary phase. An advantage of tlc is that a fresh plate is used for each analysis, avoiding "carry-over", contamination from previous experiments. Another important advantage is that many visualizing techniques are available that are not available using lc detectors. In addition, the spots or, more frequently, the band in prep tlc, can be physically cut from the plate.

Paper and tlc may be further classified as one- or two-dimensional (1D or 2D), and as either analytical or preparative. In the traditional 1D pc or tlc described above, a small spot of sample is applied, usually from a micropipet, at a point near the edge from which solvent enters the paper or plate, as described above. In 2D chromatography, after the plate or paper is removed from one solvent and allowed to dry, it is placed in another tank with a different solvent entering the paper or plate at 90° to the direction of travel of the first solvent. The result is a further resolution of components that may have had similar partitioning in the first solvent.

If large amounts of material can be separated, then one may use tlc as a form of preparative chromatography. One-dimensional tlc plates have a very high capacity, e.g. up to 20 samples and standards applied, as compared to single passes in column chromatography (hplc or gc).

Capillary electrophoresis (ce, also know as "capillary zone electrophoresis", cze) is a separations process that occurs inside a small-diameter capillary tube, based on the differential mobilities of components under the influence of an electric field. Fused silica is usually used for the capillary, since silanol groups can easily form a negatively charged surface, which contributes to electroosmotic flow. The movement of the analyte is determined by the sum of the electrophoretic velocity and the electroosometic velocity.

Capillary electrochromatography (cec) is a form of lc that uses an electric field to drive the eluent through the chromatographic column. Although the mobile phase is a liquid, the column types are similar to a gc column with packed, open-tubular, and continuous bed options.

4.1. Columns. As for gc, the column is the heart of the lc system. Columns for modern lc can be packed with a variety of materials: inert particles bonded to a liquid phase (llc); a porous gel as for sec or gpc; an ion-exchange resin as for iec; or an affinity adsorbent as for affinity chromatography (ac). Columns are generally ~5 cm in length. An important consideration is uniformity of the particles, which are generally ~5 μm in diameter for typical lc experiments.

4.2. Packings. Most packings for lc are made of chemically modified silica gel having functional groups covalently attached to the surface of the particles. Zirconia has replaced many of these original stationary phases since zirconia is thermally and chemically stable over the entire pH range from 1–14.

Reversed-phase packings have covalently bonded octadecyl groups (a C-18 phase) or octyl groups (a C-8 phase) at the surface to provide a nonpolar environment. Sometimes the further reaction of these materials with other reagents to attach trimethylsilyl groups (endcapping) is attempted. This treatment is generally supposed to cover the regions of the surface that are not covered by the first treatment, eliminating interactions that may degrade the efficiency of the column. In addition to the C-18 and C-8 column packings, other species used for chemically binding to the support particles include phenyl, nitro, and amino groups.

For sec, porous polymers such as polystyrene–divinylbenzene are sometimes used, as are the usual treated and untreated silicas.

4.3. The Mobile Phase. The great power of lc to separate the components of a mixture lies in the differential solubility of the components in the mobile liquid phase and the stationary phase. In isocratic lc, the composition of the mobile phase remains constant throughout the course of the experiment. However, the effective separating power of lc can often be enhanced by changing the composition of the mobile phase during the course of the experiment. This process, known as gradient lc, is analogous to programming the column temperature in gc. Frequently the switch is from a weak solvent for a given material to a strong one. The change can be made in a single step or by slowly varying the composition of the mobile phase with time during the separation process. Most processes involve two solvents or solvent mixtures, although there are some cases in which three solvents are used. Obviously, the more solvents used, the more complex the program of mixing. An important consideration in gradient lc is the selection of a detector. The detector must be compatible with all the solvents used in the separation process.

Solvent-delivery systems ensure uniform transfer of the mobile phase to the column. These devices must give reproducibly uniform flow without pulsations, to ensure reproducible retention times and peak areas for analyses. Because of the small diameter of the particles used in modern hplc, there is a high resistance to flow through the column, and high-pressure pumps are required (see PUMPS). To obtain the best signal from a detector, it is important that the detector be insensitive to pump strokes at all flows. The pump materials must not only be able to withstand such pressures, they must not be affected by the solvents in the system. For gradient elution lc, the pump should have a small mixing (hold-up) volume. This minimizes memory effects from solvent changes.

Sample introduction onto lc columns is usually accomplished with a sampling valve. A sample loop of volume between 5 and 50 µL attached to the sampling valve is filled or partly filled with sample solution. At the time of introduction of the sample, the valve is either manually or pneumatically actuated so flow of solvent through the sample loop moves sample onto the column.

4.4. Detectors. Liquid-chromatographic detectors must be compatible with the solvent system (mobile phase) and are optimized for sensitivity, stability, and speed of response. They are designed to retain the quality of the separation. No versatile, universal detector is in use for lc, as the flame-ionization or thermal-conductivity detectors are for gc. Instead, the most common detector found in lc is the uv detector, a selective detector that measures the absorption of radiation at a specified wavelength. These devices may be set at a fixed wavelength or the wavelength may be variable. The detectors are only sensitive to materials that absorb radiation in the range of the detector. The uv detectors are relatively insensitive to temperature or flow changes, but the response can be sensitive to solvent composition, which can effect sample absorption characteristics, as in gec.

The fluorescence detector, perhaps the most sensitive commonly used detectors in lc, is limited in its utility to the detection of materials that fluoresce or have derivatives that fluoresce. These detectors find particular use in analysis of environmental and food samples, where measurements of trace quantities are required.

Electrochemical detectors sense electroreducible and electrooxidizable compounds at low concentrations. For these detectors to work efficiently, the mobile phase (solvent) must be conductive and not subject to electrochemical decomposition.

Other detectors determine bulk properties of the system of mobile phase plus sample. The most commonly used bulk-property detector is the refractive-index (RI) detector. The RI detector, the closest thing to a universal detector in lc, monitors the difference between the RI of the effluent from the column and the pure solvent. These detectors are not very good for detection of materials at low concentrations. Moreover, they are sensitive to fluctuations in temperature.

Conductivity detectors, commonly employed in ion chromatography, can be used to determine ionic materials at levels of parts per million (ppm) or parts per billion (ppb) in aqueous mobile phases. The ir detector is one that may be used in either nonselective or selective detection. Its most common use has been as a detector in sec, although it is not limited to sec. The detector is limited to use in systems in which the mobile phase is transparent at the wavelength being monitored. It is possible to obtain complete spectra, much as in some gc–ir experiments, if the flow is not very high or can be stopped momentarily.

The ms detection is ideal for qualitiatively and quantitatively determining the constituents of a sample. The lc/ms provides a mass spectrum for each chromatographic peak or can be used in a single-ion-monitoring mode to detect components of similar fragmentation structure. The lc/ms/ms provides an additional dimension to improve the qualitaive aspect of the separation by providing not only primary structural identification but also secondary ion degradation (or the mass spectrum of a mass spectrum). The major drawback to lc/ms and lc/ms/ms is the cost and maintenace of the equipment.

4.5. Affinity Chromatography. This technique, sometimes called bioselective adsoprtion, involves the use of a bioselective stationary phase placed in contact with the material to be purified, the ligate. Because of its rather selective interaction, sometimes called a lock-and-key mechanism, this method is more selective than other lc systems based on differential solubility.

4.6. Chiral Chromatography. Chiral chromatography is used for the analysis of enantiomers, and finds applications in the separations of pharmaceuticals and biochemical compounds (see BIOPOLYMERS, ANALYTICAL TECHNIQUES). There are several types of chiral stationary phases: those that use attractive interactions, metal ligands, inclusion complexes, and protein complexes. The separation of optical isomers has important ramifications, especially in biochemistry and pharmaceutical chemistry, where one form of a compound may be bioactive and the other inactive, inhibitory, or toxic.

4.7. Ion-Exchange Chromatography. In iec, the column contains a stationary phase having ionic groups such as a sulfonate or carboxylate. The charge of these groups is compensated by counterions such as sodium or potassium. The mobile phase is usually an ionic solution, e.g. sodium chloride, having ions similar to the counterions. Ionic samples are introduced into the mobile phase, and retardation in movement results from ion exchange with the stationary phase. The more the ion interacts with the exchanger, the more strongly it is retained. For cation-exchange chromatography, positively charged ions are separated. In anion-exchange chromatography, negatively charged ions in the sample interact with and bind to cationic stationary phases (see ION EXCHANGE).

Ion chromatography (ic), a variant of ion-exchange chromatography, is a technique in which a weak ion-exchange column is used for separation. After passing through the weak ion-exchange column, the eluent passes through a subsequent column called a stripper column, in which the stream, usually made acidic or basic in the ion-exchange column, is neutralized. This stream gives no conductimetric response in this condition; however, when added ions are present, such as happens when sample is passing through the stripper column, the conductivity of the solution changes and a signal is detected. Ion chromatography is a powerful technique for examining low concentrations of anions and cations. It has the advantage over selective ion-electrode analysis that it simultaneously gives information on many ions in a single experiment (see ELECTRO-ANALYTICAL TECHNIQUES).

Ion-pair chromatography (ipc), another variant of iec, is also sometimes called pic, soap chromatography, extraction chromatography, or chromatography with a liquid ion exchanger. In this technique the mobile phase consists of a solution of an aqueous buffer and an organic cosolvent containing an ion of charge opposite to the charge on the sample ion. The sample ion and the solvated ion form an ionic pair that is soluble in the stationary phase. Thus retention is determined by the ability to form the ion pair as well as the solubility of the complex in the stationary phase.

4.8. Size-Exclusion Chromatography.   In sec or gpc, the material with which the column is packed has pores of a certain range of size. Molecules or solvent-molecule complexes too large to pass through these pores pass rapidly through the column, whereas molecules or complexes of suffiently small size are retained and are the last to exit the column. Molecules of intermediate size are partially retained and elute from the chromatographic column at intermediate times. Size-exclusion chromatography is extremely useful as a tool for characterization of polymer materials because the retention mechanism is reproducible enough to give good comparative data and can also give valuable information about the distribution of sizes of molecules in a sample.

5. Supercritical-Fluid Chromatography

Supercritical-fluid chromatography, developed in the late 1960s, was not used extensively until the early 1980s. This technique is the link between gc and lc, because its mobile phase, a supercritical fluid, has physicochemical properties intermediate between a gas and a liquid (see SUPERCRITICAL FLUIDS). The physicochemical properties of the mobile phase are strong factors determining the selectivity, sensitivity toward a component, and efficiency of separation in the chromatographic process. Supercritical fluids, e.g., can be viewed as dense gases that cannot become liquid. The density of a supercritical material increases continuously with pressure at constant temperature and its solvating power increases with pressure, because the solubility of materials in a solvent usually increases with density and can be used as a powerful means of changing retention. Carbon dioxide is the mobile phase most often used in sfc.

This technique can be performed with either capillary or lc-like packed columns. Carbon dioxide is compatible with chromatographic hardware, is readily available, and is noncorrosive. The most important detector for sfc is the flame-ionization detector because the mobile phase does not give a significant background signal. Most early applications of sfc were in the separation of petroleum products. More recent applications of sfc include separations in fields as diverse as natural products, drugs, foods, pesticides, herbicides, surfactants, and polymers. These are a direct result of the advantages that sfc has over other forms of chromatography because of low operating temperature, selective detection, and sensitivity to molecular weight.

BIBLIOGRAPHY

"Chromotography, Affinity" in *ECT* 3rd ed., Vol. 6, pp. 35–54, by A. H. Nishikawa, Haffman–LaRoche, Inc.; "Chromotography" in *ECT* 4th ed., Vol. 6, pp. 207–228, by Mary A. Kaiser, Dupont Company and Cecil Dybowski, University of Delaware; "Chromotography" in *ECT* (online), posting date: December 4, 2000, by Mary A. Kaiser, Dupont Company and Cecil Dybowski, University of Delaware.

CITED PUBLICATIONS

1. G. Wilkinson, *Today*'s Chemist **29** (Dec. 1990).
2. M. S. Tswett, *Proc. Warsaw Soc. Nat. Sci.* (1903).
3. M. S. Tswett, *Ber. Deut. Bot. Ges.* **24**, 384 (1906).
4. A. J. P. Martin and R. L. M. Synge, *Biochem. J. (London)* **35**, 1358 (1941).
5. *Res. Dev.*, 19 (Jan. 1991).
6. R. L. Grob, *Modern Practice of Gas Chromatography*, 3rd ed., John Wiley & Sons, Inc., 3rd ed., New York, 1995.

GENERAL REFERENCES

S. Ahuja, *Chiral Separations by Liquid Chromatography*, ACS Books, Washington, D.C., 1991.
H. J. Cortes, *Multidimensional Chromatography: Techniques and Applications*, Marcel Dekker, New York, 1990.
J. S. Fritz, D. T. Gjerde *Ion Chromatograph*, John Wiley & Sons, Inc., New York, 2000.
N. Grinberg, *Modern Thin-Layer Chromatography*, Marcel Dekker, New York, 1990.
R. L. Grob, *Chromatographic Analysis of the Environment*, 2nd ed., Marcel Dekker, New York, 1983.
C. Horvath and J. Nikelly, *Analytical Biotechnology: Capillary Electrophoresis and Chromatography*, ACS Books, Washington, D.C., 1990.
F. G. Kitson, C. N. McEwen, Barbara S. Larsen, *Gas Chromatography and Mass Spectrometry*, Academic Press San Diego, California, 1996.
M. S. Klee, *GC Inlets—An Introduction*, Hewlett Packard, Avondale, Pa., 1990.

H. M. McNair and J. M. Miller, *Basic Gas Chromatography*, John Wiley & Sons, Inc., New York, 1997.

S. Mori and H. G. Barth, *Size Exclusion Chromatography*, Springer-Verlag, New York, 1999.

T. Provder, *Chromatography of Polymers*, American Chemical Society, Washington, D.C., 1999.

D. Rood, *A Practical Guide to Care, Maintenace, and Troubleshooting of Gas Chromatographic Systems*, John Wiley & Sons, Inc., New York, 1999.

L. R. Snyder, J. J. Kirkland, and J. L. Glajch, *Practical HPLC Method Development*, John Wiley & Sons, Inc., New York, 1997.

H. F. Walton and R. D. Rocklin, *Ion Exchange in Analytical Chemistry*, CRC Press, Boca Raton, Fla., 1990.

CECIL DYBOWSKI
University of Delaware

MARY A. KAISER
Dupont Company

CHROMATOGRAPHY, AFFINITY

1. Introduction

Affinity chromatography is a liquid chromatographic technique that uses a biologically related agent as the stationary phase (1–7). This makes use of the selective interactions that are common in biological systems, such as the binding of an enzyme with a substrate or the binding of an antibody with a foreign substance that has invaded the body. These interactions are used in affinity chromatography by immobilizing one of a pair of interacting molecules onto a solid support. This support is then placed into a column or onto a planar surface. The immobilized molecule is referred to as the *affinity ligand* and it represents the stationary phase of the chromatographic system.

The earliest known use of affinity chromatography was in 1910, when Emil Starkenstein used insoluble starch to purify the enzyme α-amylase (8). Over the course of the next 50 years there were a few other reports that described additional solid-phase ligands for biological purification. However, it was not until the development of beaded agarose supports (9) and the cyanogen bromide immobilization method (10) in the 1960's that affinity chromatography came into common use (11).

Figure 1 shows the most common scheme for performing affinity chromatography (6,7). In this approach, a sample containing the compound of interest is first injected onto an affinity column in the presence of a mobile phase that has the right pH, ionic strength, and solvent composition for solute–ligand binding. This solvent, which represents the weak mobile phase of an affinity column, is

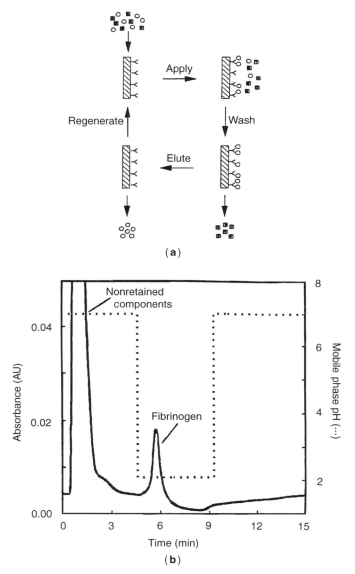

Fig. 1. A typical separation scheme for affinity chromatography. (**a**) Shows the general format that is used to apply and elute samples. The example given in (**b**) shows the use of this approach in the isolation of fibrinogen from a plasma sample, with antifibrinogen antibodies being employed as the affinity ligands and a low pH buffer being used for elution. Part **b** of this figure is reproduced with permission from Ref. 12.

referred to as the *application buffer*. As the sample passes through the column under these conditions, any compounds in the sample that are able to bind to the affinity ligand will be retained. However, due to the high selectivity of most such interactions, other substances in the sample tend to elute from the column as a nonretained peak.

After all nonretained or weakly retained substances have been washed from the column, the retained solutes are eluted by applying a solvent that displaces them from the column or promotes dissociation of the complex between each solute and affinity ligand. This second solvent, which acts as the strong mobile phase for the column, is usually referred to as the *elution buffer*. As the retained solutes are eluted from the column, they can either be collected for later use, as would be performed in a preparative application, or monitored by an on-line detector for analysis. After all of these substances have been removed from the system, the original application buffer is again applied and the affinity ligand is allowed to regenerate back to its original state prior to the application of another sample.

The wide range of ligands available for affinity chromatography makes this method a valuable tool for the purification and analysis of compounds present in complex samples. Examples that will be examined in this article include the use of both biological ligands (eg, antibodies, lectins and nucleic acids), as well as alternative ligands (eg, boronic acids, metal chelates, and biomimetic dyes) (3–7). Although most past reports on affinity chromatography have focused on its use in the isolation of biological compounds (3,4), more recent work has also used this method for chemical analysis in biochemistry, pharmaceutical sciences, clinical chemistry, and environmental testing (5–7,13–16). In addition, affinity chromatography has been used as a tool in measurements of the strength and rates of biological interactions (17,18).

2. Principles of Affinity Chromatography

2.1. Theory of Affinity Chromatography.

A number of factors affect how a compound is retained or eluted by an affinity column. These factors include the type of mobile phase that is being applied to the column, the strength of the solute–ligand interaction in this solvent, the amount of immobilized ligand present in the column, and the kinetics of solute–ligand association and dissociation. Furthermore, the type of support material that is used in the column can also play a role in determining the speed and efficiency of this chromatographic process.

For a simple affinity ligand system, many of these factors can be described by using the following reaction that shows the binding between the applied (A) and immobilized ligand (L) within the chromatographic bed.

$$A + L \underset{}{\overset{K_a}{\rightleftharpoons}} A\text{—}L \tag{1}$$

In this equation, K_a is the association equilibrium constant for the binding of A with L to form A—L, the solute–ligand complex. Based on this reaction, an expression for K_a can be written in the form shown in equation 2.

$$K_a = k_a/k_d = \{A\text{—}L\}/[A]\{L\} \tag{2}$$

In equation 2, [A] is the mobile phase concentration of A at equilibrium, while {L} and {A—L} represent the surface concentrations of the ligand and solute–ligand complex, respectively. The term k_a is the second-order association rate constant for solute–ligand binding and k_d is the first-order dissociation rate constant for the solute–ligand complex.

The above reaction can be related to the retention of solute A on an affinity column by using the solute's retention factor (k). As in any other chromatographic method, this retention factor is related experimentally to the mean elution time of a solute from the column in a given mobile phase by using the expression $k = (t_R - t_M)/t_M$, where t_R is the average retention time for the solute under these conditions and t_M is the void time of the system (ie, the observed elution time for a totally nonretained solute). The retention factor can also be shown to be directly related to the association equilibrium constant for the solute with the affinity ligand, as is demonstrated in equation 3.

$$k = K_a m_L/V_M \tag{3}$$

In this equation, m_L represents the moles of active ligand in the column, V_M is the column void volume, and the combined term m_L/V_M represents the column's phase ratio (ie, the relative amount of stationary phase versus mobile phase) (7,17,18).

According to equation 3, the retention factor for a solute on an affinity column (as well as its retention time) in the presence of a given mobile phase will depend on both the strength of the solute–ligand binding under these conditions (as represented by the association equilibrium constant, K_a) and the amount of ligand that is available for binding within the column (as represented by the binding capacity or phase ratio, m_L/V_M). Since many biologically related ligands have moderate or large equilibrium constants for their target solutes (ie, $K_a > 10^6 \ M^{-1}$), this gives these solutes strong retention and long elution times on their corresponding affinity columns under normal sample application conditions. It is for this reason that a step change to other conditions that promote lower binding is often used for the elution of solutes from affinity columns. However, for weaker binding systems ($K_a < 10^6 \ M^{-1}$), work under isocratic conditions is also possible (19–21).

2.2. General Types of Affinity Ligands. The most important item that determines the selectivity and retention of an affinity chromatographic system is the type of ligand used as the stationary phase. There are many biological agents and biological mimics that have been used for this purpose. However, all of these ligands can be placed into one of two categories: (1) high specificity ligands and (2) general, or group-specific ligands (6).

The term *high specificity ligands* refers to compounds that bind only to one or a few closely related molecules, which is used in chromatographic systems when the goal is to analyze or purify a specific solute. Typical high specificity ligands include antibodies (for binding antigens), substrates or inhibitors (for separating enzymes), and single-stranded nucleic acids (for the retention of a complementary sequence). As this suggests, most high specificity ligands tend to be biological compounds. In addition, most of these ligands have large association equilibrium constants for their solutes and high solute retention.

General, or group-specific, ligands are compounds that bind to a family or class of related molecules. These ligands are used in methods where the goal is to isolate a class of structurally similar solutes. General ligands can be of either biological or nonbiological origin. Examples include bacterial cell wall proteins (eg, proteins A and G), lectins, boronates, triazine dyes, and immobilized metal chelates. Many of these ligands have weaker binding for solutes than that observed for high specificity ligands. However, there are exceptions. For example, proteins A and G have association equilibrium constants of $10^7 \ M^{-1}$ for some types of antibodies, their target solutes. Also, some molecules that are usually considered to be high specificity ligands, such as antibodies, can be used to retain an entire class of solutes if they recognize a feature which is common to a group of closely related agents (7).

2.3. Support Materials.   Another factor to consider in affinity chromatography is the material used to hold the ligand within the column. Ideally, this support should have low nonspecific binding for sample components but should be easy to modify for ligand attachment. This material should also be stable under the flow-rate, pressure, sample, and solvent conditions to be used. In addition, the support should have sufficient surface area for immobilization, be readily available, and be simple to use in method development (22).

A wide variety of materials have been employed as supports in affinity chromatography. For instance, many carbohydrate-based materials have been used for this purpose, including starch, agarose, cross-linked agarose, cellulose, and various modified forms of cellulose. Inorganic materials like silica and glass have also been used in affinity chromatography; this first requires the modification of these substances to give them low nonspecific binding and functional groups that can be used for ligand attachment. Many types of synthetic organic-based supports have been used as well, including agarose–acrylamide copolymers, azalactone beads, dextran–acrylamide copolymers, hydroxylated polystyrene, polyacrylamide derivatives, polyethersulfone, and polymethacrylate derivatives. Most of these supports are available commercially and all can be prepared within the laboratory according to published procedures (4–7,22).

Depending on the type of support material used, affinity chromatography can be characterized as being either a low or high performance technique. In *low performance (or column) affinity chromatography*, the support is usually a large diameter, nonrigid gel. Many of the carbohydrate-based supports and synthetic organic materials fall within this category. The low back pressure of these supports means that they can be operated under gravity flow or with a peristaltic pump. This makes these gels relatively simple and inexpensive to use for affinity purification. Disadvantages of these materials include their slow mass transfer properties and their limited stability at high flow rates and pressures. These factors limit the usefulness of these supports in analytical applications, where both rapid and efficient separations are desired.

In *high performance affinity chromatography (HPAC)*, the support consists of small, rigid particles capable of withstanding the high flow rates and/or pressures that are characteristic of high performance liquid chromatography (HPLC). Supports that can be used for this include modified silica or glass, azalactone beads, and hydroxylated polystyrene media. The mechanical stability and efficiency of these supports allows them to be used with standard HPLC equipment. Although the need for more sophisticated instrumentation does make HPAC more expensive to perform than low performance affinity chromatography, the better speed and precision of this technique makes it the method of choice for analytical applications (6,7,23).

Another way affinity supports can be categorized is based on the physical form of these supports. The most common type used in affinity chromatography is the porous particulate support, which is packed within an affinity column. Examples of such supports include silica, glass beads, agarose or cellulose beads, and azalactone particles. In addition, several alternative types of affinity supports have begun to see greater interest in recent years. Examples include nonporous particulate supports, fibers and membranes, flow through particles (eg, perfusion media), expanded bed particles, and continuous bed supports (ie, monolithic media).

2.4. Immobilization Methods. A third item to consider in using affinity chromatography is the way in which the ligand is attached to the solid support, or the *immobilization method*. There are several strategies used for this purpose, as illustrated in Figure 2. The most common way for placing the ligand within the affinity column is *covalent immobilization*. This is performed by reacting functional groups on the ligand with reactive sites on the surface of the support. Several techniques are available for this, as summarized in Table 1 (24,25). For a protein or peptide, covalent immobilization generally involves coupling these molecules through free amine, carboxylic acid, or sulfhydryl residues in their structures. Immobilization of a ligand through other functional sites (eg, aldehyde groups produced by carbohydrate oxidation) is also possible.

All covalent immobilization methods involve at least two steps: (*1*) an *activation step*, in which the support is converted to a form that can be chemically

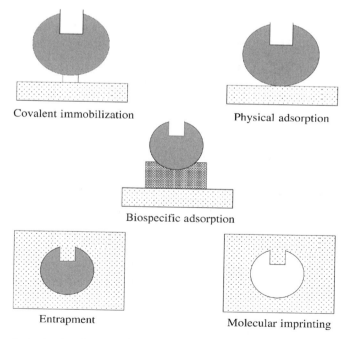

Fig. 2. General strategies that can be used to place a ligand within a column or chromatographic bed for affinity chromatography.

Table 1. **Methods for Covalent Ligand Immobilization**

Group/Compound	Immobilization technique
amines	azalactone method (for Emphaze supports)
	cyanogen bromide (CNBr) method
	N,N'-carbonyl diimidazole (CDI) method
	divinylsulfone method
	epoxy (bisoxirane) method
	ethyl dimethylaminopropyl carbodiimide (EDC) method
	fluoro methylpyridinium toluenesulfonate (FMP) method
	N-hydroxysuccinimide ester (NHS) method
	Schiff base (reductive amination) method
	tresyl chloride/tosyl chloride method
sulfhydryls	azalactone method (for Emphaze supports)
	divinylsulfone method
	epoxy (bisoxirane) method
	iodoacetyl/bromoacetyl methods
	maleimide method
	pyridyl disulfide method
	TNB–thiol method
	tresyl chloride/tosyl chloride method
carboxylates	ethyldimethylaminopropyl carbodiimide (EDC) method
hydroxyls	cyanuric chloride method
	divinylsulfone method
	epoxy (bisoxirane) method
aldehydes	hydrazide method
	Schiff base (reductive amination) method
nucleic acids	carbodiimide (CDI) method
	carboxymethyl method (for cellulose)
	cyanogen bromide (CNBr) method
	cyanuric chloride method
	diazonium (Azo) method
	epoxy (bisoxirane) method
	hydrazide method
	Schiff base (reductive amination) method
	ultraviolet (uv) irradiation (for cellulose)

attached to the ligand, and (2) a *coupling step*, in which the affinity ligand is attached to the activated support. With some techniques a third step, in which remaining activated groups are removed, is also required. The methods listed in Table 2 can be performed either in-house or by using preactivated supports available from commercial suppliers.

An alterative approach to covalent attachment is that of *physical adsorption*. In this method, the ligand is held in place through noncovalent interactions (eg, Coulombic interactions with a charged surface). This approach is relatively easy to perform but can result in a stationary phase which is unstable and uses a support that may have nonspecific interactions with sample components.

Another, more selective approach for immobilization is *biospecific adsorption*. This involves the noncovalent coupling of the ligand of interest to another secondary ligand that, in turn, is coupled to the affinity support. An example of this would be the adsorption of antibodies to an immobilized protein A or protein

G affinity column, where proteins A and G are used to capture antibodies for use in immunoaffinity chromatography.

A fourth possible approach is to use *encapsulation* or *entrapment* of the ligand within the support. One way this can be accomplished is by combining the ligand with the material used to prepare the affinity support. An example would be the entrapment of a protein by a sol–gel by preparing the sol–gel in the presence of the protein. Another example would be the trapping of large particles in the pores of the support by altering the size of the ligand particles during the entrapment process. This latter approach has been used in combination with freeze-drying to place liposomes or membrane-based particles with associated proteins into large pore-size supports such as agarose (26,27).

One final method for ligand immobilization is to make the ligand during the support's preparation. In this approach, the ligand is actually formed by binding pockets located within the support with a well-defined arrangement of functional groups. This process is known as *molecular imprinting*. In this procedure, the solute of interest is combined with a polymerization mixture that is to be used to form the affinity support. This mixture contains a cross-linking agent, monomers with functional groups that can interact with the solute, and an initiation reagent for the polymerization reaction. The resulting polymer contains pockets that were formed directly within the support and that complement the shape and arrangement of functional groups on the target solute. After the polymerization has reached completion, the imprinted solute is washed away and the now unoccupied sites can be used to isolate or retain the same solute from samples (28–30).

The correct choice of immobilization method is important in affinity chromatography since it can have a great affect on the properties of the ligand and affinity separation. For instance, if the correct immobilization procedure is not used, a decrease in ligand activity can result from multisite attachment, improper orientation and/or steric hindrance (6). *Multisite attachment* refers to the coupling of a ligand to the support through more than one functional group, which can lead to distortion and denaturation of the ligand's active region. *Improper orientation* can lead to a similar loss in ligand activity, but can be avoided by coupling the ligand through groups that are distant from its active region. *Steric hindrance* refers to the loss of ligand activity due to the presence of a nearby support or neighboring ligand molecules. Steric hindrance between neighboring ligands can be minimized by using a low ligand coverage, while steric hindrance due to the support can be reduced by adding a *spacer arm*, or tether, between the ligand and supporting material. Spacer arms are particularly important to use with small ligands that are meant to retain large solutes. Examples of spacer arms used in affinity chromatography are 6-aminocaproic acid, diaminodipropylamine, 1,6-diaminohexane, ethylenediamine, and succinic acid anhydride (24,25).

2.5. Application and Elution Conditions. Most application buffers in affinity chromatography are solvents that mimic the pH, ionic strength, and polarity experienced by the solute and ligand in their natural environment. Any cofactors or metal ions required for solute–ligand binding should also be present in this solvent. Under these conditions, the solute will probably have its highest association constant for the ligand and, therefore, its highest degree of retention on the column. The proper choice of an application buffer can also help to minimize any nonspecific binding due to undesired sample components. For example, Coulombic interactions between solutes and the support can often be decreased by altering the ionic strength and pH of the application buffer. In addition, surfactants and blocking agents [eg, Triton X-100, Tween-20, bovine serum albumin (BSA), gelatin, etc] may be added to the buffer to prevent nonspecific retention of solutes on the support or affinity ligand.

The activity of the immobilized ligand should be considered in determining how much sample can be applied to the affinity column with each use. A rough indication of the maximum possible column binding capacity can be made by assaying the total amount of ligand present. However, a better approach is to actually measure the ligand's activity. This measurement can be done by continuously applying a known concentration of solute to the affinity column (ie, breakthrough curves) or by combining the immobilized ligand with a known excess of solute and measuring the amount of free solute that remains after binding has occurred.

With some affinity systems, it is possible to see a significant fraction of non-retained solute during the application step even when the amount of injected sample is less than the known column binding capacity. This phenomenon is known as the *split-peak effect* and is caused by the presence of slow adsorption and/or mass transfer kinetics within the column (31–33). Ways in which this effect can be minimized include reducing the flow rate used for sample injection, increasing the column size, or placing a more efficient support within the column. In some cases, changing to a different immobilization method may also help by providing a ligand with more rapid binding kinetics.

The conditions used for removal of retained solutes is another item that should be considered in the design of an effective affinity separation. Even though the use of step gradient is commonly used for solute elution in affinity chromatography (as shown in Fig. 1), it is also possible to use isocratic or linear gradient methods in certain circumstances. For example, isocratic elution can be used if the association equilibrium constant for a solute with the ligand is sufficiently small to allow the solute to pass through the column in a reasonable amount of time under the application conditions. This is often used in the case of chiral separations that are performed on affinity columns and with ligands that have been selected to have weak binding affinities. This approach is referred to as *weak affinity chromatography* or *dynamic affinity chromatography* (7,19–21).

Just as the application conditions are selected to maximize specific solute–ligand interactions, the elution conditions are chosen to promote fast or gentle removal of solute from the column. The elution buffer used in affinity chromatography can be either a solvent which produces weak solute–ligand binding (ie, a small association constant) or a solvent that decreases the extent of this binding by using a competing agent that displaces solute from the column. These two approaches are known as *nonspecific elution* and *biospecific elution* (6).

Biospecific elution is the gentler of these two elution methods and is carried out under essentially the same solvent conditions as used for sample application. This makes this approach attractive for purification work, where a high recovery of active solute is desired. Biospecific elution may be performed either by adding an agent to the eluting solvent that competes with the ligand for solutes (ie, normal role elution) or by adding an agent that competes with the solute for ligand-binding sites (ie, reversed-role elution). In both cases, retained solutes are eventually eluted from the column by displacement and mass action. The main advantage of biospecific elution is its ability to gently remove analyte from the column. The main disadvantages of this method include its slow elution times, broad solute peaks, and frequent need to remove the competing agent from the eluted solute.

In *nonspecific elution* the column conditions are changed in a more drastic fashion to weaken the interactions between retained solutes and the immobilized ligand. This can be done by changing the pH, ionic strength, or polarity of the mobile phase or by adding denaturing or chaotropic agents. This results in an alteration in the structure of the solute or ligand, leading to a lower association constant and lower solute retention. Nonspecific elution tends to be much faster than biospecific elution in removing analytes from affinity columns, resulting in sharper peaks that in turn produce lower limits of detection and shorter analysis times. For these reasons, nonspecific elution is commonly used in analytical applications. This elution method can also be used in purifying solutes, but there is a greater risk of solute denaturation with this approach than there is with biospecific elution. Also, care must be taken in nonspecific elution to avoid conditions that are too harsh for the column.

3. Specific Types of Affinity Chromatography

3.1. Bioaffinity Chromatography. *Bioaffinity chromatography*, or *biospecific adsorption*, is the oldest and most common type of affinity chromatography (3–8). This refers to affinity methods that use a biological molecule as the affinity ligand. This was the first type of affinity chromatography developed and represents the most diverse category of this technique.

The earliest application of bioaffinity chromatography involved its use in enzyme purification (8), which has continued to be a major use of this technique. Some ligands used for this purpose are enzyme inhibitors, coenzymes, substrates, and cofactors. Examples include methods that use nucleotide mono-, di- and triphosphates for the purification of various kinases, the use of NAD for collecting dehydrogenases, the use of pyridoxal phosphate for the isolation of tyrosine and aspartate aminotransferases, and the use of RNA or DNA for the purification of polymerases and nucleases (3,4,34,35).

Lectins represent a class of general ligands that are common in bioaffinity chromatography. This field is sometimes referred to as *lectin affinity chromatography*. The lectins are nonimmune system proteins that have the ability to recognize and bind certain types of carbohydrate residues (36). Two lectins often used in affinity chromatography are concanavalin A, which binds to α-D-mannose and α-D-glucose residues, and wheat germ agglutinin, which binds to D-*N*-acetyl-glucosamines. Other lectins that can be employed are jackalin and lectins found in peas, peanuts, or soybeans. These ligands are used in the separation of many carbohydrate-containing compounds, such as polysaccharides, glycoproteins (eg, immunoglobulins or cell membrane proteins) and glycolipids (3–6,24).

Another useful class of bioaffinity ligands are bacterial cell wall proteins, such as protein A from *Staphylococcus aureus* and protein G from group G *streptococci* (6,24,37–39). These ligands have the ability to bind to the constant region of many types of immunoglobulins. This makes them useful in antibody purification. Protein A and protein G have their strongest binding to immunoglobulins at or near neutral pH but readily dissociate from these solutes when placed into a lower pH buffer. These two ligands differ in their ability to bind to antibodies from different species and classes. However, recombinant proteins that blend the activities of these compounds are also available.

Nucleic acids and polynucleotides can act as either general or specific ligands in bioaffinity chromatography. For instance, as high-specificity ligands they can be used to purify DNA/RNA-binding enzymes and proteins or to isolate nucleic acids that contain a sequence that is complementary to the ligand (40,41). As a group-specific ligand, an immobilized nucleic acid can be used to purify solutes that share a common nucleotide sequence. An example is the use of immobilized oligo(dT) for the isolation of nucleic acids containing poly(A) sequences.

3.2. Immunoaffinity Chromatography. The most common type of bioaffinity chromatography is that which uses an antibody or antibody-related agent as the affinity ligand. This set of methods is often referred to as *immunoaffinity chromatography (IAC)* (42–44). The high selectivity of antibody–antigen interactions and the ability to produce antibodies against a wide range of solutes has made immunoaffinity chromatography a popular tool for biological purification and analysis. Examples include methods developed for the isolation of antibodies, hormones, peptides, enzymes, recombinant proteins, receptors, viruses, and subcellular components. The strong binding constants of many antibodies requires non-specific elution for most immunoaffinity columns. However, isocratic elution methods can also be used with low affinity antibody systems (44).

Immunoaffinity chromatography was first reported by Campbell and coworkers in 1951, who used an antigen immobilized to *p*-aminobenzyl cellulose for antibody purification (45). Many current applications are still based on the use of low performance supports, particularly agarose. However, during the past decade much work has also been performed using derivatized silica, glass and perfusion media in immunoaffinity columns. The use of these supports along with an antibody or antigen ligand is referred to as *high performance immunoaffinity chromatography (HPIAC)* (43,44).

Several unique applications of immunoaffinity chromatography have appeared in recent years. One of these involves the use of affinity columns to perform immunoassays, giving rise to a technique known as a *chromatographic immunoassay* or *flow-injection immunoanalysis* (44,46). An example of a chromatographic immunoassay is given in Figure 3. Various immunoassay formats have been performed with affinity columns, including direct detection assays, immunometric assays, and competitive binding immunoassays. These assays can either be performed as stand-alone methods or as a means for detecting substances after they have been resolved by a separate chromatographic column. In this latter case, these chromatographic-based immunoassays are also referred to as *postcolumn immunodetection* (44).

Another growing use of immunoaffinity chromatography has been as a means for sample isolation and pretreatment prior to analysis by other analytical

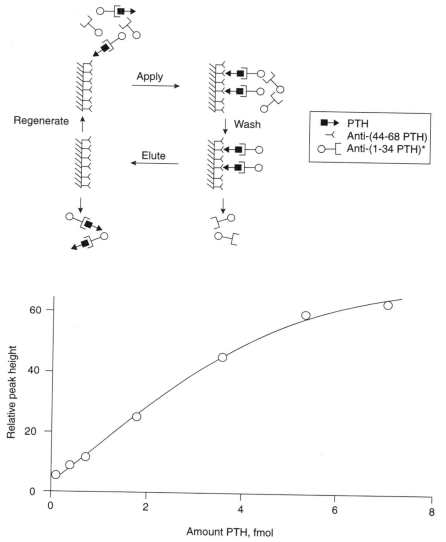

Fig. 3. An example of a chromatographic immunoassay. This particular example shows a sandwich immunoassay for measuring the hormone parathyrin (PTH) in plasma samples. The affinity column contained anti-(44-68 PTH) antibodies, while labeled anti-(1-34 PTH) antibodies were added to the sample for placing a detectable tag on the analyte. Reproduced with permission from Ref. 47.

methods, which is often referred to as *immunoextraction*. Immunoextraction can be performed as an off-line technique, in which the isolated sample components are collected from the immunoaffinity column and later transferred to a second method for measurement. But immunoextraction can also be used directly (or on-line) with some analytical methods. For instance, antibody based columns have been coupled directly with systems used in reversed-phase liquid chromatography, ion-exchange chromatography, size-exclusion chromatography, gas chromatography, capillary electrophoresis, and liquid chromatography–mass spectrometry. This produces multidimensional methods that combine the selectivity of immunoaffinity chromatography with the ability to separate structurally related compounds (as might be performed by reversed-phase liquid chromatography) or to provide structural information (as is obtained by mass spectrometry) (44).

3.3. Dye-Ligand and Biomimetic Affinity Chromatography.
Two other, related categories of affinity chromatography are the techniques of *dye-ligand affinity chromatography* and *biomimetic affinity chromatography* (48). In dye-ligand affinity chromatography, a synthetic substance like a triazine or triphenylmethane dye is used as the immobilized ligand. This approach was first reported in 1971, when Staal and co-workers used a column containing Blue Dextran to purify the enzyme pyruvate kinase (49).

Many dyes can be used as ligands in affinity chromatography. Examples are Cibacron Blue 3GA, Procion Red HE-3B, Procion Rubine MX-B, Procion Yellow H-A, and Turquoise MX-G. The structure of one of these dyes is given in Figure 4. In each case, a portion of the dye's structure interacts with a target protein by mimicking the binding of a native solute to that site. As an example, the dye Cibacron Blue 3GA binds to NAD(P)H:quinone reductase by mimicking the AMP portion of NADP+ and interacting with its associated site on the enzyme.

The selectivity of dye–ligand affinity chromatography and its use of synthetic ligands in place of natural ones has made this method an extremely popular tool for enzyme and protein purification. Well over 500 compounds have been isolated by this technique (3), including kinases, dehydrogenases, restriction endonucleases, polynucleotides, synthetases, Coenzyme A (CoA)-dependent enzymes, hydroxylases, glycolytic enzymes, phosphodiesterases, decarboxylases, clotting factors, serum lipoproteins, interferons, transferrin, and serum albumin.

Dye–ligand affinity chromatography is actually a subset of the more general technique known as biomimetic affinity chromatography. As the name of this latter method implies, it makes use of any ligand that acts as a mimic for a natural compound. This includes the use of synthetic dyes as ligands, as well as other types of agents. For instance, combinatorial chemistry and computer modeling have been used with peptide libraries to design biomimetic ligands for enzymes and other target compounds. Phage display libraries, aptamer libraries, and ribosome display have also been used for this purpose (50–52).

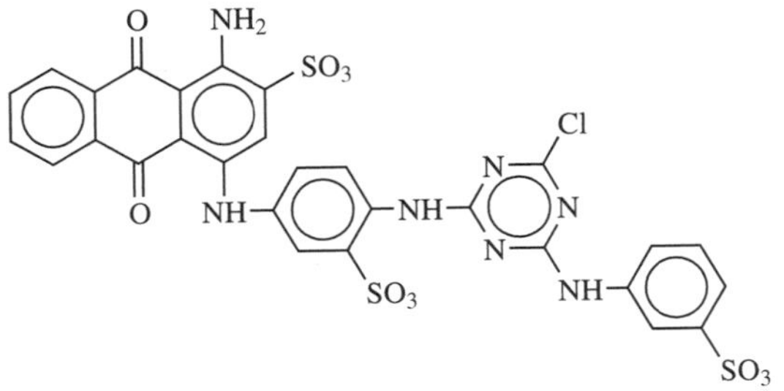

Fig. 4. Structure of Cibacron Blue G3A, a common ligand used in dye-ligand affinity chromatography.

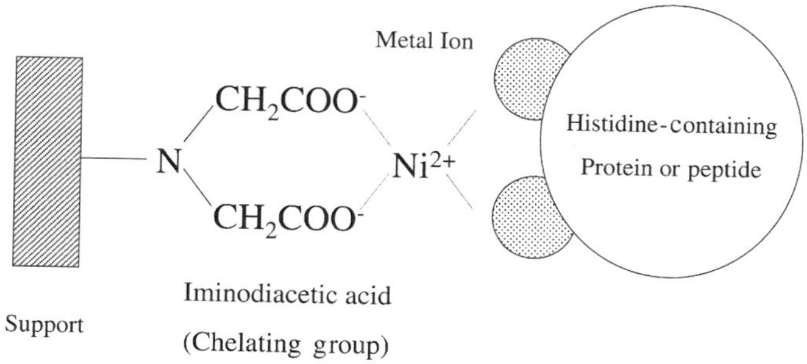

Fig. 5. Structure of Ni–IDA, a metal–chelate complex that is often employed in immobilized metal-ion affinity chromatography (IMAC) for the retention of histidine-containing proteins and peptides.

3.4. Immobilized Metal-Ion Affinity Chromatography.

Another type of affinity chromatography that uses a ligand of nonbiological origin is *immobilized metal-ion affinity chromatography (IMAC)*. This method is also known as *metal chelate chromatography* or *metal ion interaction chromatography*. In this approach, the affinity ligand is a metal ion complexed with an immobilized chelating agent (53,54). One common example is the use of Ni^{2+} ions that are complexed to a support containing iminodiacetic acid (IDA) as the chelating agent (see Fig. 5), although other types of metal ions or chelating groups can also be employed.

This type of affinity chromatography was first described by J. Porath and co-workers in 1975 (55). It was initially used to separate proteins and peptides containing electron donor groups, such as histidine, tryptophan, or cysteine residues, which can interact with the immobilized metal chelate. In more recent years, IMAC has been used for other purposes, such as the isolation of recombinant histidine-tagged proteins, studies of protein surface topography, and the isolation of phosphorylated proteins for proteomic studies.

3.5. Boronate Affinity Chromatography.

Boronic acid and its derivatives are another class of synthetic substances that have been used as affinity ligands. This makes use of the ability of such substances to form covalent bonds with compounds that contain cis-diol groups in their structure (see Fig. 6). Such a property has made boronate ligands useful for the purification and analysis of many compounds which contain sugar residues, such as polysaccharides, glycoproteins, ribonucleic acids, and catecholamines (56–58). One

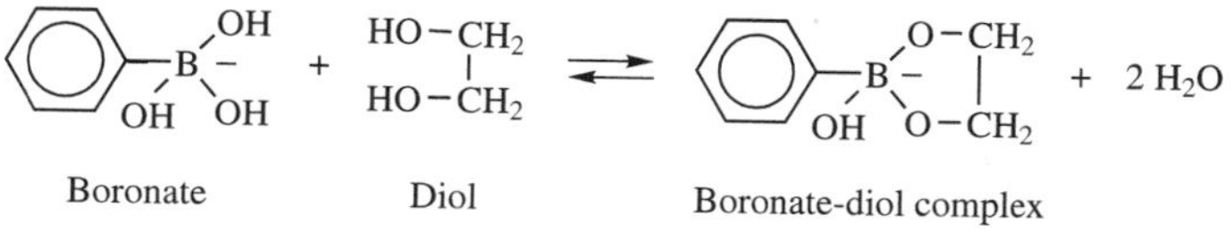

Fig. 6. Reaction of boronate with a cis-diol, illustrating the mechanism of retention in boronate affinity chromatography.

important clinical application of immobilized boronate ligands is their use in the analysis of glycosylated hemoglobin in diabetic patients (13).

   3.6. Analytical Affinity Chromatography.   Besides its use in separating molecules, affinity chromatography can also be employed as a tool for studying solute–ligand interactions. This particular application of affinity chromatography is called *analytical affinity chromatography* or *quantitative affinity chromatography* (17,59,60). Using this technique, information can be acquired regarding the stoichiometry, thermodynamics, and kinetics of biological interactions.

   Two experimental formats that are used in this field are *zonal elution* and *frontal analysis*. Both of these are illustrated in Figure 7. Zonal elution involves

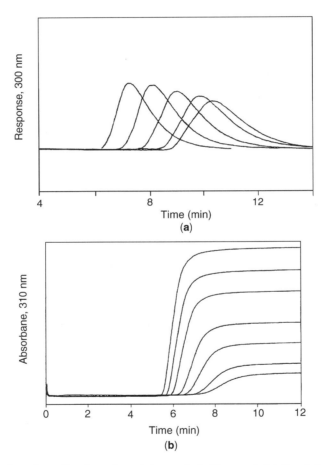

Fig. 7.   Examples of studies based on (**a**) zonal elution and (**b**) frontal analysis for the examination of solute–ligand interactions by analytical affinity chromatography. The top figure shows peaks obtained for small injections of warfarin on an immobilized human serum albumin column in the presence of increasing (from right-to-left) amounts of a competing agent in the mobile phase. The bottom figure shows frontal analysis curves obtained on an immobilized human serum albumin column for increasing (from bottom-to-top) concentrations of warfarin in the mobile phase. Reproduced with permission from Ref. 60.

the injection of a small amount of solute onto an affinity column in the presence of a mobile phase that contains a known concentration of competing agent. The equilibrium constants for binding of the ligand with the solute (and competing agent) can then be obtained by examining how the solute's retention changes with competing agent concentration. This technique was first used to study biological interactions in 1973–1974 (61,62). Since that time it has been used to examine a number of systems, such as enzyme–inhibitor binding, protein–protein interactions, and drug–protein binding (17,18,59,60).

Frontal-analysis is performed by continuously applying a known concentration of solute to an affinity column at a fixed flow-rate. The moles of analyte required to reach the mean point of the resulting breakthrough curve is then measured and used to determine the equilibrium constant for solute-ligand binding. This method was used first in 1975 to examine the interactions of trypsin with various peptide ligands (63). One advantage of this approach over zonal elution is that it simultaneously provides information on both the equilibrium constants and number of active sites involved in analyte-ligand binding. The main disadvantage of this method is the need for a larger quantity of solute than is required by zonal elution.

Information on the kinetics of solute–ligand interactions can also be obtained using affinity chromatography. A number of methods have been developed for this, including techniques based on band-broadening measurements, the split-peak effect, and peak decay analysis (17,31,60). These methods are generally more difficult to perform than equilibrium constant measurements but represent a powerful means of examining the rates of biological interactions.

Recently, a new approach for such measurements has become possible through the availability of flow-through biosensors. *Surface plasmon resonance* is one detection scheme that has been used for this purpose. In these devices, an affinity ligand is immobilized or adsorbed at the sensor's surface, while the solute of interest is applied to the surface in a flow stream of the desired buffer. Changes in the optical properties of this surface are then monitored as the solute binds to the ligand. In this approach, the rate of solute–ligand association is measured as the solute-containing buffer is applied to the system, and rate of solute release is examined when only buffer is applied to wash the surface. These results are then analyzed to determine the association and dissociation rate constants for the system, which are then combined to give the equilibrium constant for solute–ligand binding (64,65).

3.7. Miscellaneous Methods. Other methods that are related to affinity chromatography include *hydrophobic interaction chromatography (HIC)* and *thiophilic adsorption*. Hydrophobic interaction chromatography is based on the interactions of proteins, peptides and nucleic acids with short nonpolar chains on a support. This was first described in 1972 (66,67) following work that examined the role of spacer arms on the nonspecific adsorption of affinity columns. It is commonly used as a method for the purification of proteins and peptides.

Thiophilic adsorption is also known as *covalent chromatography* or *chemisorption chromatography*. This makes use of immobilized thiol groups for solute retention. Applications of this method include the analysis of sulfhydryl-containing peptides or proteins and mercurated polynucleotides (68,69).

BIBLIOGRAPHY

1. International Union of Pure and Applied Chemistry. *Nomenclature for Chromatography*. http://wingate.merck.de/english/services/chromatographie/ iupac/chrnom.htm.
2. R. E. Majors and P. W. Carr, *LC/GC* **19**, 124 (2001).
3. I. Parikh and P. Cuatrecasas, *Chem. Eng. News* **63**, 17 (1985).
4. J. Turkova, *Affinity Chromatography*, Elsevier, Amsterdam, The Netherlands, 1978.
5. W. H. Scouten, *Affinity Chromatography: Bioselective Adsorption on Inert Matrices*, John Wiley & Sons, Inc., New York, 1981.
6. R. R. Walters, *Anal. Chem.* **57**, 1099A (1985).
7. D. S. Hage, in E. Katz, R. Eksteen, P. Shoenmakers, and N. Miller, eds., *Handbook of Liquid Chromatography*, Marcel Dekker, New York, 1998.
8. E. Starkenstein, *Biochem. Z.* **24**, 210 (1910).
9. S. Hjerten, *Arch. Biochem. Biophys.* **99**, 446 (1962).
10. R. Axen, J. Porath, and S. Ernback, *Nature (London)* **214**, 1302 (1967).
11. P. Cuatrecasas, M. Wilchek, and C. B. Anfinsen, *Proc. Natl. Acad. Sci. U.S.A.* 61 **636** (1968).
12. J. P. McConnell and D. J. Anderson, *J. Chromatogr.* **615**:67 (1993).
13. D. S. Hage, in P. R. Brown and H. Isaaq, eds., *Advances in Chromatography*, Vol. 42, Marcel Dekker, New York, 2002, Chapt. 8.
14. J. Haginaka, *J. Chromatogr. A* **906**, 253 (2001).
15. Z.-L. Zhi, *Lab Rob Automat* **11**, 83 (1999).
16. P. M. Kramer, A. Franke, and C. Standfuss-Gabisch, *Anal. Chim. Acta* **399**, 89 (1999).
17. I. M. Chaiken, ed., *Analytical Affinity Chromatography*, CRC Press, Boca Raton, FL, 1987.
18. D. S. Hage, *J. Chromatogr. B* **768**, 3 (2002).
19. D. Zopf and S. Ohlson, *Nature (London)* **346**, 87 (1990).
20. M. Wikstroem and S. Ohlson, *J. Chromatogr.* **597**, 83 (1992).
21. S. Ohlson, M. Bergstrom, L. Leickt, and D. Zopf, *Bioseparation* **7**, 101 (1998).
22. Per-Olof Larsson, in T. Kline, ed., *Handbook of Affinity Chromatography*, Marcel Dekker, New York, 1993, Chapt. 2.
23. S. Ohlson, L. Hansson, P.-O. Larsson, and K. Mosbach, *FEBS Lett.* **93**, 5 (1978).
24. G. T. Hermanson, A. K. Mallia, and P. K. Smith, *Immobilized Affinity Ligand Techniques*, Academic Press, New York, 1992.
25. P.-O. Larsson, *Methods Enzymol.* **104**, 212 (1984).
26. C.-M. Zeng, Y. Zhang, L. Lu, E. Brekkan, A. Lundqvist, and P. Lundahl, *Biochim. Biophys. Acta* **1325**, 91 (1997).
27. Q. Yang and P. Lundahl, *Biochemistry* **34**, 7289 (1995).
28. D. Kriz, O. Ramstrom, and K. Mosbach, *Anal. Chem.* **69**, 345A (1997).
29. B. Sellergren, *Molecularly Imprinted Polymers—Man-Made Mimics of Antibodies and Their Applications in Analytical Chemistry*, Elsevier, Amsterdam, The Netherlands, 2001.
30. M. Komiyama, T. Takeuchi, T. Mukawa, and H. Asanuma, *Molecular Imprinting—From Fundamentals to Applications*, Wiley-VCH, Weinheim, 2002.
31. D. S. Hage, R. R. Walters, and H. W. Hethcote, *Anal. Chem.* **58**, 274 (1986).
32. D. S. Hage and R. R. Walters, *J. Chromatogr.* **436**, 111 (1988).
33. J. G. Rollag and D. S. Hage, *J. Chromatogr. A* **795**, 185 (1998).
34. M. Wilchek, T. Miron, and J. Kohn, *Methods Enzymol.* **104**, 3 (1984).
35. F. Friedberg and A. R. Rhoads, in T. Kline, ed., *Handbook of Affinity Chromatography*, Marcel Dekker, New York, 1993, Chapt. 4.
36. I. E. Liener, N. Sharon, and I. J. Goldstein, *The Lectins: Properties, Functions and Applications in Biology and Medicine*, Academic Press, London, 1986.

37. R. Lindmark, C. Biriell and J. Sjoquist, *Scand. J. Immunol.* **14**, 409 (1981).
38. P. L. Ey, S. J. Prowse, and C. R. Jenkin, *Immunochem.* **15**, 429 (1978).
39. L. Bjorck and G. Kronvall, *J. Immunol.* **133**, 969 (1984).
40. H. Potuzak and P. D. G. Dean, *FEBS Lett.* **88**, 161 (1978).
41. A. Weissbach and M. Poonian, *Methods Enzymol.* **34**, 463 (1974).
42. G. J. Calton, *Methods Enzymol.* **104**, 381 (1984).
43. T. M. Phillips, *LC Mag.* **3**, 962 (1985).
44. D. S. Hage, *J. Chromatogr. B* **715**, 3 (1998).
45. D. H. Campbell, E. Luescher, and L. S. Lerman, *Proc. Natl. Acad. Sci. U.S.A.* **37**, 575 (1951).
46. D. S. Hage and M. A. Nelson, *Anal. Chem.* **73**, 198A (2001).
47. D. S. Hage and P. C. Kao, *Anal. Chem.* **63**, 586 (1991).
48. N. E. Labrou and Y. D. Clonis, in M. A. Vijayalakshmi, ed., *Theory and Practice of Biochromatography*, Taylor & Francis, London, 2002, p. 335.
49. G. Staal, J. Koster, H. Kamp, L. Van Milligen-Boersma, and C. Veeger, *Biochem. Biophys. Acta* **227**, 86 (1971).
50. C. R. Lowe, *Curr. Opin. Chem. Biol.* **5**, 248 (2001).
51. T. S. Roming, C. Bell, and D. W. Drolet, *J. Chromatogr. B* **731**, 275 (1999).
52. P. Y. Huang and R. G. Carbonell, *Biotechnol. Bioeng.* **63**, 633 (2000).
53. G. S. Chaga, *J. Biochem. Biophys. Methods* **49**, 313 (2001).
54. J. J. Winzerling, P. Berna, and J. Porath, *Methods* **4**, 4 (1992).
55. J. Porath, J. Carlsson, I. Olsson, and B. Belfrage, *Nature (London)* **258**, 598 (1975).
56. A. Bergold and W. H. Scouten, in W. H. Scouten, ed., *Solid Phase Biochemistry*, John Wiley & Sons, Inc., New York, 1983, p. 149.
57. M. J. Benes, A. Stambergova, and W. H. Scouten, in T. T. Ngo, ed., *Molecular Interactions in Bioseparations*, Plenum Press, New York, 1993, p. 313.
58. X-C. Liu and W. H. Scouten, in P. Bailon, G. K. Ehrlich, W.-J. Fung and W. Berthold, eds., *Affinity Chromatography*, Humana Press, Totowa, 2000, pp. 119–128.
59. D. J. Winzor and C. M. Jackson, in T. Kline, ed., *Handbook of Affinity Chromatography*, Marcel Dekker, New York, 1993, p. 258.
60. D. S. Hage, *J. Chromatogr. A* **906**, 459 (2001).
61. P. Andrews, B. J. Kitchen, and D. Winzor, *Biochem. J.* **135**, 897 (1973).
62. B. M. Dunn and I. M. Chaiken, *Proc. Natl. Acad. Sci. U.S.A.* **71**, 2382 (1974).
63. K.-I. Kasai and S.-I. Ishii, *J. Biochem.* **77**, 261 (1975).
64. J. Homola, S. S. Yee, and D. G. Myszka, in C. A. Ligler, ed., *Optical Biosensors: Present and Future*, Elsevier, Amsterdam, The Netherlands, 2002, p. 207.
65. R. L. Rich and D. G. Myszka, *Curr. Opin. Biotechnol.* **11**, 54 (2000).
66. R. J. Yon, *Biochem. J.* **126**, 765 (1972).
67. S. Shaltiel, *Methods Enzymol.* **104**, 69 (1984).
68. K. Brocklehurst, J. Carlsson, M. P. J. Kierstan, and E. M. Crook, *Biochem. J.* **133**, 573 (1973).
69. P. Mohr and K. Pommerening, *Affinity Chromatography: Practical and Theoretical Aspects*, Marcel Dekker, New York, 1985.

DAVID S. HAGE
University of Nebraska

CHROMATOGRAPHY, GAS

1. Introduction

Gas chromatography (GC) is a physical method of separation in which compounds are separated using a moving gaseous phase (mobile phase) passing over or through a non-moving liquid or solid phase (stationary phase) (1). GC was first proposed in the Nobel Prize winning work of A.J.P Martin in 1941 (2) and was first developed as an instrument by Martin and James in 1952 (3). GC may be described as a form of column chromatography in that both the mobile and stationary phases are contained within a tube (column) and that the mobile phase is driven through the tube by a pressure drop between the two ends of the tube. Initially, GC was performed using packed columns, with the stationary phase consisting of solid particles packed into the column. In 1956, Golay (4) developed capillary columns, in which the stationary phase consists of a coating on the walls of a capillary tube. In this article, the technology of gas chromatography is briefly described, with reference to the original developments and to additional details that may be found in the chemical literature.

Much of the theoretical development in GC that forms the basis of current ideas on the development of GC analytical methods occurred in the 1950s and 1960s. Many current "hot topics" in GC were, in fact, proposed by researchers in these early days, including temperature programming (5), rapid separations (6), and novel stationary phases (7). The 1970s saw improvements in instrumentation, including pneumatic systems and electronic data systems. The 1970s culminated with the introduction of fused silica capillary columns in 1979 (8), which revolutionized GC analysis by making high resolution capillary columns accessible to routine users. In the 1980s, stationary phases, pneumatics, inlets and detectors were optimized for capillary columns, as the demand for these increased with increasing sales of capillary column systems. Capillary columns began to supplant packed columns for many routine applications and data systems evolved form simple chart recorders to computer-based data and instrument control systems. The 1990s saw the introduction of electronic control of the pneumatics, which provided microprocessor-controlled pneumatics and again revolutionized GC by allowing extremely precise control of gas flows, a critical parameter in GC analysis. Improvements in column, inlet and detector technologies have followed, with renewed interest in the development of rapid GC, improvements in analytical sensitivity, novel and specialty stationary phases and data systems.

In 2001, GC is a high resolution, sensitive and relatively easy to use separation technique. Samples for GC must be volatile under conditions readily achieved in GC instruments, typically temperatures <350°C. They also are typically gases, solutes dissolved in an organic solvent, or sampled from head-space and must provide a signal from a GC detector. This article describes the technology of GC, with a focus on instrumentation, stationary phases, applications, and theory, as needed to describe the technological developments in a small space. For further details, this article is heavily referenced, plus there are numerous texts and journals dedicated or focused on GC. References (9–18) include a few of the more important resources. Additional resources, especially on the most current state of the art, may be found in the annual Fundamental and Applications Reviews issues of the journal *Analytical Chemistry*, published by the American Chemical Society. Further, several research journals, including the *Journal of Chromatography A and B* (Elsevier), *Journal of Separation Science, Journal of Microcolumn Separations and Journal of High Resolution Chromatography* (Wiley), *Journal of Chromatographic Science* (Preston Publications), *Chromatographia* (Vieweg), and *LC-GC* (Advanstar) have strong emphasis in GC. A thorough discussion of GC instrumentation, including an overview and descriptions of modern inlet systems, columns, detectors and data systems is included, along with discussion of environmental, industrial, pharmaceutical, clinical, and forensic applications.

The data provided by an experiment in GC are called a chromatogram. A typical chromatogram is shown in Figure 1 (19). This figure shows the separation of a homologous series of *n*-alcohols at a temperature of 90°C. There are a number of important pieces of information that are generated by analysis of every chromatogram. First, the retention time, indicated by the time elapsed from the point of injection to the maximum of a peak, is a physical property of the compound under the conditions of the experiment. Retention times, although not unique (many compounds may have the same retention time), are used for

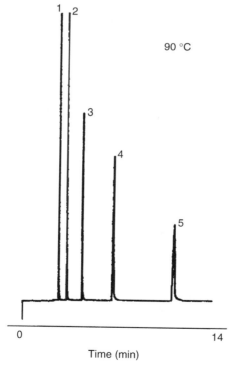

Fig. 1. Typical chromatogram of alcohols obtained from a GC analysis. 1 = *n*-Butanol, 2 = *n*-pentanol, 3 = *n*-hexanol, 4 = *n*-heptanol, 5 = *n*-octanol. [Reprinted from (19). Used with permission from the *Journal of Chemical Education*. © 1996, Division of Chemical Education, Inc.]

qualitative analysis by the matching of retention times of unknowns with those of known standards. The peak height, or peak area is related to the mass or concentration of the analyte present and is used for quantitative analysis. The gas hold-up time, defined as the retention time of a nonretained substance, is another important piece of information. The gas hold-up time is typically measured by injecting a small molecule gas, such as methane and recording the retention time.

There are several additional quantities that are commonly calculated from chromatograms. These provide the basic connection between the results generated from chromatographic data and terms familiar to most chemists. First, the retention time (t_R) for a compound, defined as the time elapsed from the point of injection to the maximum signal generated during peak elution can be divided into the time that the analyte spends sorbed in the stationary phase (t_R') and the time the analyte spends moving through the mobile phase (t_m).

$$t_R = t_R' + t_M$$

t_R', the time spent sorbed (not moving) in the stationary phase is termed the adjusted retention time and t_M, the time spent moving in the mobile phase is termed the gas hold up time. In order to relate retention times to the chemical equilibrium expressions that generate separation, the retention factor is defined as

$$k = \frac{t_R'}{t_m}$$

The retention factor, k, has an important context other than being the simple ratio of the adjusted retention time and the gas hold-up time. It also represents the ratio of mass of analyte sorbed in the stationary phase to the mass of analyte vapor in the mobile phase, at any moment that the analyte is in the column. This allows connection to a classical equilibrium constant for the phase transition between the stationary phase and the mobile phase.

$$K = k\beta$$

K is the equilibrium constant for the phase transition, k is the retention factor, defined above, and β is the ratio of mobile phase volume-to-stationary phase volume, termed the phase ratio.

If this analysis is viewed in reverse, it is seen that the retention time in GC is determined by the chemistry of the analyte-stationary phase interactions, the analyte vapor pressure, and the flow rate of the carrier gas. Temperature, carrier gas flow rate, and stationary phase composition are therefore the main variables that affect retention times. Note that while retention time is a physical property of a compound, it is not a unique property; many compounds may have the same retention time on a given column under a given set of conditions. A more thorough review of the basic theory behind retention times may be found in the texts already referenced (9–18) and in the paper by Snow (19).

In GC, column temperature is either maintained constant (isothermal GC) or the temperature is ramped (usually linearly) from a low value to a high value during the run (temperature programmed GC). Isothermal GC is much simpler, both instrumentally and conceptually, so it is often used in teaching and in process control environments, when method simplicity outweighs the need for high resolution. Due to the high thermal mass of the column and ovens used in packed column GC, isothermal conditions are often used in packed column methods. Temperature programming is most often employed with capillary columns, as they have low thermal mass and therefore, reach temperature equilibrium quickly. Also, temperature programming provides an excellent combination of improved resolution and analysis of compounds with a wide range of vapor pressures. A comparison of isothermal and temperature programmed chromatograms for a normal alkane standard is shown in Figure 2. Not only are the peaks much sharper in the temperature-programmed analysis, but there are more of them, indicating a much higher resolution separation for a wider range of analytes. The theory of temperature programming is discussed in detail in the classical text by Harris and Habgood (20) and is reviewed in the texts already referenced.

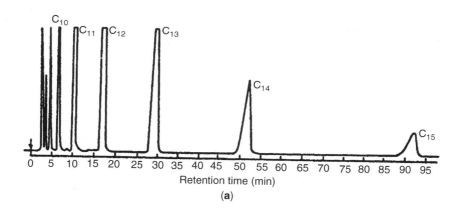

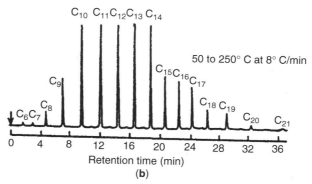

Fig. 2. Comparison of isothermal and temperature programmed chromatograms for a sample of normal alkanes. (**a**) Isothermal analysis. (**b**) Temperature programmed analysis. [Reprinted from Ref. 12, p. 145].

The peak widths are the other feature of note in the chromatogram shown in Figure 1. If it is assumed that all chromatographic peaks begin moving through the column as very sharp square bands and that they emerge as the Gaussian-appearing bands shown in the chromatogram, then random spreading mechanisms must be affecting the distribution of analyte molecules in the column. Numerous investigators have studied band broadening since the inception of GC, with the classical work being done by Van Deemter for packed columns (22) and Golay (4) for capillary columns. Their basic theories are still in use today as the most commonly applied explanations for band broadening in GC. The Van Deemter equation describes the rate of band broadening in a packed GC column. Its general form is given by

$$H = A + \frac{B}{\mu} + C\mu$$

and the Golay equation, for capillary column GC is given by

$$H = \frac{B}{\mu} + (C_S + C_M)\mu$$

In both equations, H represents the height equivalent to a theoretical plate, which roughly measures the length of column required to generate a single phase transition between the stationary phase and mobile phase, and idea drawn from the theoretical place concept used in fractional distillation. The parameter H is also a measure of the rate of band broadening; a larger value for H indicates more rapid band broadening, leading to wider peaks. Thus, minimizing H is a very important aspect of chromatographic method development. Also, in both equations, μ represents the average linear carrier gas velocity in the column, a measure of the carrier gas flow rate.

The "A term" represents band broadening due to multipath effects in packed columns. It arises from the fact that every molecule traveling through a packed bed will take a slightly different path through the particles. Thus, some analyte molecules will require more time than others to reach the column outlet. The multipath effect is most affected by the particle diameter and the quality of the packing process in a packed column. It is noted that, since a capillary column is an open tube, with no obstructions, that this term is not considered in capillary GC.

The "B term" in both equations represents molecular diffusion in the mobile phase. Given time, all populations of molecules will diffuse in the mobile phase. To minimize this term, molecules should be passed as quickly as possible through the column, using a high flow rate, along with a relatively high molecular weight carrier gas, such as nitrogen.

The "C terms" shown in both equations relate to mass transfer that occurs between the individual molecules in both the mobile phase (C_M) and the stationary phase (C_S). In the packed column equation, there is a single C term, as there is a far larger mass of stationary phase in a packed column, so the mobile phase term is neglected. The main considerations in the stationary phase term include the analyte retention factor, diffusion coefficient of the analyte in the stationary

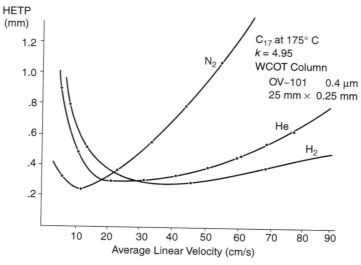

Fig. 3. Van Deemter plot showing height equivalent to a theoretical plate versus average linear carrier gas velocity. [Reprinted with permission from Ref. 24.]

phase, and stationary phase coating thickness for packed columns, plus, column inside diameter, carrier gas viscosity, and diffusion coefficient of the analyte in the carrier gas for capillary columns. Generally, low viscosity liquids, coated in thin films on the particles or capillary wall are used to minimize band broadening. For capillary columns, low molecular weigh carrier gases, such as hydrogen or helium are preferred.

Figure 3 shows a "Van Deemter plot" of height equivalent to a theoretical plate versus average linear carrier gas velocity for the three common carrier gases on a capillary column (24). It is seen that, for capillary columns, at practical linear gas velocities, the mass transfer terms dominate the rate of band broadening. Thus, for capillary columns, at practical carrier gas flow rates, hydrogen provides the best efficiency, followed by helium, then nitrogen. In the United States, helium is used most often, due to the potential safety concerns with hydrogen, while hydrogen or nitrogen are often used elsewhere. Each curve shows a minimum, which gives the optimum average linear carrier gas velocity. In practice, GC work is generally performed at higher velocities than optimum, for practical reasons. The curve shape shows the combination of the three terms, with molecular diffusion dominating at low carrier gas velocities and with mass transfer dominating at high velocities.

2. Overview of Instrumentation

A schematic diagram of a modern instrument for GC is shown in Figure 4. A GC system consists of a carrier gas supply, pneumatics, and gas scrubbers, an instrument consisting of three separately controlled heated zones: inlet, column oven and detector, and a data collection and processing system. All of these can be

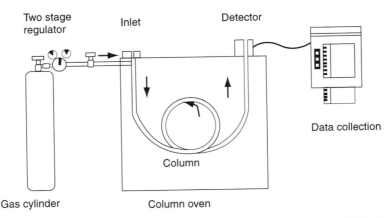

Fig. 4. Schematic diagram of a GC. (Courtesy of Professor Harold McNair.)

microprocessor controlled and generally use solid-state pneumatics and controls. Older GC's built prior to ~1995, employ a combination of digital and analog controls, while the oldest systems are mostly manually controlled. Modern GC, performed with capillary columns, requires that all ancillary equipment, such as gas supplies and equipment, syringes and devices for delivering samples and the samples be as free from contaminants as possible. This article will not address sample preparation and "cleaning-up" directly, but information on sample preparation techniques can be found in the textbooks and journals already referenced (9–18).

In capillary GC, helium is the most commonly employed carrier gas, with hydrogen used in cases where higher resolution is needed, or when the cost of helium is prohibitive. With packed columns, nitrogen is also used. The main requirements for the carrier gas are that it be of high purity and free of impurities such as water, hydrocarbons, and oxygen.

3. Inlet Systems

The ability to transfer the analyte sample into a moving gas stream at elevated pressure, without causing the system to leak is an important consideration in instrument design. Furthermore, the analyte must be transferred quantitatively, without losses or contamination and without decomposition. These requirements make the inlet and sampling system perhaps the most difficult part of the system to use and to understand for the average user. The common techniques for injecting samples into both packed and capillary GC are described there. These include simple flash vaporizers and direct inlets used with packed columns and splitters, splitless techniques, on column and programmed temperature inlets for capillary columns. Also, there are myriad on-line sampling techniques for both liquid and nonliquid samples. Some of these will be described briefly here, but the basic techniques are the focus of this chapter. For the basic techniques, there are two especially informative texts, both authored by K. Grob (25–26).

Most commonly, the injection device used in GC and with the inlets described here, is a small-volume analytical syringe. These syringes are generally composed of glass, with stainless steel plungers, and deliver volumes ranging from 0.1 to 10 μL of liquid sample, or 1 to 100 μL of gas. If gaseous samples are injected, a gas-tight syringe (1–5 mL) is generally used. Syringes should be checked often for leaks and for poor plunger performance. They generally are accurate to +/− 5% and precise to +/− 0.1% when using an autoinjector. There are several needle configurations used for specific applications, including pointed tips for manual injections, blunt tips for autoinjections and side hole tips for injection of especially labile analytes.

In order to prevent leaks during injection, the syringe must be passed through a polymeric septum or through a sampling valve. Septa come in a variety of materials and the choice of a proper septum can be critical in trace analysis. Septa are specially designed for specific inlets and it is important to use the proper septum for the inlet vendor and type. Also, especially for capillary GC work low bleed, high temperature septa should be used to prevent contamination and "ghost peaks" on chromatograms. Septa should be replaced frequently; they typically last for 30–50 injections and will leak if not replaced frequently.

3.1. Direct Inlet. Packed column systems generally employ a simple inlet called a direct inlet or a flash vaporizer. This inlet is heated to enhance rapid vaporization of the injected sample and is pressurized to enhance rapid transfer of the sample to the stationary phase. A typical packed column inlet is shown in Figure 5. The major advantage of this inlet is that a syringe needle will easily fit within a $\frac{1}{4}$- or $\frac{1}{8}$-in. outside-diameter packed column. In this system, the column is loosely fitted into a sleeve and is sealed with a compression fitting. Carrier gas flows into the inlet, around the outside of the column and into the column end, so that its temperature is equilibrated. When a sample is injected, the syringe needle pierces the septum and the sample is ejected directly onto the

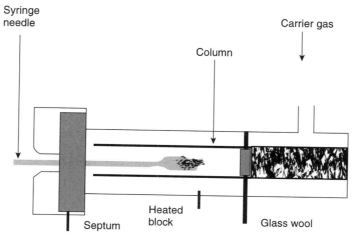

Fig. 5. Diagram of a direct injection onto a packed column (21). (Courtesy of Professor Harold McNair.)

stationary phase. The main advantage of this inlet is that all of the sample material ejected from the syringe reaches the column (see Fig. 5).

3.2. Split Inlet.   In capillary GC, there are two fundamental problems with sample injection. First, the inside diameter of most capillary columns is too small to accommodate a typical syringe. Second, the small mass of stationary phase present in a capillary column can be very easily overloaded by a 1-µL liquid sample. Thus, along with the development of the capillary column in the late 1950s, a new inlet system, the inlet splitter, was conceived. In this inlet, the sample is injected by syringe into a pressurized, heated glass sleeve. The sample vaporizes and mixes homogeneously with the carrier gas stream. Finally, the mixture is passed to two possible exits: the capillary column and a larger diameter purge vent. The purge vent exit is controlled by a needle valve that controls the split ratio, which is the ratio of the amount of the vapor mixture transferred to the vent (waste) and the amount transferred to the column. Typical split ratios range from 10:1–100:1. For example, a split ratio of 50 indicates that, for the injected sample, 50 parts is ejected out the purge vent to waste and 1 part is transferred to the capillary column. A split inlet also includes a septum purge valve that provides a small (3–5 mL/min) flow of carrier gas underneath the septum to reduce contamination. A diagram of a split inlet is shown in Figure 6.

Although first developed in the 1950s, split inlets are in very common use today. The main advantage of split injection lies in simplicity; there are three main variables: inlet pressure, inlet temperature, and split ratio. Split injection is also a very rapid technique, requiring only a few hundred milliseconds for the entire injection process to complete. This results in very sharp chromatographic peaks, necessary for high resolution separations and for good detector sensitivity. The main disadvantages of split injection are in the low final mass of sample

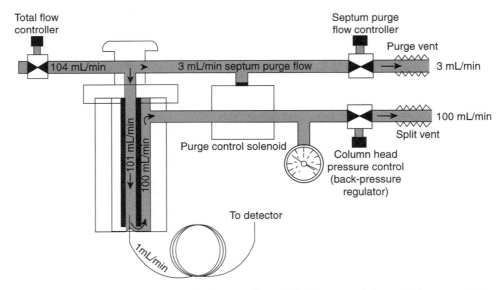

Fig. 6.   Schematic of a split inlet for capillary GC. [Reprinted from Ref. 11, p. 485].

reaching the column and in the potential for contamination or reaction of the sample components in the inlet. GC analytical sensitivity and detection limits are relatively poor (concentration detection limits ~1 ppm) when split injection is used, since most of the injected sample is transferred to the split purge, rather than to the column. Split injection also suffers from the potential reaction of analytes with the inlet components themselves. Glass sleeves should be checked for contamination and replaced often, depending upon the type and level of contaminants in the samples. Further, there are hot metal components within most split inlets that may react with organic analytes. A more complete description of the issues involved with split injection is found in the text by K. Grob (27) and in the chapter by Klee (28).

3.3. Splitless Inlets.

In 1968, K. Grob, Sr., when working on routine analysis using GC with split injection, began an analysis with the purge valve fully closed. Closing the purge valve when the instrument was not being used was common practice for reducing carrier gas usage. When the mistake was realized, shortly after beginning the analysis, the valve was opened, forcing a large flow of carrier gas through the inlet and mostly out through the purge vent. The expected result was a contaminated column and a ruined analysis. The observed result, shown in Figure 7 (29), was a chromatogram showing all analytes with very strong responses. Since this injection was performed without splitting of the injected sample vapor, it was termed "splitless" injection and has become the most commonly employed injection technique for trace analysis by GC over the past 30 years.

A schematic diagram of a modern splitless inlet is shown in Figure 8, with the "purge" shown both closed, for splitless operation and open for splitting. Note the similarity between the split and splitless inlets. In fact, on most GCs, they use the same hardware, with the difference between the configurations being the position of an electronic solenoid valve on the purge vent line. A splitless injection is begun with the purge vent closed. The electronic pneumatic controller will maintain a constant head pressure, and therefore a constant column flow. Following injection of the sample, the only outlet from the glass sleeve is the column. After allowing most of the sample to enter the column has elapsed (typically 30–45 s), the electronic solenoid valve is moved, opening the split purge. The electronic pneumatic controller will then send a large flow of carrier gas through the inlet in order to maintain the pressure with this additional outlet. This large flow, the bulk of which exits through the split purge vent, has the effect of cleaning residual solvent and sample from the inlet, causing the solvent peak to be very sharp. Unlike split injection, in which the bulk of the injected sample is lost out the split purge vent, 90–95% of the injected sample is passed to the column.

There are a large number of processes that are involved in a splitless injection, many of which are not readily obvious. Note that while the injection process may require up to 1 min to complete, the chromatographic peaks are generally very sharp, so some type of band focusing must occur. There are several processes that contribute to this focusing. These include cold trapping, solvent effects, and the use of a retention gap. Splitless injections are always associated with temperature programmed analysis, so while the inlet is heated, the column is usually relatively cool. First, high molecular weight compounds will be

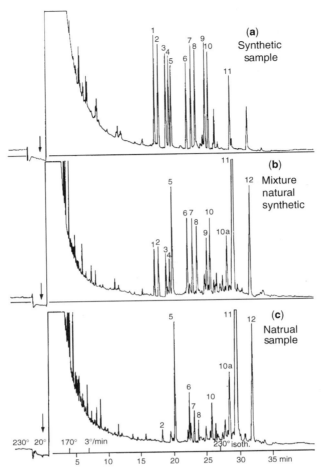

Fig. 7. First application of splitless injection for the analysis of steroids. (**a**) synthetic sample. (**b**) Mixture of natural and synthetic samples. (**c**) Natural sample. Experimental details may be found in (29). [Reprinted From the *Journal of Chrometographic Science* (29) by permission of Preston Publications. A Division of Preston Industries, Inc.].

transferred through the hot inlet, but will immediately condense as they are sorbed on the cool stationary phase. As the first molecules that reach the stationary phase stop, the remaining molecules "catch up" generating a sharp peak by concentrating the entire population of molecules into a small sharp band.

Second, solvent effects play a strong role, especially sharpening peaks representing compounds that are more volatile and are not effectively cold trapped. Solvent effects are shown schematically in Figure 9. First, the injected sample spreads out over a significant length (>1 m) of the capillary column immediately following the injection. The large amount of solvent acts like a thick stationary phase coating that dissolves the analyte components and spreads them throughout the length of the plug. If this broad band of analyte molecules is not focused further, it will be very broad, resulting in broad,

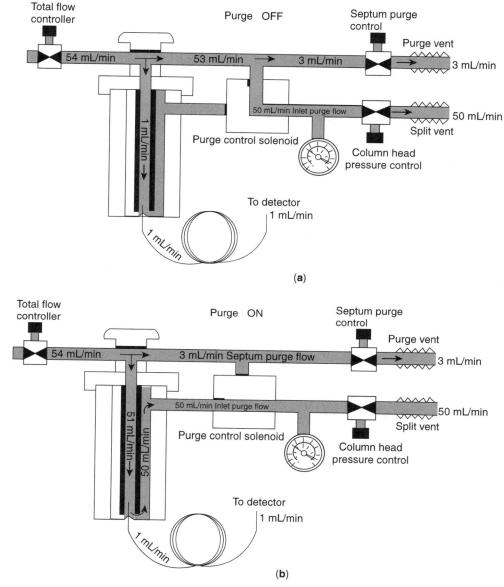

Fig. 8. Schematic diagrams of a splitless inlet for capillary GC with the inlet shown in the "Purge OFF" and "Purge ON" positions. Note the configuration of the solenoid valve. [Reprinted from Ref. 11, p. 489].

misshapen peaks. This phenomenon is termed "band broadening in space". Solvent effect focusing occurs when the flow of carrier gas causes this solvent plug to evaporate. Eventually, the solvent evaporates away, leaving the analyte behind, focused into a very sharp band. As the column temperature is increased, the focused analyte bands begin to migrate along the column and to separate.

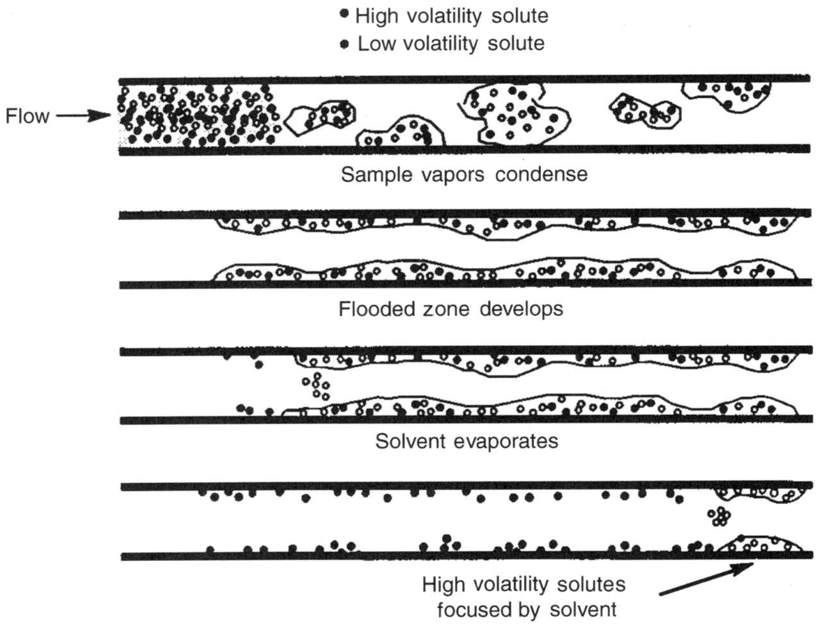

Fig. 9. Diagrammatic representation of solvent effects in splitless injection. [Reprinted with permission from Ref. 11, p. 478].

Thorough descriptions of splitless injection and its many principles are found in the chapter by Klee (30) and the text by Grob (31).

3.4. On-Column Inlet. On-column injection is receiving increased attention recently, as syringes and autoinjectors have been improved to accommodate the delicate handling required. A schematic of an on-column inlet is shown in Figure 10. Note that the column extends all the way into the inlet and that the syringe must be guided into the column by the inlet fittings. Also note the low thermal mass, so that the inlet may be temperature programmed along with the column, to ensure that the analytes elute. A syringe with a specially tapered needle is used to inject liquid sample directly into the column. During injection, the column oven and inlet are maintained below the boiling point of the sample solvent; the inlet temperature is usually increased to follow the column oven during a temperature-programmed analysis. The main advantage of on-column injection is that the entire injected sample reaches the column without the potential degradation that comes from the other hot injection techniques. The main disadvantage of on-column injection is also that the entire sample reaches the column, including any and all matrix and nonvolatile components that may be present. Column fouling and maintenance are often increased dramatically when using on-column injection techniques. Grob provides an excellent review of on-column injection (26).

A modified on-column inlet has been used for large volume injections, sometimes of up to hundreds of microliters (remember that ~1 µL is typically injected), which allows an analogous increase in analytical sensitivity and low-

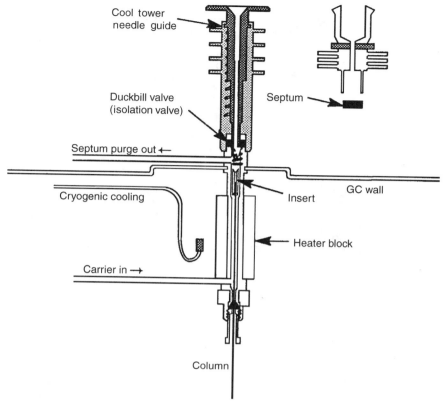

Fig. 10. Diagram of a cool-on-column inlet. [Reprinted from (11), p. 494].

ering of detection limits. The configuration of this system is shown in Figure 11. Note the addition of the retention gap, which allows physical space for the injected sample, the retaining precolumn, which helps to retain the more volatile analytes and the solvent vapor exit valve. The sample is injected with the vapor exit valve open to allow escape of the solvent vapors. When ~95% of the solvent vapor has escaped (the timing can be calculated using the vapor pressure of the solvent and the instrumental conditions) the valve is closed, directing the remaining material to the column. In both types of on-column injection presented here, the inlet is maintained at a temperature below the boiling point of the sample solvent, as the liquid must be injected onto the column without rapid evaporation, in contrast to split and splitless injections. A representative sampling of applications of large volume injection using the solvent vapor exit device is given in (32–34).

3.5. Programmed Temperature Vaporization. Temperature programmed injection has also been used in combination with the classical split and splitless injections described above. Developed in 1979 and promoted by Schomburg, the programmed temperature vaporization (PTV) inlet allows for the at-once injection of sample volumes up to 100 μL (35). The PTV inlet design is based on the classical splitless inlet, except that it has a low thermal mass to

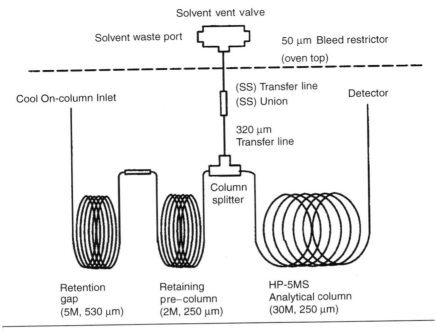

Fig. 11.   Diagram of a cool-on-column inlet configured for solvent vapor exit large volume injection. Courtesy of Agilent Technologies.

allow for rapid heating and cooling. There are a number of modes in which it can operate, including hot split and splitless, which are the same as the classical split and splitless techniques, and cold split and splitless, which involve a cool inlet during injection, which is temperature programmed to pass the injected material into the column. In a cold splitless large volume injection, the glass liner is packed with an inert, high surface area material to accommodate a large volume of injected liquid. During the injection, the inlet is cooled with the purge vent open to allow solvent vapor to escape, while analytes are trapped on the liner packing. After ~95% of the solvent vapor has evaporated, the purge vent is closed, the inlet is temperature programmed and the analytes are transferred to the column in splitless mode. Following transfer, the purge vent is opened again to clean any residual material out of the inlet. The PTV large volume injection is seeing increased attention, especially in environmental analysis, in situations where increased sensitivity is needed. Representative applications include analysis of very low levels of pesticide residues (36), constituents of landfill gases (37), and petrochemicals (38). A schematic of a PTV inlet is shown in Figure 12, which is very similar to a classical splitless inlet, except that it has low thermal mass and the capability for rapid heating and cooling.

4. Columns and Liquid Phases

In GC, the separation occurs in the column, in which the gaseous mobile phase passes over a solid or liquid stationary phase consisting of solid particles, or solid

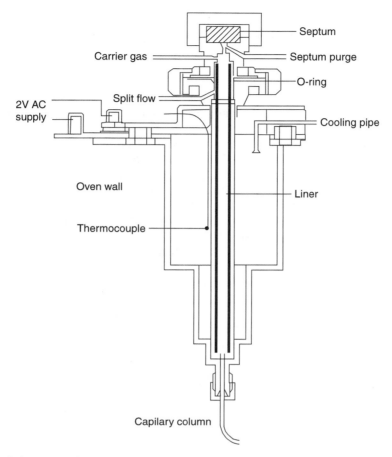

Fig. 12. Schematic of a programmed temperature vaporization inlet. (Courtesy, ATAS GL International.)

particles coated with a liquid, or consisting of a liquid or solid material coated onto the walls of a capillary tube. There is a huge variety of materials that have been employed as stationary phases over the years. With packed columns, separation efficiency is relatively low, so there are a huge number of stationary phases available, to take advantage of the myriad available surface chemistries. Inherently, capillary columns have much higher separation power than packed columns, so there is less need for a wide variety of stationary phase chemistries. Traditionally, there have been fewer capillary GC stationary phases available, although recently, specialty phases, for specific applications, have become available. In this section, the technology involved in using packed and capillary columns will be described, along with a summary of common stationary phases.

4.1. Packed Column Instrumentation. The original GC's were outfitted with packed columns. A packed column typically consists of a $\frac{1}{4}$- or 1/8-in. outside-diameter stainless steel or glass tube with length of 3−12 ft. The diameters of packed columns are generally determined by the availability of tubing and fittings from the suppliers of such equipment. The pressure drop that can be accommodated by a gas chromatograph limits the length of a packed column. Tubing materials commonly include glass, which is least reactive, but often more difficult to work with, stainless steel, which is robust but potentially reactive with organic analytes and copper, which is easy to work and install into the small ovens used in GC, but is very reactive toward organic compounds. When purchased from vendors, packed columns are generally custom configured to fit properly into the major manufacturers' GCs. Some laboratories also readily make packed columns in-house. Packed columns have a relatively large thermal mass, so temperature equilibration is a major factor in the development of methods. Commonly, to avoid this problem, packed column GCs are operated isothermally, so that temperature equilibrium and reproducibility is maintained. These constraints, taken together, leave a great deal of emphasis on the choice of stationary phase in packed column methods, which is the reason that there are a huge number (hundreds) of these available.

4.2. Capillary Column Instrumentation. In contrast to packed columns, capillary columns, also called "wall coated open tubular" columns, are available in a huge variety of lengths and inside diameters, with relatively few (dozens, rather than hundreds) stationary phases available. Generally, capillary columns vary in length from 10 to 100 m, inside diameters of 0.10 to 0.53 mm, and liquid phase coating thickness 0.1 to 5.0 μm. Since they are open tubes, capillary columns do not share the pressure drop limitations with packed columns, allowing for very long lengths. However, the relatively small inside diameter places limitations on the carrier gas flow rate, injection system and sample capacity. Capillary column instrumentation is therefore more complex and expensive (a factor of 2–5) than packed column instrumentation, with the main differences being in the inlet systems, described earlier in this article. There are also minor differences in the operation of the common detectors flame ionization detector (FID), thermal conductivity detector (TCD), electron capture detector (ECD), mass selective (MS), between capillary and packed column instruments. Capillary columns themselves are generally manufactured from polyimide coated fused silica tubing, which imparts flexibility and ease of handling. The long length necessitates coiling in to a cage for easy handling, so the user should beware that the coils may uncoil rapidly and should wear eye protection at all times when handling capillary columns. For many years, glass was the most common material used for capillary columns, but was not flexible and required expensive drawing machines to obtain the necessary inside diameters. Stainless steel and aluminum clad fused silica have been used for special high temperature applications.

4.3. Stationary Phases. In both packed and capillary GC, the stationary phase may be either a liquid or solid. In a capillary column, the stationary phase is coated or chemically bound onto the capillary wall; in a packed column, the stationary phase consists of either solid particles or liquid coated solid particles. As discussed previously, there are myriad stationary phases available for GC, as a perusal of manufacturers' literature (several WWW sites for larger manufacturers are provided here; these also include extensive application notes and educational brochures and publications) shows (39–42). Snow (43) recently reviewed recent trends and developments in liquid phases for GC. The most

Table 1. **Commonly Used Stationary Phase Materials and Applications**[a]

Stationary phase temperature range	Application
100% methyl polysiloxane – 60–350°C.	alkaloids, amines, drugs, FAME, hydrocarbons, petroleum products, phenols, solvents, waxes, general purposes
5% phenyl–95% dimethyl polysiloxane –60–350°C	alcohols, alkaloids, aromatic hydrocarbons, flavors, fuels, halogenates, herbicides, pesticides, petroleum products, solvents, waxes, general purposes
50% phenyl–50% methyl polysiloxane –60–350°C	alcohols, drugs, herbicides, pesticides, phenols, steroids, sugars
14% cyanopropylmethyl–86% dimethyl polysiloxane 0–250°C	alcohols, aroclors, alcohol acetates, drugs fragrances, pesticides
50% cyanopropylmethyl–50% phenyl polysiloxane 0–250°C	carbohydrates, FAME
trifluoropropyl polysiloxane 0–275°C	drugs, environmental samples, ketones, nitro-aromatics
polyethylene glycol 60–225°C	alcohols, flavors, fragrances, FAME, amines, acids

[a] Reprinted with permission from Ref. 43.

commonly used stationary phases and their applications are summarized below in Table 1.

A potentially confusing problem in working with stationary phases for GC is that each manufacturer uses a different name or designation to describe materials that are, for most purposes equivalent. Table 2 provides a summary of these designations for the common stationary phases and several major manufacturers.

As evidenced in the manufacturers' literature, there are also numerous variations on these common chemistries. An emphasis today is on the manufacture of specialty stationary phases, tailored to specific applications or compendial methods. A summary of some of these specialty applications is shown in Table 3.

Finally, there are numerous recent developments in the use of cyclodextrins for chiral separations (44,45), liquid crystals (46,47), crown ethers (48,49) and sol-gels (50,51) as new stationary phase materials with specific, beneficial properties for difficult separations.

5. Detectors

The purpose of the detector is to sense analytes as they elute from the column and record that information in the form of a chromatogram. The signals generated by the detector are received and recorded by a data collection device, such as a chart recorder, electronic integrator, computer data station, or central data collection system. The collected data is plotted as intensity versus time, as described previously in Figure 1 and the accompanying discussion. A summary of the capabilities of the three most common GC detectors, FID, TCD, and ECD is provided in Table 4.

Table 2. **Cross-Reference of Manufacturers' Designations for Common Stationary Phases**[a]

Stationary phase	Restek	J + W	Supelco	HP	SGE	Chrompack	Quadrex
100% polydimethyl siloxane	Rtx-1	DB-1	SPB-1, SP-2100	HP-1, Ultra-1	BP-1	CP-Sil 5 CB	007-1
95% dimethyl–5% phenyl polysiloxane	Rtx-5, XTI-5	DB-5	SPB-5	HP-5, Ultra-2	BP-5	CP Sil 8 CB	007-2
80% dimethyl–20% phenyl polysiloxane	Rtx-20		SPB-20				007-7
65% dimethyl–35% phenyl polysiloxane	Rtx-35						007-11
14% cyanopropyl phenyl–86% dimethyl polysiloxane	Rtx-1701	DB-1701			BP-10	CP Sil 19 CB	007-1701
50% methyl-50% phenyl polysiloxane	Rtx-50	DB-17	SP-2250	HP-17		CP Sil 43 CB	007-17
trifluoropropyl-methyl polysiloxane	Rtx-200	DB-210					
50% cyanopropyl phenyl–50% dimethyl polysiloxane	Rtx-225	DB-225	SP-2300	HP-225	BP-225		007-225
carbowax PEG	STABILWAX	DB-WAX	SUPELCOWAX-10	HP-20M	BP-20	CP Wax 52 CB	007-CW
carbowax PEG for amines	STABILWAX-DB	CAM					
carbowax PEG for acids	STABILWAX-DA	DB-FFAP	NUKOL, SP-1000	HP-FFAP	BP-21	CP Wax 58 CB	FFAP

[a] Reprinted with permission from Ref. 43.

426

Table 3. **Specialty Applications for Capillary GC Columns**[a]

Application	Description
thermally stable modified form of common phases; low bleed; highly inert	for GC/MS, ECD, other highly sensitive analyses
35% phenyl polysiloxane	conformational analysis
bonded poly(ethylene glycol)	fatty acid methyl esters
bis(cyanopropyl) polysiloxane	positional and geometric isomers of polysiloxanes
base modified polysiloxanes	amines; basic analytes
carbowax amine	primary, secondary, tertiary amines
6% cyanopropyl phenyl, 94% polydimethyl siloxane	USP and EP volatile organic contaminants methods

[a] Reprinted with permission from Ref. 43.

The ideal detector would be both universal, meaning that it is able to detect all compounds that elute from the column, and sensitive. However, in reality detectors are often either universal or selective. A selective detector is capable of only detecting certain types of compounds, and this selectivity is often why the detector has a high sensitivity. Analysts frequently trade selectivity for sensitivity. For these reasons, over 60 detectors have been developed and described in literature. Currently a handful of detectors are most commonly used FID, TC, MSD, ECD (52).

5.1. Flame Ionization Detector. The FID employs an ionization detection method invented specifically for GC. It was first introduced in 1958 (53,54) and has since become the most widely used detector. This detector has good

Table 4. **Summary of Characteristics—Flame Ionization, Thermal Conductivity, and Electron Capture Detectors**

	FID	TCD	ECD
limit of detection	10^{-11} g (50 ppb)	10^{-9} g (10 ppm)	10^{-14} g/s for sulfur hexafluoride, an ideal compound for ECD, the LOD is unique per compound
applications	nearly universal for organics, no fixed gases	universal	halogenated material, especially pesticide residues
linear range	10^6	10^4	10^2 in direct current (dc) mode
temperature limits	~400°C	~400°C	limited by radioactive source used: 400°C (^{63}Ni)
other	highly stable, easy to operate, conventional amplifier required	requires good temperature control otherwise stable, easy to operate, no amplification needed He carrier gas used for optimum performance	radioactive source needed one of the most easily contaminated detectors, needs ultrapure dry gases (must be free from O_2, H_2O) and clean samples

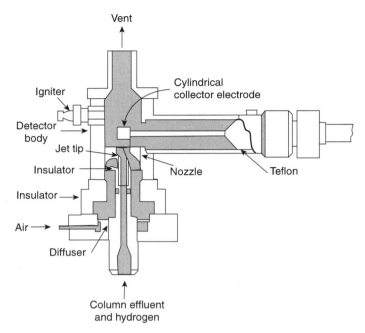

Fig. 13.   Schematic of FID. [Reprinted with permission from Ref. 12, p. 115.]

sensitivity with a minimum detectable quantity (MDQ) of ~50 ppb and wide linear range (10^6). The main advantages are its simple design, affordability and reliability. Occasionally this detector is classed as a universal detector, but in fact the FID is only able to detect organic analytes, and will not detect compounds such as water, hydrogen, helium, nitrogen, carbon monoxide, and carbon dioxide. The inability to detect these compounds is rarely an issue and when these compounds are to be detected, another method of detection, such as thermal conductivity, must be used.

A schematic for a typical FID is shown in Figure 13. A diffusion flame is used in the FID, which is to say that the detector uses two gases, hydrogen and air, and it is the rate of diffusion of the two gases that controls the burn rate of the flame. The analytes are then introduced from the column into the flame jet, where they are ionized. A voltage (300 V) is applied across the flame to produce a current that is amplified by an electrometer to pick up the signals from the analytes. This signal generates the chromatogram.

5.2. Thermal Conductivity.   Thermal conductivity detection is another commonly used detection method. This detector is a universal detector and is frequently used with packed columns and for inorganic analytes that are not detected by FID. The general characteristics of the TCD are as follows: MDQ of detection ~10 ppm, universal detection, linear for four orders of magnitude, good stability, and an upper temperature limit of 400°C.

The TCD operates on the principle that a hot body (the filament) will lose heat at a rate that is proportional to the surrounding gas and this heat loss can be used to detect the elution of analytes from the column. Since any analyte,

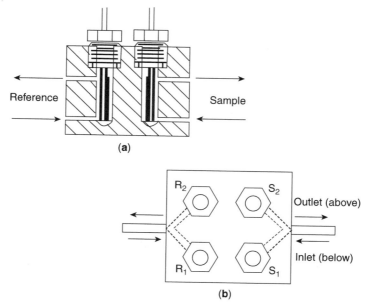

Fig. 14. Schematic of thermal conductivity detector typical four-filament tcd cell. (**a**) Side view. (**b**) Top view. [Reprinted from Ref. 12, p. 117].

except for the carrier gas itself, that passes through the detector will change the rate of heat loss this detector is truly universal. A schematic of a conventional four-filament TCD cell that is commonly used with packed columns is presented in Figure 14. Since resistance is a function of filament temperature any changes in the thermal conductivity of the gas that passes by the filament changes the resistance. The difference in resistance between a reference filament (R_1 and R_2), which is in pure carrier gas, and the sample filament (S_1 and S_2), which is in carrier gas and analyte, is used to detect the presence of the analyte as it elutes from the column. The detectors use a Wheatstone bridge circuit, which generates a signal when there is a difference between the output signals of the reference and sample filaments. For a TCD to be effective the thermal conductivity of the carrier gas must be significantly different than that of the analytes, which is not difficult to achieve since hydrogen and helium, which have the highest TC values, are commonly used as the carrier gas (55).

An important advantage of the TCD is the ability to detect air, which is not retained by most GC columns. The detection of air is useful in determining the void volume of a column. Traditionally the TCD was considerably less sensitive than the FID. However, recently the TCD has been adapted to capillary columns. The cell volume has been reduced to be more compatible with capillary GC columns and the use of a reference cell has been eliminated in some designs. Since a single cell is used, the gas flow to the detector is rapidly oscillated between the carrier gas and a reference gas. The capillary design of the TCD is reported to have sensitivities that approach that of the FID. Recently, TCD performance has improved, through the development of smaller flow cells.

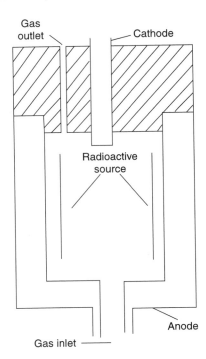

Fig. 15.   Pin-cup ECD. [Reprinted from Ref. 11, p. 293.]

5.3. Electron Capture.   The invention of ECD is attributed to Lovelock (56), who first published on the device in 1961. This selective detector is very sensitive toward compounds that can capture electrons. These are typically halogenated or nitrogen containing compounds. Its primary use is in pesticide analysis. For standard capillary column ECD the MDQ is 10 pg, the response is very selective, the linear range is three or four orders of magnitude and the detector is stable, although it is sensitive to contamination from traces of oxygen or water and requires extreme care in use and maintenance for optimal results.

A schematic of an ECD is presented in Figure 15. This detector uses a radioactive source, usually ^{63}Ni, to ionize the carrier gas (helium with nitrogen make up gas, or nitrogen, or a mixture of argon and methane), causing a high standing current. Electronegative analytes such as halogen-containing compounds, entering the detector cause a decrease in standing current as the analytes capture the free electrons. The standing current is kept constant by pumping electrons into the detector system. The decrease in electrons is measured as the analyte signal.

This is a quantitative detector, as the extent of electron capture is proportional to the analyte concentration. The ECDs are straightforward to use, but do require extra care in maintaining a clean system. Only very high quality nitrogen or argon/methane gas should be used as the makeup gas. If the detector is well maintained, conventional ECDs can easily detect picograms of analyte and micro ECD can detect as little as 4 fg of material (57). The detection limit is very dependent on the analyte's ability to capture electrons and thus the sensi-

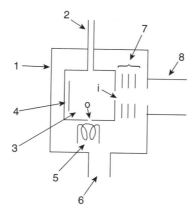

Fig. 16. Mass spectrometer ionization chamber (1) housing for ionization chamber; (2) sample inlet; (3) ionization chamber; (4) ion repeller; (5) cathode-e- emitter; (6) high vacuum system; (7) ion-beam collimator; (8) analyzer tube. Represented with permission from Ref. 59.

tivity of an ECD can drastically different from one analyte to another. For example, perflourobutene, C_4F_8, has a relative response of 1.3×10^5:1 with reference to perfluoropropane, C_3F_8 (58).

5.4. Mass Spectrometer. Perhaps the most useful detector for GC is mass spectrometry (MS). This detector provides both quantitative and qualitative analysis. State of the art bench-top GC MS systems are capable of unit mass resolution. The GC/MS system uses a narrow-bore GC capillary column; and by using a low flowrate and a vacuum pump, the column can be inserted directly into the MS.

The ionization chamber for a typical bench top mass spectrometer is diagrammed in Figure 16. The mass spectrometer ionizes the incoming sample and presents either the total ion chromatogram (TIC), or it will scan for only certain specified ions in selected ion monitoring (SIM) mode. The TIC of a sample contains all data necessary for compound identification and can be used to compare the mass spectrum of each individual peak in the chromatogram with reference spectra in a computer-based library. The SIM only monitors for one or a few ions in a sample, and so SIM data can be used to identify compounds with previously determined reference spectra.

Complete details on mass spectral interpretation and how this is applied to structure elucidation and chemical analysis may be found in the classical text by McLafferty (60) and in the more recent text edited by Busch (61).

5.5. Other Selective Detectors. Many more types of detectors exist than what has been discussed above. This section will discuss two detectors used for specific atoms, as well as infrared (ir) and ultraviolet (uv) detection.

The nitrogen-phosphorous detector (NPD) is another ionization detector that was invented for use with GC. As the name suggests, this detector is selective for nitrogen- and phosphorous-containing compounds. The detector has no flame; the NPD uses a rubidium silicate bead heated by a platinum resistance wire, and combustion is not supported because the hydrogen flow is very low.

The NPD is used in pharmaceutical labs for nitrogen-containing drugs, as it is the most sensitive detector available for nitrogen and phosphorous. Other applications of NPD are for the analysis of nitrogen- or phosphorous-containing pesticide residues, carcinogens, and amines.

The flame photometric detector (FPD), invented by Brody and Chaney in 1966 (62), is primarily used for detection of sulfur and phosphorous. A hydrogen-rich flame is used to burn the samples, which form chemiluminescent species that emit light at 394 nm for sulfur and 526 nm for phosphorous. Applications for FPD include pesticide residue analysis, air pollution studies, and petroleum analysis. Infrared spectrophotometers have been successfully attached to a GC and used as a specific detector. Discussions of these and many other selective detectors can be found in the previously referenced textbooks (9–17) and in the chapter by Henrich (63).

6. Data Collection and Handling

The purpose of a chromatographic data system is to collect analogue data from an analytical instrument and convert it to digital data. This is accomplished by an analogue-to-digital converter (ADC). The important characteristics of ADCs are speed and accuracy of conversion. Since capillary columns generate sharp peaks that elute in seconds, the major requirement of a data system is a rapid sampling rate which is easily accomplished by computer technology. The three basic types of data collection currently used in GC are integrators, dedicated computer based instrument(s) data systems and multi user server networked systems. The function of all of theses systems is to collect data from the instrument and provide the analysts with a means of interpreting the data generated by the instrument. In the distant past, chart recorders were used, but today they are rare and typically found on some educational chromatographic systems.

6.1. Integrators and Recorders. Until the early 1980s, most data was collected and stored using classical strip chart recorders and plotters, when digital computers and integrators supplanted them. The key advantage of a strip chart recorder was simplicity, however, the myriad disadvantages, especially the difficulties in postprocessing of quantitative data, led to their demise. In the 1980s, the primary data systems were digital electronic integrators, which combined strip chart recording with rudimentary computing capability. Chromatograms could be stored electronically and reprocessed, and the instrument could be controlled through the data system, although procedures for these operations tended to be tedious. Integrators are still in existence however they have been mostly replaced by dedicated and server based computer systems.

6.2. Computer-Based Data Systems. In the 1990s, stand-alone and networked computers, usually built around PC platforms, replaced many of the older integrators. These computer-based systems including servers are also capable of instrument control, data collection, and archiving, generating detailed reports and documents, generating system suitability results and interfacing with laboratory information management systems (LIMS). In many industries, strict data security is also a requirement.

Stand-alone computer based systems are typically found in smaller laboratories where only a few instrument reside. The disadvantage of these systems is the need for the user to maintain backups of the data systems and the advantage is they are generally more flexible with regards to updating the software systems which control them. Server based systems are typically used in large well-regulated corporations where the data from all the instruments is collected and stored in one location. These systems can support hundreds of instruments simultaneously. While the details of computer-based data collection are beyond the scope of this article, an excellent introduction can be found in the chapter by McDowall (64).

6.3. Regulatory Issues. In many industries, especially pharmaceutical, forensic, environmental and clinical analysis, data systems and instruments must be validated to assure that they are operating within established norms and procedures. Typically, instrument and data system vendors provide assistance with validation, although it is wise for laboratories to have internal standard operating procedures in place. Validation requirements may also go beyond instrumental and scientific concerns, to include data storage, retrieval, and security. The validation process is often lengthy and labor intensive.

For example, when using many of these data systems, the analyst must set the operating parameters for data collection and analysis and, in general data manipulation can be done after data collection. Suppose that the analyst can set the data collection rate or peak width. This determines the number of data points collected per unit time. If too few points are collected, the apex of the peak can be missed, which leads to inaccurate quantitation. If too many points are collected the analyst can data bunch or average the data collected. The problem with collecting too many data points is rapid filling of data storage space. If there is no standard operating procedure or validation in place, then two analysts, who may be running the same nominal procedure on the same samples may obtain dramatically different analytical results. A useful introduction to validation and method transfer issues in the pharmaceutical industry is found in the chapter by Crowther and co-workers (65).

7. Multidimensional GC

In order to dramatically increase separation power, multidimensional GC, employing two columns, has been developed. In multidimensional GC, the column effluent from the first column, as it elutes, is transferred to a second column, typically with another stationary phase chemistry, for further separation. This affords tremendous separation power, as, with two dimensions, chromatograms have space for thousands of peaks. Multidimensional GC is most often employed in the petroleum industry, and sometimes for toxicology and environmental problems. There are two common instrument configurations: traditional multidimensional GC, in which the effluents represented by single peaks from the first column are collected, trapped, then transferred to the second column, and comprehensive two dimensional GC, in which the column effluent is trapped continuously and transferred to the second column at regular intervals, using a trap combined with a switching valve. Due to the complexity of the valve and switching systems involved, multidimensional GC is not used by many routine analysis laboratories, but is used for specialized applications requiring especially strong separation power.

7.1. Traditional Multi-dimensional GC.   In traditional multidimensional GC, column effluent representing one or more peaks in a separation is trapped and transferred to a second column, which is generally termed "heartcutting" to represent the interesting portion of the chromatogram being further analyzed. Two-dimensional GC systems often consist of two ovens so that the two columns can be temperature programmed independently. They are often much more complex than traditional one-dimensional GCs, and are generally not in routine use. As an example, a two-dimensional separation of fire debris, in which several heart cuts were sampled and analyzed is shown in Figure 17 (66). The ability to analyze very complex mixtures, which is the main advantage of two-dimensional GC, is easily seen. Other typical applications of two-dimensional GC include chiral separations (67,68) urinary acids and (69) bornane congeners (70). Bertsch (71) provided an especially thorough description of theory and application of multidimensional GC.

7.2. Comprehensive Two-Dimensional GC.   Comprehensive two-dimensional GC, in which the effluent from a traditional column is continuously sampled into a short, narrow bore, thin film second column, was originally proposed by Schomburg (72) and was developed and promoted by Phillips and co-worker (73,74). The continuous use of the second dimension column generates tremendously high peak capacity. Notable applications of comprehensive two-dimensional GC include complex petroleum analysis (75,76) and pesticides from biological samples (77). A typical two-dimensional chromatogram showing the separation of pesticides extracted from serum is shown in Figure 18. In this two-dimensional chromatogram, viewed "from above" as a contour plot, the x-axis represents separation an the traditional column and the y-axis represents separation on the short second column. It is seen that several compounds (shown as peaks lined up in the y direction) that would have coeluted if separated just on the traditional column, are well resolved on the second column. The tremendous peak capacity is evidenced by the large amount of "blank space" seen in the two-dimensional space.

8. Fast and Micro-GC

Obtaining faster separations has been an interest of chromatographers since the pioneering work in the 1950s (6). Separations in a matter of seconds were first shown in the early 1960s, although the routine use of fast separations was not seen until the 1990s. The instrumental requirements for obtaining fast separations are more stringent than for traditional GC, as the columns are very short, the gas flows are high and require very precise control and the chromatographic peaks elute very quickly, requiring careful detector choices and optimization. Along with the drive toward faster separations in traditional bench-top systems, there has been a move toward smaller systems that are field portable. Systems for both applications have become commercially available in the late 1990s (78).

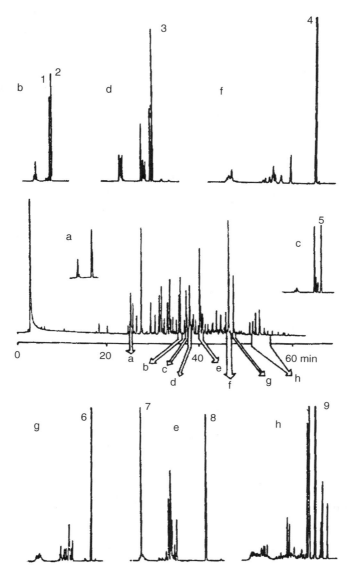

Fig. 17. Use of heart-cutting for the identification of target compounds in 90% evaporated gasoline. 1 = 1,2,4,5-Tetramethylbenzene, 2 = 1,2,3,5-tetramethylbenzene, 3 = 4-methylindane, 4 = 2-methylnaphthalene, 5 = 5-methylindane, 6 = 1-methylnaphthalene, 7 = dodecane, 8 = naphthalene, 9 = 1,3-dimethylnaphthalene. [Reprinted with permission from (65).]

The main limiting factor in developing both faster and smaller systems is the bulk of the column, column oven, inlet, detector, and fittings that accompany them. As miniaturized electronics and microprocessor-controlled pneumatics have become available, this problem has been significantly reduced. While most of these systems employ traditional inlets and detectors for ease of sample

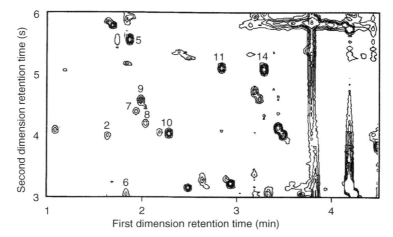

Fig. 18. Comprehensive two-dimensional GC chromatogram of supercritical fluid extract of spiked human serum. 1 = dicamba, 2 = trifluralin, 3 = dichloran, 4 = phorate, 5 = pentachlorophenol, 6 = atrazine, 7 = fonofos, 8 = diazinon, 9 = chlorothalonil, 10 = terbufos, 11 = alachlor, 12 = matalaxyl, 13 = malathion, 14 = metalochlor, 15 = DCPA, 16 = captan, 17 = folpet, 18 = heptadecanoic acid. [Reprinted with permission from Ref. 77. Copyright 1994 America Chemical Society.].

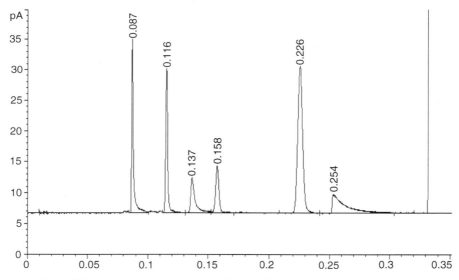

Fig. 19. Fast GC separation of residual solvents from a pharmaceutical analysis. inlet: split 100:1, 10 psi, 300°c; Column: HP-5MS, 5 m × 0.25 mm × 0.25 μm; TP: 40°C/1 min, 200°C/min to 80°C. Detector: FID, 300°C, 100 Hz. 1. Ethanol, 2. ethyl acetate 3. 1-butanol 4. Heptane 5. Toluene 6. Dimethylformide (DMF) 7. Dilution: dimethyl sulfoxide (DMSO) (80).

handling, the main modifications have been to the column and oven configuration. For fast GC, metal sheathed, resistively heated columns have been used to provide very rapid (up to 1200°C/min) temperature programming of a 5 m column. The specially sheathed column and an external controller can be added to many commercial GCs. A representative fast separation of several solvents is shown in Figure 19. An additional approach to faster and shorter columns involves wrapping a column around a heated metal rod, which allows a very small oven (79).

9. Sample Preparation

Almost all GC-based analytical methods in use today also involve some form of sample pretreatment prior to the injection and analysis. While it is beyond the scope of this article to describe sample preparation in detail, the reader is directed to a number of references for more information. Many of the standard referenced texts (9–18) and journals described above include sections or research papers dealing with sample preparation. Some of the newer techniques for which texts are available, that are often employed on-line with analysis include pyrolysys (81), static and dynamic head-space, (82) supercritical fluid extraction (83), solid-phase extraction (84), and solid-phase microextraction (85,86).

10. Conclusions

Gas chromatography is considered by many researchers to be a mature technique. Developed 50 years ago, it remains one of the most widely used instrumental techniques in analytical chemistry today. Recent advances in inlet, column, detector and data system technology have made GC straightforward and cost effective to use for myriad routine applications. Recent attention in the field has focused on novel column technologies, fast separations, selective detectors, especially MS, and sampling techniques.

11. Acknowledgments

The authors gratefully acknowledge the assistance of Ms. Rebecca Polewczak, Department of Chemistry, Clarkson University, with the editing and assembly of this manuscript.

BIBLIOGRAPHY

1. L. S. Ettre, *Pure Appl Chem.* **65**, 819 (1993).
2. A. J. P. Martin, and R. L. M. Synge, *Biochem J.* **35**, 1358 (1941).
3. A. T. James and A. J. P. Martin, *Biochem. J.* **50**, 679 (1952).
4. M. J. E. Golay, in V. J. Coates, H. J. Noebels, and I. S. Fagerson eds., *Gas Chromatography (1957 Lansing Symposium)*, Academic Press, New York, 1958, p. 1.

5. S. Dal Nogare and W. E. Langolis, *Anal. Chem.* **32**, 767 (1960).
6. S. Dal Nogare and J. Chiu, *Anal. Chem.* **34**, 890 (1962).
7. W. O. McReynolds, *J. Chromatogr. Sci.* **8**, 685 (1970).
8. R. Dandeneau and E. H. Zerenner, *J. High Resolut. Chromatogr. Chromatogr. Commun.* **2**, 351 (1979).
9. J. M. Miller, *Chromatography Concepts and Contrasts*, John Wiley & Sons, Inc., New York, 1988.
10. D. Rood, *A Practical Guide to the Care and Maintenance of Capillary Gas Chromatography Systems*, Wiley-VCH, Heidelberg, 1991.
11. R. L. Grob, ed., *Modern Practice of Gas Chromaatography*, 3rd ed., John Wiley & Sons, Inc., New York, 1995.
12. H. M. McNair and J. M. Miller, *Basic Gas Chromatography*, John Wiley & Sons, Inc., New York, 1997.
13. L. S. Ettre and J. V. Hinshaw, *Basic Relationships of Gas Chromatography*, Advanstar, Cleveland, 1993.
14. J. V. Hinshaw and L. S. Ettre, *Introduction to Open Tubular Column Gas Chromatography*, Advanstar, Cleveland, 1993.
15. M. L. Lee, F. J. Yang and K. D. Bartle, *Open Tubular Column Gas Chromatography Theory and Practice*, John Wiley & Sons, Inc., New York, 1984.
16. J. C. Giddings, *Dynamics of Chromatography, Part 1 Principles and Theory*, Dekker, New York, 1965.
17. J. C. Giddings, *Unified Separation Science*, John Wiley & Sons., Inc., New York, 1992.
18. M. McMaster and C. McMaster, *GC/MS A Practical Users' Guide*, Wiley-VCH, New York, 1998.
19. N. H. Snow, *J. Chem. Educ.* **73**, 592 (1996).
20. W. Harris and H. Habgood, *Programmed Temperature Gas Chromatography*, John Wiley & Sons, Inc., New York, 1967.
21. H. McNair, personal communication, 1999.
22. J. J. Van Deemter, F. J. Zuiderweg and A. Klinkenberg, *Chem. Eng. Sci.*, **5**, 271 (1956).
23. M. J. E. Golay in D. H. Desty, ed., *Gas Chromatography, 1958 (Amsterdam Symposium)*, Butterworths, London, 1958, pp. 36–55.
24. K. J. Hyver, in K. J. Hyver, ed., *High Resolution Gas Chromatography* 3rd ed., Hewlett-Packard, Wilmington, Del., 1989, pp. 1–16.
25. K. Grob, *Split and Splitless Injection in Capillary GC*, 3rd. ed., Wiley-VCH, Heidelberg, 1993.
26. K. Grob, *On-column Injection in Capillary Gas Chromatography*, Wiley-VCH, Heidelberg, 1987.
27. K. Grob, in Ref. 25, pp. 1–216.
28. M. Klee, in Ref. 11, pp. 485–487.
29. K. Grob and G. Grob, *J. Chromatogr. Sci.* **7**, 584 (1969).
30. M. Klee, in Ref. 11, pp. 488–493.
31. K. Grob, in Ref. 25, pp. 217–400.
32. E. Boselli, K. Grob, and G. Lercker, *J. High Resolut. Chromatogr.* **22**, 327 (1999).
33. B. Grolimund, E. Boselli, and G. Lercker, *J. High Resolut. Chromatogr.* **21**, 378 (1998).
34. E. Boselli, B. Grolimund, and R. Amado, *J. High Resolut. Chromatogr.* **21**, 355 (1998).
35. G. Schomburg, *Proceedings of the 4th International Symposium on Capillar Chromatography*, Huthig, Heidelberg, 1981, p. 921A.
36. M. Godula, J. Hajslova, K. Mastouska, and J. Krivankova, *J. Separation Sci.* **24**, 355 (2001).

37. S. Junyapoon, K. D. Bartle, A. B. Ross, M. Cooke, and B. F. Smethurst, *Int. J. Env. Anal. Chem.* **77**, 337 (2000).
38. T. P. Lynch, J. S. Lancaster, and P. G. McDowell, *J. High Resolut. Chromatogr.* **23**, 479 (2000).
39. J and W Scientific WWW Site: *http://www.jandw.com* or *http://www.chem.agilent. com/cag/cabu/jandw.htm*, Agilent Technologies/J and W Scientific, 2001.
40. Chromapck WWW Site: *http://www.chrompack.com* or *http://www.varianinc.com/ cgi-bin/nav?varinc/docs/chrompack/*, Chrompack/Varian, 2001.
41. Supelco WWW Site: *http://www.supelco.com* or *http://www.sigma-aldrich.com/ saws.nsf/SupProducts?OpenFrameset*, Sigma-Aldrich-Supelco, 2001.
42. Restek WWW Site: *http://www.restekcorp.com*, Restek Corporation, 2001.
43. N. H. Snow, in R. A. Myers, ed., *Encyclopedia of Analytical Chemistry*, John Wiley & Sons, Inc., Chichester, 2000, p. 10680.
44. W. A. Konig, *Gas Chromatographic Enentiomer Separation with Modified Cyclodextrins*, Wiley-VCH, Heidelberg, 1992.
45. Z. Juvanez and J. Petersson, *J. Microcolumn Sep.* **8**(2), 99 (1996).
46. F. Perez, P. Berdague, J. Cortieu, J. P. Bayle, S. Boudah, and M. H. Guermouche, *J. Chromatogr. A* **746**(2), 247 (1996).
47. K. P. Naikwandi, I. A. Albrecht, F. W. Karasek, and H. Gohda, *Organohalogen Compd.* **19**, 139 (1994).
48. X. Zhou, Y. Hui, W. Caiyang, and C. Yuanyin, *Sepu* **12**(6), 404 (1994).
49. Z. Zhou, Z. Yongchang, X. Minggui, and C. Ye, *Sichuan Daxue Xuebao Ziran Kexueban* **32**(1), 74 (1995).
50. D. Wang, S. L. Chang, and A. Malik, *Anal. Chem.* **69**(22), 4566 (1997).
51. A. Malik, S. L. Reese, M. L. Lee, *Chromatographia* **46**, 79 (1997).
52. L. H. Henrich, in Ref. 11, pp. 265–322.
53. I. G. McWilliam and R. A. Dawar, *Nature (London)* **181**, 760 (1958).
54. J. Horley, W. Nel and V. Pretorius, *Nature (London)* **181**, 177 (1958).
55. A. E. Lawson, Jr. and J. M. Miller, *J. Chromatogr. Sci.* **4**, 273 (1966).
56. J. E. Lovelock, *Anal. Chem.* **33**, 162 (1961).
57. D. D. Nixon, M. Abdel-Rahman, W. D. Snyder, and W. H. Wilson, *LC-GC* **18**, 268 (1998).
58. E. D. Pellizzari, *J. Chromatogr.* **98**, 223 (1974).
59. M. J. O. Brien, in R. L. Grub, ed., *Modern Practice of Gas Chromatography*, 2nd ed., John Wiley & Sons, Inc., New York, 1985. p. 251.
60. F. W. McLafferty and F. Tureck, *Interpretation of Mass Spectra*, 4th ed., University Science Books, Mill Valley, Calif., 1993.
61. R. M. Smith and K. L. Busch, *Understanding Mass Spectra A Basic Approach*, John Wiley & Sons, Inc., New York, 1999.
62. S. S. Brody and J. E. Chaney, *J. Gas Chromatogr* **4**, 42 (1966).
63. L. H. Henrich in Ref. 11, pp. 305–318.
64. R. D. McDowall, in J. M. Miller and J. B. Crowther, eds., *Analytical Chemistry in a GMP Environment A Practical Guide*, John Wiley & Sons, Inc., New York, 2000, pp. 395–422.
65. J. B. Crowther, M. I. Jimidar, N. Niemeijer, and P. Salomons, in J. M. Miller and J. B. Crowther, eds., *Analytical Chemistry in a GMP Environment A Practical Guide*, John Wiley & Sons, Inc., New York, 2000, pp. 423–458.
66. A. Jayatilaka and C. Poole *Chromatographia* **39**, 200 (1994).
67. M. Heil, F. Podebrad, T. Beck, A. Mosandl, A. Sewell, and H. Bohles, *J. Chromatogr. B* **714**, 119 (1998).
68. F. Podebrad, M. Heil, S. Leib, B. Geler, T. Beck, A. Mosandl, A. Sewell, and H. Böhles, *J. High Resolut Chromatogr.* **20**, 355 (1997).

69. D. A. Stopher and R. Gage, *J. Chromatogr. B.* **691**, 441 (1997).

70. H.-J. DeGeus, R. Baycan-Keller, J. Oehme, J. DeBoer, and U. Th. Brinkman, *J. High Resolution Chromatogr.* **21**, 39 (1998).

71. W. Bertsch, *J. High Resolut. Chromatogr.* **23**, 167 (2000).

72. G. Schomburg, *J. High Resolut. Chromatogr. Chromatogr. Commun.* **2**, 461 (1979).

73. Z. Liu and J. B. Phillips, *J. Microcol. Sept.* **1**, 159 (1989).

74. Z. Liu and J. B. Phillips, *J. Chromatogr. Sci.* **29**, 227 (1991).

75. G. S. Frysinger and R. B. Gaines, *J. High Resolut. Chromatogr.* **22**, 251 (1999).

76. R. B. Gaines, G. S. Frysinger, M. S. Hendrick-Smith, and J. D. Stuart, *Env. Sci. Technol.* **33**, 2108 (1999).

77. Z. Liu, S. A. Sirimanne, D. G. Patterson, Jr., L. L. Needham, and J. B. Phillips, *Anal. Chem.* **66**, 3086 (1994).

78. M. van Lieshout, R. Derks, and C. Cramers, *J. High Resolut. Chromatogr.* **21**, 583 (1998).

79. E. B. Overton, K. R. Carney, N. Roques, and H. P. Dharmasena, *Field Anal. Chem. Technol.* **5**, 97 (2001).

80. N. H. Snow, P. Tavlarakis, and H. T. Rasmussen, *Presented at the 24th Intl Symp. Capillar. Chromatogr.*, *http://www.meetingabstracts.com*, 2001.

81. D. J. Skahan and C. W. Amoss, in Ref. 11, pp. 640–648.

82. B. Kolb and L. S. Ettre, *Static Headspace Gas Chromatography Theory and Practice*, John Wiley & Sons, Inc., New York, 1997.

83. L. T. Taylor, *Supercritical Fluid Extraction*, John Wiley & Sons, Inc., New York, 1996.

84. J. S. Fritz, *Analytical Solid Phase Extraction*, John Wiley & Sons, Inc., New York, 1999.

85. J. Pawliszyn, *Solid Phase Microextraction Theory and Practice*, John Wiley & Sons, Inc., New York, 1998.

86. J. Pawliszyn and R. M. Smith, eds, *Applications of Solid Phase Micro-extraction (RSC Monographs)*, Springer-Verlag, London, 1998.

NICHOLAS H. SNOW
Seton Hall University

GREGORY C. SLACK
Clarkson University

CHROMATOGRAPHY, LIQUID

1. Introduction

Liquid chromatography involves the separation of compounds by differential migration as a liquid mobile phase flows over a solid stationary phase. The mode of separation varies depending on the mobile and stationary phases. (see CHROMATOGRAPHY) In HPLC, small stationary-phase particle sizes and highly controlled conditions are used to achieve high resolutions. A representative separation might involve small (5 μm diameter) particles of chemically modified silica uniformly packed into a 250 mm long by 4.6 mm i.d. stainless steel column. A mobile phase is pumped through the column at 1 mL/min at high pressure (several thousand psi). Sample injection occurs at one end of the column and the separated components are detected at the other end [eg, by ultraviolet (uv) absorbance].

For analytical applications very small amounts of material are generally added to the column. Preparative HPLC can also be used to isolate pure compounds from mixtures and this generally involves adding larger amounts of material to the column. For preparative separations columns of larger diameter (eg, 10 and 20 mm) are employed and for industrial-scale separations even larger columns are available. This article will, however, concentrate on analytical uses of HPLC.

2. Equipment

A representative HPLC instrument consists of a mobile-phase reservoir, a high pressure pump, an injection device, a separation column, a detector, and a data system (Fig. 1). The equipment can be modular, with parts from different manufacturers connected together, or an integrated system from one manufacturer.

Computers are almost universally employed with modern equipment and can be used to control the pump, detector, and robotic sample preparation equipment as well as interpret the output from the detector. In large enterprises, eg, pharmaceutical companies, the systems are highly automated and are coupled together by Laboratory Information Management Systems (LIMS). The source of each sample is recorded as are the chromatographic conditions and the

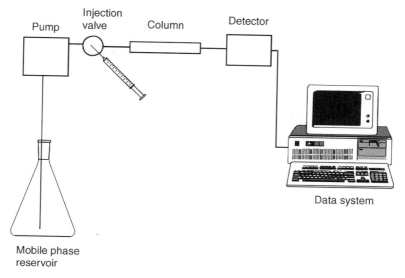

Fig. 1. Diagram of a basic HPLC instrument.

resultant chromatogram. At any time in the future, the chromatogram and related data can be retrieved for quality control purposes. The desired results are automatically calculated and presented to the operator.

The mobile-phase reservoir can be as simple as a conical flask. Some provision should be made for degassing the mobile phase by filtration, sonication, application of a vacuum, sparging with helium, online membrane degassing, or some combination of these methods. Degassing the mobile phase prevents bubbles forming in the system from outgassing and interfering with the separation. A filter should be used to prevent particulates being drawn into the pump.

The pump should provide a constant flow of mobile phase at $\sim$0.5–2 mL/min for analytical separations (with higher flow rates being used for preparative separations) at high pressure (up to 6000 psi). Ideally, pressure variations caused by the action of the pump should be as low as possible to minimize baseline noise. Gradients, in which the relative proportions of, eg, the organic and aqueous components of the mobile phase are varied during the course of the run, are commonly used. The components of the mobile phase can be mixed either before the high pressure pump (low pressure mixing) or after passing through two high pressure pumps (high pressure mixing). Gradients are generally linear (ie, the mobile phase composition changes in a linear fashion with time) but the same separation may contain gradients with different slopes. In some cases, nonlinear gradients (eg, concave and convex) are used. Gradients are particularly useful for separating compounds of greatly different polarity but may be difficult to transfer between instruments from different manufacturers because of differences in dwell (mixing) volume. The need to reequilibrate the system after the separation adds to the total run time.

The injection device can be a high pressure switching valve with a loop of narrow-bore tubing that can be wholly or partially filled with the sample. Operation of the valve brings the sample loop into the circuit between the pump and the column and the sample is carried onto the column by the mobile phase. Automated sample injectors of various designs are also available.

The separation column and detector are the heart of the system. The separation column, in which the actual separation of the sample into its components takes place, is typically a stainless steel tube 50–250 mm in length and 1–4.6 mm in internal diameter packed with small particles of modified or unmodified silica through which the liquid is pumped at a typical rate and pressure of 0.8–2 ml/min and 1000–3000 psi, respectively. The properties of the stationary and mobile phase (ie, polarity, size of particles, charge, etc) determine the mechanism of separation; thus, these components should be specifically selected to suit the nature of the sample.

The detector produces a signal relative to the concentration of the analyte. The separated compounds are presented as a set of chromatographic peaks—the chromatogram—in which the elution time is typical of a specific compound, and hence allows for its identification, and the size of the peak is relative to the concentration of the compound. Each of the various available detectors measures a different physical property of the analyte, such as the uv absorbancy of a compound or the intensity of the light emitted by a fluorescent compound, and converts it to an electrical signal that can be further processed to obtain the chromatogram.

Chromatographic systems, ie, the combination of column and mobile phase, as well as various detectors are further discussed below.

Other components that are useful include a pressure gauge, flow meter, and column heater. Most separations take place in a satisfactory manner at ambient temperature but a thermostatted column compartment will remove a possible source of inconsistency. In addition, the reduced viscosity of the warmed mobile phase will reduce the required pressure. Generally, temperatures range from 30–70°C although subambient temperatures have been used. Temperature gradients are not used in HPLC.

Generally, all components in contact with the mobile phase are made of stainless steel but some compounds may adhere to stainless steel and so it may be necessary to use components made from inert materials such as titanium, glass, or polymer [eg, PEEK (polyetheretherketone)]. To prevent the dilution of the injected sample with mobile phase (which leads to broadened peaks) it is desirable to keep the dead volume after the injector to a minimum and so narrow bore tubing, eg, 1/16 in. o.d. × 0.010 in. i.d. (1.59 mm o.d. × 0.254 mm i.d.), should be used for all connections. For similar reasons, the volume of sample injected should be kept to a minimum (≤ 25 μL). The best medium for the sample is generally the mobile phase or a weaker solvent. If a gradient system is used, the sample solvent should correspond to (or be weaker than) the initial composition of the mobile phase.

It is good practice to employ a filter before the analytical column. Guard columns, packed with a small amount of material similar, or identical, to that in the analytical column, are frequently used. Insoluble or highly retained contaminants will accumulate on the guard column rather than the main column and can be easily removed by changing the (cheaper) guard column. Sometimes aggressive mobile phases that may attack the stationary phase are passed through a column packed with low grade and less expensive chromatographic material (of the same type as in the analytical column) situated between the pump and the injector. The mobile phase attacks this sacrificial precolumn rather than the expensive analytical column. Since the precolumn is situated before the injector the sample never passes through it.

3. Sample Preparation

In some cases, eg, a solution of a bulk drug substance, sample preparation is as simple as filtering the sample before injection. However, in many cases complex sample preparation procedures are required to obtain reproducible results. Consider the analysis of a drug at a low concentration (in the ng/mL range) in blood plasma. Plasma is a complex mixture containing numerous compounds, such as proteins, and injection of an unprocessed sample would result in a large off-scale peak that would swamp any peaks attributable to the drug. Additionally, the HPLC column would rapidly become plugged and useless. To avoid such undesirable results, a variety of approaches have been developed and some of these are described below. All of these procedures can be automated using robotic equipment and this is desirable for long production runs. However, due to the effort involved in developing an automated procedure it is generally easier to use manual procedures when the number of samples to be processed is not large. Because of the resources required to set up an automated procedure the Food and Drug Administration (FDA) in the United States prefers that manual methods be submitted for validation even though automated procedures will eventually be used by the drug company (1).

3.1. Liquid–Liquid Extraction. Liquid–liquid extraction generally involves the extraction of an aqueous phase (eg, urine, plasma, and serum) with an organic solvent (2). Separation, generally facilitated by the use of a centrifuge, results in an organic layer containing the drug and an aqueous layer containing most of the potentially interfering compounds. The extraction can be repeated several times. The layers can be separated using a pipet or the aqueous layer can be frozen using a dry ice–acetone bath and the organic layer (which does not freeze) poured off. Freezing the aqueous layer is particularly helpful in removing traces of water that may interfere with the subsequent processing. The combined organic layers are generally evaporated to dryness under a stream of nitrogen or under reduced pressure and the residue reconstituted with the HPLC mobile phase prior to injection.

More than one extraction can also be used to clean up the sample. For example, basic analytes can be reextracted from the organic layer with aqueous acid. Basification of the aqueous layer and extraction with an organic solvent will result in a cleaner sample.

An increase in sensitivity can be obtained by reconstitution with a volume of mobile phase less than that of the original matrix. For example, 1 mL of plasma can be extracted and the residue reconstituted with 100 µL of mobile phase, leading to a 10-fold concentration.

The usual organic solvents can be used for extraction, eg, dichloromethane or heptane. Since most of the organic phase will eventually be vaporized toxic solvents such as benzene or chloroform should be avoided. Hexane is surprisingly toxic and should not be used (3). Also, appropriate safety precautions should be taken, eg, all procedures should be carried out in a properly functioning chemical fume hood, not in a laminar flow cabinet or on the bench. An extraction solvent combination that is sometimes used is a mixture of a nonpolar hydrocarbon with a higher alcohol, eg, heptane:isoamyl alcohol 99:1 (v/v). A number of procedures can be used for agitating the extraction mixture such as using a vibrating mixer (vortex mixer), rotating the tube, gently shaking the tube using an orbital mixer, or shaking the tube using a wrist action shaker. Frequently, small polypropylene tubes of ~1–2 mL are used for the vortex mixer and larger glass tubes (~10 mL) for the other agitators. It goes without saying that all tubes should be securely capped! Although it might be thought that more is better when it comes to agitation, excessive agitation can lead to intractable emulsions. The most effective combination of extraction solvent and agitation method should be established by a close examination of the available literature (4) and by experimentation.

3.2. Precipitation. A variation of liquid–liquid extraction is to use a water-miscible organic solvent such as acetonitrile or methanol. For example, mixing acetonitrile with plasma causes the proteins in the plasma to precipitate. Centrifuging the mixture causes the proteins to collect in the bottom of the tube and the aqueous/organic mixture may be removed, evaporated, and reconstituted as described above.

3.3. Ultrafiltration.
Plasma proteins may also be removed using a membrane filter having very small pores (eg, 0.2 μm) made of materials such as cellulose acetate, poly(tetrafluoroethylene) (PTFE), or polysulfone. The resulting filtrate is clean enough to be injected directly on an HPLC column. Low cost disposable equipment has been developed to make this procedure attractive. The plasma sample is placed in a plastic tube with a filter in the bottom and this tube is placed inside a plastic centrifuge tube. A centrifuge provides the force required to filter the plasma through the membrane and into the bottom of the tube. Small tubes of ∼0.5-mL capacity are frequently used although larger tubes are also available.

3.4. Solid-Phase Extraction.
Solid-phase extraction (SPE) is a technique whereby a crude chromatographic separation is used to effect an initial purification and produce a sample that is clean enough for injection onto the analytical HPLC column. Low cost disposable SPE cartridges are generally used although they can also be constructed by the analyst, eg, in a disposable Pasteur pipet. The ready-made SPE cartridges typically consist of 0.1–1 g of chromatographic material such as silica, C18, C8, phenyl, ion-exchange (see below) in a plastic body. The commercial cartridges may be cleaned and reused but this may prove to be a false economy if residual contaminants interfere with later separations. Note that there is no requirement that the same chromatographic material be used in the SPE cartridges and in the analytical HPLC column.

Immediately before use the SPE cartridges should be conditioned so as to remove manufacturing impurities and wet the chromatographic material. Reversed-phase cartridges (such as C18) are typically conditioned with methanol followed by water or buffer and normal phase cartridges may be conditioned with an organic solvent such as heptane. The sample (eg, plasma or urine) is added to the cartridge and, after it has been absorbed, the cartridge is washed. For example, a C18 cartridge may be washed with water and with aqueous solutions containing low percentages of methanol, and then the compounds of interest may be eluted with pure methanol. Interfering compounds such as plasma proteins are eliminated with the washes and the compounds of interest are obtained in a relatively clean methanol solution. The methanol eluate can be evaporated and reconstituted as described above. SPE cartridges of other types, such as silica or ion-exchange resin, would be washed and eluted with different solutions but the principle remains the same: Interfering compounds (in relatively large amounts) are removed with the washes and the compounds of interest (in relatively small amounts) are eluted in a relatively clean solution.

Generally, it is important not to let the SPE cartridge run dry between steps although some procedures call for all traces of liquid to be removed using a vacuum immediately before the final elution step. It is also important not to run liquid through the SPE cartridge at too fast a rate. Excessively fast flow rates will cause loss of resolution. In most applications, flow rates of 0.5–1 mL/min are satisfactory. Flow rates such as these can frequently be achieved using gravity alone. Otherwise a gentle vacuum may be applied.

Solid-phase extraction can also be used to concentrate large amounts of relatively dilute solutions. For example, a large volume (eg, 50 mL) of polluted wastewater may be run through a C18 cartridge and the water will pass straight through leaving the nonpolar pollutants trapped on the cartridge. The compounds to be analyzed can then be eluted with a small amount (eg, 1 mL) of methanol. In this example, the compounds to be analyzed have been concentrated by a factor of 50, more if the methanol is evaporated and the residue reconstituted in a smaller volume.

3.5. Column Switching. Column-switching techniques are closely related to solid-phase extraction and in some cases the distinction may become blurred. There are many variations but the simplest involves two columns, two pumps, and a switching valve.

In the example shown in Fig. 2, with the switching valve in the "load" position, the solution to be analyzed (eg, plasma) is injected onto column A and eluted to waste with a "weak" mobile phase, eg, water. When the plasma proteins and other interfering species have been washed off column A the switching valve is turned to the "analyze" position (counterclockwise in Fig. 2) and a "strong" mobile phase, eg, methanol:water 50:50, is used to elute the compounds of interest from column A onto column B. Chromatographic separation takes place on column B and the compounds of interest, eg, drugs and metabolites, are detected in the effluent. Although not strictly necessary, automation is obviously very useful with column-switching techniques. The cleanup column A may require periodic replacement but the chromatographic column B should have a normal life. Many variations of this basic technique have been reported. Since the compounds of interest tend to stay at the top of column A it is common practice to reverse the flow of the "strong" mobile phase and backflush these compounds from column A onto column B. In this way, the compounds of interest are dissolved in the minimum amount of mobile phase as they pass onto column B. Once the analytes have moved from column A to column B, column A can be cleaned with a strong

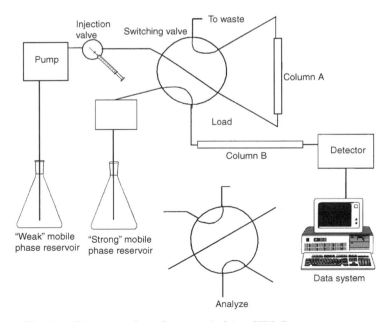

Fig. 2. Diagram of a column-switching HPLC arrangement.

solvent and reequilibrated. Different stationary phases can be used for columns A and B. Generally, the analytical column (B) is a conventional column. The initial column (A) can be a short analytical column, a commercial precolumn, or a specialized column developed for biological samples, such as an internal-surface reversed phase (ISRP) column. ISRP columns, also known as restricted internal access or Pinkerton columns, contain packing material that has narrow pores coated with reversed-phase material such as C_{18}. Large protein molecules cannot penetrate the pores and pass straight through the column. The small analyte molecules enter the pores, are separated in a conventional fashion, and elute later (5).

Column-switching techniques can be used to accomplish achiral separation of a racemic mixture of a drug from contaminants on one column, and then chiral separation of the two enantiomers on another column. For example, the enantiomers of nilvadipine can be separated in this way (6).

Just as solid-phase extraction can be used to concentrate large amounts of relatively dilute solutions so can column-switching techniques. For example, a large volume of polluted waste water may be run through a short reversed-phase column and the water will pass straight through leaving the nonpolar pollutants trapped on the cartridge. The compounds to be analyzed can then be eluted with the mobile phase onto the analytical column (7). Blood plasma can be dialyzed using a semipermeable membrane and the dialysate passed through a short column. The drug molecules pass through the membrane into the dialysate and are then trapped on the column. When dialysis is complete the compounds, eg, nonsteroidal anti-inflammatory drugs (8), are eluted from the trapping column onto the analytical column by the mobile phase and chromatographed in the usual manner.

In some cases, more than two columns are used in a column-switching setup. One group has described a complex method for the analysis of enprostil in plasma that involves solid-phase extraction, derivatization, and a column-switching apparatus with five separate columns (9).

Column-switching generally involves automation and complex plumbing and so is best suited to situations where many similar samples must be analyzed. Modern automation allows such systems to run with a minimum of attention once they have been set up.

3.6. Derivatization. In HPLC, derivatization is generally used to improve the detectability of the compounds of interest although it may also be used to improve their chromatographic properties (10,11). The derivatization reaction usually involves a reaction that is simple and irreversible, eg, reaction of an acyl halide with an alcohol or amine to give an ester or amide respectively. Thus alcohols or amines that do not have appreciable uv chromophores can be coupled to molecules with a high uv absorbance. Similarly, nonfluorescent analytes can be coupled to highly fluorescent molecules to achieve high sensitivity and selectivity. Although the chemistry of the derivatization reaction tends to be fairly simple the reagents can be quite sophisticated molecules and they can be tailored to have desirable properties, such as fluorescence at particular wavelengths.

Ideally, derivatization reactions should proceed rapidly, quantitatively, reproducibly, and irreversibly at room temperature but sometimes heating may be necessary. In some cases, particularly with fluorescent reagents, it is advantageous to remove excess reagent with a washing step before injecting the sample into the chromatographic system. Chiral separations can be achieved by using chiral derivatization reagents (11,12).

Complexing agents can be used to make metal complexes that may be easily chromatographed and detected. (12).

The examples discussed above apply to the derivatization of molecules before they are chromatographed, ie, precolumn derivatization. However, post-column reactions can also be used to derivatize molecules after they have been chromatographed. This technique has its own advantages and disadvantages and will be discussed below. Generally, precolumn derivatization is used to improve the separation characteristics and the efficiency of detection of the analytes while postcolumn derivatization is only used to improve the efficiency of detection.

3.7. Detectors.   Refractive index and evaporative light-scattering detectors are so-called universal detectors, ie, they can detect all analytes, but they have some limitations and suffer from limited sensitivity. Refractive index detectors monitor the refractive index of the eluting mobile phase. The presence of an analyte will change the refractive index and produce an output signal proportional to the amount of analyte. In an evaporative light-scattering detector, the eluting mobile phase is nebulized and evaporated as it passes through a drift tube. Particles of analyte are detected in a light scattering cell.

3.8. UV Detectors.   The most commonly used detectors detect eluting compounds by their uv absorbance. These detectors are sensitive, robust, and of relatively low cost. Although uv detectors are not universal detectors most compounds of interest can be detected. Other compounds that have essentially no uv chromophore, such as aliphatic alcohols, can be detected after derivatization. Generally a wavelength between 200 and 500 nm is used. Normally, uv absorbance increases as the wavelength decreases and so greater sensitivity can be achieved at lower wavelengths but problems may arise due to noise and mobile phase absorbance. Although sensitivity is generally less at higher wavelengths selectivity may be increased because the absorbance of the interfering compounds may decrease more than the absorbance of the analyte as the wavelength increases.

Conventional uv detectors employ a uv source, typically a deuterium lamp for uv wavelengths (<400 nm) and a tungsten lamp for visible wavelengths (>400 nm), and a monochromator to make sure that light of only one wavelength shines through the flow cell. Generally, the wavelength can be varied over a wide range (eg, 190–700 nm) but some low cost single wavelength detectors are available, eg, operating at 254 nm. To minimize peak broadening it is desirable to keep the volume of the flow cell as small as possible. However, this will lead to reduced sensitivity. To counteract the loss of sensitivity flow cells are generally arranged so that the light shines along the column of liquid rather than across it. The light that emerges from the flow cell is detected by a diode or phototube and the signal is amplified and passed to the data handling system.

In diode array detectors (DADs) [also known as photodiode array detectors (PDAs)], wavelengths between, say, 190 and 600–900 nm are continuously recorded. Light of all wavelengths passes through the flow cell and the light that emerges is dispersed by a prism or grating to provide an arc of different wavelengths. Photodiodes are situated at intervals along this arc. Each photodiode monitors light of a very narrow band of wavelengths (typically 1–4 nm). After a run a chromatogram recorded at any given wavelength can be displayed. The uv spectrum of any given peak can be obtained and this may help with compound identification. By comparing the ratio between two different wavelengths, peak purity can be assessed although this procedure is not necessarily conclusive. A pure peak will give a square wave because the ratio of the absorbance at the two wavelengths is always the same, regardless of the peak height. Two compounds eluting together will give a more complicated result because the ratio will vary if the compounds do not elute at exactly the same time.

3.9. Fluorescence Detectors.

In a fluorescence detector, the mobile phase is illuminated with a beam of light and the light emitted by a fluorescent compound is picked up and quantitated by a detector placed at right angles to the light beam. Monochromators or filters may be used to set the excitation and emission wavelengths. Greater sensitivity may be obtained by using a laser as the excitation source but then the wavelength is not readily changed. When monochromators are used the excitation and emission wavelengths may be changed during the course of the run. Fluorescence detectors offer great sensitivity and selectivity. Some compounds exhibit native fluorescence but it is frequently necessary to make a fluorescent derivative

3.10. Electrochemical Detectors.

Electrochemical detectors also offer great sensitivity and selectivity. However, only a limited range of compounds can be detected using this technique. Derivatization can be used to make the compounds of interest electroactive, ie, detectable using these detectors, but this is not commonly done. Generally, a working electrode is held at a fixed potential relative to a reference electrode and the current is monitored. When a compound that can be oxidized or reduced arrives at the electrode the change in current provides the signal. More generally (but not exclusively), oxidative conditions are used. Many different conditions have been reported for electrochemical detection. The most common electrode material is glassy carbon but many others, such as gold or platinum, have been reported. Electrode fouling and mobile-phase interference can be particular problems with this type of detector. Use of high purity solvents can help. Some methods use a guard cell before the injector to oxidize any electroactive compounds in the mobile phase. Recycling the mobile phase may also help maintain baseline stability.

One problem with electrochemical detectors is the fouling of the electrode surface by reaction products and intermediates. Fouled electrodes can be cleaned mechanically but this is a lengthy and tedious process. To get around this problem, pulsed electrochemical detectors have been developed. A series of rapid positive and negative pulses are used to clean the electrode. After the initial detection period at, eg, −50 to 200 mV a large positive potential (eg, +600 mV) is applied to the electrode. This potential causes a layer of oxide to form on the electrode and at the same time any fouling is desorbed. A large negative potential (eg, −600 mV) is then applied to remove the oxide layer. The "clean" electrode is then ready for the next detection cycle. Typically, one cycle will take <1 s. Numerous variations have been reported (13).

3.11. Mass Spectrometric Detectors. Historically, mass spectrometric (MS) detectors have been difficult to use and have lacked reliability. However, great advances have been made in recent years and LC–MS is now the method of choice for developmental pharmaceutical work, although production work and quality control generally use uv detectors. There are many different types of MS detectors and operating conditions tend to be instrument specific. Commonly used buffers, such as phosphate, that leave a residue on evaporation may not be used. Buffers that are used for LC–MS include ammonium formate and trifluoroacetic acid which evaporate completely. Gradients may be used for LC–MS.

In MS, detection ions are generated at the source where the eluate enters the detector. There are many different sources such as atmospheric pressure chemical ionization (APCI), electrospray ionization (ESI), fast-atom bombardment (FAB), thermospray (TSP), and matrix-assisted laser desorption (MALDI). The ions are separated on the basis of their mass in the mass analyzer. Common mass analyzers are the quadrupole, ion-trap, and time-of-flight (TOF) instruments. The ions are detected by an electron multiplier.

Compounds that co-elute from the chromatographic column can be readily distinguished by monitoring different ions. One of the great attractions of LC–MS is that run times may be shortened since peaks that would be merged if a uv detector was used can readily be distinguished by monitoring the different ions. Some important applications will be discussed below.

3.12. Other Detectors. Many other detectors have been described in the literature but they are not in widespread use. Conductivity detectors are useful for detecting inorganic ions in ion-exchange chromatography. Other detectors using principles such as infrared (ir), neclear magnetic resonance (nmr), radioactivity, polarimetry, and viscometry have specialized applications. Chemiluminescence detectors have been reported in the literature with increasing frequency (14).

3.13. Postcolumn Reaction Detection. Postcolumn reaction detection involves the derivatization of compounds as they elute from the chromatographic column after separation. In many cases, the chemistry is the same as that used for precolumn derivatization. For example, putrescine [110-60-1] reacts with o-phthalaldehyde and 2-mercaptoethanol to form a fluorescent derivative. This can take place precolumn (15) or postcolumn (16).

Since chromatographic separation does not occur once the compound has eluted from the HPLC column, postcolumn reaction detection can involve the use of reactions that do not necessarily lead to a single well-defined derivative. For example, ampicillin [69-53-4] in the effluent from an HPLC column reacts with a solution of sodium hypochlorite pumped at 0.2 mL/min to give, probably, a mixture of compounds. This mixture flows through a reaction coil to the detector. However, since no chromatographic separation will take place in the reaction coil it is unimportant that a mixture of compounds is produced by the reaction. A single peak that is proportional to the amount of ampicillin is produced (17).

Another technique that may lead to a mixture of products is postcolumn photochemical derivatization. The column effluent flows through a length of (typically) narrow-bore PTFE tubing that is illuminated by uv light. The uv light causes photochemical reactions to occur that produce compounds with

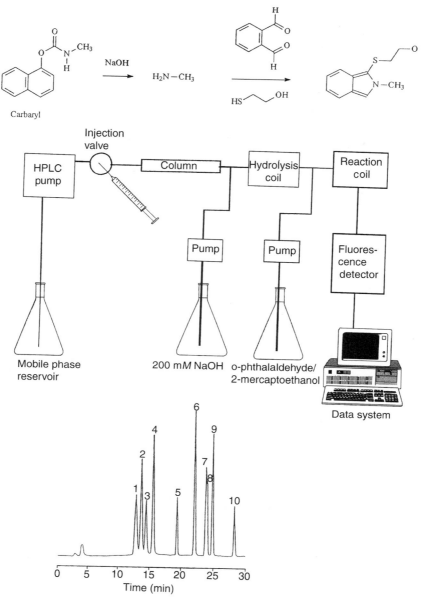

Fig. 3. Postcolumn reaction detection. Solution containing the insecticides aldicarb sulfoxide (*1*), aldicarb sulfone [1646-88-4] (*2*), oxamyl [23135-22-0] (*3*), methomyl [16752-77-5] (*4*), 3-hydroxycarbofuran [16655-82-6] (*5*), aldicarb [116-06-3] (*6*), propoxur [114-26-1] (*7*), carbofuran [1563-66-2] (*8*), carbaryl [63-25-2] (*9*), and methiocarb [2032-65-7] (*10*). Conditions: 50 × 4.6 mm i.d. Whatman pellicular ODS guard column, 250 × 4.6 mm i.d. 5 μm Apex ODS column; mobile phase methanol/water from 10:90 to 90:10 over 23 min, return to initial conditions over 4 min, reequilibrate at initial conditions for 10 min; flow rate 1.0 mL/min; detection as described in the text. Reproduced with permission from Ref. 20.

stronger uv chromophores or compounds that are fluorescent. Hence, sensitivity is increased (18). Knitting or crocheting the reaction coil leads to sharper peaks (19). Postcolumn photochemical derivatization is generally carried out empirically and the species responsible for the increased uv absorbance or fluorescence are not usually determined.

Quite sophisticated reactions can take place postcolumn. For example, N-methyl carbamate insecticides (eg, carbaryl [63-25-2]) can be separated on an HPLC column. The column effluent mixes with 200-mM NaOH pumped at 0.8 mL/min and this mixture flows through a 1 mL coil at 95°C. In the coil, the carbamate hydrolyses to methylamine and the corresponding alcohol. A reagent containing 500-mg/L o-phthalaldehyde and 1 mL/L 2-mercaptoethanol in 50-mM sodium tetraborate is added to the reaction stream at 0.8 mL/min. This mixture flows through a 0.5-mL coil at room temperature where the methylamine reacts to form a fluorescent derivative that is then detected by fluorescence (excitation 340 nm, emission 455 nm) (Fig. 3).

Postcolumn reaction procedures can take a lot of effort to set up and may require some maintenance but they can avoid the use of labor-intensive sample preparation and may provide superior sensitivity and selectivity.

4. Chromatographic Systems

4.1. Normal Phase.   As originally developed, LC involved an unmodified polar solid stationary phase, such as alumina or silica, and a nonpolar liquid mobile phase, such as octane. Since this was the first type of LC to be developed it is referred to as "normal phase". Separation is achieved by the adsorption of analyte molecules on the surface of the stationary phase. Less polar (or hydrophobic) molecules are more weakly adsorbed, and hence elute quicker, than more polar (or hydrophilic) molecules.

As currently practiced normal-phase chromatography involving unmodified stationary phases uses silica particles of ~5−10 μm with a mobile phase consisting of organic solvents such as hexane, octane, dichloromethane, or methanol. Small amounts of acids or bases can be added to suppress tailing. It is important to control the amount of water in the mobile phase at a consistent level. Too little water will cause the stationary phase to dry out and separation efficiency will be poor. Too much water will saturate the stationary phase and the compounds will not be retained on the column.

Normal-phase chromatography may also take place using some kinds of modified silica stationary phases, eg, cyano, diol, or amino. Again, these polar stationary phases are used with relatively nonpolar mobile phases. For these modified stationary phases, control of the mobile phase water content is not critical. Confusingly, these modified stationary phases may also be used in a reversed-phase mode with polar mobile phases.

Currently normal-phase chromatography is not used extensively because of the expense of obtaining (and disposing of) high purity, possibly toxic organic solvents and the difficulty of maintaining the correct water level. However, normal-phase chromatography does have some advantages and in some cases separations may be obtained that are not feasible in any other way. Normal-

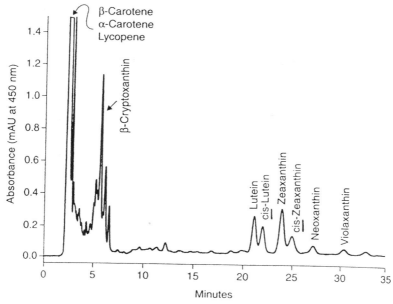

Fig. 4. A normal-phase separation. Food reference material carotenoids. Conditions: 250 × 4.6 mm i.d. 5-μm Lichrosorb Si60 column; mobile-phase hexane:dioxane/isopropanol/triethylamine 80:20:0.15:0.02; flow rate 1.0 mL/min; detector uv 450 nm. Reprinted with permission from Ref. 21.

phase chromatography is particularly useful for compounds that are unstable in aqueous solutions. It is also useful for preparative chromatography since the mobile phase is readily evaporated.

A number of good normal-phase methods using unmodified silica are found in the literature, eg, for the separation of vitamins, prostaglandins, and retinoids (Fig. 4).

4.2. Reversed Phase. By far, the most common HPLC technique is reversed-phase HPLC. In reversed-phase HPLC, stationary phases consisting of chemically modified silica are used with polar mobile phases, eg, methanol: water 50:50. Typically, these chemically modified stationary phases have nonpolar long-chain hydrocarbon groups (eg, an aliphatic chain containing 18 carbons, designated as C_{18}) bonded to the surface but many variations have been developed, particularly in recent years. Since these chemically modified stationary phases became available after unmodified silica normal-phase systems had become widely used these systems are called "reversed phase". The long chain organic groups behave like an organic liquid and molecules of the analytes partition between this nonpolar stationary phase and the polar mobile phase. Less polar (or hydrophobic) molecules spend more time in the stationary phase, and hence elute slower, than more polar (or hydrophilic) molecules. This is the reverse of the situation with normal-phase chromatography (Fig. 5).

A wide variety of stationary phases of this type is available, eg, C_1, C_4, C_8, C_{18}. Sample retention normally increases as the chain length increases.

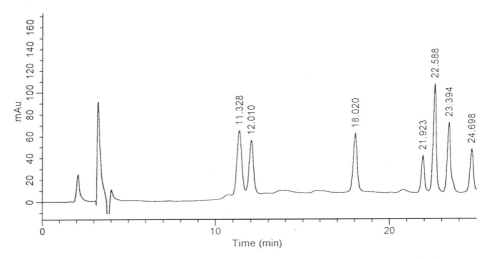

Fig. 5. A reversed-phase gradient separation. Determination of ginsenosides in ginseng root powder extract. Conditions: 250 × 4.6 mm 5 μm Phenomenex Hypersil phenyl; mobile-phase acetonitrile:water from 20:80 to 40:60 over 20 min, to 100:0 over 10 min, reequilibrate at initial conditions for 5 min; flow rate 1.0 mL/min; detector uv 202 nm.

However, there tends to be little difference between the longer chains, such as C_8 and C_{18} (22). Different brands of the same type of stationary phase (eg, C_{18}) are not necessarily interchangeable and may exhibit, in fact, considerable differences.

As the polarity of the mobile phase decreases, ie, with increasing organic content, compounds are eluted more rapidly. The polarity of the mobile phase is adjusted so that compounds do not elute too rapidly, eg, with the unretained peak, or with inconveniently long retention times. If two compounds elute as overlapping peaks the polarity of the mobile phase may be increased and the peaks separated at longer retention times. Compounds of widely different polarity may conveniently be separated by using a mobile phase gradient. During the course of the run the polarity of the mobile phase is decreased, eg, from water:methanol 90:10 to water:methanol 10:90. In this way polar compounds are easily separated from the unretained peak and later on less-polar compounds elute at convenient retention times as the polarity of the mobile phase decreases. However, gradient elution requires a reequilibration period at the initial mobile phase composition when the separation has ended so gradient elution may result in a less efficient use of instrument time than nongradient (isocratic) elution.

Different types of columns, and even columns of the same class from different manufacturers, exhibit different selectivities and this can be used to achieve optimum resolution of all compounds of interest. A more convenient way to change the selectivity of the system may be to change the composition of the mobile phase (23). Retention times depend on the exact composition of the mobile phase and not just on its overall polarity (24) (Fig. 6).

4.3. Ion-Pair Chromatography. It is difficult to chromatograph ionic compounds by reversed-phase HPLC because the polar ionic compounds prefer to stay in the polar mobile phase. However, addition of an ion-pair reagent

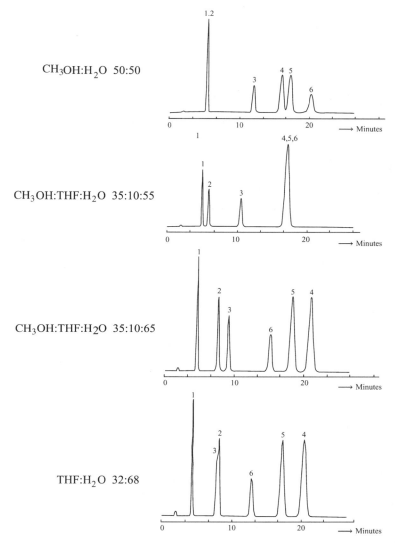

Fig. 6. An example of mobile-phase selectivity. A standard mixture of benzyl alcohol [100-51-6] (*1*), phenol [108-95-2] (*2*), 3-phenylpropanol [122-97-4] (*3*), 2,4-dimethylphenol [165-67-93] (*4*), benzene [71-41-2] (*5*), and dimethyl phthalate [131-11-33] (*6*) is separated using the mobile phases shown. The column was a 300 × 4.6-mm Nucleosil 10-RP18 column with a mobile phase flow rate of 1.5 mL/min. Reprinted with permission of Elsevier Science from Ref. 24.

such as a weak acid (eg, heptanesulfonic acid (as the sodium salt) or a weak base (eg, tetrabutylammonium phosphate) leads to the formation of neutral ion-pair between the analyte and the reagent. This neutral ion-pair exhibits the chromatographic behavior of a nonionic organic molecule and may be chromatographed by reversed-phase HPLC like neutral molecules (25). However, ion-exchange mechanisms may also be in effect (26).

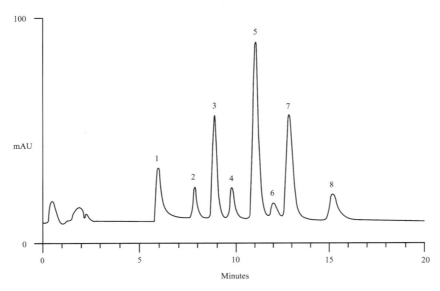

Fig. 7.   An ion-exchange chromatogram. Solution containing Fe^{3+} (*1*), Cu^{2+} (*2*), Ni^{2+} (*3*), Zn^{2+} (*4*), Co^{2+} (*5*), Cd^{2+} (*6*), Mn^{2+} (*7*), and Fe^{2+} (*8*). Conditions: 250×4 mm i.d. IonPac CS5A column (Dionex) operated in anion-exchange mode; mobile phase pH 4.2 buffer containing 7 mM pyridine-2,6-dicarboxylic acid, 66 mM potassium hydroxide, 5.6 mM potassium sulfate, and 74 mM formic acid; flow rate 1.0 mL/min; detection at 530 nm following postcolumn reaction with a solution containing 4-(2-pyridylazo)resorcinol pumped at 0.6 mL/min. Reprinted with permission of Elsevier Science from Ref. 27.

4.4. Ion-Exchange Chromatography.

In ion-exchange chromatography (Fig. 7), charged groups are covalently bound to the stationary phase. The charged ionic analyte is in competition with a counterion of the same charge in the mobile phase. Analytes that interact strongly with the stationary phase will be more retained (ie, elute later) than analytes that interact weakly.

In cation exchange, a positively charged species, A^+ (eg, a protonated base) interacts with a stationary phase containing a negatively charged group, G^- (eg, sulfonate and carboxylate). The counterion in the mobile phase might be an alkali metal such as potassium. In anion exchange a negatively charged species, B^- (eg, an ionized acid) interacts with a stationary-phase containing a positively charged group, G^+ (eg, quaternary ammonium). The counterion in the mobile phase might be a halogen such as chlorine. Thus (see Fig. 8).

4.5. Size-Exclusion Chromatography.

Size-exclusion chromatography (SEC) is also known at gel-permeation chromatography (GPC) and involves

$$A^+ \; + \; G^-K^+ \; \rightleftharpoons \; G^-A^+ \; + \; K^+$$

$$B^- \; + \; G^+Cl^- \; \rightleftharpoons \; G^+B^- \; + \; Cl^-$$

Fig. 8.   Ion-exchange chromatography.

the separation of analytes by size. The column is packed with material that contains pores of a tightly controlled size. Large molecules cannot penetrate the pores and elute first while very small molecules fully penetrate the pores and elute last. Molecules of an intermediate size partially penetrate the pores and elute somewhere in the middle. Columns with smaller pore sizes (eg, 125 Å) are useful for separating molecules of relatively small size while columns with larger pore sizes (eg, 1000 Å) are useful for larger molecules. SEC is particularly useful for synthetic polymers where the different molecules can be separated and for biological molecules such as proteins or oligonucleotides.

4.6. Chiral Separations. The two main ways of effecting chiral separations ie, separating the enantiomers of optically active compounds, are derivatization and the use of columns containing chiral stationary phases. Chiral mobile phase additives have been used but such techniques are less common, perhaps because large quantities of expensive additives may be required. In contrast, chiral additives are commonly used in capillary electrophoresis where the quantity of "mobile phase" used in the course of a day is much less.

Derivatization is an indirect method of chiral analysis. Reaction of a chiral reagent (consisting of only one enantiomer) with a racemic mixture of two enantiomers generates two diastereomers that have different physical properties and can be readily separated using reversed-phase HPLC. The advantages are that conventional HPLC equipment can be used. The disadvantages are increased sample preparation time and effort and the possibility that racemization or different reaction yields may occur during derivatization. This does not happen in every case but examples have been reported (28).

The use of a chiral column is a direct method of chiral analysis. A large variety of chiral columns are now available and many different types of compounds may be separated. In all cases, chiral molecules are bonded to the stationary phase. As the racemate moves down the column the enantiomers interact to different extents with the chiral stationary phase, and hence the enantiomers are separated. In some cases, stationary phases that are chemically identical but of opposite chirality are available. Changing the chirality of the stationary-phase changes the order of elution of the analyte enantiomers. This can be important in the analysis of trace amounts of one enantiomer in the presence of large amounts of the other enantiomer, eg, in pharmaceutical applications. The order of elution should be arranged so that the trace enantiomer elutes first since it is much easier to detect small peaks in the baseline before the large peak than in the tail of the large peak.

5. Applications

This section focuses on pharmaceutical applications of HPLC as the field in which this separation method is used most. However, other applications are mentioned and the reader is invited to look into the provided list of references to learn more about additional areas of aplication.

5.1. Pharmaceutical Applications. HPLC remains the most important chromatographic method for pharmaceutical applications. Immunoassays are also widely employed but they are specific for a single drug. Microbiological methods are also used for antibiotics but have problems of specificity since active metabolites or other antibiotics will also give a response. Because of its specificity HPLC remains the "gold standard" for drug analysis (29). Although an increasing number of papers have been published on the use of capillary electrophoresis for the analysis of drugs these papers have, so far, been mostly confined to the chromatography literature. There are only a few descriptions of the use of capillary electrophoresis in clinical studies. Gas chromatography (GC) is used in some cases, particularly in forensic investigations and for volatile compounds.

Pharmaceutical applications can be divided into two broad categories: analysis of the drug substance or dosage form and the analysis of drugs in bodily fluids (eg, blood or urine). The drug substance is a chemical substance that contains the active ingredient (eg, aspirin or ibuprofen). The drug substance is often formulated with excipeitns (inactive ingredients) and made into a dosage form (eg, tablets or capsules) that can be consumed by the patient.

Regulations.   In the United States a drug is required by the Federal Food, Drug, and Cosmetic Act and the drug regulations to meet its specifications by appropriate laboratory testing. The quality of a drug is typically controlled by a drug substance specification and a drug product specification, each of which may include test attributes such as appearance, assay, impurities, or water content. The number and type of these tests will vary for each drug and, in the case of a new drug, are established through the approval of a marketing application by the regulatory agency (ie, FDA).

In the United States the drug standards and analytical procedures found in the U.S. Pharmacopeia (30) are recognized as being official (31) and may be relied upon in legal proceedings. Additionally, analytical procedures, as well as standards, for many drugs are found in other pharmacopeias such as the European, Japanese, or British Pharmacopoeia.

In current practice, the specification for a particular aspect of the quality of a drug consists of three parts: attribute, analytical procedure, and acceptance criterion. For example, the attribute might be the assay, the procedure could be a fully described HPLC method, and the acceptance criterion could be a numerical value, eg, 90.0–110.0% of the amount stated on the label. In this case, the HPLC method should be fully validated. Validation is a complex testing process that ensures that the method gives accurate and reliable results and will continue to do so in the future.

Drug Substance.   In the case of the drug substance, it is generally desirable to assay the amount of the drug that is present and determine the amounts of process impurities (compounds produced during the synthesis of the drug substance) and degradants (compounds produced as the drug ages). Impurities and degradants together are termed related substances (see Fig. 9). Frequently, the same analytical procedure is used to determine the drug and the related substances but separate procedures are sometimes used. If the related substances have widely different chromatographic properties a gradient or even two separate chromatographic systems may be used. The chiral purity of a drug is frequently determined using a different chromatographic system from that used to determine assay or related substances. Assay and impurities are determined when the drug substance is first manufactured. Selected batches are stored under controlled conditions and tested periodically to assess the stability of the

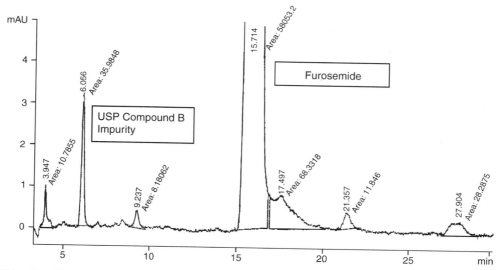

Fig. 9. Chromatogram of furosemide [54-31-9] and its impurities including USP Compound B. Conditions: 250×4.6 mm Phenomenex Luna (2) C-18 column; mobile phase THF:H_2O:acetic acid 35:65:0.1; flow rate 1.0 mL/min; detector uv 272 nm.

drug substance. If at all possible stability-indicating HPLC methods should be used. A stability-indicating assay is a validated quantitative analytical procedure that can detect the changes with time in the pertinent properties of the drug substance and drug product. A stability-indicating assay accurately measures the active ingredients, without interference from process impurities, degradation products, excipients, or other potential impurities. HPLC procedures can also be used to control the manufacturing process, eg, by determining when a reaction is complete. These tests are termed process tests or in-process controls.

Drug Product. HPLC methods developed for the analysis of the drug product tend to be similar to those developed for the analysis of the drug substance (see Fig. 10). Indeed, in many cases they are virtually identical. The main difference is that the drug product contains excipients such as microcrystalline cellulose, lactose, magnesium stearate, or gelatin. That contribute to various aspects of the drug product such as disintegration, dissolution, bioavailability, or stability. It is important that the HPLC procedures be such that the excipients do not produce interfering peaks in the chromatogram (32). It may be necessary to test for certain excipients such as preservatives, which may sometimes be done using the same HPLC procedure as used for the assay. The methods should be stability indicating.

Stability testing is also carried out for the drug product and HPLC may be used to determine assay and impurity levels as a function of time. Generally, the same HPLC procedure is used when the drug product is tested at the time of manufacture and when it is tested after storage under controlled conditions.

Drugs in Biological Fluids. Many studies involve the determination of drugs in biological fluids (4,33) (see Fig. 11). Although some procedures allow the

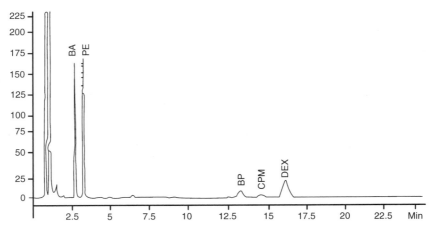

Fig. 10.   Chromatogram of Children's Tylenol Flu suspension liquid. Peaks are benzoic acid (BA, a preservative) [65-85-0], pseudoephedrine [90-82-4] (PE, a nasal decongestant), butylparaben (BP, a preservative) [94-26-8], chlorpheniramine (CPM, an antihistamine) [132-22-9], and dextromethorphan [125-69-9] (DEX, an antitussive). Conditions: 75 × 4.6 mm 3-μm Phenomenex Luna C8(2) column; mobile-phase acetonitrile:methanol:THF: buffer 5:42:2:51, buffer was 100 mM pH 2.1 sodium phosphate buffer containing 50 mM sodium 1-octanesulfonate; flow rate 1 mL/min; detector UV 214 nm. Reprinted with permission of Elsevier Science from Ref. 32.

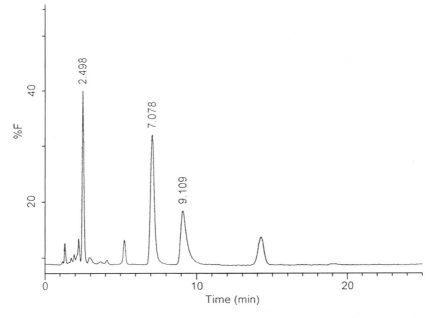

Fig. 11.   Determination of metoprolol in human plasma after sample preparation using solid-phase extraction. Peaks are α-hydroxymetoprolol [563920-16-67] (2.498 min), metoprolol [37350-58-6] (7.7078 min), and dextrorphan [125-23-5] (9.109 min (internal standard)). Conditions: 150 × 4.6 mm Metachem C4 column; mobile phase acetonitrile: THF:20 mM pH 3.0 phosphate buffer 15:2:83; flow rate 1.75 mL/min; detector fluorescence 228 nm (excitation) 320 nm (emission).

direct injection of biological fluids, sample preparation is generally critical. The concentrations of drugs in biological fluids are generally very low, in the order of nanograms per milliliter or lower, so much effort has been expended in increasing the sensitivity and selectivity of HPLC detectors. Additionally, the sensitivity and selectivity of the analytical system can be increased by the use of a derivatization procedure and by use of an appropriate sample preparation procedure. These issues have been discussed above.

The main biological fluids that are investigated are blood and urine. Some procedures describe the analysis of whole blood but generally the blood is processed to produce plasma (by immediately centrifuging whole blood) or serum (by allowing the drug to clot then centrifuging). Generally, the plasma or serum still contain the drug and are easier to work with. Analyses of drugs in many other biological matrices have been reported, however. Some examples are tissue of various kinds, feces, hair, saliva, sweat, semen, cerebrospinal fluid, and nails. Although some matrices are "cleaner" than others a sample preparation step is generally required. For the same drug the procedure may vary depending on the matrix. Many drugs are excreted in the urine as a glucuronidate or sulfate. If desired, an enzyme preparation can be used to regenerate the parent drug prior to HPLC analysis.

Measurement of drug levels in biological fluids can be used to determine the pharmacokinetics of a drug, ie, the variation of drug level with time. This yields much important information such as how frequently the drug needs to be dosed or how drugs may interact with one another. Various metabolites may also be identified and the racemization of chiral drugs in the body may be investigated.

Drugs such as antibiotics are frequently administered to animals that are to be used for food. Procedures for the analysis of drugs in tissue can be used to make sure that the drug has completely disappeared before the animal is slaughtered for human consumption.

Cassette Dosing. Historically, drug discovery proceeded by the synthesis of one compound at a time. Compounds were then screened and analytical systems devised for those molecules that showed promise. The advent of combinatorial chemistry and high-throughput techniques has greatly increased the number of compounds that may be synthesized and screened. Now a bottleneck has become the difficulty of devising analytical methods to obtain pharmacokinetic and metabolite information for the greatly increased number of compounds. Additionally, companies are coming under increasing pressure to reduce the number of animals used in testing. To help overcome these problems, a technique called cassette dosing or "N-in-1" dosing has been developed. A number of compounds (eg, 10) are administered to one animal. Blood samples are taken, processed using standard automated techniques, and samples injected into an HPLC instrument equipped with an MS detector (39). A "universal" isocratic or gradient mobile phase is used that will chromatograph most compounds. Different compounds are distinguished by scanning for ions of different mass. Before conducting the experiment, computer programs are used to make sure that different compounds are not likely to give rise to metabolites that have the same mass. In this way, the pharmacokinetics of many compounds can be assessed rapidly using a smaller number of experimental animals and with shortened assay times (34–36). However, this technique is controversial because drug–drug interactions can invalidate the results, eg, if one of the compounds is a potent inhibitor of a drug-metabolizing enzyme (37). A way to avoid the problems of drug–drug interactions is to dose one compound per animal then combine plasma samples from a number of animals and analyze the composite sample (38). However, this technique yields no reduction in the number of animals required and, as plasma samples are combined, the concentration of each drug decreases.

In a similar fashion, the power of HPLC–MS can also be used to eliminate bottlenecks in the investigation of the metabolites produced by isolated enzyme preparations acting on various substrates such as new drug candidates. In each reaction, a sample of the enzyme preparation (eg, liver microsomes) metabolizes only one compound but each reaction involves a different compound. Although it is easy to run many reactions at the same time, the expense of the analytical equipment usually causes analyses to proceed sequentially. However, when HPLC–MS is used a number of these reaction mixtures can be combined before analysis. The different compounds are then readily resolved and the extent of the metabolism of each substrate determined. Since only one compound is involved in each reaction there is no question of drug–drug interaction (40) (see Fig. 12).

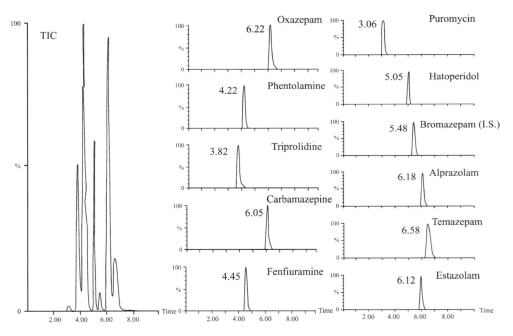

Fig. 12. Multiple compound determination. A sample of chimpanzee plasma was injected into a column-switching HPLC–MS–MS instrument. The column was a 150 × 2.0-mm Symmetry C18 (Waters) and the mobile phase was an acetonitrile/0.1% formic acid gradient pumped at 0.4 mL/min. The total ion chromatogram shows that the compounds overlap and would be hard to distinguish using a conventional (eg, uv) detector. However, the MS detector in multiple-reaction-monitoring mode can be used to monitor the distinctive ions of each compound and provide complete resolution as shown. Reprinted with permission of the American Chemical Society from Ref. 39.

Table 1. **Some Probe Compounds and the HPLC Conditions Associated with Them**[a,b]

Enzyme	Substrate	CAS Registry number	Metabolite	CAS Registry number	Column	Mobile phase	Detector
CYP1A1/2	phenacetin	[62-44-2]	acetaminophen	[103-90-2]	Zorbax phenyl	CH_3OH:pH 3.5 buffer gradient	uv 254
CYP2A6	coumarin	[91-64-5]	7-hydroxy-coumarin	[93-35-6]	NovaPak C18	CH_3OH:1% acetic acid 35:65	F ex 371 em 454uv
CYP2C8/9	tolbutamide	[64-77-7]	4-hydroxy-tolbutamide	[5719-85-7]	NovaPak C18	CH_3CN:pH 4.3 buffer gradient	uv 237
CYP2C19	S-mephenytoin	[58-12-4]	4-hydroxy-mephenytoin	[61837-65-8]	NovaPak C18	CH_3OH/pH 5 buffer gradient	uv 204
CYP2D6	dextromethorphan	[125-69-9]	dextrorphan	[125-73-5]	Zorbax C18	CH_3CN:1 mM perchloric acid 25:75	F ex 270 em 312
CYP2E1	chlorzoxazone	[95-25-0]	6-hydroxy-chlorzoxazone	[1750-45-4]	Zorbax C18	CH_3CN:0.05% phosphoric acid gradient	uv 287
CYP3A4	testosterone	[58-22-0]	6β-hydroxy-testosterone	[62-99-9]	Supelcosil LC18	CH_3OH:CH_3CN:H_2O gradient	uv 254

[a] Ref. 42.

[b] The flow rate was 1 mL/min except for phenacetin [62-44-2] (1.5 mL/min) and tolbutamide [64-77-7] (2 mL/min). Abbreviations:CH_3OH, methanol; uv, ultraviolet detector (with wavelength in nm); F, fluorescence detector (with excitation and emission wavelengths in nm); and CH_3CN, acetonitrile.

Hepatic Drug Metabolism.   Many drugs and other xenobiotic compounds (such as industrial compounds and nonmedicinal natural products) are metabolized by enzymes in the liver. Enzyme activity may be measured by using probe compounds that are predominantly metabolized by a single enzyme (41). HPLC is used to measure the extent of metabolism of these probe compounds, and hence the extent and type of enzyme activity (see Table 1).

Much useful information can be derived from the use of these probe compounds. For example, inhibitors of various enzymes can be determined because they will slow the metabolism of the probe compound. Drug interactions can then be predicted since a drug that inhibits a particular enzyme will lead to increased blood levels of other drugs that are metabolized by that enzyme. Equally, an inducer of a particular enzyme will lead to decreased levels of compounds that are metabolized by that enzyme.

In an interesting extension of the cassette-dosing strategy five probe compounds (coumarin [91-64-5], midazolam [59467-70-83], tolbutamide [64-77-7], dextromethorphan [125-69-9], and chlorzoxazone [95-25-0]) and a test compound were added to a human liver microsomal preparation. Injection of the reaction mixture into an LC–MS instrument using a short 4×2-mm Phenomenex C18 column with an acetonitrile/water/formic acid gradient led to the determination of the level of the metabolite of each probe compounds and hence the extent of the inhibitory effect of the test compound on each enzyme. Since the probe compounds do not produce metabolites with interfering ions the high resolving power of the LC–MS can generate this information in one short (2.5 min) run rather than five longer individual HPLC assays (43).

5.2. Nonpharmaceutical Clinical Applications.   It is frequently desirable to measure compounds other than drugs in biological fluids. Generally, the procedures are similar to those used to measure drugs in biological fluids. In some cases, these compounds may indicate the presence of certain diseases. For example, measurement of cysteine in plasma by HPLC (after derivatization) can show the metabolic disorder cystinuria (44). HPLC methods have been developed for many other clinically relevant analytes such as amino acids, lipids, or proteins (45).

5.3. Environmental Applications.   Many methods have been published in the literature for the analysis of pollutants in the environment (46). Generally, the main problem is that of sensitivity and so solid-phase extraction is used extensively. The U.S. Environmental Protection Agency (EPA) has published a number of methods. A list of all EPA test methods is available at http://www. epa.gov/epahome/index/ and methods for organic chemical analysis are found at http://www.epa.gov/ostwater/Tools/guide/methods.html.

5.4. Other Applications.   HPLC has found applications in many other areas. Some examples are shown below

Art conservation	S. L. Vallance, *Analyst* **122**, 75R (1997)
Color additives	W. Horwitz, ed., Official Methods of Analysis of AOAC International, 17th ed., AOAC International, Arlington, Va., 2000
Explosives	W. Horwitz, ed., Official Methods of Analysis of AOAC International, 17th ed., AOAC International, Arlington, Va., 2000

Fertilizers	W. Horwitz, ed., Official Methods of Analysis of AOAC International, 17th ed., AOAC International, Arlington, Va., 2000
Food	Current Protocols in Food Analytical Chemistry, John Wiley & Sons, Inc., New York
Food	W. Horwitz, ed., Official Methods of Analysis of AOAC International, 17th ed., AOAC International, Arlington, Va., 2000
Food additives	W. Horwitz, ed., Official Methods of Analysis of AOAC International, 17th ed., AOAC International, Arlington, Va., 2000
Forensic science	T.A. Brettell, K. Inman, N. Rudin, and R. Saferstein, *Anal. Chem.*, **73**, 2735 (2001)
Industrial processes	J. Workman, Jr., D. J. Veltkamp, S. Doherty, B. B. Anderson, K. E. Creasy, M. Koch, J. F. Tatera, A. L. Robinson, L. Bond, L. W. Burgess, G. N. Bokerman, A. H. Ullman, G. P. Darsey, F. Mozayeni, J. A. Bamberger, and M. S. Greenwood, *Anal. Chem.*, **71**, 121R (1999)
Infant formula	W. Horwitz, ed., Official Methods of Analysis of AOAC International, 17th ed., AOAC International, Arlington, Va., 2000
Molecular biology	Current Protocols in Molecular Biology, John Wiley & Sons, Inc., New York
Natural toxins	W. Horwitz, ed., Official Methods of Analysis of AOAC International, 17th ed., AOAC International, Arlington, Va., 2000
Pesticides	W. Horwitz, ed., Official Methods of Analysis of AOAC International, 17th ed., AOAC International, Arlington, Va., 2000
Petroleum	C. T. Mansfield, B. N. Barman, J. V. Thomas, A. K. Mehrota, and J. M. McCann, *Anal. Chem.* **71**, 81R (1999)
Polymers	P. B. Smith, A. J. Pasztor, Jr., M. L. McKelvy, D. M. Meunier, S. W. Froelicher, and F. C.-Y. Wang, *Anal. Chem.* **71**, 61R (1999)
Vitamins	W. Horwitz, ed., Official Methods of Analysis of AOAC International, 17th ed., AOAC International, Arlington, Va., 2000
Water quality	S. D. Richardson, *Anal. Chem.* **73**, 2719 (2001)

6. Acknowledgments

The opinions expressed in this article are those of the author and do not necessarily reflect the views or policies of the FDA.

BIBLIOGRAPHY

1. *FDA Guideline for Submitting Samples and Analytical Data for Methods Validation*, Appendix C, February 1987.

2. L. R. Snyder, J. J. Kirkland, and J. L. Glajch. *Practical HPLC Method Development*, 2nd ed., John Wiley & Sons, Inc., New York, 1997, pp. 110–119.

3. V. Meyer, *Anal. Chem.* **69**, 18A (1997).

4. G. Lunn and N. R. Schmuff, *HPLC Methods for Pharmaceutical Analysis*, Vols. 1–4, John Wiley & Sons, Inc., New York, 1997–2000.

5. T. C. Pinkerton, *J.Chromatogr.* **544**, 13 (1991).

6. A. Shibukawa, C. Nakao, T. Sawada, A. Terakita, N. Morokoshi, and T. Nakagawa, *J. Pharm. Sci.* **83**, 868 (1994).

7. E. Pocurull, R. M. Marcé, and F. Borrull, *Chromatographia* **41**, 521 (1995).

8. R. Herráez-Hernández, N. C. Van der Merbel, and U. A. T. Brinkman, *J. Chromatogr. B* **666**, 127 (1995).

9. C. H. Kiang, T. Nolan, B. L. Huang, and C. P. Lee, *J. Chromatogr.* **567**, 195 (1991).

10. K. Blau and J. Halket, eds., *Handbook of Derivatives for Chromatography*, 2nd ed., John Wiley & Sons, Inc., Chichester, U.K., 1993.

11. G. Lunn and L. C. Hellwig, *Handbook of Derivatization Reactions for HPLC*, John Wiley & Sons, Inc., New York, 1998.

12. M. Y. Khuhawar and S. N. Lanjwani. *J. Chromatogr. A* **740**, 296 (1996).

13. W. R. LaCourse and C. O. Dasenbrock, *Adv.Chromatogr.* **38**, 189 (1998).

14. A. C. Calokerinos, N. T. Deftereos, and W. R. G. Baeyens, *J. Pharm. Biomed. Anal.* **13**, 1063 (1995).

15. O. Busto, J. Guasch, and F. Borrull, *J. Chromatogr. A* **718**, 309 (1995).

16. T. Takagi, T. G. Chung, and A. Saito, *J. Chromatogr.* **272**, 279 (1983).

17. J. Haginaka, J. Wakai, H. Yasuda, T. Uno, K. Takahashi, and T. Katagi, *J. Chromatogr.* **400**, 101 (1987).

18. A. M. Di Pietra, R. Gatti, V. Andrisano, and V. Cavrini, *J. Chromatogr. A* **729**, 355 (1996).

19. M. Uihlein and E. Schwab, *Chromatographia* **15**, 140 (1982).

20. D. Chaput, *J. Assoc. Off. Anal. Chem.* **71**, 542 (1988).

21. R. E. Wrolstad, ed., *Current Protocols in Food Analytical Chemistry*, John Wiley & Sons, Inc., New York, 2001, p. F2.3.12.

22. L. R. Snyder, J. J. Kirkland, and J. L. Glajch. *Practical HPLC Method Development*, 2nd ed., John Wiley & Sons, Inc., New York, 1997, p. 193.

23. P. C. Sadek, *The HPLC Solvent Guide*, John Wiley & Sons, Inc., New York, 1996.

24. P. J. Schoenmakers, H. A. H. Billiet, and L. de Galan. *J. Chromatogr.* **218**, 261 (1981).

25. V. R. Meyer, *Practical High-Performance liquid Chromatography*, 3rd ed., John Wiley & Sons, Inc., Chichester, U.K., 1998, pp. 184–190.

26. L. R. Snyder, J. J. Kirkland, and J. L. Glajch, *Practical HPLC Method Development*, 2nd ed., John Wiley & Sons, Inc., New York, 1997, pp. 317–341.

27. N. Cardellicchio, P. Ragone, S. Cavalli, and J. Riviello, *J. Chromatogr. A* **770**, 185 (1997).

28. M. Vakily and F. Jamali, *J. Pharm. Sci.* **85**, 638 (1996).

29. P. M. Jones and K. Brune, *Clin. Chem.* **39**, 168 (1993).

30. *United States Pharmacopeia*, 24th Revision/National Formulary XVIII, United States Pharmacopeial Convention, Inc., Rockville, Md., 2000.

31. Federal Food, Drug, and Cosmetic Act, Section 501 (b); Title 21 Code of Federal Regulations 211.194(a)(2) and 314.50(d)(1).

32. M. D. Paciolla, S. A. Jansen, S. A. Martellucci, and A. A. Osei, *J. Pharm. Biomed. Anal.* **26**, 143 (2001).

33. R. S. Plumb, G. J. Dear, D. N. Mallett, D. M. Higton, S. Pleasance, and R. A. Biddlecombe, *Xenobiotica* **31**, 599 (2001).

34. J. Berman, K. Halm, K. Adkison, and J. Shaffer. *J. Med. Chem.* **40**, 827 (1997).

35. J.-T. Wu, H. Zeng, M. Qian, B. L. Brogdon, and S. E. Unger. *Anal. Chem.* **72**, 61 (2000).
36. J. E. Shaffer, K. K. Adkison, K. Halm, K. Hedeen, and J. Berman, *J. Pharm. Sci.* **88**, 313, (1999).
37. R. E. White and P. Manitpisitkul., *Drug Metab. Dispos.* **29**, 957 (2001).
38. Z. Cai, C. Han, S. Harrelson, E. Fung, and A. K. Sinhababu, *Rapid Commun. Mass Spectrom.* **15**, 546 (2001).
39. J.-T. Wu, H. Zeng, M. Qian, B. L. Brogdon, and S. E. Unger. *Anal. Chem.* **72**, 61 (2000).
40. Z. Cai, A. K. Sinhababu, and S. Harrelson, *Rapid Commun. Mass Spectrom.* **14**, 1637 (2000).
41. S. A. Wrighton, M. Vandenbranden, J. C. Stevens, L. A. Shipley, and B. J. Ring, *Drug Metab. Rev.* **25**, 453 (1993).
42. N. Chauret, A. Gauthier, J. Martin, and D. A. Nicoll-Griffith, *Drug Metab. Dispos.* **25**, 1130 (1997).
43. H.-Z. Bu, L. Magis, K. Knuth, and P. Teitelbaum, *Rapid Commun. Mass Spectrom.* **15**, 741 (2001).
44. A. Pastore, R. Massoud, C. Motti, A. Lo Russo, G. Fucci, C. Cortese, and G. Federici, *Clin.Chem.* **44**, 825 (1998).
45. D. J. Anderson, *Anal. Chem.* **71**, 314R (1999).
46. R. E. Clement and P. W. Yang, *Anal. Chem.* **73**, 2761 (2001).

GENERAL REFERENCES

R. L. Cunico, K. M. Gooding, and T. Wehr, *Basic HPLC and CE of Biomolecules*, Bay Bio-analytical Laboratory, Hercules, Calif., 1998.

V. R. Meyer, *Practical High-Performance Liquid Chromatography*, 3rd ed., John Wiley & Sons, Inc., Chichester, U.K., 1998.

U. D. Neue, *HPLC Columns : Theory, Technology, and Practice*, John Wiley & Sons, Inc., New York, 1997.

P. C. Sadek, *Troubleshooting HPLC Systems: A Bench Manual*, John Wiley & Sons, Inc., New York, 2000.

L. R. Snyder and J. J. Kirkland, *Introduction to Modern Liquid Chromatography*, 2nd ed., John Wiley & Sons, Inc., New York, 1979.

L. R. Snyder, J. J. Kirkland, and J. L. Glajch, *Practical HPLC Method Development*, 2nd ed., John Wiley & Sons, Inc., New York, 1997.

J. Swadesh, ed., *HPLC: Practical and Industry Chromatography*, 2nd ed., CRC Press, Boca Raton, Fla., 2000.

Q. A. Xu and L. A. Trissel, *Stability-Indicating HPLC Methods for Drug Analysis*, American Pharmaceutical Association, Washington, D.C., 1999.

Analyst http://www.rsc.org
Analytica Chimica Acta http://www.elsevier.nl
Analytical Chemistry http://pubs.acs.org
Analytical Letters http://www.dekker.com
Biomedical Chromatography http://www.wiley.com
Chromatographia http://www.elsevier.nl
Journal of AOAC International http://www.aoac.org
Journal of Analytical Toxicology http://www.jatox.com/
Journal of Chromatographic Science http://www.j-chrom-sci.com/
Journal of Chromatography A http://www.elsevier.nl

GEORGE LUNN
Center for Drug Evaluation and Research,
Food and Drug Administration

CHROMIUM AND CHROMIUM ALLOYS

1. Introduction

Chromium [7440-47-3] is one of the "newer" elements, celebrating the 200th anniversary of its discovery in 1997. It was about 1760 when the chromium-bearing mineral crocoite from deposits in the Ural Mountains was recognized in Europe. However, it was not until 1797/98 that chromium was isolated by Nicolas-Louis Vauquelin, a professor of chemistry at the Paris Ecole des Mines. Chromium was discovered later than other metals because it does not appear terrestrially as a native metal, and it is strongly bonded in the minerals in which it occurs. The wide variety of colorful compounds derivable from crocoite led Vauquelin to name the newly discovered element chromium, a name derived from *chroma*, the Greek word for color. Crocoite, also called *Siberian red lead*, was found to produce a yellow pigment that became popular. Thus paint became the first commercial application of chromium. Chromium was soon discovered in chromite, a much more common mineral, also from the Ural Mountains.

Chromium is used primarily in the metallurgical industry as an alloying element in steel. Chromium confers properties on the alloy that are not achievable with base metals alone. The most common use of chromium is with iron to make stainless steel, an iron–chromium alloy. Chromium confers oxidation resistance to stainless steel, making it "stainless." Stainless steel, in addition to being commonly found in home and commercial kitchens, is an important engineering alloy used throughout industry in machinery, containers, and pipes (see STEEL). Chromium is also used in chemicals for a variety of purposes. Chromite, the mineral from which chromium is extracted for use in the metallurgical and chemical industries, is used directly by the refractory industry to produce heat-, spalling-, corrosion-, and abrasion-resistant bricks for metallurgical and high-temperature industrial mineral processing applications. Chromite is not mined domestically; thus, the United States is 100% dependent on imports to meet domestic chromite demand. Domestic chromium demand is met by import of chromite ore; chromium ferroalloys, chromium metal, or chromite ore; and chemicals; and by recycling.

Chromium has a wide range of uses in metals, chemicals, and refractories. It is one of the nation's most important strategic and critical materials. The use of chromium to produce stainless steel and nonferrous alloys are two of its more important applications. Other applications are in alloy steel, plating of metals, pigments, leather processing, catalysts, surface treatments, and refractories.

The major commercially traded forms of chromium materials are chromite ore [1308-31-2] and ferrochromium [11114-46-8]. The United States was a significant world chromite ore producer before 1900. However, since that time, U.S. production has declined to nil. Ferrochromium is a product of smelting chromite ore in an electric-arc furnace. Ferrochromium is the major form of chromium used by the metallurgical industry. Historically, ferrochromium smelters developed in major steel-producing centers of the United States, Europe, and Japan. Since about 1970, the net effect of vertical integration in chromite producing nations' industry and the concomitant rationalization in developed steel-producing centers has resulted in the migration of ferrochromium production to chromite-producing countries. This trend is expected to continue. Thus the United States is a chromium importing nation. Chromium is subsequently exported from the United States in stainless-steel products.

Stainless steel was invented in the early 1900s. Soon thereafter electric furnaces evolved that could smelt chromite into ferrochromium. Before about 1960, ferrous alloys required the addition of as little carbon as possible because carbon could not be efficiently removed from molten steel. Thus, the production of low-carbon, high-chromium alloys (typically less than 0.1% carbon and more than 65% chromium) was the common practice. To make this ferrochromium, high chromium:iron ores were required (ratios greater than about 2:1).

Since 1960, major changes have occurred in the chromium industry because of changes in steelmaking technology. The development of ladle refining techniques (ie, processes that permit the chemical modification of liquid metal) such as argon–oxygen decarburization, permitted the steel industry to shift from the more costly low-carbon ferrochromium to the less costly high-carbon ferrochromium. This shift in ferrochromium grade has been accompanied by a shift in quantity of production among ferrochromium-producing countries. Since the 1970s, chromite ore–producing countries have developed their own ferrochromium production capacities. As a result, ferrochromium production has moved from the major stainless-steel-producing centers, Japan, the United States, and western Europe, to chromite-producing countries, Finland, India, the Republic of South Africa, Turkey, and Zimbabwe. With the exception of Japan, only minor ferrochromium production remains in the major stainless-steel-producing countries. In particular, the Republic of South Africa, whose ores have a chromium:iron ratio of about 1.5:1, has increased its high-carbon ferrochromium production dramatically. Significant, but declining, quantities of ferrochromium continue to be produced in Japan.

Because the United States has no chromite ore reserves and a limited reserve base, domestic supply has been a concern during every national military emergency since World War I. World chromite resources, mining capacity, and ferrochromium production capacity are concentrated in the Eastern Hemisphere. The National Defense Stockpile (NDS) contains chromium in various forms, including chromite ore, chromium ferroalloys, and chromium metal in recognition of that material is being disposed of the vulnerability of long supply routes during a military emergency.

The terms *chromium* and *chrome*, as used in the chemical industry, are synonymous. Similarly, the terms *dichromate* and *bichromate* are used interchangeably in the chemical industry.

Hargreaves and co-workers (1) reviewed the world minerals industry by mineral, country, and mining company. The composite world rank of chromium among most globally important strategic investment mineral commodities was found to be 10th out of 36. The composite world rank is an indexed composite of five factors: output by value, population and gross domestic product, resource demand, mineral reserve base, and country investment risk.

2. Occurrence

Many minerals contain chromium as a major element [see Table 1 (2)], and many minerals contain tens of percent chromium. However, only the mineral chromite occurs in large enough quantities to be a commercial source of chromium. Chromite can be found in many different rock types, but the host rocks for economically important chromite deposits are called *peridotite* and *norite*. These are distinctive rocks composed mainly of the minerals olivine and pyroxene (peridotite) and pyroxene and plagioclase (norite). These rocks occur primarily in two geologic settings: *layered intrusions*, which are large bodies of layered igneous rock that cooled very slowly in large underground chambers of molten rock; and *ophiolites*, which are large pieces of the oceanic crust and mantle that have been thrust over continental rocks by the same tectonic forces that cause continental drift. Because chromite deposits in layered intrusions tend to be tabular in form they are known as *stratiform deposits*, whereas those in ophiolites are typically podlike or irregular in form, are known as *podiform deposits* (see also MINERAL RECOVERY AND PROCESSING). Other sources of chromite are beach sands derived from chromite-containing rocks and laterites that are weathering products of peridotite. Laterites are more widely known as sources of nickel and cobalt. Beach sands and laterites historically have been a minor source of chromite.

Table 2 shows the reserves and resources of chromite worldwide.

The identified world resources of chromite are sufficient to meet conceivable demand for centuries. Current world demand is about 12 million metric tons per year. Reserves are that part of identified resources that are currently economic. The reserve base, which includes reserves, is that part of identified resources that are economic now and also may become economic with existing technology, depending on economic conditions and price of chromite.

2.1. Stratiform Deposits.   Most of the world's chromite resources occur as stratiform deposits in layered intrusions. The Bushveld Complex in South Africa contains over 8.5 billion tons of chromite while the remainder of the world's economic and subeconomic deposits have a little over 2.5 billion tons,

Table 1. **Terrestrial Minerals Containing Chromium as a Major Constituent**[a]

Name	General formula	Wt % Cr
Barbertonite	$Mg_6Cr_2(CO_3)(OH)_{16} \cdot 4H_2O$	16
Bentorite	$Ca_6(Cr,Al)_2(SO_4)_3(OH)_{12} \cdot 26H_2O$	5
Bracewellite[b]	$CrO(OH)$	61
Brezinaite	Cr_3S_4	47–50
Carlsbursite	CrN	79
Caswellsilverite	$NaCrS_2$	37
Chromian diopside	$Ca(Mg,Fe,Cr)Si_2O_6$	0.1–8
Chromian geikielite	$(Mg,Fe^{2+},Cr,Fe^{3+})(Ti,Cr,Fe^{3+})O_3$	0.5–8.5
Chromian garnet	$(Cr,Mg)_3(Al,Cr)_2(SiO_4)_3$	0.1–13
Chromite	$(Mg,Fe^{2+})(Cr,Al,Fe^{3+})_2O_4$	10–54
Chromatite	$CaCrO_4$	33
Chromian clinochlore	$(Mg,Fe^{2+})(Al,Cr)_2(Al_2,Si_2)O_{10}(OH)_8$	0.5–12
Cochromite	$(Co,Ni,Fe^{2+})(Al,Cr)_2O_4$	34–37
Crocoite	$PbCrO_4$	16
Daubreelite	$Fe^{2+}Cr_2S_4$	36
Deanesmithite	$Hg_2^{1+}Hg_3^{2+}Cr^{6+}O_5S_2$	4.3
Dietzeite	$Ca_2(IO_3)_2(CrO_4)$	10
Donathite	$(Mg,Fe^{2+})(Cr,Fe^{3+})_2O_4$	28–30
Edoylerite	$Hg_3^{2+}Cr^{6+}O_4S_2$	6.6
Embreyite	$Pb_5(CrO_4)(PO_4)_2 \cdot H_2O$	7
Eskolaite	Cr_2O_3	44–68
Fornacite	$(Pb,Cu)_3[(Cr,As)O_4]_2(OH)$	6
Fuchsite	$K(Al,Cr)_2(AlSi_3)O_{10}(OH)_2$	0.5–6
Georgeericksenite	$Na_6CaMg(IO_3)_6(CrO_4)_2(H_2O)_{12}$	5
Grimaldiite[b]	$CrO(OH)$	61
Guyanaite[b]	$CrO(OH)$	61
Heideite	$(Fe,Cr)_{1+x}(Ti,Fe)_2S_4$	0.1–18
Hemihedrite	$Pb_{10}Zn(CrO_4)_6(SiO_4)_2F_2$	13–14
Iranite	$Pb_{10}Cu(CrO_4)_6(SiO_4)_2(F,OH)_2$	10
Knorringite	$Mg_3Cr_2(SiO_4)_3$	12–23
Krinovite	$NaMg_2CrSi_3O_{10}$	14
Lopezite	$K_2Cr_2O_7$	35
Loveringite	$(Ca,Ce)(Ti,Fe^{3+},Cr,Mg)_{21}O_{38}$	0.5–10
Macquartite	$Pb_3Cu(CrO_4)SiO_3(OH)_4 \cdot 2H_2O$	6
Manganochromite	$(Mn,Fe^{2+})(Cr,V)_2O_4$	41–62
Mariposite	$K(Al,Cr)_2(Si_{3+x}Al_{1-y})O_{10}(OH)_2$	0.5–6
McConnellite	$CuCrO_2$	35
Mountkeithite	$(Mg,Ni)_{11}(Fe^{3+},Cr,Ni)_3(OH)_{24}(CO_3,SO_4)_{3.5}$ $(Mg,Ni)_2(SO_4)_2 \cdot 11H_2O$	2.2–6
Nichromite	$(Ni,CoFe^{2+})(Cr,Fe^{3+},Al)_2O$-$_4$	31–37
Phoenicochroite	$Pb_2(CrO_4)O$	8–10
Redingtonite	$(Fe^{2+},Mg,Ni)(Cr,Al)_2(SO_4)_4 \cdot 22H_2O$	0.5–3
Redledgeite	$Mg_4Cr_6Ti_{23}Si_2O_{61}(OH)_4$	11
Rilandite	$(Cr,Al)_6SiO_{11} \cdot 5H_2O$	33
Santanaite	$9PbO \cdot 2PbO_2 \cdot CrO_3$	2
Schreyerite	$(V,Cr,Al)_2Ti_3O_9$	0.7–3.6
Shuiskite	$Ca_2(Mg,Al,Fe)(Cr,Al)_2[(Si,Al)O_4](Si_2O_7)(OH)_2 \cdot H_2O$	10–17
Stichtite	$Mg_6Cr_2(CO_3)(OH)_{16} \cdot 4H_2O$	6–19
Tarapacaite	K_2CrO_4	27
Ureyite	$NaCrSi_2O_6$	23
Uvarovite	$Ca_3Cr_2(SiO_4)_3$	21
Vauquelinite	$Pb_2Cu(CrO_4)(PO_4)(OH)$	7
Vuorelaineite	$(Mn,Fe,^{2+})(V,Cr)_2O_4$	3.2–21
Wattersite	$Hg_4^{1+}Hg^{2+}Cr^{6+}O_6$	4.5
Yedlinite	$Pb_6CrCl_6(O,OH)_8$	4

[a] Modified from Lipin (2).
[b] Different crystal structures.

Table 2. **Chromite Reserves, Reserve Base, and Identified Resources in 1998**[a]

Country	Deposit type	Reserves	Reserve base	Identified resources
Albania	podiform	1,890	1,890	7,980
Australia	stratiform	0	56	1,830
Brazil	stratiform	4,450	7,140	9,060
Canada	stratiform	0	1,600	3,840
China	podiform	2,500	3,000	10,000
Cuba	podiform	739	739	1,970
Finland	stratiform	12,530	37,900	37,900
Greece	podiform	NA	380	785
Greenland	stratiform	0	0	26,000
India	stratiform	8,210	20,500	37,400
Indonesia	laterite	235	235	235
Iran	podiform	745	745	17,700
Japan	podiform	33	60	69
Kazakstan	podiform	126,000	126,000	301,000
Madagascar	stratiform	2,120	2,120	2,120
Macedonia	podiform	NA	NA	NA
Oman	podiform	NA	301	602
Papua New Guinea	laterite	0	0	2,890
Philippines	podiform	2,260	2,260	2,260
Russia	podiform	1,230	140,000	140,000
South Africa	stratiform	933,000	1,700,000	2,970,000
Sudan	podiform	513	513	513
Turkey	podiform	2,450	6,040	6,770
USA	stratiform	0	3,100	35,000
United Arab Emirates	podiform	64	64	64
Venezuela	podiform	0	0	713
Zimbabwe	stratiform	43,500	285,000	285,000
Total		*1,140,000*	*2,350,000*	*3,910,000*

[a] Values are in 1000 t, gross weight normalized to 45% Cr_2O_3 content. (*Note*: Deposit type is predominant deposit type. In many countries, more than one deposit type occurs. Reserves are economically recoverable, demonstrated resources. Reserve base is economic, marginally economic, and, possibly, some currently uneconomic, demonstrated resources. Identified resources are resources whose location, grade, quality, and quantity are known or estimated from specific geologic evidence.) Data and total rounded independently; NA = not available.

and about half of that tonnage is in the Great Dyke in Zimbabwe, another layered intrusion (3). Clearly, chromite resources in layered intrusions are not evenly distributed worldwide. Other layered intrusions that produce or have produced chromite are Stillwater Complex, Montana, USA; Kemi Complex, Finland; Orissa Complex, India; Goias, Brazil; Andriamena, Befandriana, and Ranomena, Madagascar; and Mashaba, Zimbabwe.

Stratiform deposits are not evenly distributed over geologic time, either. While intrusions of the type of rock that carry chromite deposits appear over the spectrum of geologic time, only those of Precambrian age (older than ~540 million yr) are known to carry economic chromite deposits; the youngest of these deposits is the Bushveld, at about 2.066 billion yr. A possible exception to this might be the deposits in the central Ural Mountains, which may be a disrupted layered complex of Early Silurian age (about 440 million yr old).

2.2. Podiform Deposits.

Although resources and reserves of podiform deposits are quite small compared to stratiform deposits, podiform deposits have been, and continue to be, important sources of chromite. This is because many of these deposits are large and rich enough to be economic. In addition, before certain advances in metallurgy, the composition of the chromite produced from podiform deposits was more suited for the metallurgical uses of chromite.

As stated above, podiform deposits occur in *ophiolites*, which are pieces of the oceanic crust and mantle thrust up over continental rocks. Many different rock types occur in an ophiolite, but the stratigraphically lowest of these is peridotite, which is the host for podiform chromite deposits. Podiform deposits are found in many places in the world and throughout geologic time. The most important historic sources of chromite from podiform deposits are Kempersai, Kazakstan; Perm district, Russia; Zambalas, Philippines; four districts, Albania; six districts, Turkey; Selukwe, Zimbabwe; New Caledonia; Troodos, Cyprus; and Vourinos, Greece.

Other production has come from the Appalachians in the United States, Australia, China, Cuba, the former Yugoslavia, Iran, New Guinea, Oman, Pakistan, Sudan, the coastal ranges in California and Oregon, the Shetland Islands in Scotland, and Vietnam.

Podiform and stratiform deposits have different chemical characteristics, which has determined how they are used. Industry has classified chromite ore as high-chromium, high-iron, and high-aluminum. Table 3 summarizes the relationship between these classifications, and major use. Table 4 (4) summarizes the range of chemical contents of chromite ores.

2.3. Beach Sands.

Beach sands that contain chromite exist as a result of erosion. Chromite mined from hard rock deposits, either stratiform or podiform, are concentrations in the rock commonly at least 15 vol% chromite up to 100% massive chromite. Some of them are many millions of tons in size. However, all peridotites, even those that do not contain economic concentrations of chromite, contain chromite at low levels, between 1 and 5 vol% of the rock. In addition, peridotite can occur over many hundreds of square kilometers in ophiolites. The fact that chromite is ubiquitous in peridotite at low levels and peridotite can occur over large areas allows for the possibility of streams moving through peridotite to erode the rock and deposit chromite downstream. In addition, the fact that chromite is the most dense mineral in peridotite means that wave action will naturally concentrate the mineral in a beach environment.

Table 3. Classification of Chromite by Composition, Type of Deposit, and Principal Uses

Class of ore	Composition, wt%	Type of deposit	Major use
high-Cr	Cr_2O_3 46–55% Cr/Fe>2:1	podiform and stratiform	metallurgical
high-Fe	Cr_2O_3 42–46% Cr/Fe<2:1	stratiform	metallurgical and chemical
high-Al	Cr_2O_3 33–38% Al_2O_3 22–34%	podiform	refractory

Table 4. **Range of Commercially Available Chromite Ore Chemical Characteristics Based on a Composite of Sources (in Percent)**[a,b]

Chemical compound	Observed range of values	
	Lower	Upper
Cr_2O_3	30	57
SiO_2	0.98	18
Fe	9	19.6
Al_2O_3	6	22
MgO	8	28
P	0.002	0.01

[a] Reference 4; Although the composition limits shown here come from commercially available material, they do not as a group represent any specific material.
[b] Chromium:iron ratio varies from 1.4 to 4.2.

Such is the case in Oregon, where beach sands were mined during Word War II. Since 1988, some attempts have been made to mine sands on the island of Palawan in the Philippines. Other sand or placer chromite deposits occur in Indonesia, Papua New Guinea, Vietnam, and Zimbabwe.

2.4. Laterites. Laterite forms as the result of weathering of peridotite in a tropical or a forested, warm temperate climate. Laterite is a thick red soil derived from the rock below. It is red because of the high concentration of iron. The process of laterization leaches out most of the silicate minerals in the rock, leaving higher concentrations of elements that can fit in the structures of nonsilicate minerals. Thus laterites concentrate elements such as iron, nickel, cobalt, and chromium. In some laterites chromite is concentrated to economic concentrations. This is the case in Indonesia where chromite is being mined.

2.5. Chromite. The mineral chromite is jet black in color, has a submetallic luster, yields a brown streak, is generally opaque in thin section, and has no cleavage. The density ranges from about 3.8 to 4.9 g/cm^3 and has a Vickers hardness number between 5 and 6. Chromite is a solid solution mineral of the spinel group, has cubic symmetry and a closely packed crystal lattice, hence the high density of the minerals of the spinel group. The six end-member compositions that combine to form chromite (see Fig. 1) are hercynite ($FeAl_2O_4$), spinel ($MgAl_2O_4$), Fe–chromite ($FeCr_2O_4$), picrochromite ($MgCr_2O_4$), magnetite (Fe_3O_4), and magnesioferrite ($MgFe_2O_4$). Thus, the general formula is (Mg, Fe)(Cr, Al)$_2O_4$. At high temperatures ($>1200°C$) and low oxygen fugacity, the conditions under which chromite first forms, there is complete solid solution between Mg and Fe and between Cr and Al. Other elements found in lesser amounts are Ti, Zn, Ni, V, Mn, and Co. There are no formal rules for naming chromite; however, most geologists and people in the industry use the term *chromite* when the Cr_2O_3 content rises above 15 wt%. Because chromite is a solid solution, it has no fixed composition.

2.6. Terrestrial Chromium Abundance. Chromium is the 18th most abundant element in the earth's upper crust at 35 ppm (5). Chromium is most concentrated in rocks that constitute the upper mantle, from which crustal

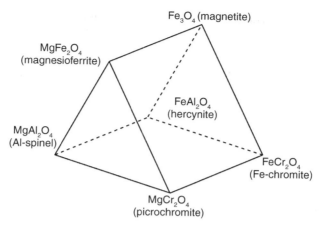

Fig. 1. Composition diagram for the spinel minerals. Each corner represents an end-member composition in a complex solid solution that involves all the end members. As a result, the corners represent the purest forms of the named minerals of this group. A naturally occurring chromite mineral composition would lie within the bounds of the prism defined by these end members.

rocks are evolved. Upper-mantle rocks are almost exclusively peridotite in which the average chromium content is 3000 ppm (6). The granitelike compositions that dominate the upper crust, and erode to form sedimentary rocks, tend to exclude chromium. The lower crust, which contains rocks that are somewhat closer in composition to the upper mantle, contains ~235 ppm (5) and the concentration in the overall crust of the Earth is 100 ppm (7). Seawater contains 2×10^{-10} g of chromium per gram of water, while rivers average 1×10^{-9} g of chromium per gram of water.

3. Properties

The data in this properties section are taken from the Ref. 8. The chemical symbol for chromium is Cr, and it has an atomic weight of 51.966 and atomic number of 24. Its melting point is 1907°C and its boiling point is 2671°C. At 20°C the specific gravity is 7.18–7.20 g/cm^3.

Chromium is one of the so-called transition elements, meaning that it has valence electrons in two shells instead of one. Chromium exists as a metal and in three valence states, 2^+, 3^+, and 6^+, when combined with other elements—all of which can occur naturally. However, the di- and trivalent forms are most prevalent in nature. The electron shell configuration of chromium is 2, 8, 13, 1, and the filling orbital is $3d5$.

Chromium is a steel gray metal, has cubic symmetry, and is very hard. It is soluble in H_2SO_4, HCl, HNO_3, and aqua regia. Chromium resists corrosion and oxidation. When used in steel at greater that 10 wt% it forms a stable oxide surface layer, which makes it particularly useful in making stainless steel and other specialty steels to ward off the corrosive effects of water. The ability of chromium to resist corrosion and accept a high polish has made it almost ubiquitous as a coating on household water faucets. See Table 5 for physical properties of chromium.

Chromium has four naturally occurring nonradiogenic isotopes. Their symbols, percent abundance, and atomic mass are as follows: $_{24}Cr^{50}$, 4.35%, 49.946046; $_{24}Cr^{52}$, 83.79%, 51.940509; $_{24}Cr^{53}$, 9.50%, 52.940651; $_{24}Cr^{54}$, 2.36%, 53.938882.

Table 5. **Physical Properties of Chromium**

Property	Value
at no.	24
at wt	51.996
isotopes	
mass	50 52 53 54
relative abundance, %	4.318 3.769 .55 2.38
crystal structure	bcc
lattice parameter, a_o, nm	0.2888–0.2884
density at 20°C, g/mL	7.19
mp, °C	1875
bp, °C	2680
vapor pressure, at 1610°C, Pa[a]	130
heat of fusion, kJ/mol[b]	14.6
latent heat of vaporization at bp, kJ/mol[b]	305
specific heat at 25°C, J/(mol · K)[b]	23.9
linear coefficient of thermal expansion at 20°C, K^{-1}	6.2×10^{-6}
thermal conductivity at 20°C, W/(m · K)	91
electrical resistivity at 20°C, $\mu\Omega \cdot m$	0.129
superconducting transition temperature, K	0.08
antiferromagnetic (Néel) transition temperature, K	311
specific magnetic susceptibility at 20°C	3.6×10^{-6}
total emissivity at 100°C[c]	0.08
reflectivity, R, %	
at 30 nm	67
at 50 nm	70
at 100 nm	63
at 400 nm	88
refractive index, α, for $\lambda = 257$–608 nm	1.64–3.28
standard electrode potential, V	
$Cr \longrightarrow Cr^{3+} + 3\,e^-$	−0.74
$Cr^{4+} \longrightarrow Cr^{6+} + 2\,e^-$	+0.95
ionization potential, V	
1st	6.74
2nd	16.6
half-life of ^{51}Cr isotope, days	27.8
thermal neutron scattering cross section, m^2	6.1×10^{-28}
elastic modulus, GPa[d]	250
compressibility[e] at 10–60 TPa	70×10^{-3}
hardness (Knoop value)	1.160
electron affinity of Cr–Cr bond, eV	0.666

[a] To convert Pa to mm Hg, multiply by 0.0075.
[b] To convert J to cal, divide by 4.184.
[c] Nonoxidizing atmosphere.
[d] To convert GPa to psi, multiply by 145,000.
[e] 99% Cr; to convert TPa to megabars, multiply by 10.

Table 6. **Standard Thermodynamic Properties of Chromium at 298.15 K**

Phase	Enthalpy of formation, kJ mol^{-1}	Gibbs energy of formation, kJ mol^{-1}	Entropy, J mol^{-1} K^{-1}	Heat capacity, J mol^{-1} K^{-1}
solid	0.0		23.8	23.4
liquid				
gas	396.6	351.8	174.5	20.8

3.1. Thermodynamic Properties. The thermodynamic properties of chromium are shown in Table 6. At various temperatures the heat capacities are (in J mol^{-1} K^{-1}): at 200 K, 19.86; at 250 K, 22.30; at 300 K, 23.47; at 350 K, 24.39; at 400 K, 25.23; at 500 K, 26.63; at 600 K, 27.72.

3.2. Other Properties. The following is a list of other properties of chromium.

- Thermal conductivity (in watts cm^{-1} K^{-1}) at selected temperatures is 0.402 at 1 K, 0. 3.85 at 10 K, at 100 K, 1.11 at 200 K, 0.937 at 300 K, 0.860 at 500 K, 0.654 at 1000 K, 0.556 at 1600 K, and 0.494 at 2000 K.
- Magnetic Susceptibility at 273 K is 180×10^{-6} cgs, at 1713 K is 224×10^{-6} cgs.
- Ionic radii (in nanometers) are as follows: Cr^{2+} (6 coordination) is 0.73; Cr^{3+} (6 coordination) is 0.62; Cr^{6+} (4 coordination) is 0.260.
- The elastic constant of a single crystal of chromium are as follows (in units of 10^{11} Newtons M^{-2}): $C_{11}=3.398$; $C_{12}=0.586$; $C_{44}=0.990$.
- Electrical resistivity at various temperatures (in 10^{-8} ohm m): 100 K = 1.6; 200 K = 7.7; 273 K = 7.7; 298 K = 12.6; 400 K = 15.8; 600 K = 24.7; 800 K = 34.6; 900 K = 39.9.

4. Mining and Processing

Exploration has nontechnologic aspects, including general and commodity-specific economic factors, and politics. Exploration requires investment, which, in turn, requires economic decisions. The first step in the exploration process is deposit discovery and description, followed by cross-correlation of deposit information. Deposit descriptions covering many important and less well-known deposits have been published and analyzed. Government and academic geologists have correlated geologic aspects of chromite deposits, defining the general geologic conditions that are consistent with known chromite deposits.

Both stratiform and podiform deposits are associated with ultramafic rocks even though the origins of these two types of chromite deposits differ. For stratiform deposits, the regular layering can be used to locate chromite deposits concealed by faulting or segmentation. Podiform deposits cannot be reliably inferred. So far, no consistently reliable geophysical or geochemical exploration technique has been found for podiform deposits. Without chromite-specific physical indicators, the traditional methods of ore body location, outcrop analysis, trenching, and drilling remain the most reliable way to locate chromite deposits. Drilling and drifting are used to locate or extend underground deposits. When an ore body has been located, structural analysis may be used to locate deposit extensions if they exist.

4.1. Beneficiation.   A wide variety of mining technology is applied to the surface and subsurface mining of chromite ore. Most ore comes from large mechanized mines. However, small labor-intensive mining operations contribute to world supply. Recovery includes surface and underground mining using unmechanized to mechanized methods.

Beneficiation to marketable chromite products varies from hand sorting to gravimetric and electromagnetic separation methods. The amount of beneficiation required and the techniques used depend on the ore source and end-use requirements. When the chromite is clean and massive, only hand sorting of coarse material and gravity separation of fine material may be required. When the ore is lumpy and mixed with other minerals, heavy-media separation may be used. When the chromite mineral occurs in fine grains intermixed with other minerals, crushing may be used in conjunction with gravity separation and magnetic separation. Processing of chromite to produce chromium products for the refractory, chemical, and metallurgical markets includes crushing and grinding and size sorting by pneumatic and hydraulic methods, kiln roasting, and electric furnace smelting (see Fig. 2). Labeling of material as it moves from the earth to the consumer is not uniform. The terms *chromite* and *chromite ore* are used here to refer to material in the ground, run-of-mine ore (ie, material removed from the ground), or material supplied to the marketplace. For the purpose of trade, imports are called *chromite ore* and *concentrate made therefrom*. This description is frequently abbreviated to chromite ore and concentrate, chromite ore, or simply chromite. Some sources use chromite ore to refer to material in the ground

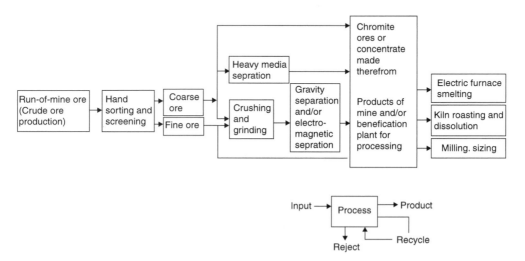

Fig. 2.   Chromite material flow process from the mine to conversion to industrial products. Not shown are reject and recycle fractions that are associated with most processes as shown in box at lower right of the figure.

and material removed from the ground before processing. The term *chromite products* is then used to refer to material supplied to the marketplace. Historically, mining operations supplied minimally processed material. Beneficiation and processing shown in Figure 2 may be carried out at the mine site, at a plant that serves several mines in one geographic area, or at a plant associated with end users. This variety in processing further complicates labeling of material. Today, quality control leads consumers to seek chromite supplies that do not vary significantly in physical or chemical properties over time. As a result, chromite ore is typically beneficiated to produce a physically and chemically uniform product before it reaches the marketplace.

Mining methods are carefully chosen to meet the characteristics of a deposit, including the ore and its environment. Since both small and large, podiform and stratiform, high-grade and low-grade, subsurface and near surface, massive and disseminated chromite deposits are exploited, a variety of mining methods are used. Since, typically, surface mining is less expensive than underground mining and ore bodies are found by their outcrops, surface mining at an outcrop precedes underground mining.

The purpose of beneficiation is to increase desirable ore attributes and decrease undesirable ones. For example, depending on end use, increasing chromic oxide content, chromium:iron ratio, or alumina content is desirable. Reducing silica or other minerals associated with chromite is desirable. Depending on end use, certain sizes may be selected or rejected. The techniques used to accomplish these tasks depend on the physical properties and sizes of the minerals present. Beneficiation does not change the chemical characteristics of the chromite mineral. However, since chromite ore is a mixture of minerals, the characteristics of the ore can be changed by altering its mineral mix. A deposit producing lumpy ore in which the chromite is easily distinguished may require only hand sorting and screening. When the chromite cannot easily be distinguished visually from associated minerals and the ore is lumpy, then heavy-media separation can be used. A deposit that yields an ore of chromite thoroughly intermixed with other minerals, however, may require milling and sizing followed by gravimetric and/or electromagnetic separation methods to produce marketable chromite products.

Beneficiation may also be selected to process tailings once sufficient quantities have been stockpiled and the technology of beneficiation and processing has been established.

4.2. Ferrochromium. The smelting of chromite ore to produce ferrochromium requires electric-arc furnace technology. Early electric furnaces having power ratings in the kilovolt-ampere range have developed into modern furnaces having power ratings of about 50 MV·A. Closed and partially closed electric-arc furnaces replaced open furnaces in the 1970s to improve pollution control, efficient furnace operation, and safety.

Ferrochromium is produced from chromite ore by smelting a mixture of the ore, flux materials (eg, quartz, dolomite, limestone, and aluminosilicates), and a carbonaceous reductant (wood, coke, or charcoal) in an electric-arc furnace. If the ore is lumpy, it can be fed directly into the furnace. However, if the ore is not lumpy, it must be agglomerated before it is fed into the furnace. Efficient operations recover chromium lost to furnace fume by collecting and remelting the dust

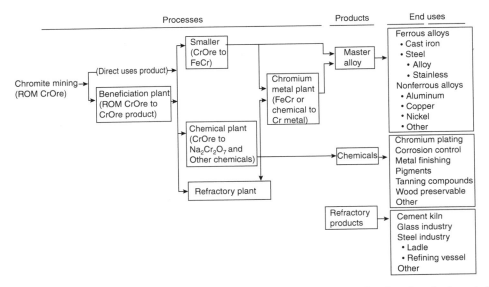

Fig. 3. Chromium material flow from mining through processing by the chemical, metallurgical, and refractory industries, to primary industrial products and the end uses in which those primary industrial products are used. CrOre = chromite ore; FeCr = ferrochromium; $Na_2Cr_2O_7$ = sodium dichromate and ROM = run-of-mine. Mining includes screening and hand sorting. Benefication includes crushing, grinding, and separation techniques including gravimetric, heavy media, magnetic, and spiral. Master alloy is an alloy used as a feed stock to produce other alloys.

and recover chromium lost to slag by crushing and beneficiating the slag. The chromium content of the ferrochromium is determined by the chromium:iron ratio of the chromite ore.

The shift from high-chromium, low-carbon ferrochromium to low-chromium, high-carbon ferrochromium, commonly called *charge-grade ferrochromium*, permitted the use of low-chromium:iron ore for smelting to ferrochromium. The Republic of South Africa is the most abundant and low-cost source of such ore. Unfortunately, this ore is friable (breaks easily into small pieces), and the finer fractions of such ore are blown out of a furnace before it can be smelted. Agglomeration technology has been developed to permit the use of fine chromite ore in the electric arc furnace. Both briquetting and pelletizing are practiced. Efficient production technology uses prereduced and preheated pelletized furnace feed. Industry is developing new production technologies using high-temperature plasmas or using alternatives to electrical power supply. The new production technologies are expected to be more cost-competitive than traditional production technology under some conditions. Advanced smelting technologies that use abundant friable ore have been and are being developed. Plasma processes, including both transferred and nontransferred arc processes, have been applied to ferrochromium production. The kiln roasting prereduction process is being applied to ferrochromium production.

4.3. Chromium Metal Production. Chromium metal is produced primarily through one of two production processes: electrodeposition process to produce electrolytic chromium metal and the reduction of chromic oxide with aluminum powder to produce aluminothermic chromium metal. The aluminothermic reduction process is more widely used and more easily installed or expanded. A wide variety of variations of reductants for the exothermic reduction process and of feed materials and electrolytes for the electrowinning process resulted in the current commercial production processes: aluminothermic reduction of chromic oxide to produce aluminothermic chromium metal and electrolytic deposition from a chromium–alum electrolyte made from high-carbon ferrochromium to produce electrolytic chromium metal. The aluminothermic process was used first. The electrolytic process was developed to provide a higher-purity chromium metal than could be obtained by the aluminothermic process. The processes from chromite ore mining to chromium metal product are shown in Figure 3.

Commercial grades of chromium metal have been produced in the United States in bulk quantities by Eramet Marietta, Inc. using the electrolytic process and by Shieldalloy Corp. using the aluminothermic process. Shieldalloy suspended production in 1990, leaving Eramet as the sole U.S. producer.

5. Sources and Supply

5.1. Availability. The U.S. Geological Survey conducted an inventory of chromium resources (9). They found that, by far, the world's major resources are centered in the Bushveld Complex deposit in Republic of South Africa and in the Great Dyke deposit in Zimbabwe. Other significant deposits were identified in the Cuttack district of Orissa State in India and in the Kempirsai district of Kazakstan.

The former U.S. Bureau of Mines (USBM) studied the availability of chromium (10). The USBM analyzed for the simultaneous availability of chromium contained in chromium ferroalloy products and in exportable chromite products (metallurgical, chemical, refractory, and foundry sands) in 10 market economy countries (MECs).

A total of about 874 million tons gross weight of *in situ* material containing about 203 million tons of chromium was analyzed. Extraction and beneficiation of this material was estimated to result in about 475 million tons of chromite products, of which 289 million tons would be available for export, and the remaining 187 million tons would be smelted in the country in which it was mined to produce about 80 million tons of chromium ferroalloys. The chromium ferroalloys would then be available for use in the country of production or for export. The 80 million tons of chromium ferroalloy included about 74 million tons of high-carbon ferrochromium, 4 million tons of low-carbon ferrochromium, and 2 million tons of ferrochromium–silicon.

The countries of South Africa and Zimbabwe held about 80% of the *in situ* contained chromium. India and Finland accounted for an additional 11% of the contained chromium; another 8% of the contained chromium was fairly evenly divided among Brazil, the Philippines, Turkey, and the United States; the remainder was in Greece and Madagascar.

Table 7. **Availability Based on Cost of Production of Chromite and Chromium Ferroalloys from 10 Market Economy Countries**[a]

Chromium materials	Quantity available, $\times 10^6$ tons, gross weight	Cost,[b] $/t	
		Weighted average	Range
chromite			
chemical grade	64.3	53	35–174
foundry sand grade	16.4	49	39–83
refractory grade	26.8	87	54–180
metallurgical grade			
primary product	145.4	101	42–705
secondary product	35.6	54	33–117
Subtotal	*181.0*	*92*	*33–705*
refractory grade	26.8	87	54–180
Total	*288.5*		
chromium ferroalloys			
ferrochromium			
high-carbon ferrochromium	74.3	473	417–1,286
low-carbon ferrochromium	3.9	937	635–1,309
ferrochromium-silicon	2.0	737	578–814
Total	*80.2*		

[a] Ref. 10.
[b] Cost of production for 0% discounted cash flow rate of return in Jan. 1989 dollars per metric ton, gross weight, of product.

Based on Cr_2O_3 content of *in situ* chromite ore, the 10 MECs split into two groups: a high-grade group and a low-grade group. The *high-grade group*, those with ore grades ranging from 33.96 to 43.01% Cr_2O_3, included the countries of India, Madagascar, South Africa, Turkey, and Zimbabwe. The *low-grade group*, those with ore grades ranging from 9.16 to 26.65% Cr_2O_3, included the countries Brazil, Finland, Greece, the Philippines, and the United States. The grade differences between the two groups resulted in a wide disparity in the respective weighted average Cr_2O_3 contents. The high-grade group averaged 38.76% Cr_2O_3; the low-grade group, only 15.90% Cr_2O_3.

Table 7 shows the results of the USBM analysis. Chromium material costs were calculated on a weighted average basis, free-on-board (fob) ship at the port of export. The chromite cost shown in Table 7 includes mining and beneficiating the ore (including mine capital and operating costs and taxes) and transportation of ore and products to port facilities. Chromium ferroalloy cost includes chromite ore but excludes smelter capital cost. The USBM analysis shows that, measured on the basis weight of production from South Africa and Zimbabwe to weight of production from all sources at the respective sources break even cost as measured in their study, South Africa and Zimbabwe could produce about 78% of metallurgical chromite ore, 93% of chemical chromite ore, 85% of refractory chromite ore, and 93% of foundry chromite ore. South Africa and Zimbabwe could also produce about 69% of the high-carbon ferrochromium, 89% of the low-carbon ferrochromium, and 100% of ferrochromium–silicon that could be produced at their respective estimated breakeven cost. The product breakdown between chromium ferroalloy and chromite and among the grades within those product categories was based on mine and smelter production capacities and known operating relationships circa the 1987/88 time period.

5.2. Strategic Considerations.   There is no production of chromite ore in the United States; primary consumption of chromium by U.S. industry is by companies that use chromite ore to produce chromium chemicals and chromite refractories and by chromium metal producers that use ferrochromium. World reserves of chromite ore are abundant, ensuring adequate long-term supply. However, major supply sources are few and remote from the United States, making supply vulnerable to disruption. The problem for the United States is one of national security. Ferrochromium is essential to production of stainless and some alloy steel, which are, in turn, essential to both the domestic economy and to the production of military hardware.

It has been the policy of the Federal Government of the United States to maintain a National Defense Stockpile of critical and strategic materials for use in the event of a national defense emergency. The U.S. government has maintained a stockpile since World War I. Industrial mobilizations resulting from World War I (1914–1918), World War II (1939–1945), and the Korean War (1950–1952), along with politically motivated peacetime supply embargoes of the former Union of Soviet Socialist Republics against the United States as a result of the Berlin crisis (1949–1950) and of the United States against Rhodesia as a result of United Nations actions (1966–1972), caused national defense planners to acquire and maintain a stockpile. The United States also implemented trade sanctions against South Africa (1986–1994); however, chromium materials were exempt for those sanctions.

Critical and strategic materials were stockpiled. *Critical materials* are essential in a national security emergency because of their important end uses. *Strategic materials* are potentially in short supply during a national emergency. The Defense Logistics Agency, the manager of the National Defense Stockpile, defines "strategic and critical materials" as materials that (*1*) would be needed to supply the military, industrial, and essential civilian needs of the United States during a national emergency and (*2*) are not found or produced in the United States in sufficient quantities to meet such need (11). Critical and strategic materials for the purpose of inclusion in the National Defense Stockpile are materials Congress directs the administration to include in the stockpile by act of Congress. So, pragmatically, critical and strategic materials are those defined to be so by Congress.

The Defense Logistics Agency, Department of Defense, is currently responsible for National Defense Stockpile operations (12). Chromium materials included in the National Defense Stockpile (NDS) are chromite ore (metallurgical, chemical, and refractory grades), chromium ferroalloys (high- and low-carbon ferrochromium and ferrochromium-silicon), and chromium metal. The purpose of the NDS is to supply military, essential civilian, and basic industrial needs of the United States during a national defense emergency, and by law the stockpile cannot be used for economic or budgetary purposes.

Changes in industrial capacity and new manufacturing and technological developments have rendered selected chromium materials in the NDS inventory obsolete, either in quality or form or both, and in need of upgrading. Subsequent

Table 8. **U.S. Government Stockpile Year-End Inventories, (metric tons, gross weight)**[a]

Material	1995	1996	1997
chromite			
chemical	219,914	219,914	217,110
metallurgical	772,587	644,957	564,799
refractory	328,107	321,966	309,406
chromium ferroalloys			
ferrochromium-silicon	52,941	52,687	52,688
high-carbon ferrochromium	737,694	717,627	689,226
low-carbon ferrochromium	282,735	282,735	282,735
chromium metal			
aluminothermic	2,667	2,667	2,667
electrolytic	5,018	5,054	5,054

[a]Inventories includes specification- and nonspecification-grade materials.

to legislative mandate, DLA began modernizing chromium materials in the NDS by converting chromite ore to high-carbon ferrochromium (1984–1994) and nonspecification-grade low-carbon ferrochromium into chromium metal (1989–1994).

As result of the dissolution of the Soviet Union in 1991, NDS planners have reduced material goals and implemented inventory reduction programs. Material disposal from the NDS takes the form both of direct sales and material used in payment for service. Table 8 shows National Defense Stockpile inventory levels (for 1995–1997).

In addition to private and government stocks, there exists a large unreported inventory of chromium contained in products, trader stocks, and scrap. The amount of these stocks varies with demand and material price. Under price pressures resulting from primary chromium shortages, recycling of consumer materials could add to the supply.

5.3. Prices. Chromium materials are not traded in open-market exchanges such as gold, silver, nickel, and some other metals. As a result, chromite ore, chromium ferroalloys, and chromium metal do not have an easily identifiable price. The price of these chromium materials is usually negotiated between buyer and seller and is known only to them. Price speculation is, of course, a very popular activity because of the great impact of prices on both producers and consumers. As a service to their readers, some periodicals report a composite price based on surveys of sellers and buyers. Included among these are American Metal Market, Industrial Minerals, Metal Bulletin, Metals Price Report, Platt's Metals Week, and Ryan's Notes. Unfortunately, the volume of trade at the reported price is unknown.

When material is imported into the United States, its value at the port of export is declared for the purpose of tax collection. This is called the free on board (fob) value. Using this value, a value history for chromium materials by import category was constructed. This value history averages reported import values over sources of supply weighted by quantity of material supplied. Using reported prices, an annual average price for chromium materials has been generated using sources that report prices in the United States. Since reported prices are

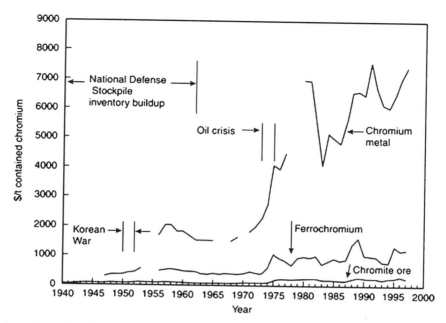

Fig. 4. Composite chromite ore, ferrochromium, and chromium metal value. The value of chromium metal exceeds that of ferrochromium by a factor of 5 and that of chromite ore by a factor of 25.

source sensitive, an annual average including all sources was calculated. Chromite ore price reported in dollars per metric ton, gross weight, can vary by nearly a factor of 2 depending on the origin and quality of the material. Ferrochromium price reported in dollars per metric ton of contained chromium shows similar variation based on material grade. Figure 4 shows the value history of chromium materials in units of dollars per metric ton of contained chromium. The value of chromite was relatively stable through about 1970, when the value started to rise. The value of chromite declined from 1982 to 1988, the time period during which steel production in general and stainless steel production in particular was weak. Strong stainless steel recovery in the 1988–1990 time period resulted in short supply of ferrochromium. The shortage nearly doubled the price of ferrochromium and stimulated capacity expansion in that industry, primarily in South Africa. The price of chromite ore rose following that of ferrochromium. Additional chromite production capacity was added to meet the anticipated additional demand from added ferrochromium production capacity, primarily in South Africa and India. When adjusted for inflation, each material is found to be less expensive today than it has been in the past. Price peaks for ferrochromium correspond to increases in world stainless steel production. The increases resulted in demand for ferrochromium in excess of material available from active production capacity and stocks.

Ferrochromium values show greater variation than those of chromite ore. Before the mid-1970s, the value of various ferrochromium grades were tightly grouped compared to the post mid-1970s time period, when the value of low-carbon ferrochromium was about double that of high-carbon ferrochromium except for a couple of years when ferrochromium shortages drove the prices together. This value differentiation shows the advantage of post-melting refining technology—it permits the use of lower cost materials. The figure also shows that reported price generally exceeded fob import value.

Figure 4 shows the value relationship among chromium materials. These values show that as chromite ore is processed to ferrochromium and to chromium metal, the added value is quite large. On a per unit scale of contained chromium basis for recent years, the value of ferrochromium is about five times that of chromite ore; the value of chromium metal is about 30 times that of chromite ore. Variations of the value of ore are shown to follow those of ferrochromium, indicating values of chromite ore change in response to demand, with ferrochromium value first to reflect demand changes. (Because chromite ore price changes may lag those of ferrochromium by only a few weeks, the tables may show peak average annual values occurring in the same year.)

5.4. Trade. The United States is 100% import dependent for chromite ore. Chromium import dependence is lessened by the supply of some chromium through recycling. The United States imports chromite ore, chromium ferroalloys, chromium chemicals, and chromium metal. Chromium ferroalloys, metal, and chemicals and chromite-containing refractories are manufactured in the United States. These materials are also exported from the United States, but in quantities smaller than those imported (except for chemicals). The United States is a major world chromium chemical producer.

The harmonized tariff schedule categories distinguish chromium-containing materials from chromium-free materials well except for chromite-containing refractories, which are included with chromite-free materials. The change from

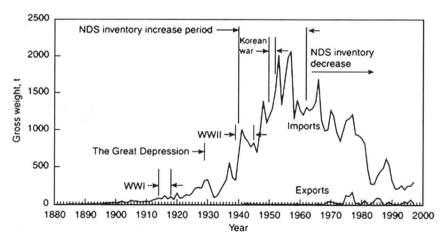

Fig. 5. U.S. chromite ore trade. U.S. chromite ore imports have exceeded exports by a large amount since such trade has been reported. At the end of World War II, U.S. chromite ore imports represented most of world chromite ore production. Part of the imports between 1939 and 1962 went into the National Defense stockpile, which reached its peak chromite ore inventory of about 8 million tons in 1962. Between 1945 and 1962, the United States imported nearly 24 million tons of chromite ore. The post-1965 decline in U.S. chromite ore imports results from declining chromite ore use in the metallurgical and refractory industries.

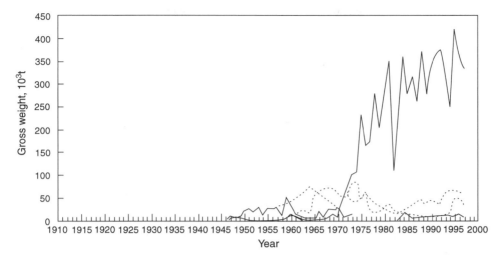

Fig. 6. U.S. chromium ferroalloy trade. Since the mid-1970s, high-carbon ferrochromium imports have dominated chromium trade. The introduction of postmelting refining processes in the steel industry after 1960 followed by rationalization of most of domestic ferrochromium production industry and strong growth in the ferrochromium consumption industry resulted in increased imports of high-carbon ferrochromium. ▬▬, High-carbon ferrochromium imports; − −, low-carbon ferrochromium imports; - - - -, ferrochromiumsilcon imports; and ▬▬, exports, composite of various grades.

the Tariff Schedule of the United States to the Harmonized Tariff Schedule of the United States resulted in many category changes. As a result, comparison of statistics across the 1988/89 boundary may result in the comparison of inconsistent materials.

Figure 5 shows that Chromite ore imports have always greatly exceeded exports. Imports grew rapidly after about 1939, when the NDS-enabling legislation was passed, and peaked in the 1950s, at about the time that the NDS reached its peak chromite ore inventory. Figure 6 shows that chromium ferroalloy imports greatly exceed exports and that high-carbon ferrochromium imports have dominated chromium ferroalloy imports since the mid-1970s, when postmelting refining technology permitted high-carbon ferrochromium to replace low-carbon ferrochromium in the production of steel. The figure shows that high-carbon ferrochromium imports have been growing since the mid-1970s. Figure 7 shows that chromium metal imports greatly exceed exports and that both imports and exports are growing. As a source of chromium to U.S. industry, chromite ore and ferrochromium imports dominate. Chromite ore was the predominant source of chromium until the 1980s, when high-carbon ferrochromium displaced chromite ore.

6. Manufacturing and Production

Chromite ore mining and chromium material manufacturing is an international industry. The major industries associated with chromium are chemical,

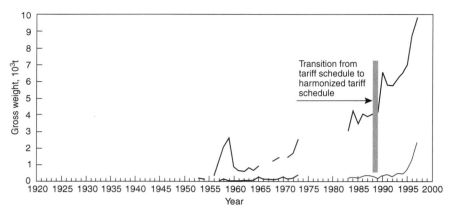

Fig. 7. U.S. chromium metal trade. Chromium metal trade is dominated by exports. The United States is both a major world chromium metal producer and consumer. ▬▬, imports; ▬, exports.

metallurgical, mining, and refractory. Mining is, of course, the first to process chromium in the form of chromite ore. The chemical industry processes chromite ore by kiln roasting to produce sodium dichromate initially and then other chromium chemicals. The metallurgical industry processes chromite ore mostly by electric-arc furnace smelting to produce ferrochromium. It also processes chromic oxide from the chemical industry and ferrochromium from the metallurgical industry into chromium metal. Ferrochromium and chromium metal are then incorporated into ferrous and nonferrous alloys. The refractory industry processes chromite ore into chromite-containing refractory materials. It also processes chromic oxide [1308-38-9] from the chemical industry into refractory materials (see also REFRACTORIES).

Production data are available for chromite ore, ferrochromium, and stainless steel by country because there are usually sufficient numbers of producers per country to maintain confidentiality of data about plants or companies. Chromium chemical and metal production and chromite-containing refractory producers are substantially fewer in number than are chromite ore, ferrochromium, and alloy producers, making it difficult to report those industries' products by country while maintaining confidentiality. Since nations are the largest grouping for which data are collected and published, production from these industries is simply not publicly available.

World chromite ore production is shown in Figure 8. The figure shows that world production has been dominated by South Africa and Kazakstan (reported as USSR before 1991) with a large number of smaller producers grouped close together. About 15 countries make up the "Other" category. Production from the most recent years indicate that South Africa, Turkey, and India are developing their chromite production potential. The decline in production from Kazakstan was substantial.

Figure 8 shows that there have been two major (i.e., >1 million tons per year) chromite ore producers over the time period shown: South Africa and Kazakhstan. (Kazakstani production was the larger share of USSR production,

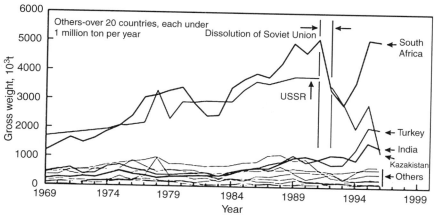

Fig. 8. World chromite ore production by country. Over most of the time period shown, South Africa and the former Soviet Union, most of which was from Kazakhstan, dominated world production. Since the dissolution of the Soviet Union in 1991, production from Kazakhstan has been displaced by increased production from India and Turkey. In any given year there are over 20 producers of chromite ore, most of which produce under one million tons per year. An immediate effect of the dissolution of the former Soviet Union was to stimulate exports of chromite ore to Western consumers traditionally supplied from Western sources. The impact was a major reduction in production by South Africa, the traditional supplier of chromium to Western consumers.

1969–1991). The figure also shows that when former Soviet markets merged with Western markets after 1991, it was South African production that declined to accommodate new chromite ore supply. When the Eastern and Western markets merged, production and capacity in the East was unchanged while Eastern demand declined rapidly. Political change in Kazakhstan appears to have negatively affected chromite ore production because, except for a one-year recovery in 1995, Kazakhstani production has declined since 1991. South Africa, on the other hand, appears to have benefited from political change over the same time period because South African, as well as Turkish and Indian production has increased. Both India and Turkey have also experienced changes in their national political paradigms in the same time period; but, with lesser impact on their national economies than have been experienced in Kazakhstan and South Africa. Increasing demand for chromium and declining production in Kazakhstan has resulted in both India and Turkey joining the major producer category in 1989 and 1994, respectively. On average over the 5-yr time period 1992–1996, the major producing countries accounted for about 80% of production.

World ferrochromium production is shown in Figure 9. The figure shows that, over most of the time period, there have been two major producers (>500,000 t/yr). However, since 1991, South Africa has stood alone as the world's largest ferrochromium producer with production over double that of the next-largest producer. Other moderate producers (200,000 to 500,000 t/yr) include China, Finland, Japan, Kazakhstan, and Zimbabwe, of which China and Japan do not have domestic ore supplies to support their ferrochromium production. On average during 1992–1996, South Africa accounted for about 30% of world

Fig. 9. World production of ferrochromium by country. World production of ferrochromium has been dominated by South Africa and the former Soviet Union, most of which came from Kazakhstan, over the time period shown. Since the dissolution of the Soviet Union in 1991, production from Kazakhstan has been displaced by increased production from China and India. In any given year there are over 20 producers of ferrochromium most of which produce under about 200,000 t/yr. An immediate effect of the dissolution of the former Soviet Union was to stimulate exports of ferrochromium to Western consumers traditionally supplied from Western sources. The impact was a major reduction in production by South Africa, the traditional supplier of chromium to Western consumers. Others includes over 15 producers each under 200,000 t/yr.

production while the moderate producers accounted for about 55%. Compared to the chromite ore industry, moderate-size producers account for a grater share of production in the ferrochromium industry.

Chromium chemical production is geographically concentrated in developed economy countries. Major producing countries where large plants (capacity in excess of 100,000 tons per year of sodium dichromate) operate include Kazakhstan, Russia, the United Kingdom, and the United States. Moderate-sized production facilities are located in Brazil, China, Japan, Romania, South Africa, and Turkey. Small-scale local producers operate in China and India.

6.1. Chromium and Chromite. Chromite is used in the metallurgical, chemical, and refractory industries. In the metallurgical industry, chromite is processed into ferrochromium or chromium metal, then is used as an alloying metal to make a variety of ferrous and nonferrous alloys. The major end use is in stainless steel, a ferrous alloy made resistant to oxidation and corrosion by the addition of chromium. Chromite is used in the chemical industry to make sodium dichromate which is both a chemical industry product and an intermediate product used to make other chromium chemicals. Chromium chemicals find a wide variety of end uses, including pigments, and plating and surface finishing chemicals. Chromite is used in the refractory industry to produce refractory materials, including shapes, plastics, and foundry sands. These refractory materials are then used in the production of ferrous and nonferrous alloys, glass, and cement. Chromite is useful in the refractory industry because it retains its physical properties at high temperatures and is chemically inert.

6.2. Chromite Consumption. Reported chromite consumption in the United States over the 5-year period from 1993 to 1997 averaged about

328,000 tons annually, indicating a decline from the 1970s, when annual production regularly exceeded 1 million tons annually. Virtually all of this chromite was imported. The chromite was used to make chromium ferroalloys and chemicals, and chromite refractory materials, including casting sand. The major reason for declining domestic chromite use is the shift from domestic to foreign ferrochromium supply as the source of chromium units for the metallurgical industry. Contributing to reduced chromite ore consumption is the decline in chromite-containing refractory use.

6.3. Metallurgical.   The metallurgical industry consumed chromite ore to produce chromium ferroalloys and metal.

6.4. Refractory.   Refractory materials resist degradation when exposed to heat. Chromite is a refractory material. Unlike the chemical and metallurgical industries, where chromite is processed to extract its chromium content, chromite is used chemically unmodified in the refractory industry. Chromic oxide, a chemical industry product, is also used to make refractories for the glass industry. Chromic oxide refractories are used in glass contact areas of glass melting furnaces to achieve long furnace life.

Refractories are broadly categorized according to their material composition into clay and nonclay refractories. The predominant nonclay refractory material is silica. Nonclay refractory materials also include alumina (bauxite), carbon (graphite), chromite, dolomite, forsterite, magnesia (magnesite), mullite, pyrophylite, silicon carbide, and zirconia (zircon). Basic refractories are a type of nonclay refractory, so called because they behave chemically as bases. Basic refractories are made of chromite, dolomite, magnesite, or various combinations of magnesite and chromite. In the refractory industry, chromite-containing refractories are called chrome refractories. Chrome–magnesite refractories are those in which more chromite than magnesite is used. Magnesite–chrome refractories are those in which more magnesite than chromite is used. The terms chrome and magnesite are used in association with refractories to indicate that the refractory was made with chromite ore and magnesia.

Refractories are further categorized by the form in which they are supplied as shaped or unshaped. *Shaped refractories* are manufactured to fit together to form a desired geometric structure, like building blocks. *Unshaped refractories* include mortars (materials used to hold shaped refractories together), plastics (materials that may be formed into whatever shape is desired), and gunning (material that may be sprayed onto a surface). In the refractory industry, the term *monolithics* is commonly used to describe refractories that are not shaped. The units used to report shipments of shaped and unshaped refractories differ. Shaped refractories have been reported in thousand 9-in. (22.86-cm) brick equivalents; unshaped refractories, in tons. A 9-in. (22.86-cm) brick equivalent is a solid volume of 0.165919 m^3, used in the United States as a "standard unit" for refractory bricks.

Chromite-containing refractory producers are shown in Table 9 along with refractory industry products and the end users and uses of those products. The major end users for chromite refractories are in the cement, copper, glass, nickel, and steel industries. Basic refractories are used in copper and nickel furnaces. In the glass industry, chromite refractories are used in glass tank regenerators and chromic oxide refractories are used in melting furnaces for the production of

Table 9. U.S. Chromite and Chromic Oxide Refractory Producers, Products, and End-Use Markets

Producers in 1998	Products	End users and uses

General Refractories Co.
U.S. Refractories Division
600 Grant Street Room 3000
Pittsburgh, PA 15219
	bricks and shapes
	mortar
	plastic
	gunning
		steel industry
		  AOD vessels
		    barrel
		    trunnion and tuyere area
		    bottom
		  electric-arc furnaces
		    sidewall

Harbison-Walker
  Refractories
a subsidiary of Global
  Industrial Technologies
One Gateway Center
Pittsburgh, PA 15222
		slagline
		
		  steel ladles
		
		  ladle metallurgical furnaces
		    slagline
		    sidewall and bottom
		  vacuum degassers
		    sidewall

National Refractories and
  Minerals Corp.
1825 Rutan Drive
Livermore, CA 94550
		snorkel
		  open-hearth furnace
		    backwalls
		    endwalls
		    lower walls

North American Refractories Co. Ltd.
500 Halle Building
1228 Euclid Avenue
Cleveland, OH 44115
		roofs
		    furnace ports
		copper–nickel industry
		  electric-arc furnaces
		    slagline
		    bottom
		  flash furnaces
		    roof and sidewall
		    bottom

Corhart Refractories
RR 6, Box 82
Buckhannon, WV 26201-8815
Phone: (304)-473-1239 (voice)
    (304)-473-1287 (fax)
		Pierce–Smith converters
		    tuyere areas
		    barrel and endwall
		  Top-blown rotary converters
		    upper cone
		    bottom and barrel
		  anode furnace
		    barrel and endwall
		  fire refining and secondary
		  furnaces
		    bottom
		    sidewall and endwall
		cement–lime industry
		  rotary kilns
		    burning zone
		    upper transition zone
		    lower transition zone
		glass industry
		  glass furnace regenerators
		    checker
		    wall
		    crown
		  fiberglass furnace
		    melting furnace

reinforcing glass fibers and textiles. In the cement industry, chromite refractories are used primarily in the transition zones of cement kilns. Basic refractories are typically used in open-hearth and electric-arc steelmaking furnaces.

Chromite refractories were used heavily in steel production using the open-hearth furnace method. Contemporary steelmaking processes that use the basic oxygen furnace or the electric-arc furnace use much less chromite-containing refractories. Whereas open-hearth furnaces used about 30 kg of refractories per ton of steel, the basic oxygen furnace uses about 1 kg/t and the electric-arc furnace uses in the range of 1–2 kg/t. As a result, the steel industry demand for basic refractories has declined dramatically as open-hearth furnace steelmaking has been phased out.

The general decline in refractory use results, at least in part, from the more cost-efficient use of refractories. Longer lasting refractories result in lower labor cost to change the refractories and higher production equipment availability because of less down time for relining. A specific reason for the decline in chromite-containing refractory use results from changes in steel industry production practice. The major end use for basic chrome refractories was in the production of steel in open-hearth furnaces. As steel production technology has shifted away from open-hearth furnace steel making, chromite refractory use has declined. Steel is no longer produced in open hearth furnaces in the United States.

6.5. Foundry Sand. Foundry sand use of chromite is a modern application. Sand is used to contain molten metal in a desired shape until the metal has solidified. Sand used in the foundry industry is washed, graded, and dried. Since silica sand is common and inexpensive, it is the most commonly used mineral. However, when physical or chemical conditions dictate, other sands are chosen, such as zircon, olivine, or chromite. Chromite foundry sand is used in the ferrous and copper casting industries.

Casting sands are defined by function and by processing. Mold and core sands are designed for the exterior and interior of a casting, respectively. Facing sand is used on the surface of a core or mold. Flour or paint may be applied to the facing sand. As indicated by its name, flour is finer in size than sand. Before casting, sand is naturally or chemically bonded. There are a variety of methods for bonding sand before casting. Chromite sand is compatible with the commonly used methods. After casting, foundry sand is reclaimed.

Chromite sand is compatible with steel castings. It is typically used as facing sand in heavy section (>4 t) casting and enjoys a technical advantage over silica sand in casting austenitic manganese steel. Chromite sand does not react with the manganese in the steel. Chromite and zircon, each having a higher melting point than silica, are chosen when casting temperatures exceed those acceptable for silica sand. U.S. foundry sand producers, products, and end-use industries are shown in Table 10. Chromite sand is also used in copper-base nonferrous casting.

Chromite sand casting was developed in South Africa, where chromite fines are readily available as an inexpensive grade of chromite associated with chemical, refractory, and metallurgical chromite production. After satisfactory results in South Africa in the late 1950s, use expanded in the 1960s to include the United Kingdom followed closely by the United States. Use of chromite sand was facilitated first by a shortage of zircon sand supply and then by the higher

Table 10. **U.S. Foundry Sand Producers, Products, and End Uses**

Producers	Products	End Uses
American Colloid Co.	chromite flour	architectural brick
1500 W. Shure Drive	chromite sand	brake shoes
Arlington Heights, IL 6004-1434		casting facing sand
		ceramic
American Minerals Inc.		colorant
901 East Eighty Avenue, Suite #200		glass
King of Prussia, PA 19406		mold coating

price of zircon sand. The foundry characteristics that make chromite sand desirable include good thermal stability, good chill properties, good moisture resistance, resistance to metal penetration, high degree of refraction, and imperviousness to chemical reaction. Its disadvantages, compared with the zircon sand it replaces, include higher thermal expansion, occasional presence of hydrous mineral impurities, and different bonding practice with some binders.

Chromite sand for U.S. foundry use was estimated to have increased from about 20,000 t in 1965 to about 36 000 t in 1971. Industry sources estimate U.S. chromite foundry sand use in 1989 and 1990 to have been about 40,000 t annually.

Reclamation is an integral part of the foundry industry. It includes mechanical, pneumatic, wet, and thermal processes, and combinations thereof. Using these processes, as much as 90% of chemically bonded foundry sand (average over all minerals used) can be reclaimed (13). Chromite sand is adaptable to these processes. After casting, chromite sand, typically used as facing sand, becomes mixed with the bulk sand (silica). Since chromite sand has a size distribution similar to that of silica sand, mechanical separation is not applicable. Hydraulic spiral separation and magnetic separation are effective at separating chromite sand from silica and zircon sand. Silica and zircon sands are nonmagnetic. Some chromite sand was found to degrade during use. However, degraded sand tends to adhere to the castings, so it does not become part of the reclaimed sand. Reclaimed chromite sand was found to be interchangeable with new chromite sand. The actual amount of chromite sand reclaimed, like the amount used, is unknown. However, Sontz estimated that about half of the foundry industries chromite demand could be met by reclaimed chromite sand (14).

7. Shipment

Chromite ore is typically transported by trackless truck or conveyor belt from the mine face to storage or processing facilities on the mine site. From there, it is transported by truck from the mine site to the local railhead. It is then transported by rail to ports or to smelters. Smelters that do not have associated loading and unloading facilities for ships transport their product by rail to ports. Following transport by ship to consumer countries, chromium materials are typically hauled by barge, truck, or rail to end users who have no loading and unloading facilities for ships.

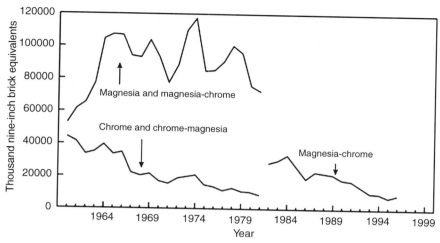

Fig. 10. U.S. chromium-containing shaped refractory shipments. Chromite ore is combined with magnesia to produce chrome–magnesia or magnesia–chrome refractories. The major end use for these refractories was in open-hearth steelmaking furnaces, which have been displaced by more efficient steelmaking methods in most of Western countries.

The historical trends of chromite-containing refractory shipments are shown in Figures 10 and 11. These figures show shipments of chromite-containing refractories since 1960. Basic refractory shipments trends shown in these figures indicate trends for chromite refractory shipments as well. Since the chrome and chrome–magnesite refractories are predominantly chromite, at least half of their content represents chromite consumed in refractories. The American Society for Testing and Materials sets specifications for the identification of chrome, chrome–magnesite, and magnesite–chrome brick. Chrome brick is identified as a refractory brick manufactured substantially or entirely of chrome ore. Chrome–magnesite and magnesite–chrome brick are classified by nominal and

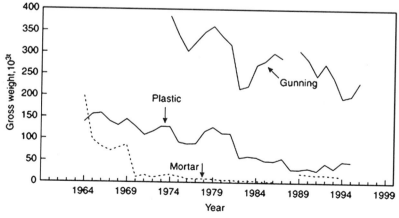

Fig. 11. U.S. chromite-containing unshaped refractory shipments.

minimum magnesia (MgO) content. Nominal MgO content ranges from 30 to 80%, so the chromite content of magnesite–chrome refractories is over 20%. However, the distribution between magnesite and magnesite–chrome within the magnesite and magnesite–chrome category is unknown.

Basic unshaped refractories may include chromite, dolomite, forsterite, magnesite, or zircon. Thus, these shipments data should be viewed as indicative of the performance of a segment of the refractory industry of which chromite is a part.

Figure 10 shows shipments of chromite-containing shaped refractories. The figure clearly shows a downward trend in chrome and chrome–magnesite refractory shipments from 1960 through 1981 when the data series was discontinued. Over the same time period the number of chrome and chrome–magnesite refractory producers fell from 9 to 7. Since 1981, the number of chromite refractory producers has fallen from 7 to 4, a trend indicative of declining production and use. Magnesite and magnesite–chrome refractories showed growth from 1960 through 1965, after which large variations in shipments occurred. The magnesite–chrome refractories shipments trend from 1982 through 1990 declined 28 to 18 million 9-in. (22.86-cm) brick equivalents. The chrome and chrome–magnesite curves are based on reported shipments. The magnesite and magnesite–chrome and the magnesite–chrome curves are composites of reported shipments data.

The trend in shipments of basic unshaped refractories is shown in Figure 11. Mortar shipments show a rapid decline from 1964 to 1970, followed by a very slow decline until 1986, when the series was discontinued. Plastic shipments show a slow decline, as does gunning. The data are not sufficiently discriminating to show trends among different materials used to make basic unshaped refractories.

In 1985, shapes started to be reported in tons, also. No specific conversion factor is applicable because the weight equivalent of the volume of shipments is dependent on the distribution of minerals used to make the bricks. From 1985 to 1990, the weight per volume of magnesite–chrome refractories averaged 4.7 kt per million 9-in. (22.86-cm) brick equivalents (kt/M9be), with a range of 4.3–5.0 kt/M9be.

8. Economic Aspects

8.1. Chromite Ore. Operations and transportation are the two major components of chromite ore cost in the market place. Operating cost includes mining (the production of run-of-mine ore) and beneficiation (the production of marketable chromite ore or concentrate from the run-of-mine ore). Mining cost is typically in the range of 70–90% of operating cost but exceeds 90% in some cases. Labor cost is the major component of mining and of beneficiation cost. Labor cost is typically in the range of 20–70% of mining cost and from 25–90% of beneficiation cost, but can be higher.

8.2. Ferrochromium. Excluding the delivered cost of chromite ore, electrical energy, other raw materials, and labor are the major components of smelting (ie, production of ferrochromium from chromite ore) cost. (Note that smelting cost excludes the cost of chromite ore feed material.) Electrical energy cost is in the range of 20–55% of smelting cost; raw materials (excluding chromite ore), 15–35%; and labor, 10–30%.

Ferrochromium production is electrical energy intensive. Charge-grade ferrochromium requires 3800–4100 kWh per ton of product, with efficiency varying with ore grade, operating conditions, and production process. Thus, ferrochromium plant location reflects a cost balance between raw materials and electrical energy supply.

8.3. Stainless Steel. Analysis of the stainless-steel industry based on historical performance and announced production capacity increases indicated that from 1987 to 1996, world annual stainless-steel production grew from 12 to 16 million tons, compound annual growth rate of 2.9%. Western stainless-steel production showed double-digit percentage growth in 1994 and 1995. Planned expansions in 1996 by nine countries (Brazil, China, India, Indonesia, Malaysia, the Republic of Korea, South Africa, Taiwan, and Thailand) were expected to add 4 million tons of crude stainless steel production capacity (3.66 million tons, rolled product) by 2000 (15).

Price for stainless steel is demand-sensitive, and an important part of it is the cost of nickel (about 70% of stainless-steel production requires nickel). Nickel availability and cost have been viewed as potential limitations to increased stainless-steel production. The discovery and development of new nickel deposits, projected to produce at nearly one-half the cost of that of currently exploited deposits, mitigate this potential limitation to stainless steel production growth.

8.4. Chromium Metal. Tosoh, the Japanese electrolytic chromium metal producer, ceased production in 1995, leading to an anticipated restructuring of the chromium metal industry. It was not until December 1996 that the company finally sold off its stocks. To a degree, Tosoh stocks have become consumer stocks. Restructuring of the chromium metal supply market started in earnest in 1997, with the remaining electrolytic producers (Russia and the United States) competing with the major aluminothermic producers (France and the United Kingdom) for the Japanese market. Both aluminothermic producers are in a position to expand production having brought new production capacity on line in 1996. The price of chromium metal was expected to increase as raw material (chromic oxide) price increases implemented in 1995 and 1996 are passed on to metal consumers. The price of low-grade chromium metal relative to ferrochromium in Japan during 1996 permitted stainless-steel producers to substitute chromium metal for ferrochromium. This substitution is expected to be curtailed as metal prices increase.

8.5. Chromite Foundry Sand. At last count, about 3100 foundries were active in the United States. These foundries tend to be small, independent operations. Chromite sand found a place in the casting industry in the 1960s when it substituted for zircon sand, which was in short supply. Since then, chromite sand has gained recognition as being technically suited to manganese steel and stainless steel casting because it produces a finish superior to that of zircon sand. Performance of the foundry industry is tied to that of the general economy, which has been strong and is expected to so continue. The automotive industry is a major demand sector for castings. Demand was good, stable, and expected to grow moderately. The use of nonmetallic materials could displace demand for metallic castings in the long term (16).

8.6. Tariffs and Taxes. Domestic producers were subject to a tax on chromium, potassium dichromate, and sodium dichromate under the Comprehensive Environmental Response, Compensation, and Liability Act of 1980 (also known as CERCLA or Superfund). The tax amounted to $4.91/t on chromium, $1.86/t on potassium dichromate, and $2.06/t on sodium dichromate. The tax expired in 1995.

Import tariffs are typically imposed to protect the domestic industry. Where there is no domestic industry, such as chromite ore production in the United States, there is no tariff. In some cases, such as ferrochromium imports to the European Community, import tariffs are used with import quotas; that is, a duty-free quota is allocated to member countries. The quotas may be revised as necessary to meet the needs of domestic consumers and producers. Only in a few cases, such as ferrochromium exports from China and certain grades of chromite ore exports from India, are export duties applied.

Chromium materials are categorized by the Harmonized Tariff System that was implemented in 1989 into the following broad areas: chromite ore, chromium ferroalloys, chromium metal, and chromium chemicals.

The trend to supply chromium in the form of ferrochromium by chromite mining countries is expected to continue. With new, efficient ferrochromium production facilities and excess capacity in chromite-producing countries, production and capacity are expected to diminish in traditional nonore-, but ferrochromium-producing countries. Production by small, less efficient producers, except where domestic industries are protected by quotas and tariffs is also expected to decline. Further upward integration of the chromium industry is expected as countries that produce chromite expand ferrochromium or stainless steel production capacity.

China has emerged as a potential major factor in the world chromium market. Because China produces only a minor amount of chromite ore, it is primarily a processor and consumer of chromium and supplies substantial quantities of ferrochromium and chromium metal to world markets. Continued industrial growth in China could result in increased demand for stainless steel there because its use is characteristic of the larger and more technologically developed economies.

For the same reasons as for China, India, too, has the potential to grow as a chromium consumer in the near future. Unlike China, India is a major chromite producing country with a vertically integrated chromium industry from chromite mining through stainless steel production; however, stainless steel production remains small.

8.7. Energy Requirements. Electric-arc furnace ferrochromium production is an electrical energy intensive process (requiring about 3500–3800 kWh per ton of ferrochromium produced) that produces similar volumes of ferrochromium and slag. However, heat recovery can reduce energy requirements. Energy-efficient processes using preheating can reduce the energy requirements to about 2500–2800 kWh ton of ferrochromium produced. Typically, all the energy required to smelt chromite is supplied in the form of electricity. Electrical energy requirements can be reduced by preheating or prereducing charge material using alternative energy sources such as coal, natural gas, or fuel oil or by recycling gases generated in the smelting furnace. Advanced production technology permits the use of liquid or gas fuel to substitute for part of the energy required. Alternative production technology is being developed that would permit nonelectrical energy sources to supply a significant fraction of the energy required to smelt chromite ore.

8.8. International Markets. In 1989, it was estimated that, on average internationally, the metallurgical industry used about 79% of chromium; the chemical industry, 13%; and the refractory industry, 8% (17). Of the chromium used in the metallurgical industry, about 60% was used in stainless steel. Thus, stainless-steel production accounted for about 50% of the chromium used internationally. In 1993, it was estimated that, on average internationally within market economy countries, the metallurgical industry used about 77% chromium; the chemical industry, 14%; and the refractory industry (including foundry sand), 9% (18). It was estimated in 1995 that 80% of chromite ore went into ferrochromium and 10% each into refractory and chemical use (19). The 80% of chromite that went into ferrochromium was estimated to have supplied 40–70% of the chromium units required by the steel (alloy plus stainless) industry. The remainder of the steel industries demand was satisfied by scrap. Of the chromium units going into the steel industry, it was estimated that 80% went into stainless steel and the remaining 20% went into alloy steel. A comparison of world production of chromite ore, ferrochromium, and stainless steel as reported in contained chromium showed that, on average, from 1992 through 1996, chromium contained in ferrochromium was about 79% of chromium contained in ore production and about 87% of that in stainless steel production. India reported its distribution of chromite ore consumption to have been: chemical, 5%; metallurgical, 88%; and refractory, 7% (20).

8.9. U.S. Markets. On average (from 1983 through 1992), U.S. chromium utilization, by end-use industry, has been as follows: metallurgical, 87%; chemical, 10%; and refractory, 3% (21). About 70% of metallurgical industry chromium use is as feed material for stainless steel production. Thus, stainless-steel production accounts for about 60% of the chromium used in the United States. The remainder of metallurgical industry use is for the production of other ferrous and nonferrous alloys. Some chemical and refractory products are used in steel production. The average chromium content of stainless steel produced in the United States from 1962 through 1983 was 17% (22). Stainless steel, by definition, contains at least 10.5% chromium but may contain as much as 36%

Table 11. World and Domestic Chromium Demand by End Use, %

	World			Domestic	
Industry	1989[a]	1990[b]	1992–1996[c]	1973–1982[d]	1983–1992[d]
chemical	13	14	—	12	10
metallurgical	79	77	79	79	87
refractory	8	9	—	9	3

[a] Ref. 17.
[b] Ref. 18.
[c] Ref. 23.
[d] Ref. 23.

chromium. For domestic demand for chromium by end use, see the two rightmost columns in Table 11 (17,18,21,23).

9. Grades, Specifications, and Quality Control

9.1. Government and Industry Organization Specifications. U.S. industry sets chemical and physical specifications for chromium materials through the American Society for Testing and Materials (ASTM). Other organizations also make specifications for chromium materials. The Defense Logistics Agency (DLA), in cooperation with the Department of Commerce, maintains purchase specifications for chromium materials contained in the National Defense Stockpile (NDS). The Treasury Department, in cooperation with the Department of Commerce and signatories to the General Agreement on Tariffs and Trade, maintains definitions of chromium materials for the purpose of recording trade and applying tariff duties. Chromium material specifications reported by ASTM are shown in Table 12.

For the purpose of trade, the U.S. government has categorized chromium materials. The import category "chromite ore and concentrates made therefrom" is subdivided by chromic oxide content as follows: containing not more than 40% chromic oxide, containing more than 40% and less than 46% chromic oxide, and containing 46% or more chromic oxide. Producers of chromite ore and concentrate typically specify chromic oxide content; chromium:iron ratio; and iron, silica, alumina, magnesia, and phosphorous contents. They also specify the size of the ore or concentrate. Typically, chromic oxide content ranges from 36 to 56%; values in the 40–50% range are most common. Chromium:iron ratios typically range from about 1.5:1 to about 4.0:1, with typical values of about 1.5:1–3.0:1. In trade, the chromite ore is also called chromium ore, chromite, chrome ore, and chrome.

Not more than 3% carbon was further subdivided in 2000 to more than 0.5% carbon but not more than 3% carbon and not more than 0.5% carbon.

The import category "chromium ferroalloys" is subdivided into ferrochromium and ferrochromium–silicon. Ferrochromium–silicon, also called *ferrosilicon–chromium* and *chromium silicide*, is not further classified. Ferrochromium is classified by its carbon content as containing not more than 3% carbon, more than 3% but not more than 4% carbon, or more than 4% carbon. Producers of ferrochromium typically classify their material as low- or high-carbon or charge-grade ferrochromium. Charge-grade ferrochromium is also called *charge chrome*. Producers of chromium ferroalloys typically specify chromium, carbon, silicon, phosphorous, and sulfur contents and material size. Ferrochromium–silicon typically contains 24–40% chromium, 38–50% silicon, and 0.05–0.1% carbon. Ferrochromium typically contains 50–75% chromium and 0.05–8% carbon. Low-carbon ferrochromium typically contains 55–75% chromium and 0.02–0.1% carbon. High-carbon ferrochromium typically contains 60–70% chromium and 6–8% carbon. Charge-grade ferrochromium typically contains 50–55% chromium and 6–8% carbon.

Table 12. Composition of Typical Chromium Ferroalloys and Chromium Metal (Composition in Percentage)[a]

Material[b]	Grade	Chromium[c]	Carbon[d]	Silicon[d]	Sulfur[d]	Phosphorus[d]	Nitrogen[d]
Ferrochromium							
High-carbon	A	51.0–56.0	6.0–8.0	6.0	0.040	0.030	0.050
	B	56.0–62.0	6.0–8.0	8.0–14.0	0.050	0.030	0.050
	C	62.0 max	6.0–8.0	3.0 max	0.050	0.030	0.050
low-carbon	A	60.0–67.0	0.025	1.0–8.0	0.025	0.030	0.12
	B	67.0–75.0	0.025	1.0	0.025	0.030	0.12
	C	67.0–75.0	0.050	1.0	0.025	0.030	0.12
	D	67.0–75.0	0.75	1.0	0.025	0.030	0.12
vacuum low-carbon	E	66.0–70.0	0.015	2.0	0.030	0.030	0.050
	G	63.0–68.0	0.050	2.0	0.030	0.030	5.0–6.0
nitrogen-bearing	—	62.0–70.0	0.050	1.0	0.025	0.030	1.0–5.0
ferrochromium–silicon	A	34.0–38.0	0.060	38.0–42.0	0.030	0.030	0.050
	B	38.0–42.0	0.050	41.0–45.0	0.030	0.030	0.050
chromium metal	A	99.0	0.050	0.15	0.030	0.010	0.050
	B	99.4	0.050	0.10	0.010	0.010	0.020

[a] 1996 Annual Book of ASTM Standards.
[b] In addition to the chemical specifications listed here, American Society for Testing Materials (ASTM) lists supplementary chemical requirements and standard sized and tolerances.
[c] Minimum, except where range of values indicating minimum and maximum appears or where noted otherwise.
[d] Maximum, except where range of values indicating minimum and maximum appears.

The Harmonized Tariff Schedule of the U.S. names and numbers for chromite ore and concentrate, chromium ferroalloys and metal, and chromium chemicals and pigments are shown in the "Economic Aspects" section of this article.

9.2. Commercial Specifications. Domestic and foreign companies supply imported chromite ore, chromium ferroalloys, chromium metal, chromium chemicals, and chromium containing refractories to U.S. consumers. Chemical specification of these materials varies among consumers and producers. Typically, consumers do not reveal detailed specification. However, producers do make typical specifications available to prospective customers. Typical chemical specifications for a variety of chromite ores, ferrochromium, and chromium metal products available to U.S. consumers have been assembled here. The chemical specifications of several chromite ores are shown in Table 13; (21) ferrochromium, in Table 14; chromium metal, in Table 15.

Domestically produced chromium chemicals and chromium-containing refractories are supplied to U.S. consumers. The chemical and physical specifications of these materials vary among consumers and producers. Consumers rarely reveal detailed purchase specifications. However, producers make typical specifications available to prospective customers. Typical chemical and physical specifications for a variety of chromium chemical and refractory products are available from U.S. producers and suppliers to the U.S. market. See the "Manufacturing, Production, and Shipments" section of this article for more details.

9.3. By-Products and Coproducts. Chromite ore is a by-product only of platinum mining of the UG-2 layer of the Bushveld complex. No coproducts or by-products are associated with chromite mining operations. Here, by-product or coproduct is assumed to mean a mineral product that is different from the primary product and not different grades of the primary mineral product. A single mining operation is likely to produce more than one grade of its product. Grades of chromite products are distinguished by ore size and chemistry.

Chromite recently became a by-product of platinum mining in South Africa. MINTEK, South Africa's mining research organization, has demonstrated smelting of beneficiated chromite-containing waste material from certain platinum mines. Platinum has been mined from the Merensky Reef, a chromite-free seam of the Bushveld Complex. As those platinum mines deplete their reserves, platinum mining is expected to move to the chromite-containing UG-2 seam, generating more chromite-containing tailings. UG-2 seam platinum mining has started. MINTEK has also demonstrated the feasibility of recovering platinum from tailings resulting from chromite mining of the LG-6 chromitite layer. The by-product chromite ore yields a ferrochromium of under 50% chromium content produced and used in South Africa for stainless-steel production.

10. Analytical Methods

Procedures for the analysis for total chromium of chromite ore and ferrochromium slags were developed at the Albany Research Center of the Department of Energy (formerly of the U.S. Bureau of Mines) as part of the study of domestic chromite ore deposits and the processing of that ore. They developed procedures to analyze for chromium content in the range of a fraction of a percent to 30% and

Table 13. **Chromite Ore Chemical Specifications**[a,b]

Company and Grade	Cr_2O_3	Fe_2O_3	FeO	SiO_2	Al_2O_3	MgO	MnO	TiO_2	T_2O_5	V_2O_5	CaO	P	S	Cr:Fe
Albchrome Ltd.														
40–42% Cr_2O_3	40–42	—	11.8	11.5	8	23.5	—	0.160	—	0.110	0.31	—	—	3
36–38% Cr_2O_3	36–38	—	12.5	15	7	23	—	0.160	—	0.020	0.50	—	—	2.7
30–34% Cr_2O_3	30–34	—	10.0	18	7	27	—	0.160	—	0.060	0.15	—	—	2.6
concentrates 48–50%	48–50	—	13.93	7	9.37	17.94	—	0.160	—	0.060	0.14	—	—	3
concentrates 45–47%	45–47	—	13.93	9	9.37	18.2	—	0.160	—	0.060	0.14	—	—	3
Advanced Mining Works Co. Ltd.														
metallurgical	50.1	—	10.26	7.35	6.24	17.96	—	—	—	—		0.003	0.001	3.341
Benguet Corporation-Masinloc Chromite Operation														
concentrates	32	—	11	5.5	27.5	18	—	—	—	—	0.45	—	—	1.9
refractory	32	—	11	5.5	27.5	18	—	—	—	—	0.45	—	—	1.8
foundry sand	31	—	14.19	4.4	27	16	—	—	—	—	0.70	—	—	1.9
Bayer (Pty.) Ltd.														
metallurgical	40.1	—	23.4	5.72	15.84	10.7	—	—	—	—	0.57	0.004	0.005	1.52
chemical	46.05	—	25.79	1.10	14.80	9.75	—	0.62	—	—	0.25	0.003	0.001	1.57
foundry sand	46.50	—	25.8	0.55	14.50	10.10	—	0.60	—	—	0.20	0.003	0.003	1.56
Bilfer Madencilik A.S. (Bilfer Mining Inc.)														
concentrates refractory	53	—	18	1.7	16	17	—	—	—	—	0.12	<0.007	—	>2.5
metallurgical	48	—	15	5	19	18	—	—	—	—	0.5	<0.007	—	>2.7
metallurgical high alumina	38–40	—	15	8–9	18	18	—	—	—	—	0.5	<0.007	—	>2.5
standard grade	34–42	—	14	12–15	8	21	—	—	—	—	0.6	<0.007	—	>2.8
refractory fines	36–40	—	15	6–9	18	17	—	—	—	—	0.5	<0.007	—	>2.5
lumps	38–41	—	—	4–6	—	—	—	—	—	—	—	—	—	—
foundry sand	52–54	—	19	1.2	15	16	—	—	—	—	0.05	—	—	—
Birlik Madencilik Dis Tic. Insaat San. ve Tic. A.S.														
metallurgical	32–46	—		8–13	6–10	16–24	—	—	—	—	—	0.005–0.008	0.005–0.007	2.2–3.0

503

Table 13 (*Continued*)

Company and Grade	Cr_2O_3	Fe_2O_3	FeO	SiO_2	Al_2O_3	MgO	MnO	TiO_2	T_2O_5	V_2O_5	CaO	P	S	Cr:Fe
chemical	40–48	—		8–10	7–10	18–21	—	—	—	—	—	0.005–0.008	0.005–0.007	2.4–2.8
refractory														
lump and fines	44–56	—		3–4	9–11	16–18	—	—	—	—	—	0.005–0.008	0.005–0.007	2.8–3.3
concentrate	48–50	—		5–7	7–10	16–18	—	—	—	—	—	0.005–0.008	0.005–0.007	2.4–2.6
chrome ore briquettes	42–48	—		6–7	7–10	16–19	—	—	—	—	—	0.005–0.008	0.005–0.007	2.4
Blue Nile Mines Co. Ltd.														
metallurgical	48–56	—	9–18	0.6–10	7–12	15–17	—	—	—	—	0.1–0.3	Traces	Traces	3:1
Chromecorp Holdings Ltd.														
metallurgical														
lumpy	38	—	24	8.4	15.5	11.6	—	—	—	—	1.9	0.002	0.005	1.5
fines	44	—	24.9	2.5	15.6	10.6	—	—	—	—	1.4	0.002	0.005	1.52
chemical	>46	—	25.7	<1	15.4	10.3	—	—	—	—	0.8	0.002	0.004	1.54
Consolidated Metallurgical Industries Ltd. (CMI)														
metallurgical	45.5	—	26	2	15	11	—	0.6	—	0.3	0.3	0.003	0.01	1.55
chemical	46.3	—	26	1	15	11	—	0.6	—	0.3	0.2	0.003	0.01	1.57
Dedeman Madencilik, Turizm, San. ve Tic. A.S.														
lumpy	38–40	—	10	9–10	10–11	21–22	—	—	—	—	0.3–0.4	0.002	0.009	2.6–2.8
fines	38–40	—	9–10	9–11	10–11	21–22	—	—	—	—	0.3–0.4	0.002	0.009	2.6–2.8
high-grade														
lump	46–48	—	13.0	6	12–13	16–18	—	—	—	—	0.2–0.4	0.002	0.007	2.9–3.1
fines	48–50	—	13–14	6–7	12–13	18	—	—	—	—	0.2–0.3	0.002	0.004	2.9–3.0
concentrate	48	—	13	6	11–12	17	—	—	—	—	0.2–0.3	0.002	0.004	2.9–3.0
refractory														
hard, lumpy	50–52	—	13–14	3	15–16	16	—	—	—	—	0.1–0.3	0.002	0.007	3.1–3.2
super concentrate	54–56	—	11–12	1–1.5	14–15	16	—	—	—	—	0.2–0.3	0.002	0.007	3.2
jig grade														
pebbles	46–48	—	12–13	6	12–13	18	—	—	—	—	0.2–0.3	0.002	0.004	2.9–3.0

concentrate	48–50	—	12–13	6	12–13	18	—	—	—	0.2–0.3	0.002	0.004	3.0–3.1
Ege Metal Endüstri A.S.													
orhaneli													
concentrates	47.84	—	15.82	6.53	7.98	17.60	0.03	—	0.01	0.42	0.002	0.01	2.66
jig fines	39.69	—	13.58	9.93	7.81	20.96	0.03	—	0.01	0.61	0.002	0.01	2.57
metallurgical	39.72	—	12.43	11.27	7.48	21.67	0.03	—	0.01	0.59	0.002	0.01	2.81
chemical	45.49	—	18.48	7.37	6.43	18.79	0.03	—	0.01	0.77	0.002	0.01	2.17
eskisehir													
concentrates	48.03	—	15.54	4.95	11.38	15.30	0.03	—	0.01	0.42	0.002	0.01	2.72
jig fines	38.57	—	12.30	11.32	8.72	21.84	0.03	—	0.01	0.72	0.00	0.01	2.76
metallurgical	38.03	—	11.82	11.02	9.71	23.15	0.03	—	0.01	0.61	0.002	0.01	2.83
refractory	53.13	—	15.94	1.85	8.55	16.20	0.03	—	0.01	0.66	0.002	0.01	2.93
concentrates lump	50.91	—	15.12	3.65	9.27	16.63	0.03	—	0.01	0.74	0.002	0.01	2.96
kop													
jig fines	38.13	—	14.88	10.05	7.96	21.13	0.03	—	0.01	0.42	0.002	0.01	2.26
metallurgical	38.18	—	14.42	10.90	7.95	20.97	0.03	—	0.01	0.51	0.002	0.01	2.33
Etibank General Management													
concentrates	37–47	—	12–15	7–11	7–11	18–23	0–0.01	—	—	0.3–1	0–0.01	0–0.04	2.9
metallurgical	42–48	—	12–15	5–9	12–14	14–19	0–0.01	—	—	0.3–1	0–0.01	0–0.04	2.9
refractory	46	—	14	3–5	16	16–17	—	—	—	0.2	—	—	2.9
Faryab Mining Co.													
metallurgical	35–55	—	1.5–10	7–11	7–11	15–21	—	—	—	—	—	—	3.0–3.5
Ferro Alloys Corporation Ltd. (FACOR)													
boula													
concentrates	39–42	—	24–26	8–10	4–8	10–14	Traces	Traces	—	1.1–1.3	Traces	Traces	1.4–1.6
metallurgical	38–47	—	14–19	8–18	7–10	10–14	Traces	Traces	—	0.2–0.3	0.003–0.004	Traces	1.8–2.9
kathpal metallurgical	40–46	—	11–13	5–10	7–8	10–13	Traces	Traces	—	0.2–0.3	0.001	Traces	2.4–2.5
ostapal metallurgical	42–45	—	23–24	8–11	6–10	10–13	Traces	Traces	—	0.6–1.0	0.010	Traces	1.6–2.0
Hayri Ögelman Mining Co. Ltd.													
orhaneli concentrates	47.36	—	16.73	10.33	8.29	14.16	0.16	—	—	—	0.01	—	2.83
others	48–52	—	—	—	—	—	—	—	—	—	—	—	—
concentrates, refractory	48–53	—	14.66	4.11	7.13	17.63	—	—	—	0.83	—	—	3.19

Table 13 (Continued)

Company and Grade	Cr₂O₃	Fe₂O₃	FeO	SiO₂	Al₂O₃	MgO	MnO	TiO₂	T₂O₅	V₂O₅	CaO	P	S	Cr:Fe
Hernic Ferrochrome (Pty.) Ltd.														
lumpy/chips	38–40	—	23	9	15.5	11.8	—	0.65	—	0.45	2.2	0.003	0.004	1.48
concentrates	44–46	—	26.5	1–3	15.5	10.5	—	0.64	—	0.44	0.28	0.003	0.004	1.50
metallurgical	44–45	—	26.5	2.5	15.5	10.5	—	0.64	—	0.44	0.3	0.003	0.004	1.50
chemical	45–46	—	26.5	<1	15.5	10.5	—	0.64	—	0.44	0.26	0.003	0.004	1.52
foundry sand	45–46	—	26.5	<1	15.5	10.5	—	0.64	—	0.44	0.26	0.003	0.004	1.52
Ingessana Chromite Mines (of AMW Co. Ltd.)														
metallurgical	50%	—	—	—	—	—	—	—	—	—	—	—	—	—
Japan Chrome Industry Co. Ltd.														
refractory	33.1	—	4.0	4.1	25.8	21.4	—	—	—	—	—	—	—	—
Kraomita Malagasy (Kraoma)														
metallurgical concentrate	48	—	17–18	6	13–16	12–14	—	—	—	—	—	0.0090	—	2.4
lumpy	42	—	13–16	12–14	13–16	17–20	—	—	—	—	—	0.0070	—	2.5
friable	48	—	17	7	—	—	—	—	—	—	—	0.0090	—	2.4
fines	49	—	21	6	—	—	—	—	—	—	—	0.0070	—	2
Krominco Inc.														
metallurgical	46	—	NA	NA	NA	NA	—	NA	—	NA	NA	—	NA	2.9
concentrates	48	—	NA	NA	NA	NA	—	NA	—	NA	NA	—	NA	2.6
Magnesita S.A.														
concentrates	45.05	17.00	—	6.50	16.50	13.50	0.15	0.27	—	—	0.53	—	—	2.59
metallurgical	41.32	14.92	—	9.01	16.8	16.05	0.12	0.31	—	—	1.46	0.0109	0.0016	2.71
refractory	49.09	17.89	—	2.71	16.58	13.15	0.16	0.29	—	—	0.13	0.0048	0.0023	2.68
chemical grade	44.55	17.00	—	7.00	16.50	13.50	0.15	0.27	—	—	0.53	—	—	2.56
foundry sand	45.55	17.00	—	6.00	16.50	13.50	0.15	0.27	—	—	0.53	—	—	2.62
Outokumpu Chrome Oy														
metallurgical concentrates	44.1	—	24.3	3.5	13.6	10.8	—	0.53	—	0.21	0.4	0.0014	0.005	1.62
lumpy	36.0	—	18.3	10.9	12.4	15.0	—	0.45	—	0.18	1.4	0.003	0.003	1.76
foundry sand	46.7	—	25.8	1.5	13.9	9.2	—	0.56	—	0.21	0.1	0.0007	0.004	1.62
PT. Palmabim Mining-PT. Bituminusa														
concentrates	41–43	—	21.6–23	0.8–1.2	19–20	19–20	—	0.8–1	—	—	0.02	0.009	0.004	1.63

					Rustenburg Minerals Development Company									
concentrates metallurgical	44	—	24.2	4	14.1	—	—	0.46	—	—	0.26	0.005	0.004	1.60
	42	—	23.3	6	14.2	—	—	0.47	—	—	0.28	0.005	0.004	1.59
					Samancor Ltd.									
metallurgical	45.0	—	25.5	2.0	15.60	10.5	—	0.5	—	—	0.25	<0.003	<0.002	1.54
refractory	46.3	—	26.3	0.7	14.5	9.6	—	0.6	—	0.2	0.13	<0.001	<0.001	1.55
chemical	46.3	—	26.3	0.7	14.5	9.6	—	0.6	—	0.2	0.13	<0.001	<0.001	1.55
foundry sand	46.5	—	26.0	0.6	14.5	10.3	—	0.6	—	0.2	0.13	<0.001	<0.001	1.57
					The Orissa Mining Corporation Limited. (OMC)									
metallurgical, chemical														
friable	40–56	—	10–18	3–8	10–18	8–15	—	—	—	—	—	0.005–0.007	0.007–0.010	1.6–3.5
lumpy	40–56	—	10–18	3–8	10–16	—	—	—	—	—	—	—	—	1.6–3.6
refractory														
lumpy ore	46–56	—	10–15	3–7	10–14	—	—	—	—	—	—	0.007	—	1.8–3.6
concentrate	45–49	—	—	5–6	12–13	—	—	—	—	—	—	0.005–0.007	0.007–0.03	2.1–2.4
					The Tata Iron and Steel Co. Ltd. (TISCO): (India)									
metallurgical														
friable ore	40–58	—	9–20	1–6	10–14	9.5–14.0	—	—	—	—	0.2–0.41	0.005–0.007	0.005–0.007	1.6–3.5
lumpy ore	36–45	—	9–15	9–14	7–11	—	—	—	—	—	0.4–0.6	0.005–0.007	0.005–0.007	2.1–2.9
refractory														
lumpy	45–55	—	8.5–12.5	3–9	—	—	—	—	—	—	—	—	—	2.8–3.9
concentrate	50–54	—	10–13	1.0–2.5	11–12	—	—	—	—	—	—	0.005–0.007	0.01 max	2.4–3.5
					Velore Mining Corp.									
concentrates	43–53	—	9–23	3–5	—	—	—	—	—	—	—	—	—	1.5–2.4
metallurgical	30–45	—	10–12	9–17	11–15	—	—	—	—	—	0.8	0.004	0.01–0.03	2.4–2.8
					Zimasco (Pvt.) Ltd.									
concentrates	42–46	—	11–14	10–13	10–14	15–17	—	0.20	—	—	3.00	—	0.20–0.3	2.5–2.9

[a] Ref. 24.
[b] *Key:*—data not reported; Na, not available.

Table 14. **Chromium Ferroalloy Chemical Specifications and Physical Specifications**[a,b]

Company and grade	Cr	C	P	Si	S	Ti	Al	Size
Albchrome Ltd.								
HCFeCr	60–65	6–9	3	0.04	0.04	0.04		<10, 10–150, >150
Chelyabinsk Electrometallurgical Integrated Plant								
HCFeCr 1	55–60	9.1	1.19	0.04	0.034			
HCFeCr 2	60–65	9.06	4.84	0.04	0.05			
HCFeCr 3	>65	10.5	4.38	0.05	0.06			
LCFeCr 1	60–65	0.5	2.8	0.044	0.01			
LCFeCr 2	65–70	0.5	2.7	0.044	0.01			
FeCrSi	28–31	0.1	46–52		0.02			
Chrome Resources (Pty.) Ltd.								
ChCr	50–53	6–8	2.5–7.0	0.04	0.05			
Cia. de Ferro Ligas da Bahia - FERBASA:								
ChCr	54.69	7.59	3.35	0.026	0.014			
LCFeCr	55.86	0.044	0.93	0.030	0.002			
FeCrSi	31.95	0.060	47.87	0.030	0.002			
ChCr 1	50–55	6–8	1–5	0.020	0.04			20–100 lumps, 5–50 granular
ChCr 2	50–52	6–7	2–6	0.020	0.035			20–150
Dalmacija Ferro-Alloys Works								
HCFeCr	65	6–8	1.5	0.035	0.08			10–200
Darfo s.r.l								
HCFeCr 1	60–65	4–6	1–2	0.02	0.04			
HCFeCr 2	60–65	6–8	1–2	0.02	0.04			
Elektrowerk Weisweiler GmbH								
LCFeCr	65–82	0.5	1.5	0.03	0.01			
Elkem a/s								
HCFeCr	60–65	4–8	1–5	0.03	0.040			
Etibank General Management								
HCFeCr	62	8	1.5–4.0	0.04	0.06			10–200
LCFeCr	68–72	0.20	1.5	0.03	0.03			10–50, 10–80, 10–100, 10–200
Faryab Mining Co. - Abadan Ferroalloys Refinery								
HCFeCr	70	8	2	0.04	0.04			
Feralloys Ltd.								
ChCr	55	6–7	4	0.02	0.05			
Ferro Alloys Corporation Ltd. (FACOR)								
ChCr	55–60	6–8	4	0.025–0.03	0.03			4–150
HCFeCr 1	60–70	6–8	2	0.03–0.05	0.035			10–150
HCFeCr 2	60–70	6–8	2–4	0.03–0.05	0.035			10–150
HCFeCr 3	60–70	6–8	4–6	0.03–0.05	0.035			10–150
MCFeCr	60–70	2–4	2–4	0.03–0.05	0.035			10–150
LCFeCr 1	60–70	0.03	1.50	0.035–0.05	0.025		0.10	10–150
LCFeCr 2	60–70	0.05	1.50	0.035–0.05	0.025		0.10	10–150
LCFeCr 3	60–70	0.10	1.50	0.035–0.05	0.025		0.10	10–150
LCFeCr 4	60–70	0.20	1.50	0.035–0.05	0.025		0.10	10–150

Table 14 (*Continued*)

Company and grade	Chemical Composition, wt%							
	Cr	C	P	Si	S	Ti	Al	Size
Albchrome Ltd.								
Ferrochrome Philippines Inc.								
HCFeCr	60	8	3	0.03	0.04			10–50, 10–80, 0–10
GMR Vasavi Industries Ltd.								
HCFeCr	60–70	6–8	1.5–4.0	0.03	0.05			
Hernic Ferrochrome (Pty.) Ltd.								
ChCr	50–54	6–7	3–7	0.025	0.050			
Hi-Tech Electrothermics (P) Ltd.								
HCFeCr	60–65	4–6	2	0.05	0.05			25–150
Huta "Laziska" Ferroalloy Plant								
HCFeCr	60–75	6–10	1–5	0.05	0.04			
Indian Metals and Ferro Alloys Ltd. (IMFA)								
HCFeCr	50–75	6–8	1.5–6.0	0.020–0.05	0.025–0.05			10–150
Integrated Chrome Corp.								
ChCr	50–55	6–8	3.5	0.025	0.03			10–150
Ispat Alloys Ltd.								
HCFeCr	60–64	6–8	4	0.03–0.035	0.04			10–100
Jiangyin Ferroalloy Factory								
HCFeCr	62–72	6–10	1–3	0.04	0.04			
Jilin Ferroalloy Works								
HCFeCr	60–70	6–9	1.5–3.0	0.04	0.04			
MCFeCr	60–70	0.3–1.0	1.5–3.0	0.03	0.03			
LCFeCr	60–70	0.01–0.15	1–2	0.03	0.03			
FeCrSi	30–40	0.02–1.0	35–45	0.03	0.01			
Jindal Ferro Alloys Ltd.								
HCFeCr	60–65	6–8	4	0.03	0.05			
Klutchevsk Ferroalloy Plant								
LCFeCr	68–75	0.03	0.3	0.02	0.005			
Macalloy Corp.								
HCFeCr	68	6.2	2	0.025	0.05			
Mandsaur Ferro Alloys Ltd.								
HCFeCr	60–70	6–8	2–4	0.035	0.03			10–150
Monnet Industries Ltd.								
HCFeCr 1	60–70	6–8	2–4	0.050	0.050			10–50
HCFeCr 2	60	6–8	2–4	0.050	0.050			3–10
HCFeCr 3	60	6–8	3–6	0.045	0.035			1–3
HCFeCr 4	58–68							100/60-mesh powder
Nanjing Ferroalloy Plant								
HCFeCr 1	62.0	9.5	3.0	0.03	0.04–0.06			<15-kg lumps, <20
HCFeCr 2	52.0–60.0	10	3.0–5.0	0.04–0.06	0.04–0.06			<15-kg lumps, <20
HCFeCr 3	60.0	8.5	3.0	0.03	0.04			<15-kg lumps, <20

Table 14 (*Continued*)

Company and grade	Chemical Composition, wt%								
	Cr	C	P	Si	S	Ti	Al	Size	

Albchrome Ltd.

Nav Chrome Ltd.

HCFeCr	64	7.5	2–4	0.05	0.05			25–100	

Nava Bharat Ferro Alloys Ltd.

| HCFeCr | 64 | 6–8 | 4 | 0.04 | 0.035 | | | | |

Nippon Denko Co., Ltd.

| HCFeCr | 65–70 | 6.0 | 1.5 | 0.04 | 0.08 | | | | |

NKK Corp.

| HCFeCr | 60–65 | 8.0 | 6.0 | 0.04 | 0.04 | | | | |
| LCFeCr | 60–65 | 0.1 | 1.0 | 0.04 | 0.03 | | | | |

Oravske Ferozliatinarske Zavody (OFZ)

HCFeCr 1	68	7.5	1.0	0.03	0.04	0.04		10–50, 10–80, 10–100, 10–150, 10–250	
HCFeCr 2	68	6	0.8	0.03	0.03			10–50, 10–80, 10–100	
HCFeCr 3	67	5.6	0.8	0.025	0.015	0.02		10–50, 10–80, 10–100	
HCFeCr 4	68	8.5	1	0.03	0.04			10–50, 10–80, 10–100, 10–150, 10–250	
ChCr	62	7.5	5	0.03	0.02			10–80, 10–150, 10–100	
LCFeCr 1	69	2.5	0.8	0.03	0.01			10–50, 10–80, 10–100	
LCFeCr 2	70	0.35	0.8	0.03	0.004			10–50, 10–80, 10–100	
FeCrSi	30	0.05	49					10–40, 10–50, 10–80, 10–100, 10–150; granules 0–5, 0–7	

Outokumpu Chrome Oy

| ChCr | 52 | 6–8 | 3–5 | 0.03 | 0.05 | | | 10–150 | |

Philippine Minerals & Alloy Corp.

| HCFeCr | 60–70 | 6–8 | 1.5–3.0 | 0.05 | 0.05 | | | 10–150 | |

S. C. Ferom S.A.

| HCFeCr | 60–65 | 8 | 4 | 0.04 | 0.06–0.08 | | | 10–100 | |

Table 14 (*Continued*)

| Company and grade | Chemical Composition, wt% | | | | | | | |
	Cr	C	P	Si	S	Ti	Al	Size
Albchrome Ltd.								
MCFeCr	65	1–4	1	0.02–0.05	0.05			10–100
LCFeCr	65	0.04–0.5	0.8–2.0	0.02–0.04	0.02			10–100
FeCrSi	55–60	6	10–18	0.04	0.03			10–100
Samancor Ltd.								
ChCr 1	50–55	6–8	3–6	0.025	0.050			10 × 150, 10 × 80, 3 × 12, <3 mm, granules
ChCr 2	50–55	8–9	1–2	0.01	0.023			10 × 150, 10 × 80, 3 × 12, <3 mm
MCFeCr	52–58	1.5–4.0	0.5	0.028	0.025			Granules
LCFeCr	58–60	0.03–0.10	1–2	0.03	0.02			10 × 100, 50 × 75, 25 × 50, 3 × 10, <3 mm
Serov Ferroalloys Plant								
HCFeCr	60–65	8–9.5	2	0.025–0.03	0.04–0.06			
MCFeCr	60–65	1–2	1.5–2.0	0.02–0.03	0.002			
LCFeCr	60–65	0.02–0.50	1.5	0.02–0.03	0.002			
FeCrSi	28	0.1	52	0.03	0.002			
Showa Denko K.K.								
ChCr	50–55	8.5	3.0	0.04	0.06			<35 mm
LCFeCr 1	60–65	0.06	1.0	0.03	0.03			10–200, 10–100, or 5–50
LCFeCr 2	60–65	0.10	1.0	0.03	0.03			10–200, 10–100, or 5–50
LCFeCr 3	60–65	0.01	1.0	0.03	0.03			10–200, 10–100, or 5–50
LCFeCr 4	60–65	0.03	1.0	0.03	0.03			10–200, 10–100, or 5–50
LCFeCr 5	65–70	0.10	1.0	0.03	0.03			10–200, 10–100, or 5–50
LCFeCr 6	60–65	0.03	0.40	0.03	0.03			10–200, 10–100, or 5–50
LCFeCr 7	85–92	0.10	1.0	0.020	0.020			10–70, 2–10, <250 mm, <104 μm

Table 14 (*Continued*)

Company and grade	Chemical Composition, wt%							
	Cr	C	P	Si	S	Ti	Al	Size
Albchrome Ltd.								
LCFeCr 8	70 min	0.03	1.0	0.03	0.03			10–70, 2–10, or <250 mm
Shri Girija Smelters Ltd.								
HCFeCr 1	60–70	6–8	2–4	0.03–0.05	0.05			25–150
HCFeCr 2	60–70	6–8	2–4	0.03–0.05	0.05			10–120
Srinivasa Ferro Alloys Ltd.								
HCFeCr 1	60–70	6–8	2–4	0.03–0.05	0.05			25–150
HCFeCr 2	60–70	6–8	2–4	0.03–0.05	0.05			10–120
Standard Chrome Ltd.								
HCFeCr	65–68	7–8	2	0.02	0.02			
The Tata Iron and Steel Company Ltd. (TISCO)								
HCFeCr	64	6–8	4	0.025	0.030			
ChCr	60	6–8	4	0.025	0.030			
Tovarna Dusika Ruse - Metalurgija d.o.o.								
HCFeCr	60–70	6–8	1.5	0.03	0.06			
LCFeCr	63–70	0.05–0.10	1.5	0.02	0.01			
V. K. Ferro Alloys Private Ltd.								
HCFeCr	60–70	6–8	0.5–4.0	0.03	0.05			
Vargön Alloys AB								
HCFeCr 1	65–67	4–6	1.5	0.02	0.08	0.04		
HCFeCr 2	65–67	6–8	1.5	0.02	0.08	0.04		
ChCr 1	55–60	6–8	1–3	0.025	0.05	0.5		
ChCr 2	55–60	6–8	3–6	0.025	0.05	0.5		
VBC Ferro Alloys Ltd.								
HCFeCr	60–70	6–8	2	0.03	0.05			
Yermakovsky Ferroalloy Plant								
HCFeCr	65–68	8–9	2.0	0.03–0.05	0.04–0.08			10–80, 10–50, 0–10
FeCrSi	28	0.1	45 (min)	0.03	0.02			10–80, 10–50, 0–10
Zimasco (Private) Ltd.								
HCFeCr	65	8	2.5	0.02	0.06			10–150
Zimbabwe Alloys Ltd.								
LCFeCr	64	0.06	1.2	0.025	0.01			3–100
FeCrSi	35	0.05	42	0.03	0.005			10–100

[a]*Key*: FeCrSi, ferrochromiumsilicon; HCFeCr, high-carbon ferrochromium; LCFeCr, low-carbon ferrochromium; ChCr, charge-grade ferrochromium; —, not reported.

[b]Cr is minimum except where range is specified unless noted otherwise. Al, C, Si, P, S, and Ti are maximum except where range is specified unless noted otherwise. Size in millimeters unless noted otherwise.

Table 15. **Chromium Metal Chemical Specifications**[a,b]

Element		Delachaux		Elkem (Vacuum-Grade)				Metallurg		Nippon Denko		Tula	ASTM		Polema			
		Vacuum	DDB	Powder	Plate	Pellets	Powder	Standard	Vacuum Refining	Standard	Request	Flake	Grade A	Grade B	EX	ERX-1	ERX-2	ERX-3
chromium	Cr	99.5	99.7	99.8	99.1	99.5	99.0	99.2	99.4	99.0	99.0	66.6	99.0	99.4	99.95	99.95	99.95	99.95
aluminum	Al	0.1	0.01	0.01	0.01	0.005	0.01	0.15	0.10	0.3		0.004	0.30	0.10	0.006	0.006	0.006	0.006
antimony	Sb	0.0005	0.0005										0.005	0.003				
arsenic	As	0.0001	0.0001									0.001	0.005	0.003				
barium	Ba	0.00003	0.00003									0.001						
bismuth	Bi	0.00005	0.00005											0.003				
boron	B	0.00001	0.00001								0.0006	0.003	0.001					
cadmium	Cd	0.0002	0.0002										0.005					
carbon	C	0.01	0.04	0.01											0.020	0.008	0.008	0.008
cobalt	Co		0.04		0.02	0.05	0.02	0.03	0.03	0.04	0.02	0.0010	0.050	0.050				
columbium	Nb											0.0003	0.003	0.001				
copper	Cu	0.002	0.001										0.050	0.050				
hydrogen	H	0.0001	0.0001		0.01	0.002	0.02			0.0007	0.0007	0.0003	0.01	0.003	0.003	0.003	0.003	0.003
iron	Fe	0.2	0.15	0.12	0.20	0.25	0.25	0.025	0.25	0.5	0.5		0.35	0.35	0.003	0.003	0.003	0.003
lead	Pb	0.0005	0.0005		0.003	0.001	0.01					0.002	0.003	0.001	0.008	0.008	0.008	0.012
magnesium	Mg	0.001	0.001									0.002			0.001	Trace	Trace	Trace
manganese	Mn	0.0015	0.0015		0.01	0.010	0.01						0.01	0.01				
molybdenum	Mo											0.002						
nickel	Ni											0.0003	0.050	0.01				
nitrogen	N	0.02	0.008	0.005	0.05	0.010	0.05	0.01	0.02	0.030	0.030	0.001	0.050	0.020	0.005	0.005	0.005	0.007
oxygen	O	0.1	0.045	0.045	0.50	0.05	0.60	0.15	0.1	0.046	0.046		0.50	0.10	0.020	0.005	0.007	0.010
phosphorus	P	0.002	0.001	0.001	0.005	0.002	0.005	0.005	0.001	0.05	0.003	0.334	0.010	0.010	0.600	0.005	0.008	0.020
selenium	Se	0.0002	0.0002										0.010	0.010				
silicon	Si	0.1	0.05	0.04	0.005	0.02	0.01	0.15	0.10	0.2	0.2	0.020	0.15	0.10	0.010	0.010	0.010	0.012
silver	Ag	0.00005	0.00005									0.003	0.003	0.001				
sulfur	S	0.01	0.004	0.005	0.030	0.010	0.04	0.005	0.01	0.05	0.05	0.0140	0.030	0.010	0.010	0.002	0.002	0.005
tantalum	Ta												0.050	0.003				
tellurium	Te	0.0002	0.0002															
thallium	Tl	0.0002	0.0002															
tin	Sn	0.0005	0.0005			0.001												
titanium	Ti											0.0001	0.001	0.001				
vanadium	V												0.050	0.003				
zinc	Zn	0.0005	0.0005									0.0003	0.050	0.050				
zirconium	Zi											0.0001	0.050	0.003				

[a] Ref. 2.
[b] Blank space, not reported.

513

to address the difficulty of dissolving samples that contain chromite. A procedure found to work is fusion with sodium peroxide followed by persulfate oxidation and titration with ferrous iron (26).

The Environmental Protection Agency (EPA) (27) has an interest in the analysis for chromium because it is classified as a toxic material, and waste materials are analyzed for chromium content. The required treatment of waste is affected by its leachable chromium content. EPA has published analysis methods for chromium and for hexavalent chromium. The analysis method for chromium is atomic absorption including direct aspiration (Method 7190) and furnace (Method 7191) techniques. The analysis methods for hexavalent chromium include coprecipitation (Method 7195), colorimetric (Method 7196A), chelation/extraction (Method 7197), and differential pulse polarography (Method 7198) techniques. The coprecipitation, colorimetric, and chelation/extraction techniques are recommended for extracts and groundwaters.

11. Environmental Concerns

In recognition of the development of environmental concerns about chromium worldwide and in response to a European Commission review of chromium occupational exposure limits, the International Chromium Development Association published industry guidelines on health, safety, and environment. The guidelines take account of extensive international changes and developments in legislation and regulation of chromium materials and is intended to help companies implement appropriate workplace practices and procedures for environmental protection.

Environmental concerns about chromium have resulted in a wide variety of studies to determine chemical characteristics, natural background levels, sources of environmental emission, movement of chromium in the environment, interaction of chromium with plants and animals, effect of chromium on plants and animals, measurement methods, and recovery technology. A broad review of many environmental factors and the role of chromium, among other metals, in the environment was published.

In the United States, the Environmental Protection Agency (EPA) regulates chromium releases into the environment. The Occupational Safety and Health Administration (OSHA) regulates workplace exposure.

The EPA regulates and monitors industrial impact on the environment. As part of its monitoring activity, EPA collects data on toxic chemicals. That information is made available in the Toxic Release Inventory (TRI). TRI is mandated under Title III of the Superfund Amendments and Re-authorization Act (SARA) of 1986. The Pollution Prevention Act of 1990 resulted in the addition of recycling activities to the material management categories covered under TRI reporting.

11.1. Environmental Regulations.   Chromium and chromium compounds are regulated by the EPA under the Clean Air Act (CAA), the Comprehensive Environmental Response, Compensation, and Liability Act of 1980 (also known as CERCLA or Superfund), National Primary Drinking Water Regulations (NPDWR), the Clean Water Act (CWA), and the Resource Conservation and Recovery Act (RCRA).

Effluent. Chromium in water effluents is manageable. The solubility of trivalent chromium compounds in neutral water usually results in a chromium concentration below that required by EPA for drinking water (0.1 ppm). Thus, when water is neutralized, chromium can be removed by filtration. If hexavalent chromium compounds are present, they must first be reduced to trivalent, a technically manageable operation.

Emissions. Congress enacted the Clean Air Act Amendments Law of 1990 (Public Law 101-549), completely revising the Air Toxics Program. Congress identified 189 hazardous air pollutants to be regulated. Chromium compounds—defined as any chemical substances that contain chromium as part of their structure—were included among those hazardous air pollutants. Under the revised Air Toxics Program, Congress instructed EPA to regulate hazardous air pollutants by regulating the source of those pollutants. Congress required EPA to identify pollution sources by November 1991, then to set emission standards for those sources. EPA eliminated the use of chromium chemicals in comfort cooling towers and regulated chromium releases from the electroplating and anodizing industries.

Solid Waste. EPA regulates solid waste generated by the chemical industry in the production of sodium chromate and dichromate. Chromium-containing treated residues from roasting and/or leaching of chrome ore is regulated under Subtitle D of the Resource Conservation and Recovery Act. EPA found no significant danger associated with treated residue from roasting and/or leaching of chrome ore based on waste characteristics, management practices, and damage case investigations.

11.2. Resource Conservation and Recovery Act. The Resource Conservation and Recovery Act (RCRA) brought waste from the extraction, beneficiation, and processing (smelting and refining) of ores and minerals under the regulatory control of EPA. EPA listed emissions from the production of ferrochromium–silicon (RCRA waste number K090) and ferrochromium (RCRA waste number K091) as hazardous waste. EPA regulates treated residue from roasting and leaching of chromite ore under Section D of RCRA. EPA was directed in 1988 by a court order to restrict the scope of exclusion to large volume, low hazard waste. An EPA study determined that treated residue from roasting and leaching of chromite ore does not pose an actual or potential danger to human health and the environment. EPA therefore decided to regulate treated residue from roasting and leaching of chromite ore under Section D of RCRA.

EPA regulates refractory material solid waste containing chromium. EPA determined that chromium-containing wastes exhibit toxicity. Therefore, they have established a policy that—if the extract from a representative waste sample contains chromium at a concentration greater than or equal to 5.0 mg/L (total chromium) as measured by a specified toxicity characteristics leaching procedure—it is hazardous. EPA promulgated a treatment standard for chromium-containing refractory brick wastes based on chemical stabilization. (Stabilization is a process that keeps a compound, mixture, or solution from changing its form or chemical nature.) EPA determined that some chromium-containing refractory brick wastes can be recycled as feedstock in the manufacture of refractory bricks or metal alloys.

EPA regulates the wood preservation industry. As a result of the Resource Conservation and Recovery Act (1988), EPA promulgated regulations on the wood preserving industry (1990) to control inorganic preservatives containing chromium labeled F035 by EPA. In the *Code of Federal Regulations* (*CFR*), EPA specified standards for drip pad design, operation, inspection, and closure, specifically in 40 *CFR* 262, 264, and 265. In effect, EPA required wood preservers to upgrade their drip pad or build new ones to meet EPA standards.

EPA regulates the emission of chromium from toxic waste incinerators. The Resource Conservation and Recovery Act (1976) made EPA responsible for managing hazardous waste disposal. EPA regulates particulate emissions from incinerators. However, EPA found that the particulate standard may not provide sufficient protection if a substantial fraction of the particulate emissions were regulated metals, leading EPA to promulgate separate regulations for toxic waste incinerators. Incineration is a desirable method of toxic waste disposal because organic waste is destroyed, leaving no future cost to society. EPA proposed regulation of chromium emission from devices burning hazardous waste in 1987 and promulgated regulations in 1990. Regulation involves control of chromium (contained in the waste stream) feed rates, chromium emission limits, and site-specific risk assessment. Based on field studies, the emission limits of chromium were complicated by the fact that stainless steel (a chromium-containing alloy) was used in the production and transportation processes.

Chromium leaching behavior in soil derived from the kiln roasting and leaching of chromite ore was reported. It was found that (*1*) leaching was highly sensitive to pH and that the most chromium leached out at soil pH between 4 and 12 and (*2*) the presence of organic matter in the soil reduced the amount of chromium leached out.

11.3. Clean Air Act. In 1992, EPA identified chromium electroplaters and anodizers as an area source of hazardous air pollutants that warrant regulation under Section 112 of the Clean Air Act and described that source's adverse impact. It was estimated that over 5000 facilities nationwide, which were collectively emitting about 175 tons of chromium per year, would be required by regulation to reduce their emission by 99%. The chromium electroplating industry includes hard chromium platers (usually a thick chromium coating on steel for wear resistance of hydraulic cylinders, zinc diecastings, plastic molds, and marine hardware), decorative chromium platers (usually over a nickel layer on aluminum, brass, plastic, or steel for wear and tarnish resistance of auto trim, tools, bicycles, and plumbing fixtures), and surface-treatment electroplaters or anodizers (usually a chromic acid process to produce a corrosion-resistant oxide surface on aluminum used for aircraft parts and architectural structures subject to high stress and corrosive conditions). EPA estimated that 1540 hard chromium electroplaters, 2800 decorative electroplaters, and 680 chromic acid anodizers nationwide are affected. EPA estimated that electroplaters collectively emit 175 tons of chromium per year, most of which is hexavalent and carcinogenic in humans. EPA estimated that the resulting U.S. nationwide population risk is an additional 110 cases of cancer per year resulting from that emission. EPA estimated the resulting individual risk in the proximity of particular facilities ranged from less than 2 chances per 100,000 for small chromic acid anodizing operations to 5 chances per 1000 for large hard plating operations. The regulation specifies emission limits, work practices, initial performance testing, ongoing compliance monitoring, recordkeeping, and reporting requirements. The EPA reported on chromium emissions from electroplating operations and chromium recovery from electroplating rinse waters.

In 1994, EPA banned the use chromium chemicals for industrial process water-cooling towers for corrosion inhibition. It was reported that 90% of industrial cooling-tower operators had eliminated the use of chromium chemicals in anticipation of such an EPA ban. However, the remaining 800 operations were given 18 months within which to comply with the new ruling.

11.4. Toxic Release Inventory. Under the Toxic Release Inventory program, EPA collected environmental release information since 1987 from manufacturing facilities that employ 10 or more persons and used a threshold amount of chromium contained in chromium compounds. (A *manufacturing facility* is one whose product is included in Standard Industrial Classification Division E (SIC) Codes 20–39. EPA was expanded in 1997 to cover additional SIC codes. Reporting under the new set of SIC codes was expected to start for the 1998 reporting year.) The threshold amount decreased from 1987 to 1989, after which time it remained constant. The threshold limit for a facility that manufactured or processed chromium compounds was about 34 tons of contained chromium in 1987, about 23 tons in 1988, and about 11 tons in 1989 and subsequent years. The threshold limit for facilities that otherwise used chromium compounds has been and remains about 5.4 tons. (Note that EPA has definitions for the terms manufacture, process, and otherwise use for the purpose of reporting releases.) When reporting chromium releases, a facility must add up the chromium released from all sources that exceed a de minimis amount. The de minimis amount for chromium compounds is 0.1%. Facilities report the amount of chromium released to the air, water, and earth environment; the amount of chromium recovered on site; and the amount transferred to offsite locations. The data are collectively referred to as the Toxic Release Inventory (TRI).

EPA denied a petition to remove chromium III compounds and chromic oxide in particular from the chemicals covered by the Emergency Planning and Community Right-to-Know Act of 1986, in particular from the section 313 list of toxic chemicals. The petition to remove chromium III compounds was based on the contention that chromium III compounds are considered nonhazardous wastes under the Resource Conservation and Recovery Act (RCRA). EPA denied the petition based on EPA's determination that the conversion of chromium(III) to chromium(VI) has been demonstrated to occur in soils and in water-treatment processes that use chlorine.

EPA started the 33/50 Program, a voluntary program to reduce environmental release and transfer of 17 toxic chemicals, including chromium and chromium compounds. The program is so named because its objective is the voluntary one-third reduction of chromium and chromium compound releases and transfers by 1992 and one-half reduction by 1995. Reductions are to be measured against 1988 TRI data. See the "Recycling and Disposal" section of this article for more details.

11.5. Water and Effluents. EPA promulgated its final rule on chromium contained in primary drinking water in 1991. EPA set the maximum contaminant level goal (MCLG) and the maximum contaminant level for chromium contained in primary drinking water at 0.1 mg/L. EPA identified the best available technologies to remove chromium(III) compounds to be coagulation with filtration, ion exchange, lime softening, and reverse osmosis. EPA identified the best available technologies to remove chromium(VI) compounds to be coagulation with filtration, ion exchange, and reverse osmosis. EPA concluded that chromium contained in drinking water should be minimized in recognition of its biological reactivity, including its potential for posing a carcinogenic hazard. EPA set the chromium(III) and chromium(VI) MCLG on the basis of the reference dose concept. The safe dose to which EPA refers is the National Academy of Sciences recommended daily intake of 50–200 µg per day.

The EPA published a retrospective study on effluent guidelines, leather tanning, and pollution prevention. The report found that industry met the chromium limitations by modifying the tanning process to get more chromium out of the tanning wastewater and into the leather. By changing chromium formulations, raising process temperature and time, and reducing bath water, industry increased chromium fixation from about 50% to about 90%. Recycling was also done to meet guidelines.

12. Recycling and Disposal

12.1. Recycling. Stainless steel, superalloys, and chromium metal are produced primarily in Europe, Japan, and the United States. Stainless steel represents about 1% of steel production domestically and worldwide. It is a specialized, small part of the steel market serving the need for durable, corrosion-resistant steel. Yet stainless steel accounts for about 50% of chromium demand.

U.S. apparent consumption of chromium is primary production (ie, chromium contained in domestic mine production of chromite ore) plus secondary production (ie, chromium contained in recycled scrap) plus net trade (ie, imports minus exports) in chromium materials (including chromite ore, chromium ferroalloys and metal, and selected chromium chemicals) plus domestic consumer, government, and producer stock changes of chromite ore and chromium ferroalloys and metal.

Chromium contained in stainless steel and other metal scrap is recycled. Both new and old scrap are collected by scrap processors and returned to stainless-steel manufacturers. Secondary production is calculated as chromium contained in reported stainless-steel scrap receipts.

Recycling (qv) is the only domestic supply source of chromium. Stainless-steel and superalloys are recycled, primarily for their nickel and chromium contents. As much as 50% of electric furnace stainless-steel production can result from recycled stainless-steel scrap. Advanced stainless-steel production technology such as continuous casting reduces prompt scrap generation and permits a higher product yield per unit of raw material feed (28).

Industry practice is to sort scrap for recycling. Chromium-containing stainless steel is collected, processed, and returned to stainless steel manufacturers for reuse. Processing may include changing the physical form of the scrap. Large pieces may be cut to smaller size, common sizes may be bundled for easier handling, and smaller-sized pieces may be melted and cast into larger sizes.

Some materials require cleaning or sorting before they can be recycled. Some processors melt and combine several alloys to produce master alloy castings that meet stainless steel or other alloy manufacturers' chemical requirements. Superalloy (nickel- and cobalt-based alloys used in the aerospace industry) reuse is carried out by certified recycling companies in cooperation with alloy producers and product manufacturers. Superalloy scrap that cannot be reused is recycled in other alloys. Small quantities of chromium metal waste and scrap are also traded.

The price of chromium-containing stainless-steel scrap is sensitive to the price and availability of its constituents from primary sources. Stainless steel is composed of two major categories: austenitic and ferritic stainless steel. Austenitic stainless steel requires nickel and chromium. Ferritic stainless steel requires only chromium. The price of austenitic stainless steel is driven mostly by the higher-valued nickel contained in the scrap.

Chromium recycling is expected to increase, driven by environmental regulations mainly in the industrialized countries. Stainless-steel use has been growing, so the availability of stainless steel obsolete scrap as well as the scrap generated as a result of processing that material should continue to increase.

Recycled chromium constituted about 20% of current apparent consumption. According the year 2000, secondary chromium was expected to rise to 25% of apparent consumption because of recycling growth and decline in non-recycling uses (see Figure 12).

12.2. Disposal. The Environmental Protection Agency surveys domestic industry for quantity and method of disposal (ie, releases plus transfers) and reports that information annually in the Toxics Release Inventory Public Data

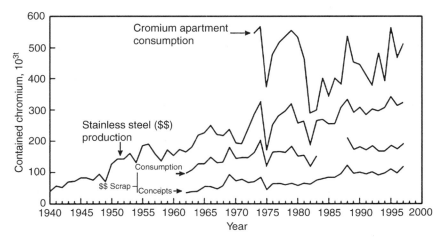

Fig. 12. U.S. chromium apparent consumption, and stainless-steel production, scrap receipts and consumption. For the purpose of calculating chromium apparent supply, secondary supply is estimated as stainless-steel scrap receipts. The trend of stainless-steel scrap receipts and consumption follow that of stainless steel production. Stainless-steel scrap consumption exceeds that of receipts by scrap generated within the consuming plant. The general trend to chromium apparent consumption is similar to that of stainless-steel production.

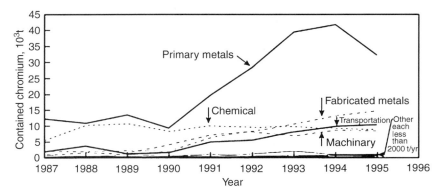

Fig. 13. Chromium disposals (ie, releases plus transfers) by industry. The major sources of chromium disposals are the same as those in the primary consuming industries: metallurgical and chemical. The major source of disposals from the metallurgical industry is primary metals, that portion that produces metal alloys in a variety of shapes for industrial consumers. Fabricated metals and machinery categories follow primary metals. Others include food, tobacco, textile, apparel, lumber, furniture, paper, printing, petroleum, rubber, leather, stone, electrical, instruments, miscellaneous other.

Release. Disposals are shown in Figure 13. Out of 20 specific categories, one—primary metals—dominates releases and transfers reflecting the fact that the major end use of chromium is in the metallurgical industry. Four other industries (fabricated metals, transportation, chemical, and machinery) account for a second tier of disposals. The remaining fifteen industries form a third tier in which each constituent accounts for less than 2.5% of total disposals. The rapid rise in disposal in fabricated and primary metals and transportation from 1990 to 1991 follows a similar rise in transfers by mode shown in Figure 14. This shows

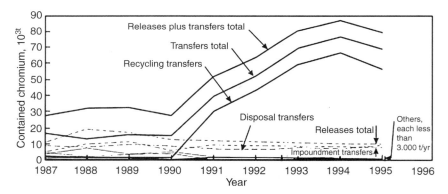

Fig. 14. Chromium releases and transfers by mode. It was in 1991 that the Environmental Protection Agency introduced "recycling transfers" as a material management category. Before 1991, "releases and transfers" were of comparable magnitude. Since 1991, the large volume of chromium contained in alloys transferred for recycling has both raised disposal (releases plus transfers) appreciably and caused both "transfers" and "disposals" to be dominated by recycling transfers. The next-largest disposal categories are "disposal transfers" and "impoundment transfers". "total releases" have declined slowly yet steadily since 1988.

that transfers exceed releases, that the difference increased from 1991 to 1994 mostly as a result of increasing transfers. Releases started to decline in 1988 and have continued to do so since then. The curves clearly show that the increase in total transfers results from the increase in recycling transfers. In 1994 and 1995, recycling accounted for in excess of 70% of releases and transfers. These increases occurred because recycling was not required to be reported until 1991.

Chromium and chromium compounds are one of the 17 priority chemicals targeted by EPA as part of the 33/50 Program, a program in which industries voluntarily tried to reduce releases with respect to those of 1988 by 33% in 1992 and by 50% in 1995. When adjusted for the change in reporting requirements from 1988 to 1992, chromium releases declined by 39%; transfers, by 43%. When adjusted for the change in reporting requirements from 1988 to 1995, chromium releases declined by 45%; transfers, by 22% (24), so the goal was not quite met, but these numbers should not be taken to mean that the program was unsuccessful. The measurement is relative to the base year, 1988, without regard to the level of industrial activity in each year. More industrial activity, even with more efficient material handling, could result in greater disposals. U.S. apparent consumption of chromium, a measure of national industrial activity, in 1988 was 537,000 tons; in 1992, 378,000 tons; and in 1995, 565,000 tons (30). The 30% decline in industrial chromium-related activity between 1988 and 1992 as indicated by the change in chromium apparent consumption between those two years likely enhanced apparent decline in chromium releases while the 5% increase from 1988 to 1995 likely diminished the apparent change in chromium releases. In other words, measuring changes in releases and/or transfers from one year to another without regard for changes in industrial activity may not be a valid way to measure industry performance at reducing releases and transfers. To measure industry performance, one should measure changes in releases and/or transfers relative to material processed. Of course, from the point of view of the environment, what matters is only how much material was released or transferred.

13. Health and Safety Factors

13.1. Health and Nutrition. Chromium is a trace mineral required by the human body. As such, the National Research Council recommends a daily intake in the range of 50–200 µg. Chromium is a cofactor for insulin, a hormone that participates in carbohydrate and fat metabolism. A cofactor is a material that acts with the material. The dietary chemical form of chromium is as trivalent compounds. Because humans cannot convert trivalent chromium to hexavalent chromium, the carcinogenicity of hexavalent chromium compounds bears no relevance to the nutritional role of trivalent chromium.

13.2. Toxicity. The effect of an element on the human body depends on several factors. These factors include the chemical or class of chemical, the route of exposure, the quantity and duration of exposure, and characteristics of the exposed subject.

The chemical distinctions typically made with respect to chromium chemicals include whether the compound is synthetic or naturally occurring. Synthetic

Table 16. **Occupational Safety and Health Standards for Workplace Exposure to Airborne Chromium**[a]

Air contaminant	Limit
chromic acid and chromates	Acceptable ceiling concentration 1 mg/10 m^3
chromium metal and insoluble salts, as chromium	Limit for air contaminant 1 mg/m^3
chromium(II) and chromium(III) compounds	0.5 mg/m^3

[a]*Note*: Acceptable ceiling concentration cannot be exceeded. Limit for air contaminant shall not exceed the stated amount measured as an 8-h time-weighted average during any 8-h work shift of a 40-h work week.

chromium compounds are typically classified by their oxidation state. Trivalent and hexavalent chromium compounds are two such classifications. Exposure to chromium compounds could typically occur through one or more of three routes: skin contact, ingestion, or inhalation. Exposure can also vary in intensity (concentration of the chemical) and duration (length of time for which exposure occurs). Response to chemical exposure is dependent on such human characteristics as age, sex, general health, and chemical sensitivity. The effect of chemical exposure on the human body can be good or bad. Chromium is one of those elements that is both essential to good health and detrimental to good health. The detrimental effects of chemical exposure are classified as acutely toxic when small amounts of the chemical cause significant damage in a short time, chronically toxic when exposure over a long time causes measurable damage, and carcinogenic when exposure can result in cancer.

Chromium generally forms chemical compounds in which chromium has either the hexavalent or trivalent oxidation state. Hexavalent chromium compounds are generally recognized as toxic. Chronic occupational exposure to hexavalent chromium has been associated with an increased incidence of bronchial cancer. The toxic status of trivalent chromium compounds is not clear. However, trivalent chromium compounds are less toxic than hexavalent chromium compounds. Chemical compounds containing chromium in lower valence states are generally recognized as benign.

The Food and Drug Administration, Department of Health and Human Services, was in the process of amending its regulations to add chromium to labeling for reference daily intakes and to add chromium to the factors in determining whether a substitute food is inferior.

The Occupational Safety and Health Administration (OSHA) regulated workplace exposure to chromium metal, soluble chromium salts, insoluble chromium salts, and chromic acid and chromates. Table 16 shows the exposure limits set by OSHA.

14. Uses

In the United States, most sodium dichromate is converted to chromic acid; some, however, is used directly by several industries. Chromium was first used in pigments and tanning compounds. Chromium plating, the electrodeposition of chromium from a solution of chromic acid, started in the early 1900s. A more recent use for chromium is in wood preservation. Chromium–copper–arsenate (CCA) impregnated wood can be protected from weathering, insects, and rotting for 40 years. Today, major end-use markets for sodium dichromate—drilling mud for the oil-and-gas industry, leather tanning, metal finishing, and wood preservation—are mature markets showing slow growth. Chromium chemicals are also used to make biocides, catalysts, corrosion inhibitors, metal plating and finishing chemicals, refractories, and printing chemicals. End uses showing declining use include chromate pigments, corrosion-control agents, and water-treatment chemicals. Newer, faster-growing markets include magnetic recording media and catalysts, and represent a small part of the market. In Europe, leather tanning is a major end use. In Japan, electroplating and metal finishing are principal end uses.

A chromium chemical end use with which many people are familiar is pigments. Chromium containing pigments are broadly classified as oxides or chromates. A rainbow of colors are produced by the pigment industry using a variety of mixed metal oxides with chromium. Oxides comprise chromic oxide green, copper chrome black, and hydrated chromium oxide green. Chromates comprise chrome green, lead chromate, and molybdate orange. Chromic oxide green pigment is used in camouflage because it has desirable infrared reflectance properties. Copper–chrome pigment is used in the black coating found on outdoor grills and wood-burning stoves. Hydrated chromium oxide green finds use in cosmetics and body soap. A variety of pigments are based on lead chromate including medium chrome yellow, lemon or primrose chrome yellow, molybdate orange, and chrome orange. Medium chrome yellow pigment is used in traffic marking yellow paint found on all major streets and highways. An important use of chromium pigments is in anticorrosion coatings. Chromium pigments that are used for corrosion control include lead, zinc, and strontium chromates. Chromate metal primers are used extensively by the federal government, in both civilian and military applications.

BIBLIOGRAPHY

"Chromium and Chromium Alloys" in *ECT* 1st ed., Vol. 3, pp. 935–940, by J. J. Vetter, Diamond Alkali Co.; in *ECT* 2nd ed., Vol. 5, pp. 451–472, by F. E. Bacon, Union Carbide Corp.; in *ECT* 3rd ed., Vol. 6, pp. 54–82, by J. H. Westbrook, General Electric Co.; "Chromium and Chromium Alloys" in *ECT* 4th ed., Vol. 6, pp. 228–263, by Jack H. Westbrook, Sci-Tech Knowledge Systems; "Chromium and Chromium Alloys" in *ECT* (online), posting date: December 4, 2000, by Jack H. Westbrook, Sci-Tech Knowledge Systems.

CITED PUBLICATIONS

1. D. Hargreaves, M. Eden-Green, and J. Devaney, *World Index of Resources and Population*, Dartmouth Publishing Co., Brookfield, Vt, 1994.
2. B. R. Lipin, "Low Grade Chromium Resources," in, W. C. Shanks, ed., *Cameron Volume on Unconventional Mineral Deposits*, Society of Mining Engineers, New York, 1983.

3. *Mineral Commodity Summaries*, U.S. Geological Survey, U.S. Government Printing Office, Washington, D.C., 1998 (available through the USGPO).
4. J. F. Papp, *Am. Ceramic Soc. Bull.* **76**(6), 84–88 (June 1997).
5. S. R. Taylor and S. M. McLennen, *The Continental Crust: Its Composition and Evolution*, Blackwell Scientific Publications, Ltd., Oxford, 1985.
6. K. Sharaki, *Resource Geol.* **47**, 319–330 (1997).
7. *Handbook of Chemistry and Physics*, 66th ed., 19,
8. D. R. Lide ed., *Handbook of Chemistry and Physics* 77th ed., CRC Press, Inc. Boca Raton, Fla., 1996–1997.
9. J. H. DeYoung, Jr., M. P. Lee, and B. R. Lipin, *International Strategic Minerals Inventory Summary Report–Chromium*, U.S. Geological Survey Circular 930-B, 1984.
10. E. H. Boyle, D. J. Shields, and L. A. Wagner, *Chromium Availability in Market Economy Countries and Network Flow Model Analysis of World Chromium Supply*, U.S. Bureau of Mines Information Circular 9337, 1993.
11. U.S. Department of Defense, *Strategic and Critical Materials Report to the Congress. Operation under the Strategic and Critical Materials Stock Piling Act during the period October 1996–September 1997*, Jan. 1998, p. 22.
12. U.S. Department of Defense, *Strategic and Critical Materials Report to the Congress. Operation under the Strategic and Critical Materials Stock Piling Act during the period October 1987–March 1988*, Sept. 1988, p. i.
13. H. J. Heine, *Foundry Management Technol.* 34–41 (April 1989).
14. A. Sontz, *AFS Transaction* 1–12 (1972).
15. M. A. Moll and K. Armitage, *Steel Times Int.* **20**(3), 38, 39, 42, 43 (1996).
16. R. Bolger, *Ind. Minerals* 35, 29–39 (Dec. 1996).
17. A. Granville and E. F. Statham, *Profits from Processing: Randburg*, Republic of South Africa, Mintek, Special Publication 13, 1989, p. 10–11.
18. Reference 10, p. 28.
19. G. Price, "Chromite Ore Resources: A Supply/Demand Perspective" in *International Chromium Development Association, Conference Proceedings*, Spring Meeting, Dusseldorf, Germany, 1995.
20. Indian Bureau of Mines, in *Indian Minerals Yearbook 1995*, Vol. **2**, Indian Bureau of Mines Press, Nagpur, India, July 1997, pp. 277–288.
21. J. F. Papp, *Chromium Life Cycle Study*, U.S. Bureau of Mines Information Circular 9411, 1994.
22. J. F. Papp, *Chromium, Nickel, and Other Alloying Elements in U.S.-Produced Stainless and Heat-Resisting Steel*, U.S. Bureau of Mines Information Circular 9275, 1991.
23. J. F. Papp, *Chromium Annual Review—1996*, Mineral Industry Surveys, U.S. Geological Survey, 1997.
24. *Chromium Industry Directory*, 1st ed., International Chromium Development Association, Paris, Sept. 1996, pp. 1-1–1-194.
25. J. F. Papp, *Chromium Metal*, U.S. Bureau of Mines Information Circular 9430, 1995.
26. D. A. Baker and J. W. Siple, *Methods for the Analysis of Mineral Chromites and Ferrochrome Slag*, Bureau of Mines Information Circular 9240, 1990.
27. United States Environmental Protection Agency, *Test Methods for Evaluation Solid Waste*, Vol. IA: *Laboratory Manual Physical / Chemical Methods*, 3rd ed., pp. 7190-1–7190-5, 7191-1–7191-5, 7195-1–7195-7, 7196A-1–7196A-6, 7197-1–7197-6, 7198-1–9, Nov. 1986.
28. Staff, Division of Mineral Commodities, *Recycled Metals in the United States*, Special Publication U.S. Department of the Interior, Bureau of Mines, Oct. 1993, Library of Congress Superintendent of Documents I29.151; "Chromium" section is pp. 9–11.
29. Environmental Protection Agency, 1995 *Toxics Release Inventory Public Data Release*, EPA 745-R-97-005, Apr. 1997.
30. J. F. Papp, *Chromium in Mineral Commodity Summaries*, Jan. 1998, pp. 44–45.

GENERAL REFERENCES

T. P. Thayer and B. R. Lipin, A Geological Analysis of World Chromite Production to the Year 2000 A.D., *Proceedings of the Council of Economics*, 107th Annual Meeting, AIME, 1978, pp. 143–152.

S. Christiansen, *Metal Bulletin Monthly* (328), 64–67 (April 1998).

American Paint & Coatings Journal, 27 (Apr. 10, 1995).

P. A. Lewis, ed., *Pigment Handbook*, Vol. I, *Properties and Economics*, 2nd ed., Wiley-Interscience, Inc., New York, 1988.

Internal Revenue Code, 26 USCS 4661 (1992).

R. A. Carnes, J. J. Santoleri, and S. H. McHale, *J. Hazard. Mater.* **30**, 343–353 (1992).

P. N. Cheremisinoff, *Calculating and Reporting Toxic Chemical Releases for Pollution Control*, SciTech Publishers, 1990.

United States Environmental Protection Agency, *Toxic Chemical Release Inventory Reporting Form R and Instructions*, revised 1990 version, EPA 560/4-91-007, Jan. 1991.

W. A. Gericke (Chairman), *Industry Guidelines—Health Safety and Environment*, International Chromium Development Assoc., Paris, 1994.

E. Merian, ed., *Metals and Their Compounds in the Environment. Occurrence, Analysis, and Biological Relevance*, VCH Publishers, Inc., New York, 1991.

Federal Register **53**(203), 41288–41300 (Oct. 20, 1988).

Federal Register **56**(114), 27300–27330 (June 13, 1991).

Federal Register **55**(15), 2322 (Jan. 23, 1990).

Federal Register **57**(137), 31586, 31588–31589 (July 16, 1992).

Federal Register **60**(16), 4948–4993 (Jan. 25, 1995).

EPA, *A Guidebook on How to Comply with the Chromium Electroplating and Anodizing National Emission Standards for Hazardous Air Pollutants*, Environmental Protection Agency Report EPA-453/B-95-001, Apr. 1995.

A. R. Gavaskar, R. F. Olfenbuttel, J. A. Jones, and co-workers, *Cadmium and Chromium Recovery from Electroplating Rinsewaters*, NTIS Report EPA/600/R-94/050, Jan. 1994.

Federal Register **56**(20), 3526–3614 (Jan. 30, 1991).

Federal Register **56**(226), 58859–58862 (Nov. 22, 1991).

Federal Register **59**(173), 46339–46353 (Aug. 8, 1994).

R. W. Midgette and K. R. Boyer, *Nat. Environ. J.* **2**(2), 21–23 (Nov./Dec. 1992).

U.S. Environmental Protection Agency, *Report to Congress on Special Wastes From Mineral Processing*, Vol. I, *Summary and Recommendations*, July 1990, pp. 2, 11–12.

U.S. Environmental Protection Agency, *Toxics in the Community—National and Local Perspectives—the 1989 Toxics Release Inventory National Report*, U.S. Government Printing Office, Sept. 1991.

U.S. Environmental Protection Agency, *Toxics in the Community—National and Local Perspectives—the 1989 Toxics Release Inventory National Report*, EPA 560/4-91-014, Sept. 1991, Chapter 6, pp. 255–305.

U.S. Environmental Protection Agency, *The Facts Speak for Themselves: A Fundamentally Different Superfund Program*, Nov. 1996; Web site: http://www.epa.gov/superfund/oerr/whatissf/index.html.

U.S. Environmental Protection Agency, *Toxic Chemical Release Inventory Questions and Answers*, revised 1990 version, EPA 560/4-91-003, Jan. 1991.

U.S. Environmental Protection Agency, *Chromium Emissions from Chromium Electroplating and Chromic Acid Anodizing Operations. Background Information for Promulgated Standards*, NTIS Report EPA/453/R-94/082B, Nov. 1994.

U.S. Environmental Protection Agency, *Effluent Guidelines, Leather Tanning, and Pollution Prevention: A Retrospective Study*, EPA Report EPA 820-R-95-006, 1995.

C. H. Weng, C. P. Huang, H. E. Allen, and co-workers, *Sci. Total Environ.* **154**, 71–86 (1994).

National Research Council, *Recommended Dietary Allowances*, National Academy of Science, Washington, D.C., 1989, pp. 241–243.

D. Burrows, *Chromium: Metabolism and Toxicity*, CRC Press, Boca Raton, Fla., 1983.

Health Effects Assessment for Trivalent Chromium, Environmental Protection Agency Report EPA/540/1-86-035, Sept. 1984, 32 pp.

Health Effects Assessment for Hexavalent Chromium, Environmental Protection Agency Report EPA/540/1-86-019, Sept. 1984.

M. A. Ottoboni, *The Dose Makes the Poison*, 2nd ed., D. Van Nostrand-Reinhold, Inc., 1991.

S. R. Taylor and S. M. McLennan, *The Continental Crust: Its Composition and Evolution*, Blackwell Scientific Publications Ltd., Oxford, 1985, pp. 15–16, 67.

Integrated Risk Information System (IRIS), IRIS Substance File Chromium(VI); http://www.epa.gov/ngispgh3/iris/subst/0144.htm.

National Research Council, *Recommended Dietary Allowances*, National Academy of Science, Washington, D.C., 1989, pp. 241–243.

Federal Register **60**(249), 67164–67175 (Dec. 28, 1995).

FDA TALK PAPER, http://www.cfsan.fda.gov/~lrd/tpsupp.html, Jan. 2, 1996; accessed Apr. 9, 1998.

Code of Federal Regulations **29**, Part 1910.1000, 1997, pp. 7–20.

JOHN F. PAPP
BRUCE R. LIPIN
U.S. Geological Survey

CHROMIUM COMPOUNDS

1. Introduction

The first chromium compound was discovered in the Ural mountains of Russia, during the latter half of the eighteenth century. Crocoite [14654-05-8], a natural lead chromate, found immediate and popular use as a pigment because of its beautiful, permanent orange-red color. However, this mineral was very rare, and just before the end of the same century, chromite was identified as a chrome bearing mineral and became the primary source of chromium [7440-47-3] and its compounds (1) (see CHROMIUM AND CHROMIUM ALLOYS).

Around 1800, the attack of chromite [53293-42-8] ore by lime and alkali carbonate oxidation was developed as an economic process for the production of chromate compounds, which were primarily used for the manufacture of pigments (qv). Other commercially developed uses were: the development of mordant dyeing using chromates in 1820, chrome tanning in 1828 (2), and chromium plating in 1926 (3) (see DYES AND DYE INTERMEDIATES; ELECTROPLATING; LEATHER). In 1824, the first chromyl compounds were synthesized followed by the discovery of chromous compounds 20 years later. Organochromium compounds were produced in 1919, and chromium carbonyl was made in 1927 (1,2).

Kazakhstan and the Republic of South Africa account for more than half the world's chromite ore production. Almost all of the world's known reserves of chromium are located in the southeastern region of the continent of Africa. South Africa has 84% and Zimbabwe 11% of these reserves. The United States is completely dependent on imports for all of its chromium (4). The chromite's constitution varies with the source of the ore, and this variance can be important to processing. Typical ores are from 20 to 26 wt% Cr, from 10 to 25 wt% Fe, from 5 to 15 wt% Mg, from 2 to 10 wt% Al, and between 0.5 and 5 wt% Si. Other elements that may be present are Mn, Ca, Ti, Ni, and V. All of these elements are normally reported as oxides; iron is present as both Fe(II) and Fe(III) (5,6).

2. Properties

Chromium compounds number in the thousands and display a wide variety of colors and forms. Examples of these compounds and the corresponding physical properties are given in Table 1. More detailed and complete information on solubilities, including some solution freezing and boiling points, can be found in References 7–10, and 13. Data on the thermodynamic values for chromium compounds are found in References 7,8,10, and 13.

Chromium is able to use all of its $3d$ and $4s$ electrons to form chemical bonds. It can also display formal oxidation states ranging from Cr(−II) to Cr(VI). The most common and thus most important oxidation states are Cr(II), Cr(III), and Cr(VI). Although most commercial applications have centered around Cr(VI) compounds, environmental concerns and regulations in the early 1990s suggest that Cr(III) may become increasingly important, especially where the use of Cr(VI) demands reduction and incorporation as Cr(III) in the product.

Preparation and chemistry of chromium compounds can be found in several standard reference books and advanced texts (7,11,12,14). Standard reduction potentials for select chromium species are given in Table 2 whereas Table 3 is a summary of hydrolysis, complex formation, or other equilibrium constants for oxidation states II, III, and VI.

2.1. Low Oxidation State Chromium Compounds. Cr(0) compounds are π-bonded complexes that require electron-rich donor species such as CO and C_6H_6 to stabilize the low oxidation state. A direct synthesis of $Cr(CO)_6$, from the metal and CO, is not possible. Normally, the preparation requires an anhydrous Cr(III) salt, a reducing agent, an arene compound, carbon monoxide that may or may not be under high pressure, and an inert atmosphere (see CARBONYLS).

$$CrCl_3 + 6\,CO \xrightarrow[4^\circ C,\ 101\ kPa,\ H_2O]{C_6H_5MgBr,\ (C_2H_5)_2O} Cr(CO)_6 + 3\,HCl \qquad (1)$$

In equation 1, the Grignard reagent, C_6H_5MgBr, plays a dual role as reducing agent and the source of the arene compound (see GRIGNARD REACTION). The $Cr(CO)_6$ is recovered from an apparent phenyl chromium intermediate by the addition of water (19,20). Other routes to chromium hexacarbonyl are possible, and an excellent summary of chromium carbonyl and derivatives can be found in

Table 1. **Physical Properties of Chromium Compounds**[a]

Compound	CAS Registry number	Formula	Appearance
chromium(0) hexacarbonyl	[13007-092-6]	$Cr(CO)_6$	colorless crystals
dibenzene chromium(0)	[1271-54-1]	$(C_6H_6)_2Cr$	brown crystals
bis(biphenyl) chromium(I) iodide	[12099-17-1]	$(C_{12}H_{10})_2CrI$	orange plates
chromium(II) acetate dihydrate	[628-52-4]	$Cr_2(C_2H_3O_2)_4 \cdot 2H_2O$	red crystals
chromium(II) chloride	[10049-05-5]	$CrCl_2$	white crystals
ammonium chromium(II) sulfate hexahydrate	[25638-51-1]	$(NH_4)_2Cr(SO_4)_2 \cdot 6H_2O$	blue crystals
chromium(III) chloride	[10025-73-7]	$CrCl_3$	bright purple plates
chromium(III) acetylacetonate	[13681-82-8]	$Cr(C_5H_7O_2)_3$	red-violet crystals
potassium chromium(III) sulfate dodecahydrate	[7788-99-0]	$KCr(SO_4)_2 \cdot 12H_2O$	deep purple crystals
chromium(III) chloride hexahydrate	[10060-12-5]	$[Cr(H_2O)_4Cl_2]Cl \cdot 2H_2O$	bright green crystals
		$[Cr(H_2O)_6]Cl_3$	violet crystals
chromium(III) oxide	[1308-38-9]	Cr_2O_3	green powder or crystals
chromium(IV) oxide	[12018-01-8]	CrO_2	dark brown or black powder
chromium(IV) fluoride	[10049-11-3]	CrF_4	very dark greenish black powder
barium chromate(V)	[12345-14-1]	$Ba_3(CrO_4)_2$	black-green crystals
chromium(VI) oxide	[1333-82-0]	CrO_3	dark red crystals
chromium(VI) dioxide dichloride	[14977-61-8]	CrO_2Cl_2	cherry-red liquid
ammonium dichromate(VI)	[7789-09-5]	$(NH_4)_2Cr_2O_7$	red-orange crystals
potassium dichromate(VI)	[7778-50-9]	$K_2Cr_2O_7$	orange-red crystals
sodium dichromate(VI) dihydrate	[7789-12-0]	$Na_2Cr_2O_7 \cdot 2H_2O$	orange-red crystals
potassium chromate(VI)	[7789-00-6]	K_2CrO_4	yellow crystals
sodium chromate(VI)	[7775-11-3]	Na_2CrO_4	yellow crystals
potassium chlorochromate(VI)	[16037-50-6]	$KCrO_3Cl$	orange needles
silver chromate(VI)	[7784-01-2]	Ag_2CrO_4	maroon crystals
barium chromate(VI)	[10295-40-3]	$BaCrO_4$	pale yellow crystals
strontium chromate(VI)	[7789-06-2]	$SrCrO_4$	yellow crystals
lead chromate(VI)	[7758-97-6]	$PbCrO_4$	yellow crystals orange crystals red crystals

[a] Refs. 7–12.

reference 2. The only access to the less stable Cr(−II) and Cr(−I) oxidation states is by reduction of $Cr(CO)_6$.

The preparation of disodium pentacarbonylchromide [51233-19-3], $Na_2[Cr(CO)_5]$, is performed in solvents such as liquid ammonia, diglyme, or tetrahydrofuran. The Cr(O) in the $Cr(CO)_6$ solution is reduced to Cr(−II) by the addition of Na, sodium amalgam, Li, Ca, or Ba. If $NaBH_4$ is used as the reducing agent, then the Cr(−I) compound disodium decacarbonyldichromide [15616-67-8],

Crystal system	Density[b] g/cm^3	Mp, °C	Bp, °C	Solubility
orthorhombic	1.77_{18}	148.5	210^c	sl sol CCl_4; insol H_2O, $(C_2H_5)_2O$, C_2H_5OH, C_6H_6
cubic	1.519	284–285	sub 150^d	insol H_2O; sol C_6H_6
	1.617_{16}	178	dec	sol C_2H_5OH, C_5H_5N
monoclinic	1.79			sl sol H_2O, C_2H_5OH; sol acids
tetragonal	2.88	815	1300	sol H_2O to blue soln, absorbs O_2
monoclinic				sol H_2O to blue soln, absorbs O_2
hexagonal	2.76_{15}	877	sub 947	insol H_2O; sol H_2O+Cr(II), Zn or Mg
monoclinic	1.34	216	340	insol H_2O; sol C_6H_6
cubic	1.826_{25}	89^e	400^f	sol H_2O
triclinic or monoclinic	1.835_{25}	95		sol H_2O, green soln turning green-violet
rhombohedral		90		sol H_2O, violet soln turning green-violet
rhombohedral	5.22_{25}	2330	3000	insol H_2O; sol hot 70% $HClO_4$ dec
tetragonal	4.98^g	dec	dec 300 to Cr_2O_3	sol acids with dec to Cr(III) and Cr(VI)
amorphous	2.89	ca 277	ca 400	sol H_2O, dec; insol organic solvents
same as $Ca_3(PO_4)_2$				sl dec H_2O; sol acids with dec to Cr(III) and Cr(VI)
orthorhombic	2.7_{25}	197	dec	v sol H_2O; sol CH_3COOH, $(CH_3CO)_2O$
	1.9145_{25}	−96.5	115.8	insol H_2O, hydrolyzes; sol CS_2, CCl_4
monoclinic	2.155_{25}	dec 180		sol H_2O
triclinic	2.676_{25}	398	dec 500	sol H_2O
monoclinic	2.348_{25}	356; 84.6^f	dec 400	v sol H_2O
orthorhombic	2.732_{18}	975		sol H_2O
orthorhombic	2.723_{25}	792		sol H_2O
monoclinic	2.497_{39}	dec		sol H_2O, hydrolyzes
monoclinic	5.625_{25}			v sl sol H_2O; sol dilute acids
orthorhombic	4.498_{25}	dec		v sl sol H_2O; sol strong acids
monoclinic	3.895_{15}	dec		sl sol H_2O; sol dilute acids
orthorhombic				
tetragonal				
monoclinic	6.12_{15}	844		insol H_2O; sol strong acids

[b] Measurement taken at temperature in °C noted in subscript.
[c] Explodes.
[d] In vacuum.
[e] Incongruent.
[f] Loses all water at temperature indicated.
[g] Calculated value.

$Na_2[Cr_2(CO)_{10}]$, is produced (21,22). The coordination number for chromium in the carbonyls and most of the derivatives is six, with octahedral geometry around the metal. However, the geometry of Cr in some organochromium(0) compounds, eg, $(C_6H_6)_2Cr$, is very different. In dibenzene chromium(0), the Cr atom is sandwiched between the two centers of high electron density provided by the benzene molecules. The π-orbitals of C_6H_6 donate electrons as the π* orbitals

Table 2. **Standard Reduction Potentials for Chromium Species**[a]

Half-cell reaction	$E°$, V
$Cr^{3+} + 3\,e^- \longrightarrow Cr$	-0.74
$Cr(OH)^{2+} + H^+ + 3\,e^- \longrightarrow Cr + H_2O$	-0.58^b
$Cr^{2+} + 2\,e^- \longrightarrow Cr$	-0.91
$Cr^{3+} + e^- \longrightarrow Cr^{2+}$	-0.41
$Cr_2O_7^{2-} + 14\,H^+ + 6\,e^- \longrightarrow 2\,Cr^{3+} + 7\,H_2O$	1.33
$Cr_2O_7^{2-} + 10\,H^+ + 6\,e^- \longrightarrow 2\,Cr(OH)_2^+ + 3\,H_2O$	1.10^b
$CrO_4^{2-} + 4\,H_2O + 3\,e^- \longrightarrow Cr(OH)_3 + 5\,OH^-$	-0.13
$CrO_4^{2-} + e^- \longrightarrow CrO_4^{3-}$	0.1^c
$Cr^{6+} + e^- \longrightarrow Cr^{5+}$	$0.6^{c,d}$
$Cr^{5+} + e^- \longrightarrow Cr^{4+}$	$1.3^{c,d}$
$Cr^{4+} + e^- \longrightarrow Cr^{3+}$	$2.0^{c,d}$

[a] Ref. 12.
[b] Calculated from free energy data, Ref. 13.
[c] Ref. 13.
[d] In acid solutions.

simultaneously accept electrons. This back donation of electron density lowers the formal oxidation state of the metal.

The normal preparation of organochromium(0) compounds is indirect. First the organochromium(I) compound is formed

$$3\,CrCl_3 + AlCl_3 + 6\,C_6H_6 + 2\,Al \xrightarrow[C_6H_6]{AlCl_3} 3\,[(C_6H_6)_2Cr]^+ + 3\,AlCl_4^- \qquad (2)$$

Table 3. **Hydrolysis, Equilibrium, and Complex Formation Constants**

Reaction	$\log K$	Ref.
$Cr^{2+} + H_2O \rightleftharpoons Cr(OH)^+ + H^+$	5.3	15
$Cr^{3+} + H_2O \rightleftharpoons Cr(OH)^{2+} + H^+$	-4.2	16
$Cr^{3+} + 2\,H_2O \rightleftharpoons Cr(OH)_2^+ + 2\,H^+$	-10.4	16
$Cr^{3+} + 3\,H_2O \rightleftharpoons Cr(OH)_3 + 3\,H^+$	-18.7	16
$Cr^{3+} + 4\,H_2O \rightleftharpoons Cr(OH)_4^- + 4\,H^+$	-27.8	16
$2\,Cr^{3+} + 2\,H_2O \rightleftharpoons [Cr_2(OH)_2]^{4+} + 2\,H^+$	-5.3	16
$3\,Cr^{3+} + 4\,H_2O \rightleftharpoons [Cr_3(OH)_4]^{5+} + 4\,H^+$	-8.7	17
$4\,Cr^{3+} + 6\,H_2O \rightleftharpoons [Cr_4(OH)_6]^{6+} + 6\,H^+$	-13.9	17
$Cr^{3+} + 3\,C_2O_4^{2-} \rightleftharpoons [Cr(C_2O_4)_3]^{3-}$	15.4	16
$Cr^{3+} + H_2EDTA^{2-} \rightleftharpoons [CrEDTA]^- + 2\,H^+$	23.4^a	16
$Cr^{3+} + SO_4^{2-} \rightleftharpoons CrSO_4^+$	1.8	18
$H_2CrO_4 \rightleftharpoons HCrO_4^- + H^+$	0.61	12
$HCrO_4^- \rightleftharpoons CrO_4^{2-} + H^+$	-5.9	12
$2\,HCrO_4^- \rightleftharpoons Cr_2O_7^{2-} + H_2O$	2.2	12
$H_2CrO_4 + Cl^- \rightleftharpoons CrO_3Cl^- + H_2O$	1.0	13
$HCrO_4^- + HSO_4^- \rightleftharpoons CrSO_7^{2-} + H_2O$	0.60	13

[a] H_2EDTA^{2-} is $[(HOOCCH_2)_2NCH_2CH_2N(CH_2COO)_2]^{2-}$, dihydrogen ethylenediamine tetraacetate.

then the salt is reduced using dithionite in the presence of base (23,24)

$$2\left[(C_6H_6)_2Cr\right]^+ + S_2O^{2-}_4 + 4\,OH^- \longrightarrow 2\,(C_6H_6)_2Cr + 2\,SO_3^{2-} + 2\,H_2O \qquad (3)$$

The reductant of equation 3 can also be hypophosphite. Mixed organocarbonyl compounds of Cr(0) and other oxidation states are also possible. These mixed compounds make the preparation of highly unstable chromium hydrides, eg, tricarbonyl(η^5-2,4-cyclopentadien-1-yl)hydrochromium [36495-37-1], $C_5H_5Cr(CO)_3H$, possible (25). Equation 2 represents a typical preparation for organochromium(I) compounds. The orange–yellow dibenzene chromium(I) cation forms sparingly soluble salts with large anions, eg, $B(C_6H_5)_4^-$.

2.2. Chromium(II) Compounds. The Cr(II) salts of nonoxidizing mineral acids are prepared by the dissolution of pure electrolytic chromium metal in a deoxygenated solution of the acid. It is also possible to prepare the simple hydrated salts by reduction of oxygen-free, aqueous Cr(III) solutions using Zn or Zn amalgam, or electrolytically (2,7,12). These methods yield a solution of the blue $Cr(H_2O)_6^{2+}$ cation. The isolated salts are hydrates that are isomorphous with FE^{2+} and Mg^{2+} compounds. Examples are chromous sulfate heptahydrate [7789-05-1], $CrSO_4\cdot7H_2O$, chromous chloride hexahydrate [83082-80-8], $CrCl_2\cdot6H_2O$, and $(NH_4)_2Cr(SO_4)_2\cdot6H_2O$.

The standard reduction potential of Cr^{2+} (Table 2) shows that this ion is a strong reducing agent, and Cr(II) compounds have been used as reagents in analytical chemistry procedures (26). The reduction potential also explains why Cr(II) compounds are unstable in aqueous solutions. In the presence of air, the oxidation to Cr(III) occurs by reaction with oxygen. However, Cr(II) also reacts with water in deoxygenated solutions, depending on acidity and the anion present, to produce H_2 and Cr(III) (27,28).

The anhydrous halides, chromium(II) fluoride [10049-10-2], CrF_2, chromium(II) bromide [10049-25-9], $CrBr_2$, chromium(II) chloride [10049-05-5], $CrCl_2$, and chromium(II) iodide [13478-28-9], CrI_2, are prepared by reaction of the hydrohalide and pure Cr metal at high temperatures, or anhydrous chromium(II) acetate [15020-15-2], $Cr_2(CH_3COO)_4$, at lower temperatures, or by hydrogen reduction of the Cr(III) halide at about 500–800°C (2,12). These halides generally display a coordination number of six, have a distorted octahedral geometry, are moisture sensitive, and are easily oxidized when exposed to humid air.

When organic acids, RCOOH, are added to aqueous Cr(II) solutions, compounds having the general formula $Cr_2(RCOO)_4L_2$ where L = H_2O are formed. The dimeric red molecules contain a quadruple Cr-Cr bond and are diamagnetic (29). They are stable in dry air but rapidly oxidize under humid conditions. Each Cr atom has a coordination number of six and an octahedral geometry. The $RCOO^{2-}$ anion serves to bridge the interpenetrating octahedra. Compounds containing quadrupole Cr-Cr bonds and octahedral geometries are also obtained from XYZ ligands such as acetanilide [103-84-4], $CH_3CONHC_6H_5$, where the Y, in this case the carbonyl carbon, connects electron-rich centers X, the carbonyl oxygen, and Z, the nitrogen. The Cr-Cr bond is part of a five-membered ring system that has one Cr bonded to X and the other to Z (30,31).

Chromium(II) also forms sulfides and oxides. Chromium(II) oxide [12018-00-7], CrO, has two forms: a black pyrophoric powder produced from the action of nitric acid on chromium amalgam, and a hexagonal brown-red crystal made from reduction of Cr_2O_3 by hydrogen in molten sodium fluoride (32). Chromium(II) sulfide [12018-06-3], CrS, can be prepared upon heating equimolar quantities of pure Cr metal and pure S in a small, evacuated, sealed quartz tube at 1000°C for at least 24 hours. The reaction is not quantitative (33). The sulfide has a coordination number of six and displays a distorted octahedral geometry (34).

The Cr^{2+} ion is extensively hydrolyzed in aqueous solutions (Table 3) and is not easily complexed in this medium. However, many complexes, such as those of cyanide, bipyridine, phenanthroline, acetylacetone, and propylenediamine, have been prepared. The first three of these ligands form octahedral complexes, acetylacetone produces a square complex, and propylenediamine forms a trigonal bipyramidal complex (35).

2.3. Chromium(III) Compounds.   Chromium(III) is the most stable and most important oxidation state of the element. The $E°$ values (Table 2) show that both the oxidation of Cr(II) to Cr(III) and the reduction of Cr(VI) to Cr(III) are favored in acidic aqueous solutions. The preparation of trivalent chromium compounds from either state presents few difficulties and does not require special conditions. In basic solutions, the oxidation of Cr(II) to Cr(III) is still favored. However, the oxidation of Cr(III) to Cr(VI) by oxidants such as peroxides and hypohalites occurs with ease. The preparation of Cr(III) from Cr(VI) in basic solutions requires the use of powerful reducing agents such as hydrazine, hydrosulfite, and borohydrides, but Fe(II), thiosulfate, and sugars can be employed in acid solution. Cr(III) compounds having identical counterions but very different chemical and physical properties can be produced by controlling the conditions of synthesis.

The anhydrous halides, chromium(III) fluoride [7788-97-8], CrF_3, chromium(III) chloride [10025-73-7], $CrCl_3$, chromium(III) bromide [10031-25-1], $CrBr_3$, and chromium(III) iodide [13569-75-0], CrI_3, can be made by the reaction of Cr metal and the corresponding halogen at elevated temperatures (12,36). Other methods of synthesis for the halides are also possible (36–38). All of the halides have a layer structure and contain Cr(III) in an octahedral geometry. They are only slightly soluble in water but dissolve slowly when Cr(II) or a reducing agent such as Zn or Mg is added.

An unusual crystal arrangement is exhibited by the isomorphous compounds $CrCl_3$ and CrI_3. The close-packed cubic array of Cl or I atoms has two-thirds of the octahedral holes between every other pair of chlorine or iodine planes filled with chromium atoms. Alternate layers of the halogen compounds are held together by van der Waals' forces (39,40).

The chemistry of Cr(III) in aqueous solution is coordination chemistry (see COORDINATION COMPOUNDS). It is dominated by the formation of kinetically inert, octahedral complexes. The bonding can be described by d^2sp^3 hybridization, and literally thousands of complexes have been prepared. The kinetic inertness results from the $3d^3$ electronic configuration of the Cr^{3+} ion (41). This type of orbital charge distribution makes ligand displacement and substitution reactions very slow and allows separation, persistence, and/or isolation of Cr(III) species under thermodynamically unstable conditions.

The simple hexaaquachromium(III) ion, $Cr(H_2O)_6^{3+}$, obtained from chromium(III) nitrate [26679-46-9], $Cr(NO_3)_3 \cdot 9H_2O$, crystals, or chromium(III) perchlorate [25013-81-4], $Cr(ClO_4)_3 \cdot 9H_2O$, is stable at room temperature. This violet ion displays strong dichroism, ie, solutions are blue by reflected light and reddish blue by transmitted light. If these solutions are heated, the color changes to green indicating hydrolysis (Table 3) and the formation of basic trivalent cations. A general equation for hydrolysis of the trivalent ion Cr^{3+} is

$$x\, Cr^{3+} + y\, H_2O \longrightarrow \left[Cr_x(OH)_y \right]^{(3x-y)+} + y\, H^+ \qquad (4)$$

The term basic used with Cr^{3+} defines the hydroxyl ion's displacement of H_2O in the primary coordination sphere of Cr(III). This displacement effectively lowers the positive charge on the cation. Aqueous solutions of other Cr(III) salts that contain the hexaaqua ion also show some tendency by the anion of the salt to displace the coordinated water molecule, even without heating.

Figure 1 illustrates the complexity of the Cr(III) ion in aqueous solutions. The relative strength of anion displacement of H_2O for a select group of species follows the order perchlorate $\leq$ nitrate $<$ chloride $<$ sulfate $<$ formate $<$ acetate $<$ glycolate $<$ tartrate $<$ citrate $<$ oxalate (42). It is also possible for any anion of this series to displace the anion before it, ie, citrate can displace a coordinated tartrate or sulfate anion. These displacement reactions are kinetically slow, however, and several intermediate and combination species are possible before equilibrium is obtained.

The carboxylic acids or anions in the displacement series prevent the formation of basic complexes whenever present in large excess. This is not true of the acids of inorganic anions. Chromium(III) acetate [1066-30-4], $Cr(CH_3COO)_3$, is not isomorphous with the Cr(II) salt and shows no tendency to form Cr–Cr bonds. Rather, the structure depends on the ratio of acetate to Cr (43). The

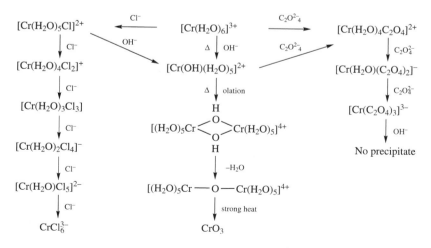

Fig. 1. Complexity of the aqueous Cr(III) ion system.

hydroxy carboxylates, eg, tartrate, may show bonding through both the alcohol and the carboxylic groups, yielding a cage-type structure (44).

The hydrolysis of Cr^{3+} (eq. 4) and the addition of less than equivalent amounts of hydroxide ion to aqueous solutions of Cr^{3+}, followed by aging, yields basic cationic polymers containing multiple chromium centers (45). The existence and formation of these polymers is of interest (17,46–49). Chromium polymers that consist of OH bridged octahedra linked via edges and/or faces, can be isolated by ion-exchange chromatography. The fraction of bridged polynuclear complexes present is proportional to the amount of hydroxide added (17,45).

When sufficient hydroxide is added to an aqueous solution of the trivalent chromium ion, the precipitation of a hydrous chromium(III) oxide, $Cr_2O_3 \cdot nH_2O$, of indefinite composition occurs. This compound is commonly misnamed as chromic or chromium(III) hydroxide [1308-14-1], $Cr(OH)_3$. A true hydroxide, chromium(III) hydroxide trihydrate [41646-40-6], $Cr(OH)_3 \cdot 3H_2O$, does exist and is prepared by the slow addition of alkali hydroxide to a cold aqueous solution of hexaaquachromium(III) ion (40). The fresh precipitate is amphoteric and dissolves in acid or in excess of hydroxide to form the metastable $Cr(OH)_4^-$. This ion decomposes upon heating to give the hydrous chromium(III) oxide. However, if the precipitate is allowed to age, it resists dissolution in excess hydroxide.

The trivalent chromium ion coordinates with almost all chelating agents (qv) and strong Lewis bases. Mixed ligand complexes of Cr(III) can yield stereoisomers. When the coordination number is 6, and J and Q are monodentate ligands, the formulas $[CrJ_4Q_2]^z$, $[CrJ_2Q_4]^z$, and $[CrJ_3Q_3]^z$, and the octahedral geometry of Cr^{3+}, allow for both cis and trans isomers. If only J is a bidentate ligand, then the formula $[CrJ_2Q_2]^z$ allows cis and trans isomers. However, if both J and Q are bidentate, then optical isomers are possible from the formulas $[CrJ_2Q]^z$, $[CrJQ_2]^z$, $[CrJ_3]^z$, and $[CrQ_3]^z$. The possible formation of binuclear complexes, eg, $[J_2Cr(OH)_2CrJ_2]^z$, allows for tartrate-type isomerism, ie, *d, l,* and *meso*. If the bidentate ligand is not symmetrical with respect to the coordination centers then structural isomers based on this asymmetry are also possible (50). The charge of the complex, z, depends on the charge of the ligands and can be positive, negative, or zero.

Chromium(III) oxide, Cr_2O_3, may be prepared by heating the hydrous chromium(III) oxide to completely remove water, as the final product of the calcination of chromium(VI) oxide, CrO_3, or by calcining chromium(III) salts that contain anions of volatile acids, eg, acetates. The Cr_2O_3 structure is isomorphous with α-alumina and α-Fe_2O_3. The best way to prepare pure chromium(III) oxide is by the decomposition of $(NH_4)_2Cr_2O_7$.

$$(NH_4)_2Cr_2O_7 \xrightarrow{200°C} Cr_2O_3 + N_2 + 4\ H_2O \qquad (5)$$

When Cr_2O_3 is introduced as an impurity into the α-Al_2O_3 lattice, as occurs in the semiprecious mineral ruby, the color is red rather than the normal green. This color anomaly is the result of ligand field splitting of the Cr(III) ion (51,52). Chromium(III) also colors other minerals (53).

Compounds that have the empirical formulas $MCrO_2$ and DCr_2O_4 where M is a monovalent and D a divalent cation, are known as chromites. These are actually mixed oxides and probably are better written as $M_2O \cdot Cr_2O_3$ and $DO \cdot Cr_2O_3$, respectively. The oxides of D are largely spinels, ie, the oxygen atoms define a close-packed cubic array having the octahedral holes occupied by the Cr(III) cation and the tetrahedral holes occupied by D (54). Chromite ore is an important member of this class of oxides.

2.4. Chromium(IV) and Chromium(V) Compounds. The formal oxidation states Cr(IV) and Cr(V) show some similarities. Both states are apparently intermediates in the reduction of Cr(VI) to Cr(III). Neither state exhibits a compound that has been isolated from aqueous media, and Cr(V) has only a transient existence in water (55). The majority of the stable compounds of both oxidation states contain either a halide, an oxide, or a mixture of these two. As of this writing, knowledge of the chemistry is limited.

Chromium(IV) fluoride [10049-11-3], CrF_4, and chromium(V) fluoride [14884-42-5], CrF_5, can be prepared by fluorinating Cr, CrF_3, or $CrCl_3$. The Cr(IV) compound is quite stable, but the fluoride of Cr(V) decomposes at 117°C and is easily hydrolyzed. The fluoride of Cr(IV) forms complexes of the type M_2CrF_6 and $DCrF_6$. These complexes are easily hydrolyzed. The K^+ salt, potassium hexafluorochromate(IV) [19652-00-7], K_2CrF_6, decomposes to potassium hexafluorochromate(III) [13822-82-7], K_3CrF_6, and CrF_5 when heated to 300°C. Although Cr(V) displays no other halides, chromium(IV) chloride [15597-88-3], $CrCl_4$, chromium(IV) bromide [51159-56-9], $CrBr_4$, and chromium(IV) iodide [23518-77-6], CrI_4, have been identified in the vapor phase of high temperature, high respective halogen vapor pressure systems.

The pure, crystalline chromium(V) oxide trifluoride can be prepared by the reaction of xenon(II) difluoride and chromium(VI) dioxide difluoride [7788-96-7], also known as chromyl fluoride, CrO_2F_2 (56):

$$XeF_2 + 2\,CrO_2F_2 \longrightarrow 2\,CrOF_3 + Xe + O_2 \tag{6}$$

Other methods of preparation, eg, the reaction of ClF_3 or BrF_3 and CrO_3, yield the oxyfluoride contaminated with reactants and side reaction products. The crystal structure of $CrOF_3$ has been found to be an infinite three-dimensional array of corner shared $CrOF_5$ octahedra (57). The species $[CrOX_4]^-$, X = F, Cl, and Br, contains the oxochromium(V) cation [23411-25-8], CrO^{3+}, exhibits a square pyramidal geometry. Compounds containing this cation are among the most stable Cr(V) compounds.

Chromium(IV) oxide [12018-01-8], CrO_2, is obtained from the hydrothermal decomposition of mixed oxides of Cr(III) and Cr(VI). A mixed oxide of the empirical formula Cr_xO_y, where the ratio of $2y$ to x is greater than 4 and less than 6, is heated at high pressure and in the presence of water to between 250 and 500°C (58). The resulting CrO_2 has an undistorted rutile structure, is ferromagnetic, and has metallic conductance. The chromium(V) oxide [12218-36-9], Cr_2O_5, is prepared by the thermal decomposition of CrO_3. It is always deficient in oxygen, giving an O to Cr mole ratio of about 2.4.

Both Cr(IV) and Cr(V) form mixed metal oxides. The blue-black and unstable tetrasodium chromate(IV) [50811-44-4], Na_4CrO_4, is formed when sodium chromite [12314-42-0], $NaCrO_2$, is heated in the presence of Na_2O to 1000°C. Compounds that have the formula D_2CrO_4 are prepared by heating the divalent metal's chromate(VI), the corresponding hydrate, and chromium(III) oxide to 1000°C. The emerald green, air-stable compounds of Sr and Ba contain the tetrahedral CrO_4^{4-} ions. Chromium(IV) mixed metal oxide species having the formulas $DCrO_3$, D_3CrO_5, and D_4CrO_6 are also known. The mixed metal oxides of chromium(V) are dark green, hygroscopic compounds of formula M_3CrO_4 or $D_3(CrO_4)_2$. The divalent cation's compound is prepared by heating a mixture of the divalent carbonate and the chromate in O_2 free nitrogen to temperatures at or above 1000°C. Both the monovalent and the divalent compounds contain the tetrahedral CrO_4^{3-} ion. Calcium chromate(V) [12205-18-4], $Ca_3(CrO_4)_2$, is isomorphous with $Ca_3(PO_4)_2$, and the other metal chromate(V) compounds show some structural similarities to the phosphates (59,60). There is also a series of compounds that have the formula $(RE)CrO_4$, where RE = La, Pr, Nd, Y, and Sm through Lu (61).

Peroxy compounds of Cr(IV) and Cr(V) are known. The chromium(IV) diperoxide adduct with ammonia [7168-85-3], $Cr(O_2)_2 \cdot 3NH_3$, crystallizes as light brown needles that are unstable and may explode, if an ammoniacal solution of ammonium perchromate is heated to 50°C, and then cooled to 0°C (62). The crystals contain Cr(IV) having a coordination number of 7 and a pentagonal bipyramidal geometry (63). Potassium tetraperoxochromate(V) [12331-76-9], $K_3Cr(O_2)_4$, is obtained as stable red-brown crystals when H_2O_2 is added to a basic solution of K_2CrO_4 maintained at 0°C (64). The geometry of Cr(V) in these crystals is dodecahedral, and its coordination number is 8 (63).

2.5. Chromium(VI) Compounds. Virtually all Cr(VI) compounds contain a Cr–O unit. The chromium(VI) fluoride [13843-28-2], CrF_6, is the only binary Cr^{6+} halide known and the sole exception. This fluoride, prepared by fluorinating Cr at high temperature and pressure, easily disproportionates to CrF_5 and F_2 at normal pressures, even at −100°C. The fluorination of chromium(VI) oxide or the reaction of KrF_2 and CrO_2F_2 in liquid HF produces chromium(VI) oxide tetrafluoride [23276-90-6], $CrOF_4$ (65). Only fluorine displays an oxyhalide having this formula.

The other Cr(VI) halides have the formula CrO_2X_2, where X = F, Cl, or Br. The mixed oxyhalides CrO_2ClY, where Y = F or Br, have been prepared but are not well characterized (66). The formula CrO_2X_2 also describes nonhalide compounds, where X = ClO_4^-, NO_3^-, SO_3F^-, N_3^-, CH_3COO^-, etc (67). Compounds containing the theoretical cation CrO_2^{2+} are commonly named chromyl. All of the chromyl compounds are easily hydrolyzed to H_2CrO_4 and HX.

The primary Cr–O bonded species is chromium(VI) oxide, CrO_3, which is better known as chromic acid [1115-74-5], the commercial and common name. This compound also has the aliases chromic trioxide and chromic acid anhydride and shows some similarity to SO_3. The crystals consist of infinite chains of vertex-shared CrO_4 tetrahedra and are obtained as an orange-red precipitate from the addition of sulfuric acid to the potassium or sodium dichromate(VI). Completely dry CrO_3 is very dark red to red purple, but the compound is deliquescent and even traces of water give the normal ruby red color. Chromium(VI) oxide is a very powerful oxidizer and contact with oxidizable organic compounds may cause fires or explosions.

Chromium(VI) oxide dissolves in water to yield the theoretical H_2CrO_4, which is only superficially similar to H_2SO_4. The two acids are about the same size, and they both have a central atom that displays a formal oxidation state of VI and a tetrahedral geometry. However, H_2CrO_4 is a very weak acid compared to sulfuric acid; H_2CrO_4 is easily reduced, but sulfuric acid is very stable. Unlike H_2SO_4, the chromium(VI) acid cannot be isolated as a pure compound, and a $HCrO_4^-$ salt analogue of $NaHSO_4$ has not been prepared. The $HCrO_4^-$ ion shows a distinct tendency to dimerize to $Cr_2O_7^{2-}$ at low total Cr(VI) concentrations, but the corresponding anion $S_2O_7^{2-}$ has not been identified in dilute aqueous sulfuric acid solutions.

The hydrolysis equilibria for H_2CrO_4 given in Table 3 are only valid in HNO_3 or $HClO_4$ solutions. Other acids yield complexes such as those shown for chloride and bisulfate ions. The exact composition of chromate(VI) anion(s) present in aqueous solution is a function of both pH and hexavalent chromium concentration (68). However, at pH values above 8, virtually all the Cr(VI) is present as the CrO_4^{2-} anion. When the pH is between 2 and 6, an equilibrium mixture of $HCrO_4^-$ and $Cr_2O_7^{2-}$ is present; when the pH is below 1, the principal species is H_2CrO_4 (68,69). At very high Cr(VI) concentrations the polychromates $Cr_3O_{10}^{2-}$ and $Cr_4O_{13}^{2-}$ may be present, but this has not been confirmed. The salts of these ions, called trichromates and tetrachromates respectively, do exist (70).

When a warm solution of $K_2Cr_2O_7$ and HCl is allowed to cool, orange needles of $KCrO_3Cl$ precipitate. The fluoride, bromide, and iodide analogues can be prepared in a similar manner. The expected oxidation of the halides by Cr(VI) is kinetically hindered allowing for the formation of CrO_3Cl^-, CrO_3Br^-, and CrO_3I^- ions. All of these compounds display a distorted octahedral geometry, hydrolyze easily, and decompose if heated (71).

The chromate(VI) salts containing the tetrahedral CrO_4^{2-} ion are a very important class of Cr(VI) compounds. Only the alkali metal, ammonium ion, and magnesium chromates show considerable water-solubility. Some cations, eg, Ag^+, Ba^{2+}, and Pb^{2+}, are so insoluble that they precipitate from acidic Cr(VI) solutions, demonstrating the labile equilibria of H_2CrO_4. Salts of colorless cations generally have a pure yellow color, but there are some useful exceptions: silver chromate(VI) [7784-01-2], Ag_2CrO_4, is a maroon color and lead chromate(VI) [15804-54-3], $PbCrO_4$, displays colors that indicate its trimorphism. The stable form is monoclinic and has and has an orange-yellow color. An unstable tetragonal orange-red form is isomorphous with, and stabilized by, $PbMoO_4$. A second unstable yellow form is orthorhombic, isomorphous with and stabilized by $PbSO_4$. The diversity shown by the lead salt is the key to its versatility as a pigment.

The dichromate(VI) salts may be obtained by the addition of acid to the chromate(VI) salts. However, they are better prepared by adding one-half the acid equivalent of a metal hydrate, oxide, or carbonate to an aqueous solution of CrO_3, then removing the water and/or CO_2. Most dichromates(VI) are water-soluble, and the salts contain water(s) of hydration. However, the normal salts of K, Cs, and Rb are anhydrous. Dichromate(VI) compounds of the colorless cations are generally orange-red. The geometry of $Cr_2O_7^{2-}$ is described as two tetrahedral CrO_4 linked by the shared odd oxygen (72).

Chromate(VI) esters and salts of organic bases are known. The esters are generally very unstable, especially those of primary alcohols. The rapid formation of chromate(VI) esters is thought to be the first step in the Cr(VI) oxidation of alcohols and aldehydes (qv) (73–75) (see ALCOHOLS, HIGHER ALIPHATIC; ALCOHOLS, POLYHYDRIC). The adduct $CrO_3 \cdot 2L$ describes the formula of virtually all the organic base salts. Examples of organic bases for the adduct L are pyridine, picolines, lutidines, and quinoline. All organic chromates(VI) are photosensitive and decompose when exposed to light.

When hydrogen peroxide is added to an acid solution of Cr(VI), a deep blue color, indicating the formation of chromium(VI) oxide diperoxide [35262-77-2], $CrO(O_2)_2$, is observed. This compound is metastable and rapidly decomposes to Cr(III) and oxygen at room temperature. The reaction sequence is unique and can be used to qualitatively confirm the presence of Cr(VI). The $CrO(O_2)_2$ species can be extracted from the aqueous solution with ether and is stable in this solvent. If pyridine is added to the ether extract, then the oxodiperoxy-(pyridine)-chromium(VI) [33361-75-0], $C_5H_5N \cdot CrO(O_2)_2$, adduct is prepared. When the acid Cr(VI) solution is at 0°C or below, the green cationic species $Cr_2(O_2)^{4+}$ and $Cr_3(O_2)_2^{5+}$ are obtained. If H_2O_2 is added to a neutral or slightly acid solution of potassium, ammonium, or thallium dichromate, the blue-violet species $[CrO(O_2)_2OH]^-$ is formed. The salts of this anion are violently explosive (63).

3. Manufacture

The primary industrial compounds of chromium made directly from chromite ore are sodium chromate, sodium dichromate, and chromic acid. Secondary chromium compounds produced in quantity include potassium dichromate, potassium chromate, and ammonium dichromate. The secondary trivalent compounds manufactured in quantity are chrome acetate, chrome nitrate, basic chrome chloride, basic chrome sulfate, and chrome oxide.

3.1. Sodium Chromate, Dichromate, and Chromic Acid.
The basic chemistry used to process chromite ore has not changed since the early nineteenth century. However, modern technologies have added many refinements to the manufacturing techniques (76,77), and plants have been adapted to meet health, safety, and environmental regulations. A generalized block flow diagram for the modern chromite ore processing plant is given in Figure 2. In the United States, chemical-grade ore from the Transvaal Region of The Republic of South Africa is employed. Historical procedures and equipment are discussed in Reference 78.

The chemical-grade ore, containing about 30% chromium, is dried, crushed, and ground in ball mills until at least 90% of its particles are less than 75 μm. It is then mixed with an excess of soda ash and, optionally, with lime and leached residue from a previous roasting operation. In American and European practice, a variety of kiln mixes have been used. Some older mixes contain up to 57 parts of lime per 100 parts of ore. However, in the 1990s manufacturers used no more than 10 parts of lime per 100 parts of the ore, and some used no lime at all (77). The roasting may be performed in one, two, or three stages, and there may be as much as three parts of leached residue per part of ore. These adaptations are responses to the variations in kiln roast and the capabilities of the furnaces used.

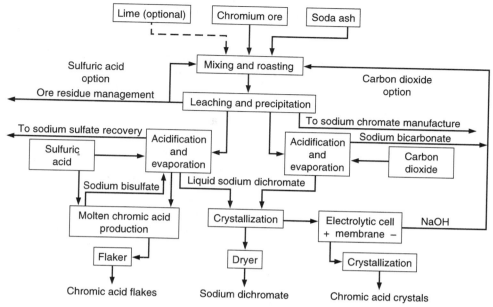

Fig. 2. Flow diagram for the production of sodium chromate, sodium dichromate, and chromic acid flake and crystals.

After thorough mixing, the mixture is roasted in a mechanical furnace, usually a rotary kiln. An oxidizing atmosphere is essential, and the basic reaction of a theoretical chromite is

$$4\,FeCr_2O_4 + 8\,Na_2CO_3 + 7\,O_2 \longrightarrow 2\,Fe_2O_3 + 8\,Na_2CrO_4 + 8\,CO_2 \qquad (7)$$

The temperature in the hottest part of the kiln is closely controlled using automatic equipment and a radiation pyrometer and generally is kept at about 1100–1150°C (see TEMPERATURE MEASUREMENT). Time of passage is about four hours, varying with the kiln mix being used. The rate of oxidation increases with temperature. However, the maximum temperature is limited by the tendency of the calcine to become sticky and form rings or balls in the kiln, by factors such as loss of Na_2O by volatilization, and by increased rate of attack on the refractory lining.

A gas-fired furnace with a revolving annular hearth also has been used to roast chrome ore (78). The mix is charged continuously at the outer edge of the hearth. A water-cooled helical screw moves it toward the inner edge where it is discharged. Mixes containing a much higher (28% Na_2CO_3) soda ash content can be handled in these furnaces. Also, the lower proportion of lime limits the formation of the suspected carcinogenic compound $Ca_3(CrO_4)_2$ (79).

Modern manufacturing processes quench the roast by continuous discharge into the leach water held in tanks equipped with agitators. At this point the pH of the leach solution is adjusted to between 8 and 9 to precipitate aluminum and silicon. The modern leaching operations are very rapid because no or little lime is used. After separation of the ore residue and precipitated impurities using rotary vacuum filters, the crude liquid sodium chromate may need to be treated to remove vanadium, if present, in a separate operation. The ore residue and precipitants are either recycled or treated to reduce hexavalent chromium to Cr(III) before disposal.

All stacks and vents attached to the process equipment must be protected to prevent environmental releases of hexavalent chromium. Electrostatic precipitators and baghouses are desirable on kiln and residue dryer stacks. Leaching operations should be hooded and stacks equipped with scrubbers (see AIR POLLUTION CONTROL METHODS). Recovered chromate values are returned to the leaching-water cycle.

Technical developments in the roasting and leaching area include refinements in pelletizing the mix fed to the kilns (80–82) and in the pre-oxidation of the ore prior to roasting (83). Both of these variants intend to increase the kiln capacity, the first through increasing the permissible fraction of soda ash in the mix, the second through increasing the effective rate of oxidation.

The neutralized, alumina-free sodium chromate solution may be marketed as a solution of 40° Bé (specific gravity = 1.38), evaporated to dryness, or crystallized to give a technical grade of sodium chromate or sodium chromate tetrahydrate [1003-82-9], $Na_2CrO_4 \cdot 4H_2O$. If the fuel for the kilns contains sulfur, the product contains sodium sulfate as an impurity. This compound is isomorphous with sodium chromate and hence difficult to separate. High purity sodium chromate must be made from purified sodium dichromate.

Sodium chromate can be converted to the dichromate by a continuous process treating with sulfuric acid, carbon dioxide, or a combination of these two (Fig. 2). Evaporation of the sodium dichromate liquor causes the precipitation of sodium sulfate and/or sodium bicarbonate, and these compounds are removed before the final sodium dichromate crystallization. The recovered sodium sulfate may be used for other purposes, and the sodium bicarbonate can replace some of the soda ash used for the roasting operation (76). The dichromate mother liquor may be returned to the evaporators, used to adjust the pH of the leach, or marketed, usually as 69% sodium dichromate solution.

Chromic acid may be produced by the reaction of sulfuric acid and sodium dichromate

$$Na_2Cr_2O_7 + 2\,H_2SO_4 \longrightarrow 2\,CrO_3 + 2\,NaHSO_4 + H_2O \qquad (8)$$

This is the sulfuric acid option of Figure 2.

Traditionally, sodium dichromate dihydrate is mixed with 66° Bé (specific gravity = 1.84) sulfuric acid in a heavy-walled cast-iron or steel reactor. The mixture is heated externally, and the reactor is provided with a sweep agitator. Water is driven off and the hydrous bisulfate melts at about 160°C. As the temperature is slowly increased, the molten bisulfate provides an excellent heat-transfer medium for melting the chromic acid at 197°C without appreciable decomposition. As soon as the chromic acid melts, the agitator is stopped and the mixture separates into a heavy layer of molten chromic acid and a light layer of molten bisulfate. The chromic acid is tapped and flaked on water cooled rolls to produce the customary commercial form. The bisulfate contains dissolved CrO_3 and soluble and insoluble chromic sulfates. Environmental considerations dictate purification and return of the bisulfate to the treating operation.

Instead of the dihydrate and sulfuric acid, 20% oleum [8014-95-7] and anhydrous sodium dichromate may be used. In this case, the reaction requires little if any external heat, and liquid chromic acid is spontaneously produced. This procedure is the basis for a continuous process (84).

Molten chromic acid decomposes at its melting point at a significant rate. The lower oxides formed impart darkness and turbidity to the water solution. Accordingly, both temperature and time are important in obtaining a quality product.

Another process depends on the addition of a large excess of sulfuric acid to a concentrated solution or slurry of sodium dichromate. Under the proper conditions, a high purity chromic acid, may be precipitated and separated (77,85).

A newer technology for the manufacture of chromic acid uses ion-exchange (qv) membranes, similar to those used in the production of chlorine and caustic soda from brine (76) (see ALKALI AND CHLORINE PRODUCTS; CHEMICALS FROM BRINE; MEMBRANE TECHNOLOGY). Sodium dichromate crystals obtained from the carbon dioxide option of Figure 2 are redissolved and sent to the anolyte compartment of the electrolytic cell. Water is loaded into the catholyte compartment, and the ion-exchange membrane separates the catholyte from the anolyte (see ELECTRO-CHEMICAL PROCESSING).

When a potential is applied across the cell, the sodium and other cations are transported across the membrane to the catholyte compartment. Sodium hydroxide is formed in the catholyte compartment, because of the rise in pH caused by the reduction of water. Any polyvalent cations are precipitated and removed. The purified NaOH may be combined with the sodium bicarbonate from the sodium dichromate process to produce soda ash for the roasting operation. In the anolyte compartment, the pH falls because of the oxidation of water. The increase in acidity results in the formation of chromic acid. When an appropriate concentration of the acid is obtained, the liquid from the anolyte is sent to the crystallizer, the crystals are removed, and the mother liquor is recycled to the anolyte compartment of the cell. The electrolysis is not allowed to completely convert sodium dichromate to chromic acid (76). Patents have been granted for more electrolytic membrane processes for chromic acid and dichromates manufacture (86).

3.2. Other Chromates and Dichromates.

The wet operations employed in the modern manufacture of the chromates and dichromates are completely enclosed and all stacks and vents equipped with scrubbers and entrainment traps to prevent contamination of the plant and its environment. The continuous process equipment that is used greatly facilitates this task. The trapped material is recycled.

Potassium and ammonium dichromates are generally made from sodium dichromate by a crystallization process involving equivalent amounts of potassium chloride or ammonium sulfate. In each case the solubility relationships are favorable so that the desired dichromate can be separated on cooling, whereas the sodium chloride or sulfate crystallizes out on boiling. For certain uses, ammonium dichromate, which is low in alkali salts, is required. This special salt may be prepared by the addition of ammonia to an aqueous solution of chromic acid. Ammonium dichromate must be dried with care, because decomposition starts at 185°C and becomes violent and self-sustaining at slightly higher temperatures.

Potassium chromate is prepared by the reaction of potassium dichromate and potassium hydroxide. Sulfates are the most difficult impurity to remove, because potassium sulfate and potassium chromate are isomorphic.

3.3. Water-Soluble Trivalent Chromium Compounds.

Most water-soluble Cr(III) compounds are produced from the reduction of sodium dichromate or chromic acid solutions. This route is less expensive than dissolving pure chromium metal, it uses high quality raw materials that are readily available, and there is more processing flexibility. Finished products from this manufacturing method are marketed as crystals, powders, and liquid concentrates.

The general method of production for aqueous trivalent compounds involves dissolving a Cr(VI) source in an acid solution of the desired anion, eg, nitric acid, in a reactor constructed of acid-resistant materials. Next, the reducing agent is added at a controlled rate until the Cr(VI) has been reduced to Cr(III). For some reducing agents it is necessary to complete the reduction at boiling or under reflux conditions. A simplified, general flow diagram for this process is given in Figure 3.

The product use determines the Cr(VI) source and limits the choice of reducing agents. High purity trivalent chromium compounds are produced from chromic acid and a variety of reducing agents that yield either the anion needed or a minimum of side reaction products. When a clean product is not required and the presence of sodium does not affect the intended application, solutions of sodium dichromate are reduced using sugars, starches, and/or other materials. Sodium-free products employ chromic acid and the same reducing agents. The reduction of Cr(VI) with sugar can be written

$$4\,Cr_2O_7^{2-} + 32\,H^+ + C_6H_{12}O_6 \longrightarrow 8\,Cr^{3+} + 6\,CO_2 + 22\,H_2O \qquad (9)$$

Although equation 9 is written as a total oxidation of sugar, this outcome is never realized. There are many intermediate oxidation products possible. Also, the actual form of chromium produced is not as simple as that shown because

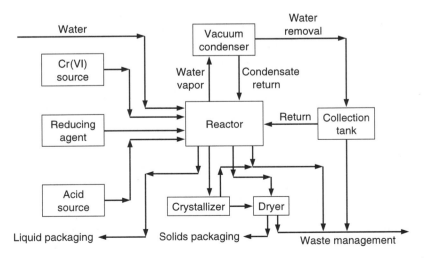

Fig. 3. Flow diagram for the manufacture of water-soluble Cr(III) compounds.

of hydrolysis, polymerization, and anion penetration. Other reducing agents are chosen to enhance the performance of the product.

The final consideration for the manufacture of Cr(III) compounds is the mole ratio of acid to Cr. This ratio determines the basicity value of the product. Basicity can also be stated as the amount of positive charge on chromium(III) neutralized by hydroxide. For example, Cr^{3+} is 0% basic, $Cr(OH)^{2+}$ is 33.3% basic, and $Cr_2(OH)_3^{3+}$ is 50% basic. The basicity value can vary continuously from 0% to 100%. It is unlikely that these formulas represent actual cationic species, but are rather simplistic images of the average charge. These formulas can be used, however, to determine the mole ratio of acid needed for manufacture of the compound. For a monovalent anion, the mole ratio of acid to Cr for 0% basicity is 3, for 33.3% basicity it is 2, and for 50% basicity it is 1.5.

Basic chrome sulfate [12336-95-7], $Cr(OH)SO_4$, is manufactured as a proprietary product under various trade names for use in leather tanning. It is generally made by reduction of sodium dichromate in the presence of sulfuric acid, and contains sodium sulfate, small amounts of organic acids if carbohydrate reducing agents are used, plus various additives. When sulfur dioxide is employed as the reducing agent, a 33.3% basic chromic sulfate is automatically obtained

$$Na_2Cr_2O_7 + 3\,SO_2 + H_2O \longrightarrow 2\,Cr(OH)SO_4 + Na_2SO_4 \qquad (10)$$

Pure sulfur dioxide is bubbled through the sodium dichromate solution in an acid-resistant tank, or sulfur burner gas is passed through a ceramic-packed tower countercurrent to descending dichromate solution. After reduction is complete, steam is bubbled through the solution to decompose any dithionate that may have formed, and to remove excess sulfur dioxide. Also, after reduction any desired additives, such as aluminum sulfate, are incorporated, and the solution is aged. It is then spray dried. Careful temperature control during drying is necessary to obtain a highly water-soluble, solid product.

The compounds are sold on a specification of chromic oxide content, 20.5–25% Cr_2O_3, and basicity, 30–58%. Solutions are also available.

4. Economic Aspects

In 2001, the total chrome ore consumption in the United States, including chemical, metallurgical (the principal use), and refractory grades, was 332,000 metric tons, having an average concentration of 42.6% Cr_2O_3. The world production of chromite in 2001 was 3,650,000 metric tons, down from the 2000 level of 4,380,000 metric tons (87).

In 2001, major producers of chromium chemicals were located in Kazakhstan, Russia, the United Kingdom and the United States. Moderate-size facilities were located in China, Japan, Romania, South Africa, and Turkey. Small-scale producers operated in China and India. Principal producers of chromium chemicals in the U.S. in 2001 were Elementis Chromium in Corpus Christi, Texas and Occidental Chemical Corp. in Castle Hayne, N. C. (87).

U.S. export data for chromium chemicals are given in Table 4. U.S. import data for chromium chemicals are given in Table 5.

Table 4. **U.S. Exports of Chromium Chemicals**[a,b]

Chromium chemicals, gross weight	2000		2001		Principal destinations, 2001
	Quantity, t	Value × 10³ $	Quantity, t	Value × 10³ $	
chromium trioxide	11,600	22,800	10,700	26,600	Canada (33%); New Zealand (10%); Australia (9%); Brazil (8%); Japan (8%); Mexico (8%); Hong Kong (5%); Germany (4%); Taiwan (4%); Korea, Republic of (3%); Thailand (2%); South Africa (1%).
chromium oxides, other	5,170	20,300	2,730	10,300	Canada (34%); Germany (18%); Belgium (11%); China (9%); Israel (3%); Japan (3%); Taiwan (3%); United Kingdom (3%); Australia (2%); Korea, Republic of (2%); Mexico (2%); Philippines (2%); Indonesia (1%).
chromium sulfates	23.5	32	13.1	200	United Kingdom (72%); Chile (18%); Hong Kong (10%).
zinc and lead chromate	287	620	158	416	Canada (7%); Chile (16%); Mexico (12%); Belgium (2%).
sodium dichromate	19,400	14,400	16,300	16,600	Canada (30%); Mexico (23%); Thailand (13%); Colombia (5%); Panama (5%); Peru (5%); Hong Kong (3%); Indonesia (3%); San Salvador (3%); Taiwan (3%); Brazil (2%); Philippines (2%).
potassium dichromate	95.4	144	18.6	44	Canada (33%); Japan (25%); New Zealand (24%); Korea, Republic of (14%); Brazil (1%); India (1%).
other chromates, dichromates, and peroxochromates	639	2,140	562	1,650	Korea, Republic of (49%); Canada (30%); Malaysia (5%); Mexico (5%); United Kingdom (5%); Saudi Arabia (3%); Japan (1%).
pigments and perparations	1,040	5,340	771	3,710	Canada (32%); Mexico (31%); Sweden (5%); Brazil (4%); Nigeria (4%); Venezuela (4%); Costa Rica (3%); Germany (2%); Guyana (2%); Trinidad and Tobago (2%); Korea, Republic of (1%); Philippines (1%).

[a] Ref. 87.
[b] Data are rounded to no more than three significant digits; may not add to totals shown.

Table 5. U.S. Imports for consumption of Chromium Compounds[a,b]

Chromium Chemicals, gross weight	2000		2001		Principal sources, 2001
	Quantity, t	Value, ×10³ $	Quantity, t	Value, ×10³ $	
chromium trioxide	8,030	13,700	10,500	17,200	Kazakhstan (57%); China (16%); Turkey (14%); United Kingdom (6%); Italy (3%); Russia (1%); South Africa (1%).
chromium oxides and hydroxides, other	3,220	12,100	2,820	10,500	Japan (31%); China (22%); Germany (22%); United Kingdom (13%); Colombia (4%); Belgium (1%); France (1%); Italy (1%); Netherlands (1%); Poland (1%); Russia (1%).
sulfates of chromium	239	227	155	151	United Kingdom (37%); Argentina (26%); Mexico (23%); Italy (14%).
chromates of lead and zinc	289	563	111	224	Norway (68%); China (20%); Japan (8%); Colombia (2%); Germany (2%).
sodium dichromate	16,900	10,500	14,800	7,760	United Kingdom (98%); China (1%); South Africa (1%).
potassium dichromate	205	392	152	322	United Kingdom (60%); Kazakhstan (26%); Netherlands (13%).
other chromates and dichromates; peroxochromates	56.9	183	110	291	Korea, Republic of (82%); United Kingdom (12%); Austria (4%); India (2%).
chromium carbide	182	2,010	267	2,900	China (45%); Germany (19%); Japan (17%); Canada (11%); United Kingdom (7%); Israel (1%).
pigments and preparations based on chromium					
chrome yellow	7,000	18,700	5,870	16,300	China (69%); Mexico (11%); China (7%); Korea, Republic of (7%); Hungary (2%); Colombia (1%); Germany (1%); Japan (1%).
molybdenum orange	1,620	7,110	1,120	5,050	Canada (90%); Colombia (4%); Mexico (3%); Germany (1%); Philippines (1%).
zinc yellow	19	21	128	98	China (62%); Czech Republic (16%); Brazil (12%); Portugal (10%).
other	1,530	6,290	1,390	4,100	China (47%); France (41%); Germany (6%); Czech Republic (3%); Japan (2%); Poland (1%).

[a] Ref. 87.
[b] Data are rounded to no more than three significant digits; may not add to totals shown.

Most of the chromium consumed worldwide is used in steel production. Production of chemicals accounted for 10%.

Sodium dichromate is the starting material for all chromium compounds. Global consumption was estimated at 660×10^3 t in 1998 (a decline in consumption). Chromium sulfide used in leather tanning was the largest consumer of sodium dichromate (37%), followed by chromic acid production (32%), and chromic oxide production (21%). Miscellaneous uses make up the remaining 10%. Chromic acid is used in the preparation of wood preservatives and in metal finishing (88).

Environmental concerns have caused a slow down in the use of sodium dichromate and is expected to show a decline of 0.3% by 2003. Environmental regulations have also had a negative impact on the use of chromium products in tanning.

5. Specifications and Shipment

Chromates and dichromates are sold in both technical and reagent grades (89,90). Chlorides and sulfates are the principal impurities. Both manufacturers' and U.S. General Services Administration (GSA) specifications exist for the technical grades (91,92) and there are also producer specifications available for some trivalent chromium compounds (93). Specifications are shown in Tables 6 and 7.

Sodium dichromate, sodium chromate, and mixtures thereof are shipped as concentrated solutions in tank cars and trucks. The chloride and sulfate contents are usually somewhat higher than in the crystalline product. Sodium dichromate is customarily shipped at a concentration of 69% $Na_2Cr_2O_7 \cdot 2H_2O$, which is close to the eutectic composition freezing at $-48.2°C$.

Chromic acid is transported in steel drums and by rail in tank cars. Multiwall paper bags, fiber drums, as well as steel drums can be employed to ship the solid chromate salts, dichromate salts, and trivalent compounds. Trivalent

Table 6. **U.S. Government Specifications for Chromium(VI) Compounds**[a]

Specification	$Na_2Cr_2O_7 \cdot 2H_2O$	CrO_3[b]	Na_2CrO_4	$K_2Cr_2O_7$[c]	$(NH_4)_2Cr_2O_7$[d]
number	O-S-595B	O-C-303D	O-S-588C	O-P-559	O-A-498B
assay,[e] %	99.0[f]	99.5	98.5	99.0	99.7
Cl⁻ wt%[g]	0.1	0.1	0.1	0.1	0.005
SO_4^{2-}, wt %[g]	0.2	0.2	1.0	0.1	0.06
H₂O insol., wt %[g]	0.2	0.1		0.1	0.02
LOD[h] at 120°C, %[g]	12.5		0.5	0.2	

[a] Ref. 91.
[b] No more than 30% may pass a 600 μm (30 mesh) U.S. Sieve screen.
[c] All must pass 2000 μm (10 mesh) U.S. Sieve screen, and no more than 25% may pass a 149 μm (100 mesh) U.S. Sieve screen.
[d] Contains 13.5% NH₃ minimally. The pH of a 20% w/v soln is 3.2.
[e] Minimum value given.
[f] Actually % $Na_2Cr_2O_7$ after drying at 120°C.

Table 7. **Specifications for Trivalent Chromium Compounds**[a]

Specification	Chrome alum	Basic chrome chloride	Chrome acetate
formula	$KCr(SO_4) \cdot 12H_2O$	$Cr_5(OH)_6Cl_9 \cdot xH_2O$[b]	$Cr(C_2H_3O_2)_3$
form	violet crystals	green powder	green liquid
Cr, wt%	10.2–10.6	29.0–33.0	11.2–11.8
basicity, %		33.0–43.0	$-5.0 - 5.0$
Cl^-, wt%		33.0–39.0	<0.05
SO_4^{2-}, wt%			<0.05
Fe, wt%	<0.01		
Cu, wt%	<0.001		
Pb, wt%	<0.005		
H_2O insolubles, wt%	<0.1	<0.25	

[a] Ref. 93.
[b] Where $8.8 \leq x \leq 12$.

chromium liquid concentrates are also available in polyethylene drums. The U.S. Department of Transportation (DOT) requires all packages having a capacity ≤416.4 L (110 gallons) to be marked with the proper shipping name and identification number of the chemical contained. The Occupational Safety and Health Administration (OSHA) requires all compounds containing chromium to be labeled as hazardous and all Cr(VI) compounds are required to contain an additional cancer hazard warning.

6. Analytical Methods

The classical wet-chemical qualitative identification of chromium is accomplished by the intense red-violet color that develops when aqueous Cr(VI) reacts with (S)-diphenylcarbazide under acidic conditions (94). This test is sensitive to 0.003 ppm Cr, and the reagent is also useful for quantitative analysis of trace quantities of Cr (96). Instrumental qualitative identification is possible using inductively coupled argon plasma–atomic emission spectroscopy (icap/aes) having a sensitivity of <10 ppb using the 205.552 nm line; using optical atomic emission spectroscopy (oaes) having an arc sensitivity of 1 ppm for the persistent emission line at 425.43 nm; and using neutron activation analysis (naa) having a sensitivity of <0.5 microgram per sample.

The methods for quantitative analysis of chromium are dependent on the concentration and the nature of the chromium containing material. There are two types of samples: water-soluble and water-insoluble. The insoluble samples, eg, ores, refractories (qv), some organochromium compounds, and some pigments, need to be converted to water-soluble forms before analyzing. This can be accomplished by fusion using sodium peroxide or dissolution in an oxidizing acid mixture. Information on preparing chromium containing samples for analysis is available (92,96–99).

6.1. Wet-Chemical Determinations. Both water-soluble and prepared insoluble samples must be treated to ensure that all the chromium is present as Cr(VI). For water-soluble Cr(III) compounds, the oxidation is easily accomplished using dilute sodium hydroxide, dilute hydrogen peroxide, and heat. Any excess peroxide can be destroyed by adding a Ni^{2+} catalyst and boiling the alkaline solution for a short time (100). Appropriate aliquot portions of the samples are acidified and chromium is found by titration either using a standard ferrous solution or a standard thiosulfate solution after addition of potassium iodide to generate an iodine equivalent. The ferrous endpoint is found either potentiometrically or by visual indicators, such as ferroin, a complex of iron(II) and o-phenanthroline, and the thiosulfate endpoint is ascertained using starch as an indicator.

To determine moderate amounts of Cr(III) and Cr(VI) in samples that have both oxidation states present, Cr(VI) is analyzed by direct titration in one sample, and the total chromium is found in a second sample after oxidation of the Cr(III). The Cr(III) concentration is determined as the difference. Trace quantities of Cr(VI) in Cr(III) compounds can be detected and analyzed by (S)-diphenylcarbazide. Trace quantities of Cr(III) in Cr(VI) may be detected and analyzed either photometrically (101) or by ion chromatography using various modes of detection (102).

6.2. Instrumental Quantitative Analysis.   Methods such as x-ray spectroscopy, oaes, and naa do not necessarily require pretreatment of samples to soluble forms. Only reliable and verified standards are needed. Other instrumental methods that can be used to determine a wide range of chromium concentrations are atomic absorption spectroscopy (aas), flame photometry, icap-aes, and direct current plasma–atomic emission spectroscopy (dcp-aes). These methods cannot distinguish the oxidation states of chromium, and speciation at trace levels usually requires a previous wet-chemical separation. However, the instrumental methods are preferred over (S)-diphenylcarbazide for trace chromium concentrations, because of the difficulty of oxidizing very small quantities of Cr(III).

Impurities in industrial chromium compounds include chloride, sulfate, insoluble matter, and trace metals. The chloride ion-selective electrode can be used to determine chloride values; sulfate is found by barium precipitation, after the reduction of Cr(VI) to Cr(III) for hexavalent chromium compounds; and a variety of methods are available to determine the specified trace metals, eg, aas and icapaes. The standard methods of organizations and agencies such as the American Society for Testing and Materials (ASTM), the American Wood Preservers' Association, the General Services Administration (GSA), and the American Leather Chemists' Association (ALCA) contain procedures for the analysis of commercially available chromium formulations, pigments, and compounds. A wider variety of tests are required for reagent chemicals (89,90).

7. Health and Safety Factors

Chromium in its hexavalent oxidation state is identified by the United States Environmental Protection Agency as one of the seventeen high-priority toxic chemicals that can cause human health problems. Hexavalent chromium (chromium VI) is found in ammonium dichromate, chromic acid, sodium dichromate, sodium chromate, and potassium dichromate. Symptoms of acute dermal exposure to hexavalent chromium are irritated skin and mucous membranes.

Table 8. **Air Standards (mg/m^3) and Classification for Various Chromium Compounds**

Substance	ACGIH	OSHA	NIOSH
chromic acid and chromates (as CrO$_3$)		ceiling 0.01	0.001 as Cr carcinogen
chromite ore processing (chromate) as Cr	0.05, A1		
chromium (II) compounds as Cr	0.5	0.5	0.5
chromium (III) compounds as Cr	0.5	0.5	0.5
chromium (VI) compounds as Cr water soluble	0.05		
chromium (VI) compounds as Cr certain water insoluble	0.01, A1		
lead chromate as Cr	0.012, A2		
strontium chromate as Cr^a	0.0005, A2		
zinc chromates as Cr	0.01, A1	ceiling 1 as CrO$_3$	0.001 carcinogen

^{a} Proposed, 1992.

Ingestion can cause serious injury or death. Deep perforating nasal ulcers know as chrome holes can result form chronic inhalation. Chromium VI is mutagenic and carcinogenic in animals and is a Classification A human carcinogen.

All chromium compounds are considered hazardous substances under CERCLA (Comprehensive Response, Compensation and Liability Act (88).

7.1. Standards, Regulations, or Guidelines of Exposure. For chromium (II) and chromium (III), the following are the exposure limits: ACGIH TLV TWA = 0.5 mg/m^3, OSHA PEL TWA = 1 mg/m^3, and NIOSH REL TWA = 0.5 mg/m^3, IDLH = 250 mg Cr(II)/m^3. ACGIH, OSHA, and NIOSH treat chromium and its compounds in slightly different ways, as shown in Table 8. It should be noted that both IARC and NTP classify certain chromium compounds as carcinogens (103). The ACGIH proposed lowering the TLV for strontium chromate to 0.0005 mg/m^3 because it is a significantly more potent carcinogen than other chromates (104). The exposure limits for the chromates (+6) is 0.1 mg/m^3 for OSHA as a ceiling value, and 0.001 mg/m^3 as a NIOSH REL. The ACGIH has a number of recommended TLVs depending upon the chromium compound.

The ACGIH biologic exposure determinants for water-soluble hexavalent chromium (the TLV is 0.05 mg/m^3) in the urine is 10 mg/g creatinine for an increase during the work shift and 30 mg/g of creatinine for the end of the shift at the end of the work week. The ACGIH noted that there is some background for those not occupationally exposed (105).

The ACGIH TLV for chromyl chloride is 0.025 mg/m^3. The NIOSH REL for chromyl chloride, which they treat as a carcinogen, is 0.001 mg/m^3 (106).

7.2. Nutrition. Chromium, in the trivalent oxidation state, is recognized as an essential trace element for human nutrition, and the recommended daily intake is 50 to 200 micrograms (107). The transport of glucose via insulin's reaction with the cell membrane, a necessary mechanism of glucose metabolism, appears to be mediated by chromium (108,109). Increased coronary disease risk, glucose intolerance, elevated serum cholesterol and elevated insulin levels have been linked to chromium deficiency (110–113). Evidence is available that suggests dietary supplements of chromium(III) may improve glucose tolerance, and there is some indication that a correlation may exist between diabetes and chromium deficient diets (114,115).

8. Environmental Concerns

The EPA has set the National Interim Primary Drinking Water Standard at 50 µg/L total chromium and the current Maximum Contamination Level (MCL) is 120 µg/L. This agency has also issued a Cr(VI) ambient water quality standard of 50 µg/L and has proposed a Maximum Contamination Level Goal (MCLG) of 0.1 µg/L (116). Industrial discharges of total Cr(VI) are regulated by National Pollutant Discharge Elimination System (NPDES) permits, specific for the area that receives the waste or discharge.

8.1. Waste Management.   All chromium-bearing waste materials in the United States are regulated by the EPA. Best practicable technology is required by the EPA to control chromium effluent in discharged industrial waters. Wastes are also subjected to designation as hazardous waste under the Resource Conservation and Recovery Act (RCRA). Chromium compounds are tracked by the EPA's Toxic Release Inventory.

In Europe, chromium is considered as cause for environmental concern. Regulations differ from country to country. Environmental issues have been a long time concern of Western Europe's leather industry. Chromium waste is significant. Sludge has been a principal concern because of the limited availability of disposal sites. An increasing number of Western European tanneries are implementing chrome-free tanning procedures.

Environmental regulations are expected to tighten with regard to air and water emissions and waste disposal.

Chromium containing solids from manufacturing and wastewater treatment sludges are classified as hazardous wastes and must be handled as such (120). These wastewater treatment sludges are F006 from electroplating wastewaters; K002, K003, K005, K006, and K008 from pigment producers' wastewaters; K086 wastewaters generated as a result of cleaning process equipment used to make chromium containing inks (qv) from pigments, driers, soaps (qv) and stabilizers; U032 wastewaters from the production of calcium chromate. These solids are characterized as D007 wastes because they exceed the Resource Conservation and Recovery Act (RCRA) threshold of 5.0 mg/L Cr as determined by the extraction procedure (EP) toxic characteristic leaching procedure (TCLP) tests and may contain both Cr(VI) and Cr(III) (121).

The EPA has established exposure levels for both Cr(III) and Cr(VI) for the general population (117,118). For exposures of short duration that constitute an insignificant fraction of the lifespan the acceptable intake subchronic (AIS) by ingestion is 979 mg/d for trivalent chromium and 1.75 mg/d for hexavalent chromium. There was insufficient data to calculate an AIS by inhalation for Cr(III), and the EPA believes this type of standard is inappropriate for hexavalent chromium (124). For lifetime exposures, an acceptable intake chronic (AIC) of 103 mg/d Cr(III) and 0.35 mg/d Cr(VI) is established for ingestion. The inhalation AIC is estimated to be 0.357 mg/d Cr(III). The EPA has calculated an inhalation cancer potency for Cr(VI) of 41 $[mg/(kg \cdot d)]^{-1}$ risk for a lifetime exposure to 1 $\mu g/m^3$ hexavalent chromium (109,117–119).

Where appropriate, the direct precipitation of hexavalent chromium with barium, and recovery of the Cr(VI) value can be employed (122). Another recycling (qv) option is ion exchange (qv), a technique that works for chromates and Cr^{3+} (123). Finally, recovery of the chromium as the metal or alloy is possible by a process similar to the manufacture of ferrochromium alloy and other metals (123).

9. Uses

9.1. Metal Finishing and Corrosion Control.
The exceptional corrosion protection provided by electroplated chromium and the protective film created by applying chromium surface conversion techniques to many active metals, has made chromium compounds valuable to the metal finishing industry. Cr(VI) compounds have dominated the formulas employed for electroplating (qv) and surface conversion, but the use of Cr(III) compounds is growing in both areas because of the health and safety problems associated with hexavalent chromium and the low toxicity of trivalent chromium (see CORROSION AND CORROSION INHIBITORS; METAL SURFACE TREATMENTS; METALLIC COATINGS).

Electroplating of Chromium. Until the middle to late 1970s, all of the commercially electroplated chromium was produced from plating baths prepared from chromic acid. Although these baths accounted for the majority of chromium electroplated products (124–127), decorative trivalent chromium baths are successfully operated in many installations (128).

Compositions and operating parameters for both Cr(VI) and Cr(III) baths are given in Table 9. Two types of trivalent baths result from different anode arrangement (129,130). The No. 1 bath uses a graphite anode that is in the bath during plating, and relies on proprietary additives combined with current

Table 9. Chemical Composition and Operating Parameters for Chromium Electroplating Baths

Parameter	Cr(III)^a No. 1	Cr(III)^a No. 2	Cr(VI)^a decorative	Cr(VI)^b functional
Cr, g/L	20–23	5–10	100–200	100–200
CrO$_3$, g/L			190–380	190–380
SO$_4^{2-}$, g/L	^{c}	^{c}	1.9–3.8	1.9–3.8
H$_3$BO$_3$, g/L	60–65	60–65		
temperature, °C	20–50	45–55	30–50	50–60
pH	2.3–2.9	3.5–3.9	<1	<1
anode type	internal	external	internal	internal
anode material	graphite	93%Pb/7%Sn	93%Pb/7%Sn	93%Pb/7%Sn
current density, A/dm^2	4–15	4–15	17.5–30	3.6–36
deposit thickness, µm	0.05–0.5	0.05–0.5	0.05–0.5	2.5–500

^{a} Ref. (128–130).
^{b} Ref. 131.
^{c} No specifications, but may be present.

density control to limit the anodic oxidation of trivalent to hexavalent chromium. Bath No. 2 employs an anode that is isolated from the plating bath by a hydrogen ion-selective membrane that allows only H^+ to pass, and therefore any anodic oxidation of Cr(III) is prevented (130). Small amounts of hexavalent chromium reduces the efficiency of trivalent baths.

Because the thickness of the plate deposited from trivalent baths is limited, these have only been employed for decorative applications. However, the bluish white deposit obtained from chromic acid baths can be closely matched by trivalent chromium baths (129).

Unlike most metals, chromium can be plated from solutions in which it is present as an anion in a high oxidation state. The deposition of chromium from chromic acid solutions also requires the presence of a catalyst anion, usually sulfate, although fluoride, fluosilicate, and mixtures of these two with sulfate have been extensively used. The amount of catalyst must be carefully regulated. Neither pure chromic acid or solutions containing excess catalysts produce a satisfactory plate. Even using carefully controlled temperature, current density, and bath composition, chromium plating is one of the most difficult electroplating operations. Throwing power and current efficiency are notably poor, making good racking procedures and good electrical practices essential.

In 1979, a viable theory to explain the mechanism of chromium electroplating from chromic acid baths was developed (132). An initial layer of polychromates, mainly $HCr_3O_{10}^-$, is formed contiguous to the outer boundary of the cathode's Helmholtz double layer. Electrons move across the Helmholtz layer by quantum mechanical tunneling to the end groups of the polychromate oriented in the direction of the double layer. Cr(VI) is reduced to Cr(III) in one-electron steps and a colloidal film of chromic dichromate is produced. Chromous dichromate is formed in the film by the same tunneling mechanism, and the Cr(II) forms a complex with sulfate. Bright chromium deposits are obtained from this complex.

Decorative chromium plating, 0.2–0.5 μm deposit thickness, is widely used for automobile body parts, appliances, plumbing fixtures, and many other products. It is customarily applied over a nonferrous base in the plating of steel plates. To obtain the necessary corrosion resistance, the nature of the undercoat and the porosity and stresses of the chromium are all carefully controlled. Thus microcracked, microporous, crack-free, or conventional chromium may be plated over duplex and triplex nickel undercoats.

Functional or hard chromium plating (125,131) is a successful way of protecting a variety of industrial devices from wear and friction. The most important examples are cylinder liners and piston rings for internal combustion engines. Functional chromium deposits must be applied to hard substrates, such as steel, and are applied in a wide variety of thicknesses ranging from 2.5 to 500 μm.

Black and colored plates can also be obtained from chromic acid baths. The plates are mostly oxides (133). Black chromium plating bath compositions are proprietary, but most do not contain sulfate. The deposit has been considered for use in solar panels because of its high absorptivity and low emissivity (131).

Chromium Surface Conversion. Converting the surface of an active metal by incorporating a barrier film of complex chromium compounds protects the metal from corrosion, provides an excellent base for subsequent painting, provides a chemical polish, and/or colors the metal. This conversion is normally accomplished by immersion, but spraying, swabbing, brushing, and electrolytic methods are also employed (134) (see METAL SURFACE TREATMENTS). The metals that benefit from chromium surface conversion are aluminum, cadmium, copper, magnesium, silver, and zinc. Zinc is the largest consumer of chromium conversion baths, and more formulations are developed for zinc than for any other metal.

The compositions of the conversion baths are proprietary and vary greatly. They may contain either hexavalent or trivalent chromium (135,136), but baths containing both Cr(III) and Cr(VI) are rare. The mechanism of film formation for hexavalent baths has been studied (137,138), and it appears that the strength of the acid and its identity, as well as time and temperature, influences the film's thickness and its final properties, eg, color. The newly prepared film is a very soft, easily damaged gel, but when allowed to age, the film slowly hardens, assumes a hydrophobic character and becomes resistant to abrasion. The film's structure can be described as a cross-linked Cr(III) polymer, that uses anion species to link chromium centers. These anions may be hydroxide, chromate, fluoride, and/or others, depending on the composition of the bath (139).

Clear-bright and blue-bright chromium conversion colors are thin films (qv) and may be obtained from both Cr(III) and Cr(VI) conversion baths. The perceived colors are actually the result of interference phenomena. Iridescent yellows, browns, bronzes, olive drabs, and blacks are only obtained from hexavalent conversion baths, and the colors are listed in the order of increasing film thickness. Generally, the thicker the film, the better the corrosion protection (see FILM DEPOSITION TECHNIQUES).

Oxide films on aluminum are produced by anodizing in a chromic acid solution. These films are heavier than those produced by chemical conversion and thinner and more impervious than those produced by the more common sulfuric acid anodizing. They impart exceptional corrosion resistance and paint adherence to aluminum and were widely used on military aircraft assemblies during World War II. The films may be dyed. A typical anodizing bath contains 50 to 100 g/L CrO_3 and is operated at 35–40°C. The newer processes use about 20 volts dc and adjust the time to obtain the desired film thickness (140).

Dichromates and chromic acid are used as sealers or after-dips to improve the corrosion resistance of various coatings on metals. For example, phosphate coatings on galvanized iron or steel as well as sulfuric acid anodic coatings on aluminum can be sealed by hexavalent chromium baths.

Chromium compounds are used in etching and bright-dipping of copper and its alloys. A typical composition for the removal of scale after heat-treating contains 30 g/L $Na_2Cr_2O_7 \cdot 2H_2O$ and 240 mL/L concentrated H_2SO_4. It is used at 50–60°C.

Chromates are used to inhibit metal corrosion in recirculating water systems. When methanol was extensively used as an antifreeze, chromates could be successfully used as a corrosion inhibitor for cooling systems in locomotive diesels and automobiles (141).

Steel immersed in dilute chromate solutions does not rust. The exact mechanism of the inhibition is not known, although it is agreed that polarization of the local anodes that serve as corrosion foci is important. In the inhibition of iron and steel corrosion a film of τ-Fe_2O_3, in which some Cr is present, appears to form. The concentration of chromate required to inhibit corrosion may range from 50 to 20,000 ppm, depending on conditions, and a pH of 8–9 is usually optimum. The inclusion of chromium compounds in formulations permits the use of such corrosive salts as zinc chloride and copper sulfate in steel cylinders.

9.2. Pigments.   Chromium pigments can be divided into chromate color pigments based on lead chromate, chromium oxide greens, and corrosion inhibiting pigments based on difficulty soluble chromate. An excellent discussion of these pigments is given in Reference 142. An older reference is also useful (143) (see PIGMENTS, INORGANIC).

Chromate Pigments Based on Lead.   Pigments based on lead can be further subdivided into primrose, lemon, and medium yellows, and chrome orange, molybdate orange, [12709-98-7] and normal lead silicochromates. Although earlier emphasis was on pure lead compounds, modern pigments contain additives to improve working properties, hue, light fastness, and crystal size and shape and to maintain metastable structures (144).

The chemical composition and ASTM specifications (145) of these pigments is given in Table 10. Details for commercial procedures are not disclosed. The pigments are characterized as follows: Medium yellows are orange–yellows that are essentially pure monoclinic lead chromate. Light lemon or primrose yellows containing up to 40% lead sulfate have some or all of the lead chromate in the metastable orthorhombic form, which is stabilized by lead sulfate and other additives. The higher the orthorhombic content, the greener the shade. Chrome oranges are basic lead chromate [18454-12-1], $PbCrO_4 \cdot PbO$. Molybdate oranges are tetragonal solid solutions of lead sulfate, lead chromate, and lead molybdate. An aging step is required in precipitation to permit development of the orange tetragonal form. Lead silicochromate, essentially medium chrome yellow precipitated on silica, has been developed for use in traffic paints where the silica gives better abrasion resistance. Chrome green, not to be confused with chromic oxide green, is a mixture of a light chrome yellow, ie, lemon or primrose, and a blue, usually iron blue. The pigment may be produced by grinding, mixing in suspension, or precipitating the yellow on the blue. The last method is the preferred. The first is hazardous because the pigment, containing both oxidizing (chromate) and reducing (ferrocyanide) components, may undergo spontaneous combustion. Phthalocyanine blues have replaced iron blues to some extent and organic greens and chromic oxide have displaced chrome green, which has poor acid and alkali resistance.

Chromium Oxide Greens.   The chromium oxide green pigments comprise both the pure anhydrous oxide, Cr_2O_3, and hydrated oxide, or Guignet's green (146). The following manufacturing processes appear to be in use.

An alkali dichromate is reduced in self-sustaining dry reaction by a reducing agent such as sulfur, carbon, starch, wood flour, or ammonium chloride. For pigment use, the reducing agent is generally sulfur. When a low sulfur grade is needed in the manufacture of aluminothermic chromium, a carbonaceous reducing agent is employed:

$$Na_2Cr_2O_7 + S \longrightarrow Na_2SO_4 + Cr_2O_3 \tag{11}$$

$$Na_2Cr_2O_7 + 2C \longrightarrow Na_2CO_3 + Cr_2O_3 + CO \tag{12}$$

Table 10. **Chemical Composition and ASTM Specifications for Chromate Color Pigments**[a]

	Composition, wt%			
			Actual	
Analyte[b]	Spec[c]	Theory	Min	Max
Primrose chrome yellow, D211-67 Type I				
$PbCrO_4$	50[d]	77.3	52.0	82.7
$PbSO_4$		22.7	4.2	25.9
TFM	8.0[e]			
Lemon chrome yellow, D211-67 Type II				
$PbCrO_4$	65[d]	72.7	52.4	68.8
$PbSO_4$		27.3	17.4	39.0
TFM	10.0[e]			
Medium chrome yellow, D211-67 Type III				
$PbCrO_4$	87[d]	100	82.4	98.2
TFM	10.0[e]			
Light chrome orange, D211-67 Type IV				
$PbCrO_4$	55[d]	59.2		
PbO		40.8		
TFM	10.0[e]			
Dark chrome orange, D211-67 Type V				
$PbCrO_4$	55[d]	59.2		
PbO		40.8		
TFM	3.0[e]			
Chrome yellow for green, D211-67 Type VI				
$PbCrO_4$	75[d]			
TFM	8.0[e]			
Pure chrome green, D212-80				
$PbCrO_4$	70[d]			
Molybdate orange, D2218-67				
$PbCrO_4$	70[d]	82.3		
$PbMoO_4$	8[d]	14.9		
$PbSO_4$		2.8		
TFM	12[e]			

[a] Ref. 145.
[b] TFM = total foreign materials or total of all substances that are not insoluble lead compounds.
[c] Spec = specification
[d] Value is minimum.
[e] Value is maximum.

The mixture is ignited with an excess of reducing agent in a reverberatory furnace or small kiln, transferred to leaching tanks, filtered, washed, dried, and pulverized. The product is 99 + % Cr_2O_3, and the metallurgical grades contain less than 0.005% of sulfur.

Chromate–dichromate solutions are reduced by sulfur in a boiling alkaline suspension (147).

$$2\,Na_2CrO_4 + Na_2Cr_2O_7 + 6\,S + 2\,x\,H_2O \longrightarrow 2\,Cr_2O_3 \cdot x\,H_2O + 3\,Na_2S_2O_3 \quad (13)$$

Excess NaOH is used to start the reaction and not over 35% of the chromium is added as dichromate. At the end of the reaction, the thiosulfate is removed by filtration and recovered. The hydrous oxide slurry is then acidified to pH 3–4 and washed free of sodium salts. On calcination at 1200–1300°C, a fluffy pigment oxide is obtained, which may be densified and strengthened by grinding. The shade can be varied by changes in the chromate:dichromate ratio, and by additives.

A dichromate or chromate solution is reduced under pressure to produce a hydrous oxide, which is filtered, washed, and calcined at 1000°C. The calcined oxide is washed to remove sodium chromate, dried, and ground. Sulfur, glucose, sulfite, and reducing gases may be used as reducing agent, and temperatures may reach 210°C and pressures 4–5 MPa (600–700 psi).

A number of manufacturers around the world are using the decomposition of ammonium dichromate to produce chrome oxide (eq. 5) (78). Generally, an excess of finely ground ammonium sulfate is mixed with sodium dichromate, and the dry mixture is heated to form chrome oxide and sodium sulfate, evolving nitrogen and steam.

$$(NH_4)_2SO_4 + Na_2Cr_2O_7 \longrightarrow Cr_2O_3 + Na_2SO_4 + N_2 + 4\,H_2O \qquad (14)$$

This is a favorable process because the side reaction products, nitrogen and water, are not pollutants and the sodium sulfate can be recovered and sold. Also, all of the wash water used to remove the sodium sulfate from the chrome oxide can be recycled.

Chromic oxide green is the most stable green pigment known. It is used where chemical and heat resistance are required and is a valuable ceramic color (see COLORANTS FOR CERAMICS). It is used in coloring cement (qv) and granulated rock for asphalt (qv) roofing. An interesting application is in camouflage paints, as the infrared reflectance of chromic oxide resembles green foliage. A minor use is in the coloring of synthetic gem stones (see GEMS). Ruby, emerald, and the dichroic alexandrite all owe their color to chromic oxide (53).

Guignet's green, or hydrated chromic oxide green, is not a true hydrate, but a hydrous oxide, $Cr_2O_3 \cdot xH_2O$, in which x is about 2. It is obtained from the production of hydrous oxide at elevated temperature, and sometimes pressure, in a borax or boric acid melt. Although Guignet's green is permanent, it does not withstand use in ceramics. It has poor tinting strength but is a very clean, transparent, bluish green. It is used in cosmetics (qv) and metallic automotive finishes (see COLORANTS FOR FOOD, DRUGS, COSMETICS, AND MEDICAL DEVICES).

Corrosion Inhibiting Pigments. Pigments inhibiting corrosion derive effectiveness from the low solubility of chromate. The principal pigment of this group is zinc chromate or zinc yellow. Others include zinc tetroxychromate, basic lead silicochromate, strontium chromate, and barium potassium chromate (148). The chemical composition and ASTM specifications of some of these pigments are shown in Table 11.

Zinc yellow became an important corrosion-inhibiting pigment for aircraft during World War II. However, the war production rate of 11,000 t/yr has not since been reached. Now, zinc yellow is widely used for corrosion inhibition on

Table 11. **Chemical Compositions and Analytical Specifications for Chromate Corrosion Inhibiting Pigments**

Analyte[a]	Composition, wt %		
	Spec[b]	Theory	Typical
Basic lead silicochromate D 1648–81 Type 1			
CrO_3	5.1–5.7		5.4
PbO	46.0–49.0		47.0
SiO_2	45.5–48.5		47.0
SSD, μm	<8.5		
Basic lead silicochromate D 1648–81 Type 2			
CrO_3	6.3–7.2		
PbO	42.5–46.0		
SiO_2	47.5–50.5		
SSD, μm	<2.0		
Strontium chromate D 1649–82			
CrO_3	41[c]		
SrO	41[c]		
SO_3	0.2[d]		
Zinc yellow (zinc chromate) D 478–49 Type I			
CrO_3	41[c]	45.8	45.0
ZnO	35–40	37.2	36.0
K_2O	13[d]	10.8	10.0
SO_3	0.2[d]		0.05
Cl	0.1[d]		
Zinc yellow (zinc chromate) D 478–49 Type II			
CrO_3	41[c]	45.8	45.0
ZnO	35–40	37.2	36.0
K_2O	13[d]	10.8	10.0
SO_3	3.0[d]		1.0
Cl	0.8[d]		
[e]	1.0[d]		
Zinc tetroxychromate			
CrO_3			17.0
ZnO			71.0
H_2O			10.0

[a] SSD = selective surface diameter determined by ASTM D1366.
[b] Ref. 145.
[c] Value is minimum.
[d] Value is maximum.
[e] When SO_3 and Cl are below maximum, the expression $[(\%SO_3/3) + (\%Cl/0.8)]$ must be used to determine conformance to specifications.

auto bodies, light metals, and steel, and in combination with red lead and ferric oxide for structural steel painting.

Zinc yellow is not a normal zinc chromate, having the empirical formula $K_2O \cdot 4ZnO \cdot 4CrO_3 \cdot 3H_2O$ [12433-50-0]. It belongs to the group of salts having the general formula $M(I)_2O \cdot 4M(II)O \cdot 4CrO_3 \cdot 3H_2O$ (149). The sodium zinc salt has occasionally been used as a pigment. The sodium copper salt has been tested as an antifouling marine pigment and is an ingredient of dips for auto bodies (see COATINGS, MARINE).

Zinc yellow is made by a variety of processes, all based on the reaction of zinc compounds, chromates, and potassium salts in aqueous solution. If products free of chloride and especially sulfate are desired, they are excluded from the system. In one process, for example, zinc oxide is swollen with potassium hydroxide and the chromates are added as a solution of potassium tetrachromate [12422-53-6] (150).

$$4\,ZnO + K_2Cr_4O_{13} + 3\,H_2O \longrightarrow K_2O \cdot 4\,ZnO \cdot 4\,CrO_3 \cdot 3\,H_2O \qquad (15)$$

The final pH is 6.0–6.6. Care must be taken in washing to avoid hydrolysis and loss of chromate.

Zinc tetroxychromate [13530-65-9], approximately $4ZnO \cdot ZnCrO_4 \cdot xH_2O$, has a somewhat lower chromate solubility than zinc yellow and has been used in wash primers.

Strontium chromate [12677-00-8], $SrCrO_4$, is used increasingly despite its high cost. It works well on light metals, and is compatible with some latex emulsions where zinc compounds cause coagulation (see LATEX TECHNOLOGY). It is also an ingredient of some proprietary formulations for chrome plating.

Basic lead silicochromate [11113-70-5] (National Lead Co. designation Pigment M-50) is a composite in which basic lead chromate, ie, chrome orange, is precipitated onto a lead silicate–silica base. It does not have an appreciable chromate solubility and depends on lead oxide for its effectiveness.

Table 12. **Chemical Composition and Specifications for Wood Preservatives**[a]

Type	Component	Composition, wt%		
		Optimum	Min	Max
	Acid copper chromate (ACC)			
	CrO_3	68.2	63.3	
	CuO	31.8	28.0	
	Chromated copper arsenate (CCA)			
type A	CrO_3	65.5	59.4	69.3
	CuO	18.1	16.0	20.9
	As_2O_5	16.4	14.7	19.7
type B	CrO_3	35.3	33.0	38.0
	CuO	19.6	18.0	22.0
	As_2O_5	45.1	42.0	48.0
type C	CrO_3	47.5	44.5	50.5
	CuO	18.5	17.0	21.0
	As_2O_5	34.0	30.0	38.0
	Chromated zinc chloride (CZC)			
	CrO_3	20	19	
	ZnO	80	76	
	Fluor chrome arsenate phenol (FCAP)			
	CrO_3	37	33	41
	As_2O_5	25	22	28
	F	22	20	24
	DNP^b	16	14	18

[a] Ref. 155.
[b] DNP = dinitrophenol.

9.3. Leather Tanning and Textiles. Although chromium(VI) compounds are the most important commercially, the bulk of the applications in the textile and tanning industries depend on the ability of Cr(III) to form stable complexes with proteins, cellulosic materials, dyestuffs, and various synthetic polymers. The chemistry is complex and not well understood in many cases, but a common denominator is the coordinating ability of chromium(III) (see LEATHER; TEXTILES).

The chrome tanning is one step in a complicated series of leather operations leading from the raw hide to the finished products. Chrome tanning is the most important tannage for all hides except heavy cattle hides, which are usually vegetable tanned. In heavy shoe uppers and soles, a chrome tanned leather is frequently given a vegetable retan to produce chrome retan leather.

Sodium dichromate and various chromic salts are employed in the textile industry (151,152). The former is used as an oxidant and as a source of chromium, for example, to dye wool and synthetics with mordant acid dyes, oxidize vat dyes and indigosol dyes on wool, aftertreat direct dyes and sulfur dyes on cotton to improve washfastness, and oxidize dyed wool. Premetallized dyes are also employed. These are hydroxyazo or azomethine dyes in which chromium or other metals are combined in the dye (see AZINE DYES; AZO DYES).

Acid Black 63 [32517-36-5] (CI 12195) is a typical premetallized dye. The commercial product contains some of the 1:2 chelate shown.

Another use of chromium compounds is in the production of water- and oil-resistant coatings on textiles, plastic, and fiber glass. Trade names are Quilon, Volan, and Scotchgard (153,154) (see WATERPROOFING AND WATER/OIL REPELLANCY).

9.4. Wood Preservation. The use of chromium compounds in wood preservation is largely because of the excellent results achieved by chromated copper arsenate (CCA), available in three modifications under a variety of trade names. The treated wood (qv) is free from bleeding, has an attractive olive-green color, and is paintable. CCA is widely used, especially in treating utility poles, building lumber, and wood foundations. About 62% of all the chromic acid produced in the United States is consumed by the wood preservation industry (77) (see BUILDING MATERIALS, SURVEY).

Chromium compounds are also used in fire-retardant formulations where their function is to prevent leaching of the fire retardant from the wood and corrosion of the equipment employed.

Chromium-containing wood preservatives and their chemical compositions are listed in Table 12 (155). Chromium compounds have a triple function in wood preservation (156). Most importantly, after impregnation of the wood the Cr(VI) compounds used in the formulations react with the wood extractives and the other preservative salts to produce relatively insoluble complexes from which preservative leaches only very slowly. This mechanism has been studied in the laboratory (157) and the field (158). Finally, although most of the chromium is reduced to chromium(III), there is probably some slight contribution of the chromium(VI) to the preservative value (159).

9.5. Drilling Muds in the Petroleum and Natural Gas Industry.
Since 1941, chromium chemicals have been used in the drilling of wells to combat fatigue corrosion cracking of drill strings, with about one metric ton of sodium chromate being used annually for an average West Texas well. Other early uses were in gas-condensate wells in Louisiana and East Texas.

However, the petroleum (qv) industry has turned to proprietary drilling-mud formulations, specially designed to suit the aqueous environment and rock strata in which the well is located (see DRILLING FLUIDS AND OTHER OIL RECOVERY CHEMICALS). In addition to heavy minerals, such as barite, and both soluble and difficulty soluble chromates for corrosion control, many of these formulations contain chromium lignosulfonates. The latter Cr(III) compounds are prepared like a tanning formula from sodium dichromate, using lignosulfonate waste from sulfite pulp (qv) mills as the reducing agent. This use amounts to about 4% of the total chromium compound consumption (160,161).

Acrylamide−polymer/Cr(III)carboxylate gel technology has been developed and field tested in Wyoming's Big Horn Basin (162,163). These gels economically enhance oil recovery from wells that suffer fracture conformance problems. The Cr(III) gel technology was successful in both sandstone and carbonate formations, and was insensitive to H_2S, high saline, and hard waters (163).

9.6. Miscellaneous Uses.
A large number of chromium compounds have been sold in small quantities for a variety of uses, some of which are described in Table 13 (185−187).

Catalysts. A more important minor use of chromium compounds is in the manufacture of catalysts (Table 13). Chromium catalysts are used in a great variety of reactions, including hydrogenations, oxidations, and polymerizations (179−181). Most of the details are proprietary and many patents are available.

Chromia−alumina catalysts are prepared by impregnating τ-alumina shapes with a solution of chromic acid, ammonium dichromate, or chromic nitrate, followed by gentle calcination. Zinc and copper chromites are prepared by coprecipitation and ignition, or by thermal decomposition of zinc or copper chromates, or organic amine complexes thereof. Many catalysts have spinel-like structures (188−191).

Photosensitive Reactions. The reduction of chromium(VI) by organic compounds is highly photosensitive, and this property is used in photosensitive dichromate-colloid systems.

A dichromate-colloid system is applied to a metal pringing plate (192). This soluble material is exposed to an image, and, where light strikes, the photochemical reaction reduces the dichromate. The chromium(III) produced forms an insoluble complex with the colloid in a reaction similar to that of dye mordanting or leather tanning. The unreacted colloid is washed off exposing bare metal that can be etched. Some of the colloids used are shellac, glue, albumin, casein, gum

Table 13. Chromium Compounds Properties and Uses

Name	CAS Registry number	Molecular formula	Properties	Uses
		Cr(VI) compounds		
ammonium chromate	[7788-98-9]	$(NH_4)_2CrO_4$	yellow crystals, 1.91, sol H_2O	textile printing, photography, dye mordant for wool, analytical reagent
barium chromate	[10294-40-3]	$BaCrO_4$	a	pyrotechnics, high temp. batteries, pigment for glasses and ceramics[b,c,d]
barium dichromate	[10031-16-0]	$BaCr_2O_7 \cdot 2H_2$	bright red-yellow needles, $-2H_2O$ at 120°C, dec in H_2O	ceramics
barium potassium chromate	[27133-66-0]	$K_2Ba(CrO_4)_2$	yellow crystals, 3.65	corrosion inhibiting pigment[e]
cadmium chromate	[14312-00-6]	$CdCrO_4$	yellow crystals, insol H_2O	catalysts, pigments
cadmium dichromate	[69239-51-6]	$CdCr_2O_7 \cdot H_2O$	orange crystals, sol H_2O	metal finishing
calcium chromate	[13765-19-0]	$CaCrO_4$	yellow crystals, sl sol H_2O	metal primers, high temp. batteries, corrosion inhibitor[b,c]
calcium dichromate	[14307-33-6]	$CaCr_2O_7 \cdot 4.5H_2O$	orange crystals, 2.136, sol H_2O	metal finishing, catalyst, corrosion inhibitor[f]
cesium chromate	[13454-78-9]	Cs_2CrO_4	yellow crystals, 4.237, sol H_2O	electronics
chromic chromate	[11056-30-7]	variable	brown, amorphous, and hydrated	catalysts, mordants
chromyl chloride	[14977-61-8]	CrO_2Cl_2	a	Etard reaction, oxidation of organics, catalyst-polymerization of olefins[g]
cobalt chromate	[13455-25-9]	$CoCrO_4$	gray-black crystals, insol H_2O, sol acids	ceramics[h]
copper chromate, basic	[12433-14-6]	$4CuO \cdot CrO_3 \cdot xH_2O$	brown, amorphous	fungicides, catalysts
copper dichromate	[13675-47-3]	$CuCr_2O_7 \cdot 2H_2O$	black crystals, 2.283, very sol H_2O, sol acids and NH_4OH	catalysts, wood preservatives
copper sodium chromate	[68399-60-0]	$Na_2O \cdot 4CuO \cdot 4CrO_2 \cdot 3H_2O$	maroon crystals, 3.57, sl sol H_2O	antifouling pigment[i]
lithium chromate	[7789-01-7]	$Li_2CrO_4 \cdot 2H_2O$	yellow crystals, 2.149, sol H_2O, transition to anhydrous at 74.6°C	corrosion inhibitor esp. in air-conditioner and nuclear reactors[j]

Table 13 (*Continued*)

Name	CAS Registry number	Molecular formula	Properties	Uses
lithium dichromate	[10022-48-7]	$Li_2Cr_2O_7 \cdot 2H_2O$	orange-red crystals, 2.34, very sol H_2O	corrosion inhibitor[k]
magnesium chromate	[16569-85-0]	$MgCrO_4 \cdot 5H_2O$	yellow crystals, 1.954, sol H_2O, turns to 7 H_2O at 17.2°C	corrosion inhibitor in gas turbines, refractories
magnesium dichromate	[34448-20-9]	$MgCr_2O_7 \cdot 6H_2O$	orange-red crystals, 2.002, sol H_2O, tr to 5H_2O at 48.5°C	catalyst, refractories[l]
mercuric chromate	[13444-75-2]	$HgCrO_4$	red crystals, slightly sol	antifouling formulations
mercurous chromate	[13465-34-4]	Hg_2CrO_4	red crystals, very sl sol	antifouling formulations
morpholine chromate	[36969-05-8]	$(C_4H_{10}NO)_2CrO_4$	yellow oily material	vapor-phase corrosion inhibitor in catalysts
nickel chromate	[14721-18-7]	$NiCrO_4$	maroon to black crystals, very sl sol	catalyst
pyridine–chromic acid adduct	[26412-88-4]	$CrO_3 \cdot 2C_5H_5N$	dark red crystals, explodes on warming	research oxidant
pyridine dichromate	[20039-37-6]	$(C_5H_5NH)_2Cr_2O_7$	orange crystals	photosensitizer in photoengraving
silver chromate	[7784-01-2]	Ag_2CrO_4	[a]	catalyst
strontium chromate	[7789-06-2]	$SrCrO_4$	[a]	corrosion-inhibiting pigment, plating additive
tetramminecopper(II) chromate	[13870-96-7]	$Cu(NH_3)_4CrO_4$	dark green needles	catalyst, gas absorbant
zinc sodium chromate	[68399-59-7]	$Na_2O \cdot 4ZnO \cdot 4CrO_3 \cdot 3H_2O$	yellow crystals, sl sol 3.24	corrosion-inhibiting pigment
Cr(III) compounds				
ammonium tetrathiocyanato diamminechromate(III)	[13573-16-5]	$NH_4(NH_3)_2Cr(SCN)_4$	red crystals	known as Reinecke's salt, analytical reagent for amines and alkaloids
basic chrome acetate	[39430-51-8]	$Cr_3(OH)_2(C_2H_3O_2)_7 \cdot xH_2O$	blue-green powder, sol H_2O	oil drilling muds, textile dye mordant, catalyst for organic oxidations[m,n]
basic chrome chloride	[50925-66-1]	$Cr_5(OH)_6Cl_9 \cdot xH_2O$	available as a green powder sol H_2O, hygroscopic	textile dye mordant, release adhesives, polymerization cross-linking agent[n]

Name	CAS Registry No.	Formula	Properties	Uses
basic chrome formate	[73246-98-7]	$Cr(OH)(OOCH)_2 \cdot 4H_2O$	green needles, example of rare crystalline basic Cr(III) salt, sol H_2O usually sold as a solution [a]	skein printing of cotton tanning
chromic acetate	[1066-30-4]	$Cr(C_2H_3O_2)_3$		printing and dyeing textiles[o]
chromic acetylacetonate	[13681-82-8]	$Cr(C_5H_7O_2)_3$		preparation Cr complexes, catalysts, antiknock compounds Cr electrowinning salt
chromic ammonium sulfate	[10022-47-6]	$NH_4Cr(SO_4)_2 \cdot 12H_2O$	violet crystals, 1.72, mp 94°C, sol water [a]	
chromic chloride	[10025-73-7]	$CrCl_3$	available as a solution	chromizing, Cr metal organochromium compounds[p] Cr plating, in catalysts chromizing
chromic fluoborate	[27519-39-7]	$Cr(BF_4)_3$		
chromic fluoride	[7788-97-8]	CrF_3	green crystals, 3.78, insol H_2O	
chromic hydroxy dichloride	[14982-80-0]	$Cr(OH)Cl_2$	sold as a water solution, can be made as isopropanol solution	manufacturing Quilon, Volan, and Scotchgard[q]
chromic naphthenate	[61788-69-0]	no definite formula	sold as soln in petroleum solvents	textile preservative
chromic nitrate	[26679-46-9]	$Cr(NO_3)_2 \cdot 9H_2O$	violet crystals, 1.80, mp 66.3°C, sol in H_2O	catalysts, textiles, manufacturing CrO_2[r]
chromic phosphate	[27096-04-4]	$CrPO_4$	green powder, also available as a solution in H_3PO_4	pigments, phosphate coatings, wash primers
chromic potassium oxalate	[15275-09-9]	$K_3[Cr(C_2O_4)_3] \cdot 3H_2O$	violet crystals [a]	dye mordant
chromic potassium sulfate	[7788-99-0]	$KCr(SO_4)_2 \cdot 12H_2O$	also available as a green powder with <12 H_2O[n]	hardening photographic emulsions, dietary supplement
chromic sulfate	[15005-90-0]	$Cr_2(SO_4)_3 \cdot xH_2O$	green amorphous powder	insolubilizing gelatin
cobalt chromite	[12016-69-2]	$CoCr_2O_4$	turquoise blue crystals, spinel	ceramics, catalysts
copper chromite	[12018-10-9]	$CuCr_2O_4$	black crystals, distorted spinel	catalysts esp. automobile exhaust[t]
magnesium chromite	[12053-26-8]	$MgCr_2O_4$	brown crystals, spinel, 4.415	refractory catalyst[s]
zinc chromite	[12018-19-8]	$ZnCr_2O_4$	green crystals, spinel, 5.30	

Other oxidation states

Name	CAS Registry No.	Formula	Properties	Uses
chromium(0) hexacarbonyl	[13007-92-6]	$Cr(CO)_6$	[a]	synthesis of organo-chromium and hydride compounds, preparation of CrO[t]
dicumene chromium(0)	[12001-89-7]	$[(CH_3)_2CHC_6H_5]_2Cr$	estd bp 300°C, explodes at 210°C	preparation of Cr carbides by vapor deposition[u]

Table 13 (*Continued*)

Name	CAS Registry number	Molecular formula	Properties	Uses
chromium(II) chloride	[10049-05-5]	$CrCl_2$	[a]	chromizing, preparation of Cr metal[v]
chromium(IV) oxide	[12018-01-8]	CrO_2	[a]	magnetic tapes[w]
calcium chromate(V)	[12205-18-4]	$Ca_3(CrO_4)_2$	green crystals, similar to $Ba_3(CrO_4)_2$[a]	corrosion inhibiting pigment, suspect carcinogen[x]

[a] See Table 1.
[b] Ref. 164.
[c] Ref. 165.
[d] Ref.166.
[e] Ref. 167.
[f] Ref. 168; Ref. 169.
[g] Ref. 170; Ref. 171.
[h] Ref. 172.
[i] Ref. 173.
[j] Ref. 174.
[k] Ref. 175.
[l] Ref. 176.
[m] Ref. 93.
[n] Ref. 162.
[o] Ref. 177.
[p] Ref. 178.
[q] Ref. 153; Ref. 154.
[r] Ref. 58.
[s] Ref. 179; Ref. 180; Ref. 181.
[t] Ref. 24; Ref. 182.
[u] Ref. 183.
[v] Ref. 184.
[w] Ref. 5; Ref. 185.
[x] Ref. 59.

arabic, and gelatin. The newer technology employs more consistent and readily controlled synthetic materials, such as poly(vinyl alcohol).

Batteries. The shelf life of dry batteries (qv) is increased from 50 to 80% by the use of a few grams of zinc chromate or dichromate near the zinc anode. This polarizes the anode on open circuit but does not interfere with current delivery.

Since World War II, the U.S. space program and the military have used small amounts of insoluble chromates, largely barium and calcium chromates, as activators and depolarizers in fused-salt batteries (165). The National Aeronautics and Space Administration (NASA) has also used chromium(III) chloride as an electrolyte for redox energy storage cells (194).

Magnetic Tapes. Chromium dioxide, CrO_2, is used as a ferromagnetic material in high fidelity magnetic tapes (qv). Chromium dioxide has several technical advantages over the magnetic iron oxides generally used (58,195).

Reagent-Grade Chemicals. Potassium dichromate is an important analytical standard, and other chromium chemicals, in reagent grades, find considerable laboratory use (89,90). This use, though small, is most important in wet analyses.

Alloys. A substantial amount of chromic oxide is used in the manufacture of chromium metal and aluminum–chromium master alloys.

Other. Recent patents describe a method for reducing cholesterol levels using proanthocyanidin and niacin-bound chromium complex (196); chromium picolinate compositions used for supplementing dietary chromium, lowering blood glucose levels and serum lipid levels and increasing lean body mass (197); heat treatable coated article with a chromium nitride ir-reflecting layer that enables the coated article to have good corrosion resistance to acid, good mechanical performance, and good color stability (198).

BIBLIOGRAPHY

"Chromium Compounds" in *ECT* 1st ed., Vol. 3, pp. 941–995, by J. J. Vetter, Diamond Alkalai Co., and C. Mueller, General Aniline & Film Corp.; in *ECT* 2nd ed., Vol. 5, pp. 473–516, by W. H. Hartford and R. L. Copson, Allied Chemical Corp.; in *ECT* 3rd ed., Vol. 6, pp. 82–120, by W. H. Hartford, Belmont Abbey College; in *ECT* 4th ed., Vol. 6, pp. 263–311, by Billie J. Page and Gary W. Loar, McGean-Rohco, Inc.; "Chromium Compounds" in *ECT* (online), posting date: December 4, 2000, by Billie J. Page and Gary W. Loar, McGean-Rohco, Inc.

CITED PUBLICATIONS

1. M. J. Udy, ed., *Chromium*, Vol. **1**, Reinhold Publishing Co., New York, 1956, pp. 1–6.
2. C. L. Rollinson, in J. C. Bailer, Jr., H. J. Emeléus, R. Nyholm, and A. F. Trotman-Dickenson, eds., *Comprehensive Inorganic Chemistry*, Vol. **3**, Pergamon Press, Oxford, U.K., 1973, 624–625.
3. U.S. Pat. 1,591,188 (Apr. 20, 1926), C. G. Fink (to United Chromium).
4. *Strategic Minerals—Extent of U.S. Reliance on South Africa*, U.S. General Accounting Office (GAO), Report to Congressional Requesters, GAO/NSIAD-88-201, Gaithersburg, Md., June 1988.

5. W. H. Hartford, in I. M. Kolthoff and P. J. Elving, eds., *Analytical Chemistry of the Elements*, Vol. **8**, Part II, Wiley-Interscience, New York, 1963, p. 278.
6. Ref. 2, p. 625.
7. Ref. 1, pp. 113–250.
8. J. A. Dean, ed., *Lange's* Handbook of Chemistry, 13th ed., McGraw-Hill Book Co., New York, 1985.
9. W. F. Linke, ed., *Solubilities*, 4th ed., Vol. **1**, D. Van Nostrand Co., Princeton, N.J., 1958.
10. R. C. Weast, ed., *CRC Handbook of Chemistry and Physics*, 65th ed., CRC Press Inc., Boca Raton, Fla., 1985.
11. F. Hein and S. Herzog, in G. Brauer, ed., *Handbook of Preparative Inorganic Chemistry*, Vol. **2**, 2nd ed., Academic Press, Inc., New York, 1965, 1334–1399.
12. F. A. Cotton and G. Wilkinson, *Advanced Inorganic Chemistry*, 5th ed., John Wiley & Sons, Inc., New York, 1988, pp. 679–697.
13. I. Dellin, F. M. Hall, and L. G. Hepler, *Chem. Rev.* **76**, 283,292 (1976).
14. "Chrom" in *Gmelins Handbuch der Anorganischen Chemie*, 8th ed., System no. 52, 1963–1965.
15. I. Nagypal and co-workers, *J. Chem. Soc. Dalton Trans.* 1335 (1983).
16. J. Kragten, *Atlas of Metal Ligand Equilibria in Aqueous Solution*, Ellis Horwood Ltd., Coll House, UK, 1978, 214–222.
17. H. Stünzi, L. Spiccia, F. P. Rotzinger, and W. Marty, *Inorg. Chem.* **28**, 66 (1989).
18. N. Tanaka, K. Ogino-Ebata, and G. Sato, *Bull. Chem. Soc. Japan* **31**, 366 (1966).
19. B. B. Owen, J. English, Jr., H. G. Cassidy, and C. V. Dundon, in L. F. Audrieth, ed., *Inorganic Syntheses*, Vol. **3**, McGraw-Hill Book Co., New York, 1950, pp. 156–160.
20. Ref. 11, p. 1741.
21. R. B. King, *Advances in Organometallic Chemistry*, Vol. **2**,. Academic Press, Inc., New York, 1964, p. 182.
22. F. Calderazzo, R. Ercoli, and G. Natta, *Organic Syntheses via Metal Carbonyls*, Vol. **1**, Wiley-Interscience, New York, 1968, p. 147.
23. Ref. 11, p. 1395.
24. Ref. 2, p. 648.
25. Ref. 2, p. 655.
26. I. M. Kolthoff and R. Belcher, *Volumetric Analysis*, Vol. **3**, Interscience Publishers, Inc., New York, 1957, 630–631.
27. Ref. 12, p. 684.
28. R. W. Kolaczkowski and R. A. Plane, *Inorg. Chem.* **3**, 322 (1964).
29. Ref. 12, pp. 685–686.
30. A. Bino, F. A. Cotton, and W. Kaim, *J. Am. Chem. Soc.* **101**, 2506 (1979).
31. J. J. H. Edema, S. Gambarotta, F. van Bolhuis, and A. L. Spek, *J. Am. Chem. Soc.* **111**, 2142 (1989).
32. Ref. 2, p. 660.
33. Ref. 11, p. 1346.
34. Ref. 12, p. 680.
35. Ref. 12, pp. 684–685.
36. B. J. Sturm, *Inorg. Chem.* **1**, 665 (1962).
37. G. W. Watt, P. S. Gentile, and E. P. Helvenston, *J. Am. Chem. Soc.* **77**, 2752 (1955).
38. G. B. Heisig, B. Fowkes, and R. Hedin, *Inorganic Syntheses*, Vol. **2**, McGraw-Hill Book Co., New York, 1946, p. 193.
39. Ref. 2, p. 664.
40. Ref. 12, p. 682.
41. F. Baslo and R. G. Pearson, *Mechanisms of Inorganic Reactions*, 2nd ed., John Wiley & Sons, Inc., New York, 1967, pp. 141–145.

42. Ref. 2, p. 679.
43. J. E. Tackett, *Appl. Spectros.* **43**, 490 (1989).
44. G. L. Robbins and R. E. Tapscott, *Inorg. Chem.* **15**, 154 (1976).
45. J. A. Laswick and R. A. Plane, *J. Am. Chem. Soc.* **81**, 3564 (1959).
46. L. Spiccia and W. Marty, *Inorg. Chem.* **25**, 266 (1986).
47. D. Rai, B. M. Sass, and D. A. Moore, *Inorg. Chem.* **26**, 345 (1987).
48. L. Spiccia, H. Stoeckli-Evans, W. Marty, and R. Giovanoli, *Inorg. Chem.* **26**, 474 (1987).
49. L. Spiccia, *Inorg. Chem.* **27**, 432 (1988).
50. Ref. 2, pp. 670–672.
51. Ref. 12, pp. 689–690.
52. B. M. Lorffler and R. G. Burns, *Am. Sci.* **64**, 636 (1977).
53. W. H. Hartford, *Rocks Miner.* **52**, 169 (1977).
54. Ref. 2, p. 666.
55. D. M. L. Goodgame and A. M. Joy, *Inorg. Chim. Acta* **135**, 115 (1987).
56. Ref. 12, p. 692.
57. G. L. Gard, *Inorg. Chem.* **25**, 426 (1986).
58. U.S. Pat. 3,117,093 (Jan. 7, 1964), P. Arthur, Jr., and J. N. Ingram (to E. I. du Pont de Nemours & Co., Inc.).
59. W. H. Hartford, in D. M. Serrone, ed., *Proceedings Chromium Symposium 1986: An Update*, Industrial Health Foundation, Inc., Pittsburgh, Pa., 1986, p. 9.
60. R. Scholder and H. Suchy, *Z. Anorg. Allgem. Chem.* **308**, 295 (1961).
61. H. Schwartz, *Z. Anorg. Allgem. Chem.* **323**, 275 (1963).
62. Ref. 11, pp. 1392–1393.
63. Ref. 12, p. 696.
64. Ref. 11, pp. 1391–1392.
65. K. O. Christe and co-workers, *Inorg. Chem.* **25**, 2163 (1986).
66. Ref. 12, p. 694.
67. K. B. Wiberg, in K. B. Wiberg, ed., *Oxidation in Organic Chemistry, Part A*, Academic Press, Inc., New York, 1965, pp. 69–184.
68. T. Shen-yang and L. Ke-an, *Talanta* **33**, 775 (1986).
69. Ref. 12, p. 693.
70. Ref. 1, p. 137.
71. Ref. 2, p. 695.
72. P. Löfgren and K. Waltersson, *Acta Chem. Scand.* **25**, 35 (1971).
73. K. B. Wiberg and W. H. Richardson, *J. Am. Chem. Soc.* **84**, 2800 (1962).
74. K. B. Wiberg and P. A. Lepse, *J. Am. Chem. Soc.* **86**, 2612 (1964).
75. F. H. Westheimer, *Chem. Rev.* **45**, 419 (1949).
76. R. J. Barnhart, *AESF Second Chromium Colloquium*, American Electroplaters and Surface Finishers Society, Orlando, Fla., 1990, Session III, A. p. 1.
77. R. J. Barnhart, private communication, Mar. 16, 1992.
78. F. Ullman, *Enzyklopädie der Technischen Chemie*, Vol. **3**, Urban and Schwarzenberg, Berlin, Germany, 1929, 400–433.; F. McBerty and Wilcoxon, *FIAT Rev. Ger. Sci. PB22627, Final Report No. 796*, 1946; Ref. 1, pp. 262–282.
79. Ref. 59, p. 21.
80. U.S. Pat. 3,095,266 (June 25, 1963), W. B. Lauder and W. H. Hartford (to Allied Chemical Corp.).
81. U.S. Pat. 3,853,059 (Dec. 3, 1974), C. P. Bruen, W. W. Low, and E. W. Smalley (to Allied Chemical Corp.).
82. S. African Pat. 74-03,604 (Apr. 28, 1975), C. P. Bruen and co-workers (to Allied Chemical Corp.).
83. U.S. Pat. 3,816,095 (June 11, 1974), C. P. Bruen, W. W. Low, and E. W. Smalley (to Allied Chemical Corp.).

84. U.S. Pat. 1,873,589 (Aug. 23, 1932), P. R. Hines (to Harshaw Chemical Co.).

85. U.S. Pat. 3,065,055 (Nov. 20, 1962), T. S. Perrin and R. E. Banner (to Diamond Alkali Co.); U.S. Pat. 3,607,026 (Sept. 21, 1971), T. S. Perrin, R. E. Banner, and J. O. Brandstaetter (to Diamond Alkali Co.).

86. U.S. Pat. 5,094,729 (Mar. 10, 1992), H. Klotz, H. D. Pinter, R. Weber, H. Block, and N. Lönhoff (to Bayer Aktiengesellschaft); U.S. Pat. 5,096,547 (Mar. 17, 1992) H. Klotz, R. Weber, and W. Ohlendorf (to Bayer Aktiengesellschaft).

87. J. F. Papp, "Chromium" *Minerals Yearbook*, U.S. Geological Survey, Reston, Va., 2001.

88. "Inorganic Chromium Compounds," *Chemical Economics Handbook*, SRI, Menlo Park, Calif., 2000.

89. *Reagent Chemicals*, 6th ed., American Chemical Society, Washington, D.C., 1981, pp. 185–190, 430–431, 434–435.

90. J. Rosin, *Reagent Chemicals and Standards*, 5th ed., D. Van Nostrand Reinhold Co., New York, 1967.

91. *Federal Specifications* O-C-303D, Aug. 13, 1985; O-A-498B, United States General Services Administration (GSA), July 11, 1985; O-S-588C, June 20, 1985; O-S-595B, May 15, 1967; O-P-559, May 26, 1952.

92. W. H. Hartford, in F. D. Snell and L. C. Ettre, eds., *Encyclopedia of Industrial Chemical Analysis*, Vol. **9**, John Wiley & Sons, Inc., New York, 1970, pp. 680–709.

93. *Chrome*, McGean Division of McGean-Rohco, Inc., Cleveland, Ohio, 1991.

94. F. Feigl, *Spot Tests*, 4th English ed., Elsevier Publishing Co., Amsterdam, The Netherlands, 1954, pp. 159–162.

95. F. D. Snell, *Photometric and Fluorometric Methods of Analysis of Metals*, John Wiley & Sons, Inc., New York, 1978, pp. 714–729.

96. E. D. Olsen and C. C. Foreback, in F. D. Snell and L. C. Ettre, in Ref. 93, Vol. **9**, pp. 632–680.

97. W. H. Hartford, in Ref. 92, Vol. 9, pp. 176–213.

98. W. H. Hartford, in I. M. Kolthoff and P. J. Elving, eds., *Treatise on Analytical Chemistry*, Part II, Vol. **8**, John Wiley & Sons, Inc., New York, 1963, pp. 273–369.

99. R. Bock, *A Handbook of Decomposition Methods in Analytical Chemistry*, John Wiley & Sons, Inc., New York, 1979.

100. Ref. 26, pp. 334–335.

101. Ref. 95, pp. 729–732.

102. V. D. Lewis, S. H. Nam, and I. T. Urasa, *J. Chromatogr. Sci.* **27**, 489 (1989).

103. National Toxicology Program (NTP), *Sixth Annual Report on Carcinogens*, NTP, Research Triangle Park, NC, 1991.

104. TLV Committee, *Appl. Occup. Environ. Hyg.* **7**, 135 (1992).

105. American Conference of Governmental Industrial Hygienists (AGCIH), *Threshold Limit Values and Biological Exposure Indices, 1992–1993*, ACGIH, Cincinnati, OH, 1992.

106. S. Langård in E. Bingham, B. Cohrssen, and C. H. Powell, eds., *Patty's* Toxicology, 5th ed., Vol. 3, John Wiley & Sons, Inc., New York, 2001.

107. W. Mertz, *Science* **212**, 1332 (1981).

108. W. Mertz, *Physiol. Ref.* **49**, 163 (1969).

109. *Health Assessment Document for Chromium*, EPA-600/8-83-014F, United States Environmental Protection Agency (EPA), 1984.

110. W. Mertz, *Contemp. Nut.* **7**(3), 2 (1982).

111. K. N. Jeejeebhoy and co-workers, *Am. J. Clin. Nutr.* **30**, 531 (1977).

112. H. Freund, S. Atamian and J. E. Fischer, *J. Am. Medical Assoc.* **241**, 496 (1979).

113. R. A. Anderson, *Sci. Total Environ.* **17**, 13 (1981).

114. R. S. Anderson, in Ref. 59, p. 238.

115. J. A. Fisher, *The Chromium Program*, Harper and Row Publishers, New York, 1990.
116. *Ambient Water Quality Criteria for Chromium*, EPA 440/5-80-35, United States Environmental Protection Agency (EPA), 1980; *Fed. Reg.* (FR), cited as Volume FR Number: 56 FR 3526; 56 FR 30266.
117. *Health Effects Assessment for Hexavalent Chromium*, EPA/540/1-86-019, United States Environmental Protection Agency (EPA), Sept. 1984; *Toxicological Profile for Chromium*, Agency for Toxic Substances and Disease Registry (ASTDR), ASTDR/TP-88/10, 1989.
118. *Health Effects Assessment for Trivalent Chromium*, EPA/540/1-86-035, United States Environmental Protection Agency (EPA), Sept. 1984.
119. A. Yassi and E. Nieboer, in J. O. Nriagu and E. Nriagu, eds., *Chromium in the Natural and Human Environments*, John Wiley & Sons, Inc., New York, 1988, pp. 443–496.
120. 52 CFR 8140.
121. 40 CFR 261.4; 6.
122. U.S. Pat. 3,552,917 (Jan. 5, 1971), C. O. Weiss (to M&T Chemicals, Inc.); U.S. Pat. 3,728,273 (Apr. 17, 1973), C. P. Bruen and C. A. Wamser (to Allied Chemical Corp.).
123. *Treatment Technology Background Document*, United States Environmental Protection Agency (EPA), 1989.
124. W. H. Hartford in A. J. Bard, ed., *Applied Electrochemistry of the Elements*, Marcel Dekker, New York, 1977.
125. G. Dubpernell in F. A. Lowenheim, ed., *Modern Electroplating*, 3rd ed., Wiley-Interscience, New York, 1974, pp. 87–151.
126. J. M. Hosdowich, in M. J. Udy, ed., *Chromium*, Vol. II, Reinhold Publishing Corp., New York, 1958, 65–91.
127. F. A. Lowenheim and M. R. Moran, *Faith, Keyes, and Clark's Industrial Chemicals*, 4th ed., Wiley-Interscience, New York, 1975, pp. 716–721.
128. V. Opaskar and D. Crawford, *Metal Finishing* **89**(1), 49 (1989).
129. D. L. Snyder, in P. H. Langdon, ed., *Metal Finishing 59th Guidebook and Directory*, Metals and Plastics Publications, Inc., Hackensack, N.J., 1991, 179–187.
130. D. Smart, T. E. Such, and S. J. Wake, *Bull. Inst. Met. Finish.* **61**, 105 (1983).
131. K. R. Newby, in Ref. 128, 173, pp. 188–196.
132. J. P. Hoare, *J. Electrochem. Soc.: Electrochem. Sci. and Technol.* **126**, 190 (1979).
133. P. Caokan, G. Barnafoldi, and I. Royik, *Bull. Doc. Cent. Inform. Chrome Dur.*, Paris, France, June 1971, pp. 11–A50.
134. F. W. Eppensteiner and M. R. Jenkins, in Ref. 173, 418–431.
135. U.S. Pat. 4,171,231 (Oct. 16, 1979), C. V. Bishop, T. J. Foley, and J. M. Frank (to R. O. Hull & Co., Inc.).
136. U.S. Pat. 4,705,576 (Nov. 10, 1987), K. Klos, K. Lindemann, and W. Birnstiel (to Elektro-Brite GmbH).
137. G. Jarrett, *Met. Finish.* **65**(3), 90 (1967).
138. L. F. G. Williams, *Surf. Technol.* **4**, 355 (1976).
139. C. V. Bishop, D. M. Burdt, and K. R. Romer, *Galvanotechnik* **71**, 1199 (1980).
140. D. C. Montgomery and C. A. Grubbs, in Ref. 173, 409–416.
141. Ref. 1, pp. 406–422.
142. T. C. Patton, *Pigment Handbook*, Vol. 1, Wiley-Interscience, New York, 1973.
143. C. H. Love, *Important Inorganic Pigments*, Hobart, Washington, D.C., 1947.
144. Ref. 143, pp. 357–389.
145. *1984 Annual Book of ASTM Standards*, Section 6, Vol. 6.02, American Society for Testing and Materials (ASTM), Philadelphia, Pa., 1984.
146. Ref. 143, pp. 351–357.

147. U.S. Pat. 2,246,907 (July 30, 1940), O. F. Tarr and L. G. Tubbs (to Mutual Chemical Company of America).
148. Ref. 143, pp. 843–861.
149. W. H. Hartford, *J. Am. Chem. Soc.* **72**, 1286 (1950).
150. U.S. Pat. 2,415,394 (Feb. 4, 1947), O. F. Tarr and M. D (to Mutual Chemical Company of America).
151. Ref. 1, pp. 283–301.
152. H. A. Lubs, *The Chemistry of Synthetic Dyes and Pigments*, Robert E. Krieger Publishing Co., Huntington, N.Y., 1972, pp. 153, 160, 161, 247, 258, 284, 426.
153. U.S. Pat. 2,662,835 (Dec. 16, 1953), T. S. Reid (to Minnesota Mining & Manufacturing Co.).
154. U.S. Pat. 2,683,156 (July 6, 1954), R. F. Iler (to E. I. du Pont de Nemours & Co., Inc.).
155. *American Wood-Preserver's* Association, Standards 1984, Section 2, Standard No. P5-83, American Wood-Preserver's Association (AWPA), Stevensville, Md., 1984.
156. W. H. Hartford, in D. D. Nicholas, ed., *Wood Deterioration and its Prevention by Preservative Treatments*, Vol. 2, Syracuse University Press, Syracuse, N.Y., 1973, 1–120.
157. S. E. Dahlgren and W. H. Hartford, *Holzforschung* **26**(2), 62 (1972); **26**(3), 105 (1972); **26**(4), 142 (1972); **28**(2), 58 (1974); **29**(3), 84 (1975); **29**(4), 130 (1975).
158. R. D. Arsenault, *Proc. Am. Wood Preserver's Assoc.* **71**, 126 (1975).
159. W. H. Hartford, *Proc. Am. Wood Preserver's Assoc.* **72**, 172 (1976).
160. W. F. Rogers, *Composition and Properties of Oil Well Drilling Fluids*, 3rd ed., Gulf Publishing Co., Houston, Tex. 1963, pp. 420–422.
161. W. G. Skelly and D. E. Dieball, *J. Soc. Petrol. Engineers* **10**(2), 140 (1970).
162. U.S. Pat. 4,683,949 (Aug. 4, 1987), R. D. Sydansk and P. A. Argabright (to Marathon Oil Co.).
163. R. D. Sydansk and P. E. Moore, *Enhanced Oil Recovery Symposium*, Casper, Wyo., May 3–4, 1988.
164. R. H. van Domelyn and R. D. Wehrle, *Proceedings of the 9th Intersociety Energy Conservation Engineering Conference*, American Society of Mechanical Engineers, New York, 1975, pp. 665–670.
165. F. Tepper, in Ref. 164, pp. 671–677.
166. R. H. Comyn, M. L. Couch, and R. E. McIntyre, *Report TR-635*, Diamond Ordinance Fuze Laboratories, 1958.
167. M. L. Kastens and M. J. Prigotsky, *Ind. Eng. Chem.* **41**, 2376 (1949).
168. W. H. Hartford, K. A. Lane, and W. A. Meyer, Jr., *J. Am. Chem. Soc.* **72**, 3353 (1950).
169. S. Budavari, ed., *The Merck Index*, Merck and Company, Rahway, N.J., 1989, p. 253.
170. W. H. Hartford and M. Darrin, *Chem. Rev.* **58**, 1 (1958).
171. Ref. 164, p. 349.
172. U.S. Pat. 3,824,160 (July 30, 1974), W. H. Hartford (to Allied Chemical Corp.).
173. W. H. Hartford, K. A. Lane, and W. A. Meyer, Jr., *J. Am. Chem. Soc.* **72**, 1286 (1950).
174. U.S. Pat. 2,764,553 (Sept. 25, 1956), W. H. Hartford (to Allied Chemical Corp.).
175. W. H. Hartford and K. A. Lane, *J. Am. Chem. Soc.* **70**, 647 (1948).
176. R. L. Costa and W. H. Hartford, *J. Am. Chem. Soc.* **80**, 1809 (1848).
177. Ref. 1, pp. 219–221.
178. U.S. Pat. 3,305,303 (Feb. 21, 1967), W. H. Hartford and E. B. Hoyt (to Allied Chemical Corp.).
179. U.S. Pat. 3,532,457 (Oct. 6, 1970), K. H. Koepernik (to KaliChemie A. G.).
180. O. F. Joklik, *Chem. Eng.* **80**(23), 49 (1973).
181. U.S. Pat. 3,007,905 (Nov. 7, 1961), G. C. Bailey (to Phillips Petroleum).
182. A. F. Wells, *Structural Inorganic Chemistry*, 3rd, ed., Oxford University Press, London, UK, 1975, p. 76.

183. W. H. Metzger, Jr., *Plating* **49**, 1176 (1962).
184. U.S. Pat. 3,414,428 (Dec. 3, 1968) and 3,497,316 (Feb. 24, 1970), W. R. Kelly and W. B. Lauder (to Allied Chemical Corp.).
185. *Chemical Week 1991 Buyer's* Guide, Vol. **147**, No. 18, Chemical Week Associates, New York, 1990.
186. *Chem Sources-USA 1988*, 28th ed., Directories Publishing Co., Ormond Beach, Fla., 1988.
187. *1991 OPD Chemical Buyers Directory*, 78th ed., Schnell Publishing Co., New York, 1990.
188. Ref. 11, pp. 1672–1674.
189. F. Hanic, I. Horváth, G. Plesch , and L. Gáliková, *J. Solid State Chem.* **59**, 190 (1985).
190. F. Hanic, I. Horváth, and G. Plesch, *Thermochimica Acta* **145**, 19 (1989).
191. Ref. 182, pp. 489–498.
192. Ref. 1, pp. 385–405.
193. R. H. Domelyn and R. D. Wehrle, in Ref. 164, pp. 665–670.
194. D. A. Johnson and M. A. Reid, *J. Electrochem. Soc.: Electrochem. Sci. Technol.* **132**(5), 1058 (1985).
195. U.S. Pat. 2,885,365 (May 5, 1959), A. L. Oppegard (to E. I. du Pont de Nemours & Co., Inc.).
196. U.S. Pat. 6,500,469 (Dec. 31, 2002), H. G. Preuss and D. Bagchi (to Dry Creek Nutrition, Inc.).
197. U.S. Pat. Appl. 20020197340 (Dec. 26, 2002), J. de la Harpe and co-workers.
198. U.S. Pat. Appl. 20030180546 (Sept. 25, 2003), G. Stachowiak, G. Neuman, and H. Wang (to Guardian Industries Corp.).

BILLIE J. PAGE
GARY W. LOAR
McGean-Rohco, Inc.

CHROMOGENIC MATERIALS, ELECTROCHROMIC

1. Introduction

A number of inorganic and organic materials exhibit redox states (reduced and/or oxidized forms) with distinct uv-visible (electronic) absorption bands. When electrochemical switching of these redox states is reversible and gives rise to different colors (ie, new or different visible region bands), the material is described as being electrochromic (1). The optical change is effected by a small electric current at low d-c potential.

The potential is usually on the order of 1V, and the electrochromic material sometimes exhibits good open-circuit memory. Unlike the well-known electrolytic coloration in alkali halide crystals, the electrochromic optical density change is often appreciable at ordinary temperatures. Where more than two redox states are electrochemically accessible in a given electrolyte solution, the electrochromic material may exhibit several colors and be termed polyelectrochromic (2).

By virtue of their numerous academic and commercial applications, electrochromic materials are currently attracting a great deal of interest.

Electrochromic antiglare car rearview mirrors have already been commercialized. Other proposed applications of electrochromic materials including their use in controllable light, reflective or light-transmissive devices for optical information and storage, sunglasses, protective eyewear for the military, controllable aircraft canopies, glare-reduction systems for offices and "smart windows" for use in cars and in buildings (3–5).

In the field of electrochromic data displays, present devices have insufficiently fast response times to be considered for such applications and cycle lives are probably also too low.

Accordingly the most exciting and attractive roles presently envisaged involve long-term display of information, such as at transport termini, reusable price labels, and advertising boards.

A large number of recent reviews on various categories of electrochromic materials and their applications have been published (2,5,6–9).

The major classes of electrochromic materials are metal oxides, Prussian blue systems, viologens, conducting polymers, transition metal and lanthanide coordination complexes metallopolymers and metal phthalocyanines. Although the latter two classes of metal coordination complexes might be considered as inorganic, they are included here because, mainly, the exhibited colors are a result of transitions that involve organic ligands. Table 1 characterizes the various classes of electrochromic materials and their main applications.

Table 1. Summary of Applications of Electrochromic Materials[a]

Material class	Examples	Possible uses
transition metal oxides	WO_3, MoO_3, V_2O_5, Nb_2O_5, $Ir(OH)_3$ and NiO_xH_y	smart windows, thermal control of satellites and electrochromic writing paper
Prussian blue systems	$[Fe^{III}Fe^{II}(CN)_6]^-$ Prussian blue; $[Fe^{III}Fe^{III}(CN)_6]^-$ Prussian brown; $[Fe^{III}_3\{Fe^{III}(CN)_6\}_2\{Fe^{II}(CN)_6\}]^-$ Prussian green and $[Fe^{II}Fe^{II}(CN)_6]^{2-}$, Prussian white	displays
viologens	1,1'-disubstituted-4,4'-bipyridinium salts[b]	car rearview mirrors and displays
conducting polymers	polypyrrole, polythiophene, polyaniline etc	smart windows and displays
transition metal and lanthanide coordination complexes and metallopolymers	metal hydrides, nitrosyl and oxo molybdenum complexes, poly-$[Ru^{II}(vbpy)_2(py)_2]Cl_2$	switchable mirrors, near-infrared switching
metal phthalocyanines	$[Lu(Pc)_2]$	electrochromic displays

[a] Ref. 1.
[b] Already used.

Coloration occurs both cathodically and anodically, as well as in both organic and inorganic materials. Compounds of all types may be classified within one or the other of two general groups based on the nature of charge balancing.

In one group, an electrolyte separates a cathode–anode pair, one or both of which may be chromogenically active. Typically the chromogenic material is a thin film on the cathode or anode.

As charge neutrality must be preserved, and the electrochromic cathode or anode is a solid, insertion–extraction of ions, often H^+ or alkaline metal, accompanies reduction-oxidation within the electrode surface layer. Insertion/extraction in the cathode or anode is the distinguishing feature of the group.

The second group is best described by referring to the viologens, a family of halides of quaternary ammonium bases derived from the 4,4'-bipyridinium structure.

Viologens are recognized as the first important organic electrochromic materials (10,11). Some of these color deeply within solution by simple reduction; others are distinguished by their deep coloration when electrodeposited from solution onto a cathode. These colorations are characteristic of the noninsertion group, although incidental insertion may accompany electrodeposition.

Members of the ion-extraction group as inorganic or organic thin films, especially the former, have attracted the widest interest most recently. Tungsten trioxide was the earliest exploited inorganic compound (12), even before the mechanism of its electrochromic response was understood (13). It is still the best known of the important ion-insertion/extraction group.

Much of the earliest published work followed from research on displays, but opportunities for switchable mirrors and windows have been highlighted as well.

With one noteworthy exception, however, there has been no remarkable commercial success because the technology involves many complex scientific and engineering principles. Also, the competing liquid crystal technology has evolved successfully in some displays applications. The one commercial exception is an electrochromic automotive rearview mirror, which has been gaining popularity since 1988. The mirror contains an encapsulated solution of viologen, that undergoes optical switching without electrodeposition (14).

2. Oxidation–Reduction in the Noninsertion/Extraction Group

The best known examples in this group are organic dyes, and the vehicle in which the oxidation–reduction takes place is in general, a liquid electrolyte. For displays, however, it is preferred that color is not developed within the liquid itself, but rather by electrodeposition. Otherwise there is drifting of the coloration and poor memory, which are especially troublesome for displaying information with high resolution.

Earlier materials, which depended on oxidation–reduction in solution or on a pH indicating effect at an electrode, were abandoned (15). The drifting, however, has proven to be acceptable for a mirror (14).

2.1. Organic Compounds. *Viologens.* Viologens typically require a very low charge density of $2mc/cm^2$ to develop sufficient contrast for display applications. They are the only compounds of the group that have been studied extensively (16). The best known viologen is 1,1'-diheptyl-4,4'-bipyridinium dibromide [6159-05-3].

$$R = C_7H_{15}$$

A cell is caused to switch from clear to bluish-purple when the divalent cation is univalently reduced in aqueous KBr solution and electrodeposited on the cathode as the bromide (10,11). The electrochromic response is visible with an applied potential more negative than $-0-66V$ against the SCE. The peak of visible absorption is at 545 nm. The added electron is delocalized on the monovalent radical ion to which it is reduced (11). Various other viologen compounds have been mentioned (17,18).

The three common viologen redox states obtained by two successive electron transfer reactions could be represented in equation 1.

$$\tag{1}$$

The dicationic form (A) is the most stable and is colorless when pure (unless optical charge transform with the counter anion occurs) (5). The developed colors are also depending on the substituents (19).

However, even with the flexibility for molecular synthesis with organic materials, the experience with viologens typifies two important difficulties.

Cycling instability associated with the onset of side reactions remains a consistent problem for displays, for which a lifetime $\geq 10^7$ cycles is generally required. Viologens suffer especially from a so-called recrystallization of the electrodeposit which impairs erasure of darkened state (20).

Efforts to alter the molecular structure of the radical ion halide salt have been without far-reaching success for display devices.

Besides this, susceptibility to degradative oxidation and photooxidation requires sealing out oxygen and minimizing exposure to uv frequencies of light. Clearly, this susceptibility to oxidation is an especially serious technical hurdle for switchable mirrors and windows exposed to outdoor conditions. When viologens are embedded in appropriate polymer matrices (eg, in poly(N-vinyl-2-pyrrolidone), they exhibit a photochromic behavior (21).

Use of Polymers Systems. A polymer electrochromic device (22) has been made, however the penalty for polymerization is a loss in device speed. Methyl viologen dichloride was dissolved in hydrated poly(2-acrylamido-2-methylpropanesulfonic acid), producing a tacky polymer electrolyte (23,24).

Poly(2-acrylamido-2-methylpropanesulfonic acid) was also used to immobilize methylene blue (22), which oxidizes from the colorless to neutral molecule by a one-electron transfer.

Chemical or electrochemical oxidation of many poly-conjugated compounds such as pyrrole, thiophene, aniline, furan, carbazole, azulene and indole produces polymer films of polypyrrole, polythiophene, or polyaniline etc. doped with counter-ions. The first stage of the polymerization mechanism involves radical cations. The doped polymers, which can adhere to the electrode surfaces are highly conducting, while the undoped, neutral forms are insulating. The oxidized (doped) and reduced (undoped) states of these polymers exhibit different colors, and electroactive conducting polymers are all potentially electrochromic as thin films (5,19).

For example polypyrrole as a thin film exhibits yellow-green (undoped) and blue-violet (doped) electrochromism (eq. 2) (5).

$$
\underset{\text{insulating}}{\left[\!\!\left[\text{pyrrole}\right]\!\!\right]_n} \underset{\text{undoping}}{\overset{\text{p-doping}}{\rightleftharpoons}} \underset{\text{conductive}}{\left[\!\!\left[\text{pyrrole}\right]^{x+}\!\!\right]_n} + nx\,e \tag{2}
$$

Polythiophenes (25) are of particular interest as electrochromic materials due to their chemical stability, ease of synthesis and processability. Polythiophene thin films are blue ($\lambda_{\max} = 730$ nm) in their doped (oxidized) state and red ($\lambda_{\max} = 470$ nm) in their undoped form. Tuning of color states is possible by suitable choice of the thiophene monomer, and this represents a major advantage of using conducting polymer for electrochromic applications. Suitable modifications to the monomer can significantly alter special properties. For example the colors available with polymer films prepared from 3-methylthiophene-based oligomer are strongly dependent on the relative position of methyl groups on the polymer back bone (26).

Colors available include pale blue, blue and violet in the oxidized form, and purple, yellow, red and orange in the reduced form. The color variations have been ascribed to changes in the effective conjugation length of the polymer chain.

Alkoxy-substituted polythiophenes are currently being intensively investigated for their electrochromic properties (27–34). Materials based on poly(3,4-(ethylenedioxy)thiophene) (PEDOT) have a band gap lower than polythiophene and alkyl-substituted polythiophenes, owing to the presence of the two electron-donating oxygen atoms adjacent to the thiophene unit. For example the band gap of PEDOT ($E_g = 1.6$ eV) is 0.5 eV lower than polythiophene which results in an absorbance maximum in the near infra-red region.

Generally the poly(3,4-alkylene dioxy thiophene)s (PXDOTS) are in a class of conducting and electroactive polymers that can exhibit high and quite stable conductivities, a high degree of optical transparency as a conductor, and the ability to be rapidly switched between conducting doped and insulating neutral states.

PXDOTS have attracted attention across academia and industry. In a recent review (35), fundamental aspects were investigated ranging from the electrochemical synthesis of PXDOTS, a variety of *in-situ* characterization techniques, the broad array of properties accessibles and morphological aspect.

Finally two electrochemically driven applications, specifically electrochromism and chemical sensors of PXDOTS, are discussed.

Polyenes and oligomers of poly(*p*-phenylene vinylene) with well-defined length of the conjugated system were prepared by straightforward synthetic procedures based upon organometallic derivatives of silicon or boron (36). Fluorinated oligo(*p*-phenylenes) containing up to five aromatic rings were easily obtained by homocoupling reaction promoted by copper(I) thiophenecarboxylate.

Metallopolymers.   Transition metal coordination complexes of organic ligands are potentially useful electrochromic materials because of their intense coloration and redox reactivity. Chromophoric properties typically arise from low-enegy metal to ligand charge transfer (MLCT), intervalence CT, intraligand excitation, and related visible region electronic transitions. Because these transitions involve valence electrons, chromophoric characteristics are altered or eliminated upon oxidation or reduction of the complex (5). While these spectroscopic and redox properties alone would be sufficient for direct use of transition metal complexes in solution phase ECDS (electrochromic devices), polymers systems have also been investigated which have potential use in all-solid-state systems. Many schemes have been described for the preparation of thin-film "metallopolymers" (37) including both the reductive and oxidative electropolymerization of suitable polypyridyl complexes. Spatial electrochromism has been demonstrated in metallopolymer films (38). Photolysis of orange poly-[$Ru^{II}(vbpy)_2(Py)_2$]Cl_2 thin films (vbpy = 4-vinyl-4-methyl-2,2′-bipyridine) on ITO glass in the presence of chloride ions leads to photochemical loss of the photolabile pyridine ligands and squential formation of poly[$Ru^{II}(vbpy)_2(py)Cl$]Cl (red) and poly[$Ru^{II}(vbpy)_2Cl_2$] (purple).

By suitable choice of metal, the color of such metallopolymer films in the M(II) redox state may be selected (eg, M = Fe, red; M = Ru, orange; M = O_S, green). Electrochromicity results from loss of the metal-to-ligand charge transfer absorption band on switching between the M(II) and M(III) redox states (5).

Metal Phthalocyanines.   Phthalocyanines (Pc) are tetraazatetrabenzo-derivatives of porphyrins with highly delocalized *II*-electron systems.

M= metal or H_2

Metallophthalocyanines are important industral pigments used primarily in inks and coloring plastics and metal surfaces (39). The water soluble sulfonate derivatives are used as dyestuffs for clothing. The purity and depth of the color of metallophthalocyanines arise from the unique property of having an isolated, single band located in the far red end of the visible spectrum near 670 nm, with ε often exceeding 10^5 dm^{3}mol^{-1}cm^{-1}. The next most energetic set of transitions is generally much less intense, lying just to the blue of the visible region near 340 nm. Introduction of additional bands around 500 nm, eg, from charge transfer transitions between the metal and the phthalocyanine ring, allows tuning of the hue (39).

Polyelectrochromism of bis (phthalocyaninato)lutecium(III) ([Lu(Pc)$_2$]) thin films was first reported in 1970, and since that time this complex has received most attention, although numerous other (mainly rare earth) metallophthalocyanines have been investigated for their electrochromic properties (3).

In addition to the familiar applications in the area of dyestuffs, the metallophthalocyanines have been intensely investigated in many fields including catalysis, liquid crystals, gas sensors, electronic conductivity, photosensitizers, nonlinear optics and electrochromism (39).

2.2. Inorganic Electrodeposition. From a comprehensive analysis (40,41) of a variety of electrodepositable metals, the reversible cathodic electroplating of silver has been determined to be the best method. A preferred aqueous solution for light shutters and displays contains 3.0–3.5M Ag(I) and 7 M Ha(I). The complementary I$_3^-$ anodic oxidation product contributes to the change in optical density. For the highest speeds and contrast intended at the time of analysis, several plating cells were required, back to back. Dependence on liquid-state electrolytes presents practical problems in cell assembly and scaling, so the solid-state is often prefered.

3. Insertion/Extraction Compounds

The seminal work on these materials began at American Cyanamid Co. in the 1960s (12,42), although these workers did not develop the ion-insertion/extraction model that has become widely accepted (13). Numerous patents were granted to American Cyanamid Co. as a result of its display-oriented work. Much of what others have written in the open literature either confirms or adds to what these patents disclosed. Important papers (23) about cathodic WO$_3^-$-based insertion devices and others (24,43,44) summarize this activity. The so-called amorphous, or poorly crystallized, tungsten oxide thin film, which developed as the most important material, is, like the viologens, of great interest because of its reversible clear-to-deep-blue coloration in transmission. Coloration efficiency is high.

An important way to assess the many insertion/extraction films known is to compare spectral coloration efficiencies, CE(λ) for the visible region.

$$CE(\lambda) = \Delta OD(\lambda)/q \qquad (3)$$

where $\Delta OD(\lambda)$ represents the change in single-pass, transmitted optical density at the wavelength of interest λ, because of a transfer of charge q, as C/cm^2. Adherence to Lambert's law must either be assumed or tested to avoid pitfalls with thin films; cathodic coloration is well described by $CE(\lambda)$ at moderate values of q for many inorganic amorphous films (45). The coloration efficiency is determined by spectroelectrochemistry, using a cell which employs a cathode–anode pair in a liquid electrolyte. The cell is operated such that only the electrode of interest is in the light path of a spectrophotometer. For example, a 1 N $LiClO_4$–propylene carbonate electrolyte solution was used with a Li anode counter electrode strip to measure the optical density change in cathodic tungsten oxide and other films during galvanostatic switching (46). Li^+ was alternately inserted into and extracted from the films. The films were deposited on conductive glass by reactive RF sputtering. Others have used similar techniques to obtain electrochromic cycling data, and propylene carbonate has been commonly used as well. Often films have been deposited by other common methods in vacuum and sometimes by anodization of the metal.

3.1. Cathodically Colored Inorganic Films. The generalized cathodic, monovalent ion-insertion reaction for inorganic thin films is

$$M_nO_{my}H_2O + xe^- + xJ^+ \rightarrow J_xM_nO_{my} \cdot H_2O \tag{4}$$

where M is a multivalent cation of the electrochromic oxide with valence $2m/n$; both m and n are taken as integers here, though that may not always be so. J^+ is the ion being inserted, and usually $0 < x < 1$. The y moles of H_2O indicate that these materials are generally variably hydrated, depending on preparation technique. Hydration and porosity are required for rapid coloration (13,47–49). On the other hand, for amorphous tungsten oxide thin films, porosity and water content are known to be associated with accelerated dissolution in acid aqueous electrolyte (13,43,48,50). It is noteworthy that single-crystal WO_3 is essentially insoluble, while otherwise interesting MoO_3 films are soluble.

Dissolution of amorphous tungsten oxide films in sealed capsules has been reported to be 2.0–2.5 nm/d at 50°C in 10:1 glycerol/H_2SO_4 (43). Dissolution is much easier in water–H_2SO_4, an even earlier electrolyte choice. With more recently and successfully developed sulfonic acid-functionalized polymer electrolytes, cell stability depends on minimizing the water content of the polymer (24). As with the film itself, however, some minimum water content is necessary in the electrolyte for achieving rapid electrochromic response. A balance must be struck for the water content of both film and electrolyte. One other critical issue with proton-based electrolytes, which applies when they are used with any ion-insertion/extraction film, is the likelihood for the water-containing electrolyte to contribute to electrochemical H_2 or O_2 gas evolution. Because of these collective problems, most of the recently published work with amorphous tungsten oxide has been done with alkali-ion insertion from nonaqueous electrolytes. Generally, however, these electrolytes have relatively low ion conductivity.

Table 2 shows ion-insertion data for some of the better known vacuum-deposited thin films of this class. These grow as amorphous (α) or poorly crystallized films at indicated low substrate temperatures. The coloration efficiency data illustrate one reason why amorphous tungsten oxide remains of high interest, despite the constraints already mentioned. All of the alternatives, including polycrystalline c-WO_3, have lesser coloration efficiencies in the all important visible region. In the near infrared, however, c-WO_3 exhibits a relatively high

Table 2. **Some Cathodically Colored, Inorganic Insertion[a]/Extraction Films[b]**

Film	CAS Registry number	Growth parameter[c]	$CE(\lambda),^d$ cm^2/C	λ, nm
α-WO$_3$	[1314-35-8]	ambient	$30 < CE < 125$	400–700
α-WO$_3^e$		240–660 nm	$30 < CE < 50$	633
α-WO$_3^f$		150°C	55	633
		1500 nm		
c-WO$_3$	[1314-35-8]	310°C	$5 < CE < 50$	400–700
c-MoO$_3$	[1313-27-5]	351°C	$30 < CE < 90$	500–700
α-Nb$_2$O$_5$	[1313-96-8]	40°C	<12	400–700
α-TiO$_2$	[13463-67-7]	40°C	<5	400–700
α-Ta$_2$O$_5$	[1314-61-0]	40°C	<5	400–700

[a] Insertion ion is Li$^+$ unless otherwise noted.
[b] Ref. 46, unless otherwise noted.
[c] Besides grow temperature, the column includes films thickness when known.
[d] Given as a value at a λ or bracketed over the indicated λ range.
[e] Also H$^+$ as insertion ion; Ref. 51.
[f] Ref. 52.

reflectance when darkened (53,54). The data in Table 2 have been used to propose an all solid-state window based on Li$^+$ insertion/extraction in cathodically coloring WO$_3$ (46). A low coloration, insertion/extraction oxide film such as α-Nb$_2$O$_5$ is deposited on conductive glass and serves as the counter electrode. A transparent thin film such as LiAlF$_4$ overlies it and is the nonliquid electrolyte. Deposited first on top of the electrolyte is a Li$^+$-precharged WO$_3$ film, and then a conductive and transparent film. During cycling, the α-Nb$_2$O$_5$ does not contribute much coloration in the WO$_3$-bleach cycle, as desired, when it assumes a cathodic role. As ideal as this seems in principle, however, it has not yet been reduced to a practical art.

By way of contrast, using proton conductivity, at least two other workable solid-state designs have actually been demonstrated at scale. The function of neither one is dictated by the coloration efficiency of a counter electrode. These use sulfonic acid-functionalized polymer electrolytes and depend on unique counter electrodes. One has a high surface area carbon paper counter electrode for reversible proton storage in a display configuration (23). The other has a very fine, reversibly oxidized copper grid that permits vision through a large-area transparency (55,56).

3.2. Anodically Colored Inorganic Films.

The important electrochromic films of this class have been discussed in the open literature since 1978 and include Prussian blue (PB) (57) and the highly hydrated (h) oxides of iridium (59) and nickel (60). Data for these are shown in Table 3. Of lesser significance are the hydrous oxides of rhodium and cobalt (61,66,67). Like the cathodically coloring insertion films, the anodically coloring films depend, for useful darkening and bleaching rates, on having open porosity and hydration. Various colors have been reported qualitatively, though full coloration is limited sometimes by the onset of O$_2$ evolution (61). As Table 3 shows, Prussian blue is especially interesting because of its relatively high coloration efficiency. Except for Prussian blue, a common method of film preparation is potential cycling (pc) on either a bulk

Table 3. **Some Anodically Colored, Inorganic Insertion/Extraction Films**

Film	CAS Registry number	Growth method	Insertion ion	CE(λ), cm^2/C	λ, nm	References
h-IrO$_x$	[12645-46-4]	pc^a	H^+/OH^-	15	633	60
h-IrO$_x$		rs	H^+/OH^-	18,12	633,600	61,62
h-NiO$_x$	[11099-02-8]	ged	H^+/OH^-	~50	~440 peak	63,64
Prussian blue	[12240-15-1]^b	ged	K^+	68	633	65

^{a}~180 nm.
^{b}CI Pigment Blue 27.

metal or a conductive film. Reactive sputtering (rs) and galvanostatic electrodeposition (ged) have also been used for these electrochromic films. In the case of Prussian blue, the film has been grown by electroless reduction, by the sacrificial anode method or by galvanostatic electrodeposition (68–72).

Despite the considerable progress made in the few years in which anodic insertion/extraction films have been known, neither film compositions, film properties, nor electrochemical reactions are sufficiently well characterized. There have been disagreements, as indicated for h-IrO$_x$ and h-NiO$_x$ in Table 3, as to whether H^+ is being extracted or OH^- inserted during coloration. The general problem is best illustrated by the important example of Prussian blue. Early work (68–71) resulted in two different sets of equations for electrochromic reduction:

$$KFe^{3+}[Fe^{2+}(CN)_6] + e^- + K^+ \rightarrow K_2Fe^{2+}[Fe^{2+}(CN)_6] \qquad (5)$$

$$Fe_4^{3+}[Fe^{2+}(CN)_6]_3 + 4\ e^- + 4K^+ \rightarrow K_4Fe^{2+}[Fe_4^{2+}(CN)_6]_3 \qquad (6)$$

The compounds $KFe^{3+}[Fe^{2+}(CN)_6]$ [25869-98-1] and $Fe^{3+}_4[Fe^{2+}(CN)_6]_3$ [14038-43-8] are both called Prussian blue. The first is known as the water-soluble form, though actually it only peptizes easily, and the second as the insoluble form. The reduced compounds $K_2Fe^{2+}[Fe^{2+}(CN)_6]$ and $K_4Fe^{2+}_4[Fe^{2+}(CN)_6]_3$ are known as Everitt's salt [15362-86-4] and Prussian white [81681-39-2] respectively. A similar lack of specificity occurs when Prussian blue is oxidized to Berlin green [14433-93-3]. This has led to propositions that film composition depends on the K^+ concentration of the growth solution (71) and on cycling (73,75). A film that is in the insoluble form initially, prior to cycling, is said thereafter to develop an intermediate K^+ content between the two forms of Prussian blue. The radius of the hydrated insertion ion is believed to be the key factor that dictates the reversibility of ion injection/extraction for the zeolite structure of Prussian blue. Reversibility has been shown to be best for K^+, Rb^+, Cs^+ and NH_4^+; hydrated radii were determined to be in the range 0.118–0.125 nm (70).

3.3. Doped/Undoped Organic Films. This class of electrochromic materials is probably the youngest and least thoroughly explored from a practical viewpoint. There has been more interest generally in the very high, metal-like conductivity of the oxidized state of some of its members than in the insertion/extraction electrochromism accompanying oxidation–reduction. Of interest have been applications for lightweight and moldable batteries and also for antistatic and electromagnetic shielding. On the other hand, there is not enough

Table 4. **Some Organic Insertion/Extraction Films**

Film	CAS Registry number	Growth method[a]	Possible dopants	Oxidizing color shift[b]	References
polyaniline	[25233-30-1]	ep	H^+, Br^-	lt yel–grn–bl	77–81
polypyrrole	[30604-81-0]	ep	ClO_4^-, BF_4^-, Li^+	yel/grn–gr/br	75,76,82–86
polythiophene	[25233-34-5]	ep	ClO_4^-, BF_4^-	red–bl	76,86,89
poly(isothia-naphthene)	[91201-85-3]	ep	ClO_4^-, BF_4^-,	bl/blk–lt grn/yel	76,90,91
TTF-function-alized polymer		spin-cast	ClO_4^-, BF_4^-,	orange–brown	92,93
TNF-MN	[1172-02-7]	vacuum	K^+	grn–transparent	94

[a] ep = electropolymerization.
[b] lt = light; yel = yellow; grn = green; bl = blue; gr = gray; br = brown; blk = black.

reported in the open literature to permit good comparisons of coloration efficiencies. Also, although the films themselves are solid state, almost all electrochromic work has been done with liquid electrolytes. This suggests that research and development are still at a fundamental stage. Nevertheless, two excellent reviews (75,76) do emphasize optical properties and color switching. Some of this is summarized in Table 4, though the dopant column should be considered with care. Analytical work using the quartz–crystal microbalance (81,84,85) suggests that charge balancing may sometimes involve both anions and cations in complicated ways. The first four materials in Table 4 are best known for the e^- conductivity that is associated with a conjugated π-electron structure. These are usually deposited by electropolymerization. Tetrathiafulvalene [31366-25-3] (TTF) and 2,4,7-trinitro-9-fluorenylidene malononitrile [1172-02-7] (TNF-MN) demonstrate some variety in the growth method for this electrochromic class. TNF-MN is also different because coloration only occurs in the reduced state. Generally, the whole class is characterized by a wide variety of transmitted colors in both the oxidized and nonoxidized forms. Some of the colors reported are shown in Table 4, though these colors are known only qualitatively and without specification of viewing conditions. They are only indicated here to show the breadth possible. Even wider color variety is possible with structural substitutions (75). Some degree of predictability may be possible because solid-state colors seem to be similar to those reported for solutions (76,94).

The same color variety is not typical with inorganic insertion/extraction materials; blue is a common transmitted color. However, rare-earth diphthalocyanine complexes have been discussed, and these exhibit a wide variety of colors as a function of potential (95–97). Lutetium diphthalocyanine has been studied the most. It is an ion-insertion/extraction material that does not fit into any one of the groups herein but has been classed with the organics in reviews. Films of this complex, and also erbium diphthalocyanine, have been prepared successfully by vacuum sublimation and even embodied in solid-state cells (98,99).

There is tangible support for ion insertion/extraction in some materials, such as polypyrrole, polyaniline, and Prussian blue, from analyses with the mass-sensitive quartz-crystal microbalance (81,84,85,100). This relatively new technique is developing as an important one for electrochromic materials generally. It increases the standard electrochemical analyses, especially cyclic voltammetry, that have been effectively used up to now. Its further use, for charge-balancing ion insertion/extraction, should give growth to the analytical technique, and also help speed along developments in electrochromism, which is still a young science.

4. Sol-Gel Systems

Gelling an electrolyte layer, which includes an electrochromic layer, has been suggested as a means of eliminating hydrostatic pressure concerns (101–103). However, not everything that is termed a gel is self-supporting, free-standing, and capable of eliminating hydrostatic pressure. Figure 1 provides a qualitative comparison of polymer thickener, polymer gels, purepolymer electrolytes and their variation viscosity. The viscosity as a function of polymer content is highly dependent on crosslink density. In the illustration, one starts with a low viscosity liquid and adds some soluble uncrosslinked polymer thickeners to increase viscosity.

    If the crosslinking is provided by chemical bond formation the resulting gel can be termed a chemical gel by contrast to a physical gel.

5. Applications

The increase in the interaction between humans and machine has made display devices indispensable for visual communication. The information which is to be communicated from a machine can be often in the form of color images. Electrochromic display device (ECD) is one of the most powerful candidate for

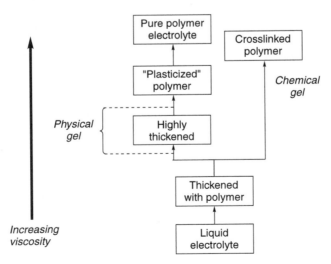

Fig. 1. Illustration of polymer thickeners, polymer gels, and polymer electrolytes as a function of viscosity and polymer crosslinking (101).

this purpose and has various merits such as multicolor, high contrast, optical memory and no visual dependence on viewing angle. A large number of electronic materials are avoidable from almost all branches of synthetic chemistry. The most important examples from major classes of electrochromic materials namely transitions metal oxides, Prussian blue, phthalocyanines, viologens, fullerenes, dyes and conducting polymers (including gels) are described (104).

In the recent years, a lot of attention has been paid to understanding the physical and chemical properties of various electrochromic materials, particularly on conducting polymer (CPs) the focus is not only on basic research, but also on commercial aspects (104,105). It has brought various electrochromic materials and devices actually into the market. In the automobile sector rear view mirrors and several parts such as sunroofs and visors are under prototype production.

Alphanumeric displays and electrochromic mirrors have also been produced. Widespread applications of ECDs (electrochromic displays) particularly for architectural applications, depend on reducing costs, increasing device lifetime, and overcoming the problem of ECD degradation. Further existing smart windows require an external power source for their operation. Photo-electrochromic systems which change color electrochemically, but only after being illuminated may be more appropriate candidates for smart windows (105,106).

Windows are divided into three classes : battery-like, with open circuit minory and solution phase and hybrid designs, which are self erasing and require a maintenance current during coloration.

Construction of windows includes all thin films state structures and devices containing polymer or gel electrolytes (6).

The study and development of photo-electrochromic materials and devices for large area window applications are in progress. Commercial production of all plastic electrochromic devices, smart windows, for monitoring time temperature application has been already achieved.

Electrochromic coatings, thin films such as WO_3 that change their optical absorbance or reflectance as a function of injected ions (typically H^+ or Li^+ species) is an area of research and development that has received considerable attention from academia, industry and government laboratories, involving manufacturing challenge (108–110).

The production of moving electrochromic images (such as moving pixels and alphanumeric displays etc) is the next step in the research and development of electrochromic materials and hence needs more efforts to be done in that direction. Recent interest for electrochromic devices for multispectral energy modulation by reflectance and absorbance has extenced the working definition. Electrochromic devices are now being studied for modulation of radiation in the near infrared, thermal infrared and micro wave regions and "color" can mean response of detectors of these wavelengths, not just the human eye (104).

BIBLIOGRAPHY

"Chromogenic Materials, Electrochromic and Thermochromic" in *ECT* 3rd ed., Vol. 6, pp. 129–142, by J. H. Day, Ohio University; "Chromogenic Materials" in *ECT* 4th ed.,

Vol. 6, pp. 312–321, by Charles B. Greenberg, PPG Industries, Inc.; "Chromogenic Materials, Electrochromic" in *ECT* (online), posting date: December 4, 2000, by Charles B. Greenberg, PPG Industries, Inc.

CITED PUBLICATIONS

1. N. M. Rowley and R. J. Mortimer, *Sciences Progress* **85**, 243 (2002).
2. R. J. Mortimer, *Electrochim. Acta* **44**, 2971 (1999).
3. P. M. S. Monk, R. J. Mortimer, and D. R. Rosseinsky in *Electrochromism: Fundamentals and Applications*, VCH, Weinheim, 1995.
4. M. Green, *Chem. Ind.* **17**, 641 (1996).
5. R. J. Mortimer, *Chem. Soc. Rev.* **26**, 147 (1997).
6. R. D. Rauh, *Electrochim. Acta.* **44**, 3165 (1999).
7. D. R. Rosseinsky and R. J. Mortimer, *Adv. Mater.* **13**, 783 (2001).
8. C. G. Granqvist, *Sol. Energy Mater. Sol. Cells.* **60**, 201 (2000).
9. C. G. Granqvist, *Electrochim. Acta* **44**, 3005 (1999).
10. C. J. Schoot, J. J. Ponjée, H. T. VanDam, R. A. Van Dorm, and P. T. Bolwijn, *Appl. Phys. Lett.* **23**, 64 (1973).
11. H. T. Van Dam, and J. J. Ponjée, *J. Electrochem. Soc.* **121**, 1555 (1974).
12. S. K. Deb, *Appl. Opt. Suppl.* **3**, 192 (1969). U. S. Pat. 3,521,941 (July 28, 1970) S. K. Deb and R. Shaw (to American Cyanamid Co).
13. B. W. Faughman and R. S. Crandall in *Topics in Applied Physics* **40**, 181 (1980).
14. U.S. Pat. 4,902,108 (Feb. 20, 1990), H. J. Byker (to Gentex Corp.).
15. I. F. Chang, in A. R. Kmetz and F. K. Von Willisem, eds., *Nonemissive Electrooptic Displays*, Plenum Press, New York, 1976, p. 155.
16. T. Oi, *Ann. Rev. Mater. Sci.* **16**, 185 (1986).
17. C. M. Lampert, *Sol. Energy Mater.* **11**, 1 (1984); U.S. DOE Contract W-7405-ENG-48, Springfield, Va., Oct. 1980.
18. A. Donnadieu, *Mater. Sci. Eng.* **B3**, 185 (1989).
19. P. M. S. Monk, R. J. Mortimer, and D. R. Rosseinsky, *Chem. Bev.* **31**, 380 (1995).
20. J. A. Barltrop and A. C. Jackson, *J. Chem. Soc. Perkin Trans. II*, 367 (1984).
21. M. Nanasauva, in J. C. Crano and R. J. Guglielmetti, eds., *Organic Photochromic and Thermochromic Compounds*, Vol. 1, Kluwer Academic/Plenum Publishers, New York, 1999, chapt. 9.
22. J. M. Calvert, T. J. Manuccia, and R. J. Nowoks, *J. Electrochem. Soc.* **133**, 951 (1986).
23. R. D. Giglia and G. Hoacke, *S. I. D. Dig.* **12**, 76 (1981); *Proc. S. I. D.* **23**, 955 (1982).
24. J.-P. Randin, *J. Electrochem. Soc.* **129**, 1215 (1982).
25. J. Roncali, *Chem. Rev.* **92**, 711 (1992).
26. M. Mastragostino, in B. Scrosati, ed., *Applications of Electroactive Polymer*, Chapman and Hall, London, 1993, Ch. 7, p. 244.
27. M. Dietrich, J. Heinze, G. Heywang, and F. Jonas, *J. Electroanal. Chem.* **369**, 87 (1994).
28. Q. Pei, G. Zuccarello, M. Ahlskog, and O. Inganäs, *Polymer* **35**, 1347 (1994).
29. G. A. Sotzing and J. R. Reynolds, *J. Chem. Soc. Chem. Commun.* **703** (1995).
30. A. Kumar and J. R. Reynolds, *Macromolecules* **29**, 7629 (1996).
31. B. Sankaran and J. R. Reynolds, *Macromolecules* **30**, 2582 (1997).
32. G. A. Sotzing and J. R. Reynolds, *Adv. Mater.* **9**, 795 (1997).
33. J. A. Irvin and J. R. Reynolds, *Polymer* **39**, 2339 (1998).
34. S. A. Sapp, G. A. Sotzing and J. R. Reynolds, *Chem. Mater.* **10**, 2101 (1998).
35. L. "Bert" Groenendaal, G. Zotti, P. H. Aubert, S. Waybright, and J. R. Reynolds, *Advanced Materials* **15**, 855 (2003).

36. F. Naso, F. Babudri and G. M. Farinola, *Pure Appl. Chem.* **71**, 1485 (1999).
37. R. J. Mortimer, in R. G. Compton and G. Hancock, eds., *Research in Chemical Kinetics*, Vol. 2, Elsevier, Amsterdam, 1994, pp. 261–311.
38. R. M. Leasure, W. Ou, R. W. Linton, and T. J. Meyer, *Chem. Mater.* **8**, 264 (1996).
39. C. C. Leznoff and A. B. P. Lever, eds., *Phthalocyanines: Properties and Applications*, John Wiley & Sons, New York, vol. 1 (1989); vol. 2 (1993); vol. 4 (1996).
40. S. Zaromb, *J. Electrochem. Soc.* **903**, 912 (1962).
41. J. Mantell and S. Zaromb, *J. Electrochem. Soc.* **109**, 992 (1962).
42. S./K. Deb, *Philos. Mag.* **27**, 801 (1973).
43. J. P. Randin, *J. Electron. Mater.* **7**, 47 (1978).
44. J. P. Randin, *J. Electrochem. Soc.* **129**, 2349 (1982).
45. S. F. Cogan, N. M. Nguyem, S. J. Perrotti, and R. D. Rauh, *J. Appl. Phys.* **66**, 1333 (1989).
46. S. F. Cogan, E. J. Anderson, T. D. Plante, and R. D. Rauh, *Proceedings of the SPIE* **562**, 23 (1985).
47. R. Hunditch, *Electron. Lett.* **11**, 142 (1975).
48. B. Reichman and A. J. Bard, *J. Electrochem. Soc.* **126**, 583 (1979).
49. P. Schlotter and L. Pickelmann, *J. Electron. Mater.* **11**, 207 (1982).
50. T. C. Arnoldusen, *J. Electrochem. Soc.* **128**, 117 (1981).
51. O. Bohnke, C. Bohnke, G. Robert, and B. Casquille, *Solid State Ionics* **6**, 121 (1982).
52. P. Schlotter, *Sol. Energy Mater.* **16**, 39 (1987).
53. O. F. Schirmer, V. Wittwer, G. Baur, and G. Brandt, *J. Electrochem. Soc.* **124**, 749 (1977).
54. R. B. Goldmer and R. D. Rauh, *Sol. Energy Mater.* **11**, 177 (1984).
55. C. B. Greenberg in S. A Marolo, ed., *Proceeding of the 15ᵗʰ Conference on Aerospace Transparent Material and Enclosures, II*, WRDC-TR-89-4044, Wright-Patteson AFB, Dayton, Ohio, p. 1124, (1989); U. S. Pat. 4,768,885 (Sept. 6, 1988), C. B. Greenberg and D. E. Singleton (to PPG Industries. Inc.).
56. K. C. Ho, D. E. Singleton, and C. B. Greenberg, *J. Electrochem. Soc.* **137**, 3858 (1990); M. K. Carpenter and D. A. Corrigan, eds., *Proceeding of the Symposium on Electrochromic Materials*, The Electrochemical Society, Inc., Pennington, N. J., 1990, p. 349.
57. V. D. Neff, *J. Electrochem. Soc.* **125**, 886 (1978).
58. S. Gottesfeld, J. D. E. Mc Intyre, G. Beniand, and J. L. Shay, *Appl. Phys. Lett.* **33**, 208 (1978); D. N. Buckly and L. D. Burke, *J. Chem. Soc. Faraday Trans. I* **71**, 1447 (1975).
59. L. D. Burke and D. P. Whelan, *J. Electroanal. Chem.* **109**, 385 (1980).
60. G. Beni and J. L. Shay, *Adv. Image Pickup Display* **5**, 83 (1982).
61. W. C. Dautremont-Smith, *Displays* **3**, 367 (1982).
62. S. F. Cogan, T. D. Plante, R. S. Mac Fadden, and R. D. Rauh, *Proceeding of the SPIE* **692**, 32 (1985).
63. M. C. Carpenter, R. S. Conell, and D. A. Corrigan, *Sol. Energy Mater.* **16**, 333 (1987).
64. S. Morisaki, K. Kawakami, and N. Baba, *Jpn. J. Appl. Phys.* **27**, 34 (1988).
65. H. Tada, Y. Bits, K. Fujino, and H. Kawahara, *Sol. Energy Mater.* **16**, 509 (1987).
66. L. D. Burke and E. J. M. O'Suliivan, *J. Electroanal. Chem.* **93**, 11 (1978).
67. L. D. Burke and O. J. Murphy, *J. Electroanal. Chem.* **109**, 373 (1980).
68. D. Ellir, M. Eckhoff, and V. D. Neff, *J. Phys. Chem.* **85**, 1225 (1981).
69. K. Itaya, H. Akahoshi, and S. Toshima, *J. Electrochem. Soc.* **129**, 1498 (1982).
70. K. Itaya, T. Ataka and S. Toshima, *J. Am. Chem. Soc.* **104**, 4767 (1982).
71. K. Itaya, I. Uchida, and V. D. Neff, *Acc. Chem. Res.* **19**, 162 (1986).
72. Y. Yano, N. Kihugasa, H. Yoshida, K. Fujino, and H. Kawahara in ref. 55, p. 125.

73. R. J. Mortimer and D. R. Rosseinsky, *J. Chem. Soc. Dalton Trans.*, 2059 (1984).
74. C. A. Lundgren and R. W. Murray, *Inorg. Chem.* **27**, 933 (1988).
75. M. Gazard, in T. A. Skotheim, ed., *Handbook of Condusting Polymers*, Vol. 1, Marcel Dekker, Inc., New York, 1986, p. 673.
76. A. O. Patil, A. J. Heeger, and F. Windl, *Chem. Rev.* **88**, 183 (1988).
77. T. Kobayashi, H. Yoneyama, and H. Tamura, *J. Electroanal. Chem.* **177**, 281, 293 (1984); **161**, 419 (1984).
78. A. G. MacDiarmid and coworkers, *Mol. Cryst. Liq. Cryst.* **121**, 173 (1985).
79. P. M. McManus, S. C. Yang, and R. J. Cushman, *J. Chem. Soc., Chem. Commun.*, 1556 (1985).
80. R. J. Cushman, P. M. McManus, and S. C. Yang, *J. Electroanal. Chem.* **291**, 335 (1986).
81. D. Orata and D. A. Buttry, *J. Am. Chem. Soc.* **109**, 3574 (1987).
82. A. F. Diaz, J. J. Castillo, J. A. Logan, and W. Y. Lee, *J. Electroanal. Chem.* **129**, 115 (1981).
83. A. F. Diaz and K. K. Kanazawa, in J. S. Miller, ed., *Extended Linear Chain Compounds*, Vol. 3, Plenum Press, New York, p. 417, 1983.
84. J. H. Kaufman, K. K. Kanazawa, and G. B. Street, *Phys. Rev. Lett* **53**, 2461 (1984).
85. K. Naoi, M. M. Lien, and W. H. Smyrl, *J. Electroanal. Chem.* **272**, 273 (1989).
86. K. Kaneto, K. Yoshino, and Y. Inuishi, *Jpn. J. Appl. Phys.* **22**, L412 (1983).
87. T.-C. Chung, J. H. Kaufman, A. J. Heeger, and F. Wudl, *Phys. Rev. B* **30**, 702 (1984).
88. F. Garnier, G. Tourillon, M. Gazard, and J. C. Dubois, *J. Electroanal. Chem.* **148**, 299 (1983).
89. K. Kaneto, H. Agawa, and K. Yoshino, *J. Appl. Phys.* **61**, 1197 (1987).
90. M. Kobayashi, N. Colaneri, M. Boysel, F. Wudl, and A. J. Heeger, *J. Chem. Phys.* **82**, 5717 (1985).
91. M. Colaneri, M. Kobayashi, A. J. Heeger, and F. Wudl, *Synth. Met.* **14**, 45 (1986).
92. F. B. Kaufman, A. H. Schroeder, E. M. Engler, S. R. Kramer, and J. Q. Chambers, *J. Am. Chem. Soc.* **102**, 483 (1980).
93. F. B. Kaufman, A. H. Schroeder, E. M. Engler, and V. V. Patel, *Appl. Phys. Lett.* **36**, 422 (1980).
94. A. Yasuda and J. Seto, *J. Electroanal. Chem.* **247**, 193 (1988); Ref. 37, p. 192.
95. P. N. Moskalev and I. S. Kirin, *Opt. Spectros.* **29**, 220 (1970); *Russ. J. Inorg. Chem.* **16**, 57 (1971); *Russ. J. Phys. Chem.* **46**, 1019 (1972).
96. G. A. Corker, B. Grant, and N. J. Clecak, *J. Electrochem. Soc.* **126**, 1339 (1979).
97. M. M. Nicholson and F. A. Pizzarello, *J. Electrochem. Soc.* **126**, 1490 (1979).
98. N. Egashira and H. Kokado, *Jpn. J. Appl. Phys.* **25**, L462 (1986).
99. M. Starke, I. Androsch, and C. Hamann, *Phys. Stat. Sol. (A)*, **120**, K95 (1990).
100. B. J. Feldman and O. R. Melroy, *J. Electroanal. Chem.* **234**, 213 (1987).
101. H. J. Byker, *Electrochemica Acta* **46**, 205 (2001).
102. M. A. Acgerter, *Solar Energy Mater. Solar Cells* **68**, 401 (2001).
103. J. Livage and D. Ganguli, *Solar Energy Mater. Solar Cells* **68**, 365 (2001).
104. P. R. Somani and S. Radhakinshinam, *Materials Chem Phys.* **77**, 117 (2002).
105. Y. Shirota, *J. Mater. Chem.* **10**, 1 (2000).
106. A. Azems and C. G. Grangvist, *J. Solid State Electrochem.* **7**, 64 (2003).
107. K. H. Heckner and A. Kraft, *Solid State Ionics* **152–153**, 899 (2002).
108. N. A. O'Brien, J. Gordan, H. Mathew, and B. P. Hichwa, *Thin Solid Films* **345**, 312 (1999).
109. K. Bange, *Solar Energy Materials and Solar Cells.* **58**, 1 (1999).
110. C. Sella, M. Maaza, O. Nemraoui, J. Lafont, N. Renard, and Y. Sampeur, *Surface Coatings Technol.* **98**, 1477 (1998).

GENERAL REFERENCES

N. Baba, M. Yamana, and H. Yamamoto, eds., *Electrochromic Display*, Sangyo Tosho Co. Ltd., Tokyo, 1991.

N. M. Rowley and R. J. Mortimer, "New Electrochromic Materials", *Sciences Progress* **85**, 243–262 (2002).

R. J. Mortimer, "Organic Electrochromic Materials", *Electrochimica Acta* **44**, 2971–2981 (1999).

P. R. Somami and S. Radhakrishman, "Electrochromic Materials and Devices : Present and Future," *Materials Chemistry and Physics* **77**, 117–133 (2002).

A. SAMAT
R. GUGLIELMETTI
Université de la Mediterannée

CHROMOGENIC MATERIALS, PHOTOCHROMIC

1. Introduction

Photochromism can be simply defined as a light-induced reversible change of color. References to photochromism have found dating back to the 1800s (1,2).

Fritzche reported in 1867 (3) the bleaching of an orange-colored solution of tetracene (**1**) in daylight and regeneration in the dark.

Later, ter Meer (4) found a change of color of the potassium salt of dinitroethane in the solid state (yellow in the dark, red in the daylight). Another early example was published by Phipson and Oor (5) who noted that a post gate painted with a zinc pigment (probably some kind of lithopone) appeared black all day and white all night. In 1899 Markwald studied the reversible change of color of 2,3,4,4-tetrachloro naphthalene-1(4*H*)-one(βTCDHN) (**2**) in the solid state (6)

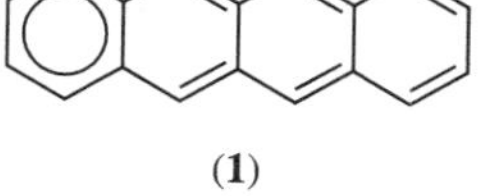

(**1**) 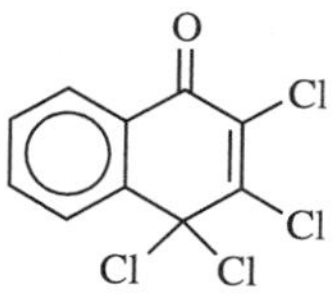　　　　　(**2**)

Interest in photochromism was continuous, but limited until the 1940–1960 period when there was an increase in mechanistic and synthetic work, particularly by the Israeli research group of Hirshberg and Fischer. In 1950 Hirshberg (7) suggested the term photochromism [(from the Greek: photos (light) and chroma (color))] to describe the phenomenon.

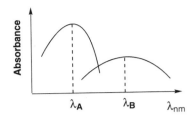

Fig. 1. Absorbance of A and B form.

However the phenomenon is not limited to colored compounds. It applies to systems absorbing from the far uv to the ir and to very rapid or very slow reactions.

Photochromism concerns the reversible transformation of a chemical species (pigment) between two forms A and B having different absorption spectra and different molecular structure (see also Fig. 1).

$$A \underset{h\nu_2 \text{ and (or) } \Delta}{\overset{h\nu_1}{\rightleftharpoons}} B$$

The thermodynamically stable form A generally absorbs in the uv-region (200–400 nm) and is transformed by irradiation ($h\gamma_1$) into the form B which absorbs in the visible region (400–750 nm). The reverse reaction can occur thermally (Δ) or photochemically ($h\gamma_2$). Some systems can function both thermally and photochemically.

The most general phenomenon is referred to as positive photochromism when the colored form B has $\lambda_{max}(B) > \lambda_{max}(A)$. Photochromism is called negative or inverse when $\lambda max(A) > \lambda max(B)$.

The most important application of this phenomenon is in variable optical transmission materials such as lenses that darken in the sun and return to their initial transparency in diffuse light. The first commercially glasses were made of glass lenses impregnated with inorganic (mainly silver) salts. In recent years organic photochromic lenses, which are lighter and therefore more confortable to wear despite their limited lifetime, have made an important jump in the world market.

Organic pigments are also more reactive, more photocolorable, and compatible with polymer matrices (8).

The photochromic phenomenon is generally observed in solution or in polymer matrix (viscous solution) and sometime in the solid state depending on the system.

Fatigue (or photodegradation) is defined as a loss in photochromic behavior, as a result of the existence of side reactions that decrease the concentration of the active species (as A or B forms) or lead to the formation of by-products that inhibit the photochemical formation of B.

$$A \underset{h\nu_2 \text{ and (or) } \Delta}{\overset{h\nu_1}{\rightleftharpoons}} B \longrightarrow \underset{\substack{\text{degradation} \\ \text{products}}}{C}$$

Light and the oxygen are the two main causes of fatigue (9). The inhibition can result from quenching of the excited states of A (singlet or triplet) or from screening of the actinic light. Therefore, the reversibility of the phenomenon is not totally possible.

Photochromic pigments are generally used as embedded into supports such as polymer matrices, liquid crystals, or other materials and constitute photochromic systems.

These photochromic systems can be divided into two broad categories: organic and inorganic. The two types are quite different in their behavior and observable mechanisms and their characteristics are discussed giving different examples. It is noteworthy that the most important development of these organic systems has occurred during the twenty last years.

2. Inorganic Photochromic Systems

2.1. Silver Halide-Containing Glasses.
The most important examples of inorganic systems are those containing silver halide crystallites dispersed throughout a glass matrix. The first description of photochromic silver halide-containing glasses appeared in 1964 (10,11). In general, these systems are characterized by broad absorption of visible light by the colored species and excellent resistance to fatigue.

The principle behind the generation of a photochromic glass with silver halide is the controlled formation of silver halide particles or crystallites suspended throughout the glass matrix (12–14). The formation of crystallites of the correct size and concentration is the key to a useful photochromic system. The general procedure involves the initial melting of a glass-forming mixture which is then cooled to a solid glass shape. Rapid cooling to room temperature results in a nonphotochromic glass. Holding the solid at a temperature in the range of 500–600°C for several minutes to hours causes the nucleation and growth of silver halide crystallites, the active photochromic species. Again, the size of the crystallites is important. With a size of less than 10 nm, significant darkening upon exposure to sunlight is not achieved. Above 20 nm, the scattering of visible light becomes a problem, leading to haziness. Also, with the larger particles the rate of thermal fading slows to an unacceptable rate.

Copper (I) (cuprous) ion serves as a catalyst for both the photochemical darkening and thermal fading reactions (10). Therefore, a small amount of cuprous ion is normally added to the glass batch.

The darkening reaction involves the formation of silver metal within the silver halide particles containing traces of cuprous halide. With the formation of metallic silver, cuprous ions are oxidized to cupric ions (10,13). The thermal or photochemical (optical bleaching) reversion to the colorless or bleached state corresponds to the reoxidation of silver to silver ion and the reduction of cupric ion to reform cuprous ion.

One of the most important characteristics of the inorganic glass matrix for a photochromic system is the temperature dependence of the solubility of silver halides (13). It is required that the solubility of silver halides be high at the temperature used to melt the glass mixture, and relatively low at the intermediate

Table 1. **Composition of a Silver Halide Photochromic Glass**[a]

Component	Wt%
SiO_2	55.8
B_2O_3	18.0
Al_2O_3	6.48
Li_2O	1.88
Na_2O	4.04
K_2O	5.76
ZrO_2	4.89
TiO_2	2.17
CuO	0.011
Ag	0.24
Cl	0.20
Br	0.13

[a] Ref. 15.

temperature at which the silver halide crystallites are formed. The approximate composition of a typical photochromic glass system that is marketed as Photogray Extra by Corning Inc., Corning, New York, is as given in Table 1 (15).

The color of the darkened state is controlled by the size and shape of the minute silver specks formed during photochemical reduction, but the relationship is not well understood. Since the shapes of the silver particles vary considerably throughout the matrix, a broad absorption over the visible range results (13,14). The color can be modified from gray to brown by changing the heat treatment for silver halide nucleation and growth and thus changing the size/shape distribution of the crystallites (16). The color can also be shifted from gray to brown by the addition of trace amounts of palladium or gold (3–4 ppm) to the glass batch (15).

An alternative to the uniform distribution of silver halide throughout the glass is the diffusion of silver ion into the surface of the glass. This has been accomplished by immersion of the glass article into a silver-containing fused salt, for example, silver nitrate plus sodium nitrate, molar ratio = 17:83 (17). Heat treatment to allow crystallite formation is still essential and copper oxide is added to the glass batch to catalyze the photochromic reactions. In general, photochromic glasses formed in this manner are not as active (do not get as dark) as the systems containing thoroughly dispersed silver, or have a slower thermal fade rate.

Thin films of photochromic glass containing silver halide have been produced by simultaneous vacuum deposition of silicon monoxide, lead silicate, aluminum chloride, copper (I) chloride, and silver halides (18). Again, heat treatment (120°C for several hours) after vacuum deposition results in the formation of photochromic silver halide crystallites. Photochemical darkening and thermal fade rates are much slower than those of the standard dispersed systems.

2.2. Other Inorganic Metal Salt Systems. An effective silver-free photochromic system can be obtained by the dispersion of crystallites containing cadmium halide and copper halide throughout an inorganic glass matrix (19,20).

Table 2. **Composition of Copper/Cadmium Halide Photochromic Glass**[a]

Component	Wt%
SiO_2	57.2–58.6
Al_2O_3	10.0–10.1
B_2O_3	21.5–22.7
Na_2O	8.1–8.2
F	1.05–1.14
Cl	0.43–0.45
CdO	0.43–0.45
SnO	0.20
CuO	0.12–0.15

[a] Ref. 20.

Heat treatment of the solid glass is again required to allow the formation of the metal halide crystallites. The mechanism of the darkening reaction is apparently the formation of colloidal copper metal particles by disproportionation of cuprous ion (20). The color of the darkened state, ranging from yellow-brown to green, is controlled by the type of heat treatment used for crystallite formation.

A typical range of glass compositions for this type of photochromic system is given in Table 2 (20).

Another inorganic photochromic glass system was prepared by the addition of europium (II) or cerium(III) to a soda–silica glass with an approximate composition of Na_2O–$2.5SiO_2$ (21). The concentration of the rare-earth ion was low (100 ppm). With europium (II)-doped glass, the photochemical darkening resulted in an amethyst color that faded rapidly thermally. These glasses were subject to fairly rapid fatigue, losing all photochromic behavior after a 20-h exposure to uv radiation centered at 332.5 nm. This was probably the result of the oxidation of Eu(II) to Eu(III). Interestingly, the photochromic behavior could be recovered by exposure of the exhausted glass to high energy uv light at 213.7 nm.

Photochromic silver–copper halide films were produced by vacuum evaporation and deposition of a mixture of the components onto a silicate glass substrate (22). The molar ratio of the components was approximately 9:1 (Ag:Cu) and film thicknesses were in the range of 0.45–2.05 μm. Coloration rate upon uv exposure was high but thermal fade rates were very slow when compared with standard silver halide glass photochromic systems.

Simultaneous deposition of cadmium chloride and copper chloride by vacuum evaporation onto fused silica or optical glass resulted in photochromic thin films (23). The thickness ranged from 0.25 to 1.3 μm.

Thin films of photochromic silver complex oxides were prepared by anodic oxidation of silver metal films (24). Complex oxides, such as Ag_2VO_4, Ag_4SiO_4, and Ag_2PO_4, darkened by exposure to visible light, but required heating to 150–250°C for thermal bleaching.

2.3. Polyoxometalates. Photochromism of materials in which electron transfer and energy transfer within polyoxometalate and related metal oxide solids have been described (25). Structural characterization of both oxidized and reduced polyoxometalates is discussed in connection with the coloration mechanism.

2.4. Sols–Gels. Research on doping a photoactive organic phase into a transparent inorganic host matrix at molecular level to obtain optical composite has intensified in recent years.

The success of low temperature synthesizing sol-gel derived gel glasses doped with photoactive organics led to some new application opportunities in non-linear optics, the solid state tuneable laser, the visible display, and photochemical hole burning for example.... The recent development of preparation techniques and characterization methods to determine the structures and properties of the organic-inorganic hybrid optical functional materials, as well as the applications in luminescence, lasers, wave guide, wavelength division multiplexing, photochromism are reviewed in Ref. 26.

The new sol-gel derived materials for optics, organically modified metal oxide matrices have widely demonstrated their great potential (27). Most of the work performed in this area was concentrated on embedding organic or organo-metallic chromophores in an oxide network to make optical devices. The main advantages of the use of hybrid organic–inorganic nanocomposites result from their high versatility in offering a wide range of possibilities to fabricate tailor-made materials in terms of their chemical and physical properties, and macroscopic shape molding. Such materials emerging in this field are known as sol-gel photonics. There were some striking examples of the use of room-temperature processed hybrids to design materials with emission, absorption, second order nonlinear optical, and photochromic properties.

Ref. 28 contains a review on the design, synthesis, and some optical properties of hybrid organic–inorganic nano composites materials. Their properties depend on the chemical nature of the components, but also on their synergy.

The interface in these nanocomposites is important and one key point of their synthesis is the control of this interface. These nanocomposites can be obtained by hydrolysis and condensation reactions of organically functionalized alkoxide precursors. Striking examples of hybrids made from modified silicon, tin and transition metal alkoxides are presented. Some optical properties (photochromic, luminescence, NLO) of siloxane based hybrids are also discussed in this review (28).

3. Organic Photochromic Systems

The organic photochromic molecular systems are numerous and it is necessary to classify them in different families based upon the mechanism of the photochromic reaction. Their photochemical or thermal reverse reaction lalso links them to different types of applications.

The most prevalent organic photochromic systems involve unimolecular reactions: the most common photochromic molecules have a coloress or pale yellow form A and a colored form B (orange, red, blue, purple or green).

Other systems are bimolecular such as those involving photocycloaddition reactions.

3.1. Photochromism Based on Geometric Isomerism. The simplest example of a photochromic reaction involving a reversible cis-trans isomerization is azobenzene [103-33-3] (29).

trans (anti)   *cis* (syn)

This easy reaction produces a modest change in the absorption of visible light, largely because the visible absorption band of *cis*-azobenzene [1080-16-6] has a larger extinction coefficient than that of *trans*-azobenzene [17082-12-1]. Several studies have examined the physical property changes that occur upon photolysis of polymer systems in which the azobenzene structure is part of the polymer backbone (30).

The cis-trans isomerization of stilbenes is another photochromic reaction of the same type (31). Although the absorption bands of the stilbene isomers occur at nearly identical wavelengths, the extinction coefficient of the lowest energy band of *cis*-stilbene [645-49-8] is generally less important than that of *trans*-stilbene [103-30-0].

trans   *cis*

3.2. Photochromism Based on Tautomerism or Hydrogen Transfer. Several substituted anils (or imines or Schiff bases) of salicylaldehydes are photochromic, but only in the crystalline state. The photochromic mechanism involves an hydrogen transfer and a geometric isomerization (32). An example is the *N*-salicylidene-2-chloroaniline [3172-42-7].

Phototropic photochromism of 3-benzoyl-2-benzyl chromones (33–35) (X = O), thiochromones (X = S) (35), or quinolones (X = N-R) (36) is another example:

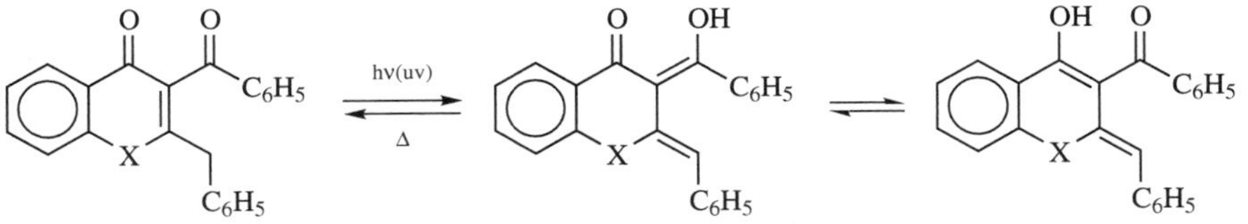

photoenol

The main limitation in this category is the weaker photosensibility (10 to 10^2 times) toward conventional photochromic pigments involving electrocyclization such as spiropyrans, spirooxazines, or naphthopyranes.

Some selectively substituted quinolones, such as 3-(2-benzylbenzoyl)-1,2-dimethyl-4(1H) quinolones (**3**) after irradiation in degassed toluene, lead to the formation of (5aS, 6R)-12-hydroxy-5,5a-dimethyl-6-phenyl-5a,6-dihydrobenzo[b]-acridin-11(5H)-one (**4**) (37).

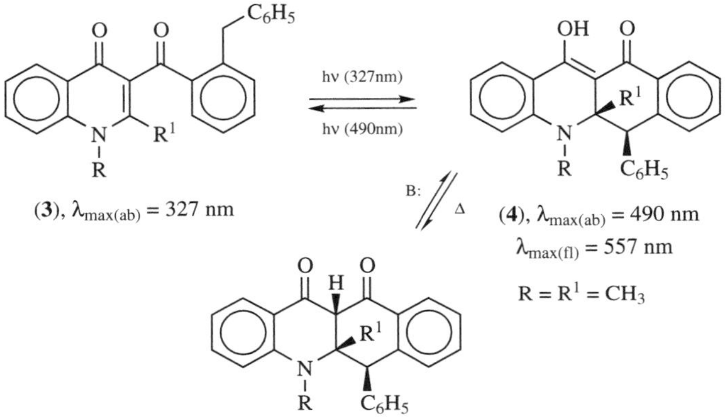

(**3**), $\lambda_{max(ab)}$ = 327 nm

(**4**), $\lambda_{max(ab)}$ = 490 nm
$\lambda_{max(fl)}$ = 557 nm

R = R^1 = CH$_3$

(**5**), $\lambda_{max(ab)}$ = 383 nm
$\lambda_{max(fl)}$ = 443 nm

Upon irradiation, derivative (**4**) (λ_{max} = 490 nm) can be converted back to (**3**). In the dark, derivative (**4**) is slowly converted to (5aS, 6R, 11aS)-5, 5a-dimethyl-6-phenyl-5a, 11a-dihydrobenzo[b]acridine-11,12(5H,6H)-dione (**5**) (λ_{max} = 386 nm). In the presence of catalytic amounts of a base, the equilibrium, (**4–5**), is established rapidly and irradiation of such solutions at 490 nm leads to the recovery of (**3**). Both (**4**) and (**5**) are fluorescent (λmax = 557 and 443 nm in toluene, respectively). No transients are detected and two isosbetic points are observed before the photostationary state is reached. Similar behavior is observed in acetonitrile, however, formation of the colored product (**4**) and its dark conversion into (**5**) proceeded more rapidly.

In a proposed mechanism, the initial photoenol or a biradical (formed upon irradiation, which is in fact a triplet state of the photoenol) can undergo further reversible cyclization owing to delocalization of the spin over the heterocyclic moiety.

This is a novel photoreversible photochromic system involving hydrogen transfer and photocyclization sequences.

Photochromic quinones exhibit photochemical migration of different hydrogen (**6b**), aryl (**6a**), or acyl groups (**6c**) (38).

(**6a**) X = O, R = Ph
(**6b**) X = CR^{1}R^2, R = H
(**6c**) X = O, R = $\overset{\text{O}}{\underset{}{\|}}$—R′

Quinones are a class of organic photochromic compounds that have been known as photochromic substances quite recently as compared to other compounds (39). They were discovered during the synthesis and studies of the properties of anthraquinones derivatives (40–42). It was determined that anthraquinone derivatives with an aryloxy group at the *peri*-position relative to the carbonyl group change color reversibly in the presence of uv light. Their photochromism is explained by the reversible photoinduced para-ana-quinoïd transformation due to photochemical migration of different hydrogen, aryl and acyl groups (43).

An example of ring-chain tautomerism and hydrogen transfer is perimidi-nespirocyclohexadienones (44), a novel type of heterocyclic photochrome.

When uv irradiated, an octane solution of 2,3-dihydro-2-spiro-4′-(2′,6′-di-*tert*-butylcyclohexadien-2′,5′-one) perimidine (**7a**) (R = H, Allyl) changes its color from yellowish to deep blue. The initial spectrum is slowly restored at room temperature (the effective life time of the colored form at room temperature is about 10^4s). No concentration dependence of the rate of the dark reaction was observed, which is in agreement with the intramolecular nature of the reactions (see Fig. 2).

The mechanism of the photochromic reaction involves cleavage of a C-N bond in the first singlet excited state of (**7a**) followed by the conformational rearrangement of the thus formed zwitterionic (biradical) intermediate (8) that precedes the final step of the intramolecular hydrogen transfer.

3.3. Photochromism Based on Dissociation Processes.

Both heterolytic and homolytic dissociation processes can result in the generation of photochromic systems. An example of an heterolytic process is the reversible formation of triaryl methyl cations by photolysis of triarylmethylchloride in acetonitrile (45).

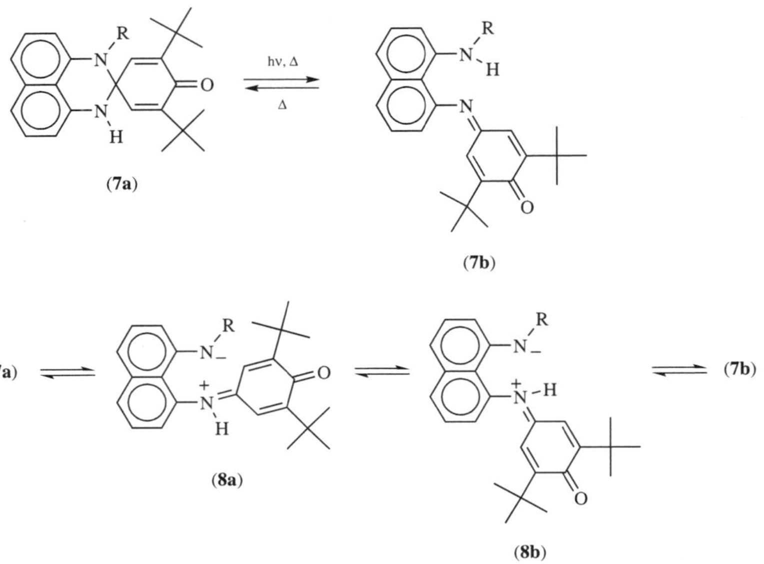

(7a)

(7b)

(7a) ⇌ **(8a)** ⇌ **(8b)** ⇌ **(7b)**

Fig. 2. Mechanism of a photochromic reaction involving perimidinespirocyclohexa-dienones.

The classical example of a photochromic process involving an homolytic dissociation is the formation of a red-purple free radical by photolysis of bis (2,4,5-triphenylimidazole) (46).

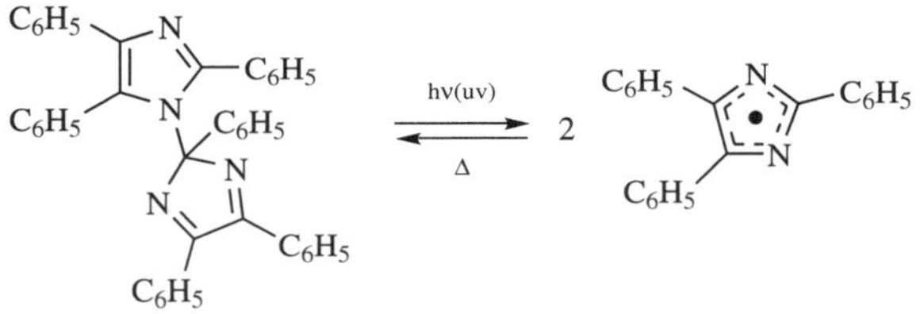

3.4. Photochromism Based on Triplet Formation.

Upon absorption of light, many polycyclic aromatic hydrocarbons and their heterocyclic analogues undergo transitions to their triplet state which has an absorption spectrum different from that of the ground state (47). In rigid glasses and some plastics, the triplet state which may absorb in the visible has a lifetime of up to 20 seconds (stabilization by the medium).

An example of such a polycyclic system is 1,2,5,6-dibenzacridine [226-36-8] which in a rigid matrix, absorbs uv irradiation to form a triplet state absorbing strongly in the visible with a λ_{max} at approximately 550 nm.

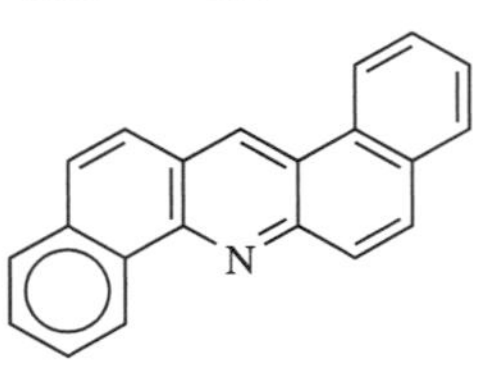

3.5. Photochromism Based on Redox Reaction.

Although the exact mechanism of the reversible electron transfer is often not defined, several viologen salts (pyridinium ions) exhibit a photochromic response to uv irradiation in the crystalline state or in a polar polymer matrix (48). An example is poly (N-vinyl-2-pyrrolidone) [9003-39-8] (49).

blue color (reduced form)

The reduced form (cation radical) is blue with a visible absorption maximum at 610 nm. The rate of the reoxidation of the reduced form is usually, but not always, strongly dependent on the presence of oxygen. The use of an electron-withdrawing counter ion leads to a photochromic system that is highly reversible in an inert atmosphere. An anion that has been used successfully is tetra-bis[3,5-di(trifluoromethyl)phenyl]borate [79230-20-9] (50).

3.6. Photochromism Based on Electrocyclic Reactions.

The most common general class of photochromic systems involves electrocyclic reversible reactions. Within this general class the most well-studied compounds are the indolinospiropyrans, the indolinospirooxazines, the naphthopyrans (or chromenes), the diaryl or heteroarylethenes, and fulgides. The most important applications are in the area of variable optical transmission materials (51,52), optical memories and switches (53,54).

Spiropyrans. Nitro-substituted indolinospirobenzopyrans or indolinospironaphthopyrans are photochromic when dissolved in organic solvents or polymer matrices (55,56). Absorption of uv radiation results in the colorless spiro compound being transformed into the colored, ring-opened species. This colored species is often called a photomerocyanine because of its structural similarity to the merocyanine dyes. Removal of the ultraviolet light source results in thermal reversion to the spiro compound.

colorless　　　　　　　　　　　　　　　　colored

The nitro-spiropyrans present a good "colorability" (good quantum yield, high extinction coefficient absorption), but they are susceptible to fatigue which has limited their application.

Indolinospirooxazines. Indolinospirooxazines exhibit photochromism by way of a mechanism that is very similar to that of the spiropyrans (57,58).

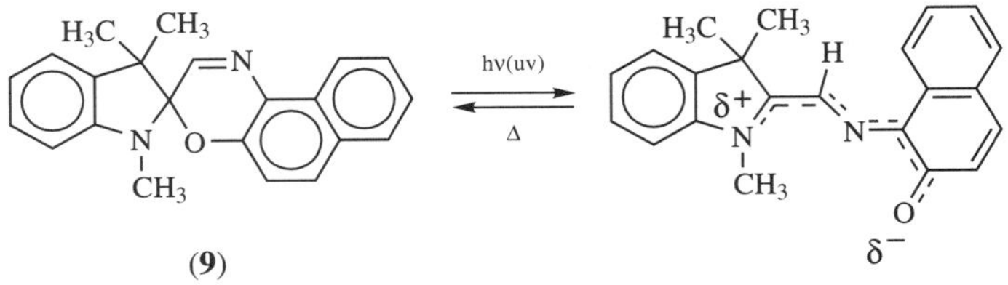

(9)

The spirooxazines however are much more resistant to fatigue and so could be used in making sunglasses.

As measured by the lifetime for photodegradation, indolinospironaphthoxazines such as (**9**) are 10 to 10^2 more stable than indolinospirobenzopyrans (59,60).

Benzo-and Naphthopyranes (or Chromenes). Intense research efforts by several groups interested in the commercialization of photochromic plastic ophthalmic lenses have, through structural modifications, dramatically enhanced the photochromic properties of benzo- and naphtho-pyrans [3,3-diaryl-3*H*-naphtho[2,1-b]pyrans (**10**) and 2,2-diaryl-2*H*-naphtho[1,2-b]pyrans (**11**)] (61).

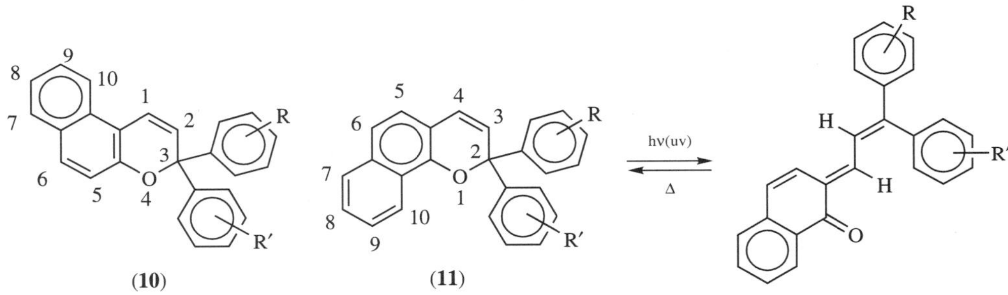

(**10**) (**11**)

closed form trans-quinoidal open form

Many patents and publications cover this field of investigation and the photochromic properties depend greatly on the type of annellation, (and the possibility of hetero annellation), and the nature and the position of organic or organometallic substituents (62). In the present examples the di-aryl substitution on tetrahedral Csp3 constitutes a main structural advantage for the application of such molecules. Indeed with this kind of substitution, the main photochromic parameters (kinetics, absorption wavelength, colorability, fatigue) are greatly improved. Recently (63) new biphotochromic systems linked by a vinyl bridge have been synthesized from functionalized spirooxazine and chromene fragments using the Wittig reaction. The photochromic behavior of the above-mentioned compounds, which were thoroughly studied later (64–68), proved to be very complex. For example, not only successive opening of both oxazine or chromene rings, but also the $Z \rightarrow E$ isomerization and electrocyclization to form compound (**15**) were observed for compound (**12**) obtained in the Z configuration.

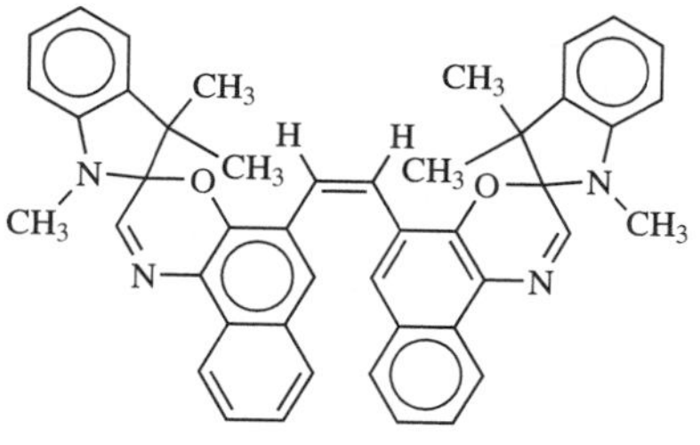

(12)

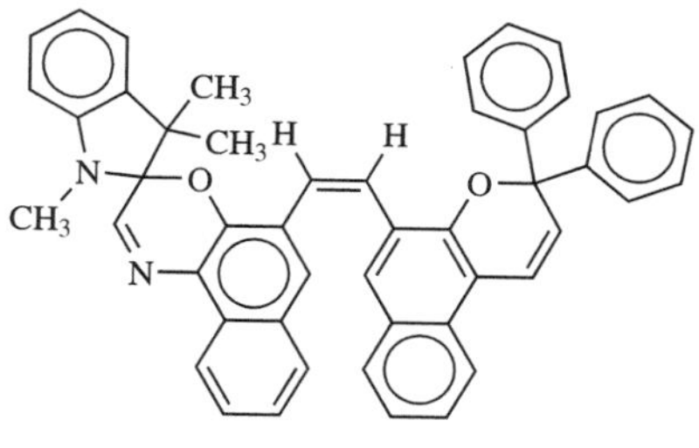

(13)

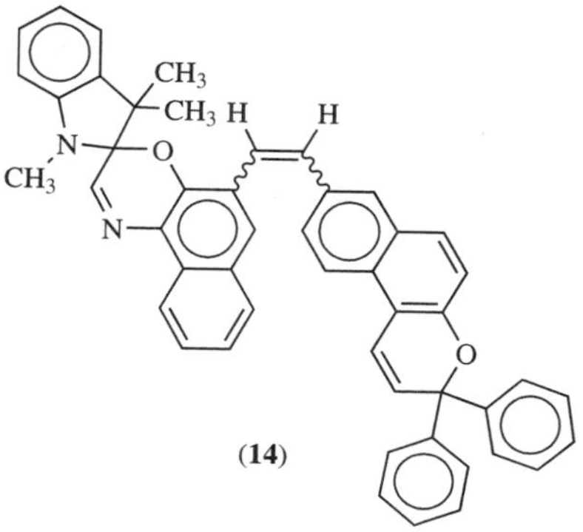

(14)

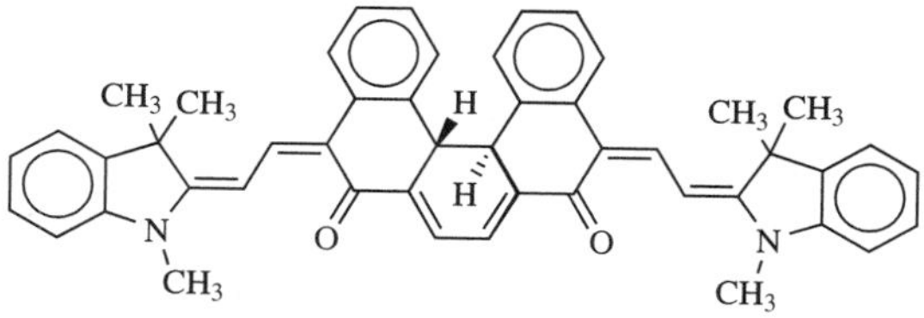

(15)

Heterocyclic, Fulgides, and Fulgimides. Another class of photochromic compounds, which operate through an electrocyclic mechanism, is the fulgides and fulgimides (69), although, with this class, the colored species is formed through a ring formation rather ring-opening (70,71). The reversion to the color-ess species does not occur thermally at ambient temperature but can efficiently be driven photochemically with visible light.

The name "fulgides" (from the Latin *fulgere*, to glisten) was given by their first investigator, Stobbe (72) because they were isolated as fine glittering crystals. Their photochromic properties have been studied extensively by H. G. Heller and co-workers (73).

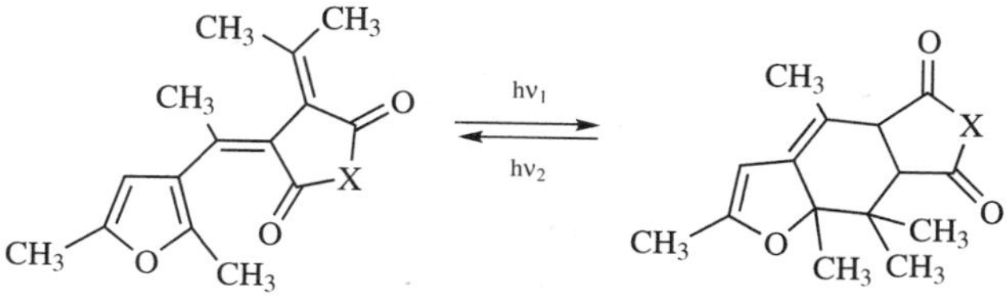

open form

closed form

X = O (fulgides)
X = NR (fulgimides)

Judiciously substituted (for example they do not have a labile hydrogen on the site of cyclization), fulgides have both high quantum yields for ring closure and opening and good fatigue resistance.

Stilbenes, Heteroarylethenes and Related Compounds. The photochemical ring closure of certain stilbenes, [eg, the highly methyl substituted compound (**16**)] and their heterocyclic analogues is the basis for another class of more sophisticated photochromic compounds (74–76).

(**16**)

By changing the substituents on the ethylenic linkage and replacing phenyl rings for heteroaromatic rings, photochromic systems (that are thermally reversible) are transformed into systems that are thermally irreversible, but photochemically reversible.

The transition between the benzothiophene-derivative isomers (**17**) and (**18**) is an example.

(**17**)   (**18**)

M. Irie and co-workers (77), in Japan, have intensively studied such photochromic systems. They have developed synthesis, investigated physico-chemical properties and applications in the area of optical memories and switches. These systems have a good resistance to fatigue and the colored forms absorb from 425 to 825 nm. The photochromism of diarylethenes molecules as single crystal was also studied (78).

Dihydroindolizines. The dihydroindolizines are photochromic compounds that undergo a photochemical ring opening reaction to form a colored zwitterionic species (79–81).

In general, this class of compounds involves complex methods of synthesis, and as in other photochromic systems, the colored species present a variety of colors depending on the substitution pattern.

The colored zwitterion reverts to the uncolored or slightly colored starting material either thermally or photochemically in the example below.

Some molecules have been linked to polymers or to inorganic salts giving supramolecular systems (80).

3.7. Photochromism Based on Cycloaddition Reactions Involving a Bimolecular Mechanism.
The photochemically reversible formation of endoperoxides is an example of this type of system (82,83) involving $(4n + 2)$ electrons. The reaction with singlet oxygen with some polycyclic aromatic compounds to form the endoperoxide and the photochemical elimination of singlet oxygen can be realized quantitatively.

colorless   colored (red)

The reaction is also accompanied by a rather drastic color change because of the disruption of the polycyclic chromophore during endoperoxide formation. A good example is the reversible formation of the endoperoxide of dibenzo(a, j) perylene-8, 16-dione (83).

In this case, the endoperoxide is colorless and the parent compound is red.

Another system concerns the dimerization of anthracenyl hydrocarbons involving $4n$ electrons [$(4 + 4)$ cycloaddition] (84).

4. Uses

Although the proposed applications for photochromic systems are numerous few have received broad as commercial use.

4.1. Variable Optical Transmission Materials.
The convenience of having lenses that darken automatically upon exposure of sunlight has proven appealing to spectacle wearers (85). By far, the most successful commercial applications is the use of photochromic silver halide-containing glasses in prescription eyewear.

The desire for plastic lenses in the ophthalmic market has also accelerated the research on organic photochromic pigments that are more compatible with polymer matrices and more photoresponsive (51,52).

In order to achieve an organic photochromic system for the eyewear market, two primary problems had to be solved. The first problem of fatigue resistance was alleviated with the discovery of indolinospirooxazines and naphthopyranes which are inherently more fatigue-resistant than other systems such as spiropyrans. The photostability of the spirooxazines can be improved even further by protecting them from oxygen (86,87) or by the addition of chemical stabilizers, specifically nickel complexes (88) or hindered amine light stabilizers (89). The second problem is caused by the relatively narrow absorption bands of activated organic photochromic molecules in the visible range of light. This problem can be handled by mixing compatible compounds such as indolinospirobenzoxazines or indolinospironaphthoxazines (57) and naphthopyranes (61) giving a much broadened visible light absorption (90).

Photochromic lenses for eyewear serve as variable density optical filters. Other applications for photochromic light filters have been proposed including glazing applications for solar attenuation, variable transmission camera lenses, and shields for protection against the light flash from a nuclear explosion.

Besides the use of photochromic systems in light filters, their color development has also received considerable attention. For example, the introduction of photochromic components into product labels, tickets, credit cards, etc, aids in the verification of authenticity (91,92). The active components are invisible until activated with an uv light source and then they are detected easily.

The color development of photochromic compounds can also be used as a diagnostic tool. The temperature dependence of the fading of 6-nitroindolinospiropyran served as the basis for a nondestructive inspection technique for honeycomb aerospace structures (93). One surface of the structure to be examined was covered with a paint containing the photochromic compound and activated to a violet color with uv-light. The other side of the structure was then heated, the transfer of heat through the honeycomb structure caused bleaching of the temperature-dependent photochromic compound. Defects in the honeycomb where heat transfer was inhibited, could be detected as darker areas.

Photochromic compounds that can be thermally faded have also been used in engineering studies to visualize flows in dynamic fluid systems (94−96).

Most photochromic compounds undergo large structural changes while being transformed from the uncolored to the colored form. This property has been used to examine the pore size of polymers by using the relationship of pore size and the kinetics of the photochromic response (97).

4.2. Optical Memories and Switches. The molecular property changes of photochromic molecules and especially the photoreversible ones can be applied to various photonic devices such as erasable optical memory media and photo-optical switch components.

The erasable memory media developed so far have been inorganic materials which use the magneto-optic effect or phase change as the basis for optical recording. But during the last years, the worldwide acceptance of CD-R (Compact Disk-Recordable) which uses organic dyes as the memory medium, has changed the situation and has given an impetus to find photochromic materials that change their refractive index by photoinduction.

Fulgides (98) and diarylethenes (99) are the main families involved in these applications. Nevertheless spiropyrans and spirooxazines could be used in association with polymers (PMMA or PVK) or liquid crystals (100).

The chirality is also used for chiroptical molecular switches in overcrowded alkenes, diarylethenes, binphthyl derivatives, fulgides, spiropyrans, photochromic polymers, polymer liquid crystals. (101)

4.3. Miscellaneous Applications. Azobenzenes, spiropyrans, fulgides, diarylethenes, anils coupled to polymers are used in the determination of linear and nonlinear optical properties of photochromic molecules and materials (102).

Photochromic materials can be used also for holographic recording (100).

The photoalignment control of liquid crystals could be realized through reversible photochromic molecules in three-dimensional networks (103).

Bacteriorhodopsin, a photochromic retinal protein having a biological function for the conversion of light energy into chemical energy, is used in data storage applications (104).

BIBLIOGRAPHY

"Chromogenic Materials, Photochromic" in *ECT* 3rd ed., Vol. 6, pp. 121–128, by R. J. Araujo, Corning Glass Works; *ECT* 4th ed., Vol. 6, pp. 321–332, by J.C. Crano, PPG Industries, Inc.; "Chromogenic Materials, Photochromic" in *ECT* (online), posting date: December 4, 2000, by John C. Crano, PPG Industries, Inc.

CITED PUBLICATIONS

1. H. Dürr and H. Bouas-Laurent eds, *Photochromism: Molecules and Systems*, Elsevier, Amsterdam, The Netherlands, 1990.
2. H. Bouas-Laurent and H. Dürr, *Pure Appl. Chem.* **73**, 639 (2001).
3. J. Fritzsche, *Compt. Rend. Acad. Sci.* Paris, **69**, 1035 (1867).
4. E. ter Meer, *Ann. Chem.* **181**, 1 (1876).
5. T. L. Phipson, *Chem. News* **43**, 283 (1881); J. B. Orr, *Chem. News* **44**, 12 (1881).
6. W. Markwald, *Z. Phys. Chem.* **30**, 140 (1899).
7. Y. Hirshberg, *Compt. Rend. Acad. Sci.* Paris, **231**, 903 (1950).
8. C. B. Mc Ardlke, ed., *Applied Photochromic Polymer Systems*, Blackie, New York, 1992.
9. V. Malatesta in J. C. Crano and R. J. Guglielmetti, eds., *Organic Photochromic and Thermochromic Compounds*, Vol. 2, Kluwer Academic/Plenum Publishers, New York, 1999, Chapt. 3, pp. 65–164.
10. W. H. Armistead and S. D. Stookey, *Science* **144**, 150 (1964).
11. U.S. Pat. 3,208,860 (Sept. 28, 1965), W. H. Armistead and S. D. Stookey (to Corning).
12. J. P. Smith, *J. Photogr. Sci.* **18**, 41 (1970).
13. R. J. Araujo, *Contemp. Phys.* **21**, 77 (1980).
14. H. J. Hoffmann, in C. M. Lampert and C. G. Granqvist eds., *Large-Area Chromogenics: Materials and Devices for Transmittance Control*, Vol. IS 4, SPIE Institutes for Advanced Optical Technologies, Bellington, Wash., 1990, p. 86.
15. U.S. Pat. 4,251,278 (Feb. 17, 1981), G. B. Hares (to Corning).
16. U.S. Pat. 4,043,781 (Aug. 23, 1977), C. V. DeMunn, D. J. Kerko, R. A. Westwig, and D. B. Wrisley, Jr. (to Corning).
17. H. M. Garfinkel, *Appl. Opt.* **7**, 789 (1968).
18. M. Mizuhashi and S. Furuuchi, *Thin Solid Films* **30**, 259 (1975).
19. U.S. Pat. 3,325,299 (June 13, 1967), R. J. Araujo (to Corning).
20. D. M. Trotter, Jr., J. W. H. Schreurs, and P. A. Tick, *J. Appl. Phys.* **53**, 4657 (1982).
21. A. J. Cohen, *Science* **137**, 981 (1962).
22. A. F. Perveyev and A. V. Mikhaylov, *Sov. J. Opt. Technol.* **39**, 117 (1972).

23. H. Marquez, J. Ma Rincon, and L. E. Celeya, *Appl. Opt.* **29**, 3699 (1990).

24. T. H. Hirono, T. Yamada, and T. Nishi, *J. Appl. Phys.* **59**, 948 (1986).

25. T. Yamase, *Chem. Rev.* **98**, 307 (1998).

26. G. Qian and M. Wang, *Guisuanyan Xuebao* (Chinese Review) **29**(6), 596 (2001).

27. B. Lebeau and C. Sanchez, *Current Opinion in Solid State and Materials Science* **4**, 11 (1999).

28. C. Sanchez, F. Ribot and B. Lebeau, *J. Mat. Chem.* **9**(1), 35 (1999).

29. H. Rau, in H. Durr and H. Bouas-Laurent, eds., *Photochromims: Molecules and Systems*, Elsevier, Amsterdam, The Netherlands, 1990, Chapt. 4, pp. 165–192.

30. G. S. Kumar and D. C. Neckers, *Chem. Rev.* **89**, 1915 (1989).

31. J. Saltiel and Y.-P. Sun, in Ref. 29, p. 64.

32. M. D. Cohen and G. M. Schmidt, *J. Phys. Chem.* **66**, 2442 (1962).

33. E. Hadjoudis in Ref. 29, Chapt. 17, pp. 685–712.

34. K. R. Huffman, M. Loy and E. F. Ullman, *J. Am. Chem. Soc.* **87**, 5417 (1965).

35. V. Rossollin, Ph. D Thesis Marseille-Luminy (France), 2001.

36. M. Vales, Ph. D Thesis Marseille-Luminy (France), 2002.

37. V. Lokshin, M. Vales, A. Samat, G. Pepe, A. Metelitsa, and V. Khodorkovsky, *J. Chem. Soc. Chem. Comm.*, 2080 (2003).

38. V. A. Barachevsky in Ref. 9, Vol. 1, Chapt. 7, pp. 267–314.

39. V. A. Barachevsky, G. I. Lashkov, and V. A. Tsekhomsky, in *Photochromism and its Applications*, Khimiya, Moscow, 1977.

40. Yu. E. Gerasimenko and N. T. Poteleshenko, *Zh. Vkho. im. Mendeleeva.* **16**, 105 (1971).

41. Yu. E. Gerasimenko and N. T. Poteleshenko, *Zh Org. Khim.* **7**, 2413 (1971).

42. V. N. Kostylev, B. E. Zaitsev, V. A. Barachevsky, N. T. Poteleshenko, and Yu. E. Gerasimenko, *Opt. Spectr.* **30**, 86 (1971).

43. N. P. Gritsan and L. S. Klimenko, *Mol. Cryst. Liq. Cryst.* **246**, 103 (1994).

44. V. I. Minkin, V. N. Komissarov and V. A. Kharlanov in Ref. 9, Vol. 1, Chapt. 8, pp. 315–340.

45. L. E. Manring and K. S. Peters, *J. Phys. Chem.* **88**, 3516 (1984).

46. T. Hayashi and K. Maeda, *Bull. Chem. Soc. Jpn.* **33**, 565 (1960).

47. J. L. Kropp and M. W. Windsor, *U.S. Air Force Technical Report*, AFML-TR-68-220, Washington, D.C., Aug. 1968.

48. M. Nanasawa in Ref. 9, Vol. 1, Chapt. 9, pp. 341–369.

49. H. Kamogawa and T. Suzuki, *Bull. Chem. Soc. Jpn.* **60**, 794 (1987).

50. T. Nagamura, K. Sakai, and T. Ogawa, *J. Chem. Soc., Chem. Commun.*, 1035 (1988).

51. J. C. Crano, T. Flood, D. Knowles, A. Kumar, and B. Van Gemert, *Pure Appl. Chem.* **68**, 1395 (1996).

52. J. C. Crano, W. S. Kwak and C. N. Welch in C. B. Mc. Ardle, ed., *Applied Photochromic Polymer Systems*, Blackie, Glasgow and London, 1992.

53. M. Irie, *Mol. Cryst. Liq. Cryst.* **227**, 263 (1993).

54. M. Irie, *Jpn. J. Appl. Phys.* **28**, 215 (1989).

55. R. Guglielmetti, in Ref. 1, Chapt. 8, pp. 314–466.

56. R. C. Bertelson in Ref. 9, Vol. 1, Chapt. 1, pp. 1–83.

57. S. Maeda in Ref. 9, Vol. 1, Chapt. 2, pp. 85–109.

58. V. Lokshin, A. Samat, and A. V. Metelitsa, *Russian Chem. Rev.* **71**, 893 (2002).

59. N. Y. C. Chu, *Proceedings of the 10th IUPAC Symposium on Photochemistry*, Interlaken, Switzerland, 1984.

60. R. Dubest, P. Levoir, J. J. Meyer, J. Aubard, G. Baillet, G. Giusti, and R. Guglielmetti, *Rev. Sci. Instr.* **64**, 1803 (1993).

61. B. Van Gemert in Ref. 9, Vol. 1, Chapt. 3, pp. 111–140.

62. S. Anguille, P. Brun, R. Guglielmetti, Y. Strokach, A. Ignatine, V. Barachevsky, and M. Alfimov, *J. Chem. Soc. Perkin Trans. 2* **4**, 639 (2001).
63. A. Samat, V. Lokshin, K. Chamontin, D. Levi, G. Pèpe, and R. Guglielmetti, *Tetrahedron* **57**, 7349 (2001).
64. F. Ortica, D. Levi, P. Brun, R. Guglielmetti, U. Mazzucato, and G. Favaro, *J. Photochem. Photobiol. A* **138**, 123 and **139**, 133 (2001).
65. G. Favaro, D. Levi, F. Ortica, A. Samat, R. Guglielmetti, and U. Mazzucato, *J. Photochem. Photobiol. A* **149**, 91 (2002).
66. J. Berthet, S. Delbaere, V. Lokshin, C. Bochu, A. Samat, R. Guglielmetti, and G. Vermeersch, *Photochem. Photobiol. Sci.* **1**, 333 (2002).
67. J. Berthet, S. Delbaere, D. Levi, A. Samat, R. Guglielmetti, and G. Vermeersch, *Photochem. Photobiol. Sci.* **1**, 665 (2002).
68. J. Berthet, S. Delbaere, D. Levi, P. Brun, R. Guglielmetti, and G. Vermeersch, *J. Chem. Soc. Perkin Trans.* **2**, 2118 (2002).
69. M. G. Fan, L. Yu and W. Zhao in Ref. 9, Vol. 1, Chapt. 4, pp. 141–206.
70. H. G. Heller, *IEE Proc.* **130**(5), 209 (1983).
71. J. Whittal, in Ref. 1, Chapt. 9, pp. 467–492.
72. H. Stobbe, *Berichte* **37**, 2236 (1904) and **40**, 3372 (1907).
73. H. G. Heller, in W. M. Horspool and Pill-soon Song, eds., *Handbook of Organic Photochemistry and Photobiology*, CRC, Boca Raton, Fla., 1995, chapt. 13.
74. S. Nakamura and M. Irie, *J. Org. Chem.* **53**, 6136 (1988).
75. Y. Nakayama, K. Hayashi, and M. Irie, *J. Org. Chem.* **55**, 2592 (1990).
76. K. Uchida, Y. Nakayama, and M. Irie, *Bull, Chem. Soc. Jpn.* **63**, 1311 (1990).
77. M. Irie, in Ref. 9, Vol. 1, Chapt. 5, pp. 207–222.
78. S. Yamamoto, K. Matsuda and M. Irie, *Angew. Chem. Int. Ed.* **42**, 1636 (2003).
79. H. Dürr, *Angew. Chem. Int. Ed.* **28**, 413 (1989).
80. H. Dürr in Ref. 9, Vol. 1, Chapt. 6, pp. 223–266.
81. H. Dürr in Ref. 1, Chapt. 6, pp. 210–269.
82. H. D. Brauer and R. Schmidt, *Photochem. Photobiol.* **37**, 587 (1983).
83. Ger. Offen. 2,910,668 (Sept. 25, 1980), H. D. Brauer, R. Schmidt, and W. Drews.
84. H. Bouas–Laurent and J. P. Desvergne in Ref. 1, Chapt. 14, pp. 561–622.
85. OMA (Optical Manufacturers Association) *National Consumer Eyewear Study VI*, Falls Church, Va., 1990.
86. U.S. Pat. 4,166,043 (Aug. 28, 1979), D. R. Uhlmann, E. Snitzer, R. J. Hovey, N. Y. C. Chu, and J. T. Fournier (to American Optical).
87. U.S. Pat. 4,367,170 (Jan. 4, 1983), D. R. Uhlmann, E. Snitzer, R. J. Hovey, and N. Y. C. Chu (to American Optical).
88. U.S. Pat. 4,440,672 (Apr. 3, 1984), N. Y. C. Chu (to American Optical).
89. U.S. Pat. 4,720,356 (Jan. 19, 1988), N. Y. C. Chu (to American Optical).
90. U.S. Pat. 4,968,454 (Nov. 6, 1990), J. C. Crano, P. L. Kwiatkowski, and R. J. Hurditch (to PPG Industries).
91. Eur. Pat. Appl. 328,320 A1 (Aug. 16, 1989), P. Wright (to Courtaulds).
92. PCT Int. Appl. WO 90/06539 A1 (June 14, 1990), S. Wallace (to Traqson Ltd.).
93. S. Allinikov, *U.S. Air Force Technical Report*, AFML-TR-70-246, Washington, D.C., Dec. 1990.
94. R. E. Falco and C. C. Chu, *Proc. SPIE (Int. Conf. Photomech. Spec. Met.)* **814**(2), 706 (1988).
95. Brit. Pat. Appl. 2,209,751 A (May 24, 1989), C. Trundle (to Plessey Co.).
96. V. Croquette, P. Le Gal, A. Pocheau, and R. Guglielmetti, *Europhys. Lett.* **1**, 393 (1986).
97. W.-C. Yu, C. S. P. Sung, and R. E. Robertson, *Macromolecules* **21**, 355 (1988).
98. Y. Yokoyama, *Chem. Rev.* **100**, 1717 (2000).

99. M. Irie, in S. Kawata, M. Ohtsu and M. Irie, eds., *Nano-Optics*, Springer, Berlin, 2002, pp. 137–150.
100. G. Berkovic, V. Krongauz, and V. Weiss, *Chem. Rev.* **100**, 1741 (2000).
101. B. L. Feringa, R. A. Van Delden, N. Koumura, and E. M. Geertseme, *Chem. Rev.* **100**, 1789 (2000).
102. J. A. Delaire and K. Nakatani, *Chem. Rev.* **100**, 1817 (2000).
103. K. Ichimura, *Chem. Rev.* **100**, 1847 (2000).
104. N. Hampp, *Chem. Rev.* **100**, 1755 (2000).

A. Samat
R. Guglielmetti
Université de la Mediterannée

CHROMOGENIC MATERIALS, PIEZOCHROMIC

1. Introduction

In its most general sense piezochromism is the change in color of a solid under compression. There are three aspects of the phenomenon.

The first is, in a sense, trivial, but it is very general. The color of a solid results from the absorption of light in selected regions of the visible spectrum by excitation of an electron from the ground electronic state to a higher level. If the two electronic energy levels are perturbed differently by pressure, compression results in a color change. This is the basic definition of pressure tuning spectroscopy. Examples include, among others, increased splitting of the *d* orbitals of transition-metal ions in complexes with pressure, the shift to higher energy of a color center (a vacancy containing an electron) in an alkali halide or glass environment, and a change in the relative energy of bonding and antibonding orbitals as pressure increases. This last phenomenon depends on the relative importance of intra- and intermolecular interactions. The compression of a bond increases the difference in energy between bonding and antibonding orbitals. On the other hand, the attractive van der Waals interactions between molecules are generally stronger around an excited molecule. Where the latter interaction dominates, one observes a shift to lower energy (red shift) of these excitations. Further examples of this include the energy associated with the transfer of an electron in an electron donor–acceptor complex and the difference in energy between the top of the valence band and bottom of the conduction band, ie, the absorption edge, in insulators and semiconductors.

These excitations are widely used to characterize electronic states and excitations, to test theories about electronic phenomena, and to delineate the nature of local sites in glasses, disordered solids, intercalates, etc. However, this aspect of changing color with pressure is so general that it is not satisfactory for defining piezochromism.

A second aspect involves a discontinuous change of color when a crystalline solid undergoes a first-order phase transition from one crystal structure to another. The most obvious example is the change of the absorption edge. For example, CdS changes from yellow to deep red at 2.7 GPa (27 kbar) when the crystal structure changes from wurtzite to sodium chloride (face-centered cubic). CdSe, ZnS, ZnSe, and ZnTe undergo similar transitions with distinct color changes at pressures from 5–15 GPa (50–150 kbar). First-order phase transitions involving alterations in crystal structure only can change the electronic excitation energy associated with almost any kind of electronic process provided the two electronic states interact differently with the changing environment. However, for most molecular crystals at modest pressures, the coupling between the electronic states of the molecule and the lattice modes is rather small so these perturbations are usually not large, with few exceptions.

The phenomenon of most interest is a change in color of a solid as a result of a change in the molecular geometry of the molecules that make up the solid. The color change takes place because the change in geometry alters the relative energy of different electronic orbitals, and therefore the electronic absorption spectrum. Frequently it rearranges the order of these orbitals or provides new combinations of atomic orbitals because of symmetry changes. The rearrangements may be discontinuous at a given pressure, may occur over a modest range of pressures, or may occur gradually over the whole range of available pressure as for chemical equilibria in solution (1,2). A few examples, together with the principles or generalizations that arise from them, are discussed.

Piezochromism has been observed in a wide variety of materials. Three classes which illustrate well some of the generalizations that have been developed are organic molecules in crystals and polymer films, metal cluster compounds, and organometallic complexes of Cu(II).

2. Principal Piezochromic Systems

2.1. Organic Molecules in Crystals and Polymer Films.

The prototypes of piezochromic organic molecules are the salicylidene anils, eg, *N*-salicylidene-2-chloroaniline (3–5). At ambient pressure in the crystalline state they are either photochromic or thermochromic, but never both, depending on the side groups on the aromatic rings. When dissolved in a polymer film they are photochromic. The ground state has an OH group opposite a nitrogen on the adjacent ring, thus it is called the "enol" form. When heated, the thermochromic compounds exhibit an absorption in the visible spectrum which corresponds to a transfer of the H from O to N without other change of molecular geometry. This is the "*cis*-keto" form. The photochromic molecules, upon irradiation at low temperature, develop an absorption assigned to the "*trans*-keto" form.

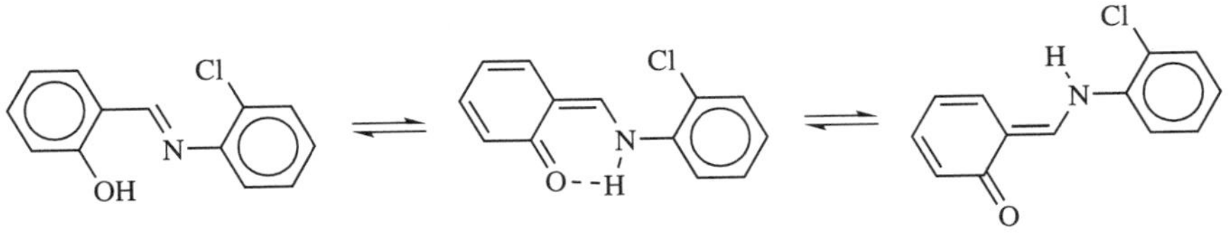

enol ("OH form")   *cis*-keto ("NH form")   *trans*-keto ("NH form″")

With increasing pressure at 25°C the *cis*-keto form is stabilized *vis á vis* both the enol and *trans*-keto isomers so that both thermochromic and photochromic materials exhibit the same type of piezochromism, but in different degrees. The conversion increases continuously with pressure, much like the changing of chemical equilibrium in liquid solution with pressure.

Two principles are illustrated here. In this case, increasing pressure and temperature favor the same process. With increasing temperature the possibility of crossing a barrier of a given height increases. With increasing pressure the potentials well-associated with the two states are perturbed differently. This perturbation can either augment or oppose the effect of increasing temperature, ie, temperature and pressure are not in general conjugate variables as is frequently assumed. Change in bond length or molecular geometry is induced by hydrogen transfer. This illustrates the point that changes in intermolecular interaction and packing are more likely to determine the amount of piezochromic or other reaction introduced by pressure than are differences in molecular geometry. Other piezochromic reactions have been observed in spiropyrans and bianthrones. The changes in molecular configuration are different, but no further principles are derived.

Other piezochromic reactions have been observed in spiropyrans and bianthrones. The changes in molecular configuration are different, but no further principles are derived.

Photogenerated and cryotrapped unstable radical species of photo- thermo-, and piezochromic hexaarylbiimidazolyl (HABI) derivatives are characterized by *in situ* X-ray cristallography. Low-temperature photolysis of *o*-Cl-HABI (2-chlorohexaarylbiimidazole) crystals generate radical pairs while retaining crystallinity. On the other hand, a pair of HABI transforms into a complex of two lophyl radicals and a piezodimer upon the photolysis (6). The piezodimer is only stable below −20°C (7), the structure of which was studied by ^{1}H nmr spectroscopy (8).

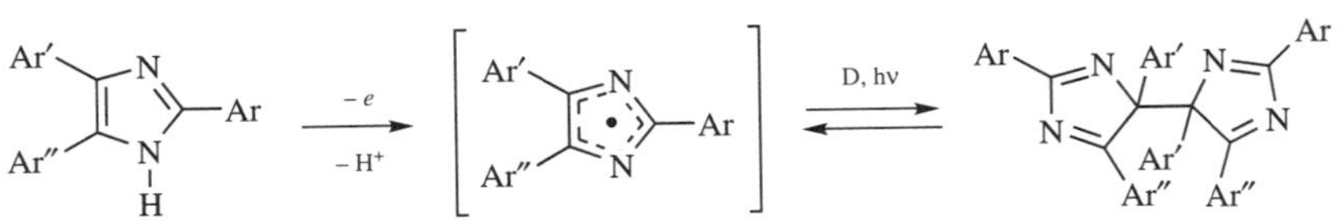

Conjugated polymers can be used as detector or transducers for chemical or physical information. Thermochromism, solvatochromism, piezochromism, photochromism, biochromism for optical transducers are discussed together with conformational and electrostatic effects for electrochemical transducers (9).

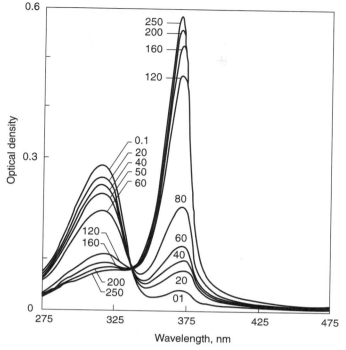

Fig. 1. Variable-pressure solid-state uv absorption spectra of poly(di-*n*-hexylsilylene) at 46°C (pressure units in MPa) (10).

The polysilylenes exhibit complex solid-state structures and electronic properties. The solid-state conformational structure of symmetrically substituted poly(di-*n*-alkylsilylenes) are described (10) (Fig. 1).

Comparisons between the different structures and their respective uv absorption characteristics demonstrate that the relationship is complex and aspects of molecular geometry beyond chain conformation must be considered. The thermochromic and piezochromic behaviors of polysilylenes are tightly linked. The all-*trans* conformation of the silicon backbone is associated with both phenomena (11,12). Polysilylene copolymers present a well-ordered structure and exhibit absorption characteristics similar to the crystalline phases of the corresponding homopolymers.

2,2,4,6-Tetraphenyldihydro-1,3,5-triazine, 2,4,4,6-tetraphenyl-1,4-dihydropyridine and related heterocycles exhibit photochromism, thermochromism, and piezochromism in the solid state (13).

A series of 3-acridinol derivatives having a hydroxyl group at 3-position of a acridine nucleus showed an absorption at about 460 nm in protic solvents and piezochromism. AM1 and PPP molecular calculations clarified that in protic solvents the lactam form of acridinol is more stable rather than the lactim form. It was proved that the piezochromic phenomenom of 9-phenyl and 9-methyl-3-acridinols occur with an intermolecular hydrogen transfer from oxygen to nitrogen under applied pressure by means of x-ray crystallography and ir analysis (14).

Application of external pressure to solutions of betaines (1) and (2) produces in all solvents used a hypsochromic shift of their long-wavelength CT absorption band (15–19).

(1) $R' = R'' = H$, $X = O$
(2) $R' = R'' = CH(CH_3)_2$, $X = O$
(3) $R' = SO_2CH_3$, $R'' = H$, $X = O$
(4) $R' = R'' = H$, $X = S$

For example, the uv/vis absorption maximum of (1) in ethanol is shifted hypsochromically by -27 nm (from $\lambda_{max} = 547$ to 520 nm) by increasing the external pressure up to 10 kbar (9869 atm) (118). This corresponds to an increase in the $E_T(30)$-value of ethanol by 2.7 kcal/mol. Thus, solvent polarity is pressure dependent. The reason for this new type of piezochromism ("negative piezo-solvatochromism") lies again in the differential solvent-mediated stabilization of the dipolar electronic betaine ground state, relative to its less dipolar excited state, with increasing external pressure. The range of piezochromism is different from solvent to solvent. The pressure-dependent absorption energies of (1) and (2), dissolved in one solvent under study, correlate well with the dielectric function $(\varepsilon - 1)/(\varepsilon + 2)$ of this solvent, which also increases with compression. This seems to indicate that only nonspecific betaine-solvent interactions that occur are, to first order, constant with pressure-at least for the alcoholic HBD solvents used (18).

2.2. Coordination Compounds and Metal Cluster Compounds.
Compounds with metal–metal bonds stabilized by appropriate ligands constitute a second class of materials where a number of cases of piezochromic behavior have been observed (1,20–22). Compounds involving a Re–Re bond stabilized either by eight halides or by bridging (bidentate) ligands, such as the pivalate ion, are two well-established cases. Octahalodirhenates, $Re_2X^{2-}_8$ (X = Cl, Br, I), in crystals with a number of counterions, exhibit an absorption peak which corresponds to an excitation from a bonding to antibonding orbital with angular momentum two around the bond (δ orbital). When pressure is applied to the iodide a new peak grows in at a lower energy. It is associated with the rotation

of the iodides on opposing Re ions from an eclipsed (directly opposite) to a staggered (not necessarily 45°) position. The reduced repulsion between opposing I^- ions more than compensates for the weakened metal–metal bond. The amount of conversion is continuous with increasing pressure.

In the case of the bridged complexes, the process involves changing from a bidentate to a monodentate configuration. For these systems the mode of transformation is variable. In close-packed crystals the rearrangement is a first-order process, ie, it occurs discontinuously at a fixed pressure. For slightly less close-packed crystals the transformation occurs over some range of pressure, eg, 2–3 GPa (20–30 kbar). In the language of physics the process corresponds to a higher order phase transition. When the molecules are dissolved in polymer films the reaction is continuous (stochastic) over the entire range of available pressure. These studies illustrate the importance of the environment on the extent of cooperativity involved in pressure-induced molecular geometry changes and thus on piezochromism.

Complexes involving larger metal clusters, eg, Au_9 or Au_2Rh_4, also undergo piezochromic rearrangements with rather dramatic changes in the absorption spectrum, and well-defined changes in molecular structure (22,23).

2.3. Organometallic Complexes of Cu(II).

Complexes of Cu(II) occur in a wide variety of distorted geometries. The d^9 configuration is stabilized by distortions from a high to a slightly lower symmetry, ie, the Jahn-Teller effect (1,20,24,25). Cu(II) complexed to organic molecules such as ethylenediamine derivatives lie in a square planar configuration of nitrogens from the organic molecules. If the axial ligands are far off or not aligned the Cu(II) retains this four-coordinate geometry in the solid state. When these axial ligands are sufficiently close the system becomes six coordinated with definite changes in the ordering of the electronic orbitals and thus in the visible spectrum. In the crystalline state the rearrangement from four to six coordinated is apparently a first-order transition, whereas when dissolved in a polymeric matrix, the transformation occurs over a range of 3–5 GPa (30–50 kbar). This is then another example of the importance of the environment on the pressure dependence of the rearrangement.

Cu(II) complexed to four Cl^- ions can adopt arrangements from tetragonal to square planar. The latter arrangement gives maximum Cu–Cl bonding; the former minimizes the Cl–Cl repulsion. In practice, complexes occur near both extremes and with almost all intermediate arrangements. The geometry of a given complex is determined by the counterion. Strongly hydrogen bonding counterions draw off electron density from the Cl^- and favor square planar symmetry. With pressure one finds that both nearly square planar and nearly tetrahedral complexes distort toward an intermediate symmetry with clear-cut changes in the electronic absorption spectrum. These transformations occur over relatively short ranges of pressure ∼1.5–2.5 GPa (15–25 kbar). The principle illustrated here is that at high compression the economy of best geometric packing overcomes the weaker van der Waals and hydrogen bonding forces.

Piezochromic effects have been observed in a variety of other Cu(II) complexes. In some cases it can be shown that the structures of a series of related complexes follow a reaction pathway with the structure of one complex at, eg, 8 GPa (80 kbar) corresponding to that of a related complex at, for example,

2 GPa (20 kbar). The changes in color of the complex, of course, follow the same sequence.

3. General Aspects and Conclusion

In general there appear to be two modes by which pressure induced isomerizations and consequently, piezochromism, can occur. There can be a transformation from one distinct conformation to another as a result of the relative stabilization of the potential well associated with conformation B with respect to that of A. This process may be either an equilibrium process, if the energy barrier is small, or involve a first- or higher-order phase transition depending on the extent of cooperativity demanded by the type of transformation and the environment. Such transformations are characterized by a change in the electronic character of the ground state. If there is no distinct change in the electronic character of the ground state, a series of compounds may follow each other along a reaction pathway as they undergo a similar piezochromic transformation over different pressure ranges.

In the foregoing discussion polymers have been used as a medium for small molecules in comparison with the crystalline state. It has also been observed that there are changes in polymer geometry and various rotational motions introduced by pressure (26–30). These are at times reflected in the absorption spectrum (usually in the ultraviolet) or in the emission spectrum and are a form of piezochromism.

The examples of piezochromism discussed so far involve rather well-established changes of molecular geometry. There are examples of pressure-induced changes in the electronic ground state with resultant changes in the electronic absorption spectrum where the changes in molecular geometry are not well established. One example involves intramolecular or intracomplex charge transfer (31–36). At ambient pressure this process takes place by optical excitation. At sufficiently high pressure the charge-transferred state may be stabilized sufficiently to become the ground state of the system with a different electronic absorption spectrum, but where the geometry changes are not well defined. In most transition-metal complexes the metallic ion exists in the state of maximum multiplicity "high spin" according to Hund's rule. With compression, the splitting of the d states may become sufficient to establish a spin paired "low spin" ground state with resultant changes in the electronic absorption spectrum (37–39). Again, the changes in molecular geometry are not well established.

Pressure can also induce a change in the spin state of a transition-metal ion in a molecule or crystal with resultant change in the spectrum. The usual change observed is from high to low spin, but the inverse transition has been observed in some cases.

The systematic study of piezochromism is a relatively new field. It is clear that, even within the restricted definition used here, many more systems will be found which exhibit piezochromic behavior. It is quite possible to find a variety of potential applications of this phenomenon. Many of them center around the estimation of the pressure or stress in some kind of restricted or localized geometry, eg, under a localized impact or shock in a crystal or polymer film, in such a film under tension or compression, or at the interface between bearings. More generally it conveys some basic information about inter- and intramolecular interactions that is useful in understanding processes at atmospheric pressure as well as under compression.

Good thin piezochromism film (40) of rare earth compounds, with controlled stoichiometry can be grown by co-evaporation of the elements (eg, SmS) (41,42).

BIBLIOGRAPHY

"Chromogenic Materials, Piezochromic," in *ECT* 4th ed., Vol. 6, pp. 332–337, by H. G. Drickamer, University of Illinois; "Chromogenic Materials, Piezochromic" in *ECT* (online), posting date: December 4, 2000, by H. G. Drickamer, University of Illinois.

CITED PUBLICATIONS

1. H. G. Drickamer, in R. Pucci and J. Picatto, eds., *Molecular Systems under High Pressure*, North Holland Press, New York, 1991, p. 91 and references therein.
2. J. Du, R. Li, C. Chem and Z. Zhou, *Grongneng Cailiao* **30**, 133 (1999).
3. D. L. Fanselow and H. G. Drickamer, *J. Chem. Phys.* **61**, 4567 (1974).
4. E. N. Hochert and H. G. Drickamer, *J. Chem. Phys.* **67**, 6168 (1977).
5. Z. A. Dreger and H. G. Drickamer, *Chem. Phys. Lett.* **179**, 199 (1990).
6. M. Kawano, *Nypon Kessho Galskaishi* **43**, 164 (2001).
7. D. M. White and J. Sinnemberg, *J. Am. Chem. Soc.* **88**, 3825 (1966).
8. H. Tamino, T. Kondo, K. Okada and T. Goto, *Bull. Chem. Soc. Jpn.* **45**, 1474 (1972).
9. M. Leclerc, *Advanced Materials* **11**, 1491 (1999).
10. F. C. Schilling, A. J. Lovinger, D. Davis, F. A. Bovey, and J. M. Zeigler, *J. Inorg. Organomet. Polymers* **2**, 47 (1992).
11. F. C. Schilling, F. A. Bovey, D. Davis, A. J. Lovinger, R. B. Macgregor, Jr., C. A. Walsh, and J. M. Zeigler, *Macromolecules* **22**, 4645 (1989).
12. J. F. Rabolt, K. Song, H. Kuzmany, R. Sooriyakumaran, G. Fickes, and R. D. Miller, *Macromolecules* **23**, 3870 (1990).
13. K. Maeda, *React. Mol. Cryst.* 248, 273 (1993).
14. K. Natsukawa, H. Nakazumi, and M. Matsui, *Senryo to Yakuhim* **40**, 300 (1995).
15. C. Reichardt, *Chem. Soc. Rev.* 147 (1992).
16. K. Tamura and T. Imoto, *Bull. Chem. Soc. Jpn.* **48**, 369 (1975).
17. J. V. Ionanne, D. A. Palmer, and H. Kelm, *Bull. Chem. Soc. Jpn.* **51**, 463 (1978).
18. W. S. Hammack, D. M. Hendrickson, and H. G. Drickamer, *J. Phys. Chem.* **93**, 3483 (1989).
19. Y. Ikushima, N. Saito, and M. Arai, *J. Phys. Chem.* **96**, 2293 (1992).
20. H. G. Drickamer and K. L. Bray, *Intern. Rev. of Phys. Chem.* **8**, 41 (1989) and references therein.
21. J. K. Grey and I. S. Butler, *Coordination. Chem. Rev.*, 219, 713 (2001).
22. J. L. Coffer, J. R. Shapley, and H. G. Drickamer, *Inorg. Chem.* **29**, 3900 (1990).
23. K. L. Bray, H. G. Drickamer, D. M. P. Mingos, M. J. Watson, and J. R. Shapley, *Inorg. Chem.* **30**, 864 (1991).
24. H. G. Drickamer and K. L. Bray, *Acc. Chem. Res.* **23**, 55 (1991) and references therein.
25. B. Scott and R. D. Willett, *J. Am. Chem. Soc.* **113**, 5253 (1991).
26. K. Song, R. D. Miller, G. M. Wallraff, and J. F. Rabolt, *Macromolecules* **24**, 4084 (1991).

27. J. F. Rabolt and co-workers, *Polym. Prep.* **31**, 262 (1991).

28. R. A. Nallicheri and M. E. Rubner, *Macromolecules* **24**, 517 (1991).

29. K. Song, H. Kuzmany, G. M. Wallraff, R. D. Miller, and J. F. Rabolt, *Macromolecules* **23**, 3870 (1990).

30. G. Cryssomallis and H. G. Drickamer, *J. Chem. Phys.* **71**, 4817 (1979).

31. W. H. Bentley and H. G. Drickamer, *J. Chem. Phys.* **42**, 1573 (1965).

32. V. C. Bastron and H. G. Drickamer, *J. Solid State Chem.* **3**, 550 (1971).

33. R. B. Ali, P. Banerjee, J. Burgess, and A. E. Smith, *Trans. Met. Chem.* **13**, 106 (1988).

34. R. B. Ali, J. Burgess, and Guardano, *Trans. Met. Chem.* **13**, 126 (1988).

35. J. Burgess, *Spectrochim Acta* **45A**, 159 (1989).

36. M. Matsui, K. Shibata, and H. Muramatsa, *Bull. Chem. Soc. Jpn.* **63**, 1845 (1990).

37. Y. Kitamara, T. Ito, and M. Kato, *Inorg. Chem.* **23**, 3826 (1984).

38. H. T. Macholt, R. Van Eldik, H. Kelm, and H. Elias, *Inorg. Chim. Acta.* **104**, 115 (1985).

39. K. L. Bray and H. G. Drickamer, *J. Phys. Chem.* **94**, 7037 (1990).

40. C. B. Greenberg, *Thin Films Solid* **251**, 81 (1994).

41. R. Suryanarayanam and G. Brun, *Thin Films Solid* **35**, 263 (1976).

42. R. Suryanarayanam, *Phys. Status Solii B* **85**, 9 (1978).

A. Samat
R. Guglielmetti
Université de la Mediterannée

CHROMOGENIC MATERIALS, THERMOCHROMIC

1. Introduction

Thermochromism is the reversible change in the spectral properties of a substance that accompanies heating and cooling. Strictly speaking, the meaning of the word specifies a visible color change; however, thermochromism has come to also include some cases for which the spectral transition is either better observed outside of the visible region or not observed in the visible at all. Primarily, thermochromism occurs in solid or liquid phase, but it also describes a thermally dependent equilibrium between brown nitrogen dioxide [10102-44-0], NO_2, and colorless dinitrogen tetroxide [10544-72-6], N_2O_4, a rare example in the gas phase (1).

Thermochromism could be defined according to J.H. Day in his review (2) as "an easily noticeable reversible color change in the temperature range limited by the boiling point of each liquid, the boiling point of the solvent in the case of solution or the melting point for solids". Other important contributions were published successively (3–5).

There are many materials, especially organic and metal-organic materials, which exhibit true thermochromism, with a variety of sometimes debatable structural transition mechanisms; it is difficult to summarize the whole with any continuity. For this reason, an effort is made to delineate the scope of the field by listing several thermochromic transitions (Table 1). Selected thermochromic material examples are accompanied in each instance by the corresponding transition stimulus for that case. Characteristically sharp transition temperatures, T_t, are indicated where appropriate. At the other extreme are examples of comparatively gradual transitions, associated for example with an equilibrium or a changing bandwidth. The sharpness of the transition is one aspect by which the several mechanisms could be classified. On the other hand, it is useful also to group materials into metal-complex, inorganic, and organic classes. In this way, the variety of thermochromic changes in each of the three material classes can easily be realized.

Thermochromic compounds such as Ag_2HgI_4 and Cu_2HgI_4 have long been known (17). These compounds color reversibly, exhibiting the discontinuous red shift of a charge transfer band edge during heating (18). As for VO_2, the characteristic hysteresis of reflectance in the visible suggests application for infrared image recording (19).

2. Metal Complexes, Spectral Transitions

Crystal field theory, which is simpler to present than the more comprehensive molecular orbital treatment, has been used to describe $d–d$ orbital excitations of transition-metal ions and the effect on these excitations of ligand coordination geometry and field strength. Absorption bands in the visible region arise in energy states made nondegenerate by the crystal field. Color changes, such as induced by heating or cooling, are therefore a direct indication of change in the surrounding environment of the metal ion. Crystal field theory, even though it does not include charge-transfer processes, has proven to be qualitatively adequate for the $3d$ orbital transition metals because the $3d$ states are not well shielded from ligand field effects. Crystal field splittings are on the order of 10,000 cm^{-1}. The lanthanide $4f$ orbitals are better shielded and have crystal field splittings only on the order of 100 cm^{-1}. It is not surprising, therefore, that many good examples of thermochromism are to be found in $3d$ transition-metal complexes and other $3d$ metal compounds.

A simple and well-known example, from Table 1, is the case of anhydrous cobalt chloride $CoCl_2$, in alcohol solutions (1,6). At room temperature, Co^{2+} is predominately tetrahedrally coordinated, as dichlorobis(ethanol) cobalt(II), $Co(C_2H_5OH)_2Cl_2$, in the case of ethyl alcohol solution, and is colored blue. The tetrahedral absorption band peaks at about 660 nm. With cooling, this band shrinks in intensity. The solution begins to acquire a pink color indicative of the growing dominance of a weak octahedral coordination band, perhaps assignable to chloropentakis(ethanol) cobalt(II), $[Co(C_2H_5OH)_5Cl]^+$ (1). The temperature-dependent equilibrium in an ethyl alcohol–water solution is a classroom demonstration of a "thermometer" for observations below and above room temperature. By analogy, the proportioning between the blue tetrahedral and pink octahedral coordinations of Co^{2+} is also evident in molten inorganic glasses (7). This is clearly not associated with oxidation to Co^{3+} because the latter is

Table 1. Some Typical Thermochromic Compounds and Their Transitions

Thermochromic material	CAS Registry number	Thermochromic transition[a]	T_t^b °C	References
Co^{2+} solutions and glasses		equilibrium shift, two coordinations		1,6,7
$[(C_2H_5)_2NH_2]_2CuCl_4$	[52003-09-5],[52003-08-4]	square planar to tetrahedral	50	8–10
$[(CH_3)_2NH_2]_3CuCl_5$	[52003-06-2]	variation in bandwidth		10
$Cu_4I_4(Py)_4$	[62121-41-9]	fluorescence variations		11
$Al_{2-x}Cr_xO_3$(ruby)	[12174-49-1]	lattice expansion/contraction		1,12,13
VO_2	[12036-21-4]	monoclinic/tetragonal	68	14–16
Cu_2HgI_4	[13876-85-2]	order/disorder	68	17–19
di-β-naphthospiropyran	[178-10-9]	close/open spiro ring		2
poly(xylylviologendibromide)	[38815-69-9]	hydration/dehydration	100	20
ETCD polydiacetylene[c]	[63809-82-5]	side group rearrangement	~115	21,22

[a] When applicable, expressed as a change upon heating; various colors have been reported, often qualitatively.

[b] Transition temperatures for sharp transitions.

[c] Urethane-substituted polymer of where R = $(CH_2)_4OCONHCH_2CH_3$.

616

unstable at the high temperatures of glass melting. Cobalt-containing glasses that are pink at room temperature become blue with heating. This color is likely to be retained with rapid quenching. However, if the glass is cooled slowly, equilibrium is more nearly approached in favor of octahedral coordination, although the change from blue may be slow and difficult to detect by the eye in any case. This sort of kinetic problem is a general one that sometimes makes difficult the distinction between processes that are reversible or irreversible.

Other Co^{2+} halide solutions have been reviewed (1), as have equilibria involving complexes and chelates of Cu^{2+} and Ni^{2+} in solutions. Good thermochromism is well known, for example, in aqueous solutions of copper chloride, $CuCl_2$, and it has also been demonstrated in solutions of $Ni(ClO_4)_2$ (23). Good thermochromism is not restricted to transition-metal ions in solution. The example of $[(C_2H_5)_2NH_2]_2CuCl_4$ in Table 1 is solid-state and it has been reported to undergo a particularly pronounced, discontinuous, first-order, reversible, green-to-yellow color transition with heating (8–10). This is associated with a change in coordination geometry for the $CuCl_4^{2-}$ anion. Similar examples abound and many other compounds have been described in depth in an extensive overview about Cu^{2+} and Ni^{2+} salts that show discontinuous thermochromism (9). Various compounds containing *N,N*-diethylethylenediamine (24–28) or the isopropylammonium ion (10,29–31) are included.

The solid-state, transition-metal example in Table 1 of $[(CH_3)_2NH_2]_3CuCl_5$ illustrates another form of thermochromism: the color shifts gradually and continuously because of changes in bandwidth with either heating or cooling (10). It is not unique, as this behavior has been mentioned for the class of compounds $(RNH_3)_2CuCl_4$, where R = alkyl group (10), and also for compounds of the form $M(N,N\text{-diethylethylenediamine})_2(X)_2$, where $M = Cu^{2+}$ or Ni^{2+} and X is an anion that does not disorder easily (9). With easily disordered anions such as ClO_4^- and BF_4^-, thermochromism occurs discontinuously at a reasonably well-defined T_t.

Halide complexes of Cu^+ with nitrogen base ligands are known to exhibit another form of reversible spectral change known as fluorescence thermochromism. The example of $Cu_4I_4(Py)_4$ from Table 1 is typical and shows red shifting in the visible emission spectrum while the sample is both cooled and irradiated with a 364 nm ultraviolet source (11).

Some organometallic derivatives containing Sb–Sb or Bi–Bi bonds exhibit drastic color changes on melting or dissolution in organic solvents. A representative example is tetramethyl distibane, a compound that is yellow as a liquid and in solution. The crystals of this distribane are bright red at temperatures close to the melting point of 18 °C.

On further cooling a blue shine is observed. At liquid nitrogen temperature the solid becomes yellow again. Similar color changes have been observed for several distribanes dibismuthanes and related compounds. The organic substituents have a strong influence on the development of color. Diarsanes do not show similar phenomena but have been investigated for a better understanding of the thermochromic properties (32).

3. Transitions in Inorganic Compounds

There are not many oxides and sulfides that may be classified as truly thermochromic; again, however, compounds of transition metals dominate. Ruby exhibits a well-known, reversible, ligand-field thermochromism at different temperatures depending on the concentration of Cr^{3+} in the Al_2O_3 lattice (1,12,13). This is a manifestation of change in the ligand field strength as dependent on lattice expansion/contraction. A whole family of oxides is known to undergo reversible, nonmetal-metal, thermoresistive transitions with heating (33–35). The shifts in band structure have frequently been debated. Sometimes these transitions are associated with symmetry changes. VO_2 is one of the best known of these compounds because its transition, which has as much as a fivefold switch in resistivity, occurs close to room temperature with a large, free-carrier, infrared change (14–16). There is also a smaller effect in the visible, as shown for easily grown thin films (16,36). Doping has been found to shift the resistivity transition temperature up and down (37–39), but this does not seem to have any great effect on the spectral range of switching for thin films made in various ways (40). This has discouraged interest in VO_2 films for large-area transparencies. Exploratory work on VO_2 films for optical storage and laser switching applications has also been discussed (41–43).

Another oxide that exhibits a nonmetal–metal transition is V_2O_3. It undergoes a symmetry change at about $-123°C$ (monoclinic/rhombohedral) with as much as a 10^7 fold change in resistivity (44,45). Fe_3O_4 undergoes a small-order symmetry change at about $-154°C$ (46,47). It is orthorhombic below this temperature but a cubic spinel above it. Also, some of the Magnéli phases of vanadium and titanium, of the form $M_nO_{2n}-1$ (M is the metal ion and $n = 3,4,5...$), have been observed to undergo relatively sharp thermal changes, especially as single crystals (33,35). However, except for VO_2, no remarkable spectral switching has been mentioned with any other member in this category of materials.

Four sulfides that also undergo reversible nonmetal–metal transitions are shown in Table 2. In three cases spectral changes are known, although in only

Table 2. Some Sulfide Compounds with Nonmetal–Metal Transitions

Compound	CAS Registry number	Thermochromic transition[a]	Approximate T_t, °C	Spectral shift	References
Ag_2S	[21548-73-2]	monoclinic/cubic	178	ir	48–50
NiS(hex)	[16812-54-7]	antiferromagnetic/ paramagnetic	−9	ir	33,51–53
FeS(tet)	[1317-37-9]	tetragonal/hexagonal	157		33,53,54
$Sm_{1-x}Ln_x^{3+}S^b$		Sm^{2+}/Sm^{3+}	[c]	visible	56,57

[a] Expressed as a change upon heating.
[b] Ln = Ce, Pr, Nd, Gd, Tb, Dy, Ho, Er, Tm, or Y.
[c] Ln_x^{3+}-dependent.

one case involves the visible region. With the $Sm_{1-x}Ln_x{}^{3+}S$ series of compounds a dramatic color change occurs at T_t with cooling. It is associated with expansion in the lattice without change in the cubic structure. Qualitatively, the sulfide is black below the transition and metallic yellow above it. This remarkable example derives its behavior from samarium sulfide, SmS, which undergoes a like black-to-yellow color transition when it is rubbed or taken to a relatively low applied pressure of 650 MPa (6400 atm). A large infrared spectral change accompanies the visible change. The high pressure form of SmS has been made stable at room temperature and atmospheric pressure by doping with Ln^{3+}. Doping promotes a shift from Sm^{2+} to Sm^{3+} by $4f$ electron delocalization, so that the sulfide can be switched back and forth below room temperature.

4. Organic and Polymer Compounds

Simple organic molecules tend to be colorless with electronic transitions in ultraviolet light, whereas visible absorption, or color, is usually associated with electronic excitations in extended and conjugated structures. Color is influenced considerably by the extent of conjugation, as well as by the molecular environment imparted by substituents. So, thermochromism arises from critical, thermally induced changes in the existing structure. Thousands of thermochromic organic examples are known (12). Three of these are in Table 1. The di-β-naphthospiropyran example represents a well-known and extensively reviewed family of compounds that develop color at the onset of a thermally induced ring opening (2). Heterolytic bond cleavage in the molecule results in polar or ionic resonance structures, with conjugation, and it is these structural characteristic that have been associated with the appearance of color. The second organic example in Table 1, poly(xylylviologen dibromide) , is conjugated between pyridinium rings. It is characterized by charge-transfer interactions with counterions. The charge-transfer energy levels in the solid state are sensitive to the molecular environment so that thermochromism occurs when the polymer is subjected to hydration/dehydration sequences (20). Similarly, for urethane-substituted ETCD polydiacetylene (Table 1), thermally induced transitions in the conformation of the unsaturated backbone have been associated with restructuring of side-group substituents (21,22). The occurrence is manifested as a change in color.

Certain poly(di-n-alkylsilanes) and germanes, when in the solid state in particular, also exhibit large spectral changes that have been associated with side-chain influence on the backbone, but these (the polymer being saturated) occur at uv rather than visible wavelengths (58–60). Poly(di-n-hexylsilane) is an example. Initially, crystallization of the n-hexyl substituent groups locks the backbone into a configuration that is characterized by a red-shifted absorption band at 374 nm. With heating through about 41°C, there is a reversible relaxation to the higher energy 317 nm band associated with disordering.

The subject of thermochromism in organic and polymer compounds has been reviewed in some depth (2,12,21), and these expansive overviews should be used by readers with deeper and more particular interest in the subject. Many more examples can be found in the reviews that further illustrate the pattern of association between thermochromism and molecular restructuring of one kind or another. The specific assignment of structures is still open to debate in many cases, and there are still not many actual commercial uses for these or any of the other thermally reversible materials discussed herein. Temperature indicators have been mentioned, though perhaps as much or more for irreversible materials.

4.1. Spiropyranes and Spirooxazines. The thermochromism of spiro-pyranes discovered in 1926, has been extensively studied. Nearly every known compound of this class leads to deep color on melting (generally, red, purple, or blue). However heating solutions of spiropyrans also causes coloration. Day (2) in review reports essentially thermochromic spiropyrans of the indoline and spirobipyran series. Bertelson (61) summarized the main spectroscopic and physicochemical data obtained up to 1971. Thermochromic properties of spiro-oxazines have been reported more recently (62).

The thermochromic mechanism in these classes has been assumed to involve a thermally sensitive equilibrium between the colorless spiroheterocyclic form (SP) and the quasi-planar open merocyanine-like structure (MC) obtained after the breaking of the C=O bond. For both these classes of spiroheterocyclic compounds, it seems certain that the most thermally stable photoinduced-colored form and the species formed thermally are spectroscopically and kinetically indistinguishable (eq. 1).

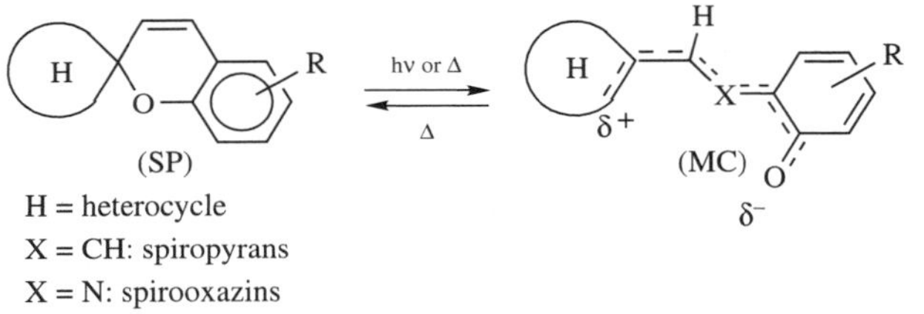

(SP)

H = heterocycle
X = CH: spiropyrans
X = N: spirooxazins

(MC)

$$\tag{1}$$

Depending on the structure, different stereoisomers of the colored form can be involved, but the most stable corresponds to a E configuration (61,63–67). The electronic distribution of the open form is situated between two resonance forms, its proximity to one form or the other depending on the structure and the medium (eq. 2).

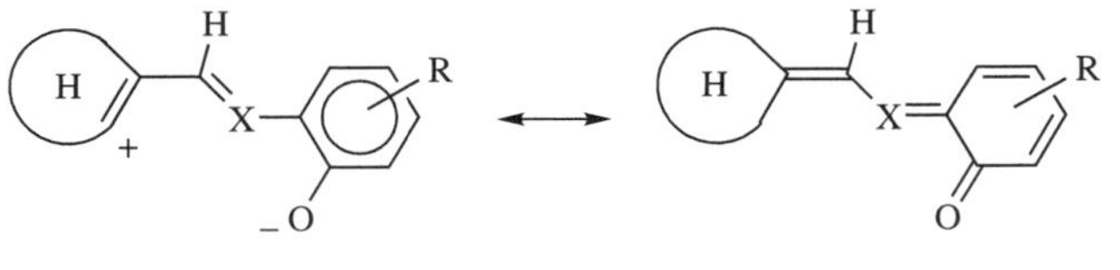

$$\tag{2}$$

For example, spiro[indoline-benzopyrans] bearing a 6-NO_2 group should have a zwitterionic tendancy (63), whereas spiro[indoline-naphthopyrans] or spiro[indoline-naphthoxazines] should have a quinoÿdal tendancy (67–73).

During the past decade, new thermochromic spiropyrans have been described, particularly by Russian teams. For example spiropyrans of the 2-oxaindane or azaindanone series with polycondensed chromene fragments exhibiting photo and thermochromic properties, have been synthesized (74,75).

The thermochomic behavior of spiropyrans of the dithiolane series are discussed in Ref. 76. These data confirm that the annellation of the benzopyran moiety favors the thermochromic properties of this class of compounds.

An important contribution (77) concerns the thermochromism of a series of spiropyrans with different types of azaheterocycles, in various solvents such as benzyl alcohol, dimethylphthalate, decalin, toluene and xylene.

Bis spiropyrans have also been studied and could be extended to conjugated bis merocyanines (78).

The mechanism of the thermal of isomeriation has been investigated through theoretical studies using semi-empirical methods (75) and *ab initio* calculations (79) in agreement with some experimental measurements of the activation energy of the ring opening of pyrans, spiropyrans, spirobipyrans or chromenes (80) using nmr spectroscopy.

The activation parameters of the isomerization reactions were determined mainly for diversely substituted thermochromic spiro[indoline-benzopyrans] (81–84).

The kinetic and thermodynamic behavior of seven spiro[indoline-oxazines] have been studied (70–72). The investigated molecules exhibit thermochromism in a polar solvent (ethanol), but not in a nonpolar solvent (methylcyclohexane). From thermodynamic data and the kinetic parameters of the bleaching reaction, the rate constants and the activation energy for the thermal breaking of the Cspiro-O bond were obtained. The activation energies of the thermal coloration reaction were found to be in the range 80–120 kJ/mol, and the reaction rates varied over a large interval (10^{-5}–10^{-2} s^{-1}) (72).

Uv-visible spectroscopy is the best method for studying the equilibrium constant Ke for the thermochromic equilibrium between the spiroform SP and the merocyanine-like form MC in different solvents ($Ke = [MC]/[SP] = k_f/k_b$). The parameters $\Delta H°$, $\Delta G°$, $\Delta S°$ and Ke were determined (5).

4.2. [2*H*]Chromenes (Benzo- or Naphthopyrans).

2-Phenylamino-2*H*-pyran (85) and benzo- or heteroannellated 2*H*-pyrans (77) have been reported to exhibit thermochromic behavior in solution. Di-β-naphthospiropyran have been investigated (2).

4.3. Other Spiroheterocyclic Compounds.

A series of novel photo- and thermochromic perimidine spirocyclohexadienones whose mechanism involved a ring-chain tautomerism with recently developed (eq. 3) (86).

$$(3)$$

Interesting zwitterionic spirocyclic compounds (87) such as 2,4,6-trinitroaryl derivatives of *o*-hydroxyaldehydes and the corresponding imines show a negative thermochromism. Indeed in this case the colored spirocyclic form is more stable than the open colorless form even in an apolar solvent (eq. 4).

$$(4)$$

4.4. Schiff Bases and Related Nitrogen-containing Compounds.

The Schiff bases of salicylaldehydes with arylamines, aminopyridines and aryl- or thienylalkylamines show thermochromism and photochromism in the solid state due to hydrogen transfer (88). At the beginning of the century, Senier and Shepheard (89) observed that many of the colored crystalline salicylideneanilines were thermochromic. In most cases, the color changed, upon heating, from yellow to orange or red.

With this in mind, extensive investigations were carried out by Cohen and Schmidt (90,91) who found that photochromism and thermochromism were mutually exclusive properties of this series of compounds and suggested that this phenomenon is related to the crystal structure of the compounds and not to the chemical nature of the ring substituents. For instance, salicylidene-2-chloroaniline (**1**) is photochromic whereas salicylidene-4-chloroaniline (**2**) is thermochromic.

(1)   **(2)**

The presence of the ortho-hydroxy group is essential for the observation of both types of chromisms, and the mechanism involves intramolecular hydrogen transfer via a six membered-ring transition state, producing enol-keto tautomeric species, with the spectra of the keto forms showing a bathochromic shift (eq. 5).

$$(5)$$

Thermochromism is restricted to planar molecules and is attributed to a shift of the tautomeric equilibrium toward the "NH" form absorbing at longer wavelengths. For a nonplanar molecule, much energy is required for hydrogen transfer in the ground state and the transfer occurs only in photochemically excited states. A subsequent Z to E isomerization leads to the photo *trans-keto* configuration (90,92).

Structural variations have been determined for many Schifft bases having an o-hydroxy group in order to investigate the problem of planarity, stack packing in the crystal and correlation with the thermochromic behavior (**3**) (**5**).

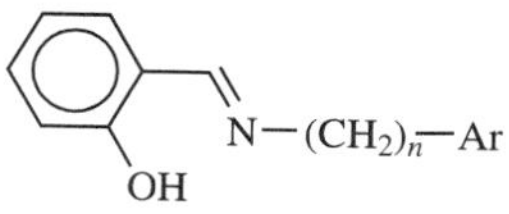

(**3**)

$n = 1, 2$
Ar = phenyl, pyridyl, thienyl

From semiempirical calculations by the CNDO/2 method, the relative energy of photo- and thermochromic salicylidenearylamines can be calculated (93).

Intensive investigations have been conducted to elucidate the nature of the mechanism of thermochromism of salicylidene Schiff bases.

Different techniques or methods suitable for the study of the tautomeric equilibrium between the enol form and the (Z)-keto form have been used, including X-ray diffraction, ^{1}H and ^{13}C nmr, infrared (ir) and Raman spectroscopy and theoretical calculations.

The temperature dependence of the ^{14}N NQR frequencies has been measured (94) for a series of seven thermochromic salicylideneanilines or salicylidene benzylamines and one photochromic compound. The energy difference (ΔE) between the two forms varies from 56 meV to 104 meV depending on the structure and the substitution. ΔE is larger for nonplanar photochromic compounds than for planar thermochromic ones. In the latter case, there is an equilibrium between the enol and (Z)-keto isomers, while for photochromic compound, at 300 K, only the enol form exists.

Thermochromic N-bis(salicylidene) diamines (**4**) and thermochromic N,N′-bis(salicylidene)diamines (**5**) (95,96) have been also synthesized and their physico-chemical behavior studied.

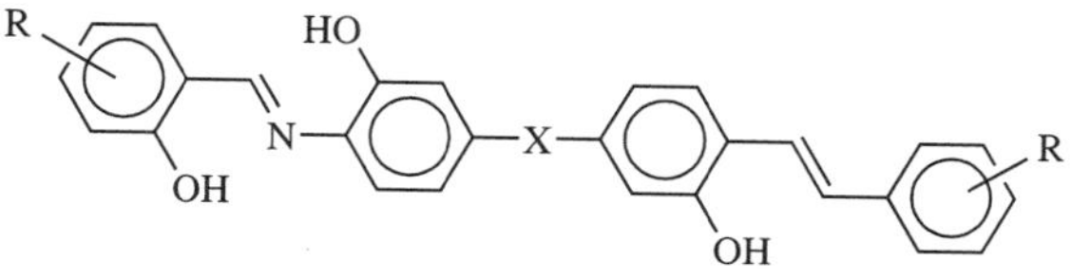

X = CH$_2$, SO$_2$

(**4**)

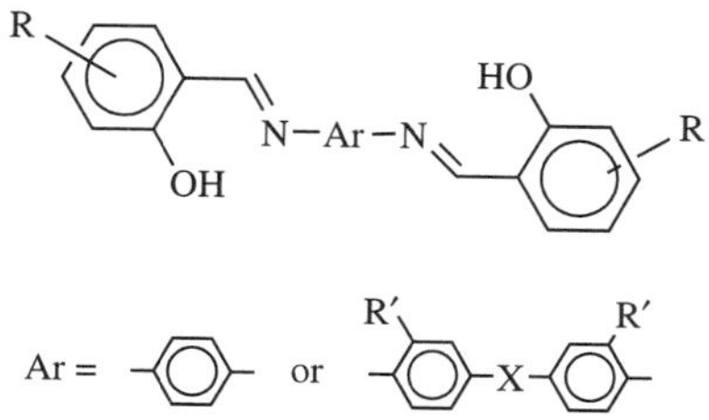

Ar =   or

with R′ = H, CH$_3$ and X = ——, CH$_2$, O

(**5**)

4.5. Bianthrones and Other Overcrowded Ethenes. Bianthrones and related bianthrylidene systems undergo a reversible color change induced by light (photochromism), temperature (thermochromism) or pressure (piezochromism).

Since the original description of thermochromism of bianthrones at the beginning of the century, this phenomenom has evoked considerable interest and has been the subject of numerous reviews, the most recent was published in 1988 (97).

The thermochromism of bianthrone in solution has been shown to result from a thermal equilibrium between two distinct and interconvertible isomeric species A→B. Species A exists at room temperature as a yellow form ($\lambda_{max} \cong 380$ nm); upon heating of solutions of A, a significant fraction is converted to the green B form ($\lambda_{max} \cong 680$ nm), whose enthalpy is 12.5 kJ mol^{-1} greater than that of the A form (eq. 6) (98).

$$\tag{6}$$

(E)   (Z)

In the A form, the severe repulsion between the two anthrone moieties is avoided by the adoption of a folded conformation (Fig. 1). This feature has been confirmed by X-ray crystallographic investigations and dynamic nmr studies (98,99). Among the various proposed structures for the thermoinduced B species (supposedly identical to the photochromic species (97)) the twisted conformation has gathered a variety of experimental and theoretical support (100,101).

The experimental value of the activation energy for the A → B transformation is close to 84 kJ mol^{-1} (97,98).

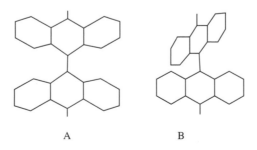

A   B

Fig. 1.   Representation of the folded A form and of the twisted B form of bianthrone (5).

The thermochromism is prevented by bulky substituents at the 1,1'- and 8,8'-positions. In these cases, potential barriers that must be surmounted on the thermal path leading to the B form are too high, and the B form is therefore unobservable.

Besides bianthrone derivatives, other sterically overcrowded alkenes (101) have generated considerable interest owing to their intriguing thermochromic (and photochromic) properties (97,102), such as dixanthenylidenes, bithioxanthylidenes, 9,9'-fluorenylidene anthrones, 9-diphenylmethyleneanthrones, xanthylideneanthrones, and [2-(thioxanthen-9-ylidene) indane-1,3-dione (103).

4.6. Miscellaneous Compounds. Some indano[1,2-b]aziridines are reported to exhibit thermo-(and photo-)chromism, eq. 7 is an example (103,105).

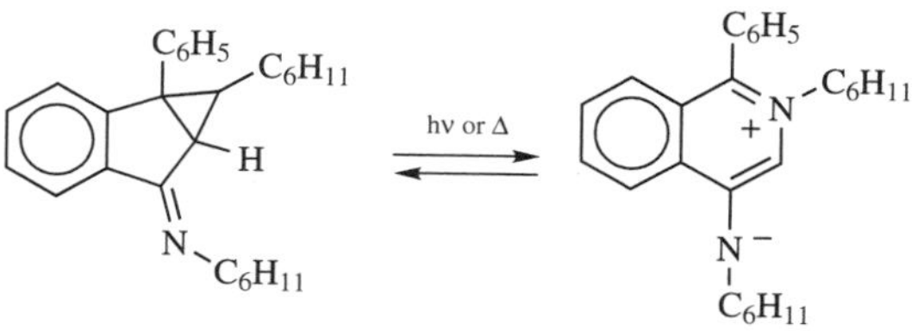

$$(7)$$

A reversible color change of 4,6,7-tri(alkoxy-substituted phenyl)-1,2,5-thiadiazolo[3,4-c]pyridines is observed in the solid state (**6**) (106).

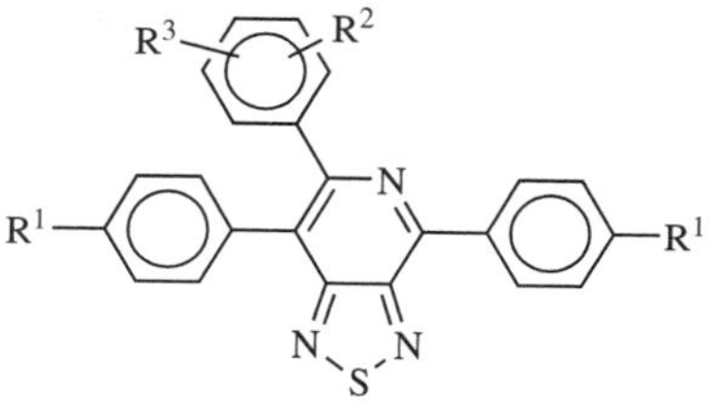

(**6**)

The polysubstituted semibullvalene (**7**) and barbaralane (**8**) skelettons exhibit thermochromism in solution and in solid state (107–109).

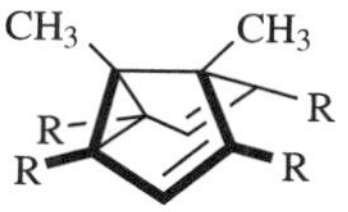

(**7**) 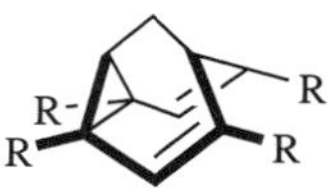   (**8**)

Thermochromism of hindered amino-substituted cyclohexadiones due to C—N bond cleavage has been described by Russian authors (Fig. 2) (110,111).

In the case of Mannich bases exhibiting thermo-(and photo-)chromism the proposed mechanism is quite similar (112,113).

A solution of pyridinium N-phenoxide betaine in ethanol is blue-violet ($\lambda_{max} = 568$ nm) at $+78°C$ and red-colored ($\lambda_{max} = 513$ nm) at $-78°C$. This corresponds to a hypsochromic shift of the charge transfer band of the betaine by -55 nm with decreasing temperature ($\Delta T = 156°C$). This temperature

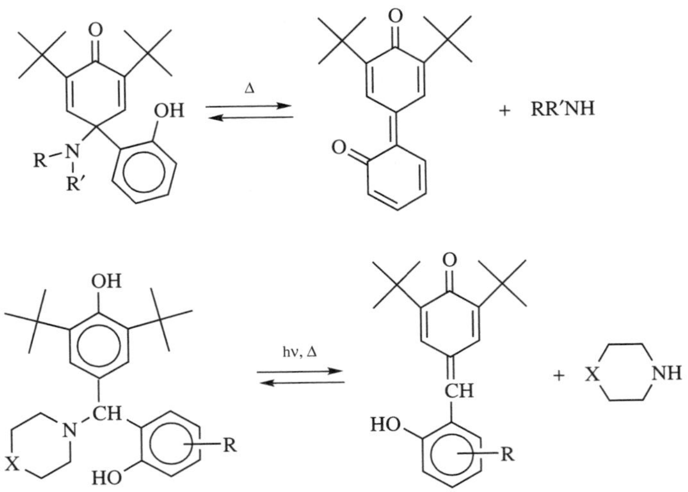

Fig. 2. Thermochromism of hindered amino-substituted cyclohexadiones. $X = O, CH_2$; $R = $ benzo, Cl, Br.

dependent behavior of the betaine represents a new type of thermochromism ("negative thermo-solvatochromism"). It is obviously caused by the increased differential stabilization of the dipolar electronic betaine ground state, relative to its less polar excited state, with decreasing solution temperature (**9**) (114).

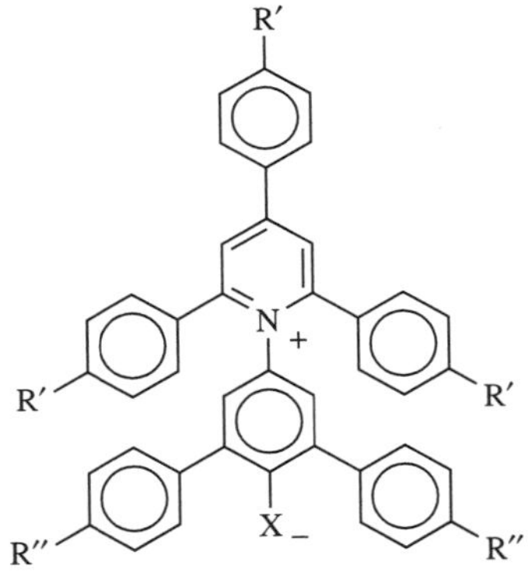

$$R^1 = R^2 = H, X = O$$

(**9**)

5. Uses

The optical properties of thermochromic coatings are strongly temperature dependent. In recent years, it has been shown that some transition metal oxide and related compounds undergo an abrupt reversible transition from a metallic to a semiconductor state with decreasing temperature (115). The material showing maximum promise is vanadium dioxide, VO_2, with a transition temperature at 68°C (116), ie, in a convenient range for technical applications (117).

Films of vanadium oxide were produced by reactive RF magnetron sputtering. For the major technological applications, the undoped VO_2 has a high transition temperature. Addition of a dopant such as tungsten or molybdenum lowers the transition temperature.

It has been reported (118) that in the tungsten-doped VO_2 films, the transition temperature decreases linearly with the concentration of tungsten in the film (-23°C per at % of W). For example, one can reach a transition temperature of 0°C with 3% of W in a $V_{1-x}W_xO_2$ mixed oxide.

Commercialy, only two types of thermochromic systems have been successfully applied to textiles: the liquid crystal type and the molecular rearrangement type. Both depend on microencapsulation involving polymer binders for their success. At present more research needs to be carried out with respect to the formulation, encapsulation, and application of thermochromomic systems (119).

Thermochromic coatings that reduce the transmission of solar energy as the temperature rises can prevent overheating and find application in the thermal control of building and housing sectors, satellites, and spatial equipment (116).

The scientific and technological challenges of large-area electrochromic and thermochromic devices are great, involving improvements such as low cost, high uniformity, high rate and large area thin-film deposition techniques (DC and RF sputtering, chemical vapor deposition [CVD, plasma-enhanced CVD and dip-coating]). Sputtering techniques are particularly interesting. The remaining problems of uniformity of the films and reproducibility are expected to be solved by accurately monitoring the substrate temperature and the partial pressure of oxygen in reactive sputtering (115).

Very few new thermochromic molecular systems have been reported during the past 15 years.

BIBLIOGRAPHY

"Chromogenic Materials, Electrochromic and Thermochromic" in *ECT* 3rd ed., Vol. 6, pp. 129–142, by J. H. Day. Ohio University; "Chromogenic Materials, Thermochromic" in *ECT* 4th ed., Vol. 6, pp. 337–343, by Charles B. Greenberg, PPG Industries, Inc.; "Chromogenic Materials, Thermochromic" in *ECT* (online), posting date: December 4, 2000, by Charles B. Greenberg, PPG Industries, Inc.

CITED PUBLICATIONS

1. K. Sone and Y. Fukuda, *Inorganic Thermochromism*, Vol. 10, Springer-Verlag, New York, 1987, pp. 2, 13.
2. J. H. Day, *Chem. Rev.* **63**, 65 (1963).
3. J. Vitry, *Chim. Ind. Genie Chim.* **102**, 1333 (1969).
4. M. M. Sidky, *Chem. Stosow* **27**, 165 (1983).
5. A. Samat and V. Lokshin in J. C. Crano and R. J Guglielmetti, eds., *Organic Photochromic and Thermochromic Compounds*, Vol. 2, Kluwer Academic/Plenum Publishers, New York, 1999, Chapt. 10.

6. W. C. Nieuwpoort, G. A. Wesselink, and E. H. A. M. Van der Wee, *Rec. Trav. Chim. Pays-Bas* **85**, 397 (1966).
7. W. A. Weyl, *Coloured Glasses*, The Society of Glass Technology, Sheffield, UK, 1951, p. 179.
8. D. R. Bloomquist, M. R. Pressprich, and R. D. Willett, *J. Am. Chem. Soc.* **110**, 7391 (1988).
9. D. R. Bloomquist and R. D. Willett, *Coord. Chem. Rev.* **47**, 125 (1982).
10. R. D. Willett, J. A. Haugen, J. Lebsack, and J. Morrey, *Inorg. Chem.* **13**, 2510 (1974).
11. H. D. Hardt and A. Pierre, *Inorg. Chim. Acta* **25**, L59 (1977).
12. K. Nassau, *The Physics and Chemistry of Color*, John Wiley & Sons Inc., New York, 1983, pp. 77, 109.
13. D. S. McClure, *J. Chem. Phys.* **36**, 2757 (1962).
14. F. J. Morin, *Phys. Rev. Lett.* **3**, 34 (1959).
15. A. S. Barker, Jr., H. W. Verleur, and H. J. Guggenheim, *Phys. Rev. Lett.* **17**, 1286 (1966).
16. H. W. Verleur, A. S. Barker, Jr., and C. N. Berglund, *Phys. Rev.* **172**, 172 (1968).
17. J. H. Day, *Chem. Rev.* **68**, 649 (1968).
18. H.-R. C. Jaw, M. A. Mooney, T. Novinson, W. C. Kaska, and J. I. Zink, *Inorg. Chem.* **26**, 1387 (1987).
19. J. S. Chivian, R. N. Claytor, D. D. Eden, and R. B. Hemphill, *Appl. Opt.* **11**, 2649 (1972).
20. J. S. Moore and S. I. Stupp, *Macromolecules* **19**, 1815 (1986).
21. D. N. Batchelder, *Contemp. Phys.* **29**, 3 (1988).
22. M. F. Rubner, D. J. Sandman, and C. Velazquez, *Macromolecules* **20**, 1296 (1987).
23. T. R. Griffiths and R. K. Scarrow, *Trans. Farad. Soc.* **65**, 1727 (1969).
24. W. E. Hatfield, T. S. Piper, and U. Klabunde, *Inorg. Chem.* **2**, 629 (1963).
25. H. Yokoi, M. Sai, and T. Isobe, *Bull. Chem. Soc. Jpn.* **42**, 2232 (1969).
26. A. B. P. Lever, E. Mantovani, and J. C. Donini, *Inorg. Chem.* **10**, 2424 (1971).
27. L. Fabbrizzi, M. Micheloni, and P. Paoletti, *Inorg. Chem.* **13**, 3019 (1974).
28. J. R. Ferraro, L. J. Basile, L. R. Garcia-Ineguez, P. Paoletti, and L. Fabbrizzi, *Inorg. Chem.* **15**, 2342 (1976).
29. S. A. Roberts, D. R. Bloomquist, R. D. Willett, and H. W. Dodgen, *J. Am. Chem. Soc.* **103**, 2603 (1981).
30. D. R. Bloomquist, R. D. Willett, and H. W. Dodgen, *J. Am. Chem. Soc.* **103**, 2610 (1981).
31. D. R. Bloomquist and R. D. Willett, *J. Am. Chem. Soc.* **103**, 2615 (1981).
32. H. J. Breuning in S. Patai, ed., *The Chemistry of Organic Arsenic, Antimony and Bismuth Compounds*, John Wiley & Sons, Inc., New York, 1994, pp. 441–456.
33. D. Adler, *Rev. Mod. Phys.* **40**, 714 (1968).
34. D. Adler in J. I. Budnick and M. P. Kawatra, eds., *Conference on Dynamical Aspects of Critical Phenomena*, Gordon and Breach, London, 1972, p. 392.
35. J. M. Honig and L. L. Van Zandt, in R. A. Huggins, ed., *Annual Review of Material Science*, Vol. 5, Annual Reviews Inc., Palo Alto, 1975, p. 225.
36. C. B. Greenberg, in S. A. Marolo, ed., *Proceedings of the 15th Conference on Aerospace Transparent Materials and Enclosures II*, WRDC-TR-89-4044, Wright-Patterson AFB, Dayton, Ohio, 1989, p. 1124.
37. M. Nygren and M. Israelsson, *Mater. Res. Bull.* **4**, 881 (1969).
38. T. Horlin, T. Niklewski, and M. Nygren, *Mater. Res. Bull.* **7**, 1515 (1972).
39. J. M. Reyes, G. F. Lynch, M. Sayer, S. L. McBride, and T. S. Hutchinson, *J. Can. Ceram. Soc.* **41**, 69 (1972).
40. C. B. Greenberg, *Thin Solid Films* **110**, 73 (1983).
41. W. R. Roach, *Appl. Phys. Lett.* **19**, 453 (1971).

42. A. W. Smith, *Appl. Phys. Lett.* **23**, 437 (1973).
43. I. Balberg and S. Trokman, *J. Appl. Phys.* **46**, 2111 (1975).
44. M. Foëx, *Compt. Rend.* **223**, 1126 (1946).
45. J. Feinleib and W. Paul, *Phys. Rev.* **155**, 841 (1967).
46. E. J. W. Verwey, *Nature* **144**, 327 (1939).
47. P. A. Miles, W. B. Westphal, and A. von Hippel, *Rev. Mod. Phys.* **29**, 279 (1957).
48. M. H. Hebb, *J. Chem. Phys.* **20**, 185 (1952).
49. P. Brüesch and J. Wullschleger, *Solid State Commun.* **13**, 9 (1973).
50. T.-Y. Hsu, H. Buhay, and N. P. Murarka, *Proceedings of the SPIE*, **259**, 38 (1980).
51. J. T. Sparks and T. Komoto, *J. Appl. Phys.* **34**, 1191 (1963); *Phys. Letters* **25A**, 398 (1967); *Rev. Mod. Phys.* **40**, 752 (1968).
52. A. S. Barker, Jr. and J. P. Remeika, *Phys. Rev. B* **10**, 987 (1974).
53. T. Ohtani, *J. Phys. Soc. Jpn.* **37**, 701 (1974).
54. M. Murakami, *J. Phys. Soc. Jpn.* **16**, 187 (1961).
55. E. F. Bertaut, P. Burlet, and J. Chappert, *Solid State Commun.* **3**, 335 (1965).
56. A. Jayaraman, E. Bucher, P. D. Dernier, and L. D. Longinotti, *Phys. Rev. Lett.* **31**, 700 (1973).
57. A Jayaraman, P. D. Dernier, and L. D. Longinotti, *Phys. Rev. B* **11**, 2783 (1975); *High Temp.-High Press.* **7**, 1 (1975).
58. R. D. Miller, D. Hofer, J. Rabolt, and G. N. Fickes, *J. Am. Chem. Soc.* **107**, 2172 (1985).
59. J. F. Rabolt, D. Hofer, R. D. Miller, and G. N. Fickes, *Macromolecules* **19**, 611 (1986).
60. R. D. Miller and R. Sooriyakumaran, *J. Polym. Sci., Part A: Polym. Chem.* **25**, 111 (1987).
61. R. C. Bertelson in G. H. Brown, ed., *Photochromism*, John Wiley & Sons, Inc., New York, 1971, pp. 45–431.
62. N. Y. C. Chu, *Can. J. Chem.* **61**, 300 (1983).
63. R. Guglielmetti, in H. Dürr and H. Bouas-Laurent eds., *Photochromism : Molecules and Systems*, pp. 314–466, Elsevier, Amsterdam, The Netherlands, 1990.
64. S. Nakamura, K. Uchida, A. Murakami, and M. Irie, *J. Org. Chem.* **58**, 5543 (1993).
65. V. Malatesta, G. Ranghino, U. Romano, and P. Allegrini, *Int. J. Quantum Chem.* **42**, 879 (1992).
66. H. Pommier, A. Samat, R. Guglielmetti, M. Rajzmann, and G. Pèpe, *Mol. Cryst. Liq. Cryst.* 246, 241 (1994).
67. P. Lareginie, V. Lokshin, A. Samat, R. Guglielmetti, and G. Pepe, *J. Chem. Soc. Perkin Trans.* 2 **107** (1996).
68. J. L. Pozzo, A. Samat, R. Guglielmetti, and D. Dekeukeleire, *J. Chem. Soc. Perkins Trans* 2 **1327** (1993).
69. G. Favaro, F. Masetti, U. Mazzucato, G. Ottavi, P. Allegrini, and V. Malatesta, *J. Chem. Soc. Faraday Trans.* **90**, 333 (1994).
70. G. Favaro, F. Ortica, and V. Malatesta, *J. Chem. Soc. Faraday Trans.* **91**, 4099 (1995).
71. G. Favaro, V. Malatesta, U. Mazzucato, G. Ottavi, and A. Romani, *J. Photochem. Photobiol. A* **87**, 235 (1995).
72. G. Favaro, V. Malatesta, U. Mazzucato, C. Miliani, and G. Ottavi, *Proc. Indian Acad. Sci. Chem. Sci.* **6**, 659 (1995).
73. Y. S. Lee, J. G Kim, Y. D. Huh, and M. K. Kim, *J. Korean. Chem. Soc.* **38**, 864 (1994).
74. S. M. Aldoshin, V. A. Lokshin, A. N. Rezonov, N. V. Volbushko, N. E. Shelepin, M. I. Knyazhanskii, L. O. Atovmyan, and V. I. Minkin, *Khim. Geterotsikl. Soedin* **144** (1987).
75. S. P. Makarov, B. Ya. Simkin, and V. I. Minkin, *Chem. Heterocycl. Compd.* 140 (1988).

76. T. B. Krasieva, Ya. N. Malkin, V. A. Lokshin, S. V. Mel'nichuk, S. A. Tikhomirov, and V. A. Kuz'min, *Izv. Akad. Nauk. SSSR. Ser. Khim.* 2504 (1989).
77. B. Hellrung and H. Balli, *Helv. Chim. Acta.* **72**, 1583 (1989).
78. Ya. N. Malkin, V. A. Kuz'min, G. G. Dyadyusha, A. M. Boguslavskaya, and F. A. Mikhailenko, *Bull. Acad. Sci. USSR* **25**, 526 (1976).
79. P. N. Day, Zh. Wang and R. Pachter, *J. Phys. Chem.* **99**, 9730 (1995).
80. A. Mannschreck, K. Lorenz, and M. Schinabeck in J. C. Crano and R. J. Guglielmetti, eds., *Organic Photochromic and Thermochromic Compounds*, Vol. 2, Kluwer Academic/Plenum Publishers, New York, 1999, Chapt. 6.
81. Y. Sueishi, M. Ohcho, and N. Nishimura, *Bull. Chem. Soc. Jpn.* **58**, 2608 (1985).
82. Y. Sueishi, M. Ohcho, S. Yamamoto, and N. Nishimura, *Bull. Chem. Soc. Jpn.* **59**, 3666 (1986).
83. N. Nishimura, J. Miyake, and Y. Sueishi, *Bull. Chem. Soc. Jpn.* **62**, 1777 (1989).
84. Y. Sueishi and N. Nishimura, *J. Phys. Org. Chem.* **8**, 335 (1995).
85. B. Ya. Simkin, S. P. Makarov, M. G. Furmanova, K. Sh. Karaev, and V. I. Minkin, *Khim. Geterotsikl. Soedin.*, 747 (1984).
86. V. I. Minkin, V. N. Komissarov, and V. A. Kharlanov, in Ref. 81, Vol. 1, Chapt. 8.
87. L. L. Oleknovitch, I. E. Mikhailov, V. I. Minkin, N. G. Furmanova, O. E. Kompan, Yu. T. Strutchkov, and A. V. Lukash, *Zh. Org. Khimi.* **18**, 484 (1982).
88. E. Hadjoudis in H. Durr and H. Bouas-Laurent, eds. *Photochromism : Molecules and Systems*, Elsevier, Amsterdam, The Netherlands, 1990, pp. 685–712.
89. A. Senier and F. G. Shepheard, *J. Chem. Soc.* **95**, 1943 (1909).
90. M. D. Cohen and G. M. Schmidt, *J. Phys. Chem.* **56**, 2442 (1962).
91. M. D. Cohen, G. M. Schmidt, and S. Flavian, *J. Chem. Soc.* 2041 (1964).
92. E. Hadjoudis, *J. Photochem.* **17**, 355 (1981).
93. S. M. Aldoshin and L. O. Atovmyan, *Khim. Fiz.* **3**, 915 (1984).
94. E. Hadjoudis, F. Milia, J. Seliger, V. Zagar and R. Blinc, *Chem. Phys.* **156**, 149 (1991).
95. N. Hoshino, T. Inabe, M. Hoshino, T. Mitani, and Y. Maruyama, *Bull. Chem. Soc. Jpn.* **61**, 4207 (1988).
96. T. Inabe, M. Hoshino, T. Mitani, and Y. Maruyama, *Bull. Chem. Soc. Jpn.* **62**, 2245 (1989).
97. K. A. Muszkat in S. Patai and Z. Rappoport, eds., *Chemistry of Quinoïd Compounds*, Vol. 2, John Wiley & Sons, Inc., Chichester, 1988, pp. 203–224.
98. Y. Tapuhi, O. Kalisky, and I. Agranat, *J. Org. Chem.* **44**, 1949 (1979).
99. I. Agranat and Y. Tapuhi, *J. Org. Chem.* **44**, 1941 (1979).
100. R. Korenstein, K. A. Muszkat, and S. Sharafy-Ozeri, *J. Am. Chem. Soc.* **95**, 6177 (1973).
101. G. Shoham, S. Cohen, M. R. Suissa, and I. Agranat in J. J. Stezowski, J. L. Huang, and M. C. Shao, eds., *Molecular Structure, Chemical Reactivity and Biological Activity*, Oxford Univ. Press., Oxford, 1988, pp. 290–312.
102. H. Dürr, *Angew. Chem. Int. Ed. Engl.* **28**, 413 (1989).
103. J. J. Stezowski, P. U. Biedermann, T. Hildenbrand, J. A. Dursch, C. J. Eckhardt, and I. Agranat, *J. Chem. Soc. Chem. Commun.* 213 (1993).
104. J. W. Lown and K. Matsukato, *J. Chem. Soc.* **692** (1970).
105. A. Padwa and E. Vega, *J. Org. Chem.* **40**, 175 (1975).
106. S. Mataka, K. Takahashi, M. Tashiro, W. H. Lin, I. Iwasaki, T. Tsutsui, S. Saito, S. Akiyama, and T. Yonemitsu, *J. Heterocycl. Chem.* **26**, 215 (1989).
107. H. Quast, K. Knoll, E. M. Peters, K. Peters, and H. G. Von Schnering, *Chem. Ber.* **126**, 1047 (1993).
108. H. Quast, E. Geissler, T. Herkert, K. Knoll, E. M. Peters, K. Peters, and H. G. Von Schnering, *Chem. Ber.* **126**, 1465 (1993).

109. H. Quast, T. Herkert, A. Witzel, E. M. Peters, K. Peters, and H. G. Von Schnering, *Chem. Ber.* **127**, 921 (1994).

110. V. N. Komissarov, L. Yu Ukhin, V. A. Kharlanov, L. V. Vetoshkina, L. E. Konstanti-novskii, S. M. Aldoshin, O. S. Filipenko, M. A. Movozhilova, and L. O. Atovmyan, *Izv. Akad. Nauk, SSSR. Ser. Khim.*, 1121 (1991).

111. S. M. Aldoshin, O. S. Filipenko, M. A. Movozhilova, L. O. Atovmyan, V. N. Komis-sarov, and L. Yu Ukhin, *Izv. Akad. Nauk, SSSR, Ser. Khim.* 1808 (1991).

112. V. N. Komissarov, L. Yu Ukhin, V. A. Kharlanov, V. A. Lokshin, E. Yu. Bulgarevich, V. I. Minkin, O. S. Filipenko, M. A. Novozhilova, S. M. Aldoshin, and L. O. Atovm-yan, *Izv. Akad. Nauk, SSSR, Ser. Khim.* 2389 (1992).

113. V. N. Komissarov, V. A. Kharlanov, L. Yu Ukhin, E. Yu. Bulgarevich, and V. I. Minkin, *Zh. Org. Khim.* **28**, 513 (1992).

114. C. Reichardt, *Chem. Soc. Rev.* 147 (1992).

115. C. Sella, M. Maaza, O. Nemraoui, J. Lafait, N. Renard, and Y. Sampeur, *Surface and Coating Technology* **98**, 1477 (1998).

116. C. G. Granqvist, *Critical Rev. Solid State and Mat. Sci.* **16**, 291 (1990).

117. C. B. Greenberg, *Thin Solid Films* **251**, 81 (1994).

118. P. Jin, K. Yoshimura and S. Tanemura in *Conference Proceedings of Window Inno-vations'95*, Toronto, June 1995, p. 481 5–6.

119. D. Aitken, S. M. Burkinshaw, J. Griffiths, and A. D. Towns, *Rev. Prog. Coloration* **26**, 1 (1996).

A. Samat
R. Guglielmetti
Université de la Mediterannée

CITRIC ACID

1. Introduction

Citric acid [77-92-9] (2-hydroxy-1,2,3-propanetricarboxylic acid), is a natural component and common metabolite of plants and animals. It is the most versatile and widely used organic acid in foods, beverages, and pharmaceuticals.

$$
\begin{array}{l}
CH_2-COOH \\
| \\
HO-C-COOH \\
| \\
CH_2-COOH
\end{array}
$$

Because of its functionality and environmental acceptability, citric acid and its salts (primarily sodium and potassium) are used in many industrial applica-tions for chelation, buffering, pH adjustment, and derivatization. These uses include laundry detergents, shampoos, cosmetics, enhanced oil recovery, and chemical cleaning.

Citric acid specifications are defined in a number of compendia including *Food Chemicals Codex* (FCC), *United States Pharmacopoeia* (USP), *British Phar-macopoeia* (BP), *European Pharmacopoeia* (EP), and *Japanese Pharmacopoeia* (JP).

Historically, about AD 1200, the alchemist Vincentius Bellovacensis recognized that lemon and lime juices contained an acid substance. In 1784 Scheele first isolated crystalline citric acid from lemon juice. In 1834 Liebig recognized citric acid as a hydroxy tribasic acid, and in 1893 Wehmer indicated that certain fungi produce citric acid when grown on sugar solutions. The microbial fermentation of a carbohydrate substrate is virtually the exclusive commercial procedure to produce citric acid.

2. Occurrence

Citric acid occurs widely in the plant and animal kingdoms (1). It is found most abundantly in the fruits of the citrus species, but is also present as the free acid or as a salt in the fruit, seeds, or juices of a wide variety of flowers and plants. The citrate ion occurs in all animal tissues and fluids (1). The total circulating citric acid in the serum of humans is approximately 1 mg/kg body weight. Normal daily excretion in human urine is 0.2–1.0 g. This natural occurrence of citric acid is described in Table 1.

2.1. Physiological Role of Citric Acid. Citric acid occurs in the terminal oxidative metabolic system of virtually all organisms. This oxidative metabolic system (Fig. 1), variously called the Krebs cycle (for its discoverer, H. A. Krebs), the tricarboxylic acid cycle, or the citric acid cycle, is a metabolic cycle involving the conversion of acetate derived from carbohydrates, fats, or proteins to carbon dioxide and water. This cycle releases energy necessary for an organism's growth, movement, luminescence, chemosynthesis, and reproduction.

Table 1. Natural Occurrence of Citric Acid

Fruits and vegetables		Animal tissues and fluids	
Plant	Citric acid, wt%	Location	Citric acid, ppm
lemons	4.0–8.0	human whole blood	15
grapefruit	1.2–2.1	human blood plasma	25
tangerines	0.9–1.2	red blood cells	10
oranges	0.6–1.0	human milk	500–1250
currants		urine	100–750
black	1.5–3.0	semen	2000–4000
red	0.7–1.3	thyroid gland	750–900
raspberries	1.0–1.3	kidney	20
strawberries	0.6–0.8	bone	7500
apples	0.008	saliva	4–24
potatoes	0.3–0.5	sweat	1–2
tomatoes	0.25	tears	5–7
asparagus	0.08–0.2		
turnips	0.05–1.1		
peas	0.05		
corn kernels	0.02		
lettuce	0.016		
eggplant	0.01		

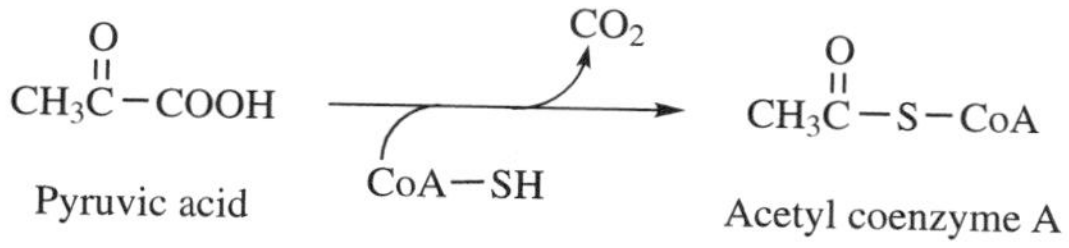

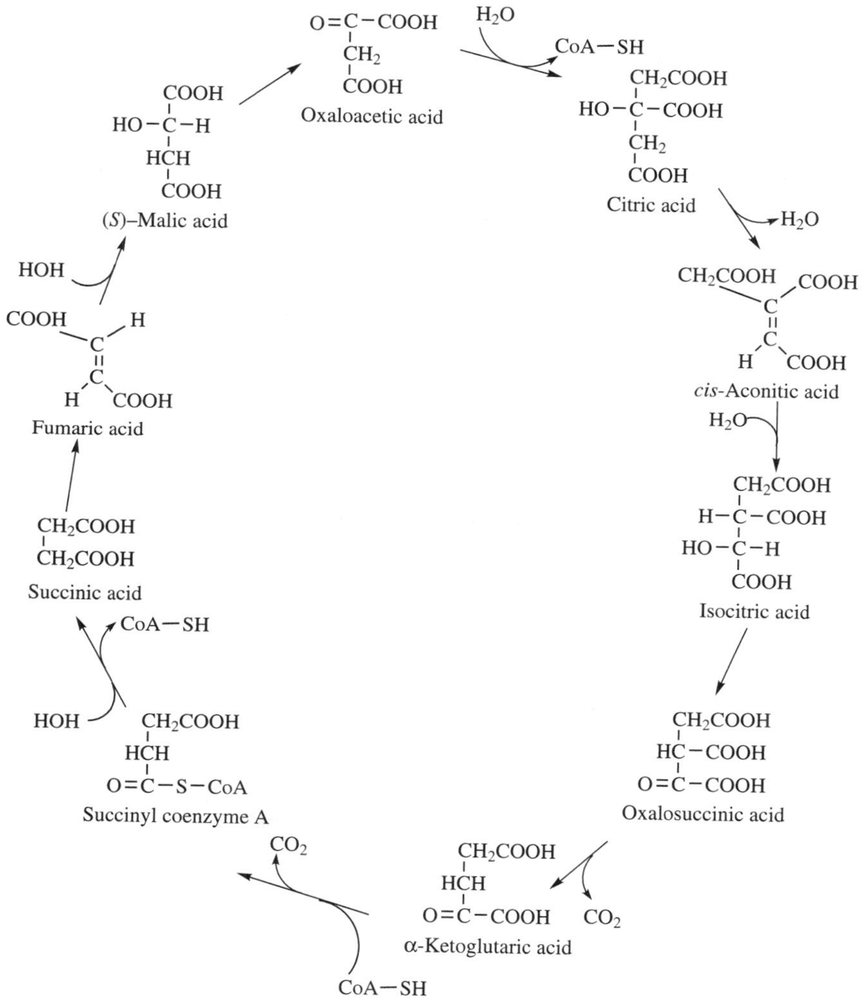

Fig. 1. Krebs (citric acid) cycle. Coenzyme A is represented CoA–SH. The cycle begins with the combination of acetyl coenzyme A and oxaloacetic acid to form citric acid.

The cycle also provides the carbon-containing materials from which cells synthesize amino acids and fats. Many yeasts, molds, and bacteria conduct the citric acid cycle, and can be selected for their ability to maximize citric acid production in the process. This is the basis for the efficient commercial fermentation processes used today to produce citric acid.

3. Physical Properties

Citric acid, anhydrous, crystallizes from hot aqueous solutions as colorless translucent crystals or white crystalline powder. Its crystal form is monoclinic holohedra. Citric acid is deliquescent in moist air and is optically inactive. Some physical properties are given in Table 2 (2,3). The solubility of citric acid in water and some organic solvents is given in Table 3. The pH and specific gravity of aqueous solutions of citric acid are shown in Table 4.

Aqueous solutions of citric acid make excellent buffer systems when partially neutralized because citric acid is a weak acid and has three carboxyl groups, hence three pK_as. At $20°C pK_1 = 3.14$, $pK_2 = 4.77$, and $pK_3 = 6.39$ (1). The buffer range for citrate solutions is pH 2.5 to 6.5. Buffer systems can be made using a solution of citric acid and sodium citrate or by neutralizing a solution of citric acid with a base such as sodium hydroxide. In Table 5 stock

Table 2. **Physical Properties of Citric Acid, Anhydrous**

Property	Value
molecular formula	$C_6H_8O_7$
mol wt	192.13
gram equivalent weight	64.04
melting point, °C	153
thermal decomposition temp., °C	175
density, g/mL	1.665
heat of combustion,[a] MJ/mol[b]	1.96
heat of solution, J/g[b]	117

[a] At 25°C.
[b] To convert J to cal, divide by 4.184.

Table 3. **Solubility[a] of Citric Acid, Anhydrous**

Temperature, °C	g/100 g satd soln
In water	
10	54.0
20	59.2
30	64.3
40	68.6
50	70.9
60	73.5
70	76.2
80	78.8
90	81.4
100	84.0
In organic solvents at 25°C	
amyl acetate	4.2
diethyl ether[b]	1.0
ethyl alcohol[b]	38.3

[a] Ref. 4.
[b] Absolute.

Table 4. **pH and Specific Gravity of Aqueous Citric Acid Solutions**

Concentration, % w/w	pH	Specific gravity at 25°C
0.1	2.8	
0.5	2.4	
1.0	2.2	
5.0	1.9	
10.0	1.7	1.035
20.0		1.084
30.0	1.2	1.131
40.0		1.182
50.0	0.8	1.243
60.0		1.294

Table 5. **Citric Acid Buffer Solutions**

0.1 M Citric acid, mL	0.1 M Sodium citrate, mL	Buffer solution pH
46.5	3.5	3.0
33.0	17.0	4.0
20.5	29.5	5.0
9.5	41.5	6.0

solutions of 0.1 M (0.33 N) citric acid are combined with 0.1 M (0.33 N) sodium citrate to make a typical buffer solution.

Citric acid monohydrate [5949-29-1] has a molecular weight of 210.14 and crystallizes from cold aqueous solutions. When gently heated, the crystals lose their water of hydration at 70–75°C and melt in the range of 135–152°C. Rapid heating causes dehydration at 100°C to form crystals that melt sharply at 153°C. Citric acid monohydrate is available in limited commercial quantities since most applications now call for the anhydrous form.

4. Chemical Properties

Citric acid undergoes most of the reactions typical of organic hydroxy polycarboxylates. Reactions such as esterification, salt formation and anhydride reactions can be easily perfomed. However, the tertiary hydroxyl group does not undergo all the common reactions. If a reaction requires more strenous conditions, dehydration to aconitic acid is promoted.

4.1. Decomposition. When heated above 175°C, citric acid decomposes to form aconitic acid [499-12-7], citraconic acid [498-23-7], itaconic acid [97-65-4], acetonedicarboxylic acid [542-05-2], carbon dioxide, and water, as shown in Figure 2.

4.2. Esterification. Citric acid is easily esterified with many alcohols under azeotropic conditions in the presence of a catalyst such as sulfuric acid,

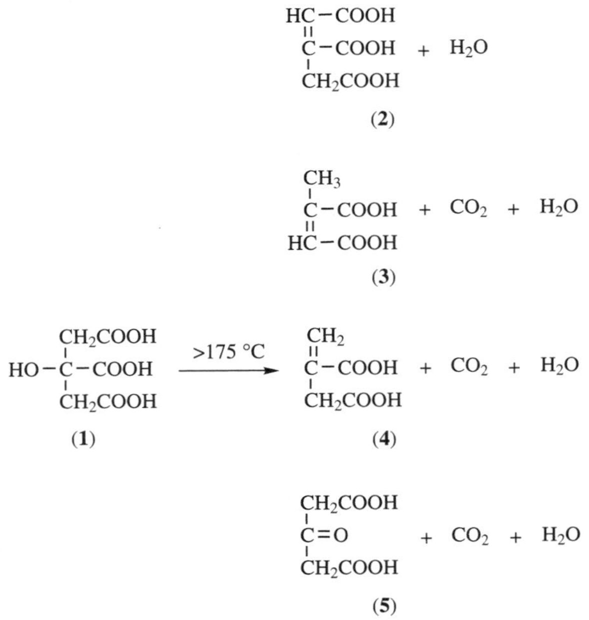

$$
\begin{array}{l}
HC-COOH \\
\;\;\|\\
\;\;C-COOH \;\;\; + \;\; H_2O \\
\;\;\;\;\;| \\
CH_2COOH
\end{array}
$$

(**2**)

$$
\begin{array}{l}
CH_3 \\
\;\;| \\
\;\;C-COOH \;\;\; + \;\; CO_2 \;\; + \;\; H_2O \\
\;\;\|\\
HC-COOH
\end{array}
$$

(**3**)

$$
\begin{array}{l}
CH_2COOH \\
\;\;\;| \\
HO-C-COOH \\
\;\;\;| \\
CH_2COOH
\end{array}
\xrightarrow{>175\,°C}
\begin{array}{l}
CH_2 \\
\;\|\\
\;\;C-COOH \;\;\; + \;\; CO_2 \;\; + \;\; H_2O \\
\;\;| \\
CH_2COOH
\end{array}
$$

(**1**)   (**4**)

$$
\begin{array}{l}
CH_2COOH \\
\;\;| \\
\;\;C=O \;\;\;\;\; + \;\; CO_2 \;\; + \;\; H_2O \\
\;\;| \\
CH_2COOH
\end{array}
$$

(**5**)

Fig. 2. Thermal decomposition of citric acid (**1**) to aconitric acid (**2**), citraconic acid (**3**), itaconic acid (**4**), and oxidation to acetonedicarboxylic acid (**5**).

p-toluenesulfonic acid, or sulfonic acid-type ion-exchange resin. Alcohols boiling above 150°C esterify citric acid without a catalyst (5–8).

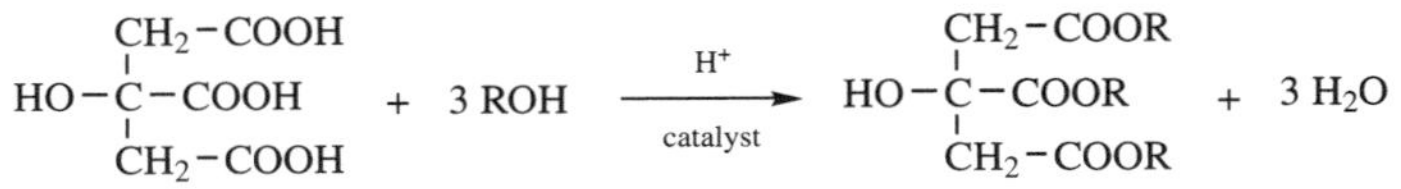

$$
\begin{array}{l}
CH_2-COOH \\
\;\;\;| \\
HO-C-COOH \\
\;\;\;| \\
CH_2-COOH
\end{array}
\;\; + \;\; 3\,ROH \;\;
\xrightarrow[\text{catalyst}]{H^+}
\begin{array}{l}
CH_2-COOR \\
\;\;\;| \\
HO-C-COOR \\
\;\;\;| \\
CH_2-COOR
\end{array}
\;\; + \;\; 3\,H_2O
$$

Alcohols typically used in citric acid esterification are methyl, ethyl, butyl, and allyl alcohols.

4.3. Oxidation.   Citric acid is easily oxidized by a variety of oxidizing agents such as peroxides, hypochlorite, persulfate, permanganate, periodate, hypobromite, chromate, manganese dioxide, and nitric acid. The products of oxidation are usually acetonedicarboxylic acid (**5**), oxalic acid (**6**), carbon dioxide, and water, depending on the conditions used (5).

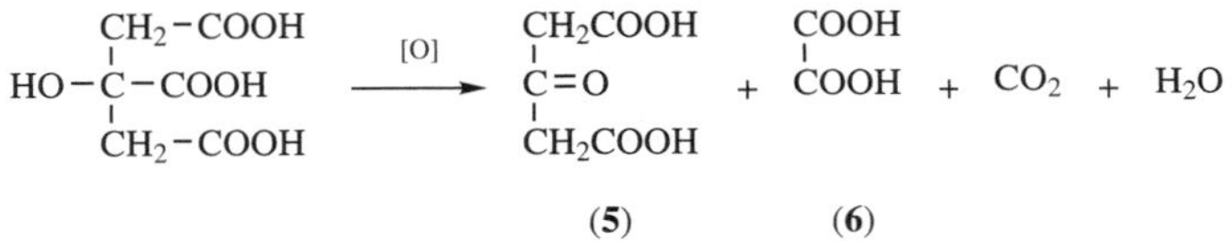

$$
\begin{array}{l}
CH_2-COOH \\
\;\;\;| \\
HO-C-COOH \\
\;\;\;| \\
CH_2-COOH
\end{array}
\xrightarrow{[O]}
\begin{array}{l}
CH_2COOH \\
\;\;| \\
\;\;C=O \\
\;\;| \\
CH_2COOH
\end{array}
\;\; + \;\;
\begin{array}{l}
COOH \\
\;| \\
COOH
\end{array}
\;\; + \;\; CO_2 \;\; + \;\; H_2O
$$

(**5**)   (**6**)

4.4. Reduction.   The hydrogenation of citric acid yields 1,2,3-propanetricarboxylic acid [99-14-9]. (5)

$$
\begin{array}{c}
\text{CH}_2-\text{COOH} \\
| \\
\text{HO}-\text{C}-\text{COOH} \\
| \\
\text{CH}_2-\text{COOH}
\end{array}
\quad \xrightarrow{\text{[H]}} \quad
\begin{array}{c}
\text{CH}_2-\text{COOH} \\
| \\
\text{CH}-\text{COOH} \\
| \\
\text{CH}_2-\text{COOH}
\end{array}
\quad + \text{ H}_2\text{O}
$$

Catalytic hydrogenation is more difficult. Hydrogenolysis yields carbon dioxide, water, methane, formic acid, acetic acid and a small amount of methyl succinic acid (5).

Hydrogenation of trisodium citrate over a Ni catalyst at 8.6 MPa (85 atm) and a temperature of 220–230°C results in hydrogenolysis fragments.

4.5. Salt Formation. Citric acid forms mono-, di-, and tribasic salts with many cations such as alkalies, ammonia, and amines. Salts may be prepared by direct neutralization of a solution of citric acid in water using the appropriate base, or by double decomposition using a citrate salt and a soluble metal salt.

Trisodium citrate is more widely used than any of the other salts of citric acid. It is generally made by neutralization of a water solution of citric acid using sodium hydroxide. The neutralization reaction is highly exothermic giving off 1109 J/g of citric acid. To conserve energy, the heat evolved can be used in the sodium citrate concentration and crystallization steps.

$$
\begin{array}{c}
\text{CH}_2-\text{COOH} \\
| \\
\text{HO}-\text{C}-\text{COOH} \\
| \\
\text{CH}_2-\text{COOH}
\end{array}
\quad + \text{ 3 NaOH} \quad \longrightarrow \quad
\begin{array}{c}
\text{CH}_2-\text{COONa} \\
| \\
\text{HO}-\text{C}-\text{COONa} \cdot 2\,\text{H}_2\text{O} \\
| \\
\text{CH}_2-\text{COONa}
\end{array}
\quad + \text{ H}_2\text{O}
$$

Other sources of sodium ion that are used to make sodium citrate are sodium carbonate and sodium bicarbonate. These reactions evolve large volumes of carbon dioxide gas, resulting in much foaming but less exotherm.

The mono- and disodium citrate salts are made by limiting the amount of sodium available by using only one mole of base for each mole of citric acid for the monosodium citrate and two moles for the disodium citrate. The result is primarily the mono or disalt with small amounts of the other forms and citric acid being present. Other salts that have been offered commercially are shown in Table 6.

4.6. Chelate Formation. Citric acid complexes with many multivalent metal ions to form chelates (9,10). This important chemical property makes citric acid and citrates useful in controlling metal contamination that can affect the color, stability, or appearance of a product or the efficiency of a process.

Citric acid, with its one hydroxyl and three carboxyl groups, is a multidentate ligand. Two or more of these sites are utilized to form a ring structure. The normal molar ratio of metal-to-ligand is 1:1. With some metal ions, under certain conditions, more than one ring can be formed allowing a higher metal-to-ligand ratio (see CHELATING AGENTS).

When a metal ion is chelated by a ligand such as citric acid, it is no longer free to undergo many of its chemical reactions. A metal ion that is normally colored may, in the presence of citrate, have little or no color. Under pH conditions that may precipitate a metal hydroxide, the citrate complex may be soluble. Organic molecules that are catalytically decomposed in the presence of metal ions can be made stable by chelating the metal ions with citric acid.

Table 6. **Salts of Citric Acid**

Salt	CAS Registry number	Molecular formula
ammonium citrate	[3012-65-5]	$(NH_4)_2HC_6H_6O_7$
calcium citrate	[813-94-5]	$Ca_3(C_6H_5O_7)_2$
calcium citrate tetrahydrate	[5785-44-4]	$Ca_3(C_6H_5O_7)_2 \cdot 4H_2O$
cobalt citrate	[866-81-9]	$Co_3(C_6H_5O_7)_2$
copper citrate	[866-82-0]	$C_6H_8O_7 \cdot 2Cu$
ferric ammonium citrate	[1185-57-5]	$C_6H_8O_7xFe \cdot xNH_3$
ferric citrate	[2338-05-8]	$C_6H_8O_7 \cdot xFe$
lead citrate	[512-26-5]	$Pb_3(C_6H_5O_7)_2$
lithium citrate	[919-16-4]	$Li_3C_6H_5O_7$
magnesium citrate	[3344-18-1]	$Mg_3(C_6H_5O_7)_2$
manganese citrate	[5968-88-7]	$MnC_6H_5O_7$
nickel citrate	[6018-92-4]	$Ni_3(C_6H_5O_7)_2$
potassium citrate	[866-84-2]	$K_3C_6H_5O_7$
potassium citrate hydrate	[6100-05-6]	$K_3C_6H_5O_7 \cdot H_2O$
sodium citrate dihydrate	[6132-04-3]	$Na_3C_6H_5O_7 \cdot 2H_2O$
zinc citrate	[546-46-3]	$Zn_3(C_6H_5O_7)_2$

Chelation is an equilibrium reaction. There are always some free-metal ions present as well as chelated metal ions. In a system where a metal salt is being reduced, such as in metal plating, the rate of the reaction forming the metal can be controlled by using the metal citrate chelate.

The log function of the ratio of chelated metal ions to free-metal ions is expressed as the stability constant or formation constant as shown in Table 7. The higher the stability constant the greater the percentage of metal ions that are chelated (11).

Stability constants are measured at their optimum pH. Conditional stability constants are measured at a specific pH. In general, stability constants for metal citrates are very low below pH 2–3, high at pH 3–10, and low above pH 10–12.

4.7. Corrosion. Aqueous solutions of citric acid are mildly corrosive toward carbon steels. At elevated temperatures, 304 stainless steel is corroded

Table 7. **Stability Constants for Metal Citrates**

Metal	Valence	$\log K$
Fe	+3	12.5
Al	+3	7.00
Pb	+2	6.50
Ni	+2	5.11
Co	+2	4.80
Zn	+2	4.71
Ca	+2	4.68
Cu	+2	4.35
Cd	+2	3.98
Mn	+2	3.67
Mg	+2	3.29
Fe	+2	3.08
Ba	+2	2.98

by citric acid, but 316 stainless steel is resistant to corrosion. Many aluminum, copper, and nickel alloys are mildly corroded by citric acid. In general, glass and plastics such as fiber glass reinforced polyester, polyethylene, polypropylene, poly(vinyl chloride), and cross-linked poly(vinyl chloride) are not corroded by citric acid.

5. Manufacturing and Processing

Historically, citric acid was isolated by crystallization from lemon juice and later was recognized as a microbial metabolite. This work led to the development of commercial fermentation technology (12). The basic raw materials for making citric acid include Molasses (mainly beet), sucrose, dextrose (mainly from corn, wheat or tapioca) and unrefined sweet potato.

5.1. Fermentation. Several microorganisms (yeasts and molds) have been identified as citric acid accumulators (i.e., *Yarrowia sp., Candida sp.* and *Aspergillus niger*). However, the most common microorganisms used for the commercial production for citric acid is *Aspergillus niger*. Commercial scale fermentation was begun in 1923 utilizing certain strains of *Aspergillus niger* to produce citric acid on the surface of a sucrose and salt solution. Surface or tray fermentation is still used for production of citric acid in areas of the world that are agriculturally rich and still accounts for a reasonable proportion of worldwide citric acid production. Surface fermentation is more labor intensive and is easier to maintain aseptically in industrial operations. Nowadays, the preferred method for large-volume industrial production is a submerged process known as deep tank fermentation (13–21).

In the deep tank submerged process, *Aspergillus niger* mold spores are grown under controlled aseptic conditions on a test-tube slant and transferred to a seed tank or inoculum which is added to a fermentor along with pasteurized syrup. The pH is adjusted and nutrients added. Sterile air is sparged into the fermentor while the sugar is converted to citric acid. The complete fermentation cycle can take as long as 7 days or even less if yeasts are used.

Several patents have recently been granted for production of citric acid/ citrates by fermentation. One is a process for continuous production of citric acid/ citrate comprising the steps of continuously feeding to a fermentor containing yeasts, oxygen, sugar (especially glucose), and an ammonium compound. Sugar continuously fed to the fermentor is transformed to citric acid and the product is withdrawn continuously from the fermentor (22).

Another uses an aqueous solution of raw material containing citric acid by means of fermentation. Second, third, and fourth solutions are obtained by cell separation, filtration, and ion exchangers. A final solution is obtained containing citric acid/ citrate in water (23).

5.2. Recovery. Citric acid fermentation broth is generally separated from the biomass using filtration or centrifugation. The citric acid is usually purified using either a lime-sulfuric acid method or a liquid extraction process (24). Choice between these two methods is dictated in part by the fermentation feedstock. Lime-sulfuric extraction is more traditional, so it is used in many of the older plants. It is still used in newer production facilities where crude feedstocks

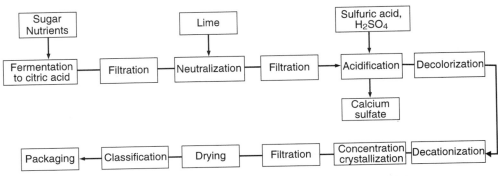

Fig. 3.   Lime-sulfuric recovery process for citric acid.

are still used. Solvent extraction is currently viable where pure substrate is used for fermentation.

Lime-Sulfuric.   Recovery of citric acid by calcium salt precipitation is shown in Figure 3. Although the chemistry is straightforward, the engineering principles, separation techniques, and unit operations employed result in a complex commercial process. The fermentation broth, which has been separated from the insoluble biomass, is treated with a calcium hydroxide (lime) slurry to precipitate calcium citrate. After sufficient reaction time, the calcium citrate slurry is filtered and the filter cake washed free of soluble impurities. The clean calcium citrate cake is reslurried and acidified with sulfuric acid, converting the calcium citrate to soluble citric acid and insoluble calcium sulfate. Both the calcium citrate and calcium sulfate reactions are generally performed in agitated reaction vessels made of 316 stainless steel and filtered on commercially available filtration equipment.

The citric acid solution is deionized at this stage to remove trace amounts of residual calcium, iron, other cationic impurities, and to improve crystallization. In some processes, trace-impurity removal and decolorization are accomplished with the aid of adsorptive carbon.

The aqueous citric acid solution is concentrated in a series of crystallization steps to achieve the physical separation of citric acid from remaining impurities. Standard evaporation, crystallization, and filtration equipment can be employed in this operation. The choice of crystallizer temperature dictates the formation of anhydrous or monohydrate citric acid. Above 37°C, the transition point, anhydrous citric forms whereas below 37°C, the monohydrate will crystallize. The finished citric acid is dried and classified. Because anhydrous citric acid is hygroscopic, to protect against caking care must be taken to avoid handling, packaging, and storage of the crystals in areas of high temperature and high humidity.

Liquid Extraction.   The recovery process, shown in Figure 4, was developed in the 1970s and involves the extraction of citric acid from fermentation broth using a mixture of trilaurylamine,n-octanol, and C_{10} or C_{11} hydrocarbon, followed by re-extraction of the citric acid from the solvent phase into water (25). Efficient citric acid extraction is achieved through a series of countercurrent steps that ensure intimate contact of the aqueous and nonaqueous phases. When transfer of the citric acid to the solvent phase is complete, the citric acid is

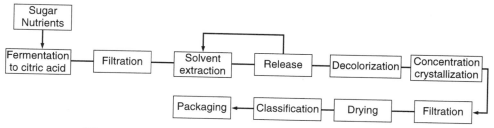

Fig. 4. Liquid extraction recovery process for citric acid.

re-extracted into water, again using a multistage countercurrent system. The two steps differ mainly in the temperature at which they are performed.

The final processing steps are a wash of the aqueous citric acid solution by the hydrocarbon solvent, followed by passage of the acid solution through granular activated carbon columns. Effluent from the carbon columns is processed through a conventional sequence of evaporation/crystallization/drying and packaging steps to complete the manufacturing process.

Citric acid is also commercially available as a 50% w/w solution made either by dissolving crystalline citric acid in water, or a combination or crystalline citric acid, and one of the citric acid process streams. There are several grades of citric acid solutions available, each made according to quality which is measured by color and trace impurities. The citric acid content of each grade can be identical, 50% w/w, which is near the solubility limit.

Several methods of recovery of citric acid from a medium in which it is contained have been reported in ref. 26. The preferred method involves contact of citric acid with a solid-phase free base resin having a tertiary amine group to absorb the citric acid.

5.3. By-Products. The biomass from the fungal fermentation process is called mycellium and can be used as a supplement for animal feed since it contains digestable nutrients (27,28). The lime–sulfuric purification and recovery process results in large quantities of calcium sulfate cake, which is usually disposed of into a landfill but can find limited use in making plaster, cement, wallboard, or as an agricultural soil conditioner. The liquid extraction purification and recovery process has the advantage of little solid by-products.

5.4. Energy. In recent years the concern for energy conservation has resulted in many innovative process improvements to make the manufacture of citric acid more efficient. For example, heat produced by the exotherm of the neutralization of citric acid with lime is used in another part of the process where heat is required, such as the evaporation/crystallization step.

5.5. Chemical Synthesis. The chemical synthesis of citric acid was reported in 1880 (29). Since then, many different synthetic routes have been investigated, reported, and patented (30–38). However, none of these have proven to be commercially feasible.

6. Shipment and Storage

Crystalline citric acid, anhydrous, can be stored in dry form without difficulty, although conditions of high humidity and elevated temperatures should be

Table 8. **Citric Acid**^{a} **Corrosion Rates**^{b}

Material	Temperature, °C	Corrosion rate, mm/yr
316 stainless steel	25	0.03
	50	0.03
304 stainless steel	25	0.03
	50	0.23
carbon steel	25	4.6
	50	32.8

^{a} 50 wt% solution.
^{b} Ref. 37.

avoided to prevent caking. Storage should be in tight containers to prevent exposure to moist air. Several granulations are commercially available with the larger particle sizes having less tendency toward caking.

Liquid citric acid is commercially available in a variety of concentrations with 50% w/w being most common. Grades are available that vary in appearance, purity, and color. Packaging is usually in drums, tank trucks, or rail cars. Liquid citric acid should be kept above 0°C to prevent crystallization.

Solutions of citric acid are corrosive to normal concrete, aluminum, carbon steel, copper, copper alloys, and should not be used with nylon, polycarbonates, polyamides, polyimides, or acrylics.

Recommended materials of construction for pipes, tanks, and pumps handling citric acid solutions are 316 stainless steel, fiber glass-reinforced-polyester, polyethylene, polypropylene, and poly(vinyl chloride). At elevated temperatures, 304 stainless steel is not recommended (Table 8).

Although not as corrosive as the acid, the sodium and potassium salts of citric acid should be handled in the same type of equipment as the acid to avoid corrosion problems.

7. Economic Aspects

Citric acid is manufactured in over 20 countries with 1998 worldwide production estimated at approximately 879,000 MT, distributed as shown in Figure 5. Most of this production is used for foods and beverages; however, industrial applications, eg, detergents, metal cleaning, of citric acid are becoming more important on a worldwide basis.

It was estimated that 1990 U.S. citric acid and citrate salt consumption was 152,000 t. Citric acid represents approximately 90% of this volume. This citric acid/citrate use and its historical distribution in various markets is described in Table 9.

In the 1980s, citric acid was produced mainly by pharmaceutical companies. However, in 1990–1991, when corn wet miling companies entered the market, back integration with a source of fermentation feedstock (dextrose, sugar, or molasses) became important. This marked the movement of citric acid from a pharmaceutical company product to a commodity chemical product.

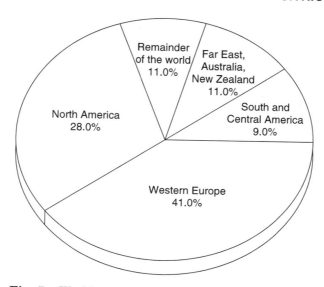

Fig. 5. Worldwide citric acid production in 1998 (40).

The list price for citric acid in the U.S. for 1998 was approximately 1.65 USD/g (40).

8. Specifications, Standards, and Quality Control

Since citric acid is produced and sold throughout the world, it must meet the criteria of a variety of food and drug compendia (41–45).

9. Analytical and Test Methods

Aqueous titration with 1 N NaOH remains the official method for assaying citric acid (41,42). Although not citrate-specific, the procedure is satisfactory in the absence of interfering substances. Low concentrations of citric acid can be determined by a spectrophotometic method based on the Furth and Herrmann reaction with pyridine and acetic anhydride (PAA) (46). This PAA method is

Table 9. **U.S. Citric Acid/Citrate Distribution by End Use,** $\%^a$

Product	1998	1996	1994
beverages	48	46	47
foods	23	22	23
pharmaceuticals and cosmetics	18	19	20
household detergents and cleaning products	6	6	6
misc. nonfood	5	7	4
Total annual consumption	*233,000*	*206,000*	*203,000*

^{a} Ref. 40.

citrate-specific at 420 nm and is sensitive for citrate ions at concentrations down to 5 ppm. The PAA method can be used to quantify citrate in foods, beverages, and industrial products such as detergents.

An enzymatic method (47), which is specific for the citrate moiety, can be used as a combined assay and identification test for citric acid and its common salts down to 20 ppm.

A high performance liquid chromatography (HPLC) method to determine citric acid and other organic acids has been developed (48). The method is an isocratic system using sulfuric acid to elute organic acids onto a specific hplc column. The method is sensitive for citric acid down to ppm levels and is capable of quantifying citric acid in clear aqueous systems.

10. Health and Safety Factors

Citric acid, as well as its common sodium and potassium salt forms, are Generally Recognized As Safe (GRAS) by the U.S. Food and Drug Administration as Multiple Purpose Food Substances (49). Citric acid is also approved by the Joint FAO/WHO Expert Committee on Food Additives for use in foods without limitation (50). The use of citric acid and certain of its salts and esters has been evaluated by a Special Committee on GRAS Substances (SCOGS) of the Federation of American Societies for Experimental Biology under contract with the FDA (51). The evaluation was based largely on two scientific literature reviews prepared for the FDA that summarize the world's applicable scientific literature from 1920 through 1973 (1). Citric acid, sodium and potassium citrates are not listed under the Safe Drinking Water and Toxic Enforcement Act of 1986 (California proposition 65) that lists chemicals known to cause cancer or reproductive toxicity. These compounds are not listed by the National Toxicology Program (NTP) or the International Agency for Research on Cancer (IARC) or regulated as carcinogens by the U.S. Occupational Safety and Health Administration (OSHA) (52,53).

Tests have shown that citric acid is not corrosive to skin but is a skin and ocular irritant (54). For these reasons it is recommended that individuals use appropriate personal protection to cover the hands, skin, eyes, nose, and mouth when in direct contact with citric acid solutions or powders. This product is not hazardous under the criteria of the OSHA Hazard Communication Standard (48).

11. Environmental Considerations

Citric acid is biodegraded readily by many organisms under aerobic and anaerobic wastewater treatment conditions and in the natural environment (55). Under United States Environmental Protection Agency legislation, citric acid and its sodium and potassium salts are reported on the Toxic Substance Control Act (TSCA) inventory. They are exempt from the Emergency Planning and Community Right-to-Know Act (SARA Title III) and contain no section 313 toxic chemical although they may contain up to the FCC limits for arsenic (1 ppm), lead

(0.5 ppm), and heavy metals (5 ppm as lead). Under the U. S. Resource Conservation and Recovery Act (RCRA), if discarded in its purchased form, this product would not be a hazardous waste either by listing or by characteristic (56).

12. Uses

Citric acid is utilized in a large variety of food and industrial applications because of its unique combination of properties. It is used as an acid to adjust pH, a buffer to control or maintain pH, in a wide range (2.5–6.5), a chelator to form stable complexes with multivalent metal ions, and a dispersing agent to stabilize emulsions and other multiphase systems (see DISPERSANTS). In addition, it has a pleasant, clean, tart taste making it in the acidulant of choice in food and beverage products applications.

12.1. Food Uses. *Beverages.* Citric acid, sodium citrate, and potassium citrate are used extensively in carbonated and non-carbonated beverages (57). Juice-added beverages, low calorie beverages, and thirst quenchers, also known as isotonic drinks, use citric acid alone and in combination with citrate salts for flavoring and buffering properties and to increase the effectiveness of antimicrobial preservatives. The high solubility of citric acid is important in beverage syrups. Citric acid also combines well with other acids (i.e. malic acids) to give unique flavor profiles. The amount of acid used depends on the flavor desired in the product as well as taste evaluation and customer preference. Citric acid and its salts are also used to stabilize the pH in reduced-calorie drinks that contain acid-sensitive sweeteners. The acid concentration of most fruit-flavored carbonated beverages (qv) falls in the range of 0.10–0.25% w/w.

Citric acid and its salts are used in dry beverage mixes, convenience teas, New age drinks and cocktail mixes for pH control and flavor, and are used in wine coolers at 0.10–0.55%, combining well with fruity and light flavors. Because citric acid is derived from a natural fermentation process, its use causes no problems in labeling claims. Therefore, citric acid is ideal for the marketing of health-oriented drinks.

Jams, Jellies, and Preserves. Citric acid is used in jams and jellies to provide tartness and to adjust the product pH for optimum gelation (58).

Candy. Citric acid is added in candy for tartness (59,60). To suppress the inversion of sucrose, it should be added after the cook, at levels from 0.5 to 2.0%. The pH of pectin gel candies is adjusted with citric acid for maximum gel strength. Sodium and potassium citrate are also used to reduce and control inversion and to regulate tartness.

Salads. A combination of citric acid and ascorbic acid is used as an alternative to sulfites in prevention of enzymatic browning in fresh prepared vegetables (61).

Frozen Food. The chelating and acidic properties of citric acid enable it to optimize the stability of frozen food products by enhancing the action of antioxidants and inactivating naturally present enzymes which could cause undesirable browning and loss of firmness (62,63).

Citric acid also inhibits color and flavor deterioration in frozen fruit. Here again the function is to inhibit enzymatic and trace metal-catalyzed oxidation.

Canned Fruits and Vegetables. The use of citric acid to bring the pH below 4.6 can reduce heat treatment requirements in canned fruits and vegetables. In addition, citric acid chelates trace metals to prevent enzymatic oxidation and color degradation, and enhances the flavor, especially of canned fruits.

Fats and Oils. The oxidation of fats and oils in food products can be prevented by the addition of citric acid to chelate the trace metals that catalyze the oxidation. Citric acid is also used in the bleaching clays and the degumming process during oil refining to remove chlorophyll and phospholipids (64–68).

Confections and Desserts. Citric acid andsodium citrate are utilized in the confection industry to optimize gel-setting characteristics, provide tartness, and enhance flavor. Citric acid acts as a flavor adjunct in compressed candy, tablets, cream candies and startch-based jellies.

Pasteurized Process Cheese. Sodium citrate is used in pasteurized process and sliced cheese as an emulsifying salt to stabilize the water and oil emulsion and improve process cheese body and texture (69).

Dairy Products. Sodium citrate is an important stabilizer used in whipping cream and vegetable-based dairy substitutes. Addition of sodium citrates to ice cream, ice milk, and frozen desserts before pasteurization and homogenization reduces the viscosity of the mix, making it easier to whip.

Seafood. Citric acid is used in combination with other preservatives/antioxidants to lower the pH to retard microbial growth, which can lead to spoilage, formation of off-flavors, and colors on fish and other seafood products. It is also used to chelate trace metals that catalyze oxidation.

Meat Products. Citric acid is used in cured meat products to increase the effectiveness of the antioxidant preservatives, as a processing aid, and a texture modifier. It is often encapsulated and released at a specific temperature from a controlled release matrix. Carcass rinse solutions contain citric acid to adjust the pH and enhance antimicrobial properties.

Citric acid or sodium citrate can be used as a replacement of up to 50% of cure accelerators such as ascorbic acid, erythorbic acid, sodium ascorbate, and sodium erythorbate in the preparation of cured, dried meat and poultry products.

12.2. Medical Uses. Citric acid and citrate salts are used to buffer a wide range of pharmaceuticals at their optimum pH for stability and effectiveness (70–78). Effervescent formulations use citric acid and bicarbonate to provide rapid dissolution of active ingredients and improve palatability. Citrates are used to chelate trace metal ions, preventing degradation of ingredients. Citrates are used to prevent the coagulation of both human and animal blood in plasma and blood fractionation. Calcium and ferric ammonium citrates are used in mineral supplements. A patent derivatives a chewing gum with fluoride and citric acid for improved dental health (79).

12.3. Agricultural Use. Citric acid and its ammonium salts are used to form soluble chelates of iron, copper, magnesium, manganese, and zinc micronutrients in liquid fertilizers (80–86). Citric acid and citrate salts are used in animal feeds to form soluble, easily digestible chelates of essential metal nutrients, enhance feed flavor to increase food uptake, control gastric pH and improve feed efficiency.

12.4. Industrial Uses. *Laundry Detergents.* Sodium citrate is used in both liquid and powder laundry detergents as a builder(87–104). In many detergent applications, builder systems containing citrates are used as environmentally acceptable replacements for phosphates. Citrates chelate water hardness ions, disperse soil, and are used as processing aids. High water solubility and performance at both low and high pH and low and high temperatures are the keys to citrate use in detergents. In powder detergents citrates are used as auxiliary co-builders usually with zeolites and carbonates (see DETERGENCY).

Hard Surface Cleaners. Citric acid and sodium citrate are used in hard surface cleaners as acidifying and chelating agents for dissolving hard water deposits and as a builder to increase the efficacy of the surfactants. In carpet cleaning, citric acid stabilizes dyes and removes carpet yellowing, especially in stain-resistant carpets.

Reverse Osmosis Membrane Cleaning. Citric acid solutions are used to remove iron, calcium, and other cations that foul cellulose acetate and other membranes in reverse osmosis and electrodialysis systems. Citric acid solutions can solubilize and remove these cations without damaging the membranes (105–107).

Metal Cleaning. Citric acid, partially neutralized to ~pH 3.5 with ammonia or triethanolamine, is used to clean metal oxides from the water side of steam boilers and nuclear reactors with a two-step single fill operation (108–126). The resulting surface is clean and passivated. This process has a low corrosion rate and is used for both pre-operational mill scale removal and operational cleaning to restore heat-transfer efficiency.

High-pressure sprays of heated neutralized citric acid solutions replace sandblasting techniques to clean stainless steel equipment and areas not easily accessible such as ship bilges.

Petroleum. Citric acid is added to hydrochloric acid solutions in acidizing limestone formations. Citric acid prevents the formation of ferric hydroxide gel in the spent acid solution by chelating the ferric ions present. Formation of the gel would plug the pores, preventing the flow of oil to the producer well (127–131).

A clear solution of aluminum citrate neutralized to pH 7 is used for *in situ* gelling of polymers in polymer flooding and well stimulation in enhanced oil recovery techniques (132–136). The citrate chelate maintains aluminum ion solubility and controls the rate of release of the aluminum cross-linker.

Citric acid is used to chelate vanadium catalyst in a process for removing hydrogen sulfide from natural and refinery gas and forming elemental sulfur, a valuable product (137).

Flue Gas Desulfurization. Citric acid can be used to buffer systems that can scrub sulfur dioxide from flue gas produced by large coal and gas-fired boilers generating steam for electrical power (138–147). The optimum pH for sulfur dioxide absorption is pH 4.5, which is where citrate has buffer capacity. Sulfur dioxide is the primary contributor to acid rain, which can cause environmental damage.

Mineral and Pigment Slurries. Citric acid can be used as a dispersing agent in slurries of ores, rocks, clays, and pigments during refining and transport. Citric acid controls swelling of clays and reduces pumping viscosity by contributing thixotropic properties to the dispersions (148–156).

Electrodeposition of Metals. Citric acid and its salts are used as sequestrants to control deposition rates in both electroplating and electroless plating of metals (157–175). The addition of citric acid to an electroless nickel plating bath results in a smooth, hard, nonporous metal finish.

Concrete, Mortar, and Plaster. Citric acid and citrate salts are used as admixtures in concrete, mortar, and plaster formulations to retard setting times and reduce the amount of water required to make a workable mixture (176–184). The citrate ion slows the hydration of Portland cement and acts as a dispersant, reducing the viscosity of the system (185). At levels below 0.1%, citrates accelerate the setting rate while at 0.2–0.4% the set rate is retarded. High early strength and improved frost resistance have been reported when adding citrate to concrete, mortar, and plaster.

Textiles. Citric acid acts as a buffer in the manufacture of glyoxal resins which are used to give textiles a high quality durable-press finish (see AMINO RESINS). It has been reported to increase the soil-release property of cotton with wrinkle-resistant finishes and is used as a buffer, a chelating agent, and a non-volatile acid to adjust pH in disperse dying operations (186–197).

Plastics. Citric acid and bicarbonate are used as an effervescent blowing agent to foam polystyrene for insulated food and beverage containers replacing blowing agents such as chlorinated fluorocarbons (198–210).

Citric acid is used as a chelating agent in catalyst systems for making resins, and citrate esters are used as plasticizers (qv) inPVC film, especially in food packaging (211).

Paper. Citric acid is added to the pulp slurry prior to bleaching to sequester metal ions and prevent discoloration (212–215). Citrates are used in cigarette paper to control the burning rate to match that for tobacco.

Tobacco. Citric acid is a natural constituent of the tobacco leaf and during tobacco processing additional citric acid is added to enhance the flavor and to effect more complete combustion of tobaccos (216).

Cosmetics and Toiletries. Citric acid and bicarbonate are used in effervescent type denture cleansers to provide agitation by reacting to form carbon dioxide gas. Citric acid is added to cosmetic formulations to adjust the pH, act as a buffer, and chelate metal ions preventing formulation discoloration and decomposition (91,217–222). The use of active such in creams to alleviate signs of dermatological aging and wrickling is discussed in Ref. 221.

Refractories and Molds. Citric acid is used as a binder for refractory cements, imparting volume stability and strength in ceramic materials for electrical condensers, foundry and glassmaking molds, and sand molds for metal castings (223–227).

13. Derivatives

13.1. Salts. The trisodium citrate salt is made by dissolving citric acid in water at a concentration of 50% w/w or higher. A 50% solution of sodium hydroxide is carefully added to pH 8.0–8.5. The reaction is exothermic and cooling is necessary to prevent boiling. The hot solution can be treated with activated carbon to remove impurities before evaporating and crystallizing. The concentrated slurry is filtered to separate the sodium citrate dihydrate [6132-04-3], which is washed with water, dried in a hot air dryer, classified, and packaged in bags, drums, or large sacks. The tripotassium salt of citric acid is made in a similar manner using potassium hydroxide. The product crystallizes as the monohydrate [6100-05-6].

Ammonium salts of citric acid are made by adding either aqueous or anhydrous ammonia to citric acid dissolved in water. They are usually used in the liquid form rather than isolated as a dry product. Citric acid salts are listed in Table 5. Solubility data is as follows (1).

Citrate salt	Water solubility, wt%
diammonium citrate	50
calcium citrate tetrahydrate	0.10
ferric ammonium citrate	very soluble
potassium citrate monohydrate	60
sodium citrate dihydrate	42

13.2. Esters. The significant esters of citric acid are trimethyl citrate, triethyl citrate, tributyl citrate, and acetylated triethyl- and tributyl citrate. Many other esters are available but have not been used on a commercial scale. Citric acid esters are made under azeotropic conditions with a solvent, a catalyst, and the appropriate alcohol.

Catalysts used are usually acids such as sulfuric acid, p-toluenesulfonic acid, sulfonic acid ion-exchange resins, and others. The water from the reaction of the citric acid and the alcohol is continuously removed as the azeotrope until no more water is formed. At this point, the reaction is usually complete and the solvent and any excess alcohol is distilled off under mild vacuum. The catalyst is neutralized using carbonate or sodium hydroxide, leaving a crude product. If a pure product is desired, the ester can be distilled under high vacuum.

The properties of citric acid esters are described in Table 10.

Citric acid esters are used as plasticizers in plastics such as poly(vinyl chloride), poly(vinylidene chloride), poly(vinyl acetate), poly(vinyl butyral), polypropylene, chlorinated rubber, ethylcellulose, and cellulose nitrate. Most citrate esters are nontoxic and are acceptable by the FDA for use in food-contact

Table 10. **Properties of Citric Acid Esters**

Name	CAS Registry number	Molecular weight	Density,[a] g/mL	Bp[b], °C
triethyl citrate	[77-93-0]	276.29	1.136	126–127
tri-n-butyl citrate	[77-94-1]	360.43	1.042	169–170
tricyclohexyl citrate	[4132-10-9]	438.57	1.7	57[c]
acetyl triethyl citrate	[77-89-4]	318.31	1.135	131–132
acetyl tri-n-butyl citrate	[77-90-7]	402.46	1.046	172–174
acetyl tri-2-ethylhexyl citrate	[144-15-0]	570.81	0.983	225

[a] At 25°C.
[b] At 133 Pa = 1 mm Hg.
[c] Melting point.

packaging and for flavor in certain foods. As a plasticizer, citrate esters provide good heat and light stability and excellent flexibility at low temperatures. Triethyl citrate, tri-*n*-butyl citrate, isopropyl citrate, and stearyl citrate are considered GRAS for use as food ingredients (228–232).

BIBLIOGRAPHY

"Citric Acid" in *ECT* 1st ed., Vol. 4, pp. 8–23, by G. B. Stone, Chas. Pfizer & Co., Inc.; in *ECT* 2nd ed., Vol. 5, pp. 524–540, by L. B. Lockwood and W. E. Irwin, Miles Chemical Co.; in *ECT* 3rd ed., Vol. 6, pp. 150–179, by E. F. Bouchard and E. G. Merritt, Pfizer Inc.; in *ECT* 4th ed., Vol. 6, pp. 354–380, by G. Blair and P. Staal, Haarman & Reimer Corporation; "Citric Acid" in *ECT* (online), posting date: December 4, 2000, by G. Blair and P. Staal, Haarmann & Reimer Corporation.

CITED PUBLICATIONS

1. J. A. Dean, *Lange's* Handbook of Chemistry, 12th ed., McGraw-Hill Book Co., Inc., New York, 1979.
2. *The Merck Index*, 11th ed., Merck & Co., Rahway, N.J. 1989.
3. R. C. Weast, *CRC Handbook of Chemistry and Physics*, 69th ed., CRC Press, Boca Raton, Fla., 1988, 1989, p. 163.
4. *Perry's* Chemical Engineering Handbook, 6th ed., McGraw-Hill Book Co., Inc., New York, 1984.
5. A. Seidell, *Solubilities of Inorganic and Organic Compounds*, 3rd ed.,Vol. 2, D. Van Nostrand Co., Inc., New York, 1941, 427–429.
6. G. T. Blair and M. F. Zienty, *Citric Acid: Properties and Reactions*, Miles Laboratories Inc., 1979.
7. U.S. Pat. 3,997,596 (Dec. 14, 1976), F. Smeets (to Citrex SA, Belgium).
8. *Citrest-Citric Acid Fatty Esters*, technical information, Cyclo Chemicals Corp., Miami, Fla.
9. C. J. Knuth and A. Bavley, *Plast. Technol.* **3**, 555 (1957).
10. P. W. Staal, *Chelation*, Technical Bulletin A-1014, Haarmann & Reimer Corp., Springfield, N.J., 1989.
11. M. K. Musho, *Citric Acid Chelation Chemistry*, Technical Bulletin A-1013, Haarmann & Reimer Corp., Springfield, N.J., 1989.
12. *Scientific Literature Review on GRAS Food Ingredients-Citrates*, PB-223 850, National Technical Information Service, Springfield, Va., Apr. 1973; *Scientific Literature Review on GRAS Food Ingredients-Citric Acid*, PB-241 967, National Technical Information Service, Springfield, Va., Oct. 1974.
13. G. T. Austin, *Shreve's* Chemical Process Industries, 5th ed., McGraw-Hill Book Co., Inc., New York, 1984.
14. L. M. Miall, in A. H. Rose, ed., *Primary Products of Metabolism*, Academic Press, Inc., New York, 1978.
15. J. N. Currie, *J. Biol. Chem.* **31**, 15 (1917).
16. H. Amelung, *Chem. Ztg.* **54**, 118 (1930).
17. L. H. C. Perquin, *Bijdrage Tot De Kennis Der Oxydative Dissimilatic Van Aspergillus niger van Tieghem*, Meinema, Delft, The Netherlands, 1938.
18. S. M. Martin and W. R. Waters, *Ind. Eng. Chem.* **44**, 2229 (1952).

19. D. S. Clark, *Can. J. Microbiol.* **8**, 133 (1962).

20. Brit. Pat. 653,808 (Mar. 23, 1951), R. L. Snell and L. B. Schweiger (to Miles Laboratories Inc.).

21. U.S. Pat. 3,285,831 (Nov. 15, 1966), E. J. Swarthout (to Miles Laboratories Inc.).

22. U.S. Pat. 6,303,351 (Oct. 16, 2001), S. Anastassiadis, A. Aivasidis, and C. Wandrey.

23. U.S. Pat. 6,087,139 (July 11, 2000), R. Boensch and co-workers(to Metallgesellschaft Aktiengesellschaft, Frankfurt am Main).

24. Brit. Pat. 1,145,520 (Mar. 19, 1969), M. A. Batti (to Miles Laboratories Inc.).

25. *Ullmanns Encyklopädie der Technischen Chemie*, 4th ed., Vol. 9, Urban & Schwarzenberg, Munich, Berlin, Germany, 1975, pp. 624–636.

26. U.S. Pat. 6,137,004 (Oct. 24, 2000), D. W. McQuigg and co-workers(to Reilly Industries, Inc.).

27. *Code of Federal Regulations*, Title 21,§173.280, U.S. Government Printing Office, Washington, D.C., 1990.

28. *The Association of American Feed Control Officials*, Official Publication, 1977, Section 36, pp. 87–88.

29. H. C. DeRoo, *Conn. Agri. Exp. St. Bull.* 750 (1975).

30. E. Grimoux and P. Adam, *C. R. Acad. Sci. Paris* **90**, 1252 (1880); *Bull. Soc. Chim. Fr.* **36**, 18 (1881).

31. H. V. Pechmann and M. Dunschmann, *Ann. Chem.* **261**, 162 (1891); A. Haller and A. Held, *Ann. Chim. Phys.* **23**, 175 (1891).

32. W. T. Lawrence, *J. Chem. Soc.* **71**, 457 (1897); E. Ferrario, *Gazz. Chim. Ital.* **38**, 99 (1908).

33. E. Baur, *Chem. Ber.* **46**, 852 (1913).

34. H. Franzen and F. Schmitt, *Chem. Ber.* **58**, 222 (1925).

35. H. O. L. Fischer and G. Dangschat, *Helv. Chim. Acta* **17**, 1196 (1934); E. Baer, J. M. Grosheintz, and H. O. L. Fischer, *J. Am. Chem. Soc.* **61**, 2607 (1939).

36. F. Knopp and C. Martius, in Hoppe-Seyler, ed., *Z. Physiol. Chem.* **242**, 204 (1936); C. Martius, *Ibid.*, 279, 96 (1943).

37. A. M. Gakhokidze and A. P. Guntsadze, *J. Gen. Chem. USSR* **17**, 1642 (1947).

38. P. E. Wilcox, C. Heidelberger, and V. R. Potter, *J. Am. Chem. Soc.* **62**, 5019 (1950).

39. M. Taniyama, *Toho-Reiyon Kenkyu Hokoku* **1**, 40 (1954).

40. T. Esker, H. Jansherkar, Y. Sakuma, CEH Marketing Research Report, "Citric Acid", *Chemical Economics Handbook*, SRI, Menlo Park, Calif., 1999.

41. *Food Chemicals Codex*, 3rd ed., 3rd Suppl., National Academy Press, Washington, D.C., 1992.

42. *United States Pharmacopoeia XXII*, United States Pharmacopoeial Convention, Inc., Rockville, Md., 1990.

43. *British Pharmacopoeia*, British Pharmacopoeia Commission, London, 1988.

44. *The Pharmacopoeia of Japan*, The Society of Japanese Pharmacopoeia, Tokyo, Japan, 1987.

45. *European Pharmacopoeia*, 2nd ed., 1986.

46. C. G. Hartford, *Anal. Chem.* **34**, 426 (1962).

47. J. A. Taraborelli and R. P. Upton, *J. Am. Oil Chem. Soc.* **52**, 248 (1975).

48. G. D. Guerrand and co-workers, "Organic Acid Analysis," *J. Clin. Microbiol.* **16**, 355 (1982).

49. *Code of Federal Regulations*, Title 21,§181.27, 182.1033, 182.6033, Title 29, §1910.1200 U.S. Government Printing Office, Washington, D.C., 1990.

50. *FAO Nutrition Meetings Report*, Series No. 40 A,B,C, Food and Agriculture Organization of the United Nations World Health Organization, New York, 1967, p. 134.

51. *Tentative Evaluation of the Health Aspects of Citric Acid, Sodium Citrate, Potassium Citrate, Calcium Citrate, Triethyl Citrate, Isopropyl Citrate, and Stearyl Citrate as Food Ingredients*, PB280 954, National Technical Information Service, Springfield, Va., 1977.

52. Office of Environmental Health Hazard Assessment, California Protection Agency, Safe drinking water and toxic enforcement act of 1986. Proposition 65. Chemical listed effective August 20, 1999 as known to cause cancer or reproductive toxicity (1999).

53. Code of Federal Regulations, 40 CFR 700-799. Toxic Substances Control Act(1977)

54. Internal Reports: *Skin Corrosion Potential of Liquid Citric Acid 50%*, Jan. 19, 1979, *Dermal Irritation of Citric Acid in the Rabbit*, Dec. 20, 1990, *Ocular Irritation of Citric Acid in the Rabbit*, May 14, 1991, Miles Inc., Elkhart, Ind.

55. *Ecological Effects of Non-Phosphate Detergent Builders-Final Report on Organic Builders Other than NTA*, International Joint Commission, Windsor, Ontario, Canada, July 21, 1980.

56. United States Enviornmental Protection Agency. Title III List of Lists. Consolidated List of Chemicals Subject to the Emergency Planning and Community Right-To-Know Act (EPCRA) and Section 112(r) of the Clean Air Act, as Amended, 1999.

57. *The Story of Soft Drinks*, National Soft Drink Association, Washington, D.C., 1982.

58. *Preservers Handbook*, Sunkist Growers, Inc., Ontario, Calif., 1964.

59. C. D. Barnett, *The Science & Art of Candy Manufacturing*, Harcourt Brace Jovanovich Publications, Duluth, Minn., 1978, p. 58.

60. A. F. Porter, *Spice Up Your Candy With Citric Acid*, Candy Industry, Aug. 1985.

61. N. A. Eskin, H. M. Henderson, and R. J. Townsend, *Biochemistry of Foods*, Academic Press, New York, 1981; *Code of Federal Regulations*, Title 21, Part 182, Subpart D, U.S. Government Printing Office, Washington, D.C., 1990.

62. K. H. Moledina and co-workers, *J. Food Sci.* **42**, 759 (1977).

63. C. E. Wells, D. C. Martin, and D. A. Tichenor, *J. Am. Dietetic Assoc.* **61**, 665 (1972).

64. L. L. Diosady, P. Sleggs, and T. Kaji, *JAOCS* **59**(7), 313–316 (1982).

65. A. Smiles, Y. Kakuda, and B. MacDonald, *JAOCS* **65**,(7), 1151–1155 (1988).

66. K. S. Law and K. G. Berger, *Citric Acid in the Processing of Oils and Fats*, No. 11, Porim Technology, Palm Oil Research Institute of Maylasia, July 1984.

67. D. D. Brooks and co-workers, "The Synergistic Effect of Neutral Bleaching Clay and Citric Acid: Chlorophyll Removal," paper presented at the *79th Annual AOCS Meeting*, Phoenix, Ariz. 1988.

68. S. K. Brophy and co-workers, "Chlorophyll Removal from Canola Oil: A New Concept," paper presented at the *80th Annual AOCS Meeting*, Cincinnati, Ohio, 1989.

69. *Code of Federal Regulations*, Title 21, Part 133, U.S. Government Printing Office, Washington, D.C., 1977.

70. U.S. Pat. 2,999,293 (Sept. 12, 1961), J. White and R. Kolb (to Warner Lambert Pharmaceutical Co.).

71. *Remingtons' Pharmaceutical Sciences*, 15th ed., Mack Publishing Co. Inc., Easton, Pa., 1975, p. 1574.

72. *The National Formulary*, 14th ed., American Pharmaceutical Association, Washington, D.C., 1975, pp. 389–390.

73. U.S. Pat. 3,956,156 (May 11, 1976), A. N. Osband, F. W. Gray, and J. C. Jervert (to Colgate-Palmolive Co.).

74. *Handbook of Non-Prescription Drugs*, 5th ed., American Pharmaceutical Association, Washington, D.C., 1977, pp. 3–17.

75. D. Entriken and C. Becker, *J. Am. Pharm. Assoc.* **43**, 693 (1954).

76. S. Bhattacharya and co-workers, *J. Indian Chem. Soc.* **31**, 231 (1954).

77. *USP XIX*, The United States Pharmacopeial Convention Inc., Rockville, Md., 1975, pp. 33–35;*USP XXII*, 1990, 1989,p. 103.

78. *Code of Federal Regulations*, Title 21,§182.1195, 182.5195, 182.6195 (calcium citrate); 182.1625, 182.6625 (potassium citrate), 182.5449 (manganese citrate), U.S. Government Printing Office, Washington, D.C., 1990.

79. U.S. Pat. 5,698,215 (Dec. 16, 1977), T. Kallilis andA. A. Kaputo

80. U.S. Pat. 2,813,014 (Nov. 12, 1957), J. R. Allison and C. A. Hewitt (to Leffingwell Chem. Co.).

81. Brit. Pat. 827,521 (Feb. 3, 1960), I. S. Perold (to Union of South Africa, Dir. of Tech. Service, Dept. of Agriculture).

82. U.S. Pat. 3,869,272 (Mar. 4, 1975), R. J. Windgasen (to Standard Oil Co.).

83. A. Marchesini, P. Sequi, and G. A. Lanzani, *Agrochimica* **10**(2), 183 (1966).

84. F. L. Daniel, P. O. Ramaswani, and T. P. Mahadevan, *Madras Agr. J.* **55**(1), 31 (1968).

85. M. S. Omran and co-workers, *Egypt. J. Soil, Sci.* **27**(1), 31–42 (1987).

86. J. J. Mortvedt, *Solutions*, 64–79 (May–June 1979).

87. *Environmental Impact of Citrates*, Information Sheet No. 2030, Pfizer Chemicals Division, New York, 1974.

88. D. L. Muck and H. L. Gewanter, "The Detergent Building Properties of Trisodium Citrate," paper presented at the *American Oil Chemist Society Meeting*, Apr. 1972.

89. U.S. Pat. 4,028,262 (June 7, 1977), B.-D. Cheng (to Colgate-Palmolive Co.).

90. U.S. Pat. 4,013,577 (Mar. 22, 1997) (to Colgate-Palmolive Co.).

91. U.S. Pat. 4,009,114 (Feb. 22, 1977), J. A. Yurke (to Colgate-Palmolive Co.).

92. Brit. Pat. 1,477,775 (June 26, 1977) (to Hoechst AG).

93. Brit. Pat. 1,427,071 (Mar. 3, 1976), M. Filcek and co-workers (to Benckiser GmbH).

94. U.S. Pat. 3,985,669 (Oct. 12, 1976), H. K. Krummel and T. W. Gault (to The Procter & Gamble Co.).

95. U.S. Pat. 4,021,377 (May 3, 1977), P. J. Borchert and J. L. Neff (to Miles Laboratories, Inc.).

96. Belg. Pat. 848,533 (May 20, 1977) (to Henkel & Cie, GmbH).

97. U.S. Pat. 4,024,078 (May 7, 1977), A. Gilbert and J. W. Schuette (to The Procter & Gamble Co.).

98. *Citrosol-50 T, W and E*, Information Sheet No. 626, Pfizer Chemicals Division, New York, 1974.

99. U.S. Pat. 3,968,048 (July 6, 1976), J. A. Bolan (to The Drackett Co.).

100. U.S. Pat. 3,920,564 (Nov. 18, 1975), J. J. Greecsek (to Colgate-Palmolive Co.).

101. U.S. Pat. 4,379,080 (Apr. 5, 1983), A. P. Murphy (to The Procter & Gamble Co.).

102. U.S. Pat. 4,490,271 (Dec. 25, 1984), G. L. Spadini, A. L. Larabee, and D. K. K. Liu (to The Procter & Gamble Co.).

103. U.S. Pat. 4,605,509 (Aug. 12, 1986), J. M. Corkill, B. L. Madison, M. E. Burns (to The Procter & Gamble Co.).

104. U.S. Pat. 4,965,013 (Oct. 23, 1990), K. L. Pratt (to Miles Inc.).

105. Jpn. Pat. 76 18,280 (Feb. 13, 1976), T. Mizumoto and co-workers (to Ebara-Infilco Co., Ltd.).

106. Jpn. Pat. 75 153,778 (Dec. 11, 1975), T. Mizumoto and co-workers (to Ebara-Infilco Co., Ltd.).

107. K. J. McNulty and co-workers, *Laboratory and Field Evaluation of NS-100 Reverse Osmosis Membrane*, EPA-600/2-80-059, Industrial Environmental Research Laboratory, EPA, Cincinnati, Ohio, Apr. 1980.

108. W. J. Blume, *Mater. Perform.* **16**(3), 15 (1977).
109. U.S. Pat. 3,806,366 (Apr. 23, 1974), D. B. Cofer and co-workers (to Southwire Co.).
110. U.S. Pat. 3,664,870 (May 23, 1972), A. W. Oberhofer and co-workers (to Nalco Chemical Co.).
111. U.S. Pat. 3,072,502 (Jan. 8, 1963), S. Alfano (to Pfizer Inc.); 3,248,269 (Apr. 26, 1966), W. E. Bell (to Pfizer Inc.).
112. L. D. Martin and W. P. Banks, "Electrochemical Investigation of Passivating Systems," *Proceedings of the 35th International Water Conference*, Pittsburgh, Pa., Oct. 29–31, 1974.
113. G. W. Bradley and co-workers, "Investigation of Ammonium Citrate Cleaning Solvents," *Proceedings of the 36th International Water Conference*, Pittsburgh, Pa., Nov. 4–6, 1975.
114. U.S. Pat. 3,003,898 (Oct. 10, 1961), C. F. Reich (to The Dow Chemical Company).
115. U.S. Pat. 3,496,017 (Feb. 17, 1970), R. D. Weed (to U.S. Atomic Energy Commission).
116. U.S. Pat. 3,013,909 (Dec. 19, 1961), G. P. Pancer and J. L. Zegger (to U.S. Atomic Energy Commission).
117. McCollum and Logan, *Electrolytic Corrosion of Iron in Soils*, Technical Paper No. 25, Bureau of Standards, Washington, D.C., 1913, p. 7.
118. N. Hall and Hogaboom, "Metal Finishing," *21st Annual Guidebook Directory*, Westwood, N.J., 1953.
119. U.S. Pat. 2,558,167 (June 26, 1951), A. J. Beghin, P. F. Hamberg, and H. E. Smith (to Insl-X Corp.).
120. E. C. Wackenhuth, L. W. Lamb, and J. P. Engle, *Power Eng.*, 68 (Nov. 1973).
121. Jpn. Kokai 75 47,457 (Apr. 26, 1975), Y. Kudo and co-workers (to Mitsubishi Heavy Industries, Ltd.).
122. S. Arrington and G. Bradley, "Service Water System Cleaning with Ammoniated Citric Acid," paper presented at *Corrosion 87-NACE*, No. 387, San Francisco, Calif., 1987.
123. U.S. Pat. 4,190,463 (Feb. 26, 1980), R. I. Kaplan (to Nalco Chemical Co.).
124. U.S. Pat. 4,540,443 (Sept. 10, 1985), A. G. Barber (to Union Carbide Corp.).
125. C. A. Poulos, *Mater. Perform.* **23**(8), 19–21 (1984).
126. J. R. Gatewood and co-workers, *Mater. Perform.* **18**(7), 9–14 (1979).
127. R. T. Johansen, J. P. Powell, and H. N. Dunning, *U.S. Bur. Mines Inform. Circ.*, 7797 (1957).
128. *Pfizer Products for Petroleum Production*, Technical Bulletin No. 97, Pfizer Inc., New York, 1961.
129. U.S. Pat. 3,335,793 (Aug. 15, 1967), J. W. Biles and J. A. King (to Cities Service Oil Co.).
130. U.S. Pat. 3,402,137 (Sept. 17, 1968), P. W. Fischer and J. P. Gallus (to Union Oil Co. of California).
131. U.S. Pat. 3,732,927 (May 15, 1973), E. A. Richardson (to Shell Oil Co.).
132. U.S. Pat. 4,447,364 (May 8, 1984), P. W. Staal (to Miles Laboratories, Inc.).
133. C. W. Crowe, *J. Petrol. Technol.* 691–695, (Apr. 1985).
134. U.S. Pat. 4,151,098 (Apr. 24, 1979), W. R. Dill, J. A. Knox (to Halliburton Co.).
135. U.S. Pat. 3,952,806 (Apr. 27, 1976), J. C. Trantham (to Phillips Petroleum Co.).
136. J. C. Mack and M. L. Duvall, "Performance and Economics of Minnelusa Polymer Floods," paper presented at the *1984 Rocky Mountain Regional Meeting of Society of Petroleum Engineers*, Casper, Wyo., May 21–23, 1984.
137. U.S. Pat. 4,432,962 (Feb. 21, 1984), H. W. Gowdy and D. M. Fenton (to Union Oil of California).
138. D. R. George, L. Crocker, and J. B. Rosenbaum, *Min. Eng.* **22**(1), 75 (1970).

139. J. B. Rosenbaum, D. R. George, and L. Crocker, "The Citrate Process for Removing SO2 and Recovering Sulfur from Waste Gases," paper presented at the *AIME Environmental Quality Conference*, Washington, D.C., June 7–9, 1971.

140. J. B. Rosenbaum and co-workers, *Sulfur Dioxide Emission Control by Hydrogen Sulfide Reaction in Aqueous Solution—The Citrate System*, PB221914/5, National Technical Information Service, Springfield, Va., 1973.

141. L. Korosy and co-workers, *Adv. Chem. Ser.* **139**, 192 (1975).

142. U.S. Pat. 3,757,488 (Sept. 11, 1973), R. R. Austin and A. L. Vincent (to International Telephone and Telegraph Corp.).

143. U.S. Pat. 3,933,994 (Jan. 20, 1976), G. L. Rounds (to Kaiser Steel Corp.).

144. Ger. Pat. 2,432,749 (Jan. 30, 1975), W. J. Balfanz, R. M. DePirro, and L. P. Van Brocklin (to Stauffer Chemical Co.).

145. T. Wasag, J. Galka, and M. Fraczak, *Ochr. Powietrza* **9**(3), 72 (1975).

146. P. M. Bever and G. E. Klinzing, *Environ. Prog.* **4**(1)1–6 (Feb. 1985).

147. B. K. Dutta and co-workers, *Ind. Eng. Chem. Res.* **26**(7), 1291–1296 (1987).

148. U.S. Pat. 4,042,666 (Aug. 16, 1977), H. L. Rice and R. A. Wilkins (to Petrochemicals Co., Inc.).

149. U.S. Pat. 3,663,284 (May 16, 1972), D. J. Stanicoff and H. J. Witt (to Marine Colloids, Inc.).

150. U.S. Pat. 2,952,580 (Sept. 13, 1960), J. Frasch.

151. U.S. Pat. 3,029,153 (Apr. 10, 1962) K. L. Hackley, (to Champion Papers, Inc.).

152. U.S. Pat. 3,245,816 (Apr. 12, 1966), H. C. Schwaibe (to Mead Corp.).

153. V. Laskova, *Pap. Cellul.* **30**(7–8), 173 (1975).

154. U.S. Pat. 2,336,728 (Dec. 14, 1943), H. W. Hall (to the Dicalite Co.).

155. U.S. Pat. 4,144,083 (Mar. 13, 1979) W. F. Abercrombie, Jr. (to J. M. Huber Corp.).

156. U.S. Pat. 4,309,222 (Jan. 5, 1982) H. L. Hoyt, 4th (to Pfizer, Inc.).

157. C. W. Smith and C. B. Munton, *Met. Finish.* **39**, 415 (1941).

158. A. W. Hothersall, "The Adhesion of Electrodeposited Nickel to Brass," paper presented to *Electroplaters'* and Depositors' Technical Society, Northampton Polytechnic Institute, London, UK, May 18, 1932.

159. Fr. Pat. 813,548 (June 3, 1937) (to The Mound Co., Ltd.).

160. U.S. Pat. 2,474,092 (June 21, 1949), A. W. Liger (to Battelle Development Corp.).

161. C. W. Fleetwood and L. F. Yntema, *Ind. Eng. Chem.* **27**, 340 (1935).

162. J. Kashima and F. Fuhushima, *J. Electrochem. Assoc. Jpn.* **15**, 33 (1947).

163. W. E. Clark and M. L. Hold, *J. Electrochem. Soc.* **94**, 244 (1948).

164. N. N. S. Siddhanta, *J. Indian Chem. Soc. Ind News Ed.* **14**, 6 (1951).

165. C. N. Shen and H. P. Chung, *Chin. Sci.* **2**, 329 (1951).

166. K. G. Sodeberg and H. L. Pinkerton, *Plating* **37**, 254 (1930).

167. W. E. Brodt and L. R. Taylor, *Trans. Electrochem. Soc.* **73** (1938).

168. L. F. Yntema, *J. Am. Chem. Soc.* **54**, 3775 (1932).

169. U.S. Pat. 2,599,178 (June 3, 1952), M. L. Holt and H. J. Seim (to Wisconsin Aluminum Research Foundation).

170. Can. Pat. 443,256 (July 29, 1947), E. M. Wise and R. F. Vines (to The International Nickel Company of Canada, Ltd.).

171. L. E. Netherton and M. L. Holt, *J. Electrochem. Soc.* **95**, 324 (1949).

172. *Ibid.* 98, 106 (1951).

173. Ital. Pat. 444,078 (Jan. 12, 1949), A. Sacco and M. Gandusi.

174. K. S. Rajam, *Met. Finish.*, 41–45 (Oct. 1990).

175. U.S. Pat. 4,371,573 (Feb. 1, 1983), H. Januschkowetz and H. Laub (to Siemens AG).

176. *J. Am. Concr. Inst.* **60**(11), (Nov. 1963).

177. U.S. Pat. 2,174,051 (Sept. 26, 1939), K. Winkler.

178. U.S. Pat. 3,656,985 (Apr. 18, 1972), B. Bonnel and C. Hovasse (to Progil, France).

179. U.S. Pat. 2,542,364 (Feb. 20, 1951), F. A. Schenker and A. Ammann (to Kaspar Winkler Cie).

180. F. Tamas, NASNRC Publ. **1389**, 392 (1966).

181. F. Tamas and G. Liptay, *Proc. Conf. Silicate Ind.* 8,299 (1965).

182. E. C. Combe and D. C. Smith, *J. Appl. Chem. (London)* **16**(3), 73 (1966).

183. *Pfizer Organic Chelating Agents*, Technical Bulletin No. 32, Pfizer Inc., Chemicals Division, New York, 1972.

184. U.S. Pat. 4,004,066 (Jan. 18, 1977), A. J. DeArdo (to Aluminum Company of America).

185. N. B. Singh and co-workers, *Cem. Concr. Res.* **16**, 911–920 (1986).

186. U.S. Pat. 3,754,860 (Aug. 28, 1973), J. G. Frick Jr., and co-workers (to U.S. Secretary of Agriculture).

187. U.S. Pat. 3,212,928 (Oct. 19, 1965), H. R. Hushebeck (to Joseph Bancroft and Sons Co.).

188. D. D. Gagliardi and F. B. Shippee, *Am. Dyestuff Rep.* **52**, 300 (1963).

189. L. Benisek, *J. Soc. Dyers Colour.*, 277 (Aug. 1971).

190. "Mordanting of Wool: A Dyeing Technique to Increase the Flame Resistance of Wool Shag Carpets," *Wool Facts*, Vol. 1, No. 1, Wool Bureau Inc., New York, privately presented, Wool Bureau Technical Center, Woodbury, N.Y., 1971.

191. "Mordanting Wool with Zirconium," *Wool Facts*, Vol. 1, No. 5, Wool Bureau, Inc., New York. 1971.

192. R. R. Haynes, J. H. Mathews, and G. A. Heath, *Text. Chem. Color.* **1**(3), 16/74 (1969).

193. U.S. Pat. 2,720,441 (Oct. 11, 1955), J. G. Wallace (to E. I. du Pont de Nemours & Co., Inc.).

194. U.S. Pat. 2,898,179 (Aug. 4, 1959), C. J. Rogers (to E. I. du Pont de Nemours & Co., Inc.).

195. *Replacement of Phosphates with Citric Acid in Nylon Carpet Dyeing*, Information Sheet No. 2025, Pfizer Chemicals Division, New York, 1973.

196. B. J. Harper and co-workers, *Text. Chem. Color.* **3**(5), 65/127–70/132 (May 1971).

197. B. A. Kottes Andrews and R. M. Reinhardt, "How Mixed Catalysts Differ," paper presented at the *Fourth Annual Natural Fibers Textile Conference*, New Orleans, La., Sept. 14–16, 1981.

198. Neth. Appl. 6,605,358 (Oct. 24, 1966) (to Koppers Co.).

199. U.S. Pat. 2,950,263 (Aug. 23, 1960), W. Abbotson, R. Hurd, and H. Jackson (to Imperial Chemical Ind., Ltd.).

200. U.S. Pat. 3,185,588 (May 25, 1965), J. Y. Resnick (to Int Res. & Dev. Co., New York).

201. U.S. Pat. 4,016,110 (Apr. 5, 1977), W. E. Cohrs and co-workers (to The Dow Chemical Company).

202. V. G. Kharakhash, *Plast. Massy* **6**, 40 (1971).

203. U.S. Pat. 3,523,988 (Aug. 11, 1970), Z. M. Roehr, R. Berger, and P. A. Plasse (to Roehr Metals & Plastics Co.).

204. U.S. Pat. 3,482,006 (Dec. 2, 1969), F. A. Carlson (to Mobil Oil Corp.).

205. U.S. Pat. 3,069,367 (Dec. 18, 1962), R. D. Beaulieu, P. N. Speros, and D. A. Popielski (to Monsanto Chemical Co.).

206. Ger. Pat. 11,144,911 (Mar. 7, 1963), F. Stastny, B. Ikert, and E. F. v. Behr (to Badische Anilin-und Soda-Fabrik).

207. U.S. Pat. 3,660,534 (May 25, 1972), F. E. Carrock and K. W. Ackerman (to Dart Industries).

208. B. C. Mitra and S. R. Palit, *J. Indian Chem. Soc.* **50**(2), 141 (1973).

209. Jpn. Pat. 72 28,766 (July 29, 1972), K. Okuno, K. Itagaki, and H. Takashi (to Mitsubishi Chemical Industries Co.).

210. J. Jarusek and J. Mleziva, *Chem. Prumyst* **16**(11), 671 (1966).

211. E. H. Hull and K. K. Mathur, *Mod. Plast.*, 66–70 (May 1984).

212. U.S. Pat. 3,674,619 (July 4, 1972), H. I. Scher and I. S. Ungar (to Esso Research Engineering Co.).

213. Ger. Pat. 1,269,874 (June 6, 1968), R. H. McKillip and R. Henderson (to Olin Mathieson Chemical Corp.).

214. Brit. Pat. 1,079,762 (June 4, 1969), I. R. Horne and R. N. Lewis (to Bakelite Ltd.).

215. *Ind. Eng. Chem.* **53**(1), 28A (1961).

216. C. O. Jensen, *Chemical Changes During the Curing of Cigar Leaf Tobacco*, The Pennsylvania State College, State College, Pa., Sept. 1951.

217. L. MacDonald, *Am. Perfum.* **76**(7), 22 (1961).

218. M. Ash and I. Ash, *Formulary of Cosmetic Preparations*, Chemical Publishing Co. Inc., New York, 1977.

219. L. Smith and M. Weinstein, *Household Pers. Prod. Ind.* **14**(10), 54 (1977).

220. U.S. Pat. 3,718,236 (Feb. 27, 1973), E. M. Reyner and M. E. Reyner.

221. U.S. Pat. 5,674,903 (Oct. 7, 1997), U.S. Pat. 5,547,988 (Aug. 20, 1996), R. J. Yu and E. J. van Scott.(to Tristrata Technology, Inc.);

222. E. W. Flick, *Cosmetic and Toiletry Formulations*, Noyes Publications, Park Ridge, N.J., 1984.

223. U.S. Pat. 3,333,972 (Aug. 1, 1967), J. T. Elmer and B. G. Atlman (to Kaiser Aluminum & Chemical Corp.).

224. Aust. Pat. 259,442 (Jan. 10, 1968), T. Chvatal.

225. Neth. Appl. 6,409,880 (Mar. 1, 1965), (to Sprague Electric Co.).

226. Fr. Appl. 2,172,160 (Nov. 2, 1973), F. Sembera.

227. Jpn. Pat. 75 03,020 (May 15, 1973), S. Sugiyama, Y. Heima, and M. Sugi (to Toshiba Mach. Co. Ltd.).

228. *Code of Federal Regulations*, Title 21, §175.105, 175.300, 175.320, 175,380, 175.390, 176.170, 177.1210, 178.3910, 181.27(acetyl triethyl citrate and acetyl tri-*n*-butyl citrate); 175.105, 175.300, 175.320, 175.380, 175.390, 176.170, 177.1210, 181.27 (triethyl citrate), 175.105 (tri-*n*-butyl citrate), U.S. Government Printing Office, Washington, D.C., 1990.

229. *Code of Federal Regulations*, Title 21,§182.1911(triethyl citrate), U.S. Government Printing Office, Washington, D.C., 1990.

230. R. L. Hall and B. L. Oser, *Food Technol.* **19**(2), 151 (1965).

231. *Code of Federal Regulations*, Title 21, §182.6386(isopropyl citrate), 182.6511 (mono-isopropyl citrate), U.S. Government Printing Office, Washington, D.C., 1990.

232. *Code of Federal Regulations*, Title 21, §182.6851, U.S. Government Printing Office, Washington, D.C., 1990.

REBECCA LOPEZ-GARCIA
Tate & Lyle

CLAYS, SURVEY

1. Introduction

The term "clay" is somewhat ambiguous unless specifically defined because it is used in three ways: (*1*) as a diverse group of fine-grained minerals, (*2*) as a rock term, and (*3*) as a particle size term. Clay is a natural earthy, fine-grained material comprised largely of a group of crystalline minerals known as the clay minerals. These minerals are hydrous silicates composed mainly of silica, alumina, and water. Several of the clay minerals also contain appreciable quantities of iron, alkalies, and alkaline earths. Many definitions include the statement that clay is plastic when wet, which is true because most clays do have this property but some clays are not plastic, eg, flint clays and halloysite (1). As a rock term many authors use the term clay for any fine-grained natural earthy argillaceous material (2) that would include shale, argillite, and some argillaceous soils. As a particle size term, clay is used for the category that includes the smallest particles. Soil scientists and mineralogists generally use 2 μ as the maximum particle size. On the widely used Wentworth scale, clay is any material finer than 4 μ (3). Even though there is no universally accepted definition of the term "clay", geologists, agronomists, ceramists, engineers, and others who use the term understand its meaning.

Because of the extremely fine particle size of clays and clay minerals, they require special techniques for identification. The optical microscope generally cannot resolve particles below 5 μ. The most useful instrument for identification and semiquantification of clay minerals is X-ray diffraction (4). Other methods of identification used are, differential thermal analysis (5) electron microscopy (6), and infrared (ir) spectroscopy (7). Identification and quantification of the clay and non-clay minerals present in a clay material are important because the uses and engineering properties are controlled by the specific clay minerals present and the type and quantity of non-clay mineral present. As pointed out by Grim (1), the uses and properties of clay are dependent on the clay mineral composition, non-clay mineral composition, presence of organic material, the type and amount of exchangeable ions and soluble salts, and texture. The major clay mineral groups are kaolins, smectites, illites, chlorites, and palygoskite-sepiolite.

Clay is an abundant natural raw material that has an amazing variety of uses and physical properties. Clays are among the leading industrial minerals in both tonnage produced and total value. Clays and clay minerals are important to industry, agriculture, geology, environmental applications, and construction. Clay minerals are important indicators in petroleum and metallic ore exploration and in reconstructing the geological history of deposits. However, clay minerals may be deleterious in aggregates and in oil reservoirs.

There are several scientific groups whose work is devoted to the study of clays and clay minerals. These include AIPEA (Association Internationale pour le Etude des Argile), Clay Minerals Group of the Mineralogical Society of Great Britain, CMS (The Clay Minerals Society), European Clay Minerals Society, Clay Minerals Society of Japan, Australian Clay Minerals Society, and Czech National Clay Group. Major publications include the *Proceedings of International Clay*

Conferences published by AIPEA; *Clay Minerals* published by the Mineralogical Society of Great Britain; *Clays and Clay Minerals* published by CMS: *Clay Mineral Science*, published by the Clay Society of Japan; and *Applied Clay Science* published by Elsevier. CMS sponsors an annual clay conference; The European Clay Minerals Society sponsors a meeting every 2 years; and AIPEA holds an international clay conference every 4 years.

2. Occurrence and Geology of Major Clay Deposits

Some clay deposits are comprised of relatively pure concentrations of a particular clay mineral and others are mixtures of clay minerals. Kaolins, smectites, and palygorskite-sepiolite can occur in relatively pure concentrations whereas illite and chlorite usually occur in mixtures of clay minerals and non-clay minerals such as shales, which are the most common sedimentary rock. Each of these will be discussed under the headings kaolins, smectites, palygorskite-sepiolite, and common clays.

2.1. Kaolins. Kaolins are hydrated aluminum silicates. Kaolin is a mineral group consisting of the minerals kaolinite, dickite, nacrite, and halloysite. Kaolinite is the most common of the kaolin group minerals. The term kaolin is derived from the Chinese word kauling, which in the Chinese language means high ridge. At Jauchian Fu in China, kaolin was mined for centuries from a hill or ridge. The term china clay is synonymous with kaolin and has two connotations. One refers to the use of kaolin in fine china ceramic products and the other makes reference to the fact that kaolin was first mined in China and thus is called china clay.

Kaolin, in addition to being a group mineral name, is a rock term and is used for any rock that is comprised predominantly of one of the kaolin minerals. Kaolin deposits are classed as primary or secondary (8). Primary or residual kaolins are those that have formed by the alteration of aluminous crystalline rocks such as granite and remain in the place where they formed. The alteration results from surficial weathering, groundwater movement below the surface, or action of hydrothermal fluids. Secondary kaolins are sedimentary rocks that were eroded, transported, and deposited as beds or lenses in association with other sedimentary rocks. Most secondary or sedimentary kaolins were deposited in deltaic environments.

The most common parent minerals from which kaolin minerals form are feldspars and muscovite. The transformation of potassium feldspar into kaolinite results from weathering or hydrothermal alteration, which leaches potassium and silica according to the equation:

$$\underset{\text{Potash Feldspar}}{2KAlSi_3O_8} + \underset{\text{Water}}{3H_2O} \rightarrow Al_2Si_2O_5(OH)_4 + \underset{\text{Silica}}{4SiO_2} + \underset{\text{Potash}}{2K(OH)}$$

The potassium must be flushed out of the system in order for kaolinite to form, otherwise the mineral illite may form. Granites and rhyolites (Fig. 1) weather readily to kaolinite and quartz under favorable conditions of high rainfall, rapid drainage, temperate to tropical climate, a low water table, and

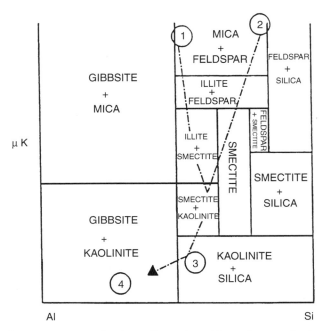

Fig. 1. Phase diagram on weathering of granite. Alteration products of mica (1) and feldspar (2) are smectite and kaolinite and (3) to (4) results from intense weathering in tropical climates.

adequate water movement to leach the soluble components. The more immobile components are alumina and silica, whereas alkalies and alkaline earths are mobile (9). Plagioclose feldspars are relatively unstable and alter before either potash feldspar or muscovite (10).

Kaolinite is the dominant kaolin mineral in secondary or sedimentary deposits. Dickite, nacrite, and halloysite are more commonly found in hydrothermal deposits (8). Kaolin deposits occur on every continent but relatively few are mined and processed for industrial use. Some of the more important and largest deposits are described and discussed below.

The largest kaolin-producing district in the world is located in Georgia and South Carolina in the United States. These deposits are late Cretaceous and early Tertiary in age (11) and occur in lenses and beds in relatively coarse sands. The kaolin was derived from granitic rocks, phyllites, and schists on the Piedmont Plateau (12). The deposits are concentrated in a belt ~30 km wide extending from central Georgia northeast into South Carolina (Fig. 2). The deposits are located south of the fall line, which is the boundary between the crystalline rocks of the Piedmont Plateau and the Coastal Plain sedimentary rocks.

These mineable deposits range in thickness from 3 to 15 m and extend laterally from a few hundred meters to kilometers in length and width. The Cretaceous age kaolins are relatively coarse with a particle size ranging between 55 and 70% finer than two micrometers whereas the early Tertiary kaolins range from 85 to 95% finer than 2 μ. The relatively coarse Cretaceous kaolins are called soft

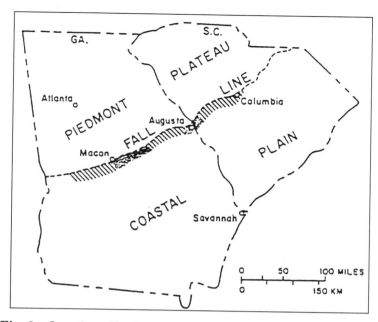

Fig. 2.　Location of kaolin deposits in Georgia and South Carolina.

kaolin and the fine Tertiary kaolins are called hard kaolin (8). The Georgia and South Carolina kaolins contain ~85–95% kaolinite. Other minerals present in these Georgia and South Carolina kaolins are quartz, muscovite, biotite, smectite, ilmenite, anatase, rutile, leucoxene, hematite, goethite, and trace amounts of zircon, tourmaline, kyanite, and graphite.

Another region where large sedimentary deposits of kaolin occur is in the lower Amazon area of Brazil (13). A large deposit is mined adjacent to Jari River on the border of the states of Para and Amapa (13). This deposit is fine in particle size similar to the Tertiary deposits in Georgia and South Carolina. This deposit is 40 m thick and extends over an area of 60 km^2. This fine particle kaolin was derived from crystalline rocks on the Guyana shield and is Pliocene in age. Kaolinite is the major mineral present ranging between 80 and 98%. Other minerals present are quartz, gibbsite, rutile, anatase, tourmaline, zircon, and goethite.

Another district in Brazil where kaolin deposits occur is in the Rio Capim basin located south of the city of Belem, which is the major port city on the lower Amazon River. These Capim kaolins are relatively coarse, similar to the Cretaceous kaolins in Georgia. These kaolin deposits are associated with low lying plateaus in the Rio Capim basin and are believed to be Tertiary in age. These coarse kaolins are excellent in quality and the kaolinite content ranges from 65 to 80% with quartz as the predominant non-clay mineral present. Minor quantities of anatase, rutile, zircon, tourmaline, and goethite are also present. It is believed that the Rio Capim kaolins were derived from granites and gneisses northeast of the Capim Basin. The Capim kaolin deposits are in the state of Para (Fig. 3).

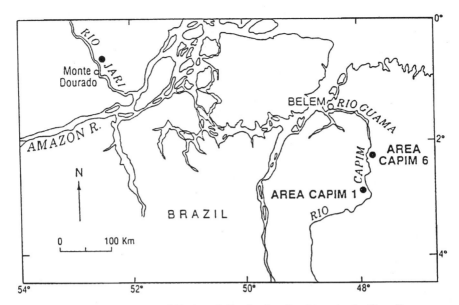

Fig. 3.   Location of Jari and Capim kaolin deposits in Brazil.

Large primary kaolin deposits occur in the Cornwall district of southwest England. These kaolins were formed by the alteration of large granite bodies by warm acidic hydrothermal solutions migrating upward along faults, fractures, and joints (14). These solutions altered the feldspars in the granite leaving quartz, mica, and tourmaline relatively unaltered. Superimposed on the hydro-thermally altered granite is additional alteration by surface weathering (15). Drilling has shown that kaolinization exists at depths >250 m. The kaolinite content of these altered granites ranges between 10 and 20%. The kaolinite is recovered by hydraulic methods that washes the kaolinite from the soft granite leaving quartz, mica, unaltered feldspar, and other minerals in the residue.

The Czech Republic is the principal producer of kaolin in Eastern Europe. The majority of the deposits are located in the vicinity of Karlovy-Vary, Pilzen, and Podborany. The kaolin primarily occurs as a weathering crust on crystalline rocks of the Bohemian Massif (16). The crystalline rocks where the kaolin deposits occur are granites and gneisses in which the feldspar is almost completely kaolinized. The major non-clay minerals are quartz and mica.

Both primary and secondary deposits are mined in eastern Germany. Near Dresden are two large primary deposits at Caminau and Kemmlitz (17). These kaolins are similar to those in the Czech Republic and are the result of the weathering of the Bohemian Massif, which extends into this area of Germany. The largest secondary deposit is in Bavaria near the village of Hirschau. Kaolin is the altered product of feldspar contained in a Triassic age arkose (18). The deposit extends for ~15 km and ranges in thickness from 3 to 8 m. The kaolinite content in this arkosic sand is ~10% ± 5, with quartz being the dominant non-clay mineral present.

Kaolin occurs in Spain in two areas, a primary kaolin is mined near Galicia in northwest Spain and a secondary kaolinitic sand is mined in the state of Guadalajara east of Madrid (19). This secondary kaolin is a minor constituent (10–20%) of a friable white sand used for making glass.

Near Proyanovski in Ukraine there is a large primary kaolin deposit (13). The major non-clay minerals are quartz and mica. There is also some smectite in the lower portion of the deposit. In some areas of the deposit it is 30 m thick.

Both primary and secondary deposits of kaolin occur in Australia (20). The largest primary deposits are near Pittong, north of Melbourne, where the kaolin is a residual weathering product of granite. The major non-clay minerals are quartz and muscovite. The sedimentary deposits of probable early Tertiary age occur below the extensive bauxite deposits on the Cape York Peninsula in northeastern Australia. The kaolinitic sand, which contains from 40 to 60% quartz, is located on the western margin of the Cape York Peninsula near Weipa and Skardon River.

Several kaolin deposits are mined in China. Both primary and secondary deposits occur mainly in south China (21). The largest primary kaolin deposit is located in the Suzhou district southwest of Shanghai and is the result of both hydrothermal and weathering alteration of granites and other acid igneous rocks. This kaolin contains both kaolinite and halloysite along with non-clay minerals quartz, mica, alunite, smectite, and pyrite. Another primary kaolin deposit is located in the Longyan region in southern Fujian Province. This residual weathered kaolin is comprised of kaolinte and halloysite along with some illite and small amounts of quartz, feldspar, and mica. A large secondary kaolinitic sand deposit is located at Maoming in southwestern Guanzhou province (22). The kaolinitic sand is comprised of about 80% quartz and 20% kaolinite.

Other kaolin deposits of local importance are located in Argentina, Brazil, Chile, Egypt, Indonesia, India, Korea, Mexico, Portugal, Tanzania, South Africa, Suriname, and Venezuela (11,23).

Halloysite, a kaolin group mineral, has an elongate, tubular shape and has a higher degree of hydration than kaolinite. The only halloysite deposit currently mined is located in New Zealand ~240 km north of Auckland. The deposits were formed by hydrothermal alteration of rhyolite volcanics of Pliocene or Pleistocene age (24). The other kaolin group minerals dickite and nacrite are relatively rare. There is one hydrothermal deposit in Japan with lenses of pure dickite at the Shokozan mine in the Hiroshima prefecture (25). Relatively pure nacrite lenses were reported in one deposit in Nyarit, Mexico (26). Generally, halloysite, dickite, and nacrite occur in minor quantities mainly in hydrothermal deposits although they can also occur in some secondary deposits.

There are special types of kaolinitic clays that are used primarily for ceramics. One of these types is ball clays. Ball clays are secondary and are characterized by the presence of organic matter, high plasticity, high dry and fired strength, long vitrification ranges, and light color when fired. Kaolinite is the principal mineral constituent of ball clay and typically comprises 70% or more of the minerals present. Other minerals commonly present are quartz, illite, smectite and feldspar, as well as lignitic material. The three largest areas of ball clay occurrences are in western Tennessee and Kentucky in the United

States (27), southwest England near Devonshire (28) and Germany (18). A second type of secondary kaolinitic clays used for ceramics is called refractory clays or fire clays. These kaolin clays are used in the manufacture of products requiring resistance to high temperatures (27). Major areas of occurrence of refractory clays are Missouri, Kentucky, Ohio, and Pennsylvania in the United States (27), Australia (20), Argentina, Mexico, China, and India. The refractory clays in Missouri and Kentucky are called flint clays, which are dense, brittle, fine particle size kaolinite (29). Flint clays also occur in Australia and China.

2.2. Smectites.   Smectite is the name for a group of sodium, calcium, magnesium, iron, and lithium aluminum silcates. The group includes the specific clay minerals montmorillonite, saponite (magnesium smectite), nontronite (iron smectite), beidellite and hectorite (lithium smectite). Bentonite is the rock in which these smectite minerals are usually the dominant constituent. The term bentonite was first suggested by Knight (30) and is the term used today to describe the industrial minerals in which a sodium or calcium montmorillonite is the major mineral constituent. The term bentonite was defined by Ross and Shannon (31), who restricted the term to a clay material altered from a glassy igneous material usually volcanic tuff or glass. Wright (32) suggested that bentonite was any clay dominantly comprised of a smectite clay mineral and whose physical properties are dependent on this clay mineral. Grim and Guven (33) used Wright's definition and today some authors use Wright's definition and others use Ross and Shannon's.

Bentonites in which the smectite sodium montmorillonite is the major mineral component normally have a high swelling capacity (34). The best quality and largest sodium bentonite deposits are located in South Dakota, Wyoming, and Montana (35). These high swelling clays are sometimes referred to as Western or Wyoming bentonites. The bentonite beds are Cretaceous age and are in the Mowry formation (35). There are several bentonite layers in the Mowry and the thickest and most extensive is called The Clay Spur Bentonite Bed (Fig. 4), which extends west from Belle Fourche, S.D. across Wyoming to Greybull and then north across Montana into Alberta, Canada. A thickness of 4 m has been reported (36). The sodium bentonite in the Mowry formation was formed by the alteration of volcanic ash. The major mineral is sodium montmorillonite and the non-clay minerals are quartz, opal CT, zeolite, feldspar, and mica (31). Locally secondary calcite and gypsum may also be present.

Bentonites in which calcium montmorillonite is the major mineral component commonly have a low swelling capacity (34). These calcium bentonites are sometimes referred to as southern or sub-bentonites. Large deposits of calcium bentonites occur in Mississippi and Texas (33). In Mississippi the calcium bentonite is in the Eutaw formation of Upper Cretaceous age (37). It is mined in Itawamba and Monroe Counties where it is as much as 4 m thick. Quartz, feldspar, and mica are the major non-clay minerals present. It is presumed that this bentonite is an altered volcanic ash that was deposited from major volcanic eruptions in early Cretaceous time in central Mississippi. Another calcium bentonite is mined in Tippah County near Ripley from the Paleocene Porters Creek formation for its absorbent properties (38). The Texas calcium bentonites occur in a belt, which parallels the present Gulf Coast in late Cretaceous and Tertiary sediments. They are best developed in the Jackson and Gueydan formations of

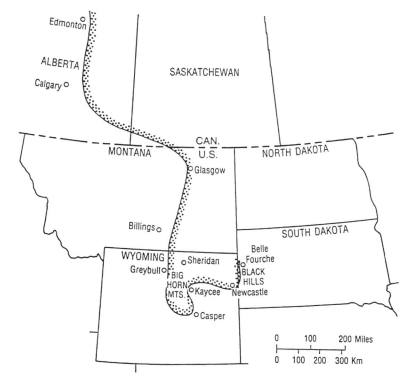

Fig. 4. Location of the clay spur bentonite bed.

Tertiary age in Gonzales and Fayette Counties, which are directly east of San Antonio (39). The calcium bentonites in these two counties range from 0.5 to 3 m in thickness. The color of these bentonites varies from white to yellowish green to dark brown almost chocolate colored. Some of the white bentonite in Gonzales County contains kaolinite (40). The major non-clay minerals are quartz, opal CT, feldspar, and mica. These bentonites are reworked volcanic ash that was deposited in lagoons and small depressions along the ancient Tertiary shoreline.

Other calcium bentonites in the United States occur in Alabama, Arizona, and Nevada. The best known and most extensively mined deposit is near Cheto, Arizona. At this location, a 5 m thick ash bed in the Pliocene Bidahochi formation has altered to calcium montmorillonite (41). This bentonite is very light gray and contains a small percentage of kaolinite. Quartz, mica, feldspar, and opal CT are the non-clay impurities. In Nevada, there is a calcium bentonite in the Amargosa Valley ~100 miles north of Las Vegas (42). This bentonite is Pliocene in age and is white to light gray. The non-clay minerals in this clay include quartz, mica, zeolite, opal CT, calcite, and dolomite. A calcium bentonite is mined at Sandy Ridge in Alabama from the Ripley formation (43).

Calcium bentonites are produced in many other countries including England, Germany, Italy, Greece, Hungary, Republic of Georgia, India, Argentina,

Japan, China, Malaysia, and Brazil. The more important of these are the Redhill Surrey deposits in England of Upper Cretaceous age (44); the German bentonites north of Munich (45); the calcium bentonite on the Island of Sardinia in Italy of Miocene age (46); the Pliocene calcium bentonite on the island of Milos in Greece (33); the Askana bentonite deposit in the Republic of Georgia near Tbilisi (33); the bentonite deposit near Neuquen in Argentina (13); and the lower tertiary age bentonite in The Barmer district of Rajasthan near Akli in India (47).

Some less common bentonites including hectorite and saponite are mined in the United States. Hectorite, a lithium montmorillonite is mined near Hector, California and is hydrothermal in origin. It was formed where siliceous lithium and fluorine solutions reacted with carbonates (48). Saponite, the magnesium montmorillonite, is mined in the Ash Meadows area in the Amargosa Valley in Nevada. The saponite is Pliocene in age (42). The most comprehensive description of many of the bentonite deposits of the world may be found in the book by Grim and Guven (33).

2.3. Palygorskite and Sepiolite.   Hormite is a group name that has been used for the minerals palygorskite and sepiolite. However, the International Nomenclature Committee (INC) has not yet accepted this name. Attapulgite and palygorskite are synonymous but the name palygorskite was used first so the INC discourages the uses of attapulgite as a mineral name. Palygorskite and sepiolite are hydrated magnesium aluminum silicates that have an elongate shape. Because of their large surface area, they are sometimes referred to as sorptive clays called Fuller's earth. Fuller's earth is a term used for clays and other fine particle size earthy materials suitable for use as sorbent clays and bleaching earths. The term was first used for the earthy material used in cleansing and fulling wool, thereby removing the lanolin and dirt, so this clay acquired the name Fuller's earth (49).

Palygorskite deposits are relatively few in number. Deposits are located in the United States near the Georgia–Florida border in the vicinity of Quincy, Florida and Attapulgus, Georgia; in Senegal east of Dakar; in China at Mingguang in Anhui Province and in Xuyi in Jiangsu Province, and in Ukraine (18). Sepiolite deposits are located in Spain, Somalia, and Turkey. Sepiolite is very similar to palygorskite but normally has slightly higher magnesium content and a somewhat different crystal structure.

The palygorskites in Southeastern United States are Miocene age and were deposited in marine lagoons (50). The deposits range in thickness from 0.5 to 3 m or more and are gray or tannish brown in color. These deposits are mixtures of palygorskite and smectite with the palygorskite content highest in the southern part of the district where the deposits are predominantly palygorskite. In the northern part of the district near Meigs, Georgia (51), the deposits are approximately evenly both smectite and palygorskite. Minerals other than palygorskite include smectite, quartz, and dolomite.

The Senegal palygorskite (52) deposit is very large extending from Theis south-southwest ~100 km to the Senegal-Barrie border. The deposit is ~50 km west of Dakar and is early Eocene in age and ranges in thickness from 1.5 to 6 m. The color of this clay is light tan to greenish gray. Smectite is a common constituent but is normally present in small amounts. The major impurities are quartz and dolomite.

In China, palygorskite deposits are located near Mingguang in Anhui Province and at Xuyi in Jiangsu Province. At Mingguang, the palygorskite occurs directly under a basalt and ranges from 4 to 6 m in thickness (53). Near Xuyi the palygorskite is very near the surface under a smectite-rich layer. The thickness of the palygoskite is ~6 m. These palygorskites are Middle Miocene in age and the deposits were formed from volcanic ash that fell into a lacustrine environment and altered to palygorskite (53). The palygorskite is exceptionally pure but contains some quartz and smectite impurities.

The palygoskite deposit in Ukraine is in the central part of the Ukrainian crystalline massif along the borders of the Cherkassay and Kiev regions. The palygorskite bed occurs in the middle section of a bentonite and is Lower Miocene in age (54). The thickness of the palygorskite layer is ~2 m. Other mineral constituents are smectite, mica, and quartz.

Sepiolite is mined in Spain near Toledo. The deposit is Tertiary in age and was formed in a lacustrine environment (55). The color of this clay is light tan and ranges in thickness from 1 to 5 m.

In Turkey, sepiolite deposits are found in the vicinity of Eskisehar in Anatolia (56). The sepiolite is white and is interbedded with palygorskite layers.

Some very large deposits of sepiolite occur in Somalia. They are Miocene in age but at present are only used by artists for carving various articles. Sepiolite is mined in small quantities in the Ash Meadows area in the Amargosa Valley in Nevada (57). The sepiolite occurs in 0.3–1 m stringers in saponite clays of lacustrine origin. The sepiolite is brown in color and is Pliocene-Pleistocene in age (42).

2.4. Common Clay.   Common clay includes a variety of clays and shales that are fine in particle size. The clay mineral composition of these common clays is mixed but usually illite and chorite are the most common clay minerals present. Kaolinite and smectites are usually present in smaller quantities. Illite is a hydrated potassium iron aluminum silicate and chlorite is a hydrated magnesium iron aluminum silicate. These common clays are used in many structural clay products such as brick, tile, pottery, stoneware, etc (58).

Common clay occurs in rocks ranging in age from Precambrian to Holocene. They include soils, glacial clays, alluvium, loess, shale, schist, slate, and argillite. The mineral composition of these clays and shales is quite variable. In addition to the clay minerals, quartz is the most common constituent along with mica, feldspar, and many other detrital minerals. Common clays are present in every state in the United States and in every country in the world.

3. Structure and Composition of Clay Minerals

Clay minerals are phyllosilicates, which are sheet structures basically composed of silica tetrahedral layers and alumina octahedral layers. Each clay mineral has a different combination of these layers. Chemical substitutions of aluminum for silicon, iron and/or magnesium for aluminum, etc, generates positive or negative charges in the structure. The clay minerals are crystalline and their identification is determined by X-ray diffraction techniques. Grim (2) proposed a classification of four types. (1) Two layer types consisting of sheet structures composed

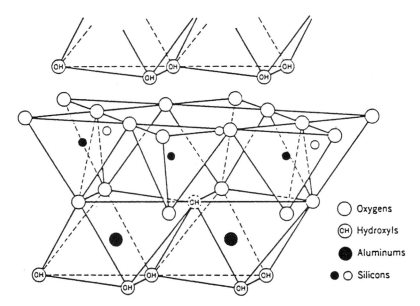

Fig. 5.   Diagrammatic sketch of the structure of kaolinite.

of one layer of silica tetrahedrons and one layer of alumina octahedron (termed "1:1 layer types"). (*2*) Three-layer types consisting of sheet structures composed of two layers of silica tetrahedrons and one central octahedral layer (termed "2:1 layer types"). (*3*) Regular mixed-layer types, which are ordered stacking of alternate silica tetrahedral and alumina octahedrals layers. (*4*) Chain structure type, which are hornblend-like chains of silica tetrahedrons linked together by octahedral groups of oxygen and hydroxyls containing aluminum and magnesium ions.

The kaolin minerals (59) are hydrous aluminum silicates with the composition of $2H_2O \cdot Al_2O_3 \cdot 2Si\ O_2$. Kaolinite is the most common of the kaolin minerals. The structure of kaolinite is a single silica tetrahedral sheet and a single alumina octahedral sheet combined to form the 1:1 kaolinite unit layer (Fig. 5). These unit layers are stacked and are held together by hydrogen bonding. Variations in the orientation of the unit layers in stacking cause the differentiation between kaolinite, dickite, and nacrite (60). Figure 6 is a scanning electron micrograph of kaolinite. Halloysite is an elongate kaolin mineral that has a layer of water between the unit layers so the composition includes $4H_2O$ instead of $2H_2O$. Bates and co-workers (61) proposed that halloysite consists of curved sheets of kaolinite unit layers that cause the elongate tubes as shown in Figure 7.

Smectite structures are comprised of two silica tetrahedral sheets with a central octahedral sheet (Fig. 8). The structure has an unbalanced charge because of the substitution of aluminum for silicon in the tetrahedral sheet and iron and magnesium for aluminum in the octahedral sheet. In order to balance this negative charge, cations accompanied by water molecules enter between the 2:1 layers (Fig. 8). Sodium montmorillonite is a smectite that has sodium ions and water molecules in the interlayer and calcium montmorillonite has calcium ions and water molecules in the interlayer. Figure 9 shows a scan-

Fig. 6.   Scanning electron micrograph of kaolinite from Georgia.

ning electron micrograph of sodium montmorillonite. Nontronite has iron ions in the structure, saponite has magnesium ions in the structure, and hectorite has lithium and magnesium ions in the structure. Smectites expand when water and other polar molecules enter between the layers.

Fig. 7.   Electron micrograph of halloysite.

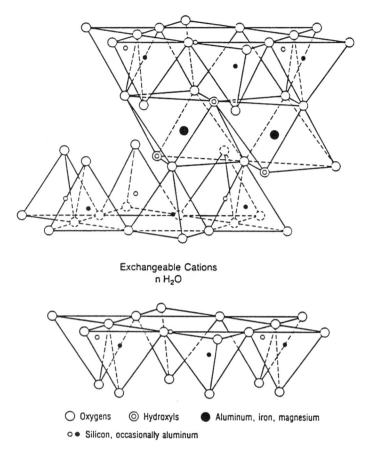

Exchangeable Cations
n H₂O

○ Oxygens   ◎ Hydroxyls   ● Aluminum, iron, magnesium

○ ● Silicon, occasionally aluminum

Fig. 8.   Diagrammatic sketch of the structure of smectite.

Illite is a clay mineral akin to mica (62). The basic structure is a 2:1 layer similar to smectite except that more aluminum ions replace silicon in the tetrahedral sheet, which results in a higher charge deficiency, balanced by potassium ions (Fig. 10). These large diameter potassium ions act as a bridge between the 2:1 layers and so strongly bind them together that illite is nonexpandable. The composition of illite is a potassium aluminum silicate with the general structural formula of $(OH)_4 K_2 (Si_6 AL_2) Al_4 O_{20}$. In this structure, iron and magnesium can substitute for aluminum in the octahedral sheet.

Chlorite is a common clay mineral in shales and the structure consists of alternate silica tetrahedral layers and aluminum octahedral layers (Fig. 11). The octahedral layer has considerable substitution of iron and magnesium for aluminum (63). However, the 2:1 mica sheet octahedral layer may have a different composition than the octahedral brucite layer between the 2:1 mica sheets.

Palygorskite and sepiolite have a chain-like structure, (Fig. 12) that consists of inverted ribbons of silica tetrahedral ribbons linked together by aluminum and magnesium octahedrals. The structural formula of palygorskite is

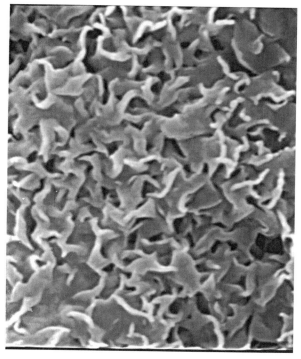

Fig. 9. Electron micrograph of sodium montmorillonite.

$(OH_2)_4 (OH)_2 Mg_5 Si_8 O_{20} \cdot 4 H_2O$. The difference between palygorskite and sepiolite is that sepiolite has a higher content of magnesium and paylgorskite has more aluminum as well as a slightly different crystal structure. Figure 13 is an electron micrograph of palygorskite.

Because most of these layer silicates are all made up of 1:1 and 2:1 layers, there are many possible unit layer mixtures of illite, chlorite, and smectite compositions. These mixed-layer clays are relatively common in occurrence. Illite/smectite, illite/chlorite, chlorite/smectite, and kaolinite/smectite have been described. Mixed-layer clays with both random and regular layering were studied by Reynolds (64) and Moore and Reynolds (4).

4. Mining and Processing

Clays of all types are mined principally by open pit and there are very few underground mines. An understanding of the geology and the origin of clays plays an important role in clay exploration. After a clay deposit is discovered, it must be evaluated to determine thickness, quality and quantity. This may be accomplished by testing either core or auger drilled clay (11). The spacing of the drill holes depends on the geologic and surface conditions associated with the

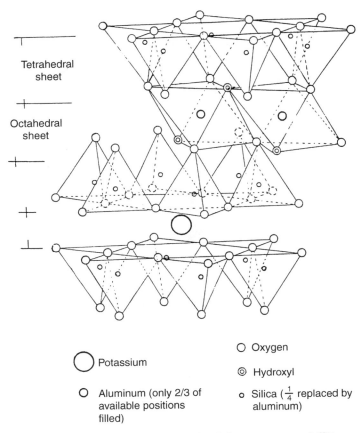

Fig. 10. Diagrammatic sketch of the structure of illite.

particular clay deposit. A drilling pattern used to test a sedimentary clay deposit is very different from the pattern used to evaluate a hydrothermal clay deposit.

Testing the core or auger samples is an important second step in determining the quality of a particular clay deposit. The tests performed to evaluate a kaolin deposit are much different than those performed on a smectite or palygorskite clay deposit. The tests are related to the applications of the clay and to determine the type of processing that will be required to make the final product. General tests that are performed on clays are (1) mineralogy, (2) percent grit (plus 325 mesh or 44 micrometers), and (3) color. Special tests for kaolins are (1) particle size distribution, (2) brightness, (3) low and high shear viscosity, and (4) leach response to improve the brightness. Other special tests also may be performed such as magnetic separation, flotation, selective flocculation, and abrasion. Many of these test procedures have been previously described (65). Special tests for bentonites include ion exchange capacity, viscosity, swelling capacity in water, foundry tests, surface area, water and oil absorption, and bulk density. Special tests for paylgorskite and sepiolite include surface area, exchange capacity, viscosity, water and oil absorption, and bulk density. After drilling and testing a mining plan can be designed for obtaining the maximum quantity of clay.

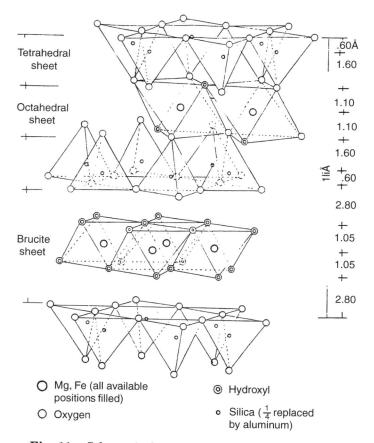

Tetrahedral
sheet

Octahedral
sheet

Brucite
sheet

.60Å
1.60
1.10
1.10
1.60
.60
2.80
1.05
1.05
2.80

11Å

○ Mg, Fe (all available positions filled)

○ Oxygen

◎ Hydroxyl

∘ Silica ($\frac{1}{4}$ replaced by aluminum)

Fig. 11. Schematic diagram of the chlorite structure.

4.1. Kaolin.

Once drilling and testing have determined the quality and quantity of a deposit, the processing that is required to produce a saleable product is determined. Either a dry or wet process accomplishes beneficiation. The higher quality grades of kaolin that are used in the paper, paint, and plastics industries are prepared by wet processing because the product is more uniform, has better brightness and color, and is relatively free of impurities.

Dry Process. The dry process (66) is simple and yields a lower cost and lower quality product than the wet process. In the dry process, the properties of the finished kaolin reflect the quality of the crude kaolin. The general dry process flow sheet is shown on Figure 14. The kaolin is mined by using drag lines, shovels or front end loaders, back hoes and transported in trucks to the processing plant. There it is shredded or crushed to about egg size. After crushing, it is dried, commonly with a rotary drier. After drying, the kaolin is pulverized in a roller mill, hammer mill, disk grinder, or some other grinding device. Commonly heat is applied during the grinding to further reduce the moisture content. The pulverized kaolin is then classified to separate the fine and coarse particles. The finished product can be loaded in bulk bags and shipped by railcar or truck.

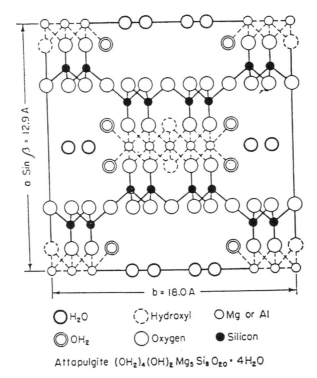

Attapulgite $(OH_2)_4(OH)_2 Mg_5 Si_8 O_{20} \cdot 4H_2O$

Fig. 12. Schematic diagram of the structure of palygorskite, (attapulgite).

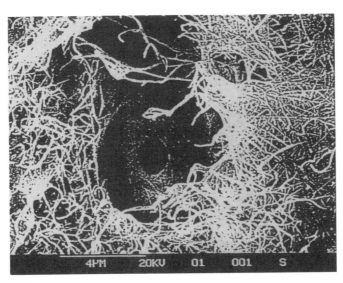

Fig. 13. Elongate particles of palygorskite from Mingguang, China.

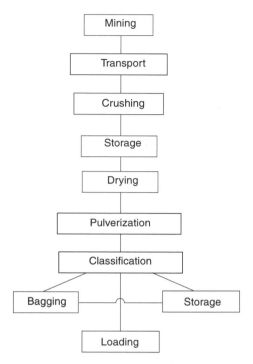

Fig. 14.   General flow sheet for dry processing kaolin.

Wet Process.   The general wet process flow sheet for beneficiating kaolin is shown on Figure 15. The kaolin is mined with shovels, draglines, motorized scrapers, or front-end loaders and can be either transported to the processing plant or fed into a stationary or mobile blunger. The blunger separates the kaolin into small particles, which are mixed with water and a dispersing chemical to form a clay–water slurry. The dispersing chemical can be sodium polyphosphate, sodium silicate, or sodium polyacrylate, each of which can be blended with soda ash to bring the slurry to a neutral pH that optimizes dispersion and reduces cost. The percent solids of the slurry is normally between 30 and 40, but in some special circumstances it can be in excess of 60 and as high as 70 solids. This clay–water slurry is pumped from the blunger to rake classifiers or hydrocyclones and screens to remove the grit (material >325 mesh). The grit removed from the kaolin slurry is discarded into waste ponds or into mined-out areas. The degritted slurry is collected into large storage tanks with agitators and is then pumped to the processing plant, which may be several miles away. The kaolin slurry is collected in large storage tanks at the plant before it is processed. Although each kaolin producer may process the kaolin slightly differently, the steps in the processing are generally as shown in Figure 15. The first step is to separate the kaolin particles into a coarse and fine fraction through continuous centrifuges. The degritted slurry is then passed through a high gradient magnetic separator prior to centrifugation to upgrade the crude clay by removing iron and titanium minerals. High gradient magnetic separation (HGMS) has

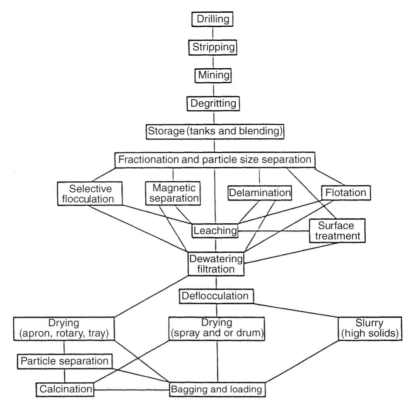

Fig. 15.   Generalized flow sheet for wet processing kaolin.

become a standard processing technique in the kaolin industry. The HGMS process (67) uses a canister filled with a fine stainless steel wool that removes iron, titanium, and some mica minerals from the kaolin slurry as it passes through the canister (Fig. 16). The process can be used to upgrade marginal low brightness crude kaolins and to produce high brightness coating clays. It also reduces the amount of chemicals that are needed to leach out the iron in a subsequent processing step. The development of HGMS has dramatically increased the usable reserves of kaolin in Georgia. Superconducting cryogenic magnets with field strengths up to 5 tesla are now being used in addition to the 2-tesla electromagnets. The superconducting magnets use very little power and are effective in removing ultrafine paramagnetic minerals (68). The coarse fraction can be used as a filler in paper, plastics, paint, and adhesives, as a casting clay for ceramics, or as a feed to delaminators. Delamination, a subsequent processing step to produce a special paper coating grade, is described in a following section. The fine particle size fraction is used in paper coatings, high gloss paints, inks, special ceramics, and rubber. As shown in Figure 15, there are many different steps that can be taken after the clay is fractionated. Normally, the coarse kaolin takes one of two routes. (*1*) It can go directly to the leaching department,

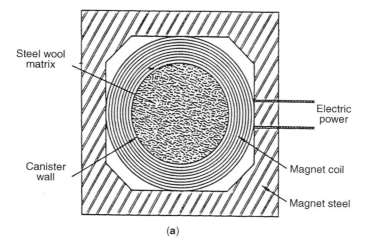

(a)

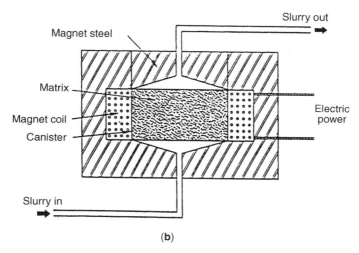

(b)

Fig. 16. Diagrammatic sketch of a high gradient magnetic separator. (**a**) Top view, (**b**) Side view section through magnet.

where it can be chemically treated to solubilize some of the iron if the brightness needs to be upgraded, or it can be flocculated so that the dewatering step is facilitated. (*2*) It can go through magnetic separation and into the delaminators or directly to the delaminators and then to the magnetic separator or to the delaminators without any magnetic separation step. In the leaching operation, the kaolin slurry is acidified with sulfuric acid to a pH of ~3.0 to solubilize some of the colloidal iron. Sodium dithionite, a strong reducing agent, is then added to reduce the iron to iron(II) state, which is more soluble and forms a clear iron sulfate that is removed with the filtrate in the dewatering step. In some cases, the iron in the kaolin is in a reduced state rather than the normal oxidized state. Usually, such kaolins are slightly gray. These gray kaolins are treated with a strong oxidizing agent, such as ozone or sodium hypochlorite, prior to acidification and the normal reduction leach procedure. To facilitate flocculation prior to dewatering, in many cases alum is added along with the sulfuric acid. At this point in the process, the kaolin slurry is between 20 and 30% solids and is flocculated. This flocculated slip is then dewatered, which raises the solids to >60%. The dewatering is accomplished by large rotary vacuum filters or plate and frame filter presses. The dewatered filter cake is then prepared for drying by using rotary, apron, drum, or spray dryers. Most of the clay that is shipped in dry bulk form is spray dried in a dispersed state. The usual process is to take the filter cake, which is flocced, and disperse it by adding a long- chain sodium polyphosphate or sodium polyacrylate.

This dispersed high solids slurry (>60%) is fed into the spray dryer, where it is sprayed or atomized and dried to meet the final moisture specifications, which normally range from <1% to as high as 6%. This dispersed clay slurry can also be dried by rotary drum dryers for some special uses that require the flake form. Slurry clay, which is normally shipped in tank cars or trucks at 70% solids, is produced by evaporaters or mixing the dispersed clay from the filters with enough spray-dried clay to bring the final solids to 70%. About 80% of all shipments of coating and fine filler clays to the paper industry is now being shipped in slurry form. If the filter cake is to be dried with rotary or apron dryers, then the filter cake is extruded into noodles ~3/8 in. (1 cm) in diameter and fed into the rotary dryer or onto an apron dryer. The final moisture from these dryers is generally ~6%. This type of clay can be shipped in lump form, or if necessary in powder form, using hammer mills swept with hot air. The final moisture of this disintegrated kaolin can be <0.5%. It is commonly referred to in the kaolin industry as "acid clay" because the pH is ~4.5. The wet processes are used on the sedimentary kaolins in Georgia and Brazil.

Primary kaolins such as the English china clays produced in the Cornwall district of southwest England are wet processed in much the same manner as the Georgia kaolins except for the mining methods. The kaolin content of the weathered granites ranges from 10 to 20%. Mining uses high pressure hydraulic monitors that play a stream of water on the mine face, washing out the fine particle kaolin and leaving the coarse quartz and mica residue. The fine kaolin is suspended and is transported in small rivulets to a collecting basin in the bottom of the open pit. This slurry is pumped to large thickeners, where the clay slurry is concentrated, and then processed generally following Figure 15.

Special Processes.   Several special processes are used to produce unique and special quality grades of kaolin. One of these special processes is delamination, which is a process that takes a large kaolin stack and separates it into several thin, large-diameter plates (Fig. 17). These thin, large-diameter plates have excellent covering power on rough base sheets of paper and are also used to produce high quality lightweight coatings. The process of delaminating involves attrition mills into which fine media, such as glass beads or nylon pellets, are placed along with the coarse kaolin stacks and intensely agitated. The fine media impact the kaolin stacks, separating them into thin plates (69). The brightness and whiteness of the delaminated kaolin is very good; the clean newly separated basal plane surfaces are white because they have been protected from ground water and iron staining. The delamination process has enabled the

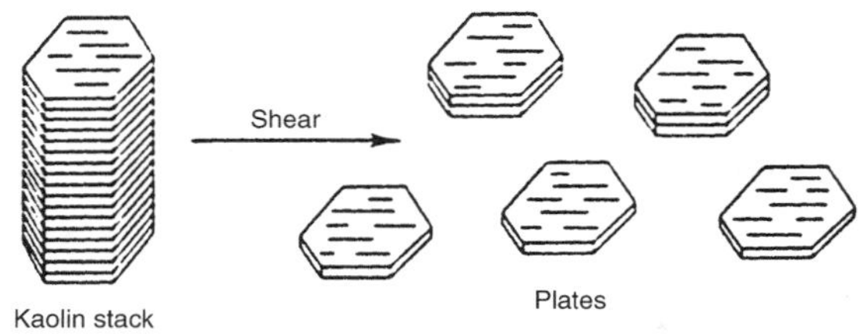

Fig. 17. Diagrammatic representation of delamination of a kaolin stack into large thin plates.

kaolin industry to convert low priced coarse clay into a higher value special coating and filler clay. Another special process is ultraflotation, which is a flotation process to remove iron and titanium contaminants to make a 90% brightness kaolin product (70). Selective flocculation (71–73) is another special process that produces a 90% brightness kaolin product by selectively removing iron and titanium impurities.

Calcining is another special treatment that is used to produce special grades. One grade is thermally heated to a temperature just above the point where the structural hydroxyl groups are driven out as water vapor, which is between 650 and 700°C. This procedure produces a bulky product that is used as paper coating additive to enhance resiliency and opacity in low basis weight sheets. A second grade is thermally heated to 1000–1050°C. By proper selection of the feed kaolin and careful control of the calcination and final processing (74), the abrasiveness of the calcined product can be reduced to acceptable levels. The brightness of this fully calcined, fine-particle kaolin is 92–95%, depending on the feed material. It is used as an extender for titanium dioxide in paper coating and filling and in paint and plastic formulations.

Another special process is *surface treatment*. Kaolin is hydrophilic and can be easily dispersed in water. Because of the nature of the kaolinite surface, it can be chemically modified to produce hydrophobic and organophilic characteristics. Generally, an ionic or a polar nonionic surfactant is used as a surface-treating agent. These surface modified kaolins (75), are used in paint (1), plastics (76), and rubber (77).

4.2. Smectite. Virtually all bentonite, which is comprised mainly of a smectite mineral, is surface-mined. Bulldozers and motorized scrapers are most commonly used for removing overburden. In a typical open pit, the overburden is removed in panels. The exposed bentonite is then mined by loading the material into trucks with draglines, shovels, or front-end loaders. Once the bentonite is removed, the overburden from the next panel is shifted to the mined-out panel.

Bentonite beneficiation and processing involves relatively simple milling techniques that includes crushing or shredding, drying, and grinding and screening to suitable sizes (Fig. 18). The high swelling Wyoming or sodium bentonites have ~25% moisture. The processed bentonite generally contains only 7–8% moisture, although because bentonite is hygroscopic, it may contain considerably more moisture when used. The raw bentonite is passed through some sort of crushing or shredding device to break up the large chunks before drying. Drying

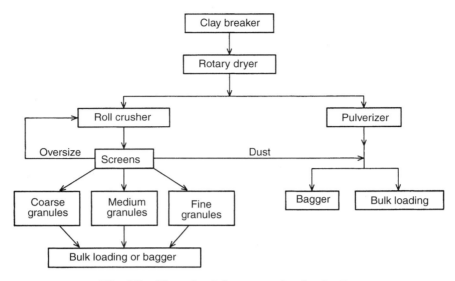

Fig. 18. Flow sheet for processing bentonite.

in most plants is accomplished by gas- or oil-fired rotary dryers, but in one bentonite plant in Wyoming, a fluidized-bed dryer is used. The bentonite properties
can be seriously affected by overdrying, so the drying temperature must be carefully controlled. The dried bentonite is ground and sized in several ways. Granular bentonite is cracked by using roll crushers and screened to select the proper
size range granules. Most powdered bentonite is ground with roll or hammer
mills to ~90% finer than 200 mesh. The bentonite is shipped mainly in bags,
but some is shipped in bulk.

Special processing is used in some plants to produce special products.
Extruders can be used to pug the clay and mix it with additives that may improve
the viscosity and dispersion properties. Some hectorite and Southern bentonites
are beneficiated wet. In this process, the bentonite is dispersed in water and
degritted by centrifuges, hydrocyclones, settling devices, or screens. The slurry
is either filtered and then flash-dried or sent directly to rotary drum or spray
dryers.

Organic-clad bentonite is bentonite that is processed to replace the inorganic exchangeable ions with alkylamine cations, which produces a hydrophobic
clay (78). These organic-clad bentonites are used in paints, greases, oil-base
drilling muds, and to gel organic liquids.

Acid-activated bentonites are special sorptive clays used for bleaching
edible oils and decolorizing special lubricating oils. The process flow sheet for
acid activation is shown in Figure 19. The process involves slurrying the bentonite in warm water, removing the grit by hydrocylcones, reacting the clay with
either sulfuric or hydrochloric acid at elevated temperature, dewatering in plate
and frame filter presses, and flash drying. A dry activation process is also used
and involves adding a concentrated sulfuric acid to the clay in a blender and pulverizing and drying the acid−clay mixture to ensure uniform distribution of the

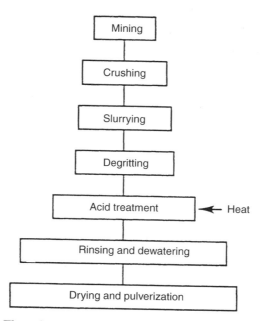

Fig. 19.   Flow sheet for the production of acid activated bentonite.

acid. Aluminum, magnesium, and iron octahedral ions are partially removed, resulting in a highly charged particle saturated with hydrogen ions, which makes for a good bleaching and decolorizing clay.

4.3. Palygorskite-Sepiolite.   The mining of palygorskite clays is no different than mining bentonites. The processing is also similar (Fig. 20). The clay is shredded or crushed, dried, screened, air-classified, and packaged. Some of this clay is thermally treated to harden the particles to prevent disintegration of the granules during bagging, transport, and handling prior to arrival to the customer. The temperature of this thermal treatment is <400°C so that no structural modification results, but all of the absorbed water is driven from the surface and interior of the particles.

When palygorskite (attapulgite) is to be used for salt-water drilling mud, the clay is extruded at high pressure through small orifices. This shears apart the bundles of palygorskite into individual elongate particles, which produces a higher viscosity. Extrusion with magnesium also produces a gelling quality product.

5. Environmental Considerations

Almost all clays are surface mined, so the industry is required to reclaim the disturbed land in most countries. Common practice is to open a cut and then deposit the overburden from the following panels or cuts into the mined out areas. The land is leveled or sloped to meet the governmental requirements and then planted with grasses or trees. In the processing plants the waste materials,

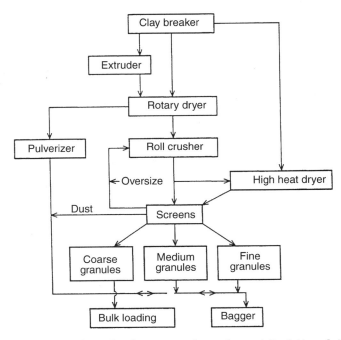

Fig. 20.   Flow sheet for dry processing palygorskite (attapulgite).

which include chemicals and separated clay particles or other minerals, are collected in impounds. The clay and other particles are flocculated with alum or other chemical flocculants, and the clear water is released into streams after adjusting the pH to 6–8.

Air quality is maintained in the processing plants by using dust collectors on the dryers and by enclosing transfer points where dry clay may cause dust. In areas such as the bagging departments, the workers may be required to wear dust masks. None of the clays, kaolins, smectites, or palygorskite-sepiolite are health hazards. Any dust, if inhaled in large quantities, can be a problem, but the industries use dust collectors and require workers to wear masks, therefore, no serious lung ailments result. The clays themselves contain no deleterious trace elements or chemicals that are hazardous to human health.

6. Production and Consumption

Clays are one of the most important and versatile industrial minerals. They are used in construction, agriculture, process industries, and environmental applications (79). They are necessary ingredients in many products including kaolins in paper, paint, ceramics and rubber, bentonite in drilling fluids, foundries, and fluid barrier applications, palygorskite-sepiolite in special drilling fluids, carriers for agricultural chemicals, suspending agents in paints and pharmaceuticals, and common clays for making bricks and other structural clay products.

The total kaolin production in the world is estimated to be $\sim 40 \times 10^6$ tons/year; bentonite production about 10×10^6 tons/year; palygorskite-sepiolite about 3×10^6 tons/year; and common clays over 100×10^6 tons/year.

Future research and development efforts will improve current clay products and develop new value added products. There are adequate clay reserves in the world so that tomorrow's markets will continue to be served.

BIBLIOGRAPHY

"General Survey" under "Clays" in *ECT* 1st ed., Vol. 4 pp. 24–38, by W. D. Keller, University of Missouri: "Survey" under "Clays" in *ECT* 2nd ed., Vol. 5, pp. 541–560, by W. D. Keller, University of Missouri; in *ECT* 3rd ed., Vol 6, pp. 190–206, by W. D. Keller, University of Missouri; in *ECT* 4th ed., Vol. 6, pp. 381–405, by T. Dombrowski, Engelhard Corporation; "Clays, Survey" in *ECT* (online), posting date: December 4, 2000, by T. Dombrowski, Engelhard Corporation.

CITED PUBLICATIONS

1. R. E. Grim, *Applied Clay Mineralogy*, McGraw-Hill Book Co., Inc., New York, 1962.
2. R. E. Grim, *Clay Mineralogy*, 2nd ed., McGraw-Hill Book Co., Inc., New York, 1968.
3. C. K. Wentworth, *J. Geol.* 30 (1922).
4. D. M. Moore and R. C. Reynolds, Jr., *X-Ray Diffraction and Identification of Clay Minerals*, Oxford University Press, 1989.
5. R. C. MacKenzie, *The Differential Thermal Investigation of Clays*, Mineralogical Society, London, 1957.
6. H. Beutelspacher and H. W. VanDerMarel, *Atlas of Electron Microscopy of Clay Minerals and Their Admixtures*, Elsevier, Amsterdam, The Netherlands, 1968.
7. H. W. VanDerMarel and H. Beutelspacher, *Atlas of Infrared Spectroscopy of Clay Minerals and Their Admixtures*, Elsevier, Amsterdam, The Netherlands, 1976.
8. H. Murray, W. Bundy, and C. Harvey, *Kaolin Genesis and Utilization*, Clay Minerals Society Special Publication, Vol. 1, 1993.
9. B. Velde, *Clay Minerals: A Physico-Chemical Explanation of Their Occurrence*, Elsevier, Amsterdam, The Netherlands, 1985.
10. H. Murray, P. Partridge, and J. Post, *Schriftemr. Geol. Wiss*, 11 (1978).
11. S. Pickering, Jr., and H. Murray, in D. Carr, ed., *Kaolin, Industrial Minerals and Rocks*, 6th ed., SME, Littleton, Clo., 1994.
12. T. Dombrowksi, Ph. D. Dissertation, Indiana University, Bloomington, Ind., 1992.
13. H. Murray, Preprint 99-135, SME Annual Mtg., Denver, Clo., 1999.
14. E. M. Durramce and co-workers, *Proceedings of the Annual Conference of the Usher Society* 1982.
15. C. M. Bristow, *Proceedings of the 8th International Kaolin Symposium Madrid-Rome*, 1977.
16. M. Kuzvart, *International Geological Congress Rep. Sess. 23rd* (1969).
17. M. Storr, 5th International Kaolin Symposium Ernst-Mortiz-Arndt University Greifswald, 1975.
18. H. Koster, 7th International Kaolin Symposium Proceedings, University of Tokyo, 1976.
19. L. Martin-Vivaldi, International Geologic Congress Rep. Sess. 23rd, 1969.

20. H. Murray and C. Harvey, Preprint 82-83, 1st International SME Mtg., Honolulu, Hawaii, 1982.
21. Z. Zheng and co-workers, *Mineral Deposits of China* **4** (1996).
22. J. Yuan and H. Murray, in Special Pub. No. 1, Clay Minerals Society, 1993.
23. H. Murray, *Paper Coating Pigments*, TAPPI, Atlanta, Ga. (1984).
24. C. Harvey and H. Murray, Special Publisher 1, Clay Minerals Society, 1993.
25. T. Sudo and S. Shimoda, *Clays and Clay Minerals of Japan*, Elsevier, The Netherlands, 1978.
26. R. Hanson, R. Zamora, and W. Keller, *Clays Clay Minerals* **29** (1981).
27. S. Patterson and H. Murray, U.S. Geol. Survey Prof. Paper 1306, 1984.
28. C. Bristow, International Geologic Congress Rep. Sess. 23rd (1969).
29. W. Keller, J. Westcott, and A. Bledsoe, Proceedings of the 2nd National Conference on Clay-Nat. Acad. Sci. 327, 1954.
30. W. Knight, *Eng. Min. J.* 66 (1898).
31. C. S. Ross and E. Shannon, *J. Am. Cer. Soc.* 9 (1926).
32. P. Wright, *J. Geol. Soc. Aust.*, 15 (1968).
33. R. E. Grim and N. Guven, *Bentonites*—Development Sedimentol 24 Elsevier, Amsterdam, The Netherlands, 1978.
34. J. Elzea and H. Murray, *Ind. Minerals and Rocks*, 6th ed., SME, 1994.
35. M. Knechtel and S. Patterson, U.S. Geol. Survey Bull. 22 (1962).
36. M. Slaughter and J. Early, Spec. Paper 83, Geological. Soc. America, 1965.
37. R. E. Grim, *Miss. State Geol. Surv. Bull* 22 (1928).
38. F. Reisch, M.S. Dissertation, Indiana University, 1991.
39. A. F. Hagner, *Am. Mineral.* 24 (1939).
40. P. Y. Chen, Ph. D. Dissertation, University of Texas, Austin, Tx., 1968.
41. R. Sloan and J. Gilbert, *Clays Clay Mineral* 15 (1966).
42. K. Papke, *Clays Clay Mineral* 17 (1969).
43. W. Monroe, *Ala Geol. Surv. Bull.* 48 (1941).
44. I. A. Cowperthwaite and co-workers, *Clay Mineral Bull.* 9 (1972).
45. R. Fahn, *Interceram* 12 (1965).
46. V. Annedda, *Resoconti Assoc. Min. Sarda* 60 (1956).
47. N. Siddiquie and D. Bahl, *Mem. Geol. Surv. India* 96 (1965).
48. L. Ames, Jr., and co-workers, *Econ. Geol.* 53 (1958).
49. R. H. S. Robertson, *Fuller's Earth*, Volturna, Kent, U.K., 1986.
50. S. Patterson, U.S. Geol. Surv. Prof. Paper 828, 1974.
51. R. Merkl, Ph. D. Dissertation, Indiana University, 1989.
52. L. Wirth, Lab de Geologie, University de Dakar, 26 (1968).
53. H. Zhou, Ph. D. Dissertation, Indiana University, 1996.
54. F. Ovcharenko and co-workers, *The Colloid Chemistry of Palygorskite*, Israel Program for Scientific Translations, Jerusalem, 1964.
55. E. Galan and co-workers, *Proc. Int. Clay Conf.* (1975).
56. Anon., *Ind. Minerals Mag.* 126 (1978).
57. H. Khoury, D. Eberl, and B. Jones, *Clays Clay Minerals* 30 (1982).
58. H. Murray, *Ind. Minerals and Rocks*, 6th ed. SME, 1994.
59. C. S. Ross and P. F. Kerr, U.S. Geol Surv. Prof. Paper 165-E, 1931.
60. S. W. Bailey, in Special Pub. 1, Clay Minerals Soc., 1993.
61. T. F. Bates and co-workers, *Am. Mineral* 35 (1950).
62. R. E. Grim and co-workers, *Am. Mineral* 22 (1937).
63. S. W. Bailey, in J. Gieseking, ed., *Soil Components*, Vol. 11, Springer, Verlag, New York, 1975.
64. R. C. Reynolds, *Monograph 5*, Mineral. Society of London, 1980.
65. H. Murray, *Acta Mineral—Petrogr.* 23 (1980).

66. H. Murray, *Interceram* 31 (1982).

67. J. Iannicelli, *IEEE Trans. Magn. Mag.* 12 (1976).

68. J. A. Wernham and T. H. Falconer, Conference on Superconducting Magnetic Separation, Imperial College, London, 1985.

69. U.S. Pat. 3,171,718 (1965), F. A. Gunn and H. H. Morris.

70. E. W. Greene and J. B. Duke, *Trans. Soc. Min. Eng. AIME* 223 (1962).

71. U.S. Pat. 3,477,809 (1969), W. M. Bundy and J. P. Berberich.

72. U.S. Pat. 3,371,988 (1968), R. N. Maymard and co-workers.

73. U.S. Pat. 3,701,417 (1972), V. V. Mercade.

74. U.S. Pat. 3,586,523 (1971), H. R. Fanselow and D. A. Jacobs.

75. W. M. Bundy and co-workers, *Proceedings of the Technical Association Pulp. Paper Industries*, 1983.

76. U.S. Pat. 3,248,314 (1966), P. G. Nahin.

77. P. W. Libby and co-workers Abstr. Papers American Chemical Society 154th (1967).

78. U.S. Pat. 2,531,440 (1950), J. W. Jordan.

79. H. Murray, *Proceedings of the 10th Int. Clay Conference*, Adelaide, Australia, 1993.

H. H. Murray
Indiana University

CLAYS, USES

1. Introduction

Clays are fine particle size materials comprised of clay minerals, which are basically hydrated aluminum silicates with associated alkali and alkaline earth elements. The clay mineral groups are kaolin, smectite, palygorskite-sepiolite, illite, chlorite, and mixed-layered clays. The properties of these clays are very different because of differences in their structure and composition. All are extremely fine and contain non-clay minerals such as quartz, feldspar, mica, calcite, dolomite, opal CT, and minor heavy minerals including ilmenite, rutile, anatase, tourmaline, zircon, kyanite, and other detrital minerals. The mineral content can be determined using X-ray diffraction techniques. Each group of clay minerals has very different applications because of differences in their physical and chemical properties.

The structure and composition of kaolins, smectites, and palygorskite-sepiolite are very different even though each has silica tetrahedral and alumina octahedral sheets as their basic building blocks. The arrangement and composition of the octahedral and tetrahedral sheets account for most of the major and minor differences in the physical and chemical properties of those clay minerals and thus on their ultimate application. Also important are the type and amount of non-clay minerals that are present. A brief summary of some of the important characteristics and properties of kaolins, smectites, and palygorskite are summarized in Table 1.

The particle size, shape, and distribution are physical properties that are intimately related to the applications of the clay minerals. Other important

Table 1. Some Important Properties of Clay Minerals that Relate to Their Application

Kaolin	Smectite	Palygorskite
1:1 layer	2:1 layer	2:1 layer inverted
white or near white	tan, olive green, gray or white	light tan or gray
little substitution	octahedral and tetrahedral substitution	octahedral substitution
minimal layer charge	high layer charge	moderate layer charge
low base exchange capacity	high base exchange capacity	moderate base exchange capacity
pseudohexagonal flakes	thin flakes and laths	elongate particles
low surface area	very high surface area	high surface area
very low absorption capacity	high absorption capacity	high absorption capacity
low viscosity	very high viscosity	high viscosity

properties are surface chemistry, area, and charge. These along with color and brightness affect many properties including low and high shear viscosity; absorption capacity and selectivity; plasticity; green, dry, and fired strength; casting rate; permeability; bond strength; and optical coating properties for paper and paint. In most every application the clays and clay minerals perform a function and are not just inert components in the system. Improved processing techniques, which have evolved over the past 40 years, have had a profound effect on the traditional and new applications. Statistical information supplied by the U.S. Geological Survey (formerly the U.S. Bureau of Mines) classifies clays into six groups: kaolin, ball clay, fireclay, bentonite, Fuller's earth, and common clay and shales. Because kaolins, eg, spill over in the figures for ball clay and fireclay and Fuller's earth includes attapulgite (palygorskite) and calcium montomorillonite, it is difficult to get correct tonnage and monetary values for the clay mineral types.

The name palygorskite and attapulgite are synonymous. The International Nomenclature Committee (INC) favors palygorskite because it predates the name attapulgite; however, attapulgite is so well ingrained in commercial literature and usage that industry continues to use the name attapulgite. The term Fuller's earth is used for any fine-grained material that is absorbent and is a naturally active bleaching earth. The origin of the term goes back to when earthy material was used in cleansing and fulling wool, thereby removing lanolin and dirt and thus the name Fuller's earth (1). Both palygorskite (attapulgite) and calcium bentonite are classified in the Government statistics as Fuller's earth.

The discussion of uses will be grouped using the following headings; kaolins, smectites, palygorskite-sepiolite, and common clays.

2. Kaolins

As mentioned in the article on Clays, Survey (see CLAYS, SURVEY) kaolin is a group mineral name for kaolinite, dickite, nacrite, and halloysite. The most common mineral in the kaolin group is kaolinite. It's physical and chemical properties and applications are discussed in this section. The uses are governed by several factors including the geological conditions under which the kaolin formed, the mineralogical composition of the kaolin deposit, and the physical and chemical properties of the kaolinite. Kaolin deposits can be sedimentary, residual or hydrothermal (2) and in almost every instance the kaolin has different properties and thus must be fully tested and evaluated to determine its utilization. Sedimentary kaolins are called secondary deposits and residual or hydrothermal kaolins are called primary deposits.

The occurrences of kaolin are numerous but commercially useable deposits are few in number. The best known and most highly utilized deposits are the primary hydrothermal kaolins in the Cornwall area of southwestern England, the secondary kaolins in Georgia and South Carolina in the United States, and the Jari and Capim kaolin deposits in the lower Amazon region of Brazil (3). Other important kaolin deposits that are regionally utilized, are the primary kaolin deposits in the Czech Republic (4); the primary kaolin deposits in Caminau and Kemmlitz in eastern Germany (5) and the secondary kaolins in Bavaria near Hirschau Germany (6); in Spain a primary kaolin near Galicia and a secondary kaolinitic sand in the state of Guadalajara (7); and a large primary kaolin deposit near Proyanoski in Ukraine (8). Other kaolin deposits used locally are located in Argentina, Australia, Brazil, China, Chile, Egypt, Indonesia, India, Korea, Mexico, Portugal, Tanzania, South Africa, Suriname, and Venezuela (3).

Kaolins are fine in particle size and are hydrophilic. With a small amount of chemical dispersant kaolin will easily disperse in water. The particle shape of kaolinite is important for many applications. The thin pseudohexagonal plates orient on coated surfaces such as on paper or in paint because of their two-dimensional nature (Fig. 1). Kaolin is one of the most versatile industrial minerals (Table 2) because it

1. It is chemically inert over a wide pH range (4–9).
2. It is white or near white in color.

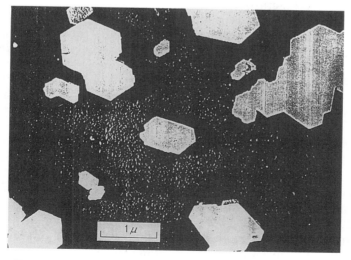

Fig. 1.   Electron micrograph of platy kaolinite particles.

Table 2. **Uses of Kaolin**

paper coating	cement	food additives
paper filling	pencil leads	bleaching
extender in paint	adhesives	fertilizers
ceramic raw material	tanning leather	plaster
filler in rubber	pharmaceuticals	filter aids
filler in plastics	enamels	cosmetics
extender in ink	pastes and glues	crayons
petroleum	insecticides	detergents
fiberglass	medicines	roofing granules
foundry bond	sizing	linoleum
dessicants	textiles	polishing compounds

3. It has good covering or hiding power when used as a pigment or extender in coatings.
4. It is soft and nonabrasive.
5. It is fine in particle size.
6. It is plastic, refractory, and fires to a white or near white color.
7. It has low conductivity of both heat and electricity.
8. It is hydrophilic and disperses readily in water.
9. Some kaolins have a very low viscosity and flow readily at 70% solids.
10. It is relatively low in cost.

2.1. Ceramic Products. A large proportion of the annual production of ball clay and fireclay and a large amount of kaolin are used in the manufacture of ceramic products. Ball clays, which are fine particle size kaolinitic clays, are used as a raw material in whiteware, sanitaryware and tile. These clays are plastic, have high green, dry, and fired strength, are relatively low in shrinkage, fire white or near white, and have good casting qualities for sanitaryware. Fireclays are used in the manufacture of refractories because of their high melting point and low shrinkage. A special type of fireclay is the high alumina clays and flint clays found in Missouri (9), Kentucky (10), and Australia (11). Flint clays are very dense and brittle and are essentially pure, extremely fine-grained kaolinite. Another use for the high alumina clays and kaolins is to calcine them at high temperature (1300°C) to produce refractory grog. The grog is used in the refractory mix to reduce shrinkage and increase the melting point of the refractory brick for molten glass and metal processing. Kaolins are widely used as an important ingredient in whiteware, tile, insulators, and sanitaryware. Relatively pure kaolins melt or fuse at a temperature of 1850°C. Kaolin is an important constituent in the manufacture of catalytic converters for automobiles because of it's plasticity, controlled shrinkage, and fired strength. Ball clays and kaolins are used as suspending material to carry the frits and coloring pigments in porcelain enamels so that the coating is uniform.

For refractory applications the pyrometric cone equivalent (PCE) is measured. The pyrometric cone measures the combined effects of temperature and time (12). The cones consist of a series of standardized unfired ceramic compositions molded into the shape of triangular pyramids. The sample of kaolin or ball clay is molded into the standard cone shape and is heated along with several standard cones so that its end point of fusion can be determined in terms of equivalent cone number. Refractory PCE values are 32–35.

Casting rate in the production of sanitaryware is important and fine-grained bodies cast more slowly than coarse ones. Viscosity of a slip must be carefully controlled because if it is too viscous the slip will not properly fill the mold or drain quickly. Therefore, viscosity of kaolins and ball clays is measured and controlled on shipments used in the casting process.

2.2. Paper. The paper industry is the largest consumer of wet processed kaolin (see CLAYS, SURVEY). Kaolin used in the paper industry has two main uses, as a filler where the kaolin is mixed with the pulp fibers and as a coating where the kaolin is mixed with water, adhesives, and various additives and coated onto the surface of the paper. The largest use of kaolin is for coating paper. As a filler kaolin improves the opacity of the paper sheet, imparts smoothness to the surface, and replaces some of the more expensive pulp fibers. In coatings on the paper sheet, the kaolin imparts opacity, brightness, a glossy finish, smoothness, and improved printing quality. For example, a coated sheet of paper in National Geographic Magazine contains ~35% by weight kaolinite.

Kaolins used by the paper industry are of three types based upon the type of processing: air-floated, water washed, and calcined (see CLAYS, SURVEY). Water washed kaolins are of higher value than air-floated because the more elaborate processing results in more uniform and higher quality products. Calcined kaolins are heated to ~1000°C, which converts the kaolinite to mullite and opal CT and results in significantly higher brightness and opacity (13). Titanium dioxide is a prime pigment that has very high brightness and opacity but is expensive. Calcined kaolin, which is much less costly, can replace a majority of the titanium dioxide with little or no loss in brightness or opacity. Calcined kaolins have brightnesses ranging between 92 and 95%.

The properties of kaolin that make it useful in the paper industry are brightness, viscosity, and particle size and shape. Air-floated kaolins are at the lower end of the brightness range for fillers used by the paper industry. The air-floated fillers are coarser and have a brightness range of 80–84%. The particle size is in the range of 50–70% finer than 2 m. They are low in cost but their uniformity and quality is not as high as the water washed filler clays. Improvement in paper sheet brightness and opacity results from the addition of kaolin filler clays. The best results are obtained using calcined clay followed by water washed kaolins and then air-floated kaolin (Fig. 2).

Coating grade kaolins are produced using a wet process (see CLAYS, SURVEY). The important qualities that are controlled for producing a finished product are brightness, particle size, and viscosity. Table 3 shows the typical kaolin grades produced in Georgia for the paper industry. The largest sources for coating grade kaolins are Georgia in the United States, Cornwall in southwestern England, and the lower Amazon region of Brazil (see CLAYS, SURVEY). Kaolins from each area have different characteristics that can be traced to the geologic origin of the crude kaolin. In the Southeastern United States, the kaolins of Cretaceous age that are mined in middle Georgia are relatively coarse and those of Tertiary age in east Georgia are very fine in particle size. In Brazil, the Jari kaolin is very fine and the Capim kaolin is much coarser. Both the Georgia

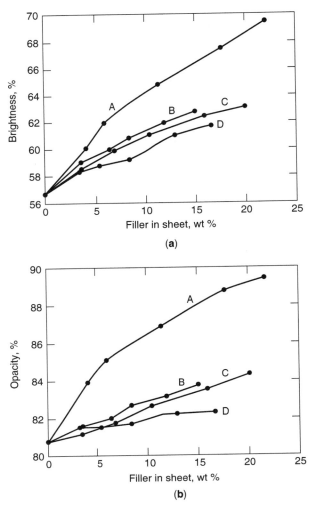

Fig. 2.   (**a**) Brightness improvement obtained by use of kaolin as a filler in paper (**b**) Opacity improvement obtained as pulp fibers are replaced with various kaolins. A, calcined clay; B, high brightness No. 1 clay; C, coarse particle water washed filler clay; and D, air-floated kaolin.

and Brazil kaolins are sedimentary and the English kaolin is primary, which gives it somewhat different characteristics.

Viscosity of the kaolin coating grade clays is a very important characteristic. The viscosity is measured at 70% solids. Two types of viscosity are measured, the Brookfield low shear viscosity and the Hercules high shear viscosity, both of which are important to the paper coater. The low shear viscosity relates to the pumpability and flow from a tank car through the coating preparation facility, and the Hercules high shear viscosity relates to runnability on the coating machine (14).

Other properties of kaolin coating clays that are important to the paper coater are dispersion, opacity, gloss and smoothness, adhesive demand, film

Table 3. **Grades, Particle Size, and Brightness of Georgia Kaolins for Paper Coating**

Grade	Particle size	Brightness
No. 3	73% < 2 µm	85–86.5
No. 2	80–82% < 2 µm	85.5–87.0
No. 1	90–92% < 2 µm	87.0–88.0
fine No. 1	95% <2 µm	86.0–87.5
delaminated	80% <2 µm	88.0–90.0
high brightness No. 2	80% <2 µm	90–91
high brightness No. 1	92% <2 µm	90–91
high brightness fine No. 1	95% <2 µm	90–92

strength, and ink receptivity. In order to obtain the maximum efficiency of a coating clay, the individual clay particles must be completely dispersed. Kaolinite is easily dispersed because it is hydrophilic. Opacity is strongly influenced by particle packing that is dependent on particle size and shape and particle size distribution (14). Gloss and smoothness are related to particle size and shape. Superior gloss is obtained with kaolins of fine particle size. Coatings made with thin plates of kaolinite that orient parallel to the paper sheet have better smoothness. Adhesive demand is related to surface area. Finer particle size kaolins require slightly higher amounts of adhesive. The amount of adhesive in the coating affects the opacity, brightness, color, and smoothness and so the least amount of adhesive that will bond the coating to the base sheet is advantageous. Film strength or pick strength is related to preferential adhesive migration into the substrate and to the preferred orientation of the kaolinite particles (15). Maximum film strength is attained with well oriented dense films. If pick occurs within the coating layer it is generally caused by insufficient adhesive, excessive adhesive migration, or poorly dispersed kaolinite. Ink receptivity is related to film permeability as influenced by void volume. Small diameter particles, randomly oriented, give excellent ink receptivity. Uniform absorbency of the ink is particularly important or the printed surface may have a mottled appearance. The interrelationship of the kaolinite and adhesive is a major control over ink receptivity and holdout (16).

Particle size is a very important parameter to control in coating grade kaolins (Fig. 3). Table 4 shows how coated paper properties change with particle size. In general, finer particle kaolins give higher gloss, opacity, and brightness. For extremely fine kaolins, however, opacity and brightness may decrease as a result of the loss of light scattering power.

Testing. The Technical Association of the Pulp and Paper Industry (TAPPI) publishes test methods that are used by both kaolin suppliers and the paper industry. These include tests for viscosity, viscosity stability, brightness, pH, particle size distribution, moisture content, screen residue, and abrasion (17). These test procedures are continually upgraded and are monitored by a TAPPI committee. Other tests that relate to use properties are carried out by the kaolin suppliers.

Bentonite (Smectite). Because of its very fine particle size and high sorption capacity, smectite is sometimes used in the deinking process to recover cellulose fibers. The deinking process generally involves heating the paper in a

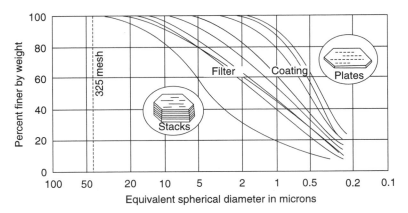

Fig. 3. Particle size distribution of coating and filler clays.

caustic soda solution to break down the ink in order to free the ink pigment. A detergent is then added to release the pigment from the paper fibers. Smectite is added, which serves to disperse the pigment particles and adsorb them. Washing then removes the smectite carrying with it the ink pigment. Sodium montmorillonite is the preferred smectite for this application. Smectite, usually sodium montmorillonite, is also used in papermaking to prevent agglomeration of pitches, tars, waxes, and resinous materials. The addition of 0.5% sodium montmorillonite, based on the dry weight of the paper stock, will prevent agglomerates so that the globules will not stick to screens, machine wires, press rolls, etc, and cause defects in the paper.

In some instances, small quantities of sodium montmorillonite have been used to increase the low shear viscosity in certain coating color formulations. One percent or less will significantly increase the low shear viscosity at 10 Rpm, particularly in high solids coating colors. Also, the addition may lower the high shear viscosity into the low dyne region. The addition of sodium montmorillonite will also retain water in the coating layer and slow down its migration into the paper substrate. The presence of small quantities of smectite, because of its high surface area and strong sorptive capacity, promotes ink receptivity and inhibits ink holdout (16). The amount of smectite used by the paper industry is very small.

Palygorskite (Attapulgite).   Palygorskite is sometimes used in specialty copy papers particularly in non-carbon required (NCR) papers to make multiple

Table 4. **Effect of Particle Size on Coated Sheet Properties**

Particle size		Coated sheet properties		
% <2 μm	Median, μm	Gloss	Brightness %	Opacity %
35	3.80	25	71.6	84.2
54	1.80	31	72.2	84.6
78	0.75	33	72.6	84.6
85	0.68	39	72.7	84.7
96	0.46	45	72.6	84.7

copies without carbon paper. The receiving surface of the copy sheet is coated with paylgorskite. The lower or transfer surface is coated with adhesive containing minute encapsulated droplets of dyestuffs. Writing or typing breaks the dyestuff gobules so that they penetrate the palygorskite coated layer that catalyzes their conversion into a colored letter or mark. This use is decreasing because of the ever increasing use of computers and copy machines.

2.3. Paint. Clays are widely used in both water and oil-based paints. In this application, clays perform several important functions. They extend the much higher cost titanium dioxide opacifying pigment, control viscosity so as to prevent pigments from settling during storage, provide thixotropy so that the paint is easily applied yet does not sag or run after application, improve gloss, promote film strength, and aid in tint retention.

Kaolin. Kaolins used in paints are very similar to those used in paper but are processed differently so as to enhance the properties of the paint. The conductivity of the kaolin is measured so as to control the amount of soluble salts present. Soluble salts are detrimental to some of the ingredients that make up the paint formulation so the kaolin is well washed to remove the soluble salts. Both hydrous and calcined kaolins are used in paint formulations. Calcined kaolins are used to extend titanium dioxide with little or no loss of hiding power.

Surface treatments of the hydrophilic kaolin are sometimes used to enhance the oleophilic properties for use in oil-based paints. For certain paint applications the Hegman fineness of grind test specification is important. This test measures the degree of dispersion. Some paint customers specify that a pigment or extender must be finer than a particular Hegman number. Kaolin in water-based paints may constitute as much as 30% of the pigment.

Bentonite. Sodium montmorillonite is used as a viscosity control and suspending agent in water-based paints. Because of their fine particle size they do not contribute to opacity so are not used as an extender. The smectites used for paint are processed to remove oversize particles. Oganoclays that are organic treated sodium montmorillonite are widely used as suspending and antisag agents in oil-based paints. The most common surface treatment chemical is hydrogenated tallow quaternary amine. Whiter grades are preferred so as to contribute as little as possible to the color of the finished paint. The American Society for Testing and Materials (ASTM) provides procedures and specifications for paint raw materials (18).

Palygorskite (Attapulgite). Attapulgite is used as a viscosity control and suspending agent. Because of the elongate particle shape and surface properties of attapulgite it does not flocculate in the presence of electrolytes. The physical hindrance to settling because of its particle shape makes it a preferred thickening and suspending agent. Fine particle size attapulgite with no oversize coarse particles is required for this use.

2.4. Plastics and Rubber. Kaolin is the only clay used in plastics and rubber as extenders and functional fillers (19). Various grades are produced specifically for use by the plastics and rubber industries: Table 5 lists the more important kaolin grades used as components in plastics and rubber. More than half the tonnage used is air-floated kaolin, which is used extensively by the rubber industry.

Table 5. **Kaolin Grades for Polymer Applications**

Kaolin type	Brightness %	Median particle size, μm	Uses	Relative cost[a]
		Air floated		
regular	73–76	0.2–1.0	rubber	L
surface treated	73–76	0.2–1.0	rubber	M
		Water washed		
coarse particle	79–82	4.8	thermosets	M
intermediate	85–87	0.6	PVC[b]	M
fine	86–88	0.4	PVC[b]	M
		Calcined		
meta kaolin	84–86	1.4	PVC[b] insulation	H
high temperature	90–92	1.4	engineering plastics, PVC	H
surface treated	90–92	1.4	engineering plastics, PVC	Ht

[a] L = low; M = moderate; H = high; Ht = highest.
[b] Poly (vinyl chloride) = PVC.

As seen from Table 5, the kaolins used are air-floated, water-washed, and calcined. The calcined meta-kaolin is used almost exclusively in PVC wire insulation because it enhances the electrical resistivity of the wire coating.

Surface treated clays shown in Table 5 are an important group of clays used in polymers. These are made from calcined kaolin, air-floated kaolin, and water-washed kaolin. Several types of chemical compounds are used to convert the hydrophilic surface of the kaolinite into an oleophilic surface that is more compatible with polymers (19). Silane treatment of calcined and hydrous kaolins raises the cost significantly, but the improvement in polymer properties justifies the extra cost. The data in Table 6 shows the effect on physical properties of adding 40% of an aminosilane treated calcined kaolin to a nylon compound. The improvement in strength properties and the marked increase in heat deflection temperature are noteworthy.

Properties that make kaolin useful in the plastics and rubber industries are color, particle size and shape, and viscosity. Clays used in the polymers industry

Table 6. **Effect of Clay Filler on Nylon Properties**[a,b]

Property	Unfilled	Filled
tensile strength, MPa[b]	82	97
tensile elongation, %	60	8
tensile modulus, MPa[b]	2900	6200
flexural strength, MPa[b]	97	159
flexural modulus, MPa[b]	2900	6200
izod impact, J/m	0.020	0.017
deflection temp. at 1.82 MPa, C	77	204

[a] Data for 40% loading of aminosilane-treated calcined kaolin in nylon-6,6.
[b] To convert MPa to psi, mutiply by 145.

are required to be white or nearly so except for applications in black compounds. The refractive index of kaolin closely matches that of most polymer compounds so kaolins contribute little to opacity and color. Particle size and shape are important because large particles may give a rougher than desired surface whereas small particles are more effective in increasing polymer hardness. Particles having a high aspect ratio, ie, platy particles, usually give greater reinforcement than more equant particles. The incorporation of a kaolin filler into a polymer may strongly affect the viscosity of the compound, especially when fine particle size clays are added. In general, finer particle clays increase viscosity more than coarse particle clays. Depending on the polymer and its application, increased viscosity may be desirable or undesirable. For example, a polymer that is too fluid to handle by some processing equipment may be rendered sufficiently viscous to be utilized by the incorporation of a kaolin filler. Surface treated kaolins because of the more oleophilic surface, give lower viscosity compounds than do untreated kaolins. Also the surface treatments generally permit much higher loading in the polymer.

Testing. Various test methods are provided by ASTM (18). These include pigment tests such as chemical analyses, presence of oversize particles, oil absorption, particle size distribution, degree of dispersion, presence of soluble components, etc. Numerous tests are also given by ASTM to determine the properties of filled and unfilled polymers. These include such properties as impact resistance, stiffness, viscosity, tear resistance, hardness, color, and electrical resistivity.

2.5. Halloysite. Halloysite is a member of the kaolin group and has a layer of water between the unit layers (see CLAYS, SURVEY). This causes the shape to be elongate tubes, which greatly changes its applications. Halloysite is produced in New Zealand and is used primarily in whiteware ceramics as an additive where it adds dry and fired strength and makes the dinnerware more translucent. Halloysite is also used as a filler in paper and as a raw material to make synthetic zeolites.

As shown in Table 2, kaolin has a large number of uses. The uses for paper, ceramics, paint, plastics, and rubber have been discussed and the remaining miscellaneous uses will be discussed in the following sections.

2.6. Inks. Kaolins are a common ingredient in a large variety of printing inks. In addition to extending the more expensive polymers present, kaolin also contributes to improved color intensity, limits ink penetration into the paper, controls rheology, and improves adhesion. Kaolin for use in inks must usually be as white as possible and free from oversize particles. Surface treated kaolins are used in oil-based inks and also to improve color acceptance.

2.7. Cracking Catalysts. Kaolins are used as a raw material to make zeolites and aluminum silicates for use as catalysts in the refining of petroleum (20). Several thousand tons are used annually for this purpose. Halloysite has also been used in the manufacture of petroleum cracking catalysts.

2.8. Chemical Raw Materials. Several tens of thousands tons annually of air-floated kaolin are used as a chemical raw materials to make fiberglass. A low iron and titanium content is required for this application. Kaolin is used in some cement plants to whiten the color, to provide silica and alumina for the cement reactions, and increase cement strength. In the production of

Portland cement, the limestone and kaolin react when heated to ~1600°C to form a calcium alumino—silica clinker. The clinker is pulverized and when water and aggregate are added, concrete is formed (21). Kaolin is used as a raw material for producing zeolites, which are an ingredient in detergents.

2.9. Special Fillers. Kaolin is used as a filler largely because of its fine particle size and white color in adhesives, pastes and glue, crayons, and linoleum (22).

2.10. Pencil Leads. Kaolin is used as an additive to graphite in making pencil leads (23). The plasticity and fine particle size are useful in extruding the graphite-kaolin mixture, and when the lead is fired, the kaolin gives the lead strength or hardness. The hardness of the lead is controlled by the percentage of kaolin in the graphite. A2H pencil contains a small amount of kaolin and the hardness increases in 3H, 4H, and 5H pencils by adding additional amounts of kaolin.

2.11. Suspensions and Diluents. Kaolins are used as suspending agents in pharmaceuticals, cosmetics, enamels, and medicines. A good example is kaopectate, which is a suspension of pectin in a kaolin slurry. Some pills use kaolin as a diluent and binder in pressing pills. Very fine particle size kaolins are best for these applications.

2.12. Fertilizers, Dessicants, and Insecticides. Kaolin is used as a carrier for certain insecticides and fertilizers and is used as a dessicant to promote flow when mixed with highly deliquescent materials. Some kaolin dessicants are coated with organic compounds to make them better flow agents.

2.13. Foundry Binders. Some plastic, fine particle kaolins are used as a binder for sand to make high temperature resistant molds for special metals.

2.14. Roofing Granules and Polishing Compounds. Roofing granules are produced by calcining coarse granular particles of white kaolin. Fine particle calcined kaolin with a top particle size of 3 μm is used in polishing compounds for automobiles, silver, copper, brass, and other metals. The hardness of calcined kaolin is ~6.5 on the Moh's scale. Because of its hardness and very fine particle size, it is an excellent polishing agent.

3. Smectites

Smectite is the group name for a number of clay minerals including sodium montmorillonite, calcium montmorillonite, saponite, nontronite, and hectorite (see CLAYS, SURVEY). The rocks in which the smectite minerals are dominant are called bentonites. Industrial quality bentonites are predominantly comprised of either sodium montmorillonite or calcium montmorillonite and to a much lesser extent, hectorite. Industrial nomenclature for these materials is sodium bentonite, calcium bentonite, and lithium bentonite. These smectite minerals and bentonites have significantly different physical properties, which dictate to a large degree their industrial uses.

The most widely used sodium bentonite in the world is located in Wyoming, South Dakota, and Montana. Important calcium bentonites are located in Mississippi and Texas in the United States, England, Germany, Italy, and Greece in Europe, South Africa, India, China, and Argentina.

Table 7. **Industrial Uses of Smectites**

drilling mud	industrial oil absorbents	beer and wine clarification
foundry binders	bleaching clays	suspension aids
iron ore pelletizing	catalysts	deinking on paper
cat litter	detergents	tape joint compounds
sealants	ceramics	emulsion stabilizers
animal feed binders	cosmetics	slurry trench excavation
paint	dessicants	adhesives
agricultural carriers	crayons	pharmaceuticals
nanoclays	medical formulations	organoclays

The uses of smectites are shown in Table 7. The largest uses by far are for drilling muds, foundry binders, iron ore pelletizing, cat litter, and sealants.

Smectites are very fine in particle size and the particles are extremely thin, which gives the material a high surface area. Smectites and particularly sodium montmorillonite have a high base exchange capacity on the order of 75–100 meq/100 g of clay. Sodium montmorillonite has a high swelling capacity of 10–15 times its dry volume. These characteristics give smectites a high degree of absorbency for many materials, and when mixed with water, the high charge and fine particle size give the fluid a very high viscosity. These properties of smectites control the many industrial uses shown in Table 7, which are described in detail in the following paragraphs. Bentonite production in the United States is 4,500,000 tons annually and the world production is ~10,000,000 tons.

3.1. Drilling Mud. Sodium montmorillonite is the major constituent of the Western bentonite, which is a high swelling clay. This high swelling property makes it a necessary ingredient in freshwater drilling muds throughout the world. Commonly, this Western or Wyoming bentonite yields 100 barrels of 15 CP drilling mud per ton (24). Only 5% by weight of this high swelling bentonite are required to produce the high viscosity, thixotropic fluid with low filter cake permeability that is necessary to meet the American Petroleum Institute (API) drilling fluid specifications. Hectorite also makes an excellent drilling mud but because it is used for more value added products, hectorite is not used for this purpose. Some calcium montmorillonites can be treated with a sodium salt such as soda ash to make a drilling mud, but neither the viscosity nor the water loss properties are as good as that of the Wyoming sodium bentonites.

3.2. Foundry Binders. The molding sands used in foundries are comprised of high silica sand and ~5–8% bentonite. The bentonite provides bonding strength and plasticity. A small amount of tempering water is added to the mixture to make it plastic. This mixture of sand, clay, and water can then be molded around a pattern shape and be cohesive enough to maintain the proper shape after the pattern is removed and molten metal is poured into the mold. The important foundry properties are green compression strength, dry compression strength, hot strength, flowability, permeability, and durability. These properties vary greatly with the amount of tempering water (24). Both sodium and calcium bentonites are used as bonding clays. Calcium bentonite has a higher green strength, lower dry strength, lower hot strength, and better flowability than sodium bentonite. Blends of sodium and calcium bentonite are commonly used to gain the optimum properties from each.

3.3. Iron Ore Pelletizing.   Sodium bentonites are used to pelletize iron ores (25). Finely pulverized ore concentrates are pelletized into marble-sized balls ~2.5cm in diameter for ease of shipping and to produce a superior furnace feed. Sodium bentonite constitutes ~0.5 wt% of the ore. The superior dry strength of the sodium bentonite makes it the preferred clay for pelletizing.

3.4. Cat Litter.   Both calcium and sodium bentonites are used in making cat litter. Because of its high absorbance quality, calcium bentonite is crushed and sized into a granular product for use as cat box filler. Calcium bentonites from Mississippi, Georgia, Illinois, and Missouri are the major sources for making the granular cat litter. Over the past several years, a new product called clumping cat litter has emerged as a preferred litter. This is made by blending high swelling sodium bentonite and calcium bentonite granules. When liquid feline waste hits this blend of granules, the sodium bentonite swells and forms a hard clump, which is easy to remove from the litter box. This has now become the highest tonnage use of sodium bentonite.

3.5. Sealants.   Sodium bentonites are used extensively for water impedance because of their high swelling capacity. The bentonite is used to impede the movement of water through earthen structures such as dams, to seal irrigation ditches, to prevent seepage of water from ponds and impounds, and to prevent water from entering basements of homes. Sodium bentonite is also used in barrier walls and liners in landfills and toxic waste dumps (26).

3.6. Slurry Trench Excavations.   High swelling sodium bentonite is used in the slurry trench or diaphragm wall method of excavation in construction in areas of unconsolidated rocks and soils (27). In this method, the trench being excavated is filled with bentonite slurry and the earth material being excavated is moved through it. A thin filter cake on the walls of the excavation prevents loss of fluid and the hydrostatic head of the slurry prevents caving and running of loose soil. This makes costly shoring unnecessary. This technique is used extensively in Europe and is becoming popular in the United States.

3.7. Absorbents.   Calcium bentonites are excellent absorbent clays. Small granules of the bentonite are prepared and used for absorbing pesticides and insecticides for protecting farm crops. The absorbed pesticide is less toxic to humans and allows for targeted emplacement. Another use is absorbing oil and grease spills in factories and gas stations. The oil or grease is absorbed on the clay granules, which are swept up and discarded. The oil-soaked granules do not present a fire hazard and also provide an anti-slip surface to an oily floor.

3.8. Suspension Aids.   Sodium bentonite is used as suspension aids in cosmetics, medical formulations, pharmaceuticals, and for use in the distribution of suspension fertilizers, which is becoming a popular method to spread fertilizer on farm crops.

3.9. Bleaching Clays.   Calcium bentonites are treated with sulfuric and/or hydrochloric acid to remove ions from the surface and from the octahedral layer to increase the charge on the clay particle. These acid activated clays are widely used to decolorize mineral, vegetable, and animal oils. Almost all edible oils are bleached using acid activated clays (28).

3.10. Organoclays.   Sodium montmorillonite and hectorite are processed so that the exchangeable sodium ions are replaced with alkylammonium cations to produce a hydrophobic surface (29). These organic clad sodium montmorillonites are used as thickeners in oil-base paints, greases, oil-base drilling muds, and to gel various organic liquids (30).

3.11. Animal Feed Binders. Both sodium and calcium bentonites are used to bind animal feed into pellets. The fine particle bentonite is very plastic and binds the feed into pellets, which are easy to package and handle. In addition, the montmorillonite acts as an absorbent for bacteria and certain enzymes, which when removed from the animal, promotes growth and health.

3.12. Wine and Beer Clarification. Both calcium and sodium bentonites are used to remove colloidal impurities such as haze-forming compounds in wine and beer. These colloidal impurities are positively charged and are attracted and coagulated by the negatively charged smectite clay.

3.13. Nanoclays. A recent development using sodium montmorillonite is the separation of the unit layers into almost unit cell thickness (1 nm or 10 Å) for use in polymer compositions called nanocomposites. These extremely thin, platy particles are exchanged using organic molecules, which will interact with the polymer to produce very strong and heat resistant composites. These nanoclays are currently being utilized in certain automotive components and in polymer food wrappings and packaging (31).

4. Desiccant

A calcium bentonite from Cheto, Arizona is processed for use as a desiccant. The bentonite is dried, crushed, and screened into small granules, which have the property of absorbing water. This particular bentonite can absorb moisture up to 80 % of its dry weight.

5. Palygorskite and Sepiolite

As mentioned in the chapter on (see CLAYS, SURVEY), palygorskite and attapulgite are names for the same mineral. Also, as previously discussed, attapulgite and calcium montmorillonites are classed as Fuller's earth. Therefore the term fuller's earth is quite often used to include palygorskite (attapulgite) and calcium montmorillonite. Palygorskite and sepiolite have a high surface area, a small octahedral layer charge, a fine particle size, and an elongate shape that gives palygorskite and sepiolite a high absorption capacity that makes these clays useful in many industrial applications. The elongate particles cause high viscosity when added to any liquid. The viscosity results from the elongate shape of the mineral and is not chemically induced so it is very stable as a suspending medium in many applications, eg, in salt water drilling fluids. The applications of these two minerals are many and diverse as shown in Table 8.

5.1. Oil Well Drilling Fluids. The drilling mud circulated through a well serves the primary function of removing rock cuttings from the hole. In addition, it lubricates the bit, prevents hole sloughing, and forms an impervious filter cake on the walls of the hole, thus preventing loss of fluid into porous formations. Of prime importance among the characteristics of a clay for drilling mud is the ability of the clay to build a suitable viscosity at a relatively low solids level, and

Table 8. Industrial Uses of Palygorskite (attapulgite) and Sepiolite

drilling fluids	agricultural carriers	ceramics
paint	industrial floor absorbents	asphalt
tape joint compounds	oil refining	adhesives
paper	cat box litter	pharmaceuticals
suspension fertilizers	animal feed binder	wax emulsion stabilizer
catalyst supports	medicinal suspension	polishes

to maintain the desired viscosity throughout the drilling of the well (32). Sodium bentonite is widely used for this purpose, but can only be used with fresh water. Brines and soluble salts encountered in many oil wells prevent the swelling of bentonite so it becomes ineffective in maintaining the desired viscosity when these salts are present. Attapulgite does not depend on swelling for viscosity and is stable in the presence of these contaminants so it is the preferred clay when brines and other salt contaminants are encountered. Also attapulgite drilling muds are more stable under the high temperatures encountered in deep drilling.

5.2. Liquid Suspension Fertilizers. Liquid fertilizers supply an increasing proportion of plant food consumed in the United States and in Europe and China. Liquid fertilizers require the complete solution of the components in order to be useful. The suspension fertilizers bypass the limitation of all components being in solution and 1 or 2% attapulgite stabilizes the suspension and prevents settling of the insoluble components. Attapulgite is the obvious choice for this application because of its highly stable colloidal properties in high concentrations of salts.

5.3. Adhesives. In the production of corrugated paper board, the starch adhesives lose viscosity under shear making it very difficult to apply a constant amount of adhesive during production (32). Attapulgite develops viscosity under shear so the incorporation of attapulgite is an effective method for counteracting the loss of viscosity of the starch. The addition of attapulgite to other types of adhesives stabilizes them so that uniformity is maintained by keeping the components uniformly distributed.

5.4. Colloidal and Suspension Applications. As mentioned in the section Oilwell Drilling Fluids, attapulgite-based suspensions are very stable in the presence of salts and electrolytes. Uses that utilize this property are paint, medicinal suspensions, pharmaceuticals, and polishes. In all these applications, the attapulgite prevents settling and separation of the components thus maintaining uniformity of viscosity and distribution of the components.

5.5. Absorbent Applications. The high surface area and high porosity of attapulgite leads to its use as a floor absorbent, agricultural chemical carrier, and cat box litter. Attapulgite is used in granular form for the removal of oil, grease, and water from garage and factory floors. The pores in attapulgite are large enough to permit rapid saturation and are small enough to hold the liquid firmly by capillary action. These same properties make it useful in absorbing liquid pesticides and herbicides onto free flowing granules, which may readily be applied in the field, and for extensive use as a granular product for use in cat boxes as an absorbent for cat wastes.

5.6. Tape Joint Compounds. Attapulgite pastes are used to fill in and smooth the joints in wall board. The attapulgite makes the paste a thixotropic gel that is easily applied and smoothed and also does not shrink when the paste dries.

5.7. Oil Refining. The mechanical and thermal stability and high surface area of granular attapulgite makes it useful as a percolation absorbent to remove high molecular weight compounds such as sulfonates, resins, and asphaltines in petroleum oils. The granular clay is heated to 400°C and then charged to the filter bed. The oil is percolated through the clay bed until the oil reaches a predetermined quality. The attapulgite can be washed with naptha and regenerated by heating to ~600°C and the clay reused. After 10 or so cycles, the clay is discarded.

5.8. Miscellaneous Additional Uses. The use of attapulgite as a coating on carbon-less copy papers has been previously described in the section Palygorskite (Attapulgite). Attapulgite and sepiolite are used as binders for animal feed as described in the section 8.11. Attapulgite is used to stabilize asphalt and wax emulsions and is also used as a plasticizing agent in various putties, mastics, and caulking compounds. Recently, a highly refined attapulgite has been used as a suspension medium for special ceramics to keep the various oxides uniformly distributed and to prevent settling.

6. Common Clays

Common clays and shales are important raw materials for structural ceramic clay products in most every country in the world. The products include bricks, roof tiles, sewer pipe, conduit tile, structural tile, flue linings, and others. The properties that are important are plasticity, shrinkage, dry and fired strength, fired color, and vitrification range. Shales are largely comprised of the clay minerals illite and chlorite along with quartz. Clays occurring under coals and lignites are generally mixtures of kaolinite and illite with some organic carbon. Some of these underclays are used for face brick and a low grade refractory brick. No figures for the production of these structural products are available, but a good estimate would be of the order of 100,000,000 tons. Some shales and slates are used as a raw material to produce lightweight aggregate. These shales must contain a material that will produce a gas at the vitrification temperature of the shale so that small gas vesicles will form and cause the shale to bloat (33).

BIBLIOGRAPHY

"Clays" in *ECT* 1st ed., Vol. 4: "Ceramic Clays," pp. 38–49, by W. W. Kriegel, North Carolina State College; "Fuller's Earth," pp. 49–53, by W. A. Johnston, Attapulgus Clay Co.; "Activated Clays," pp. 53–57, by G. A. Mickelson and R. B. Secor, Filtrol Corp.; "Papermaking, Paint, and Filler Clays," pp. 57–71, by S. C. Lyons, Georgia Kaolin Co.; "Rubbermaking Clays," pp. 71–80, by C. A. Carlton, J. M. Huber Corp.; and "Clays (Uses)" in *ECT* 2nd ed., Vol. 5, pp. 560–586, by R. E. Grim, University of Illinois; in *ECT* 3rd ed., Vol. 6, pp. 207–223, by R. E. Grim, University of Illinois; in *ECT* 4th ed., Vol. 6, pp. 405–423, by Paul Sennett, Engelhard Corporation; "Clays, Uses" in *ECT* (online), posting date: December 4, 2000, by Paul Sennett, Engelhard Corporation.

CITED PUBLICATIONS

1. R. H. S. Robertson, *Fuller's Earth*, Volturna, Kent, U.K., 1986.
2. H. Murray and W. Keller, Special Pub. 1, Clay Minerals Society, 1993.
3. S. Pickering and H. Murray, *Industrial Minerals and Rocks*, 6th ed., SME, Littleton, Cl., 1994.
4. M. Kuzvart, International Geological Congress Report Sess. 23rd, 1969.
5. M. Storr, 5th International Kaolin Sympocium, Ernst-Moritz-Arndt University Greifswald, 1975.
6. H. Koster, 7th International Kaolin Sympocium, Proceedings, University of Tokyo, 1976.
7. L. Martin-Vivaldi, International Geological Congress Report Sess. 23rd, 1969.
8. H. Murray, Preprint 99-135, SME Annual Mtg., Denver, Cl. 1999.
9. W. Keller and co-workers *Procedings 2nd National Conference on Clays*, National Acaddemy Science 327, 1954.
10. S. Patterson and H. Murray, U.S. Geological Survey Prof. Paper 1306, 1984.
11. H. Murray and C. Harvey, Preprint 82-83, 1st International SME Mtg., Honolulu, Hawaii, 1982.
12. F. H. Norton, *Refractories*, McGraw-Hill Book Co., New York, 1968.
13. H. Murray, *Appl. Clay Sci.* 5 (1991).
14. H. Murray, *Paper Coating Pigments*, TAPPI Momograph 38, Atlanta, Ga., 1976.
15. A. C. Eames, *TAPPI* 43 (1960).
16. W. Bundy and co-workers, *TAPPI* 48 (1965).
17. *TAPPI Test Methods*, Technical Association of the Pulp and Paper Industry, Atlanta, Ga., 1991.
18. *ASTM Handbook*, Philadelphia, Pa., 1992.
19. D. G. Sekutowski, in J. D. Edenbaum, ed., *Plastics Additives and Modifiers Handbook*, Van Nostrand Reinhold, New York, 1992.
20. H. Murray, *Industrial Minerals and Rocks*, 6th ed., SME, Littleton, Cl., 1994.
21. R. W. Grimshaw, *The Chemistry and Physics of Clays*, 4th ed., Wiley-Interscience, New York, 1971.
22. N. Trivedi and R. Hagemeyer, *Industrial Mineral and Rocks*, 6th ed., SME, Littleton, CO. 1994.
23. U.S. Pat. 2,986,472 (1961), H. Murray.
24. R. E. Grim and N. Guven, *Bentonites*, Devel. Sedimentol, Vol. 24, Elsevier, 1978.
25. U.S. Pat. 2,743,172 (1956), F. D. DeVaney.
26. K. S. Keith and H. H. Murray, *Industrial Minerals and Rocks* 6th ed., SME, Littleton, CO., 1994.
27. W. D. Lang, Preprint 71-H-29, SME, Littleton, Cl., 1971.
28. I. E. Odom, *Philos. Trans. R. Soc., London, Ser. A* 311 (1984).
29. U.S. Pat. 2,531,440 (1950), J. W. Jordan.
30. J. W. Jordan, *Proceedings of the 10th Conference on Clays and Clay Minerals*, Pergamon, Oxford, U.K., 1963.
31. G. W. Beall and co-workers Abstract: 33rd Annual Meeting, The Clay Minerals Society Program and Abstracts, 1996.
32. L. W. Haden, SME Preprint 72H-327, AIME, N.Y., 1972.
33. H. H Murray and J. M. Smith, *Indiana Geol. Survey, Rept. Prog.* 12 (1958).

HAYDN H. MURRAY
Indiana University

COAL

1. Introduction

The use of coal, known as the rock that burns, was recorded in China, Greece, and Italy >2000 years ago. Coal mining began in Germany around the tenth century AD and enough coal was mined in England for export in the thirteenth century. Coal mining began in the United States in ~1700.

Coal is usually a dark black color, although geologically younger deposits of brown coal have a brownish red color (see LIGNITE AND BROWN COAL). The color, luster, texture, and fracture vary with rank, type, and grade. Coal is the result of combined biological, chemical, and physical degradation of accumulated plant matter over geological ages. The relative amounts of remaining plant parts leads to different types of coal, which are sometimes termed banded, splint, nonbanded (cannel and boghead); or hard or soft; or lignite, subbituminous, bituminous, or anthracite. In Europe, the banded and splint types are generally referred to as ulmic or humic coals. Still other terms refer to the origins of the plant parts through maceral names such as vitrinite, liptinite, and inertinite. The degree of conversion of plant matter or coalification is referred to as rank. Brown coal and lignite, subbituminous coal, bituminous coal, and anthracite make up the rank series with increasing carbon content. The impurities in these coals cause differences in grade.

Coal consists primarily of carbon, hydrogen, and oxygen, and contains lesser amounts of nitrogen and sulfur and varying amounts of moisture and mineral matter. The mode of formation of coal, the variation in plant composition, the microstructure, and the variety of mineral matter indicate that there is a mixture of materials in coal. The nature of the organic species present depends on the degree of biochemical change of the original plant material, on the historic pressures and temperatures after the initial biochemical degradation, and on the finely divided mineral matter deposited either at the same time as the plant material or later. The principal types of organic compounds have resulted from the formation and condensation of polynuclear and heterocyclic ring compounds containing carbon, hydrogen, nitrogen, oxygen, and sulfur. The fraction of carbon in aromatic ring structures increases with rank.

Nearly all coal is used in combustion and coking (see COAL CONVERSION PROCESSES). At least 80% is burned directly in boilers for generation of electricity (see MAGNETOHYDRODYNAMICS; POWER GENERATION) or steam for industrial purposes. Small amounts are used for transportation, space heating, firing of ceramic products, etc. The rest is essentially pyrolyzed to produce coke, coal gas, ammonia (qv), coal tar, and light oil products from which many chemicals are produced (see FEEDSTOCKS, COAL CHEMICALS). Combustible gases and chemical intermediates are also produced by the gasification of coal (seeFUELS, SYNTHETIC), and different-carbon (qv) products are produced by various heat treatments. A small amount of coal is used in miscellaneous applications such as fillers (qv), pigments (qv), foundry material, and water (qv) filtration (qv).

In 1991, the annual coal production was ~1.1×10^9t in the United States and 4.7×10^9t worldwide, the latter essentially unchanged since 1991 (1,2).

World reserves of bituminous coal and anthracite are $\sim 5.6 \times 10^{12}$t of coal equivalent, ie, 29.3 GJ/t (12.6×10^3Btu/lb), and subbituminous and lignite are 2.9×10^{12} t of coal equivalent (see FUEL RESOURCES). For economic and environmental reasons coal consumption has been cyclic.

2. Origin of Coal

Coal evolved from partially decomposed plants in a shallow-water environment. Various chemical and physical changes occurred in two distinct stages: one biochemical and the other physicochemical (geochemical) (3–7). Because some parts of plant material are more resistant to biochemical degradation than others, optical variations in petrologically distinguishable coals resulted. The terms vitrain and clarain refer to bright coals; durain is a dull coal, and fusain is structured fossil charcoal. Exposure to pressure and heat during the geochemical stage caused the differences in degree of coalification or rank that are observable in the continuous series: peat, brown coal and lignite, subbituminous coal, bituminous coal, and anthracite. The carbon containing deposits in which the inorganic material predominates, such as in oil shale (qv) and bituminous shale, are not classified as coal.

Complete decay of plant material by oxidation and oxygen-based bacteria and fungi is prevented only in water-logged environments such as swamps in regions where there is rapid and plentiful plant growth. Peat is formed in such swamps from plant debris such as branches and twigs, bark, leaves, spores and pollen, and even tree trunks that are rapidly submerged in the swamp water. A series of coal seams have been formed from peat swamps growing in an area that has undergone repeated subsidence followed by deposition of lacustrine or marine intrusion material. Periods during which vegetation flourished and peat accumulated were followed by rapid subsidence resulting in submergence of the peat swamp and covering of the deposit with silt and sand. It has been suggested that in the United States the Dismal Swamp of Virginia and North Carolina, which is gradually being flooded by Lake Drummond, is an area undergoing active subsidence (8).

According to the autochthonous, *in situ*, theory of coal formation, peat beds and subsequently coal were formed from the accumulation of plants and plant debris in place. According to the allochthonous theory, the coal-producing peat bogs or swamps were formed from plant debris that had been transported, usually by streams or coastal currents, to the observed burial sites.

3. Biochemical Stage

The initial biochemical decomposition of plant matter depends on two factors: the ability of the different plant parts to resist attack and the existing conditions of the swamp water. Fungi and bacteria can cause complete decay of plant matter that is exposed to aerated water or to the atmosphere. The decay is less complete if the vegetation is immersed in water containing anaerobic bacteria. Under these latter conditions, the plant protoplasm, proteins, starches, and to a lesser extent the cellulose (qv) are easily digested. Lignin (qv) is more resistant. The most decay-resistant plant parts, for both anaerobic and aerobic decomposition, are the waxy protective layers, ie, cuticles, spore, and pollen walls, and the resins. Vitrain results from the partial decay of lignin and cellulose in stagnant water. The original cell structure of the parent plant tissue can be recognized in many samples.

The clarain (9) and bright attritus (8) are finely banded bright parts of coal that evolved from the residues of fine woody material such as branches, twigs, leaves, spores, bark, and pollen. In aerated waters, the plant parts were more decomposed and show a higher concentration of resins, spores, and cuticles. Dull coal, called durain (9), was formed under these conditions and occurs commonly in Pennsylvanian coals. In the United States, it is known as splint or block coal. More selective chemical and biochemical activity, probably in a drier environment, led to the formation of soft, charcoal-like fusain from woody plant material. The conversion was rapid and probably complete by the end of the peat formation stage. Cannel coal is believed to have formed in aerated water, which decomposed all but the spores and pollen. The name is derived from its quality of burning in splints with a candlelike flame. Boghead coal closely resembles cannel coal but was derived from algae instead of plant spores.

3.1. Geochemical Stage. The conversion of peat to bituminous coal is the result of the cumulative effects of temperature and pressure over a long time. The sediment covering the peat provides the pressure and insulation so that the earth's internal heat can be applied to the conversion. The temperature increase is ~2–5°C for each 100 m of depth. The changes in plant matter are termed normal coalification.

Moisture is lost and the chemical composition changes during coalification. Oxygen and hydrogen decrease and carbon increases. These compositional changes are accompanied by decreases in volatile matter and increases in calorific value. The volatile matter and calorific content are the main criteria used for commercial classification in the United States and for the International Classification.

The change in rank from bituminous coal to anthracite involves the application of significantly higher pressures, ie, as in mountain building activity, and temperatures, ie, as in volcanic activity. The more distant the coal from the disruption, the less proportionate the alteration. Tectonic plate movements involved in mountain building provide pressure for some changes to anthracite. As a general rule, the older the coal deposit, the more complete the coalification and the higher the rank of coal. Most commercial bituminous coal fields were deposited during the Pennsylvanian (~285–320 million years ago), Upper Cretaceous (~65–100 million years ago), and early Tertiary (~20–65 million years ago) ages. The lower rank coals come primarily from the Tertiary and Upper Cretaceous ages, and peat deposits are relatively recent, <1 million years old. However, age alone does not determine rank. The brown coal of the Moscow basin is not buried deeply, and although it was deposited during the Lower Carboniferous or Mississippian age (~320–360 million years ago), there was not enough heat and pressure to convert it further.

4. Coal Petrography

Careful examination of a piece of coal shows that it is usually made up of layers or bands of different materials that are distinct entities upon microscopic examination, which are distinguishable by optical characteristics (10–12). The study of the origin, composition, and technological application of these materials is called coal petrology, whereas coal petrography involves the systematic quantification of the amounts and characteristics by microscopic study. The petrology of coal may involve either a macroscopic or microscopic scale.

On the macroscopic scale, two coal classifications have been used: humic or banded coals and sapropelic or nonbanded coals. Stratification in the banded coals, which result from plant parts, is quite obvious; the nonbanded coals, which derive from algal materials and spores, are much more uniform. The physical and chemical properties of the different layers in a piece of coal or a seam can vary significantly. Therefore the relative amounts of the layers are important in determining the overall characteristics of the mined product. Coal petrography has been widely applied in cokemaking and is important in coal liquefaction programs.

If the mineral matter in the coal exceeds ∼40%, then the material is referred to as a coaly or carbonaceous shale. If the mineral matter is a finely divided clay, well dispersed in the coal, then the material may be described as a stony coal or bone coal.

4.1. Macerals. Coal parts derived from different plant parts, are referred to as macerals (13). The maceral names end in "-inite". The most abundant (∼85%) maceral in U.S. coal is vitrinite, derived from the woody tissues of plants. Another maceral, called liptinite, is derived from the waxy parts of spores and pollen, or algal remains. The liptinite macerals fluoresce under blue light permitting a subdivision based on fluorescence. A third maceral, inertinite, is thought to be derived from oxidized material or fossilized charcoal remnants of early forest fires.

A number of subdivisions of the maceral groups have been developed and documented by the International Commission on Coal Petrology (14). Table 1 lists the Stopes-Heerlen classification of higher rank coals. Periodic revisions include descriptions of the macerals, submacerals, morphology, physical properties, and chemical characteristics. Theories on the mode of formation of the macerals and their significance in commercial applications are also included in (14).

The macerals in lower rank coals, eg, lignite and subbituminous coal, are more complex and have been given a special classification. The term huminite has been applied to the macerals derived from the humification of lignocellulosic tissues. Huminite is the precursor to the vitrinite observed in higher rank coals.

The elemental composition of the three maceral groups varies. The vitrinite, which frequently is ∼85% of the sample in the United States, is similar to the parent coal. The liptinites are richer in hydrogen, whereas the inertinites are relatively deficient in hydrogen and richer in carbon. The liptinites also contain more aliphatic materials; the inertinites are richer in aromatics. The term inertinite refers to the relative chemical inertness of this material, making it especially undesirable for liquefaction processes because it tends to accumulate in recycled feedstock streams.

Table 1. **Stopes-Heerlen Classification of Maceral Groups, Macerals, and Submacerals of Higher Rank Coals**[a]

Maceral Group	Maceral Subgroup	Maceral (ICCP, 1995)	Maceral (ASTM D 2799)
Vitrinite	Telinite	Telinite	Vitrinite
		Collotelinite	
	Detrovitrinite	Vitrodetrinite	
		Collodetrinite	
	Gelovitrinite	Gelinite	
		Corpogelinite	
Liptinite			Sporinite
			Cutinite
			Resinite
			Alginite
Inertinite	Telo-Inertinite		Fusinite
			Semifusinite
			Sclerotinite
	Detro-Inertinite		Inertodetrinite
			Micrinite
	Gelo-Inertinite		Macrinite

[a] See Ref. 9.

4.2. Vitrinite Reflectance.

The amount of light reflected from a polished plane surface of a coal particle under specified illumination conditions increases with the aromaticity of the sample and the rank of the coal or maceral. Precise measurements of reflectance, expressed as a percentage, are used as an indication of coal rank.

Precise reflectance measurements are carried out using incident light having wavelength of 546 nm (green mercury line), and carefully polished coal specimens immersed in an oil having a refractive index of 1.518 at 23°C. Comparison is made to a calibrated standard, using a photomultiplier system. Coal is an anisotropic material, so the reflectance varies according to the orientation of the particle. A typical procedure involves making many measurements on the vitrinite particles in a coarsely ground sample to obtain a range of values that are then used to determine a maximum vitrinite reflectance that correlates with coal rank. Minimum values can also be correlated. Figure 1 illustrates the relationship between reflectances and the carbon content (12). The reflectance of liptinite macerals is less than that for vitrinite, and the petrographer can distinguish the two for low rank coals. However, as measurements are made on progressively higher rank coals, the reflectivities of the liptinites and vitrinites become similar and are the same for medium volatile bituminous coals. For inertinites, the distinctions between reflectivities for vitrinite persist into the anthracite range. Table 2 indicates the vitrinite reflectances for the various coal ranks (12).

4.3. Application of Coal Petrology and Petrography.

Petrographic analysis is frequently carried out for economic evaluation or to obtain geologic information. Samples are usually lumps or more coarsely ground material that have been mounted in resins and polished. Maceral analysis involves the examination of a large number (usually >500) of particles during a traverse of a

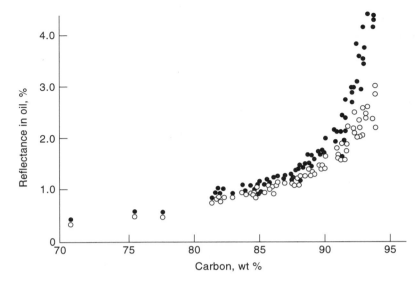

Fig. 1. Relation of vitrinite reflectance (R_o) in percent to maximum temperature (T_{max}) and effective heating time (t_{eff}) where E_A = activation energy in kJ and t_{eff} is within 15°C of T_{max}. To convert kJ to kcal, divide by 4.184 (11).

polished surface to identify the macerals at specified intervals. A volume percentage of each of the macerals present in a sample is calculated.

Seam correlations, measurements of rank and geologic history, interpretation of petroleum (qv) formation with coal deposits, prediction of coke properties, and detection of coal oxidation can be determined from petrographic analysis. Constituents of seams can be observed over considerable distances, permitting the correlation of seam profiles in coal basins. Measurements of vitrinite reflectance within a seam permit mapping of variations in thermal and tectonic histories. Figure 2 indicates the relationship of vitrinite reflectance to maximum temperatures and effective heating time in the seam (11,15).

Table 2. **Vitrinite Reflectance Limits, Taken in Oil, and ASTM Coal Rank Classes**[a]

Coal rank	Maximum reflectance, %
subbituminous	<0.47
high volatile bituminous	
C	0.47–0.57
B	0.57–0.71
A	0.71–1.10
medium volatile bituminous	1.10–1.50
low volatile bituminous	1.50–2.05
semianthracite	2.05–3.00[b]
anthracite	>3.00[b]

[a] Ref. 11.
[b] Approximate value.

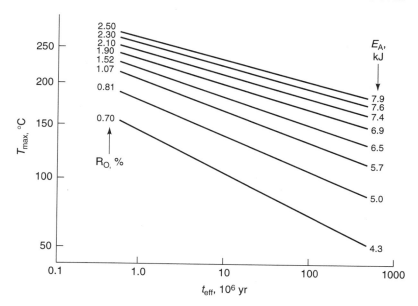

Fig. 2.   Relation of vitrinite reflectance (R_o) in percent to maximum temperature (T_{max}) and effective heating time (t_{eff}) where E_A = activation energy in kJ and t_{eff} is within 15°C of T_{max}. To convert kJ to kcal, divide by 4.184 (11).

The coking behavior of coal depends on the rank of the coal, the properties of the individual constituents, and their relative amounts (3–7,10). For some purposes, a blend of coals can be selected to achieve desired coking properties. The maceral groups behave differently on heating: vitrinite from most medium rank coal (9–33% volatile matter) has good plasticity and swelling properties and produces an excellent coke; inertinite is almost inert and does not soften on heating, and exinite becomes extremely plastic and is almost completely distilled as tar. By careful control of the petrological composition and the rank of a coal blend, behavior during carbonization can be controlled. Additionally, coking behavior can be reasonably predicted using petrography and maceral breakage (10). Oxidation reduces the coke-forming properties of a given coal and can also be detected by petrographic techniques (16).

5. Classification Systems

Prior to the nineteenth century, coal was classified according to appearance, eg, bright coal, black coal, or brown coal. A number of classification systems have since been developed. These may be divided into two types, which are complementary: scientific and commercial. Both are used in research, whereas the commercial classification is essential industrially. In the scientific category, the Seyler chart has considerable value.

5.1. The Seyler Classification.   The Seyler chart, shown in Figure 3, is based on the carbon and hydrogen content of coals determined on a dry

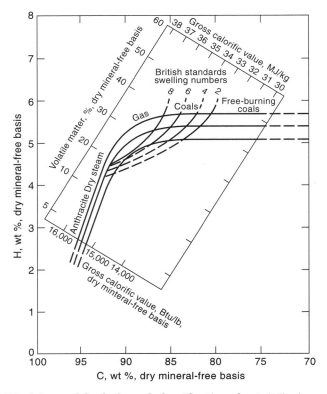

Fig. 3. Simplified form of Seyler's coal classification chart (17). An updated version of Seyler's coal classification is described in Reference6. Note that ASTM uses the free-swelling index (18).

mineral-matter-free basis (17). Points representing different coal samples lie along a broad band. The center band on the chart shows the properties of coal rich in vitrinite. The location of the band indicates the range and interrelationship of the properties. Coals above the band are richer in hydrogen, eg, cannel and boghead coals, and the liptinite macerals in the usual coals. Coals below the band are represented by the maceral inertinite. Other properties, such as moisture and swelling indexes, also fit into specific areas on this chart. The curve in the solid band represents a composition range where the properties of the coal change rapidly. Swelling indexes, coking power, and calorific values are maximized, and moisture is minimized. The lowest rank coals lie on the right side of this curve; the highest rank coals are on the left at the lower part of the band.

5.2. The ASTM Classification. The ASTM classification system was adopted in 1938 as a standard means of specification. This system is used in the United States and in many other parts of the world, and is designated D388 in the ASTM Standards (18). The higher rank coals are specified by fixed carbon ≥69%, or for volatile matter ≤31%, on a dry, mineral-free basis. Lower rank coals are classified by calorific value on the moist, mineral-matter-free basis. These parameters are given in Table 3. Calorific value depends on two

Table 3. Classification of Coals by Rank

Coals	Fixed carbon, %[a] ≥	Fixed carbon, %[a] <	Volatile matter, %[a] >	Volatile matter, %[a] ≤		Gross calorific value, kJ/kg ≥	Gross calorific value, kJ/kg <	Agglomerating character
					Anthracitic			
meta-anthracite	98			2				nonagglomerating
anthracite	92	98	2	8				
semianthracite	86	92	8	14				
					Bituminous			
low volatile	78	86	14	22				
medium volatile	69	78	22	31				commonly agglomerating[e]
high volatile A		69	31			32,500[d]		
B						30,200[d]	32,500	
C						26,700	30,200	agglomerating
						24,400	26,700	
					Subbituminous		22,100	
A						24,400	26,700	
B						22,100	24,400	
C						19,300	22,100	
					Lignitic			
A						14,600	19,300	nonagglomerating
B							14,600	

[a] Dry, mineral-matter-free basis.
[b] To convert from kJ/kg to Btu/lb, multiply by 0.4302; moist mineral-matter-free basis, ie, contains inherent moisture but not water visible on the surface.
[c] If agglomerating, classify in low volatile group of the bituminous class.
[d] Coals having 69% or more fixed carbon on the dry, mineral-matter-free basis are classified according to fixed carbon, regardless of gross calorific value.
[e] There may be nonagglomerating varieties in the groups of the bituminous class, and there are notable exceptions in high volatile C bituminous group.

711

properties: moisture absorbing capacity and the calorific value of the pure coal matter. When some overlap between bituminous and subbituminous coals occurs, it is resolved on the basis of the agglomerating properties.

5.3. National Coal Board Classification for British Coals. The classification proposed in 1946 by the U.K. Department of Scientific and Industrial Research led to the system in use by the National Coal Board for coals in the United Kingdom. There are two parameters: the quantity of volatile matter determined on a dry, mineral-matter-free basis, and the Gray-King coke-type assay, a measure of coking power as designated in the British Standards (18). This latter assay is used as a primary means of classification for lower rank coals. The classification applies to coals having <10% ash. High ash coals are cleaned before analysis by a float-sink separation to reduce the ash content <10%.

5.4. International Classification. *Hard Coal.* The amount of coal in international commerce since ~1945 necessitated an international system of coal classification and in 1956 the Coal Committee of the European Economic Community agreed on a system designated the International Classification of Hard Coal by Type (3). Volatile matter and gross calorific value on a moist, ash-free basis are among the parameters considered. Table 4 shows the various classes of the international system and gives the corresponding national names used for these coals.

A three-digit classification is employed in the international system (10), where the first digit indicates the class or rank, such that higher digits correspond to lower ranks; the second digit indicates the group indicated by caking properties such as the free-swelling index or the Roga index; and the third digit defines a subgroup based on coking properties as measured using a dilatometer or the Gray-King assay. Coals having volatile matter up to 33% are divided into classes 1–5; coals having volatile matter <33% are divided into classes 6–9. The calorific values are given for a moisture content obtained after equilibrating at 30°C and 96% rh. The nine classes are then divided into four groups as measured through either the free-swelling index (17) or the Roga index. These tests indicate properties observed when the coal is heated rapidly.

Brown Coal and Lignite. The brown coals and lignites, defined as coals having heating values that are <23,860 kJ/kg (10,260 Btu/lb, 5700 kcal/kg), are classified separately (see LIGNITE AND BROWN COAL). A four-digit code is used for classification. The first two digits (class parameter) are defined by total moisture content of freshly mined coal on an ash-free basis. The third and fourth digit are defined by the tar yield on a dry, ash-free basis.

6. Composition and Structure

The constitution of a coal involves both the elemental composition and the functional groups that are derived therefrom. The structure of the coal solid depends to a significant extent on the arrangement of the functional groups within the material.

6.1. Composition. The functional groups within coal contain the elements C, H, O, N, or S (3,4,5,19). The significant oxygen-containing groups

Table 4. The International and Corresponding National Systems of Coal Classes[a]

	International system			National classifications						
Class no.	Volatile matter, %	Calorific value, kJ/g[b,c]	Belgium	Germany	France	Italy	The Netherlands	Poland	United Kingdom	United States
0	0–3					antraciti speciali		*meta*-antracyt		*meta*-anthracite
1A	3–6.5		maigre	Anthrazit	anthracite	antraciti communi	anthraciet	antracyt	anthracite	anthracite
1B	6.5–10									
2	10–14		1/4 gras	Mager-kohle	maigre	carboni magri	mager	polantracyt chudy	dry steam	semianthracite
3	14–20		1/2 gras	Esskohle	demigras	carboni semigrassi	esskool	polkoksowy *meta*-koksowy	coking steam	low volatile bituminous
4	20–28		3/4 gras	Fettkohle	gras à courte flamme	carboni grassi corta fiamma	vetkool	*orto*-koksowy	medium volatile coking	medium volatile-bituminous
5	28–33		gras	Gaskohle	gras proprement dit	carboni grassi media fiamma				
6	>33 (33–40)	32.4–35.4				carboni da gas	gaskool	gazowo koksowy		high volatile bituminous A
7	>33 (32–44)	30.1–32.4			flambant gras	carboni grassi da vapore	gasvlam-kool	gazowy	high volatile	high volatile bituminous B
8	>33 (34–46)	25.6–30.1		Gas flamm-kohle	flambant sec	carboni secchi	vlamkool	gazowo-plomienny		high volatile bituminous C
9	>33 (36–48)	<25.6						plomienny		subbituminous

[a] Ref. 3.
[b] Calculated to standard moisture content.
[c] To convert kJ/g to Btu/lb, multiply by 430.2.

found in coals are carbonyl, hydroxyl, carboxylic acid, and methoxy. The nitrogen-containing groups include aromatic nitriles, pyridines, carbazoles, quinolines, and pyrroles (20). Sulfur is primarily found in thiols, dialkyl and aryl–alkyl thioethers, thiophene groups, and disulfides. Elemental sulfur is observed in oxidized coal (20).

The relative and absolute amounts of the various groups vary with coal rank and maceral type. The principal oxygen-containing functional groups in vitrinites of mature coals are phenolic hydroxyl and conjugated carbonyls as in quinones. Spectroscopic evidence exists for hydrogen bonding of hydroxyl and carbonyl groups. There are unconjugated carbonyl groups such as ketones in exinites. The infrared (ir) absorption bands are displaced from the normal carbonyl range for simple ketones by the conjugation in vitrinites. Interactions between the carbonyl and hydroxyl groups affect the normal reactions.

A range of quantitative organic analytical techniques may be used to determine functional group concentrations. Acetylation and O-alkylation are used to determine hydroxyl groups, whereas carbonyl groups are difficult to quantify using simple procedures. A variety of instrumental techniques has also been used to aid in the understanding of coal structure and constitution. Magnetic resonance techniques have been particularly helpful in determining relative amounts of different carbon species within a coal. Table 5 contains data obtained using these techniques (21).

Aromaticity of coal molecules increases with coal rank. Calculations based on several models indicate that the number of aromatic carbons per cluster varies from 9 for lignite to 20 for low volatile bituminous coal, and the number of attachments per cluster varies from 3 for lignite to 5 for subbituminous through medium bituminous coal. The value is 4 for low volatile bituminous (21).

Reaction of coals and mild selective oxidizing agents such as benzoquinone (20,22) causes the coals to lose much of the hydrogen content. Similarly, a palladium catalyst can cause the evolution of molecular hydrogen (23,24). These methods may give an indication of the minimum amount of hydrogen in the coal that is involved in hydroaromatic rings. This amount is close to the total nonaromatic hydrogen determined for lower rank coals. Other hydrogen determining methods involve dehydrogenation using sulfur (25) and using halogens (26). The values obtained by these last methods are somewhat lower than that of benzoquinone.

Hydrogen can be added to the aromatic structures converting them to hydroaromatic rings. The hydrogen addition and removal is generally but not entirely reversible (24).

High resolution mass spectrometry (ms) (qv) has been used with extracts of a series of coals to indicate the association of different heteroatoms (27). Various types of chromatography (qv) have also been used to identify the smaller species that can be extracted from coal.

6.2. Coal Structure. *Bonding in Macromolecules.* Conclusions regarding the chemical structure of the macromolecules within coal are generally based on experimental measurements and an understanding of structural organic chemistry (3,4,20,28). The description given herein refers to vitrinites.

Several requirements must be met in developing a structure. Not only must elementary analysis and other physical measurements be consistent, but limita-

Table 5. Carbon Structural Distribution of the Argonne Premium Coals Based on NMR Measurements[a]

Coal[c]	f_a	f'_a	f_a^C	f_a^H	f_a^N	f_a^P	f_a^S	f_a^B	f_{al}	f_{al}^H	f_{al}^*	f_{al}^O
North Dakota (L)	0.61	0.54	0.07	0.26	0.28	0.06	0.13	0.09	0.39	0.25	0.14	0.12
Wyodak (SB)	0.63	0.55	0.08	0.17	0.38	0.08	0.14	0.16	0.37	0.27	0.10	0.10
Blind Canyon (HVB)	0.65	0.64	0.01	0.22	0.42	0.07	0.15	0.20	0.35	0.22	0.13	0.04
Illinois no. 6 (HVB)	0.72	0.72	0.00	0.26	0.46	0.06	0.18	0.22	0.28	0.19	0.09	0.05
Pittsburgh (HVB)	0.72	0.72	0.00	0.27	0.45	0.06	0.17	0.22	0.28	0.13	0.15	0.03
Lewiston-Stockton (HVB)	0.75	0.75	0.00	0.27	0.48	0.05	0.21	0.22	0.25	0.14	0.11	0.04
Upper Freeport (MVB)	0.81	0.81	0.00	0.28	0.53	0.04	0.20	0.29	0.19	0.09	0.10	0.02
Pocahontas (LVB)	0.86	0.86	0.00	0.33	0.53	0.02	0.17	0.34	0.14	0.08	0.06	0.01

[a] Ref. 21.
[b] The symbols f_a and f_{al} correspond to total fraction of sp^2 and sp^3 hybridized carbon, respectively. f'_a represents the fraction of sp^2 carbon in aromatic rings; f_a^C = the fraction in carbonyls, $\delta > 165$ ppm; f_a^H = the aromatic fraction that is protonated; f_a^N = the aromatic fraction that is nonprotonated; f_a^P = the phenolic or phenolic ether carbon, $\delta = 150$–165 ppm; f_a^S, the alkylated aromatic carbon, $\delta = 135$–150 ppm; f_a^B, the aromatic bridgehead carbon; f_{al}^H represents the fraction of CH or CH_2 aliphatic carbon; f_{al}^* = the CH_3 or nonprotonated aliphatic carbon; and f_{al}^O = the aliphatic carbon bound to oxygen, $\delta = 50$–90 ppm.
[c] L = lignite; SB = subbituminous; HVB = high volatile bituminous; MVB = medium volatile bituminous; and LVB = low volatile bituminous.

tions of structural organic chemistry and stereochemistry must also be satisfied. Mathematical expressions have been developed to test the consistency of any given set of parameters used to describe the molecular structure of coal and analyses of this type have been reported (4,6,19,20,29,30).

Evidence suggests that the structure for vitrinites in bituminous coals and anthracite has the following characteristics: (1) the molecule contains a number of small aromatic nuclei or clusters, each usually having from one to four fused benzene rings. The average number of clusters in the molecule is characteristic of the coal rank. This average increases slightly to 90% carbon and then increases rapidly; (2) the aromatic clusters are partly linked together by hydroaromatic, including alicyclic, ring structures. These latter rings also contain six carbon atoms and thus, upon loss of hydrogen can become part of the cluster, increasing the average cluster size; (3) other linkages between clusters involve short groupings such as methylene, ethylene, and ether oxygen ; (4) a significant amount of hydrogen sites on the aromatic and hydroaromatic rings have been substituted by methyl and sometimes larger aliphatic groups; (5) the oxygen in vitrinites usually occurs in phenolic hydroxyl groups and ethers (19), substituting for hydrogen on the aromatic structures; (6) a small fraction of the rings, both aromatic and hydroaromatic, have oxygen substituted for a carbon atom. Some of these heterocyclic rings may be five membered; (7) a lesser amount of oxygen than that occurring in phenolic groups appears in the carbonyl moieties, ie, in quinone form on aromatic rings or as ketones attached to hydroaromatic ones. In both cases, the oxygen is apparently hydrogen bonded to adjacent hydroxyl groups: the reactivity and peak location for the characteristic ir absorptions are different from those in the typical quinones and ketones; (8) nitrogen is less abundant in vitrinites than oxygen. It is usually present as a heteroatom in a ring structure or as a nitrile (19); (9) most of the sulfur in as-mined coal, especially if the S content exceeds 2 wt%, is associated with inorganic material. However, for very low sulfur coals the organic sulfur which occurs as both aliphatic and aromatic usually exceeds the inorganic. Sulfide and disulfide groups may also link clusters (19,20); (10) a given piece of coal contains a variety of molecules having different proportions of a given structural feature and varying molecular weight. These molecules are composed of planar fused ring clusters linked to nonplanar hydroaromatic structures. The overall structure is irregular, open, and complex. Entanglement between molecules occurs, as do cross-links of hydrogen-bonded species, leading to difficulties in molecular weight determinations for extracts and to changes in properties on heating. The different molecular shapes and sizes in a piece of coal lead to irregularities in packing and hence the amorphous nature and the extensive ultrafine porosity; and (11) some of the evidence suggests that the molecular weights of extracts representing 5–50 wt% of the coal average 1000–3000. Larger transient units also exist owing to aggregation and it is probable that unextracted coal contains even larger molecular units (19).

Figure 4 gives a representation of the coal molecule (28) that correlates with products obtained from liquefaction. Heating >400°C or mild chemical oxidation changes some hydroaromatic into aromatic structures. Hydroaromatic structures appear to be involved in tar formation (31); however, the tar probably contains some of the same smaller aromatic structures also found in the coal. Tar

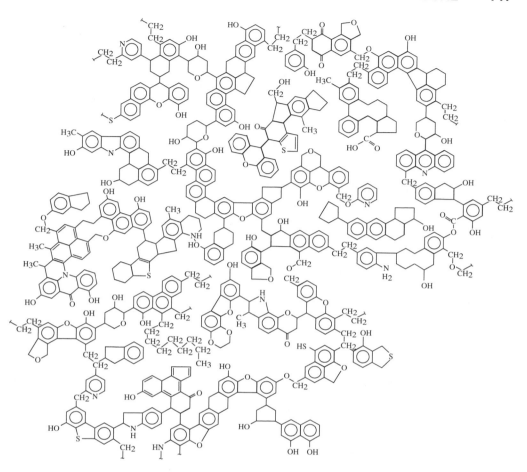

Fig. 4. Model of bituminous coal structure (28).

and gas production, including changes with rank, is consistent (3,4) with the structural model described. Most of the larger units remain in the char or coke produced by heating. The average size of the molecular units is increased by polymerization and conversion of hydroaromatic links to aromatic rings.

Bonding between Macromolecules. The macromolecules that make up the coal structure are held together by a variety of forces and bonds (32). The coal network model is one approach to describing the three-dimensional (3D) structure of the solid. Aromatic clusters are linked by a variety of connecting bonds, through oxygen, methylene or longer aliphatic groups, or disulfide bridges, and the proportions of the different functional groups change as the rank of the coal is progressively increased. For example, oxygen is diminished. Acid groups decrease early in the rank progression and other groups follow leaving ether moieties that act as cross-links between the larger clusters in the high rank coals. Another type of linkage involves hydrogen bonds which, eg, hold hydroxy and ketogroups together in the solid. A review of coal models, the mechanical properties of the network, and the glass-transition temperature corresponding to the change from a fluid to a rigid amorphous solid are available (32).

A model for coal fluidity based on a macromolecular network pyrolysis model has been developed (33). In that model, bond breaking is described as a first-order reaction having a range of activation energies. A variety of lattices have also been used to describe the bonding in coal. In turn, these structures have been used to describe devolatilization, combustion, and char formation (34). The form of the macromolecule in a liquid extract tends to be spherical as a result of surface energy considerations, hydrogen bonds, and van der Waals forces (32).

6.3. Coal Constitution.   Chemical composition studies (35,36) indicate that brown coals have a relatively high oxygen content. About two-thirds of the oxygen is bonded carboxyl, acetylatable hydroxyl, and methoxy groups. Additionally, unlike in bituminous coals, some alcoholic hydroxyl groups are believed to exist.

Anthracites.   The anthracites, which approach graphite in composition (see CARBON,GRAPHITE), are classified higher in rank, have less oxygen and hydrogen, and are less reactive than bituminous coals. Anthracites are also insoluble in organic solvents. These characteristics become more pronounced as rank increases within the anthracite group. The aromatic carbon fraction of anthracites is at least 0.9, and the number of aromatic rings per cluster is greater than that for the low volatile bituminous coals, with a value of ~10 for anthracite having 95 wt% C. There is X-ray diffraction evidence (37) to indicate that the aromatic rings are more loosely and variably assembled than those in bituminous coal clusters. The anthracites have greater optical and mechanical anisotropy than lower rank coals, and the internal pore volume and surface increase with rank after the minimum below ~90 wt% C.

6.4. Mineral Matter in Coal.   The mineral matter (7,38) in coal results from several separate processes. Some comes from the material inherent in all living matter; some from the detrital minerals deposited during the time of peat formation; and a third type from secondary minerals that crystallized from water that has percolated through the coal seams.

A variety of instrumental techniques may be used to determine mineral content. Typically the coal sample is prepared by low temperature ashing to remove the organic material. Then one or more of the techniques of X-ray diffraction, ir spectroscopy, differential thermal analysis, electron microscopy, and petrographic analysis may be employed (7).

The various clay minerals are the most common detrital mineral (see CLAYS); however, other common ones include quartz, feldspar, garnet, apatite, zircon, muscovite, epidote, biotite, augite, kyanite, rutile, staurolite, topaz, and tourmaline. The secondary minerals are generally kaolinite, calcite, and pyrite. Analyses have shown the presence of almost all elements in at least trace quantities in the mineral matter (39). Certain elements, ie, germanium, beryllium, boron, and antimony, are found primarily with the organic matter in coal, whereas zinc, cadmium, manganese, arsenic, molybdenum, and iron are found with the inorganic material. The primary elemental constituents of mineral matter in coal are aluminum, silicon, iron, calcium, magnesium, sodium, and sulfur.

The relative concentrations depend primarily on the geographical location of the coal seam, and vary from place to place within a given field. In the eastern United States, the most abundant mineral elements are silicon, aluminum, and iron and there are much lower amounts of alkali and alkaline earth elements. West of the Mississippi River the relative amounts of silicon, aluminum, and iron are much less and the alkaline earth and alkali elements are much greater.

7. Properties

Pieces of coal are mixtures of materials somewhat randomly distributed in differing amounts. The mineral matter can be readily distinguished from the organic, which is itself a mixture. Coal properties reflect the individual constituents and the relative proportions. By analogy to geologic formations, the macerals are the constituents that correspond to minerals that make up individual rocks. For coals, macerals, which tend to be consistent in their properties, represent particular classes of plant parts that have been transformed into coal (40). Most detailed chemical and physical studies of coal have been made on macerals or samples rich in a particular maceral, because maceral separation is time consuming.

The most predominant maceral group in U.S. coals is vitrinite. The other important maceral groups include inertinite, including fusinite, a dull fibrous material similar to charcoal, and the liptinite group, including sporinite, which is relatively fusible and volatile. Differences in macerals are evident over the range of coal rank, ie, from brown coal or lignite to anthracite. The definition of rank is that generally accepted as the wt% C, on a dry, mineral-free basis, in the vitrinite associated with the given coal in the seam. The range of ranks in which differences between macerals are most significant is 75–92 wt% C content of the vitrinite. These coals are bituminous.

In the United States the commercial classification of coals is based on the fixed carbon (or volatile matter) content and the moist heating value. One correlation is made by plotting the hydrogen content, on a mineral-free basis, against the corresponding carbon content. A similar plot, made using the commercial criteria of volatile matter and heating placed on axes at an appropriate angle to the % C, % H axes, forms the Seyler coal classification chart (17). Both are illustrated in Figure 3. Table 6 indicates the usual range of composition of commercial coals of increasing rank.

7.1. Physical Methods of Examination. Physical methods used to examine coals can be divided into two classes that, in the one case, yield information of a structural nature such as the size of the aromatic nuclei, ie, methods such as X-ray diffraction, molar refraction, and calorific value as a function of composition; and in the other case indicate the fraction of carbon present in aromatic form, ie, methods such as ir and nuclear magnetic resonance (nmr) spectroscopies, and density as a function of composition. Some methods used and types of information obtained from them are (41).

The scattering of X-rays (6,37,42) gives information on the average distances between the carbon atoms in coal and insight into the bonding between these atoms. Because X-ray scattering depends on the number of protons in the

Table 6. **Composition of Humic Coals**

Type of coal	Composition, wt%[a]						Calorific value, kJ/g[b]
	C	H	O	N	Moisture as found	Volatile matter	
peat	45–60	3.5–6.8	20–45	0.75–3.0	70–90	45–75	17–22
brown coals and lignites	60–75	4.5–5.5	17–35	0.75–2.1	30–50	45–60	28–30
bituminous coals	75–92	4.0–5.6	3.0–20	0.75–2.0	1.0–20	11–50	29–37
anthracites	92–95	2.9–4.0	2.0–3.0	0.5–2.0	1.5–3.5	3.5–10	36–37

[a] Dry, mineral-matter-free basis except for moisture value.
[b] To convert kJ/g to Btu/lb, multiply by 430.2.

nucleus, carbon is much more effective in scattering X-rays than hydrogen (see X-RAY TECHNOLOGY). The ultraviolet (uv) and visible spectra (6) of coal and various solvent extracts show decreasing absorption with increasing wavelength and lack features to aid in interpreting structure except for one peak ~270 nm, which is believed to result from superposition of effects from many similar species. In studies of specific features, comparisons are usually made between coal or coal-derived samples, and pure, usually aromatic, compounds indicating probable presence of particular structures or functional groups. Similar statements can be made concerning reflectance and refractive index (3,4). The derived optical anisotropy is especially evident in coals having carbon contents that exceed 80–85 wt%. Measurements perpendicular and parallel to the bedding plane give different results for optical and some other characteristics (see SPECTROSCOPY).

A significantly greater amount of information concerning functional groups such as hydroxyl can be obtained from ir absorption (3,4,6); however, this is less specific than the information obtained from an individual organic compound (Fig. 5) (6). An estimate of the relative amount of hydrogen attached to aromatic and nonaromatic structures can, however, be made by using this method (see INFRARED AND RAMAN SPECTROSCOPY). Studies may be carried out on raw coal or products derived from the coal. Physical separation is used to separate fractions of extract and aid in the deduction of the parent coal structure. A method of characterizing coal liquids in terms of 10 fractions of different functionality has been described (43).

Magnetic resonance spectra (^{1}H and ^{13}C) (6,21) also yield information on bonding for hydrogen and carbon, including the distribution between aromatic and nonaromatic structures, as well as bonding to various heteroatoms. Additional estimates may be made of hydrogen in CH, CH_2, and CH_3 groups. Developments in solid-state ^{13}C nmr spectroscopy, coupled with cross-polarization (cp), magic angle spinning (mas), and dipolar-decoupling techniques have made these estimates somewhat more quantitative than those from ir measurements (see MAGNETIC SPIN RESONANCE). Quantitation has been somewhat limited because of a fraction of the atoms that are not observed owing to paramagnetic centers or spin dynamics of the system (21) (see Table 5).

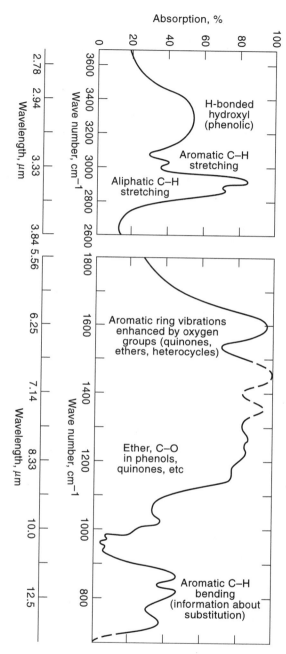

Fig. 5. Infrared spectrum of a medium rank coal where the dashed lines represent variations that occur as a result of differing maceral content or because of functional group conjugation.

Electron spin resonance (esr) (6,44) has had more limited use in coal studies. A rough estimate of the free-radical concentration or unsatisfied chemical bonds in the coal structure has been obtained as a function of coal rank and heat treatment. For example, the concentration increases from 2×10^{18} radicals/g at 80 wt% carbon to a sharp peak of $\sim 50 \times 10^{18}$ radicals/g at 95 wt % carbon content and drops almost to zero at 97 wt% carbon. The concentration of these radicals is less than that of the common functional groups such as hydroxyl. However, radical existence seems to be intrinsic to the coal molecule and may affect the reactivity of the coal as well as its absorption of uv radiation. Measurements from room temperature to 900 K indicate that the number of electron spins/g increases sharply above ~ 600 K and peaks from 773 to 850 K, with increasing values for higher rank coals. Oxidation increases the number of radicals by a factor of 3 over 8 days (44).

The other physical measurements (4,6), except for diamagnetic susceptibility (4) and possibly density (4), are primarily of interest for determining chemical structural properties of coal.

7.2. Physical Properties. Most of the physical properties discussed herein depend on the direction of measurement as compared to the bedding plane of the coal. Additionally, these properties vary according to the history of the piece of coal. Properties also vary between pieces because of coal's brittle nature and the crack and pore structure. One example concerns electrical conductivity. Absolute values of coal sample specific conductivity are not easy to determine. A more characteristic value is the energy gap for transfer of electrons between molecules, which is determined by a series of measurements over a range of temperatures and is unaffected by the presence of cracks. The velocity of sound is also dependent on continuity in the coal.

The specific electrical conductivity of dry coals is very low, specific resistance 10^{10}–10^{14} Ωcm, although it increases with rank. Coal has semiconducting properties. The conductivity tends to increase exponentially with increasing temperatures (4,6). As coals are heated to above $\sim 600°C$ the conductivity rises especially rapidly owing to rearrangements in the carbon structure, although thermal decomposition contributes somewhat below this temperature. Moisture increases conductivity of coal samples through the water film.

The dielectric constant is also affected by structural changes on strong heating. Also the value is very rank dependent, exhibiting a minimum at ~ 88 wt% C and rising rapidly for carbon contents >90 wt% (4,6,45). Polar functional groups are primarily responsible for the dielectric of lower ranks. For higher ranks the dielectric constant arises from the increase in electrical conductivity. Information on the freedom of motion of the different water molecules in the particles can be obtained from dielectric constant studies (45).

Magnetic susceptibility measurements indicate that the organic part of the coal is diamagnetic, having traces of paramagnetic behavior resulting from free radicals or unpaired electrons (6).

Density values (4,6) of coals differ considerably, even after correcting for the mineral matter, depending on the method of determination. The true density of coal matter is most accurately obtained from measuring the displacement of helium after the absorbed gases have been removed from the coal sample. Density values increase with carbon content or rank for vitrinites. Values are

1.4–1.6 g/ml >85 wt % carbon where there is a shallow minimum. A plot of density versus hydrogen content gives almost a straight-line relationship, and if the reciprocal of density is plotted, the linear relationship is improved. Values for different macerals as well as for a given maceral of different ranks are almost on the same line.

Thermal conductivity and thermal diffusivity are also dependent on pore and crack structure. Thermal conductivities for coals of different ranks at room temperature are in the range of 0.230.35 W/(m × K). The range includes the spread owing to crack variations and thermal diffusivities of $(1–2) \times 10^{-3}$ cm ^{2}/s. At 800°C, these ranges increase to 1–2W /(m × K) and $(1–5) \times 0^{-2}$ cm ^{2}/s, respectively. The increase is mainly caused by radiation across pores and cracks.

The specific heat of coal can be determined by direct measurement or from the ratio of the thermal conductivity and thermal diffusivity. The latter method gives values decreasing from 1.25J/(g × K)[0.3cal/(g × K)] at 20°C to 0.4J/ [g × K)(0.1cal /(g × K)] at 800°C. The specific heat is affected by the oxidation of the coal (46).

7.3. Ultrafine Structure. Coal contains an extensive network of ultra-fine capillaries (3,4,6,47) that pass in all directions through any particle. The smallest and most extensive passages are caused by the voids from imperfect packing of the large organic molecules. Vapors pass through these passages during adsorption, chemical reaction, or thermal decomposition. The rates of these processes depend on the diameters of the capillaries and any restrictions in them. Most of the inherent moisture in the coal is contained in these capillaries. The porous structure of the coal and products derived from it have a significant effect on the absorptive properties of these materials.

A range of approaches has been developed for studying the pore structure. For example, heat of wetting by organic liquids is one measure of the accessible surface. The use of liquids having different molecular sizes gives information about restrictions in the pores. Measurements of the apparent density in these liquids give corresponding information about the volume of capillaries. Measurement of the adsorption of gases and vapors provide information about internal volume and surface area. Pores have been classified into three size ranges: (1) micropores (<0.4–2.0 nm) measured by CO_2 adsorption at 298 K; (2) mesopores (2.0–50 nm) from N_2 adsorption at 77 K; and (3) macropores (>50 nm) from mercury porosimetry. For coals having <75 wt% C, macropores primarily determine porosity. For 76–84 wt% C, ~80% of the pore volume primarily results from micro and transitional pores. For the higher rank coals porosity is caused primarily by micropores (48).

Bituminous coals appear to have specific internal surfaces in the range of 30–100 m^2/g arising almost entirely from ultrafine capillaries of <4-nm diameters. The surface area of the very fine capillaries can be measured accurately by using methods not too far below room temperature, depending on the gas or vapor used (49). Diffusion into the particle is very slow at low temperatures. Therefore, measurements at liquid nitrogen temperature (77 K) relate to the external surfaces and macro- as well as mesopores, and may yield areas that are lower than ambient temperature measurements by factors of 100. Sorption by neon or krypton near room temperature and heat of wetting in methanol have given surface area values. The methanol method is affected, up to a factor

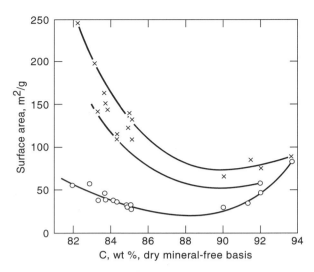

Fig. 6. Surface area of coals as estimated from (○) neon sorption at 25°C and (x) methanol heat of wetting.

of 4, by polar groups, but it is faster. Pore characteristics may also be determined by nmr measurements, as with xenon (50). Measurements of the change in the nmr chemical shift with varying Xe pressures permit the calculation of pore diameters. Total porosity volumes of bituminous and anthracite coal particles are ~10–20%, and ~3–10% are in the microphase range. There are shallow minima in plots of internal area as can be seen from Figure 6 and in plots of internal porosity against coal rank in the range of vitrinite carbon content of 86–90% (6). It is possible to use low angle scattering of X-rays to obtain a value of internal surface, but this does not distinguish between accessible capillaries and closed pores (51).

Different coals have been observed in the electron microscope when two pore-size ranges appear, one of >20 nm and the other <10 nm (52). Fine pores from 1–10 nm across have been observed using a lead impregnation procedure (53). Effectiveness of coal conversion processes depends on rapid contact of gases with the surface. Large internal surfaces are required for satisfactory rates (54).

7.4. Mechanical Properties. Mechanical properties (4,6,55) are important for a number of steps in coal preparation from mining through handling, crushing, and grinding. The properties include elasticity and strength as measured by standard laboratory tests and empirical tests for grindability and friability, and indirect measurements based on particle size distributions.

Deformation Under Load. The mechanical behavior of coal is strongly affected by the presence of cracks, as shown by the lack of proportionality between stress and strain in compression tests or between strength and rank. However, tests in triaxial compression indicate that as the confirming pressure is increased different coals tend to exhibit similar values of compressive strength perpendicular to the directions of these confining pressures. Except for anthracites, different coals exhibit small amounts of recoverable and irrecoverable strain under load.

Dynamic tests have been used to measure the variation of elastic properties with coal rank. Tests using vitrains suggested that coals were mechanically isotropic up to 92 wt% C, with anisotropy increasing above this value. Dynamic tests were used to measure internal energy losses in vibration and to study the fluidity changes of coking coals on oxidation. The Young's modulus for median rank coals has been found to be about 4 GPa (4×10^{10} dyn/cm^2) (6). Sharp increases in the Young's and shear moduli have been found in vitrinites having increase in carbon content >92 wt%.

Strength. The strength of a coal as measured in the laboratory may not be relevant to mining or size reduction problems where the applied forces are much more complex. There are indications that compressive strength, measured by compression of a disk, may give useful correlations to the ease of cutting for different kinds of equipment. Studies of the probability of survival of pieces of different size suggest that the breaking stress S should be most closely related to the linear size x rather than the area or volume of the piece. The results of a number of studies (6) indicate that S is proportional to x^{-r} where r frequently has the value 1–2.

The effects of rank on both compressive and impact strength have been studied, and usual minima were found at 20–25% dry, ash-free volatile matter (88–90 wt% carbon). Accordingly, the Hardgrove grindability index exhibits maximum values in this area.

Size Distribution Relationships. Different models have been used to describe the size distribution of particles experiencing single and multiple fractures. A model based on fracture at the site of the weakest link and a distribution of weakest links in the system gave results that could be described as well by the Rosin-Rammler relation (56). The latter is based on the concept that fracture takes place at prexisting flaws that are distributed randomly throughout the particle.

Comminution. The size reduction of coal during handling of comminution results from many ill-defined forces. Grindability and friability tests are useful indicators of size reduction for any given coal having a specific energy consumption. The Hardgrove test yields an index that varies with coal rank, moisture content, and ash and maceral distribution. The higher the grindability index, the lower the energy requirements to achieve a given size reduction. These indexes are useful in establishing capacity factors for pulverizers. Grinding is easiest for coals having 75–80% dry, ash-free fixed carbon. Optimum moisture contents have been observed for the younger coals. High moisture contents lead to difficulty in grinding, and excessive drying causes the coal particles to be tougher than the optimum dryness.

A relationship between energy consumption and size reduction would be helpful for comminution processes, but none of the many attempts to develop this have been broadly applicable. One reason is that generation of new surface is only one of many phenomena in the size reduction (qv) process. The energy requirements of a comminution system may, however, be estimated from laboratory tests for given amounts of size reduction. For pulverized coal-fired boilers 70% <74mm size (−200mesh) is frequently used. Product size distributions of reproducible forms are obtained from a range of graded coal input sizes and careful control of crushing conditions in the laboratory. The Hardgrove test gives data for comparing ease of grinding different coals (57). The efficiency of pulverizers can then be calculated from the energy requirements for each product size from a series of tests. The relationship between a particle size and energy consumption obtained from plant data is frequently expressed in terms of Kick's or Rittinger's laws or some modification such as Bond's law (58). These empirical relationships do not provide much insight into the mechanism of the grinding process.

The development of a continuous grinding index was the focus of work in the late 1970s (59). The laboratory test equipment used is similar to that for the Hardgrove test but permits classifying the product and recycling the oversize material. An improved correlation is obtained that may, however, need to be corrected for the relative sizes of the test grinding balls versus those used in commercial scale equipment. The continuous grinding index is especially useful for lower rank coals.

7.5. Properties Involving Utilization.   Coal rank is the most important single property for application of coal. Rank sets limits on many properties such as volatile matter, calorific value, and swelling and coking characteristics. Other properties of significance include grindability, ash content and composition, and sulfur content.

Combustion.   Most of the mined coal is burned to produce steam for electric power generation (qv). The calorific value determines the amount of steam that can be generated. However, the design and operation of a boiler requires consideration of a number of other properties (see FURNACES, FUEL-FIRED).

In general, high rank coals (high calorific value) are more difficult to ignite, requiring supplemental oil firing and slower burning in large furnaces to reach complete combustion. Greater reactivity makes lower rank coals better suited for cyclone burners, which carry out rapid, intense combustion to maximize carbon utilization and minimize smoke emission. The burning profile, a derivative thermogravimetric analysis of oxidation, is used to characterize coal for oxidation or combustion behavior (see COMBUSTION TECHNOLOGY). Volatile matter is important for ease of coal ignition. Because high rank coals have low volatile matter contents, they burn more slowly and with a short flame. They are primarily used for domestic heating, where heat is transferred directly from the fuel bed. For kilns, long hot flames are preferred and the coal should have medium to high volatile matter. The heating value released with the volatile matter for the various coals is given in Table 7 (6).

Table 7. Rank and Heating Value in Volatile Matter

Rank (ASTM)	Volatile matter, %	Total heat energy liberated in volatile matter, %
anthracite	<8	5–14
semianthracite	8–14	14–21
low volatile bituminous	14–22	21–28
medium volatile bituminous	22–31	28–36
high volatile A bituminous	>31	36–47

The swelling and caking properties of coal are not important for most boiler firing, such as pulverized coal-fired use. Some units, however, such as retort stokers, form coke in their normal operation. The smaller domestic heating units also require noncaking coal for satisfactory operation. For pulverized coal firing, a high Hardgrove index or grindability index is desired. The easier the coal is to grind, the lower the energy cost for pulverizing. The abrasiveness of the coal is also important because this determines the wear rate on pulverizer elements.

Coal moisture content, which affects both handling characteristics and freight costs, is most important for fine (<0.5 mm) particles. The lower rank coals have higher moisture contents. The moisture acts as a diluent, lowering flame temperatures and carrying sensible heat out with the flue gases. For pulverized coal firing the moisture content must be low to maximize grindability and avoid clogging. Thus dry run-of-mine coal having up to 30% ash may be more desirable than cleaning and drying coal.

The moisture content of peat or brown coal that is briquetted for fuel must be reduced to ~15% for satisfactory briquetting. Mechanical or natural means are used because of the cost of thermal drying. Moisture is sometimes desirable. About 8% is necessary for prevention of combustible loss from a chain-grate stoker.

Ash content is also important. Ash discharge at high temperature, as molten ash from a slagging boiler, involves substantial amounts of sensible heat. However, the higher cost of washed coal of lower (~10%) ash content does not always merit its use. Ash disposal and extra freight costs for high ash coals also affect the coal selection. The use of continuous mining equipment produces coal having ~25% ash content. The average ash content of steam coal burned in the United States is ~15%. For some applications, such as chain-grate stokers, a minimum ash content of ~7–10% is needed to protect the metal parts.

Ash fusion characteristics are important in ash deposition in boilers. Ash deposition occurring on the furnace walls is termed slagging, whereas accumulation on the superheater and other tubes is termed fouling. A variety of empirical indexes have been developed (60,61) to relate fouling and slagging to the ash chemical composition through parameters such as acidic and basic oxides content, sodium, calcium and magnesium, and sulfur.

A related property is the viscosity of coal ash. Ash viscosity affects the rate at which ash deposits may flow from the walls, and thus the requirements for ash removal equipment such as wall blowers and soot blowers. The preferred coal ash has a narrow temperature range through which it passes the plastic range, ~25,000–1,000,000 mPa × s (=cP) (62).

Some minor constituents can interfere in firing. High (>0.6%) chlorine is associated with high sodium and complex sulfate deposits that appear to be required to initiate deposition on superheater tubes, as well as initiate stress corrosion cracking of superheater tubes. Phosphorus (>0.03% of the coal) contributes to phosphate deposits where high firing temperatures are used. Sulfur forms complex sulfates, however, its most damaging effect is corrosion of the boiler's coolest parts through condensation of sulfur oxides as sulfuric acid. Control is achieved by setting flue gas temperatures above the acid dew point in the boiler areas of concern.

Sulfur content plays an important role in meeting air quality standards. In the United States, the EPA has set an emission limit for SO_2 of 516 g/10^6 kJ (490 g/ $\times 10^6$ Btu) of coal burned. To meet this, steam coals have to contain <1% sulfur. Regulations resulting from the Clean Air Act of 1991 call for reduction of the total amount of sulfur oxide emissions by 8–9 million tons annually. One-half of this reduction was required by 1995, the remainder by the year 2000. A cap on the total emissions is given, and reductions in NO_x and particulate emissions are also mandated. Credits given for reductions of emissions that exceed the amount indicated for a given plant may be sold to other facilities unable to meet the requirements. This requirement is expected to force the addition of SO_2, NO_x, and particulate removal equipment to all boilers. Technology is being developed to control SO_x and NO_x through a combination of sorbent injection into the furnace and scrubbing and/or baghouse treatment to neutralize the acid gases and catalytically convert the NO_x to nitrogen (63).

Fluidized-bed boilers have been built with sizes up to 150 mw for commercial power generation and cogeneration units. This type of technology is displacing some stoker fired units at the low capacity (<200,000 kg/h of steam) boilers and smaller pulverized coal units at the large size of the fluidized-bed range. Bubbling and circulating bed designs are used, with operating temperatures in the 816–899°C range. Sulfur oxides are controlled using dolomite or limestone injection in the bed. Higher (2:1–5:1) Ca/S ratios are needed for fluid bed units than for wet scrubbing (1:1) or spray dryers (1.2–1.5:1). Nitrous oxide emissions are higher for fluid-bed units than for the other methods, possibly because of formation and oxidation of hydrogen cyanide (64,65) (see REACTOR TECHNOLOGY).

Coke Production.   Coking coals are mainly selected on the basis of the quality and amount of coke that they produce, although gas yield is also considered. About 65–70% of the coal charged is produced as coke. The gas quality depends on the coal rank and is a maximum, measured in energy in gas per mass of coal, for coals of ~89 wt% carbon on a dry, mineral matter-free basis, or 30% volatile matter.

Coals having 18–32% volatile matter are used to produce hard metallurgical coke. Methods have been developed to blend coals having properties outside this range to produce coke. Several coals are frequently blended to improve the quality of the coke (6,66). Blending also affects the shrinkage required to remove the coke from the ovens after initial swelling. Lower rank coals having up to 40% volatile matter may be used alone or in blends at a gasmaking plant. This coke, which need not be as strong as metallurgical coke, is more reactive, and is used in the domestic market.

Coking coal is cleaned so that the coke ash content is not >10%. An upper limit of 1–2 wt% sulfur is recommended for blast furnace coke. A high sulfur content causes steel (qv) to be brittle and difficult to roll. Some coal seams have coking properties suitable for metallurgical coke, but the high sulfur prevents that application. Small amounts of phosphorus also make steel brittle, thus low phosphorus coals are needed for coke production, especially if the iron (qv) ore contains phosphorus.

Solvent Extraction.   Coal partially dissolves in a number of solvents and this property has been used to aid in characterization of coal material, because the composition of extracts is sometimes similar to the coal. A wide range of organic solvents can be used (6,67), but dissolution is never complete and usually requires heating to temperatures sufficient for some thermal degradation or solvent reaction to take place, eg, ∼400°C. Dissolution of up to 40 wt% can be achieved near room temperature and up to 90% near 400°C. At room temperature the best solvents are primary aliphatic amines, pyridine, and some higher ketones, especially when used with dimethylformamide (DMF). Above 300°C large amounts can be dissolved using phenanthrene, 1-naphthol, and some coal-derived high boiling fractions. Coals having 80–85% carbon in the vitrinite give the largest yields of extract. Very little coal having >90 wt%C dissolves. Ultrasonic enhancement of extraction increases the yield of product by about 2.5 times the nonirradiated material (68) (see ULTRASONICS). The increase occurs only in solvent mixtures that significantly swell coals that range in rank from lignite to high volatile bituminous coal.

When the concentration of dissolved coal exceeds ∼5% of the solution by weight, the extracted material resembles the parent coal in composition and some properties. The extract consists of the smaller molecules within the range of the parent coal. Recovered extract is relatively nonvolatile and high melting. A kinetic study of coal dissolution indicated increasing heats of activation for increasing amounts of dissolved coal (69).

Gasification. Many of the coal selection criteria for combustion apply to gasification, which is typically a form of partial oxidation. Gasifiers are primarily described as fixed bed, fluidized bed, entrained, or rotating bed (70). The fixed bed involves an upward flow of reaction gas through a relatively stationary bed of hot coal. The gas velocity is slow enough to avoid blowing the coal out of the bed. The fluidized bed operates at higher gas velocities than the fixed bed and utilizes somewhat smaller particles. The entrained bed operates with parallel flows of reaction gas and finely pulverized coal particles to minimize reaction time and maximize throughput of product. The rotating bed is similar to a kiln that operates with the coal entering at the upper part of the inclined kiln. Rotation avoids clinkering and exposes fresh surfaces to enhance completion of the reaction. The range of coals that may be used vary from one gasifier type to another. Entrained flow gasifiers are able to handle the widest range of raw coals. Fixed-bed gasifiers require mildly caking or noncaking feedstocks for normal operation.

The Lurgi fixed-bed gasifier operates using lump coal of a noncaking type having an ash composition chosen to avoid a sticky, partly fused ash in the reactor. A slagging version of this gasifier has been tested in Westfield, Scotland. Other fixed-bed gasifiers have similar coal requirements.

The Shell-Koppers-Totzek gasifier is an entrained-bed type. It can gasify lignite and subbituminous or bituminous coal. The coal is fed as a pulverized fuel, usually ground to 70% <74 mm (−200 mesh) as used for pulverized coal fired boilers. Residence times are only a few seconds, therefore coal reactivity is important. The gasifier operates at >1650°C, so that coal ash flows out of the gasifier as a molten slag. Coal ash composition must permit continuing molten ash flow.

Fluidized-bed gasifiers typically require a coal feed of particles near 2–3 mm diameter. Caking coals are to be avoided because they usually agglomerate in the bed. This can be avoided using a pretreatment consisting of a surface oxidation with air in a fluidized bed. A useful flue gas is produced. Examples of this type include the commercially available Winkler, and the U-Gas technology developed at the Institute of Gas Technology in Chicago.

8. Chemistry

Coal reactions, which on heating are important to the production of coke and synthetic fuels, are complicated by its structure.

Mature (>75 wt % C) coals are built of assemblages of polynuclear ring systems connected by a variety of functional groups and hydrogen-bonded cross-links (Fig. 4) (3,4,7,21). The ring systems themselves contain many functional groups. These polynuclear coal molecules differ one from another to some extent in the coal matter. For bituminous coal, a tarlike material occupies some of the interstices between the molecules. Generally, coal materials are nonvolatile except for some moisture, light hydrocarbons, and contained carbon dioxide. The volatile matter produced on carbonization reflects decomposition of parts of the molecule and the release of moisture. Rate of heating affects the volatile matter content such that faster rates give higher volatile matter yields.

Coal composition is denoted by rank. Carbon content increases and oxygen content decreases with increasing rank. Table 8 gives a listing of the empirical formulas in terms of hydrogen, oxygen, nitrogen, and organic sulfur per 100 carbon atoms for a set of eight premium coal samples.

The surface of coal particles undergoes air oxidation, a process that may initiate spontaneous combustion in storage piles or weathering with a loss of heating and coking value during storage. Combustion produces oxides of sulfur and nitrogen as well as carbon dioxide and water vapor. The SO_x results from oxidation of both organic sulfur and inorganic forms such as pyrite. Nitrogen

Table 8. Empirical Composition of Argonne Premium Coal Samples[a]

Coal		Composition, atoms/100 atoms C			
Type	From	H	O	N	S
Nonbituminous coals					
lignite	Beulah-Zap	79.5	20.9	1.4	0.4
subbituminous	Wyodak-Anderson	85.6	18.0	1.3	0.2
Bituminous coals					
high volatile 1[b]		77.3	13.1	1.5	1.2
high volatile 2[c]		85.7	10.8	1.7	0.2
high volatile 3[d]		76.3	8.9	1.6	0.3
high volatile 4[e]		76.7	8.0	1.7	0.4
medium volatile	Upper Freeport	66.0	6.6	1.6	0.3
low volatile[f]		58.5	2.0	1.2	0.2

[a] Ref. 71.
[b] Illinois no. 6 coal.
[c] Blind Canyon coal.
[d] Lewiston-Stockton coal.
[e] Pittsburgh coal.
[f] Pocahontas no. 3 coal.

oxides are formed primarily from the nitrogen in the coal during high temperature combustion, rather than from the air used for combustion.

Partial oxidation as carried out in gasification produces carbon monoxide, hydrogen gas, carbon dioxide, and water vapor. The carbon dioxide reacts with hot carbon from the coal to produce carbon monoxide, and steam reacts with the carbon to produce carbon monoxide and hydrogen. The hydrogen can react with carbon through direct hydrogen gasification:

$$C + 2\,H_2 \rightarrow CH_4$$

at high hydrogen pressure, frequently 6.9 MPa (1000 psi) and moderate temperatures of 650–700°C. Methane may also be produced from

$$CO + 3\,H_2 \rightarrow CH_4 + H_2O$$

in a nickel-catalyzed reactor. This latter reaction is highly exothermic and is used to provide steam for the process. The correct 3:1 ratio of hydrogen to carbon monoxide is achieved using the water-gas-shift reaction:

$$CO + H_2O \rightarrow H_2 + CO_2$$

A mixture of CO and H_2, called synthesis gas, may also be used in other catalytic reactors to make methanol (qv) or hydrocarbons (qv):

$$CO + 2\,H_2 \rightarrow CH_3OH$$

$$n\,CO + 2n\,H_2 \rightarrow (CH_2)_n + n\,H_2O$$

Surface oxidation short of combustion, or using nitric acid or potassium permanganate solutions, produces regenerated humic acids similar to those extracted from peat or soil. Further oxidation produces aromatic acids and oxalic acid, but at least one-half of the carbon forms carbon dioxide.

Treatment with hydrogen at 400°C and 12.4 MPa (1800 psi) increases the coking power of some coal and produces a change that resembles an increase in rank. Hydrogenation using an appropriate solvent liquefies coal. Noncatalyzed processes primarily produce a tarlike solvent-refined-coal used as a boiler fuel. Catalysts and additional hydrogen were used in the H-Coal process developed by Hydrocarbon Research, Inc. to produce a higher quality liquid product. A 450 t/day plant was built in Catlettsburg, Kentucky, to demonstrate this process by making a coal-derived refinery feedstock. The reactor used a catalyst suspended in a process derived liquid or ebulated bed. Hydrogen reactions over short (0.1–2 s) times with very rapid heating produce a range of liquids such as benzene, toluene, xylene, and phenol. A less rapid heating and lower maximum temperatures permit removal of some sulfur and nitrogen from the coal (72). These efforts have not been commercialized.

Treatment of coal with chlorine or bromine results in addition and substitution reactions. At temperatures up to 600°C chlorinolysis produces carbon tetrachloride, phosgene, and thionyl chloride (73). Treatment with fluorine or chlorine trifluoride at atmospheric pressure and 300°C can produce large yields of liquid products.

Hydrolysis using aqueous alkali has been found to remove ash material including pyrite. A small pilot plant for studying this process was built at the Battelle Memorial Institute in Columbus, Ohio (74) and subsequently discontinued. Other studies have produced a variety of gases and organic compounds such as phenols, nitrogen bases, liquid hydrocarbons, and fatty acids totaling as much as 13 wt% of the coal. The products indicate that oxidation and other reactions as well as hydrolysis take place.

The pyritic sulfur in coal can undergo reaction with sulfate solutions to release elemental sulfur (see SULFUR REMOVAL AND RECOVERY). Processes to reduce the sulfur content of coal have been sought (75). The reaction of coal and sulfuric acid has been used to produce cation exchangers, but it was not very efficient and is no longer employed. Efforts have turned to the use of hot concentrated alkali in a process called Gravimelt.

Many of the products made by hydrogenation, oxidation, hydrolysis, or fluorination are of industrial importance. Concern about stable, low cost petroleum and natural gas supplies is increasing the interest in some of the coal products as upgraded fuels to meet air pollution control requirements as well as to take advantage of the greater ease of handling of the liquid or gaseous material and to utilize existing facilities such as pipelines (qv) and furnaces. A demonstration plant was built in North Dakota for conversion of coal to methane, also known as substitute natural gas (SNG) production. This plant, operated by Great Plains Gasification Associates and in use at this writing, may be converted to produce methanol instead of methane (see GAS, NATURAL). A chemistry based on the conversion of synthesis gas has been developed and applied extensively in South Africa to the production of liquid fuels and many other products. A small-scale production is used in the manufacture of photographic film materials from coal-derived synthesis gas in the Eastman Kodak plant in Kingsport, Tennessee. However, the principal production of chemicals from coal involves the by-products of coke manufacturing.

8.1. Reactions of Coal Ash.   Mineral matter impurities have an important effect on the utilization of a coal. One of the constituents of greatest concern is pyrite because of the potential for sulfur oxide generation on combustion. The highest concentrations of pyrite are associated with coal deposition under marine environments, as typified by the Illinois Basin, including parts of Illinois, Indiana, and Kentucky. Additionally, the mineral matter has a tendency to form sticky deposits in a boiler. This tendency is most pronounced using mixtures that are rich in water-soluble alkalies such as are found in the Western Plains states. Coals from North Dakota, South Dakota, Wyoming, and Montana are typically low in the sulfur-bearing constituents and therefore otherwise desirable as fuels.

Coal deposits from east of the Mississippi River generally have acidic mineral constituents, ie, they are richer in silica and alumina and tend to produce higher melting ash mixtures. These materials do not soften until >1000°C and have limited problems with deposition on the inside walls of the boiler (slagging) or on the superheater tubes inside the boiler (fouling).

Coal ash passes through many reactors without significant chemical change. High temperature, exceeding the ash-softening temperature for the coal, permits reactions of the simpler ash constituents to form more complex species. Molten ash behavior affects slagging and ash removal. Correlations of viscosity have been made with a variety of chemical parameters, and descriptions based on acid–base chemistry appear to correlate with observed effects (74). Iron may be interconverted between the Fe(II) and Fe(III) states. Significant reduction in viscosity occurs as ferrous concentrations increase.

Corrosion of boiler tubes appears to be initiated in some cases with the formation of a white layer of general composition $(Na,K)_3Al(SO_4)_3$. Conditions for initiation of the deposit are favored by coals having high alkali and sulfur contents. The white layer bonds to the tubes and permits growth of ash deposits that insulate the layer and permit further corrosion.

8.2. Plasticity of Heated Coals.   Coals having a certain range of composition associated with the bend in the Seyler diagram (Fig. 3) and having 88–90 wt% carbon soften to a liquid condition when heated (4,6). These materials are known as prime coking coals. The soft condition is somewhat reversible for a time, but does not persist for many hours at 400°C, and is not observed above ~550°C if the sample is continuously heated as in a coking process. Continuous or lengthy heating result in degradation of coal matter, releasing vapors and resulting in polymerization of the remaining material. The coal does not behave like a Newtonian fluid and only empirical measurements of plasticity can be made (see RHEOLOGICAL MEASUREMENTS). About 10–30% of the coal becomes liquid, having a melting point <200°C, and this molten material plasticizes the solid matrix remaining.

The molten part of a vitrinite is similar to the gross maceral, and a part of the maceral is converted to a form that can be melted after heating to 300–400°C. The molten material is unstable and forms a solid product (coke) >350°C at rates that increase with temperature. The decomposition of the liquid phase is rapid for lower rank noncoking coals, and less rapid for prime coking coals. The material that melts resembles coal rather than tar and, depending on rank, only a slight or moderate amount is volatile.

The fluidity of coal increases and then decreases at a given temperature. This has been interpreted in terms of reaction sequence of coal–fluid coal–semicoke. In the initial step, a part of the coal is decomposed to add to that which normally becomes fluid. In the second step, the fluid phase decomposes to volatile matter and a solid semicoke. The semicoke later fuses accompanied by evolution of additional volatile matter to form a high temperature coke.

Formation of a true coke requires that the fluid phase persist long enough during heating for the coal pieces to form a compact mass before solidification occurs from the decomposition. Too much fluidity leads to an expanded froth owing to formation of dispersed bubbles from gas evolution in the fluid coal. Excess bubble formation results in a weak coke. The porous nature of true coke is caused by the bubble formation during the fluid phase. The strength of semicoke is set by the degree of fusion during the fluid stage and the thickness of the bubble walls formed during the frothing. In the final conversion to a hard high temperature coke, additional gas evolution occurs while the solid shrinks and is subjected to thermal stresses. The strength of the resultant coke and the size of the coke pieces are strongly affected by the crack structure produced as a result of the thermal stresses. Strong large pieces of coke are desirable to support the ore burden in blast furnaces.

Several laboratory tests (3,6) are used to determine the desirability of a coal or blend of coals for making coke. These are empirical and are carried out under conditions that approach the coking process. The three properties that have been studied are swelling, plasticity, and agglomeration.

Several dilatometers have been developed to determine the swelling characteristics of coals. The sample is placed in a cylindrical chamber with a piston resting on the coal surface. The piston motion reflects coal volume changes and is recorded as a function of temperature with a constant heating rate. When the coal first softens, contraction is caused by the weight of the piston on softened coal particles that deform to fill void spaces. Swelling then takes place when the particles are fused sufficiently to resist the flow of the evolving gases. The degree of swelling depends on the rate of release of volatile matter and the plasticity of the coal. The mass stabilizes at 450–500°C as the semicoke hardens. The shapes of the curves depend on the dilatometers. Curves obtained using the Hofmann apparatus have been classified into four main types that permit distinguishing coals having the optimum softening and swelling properties for production of a strong coke (76). Types A and C, and to some extent type D, can soften so that curves like those shown in Figure 7 can be obtained.

Free-swelling tests are commonly used to measure a coal's caking characteristics. A sample of coal is packed in a crucible or tube, without compaction, and heated at a fixed rate to ~800°C. Infusible coals distill without changing appearance or state of agglomeration. The fusible coals soften, fuse, and usually swell. The profile of the resultant coke is compared to a series of reference

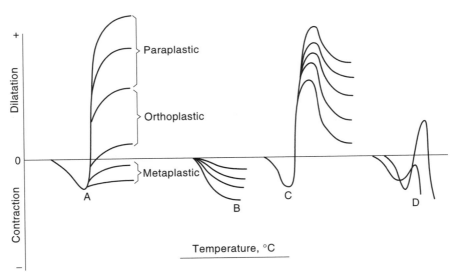

Fig. 7.   Coal classification system according to Hofmann where A = Eu plastic; B = plastic;   C = Per plastic;   and D = Fluido plastic.Courtesy of Centre d'Etudes et Recherches des Charbonnages de France and Brennstoff-Chemie.

profiles so that a swelling index can be assigned. The profiles represent indexes between 0 and 9. The best cokes come from coals having indexes between 4 and 9.

The Gray-King assay, primarily carried out in Europe and the United Kingdom, is obtained from a similar test. The coal is heated to 600°C in a horizontal tube. Standard photographs are used to compare general appearance, profile, and size of the coke mass. Before testing, the more fusible coals are mixed with varying amounts of a standard electrode carbon of carefully selected size. A nonuniform scale termed A-F and G-G9 has been developed from the coke appearance for low swelling coals or from the amount of carbon required to give a standard appearance for the high swelling coals. The U.K. National Coal Board Rank Code Numbers are partly assigned on the Gray-King assay and partly on the volatile matter. The Gray-King assay procedure can also permit evaluation of yields of tar, gas, and liquor.

Plasticity can be studied using a device known as the Gieseler plastometer. A constant torque is applied to a shaft with rabble arms imbedded in coal in a crucible heated at a fixed rate. The rate of rotation of the shaft indicates the fluidity of the coal and is plotted as a function of the coal temperature. These curve, as shown in Figure 8, has a well-defined peak for coking coals usually near 450°C. Softening occurs at 350–400°C. At a normal heating rate of 3°C/min, the fluid hardening may be complete by 500°C.

Several agglutinating and agglomerating tests that indicate the bonding ability of the fusible components and depend on the crushing strength of a coke button produced, in some cases with addition of inert material, are also used. The Roga agglutinating test, developed in Poland, provides one of the criteria of the Geneva International Classification System. The coal sample is

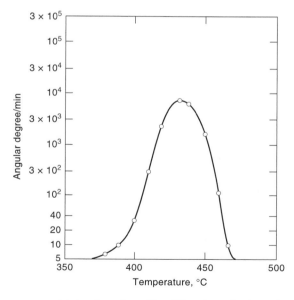

Fig. 8. Plasticity curve obtained using the Gieseler plastometer. Heating rate is 2°C/min. Courtesy of Centre d'Etudes et Recherches des Charbonnages de France.

mixed with carefully sized anthracite, compacted, and heated to 850°C in 15 min. The part of the product that passes through a 1 mm (~18 mesh) screen is weighed, and a rotating drum further degrades the product. Roga indexes from 0–70 have been determined.

A coherent plastic layer from a few millimeters to 2–3 cm thick separates the semicoke and coke from the unfused coal in the coke oven. Coking properties are assessed in Russia and some other countries by a measurement of the thickness of this plastic layer. A standardized test widely used in eastern Europe is the best known of this type (6) and involves a penetrometer used to measure the thickness of the plastic layer in a column of coal heated from the bottom. The various standard tests give results that are similar but do not give close correlations with each other.

The behavior of different polymerizing and gas-relating materials has been used to relate the plastic behavior of coal with known kinds of chemical change (3). The plastic nature of coal matter is determined by the competition between the reactions that generate the liquid phase, and those that convert it to semi-coke. In general for vitrinites the greater the heating rate, the greater the fluidity or plasticity and the dilatation. Inertinite essentially does not contribute to the plastic properties of the coal. Exinite becomes fluid when heated, but also rapidly devolatilizes instead of forming semicoke, has little value as a binder, and can increase the fluidity to an undesirable extent.

8.3. Pyrolysis of Coal.   Most coals decompose below temperatures of ~400°C (5,6), characteristic of the onset of plasticity. Moisture is released near 100°C, and traces of oil and gases appear between 100 and 400°C, depending on the coal rank. As the temperature is raised in an inert atmosphere at a rate of 1–2°C/min, the evolution of decomposition products reaches a maximum rate near 450°C, and most of the tar is produced in the range of 400–500°C. Gas evolution begins in the same range but most evolves >500°C. If the coal temperature in a single reactor exceeds 900°C, the tars can be cracked, the yields are reduced, and the products are more aromatic. Heating beyond 900°C results in minor additional weight losses but the solid matter changes its structure. The tests for volatile matter indicate loss in weight at a specified temperature in the range of 875–1050°C from a covered crucible. This weight loss represents the loss of volatile decomposition products rather than volatile components.

A predictive macromolecular network decomposition model for coal conversion based on results of analytical measurements has been developed called the functional group, depolymerization, vaporization, cross-linking (FG-DVC) model (77). Data are obtained on weight loss on heating (thermogravimetry, tg) and analysis of the evolved species by Fourier transform infrared (FTir) spectrometry. Separate experimental data on solvent swelling, solvent extraction, and Gieseler plastometry are also used in the model.

Six factors form the basis of this model: (1) the decomposition of functional group sources in the coal yield the light gas species in thermal decomposition. The amount and evolution kinetics can be measured by tg/Ftir spectrometry and the functional group changes by Ftir and nmr; (2) the decomposition of a macromolecular network yields tar and metaplast. The amount and kinetics of the tar evolution can be measured by tg/Ftir and the molecular weight by field ionization mass spectrometry (Fims). The kinetics of metaplast formation and destruction can be measured by solvent extraction, by Gieseler plastometry, and by proton magnetic resonance thermal analysis (pmrta); (3) the molecular weight distribution of the metaplast depends on the network coordination number, ie, the average number of attachments on aromatic ring clusters. The coordination number can be determined by solvent swelling and nmr; (4) the network decomposition is controlled by bridge breaking. The number of bridges broken is limited by the available donatable hydrogen; (5) the network solidification is controlled by cross-linking. The changing cross-link density can be measured by solvent swelling and nmr. Cross-linking appears to occur with evolution of both CO_2 prior to bridge breaking and CH_4 after bridge breaking. Thus low rank coals, which form a lot of CO_2, cross-link prior to bridge breaking and are thermosetting. High volatile bituminous coals, which form little CO_2, undergo significant bridge breaking prior to cross-linking and become highly fluid. Weathering, which increases the CO_2 yield, causes increased cross-linking and lowers fluidity; and (6) the evolution of tar is controlled by mass transport in which the tar molecules evaporate into the light gas species and are carried out of the coal at rates proportional to their vapor pressure and the volume of light gases. High pressures reduce the volume of light gases and hence reduces the yield of heavy molecules having low vapor pressures. These changes can be studied using field ionization mass spectrometry.

Nature and Origin of Products. Volatile matter yields decrease with increasing coal rank. For slow heating, the final weight loss depends on the maximum temperature (Fig. 9). A variety of reactions take place and increasing temperatures provide the thermal energy required to break the stronger chemical

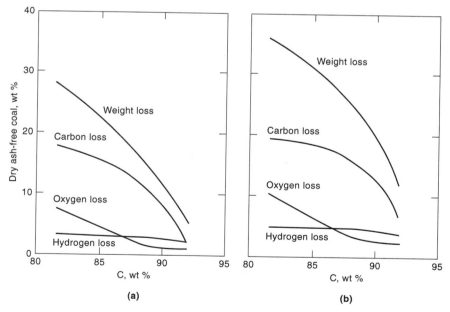

Fig. 9. Composition of volatile matter as a function of rank (bright coals) at (**a**) 500°C and (**b**) 900°C. The wt% of C is on a dry ash-free basis of unheated coal. Courtesy of Institute of Fuel.

bonds. Much decomposition takes place in a short time (apparently <1s), but detection is limited by the rate of diffusion of the volatile products through the solid. The liquids result from initial decomposition and gases from decomposition of liquid material. Very rapid heating rates produce weight losses as high as 72% at 1900°C, suggesting that the intrinsic volatile matter is limited only by the vapor pressure of the initial pyrolysis fragments, and would be expected to increase with temperature and decreasing coal rank (78).

The residual solid or char heated to 500°C contains 3–3.5 wt% H and up to 5 wt% O. On further heating to 900°C the solid contains only 0.8 wt% H and up to 0.3 wt% O. An aqueous liquor is produced that comes from the moisture in the coal as well as hydroxyl and possibly other oxygen-containing groups. Phenols in the tar are probably derived from hydroxyl aromatic groups in the coal. The total tar yield appears to be proportional to the fraction of aromatic carbon in the coal (see TAR AND PITCH). Coke oven gas is obtained from a variety of reactions that include cracking some of the tar. The hydrogen in the gas is generated after the char is heated to 400°C, but most is evolved in the conversion of the fluid coal to semicoke or coke at 550–900°C. The steam in the ovens can also produce hydrogen on reaction with hot coke.

Pyrolysis Reaction Mechanisms.   An overall picture of the pyrolysis process is generally accepted but the detailed mechanism is controversial. Information has been obtained from: the sequence of volatile material appearing in a coke plant as determined by gas chromatography; laboratory work simulating coking and minimizing secondary reactions by working in vacuum or sweeping with inert gas (79); laboratory studies using model organic compounds to determine the mechanism by which these materials are converted to coke, liquid, and gaseous products; and laboratory work with more complex materials, including specially synthesized polymers, to better provide a model of coal (4,80). Radioactive tracers have been used in the last two studies to follow the transformation to materials in the products (4). In the last study, gas generating materials were added to aid in simulating the swelling process. The dehydrogenation of coal, which can alter the distribution of products, also provides information regarding the formation mechanism (81). The mechanism of formation of metallurgical coke and its effect on coke properties has also been described (82).

The mechanism of coal pyrolysis has been discussed (77,79,82) and a table summarizing the various changes has been prepared (79). The early stages involve formation of a fluid through depolymerization and decomposition of coal organic matter containing hydrogen. Around 400–550°C aromatic and non-aromatic groups may condense after releasing hydroxyl groups. The highest yields of methane and hydrogen come from coals having 89–92 wt% C. Light hydrocarbons other than methane are released most readily <500°C; methane is released at 500°C. The highest rate for hydrogen occurs >700°C (77,79).

9. Resources

9.1. World Reserves.   Amounts of coal of some specified minimum deposit thickness and some specified maximum overburden thickness existing in the ground are termed resources. There is no economic consideration for resources, but reserves represent the portion of the resources that may be recovered economically using conventional mining equipment. The first inventory of world coal resources was made during the Twelfth International Geological Congress in Toronto in 1913. An example of the changes since 1913 can be seen from an examination of the coal resources for Canada. These were estimated to have been 1217×10^9 metric tons in 1913, based on a few observations and statistical allowance for all possible coalbeds to a minimum thickness of 0.3 m and to a maximum depth of 1220 m below the surface. In 1974, however, the estimate of solid fossil fuel resources (excluding peat) from the World Energy Conference gave the total resources as only 109×10^9 metric tons, and in 1986 the proven recoverable resources and estimated additional amount in place was given as 50×10^9 metric tons (83), <5% of the earliest figures (see FUEL RESOURCES).

Comprehensive reviews of energy sources are published by the World Energy Conference, formerly the World Power Conference at 6-year intervals (83). The 1986 survey includes reserves and also gives total resources. In 1986, the total proven reserves of recoverable solid fuels were given as 6×10^{11} metric tons. One metric ton is defined as 29.2×10^3 MJ (27.7×10^6 Btu) to provide for the variation of calorific value in different coals. The total estimated additional reserves recoverable and total estimated additional amount in place are 2.2×10^{12} and 7.7×10^{12} metric tons, respectively. These figures are about double the 1913 estimates, primarily because significantly increased reserves have been indicated for Russia.

The part of the resource that is economically recoverable varies by country. The estimates made in the survey show that the proven recoverable reserves would last ∼1200 years at the 1988 annual rate of production and that the estimated additional amount in place represent almost 1700 years at 1988 annual consumption.

In Table 9 (84), a somewhat different basis is used. The estimated total original coal resources of the world include beds 30-cm thick, and generally

Table 9. Estimated Total Original Coal Resources of the World by Continents[a]

Continent	Identified resources,[a] 10^9 t	Hypothetical[b] resources,[c] 10^9 t	Estimated total resources,[d] 10^9 t
Asia[c]	3,635[d]	6,362	9,997[e]
North America	1,727	2,272	3,999
Europe[f]	273	454	727
Africa	82	145	227
Oceania[g]	64	55	118
Central and South America	27	9	36
Totals	*5,808*	*9,297*	*15,104*

[a] Ref. 84.
[b] Original resources in the ground in beds 30 cm or more thick and generally <1299 m below surface but includes small amount between 1200 and 1800 m.
[c] Includes European Russia.
[d] Includes about 2090×10^9 metric tons in Russia.
[e] Includes ∼8600×10^9 in Russia.
[f] Includes Turkey.
[g] Australia, New Zealand, and New Caledonia.

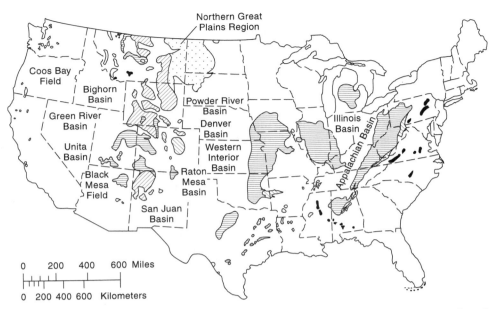

Fig. 10. Coal fields of the conterminous United States where ■ represents anthracite and semianthracite; ⊠, low volatile bituminous coal; ⇗∧, medium and high volatile bituminous coal; ⊠, subbituminous coal; and ⊡, *lignite*(84).

<1220 m below surface but also include small amounts between 1229 and 1830 m. The data from column 1 are from earlier World Power Conference Surveys, whereas the figures for hypothetical resources (col. 2) and total estimated resources (col. 3) may be less reliable. This estimate represents about one-third more than the World Energy Conference Survey.

Reserves in the United States.   Coal is widely distributed and abundant in the United States as indicated in Figures 10 and 11. A large portion of the coal fields contain lignite and subbituminous coal, however, and another portion of the coal is contained either in thin or deep beds that can be mined only with difficulty or great cost. Reserve estimates for the United States as of 1974 were on the order of 1.6×10^{12}t (84). The U.S. Geological Survey is currently completing a thorough revision of the previous estimates. This information is useful for showing the quantitative distribution of reserves, selecting appropriate areas for further exploration of development, and in planning coal-based industrial activity.

The reserves of 21 states have been classified by overburden thickness, reliability of estimates, and bed thickness. This coal represents ~60% of the total identified tonnage. Of this, 91% is <305 m from the surface, 43% is bituminous, and 58% is in beds thick enough to be mined economically.

On a uniform calorific value basis, coal constitutes 69% of the total estimated recoverable resources of fossil fuel in the United States. Petroleum and natural gas are ~7% and oil in oil shale, which is not as of this writing used as a fuel, is ~23%. The 1989 total recoverable reserves of coal are ~500 times the 1989 annual production (2), whereas the reserves of oil and gas are smaller,

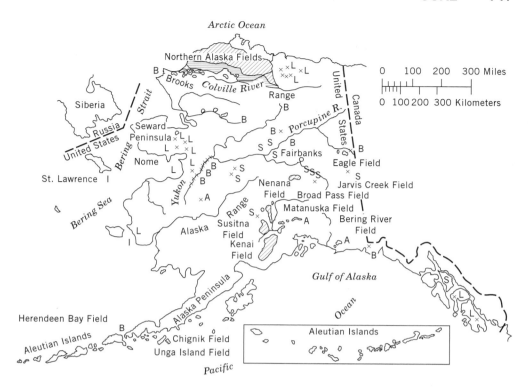

Fig. 11. Coal fields of Alaska where ⩕ represents bituminous coal; ▨, subbituminous coal and lignite; and X is an isolated occurrence of coal of unknown extent. A = anthracite; B = bituminous; S = subbituminous; and L = lignite (84).

the production and consumption rate of oil and gas in the United States is three times that of coal.

9.2. Coal Production. In 1860 world coal production was 122×10^6t / year. Production increased to 1140×10^6t in 1913, giving a 4.2% annual average rate of increase. The rate has slowed and been erratic since that time. Statistical data on world coal production from 1860 to 1960 is given in (81). World coal and lignite production rose to $\sim 4.7 \times 10^9$t in 1988 (1).

United States Coal Production. Coal production in the United States, which dates back to 1702, started in earnest ~ 1820. It has increased with fluctuations to 617×10^6t in 1976, 755×10^6t in 1982, and 1099×10^6t in 1999 (2). In 1999 the United States produced 23.2% of the world's coal supply (1). U.S. coal production by state is given in Table 10 (2). Coal shipments in 2000 totaled 1048×10^6t to United States destinations. Of these, electric utilities received 894×10^6t; coke and gas plants 28×10^6t; and industrial, commercial, retail, and transportation 65×10^6t (2).

The demand for energy is continually increasing and the highest energy consumption in the world occurs in the United States. In 1989, consumption totaled 8.6×10^{13} MJ (81.3×10^{15}Btu) or 11.7 metric tons of coal-equivalent per

Table 10. Coal Production by State, 1991, 1996–2000 (Thousand Short Tons)

Coal-producing state and region	2000	1999	1998	1997	1996	1991	Percent change 1999–2000	Average annual per-cent change 1996–2000	1991–2000
Alabama	19,324	19,504	23,013	24,468	24,637	27,269	-0.9	-5.9	-3.8
Alaska	1,641	1,565	1,344	1,450	1,481	1,436	4.8	2.6	1.5
Arizona	13,111	11,787	11,315	11,723	10,442	13,203	11.2	5.8	-.1
Arkansas	12	22	24	18	21	52	-47.3	-13.7	-15.3
California						57			
Colorado	29,137	29,989	29,631	27,449	24,886	17,834	-2.8	4.0	5.6
Illinois	33,444	40,417	39,732	41,159	46,656	60,258	-17.3	-8.0	-6.3
Indiana	27,965	34,004	36,803	35,497	29,670	31,468	-17.8	-1.5	-1.3
Iowa						344			
Kansas	201	409	341	360	232	416	-50.8	-3.5	-7.8
Kentucky	130,688	139,626	150,295	155,853	152,425	158,980	-6.4	-3.8	-2.1
Total									
Eastern	*104,901*	*110,043*	*116,654*	*120,918*	*116,951*	*117,220*	*-4.7*	*-2.7*	*-1.2*
Western	*25,787*	*29,583*	*33,641*	*34,936*	*35,474*	*41,760*	*-12.8*	*-7.7*	*-5.2*
Louisiana	3,699	2,953	3,216	3,545	3,221	3,151	25.3	3.5	1.8
Maryland	4,546	3,837	4,060	4,160	4,093	3,773	18.5	2.7	2.1
Mississippi	902	18					nm[a]		
Missouri	436	392	372	401	710	2,304	11.3	-11.4	-16.9[b,c]
Montana	38,352	41,102	42,840	41,005	37,891	38,237	-6.7	0.3	.0
New Mexico	27,323	29,156	28,597	27,025	24,067	21,518	-6.3	3.2	2.7
North Dakota	31,270	31,135	29,912	29,580	29,861	29,530	.4	1.1	.6
Ohio	22,269	22,480	28,048	29,154	28,572	30,569	0.9	-6.0	-3.4
Oklahoma	1,588	1,661	1,661	1,621	1,701	1,841	-4.4	-1.7	-1.6
Pennsylvania	74,619	76,399	81,036	76,198	67,942	65,381	-2.3	2.4	1.5
Total									
Anthracite	*4,572*	*4,753*	*5,231*	*4,678*	*4,751*	*3,445*	*-3.8*	*-.9*	*3.2*
Bituminous	*70,046*	*71,646*	*75,805*	*71,520*	*63,190*	*61,936*	*-2.2*	*2.6*	*1.4*
Tennessee	2,669	3,037	2,696	3,300	3,651	4,290	-12.1	-7.5	-5.1

742

Texas	49,498	53,072	52,583	53,328	55,164	53,825	-6.7	-2.7	0.9
Utah	26,656	26,373	26,075	26,683	27,507	21,945	1.1	9.8	2.2
Virginia	32,834	32,294	33,747	35,837	35,590	41,954	1.7	-2.0	-2.7
Washington	4,270	4,101	4,638	4,495	4,565	5,143	4.1	-1.6	-2.0
West Virginia	158,257	157,978	171,145	173,743	170,433	167,352	0.2	-1.8	9.6
Total									
Northern	37,601	38,788	44,618	42,802	45,910	52,155	-3.1	-4.9	-3.6
Southern	120,656	119,191	126,527	130,941	124,523	115,196	1.2	9.8	0.5
Total									
Wyoming	338,900	337,119	314,409	281,881	278,440	193,854	0.5	5.0	6.4
Appalachian	419,419	425,573	460,400	467,778	451,868	457,808	-1.4	-1.8	-1.0
Total [b]									
Interior Total [b]	143,531	162,530	168,374	170,863	172,848	195,418	-11.7	-4.5	-3.4
Western Total [b]	510,661	512,328	488,762	451,291	439,140	342,758	0.3	3.8	4.5
East of Miss. River	507,517	529,594	570,576	579,369	563,668	591,294	-4.2	-2.6	-1.7
West of Miss. River	566,094	570,837	546,960	510,563	500,188	404,690	-.8	3.1	3.8
U.S. Total	1,073,612	1,100,431	1,117,535	1,089,932	1,063,856	995,984	-2.4	0.2	0.8

[a] nm/Not meaningful as value is >500 %.

[b] Data round to zero.

[c] For a definition of coal-producing regions, see Appendix C.

[d] Notes: Coal production excludes silt, culm, refuse bank, slurry dam, and dredge operations except for pennsylvania anthracite. Totals may not equal sum of components due to independent rounding.

[e] Sources: Energy Information Administration, Form EIA-7A, "Coal Production Report"; State Mining Agency Coal Production Reports; and/or U.S. Department of Labor, Mine Safety and Health Administration, form 7000-2, "Quarterly Mine Employment and Coal Production Report".

Table 11. **Energy Source Usage in the United States, % Energy Basis**[a]

Source	1942	1952	1962	1972	1988	1991	1999
coal	68.7	45.2	32.5	25.6	23.5	23.1	21.4
petroleum	17.1	25.4	23.6	22.8	42.7	40.1	39.7
natural gas	9.0	23.3	38.3	44.4	23.1	24.7	23.3
hydroelectric	5.2	6.1	5.5	6.0	3.3	3.8	3.7
nuclear			0.1	1.2	7.1	8.0	8.1

[a]Ref. 2.

capita (85). World recoverable reserves were ~120 times the annual coal production in 1988 and ~10 times that for the additional reserves believed to be in place (1). Estimated coal consumption reduces the known recoverable reserves at ~1%/year. Whereas the use of bituminous coal is expected to continue to increase in terms of tonnage, the percentage of coal used in the United States has stabilized as shown in Table 11.

10. Sample Sources

Basic coal research requires a variety of coal samples of different ranks that workers may access using a minimum of effort. Coal sample banks fill this need. Moreover, over the past decades it has become evident that the quality of samples degrades from atmospheric oxidation and the degradation has limited the ability of researchers to compare results. The U.S. Department of Energy Office of Basic Energy Sciences has sponsored the Argonne Premium Coal Sample Program to permit the acquisition of ton-sized samples of each of eight different coals representing a range of coal ranks, chemical composition, geography, and maceral content (71).

The samples were collected and processed in a manner to avoid exposure to oxygen and control loss of moisture; then they were pulverized to convenient sizes for research, ie, <149 μ (−100 mesh) and <840 μ (−20 mesh); the entire ton was thoroughly mixed; and packaged in sealed glass ampules under nitrogen. These samples have been well characterized and are available in quantities that are expected to last for decades (Table 8).

Other coal sample banks are also in existence. The Penn State Sample Bank at Pennsylvania State University has the most diverse collection of samples (86). The Illinois Basin Coal Sample Program at the Illinois State Geological Survey specializes in samples from the Illinois Basin (87). The European Center for Coal Specimens, located in Eygelshoven in The Netherlands, has a significant collection of samples from the entire world (88). Each makes samples available in kilogram quantities.

11. Mining and Preparation

11.1. Mining. Coal is obtained by either surface mining of outcrops or seams near the surface or by underground mining depending on geological conditions, which may vary from thick, flat seams to thin, inclined seams that are folded and need special mining methods. Coal mining has changed from a labor intensive activity to one that has become highly mechanized. In 2001, the average output per person per hour in underground mines in the United States was 4.2 tons. For surface mines, the output was 11.2 tons per person per hour (2).

Strip or open-pit mining involves removal of overburden from shallow seams, breaking of coal by blasting or mechanical means, and loading of the coal. The development of very large excavating equipment, including draglines, bulldozers, front-end loaders, and trucks, has been responsible for significantly increased production from strip mining.

The two methods of underground mining commonly used are room-and-pillar and longwall. In room-and-pillar mining the coal is removed from two sets of corridors that advance through the mine at right angles to each other. Regularly spaced pillars, constituting about one-half of the coal seam, are left behind to support the overhead layers in the mined areas. This method is used extensively in the United States and other nations having large reserves. The pillars may later be removed, leading to probable subsidence of the surface. Longwall mining is used to permit recovery of as much of the coal as possible (87). Two parallel headings are made 100–200-m apart and at right angles to the main heading. The longwall between the two headings is then mined away from the main heading. The equipment provides a movable roof support system that advances as the coal is mined and allows the roof to collapse in a controlled manner behind it. This method also leads to subsidence of the overhead layers.

Another method used in Europe for steeply inclined seams is called horizontal mining. Horizontal shafts are cut through rock below the coal seams. Vertical connections are then made to the seam to permit coal removal.

The mechanical equipment used in room-and-pillar underground mining usually involves a series of specific operations with continuous mining equipment. Continuous miners use rotating heads equipped with bits to pick or cut through the coal without blasting and load it into a shuttle car for conveying to a belt system.

11.2. Preparation.   Coal preparation is of significant importance to the coal industry and to consumers (6,55,90). Preparation normally involves some size reduction of the mined coal and the systematic removal of some ash-forming material and very fine coal. The percentage of mined coal that is mechanically cleaned in the principal coal producing countries has risen during the past 30 years. There are a number of reasons for this. The most important in the United States is the increased use of continuous mining equipment. The nature of this operation tends to include inorganic foreign matter from the floor and ceiling of the seams, thus run-of-mine coal includes ∼25% mineral matter. The size consist of the mined coal is also smaller when produced using continuous mining equipment. The depletion of the better quality seams, which are low in ash and sulfur, in many coal fields necessitates cleaning of the remainder. Moreover, the economic need to recover the maximum amount of salable coal has led to cleaning of the finer sizes that had previously been discarded. Stringent customer demands for coal meeting definite specifications, regulations requiring the removal of pyrites to control air pollution (qv), increased freight rates, and ash disposal costs all contribute to the upsurge in coal cleaning (see COAL CONVERSION PROCESS, CLEANING AND DESULFURIZATION).

In earlier times the coal was hand-picked for removal of larger pieces of refuse, but higher labor costs have made this method uneconomical. Mechanical separation methods are used, most of which depend on the difference in density between the coal and refuse. The washability characteristics of a coal determine the extent to which the refuse may be removed. The laboratory float-and-sink analysis gives information on the percentages and quality of the coal material occupying different density ranges. From this information, graphs are constructed showing the composite quantity and coal quality that can be obtained by cleaning at different specific gravities. This information is considered with the economic factors involved in the sale of the washed coal to choose an optimum method of cleaning. Cleaning plants are usually designed to handle the output of specific mines and to clean for a specific market. The plant includes various types of cleaning methods designed to move the different fractions through several cleaning circuits to maximize recovery.

In some areas, run-of-mine coal is separated into three products: a low gravity, premium-priced coal for metallurgical or other special use, a middlings product for possible boiler firing, and a high ash refuse. The complete preparation of coal usually requires several processes.

Cleaning Methods.   Jig washing is the most widely used of all cleaning methods. A bed of coal particles is subjected to alternate upward and downward currents of water causing a moving bed of particles to stratify. The lighter clean coal particles go to the top and the heavier refuse particles to the bottom (see MINERALS RECOVERY AND PROCESSING). The heavy medium process is a simple float-and-sink one that is widely used for coarse coal cleaning. The medium is usually a suspension of pulverized magnetite, which is mixed to the desired specific gravity. This method is also used in cyclones for a wide range of coal cleaning.

By using trough washers, the coal is fed to a trough in a stream of water that carries the coal particles forward but allows heavier particles to sink and be removed. Washing tables are used for cleaning fine coal. A coal suspension in water flows across a slanted table that oscillates at right angles to the direction of flow. The heavier refuse particles settle onto the table and are trapped by riffles or bars, while the lighter coal particles are carried over the riffles in the current. If dry cleaning is used, the coal passes over a perforated, oscillating table through which air is blown. This method creates a dust problem, although it eliminates the need for drying the coal. In countries using hydraulic mining or underground dust suppression with water, there is limited opportunity for dry cleaning.

Froth flotation (qv) is the most important method for cleaning fine coal because very small particles cannot be separated by settling methods. Air is passed through a suspension of coal in water to which conditioning reagents, usually special oils, have been added. The oils are selected so that the coal particles preferentially attach themselves to the bubbles and separate from the refuse that remains in suspension.

Dewatering.   The coal leaving the cleaning plant is very wet and must be at least partially dried to reduce freight charges, meet customer requirements, and avoid freezing. Draining on screens removes substantial amounts of water from larger coal, but other dewatering (qv) methods are required for smaller sizes having larger surface areas.

Vibrating screens and centrifuges are used for dewatering. For very fine coal, such as that obtained from flotation, vacuum filtration with a disk or drum-type filter may be used. Flocculants may be added to aid filtration (see FLOCCULATING AGENTS). They are also used for cleaning wastewater and pollution control. If very low ($\leq$2–3%) moisture contents are required, thermal drying must be used. A number of dryer types are available including fluidized bed; suspension; and rotary- and cascade-dryers. All of these are expensive to operate, however (see DRYING).

Storage.   Storage of the coal may be necessary at any of the various steps in production or consumption, ie, at the mine, preparation plant, or consumer location. Electric utilities have the largest amounts of coal in storage, having stockpiles that frequently are able to meet 60–100 day normal demand thus protecting against delays, shortages, price changes, or seasonal demands (2).

For utilities, two types of storage are used. A small amount of coal in storage meets daily needs and is continually turned over. This coal is loaded into storage bins or bunkers. However, long-term reserves are carefully piled and left undisturbed except as necessary to sustain production.

Coal storage results in some deterioration of the fuel owing to air oxidation. Moreover, if inadequate care is taken, spontaneous heating and combustion may result. As the rank of coal decreases, it oxidizes more easily and must be piled more carefully. Anthracite does not usually present a problem.

The surface of the coal particles oxidizes or weathers resulting in cracks, finer particles, and reduced agglomeration all of which may destroy coking properties. If spontaneous heating takes place, the calorific value of the coal is reduced. Hot spots must be carefully dug out and used as quickly as possible. Without spontaneous heating and with good compaction, calorific value losses <1%/year have been recorded.

Coal piles are carefully constructed to exclude air or to allow adequate ventilation. The latter requires larger sizes, graded as 4 cm+ without fines, for avoiding heating by ventilation. For exclusion of air, mixed sizes provide fines to fill the gaps between larger pieces. Pockets of large sizes must not be allowed because these provide access for air. The coal should be compacted to maximize the bulk density of the coal pile.

Several approaches have been effective for storage: (*1*) large compacted layered piles where sides and top are sealed using an oil or asphalt (qv) emulsion. Four liters of oil seals 1 m^2 of coal; (*2*) large compacted layered piles where sides and top are covered first with fines to seal the pile and then with coarse coal to protect fines from wind and weather. The sides may slope at angles $\geq$30°; (*3*) piles of compacted layers in open pits having tight sides so that the air has access only at the top; (*4*) sealed bins or bunkers in which airtight storage can be provided for smaller amounts of coal for long times; and (*5*) underwater storage in concrete pits. This is expensive and rarely used but effectively prevents deterioration although it introduces other problems related to handling wet coal.

Large compacted storage piles should be located on hard surfaces and not subject to flooding. A layer of fines may be put down first to facilitate recovery. Each layer of coal should be compacted after it is deposited. The top of the pile should have a slight crown to avoid water accumulation. Excessive heights should be avoided to prevent air infiltration caused by wind. Coal removal should be done in layers followed by compacting and smoothing the surface. Piles should be limited to the same rank of coal depending on the intended use.

12. Transportation

The usual means of transporting coal are railroad, barge, truck, conveyer belt from mine to plant, and slurry pipelines (2,4) (see TRANSPORTATION). In 2000, 1038×10^6 t of coal was transported to United States destinations; of this, 65.5% was shipped by railroad, 14.3% by barge, 10.6% by truck, and 9.6% by conveyer, slurry pipeline, and other methods (2). Electric utilities consumed 80.04% of the coal transported in the United States in 2000 (2).

The unit train handles about one-half of rail transportation (71). Most unit trains consist of ~100 rail cars and are dedicated to coal haulage from the mine to the consuming plant. Almost no time is wasted at either the loading or unloading site as a result of efficient loading and unloading equipment. Diesel and diesel-electric trucks having capacities up to 320 t (2) handled the off-highway transport of coal.

A 437-km slurry pipeline, 46 cm in diameter, was started in 1970 to move coal from Arizona to southern Nevada. The coal is crushed and ground to the fineness needed for proper viscosity and settling. About 18–20% is −325 mesh (<44 μ), 35–45% is −100 mesh (<149 μ), and 0–2% is +14(<1070 μ) mesh. The solids content of the slurries has approached 70% using additives to stabilize the mixture. The slurry is dewatered with centrifuges before combustion of the coal.

For shipment in cold climates, a freezeproofing treatment using inorganic chemicals or oil spray is used. An oil spray on the inside of the coal cars is also effective at 3.8–5.7 L/car for four round trips of 1280 km each. Oil treatment has also been used for dustproofing; wind loss can be prevented by use of an asphalt emulsion on the top of rail cars (6).

Coal pipelines have been built in countries such as France (8.8 km), and Russia (61 km), and pipelines are also used for transporting limestone, copper concentrates, magnetite, and gilsonite in other parts of the world. The first coal pipeline, built in Ohio, led to freight rate reductions. The pipeline stopped operation after introduction of the unit train, used exclusively to transport coal from the mine to an electric power generation station.

Hydraulic transport is used in mines and for lifting coals to the surface in Russia, Poland, and France. Pneumatic transport of coal is used over short distances in power plants and steel mills. The longest (14.5 km) single flight conveyer belt in the world near Uniontown, Kentucky, has a capacity of 1360 t/h.

13. Economic Aspects

Table 12 gives the estimated destinations for United States bituminous coal exports in 1976, 1989, and 2000. About one-third of the bituminous coal exported from the United States in 1976 went to Europe, another third to Asia, and the remainder to North and South America. The pattern shifted to ~42% to North and South America, ~43% to Europe, 11% to Asia, and the balance to Africa (1).

Of the 1989 total, 65,128,000 t were metallurgical coal and 34,910,000 t were steam coal. Exports of coke from the United States in 1989 were 1,169,120 t, whereas anthracite exports were 745,749 t. Lignite exports were 163,628 t. In 1989 Canada produced 77,727,000 t, imported 16,160,000 t and exported 36,094,000 t. Japan is the principal recipient of Canadian exports. Selected coal exports and imports are given in Table 13.

The weighted average values for underground and surface mines decreased from $28.24/t in 1984 to $23.99/t in 1990 (2). Underground mine prices decreased from $36.66/t to $31.51/t, whereas surface mine prices dropped from $22.70/t to $18.72/t.

14. Analysis

Most countries have an official national organization, which is responsible for developing and maintaining standards for testing and analysis. The ASTM serves this purpose in the United States as does the British Standards (BS) Organization in the United Kingdom. In Geneva, the International Organization for Standardization (ISO) formed a committee (T.C.27), which is responsible for developing international standards. Each organization issues periodic updates of their standards. ASTM does this on an annual basis. The ASTM coal standard methods are each assigned a number, preceded by a D. The methods are periodically reviewed and revised. A two-digit number may follow the method number to indicate the year of the revision. BS standards of similar type also exist for most methods described. Details of an individual test may be found in the compilation for the respective source organization.

14.1. Sampling. The procedures for taking a sample, reducing the particle size of the sample, and separation of a smaller portion for later analysis are given in ASTM D2234 and D2013 (18) and BS1017. The procedures describe the minimum amount of sample needed to maintain a representative sample for analysis.

14.2. Size Analysis. ASTM and BS (18) provide a number of methods dealing with the size specifications and size analysis procedures including D197, D410, D311, and D431.

14.3. Moisture Holding Capacity. The bed or equilibrium moisture is the amount of moisture retained after equilibration at 96–97% rh at 30°C (D1412) (18). Total moisture is determined by air drying, crushing to smaller particle sizes and heating at 107°C to constant weight (D3302).

14.4. Analysis. The proximate analysis is based on determinations of volatile matter, moisture, and ash for a coal sample. Fixed carbon is then

Table 12. **U.S. Coal Exports by Destination, 1991, 1996–2000 (Thousand Short Tons)**

Continent and country of destination	2000	1999	1998	1997	1996	1991	Percent change 1999–2000	Average Annual percent change 1996–2000	1991–2000
North America									
Canada[a]	18,769	19,826	20,654	14,975	12,029	11,178	−5.3	11.8	5.9
Mexico	819	1,411	1,543	1,899	1,509	92	−41.9	−14.2	27.5
Other[b]	24	7	119	73	72	184	228.4	−23.8	−20.1
Total	*19,612*	*21,244*	*22,316*	*16,947*	*13,609*	*11,454*	*−7.7*	*9.6*	*6.1*
South America									
Argentina	204	3	324	325	304	429	NM	−9.4	−7.9
Brazil	4,536	4,442	6,475	7,455	6,540	7,052	2.1	−8.7	−4.8
Chile	53	43	51	146	574	135	21.7	−45.0	−9.9
Other[b]	30	32	185	288	87	45	−5.3	−23.4	−4.3
Total	*4,823*	*4,521*	*7,034*	*8,214*	*7,505*	*7,661*	*6.7*	*−10.5*	*−5.0*
Europe									
Belgium & Luxembourg	2,890	2,073	3,195	4,319	4,569	7,464	39.4	−10.8	−10.0
Bulgaria	919	522	989	1,114	1,387	946	75.9	−9.8	−.3
Denmark	77		274	350	1,316	4,658		−50.8	−36.6
Finland	317	233	463	662	704	530	36.2	−18.0	−5.5
France	3,044	2,522	3,192	3,398	3,852	9,509	20.7	−5.7	−11.9
Germany, FR	976	573	1,247	870[c]	1,055	1,742	70.3	−1.9	−6.2
Hungary	72								
Iceland	53	51	39	54	62	45	4.1	−3.6	1.9
Ireland	502	868	1,150	637	765	1,313	−42.1	−10.0	−10.1
Italy	3,711	4,014	5,317	7,019	9,204	11,274	−7.5	−20.3	−11.6
Netherlands	2,623	3,432	4,516	4,825	7,058	9,625	−23.5	−21.9	−13.4
Norway	130	86	93	96	85	200	50.1	11.1	−4.7
Portugal	596	745	746	1,470	1,803	1,629	−20.0	−24.2	−10.6
Romania	489	322	1,097	2,244	1,512	1,147	52.0	−24.6	−9.0
Spain	2,686	2,472	3,156	4,134	4,093	4,694	8.6	−10.0	−6.0

Sweden	708	638	757	834	1,070	1,239	10.8	−9.8	−6.0
Turkey	1,809	795	1,592	2,092	2,167	2,186	127.6	−4.4	−2.1
United Kingdom	3,294	3,162	5,947	7,185	6,196	6,171	4.2	−14.6	−6.7
Yugoslavia, FR	73	1	3	29	296	979			−25.1
Other[b]	c					169	−82.3	−84.1	−53.0
Total	*24,969*	*22,508*	*33,773*	*41,331*	*47,193*	*65,520*	*10.9*	*−14.7*	*−10.2*
Asia									
China (Taiwan)	386	1,215	1,519	2,241	2,441	4,547	−68.3	−36.9	−24.0
Israel	62	603	527	593	1,202	651	−89.8	−52.4	−23.0
Japan	4,446	4,953	7,734	7,974	10,529	12,269	−10.2	−19.4	−10.7
Korea, Republic of	1,768	2,365	2,453	3,489	3,773	3,711	−25.2	−17.3	−7.9
Other[b]	40	21	78	201	36	611	91.7	3.0	−26.1
Total	*6,702*	*9,157*	*12,311*	*14,498*	*17,980*	*21,788*	*−26.8*	*−21.9*	*−12.3*
Oceania & Australia									
Other[b]	c	c	5	1	1	c	216.2	−27.4	7.4
Total	*c*	*c*	*5*	*1*	*1*	*c*	*216.2*	*−27.4*	*7.4*
Africa									
Algeria	296	317	343	264	177	522	−6.6	13.8	−6.1
Egypt	753	260	891	1,130	1,038	769	189.8	−7.7	−.2
Morocco	909		68	142	1,650	1,013	c	−13.8	−1.2
South Africa, Rep of	424	469	1,299	987	1,320	239	−9.5	−24.7	6.6
Other[b]	1	c	8	31	c	2	d	c	−13.2
Total	*2,383*	*1,046*	*2,609*	*2,554*	*4,184*	*2,545*	*127.9*	*−13.1*	*−.7*
Grand Total	*58,489*	*58,476*	*78,048*	*83,545*	*90,473*	*108,969*	*c*	*−10.3*	*−6.7*

[a] Based on the U.S. Canada Free Trade Agreement, as of January 1990, the U.S. Department of Commerce began reporting statistics on U.S. exports to Canada based on information on imports provided monthly by the Canadian government.

[b] Includes countries with exports less than or equal to 50,000 short tons in 2000.

[c] Data round to zero.

[d] Not Meaningful as value is >500 %.

Note: Totals may not equal sum of components due to independent rounding.

Source: U.S. Department of Commerce, Bureau of the Census, "Monthly Report EM 545."

Table 13. **World Coal Supply and Disposition, 1998 (Trillon Btu)**

Region/Country	Production	Imports	Exports	Apparent consumption
North America				
Canada	1,989	526	988	1,526
Mexico	211	68	0	275
United States	23,719	313	2,076	21,660
Total	*25917.72*	*906.98*	*3064.24*	*23460.63*
Central and South America				
Argentina	7	39	5	38
Brazil	67	509	0	568
Chile	7	132	1	147
Colombia	914	0	797	157
Peru	1	17	0	16
Venezuela	228	0	201	0
Other	0	24	0	23
Total	*1223.28*	*720.37*	*1002.97*	*949.85*
Western Europe				
Austria	12	122	0	130
Belgium	8	402	35	365
Bosnia and Herzegovina	16	0	0	16
Croatia	1	8	0	10
Denmark	0	203	4	233
Finland	0	131	0	145
France	155	496	15	666
Germany	2,585	724	10	3,395
Greece	339	37	2	372
Italy	1	456	3	461
Luxembourg	0	4	0	4
Macedonia, TFYR	72	10	0	81
Netherlands	0	551	226	348
Norway	9	43	8	45
Portugal	0	131	2	128
Serbia and Montenegro	387	12	0	399
Slovenia	56	3	0	59
Spain	335	327	14	658
Sweden	0	92	2	94
Turkey	572	312	0	893
United Kingdom	1,065	617	32	1,675
Other	0	85	0	87
Total	*5614.43*	*4767.92*	*353.99*	*10262.72*
Eastern Europe & Former U.S.S.R.				
Bulgaria	279	97	(0)	373
Czech Republic	961	44	113	901
Hungary	127	32	5	153
Poland	3,158	102	722	2,569
Romania	209	97	4	312
Slovakia	48	129	1	185
Belarus	0	18	0	18
Estonia	0	13	1	13
Kazakhstan	873	39	438	474
Kyrgyzstan	6	15	0	21
Latvia	0	3	0	3
Lithuania	0	6	0	6

Table 13 (*Continued*)

Region/Country	Production	Imports	Exports	Apparent consumption
Moldova	0	9	0	11
Russia	3,998	405	483	3,970
Tajikistan	0	2	0	2
Ukraine	1,613	215	49	1,781
Uzbekistan	43	0	0	43
Other	1	0	0	1
Total	*11317.77*	*1228.52*	*1815.28*	*10834.33*
Middle East				
Iran	24	18	0	41
Israel	0	251	0	242
Other	0	6	0	6
Total	*23.62*	*275.15*	*0.31*	*289.41*
Africa				
Algeria	1	16	0	20
Botswana	24	0	0	24
Congo (Kinshasa)	2	5	0	7
Egypt	9	48	11	47
Kenya	0	3	0	3
Malawi	1	0	0	2
Mauritius	0	2	0	2
Morocco	6	88	0	94
Mozambique	1	0	0	2
Niger	4	0	0	4
Nigeria	2	0	0	2
South Africa	5,273	31	1,622	3,396
Swaziland	7	0	0	7
Tunisia	0	2	0	2
Zambia	4	0	0	4
Zimbabwe	112	0	5	107
Other	0	1	0	1
Total	*5446.98*	*198.24*	*1638.54*	*3723.00*
Far East and Oceania				
Afghanistan	0	0	0	0
Australia	6,043	0	4,147	2,015
Bangladesh	0	4	0	4
Bhutan	0	1	0	0
Burma	1	0	0	2
China	24,595	33	1,025	23,781
Hong Kong	0	182	0	200
India	5,694	248	0	5,952
Indonesia	1,546	12	1,199	376
Japan	85	3,181	69	3,194
Korea, North	1,072	58	9	1,121
Korea, South	82	1,248	0	1,330
Malaysia	10	55	0	62
Mongolia	49	3	2	50
Nepal	0	3	0	3
New Caledonia	0	4	0	4
New Zealand	67	0	32	38
Pakistan	59	26	0	85
Philippines	23	81	0	101

Table 13 (*Continued*)

Region/Country	Production	Imports	Exports	Apparent consumption
Singapore	0	0	0	0
Taiwan	2	1,017	0	1,008
Thailand	242	43	0	290
Vietnam	250	1	75	121
other	0	1	0	1
Total	*39819.50*	*6199.75*	*6559.00*	*39737.60*
World Total	*89363.31*	*14296.93*	*14434.33*	*89257.53*

calculated by difference. Volatile matter is determined empirically by measuring the weight loss when coal is heated in a covered crucible at either 950°C (ASTM D3175 or D5142) or 900°C (BS).

The ultimate analysis gives the elemental composition in terms of C, H, N, S, and O (D3176, D3177, or D4239, and D3178) (18). The C and H analyses are based on oxidation of the sample in a tube and reaction of the gaseous products with absorbents to permit calculation of C and H content. The N analyses involves a Kjeldahl determination and the S involves oxidation and detection of SO_2 by ir detection, titration with base, or conversion to an insoluble sulfate to determine total S. Oxygen is determined by difference. Oxygen in the organic material is calculated by subtracting the dry, ash-free percentage of C, H, N, and organic S from 100. Sulfur forms are determined in a separate procedure. Sulfate and pyritic sulfur are determined separately and subtracted from total S to give organic S.

Other Elements. To determine chlorine, the sample is mixed with Eschka's mixture and burned to convert the chlorine to chloride or decomposed in an oxygen bomb (18). Chloride is determined by titration (D2361) or using a chloride selective electrode (D4208) (18) (see ELECTROANALYTICAL TECHNIQUES).

Phosphorus determination involves the conversion of phosphorus to soluble phosphate by digesting the coal ash with a mixture of sulfuric, nitric, and hydrofluoric acids (18). Phosphate is precipitated as ammonium phosphomolybdate, which may be reduced to give a blue solution that is determined colorimetrically or volumetrically (D2795) (18).

14.5. Calorific Value. To determine calorific value, a sample is placed in a bomb, pressurized with oxygen, and ignited. The temperature rise in the water bath of the calorimeter surrounding the bomb is used to determine the calorific value (D2015, D3286, or D1989) (18).

14.6. Ash Fusibility. A molded cone of ash is heated in a mildly reducing atmosphere and observed using an optical pyrometer during heating. The initial deformation temperature is reached when the cone tip becomes rounded; the softening temperature is evidenced when the height of the cone is equal to twice its width; the hemispherical temperature occurs when the cone becomes a hemispherical lump; and the fluid temperature is reached when no lump remains (D1857) (18).

15. Swelling and Coking Tests

For the free swelling index which is also known as the crucible swelling number (ASTM D720), a coal sample is rapidly heated to 820°C in a covered crucible. Then the profile of the resulting char is compared to a series of standard numbered profiles (18). For the Roga index weighed amounts of coal and standard anthracite are mixed and carbonized, and the product coke is tested in a Roga drum for its resistance to abrasion (89).

For the Gray-King coke-type assay test (91,92) coal is heated in a retort tube to 600°C and the product coke is compared to a series of standard cokes. For a strongly swelling coal, enough anthracite or electrode carbon is added to the coal to suppress the swelling. This method is primarily used in Europe.

In the Audibert-Arnu dilatometer test (91), a thin cylinder of compressed powdered coal contacting a steel piston is heated at a rate not >5°C/min. The piston movement is used to calculate the percent dilation.

15.1. Hardgrove Grindability Index and Strength Tests. A specially sized coal sample is ground in a specifically designed ball and race grinding mill (D409). The index is determined from the amount of coal remaining on a 74 mm (200 mesh) screen (18). The higher the index, the easier it is to grind the coal.

The drop shatter test indicates the resistance of a coal or coke to breakage on impact (see D440). A sample is dropped in a standard way a number of times from a specified height. For the tumbler test or abrasion index (ASTM D441), the coal or coke is rotated in a drum to determine the resistance to breakage by abrasion (18).

16. Health and Safety Factors

Coal mining has been a relatively dangerous occupation (2,91–93). During the period from 1961–1967 the average fatality rate in the United States for each million person hours worked was 1.05. In the 7 years after the passage of the Federal Coal Mine Health and Safety Act of 1969, the average fatality rate decreased to 0.58, and by 1989 the rate was 0.25 (2).

The rates of occupational injuries are reported per 200,000 employee-hours which correspond to ~100 employee-years. In 1989, the total for all mines was 11.84 or 11.84%. Over the 3-year periods before and after passage of the 1969 act, the rates for underground mining were 48.60% and 40.07%, respectively. The principal causes of fatalities are falling rock from mine roofs and faces, haulage, surface accidents, machinery, and explosions. For disabling injuries the primary causes are slips and falls, handling of materials, use of hand tools, lifting and pulling, falls of roof rock, and haulage and machinery (2).

16.1. Gases and Coal Dust Explosions. Gases can be hazardous in coal mines. Methane is of greatest concern, although other gases including carbon monoxide and hydrogen sulfide may be found in some mines. Methane must be detected and controlled because mixtures of air and 5–15% of methane are explosive.

The U.S. Mine Health and Safety Act of 1969 requires that a mine be closed if there is 1.5% or more methane in the air. The use of an electrical methane detection device is required. High capacity ventilation systems are designed to sweep gases from the cutting face and out of the mine. These systems remove all gases before they become harmful.

Whereas an explosion from methane tends to be localized, it may start coal dust explosions resulting in more widespread injury and loss of life. All coal breaking operations result in formation of fine coal particles; some are controlled with water during the mining operation. Breakage associated with hauling disperses dust, and dust accumulations can be made safe by rockdusting. Powdered limestone is spread over the mine surfaces to cover the dust.

16.2. Drainage.   Some mines are located beneath subsurface streams, or the coal seams may be aquifers. These mines may become flooded if not continually pumped. In Pennsylvania anthracite mines as much as 30 tons of water may be pumped for each ton of coal mined (94).

Air or biological oxidation of pyrite leads to sulfate formation and dilute sulfuric acid in the mine drainage. This pollutes streams and the water supplies into which the mine water is drained. Means of controlling this problem are under study.

16.3. Other Hazards.   Rocks falling from the roofs of mines used to cause the largest number of accidents. Roof bolts are placed in holes drilled into the roofs of working areas to tie the layers of rock together and thus prevent rock falls. A disease called pneumoconiosis, also called black lung, results from breathing coal dust over prolonged periods of time. The coal particles coat the lungs and prevent proper breathing.

16.4. Regulations.   The U.S. Bureau of Mines, Mining Enforcement and Safety Administration (MESA) studies hazards and advises on accident prevention. MESA also administers laws dealing with safety in mines. Individual states may also have departments of mines to administer state standards.

The Federal Coal Mine Health and Safety Act set standards for mine ventilation, roof support, coal dust concentrations levels, mine inspections, and equipment. As a part of this comprehensive act, miners must receive medical examinations at employer expense, and payments are made from the U.S. government to miners who cannot work because of black lung disease.

17. Uses

17.1. Coal As Fuel.   Coal is used as a fuel for electric power generation, industrial heating and steam generation, domestic heating, railroads, and coal processing. About 87% of the world's coal production is burned to produce heat and derived forms of energy. The balance is practically all processed thermally to make coke, fuel gas, and liquid by-products. Other uses of coke and fuel gas also contribute to coal consumption for heat. In the United States, coal use for power generation has increased to 86.1% in 1988, whereas coking coal use has dropped to 4.7% and the industrial/retail market declined to 9.2% (2).

Electric Power Generation.   Coal is the primary fuel for thermal electric power generation. Since 1940 the quantity of bituminous coal consumed by electric utilities has grown substantially in each successive decade, and this growth is expected to continue for many years. Coal consumed by electric utilities increased from ~536×10^6t in 1981 to 991×10^6t in 2000 (2). The reasons for increased coal demand include availability, relative stability of decreasing coal prices, and lack of problems with spent fuel disposal as experienced in nuclear power plants (see NUCLEAR REACTOR TYPES).

The overall efficiency of electric power plants consisting of coal-fired boilers and steam turbines has plateaued at ~39%. The addition of pollutant control equipment has increased the internal power use on the stations and lowered the effective efficiency of the plant. The increased efficiencies have been achieved through use of larger units (up to 1500 MW) and higher pressures to 24.1 MPa (3500 psi) and reheat, but concerns about reliability and ability to match power generation and demand have kept plant sizes below these values. Maximum temperatures have not been increased because of the difficulties of corrosion owing to coal ash constituents, materials properties, and costs of better alloys. The advent of any future increases in efficiency depends on development of new systems of power generation, which might include fluid-bed boilers, gasification of coal to power a gas turbine having hot exhaust directed to a waste heat boiler in a combined cycle (gas turbine and steam turbine), or use of magnetohydrodynamics (qv) (see FURNACES,FUEL-FIRED).

Almost all modern large coal-fired boilers for electric power generation use pulverized coal. The cyclone furnace, built mainly for use in Germany and the United States, uses coarser pulverized coal. The ash is removed primarily as a molten slag from the combustor. Apparently, this design is no longer offered in the United States. This method of firing has not been accepted in the United Kingdom because of the higher softening temperature of the ash of the British coals. Stoker firing is generally limited to the smaller obsolete stand-by utility plants and generation plants used by industrial companies.

One significant advantage of pulverized coal boilers is the ability to use any kind of coal, including run-of-mine or uncleaned coals. However, with the advent of continuous mining equipment, the ash content frequently is ~25%, and some preparation is frequently practiced. There were 931 coal preparation plants in the United States in 1988, mainly in Kentucky, West Virginia, and Pennsylvania.

The advent of fluidized-bed boilers has enabled the size of units to go to 150 MW for commercial power generation and cogeneration in the last decade. This technology is displacing some stoker fired units at the low capacity applications and smaller pulverized coal units at the large size of the fluidized-bed range. Bubbling and circulating bed designs are used, and operating temperatures are in the 815–900°C range. Sulfur oxides are controlled using dolomite or limestone injection in the bed. Higher (2:1–5:1) calcium to sulfur ratios are needed for fluid-bed units than for wet scrubbing (1.0.:-1) or spray dryers (1.2–1.5:1). Nitrous oxide emissions are higher than for other methods, possibly as a result of formation and oxidation of hydrogen cyanide (64). Several processes are being developed to reduce nitrogen oxides emissions. For example, three pressurized fluid-bed combustors are being demonstrated under the U.S. Department of Energy's Clean Coal Technology Program (15).

Integration of coal gasification and a combination of a gas turbine for power generation and a waste heat boiler for power generation is termed integrated gasification combined cycle (IGCC). Efficiencies are currently ~42% and promise to be higher as gas turbine technology improves (63). As of 1992 five plants using this technology have been announced in the United States. The IGCC technology uses sulfur gas removal techniques that result in higher removal rates than conventional scrubbers, in part because of the improved efficiency of scrubbing the more concentrated gases.

A primary concern in coal-fired power generation is the release of air pollutants. Limits on SO_2 output, 0.52 g/MJ equivalent of coal input to a new plant, have been established. For a bituminous coal of 27.9 MJ/kg, there is thus an upper limit of 0.72% sulfur content. Relatively few coals can meet this requirement. The U.S. Department of Energy indicated recoverable reserves of 420×10^9t in 1987 (2) that were categorized by sulfur content: 33.5% had 0.6% S or less, 15.4% had between 0.61% and 0.83% S, 16.1% had between 0.84 and 1.67% S, 12.4% had between 1.68 and 2.50% S, and 22.6% had >2.5% S. The lowest sulfur coal, ~86%, is found west of the Mississippi River, mainly in Montana and Wyoming, quite distant from the electric power demand centers in the East. A trend to utilization of the western coals has developed.

Industrial Heating and Steam Generation.   The principal industrial users of coal include the iron (qv) and steel (qv) industry and the food, chemicals, paper (qv), engineering, bricks, and other clay products, and cement (qv) industries, and a group of miscellaneous consumers such as federal and local government installations, the armed services, and small industrial concerns. Most of the coal is burned directly for process heat, ie, for drying and firing kilns and furnaces, or indirectly for steam generation for process needs or for space heating, and for a small amount of electric power generation. Industrial coal usage in the United States has diminished significantly in past decades, especially among small users, because of the greater convenience in storing and handling gaseous and liquid fuels and the higher initial cost of coal-fired equipment.

Several developments are being pursued to utilize coal directly, ie, automation of controls, coal and ash handling equipment for smaller stoker and pulverized coal-fired units, design of packaged boiler units, and pollution control equipment. In the cement industry coal firing has been used, because the sulfur oxides react with some of the lime to make calcium sulfate in an acceptable amount.

17.2. Coal Processing to Synthetic Fuels and Other Products.   The primary approaches to coal processing or coal conversion are thermal decomposition, including pyrolysis or carbonization (5,6), gasification (6), and liquefaction by hydrogenation (6). The hydrogenation of coal is not currently practiced commercially.

In the United States, the Clean Coal Technology program was created to develop and demonstrate the technology needed to use coal in a more environmentally acceptable manner. Activities range from basic research and establishing integrated operation of new processes in pilot plants through demonstration with commercial-scale equipment.

High Temperature Carbonization.   High temperatures and long processing times are used in carbonizing coking coals in coke ovens or gas retorts. Besides metallurgical or gas coke the products include fuel gas, crude tar, light oils (benzene, toluene and xylene, referred to as BTX, and solvent naphtha), and ammonia gas (see COAL CONVERSION PROCESSES, CARBONIZATION).

Most coal chemicals are obtained from high temperature tar with an average yield over 5% of the coal which is carbonized. The yields in coking are ~70%

of the weight of feed coal. Tars obtained from vertical gas retorts have a much more uniform chemical composition than those from coke ovens. Two or more coals are usually blended. The conditions of carbonization vary depending on the coals used and affect the tar composition. Coal-tar chemicals include phenols, cresols, xylenols, benzene, toluene, naphthalene, and anthracene.

The largest consumer of coke is the iron and steel industry. In the United States, ~600 kg of coke is used to produce a metric ton of steel. Japanese equipment and practice reduce the requirement to 400–450 kg. Coke is also used to gas from the char in one vessel. The reducing gas converts iron oxide to iron in the upper two stages of a second vessel. Steam is converted to hydrogen and reoxidizes the iron in two stages in the lower half of the vessel.

None of these second- or third-generation processes has been commercialized, largely because of the relatively low price of available liquid and gaseous fuels.

A large commercial plant was completed in 1981 by a consortium of American Natural Gas and Peoples Gas, Light and Coke, and others for Mercer County, North Dakota. This plant has a design capacity of $3.7 \times 10^6 m^3$ (137.5×10^6 standard cubic feet [scf] of methane per day. The plant uses 14 Lurgi gasifiers and 12,700 t/day of lignite, 2,585 t/day oxygen, and 12,383 t/day of steam. The air separation plant is the largest in the hemisphere (see CYROGENICS; NITROGEN). The Phosam process is used for recovery of 113 t/day of ammonia, and the Stretford process was initially used for the recovery of 106 t/day of sulfur. The other products are used primarily as boiler fuels and include tar oil, naphtha, and crude phenol. The coal supply is Beulah-Zap lignite produced at an adjacent mine. The fine coal is removed before gasification and is sold to the neighboring electric utility (Basin Electric Power Cooperative) for use in the adjacent power plant and another plant about 48 km away. The cost of the gas was subsidized in the initial decade of operation. Future plans include the production of SNG and the use of the site for demonstration of a coal to methanol plant (95). In 1988 the ownership of the plant was transferred to the Dakota Gasification Co. of Bismarck. The agreement calls for operation of the plant until 2009 as long as revenues exceed expenses.

Processes for intermediate-Btu gas, ie, 9.3–18.6 MJ × m³ (250–500 Btu × ft³), or synthesis gas production were also developed. In the IGT U-Gas, or a similar Westinghouse process, crushed coal is fed into a fluidized-bed gasifier. Steam and oxygen enter the base of the bed. A part of these gases carry unreacted fines into a hot spout, which accelerates gasification and permits the ash to soften and particles to agglomerate. Ash agglomerates discharge below the spout. Product gases can be cleaned and pipelined as industrial fuel gas near 11.2 MJ × m³ (300 Btu × ft³). This technology has been offered by a consortium of Stone & Webster and Tampella Keeler. This group uses air-blown gasification and hot gas cleanup to lower the capital costs. The modification provides a lower heating value product.

Several plants use the Texaco partial oxidation gasifier developed as a modification of Texaco's oil consuming partial oxidation process. Pulverized coal falls through the reactor at high pressure and temperature to produce the gas which is then cleaned. The ratio of carbon monoxide to hydrogen can be adjusted by the water gas shift reaction as needed for a variety of chemical intermediates. This design was used in a plant to make chemical intermediates for the Eastman Kodak Co. in Kingsport, Tennessee. The Coolwater IGCC (integrated gasification-combined-cycle) demonstration plant in Southern California used this gasifier to provide fuel gas for boilers for electric power generation or for gas turbines for combined cycle power generation. The plant was technically successful but not able to compete economically. The scrubbing system removes a very high amount of the sulfur in the coal (96,97).

The high capital cost, ~$1500 × kW, is the principal deterrent to growth of the IGCC concept. The ability to remove up to 99% of the sulfur species from the combustion products make the IGCC an environmentally desirable option as make calcium carbide (see CARBIDES), from which acetylene is made. Synthesis gas for methanol and ammonia production is also made from gasification of coke.

Considerable research has been carried out to produce metallurgical grade coke from low rank bituminous and subbituminous coal, which is especially true in areas where coking coal reserves are becoming significantly depleted or are unavailable. The leading countries in this area of research are the United States (FMC Formcoke), Japan (Itoh process), and Germany (BFL process). These processes generally involve carbonization of crushed coal in fluidized beds, agglomerating the semicoke into conveniently sized balls with a binder, and calcining. The advantages of this technology include better heat transfer, shorter carbonizing time, continuous operation, and utilization of a much broader range of coals.

Low Temperature Carbonization. Lower temperature carbonization of lump coal at ~700°C, primarily used for production of solid smokeless fuel, gives a quantitatively and qualitatively different yield of solid, liquid, and gaseous products than does the high temperature processes.

Although a number of low temperature processes have been studied, only a few have been used commercially. These have been limited in the types of coal that are acceptable, and the by-products are less valuable than those obtained from high temperature processing. The Disco process is used in the United States to supply a limited amount of fuel to meet requirements of smoke ordinances. The British Coalite and Rexco processes produced substantial amounts of domestic smokeless fuel. Development of fluid-bed methods of carbonizing finer coal at ~400°C has been studied in the United Kingdom. A reactive char is briquetted without a binder to produce a premium open-fire smokeless fuel.

17.3. Gasification. Gasification of coal is used to provide gaseous fuels by surface and underground applications, liquid fuels by indirect liquefaction, ie, catalytic conversion of synthesis gas, and chemicals from conversion of synthesis gas. There are also applications in steelmaking (see COAL CONVERSION PROCESSES, GASIFICATION).

Gasifier Designs. A number of gasifiers are either available commercially or in various stages of development. These are described as fixed bed, fluidized bed, and entrained or rotating bed. The fixed bed involves an upward flow of reaction gas through a relatively stationary bed of hot coal. The gas velocity is slow enough to avoid blowing the coal out of the bed. The fluidized bed operates at higher gas velocities than the fixed bed and utilizes somewhat smaller particles. The entrained bed operates with parallel flows of reaction gas and finely pulverized coal particles to minimize reaction time and maximize throughput of product. The rotating bed is similar to a kiln, which operates with the coal entering at the upper part of the inclined kiln. Rotation avoids clinkering and exposes fresh surfaces to enhance completion of the reaction. The range of coals that may be used vary from one gasifier type to another with entrained flow gasifiers able to handle the widest range of raw coals. Fixed-bed gasifiers require mildly caking or noncaking feedstocks for normal operation.

The Lurgi fixed-bed gasifier operates using lump coal of a noncaking type with an ash composition chosen to avoid a sticky, partly fused ash in the reactor. A slagging version of this gasifier has been tested in Westfield, Scotland. Other fixed-bed gasifiers have similar coal requirements.

Fluidized-bed gasifiers typically require a coal feed of particles near 2–3 mm in diameter. Caking coals are to be avoided, because they usually agglomerate in the bed. This can be avoided using a pretreatment consisting of a surface oxidation with air in a fluidized bed. A useful fuel gas is produced. Examples of this type include the commercially available Winkler, and the U-Gas technology developed at the Institute of Gas Technology in Chicago. The latter is offered by a joint venture of Stone & Webster and Tampella Keeler. This system uses air-blown gasification and hot gas cleanup.

The Texaco gasifier and a similar unit developed by The Dow Chemical Company are pressurized entrained gasifiers. At the top, pulverized coal is mixed with reaction gas and is blown down into the gasifier. The reaction products leave from the side, and ash is blown down to a water pool where it is quenched. These units have operated at an Eastman Kodak facility in Kingsport, Tennessee and at the Coolwater power station in California for an integrated combined cycle power plant.

Pulverized coal is used in several entrained gasifiers and was studied in Germany before World War II. The Koppers-Totzek gasifier has been used commercially in different parts of the world. The original design used multiple (2 or 4) heads to feed coal, air or oxygen, and steam into an entrained atmospheric pressure reactor. Molten slag is discharged. The Babcock and Wilcox company also built an entrained bed gasifier for the DuPont Company at Belle, West Virginia, for chemical feedstock.

The Shell-Koppers-Totzek gasifier is also an entrained type. It can gasify lignite and subbituminous or bituminous coal. The coal is fed as a pulverized fuel, usually ground to 70% <74mm (−200 mesh) as used for pulverized coal fired boilers. Residence times are only a few seconds, therefore coal reactivity is important. The gasifier operates at >1650°C, at 2.2 MPa (22 atm) so that coal ash flows out of the gasifier as a molten slag. Coal ash composition must permit continuing molten ash flow. The joint development of the Shell Oil Co. and Koppers-Totzek led to a demonstration plant in The Netherlands having a gasifier for a 250-MW (50 cycle) integrated gasification combined cycle scheduled to begin operation in 1993. This is to be one of the first of the new generation of these plants to operate (95).

Surface Gasification Technology. Gasification of coal for fuel gas and chemical intermediate production has been developed commercially, and improvements in technology are being studied in a number of facilities. In the United States, the purpose of a number of programs has shifted from production of a substitute natural gas (methane) to electric power generation (qv) through the integrated IGCC plants. The interest in this use of coal results from the low emission levels that can be achieved and the potential for higher power generation efficiency.

Efficiencies of ~42% from natural gas to electricity have been indicated and can improve as the high temperature capabilities of turbines improve. Coal gasification would lower the overall energy efficiency but still give efficiencies greater than those with conventional coal-fired plants having typical emission control systems. Conventional power plants are able to produce electricity having heat rates of ~10 MJ × kWh and 90% SO_2 removal. The heat rates for IGCC plants are expected to be from 8 to 9.5 MJ × kWh having 99% SO_2 removal (96).

The Lurgi process (6) is the most successful complete gasification process for converting weakly caking coals as well as noncaking ones. The gasification takes place with steam and oxygen at 2–3 MPa (20–30 atm) to produce a 13.0–14.9 MJ × m^3 (350–400 Btu × ft^3) gas, which may be enriched with hydrocarbons to meet town gas specifications. The reactor is a slowly moving bed and is federal lump coal. Fine coal particles are usually removed before feeding to the gasifier.

The first commercial operation of the Lurgi process was in Germany in 1936 using brown coal. The reactor was modified to stir the coal bed to permit utilization of bituminous coal. One plant was built at the Dorsten Works of Steinkohlengas AG, and the Sasol plants were built in South Africa to provide synthesis gas for liquid fuels.

The gasifier for the 250 MW IGCC project in The Netherlands, scheduled to begin operation in 1993, is a 55 MW gas turbine with the balance of the power from a steam turbine. An Australian coal is to be used, and sulfur removal is expected to be 98.5% (96).

In the 1970s, a combined U.S. Federal government–American Gas Association program supported the development of second generation processes for making pipeline quality gas. In these processes coal is prepared, gasified, the gas is cooled, shifted if necessary to adjust the H_2/ CO ratio to ~3:1, the acid gases (H_2S and CO_2) removed, and then catalytic conversion to methane is carried out. Under this program the Institute of Gas Technology in Chicago developed the Hygas process in a 68 t/day pilot plant in which the gasifier at 6.9 MPa (1000 psi) accepts a coal slurry, dries it, goes through the first stage hydrogenation at 650–730°C, and second stage at 815–930°C before steam-oxygen gasification of the char to obtain high carbon utilization. The process also produces some benzene, toluene, and xylene, which were used in the pilot plant to make up the slurry. This process has been operated successfully using lignite, subbituminous, and bituminous coals.

The CO_2 Acceptor Process was also developed under this program by Consolidation Coal Co. in a 36-t/day pilot plant at Rapid City, South Dakota. Heat to drive the gasification process was provided by the reaction of calcined dolomite (MgO–CaO) and CO_2 produced in gasification of lignite or subbituminous coal using steam at 1 MPa (10 atm) and 815°C. The spent dolomite is regenerated at 1010°C in a separate vessel and returned to the gasifier. The process has operated successfully using lignite and subbituminous coal.

Still another process, called BI–GAS, was developed by Bituminous Coal Research in a 73-t/day pilot plant in Homer City, Pennsylvania. In this entrained-bed process, pulverized coal slurry was dried and blown into the second stage of the gasifier to contact 1205°C gases at ~6.9 MPa (1000 psi) for a few seconds residence time. Unreacted char is separated and recycled to the first stage to react with oxygen and steam at ~1650°C to produce hot gas and molten slag that is tapped.

The Synthane process was developed by the DOE at the Pittsburgh Energy Research Center. This fluidized-bed process operated at a ~6.9 MPa (1000 psi) and 980°C to gasify coal and produce some char. It used subbituminous coal. A third-generation process called Steam-Iron was also developed by the Institute of Gas Technology at a pilot plant in Chicago. This plant generates hydrogen from char produced in any gasification process. A gas producer uses air to make a reducing the Clean Air Act Amendment of 1991 phases in and the increased efficiency reduces the carbon dioxide emissions.

A similar design has been developed using a 161-MW plant by The Dow Chemical Company in its Plaquemine, Louisiana location. Destec, Inc. is a power subsidiary of The Dow Chemical Company and has joined with Public Service Of Indiana to build a new 230-MW plant near Terre Haute, Indiana. Operation is projected for 1995 (96).

Future large gasification plants, intended to produce ~7 × 10⁶m³ standard (250 million scf) of methane per day, are expected to be sited near a coal field having an adequate water supply. It is cheaper to transport energy in the form of gas through a pipeline than coal by either rail or pipeline. The process chosen is expected to utilize available coal in the most economical manner.

Underground Coal Gasification (UCG).   Underground coal gasification is intended to gasify a coal seam *in situ*, converting the coal into gas and leaving the ash underground. This approach avoids the need for mining and reactors for gasification. UCG is presently considered most interesting for deep coal or steeply sloping seams. This approach involves drilling holes to provide air or oxygen for gasification and removal of product gases and liquids (98).

A low calorific value gas, which includes nitrogen from air, could be produced for boiler or turbine use in electric power production, or an intermediate calorific value gas containing no nitrogen for an industrial fuel gas, or synthesis gas for chemical and methane production could be provided. This approach, which has been studied in Russia, Europe, Japan, and the United States, is still noncommercial in part because it is not economically competitive.

Although many environmental and safety problems can be avoided using UCG, there is some concern about groundwater contamination as a result of the process (see GROUNDWATER MONITORING).

In the United States a program, carried out near Hanna, Wyoming for the Department of Energy, examined different approaches to gasification, including use of air and oxygen. Other programs under government sponsorship included use of a longwall generator at the Morgantown, West Virginia Energy Technology Center.

Industrial testing programs have been carried out by Gulf Research and Development Co. in Western Kentucky (99) on a coal seam at a depth of 32.6 m and a thickness of 2.7 m. The coal seam was excavated for study after the gasification program. Another program using Russian technology is being carried out by Texas Utilities Services in an East Texas lignite deposit.

A joint Belgian–West German program is aimed at gasifying seams ≥ 1000 m underground and using the gases for combined cycle electrical generating plants. If initial efforts are effective, hydrogen is to be pumped into hot coal seams to make substitute natural gas from the exothermic carbon–hydrogen reaction. Work using underground coal gasification has been most extensive in Russia where the technology has been applied to produce gas for four or five electrical generation stations. An institute was established in 1933 to study this process and has primarily studied air-blown gasification that produced a gas of ~ 3.35–4.20 MJ/m^3 (90–113 Btu/ft^3) heat content. Other work has produced synthesis gas suitable for chemical production.

The chemistry of underground gasification has much in common with surface gasification; however, many of the parameters cannot be controlled because the reaction occurs in a remote site. Heat energy to drive the gasification comes primarily from carbon combustion to produce CO and then CO_2. Because many coal seams are also aquifers there is a considerable amount of water intrusion, which leads to steam generation at the expense of the reaction energy. As a result the rate of air or oxygen passage through the injection wells and seam are adjusted to maintain a low level of moisture in the product gas. The steam is beneficial for char gasification and some is consumed in the water gas shift reaction to produce H_2 and CO_2 from H_2O and CO. Some H_2 reacts with C to produce CH_4, which enhances the calorific value of the gas.

UCG is started by drilling wells to serve as injection points for oxidant and steam as well as collection points for product gases. Permeability of the coal seam is achieved by directional drilling, countercurrent combustion, electrolinking or hydraulic fracturing. Permeability is needed to provide a high rate of production with a minimum of pressure drop through the reaction zone. Low rank, ie, lignite and subbituminous, coals crack and shrink during gasification, rendering the seams more permeable. The bituminous coals swell and plug gas channels unless carefully preconditioned with preliminary oxidation to avoid this.

Liquid Fuels and Chemicals from Gasification of Coal.   Gasification of coal using steam and oxygen in different gasifiers provides varying proportions of carbon monoxide and hydrogen. Operations at increasing pressures increases the formation of methane. Because mixtures of CO and H_2 are used as the start of chemical synthesis and methane is not wanted or needed for chemical processes, the conditions favoring its formation are avoided. The product gases may then be passed over catalysts to obtain specific products. Iron-based catalysts are used to produce hydrocarbons in the Fischer-Tropsch process, or zinc or copper catalysts are used to make methanol.

The Fischer-Tropsch process has not been economical in competitive markets. The South African Sasol plant (100) has operated successfully using the Kellogg and German Arge (Ruhr Chemie Lurgi) modification of the Fischer-Tropsch process. The original plant was designed to produce 227,000 t/year of gasoline, diesel oil, solvents, and chemicals from 907,000 metric tons of noncaking high ash subbituminous coal. The Lurgi gasification process is used to make the synthesis gas. The capacity of this plant was expanded substantially with Sasol II and Sasol III commissioned in 1980 and 1983 to meet transportation fuel needs for South Africa. The combined annual production capacity of the three Sasol facilities is 8×10^6 m^3 (50 million barrels) of liquid products (97).

The success of the Sasol project is attributed to the availability of cheap coal and the reliability of the selected components. Plants using Lurgi or Koppers-Totzek gasifiers for making chemicals are located in Australia, Turkey, Greece, India, and Yugoslavia, among other countries.

A variety of pilot plants using fluid-bed gasifiers have been built in the United States, Germany, and elsewhere. The Winkler process is the only one that has been used on a large scale. It was developed in Germany in the 1920s to make synthesis gas at atmospheric pressure. Plans were being made to develop a pressurized version. Plants using bituminous coal have been built in Spain and Japan with the atmospheric pressure gasifier.

Gasification and Metallurgy.   Some interesting combinations of these technologies include direct reduction of iron ore and direct injection of coal into the blast furnace. In direct reduction, a reducing gas mixture of methane or carbon monoxide and hydrogen reduces iron ore pellets into elemental iron by reaction at 1000–1200°C. These pellets may later be used to feed steelmaking processes. In 1983, 45 plants having a capacity of 15×10^6 metric tons were in operation. Pulverized coal has been successfully injected into the tuyeres of a blast furnace of the Armco Co. in Middletown, Ohio to supplement coke (see IRON BY DIRECT REDUCTION).

17.4. Liquefaction.   Liquefaction of coal to oil was first accomplished in 1914. Hydrogen was placed with a paste of coal, heavy oil, and a small amount of iron oxide catalyst at 450° and 20 MPa (200 atm) in stirred autoclaves. This process was developed by the I. G. Farbenindustrie AG to give commercial quality gasoline as the principal product. Twelve hydrogenation plants were operated during World War II to make liquid fuels (see COAL CONVERSION PROCESSES, LIQUEFACTION).

Imperial Chemical Industries in Great Britain hydrogenated coal to produce gasoline until the start of World War II. The process then operated on creosote middle oil until 1958. As of this writing none of these plants is being used to make liquid fuels for economic reasons. The present prices of coal and hydrogen from coal have not made synthetic liquid fuels competitive. Exceptions are those cases, as in South Africa, where there is availability of cheap coal, and fuel liquids are very important.

The Pott-Broche process (101) was best known as an early industrial use of solvent extraction of coal but was ended owing to war damage. The coal was extracted at ~400°C for 1–1.5 h under a hydrogen pressure of 10–15 MPa (100–150 atm) using a coal-derived solvent. Plant capacity was only 5 t/h with an 80% yield of extract. The product contained <0.05% mineral matter and had limited use, mainly in electrodes.

Solvent extraction work was carried out by a number of organizations in the United States. Pilot plants for producing SRC were built and initially sponsored by the Southern Company Services and Electric Power Research Institute in Wilsonville, Alabama in 1973 and built with Department of Energy sponsorship near Tacoma, Washington in 1974 having capacities of 5 and 45 t/day of coal input, respectively. The Wilsonville plant was closed in 1992 after many modifications from the initial design; the Tacoma plant is closed.

In the SRC work, coal was slurried with a process-derived anthracene oil and heated to 400–455°C at 12.4–13.8 MPa (1800–2000 psi) of hydrogen for

0–1 h. A viscous liquid was extracted. The product stream contains some hydro-carbon gases, and H_2S. The residue is gasified to generate hydrogen for the process. The remaining filtrate is separated into solvent, which is recycled, and SRC, a low ash, tarlike boiler fuel.

Heating value of the product (SRC) is ∼37 MJ/kg (16,000 Btu/lb). Sulfur contents have been reduced from 2 to 7% initially to 0.9% and possibly less. Ash contents have been reduced from 8–20% to 0.17% (102). These properties permit compliance with environmental protection aging (EPA) requirements for SO_2 and particulate emissions. The SRC is primarily intended to be used as a boiler fuel in either a solid or molten form (heated to ∼315°C). The solid has a Hardgrove index of 150 (103). Boiler tests have been successfully carried out using a utility boiler.

A series of process improvements have been developed at Wilsonville to produce high quality transportation fuels. Two integrated stages of liquefaction separated the initial coal dissolution from the hydrogenation to upgrade the product. This was known as SRC-II. An intermediate step, critical solvent deashing, was added to remove mineral matter to extend the life of catalysts used in hydrogenation. Later efforts involved the use of an ebulated bed developed by Hydrocarbon Research, Inc. (HRI) and eliminated the mineral matter removal between stages. Temperatures were lowered to reduce contamination of catalysts, which were also added to the first stage. This approach has been called the integrated catalytic two-stage liquefaction process (104).

Several processes progressed to demonstration scales but have not been commercialized, primarily because of economic inability to compete with available petroleum products. The H-Coal process developed by Hydrocarbon Research, Inc. was demonstrated at Catlettsburg, Kentucky using a 545 t/day plant and Department of Energy (DOE) support. The Exxon donor solvent liquefaction process was not commercialized either.

Processes for hydrogen gasification, hydrogen pyrolysis, or coking of coal usually produce liquid coproducts. The Hygas process produces about 6% liquids as benzene, toluene, and xylene. Substitution of petroleum residuum for the coal-derived process oil has been used in studies of coal liquefaction and offers promise as a lower cost technology (104).

17.5. Bioprocessing and Biotreatment of Coal. The use of biotechnology to process coal to make gaseous and liquid fuels is an emerging field (105). Bacteria and enzymes have been studied to establish the technical feasibility of conversion. The earliest work was done on microbial decomposition of German hard coals (106). Reactors have been designed to use a variety of bacteria and fungi to break down the large molecular structure into smaller units that may be useful as intermediates (solubilization) or as liquid and gaseous fuels (conversion). Efforts have focused on lower rank coals, lignite or brown coal and subbituminous coal, because of greater reactivity. The conversion processes frequently introduce chemically combined oxygen through hydrolysis or related reactions to make the solid soluble in the reaction mixture as an initial step. Further reaction involves biological degradation of the resulting material to form gases or liquids.

The large-scale processing of coal is expected to involve plants similar to sewage treatment facilities in the handling of liquid and solid materials (see

WATER, SEWAGE). The reaction rates are substantially lower than those achieved in high temperature gasifiers and liquefaction reactors requiring much larger systems to achieve comparable coal throughput.

Biological processes are also being studied to investigate ability to remove sulfur species in order to remove potential contributors to acid rain (see AIR POLLUTION). These species include benzothiophene-type materials, which are the most difficult to remove chemically, as well as pyritic material. The pyrite may be treated to enhance the ability of flotation processes to separate the mineral from the combustible parts of the coal. Genetic engineering (qv) techniques are being applied to develop more effective species.

17.6. Other Uses. The quantity of coal used for purposes other than combustion or processing is quite small (2,6). Coal, especially anthracite, has established markets for use as purifying and filtering agents in either the natural form or converted to activated carbon (see CARBON). The latter can be prepared from bituminous coal or coke, and is used in sewage treatment, water purification, respirator absorbers, solvent recovery, and in the food industry. Some of these markets are quite profitable and new uses are continually being sought for this material.

Carbon black from oil is the main competition for the product from coal, which is used in filters. Carbon for electrodes is primarily made from petroleum coke, although pitch coke is used in Germany for this product. The pitch binder used for electrodes and other carbon products is almost always a selected coal tar pitch.

The preparation of pelletized iron ore represents a substantial market for coke and anthracite for sintering. Direct injection into the blast furnace of an auxiliary fuel, coal, or oil is now practiced to provide heat for the reduction and some of the reducing agent in place of the more expensive coke that serves these purposes. Some minor uses of coal include the use of fly ash, cinders, or even coal as a building material (see BUILDING MATERIALS, SURVEY); soil conditioners from coal by oxidizing it to humates (see SOIL STABILIZATION); and a variety of carbon and graphite products for the electrical industry, and possibly the nuclear energy program. The growth of synthetic fuels from coal should also provide substantial quantities of by-products including elemental sulfur, fertilizer as ammonia or its salts, and a range of liquid products. The availability of ammonia and straight-chain paraffins may permit future production of food from fossil fuels.

BIBLIOGRAPHY

"Coal" in *ECT* 1st ed., Vol. 4, pp. 86–134, by H. J. Rose, Bituminous Coal Research, Inc.; in *ECT* 2nd ed., Vol. 5, pp. 606–678, by I. G. C. Dryden, British Coal Utilisation Reserch Association; in *ECT* 3rd ed., Vol. 6, pp. 224–283, by K. S. Vorres, Institute of Gas Technology; in *ECT* 4th ed., Vol. 6, pp. 423–449, by K. S. Vorres, Argonne National Laboratory; "Coal" in *ECT* (online), posting date: December 4, 2000, by K. S. Vorres, Argonne National Laboratory.

CITED PUBLICATIONS

1. Energy Information Administration, http://www.eia.doe.gov/pub/international/iealf/table25.xls, accessed 2002.
2. *Mining Statistics*, National Mining Association, http://www.nma.org/, accessed 2002.
3. D. W. van Krevelen, *Coal*, Elsevier Scientific Publishing Co., Amsterdam, The Netherlands, 1961; P. G. Hatcher and D. J. Clifford, *Org. Geochem.* **27**, 251 (1997).
4. H. H. Lowry, ed., *Chemistry of Coal Utilization*, Vols. 1 and 2, John Wiley & Sons, Inc., New York, 1945.
5. *Ibid*, Suppl. Vol., 1963.
6. M. Elliott, in Ref. 4, Second Supplementary Vol., 1981; this is an exceptionally extensive source.
7. H. J. Gluskoter, N. F. Shimp, and R. R. Ruch, in Ref. 6, Chapt. 7.
8. R. Thiessen, *U.S. Bur. Mines Inform. Circ.*, 7397 (1947).
9. M. C. Stopes, *Proc. R. Soc. London Ser. B.* **90** 470 (1919); *Fuel* **14** 4 (1935); International Committee for Coal and Organic Petrology, *Vitrinite classification: ICCP 1994* (1995).
10. R. C. Neavel, in Ref. 4, Chapt. 3.
11. C. R. Ward, ed., *Coal Geology and Coal Technology*, Blackwell Scientific Publications, London, 1984. This is an excellent text on this area.
12. A. Davis, in Ref. 11, Chapt. 3.
13. M. C. Stopes, *Fuel* **14**, 4 (1935).
14. *International Handbook of Coal Petrography*, 2nd ed., International Committee for Coal Petrology, Centre National de la Recherche Scientifique, Paris, France, 1963, 252 pp.; 2nd Suppl. to 2nd ed., 1976.
15. A. Hood, C. C. M. Gutjahr, and R. L. Heacock, *Bull. Am. Assn. Petrol. Geol.* **59**, 986 (1975).
16. R. J. Gray, A. H. Rhoades, and D. T. King, *Trans. Soc. Min. Engrs. AIME* **260**, 334 (1976).
17. C. A. Seyler, *Fuel* **3**, 15, 41, 79 (1924); *Proc. S. Wales Inst. Eng.* **53**, 254, 396 (1938).
18. "Gaseous Fuels, Coal and Coke," *Annual Book of ASTM Standards*, Vol. 5.05, American Society for Testing and Materials, Philadelphia, Pa., published annually; *British Standards 1016*, parts 1–16, British Standards Institute, London, published annually.
19. I. Wender and co-authors, in Ref. 6, Chapt. 8.
20. L. M. Stock, R. Wolny, and B. Bal, *Energy Fuels* **3**, 651 (1989).
21. M. S. Solum, R. J. Pugmire, and D. M. Grant, *Energy Fuels*, **3**(2), 187 (1989).
22. M. E. Peover, *J. Chem. Soc.*, 5020 (1960).
23. R. Raymond, I. Wender, and L. Reggel, *Science* **137**, 681 (1962).
24. L. Reggel, I. Wender, and R. Raymond, *Fuel* **43**, 75 (1964).
25. B. K. Mazumdar and co-workers, *Fuel* **41**, 121 (1962).
26. B. K. Mazumdar, S. S. Choudhury, and A. Lahiri, *Fuel* **39**, 179 (1960).
27. R. E. Winans and P. H. Neill, in W. L. Orr and C. M. White, eds., *Geochemistry of Sulfur in Fossil Fuels*, ACS Symposium Series, No. 429, American Chemical Society, Washington, D. C., 1990 p. 249.
28. J. H. Shinn, *Fuel*, 1187 (1984).
29. I. G. C. Dryden, *Fuel* **37**, 444 (1958).
30. I. G. C. Dryden, *Fuel* **41**, 55 and 301 (1962).
31. A. C. Bhattacharya, B. K. Mazumdar, and A. Lahiri, *Fuel* **41**, 181 (1962); S. Ganguly and B. K. Mazumdar, *Fuel* **43**, 281 (1964).
32. T. Green, J. Kovac, D. Brenner, and J. W. Larsen, in R. A. Meyers, ed., *Coal Structure*, Academic Press, New York, 1982 p. 199.

33. P. R. Solomon, P. E. Best, Z. Z. Yu, and S. Charpenay, *Energy Fuels* **6**, 143 (1992); see also P. R. Solomon, D. G. Hamblen, R. M. Carangelo, M. A. Serio, and G. V. Deshpande, *Energy Fuels Ibid.* **2**, 405 (1988).

34. D. M. Grant, R. J. Pugmire, T. H. Fletcher, and A. R. Kerstein, *Energy Fuels* **3**, 175 (1989).

35. I. Wender, *Chem. Rev. -Cat. Sci.* **14**, 97 (1976).

36. A. L. Chaffee, G. J. Perry, R. B. Johns, and A. M. George, *Am. Chem. Soc. Adv. Chem. Ser.* **192**, Chapt. 8 (1981).

37. L. Cartz and P. B. Hirsch, *Phil. Trans. R. Soc. London Ser. A* **252**, 557 (1960).

38. R. D. Harvey and R. R. Ruch, in K. S. Vorres, ed., *Mineral Matter and Ash in Coal*, ACS Symposium Series, No. 301, American Chemical Society, Washington, D.C., 1986, Chapt. 2.

39. H. J. Gluskoter and co-workers, *Ill. State Geol. Survey Circ.*, 499 (1977).

40. W. Spackman, "What Is Coal?", *Short Course on Coal Characteristics and Coal Conversion Processes*, Pennsylvania State University, University Park, Pa., Oct. 1973, 48 pp.

41. A. G. Sharkey, Jr. and J. T. McCartney, in Ref. 6, Chapt. 4.

42. D. L. Wertz, *Energy Fuels* **4**(5), 442 (1990).

43. M. Farcasiu, *Fuel* **56**, 9 (1977).

44. M. Bakr, T. Yokono, and Y. Sanada, *Proceedings of the 1989 International Conference on Coal Science*, Oct. 23–27, Vol. 1, Tokyo, Japan, p. 217.

45. I. Chatterjee and M. Misra, *J. Microwave Power Electromagnet. Energy* **25**(4), 224 (1990).

46. R. A. MacDonald, J. E. Callanan, and K. M. McDermott, *Energy Fuels*, **1**(6), 535 (1987).

47. O. P. Mahajan, *Coal Porosity in Coal Structure*, Academic Press, New York, 1982 p. 51.

48. H. Gan, S. P. Nandi, and P. L. Walker, *Fuel* **51**, 272 (1972).

49. J. W. Larsen and P. Wernett, *Energy Fuels* **2**(5), 719 (1988).

50. P. C. Wernett, J. W. Larsen, O. Yamada, and H. J. Yue, *Energy Fuels* **4**(4), 412 (1990).

51. Z. Spitzer and L. Ulicky, *Fuel* **55**, 212 (1976).

52. J. T. McCartney, H. J. O'Donnell, and S. Ergun, *Coal Science, Advances in Chemistry Series*, Vol. 55, American Chemical Society, Washington, D.C., 1966 p. 261.

53. G. H. Taylor, in Ref. 6, p. 274.

54. W. H. Wiser, "Some Chemical Aspects of Coal Liquefaction," in Ref. 40.

55. H. F. Yancey and M. R. Geer, in J. W. Leonard and D. R. Mitchell, eds., *Coal Preparation*, 3rd ed., American Institute of Mining, Metallurgical and Petroleum Engineers, Inc., New York, 1968, pp. 3–56.

56. P. Rosin and E. Rammler, *J. Inst. Fuel* **7**, 29 (1933); J. G. Bennett, *J. Inst. Fuel* **10**, 22 (1936).

57. F. Kick, *Dinglers Polytech. J.* **247**, 1 (1883); P. von Rittinger, *Lehrbuch der Aufbereitungskunde*, Ernst and Korn, Berlin, Germany, 1867 p. 595; J. C Hower, *Minerals Metallurgical Proc.* **15**(3) 1 (1998).

58. F. C. Bond, *Min. Eng.* **4**, 484 (1952).

59. S. J. Vecci and G. F. Moore, *Power* 74 (1978).

60. R. C. Attig and A. F. Duzy, *Coal Ash Deposition Studies and Application to Boiler Design*, American Power Conference, Chicago, Ill., 1969.

61. E. C. Winegartner and B. T. Rhodes, *J. Eng. Power* **97**, 395 (1975).

62. *Steam, Its Generation and Use*, The Babcock & Wilcox Co., New York, 1972, pp. 15–4.

63. R. Smock, *Power Eng.* **95**(2), 32 (1991).

64. J. Makansi, *Power* **135**(3), 15 (1991).

65. G. A. Nelkin and R. J. Dellefield, *Mech. Eng.* **112**(9), 58 (1990).

66. J. A. Harrison, H. W. Jackman, and J. A. Simon, *Ill. State Geol. Survey Circ.*, 366 (1964); M. Diez and co-workers, *Int. J. Coal Geol.* **50** in press (2002).

67. T. Takanohashi and M. Iino, *Energy Fuels* **4**(5), 452 (1990).

68. M. G. Matturro, R. Liotta, and R. P. Reynolds, *Energy Fuels* **4**(4), 346 (1990).

69. G. R. Hill and co-workers, *Advances in Chemistry Series*, Vol. 55, American Chemical Society, Washington, D.C., 1966 p. 427.

70. Ref. 6, Chapt. 24.

71. K. S. Vorres, *Energy Fuels* **4**(5), 420 (1990).

72. D. K. Fleming, R. D. Smith, and M. R. Y. Aquino, *Preprints, Fuel Chem. Div., Am. Chem. Soc.* **22**(2), 45 (1977).

73. S. C. Spalding, Jr., J. O. Burckle, and W. L. Teiser, in Ref. 69, p. 677.

74. E. P. Stambaugh, Coal Desulfurization, *Chemical and Physical Methods*, ACS Symposium Series 64, American Chemical Society, Washington, D.C., 1977 p. 198.

75. J. W. Hamersma, M. L. Kraft, and R. A. Meyers, in Ref. 72, pp. 73, 84.

76. H. Hofmann and K. Hoehne, *Brennstoff Chem.* **35**, 202, 236, 269, 298 (1954).

77. P. R. Solomon and co-workers, *Preprints, Fuel Chem. Div., Am. Chem. Soc.* **36**(1), 267 (1991).

78. M. D. Kimber and M. D. Gray, *Combust. Flame* **11**, 360 (1967).

79. D. Fitzgerald and D. W. van Krevelen, *Fuel* **38**, 17 (1959).

80. K. Ouchi and H. Honda, *Fuel* **38**, 429 (1959).

81. B. K. Mazumdar, S. K. Chakrabartty, and A. Lahiri, *Proceedings of the Symposium on the Nature of Coal*, Central Fuel Resource Institute, Jealgora, India, 1959 p. 253; S. C. Biswas and co-workers, *Ibid.* p. 261.

82. H. Marsh and D. E. Clark, *Erdol Kohle* **39**, 113 (1986); see also *Proceedings of the Iron & Steel Society*, meeting in Toronto, Apr. 1992, and *Iron & Steel Society AIME*, in Aug. 1992.

83. *Survey of Energy Resources 1986*, World Energy Conference, Central Office, London, 1986.

84. P. Averitt, "Coal Resources of the United States, Jan. 1, 1974," *U.S. Geological Survey Bulletin*, 1975 p. 1412.

85. U. S. Dept. of Energy, Energy Information Agency, *Monthly Energy Review*, Mar. 1990.

86. *The Penn State Coal Sample Bank and Data Base*, Energy and Fuels Research Center, Pennsylvania State University, University Park, Pa., Apr. 1988.

87. R. D. Harvey and C. W. Kruse, *J. Coal Quality* **7**(4), 109 (1988).

88. *SBN Sample Catalogue*, European Center for Coal Specimens, Eygelshoven, The Netherlands, revisions issued periodically.

89. *Coal Age* **82**, 59 (1977).

90. *Coal Age* **68**, 226 (1963).

91. *International Classification of Hard Coals by Type*, United Nations Publication No. 1956 II, E. 4 E/ECE/247; E/ECE/Coal/100, 1956.

92. *Analysis and Testing of Coal And Coke*, British Standards, parts 1–16, 1957–1964 p. 1016.

93. *Coal Age* **68**, 62 (1963).

94. *World Book Encyclopedia*, Field Enterprises Educational Corp., Chicago, Ill., 1975 p. 566.

95. M. Valenti, *Mech. Eng.* **114**(1), 39 (1992).

96. R. Smock, *Power Eng.* **95**(2), 32 (1991).

97. R. D. Doctor and K. E. Wilzbach, *J. Energy Res. Technol.* **111**, 160 (1989).

98. T. F. Edgar and D. W. Gregg, "Underground Gasification of Coal," in Ref. 6.

99. D. Raemondi, P. L. Terwilliger, and L. A. Wilson, Jr., *J. Petrol. Tech.* **27**, 35 (1975).

100. *PETC Rev.*, Issue 4, Fall 1991 p. 16.
101. A. Pott and co-workers, *Fuel* **13**, 91, 125, 154 (1934).
102. *Environ. Sci. Technol.* **8**, 510 (1974).
103. W. Downs, C. L. Wagoner, and R. C. Carr, *Preparation and Burning of Solvent Refined Coal*, presented at American Power Conference, Chicago, Ill., Apr. 1969.
104. *PETC Rev.*, Issue 3, Pittsburgh Energy Technology Center, Pittsburgh, Pa., Mar. 1991.
105. D. L. Wise, *Bioprocessing and Biotreatment of Coal*, Marcel Dekker, Inc. New York, 1990, 744 pp.
106. R. M. Fakoussa, translation of the *Investigations of the Microbial Decomposition of Untreated Hard Coals; Coal as a Substrate for Microorganisms*: doctoral dissertation of R. Fakoussa, Bonn, 1981; prepared for the U.S. Department of Energy, Pittsburgh Energy Technology Center; translated by the Language Center Pittsburgh under Burns and Roe Services Corp., Pittsburgh, Pa., June 1987.

GENERAL REFERENCES

D. L. Crawford, ed., *Biotransformations of Low Rank Coals*, CRC Press Inc., Boca Raton, Fla., 1992.
D. L. Wise, ed., *Bioprocessing and Biotreatment of Coal*, Marcel Dekker, New York, 1990, 744 pp.
Fuel **70**(3), (1991) contains a series of papers by leading researchers presented at the International Conference on Coal Structure and Reactivity; chemical, physical, and petrographic aspects, Cambridge, UK, Sept. 5–7, 1990.
Fuel **70**(5), (1991) contains a series of papers presented at "Biotechnology for the Production of Clean Fuels", Aug. 27–28, 1990, Washington, D.C., pp. 569–620.
W. Francis, *Coal*, 2nd ed., Edward Arnold & Co., London, 1961.
E. Stach and co-workers, *Stach's Textbook of Coal Petrology*, 3rd ed., Gebrüder Borntraeger, Berlin, Germany, 1982, 535 pp. Excellent text on coal petrography; superseded by G. Taylor et al., 1998, *Organic Petrology*, Gebrüder Borntraeger, Berlin, Germany.
Int. J. Coal Geol. **50** (2002) contains a collection of 16 papers on subjects from the deposition of coal, to chemical and mineralogical aspects of coal, to coal utilization.

KARL S. VORRES
Argonne National Laboratory

COAL GASIFICATION

1. Introduction

Coal gasification is the process of reacting coal with oxygen, steam, and carbon dioxide to form a product gas containing hydrogen and carbon monoxide. Gasification is essentially incomplete combustion. The chemical and physical processes are quite similar; the main difference being the nature of the final products. From a processing point of view the main operating difference is that gasification consumes heat evolved during combustion. Under the reducing environment of gasification the sulfur in the coal is released as hydrogen sulfide rather than sulfur dioxide and the coal's nitrogen is converted mostly to ammonia rather than nitrogen oxides. These reduced forms of sulfur and nitrogen are easily isolated, captured, and utilized, and thus gasification is a clean coal technology with better environmental performance than coal combustion.

Depending on the type of gasifier and the operating conditions gasification can be used to produce a fuel gas suitable for any number of applications. A low heating value fuel gas is produced from an air blown gasifier for use as an industrial fuel and for power production. A medium heating value fuel gas is produced from enriched oxygen blown gasification for use as a synthesis gas in the production of chemicals such as ammonia, methanol, and transportation fuels. A high heating value gas can be produced from shifting the medium heating value product gas over catalysts to produce a substitute or synthetic natural gas (SNG).

Coal gasification is presented by first describing the chemistry of the process and the coal characteristics that affect the processes. Coal gasification processes have been tailored to adapt to the different types of coal feedstocks available. The development of gasification is then presented from an historical perspective. This leads into the discussion of the types of gasifiers most commonly used and the process improvements made to meet the changing market needs. Complete gasification systems are then described including typical system configuration, required system attributes, and environmental performance. The current status, economics of gasification technology, and future of gasification are also discussed.

2. Coal Gasification Chemistry

In a gasifier, coal undergoes a series of chemical and physical changes as shown in Figure 1.

Each of the steps is described in more detail below. As the coal is heated most of the moisture is driven out when the particle temperature is $\sim$105°C. Drying is a rapid process and can be essentially complete when the temperature reaches 300°C (1) depending on the type of coal and heating method used.

Devolatilization or pyrolysis accounts for a large percentage coal weight loss and occurs rapidly during the initial stages of coal heat up. During this process, the labile bonds between the aromatic clusters in coal are cleaved, generating

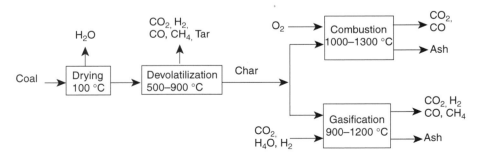

Fig. 1. Chemical and physical changes of coal.

fragments of molecular weight much smaller than coal. Fragments with low molecular weights vaporize and escape from the coal particle to constitute light gases and tar. The fragments with high molecular weight, and hence low vapor pressures, remain in the coal under typical devolatilization conditions until they reattach to the char lattice. These high molecular weight compounds plus the residual lattice are referred to as metaplast (2). During this period some coals swell to a degree, depending on its swelling index and the heating conditions (1). The metaplast further depolymerizes to yield char and volatiles.

The volatile yield and composition depends on the heating rate and final temperature. At slow-heating rates ($<1°C/s$) the volatile yield is low due to repolymerization. Then the total volatile yield will be equal to the volatile matter content determined from the ASTM Proximate Analysis, which is an analysis done at a slow heating rate. Under rapid-heating rate ($500-10^{5}°C/s$) the volatile yield is 20–40% more than that at slow-heating rates (1). At any given temperature only a certain fraction of the volatiles is released. Significant devolatilization begins when the coal temperature is about 500°C. As the temperature is increased more volatiles are released. The maximum volatile yield occurs when the temperature is >900°C, the temperature at which the American Society for Testing and Materials (ASTM) Proximate Analysis for volatile matter is conducted.

The fraction of the devolatilization gas that condenses at room temperature and pressure is called tar. It is a mixture of hydrocarbons with an average molecular weight ranging from 200–500 g/mol (2). The yield of tar depends on the coal rank; higher rank coals produce lesser amounts of tar. Higher gasifier temperature also reduces the amount of tar in the gasifier products because of increased cracking of tar into lighter gases. The amount of tar also decreases with increasing pressure and decreasing heating rates.

The devolatilization gas that does not condense at room temperature and pressure consists mainly of CO, CO_2, CH_4, H_2, and H_2O. The predominant source of CH_4 in the gasifier-product gas is the devolatilization process, and its production is favored by low temperature and high pressure. Therefore, the amount of methane in the product of moving bed gasifiers, which operates at a low temperature, is higher than that in typical fluidized bed and entrained bed gasifiers (3).

The solid product left over from devolatilization is char. During devolatilization the porosity changes from 2 to 20%, typical of coal, to >80%. The nitrogen surface area increases from $10-20$ m^2/g (coal) to $200-400$ m^2/g (1). The increased surface area increases the reactivity of char. The reactivity of char depends on properties of coal minerals, pyrolysis conditions, and gasification conditions. If the char porosity reaches a critical porosity (70–80%) the char will fragment into fine solids, which also increases the reactivity of char.

Char in an oxygen atmosphere undergoes combustion. In gasifiers partial combustion occurs in an oxygen-deficient, or reducing, atmosphere. Gasifiers use 30–50% of the oxygen theoretically required for complete combustion to carbon dioxide and water. Carbon monoxide and hydrogen are the principal products, and only a fraction of the carbon in the coal is oxidized completely to carbon dioxide. The combustion reaction is written in a general form as follows:

$$(1 + \lambda)\, C + O_2 \rightarrow 2\lambda\, CO + (1 - \lambda)\, CO_2 \qquad \Delta H^{\circ}_{298\ K} = 172.5\ \lambda - 393.5 \text{ kJ/mol}$$

Table 1. **Relative Gasification Rates at 10 kPa and 800°C**

Reaction	Relative rate
$C + O_2$	10^5
$C + H_2O$	3
$C + CO_2$	1
$C + H_2$	10^{-3}

where λ varies from 0 (pure CO_2 product) to 1 (pure CO product). The value of λ depends upon the gasification conditions and is usually close to 1. Under typical gasifier conditions this reaction rate is controlled by diffusion limitation. For a particle size of 90 μm the rate is limited by diffusion rate for temperatures >475°C. For smaller particles, diffusion rate becomes limiting only at a higher temperature (1325°C for 20 μm particles). The heat released by the partial combustion provides the bulk of the energy necessary to drive the endothermic gasification reactions.

The oxygen is rapidly consumed in the combustion zone, which occupies a small volume of the reactor. Further conversion of char occurs through the much slower, reversible gasification reactions with CO_2, H_2O, and H_2.

$$C + CO_2 \leftrightarrow 2\,CO \qquad \Delta H^\circ_{298\ K} = 172.5 \ \text{kJ/mol}$$

$$C + H_2O \leftrightarrow CO + H_2 \qquad \Delta H^\circ_{298\ K} = 131.3 \ \text{kJ/mol}$$

$$C + 2H_2 \leftrightarrow CH_4 \qquad \Delta H^\circ_{298\ K} = -74.8 \ \text{kJ/mol}$$

The rate of gasification reaction depends upon the char properties and the gasification conditions. Table 1 (4) gives the typical orders of magnitude of various reactions: Another important chemical reaction in a gasifier is the water–gas shift reaction:

$$CO + H_2O \leftrightarrow CO_2 + H_2 \qquad \Delta H^\circ_{298\ K} = -41.2 \ \text{kJ/mol}$$

Mineral matter in the coal catalyzes this gas-phase reaction. Other gas-phase reactions are the combustion of CO, H_2, and CH_4 and tar cracking.

The physical and chemical changes in the sorbent material are depicted in Figure 2

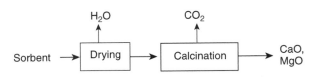

Fig. 2.   Physical and chemical changes in solvent material.

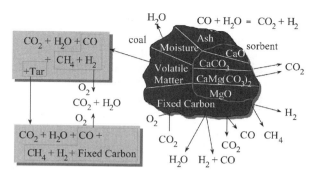

Fig. 3. Graphical representation of chemical reactions in coal, sorbent, and the gas phase during coal gasification (5).

When limestone is heated in the gasifier, initially the moisture is driven out. Further heating decomposes or calcines the limestone:

$$CaCO_3 \leftrightarrow CaO + CO_2 \qquad \Delta H^{\circ}_{298\ K} = 183\ kJ/mol$$

The calcium oxide removes some of the sulfur gases as calcium sulfide.

Trace elements, such as sulfur and nitrogen, are also involved in the gasification reactions. Sulfur in coal is converted primarily to H_2S under the reducing conditions of gasification. Approximately 5–15% of the sulfur is converted to COS, whereas the coal nitrogen is converted primarily to N_2; trace amounts of NH_3 and HCN are also formed.

The heating, drying, devolatilization, combustion, and gasification of coal in the presence of a sorbent are represented graphically in Figure 3. The reactions portrayed in Figure 3 have been incorporated into computer simulations to describe the behavior of coal during various coal gasification processes (5).

High temperature favors endothermic reactions (increases the products on the right hand side). High temperature favors reactions in which there is a reduction in the number of moles, as in the methanation reaction (2 mol of hydrogen gives 1 mol of methane). Hydrogen and carbon monoxide production increases with decreasing oxygen in the feed, with decreasing pressure, and with increasing temperature. Hydrogen production increases and carbon monoxide production decreases with increased steam rate. Methane production increases with decreasing temperature and increasing pressure. The product gas of air blown gasifiers is diluted by nitrogen. Upon heating bituminous coals become sticky and swell, which can cause problems in fixed-bed gasifiers. Such coals are more easily handled in fluidized and entrained bed gasifiers.

3. Syngas Chemistry

Whereas near-term application of coal gasification is expected to be in the production of electricity through combined cycle power generation systems, longer term applications show considerable potential for producing chemicals from coal using syngas processing (7). Products include ammonia, methanol, synthetic natural gas, and conventional transportation fuels.

The cost and availability of oil and natural gas influence the competitiveness of coal gasification, but coal is expected to continue to play an ever-increasing role as a significant resource base for both energy and chemicals.

3.1. Ammonia.   Ammonia is produced through the reaction of hydrogen and nitrogen. In a coal-to-ammonia facility, coal gasification produces the hydrogen and an air separation plant, which also provides oxygen for coal gasification, supplies the nitrogen. Because coal gasification produces a mixture of hydrogen and carbon monoxide, the CO is combined with steam in a water gas shift reactor to produce carbon dioxide and H_2. Following CO_2 removal, the hydrogen stream is fed to an ammonia synthesis reactor where it reacts with molecular nitrogen to produce ammonia.

The water gas shift reaction is used to increase the amount of hydrogen in the gas. For shifting coal-derived gas, conventional iron–chromium catalysts can be used. Because coal gas has a significantly higher concentration of carbon monoxide than is found in gas streams in conventional refineries, the catalyst must be able to withstand high thermal loads. However, potential catalyst poisons such as phenol and other hydrocarbons are not a concern in entrained-bed gasifiers.

3.2. Methanol.   Methanol is produced by stoichiometric reaction of CO and H_2. The syngas produced by coal gasification contains insufficient hydrogen for complete conversion to methanol, and partial CO shifting is required to obtain the desired concentrations of H_2, CO, and CO_2. These concentrations are expressed in terms of a stoichiometric number, $(H_2 - CO)/(H_2 + CO_2)$, which has a desired value of 2. In some cases CO_2 removal is required to achieve the stoichiometric number target. Both CO and H_2 are combined to form methanol in a catalytic methanol synthesis reactor.

The exothermic reaction

$$CO + 2H_2 \rightarrow CH_3OH \qquad \Delta H_R = -109 \text{ kJ } (-26 \text{ kcal})$$

is enhanced by high pressures and low temperatures. Catalysts used in the reactor are based on copper, zinc, or chromium oxides (8), and reactors are designed to remove the exothermic heat of reaction effectively.

Mobil Oil Corporation has developed a process on a pilot scale that can successfully convert methanol into 96-octane gasoline. Although methanol can be used directly as a transportation fuel, conversion to gasoline would eliminate the need to modify engines and would also eliminate some of the problems encountered using gasoline–methanol blends (see ALCOHOL FUELS; GASOLINE AND OTHER MOTOR FUELS)

3.3. Synthetic Natural Gas.   Another potentially very large application of coal gasification is the production of SNG. The syngas produced from coal gasification is shifted to produce a H_2/CO ratio of approximately 3:1. The carbon dioxide produced during shifting is removed, and CO and H_2 react to produce methane (CH_4), or SNG, and water in a methanation reactor.

The following reactions can occur simultaneously within a methanation reactor:

$$CO + 3H_2 \rightarrow CH_4 + H_2O$$

$$CO_2 + 4H_2 \rightarrow CH_4 + 2H_2O$$

$$CO + H_2O \rightarrow CO_2 + H_2$$

$$2CO \rightarrow CO_2 + C$$

The heat released from the CO–H_2 reaction must be removed from the system to prevent excessive temperatures, catalyst deactivation by sintering, and carbon deposition. Several reactor configurations have been developed to achieve this (9).

The tube wall reactor (TWR) system features the use of catalyst-coated tubes. The Raney nickel catalyst is flame-sprayed onto the inside surface of the tubes, and the tubes are immersed in a liquid, such as Dowtherm, which conducts the heat away. Some quantity of recycle gas, in the ratio from 0 to 5, may also be used. In fluidized-bed catalyst systems, Raney or thorium nickel catalysts operate under moderate pressure, and heat is quickly removed from the system by the off-gas stream. In the liquid-phase, methanation system developed by Chem Systems, an inert liquid is pumped upward through the reactor, operating at 2–7 MPa (300–1000 psi) and 300–350°C, at a velocity sufficient to fluidize the catalyst and remove process heat. At the same time, the coal gas is passed up through the reactor, where methanation occurs in the presence of the catalyst. It has been found that catalyst attrition is substantially reduced over that in gas-fluidized beds because of the cushioning effect of the liquid. Processes have also been developed for hydrogasification that maximize direct conversion of coal to methane. A good example is the HYGAS process, which involves the direct hydrogenation of coal in the presence of hydrogen and steam, under pressure, in two fluidized-bed stages. Additional developments have been pursued with catalysts, such as Exxon's catalytic gasification process, but these processes have not been commercialized.

A coal-to-SNG facility can be built at a coal mine-mouth location, taking advantage of low cost coal. SNG can then be pipelined to local distribution companies and distributed through the existing infrastructure. This approach is used in the Great Plains Coal Gasification Project in Beulah, North Dakota, which employs Lurgi gasifiers followed by shift and methanation steps. SNG has the advantage that it can directly displace natural gas to serve residential, industrial, and utility customers reliably.

Another technology that is being pursued for fuel utilization of coal is mild gasification. Similar to pyrolysis, mild gasification is performed at atmospheric pressure at temperatures <600°C. By drying and heating under controlled conditions, the coal is partially devolatilized and converted to gases and a solid residue. The gases can be used as fuel and partially condensed to produce a liquid fuel similar to residual fuel oil. The solid product is similar to low moisture, high heating value coal. A demonstration project for Powder River Basin coal was tested by ENCOAL in Wyoming.

3.4. Conventional Transportation Fuels. Synthesis gas produced from coal gasification or from natural gas by partial oxidation or steam reforming can be converted into a variety of transportation fuels, such as gasoline, aviation turbine fuel (see AVIATION AND OTHER GAS TURBINE FUELS), and diesel fuel. The Fischer-Tropsch process that converts synthesis gas into largely aliphatic hydrocarbons over an iron or cobalt catalyst is widely used for this application. The process was operated successfully in Germany during World War II and is used commercially at the Sasol plants in South Africa.

More recently, Shell developed proprietary technology for converting syngas into liquid hydrocarbons (10). This technology is particularly well suited for producing high quality distillate fractions and is therefore referred to as the Shell Middle Distillate Synthesis (SMDS) process. This is a modernized version of the classical Fischer-Tropsch technique. In the first step, the synthesis gas components, hydrogen and carbon monoxide, react to form predominantly long-chain paraffins that extend well into the wax range.

Underlying the Fischer-Tropsch reaction is a chain-growth mechanism. The product distribution is in accordance with Schultz-Flory polymerization kinetics and can be characterized by the probability of chain growth; the higher the probability of chain growth, the heavier the waxy product. In the development of the SMDS process, proprietary catalysts have been developed with a high selectivity toward heavier products and, therefore, with a low yield of products in the gas and gasoline range. Much attention has been paid to the selection of a reactor for this very highly exothermic process. In principle, three different types of reactors can be used for the synthesis: a fixed-bed reactor, an ebulliating or fluidized-bed reactor, and a slurry reactor.

The use of a fluidized-bed reactor is possible only when the reactants are essentially in the gaseous phase. Fluidized-beds are not suitable for middle distillate synthesis, where a heavy wax is formed. For gasoline synthesis processes like the Mobil MTG process and the Synthol process, such reactors are especially suitable when frequent or continuous regeneration of the catalyst is required. Slurry reactors and ebulliating-bed reactors comprising a three-phase system with very fine catalyst are, in principle, suitable for middle distillate and wax synthesis, but have not been applied on a commercial scale.

For the Fischer-Tropsch reaction in the first stage (heavy paraffin synthesis, or HPS) of the SMDS process, a tubular fixed-bed reactor has been chosen for its inherent simplicity in design and operation and also for its proven technology in other processes, such as methanol synthesis. The catalyst is located in the tubes, which are cooled by boiling water around them, and considerable heat can thus be removed by boiling heat transfer. The good stability of the SMDS catalyst makes it possible to use a fixed-bed reactor. In the next step, heavy paraffin cracking (HPC), the long-chain waxy paraffins are cracked to desired size under mild hydrocracking conditions using a commercial Shell catalyst. In the final step, by selection of the corresponding cut points, the product stream is split into fractions of the required specification. The products manufactured in the SMDS process are predominantly paraffinic and free of impurities such as nitrogen and sulfur.

4. Coal Characteristics Affecting Gasification

Developers of coal gasification technology have studied the impact of key coal properties on different parts of the gasification process. These tests have

Table 2. **Feed Property Extremes as Tested in SCGP-1**

Constituent	Composition, wt%	
	High	Low
ash	24.5[a] (up to 35)	0.5[b]
oxygen	16.3[c]	0.1[b]
sulfur	5.2[b]	0.3[d]
chlorine	0.41[e]	
moisture	30.7[a]	
Na_2O	3.1[d]	
K_2O	3.3[f]	
CaO	23.7[c]	0.8[b]
$CaO + MgO + Fe_2O_3$		10.2[g]
Fe_2O_3	27.8[h]	
SiO_2	58.9[i]	
Al_2O_3	32.6[j]	

[a] Texas lignite.
[b] Petroleum coke.
[c] Buckskin, Powder River Basin, Wyo.
[d] SUFCo.
[e] Pyro No. 9.
[f] Pike County.
[g] Newlands.
[h] R&F.
[i] El Cerrejon.
[j] Skyline.

provided a good understanding of the influence of coal properties and have led to the development of process and equipment options. For example, a comprehensive demonstration program conducted on the Shell Coal Gasification Process (SCGP) at the demonstration plant near Houston (SCGP-1) included 18 different feeds including petroleum coke and a very broad range of coals (11). The property extremes are shown in Table 2.

4.1. Reactivity. Reactivity is used to describe the relative degree of ease with which a coal undergoes gasification reactions. The primary property affecting the ease of conversion is the coal rank, which in turn reflects its volatile matter content, oxygen content, level of maturity, extent of aromatic ring condensation, and porosity. The lower the rank the higher the volatile matter content and the more open the pore structure. Also, lower rank coals have more heteroatoms (oxygen, nitrogen, and sulfur) within the organic structure and the aromatic structures are poorly aligned. Such an amorphous and open structure contains more active sites making reaction with oxygen and steam easier. As the rank increases the carbon lattice becomes better aligned and the porosity reduces until, in anthracite coals the carbon structure becomes less reactive developing the flat basal structures found in graphite. Reactivity varies dramatically with rank with some low rank coals being several orders of magnitude more reactive than high rank coals (6).

Other factors that have impact on reactivity are maceral distribution and the content of some catalytic mineral components. Vitrinites is the most common coal maceral derived from woody tissue. The properties and reactivity of vitrinites vary with the extent of geological maturation or coal rank. Fusinites originate from biodegraded or charred wood and are relatively unreactive C-rich macerals; while liptinites come from hydrogen-rich sources such as spores, leaf cuticles, and algal colonies. These liptinites are quite reactive.

The cause of the higher reactivity in low rank coals is the higher porosity, the larger number of active sites, and higher content and dispersion of catalytic metals such as calcium, potassium, and sodium. The organic matrix in low rank coals have carboxylic acid and other heteroatomic (O, S, and N) functional groups that have exchanged hydrogen ions with these cations producing ideal catalytic sites lowering the activation energy for gasification reactions. From Table 2, it can be seen that anthracite coal and petroleum coke that have the lowest reactivity also have low oxygen content [0.1% moisture free (MF)], whereas subbituminous coal from the Powder River basin, named Buckskin, has the highest reactivity (an oxygen content of 16.3% MF). It has been shown that the high degree of dispersion of catalytic minerals in low rank coals, like the Powder River basin coal, account for its high reactivity (12). Ion-exchangeable cations are attributed to this coal's superior performance in the Transport gasifier (13).

4.2. Moisture and Oxygen Content. The moisture present in the coal is primarily a diluent. Although steam is used for gasification, there are several sources of water for this steam and only a small amount of the total water available is actually converted in the steam-carbon reaction. Steam is introduced with air and oxygen to moderate the temperature in the combustion zone of the gasifier. Some entrained gasifiers introduce coal into the reactor with the aid of water as a coal–water slurry. In addition, all coals have moisture content. Lower rank coals, which have been exposed to geological maturation over shorter periods of time and at lower temperatures, have more moisture than higher rank coals. As a result, the feed rate of lower rank coals must be increased to obtain a gas product of equivalent heat value.

Similarly, the higher oxygen content of lower rank coals reduces the heating value of these coals relative to higher rank coals. To offset the effectively higher oxidation state of the low rank coals a higher coal feed rate is required to obtain similar product gas quality.

In addition, the integrity of the coal structure becomes weaker and the coal particle becomes more friable as the oxygen and moisture contents increase. Low rank coals can exhibit serious particle decrepitation, dusting, and even spontaneous combustion when exposed to drying conditions upon storage, crushing, and handling.

Fixed-bed gasifiers can accommodate coal with moisture contents as high as 35% as long as the ash content is <10%. Entrained and fluid beds generally prefer less moisture to aid in forming the coal water slurry and reducing the extent of solids losses via elutriation and entrainment into the product gas stream. Because of the large heat of vaporization of water, excessive moisture can affect the gasifier temperature and impede normal operations.

4.3. Caking Properties. When bituminous coals are heated to 300–350°C the particles tend to swell and agglomerate producing a consolidated cake. The handling of this caking char and the heavy tars that accompany it has been critical to the development of gasification processes. The agglomerate that forms in a fixed or fluid bed disrupts gas flow patterns and lowers thermal efficiency.

It is useful to understand the nature of this caking phenomenon. Compared to low rank coals bituminous coals have undergone a greater degree of metamorphism: the organic structure has fewer heteroatomic cross-links, the aromatics are more highly condensed and aligned, and the pore structure is less open and more microporous. Upon drying bituminous coal particles retain their strength rather than becoming weak and friable as in low rank coals. However, upon heating to the point where organic constituents begin to decompose, organic volatiles form and partially dissolve the more cross-linked features. The result is essentially a plastic melt. This plastic transformation of the bituminous coal structure is referred to as metaplast and results in a sticky, particle-agglomerating phase during the heating process. Once the volatile matter content of the coal falls below ~20% the caking tendencies are reduced.

In addition, to formation of cake, bituminous coals also produce high molecular weight tars. To handle the tars formed in coal process units, the downstream gas cleanup system must be engineered to avoid plugging and fouling of lines, heat exchangers, and filters.

4.4. Mineral Composition.   The mineral content affects gasifier performance, especially for most slagging gasifiers, because minerals melt to form of slag and provide an insulating coverage on the wall of the gasifier, which reduces the heat transferred during the gasification reaction (14). Mineral content also influences the requirements of the slag tap and the slag handling system. A related parameter is the slagging efficiency, which is the percentage of mineral solids recovered as slag out of the bottom of the gasifier relative to the total mineral solids produced by the process. As shown in 2, the feeds that were gasified at SCGP-1 had mineral producing ash contents ranging from 0.5% MF for petroleum coke to 24.5% MF for Texas lignite.

Iron sulfide, or pyrite, has a high density and relatively low melting point and is often found in the clinkers produced by temperature excursions in fluid-bed gasifiers. Fluxing agents, such as sodium and calcium, particularly when combined with the ubiquitous clays and pyrite reduce the ash melting temperatures and are often identified as the cause for the initiation of process clinkers. As a result, coal mineral composition is monitored as closely as the heating value in an attempt to avoid operational problems. These minerals are generally evaluated by analyzing the oxide residue or ash produced upon burning the coal.

Ash Melting Point/Slag Viscosity.   For gasification technologies utilizing a slagging gasifier, slag flow behavior is an important parameter. To determine slag viscosity, the viscosity of coal ash is measured in a reducing atmosphere. Coals having a wide range of ash fusion temperatures were tested at fluid temperatures ranging from 1190°C for Illinois No. 5 coal to 1500°C and higher for several Appalachian coals, such as Skyline, Robinson Creek, and Pike County. Coal ash viscosity plots versus temperature are shown in Figure 4 for a variety of coals. Slag viscosity varies over several orders of magnitude for the different coals at representative gasifier temperatures. For example, Buckskin, which is a subbituminous coal from the Powder River Basin in Wyoming, has a much lower viscosity than an Appalachian coal such as Blacksville No. 2. For coals having

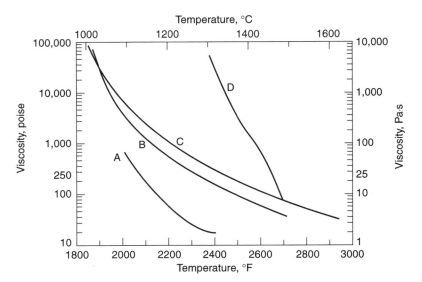

Fig. 4. Viscosity–temperature profiles of coal ash samples, where A represents Buckskin; B, pyro No. 9; C, Blacksville (Appalachia) No. 2; and D, Drayton.

high slag viscosities, slag behavior can be modified by the addition of a flux such as limestone (calcium carbonate).

Fouling Precursors. Fouling of heat transfer surfaces can result from constituents such as chlorine, sodium, potassium, and calcium. Most fouling indexes are based on experience with pulverized coal boilers. This information is often used to select conventionally high fouling coals to obtain fouling data on gasification units. In order to establish fouling indexes, the coal selection criteria in demonstration programs have included a wide range of fouling agents. As shown in 2, the calcium oxide content ranged from 0.8% for petroleum coke to 23.7% for Buckskin coal.

Corrosive Components. The primary coal properties affecting corrosion are sulfur and chlorine levels. Upon gasification these elements form acidic gases, hydrogen sulfide, and hydrogen chloride, which are responsible for corrosion of metals and other materials. The formation of these corrosive species is dependent on their content in the feed coal. The range of sulfur levels in Table 2 shows a low of 0.3% for SUFCo and a high of 5.2% for petroleum coke. Coals in the U.S. interior basin (including Illinois, Indiana, and Ohio typically have the highest sulfur levels, although the Appalachian region coals have high levels as well. R&F coal has 4.2% sulfur. Chlorine is less abundant and not uniformly distributed through the geologic basins. The highest chlorine level in this group of coals is 0.41% for Pyro No. 9 coal.

5. History

5.1. Early Coal Use and Gasification.
Coal has been used for centuries (see Table 3). The very first recorded use of coal was in China between

Table 3. **Significant Events Related to Coal**

Year AD	Event
589	First recorded use of coal in China
852	Coal first mentioned in the "Saxon Chronicle" of the Abbey of Petersborough
1180	Coal systematically mined in England
1250	Coal recognized as a commercial commodity
1316	Royal Proclamation forbidding use of coal in London due to its "noisome smell"
1609	Van Helmont identifies gas production from coal combustion.
1659	Shirley investigated "natural gas" released from a well in Lancashire England
1675	Coal was distilled for the production of tar
1780	Fontana proposes making "blue water gas" by passing steam over incandescent carbon.
1792	Murdoch lights his Scotch home with gas from coal heated in an iron retort
1803	Huge gas powered lamp installed on 40 ft. high tower on Main Street, Richmond, Virginia
1812	London Streets illuminated by the London and Westminster Gas Light and Coke Co.
1859	Drake drills first oil well near Titusville, Pennsylvania
1872	Lowe invents carbureted gasifier
1880	Development of modern day fixed bed coke ovens
1920	Fischer and Tropsch develop catalysts to convert coal synthesis gas to liquids.
1926	Rheinbraun develops fluid bed gasifier
1936	Development of the modern day entrained bed Koppers-Totzek gasifier
1950	Production of gasoline and diesel fuel using Lurgi gasifier in South Africa
1970	Clean Air Act
1973	Arab oil embargo
1983	Syngas production for chemical production using Texaco gasifier at Tennessee Eastman
1984	Production of synthetic natural gas at Dakota Gasification Plant using Lurgi gasifiers
1996	Clean coal demonstration power production plants using Texaco and E-Gas gasifiers
2001	Power shortages in California

220–589 AD. As noted in the Saxon Chronicle for the Abbey of Petersborough, coal was being used in Europe as early as 852 AD. However, the association of coal with gas production was not made until 1609. It was in Jean Baptist van Helmont's alchemy laboratory that coal was heated and the release of gas was first noted. He wrote how the coal "did belch forth a wild spirit or breath... not susceptible of being confined in vessels, nor capable of being reduced to a visible body". Helmont was likely frightened by this spirit and quite naturally called this spirit "gas" as derived from the Dutch word "Geest" for ghost (15).

To understand the development of coal conversion technologies, such as coal gasification, one must understand the factors that drove its development. Coal was the energy source that fueled the Industrial Revolution. Coal has a higher heating value than wood making it less bulky and easier to store and transport to the marketplace. It was found in concentrated seams underground and as such was systematically mined in England as early as 1180. These facts coupled with its widespread abundance in Europe made coal the preferred fuel in cities and towns.

In 1272, coal was first used in London; however, a Royal Proclamation in 1316 outlawed its use because of the "noisome smell". With the population growth around key cities the uncontrolled use of coal for home heating and the workshops created the first recorded air pollution. Again, in 1580 Queen Elizabeth forbade the use of coal in London when Parliament was in session because "the health of the knights of the shires might suffer during the abode in the Metropolis." The use of coal continued unabated so that in 1662 King Charles II raised a huge sum of 200,000 lb from a "health tax" imposed on fireplaces (15).

As the world emerged from the Middle Ages coal provided more than just the fuel for innovation. Coal gas also played a key role in innovations in developments in materials and lighting. Development of industry required better materials for tools and machines. The first record of a Blast Furnace for the reduction of iron ore to iron was in the England in 1496. It was 1709 when Abraham Darby introduced coke to the blast furnace in place of charcoal, allowing the furnace size to be increased making the production of large amounts of cheap iron possible. Prior to this time charcoal would collapse under the extra weight of the charge of iron ore in a large furnace (see Fig. 5).

5.2. Coke Manufacturing Processes. These processes were developed to convert the softer bituminous coals into a strong hard coke ideally suited for iron making. It was found that when bituminous coals are heated slowly in the absence of air and then cooled the solid coke is both hard and inherently strong. The *coking process* requires heating to temperatures approaching 1000°C. This process heat is provided either by burning a portion of the coal directly and allowing the hot products of combustion to pass through a bed of coal, or by indirectly heating the exterior walls of a vessel containing the coal. The indirect method of heating a coke oven is still the most commonly used coking process used today because it maximizes the formation and quality of the coke product. In both heating methods, a flammable coal gas is produced that were readily put to use in a variety of applications.

The first coal *gasifier* was used by Fontana in 1780 when he directed a flow of water (steam) over red-hot, or incandescent, coal that was previously heated in an air blast. This process can be simply described using two chemical reactions: char combustion for producing heat and steam gasification for producing synthesis gas, a combustible mixture of carbon monoxide and hydrogen.

Fontana called the resulting coal gas "blue water gas", because it produced a pale blue flame when burned in air (4).

5.3. Manufacture of "Blue Water Gas" and "Town Gas". At the time of the French and American Revolutions houses were lit using candles and whale oil and heated with wood fireplaces or coal burners. Textile factories and mills were poorly lit and cold places to work. Thomas Edison did not discover the light bulb until late in the nineteenth century, 1879. The first electric power plant was not built until 1882. However, as early as 1792 a Scotch engineer named Murdock, used Fontana's process to produce a fuel gas and light his house. Later, James Watt, the inventor of the steam engine, employed Murdock to light one of his foundries with the newly discovered coal gas (16).

The first recorded use of coal gas in America was to light a street lamp mounted on a 40-ft tower on Main Street in Richmond, Virginia in 1803. The

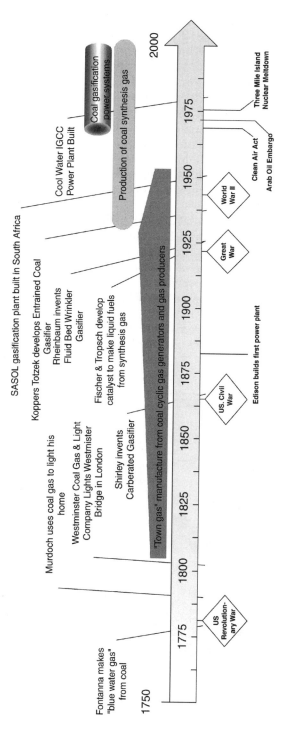

Fig. 5. Timeline for coal gasification development.

785

Westminster and London Gas Light and Coke Co. was given a royal grant to provide lighting for London streets in 1812 becoming the first company to produce
coal gas commercially. In spite of the hostilities in the War of 1812, the British
influence in America was evident from the fact that just five years later the Gas
Light Company of Baltimore became America's first gas company. In all of these
cases the fuel gas was manufactured from coal gasification. However, it was not
until 1865 that the use of coal gas for lighting became common practice.

During the nineteenth century, gas manufactured from coal was the only
source of fuel gas. Although Thomas Shirley first described a discovery of natural
gas from a well in Lancashire in 1659, but this resource was not commercialized
until the early 1900s. Coal gasification was used for lighting and for heating in
industrial processes. Later coal gas served as a rich source of chemicals and
liquid fuels as well.

Shortly after the American Civil War, by 1875, every large and medium
sized city in industrialized Europe and America developed coal gasworks and
gas distribution networks. Gasworks consisted of a large number of individual
gasifiers operating in various stages of the process: some being loaded, some
being heated under air blast, others producing gas under a steam blast, and
still others being cleaned out or emptied. Loading and unloading was typically
done by hand (see Fig. 6).

The "blue water gas" produced from *cyclic gas generators* of the type used by
Fontana were not strictly suitable for street or household lighting. The first coal

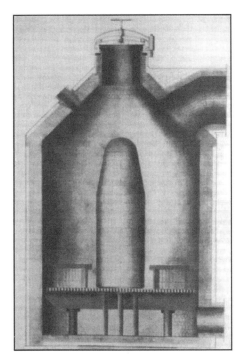

Fig. 6.   Water–gas generator with central pier for improved heat distribution (17).

gasifiers built for this purpose undoubtedly suffered several setbacks. Although the heating value was high enough to sustain combustion, the blue flame was not sufficiently bright to illuminate a street or room in a house. Early developers learned that the coal gas needed to contain components called "illuminates" to provide a bright luminous yellow flame. These components consisted of higher hydrocarbons with hydrocarbon chain lengths of 2 or more. However, a second problem was observed if too many or the wrong types of higher hydrocarbons were added; condensable species, tars and naphthalene caused fouling and plugging in the piping and transfer lines. In addition, noxious and poisonous gases were also produced. These impurities included primarily hydrogen sulfide, hydrogen cyanide, and ammonia.

The solution to produce a brighter yellow flame was to simply add "illuminates" as the product gases left the gasifier. In the *carbureted gasifier process* developed in 1872, Professor Thaddeus Lowe produced both an "oil gas" and "blue water gas". The "oil gas" contained condensable coal tars or cheap petroleum distillate oil and was sprayed onto a hot brick matrix or checker work. After the "blue water gas" was formed it was passed over this checker work and mixed with the volatile products of "oil gas" decomposition. The high temperature of the brick and reducing nature of the product gas leaving the gasifier would cause cracking and vaporization of the coal tars and oils (15).

Solutions were readily developed to separate and clean the undesirable components from the coal gas. Cooling, direct water spray, and separation methods such as filtering of the gases were usually necessary to remove condensables and impurities. The carbureted gas was commonly called "town gas" because this gas was distributed through a network of pipelines for use in lighting streets and eventually homes in many of the late nineteenth century cities in America and Europe.

As is still true today, the coal gas in those early gasifiers was produced in several different reactors depending on the desired products and end use. Coal technologies besides combustion included: *pyrolysis, coking, cyclic gas generators*, and *gas producers*. All of these processes are heated with insufficient air to convert the coal to the final products of combustion. The products of all of these processes are a solid fuel, condensables and tars, and flammable gases. Both the quality and quantity of these products depends on how the coal is heated, the gas atmosphere used during the process, and the temperature. A comparison of the various products generated from these coal conversion processes is presented in Table 4.

Pyrolysis, or heating coal in the absence of oxygen, was conducted principally for the manufacture of chemicals. This conversion process is geared toward the production of condensable products such as benzene, toluene, naphthalene, phenols, creosote, and pitch. *Coking* is another process in which coal is heated in the absence of oxygen, but unlike pyrolysis the principal product of interest is the solid coke. In *pyrolysis*, the coal beds are thin and temperatures are low and near the melting, or the decomposition, temperature of coal to promote production of liquids and gases. While in *coking* the coal beds are deep and the temperatures are much higher to promote resolidification of the decomposing coal. In both of these processes, fuel gas is formed in significant quantities and of high quality as measured by the gas heating value in terms of British thermal units

Table 4. Operating Conditions and Product Distribution for Various Coal Conversion Processes Using Bituminous Coal[a]

Product composition	Gas producers				Entrained gasifier		Coke oven
	Blue water gas	Town gas	Fixed bed	Fluid bed	Fixed bed	Fluid bed	
oxidant	steam	steam	air, steam	oxygen, steam	air, steam	oxygen, steam	none
fuel	bit. coal	bit. coal	bit. coal	brown coal	subbit. coal	bit. coal	bit. coal
exit temp, °C			142	482	1093	1093	982
Gases, %wt							
H_2	50.5	40.5	14.5	36.0	12.9	35.8	46.5
CO	38.5	34.0	25.0	44.4	23.5	50.7	6.3
CH_4	1.0	10.2	3.1	1.6	0.02	0.1	32.1
illuminates[b]		8.0		0	0	0	4.0
N_2	3.5	2.9	52.7	0.8	60.3	0.2	8.0
CO_2	6.0	3.0	4.7	15.7	3.1	13.1	2.2
char, wt%			19.5	20.0	11.8	11.4	75
gas HHV, Btu/scf	300	550	167	267	117	280	584

[a] From Refs. (15,18).
[b] Noncondensable gases consisting of hydrocarbons with carbon chains between C_2–C_4.

per standard cubic foot of gas (Btu/scf). Where possible this gas was used for process heat, but any excess gas could be sold.

There were two types of coal gasifiers in the mid-1800s *cyclic gas generators* similar to those first used by Fontana and *gas producers*. The *cyclic gas generators* produced a high quality fuel gas. This high heat value gas was made possible by collecting only the products during the steam blast and venting the products of combustion during the air blast. As we noted above the *carbureted gasifier* produced "town gas" that was particularly well suited for lighting. The original *cyclic gas generators* produced "blue water gas" that was an ideal industrial fuel used in kilns, boilers, brick ovens, and for curing materials, particularly where particulate impurities were to be avoided. However, it was readily seen that the production rate could be improved by making this cyclic process continuous.

Gas manufacturers in the early 1800s learned that they could improve both the rate and efficiency of the gas making process by introducing steam and air together into the coal. In this way, the steam was available to gasify the carbon at the same time that heat was being produced from coal combustion. The result was the ability to substantially increase the overall conversion of coal to gas. Thus, modern *gas producers* were developed. Gasifiers were batch units that were loaded by hand, and the ash or coke, often still glowing hot, had to be removed by hand usually with nothing more than a shovel. As late as 1936 the Federal Trade commission reported that there were 3800 machine fed and $\sim$1000 hand-fed *gas producers* in the United States alone (1).

The development of *gas producers* did not, however, displace the *cyclic gas generators*. By firing air and steam simultaneously the product gas in these producers now contained the unreactive nitrogen that is present in the air stream fed to the gasifier. Air is 79% N_2 and only 21% O_2 by volume. This added nitrogen is merely a diluent and plays no role in the combustion or gasification reactions. In addition, this diluent must be heated both during the gasification process and later as the fuel is burned. As a result the producer gas has only about one-half of the heating value compared to "town gas". There are several issues for using this low quality gas fuel. The large gas volume and lower heat value reduce the feasibility to economically transport producer gas for more than a short distance. Upon combustion the flame is longer, cooler, and more luminous. However, producer gas found many industrial applications to take advantage of the flame properties of this low Btu gas. For example, in glass making the relatively low temperatures and long flame are useful to avoid hot spots and nonuniformities in the glass.

By 1930, there were over 11,000 coal gasifiers operating in the United States and in the early 1930s over 11 million metric tons of coal were gasified annually. Fredersdorff and Elliott (19) described "the ideal complete gasification process as a single stage, continuous process which uses air as the oxidizing medium and converts any type of coal into a combustible or synthesis gas low in inerts". The development of the modern-day gasifier strives to meet this goal while achieving the most desirable mix of products depending on the end-use or process application. Three modern coal gas producer types are shown schematically in Figure 7, called Fixed Bed, Fluid Bed, and Entrained-Flow gasifiers.

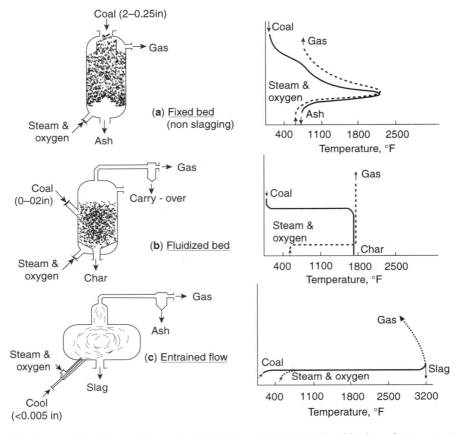

Fig. 7. Types of gasifiers: (**a**) moving-bed (dry ash), (**b**) fluidized bed, and (**c**) entrained-flow (20).

Each of these gasifiers has advantages, disadvantages, and potential for process improvement. Each type of gasifier converts certain coal feedstock over unique time–temperature profiles resulting in various product distributions. In Figure 5, a typical temperature profile is also portrayed for each reactor system, highlighting the process differences for these technologies. Typical product distributions for each are presented in Table 4 and compared to the early cyclic gas generators. The prime developments in the modern gasifiers were the variety of different coals that could be gasified, the coal throughput or capacity, and the overall coal conversion. The sacrifice made to achieve these improvements was the reduced quality of the fuel gas that was often lower, particularly when gasifying coal in air and steam. However, the composition of hydrogen and carbon monoxide produced in the newer generation of gasifier was sufficient for new applications whether they were liquid production, chemical synthesis, or power generation.

5.4. Gasification for Liquid and Chemical Feedstock. Although Drake discovered oil in Titusville, Pennsylvania in 1859, petroleum products did not make inroads into the gas manufacturing markets until the 1930s. At that time oil and natural gas began to effectively displace manufactured gas from pipeline distribution networks. However, coal gasification remained a strong industry as a result of the emerging new transportation and chemical industries.

During the first quarter of the twentieth century coal was king, providing: heat for homes and industry, fuel gas for town lighting, process gas for industry, coke for the iron industry, and fuel for steam locomotives and shipping. By 1850, with the completion of the transcontinental railroad in the United States coal-fired steam locomotives tied the nation together. By 1900, electric power was beginning to compete with the coal gas industry for public lighting. The electric power industry was also coal based, although large hydroelectric power plants from dams would be built between 1900 and 1930. With the development of the internal combustion engine and Ford's innovations in the automobile industry, it was natural for scientists to look toward coal for the liquid fuel that was needed.

During the 1930s new cheaper and cleaner fuels began to emerge—natural gas and oil industries were born at this time. However, this did not happen overnight. Even after the Second World War, 5 million tons of coal was consumed per year through the manufacture of carbureted gas.

The first liquid fuels developed from coal can be traced to Bergius in Germany in 1913. He used a liquid extraction process heating pulverized coal with oil under high pressures and with hydrogen. This is called direct hydrogenation or direct liquefaction of coal.

In 1920, two other German researchers, Fischer and Tropsch, developed a catalyst to convert the hydrogen and carbon monoxide from coal gasifier gas to hydrocarbon liquids. The Fischer-Tropsch process became known as indirect coal liquefaction since coal was first gasified before the gases were reduced to liquids.

By the mid-1930s Germany had plants to produce gasoline and oil from both direct and indirect liquefaction technologies, ~1 million and 0.5 million gal/year, respectively (21). Individual reactors in the early Fischer-Tropsch plants yielded ~5000 L of gasoline per day. Literally 100 such gasifiers were used in the first German plants (Fig. 8). But because the Bergius process plants were further developed and could be more readily scaled up to larger size, that was the technology chosen in 1939 when Germany expanded production to fuel the Nazi war machine. Direct hydrogenation reactors were eventually built to process up to 350 tons of coal per day yielding 250,000 L of gasoline. The capacity for oil and gasoline manufacture from direct liquefaction plants quadrupled during the war years in Germany. Elsewhere, however, oil resources were sufficient to meet demands and coal was not extensively used to produce liquids. After the war the technology used in the German synthetic fuels industry was studied, but considered uneconomical given the more competitive costs of natural gas and oil production.

A unique economic climate developed in South Africa, where the South African Coal, Oil, and Gas Company has operated the SASOL plant to produce liquid transportation fuels. SASOL is a coal gasification plant operating since 1955 using the Fischer-Tropsch process for the production of liquid fuels. South Africa was economically sanctioned for their social policies of Apartheid,

Fig. 8.   Fischer-Tropsch synthesis for gas generators, Ruhrchemie plant (21).

and had no known or available oil resources, but ample coal supplies. The individual reactors in these plants are 100 times larger than the plants used in Germany during the Second World War. SASOL had a capacity of nearly 2.5 million gal of oil and gasoline per day in 1982 using Lurgi fixed-bed gasifiers.

6. Gasifier Types

6.1. Fixed-Bed Gasification.
As described above, the early gasification processes were developed using a countercurrent, *fixed bed gasifier*. Within the bed the fuel is not actually fixed but in fact moves, by gravity flow, as the combusted ash is withdrawn from the gasifier. Typically, the air and steam are introduced at the bottom and travel upward through the coal bed. The coal is fed onto the top of the bed and travels downward countercurrent to the flow of gases. The gas outlet and coal feed inlets fix the upper level of the bed, while the bottom of the bed is most commonly fixed by the presence of a rotating grate.

The *fixed bed* was long considered to be the most efficient method of converting carbonaceous fuels to fuel gas. A schematic of the processes taking place within this *fixed bed gasifier* are represented in Figure 9 (22). The fuel bed is generally divided into different temperature zones corresponding to the following:

1. The topmost layer where coal is dried and preheated and volatile hydrocarbons are released.
2. The reduction, or gasification, zone where the hot char reacts with steam and carbon dioxide to produce hydrogen and carbon monoxide.
3. The oxidation zone where the residual carbon reacts with oxygen producing heat for gasification reactions.
4. The ash cooling and air preheat zone at the bottom of the gasifier.

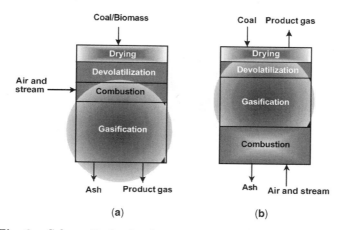

Fig. 9. Schematic for fixed- or moving-bed gas producer (22).

Fixed beds have several advantages. The flow of the hot gases up from the combustion zone preheats the coal leading to maximum heat economy. High carbon conversion is assured by plug flow of solids through the gasification and combustion zones and the relatively long residence times of the fuel in the vessel. The product gas exits relatively cool temperatures and without contamination of solids.

Fixed-Bed gasifiers have been developed to handle a variety of solid fuels. The factors important in affecting a fuels performance in a fixed-bed gasifier are particle size and size distribution, tendency for coal to melt and form an agglomerated mass, the temperature at which the ash melts and fuses, and the reactivity of the coal. The most suitable coals are uniformly sized crushed particles without tendency to agglomerate and with a minimum of fines but having reasonable mechanical strength. With such fuels the flow of gases through the bed is distributed uniformly through the bed resulting in uniform temperature distribution and stable reaction zones. Reactivity of the coal or coke feedstock affects the size of the reaction zones. Less reactive coals need larger bed depths, higher reaction temperatures, and longer residence times to achieve complete carbon conversion.

The disadvantage of the *fixed-bed gasifier* is the inability to process caking coals. These coals, generally of bituminous coal rank have a tendency to swell and agglomerate upon heating. As a result of this behavior such coals cause mal-distribution of both gas and solids flows leading to process failure. In order to use these coals a pretreatment by either preoxidizing, or more commonly coking, to eliminate any caking tendencies in the coal. The production of tars in the heating zone can also lead to fouling of the gas pipelines when the gas is used as a chemical feedstock, or synthesis gas, or when distributing this gas to public utilities. Coals having >35% moisture content lose strength upon heating and are not suitable feedstocks for fixed-bed gasifiers.

In fixed-bed gasifiers the devolatilization products exit the gasifier with the syngas, because of low temperatures and lack of oxygen in the devolatilization zone. This causes increased amount of methane in the product gas, which

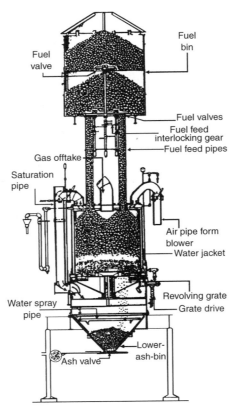

Fig. 10. Wellman-Galusha fixed-bed gasifier showing gravity coal feed system (18).

increases the heating value of the gas. But the low temperature and the countercurrent operation also allows the tar to escape, which is detrimental to the down
stream equipment.

The Wellman-Galusha is a good example of an atmospheric *fixed-bed gasifier* with automated coal feed and a rotating grate for ash removal. The development of a coal feed system without mechanical parts provided a reliable
maintenance free system as long as the coal feed size could be adequately
controlled (Fig. 10).

The more critical performance limit for coal gasifiers, however, was the size,
capacity, and coal throughput. The Lurgi gasifier was first commercialized in
1936. By 1950 the Lurgi brought significant technological advancements to
fixed-bed gasification increasing the capacity and throughput. The primary
improvements were to increase the operating pressure, temperature, and use
of oxygen rather than air (Fig. 11).

Higher pressure increases the gasification rates per unit volume, decreases
the heat losses, and reduces the number of reactor vessels to achieve a desired
gas production rate. Higher reaction rates increase a reactor's capacity and
throughput. Faster reactions also result in a shorter combustion zone. A shorter

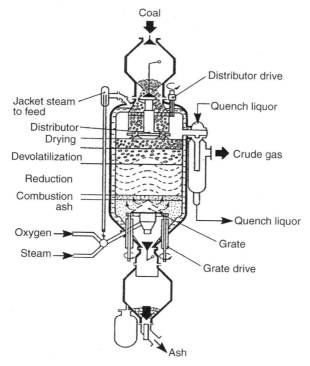

Fig. 11.   The pressurized, oxygen blown, Lurgi gasifier showing with superimposed reaction zones (20).

high temperature combustion zone reduces heat losses and thus process inefficiencies.

The use of oxygen from an air separation facility also produces higher temperature in the combustion zone that, in turn, also increases the combustion and gasification rates. Using air the maximum combustion temperatures reach 1200–1300°C, but using only oxygen the temperatures reach 1500–1800°C. The use of oxygen in the gasifier also produces a higher quality gas fuel through the elimination of the nitrogen diluent in the produced gas.

The main technological hurdle to overcome with higher operating temperatures was the fact that the mineral impurities in coal, mainly sand and clays, melt and become sticky at these higher temperatures. This molten ash is known as slag. In a slagging gasifier this molten ash is kept fluid and drained (Fig. 12). Lurgi offered the first pressurized, oxygen-blown, fixed bed, slagging gasifier in the British Gas Lurgi gasifier in 1976. The slagging gasifier technology was developed at BGCs Westfield facility in Scotland, initially on a 275-tons/day pilot plant, and subsequently, a 550-tons/day demonstration unit.

A significant efficiency advantage is gained by reducing the steam requirement to only ~15% of that required by the dry-ash Lurgi gasifier. Compared with raw gas from the dry-ash Lurgi gasifier, the raw gas from the slagging gasifier has lower H_2O, CO_2, and CH_4 and higher CO content, primarily because of the

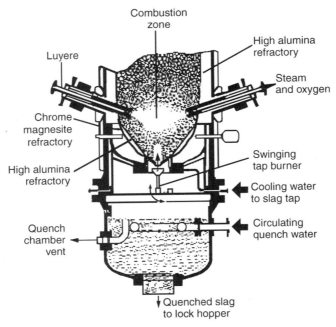

Fig. 12. The base of the British gas Lurgi slagging gasifier showing the hearth area and slag tapping (20).

lower steam consumption. Recycle of the tar and oil in the slagging gasifier increases the gas yield by reducing the net hydrocarbon liquid production to only naphtha and phenols.

6.2. Fluid-Bed Gasifiers.   Another approach developed in the 1930s to overcome the size limitations and lack of fuel flexibility in the early fixed-bed gasifiers was the fluidized-bed gasifier. Rheinbraun developed the most commercially successful fluid bed gasifier in 1926, the Winkler gasifier. In this process, the reactor vessel was designed so that the air and steam flow required for gasification was sufficient to fluidize the bed of coal, char, and ash. Fluidization occurs when the gas flow velocity lifts the particles causing the gas–solid mixture to flow like a fluid. The coal feed to this gasifier is a finer crushed coal rather than the larger nuggets used in fixed beds (Fig. 13).

Fluidized-bed gasifiers provide better mixing and uniform temperatures that allow oxygen to react with the devolatilization products. These products also undergo thermal cracking, primarily on hot char surfaces, reacting with steam and H_2. In dry fluidized-bed gasifiers, temperatures have to be maintained below the ash melting point, which leads to incomplete carbon conversion for unreactive coals. Agglomerating ash gasifiers operate at higher temperatures, near the ash softening point, which provides improved carbon conversion.

The primary advantage of the fluid-bed gasifier is the flexibility to use caking coals as well as low quality coals of high ash content. In addition, a fluid-bed gasifier is able to operate over a wide range of operating loads or outputs without significant drop in process efficiency. This fluid-bed process has a large inventory

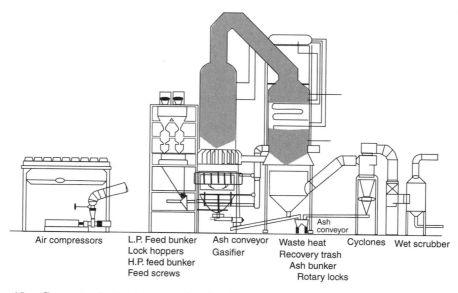

Air compressors	L.P. Feed bunker	Ash conveyor	Waste heat	Cyclones	Wet scrubber
	Lock hoppers	Gasifier	Recovery trash		
	H.P. feed bunker		Ash bunker		
	Feed screws		Rotary locks		

Fig. 13. Conceptual plant layout for the Winkler fluid bed gasifier process showing major auxilliary equipment (20).

of hot solids that stabilizes the temperature and eliminates the potential of oxygen breaking through and burning with the fuel gases in the event of an inadvertent loss of coal feed. Fluid beds also have high heat transfer rates and good solids and gas mixing. This minimizes the formation of localized hot spots that can produce molten ash agglomerates known as clinkers.

In addition, fluid-bed gasifiers can include inexpensive disposable sorbents, such as limestone, to absorb sulfur, reducing air emissions. The temperature regime is ideal for capture of hydrogen sulfide using limestone or dolomite.

One drawback, actually a result of good mixing in the fluid-bed gasifier, is the high temperature of the fuel gas at the exit of the reactor. When using cold gas clean up this high exit temperature represents a loss in process heat and thus process inefficiency. Likewise, solids drained from this type of well-mixed reactor also include a significant amount of carbon that must be utilized to avoid inefficiencies in carbon utilization. Solid particulates entrained in the product fuel gas are also high in carbon representing unreacted coal that must be recovered. Lower quality coals are more friable and result in higher loadings of such dust exacerbating the technical challenges.

In 1976, a vertical fluid bed gasifier was developed by Kellogg-Rust-Westinghouse (KRW) and the U.S. Department of Energy to overcome these drawbacks in the fluid bed gasifier. In the Kellogg Rust Westinghouse (KRW) gasifier, a central air or oxygen jet generates a hot zone at the base of the fluid bed to agglomerate ash. Unconverted char is recycled to this zone, which is hot enough to gasify the char and soften the ash. The ash particles stick together, growing in size and density, until they settle from the fluid bed and are separated from the char and removed from the bottom through dry pressure lock-hoppers. Agglomerated ash operation improves the ability of fluidized-bed processes to gasify unreactive high rank coals and caking coals efficiently. In addition, this technology offers the potential to reduce the loss of carbon and the resulting process inefficiencies. This process permits the fluid bed to operate at higher temperatures (up to 1150°C), and thus increases gasification rates and coal throughput. The development of this process has culminated with the construction of a 300 MW Clean Coal Technology plant by Siearra Pacific Power Co. at Pinon Pine power plant near Reno, Nevada in 1998.

A Transport gasifier is being developed by KBR-Haliburton, Southern Co., and the U.S. Department of Energy in further attempts to decrease the capital costs of gasification by increasing the coal through-put (23). The transport reactor achieves more rigorous mixing than bubbling or jetting fluid beds through the aggressive recirculation of char and ash: 100 kg of char is recycled into the reactor for every kilogram of coal. The Transport reactor also exhibits the plug flow characteristics of entrained reactors without as high a gas exit temperature. The unique feature of the Transport gasifier is the large inventory of hot recycle solids stabilizing the reaction zone while moderating temperatures. The process is operated at elevated pressure of 10–15 atm. Powder River basin subbituminous coals are reported to react to 90% C conversion in <20 s at only 950°C (12). Analysis conducted confirmed that this type of coal could be gasified in the Transport reactor; however, less reactive coals were found to be only partially combusted and would produce hot spots if higher operating temperatures were attempted (12). Under air-blown operations S capture with the inherent Ca -containing minerals exceeds 95%; however, oxygen-blown attempts at the 100-kg/h facility have merely replaced nitrogen with steam and a shift in equilibrium results in only 20% S capture even with added limestone. Uncertainties that remain are the separation of solids from the product gas and the ability to convert a wider variety of coals.

6.3. Entrained Flow Gasifiers.   Like fluidized beds, entrained coal gasifiers were developed to improve the gas production rate and operate with a wider range of fuel feedstocks. In an entrained gasifier the coal is introduced into air or oxygen in a dilute stream and heated to high temperatures, 1300–1475°C, over only a very short period, 2–3 s. Entrained-bed slagging gasifiers provide uniform high temperatures, resulting in complete conversion of all coals to hydrogen, carbon monoxide, and carbon dioxide, and producing no tars, oils, or phenols. As a result the throughput and capacity of the entrained reactor is the highest of all gasifiers. Coal friability does not affect operations since the coal must be pulverized for the entrained flow gasifiers. Likewise, coal swelling and agglomeration do not influence gasification performance since the particles are separated from each other in the flowing gas stream. The product stream contains no tars and very little methane because the heavy volatiles are rapidly released and cracked at the high temperature and within the short time available in the reactor.

Koppers Totzek began developing the first entrained coal gasifier in 1938 (Fig. 14). The co-current flow of gases and solids results in a high gas exit temperature. This requires gas cooling and means loss of process heat or relatively low process efficiency.

The Texaco coal gasification process (TCGP) is the most widely employed commercial entrained flow gasifier. The Texaco gasifier has been licensed several times for use in chemical manufacture. The first license was issued to Tennessee

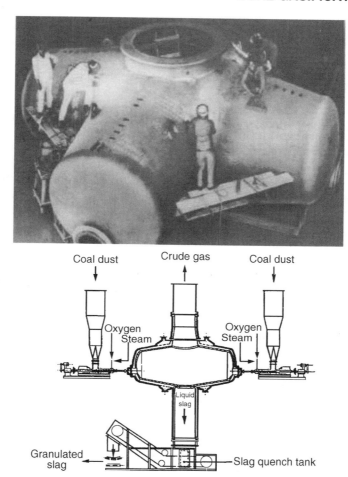

Fig. 14. Schematic Koppers-Trotzek entrained flow gasifier and photograph of gasifier with four burner heads (20).

Eastman for an 800 t/d plant, which was started up in 1983 and is used in the production of acetic anhydride. In 1984, a second Texaco coal gasification plant having a coal capacity of 1650 ton /day was built in Japan to produce ammonia. Since 1986 a third commercial plant, having a coal capacity of 800 t/d, was operated in Germany for use in oxo-chemicals manufacture.

Shell developed oxygen blown, pressurized, slagging, entrained coal gasification process. In 1993, the largest and most integrated coal gasification cycle plant in the world, a 253-MW power plant, was built in the Netherlands. The net efficiency at full load is ~43% based upon the lower heating value (LHV) of the coal. The level of sulfur removal is ~98% minimum (75 mg/m^3), and a maximum NO$_x$ emission from the gas turbine of 95 grams/gigajoule (g/GJ). By-products are slag, fly ash, sulfur, and salt. The coal consumption at full load is about 2000-ton/day (585 MW equiv) using dry ground coal. The gasification pressure is ~2.8 megapascals (MPa or 28 bar); reaction temperature, about

1500°C; steam pressures of 12.5, 4, and 0.8 MPa (125, 40, and 8 bar); and sulfur production, ~5000 tons/year.

7. Gasifier Performance

Entrained-flow gasifiers can process a wide variety of coals, and data for several SCGP-1 coals are shown. In Table 5 compositional analyses; Table 6 shows ash minerals; and Table 7 presents a summary of gasification parameters for the different feeds. These feeds range from high rank bituminous coals from Appalachia, eg, Pike County coal, to low rank Texas lignite and petroleum coke (24–31).

Operating Parameters. The primary gasifier operating parameters are coal composition, coal throughput, oxygen/coal ratio and steam/oxygen ratio. The amount of oxygen and steam fed to the gasifier depends on the coal composition. In general, low rank coals are very reactive and require less oxygen and little to no steam, whereas high rank coals are relatively unreactive, requiring more oxygen and a moderate amount of steam. Steam provides an alternative source of oxygen for the gasification reaction and helps to moderate the gasification temperature. As a source of hydrogen, steam also helps to balance the H_2/CO ratio, giving a constant syngas composition for all coals including petroleum coke. Gasifier performance is evaluated in terms of syngas production and composition, carbon conversion, and cold gas efficiency (see Table 7).

Cold Gas Efficiency (CGE). Cold gas efficiency, a key measure of the efficiency of coal gasification, represents the chemical energy in the syngas relative to the chemical energy in the incoming coal. Cold gas efficiency on a sweet gas basis is calculated as the percentage of the heating value in coal that is converted to clean product syngas after removal of H_2S and COS.

Carbon Conversion. Carbon conversion on a once-through basis is a function of the coal composition and is strongly influenced by the oxygen/coal ratio. For some coals, the level of steam in the blast also affects the conversion pattern. Another factor is fly ash recycle, which raises the carbon conversion by recycling the unconverted carbon, most of which resides on the fly ash. This results in an overall carbon conversion greater than 99%.

Gas Composition and Heating Value. In Table 7 the syngas composition for a number of feedstocks is presented. These numbers reflect the composition of the gasifier off-gas on a dry basis. The primary gas components are CO and H_2, ranging from 59 to 67% and from 25 to 31%, respectively. Generally the gas composition is constant within a fairly narrow band for all coals including petroleum coke. The moderate variation is primarily because of variation in the CO_2 concentration caused by different steam/oxygen levels in the blast and oxygen/moisture and ash-free (MAF) coal ratios. The HHV of the product syngas after removal of H_2S, COS, and CO_2 is typically 12 MJ/m^3 (300 Btu/ft^3) and does not change significantly with changes in feedstock or gasifier conditions. The product syngas, also called medium-Btu gas (MBG), makes an excellent fuel for commercial gas turbines.

Heat Balance. Mass and heat balances are calculated around the gasification block, which includes the gasifier, quench, syngas cooler, and solids removal systems. A typical heat balance for Pike County coal is shown in

Table 5. **Analyses of SCGP-1 Feedstocks[a]**

Component	Texas lignite	Pike County[b]	Pike County[c]	Dotiki	Newlands	El Cerrejon	Skyline	Robinson Creek	R
					Samples as received				
moisture	30.70	8.55	6.04	6.71	7.01	11.86	7.95	5.61	5.7
ash	16.96	6.87	10.71	8.36	14.14	7.75	8.80	7.23	13.
volatile	30.19	32.13	30.80	34.61	25.30	33.30	33.36	32.63	35.
fixed carbon	22.15	52.45	52.45	50.32	53.55	47.09	49.89	54.53	46.
total	*100.00*	*100.00*	*100.00*	*100.00*	*100.00*	*100.00*	*100.00*	*100.00*	*100*
					Dry basis samples				
ash	24.48	7.51	11.40	8.96	15.20	8.79	9.56	7.66	13.
carbon	56.22	79.40	75.24	74.47	71.49	74.71	74.44	78.31	69.
hydrogen	4.36	5.18	4.67	5.23	4.28	4.99	4.83	5.09	4.5
nitrogen	1.13	1.59	1.49	1.58	1.61	1.53	1.51	1.43	1.4
chlorine	0.08	0.20	0.12	0.23	0.09	0.02	0.02	0.16	0.1
sulfur	1.67	0.67	0.79	3.10	0.63	0.98	1.11	1.15	4.1
oxygen	12.06	5.46	6.29	6.44	6.70	8.99	8.52	6.20	6.2
total	*100.00*	*100.00*	*100.00*	*100.00*	*100.00*	*100.00*	*100.00*	*100.00*	*100*
					Other properties				
HHV[d] kJ/kg[e]	22839	32586	30749	31439	29225	30918	30523	32567	293
Hardgrove grindability	63	47	49	55	57	53	36	43	60

[a] Composition is given in units of wt% unless other units are indicated.
[b] Washed samples.
[c] Run-of-mine samples.
[d] HHV = higher heating value, on dry basis.
[e] To convert kJ to kcal, divide by 4.184.

Table 6. Ash Minerals Variability of SCGP-1 Feedstocks[a]

Component	Texas lignite	Pike County[b]	Pike County[cd]	Dotiki	Newlands	El Cerrejon	Skyline	Robinson Creek	R&D
					Ash mineral				
P_2O_5	0.16	0.06	0.31	0.25	1.36	0.20	0.30	0.28	0.3
SiO_2	48.82	52.30	53.22	51.10	50.89	58.94	51.20	50.92	471
Fe_2O_3	7.26	5.89.	728	13.31.	7.59	9.04	9.09	11.05	277
Al_2O_3	15.12	31.00	28.70	21.50	31.62	17.45	32.61	29.06	190
TiO_2	1.05	1.33	1.49	1.30	1.61	0.78	1.59	1.38	0.8
CaO	11.67	4.30	1.23	4.46	218	4.16	1.19	1.63	1.1
MgO	2.05	1.04	1.10	0.69	0.43	2.45	0.70	1.04	0.6
SO_3	10.89	1.18	0.75	4.05	0.94	3.97	0.94	1.05	0.9
K_2O	1.06	2.24	3.32	2.34	0.51	1.83	1.85	2.74	1.6
Na_2O	0.42	0.27	0.36	0.44	0.12	0.53	0.21	0.52	0.1
Total	99.49	99.61	97.86	99.44	97.23	99.36	99.66	99.67	99.8
					Ash content				
ash (MF), mean%	24.48	7.50	hAG	8.96	15.20	8.79	9.56	7.66	13.8
standard deviation	3.31	0.72	1.09	0.42	0.99	1.18	2.97	2.17	1.9
standard deviation, %	13.52	9.60	9.56	4.69	6.51	13.42	31.07	28.33	144

[a] Composition is given in units of wt%.
[b] Washed sample.
[c] Run-of-mine sample.
[d] Sample also contains 71.77 wt% V_2O_5 and 7.37 wt% NiO.

Table 7. **Summary of SCGP-1 Gasification Performance**

Parameter	Texas lignite[a]	Texas lignite[b]	Pike County[c]	Pike County[d]	Dotiki	Newlands	El Cerrejon	Skyline
coal to plant, t/d	335	248	154	175	166	171	246	179
oxyge/MAF–coal ratio	0.877	0.865	1.006	0.974	0.970	0.986	0.922	0.955
burner steam/oxygen ratio			0.141	0.108	0.128	0.089	0.141	0.122
gasifier off-gas, vol %[e]								
CO	60.59	61.82	63.08	64.43	62.05	65.30	63.41	63.17
H_2	28.20	28.01	29.81	30.14	30.33	26.90	30.78	29.24
CO_2	5.38	4.47	2.59	0.68	2.47	2.26	1.68	1.99
$H_2S + COS$	0.71	0.80	0.24	0.36	0.90	0.28	0.27	0.34
$N_2 + Ar + CH_4$	5.08	4.83	4.18	4.34	4.18	5.19	3.83	5.21
sweet syngas, ka/t	12062	10381	12213	13230	12275	12116	16707	13437
HHV[f] energy basis, GJ/h[g]	155.2	130.9	158.6	173.0	161.7	153.2	227.2	172.0
sulfur removal, %[h]	99.1	99.8	99.7	99.8	99.8	99.5	98.6	99.5
carbon conversion, %	99.7	99.4	99.9	99.1	99.9	99.7	99.6	99.9
cold gas efficienty, % HHV[f] (sweet gas basis)	78.8	80.3	80.9	83.0	80.1	80.3	83.4	82.4

[a] High ash content.
[b] Low ash content.
[c] Washed samples.
[d] Run-of-mine samples.
[e] Dry gas.
[f] HHV = higher heating value.
[g] To convert J to cal, divide by 4.184.
[h] From syngas.

803

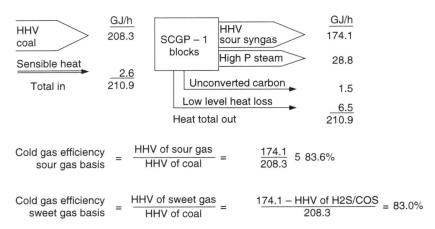

$$\text{Cold gas efficiency}\atop\text{sour gas basis} = \frac{\text{HHV of sour gas}}{\text{HHV of coal}} = \frac{174.1}{208.3}\,5\,83.6\%$$

$$\text{Cold gas efficiency}\atop\text{sweet gas basis} = \frac{\text{HHV of sweet gas}}{\text{HHV of coal}} = \frac{174.1 - \text{HHV of H2S/COS}}{208.3} = 83.0\%$$

Fig. 15. SCGP-1 heat balance for run-of-mine Pike County coal. To convert GJ/h to Btu/h, multiply by 9.48'105.

Figure 15. Input streams are HHV of the incoming coal and sensible heat of the coal, oxygen and steam blast. The output streams are HHV of the sour syngas obtained from gas analysis, HHV of unconverted carbon obtained from analysis and weights of solids, heat recovered in the steam system in the gasifier and syngas cooler, and low level heat representing unrecovered sensible heat in the syngas. The low level heat is calculated by difference, thus forcing the heat balance to 100%, and it is typically 3–4% for all feedstocks. Thus in most cases at least 95% of the energy of the feed streams, mostly heating value of the coal, is converted to usable energy in the form of syngas and high pressure steam.

8. Gasification Systems

8.1. System Configurations. The flexibility of gasification technology allows it to be integrated into a variety of system configurations to produce electrical power, thermal energy, fuels, or chemicals (Fig. 16). The heart of the system is the gasifier. It converts a carbonaceous feedstock (such as coal) in the presence of steam and oxygen (or air) at high temperatures and moderate pressure, into synthesis gas, a mixture of carbon monoxide and hydrogen (with some carbon dioxide and methane). Minerals (ash) in the feedstock separate and leave the bottom of the gasifier either as an inert glass-like slag or other marketable solid product. A small fraction of the ash becomes entrained in the syngas and requires downstream particulate removal. Any sulfur in the feedstock gets converted to hydrogen sulfide and along with ammonia, hydrogen chloride or other contaminants, needs to be removed to meet pollutant emission limits or predetermined levels necessary for further downstream processing. The cleaned synthesis gas is then combusted in a high efficiency gas turbine/generator to produce both electrical power and supply compressed air to the air separation unit that generate oxygen for the gasifier. The hot combustion gas from the turbine is sent to a heat recovery steam generator (HRSG), which in turn, drives a

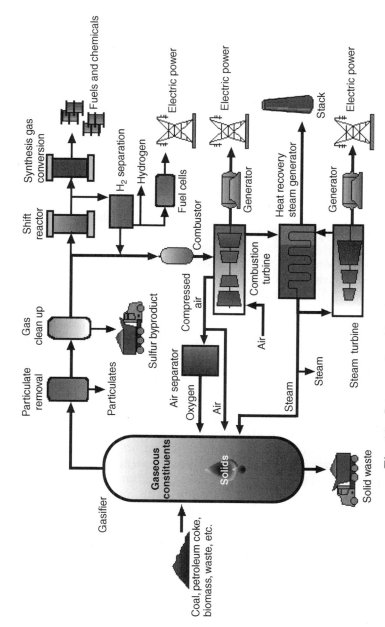

Fig. 16. Integrated gasification system configurations.

805

steam turbine/generator to produce additional electrical power. These plants are referred to as IGCC. This combined use of combustion and steam turbines significantly boosts plant efficiencies over single cycles. In this mode of operation, approximately two-thirds of the electricity is produced in the combustion gas turbine/generator. A variation of this concept, cogeneration, can produce high-grade heat or steam for specific applications.

The other significant mode that this technology can be configured into is coproduction. This term refers to the coproduction of power, fuels, or chemicals. Products can be produced by either processing the feedstock prior to gasification to remove valuable components or by converting the feedstock into synthesis gas and later into products. Although not highlighted in Figure 14, feedstocks such as coal can be extracted to produce valuable precursors for the manufacture of high strength, lightweight carbon fibers and anode coke for the manufacturing industry. The rejected carbon-containing materials are then gasified and converted to power or other products. In the second option, raw feedstocks are directly gasified and cleaned to produce synthesis gas that can be routed through the combined cycle and/or be catalytically converted into fuels or chemicals.

8.2. Attributes of Gasification Technology. Gasification has many positive attributes that make it a desirable technology for the production of power, fuels, and/or chemicals. Some of those attributes that have helped to stimulate the current market and provide for a promising future are as follows:

Fuel Flexibility. In general, gasification has the ability to utilize all carbon-containing feedstocks. In addition to primary fuels such as coal, gasification can process hazardous wastes, municipal solid waste, sewage sludge, biomass, etc, after proper preparation to produce clean synthesis gas for further processing. Because of its ability to use low-cost feedstocks, gasification is the technology of choice for many industrial applications such as the gasification of petroleum coke in refineries. The ability to cofeed opportunity fuels (low cost) gives gasification good market adaptability.

Product Flexibility. Gasification is the only technology that offers both upstream (feedstock flexibility) and downstream (product flexibility) advantages. Integrated gasification combined cycle, and gasification processes in general, is the only advanced power generation technology capable of coproducing a wide variety of commodity and premium products (eg, methanol, higher alcohols, diesel fuel, naphtha, waxes, hydrogen...) in addition to electricity, to meet future market requirements. It is this ability to produce value-added products that has made gasification economical in selected situations and will be a key driver in a deregulated power market.

Cleanup. Because gasification operates at high pressure with a reducing atmosphere, the products from the gasifier are more amenable to cleaning to reduce ultimate emissions of sulfur and nitrogen oxides as well as other pollutants than those from combustion processes. In general, the volume of the fuel gas processed in an IGCC plant for contaminant removal is typically one-third that from a conventional combustion power plant. Processing lower volumes of gas translates to lower capital cost for pollution prevention. The removal of sulfur, nitrogen, and other contaminants from the reducing gas is also much easier than from combustion systems. This results in sulfur and nitrogen oxide emissions being more than an order of magnitude less than those of conventional combustion processes. Gasification plants can also be configured to reach near-zero levels of emissions when required.

By-Product Utilization. Unlike that from combustion processes, the by-product ash and slag from the gasification technologies have also been shown to be nonhazardous. The material can be readily used for landfill without added disposal cost or can be used in construction materials or further processed to produce value-added products, leading to a zero discharge plant. Sulfur can also be readily removed and converted into elemental sulfur or sulfuric acid as a saleable product.

Efficiency. Compared to combustion systems, gasification is the most efficient and environmentally friendly technology for producing low cost electricity from solid feedstocks, and IGCC can be made to approach the efficiency and environmental friendliness of natural gas combined cycle plants. Further increases in efficiency can be achieved through integration with fuel cells and other advanced technologies. These higher efficiencies translate to lower operating costs, resource conservation, and lower carbon dioxide emissions (a global warming issue). In addition, the gasification process can be readily adapted with advanced technologies for the concentration of CO_2 with minimized impact on cost and thermal efficiency. The ability of a technology to achieve higher efficiencies and concentrate CO_2 with minimal impact on the cost of final products are major factors in technology selection for future energy plants.

System Flexibility. Gasification technology can be configured into a wide variety of systems to maximize efficiency, achieve fuel/product flexibility, or emphasize environmental performance. Although current cost for green field sites is high, gasification processes can be economically integrated into existing refineries and chemical plants. With proper integration and the use of existing infrastructure, the overall cost of a project can be significantly reduced. Through deployment in such environments, additional knowledge and experience will be gained, thereby reducing capital, operating, and maintenance costs for future facilities.

9. Environmental Performance

One advantage of modern IGCC systems is excellent environmental performance. Not only are regulatory standards met, but also emissions and effluents are well below acceptable levels (13,33,34). Regulations regarding pollutant emissions are becoming increasingly stringent. The need to reduce emissions of SO_2, NO_X, particulate matter, and other pollutants is principally governed by the Clean Air Act Amendment (CAAA) of 1990, which is a very complex law to interpret and is site specific. Some of the more major sections of the CAAA are as follows:

1. Clean Air Act (Including 1990 Amendments)

   Title I—Air Pollution Prevention and Control
   Part A—Air Quality and Emission Limitations
   Section 109—National Ambient Air Quality Standards

In July 1997 EPA promulgated new standards for particulate matter finer than 2.5 μm (PM2.5) and revised the ambient ozone standards.

Section 111—Standards of Performance for New Stationary Sources

Part C—Prevention of Significant Deterioration of Air Quality

Section 169A&B—Visibility protection for Federal Class I Areas

(EPA has issued new requirements to improve the visibility in National Parks and other Class I areas primarily through the reduction of fine particles).

Title III—Hazardous Air Pollutants

Section 3.02—Listing of 189 Hazardous Air Pollutants (HAPs)

Several HAPs are released as particles (eg, arsenic, cadmium, chromium, lead) or as acid gases (eg hydrochloric acid) during the combustion of coal.

Section 3.04—Promulgation of Maximum Achievable Control Technology standards

Section 3.11—Atmospheric Deposition to Great Lakes and Coastal Waters

Section 3.12—Specific Studies

Hazardous Air Pollutant Report to Congress

Hydrofluoric Acid and Uses

Section 3.15—Prevention of Accidental Release

Includes reporting of sulfur trioxide releases

Title IV—Acid Deposition Control

Section 404—Phase I Sulfur Dioxide Requirements

Section 405—Phase II Sulfur Dioxide Requirements

Section 407—Nitrogen Oxides Emissions Reductions Program

Another significant environmental issue is global warming. World carbon emissions are expected to reach 8 billion metric tons by 2010. Emissions from developing countries were about 60% of those from the industrialized countries in 1990, but by 2010, will surpass them with respect to carbon emissions. The sharp increase is expected to result from economic expansion, accompanied by increasing power demand, and by continued heavy reliance on coal, especially in Asia. There is worldwide concern over greenhouse gas production and general agreement that reduction would be prudent. However, reduction goals and mechanisms for reduction implementation have not yet been established.

9.1. Acid Rain Emissions. Integrated gasification combined cycle represents a superior technology for controlling SO_2 and NO_x emissions. Emissions are much lower than those from traditional coal combustion technologies (34). During gasification, the sulfur in the coal is converted to reduced sulfur compounds, primarily H_2S and a small amount of carbonyl sulfide, COS. Because the sulfur is gasified to H_2S and COS in a high pressure concentrated stream, rather than fully combusted to SO_2 in a dilute-phase flue gas stream, the sulfur content of the coal gas can be reduced to an extremely low level using well-established acid gas treating technology. The sulfur is recovered from the gasification plant as salable, elemental sulfur. A small quantity of sulfur can also be captured in the slag as sulfates.

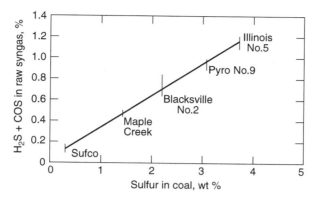

Fig. 17. Sulfur profile for SCGP-1 feedstocks. Overall sulfur removal is > 99.4% and the sulfur in the sweet syngas is < 20 ppm.

The gas treating and sulfur recovery processes employed in coal gasification have been broadly applied and operated for decades in refinery and petrochemical facilities and in natural gas sweetening plants. Operating experience from SCGP-1 (13,30) has confirmed that overall sulfur removal efficiencies of 99.4% from the raw syngas are achievable, independent of coal sulfur content for a variety of coals (Fig. 17). Modern pulverized coal (PC) plants generally have flue gas desulfurization (FGD) units capable of 95% sulfur capture. See Figure 18 for a comparison of emissions. New technologies are being developed for removing sulfur and other contaminants at high temperature. One hot-gas cleanup process uses metal oxide sorbents to remove $H_2S + COS$ from raw gas at high (>500°C) temperature and system pressure.

During coal gasification the nitrogen content of coal is converted to molecular nitrogen, N_2, ammonia, NH_3, and a small amount of hydrogen cyanide, HCN. In moving-bed gasifiers, some of the nitrogen also goes into tars and oils. The NH_3 and HCN can also be removed from the coal gas using conventional (cold) gas treating processes. Other techniques are being investigated in hot-gas

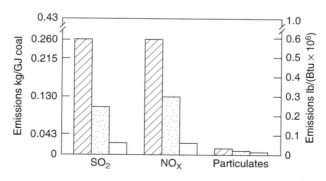

Fig. 18. Environmental emissions, where ▨ represents new source performance standards (NSPS) requirements; ▩ represents a pulverized coal (PC) plant; and □ represents SCGP-1.

cleanup technologies. After removal of HCN and NH_3, combustion of the coal gas in the gas turbine produces no fuel-based NO_x. Only a small amount of thermal NO_x is formed and this can be controlled to low levels through turbine combustor design and, if necessary, steam or nitrogen addition. Based on tests using SCGP-type coal gas fired in a full-scale GE-frame 7F combustor (35), a NO_x concentration of no >10 ppm in the gas turbine flue gas is attainable. See Figure 16 for a comparison of NO_x emissions from a PC plant equipped with low NO_x burners.

9.2. Criteria Air Pollutants. Moving-bed gasifiers produce tars, oils, phenols, and heavy hydrocarbons, the concentrations in the gas product are controlled by quenching and water scrubbing. Fluidized-bed gasifiers produce significantly lower amounts of these compounds because of higher operating temperatures. Entrained-flow gasifiers operate at even higher temperatures, typically in excess of 1650°C. SCGP-1 experience has confirmed that carbon conversions of >99.5% are easily attainable for any coal and that essentially no organic compounds heavier than methane are produced (35). Emissions of volatile organic compounds (VOC) from a IGCC plant are expected to be ~300 times lower than those from a similarly sized coal-fired steam plant equipped with low NO_x burners and an FGD unit.

The product gas after cleanup consists of primarily CO and H_2. Combustion of coal gas in high firing-temperature gas turbines converts virtually all of the CO to CO_2, and gas turbine exhaust is expected to contain no >10 ppm CO when operating at design conditions. Carbon monoxide emissions from an IGCC plant are thus expected to be around one-tenth those of a modern coal-fired plant equipped with low NO_x burners.

Particulate removal from the coal gas is effected either through a series of dry–solid and wet–solid removal steps or through the use of dry solids filters, so that the gas fed to the combustion turbine is essentially free of suspended particulates. The emissions of total suspended particulates (TSP) from an IGCC plant are about one-third those from a comparable pulverized coal plant equipped with a fabric filter and FGD unit.

9.3. Hazardous Air Pollutants. A number of the metals present in coal have the potential to be released as toxic air emissions. In moving- and fluidized-bed gasifiers, these metals are captured in the water. In entrained-bed gasifiers, a majority of these metals are captured in the slag. Because the coal ash in entrained gasifiers becomes vitrified at the high gasifier temperatures, the resultant glasslike slag encapsulates the metals in nonleachable form. In conventional treating systems using cold-gas cleanup, the small fraction of metals released to the gas phase is captured effectively in the gas cooling and gas treating steps. The combination of gas cooling and multistage gas–liquid contacting reduces very substantially the potential for airborne emissions of volatile metals such as lead, beryllium, mercury, or arsenic.

The total emissions of hazardous air pollutants from an IGCC plant having wet cleanup are expected to be at least an order of magnitude lower than those achievable from a modern coal-fired steam plant (37). Metals removal in hot-gas cleanup systems is still under development.

9.4. Water Consumption and Effluent Characterization. Another advantage of IGCC power generation is derived from lower water requirements. Because more than one half of the power generated in a IGCC plant comes from the gas turbine, the water requirement is only 70–80% of that required for a coal-fired power plant, where all of the power is generated from steam turbines.

Whereas moving-bed gasifiers require complex water-treatment systems to address tars, phenols, and metals, this complexity is mostly alleviated for fluidized-bed gasifiers and is eliminated for entrained-flow gasifiers. The exiting water streams of SCGP-1 contain no detectable amounts of volatile or semivolatile organics. The effluent from an IGCC facility can be biotreated to meet National Pollutant Discharge Elimination System (NPDES) standards (38). Biological treatment provides oxidation for the small amounts of inorganic nitrogen and sulfur species that remain in the water. Effluent from SCGP-1 has pH 7.8 and contains (see also Table 8).

Both acute and chronic toxicity testing of the treated effluent on daphnia shrimp and fathead minnows have indicated that the effluent is completely suitable for discharge into receiving waters with no adverse impact (38).

9.5. Solid By-Products. Coal gasification power generation systems do not produce any scrubber sludge, a significant advantage over both direct coal combustion processes that use limestone-stack gas scrubbers and fluidized-bed

Table 8. **The Biotreated SCGP-1 Effluent Contains Fully Oxidized Products and Very Low Concentrations of Trace Metals**

Chemical analysis	ppmwt
oil and grease	<1
phenols	<0.1
ammonia	<0.5
nitrite	<0.5
nitrate	50
total cyanide	<0.27
thiocyanate	0.1
formate	<0.1
thiosulfate	<0.1
sulfate	109
sodium	470
chloride	510
pH	7.8

Metal	ppmwt
antimony	0.01
arsenic	0.02
beryllium	<0.01
cadmium	<0.03
chromium	<0.01
copper	<0.01
lead	<0.03
mercury	<0.001
nickel	<0.02
selenium	0.35
silver	0.03
thallium	<0.001
zinc	0.03

combustion processes that use solid absorbents for sulfur capture. In coal gasification, the sulfur in the coal is recovered as bright yellow elemental sulfur for which there are several commercial applications, the largest being in the phosphate fertilizer industry (see FERTILIZERS). Elemental sulfur is a commodity traded worldwide, with 1990s prices in excess of $100/ton.

The ash in the coal is converted to slag, fly slag, or fly ash. Moving-bed and fluidized-bed gasifiers produce fly ash, which may be disposed of in a manner similar to that used for conventional power plant fly ash. In slagging gasifiers, the coal ash is mostly converted to a glasslike slag that has very low leach-ability. Environmental characterization of SCGP-1 slag and fly slag was performed for several coals using the extraction procedure (EP) toxicity tests and the toxicity characteristic leaching procedure test (TCLP), confirming that toxic trace metal concentrations in the leachate were well below Resource Conservation and Recovery Act (RCRA) requirements (39). Many of the elements, if present, were even below the detection limits. Additionally, the runoff from the slag storage area was collected and analyzed for comparison with the National Interim Primary Drinking Water Standards. The results of this comparison show that the measured values are typically much less than those allowed by the stringent national standard (39).

As part of a solids utilization program at SCGP-1, gasifier slag has been used as a principal component in concrete mixtures (Slagcrete) to make roads, pads, and storage bins. Other applications of gasifier slag and fly slag that are expected to be promising are in asphalt aggregate, Portland cement kiln feed, and lightweight aggregate (see CEMENT) (40). Compressive strength and dynamic creep tests have shown that both slag and fly slag have excellent construction properties.

9.6. CO_2 Emissions and Global Warming. The high coal-to-busbar efficiency of an IGCC system provides a significant advantage in responding to CO_2 emissions and thus to global warming concerns. High efficiency translates to lower coal consumption and lower CO_2 production per unit of electricity generated. The average existing PC unit has a heat rate of more than 10,550 kJ/kWh (10,000 Btu/kWh) on a higher heating value basis, which means that associated CO_2 emissions for a nominal 450-MW plant are well over 4 million tons/yr. The most efficient IGCC units offer heat rates of 8650 kJ/kWh (8200 Btu/kWh) and reduce CO_2 emissions by $\sim$15–20% relative to the emissions from a PC unit. Current IGCC units such as the Buggenum, The Netherlands, IGCC plant has reported even higher efficiency than this with a heat rate of 7935 Btu/kWh or 43% efficiency.

10. Economics

10.1. Gasification in the Evolving World Energy Market. Today's energy picture is colored by several recent world events. Western economic development has proved to have an enormous appetite for energy with the United States being the largest energy producer and consumer. Requirements for transportation fuel has outstripped domestic oil supplies, but a coal gas/liquid industry does not exist to offset any increases in demand. As a result secure international supplies of oil are considered critical to economic health. Thus, situations in oil-producing countries such as the Arab–Israeli wars in 1967 and 1973, the oil embargo in 1973, the Islamic revolution in Iran in 1979, the Iran–Iraq war in 1980, and the Iraqi attack of Kuwait 1990 have all resulted in keen interest from western countries and particularly responses from U.S. government. The continued development of a coal gasification industry, utilizing abundant U.S. domestic coal resources, could help to temper our responses to these threatening economic situations.

The power industry has also undergone dramatic changes over the past 50 years. The production of electric power has been dominated by coal-fired power plants. At the beginning of the twentieth century the power industry was initially developed as a public service through local, municipal, and regional power companies. By 1955 the power industry was well established. Power companies with approval of the Public Utilities Commission were able to build large power plants with scheduled costs and prices. The capacity and distribution of power was planned and orchestrated to achieve conservative margins of safety. The Clean Air and Water Legislation in the 1970s and 1980s have changed the way Electric Power is produced. Strong financial incentives and penalties were provided to develop cleaner power systems. Large public power utilities were slow to convert their existing plants to meet these new stricter restrictions in sulfur oxides, nitrogen oxides, and particulate emissions. There was public resistance to the huge costs associated with the cleanup systems required. As a result, new construction of power plants slowed in favor of retrofitting existing plants and excess capacity dropped to dangerously low levels. With deregulation of the power industry this situation continued unabated until the blackouts in California in the summer of 2001.

Nuclear power was expected to be the answer to clean and inexpensive electricity. However, scientist began to understand the long lasting hazards of radiation and costs for these power plants began to rise to meet safety regulations. Finally, a nuclear meltdown at the Three Mile Island power plant in 1979 and an explosion at Chernobyl in 1986 essentially halted further expansion of nuclear power plants. In addition, these incidents along with finding and cleanup of industrial waste sites raised the general public awareness of potential hazards and environmental concerns about pollution from all types of power plants.

To encourage development of more economic and cleaner power plants, the power industry was deregulated in the 1990s allowing independent power producers to compete with public utilities. Combined cycle plants, coal waste burners, and cogeneration facilities were now economically competitive and were built in the 1990s on a smaller scale to take advantage of special situations in various localities. However the changing nature of environmental legislation and the uncertainty that this produces have stifled the construction of new power plants. As a result, by the year 2000 the excess capacity of electric power generating facilities in the United States has fallen to <5%. The combined effects of deregulation and lack of power generating facilities and inadequate transmission lines led to California's electric power outages in the summer of 2001.

Coal power systems historically have raised steam and expanded it through a steam turbine to produce electricity. This process is limited in its efficiency because of the high heat requirement necessary to convert water from a liquid to steam. This heat of vaporization amounts to 540 cal/g of water. Nearly two-thirds of the heat needed for the steam cycle is lost to raising steam and cannot be recovered as useful power. However gas turbines, which can use synthesis gas to generate power, do not have this inherent inefficiency.

Gas turbines, originally built for jet engines, have been improved to the efficient high temperature machines of today. In the gas turbine the release of chemical energy by the production of gases and thermal expansion of those gases under pressure is more completely recovered than from a steam turbines. Fuel cells, which can use synthesis gas or hydrogen, are also being developed and promise to provide the most efficient conversion of chemical energy to electricity. These technologies will provide the driving force for the next generation of highly efficient electric power plants. Coal gasification is uniquely posed to take advantage of these new technologies based upon the reliable abundant coal supplies.

Coal, the primary fuel for electricity generation in the United States and other countries, is expected to have an increasing role in the future. Conventional coal-fired electricity generation has resulted in numerous environmental problems, notably emissions of sulfur and nitrogen compounds, both of which have been linked to acid rain, and emissions of particulates (see AIR POLLUTION). Conventional coal-combustion technologies only partially solve these problems. Modern coal gasification combined cycle power generation technologies, also known as IGCC systems, present electric power producers with important options and opportunities to improve efficiency, environmental performance, and overall cost effectiveness.

10.2. Increasing Power Demands. Possibly the single largest driver for the demand of gasification technology is electrical power. In the United States, the Energy Information Administration (EIA) predicts that electricity demand in the commercial and industrial sectors will grow by 2.0 and 1.4% per year, respectively, between 1999 and 2020 (41). This translates to upward of 700 GW of new capacity, repowering, and replacement of existing power plants. It is likely that a large majority of this power (estimated at >85%) will come for natural gas-fired turbine generators due to their low capital cost and high efficiency. However, over half of all electricity in the United States is produced by coal-fired power plants (41). A great opportunity exists to retrofit and repower many of these existing plants with clean coal technology. It has been estimated that 40,000 MW of increased electricity production is possible over the next 3 years by retrofit and repowering of the existing coal fleet with state-of-the-art technologies (42).

Internationally, the market for power generation is somewhat different. By 2020, worldwide electricity demand is expected to be more than triple to 23 trillion kWh. The greatest gains are expected in developing Asia and in Central and South America. In Asia, electric power production cannot keep pace with economic growth and electricity shortages are common. Because these nations have large indigenous coal supplies, economical and environmentally acceptable coal-fired power plants have tremendous market potential with the initial commercial units being mostly conventional technology. As environmental requirements become more stringent and more capital becomes available to those markets, opportunities will be available for advanced coal-based technologies. In contrast to the vast coal resources in Asia, Central and South America currently rely heavily on hydropower for electricity generation, and new capacity will likely be fueled by the large reserves of natural gas in the region.

There are also several niche industrial markets that require power in which gasification could play a predominant role. For example, the refining industry generates large quantities of high Btu content waste streams that could be gasified to generate power, steam, and synthesis gas. Refineries are heavy users of power and steam, are large users of hydrogen and are often located in close proximity to large chemical complexes (potential customers for both the power and the synthesis gas coproduct). One estimate places the worldwide refining market for gasification over the next 5 years as high as 4000 MW (~2500 MW of power and the remainder to synthesis gas). The near-term market is in China and Japan where there is growth in new refining capacity and an expanding need for more residual upgrading. The need to replace base-load power generation capacity creates a domestic market for gasification refinery applications ~2010.

Another niche market example is in the pulp and paper industry. Much of the black liquor and biomass-fueled steam turbine cogeneration capacity in the U.S. pulp and paper industry will need to be retired and replaced over the next 20 years. Gasification could be employed to efficiently convert the biomass waste into steam and power and to recover chemicals. In 1994, 1.2×10^{18} J (or 38,000 MW) of black liquor were consumed by the U.S. pulp and paper industry. If biomass gasifiers were coupled with the Tomlinson boilers to cogenerate power and steam from this amount of black liquor, an additional 22 billion kw-h/year (or nearly one-half of the 51 billion kWh that the total industry purchased in 1994) could be generated. Gasification must compete against FBC in this market; however, gasification has the advantage of generating synthesis gas fuel for kiln operation in addition to being able to destroy hazardous chemicals and reduce sulfur emissions.

10.3. Deregulation. In the United States, deregulation is expected to completely restructure the U.S. electric utility industry. New capacity additions will likely favor low capital-cost power generation technologies. Competition will force utilities to increase utilization rates at existing plants, maximize efficiency, and minimize operating costs. Aging plants nearing retirement may be kept open if their operating costs are proven to be competitive, and plants with high fixed costs will be forced to close. Traditional "electric only" utility generators will see their market share dwindle as energy firms (firms that produce a variety of energy products such as steam, chemicals, fuels and synthesis gas) capture an increasing volume of electricity sales. Many niche industrial applications that integrate electric power generation with industrial processes to provide gains in productivity, environmental performance, and capital utilization will become attractive for gasification technologies.

Unless fuel price or availability at specific locations dictates otherwise, natural gas-fired turbine generators are the preferred low-cost technology. The capital cost for a natural gas-fired combined cycle plant is $400–600/kW, about one-half the cost of an IGCC plant that gasifies coal. IGCC is capital intensive; it needs economies of scale to be an attractive investment option. However, IGCC costs can be improved by integrating in a synergistic way with various industrial applications. For example, gasification can operate on low cost opportunity feedstocks, it can be used to convert hazardous waste into useful products reducing or eliminating waste disposal costs, and it can coproduce power, steam, and high value products for use within the host plant or for export.

10.4. Fuel Reliability and Affordability. Numerous factors threaten the reliability and affordability of electricity throughout the United States. In the past year alone, rolling blackouts have become commonplace in California and skyrocketing electricity prices have shocked the state's $1.3 trillion dollar economy—the sixth largest on earth (43). Other parts of the country are not immune from these effects. The demographics of population growth, demand increases, transmission bottlenecks, supply gaps, and an aging infrastructure are creating reliability problems in power generation markets across the country. The likelihood of power outages continues to increase and rolling blackouts like those experienced in California could occur elsewhere. Additionally, current environmental regulatory constraints and anticipated new, more stringent regulations pose potential hurdles and longer lead times for siting and permitting of new power plants.

Fuel diversity is the best way to insure reliable and affordable electricity. A diverse fuel mix helps keep electricity prices low by insulating companies and consumers from issues related to fuel variability, fuel price fluctuations, regional shortages, regional disruptions and changes in regulatory policies. Regional reserves, fuel transportation infrastructure and fuel cost often dictate the best fuel mix for a given region. These factors have contributed to substantial regional fuel diversity throughout the United States, which is recognized by power producers as an important aspect to reliable and affordable electricity throughout the United States (44). Technologies must be available to accommodate a diverse fuel mix and coal plays a dominant role in the fuel mix throughout the United States.

Figure 19 depicts projected fuel prices through 2020 (41). The cost for coal is projected to be very stable or slightly lower throughout that period, while natural gas and fuel oil are projected to increase in cost. To evaluate the potential impact of fuel price on technology selection, a recent study was sponsored by the Department of Energy. Figure 20 illustrates the effect of increasing natural gas costs on the cost of electricity for natural gas combined cycle and coal-fired power plants.

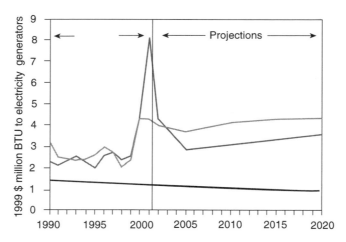

Fig. 19. Fuel price projections through 2020.

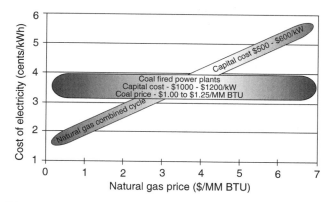

Fig. 20. Effect of natural gas price on cost of electricity.

10.5. Status of Gasification Technology. The worldwide activity in gasification technology has increased significantly over the past 30 years (Fig. 21). Nearly all of the gasification capacity through the mid-1970s can be attributed to the 19 Lurgi gasifiers operating at Sasol in South Africa. The relatively large increases in capacity in the latter part of the 1970s and the early 1980s represent the startup of 80 gasifiers associated with Sasol II and III, representing a combined increase of nearly 8300 mega Watt thermal (MWth) of synthesis gas capacity. A lesser, but notable, increase in capacity also occurred in the early 1980s with the commissioning of 14 Lurgi gasifiers at the Dakota Gasification plant in Buelah, North Dakota, adding another 1500 MWth of capacity. Following this, capacity remained relatively flat for over a decade. However, within a few short years due to deregulation and an increased need to more fully utilize the resources available, capacity increased by almost 50% and is expected to grow by nearly 60% in the next 5 years. Currently, there are >385 gasifiers in operation or under construction at 138 sites located in 22 nations in North and South America, Europe, Asia, Africa, and Australia. The largest 30 gasification plants in the world are listed in Table 9. This table also lists the principal product for each.

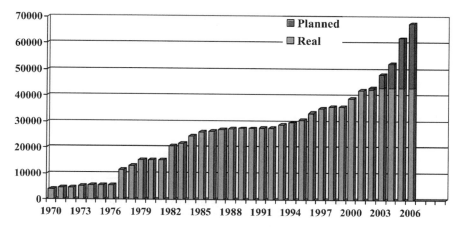

Fig. 21. Worldwide gasification plant activity.

Table 9. Thirty Largest Gasification Plants Worldwide

Plant owner	Country	Gasifier	Year	MWthOut	Fuel	Product
Sasol Chemical Ind. (Pty.) Ltd./Sasol Ltd.	South Africa	Lurgi	1977	5,090.0	bit.coal	FT liquids
Sasol Chemical Ind. (Pty.) Ltd./Sasol Ltd.	South Africa	Lurgi	1982	5,090.0	bit.coal	FT liquids
Dakota Gasification Co.	United States	Lurgi	1984	1,900.3	ligruite and ref. resid.	SNG and CO_2
SARLUX srl	Italy	Texaco	2001	1,216.7	visbreaker residue	electricity, H_2 and steam
Shell MDS (Malaysia) Sdn. Bhd.	Malaysia	Shell	1993	1,032.4	natural gas	mid-distillates
Mitteldeutsche Erdoöl-Raffinerie GmbH	Germany	Shell	1985	984.3	visbreaker residue	H_2, methanol and elec.
ISAB Energy	Italy	Texaco	2000	981.8	ROSE asphalt	electricity, H_2 and steam
Sasol Chemical Ind. (Pty.) Ltd./Sasol Ltd.	South Africa	Lurgi	1955	970.6	bit. coal	FT liquids
Global Energy, Inc.	Germany	Lurgi	1964	848.3	municipal waste	electricity and methanol
Nippon Petroleum Refining Co.	Japan	Texaco	2003	792.9	vac.residue	eletricity
Millenium (Quantum)	United States	Texaco	1979	656.2	natural gas	methanol and CO
Hydro Agri Brunsbuüttel	Germany	Shell	1978	642.5	heavy vis. residue	ammonia
Shell Nederland Raffinaderij BV	Netherlands	Shell	1977	637.3	visbreaker residue	H_2 and electricity
Sokolovska Uhelna, A.S.	Czech Republic	Lurgi	1996	636.4	liguite	electricity and steam
Global Energy, Inc.	United States	E-Gas	1995	590.6	petcoke	electricity

Company	Country	Licensor	Year	Capacity	Feedstock	Product
VEBA Chemie AG	Germany	Shell	1973	587.8	vac. residue	ammonia and methanol
Elcogas SA	Spain	PRENFLO	1997	587.8	coal and petcoke	electricity
Motiva Enterprises LLC	United States	Texaco	2001	519.5	fluid petcoke	electricity and steam
api Energia S.p.A.	Italy	Texaco	2001	496.2	visbreaker residue	electricity and steam
Chemopetrol a.s.	Czech Republic	Shell	1971	492.1	vac.residue	methanol and ammonia
Demkolec BV	Netherlands	Shell	1994	465.9	bit. coal	electricity
Ultrafertil S.A.	Brazil	Shell	1979	451.1	asphalt residue	ammonia
Tampa Electric Co.	United States	Texaco	1996	451.1	coal	electricity
Shanghai Pacific Chemical Corp.	China	GTI	1994	410.1	bit.coal	fuel gas and town gas
Gujarat National Fertilizer Co.	India	Texaco	1982	405.3	ref.residue	ammonia and methanol
Esso Singapore Pty. Ltd.	Singapore	Texaco	2001	363.6	residual oil	electricity, H_2 and steam
ExxonMobil	United States	Texaco	2000	347.2	deasphalter pitch	syngas
BASF AG	Germany	Texaco	1974	341.8	vac. resid. and fuel oil	methanol
China National Petrochem. Corp./Sinopec	China	Texaco	1988	341.8	visbreaker residue	gases
Quimigal Adubos	Portugal	Shell	1984	328.1	vac.residue	ammonia

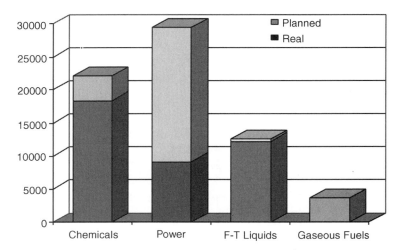

Fig. 22. Gasification by application.

A tremendous rise in capacity is expected to continue beyond 2005. Most of the capacity growth for gasification technology over the next several years appears to be power-based (Fig. 22). Both deregulation in the United States and demand growth worldwide appear to be the principal drivers. With this capacity increase, the use of synthesis gas for the production of electricity is approaching that for the production of chemicals.

To meet this need, an additional 33 plants with 48 gasifiers are expected to be constructed in the next 5 years, which would add another 18,000 MWth of synthesis gas capacity (45). A significant fraction of this expected growth will occur in the developing nations in the Pacific Rim as the need for further electrification of these nation's economies grow (Fig. 23). Western Europe will also see large generation increases where refineries need to fully utilize available

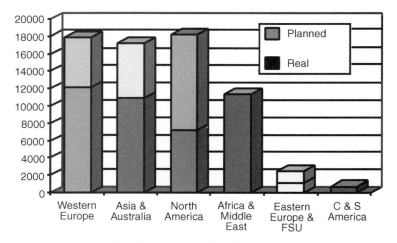

Fig. 23. Gasification by region.

feedstocks while reducing fuel oil production. Major capacity addition is expected in North America and will be concentrated in the refining industry as well. Very little or no growth is anticipated to occur in Africa or other regions of the world (see Table 9).

With regard to fuel, coal and petroleum residuals are by far the dominant feedstocks and account for >70% of the synthesis gas capacity. The major users of coal for synthesis gas production are Sasol, Dakota Gasification, and the current IGCC demonstration projects being sponsored by the U.S. Department of Energy. Natural gas is also an important feedstock for synthesis gas production, accounting for ~20% of today's capacity, and is used almost exclusively in the production of chemicals. Over the next several years, most of the growth in capacity will be from the gasification of coal and petroleum residuals, with a small fraction from petroleum coke. The growth in these feedstocks, however, will be used primarily to produce electricity. Coal, petroleum, and petcoke are the predominant feedstocks (Fig. 24) as indicated by the projected plants. No capacity additions for synthesis gas production from natural gas are projected.

10.6. Future Gasification Technology Development. *Power Generation.* For many years, government and industry have worked to develop the concept of integrating coal gasification with clean, efficient gas and steam turbines to create IGCC and various hybid power systems. In the United States, IGCC technologies are now being demonstrated as part of the U.S. Department of Energy's (DOEs) Clean Coal Technology (CCT) Demonstration Program. The significant strides in this program together with those from plants operating in the Netherlands and Spain have successfully demonstrated the performance of these coal-based power generation facilities. However, the capital cost of such advanced new plants and the risks involved with first-of-a-kind facilities and

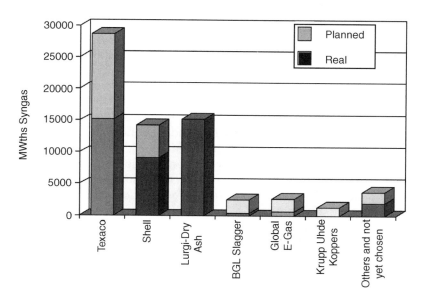

Fig. 24. Gasification by primary feed.

Table 10. **Impact of Technology Advances on IGCC Cost and Performance**

Gasifier "gas turbine" plant start-up	Destec "F" (2000)	Destec "G" (2005)	Destec "H" (2010)	Transport 2010
coal feed, t/day	4319	2449	2781	2552
oxygen feed, 95%, t/day	3592	1842	2027	N/A
no. of gasifiers trains	2	1	1	2
gas cleanup	amine	transport HGD	transport HGD	*In situ* desulf. transport/HG polisher
gas turbine (number)	GE7001Fa (2)	Westinghouse G (1)	GE "H" (1)	GE "H" (1)
gas turbine output, MWe	394	263	335	271
steam turbine output	255	141	155	140
net power, MWe	543	349	427	398
efficiency, HHV, %	40.1	45.4	49.0	49.7
total plant cost, KS, $/kW	1241	1229	1087	961
production cost, mills/k Wh	14.7	14.5	13.3	13.0
cost of electricity, mills/k Wh	38.8	37.6	33.8	31.2

technologies will likely delay the widespread application of IGCC for power generation until the next decade.

Although estimates for IGCC applications range from $1200–1500/kW, there is promise that costs may be reduced to below $1000/kW making coal-fired IGCC a more attractive alternative to gas-fired gas turbine generators. A recent study sponsored by the DOE analyzed the impact of technology improvements on overall cost and performance of an IGCC system (46). In Table 10 the impact is presented in terms of the evolutionary improvements with time in gas turbine technology (ie, F-, G-, and H-class turbines), gasification technology (Destec, renamed E-Gas, and transport gasifiers) and integration improvements. As apparent by inspecting Table 10, these evolutionary improvements have a significant impact on thermal efficiency and capital and O&M costs, ultimately reducing the cost of electricity. The nearly 10% increase in thermal efficiency due to the use of hot gas desulfurization and advanced turbines results in a 20% reduction in the emission of carbon dioxide per kWe generated (47).

The costs presented in Table 10 represent those for the first plants built with the indicated technologies and are likely to be somewhat higher than that for subsequent plants using the same technologies. Cost reductions and performance improvements are being made and will continue to be made as increased operating experience leads to improved, optimized process designs. Figure 25 provides an indication of what cost reductions might be expected from deployment of successive plants. The slope of the "learning curves" is representative of cost reductions associated with the deployment of other advanced technologies in the marketplace in the recent past. This figure shows that through successive deployment of the technology, costs may be reduced to <$1000/kWe. As shown by

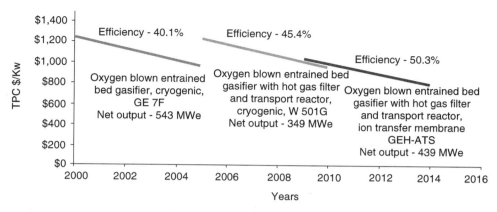

Fig. 25. Gasification cost and efficiency projections for various process improvements.

the curve on the right, potential exists for achieving a cost of about $850/kWe, that which is considered by industry to be competitive with Natural Gas Combined Cycle (NGCC).

Coproduction. In the absence of niche market applications such as the petroleum industry or in the case of very high natural gas prices, gasification for power production will likely not occur to an appreciable extent until capital costs are significantly lowered. Identifying innovative opportunities will be critical to gasification's near-term market acceptance. For example, there continues to be a high demand for peaking and intermediate load capacity. Combining coproduction of chemicals and power in a gasification facility allows the flexibility to maximize power generation during peak demand periods and maximize chemicals production during off-peak periods, making gasification more attractive to the existing power market. It also allows flexibility in geographical location based on local grid peak-shaving needs. Also, combined power and methanol plant requires less capital than separate power and methanol processes. Locating the integrated facility near a chemical complex provides a means of improving gasification economics by ensuring full utilization of the gasifier and exploiting synergies between the processes. Utilizing cofeeding of opportunity fuels or "waste" with coal while coproducing power and fuels and/or chemicals is another way to avoid waste disposal costs and diversify fuel supply mix as a hedge against volatile natural gas prices.

An analysis was conducted for the DOE that studied the economics of coproduction, and demonstrated the advantage of multiple revenue streams, ie, from sale of electric power, liquid fuels, and chemicals. In one scenario, it is assumed that the power generated by coproduction is sold at the same price as that generated by natural gas combined cycle. Thus, the price of natural gas sets the cost of power and establishes a benchmark for economic comparison. In addition, liquid fuels are valued at an assumed premium of $8/barrel over the reference world oil price of $21/barrel in 2010 as predicted by the U.S. Energy Information Agency. Based on a fixed return on equity (ROE) of 15% and a coal price of $30/ton, IGCC is competitive with natural gas at ~$3.75/million Btu. For comparison, by generating additional revenue, coproduction attains the same ROE at a lower natural gas price of about $3.25/million Btu. These figures illustrate the improved profitability of coproduction (48). Yet another study concluded that through proper integration, coproduction can offer higher process efficiencies with little added capital (49).

Technology Development Needs.   To reduce capital costs, increase efficiency and decrease emissions, technology development is required. The U.S. Government is undertaking a variety of research, development and demonstration projects with industry to accomplish those goals. Some of the general efforts being undertaken are as follows:

Gasification.   Advances in the gasifier itself to enhance efficiency, reliability, and feedstock flexibility and economics are crucial for gasification system improvements. Research is being conducted on advanced gasifiers, such as a high throughput transport gasifier, so higher performance goals can be reached and the variety of possible feedstocks can be further expanded. Advanced refractory materials and new process instrumentation are being developed to improve system reliability and availability, operational control, and overall system performance. Studies of alternative feedstocks (biomass and waste from refineries, industries, and municipalities) are being conducted to improve gasifier flexibility and utility. Data from fluid dynamic models are being used to develop and improve advanced gasification.

Gas Cleanup and Conditioning.   In the gas cleaning and conditioning area, the goal is to achieve near-zero emissions while simultaneously reducing capital and operating costs. Novel gas cleaning and conditioning technologies are being developed to reach this goal. Processes that operate at mild to high temperatures and incorporate multicontaminant control to parts per billion levels are being explored. These include a two-stage process for H_2S, trace metals, HCl, and particulates removal; membrane processes for control of H_2S, Hg, and CO; and sorbents/catalysts for NH_3 and H_2S control. Investigation of technologies for mercury removal is also underway. For removal of particulates, both ceramic and metallic filters are being assessed.

Gas Separation.   Advanced gas separation research offers the potential for substantial improvement in environmental and cost performance. It has been estimated that the cost of a conventional cryogenic air separation subsystem of an oxygen-blown gasification plant is >15% of the total plant cost. Parasitic power losses for air compression are also quite high. Development efforts are underway on oxygen ion separation membranes. This technology will also potentially enhance process efficiency as well. Improved hydrogen recovery and CO_2 removal are also important. Efforts are underway for developing high-temperature ceramic membranes for hydrogen recovery from syngas streams, as well as low-temperature approaches for hydrogen recovery and carbon dioxide removal.

Product/By Product Utilization.   Markets and applications are being assessed for utilization of gasifier bottoms as a saleable by-product. Processes for near-zero discharge of wastes are being evaluated. Post-H_2S removal processes for the by-product production of sulfur and/or sulfuric acid are being researched. A single-step sulfur removal and by-product production process is also being evaluated.

Systems Analysis and Technology Integration. Plant design and system configurations for various market applications are being assessed. Novel approaches to technology integration, flexible fuel supplies, and product yields are being evaluated to achieve improved plant economics for gasification systems.

Gasification technologies will continue to play a major role in satisfying the increasing worldwide energy demand during this cmentury. Market forces and environmental concerns are resulting in the expanded use of integrated gasification combined cycle plants and hybrid systems (combined gasification and combustion). The gasifier vastly expands the fuel base beyond natural gas to include abundant and lower cost feedstock such as coal, biomass, and agricultural, forestry, and refinery wastes. It enables the separation of pollutants from the product gas and the reduction of greenhouse gases. These facilitate the clean and efficient generation of electricity and the production of chemicals and clean liquid fuels. The capability to coproduce electricity, chemicals, and liquid fuels makes the technology economically attractive to a broad range of industrial applications. Feedstock and product flexibility, high efficiency and near zero pollutant emissions make gasification the technology of choice for electricity generation and chemicals and liquid fuel production in the twenty-first century. Table 11 provides a list of upcoming or planned gasification plants in the near future.

Table 11. **Planned Future Gasification Plants Worldwide**

Plant Owner	Country	Gasifier	Year	MWth	Fuel	Product
Global Energy, Inc./Fife Electric	United Knigdom	BGL	2002	732.5	coal and sludge	electricity
IBIL energy Systems Ltd. (IES)	India	GTI U-GAS	2002	109.1	lignite	electricity and Steam
Global Energy, Inc.	United States	BGL	2003	727.0	coal and MSW	electricity and diesel
Global Energy, Inc.	United States	BGL	2003	727.0	coal and MSW	electricity and H_2
Indian Oil Corp. Ltd.	India	Shell	2003	888.6	petcoke	H_2 and electricity
Sokolovska Uhelna, A.S.	Czech Republic	HTW	2003	787.4	lignite	electricity
Sinopec/Shell	China	Shell	2003	466.2	coal	ammonia
AGIP Raffinazione S.p.A.	Italy	Shell	2003	456.6	visbreaker resid.	electricity and H_2
Calla Energy Partners, LLC	United States	GTI U-GAS	2003	80.7	biomass	electricity
Sistems de Energia Renovavel	Brazil	TPS	2003	68.4	biomass	electricity
ATI Sulcis	Italy	Shell	2004	956.9	coal	electricity
Sinopec/Shell	China	Shell	2004	956.9	coal	electricity and syngas
Unspecified Utility Consortium	Japan	ICGRA	2004	585.4	coal	electricity
Sinopec/Shell	China	Shell	2004	478.5	coal	ammonia
Sinopec/Shell	China	Shell	2004	466.2	coal	ammonia
Waste Management & Processors, Inc.	United States	Texaco	2004	410.1	ant. culm	diesel and electricity

Table 11 (*Continued*)

Plant Owner	Country	Gasifier	Year	MWth	Fuel	Product
Koa Oil Co. Ltd.	Japan	Texaco	2004	287.1	petcoke Electricity	
Boise Cascade Corp.	United States	GTI U-GAS	2004	102.5	biomass	electricity
Port of Port Arthur/Sabine Power I, Ltd.	United States	E-GAS	2005	2,029.4	petcoke	electricity
Petronor (Repsol-YPF)/Iberdrola (PIEMSA)	Spain	Texaco	2005	1,654.1	vac. residue	electricity and H_2
TECO Power Services Corp./ Citgo/Texaco	United States	Texaco	2005	1,406.7	petcoke	elect, H_2 and steam
Eagle Energy (TECO Power Services/ Texaco)	United States	Texaco	2005	1,367.1	petcoke	electricity
Proj. IGCC Normandie (Total-FinalElf/EdF/ Texaco)	France	Texaco	2005	1,043.1	fuel oil	elect, H_2 and steam
Hindustan Petroleum Corp. Ltd.	India	Texaco	2005	649.3	petcoke	electricity
Refineria Gdanska SA	Poland	Texaco	2005	496.2	visbreaker resid.	elect, H_2 and steam
Unspecified owner	Unspec. Eur. Country	Shell	2005	478.5	residue	electricity
Agip Raffinazione S.p.A.	Italy	Texaco	2005	409.1	visbreaker resid.	electricity
Unspecified Owner	United States	Texaco	2006	2,761.4	coal	electricity
Shell Deer Park Refining Co.	Untied States	Texaco	2006	1,400.0	petcoke	elect, H_2 and steam
Dong Ting	China	Texaco	2006	1,170.8	coal	syngas
Beijing Coking	China	Texaco	2006	286.6	fuel oil	methanol
Netherlands Refining Co. BV	Netherlands	Texaco	2006	47.6	waste plastics	electricity and CO

11. Glossary

Agglomerating ash gasifier	Gasifier that maintains a sufficiently high temperature to promote ash agglomeration (eg, U-GAS)
Ash fusion temperature	Temperature at which the ash melts (°C)
ASTM	ASTM
BGC	British Gas Company
Bituminous coal	ASTM coal ranking for coal with heating value >10,500 BTU/lb and Volatile Matter content >14%

Blue Water gas	Product of cyclic gas generators that produce a blue
Caking coal	Coal that become sticky upon heating
Clean Air Act	A federal law enacted in United States in 1990 that entrusts the EPA to set limits on how much of a pollutant can be in the air anywhere in the United States.
CAAA	Clean Air Act Amendment
CCT	Clean Coal Technology Demonstration Program: A U.S. department of energy initiative.
Clean Water Act	A federal law enacted in United States in 1972 to protect the nation's waters, including lakes, rivers, aquifers and coastal areas.
Cold Gas Efficiency	The ratio of the chemical energy in the syngas relative to the chemical energy in the incoming coal.
Cogeneration	Production of high grade heat or steam in addition to electric power.
Co-production	Production of fuels or chemicals in addition to electric power.
Criteria Air Pollutants	VOC
Cyclic Gas generator	Gas generators cyclically operated in combustion (with air) and gasification (with steam) modes.
Destec gasifier	Entrained flow gasifier originally developed by Shell Oil Co.
Deregulation	Changes enacted by several U.S. states to make electric industry open to competition, giving consumers the power to choose their electricity provider.
Devolatilization	The process of volatile matter release from coal as it is heated.
Direct hydrogenation	Obtaining liquid fuels from coal by heating pulverized coal with oil under high pressures and with hydrogen.
Direct liquefaction	Same as direct hydrogenation.
Dry ash gasifier	Gasifier in which the ash temperature is maintained below ash fusion temperature.
E-Gas	Same as Destec gasifier
EIA	U.S. Energy Information Administration
ENCOAL	Mild Coal Gasification Project
Entrained flow gasifier	Gasifier in which the coal particles are entrained with the gasifying agent to react in a cocurrent flow.
EPA	U.S. EPA
EP	Extraction Procedure (toxicity tests)
FBC	Fluidized bed combustor
Fisher-Tropsch synthesis	Chemical process for producing liquid fuels from syngas
Fixed bed gasifier	A gasifier with a slowly moving packed bed of coal with a fixed height. Also known as moving bed gasifier.

Fusinite	Coal maceral derived from biodegraded or charred wood.
FGD	Flue gas desulfurization.
Fluidized bed Gasifier	A gasifier in which the bed of coal, char, and ash are fluidized.
HAPs	Hazardous Air Pollutants
HHV	Higher heating value
High rank coal	Coals with heating value greater than or equal to that of bituminous coal (>10,500 btu/lb)
HPC	Heavy paraffin cracking
HPS	Heavy paraffin synthesis
HRSG	Heat recovery steam generator
HYGAS process	Direct hydrogenation of coal in the presence of hydrogen and steam, under pressure, in two fluidized-bed stages
Hybrid power systems	Combined power systems employing both gasification to drive gas turbines and combustion steam boilers to fuel steam turbines.
IGCC	Integrated gasification combined cycle (steam and gas cycles)
Indirect liquefaction	Process of producing syngas from coal and then converting the syngas to liquid fuels
Koppers-Totzek gasifier	A commercialized entrained bed gasifier
KRW gasifier	Kellogg, Rust, Westinghouse gasifier. A fluidized bed gasifier tested in a pilot plant.
Liptinite	Coal macerals derived from hydrogen-rich sources such as spores, leaf cuticles, and algal colonies.
Low rank coal	Nonagglomerating coals with heating value less than that of bituminous coal (<10,500 btu/lb)
LHV	Lower heating value
Lurgi Gasifier	A commercialized fixed bed gasifier.
NPDES	National Pollutant Discharge Elimination System standards
NGCC	Natural Gas Combined Cycle
MAF	Moisture and ash free (analysis of coal)
MBG	Medium BTU gas (250–500 Btu/cu ft)
MF	Moisture free (analysis of coal)
Moving bed gasifier	Same as fixed-bed gasifier
MTG	Methanol to gasoline process
O&M costs	Operation and maintenance costs
Opportunity fuels	Low cost fuels readily available in certain markets (eg, biomass, municipal solid waste, sewage sludge, etc)
PC plants	Pulverized coal plants
Public Utilities Commission	State regulatory agencies for electric utility companies
Pyrolysis	Decomposition of coal with heat

Rank	A classification of coals describing the relative maturity of the coal; higher ranks being further evolved having less volatiles, higher carbon, and less hetero-atoms.
RCRA	Resource Conservation and Recovery Act establishes a framework for U.S. national programs to achieve environmentally sound management of both hazardous and nonhazardous wastes.
ROE	Return on equity
Repowering	Modification or enhancements to existing power plant to replace aging or outdated components or systems.
Retrofitting	Upgrading or improving an outdated technology using modern developments.
SASOL process	Synthetic liquid fuels plant operated by the South African Coal, Oil, and Gas Co.
SCGP	Shell coal gasification process
Slagging gasifier	A gasifier designed to melt and remove the coal minerals
Slagcrete	A concrete mixture with gasifier slag as a principal component.
SNG	Synthetic or substitute natural gas, consisting mainly of CH_4
Synthol process	A Fischer-Tropsch process developed at SASOL
SMDS	Shell Middle Distillate Synthesis process
Softening point	Temperature required for the onset of melting the coal ash
Sour syngas	Product gas containing acidic components such as H_2S
Steam gasification	Reaction of water and carbon to produce CO and hydrogen
Subbituminous coal	ASTM coal ranking for nonagglomerating coal with heating value >8,100 and <10,500 BTU/lb
sweet syngas	Product syngas with acidic components removed.
Syngas	A mixture of carbon monoxide and hydrogen (with some carbon dioxide and methane).
Tar	Complex condensable hydrocarbons produced from the heating of coal
TCLP	The standard test: toxicity characteristic leaching procedure test
TEXACO gasifier	Entrained flow gasifier commercialized by Texaco Oil Co.
TSP	Total suspended particulates
TWR	The tube wall reactor, a reactor that uses catalyst-coated tubes
Town Gas	Synthetic natural gas distributed through pipeline network to cities and town for street lighting and domestic uses.

Vitrinite   The most common coal maceral derived from woody
                                       tissue
VOC   Volatile organic compounds
Wellman-Galusha gasifier   A commercialized fixed bed gasifier
Winkler Gasifier   A commercialized fluidized bed gasifier

BIBLIOGRAPHY

"Gasification" under "Coal Chemicals and Feedstocks" in *ECT* 3rd ed., Supplement
pp. 194–215, by J. Falbe, D. C. Frohning, and B. Cornils, Ruhrchemie AG. "Coal Conversion
Processes, Gasification" in *ECT* 4th ed., Vol. 6, pp. 541–568, by Uday Mahagaokar and
A. B. Krewinghaus.; "Coal Conversion Processes, Gasification" in *ECT* (online), posting
date: December 4, 2000, by Uday Mahagaokar and A. B. Krewinghaus, Shell Development
Company.

CITED PUBLICATIONS

1. C.Y. Wen and S. Dutta, Rates of Coal Pyrolysis and Gasification Reactions, in C. Y.
   Wen and E. S. Lee, eds., *Coal Conversion Technology*, Addison-Wesley Publishing Co.,
   Reading Mass., 1979 pp. 57–170.
2. T. H. Fletcher, A. R. Kerstein, R. J. Pugmire, M. Solum, and D. Grant, *"A chemical
   percolation model for devolatilization: Summary"*, http://www.et.byu.edu/~tom/cpd/
   CPD_Summary.pdf.
3. R. Moreea-Taha, "Modeling and simulation for coal gasification", Report CCC/42, IEA
   Coal Research, London, U.K., December 2000.
4. R. F. Probstein and R. E. Hicks, *Synthetic Fuels*, McGraw-Hill Book Co., New York,
   1979, pp.1–256.
5. M. Syamlal, S. Venkatesan, and S. M. Cho, "Modeling of Coal Conversion in a Carbonizer", *Proceedings of Thirteenth Annual International Pittsburgh Coal Conference*,
   Vol. 2, in S.-H. Ciang, ed., University of Pittsburgh, Pittsburgh, Pa., September
   3–7, 1996, pp. 1309–1314.
6. D. Vamvuka, *Energy Exploration and Exploitation* **17**(6), 515 (1999).
7. G. A. Cremer, N. Hauser, and C. A. Bayens, paper presented at the *AICE Spring
   National Meeting*, Mar. 1990.
8. *Lignite to Methanol: An Engineering Evaluation of Winkler Gasification and ICI
   Methanol Gasification Route, AP-1592*, Electric Power Research Institute, Palo
   Alto, Calif., Oct. 1980.
9. P. F. H. Rudolph, "The Lurgi Process—The Route to SNG From Coal", *Fourth
   Synthetic Pipeline Gas Symposium*, Chicago, Ill., 1972.
10. J. R. Williams and G. A. Bekker, paper presented at the *Thirteenth International
    LNG/LPG Conference*, Kuala Lumpur, Malaysia, Oct. 1988.
11. J. N. Phillips, M. B. Kiszka, U. Mahagaokar, and A. B. Krewinghaus, "Shell Coal
    Gasification Project Final Report on Eighteen Diverse Feeds", *EPRI TR-100687*, 1992.
12. L. R. Radovic, P. L. Walker, Jr., and R. G. Jenkins, "Importance of Catalyst Disper-
    sion in the Gasification of Lignite Chars," *J. Catal.* **82**, 382 (1983).
13. L. J. Shadle, E. R. Monazam, and M. L. Swanson, *Ind. Eng. Chem. Res.* **40**, 2782
    (2001).
14. P. C. Richards and A. B. Krewinghaus, "Coal Flexibility of the Shell Coal Gasification
    Process", *Sixth Annual International Pittsburgh Coal Conference*, Sept. 1989.

15. L. G. Massey, in C. Y. Wen and E. S. Lee, eds., *Coal Conversion Technology*, Addison-Wesley Publishing Co., Reading Mass., 1979, pp. 313–427.

16. L. Shindman, in H. H. Lowry, ed., *Chemistry of Coal Utilization*, Volume II, John Wiley & Sons Inc. New York, 1945, pp. 1252–1286.

17. J. J. Morgan, Producers and Producer Gas, in H. H. Lowry, ed., *Chemistry of Coal Utilization*, Volume II, John Wiley & Sons, Inc. New York, 1945, pp. 1673–1750.

18. B. J. C Van Der Hoeven, in H. H. Lowry, ed., *Chemistry of Coal Utilization*, Volume II, John Wiley & Sons,, Inc., New York, 1945, pp. 1586–1672.

19. C. G. Von Fredersdorff, and M. A. Elliott, in H. H. Lowry, ed., *Chemistry of Coal Utilization, Supplementary Volume*, John Wiley & Sons, Inc., New York, 1953, pp. 892–1022.

20. D. Herbden and H. J. F. Stroud, in M. A. Elliott, ed., *Chemistry of Coal Utilization, Second Supplementary Volume*, John Wiley & Sons, New York, 1945, pp. 1602–1752.

21. H. H. Storch, in H. H. Lowry, ed., *Chemistry of Coal Utilization*, Volume II, John Wiley & Sons, New York, 1945, pp. 1751–1802.

22. E. R. Monazam and L. J. Shadle, "Predictive tool to aid design and operations of pressurized fixed bed coal gasifiers"., *Ind. Eng. Chem. Res.* **37**(1), 120 (1998).

23. J. Longanbach, E. Moorehead, G. Styles, and P. Vimalchand, *Mod. Power Systems*, (Nov. 37, 1996).

24. U. Mahagaokar and A. B. Krewinghaus, "Shell Coal Gasification Project, Gasification of SUFCo Coal at SCGP-1,*EPRI Report No. GS-6824*, Interim Report, May 1990.

25. U. Mahagaokar, A. B. Krewinghaus, and M. B. Kiszka, "Shell Coal Gasification Project: Gasification of Six Diverse Coals", EPRI Report GS-7051, Interim Report, Nov. 1990.

26. J. N. Phillips, U. Mahagaokar, and A. B. Krewinghaus, "Shell Coal Gasification Project: Gasification of Eleven Diverse Feeds," EPRI Report GS-7531, Interim Report, May 1992.

27. R. P. Jensen, U. Mahagaokar, and A. B. Krewinghaus, "SCGP—Progress in a Proven, Versatile, and Robust Technology", *Ninth EPRI Conference on Coal Gasification Power Plants*, Palo Alto, Calif., Oct. 16–19, 1990.

28. U. Mahagaokar and A. B. Krewinghaus, "Shell Coal Gasification Plant No. 1—Recent Results on Domestic Coals", Power-Gen '90 Conference, Orlando, Fla., Dec. 4–6, 1990.

29. U. Mahagaokar and A. B. Krewinghaus, "Shell Coal Gasification Process—Recent Performance Results on Drayton, Buckskin, Blacksville No. 2, and Pyro No. 9 Coals," *1990 International Joint Power Conference*, Boston, Mass., Oct. 1990.

30. U. Mahagaokar and co-workers, "Shell's SCGP-1 Test Program—Final Overall Results," *Tenth Annual EPRI Conference on Gasification Power Plants*, San Francisco, Calif., Oct. 1991.

31. J. N. Phillips and co-workers, "SCGP Recent Results on Low Ash Feedstocks Including Petroleum Coke," *Eighth Annual Pittsburgh Coal Conference*, Pittsburgh, Pa., Oct. 1991.

32. "Cool Water Coal Gasification Program," *Final Report EPRI GS-6806*, Dec. 1990.

33. D. G. Sundstrom and J. U. Bott, paper presented at the *Tenth Annual EPRI Conference on Gasification Power Plants*, San Francisco, Calif., Oct. 1991.

34. G. A. Cremer and C. A. Bayens, "Shell GCC Sets New Standards For Clean Power From Coal," *Alternate Energy 1991*, Scottsdale, Ariz., Apr. 1991.

35. R. P. Allen, R. A. Battista, and T. E. Ekstrom, "Characteristics of an Advanced Gas Turbine with Coal Derived Fuel Gases", *Ninth Annual EPRI Conference on Gasification Power Plants*, Palo Alto, Calif., Oct. 1990.

36. W. V. Bush, K. R. Loos, and P. F. Russell, "Environmental Characterization of the Shell Coal Gasification Process. I. Gaseous Effluent Streams," *Fifteenth Biennial Low-Rank Fuels Symposium*, St. Paul, Minn., May 22–25, 1989.

37. D. C. Baker, W. V. Bush, and K. R. Loos, "Determination of the Level of Hazardous Air Pollutants and Other Trace Constituents in the Syngas from the Shell Coal Gasification Process", *Conference on Managing Hazardous Air Pollutants—State of the Art*, Washington, D.C., Nov. 4–6, 1991.

38. D. C. Baker, W. V. Bush, K. R. Loos, M. W. Potter, and P. F. Russell, "Environmental Characterization of the Shell Coal Gasification Process. II. Aqueous Effluent," *Sixth Annual Pittsburgh Coal Conference*, Pittsburgh, Pa., Sept., 1989.

39. R. T. Perry and co-workers, "Environmental Characterization of the Shell Coal Gasification Process. III. Solid By-Products," *Seventh Annual International Pittsburgh Coal Conference*, Pittsburgh, Pa., 1990.

40. J. A. Salter and co-workers, "Shell Coal Gasification Process: By-Product Utilization,"*Ninth International Coal Ash Utilization Symposium*, Orlando, Fla., Jan. 1991.

41. EIA, "Annual Energy Outlook 2001", Energy Information Administration, December 2000.

42. NCC, "Increasing Electricity Availability From Coal-Fired Generation in the Near-Term", The National Coal Council, May 2001.

43. TIME, "The New Energy Crunch, *TIME M*, 37–48 (Jan. 29, 2001).

44. EEI, "Fuel Diversity — Key to Affordable and Reliable Electricity", Edison Electric Institute, April 2001.

45. D. R. Simbeck, "Report on SFA Pacific Gasification Database and World Market Report", *Proceedings 1999 Gasification Technologies Conference*, San Francisco, October 17–20, 1999.

46. Parsons Infrastructure & Technology Group Inc., Market-Based Advanced Coal Power Systems, final report, contract No. DE-AC01-94FE62747, December 1998.

47. G.J. Stiegel and R. C. Maxwell, "Gasification technologies: the path to clean, affordable energy in the 21st century, *Fuel Processing Technology* **71**, 79–97. (2001).

48. U.S. Department of Energy, National Energy Technology Laboratory, "Gasification Technologies: A Program to Deliver Clean, Secure, and Affordable Energy", November 2001.

49. D. Gray and G. Tomlinson, "Co-production of Electricity and High Quality Transportation Fuels in One Facility", *Proceedings Gasification 4: The Future, Noordwuk*, The Netherlands, April 11–13, 2000.

LAWRENCE J. SHADLE
DAVID A. BERRY
U.S. Department of Energy

MADHAVA SYAMLAL
Fluent Inc.

COAL LIQUEFACTION

1. Introduction

Liquefaction is the generic term for converting coal to fuels and chemicals (see also FUELS, SYNTHETIC, GASEOUS FUELS; FUELS, SYNTHETIC, LIQUID FUELS). Coal (qv) has been described variously, depending on the context, as "nature's dump" and "nature's storehouse." The reason is that while the primary constituents of coal are carbon and hydrogen, one also finds oxygen, sulfur, and nitrogen (generally classified as "heteroatoms"). Lesser amounts of many other elements can be detected (as inorganic oxides, or "mineral matter") as well. All methods for converting coal to fuels, and most methods of converting coal to chemicals, require both an increase in the hydrogen/carbon (H/C) ratio (from $\sim$0.2 to $\sim$2, order of magnitude, both molar) and removal of sulfur, nitrogen, and the other elements.

Coal can be converted to liquid and gaseous fuels and chemicals by two different processing routes, normally termed "direct" and "indirect." Direct liquefaction processes result in primary products (liquids or solids) of molecular weight greater than, or of the order of magnitude of, the fuels and chemicals desired. Catalysts may be used. Secondary processing is usually required to form fuels and chemicals. Some direct liquefaction schemes also involve chemical pretreatment of the coal. Other schemes involve a second feed source, generally heavy fractions of petroleum (coal–oil coprocessing), sometimes recyclable wastes (coal–waste coprocessing). In indirect liquefaction (IL) processes, on the other hand, the first step is always gasification of coal to synthesis gas ("syngas," $CO + H_2$), and this is followed by additional steps in which the syngas is catalytically recombined to form hydrocarbons and/or oxygenates.

In the 1990s, the US Department of Energy (USDOE) considered catalytic two-stage liquefaction and coal/oil and coal–waste coprocessing as the two major elements of its direct coal liquefaction (DCL) program. Major elements of the indirect coal liquefaction program were advanced Fischer-Tropsch technology for transportation fuels and processes for oxygenated fuel additives and high value chemicals. At the turn of the century, USDOE's Vision 21 Concept has as a goal the development of a suite of "modules" that can be interconnected to design a plant that takes advantage of local resources and supply local needs. The object is for the plant to be able to use one or more fuel types (coal, natural gas, biomass, petroleum coke from oil refineries, waste from municipalities) and, coupled with carbon sequestration techniques, to produce multiple products (one or more of electricity, heat, fuels, chemicals, hydrogen) at high efficiencies with no emission of greenhouse gases (1).

Reviews of coal liquefaction may be found in (2–5). Below, the processing schemes for DCL, including some of the research results on catalytic DCL since the 1990s, are discussed first followed by coprocessing schemes. This is followed by a discussion of indirect coal liquefaction.

2. Direct Coal Liquefaction

From the above, there are two chemical concerns in DCL—introduction of hydrogen to the parent coal molecule, and removal of heteroatoms (nitrogen, sulfur, and oxygen) and mineral matter. A third concern is the transport of solid and slurry-phase material. Though a physical rather than a chemical problem, it has ramifications in the economics of commercial–scale plant design.

Hydrogenation (hydroprocessing, hydrorefining, solvent refining, or hydroliquefaction) and pyrolysis are the two means used for DCL. In hydrogenation, the organic components of coal are dissolved under a moderate-to-high hydrogen pressure using a solvent, generally a coal-derived heavy aromatic material (6).

Here the primary reactions are a combination of homogeneous thermal cracking (ie, free-radical generation) and heterogeneous hydrogenation (involving hydro-aromatics in the slurry vehicle and/or the coal itself as hydrogen-transfer agents). Rapid and efficient capping of the primary free radicals generated by heating is thought to be necessary in order to prevent retrogressive reactions leading to formation of solid char (7). Other theories of coal liquefaction suggest that hydrogen can engender reactions involving scission of strong bonds in the coal macromolecule, and hence can act as an active bond-cleaving agent rather than simply a passive radical quencher (8). Typically, the object of hydrogenation is to maximize the yield of distillate fractions that can subsequently be converted into fuels. [Distillate materials comprise the naphtha fraction, typically boiling between C_5 and 420°F (215°C) plus the middle distillate fraction, typically between 300 and 700°F (150–370°C), with the exact cuts depending on product specifications.]

Pyrolysis normally involves heating in an inert or reducing atmosphere and produces char and oil, and often a low-BTU gas. The relative proportion of char to the other products can be quite high, hence the rationale for liquefaction by pyrolysis is often not production of coal-derived distillate materials but rather the solid. Hydropyrolysis (heating in the presence of hydrogen) and/or pyrolysis under conditions of rapid heating can, however, generate yields of distillate products significantly in excess of the volatile matter content of the starting coal.

2.1. Hydrogenation. *Early Work.* Bergius was awarded the Nobel Prize for chemistry in 1931 based on his pioneering work on DCL (9). The work of I. G. Farben on the Bergius process led to the development of a two-stage direct-hydrogenation liquefaction process. Here, primary coal solubilization was carried out in the first stage using added disposable catalysts in bubble-column reactors. Distillate materials were subsequently catalytically upgraded to liquid transportation fuels using supported hydrotreating/hydro-cracking catalysts in fixed-bed reactors (10). The Bergius process was significant in the years of World War II but fell into disuse due to the decreasing price of crude. In the 1960s, interest was rekindled when oil prices increased and a better understanding of the fundamentals of hydrogen donation made solvent refining more attractive as a DCL technique (6).

Solvent-Refined Coal Process. Work in the mid-1960s by the Spencer Chemical Co. (11) and during the 1970s by its successor, the Gulf Chemical Corp. (now Chevron) led to two solvent-refined coal (SRC) processing schemes: SRC-I for production of low ash solid boiler fuels and SRC-II for distillates, eg, "syncrude."

A schematic flow diagram for the SRC-I process is shown in Figure 1. Coal is first slurried in a recycle solvent, then preheated and finally reacted in a bubble column-type reactor at 450°C in the presence of gaseous hydrogen. Because of the high reactivities of the coals tested, primarily eastern U.S. high- and medium-volatile bituminous coals, no catalysts were added and the reaction was carried out at pressures as low as 6.9 MPa (1000 psig). Mean residence time in the reactor was reported to be on the order of 30 min. Solids were removed by use of either rotary pressure precoat filters or hydroclones. SRC yields, computed as the mass of SRC-I per mass of moisture- and ash-free (maf) coal exclusive of light hydrocarbon gas make, of ~60% were achieved.

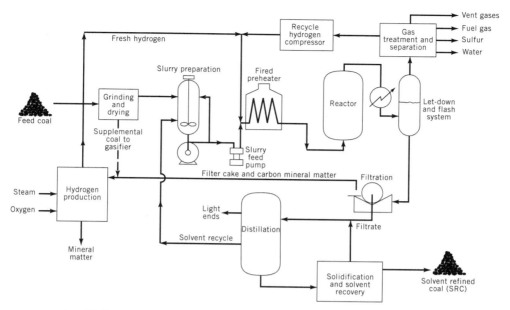

Fig. 1. SRC process. The reactor operated at 450°C and 6.9 MPa (1000 psig).

Ash removal was quite high by this process and, depending on the forms of sulfur in the parent coal (inorganic/organic), sulfur reduction was also substantial. Data for a Kentucky No. 9 high pyrite, eastern U.S. coal showed SRC-I/parent-coal weight percent ratios for total sulfur as 0.22 and for mineral matter as 0.0095. The heating-value ratio was 1.2.

The SRC-I technology was tested at a large-scale (50 t/day) demonstration plant in Fort Lewis, Washington, commissioned in 1974. This plant operated for several years, but severe problems were encountered, primarily with solvent balance and with operation and reliability of the solids-separation portion of the facility (12). The final design for this facility differs from Figure 1 in that an expanded-bed hydrocracking unit was added as a second-stage reactor to increase the yield of distillate material (13). Total solids [SRC plus two-stage liquefaction (TSL) solids] were reduced to ~27% by this modification, resulting in an increase in distillate materials. Bench-scale testing of the hydrocracking step indicated that the naphtha fraction, C_5 to 420°F (215°C), would be low in sulfur (0.01 wt%) but high in nitrogen (~0.1%). Severe hydrotreating of this material would be required before refining into gasoline via catalytic reforming. An overall thermal efficiency (defined loosely as the energy capable of being generated by combustion of product from a unit weight of feed per energy generated by combustion of a unit weight of feed) of this proposed facility was calculated to be 70%.

The SRC-II process, shown in Figure 2, was developed in order to minimize the production of solids from the SRC-I coal-processing scheme. The principal variations were incorporation of a recycle loop for the heavy ends of the primary liquefaction process and imposition of more severe conditions during

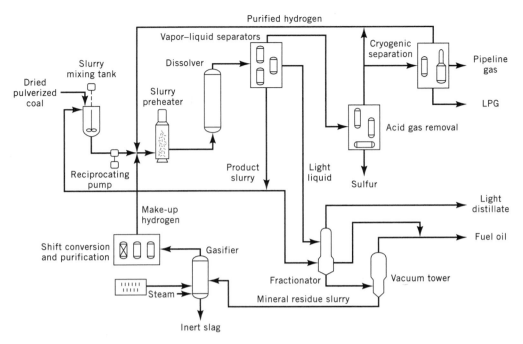

Fig. 2. SRC-II process where LPG is liquefied petroleum gas.

hydrogenation (5). It was quickly realized that minerals concentrated in this recycle stream served as heterogeneous hydrogenation catalysts that aid in the production of distillate. In particular, pyrrhotites, Fe_xS_y (where x and y are $\sim$1), nonstoichiometric iron sulfides produced by reduction of iron pyrite (FeS_2), were identified as being especially important. Pyrite was subsequently added for cases where the inherent pyrite content of the coal was low (14,15). Yields of some of the primary liquefaction products formed when pyrite is added to a slurry containing a moderately reactive but relatively low pyrite coal are given in Table 1.

A yield comparison between the products of the SRC-I and SRC-II processes is given in Table 2.

Table 1. **Yield^a for Addition of Pyrite to Coal^b**

	Pyrite addition, wt%		
Product	0.0	3.0	7.5
light hydrocarbon gases^c	16.6	17.1	17.6
naphtha	7.3	9.4	11.4
total oil	37.5	40.9	44.7
SRC	29.8	27.5	23.5
insoluble organic matter	5.9	5.3	5.2

^{a}Based on wt% maf coal.
^{b}Pittsburgh seam bituminous coal from West Virginia containing 0.9 wt% pyrite.
^{c}Up to C_4.

Table 2. **Comparison between Products of the SRC-I and SRC-II Processes**[a]

Process	SRC-I	SRC-II
Product yield, wt %		
C_1-C_4	10.5	16.1
total oil	25.9	38.9
SRC solids	42.7	21.0
insoluble organic matter	4.1	5.1
H_2[b]	−2.4	−5.6

[a]High-volatile Kentucky bituminous coal.
[b]The negative sign indicates that hydrogen is being consumed.

Changing the process configuration to SRC-II was successful in producing ~50% additional oil. However, a large increase in light hydrocarbon gas make accompanied this increase, with an attendant reduction in hydrogen utilization efficiency. Problems persisted using many coals, particularly subbituminous coals (4).

Exxon Donor Solvent Process.   A schematic flow diagram for the Exxon Donor Solvent (EDS) process is shown in Figure 3. The principal modification in

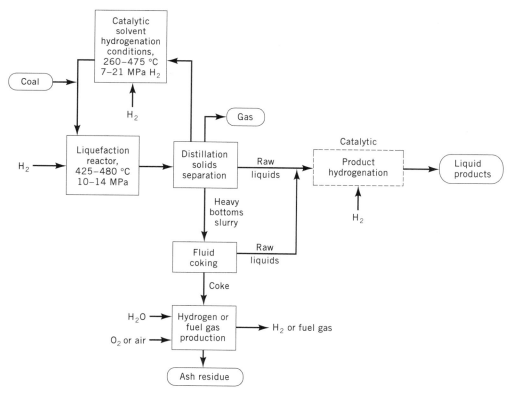

Fig. 3.   Exxon Donor Solvent process. To convert MPa to psi, multiply by 145.

this technology was the incorporation of a fixed-bed catalytic hydrogenating unit for the recycle solvent stream. This additional unit was required to keep the hydrogen donating/shuttling capacity of the recycle solvent oil at an acceptably high value (16). The use of bottoms slurry recycle to increase the solvent "make" fraction by taking advantage of the catalytic properties of minerals was also investigated, and improved yields in the bottoms recycle mode were generally reported. Recycle of this fraction was also reported to improve operability of the process dramatically, especially using low rank coals where viscosity of the bottoms stream was a significant problem (17). The primary liquefaction part of the reaction system operated at temperatures of 425–480°C and pressures of 10–14 MPa (1450–2030 psi), using mean residence times in the range of 15 min–2 h, depending on coal reactivity and process configuration. Operation at these conditions required significant advances in hardware, such as the design of a slurry let-down valve, required to reduce the pressure of the let-down slurry (15% solids) from 14 to 1 MPa at 450°C (18).

Operation of the EDS process was demonstrated in a 230 t/day unit in Baytown, Texas that started up in 1980 and was dismantled in late 1982. Exxon (now ExxonMobil) was the first to investigate the suitability of a wide range of different U.S. coals for conversion. Data on the response of a variety of coals to once-through and bottoms-recycle operations are shown in Figure 4. Figure 5 presents typical liquefaction product distributions for the system operated both with and without the Flexicoking (fluidized-bed coking) option.

H-Coal.   A significantly different scheme for DCL, developed by Hydrocarbon Research Inc. (HRI; now Hydrocarbon Technologies Inc., HTI; as of August

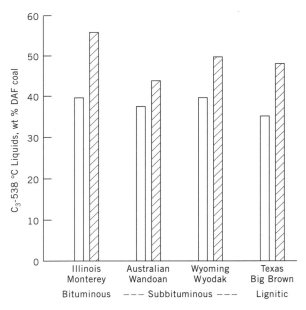

Fig. 4. Product yields for EDS process on (open) once-through basis and (shaded) bottoms-recycle basis for various types of coal. DAF = dry, ash-free. C_3–538 °C = a boiling fraction.

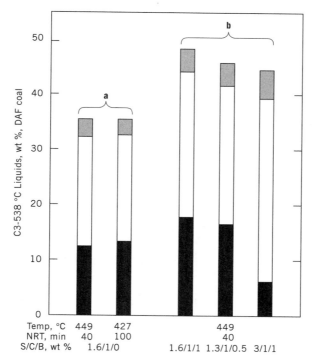

Fig. 5. Product yields for the EDS process in a 34-kg/day pilot plant **a** with, and **b** without, the Flexicoking option. Gray bars represent gas, clear bars represent naphtha, and dark bars represent oil. The pressure for the once-through process is 10.3 MPa, for the recycle 13.8 MPa. NRT, nominal residence time; S/C/B, solvent/coal/bottoms. To convert MPa to psi, multiply by 145.

2001, a wholly owned subsidiary of Headwaters Inc), was based on research and development on the H-Oil ebullated bed catalytic reactor for hydrotreating and hydrocracking heavy oil. The heart of this process is the reactor, where coal, catalyst, solvent, and hydrogen are all present in the same vessel (Fig. 6). The reactor is maintained in a "bubbling" or ebullated, ie, well-mixed, state by internal agitation coupled with the action of the gas bubbling through the fluid. This process was piloted by the then HRI and Ashland Synthetic Fuels, Inc. in a 600 t/day pilot plant adjacent to Ashland's refinery in Catlettsburg, Kentucky (12) . The process consists of slurry preparation followed by catalytic hydrogenation/hydrocracking at 450°C and 15 MPa (2200 psi) in the ebullated bed reactor (Fig. 7).

A principal focus of this project was research and development for catalysts that were tolerant of the coal-derived mineral matter in the reactor. Typical early catalysts showed rapid deactivation because of coking and loss of surface area, presumably from pore–mouth blockage by coke and metals laydown. Coke buildups of 20–25 wt% and surface area reductions from 300 m^2/g for the fresh catalyst to 25 m^2/g for the aged catalyst were reported after only 5 days on stream (19). Although one of the primary advantages of the H-Coal processing scheme was the ability to add and withdraw catalyst continuously from the reactor in

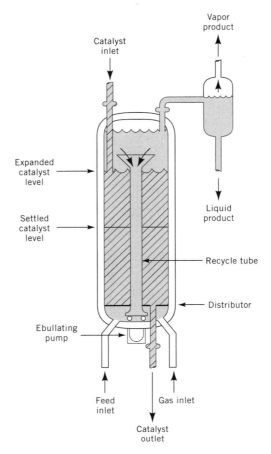

Fig. 6. H-Coal ebullating-bed reactor.

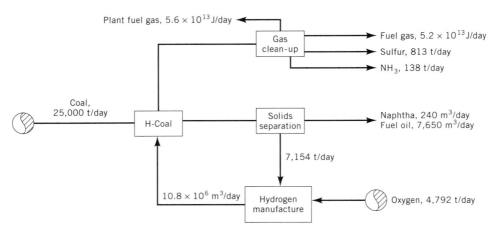

Fig. 7. H-Coal process using Illinois No. 6 coal. To convert J to Btu, multiply by 6.48×10^{-4}. To convert m^3 to bbl, multiply by 6.29. To convert m^3 to standard cubic feet (scf), multiply by 35.3.

order to maintain a stable level of activity, catalyst replacement and consumption rates were unacceptably high under these conditions.

In a later version, two reactors were used, coupled closely together, one operating at 400°C and the other at 420–425°C. The catalyst was also changed, from a cobalt–molybdenum/alumina to a nickel–molybdenum on bimodal alumina (4). Catalyst was added to both reactors, and ash was not removed between reactors. The deasher feed, now from the second reactor, was sufficently light that the deasher previously used could be replaced by a (simpler) pressure filter. In this Catalytic Two-Stage Liquefaction (CTSL) process, conversion was increased and hydrogen was more efficiently used (5,20).

Wilsonville Coal-Liquefaction Facility. Beginning in 1973, a 6 t/day coal liquefaction pilot plant was built in Wilsonville, Alabama by the Edison Electric Institute (EEI) and Southern Company Services. The Electric Power Research Institute (EPRI) assumed project sponsorship in late 1973, and USDOE became the primary sponsor in 1976. Amoco Oil Co. (now BP) joined the project in 1984. The purpose of the Wilsonville Advanced Coal Liquefaction R&D facility was to provide a flexible but reasonably large-scale pilot plant where effects of coal type and processing, ie, reactor configuration, could be tested and evaluated. Research on the Kerr–McGee critical solvent deashing (CSD) technology (also termed the Residual Oil Solvent Extraction, or ROSE, process) was also carried out, resulting in the development of alternative methods for solids removal from primary liquefaction products. Because of the role of Wilsonville in consistently demonstrating success or failure of a large number of concepts, many of whom were initialized elsewhere, particular attention should be paid to the results from this facility.

The plant began operation in 1974 in the SRC-I mode, but evolved to a two-stage operation utilizing two ebullating bed catalytic reactors (21). Initial efforts in TSL focused on catalytic upgrading of the thermal products, or Nonintegrated Two-Stage Liquefaction (NTSL). This configuration, termed nonintegrated because the coal-derived resid hydrocracking step did not interact with the primary thermal part of the plant, was excessively inefficient because of high hydrogen consumptions associated with the thermal part of the operation.

In Integrated Two-Stage Liquefaction (ITSL), a short contact-time thermal reactor was close coupled to an ebullated-bed catalytic reactor and process solvent was generated by distillation of the hydrocracked products. The thermal resid produced in the ITSL at short contact times was more reactive toward expanded-bed hydrocracking, thus permitting operation of the ebullated bed reactor at lower severity and minimizing gas make (22). Results on liquefaction of an Illinois No. 6 high-volatile bituminous coal using both the NTSL and ITSL modes at Wilsonville are shown in Table 3 (23).

Coal throughput, ie, space velocity per unit reactor volume, was substantially improved in going to the ITSL mode. The higher reactivity of the coal-derived resid permitted operation of the hydrocracker at lower temperature; this would be expected to reduce the rate of coke lay-down on catalyst, and to improve hydrogen utilization efficiency by minimizing formation of light hydrocarbon gases (higher distillate selectivity). A 35% increased yield of C_4^+ distillate was obtained.

Table 3. Operating Conditions and Yields at Wilsonville Plant[a]

parameter	NTSL 241CD Armak	ITSL 7242BC; 243JK/244B Shell 324M	RITSL 247D Shell 324M	CC–ITSL 250D Amocat IC	CTSL 250G (a) Amocat
run number / catalyst		Mode of Operation[b]			
Thermal stage					
average reactor temperature, °C	429	460;432	432	440	443
coal space velocity at temp >371°C, kg/m^3	320	690;450	430	320	320
pressure, MPa[c]	15	17;10–17	17	17	17
Catalytic stage					
average reactor temperature, °C	416	382	377	399	399
space velocity catalyst[d], h^{-1}	1.7	1.0	0.9	2.08	2.23
catalyst age resid/cat	260–387	278–441; 380–850	446–671	697–786	346–439
Yields[e]					
C$_1$–C$_3$ gas	7	4;6	6	7	8
C$_4^+$ distillate	40	54;59	62	64	63
resid	23	8;6	3	2	5
hydrogen consumption	4.2	4.9;5.1	6.1	6.1	6.4
Other					
hydrogen efficiency, C$_4^+$ distillate/H$_2$ consumed	9.5	11;11.5	10.2	10.5	9.8
distillate selectivity, C$_2$–C$_3$/C$_4^+$ distillate	0.18	0.07;0.10	0.10	0.11	0.12
energy content of feed coal reject to ash concentrate, %	20	24;20–23	22	23	16

[a]Illinois No. 6 coal.
[b]See text for term definition.
[c]To convert MPa to psi, multiply by 145.
[d]On a wt of feed per wt of catalyst basis.
[e]Wt% on a maf coal basis.

Also explored were Reconfigured Integrated Two-Stage Liquefaction (RITSL), where solvent deashing was practiced after the hydrocracking step, and Close-Coupled Integrated Two-Stage Liquefaction (CC–ITSL), where the two reactors (thermal/catalytic) were linked directly without any intervening processing steps (24,25). Typical results for these processes are also shown in Table 3. Incremental improvements in distillate yield and selectivity were realized by changing the process configuration, but at the expense of increased hydrogen consumption.

From 1985 to 1992, process development at Wilsonville focused on development of a Catalytic/Catalytic Two-Stage Liquefaction (CTSL) scheme utilizing ebulating-bed catalytic reactors in both stages. Initial work (26) indicated that distillate yields as high as 78% could be obtained by operating the first stage at low severity (399°C) and by using a large-pore bimodal NiMo catalyst having a mean micropore diameter in the 11.5–12.5-nm range. Results in the CTSL mode for three different coals are shown in Table 4. These data show the significant improvement in distillate production that can be achieved by use of catalyst in both stages, but (again) at the cost of increasing levels of hydrogen consumption.

Table 4. Operating Conditions and Yields at Wilsonville for Three Coals in CTSL Mode

Parameter	Coal type		
	Illinois No. 6	Ohio 6[a]	Wyodak
run number	253A	254G	251-IIIB
catalyst	Shell 317	Shell 317	Shell 324
First stage			
average reactor temperature, °C	432	433	441
inlet hydrogen partial pressure, MPa[c]	14.1	15.0	17.3
feed space velocity, h^{-1}	4.8	4.3	3.5
pressure, MPa[b]	17.9	18.8	17.9
catalyst age, resid/catalyst	150–350	1003–1124	760–1040
Second stage			
average reactor temperature, °C	404	421	382
space velocity, feed/catalyst, h^{-1}	4.3	4.2	2.3
catalyst age, resid/catalyst	100–250	1166–1334	371–510
Yield[c]			
C_1–C_3 gas	6	8	11
$C_4{}^+$ distillate	70	78	60
resid	−1	−1	+2
hydrogen consumption	6.8	6.9	7.7
Other			
hydrogen efficiency, $C_4{}^+$ distillate/H_2 consumed	10.3	11.3	7.8
distillate selectivity, C_1–C_3/$C_4{}^+$ distillate	0.08	0.11	0.18
energy content of feed coal rejected to ash concentrate, %	20	10	15

[a]Approximately 6% ash.
[b]To convert MPa to psi, multiply by 145.
[c]Wt% on a maf coal basis.

Block diagrams of NTSL, ITSL, RITSL CC–ITSL, and CTSL operations at Wilsonville are given in Figure 8. The Wilsonville Advanced Liquefaction R&D facility was shut down in early 1992 and was decommissioned shortly thereafter.

NEDOL.   The Japanese NEDOL process and the German Kohleoel process (below) are considered to be demonstrated and suitable for commercialization by their companies. NEDOL is similar to the EDS process. However, a catalyst is used ("natural" pyrite of mean particle size 0.7 μ) with the hydrogenated recycle solvents. A pilot plant has been reported to be successfully functioning at 150 t/day (27). A 2500-t/day plant operating at mild reaction conditions, 16.8–18.8 MPa and 450°C, has been simulated (28). The pilot-plant schematic is shown in Figure 9.

Kohleoel-Integrated Gross Oil Refining (IGOR+).   Ruhrkohle and VEBA Oel have collaborated on this process. In the primary reactor, conditions are maintained at 30 MPa and 470°C, and an iron oxide catalyst is used. Products are separated hot, and the vapor products are hydrotreated at 30 MPa and 350–420°C. The liquid from the first separator is recycled as part of the solvent. Liquid from the second separator is distilled at atmospheric pressure to yield a light oil and a medium oil. The process schematic is shown in Figure 10 (29). Process yields are shown in Table 5.

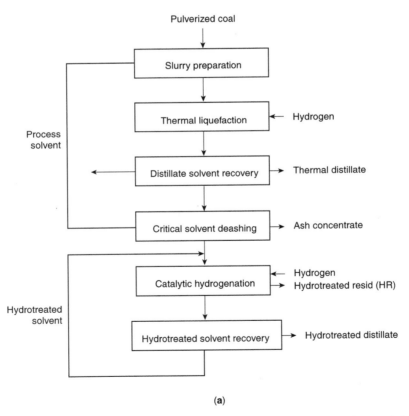

(a)

Fig. 8.   Block diagrams for (**a**) NTSL, (**b**) ITSL, (**c**) RITSL, (**d**) CC-ITSL, and (**e**) CTSL operations at Wilsonville (4).

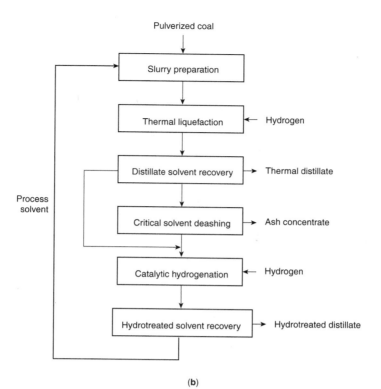

(b)

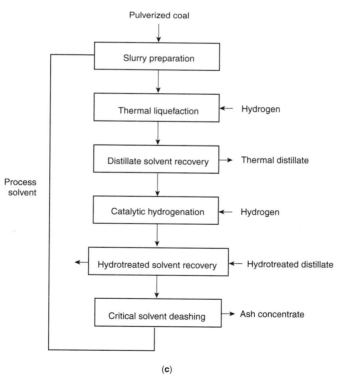

(c)

Fig. 8 (*Continued*)

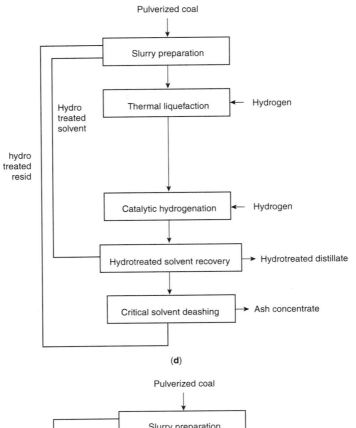

(d)

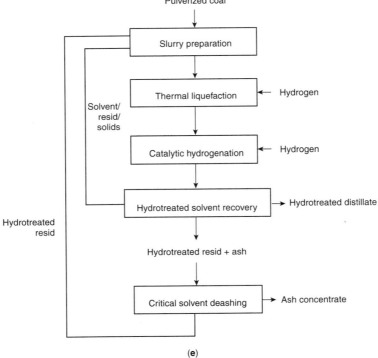

(e)

Fig. 8 *(Continued)*

846

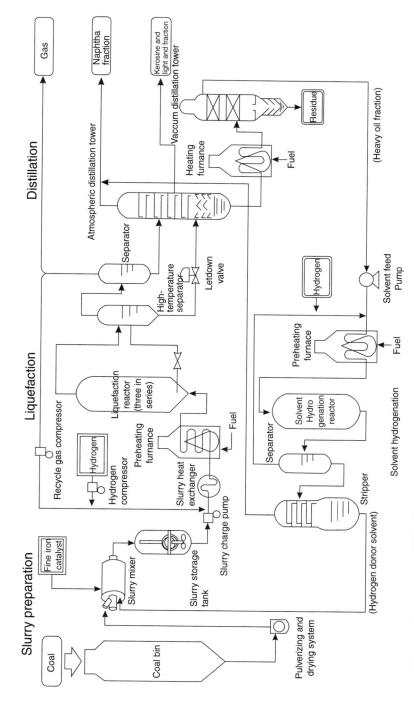

Fig. 9. Schematic of NEDOL Pilot Plant. Reprinted from K. Hirano, Ref. 27 with permission from Elsevier Science.

847

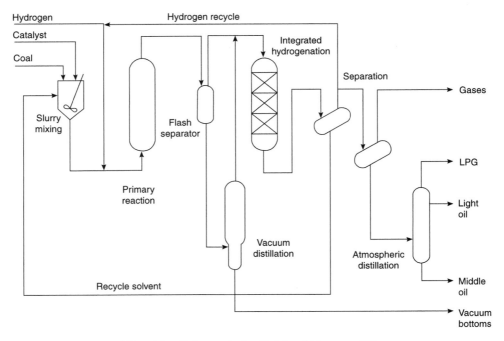

Fig. 10. Schematic for Kohleoel Process (29).

Other Processes. Other variations of catalytic and noncatalytic coal liquefaction schemes have also been developed. A 2.5-t/day pilot plant was operated for 4 years by the National Coal Board in the United Kingdom at Point of Ayr in Wales, but has since been decommissioned (29,30). The Catalytic Coal Liquids (CCL) process (31) reacted a coal–oil slurry with hydrogen over baskets containing a proprietary catalyst developed by then—Gulf Oil (now a part of Chevron), reportedly unaffected by coal ash. The Consolidation Synthetic Fuels (CSF) process used a complicated multistage operation in place of the second-stage hydrogenation/fractionation typically carried out in TSL processes (32).

Table 5. **Yields and Product Quality for the Kohleoel Process**[a]

Process yields	Yield	
hydrocarbon gases (C1–C4)	19.0	
light oil (C5–200°C)	25.3	
medium oil (200–325°C)	32.6	
unreacted coal and pitch	22.1	
Product quality	Light oil	Medium oil
hydrogen (%)	13.6	11.9
nitrogen (ppm)	39	174
oxygen (ppm)	153	84
sulfur (ppm)	12	<5
density (kg m^{-3})	772	912

[a]Prosper coal (German bituminous).

Table 6. **Effect of Emulsified Mo Catalyst on Product Distribution**[a]

Mo added, ppm	216	108	0
Product			
gases and light oil, wt %	33.3	32.6	17.6
hexane-soluble oil, wt %	22.2	26.6	5.4
asphaltenes[b]	23.4	21.1	50.3
hydrogen consumed[c]	6.1	6.1	4.4

[a]Pittsburgh No. 8 bituminous coal, 400°C, 13.7 MPa.
[b]Defined as toluene-soluble, hexane-insoluble material.
[c]kg H_2/100 kg maf coal.

The Brown Coal Liquefaction process was developed by NEDOL to handle very low rank coals with high moisture levels, such as those found in Australia (29). The China Coal Research Institute has comissioned feasibility studies, pilot plants and demonstration units for DCL from Germany, Japan and HTI, to use various Chinese coals.

Bench-Scale Research on Catalysts for DCL. Bench-scale test results are generally looked upon with skepticism because it is not clear how they relate to processes going on in industrial-scale reactors. However, Xu and co-workers (33) compared various types of bench-scale reactors and found that results similar to those from a large-scale ebullated-bed reactor could be obtained from a microautoclave reactor shaken at 400 cpm and containing a steel ball for efficient mixing.

While materials like zinc chloride were tested as catalysts for DCL (34), most catalytic research employed expensive materials such as molybdenum or cheap, disposable materials such as iron. The increase in conversion using an emulsified molybdenum catalyst is shown in Table 6. However, the costs and recovery of such material posed problems.

In the 1990s, USDOE started a consolidated program where various iron-based catalysts were used in bench-scale tests in different laboratories. Standard coals DECS-6 or DECS-17, chosen for their extremely low iron content, were used for the bench-scale tests.

Researchers at the University of Pittsburgh used finely divided sulfated iron oxides for the direct liquefaction of coal (36). Figure 11 shows results (37) from liquefaction runs using DECS-17 with a series of anion-modified iron(III) oxides and monoclinic pyrrhotite (Fe_7S_8) as catalysts (0.35 wt% Fe relative to coal). The overall conversion of coal (defined as product soluble in methylene chloride, MC, as a fraction of the initial coal) and the yield of oil (defined as MC-soluble, pentane-soluble product as a fraction of the initial coal) are both significantly smaller for the unmodified iron oxide. Using the sulfated iron oxide and the sulfated iron hydroxide results in conversions and yields comparable to modifying the iron oxide with 5% tungstate or 5% molybdate anions, and all are comparable to using pyrrhotite alone.

A series of catalysts were made at West Virginia University by preparing and disproportionating ferric sulfide (Fe_2S_3) under various conditions. The disproportionation is assumed to lead to an intimate mixture of pyrite, nonstoichiometric pyrrhotite and elemental sulfur (38). The catalysts are extremely sensitive

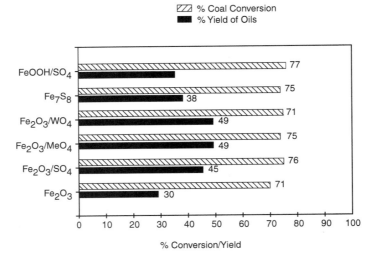

Fig. 11. Activities of iron(III) oxides, modified by small amounts of different anions. DECS- 17 coal, 400°C, 6.9 MPa (1000 psig) hydrogen pressure (cold). 300-mL stirred batch autoclave, 1200 rpm, 60 min, tetralin/coal = 3, Fe/coal = 0.35 wt%. Reprinted Ref. 37 from V. R. Pradhan, J. Hu, J. W. Tierney, and I. Wender, with permission of the American Chemical Society.

to air or oxygen. The materials can be made as small particles by preparing *in situ* on the coal surfaces (39), or in an aerosol reactor (40), or using surfactants (41). The *in situ* technique generally works best. Figure 12 shows the effect of various amounts of *in situ* impregnated catalyst on conversion [tetrahydrofuran(THF) soluble] and oil + gas yield (THF soluble, hexane soluble). Addition of second metals as sulfides sometimes leads to alloys, but with little improvement in conversion or yield (42).

Researchers at Pacific Northwest Laboratory developed techniques (43) for generating iron-based catalysts in large quantities using flow-through techniques termed Rapid Thermal Decomposition of Solutions (RTDS) and Modified Reverse Micelle (MRM) processes. In RTDS, precursors are exposed briefly to conditions of high temperature and high pressure to initiate nucleation of iron oxide and hydroxide. In MRM, high loadings of iron-bearing salts in water-in-oil microemulsions are precipitated by changing the conditions of the microemulsions. Various catalysts obtained using these methods were tested using the model compound naphthyl bibenzylmethane:

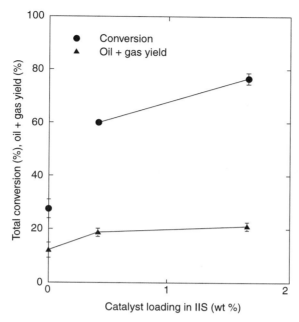

Fig. 12. Effect of loading of *in situ* impregnated ferric-sulfide-based catalyst. DECS-6 coal, 400°C, 6.9 MPa (1000 psig) hydrogen pressure (cold), 57-mL tubing-bomb reactor, vertical agitation at 5 Hz, 30 min, 3-g coal, 5-mL hexadecane. Reprinted from Ref. 39 with permission.

Here the bonds marked by arrows are analogous to those that would be cleaved in initial DCL reactions ("primary" DCL), while cleavage of the other bonds represent other, later, significant reactions in DCL. Results for some of these tests are shown in Table 7.

A good snapshot of work on iron-based catalysts in laboratories supported by USDOE as well as in other laboratories can be found in (44). However, Table 7 and Figures 11 and 12 demonstrate the difficulty in comparing catalyst performance by comparing results from different laboratories using not only different reactors, but also different solvents, coals, conditions, and even analytical techniques. To solve this problem, USDOE commissioned Sandia National Laboratories to test catalysts from different laboratories using an identical set of conditions and reagents and using a statistical design of experiments. To the authors' knowledge, the study was never completed. However, Table 8 shows results of a partial report (45).

2.2. Pyrolysis and Hydropyrolysis. The second category of DCL aimed at producing distillate materials from coal is pyrolysis and hydropyrolysis. Here a solvent is typically not used, and neither is a catalyst. Pyrolysis processes are burdened with poor liquid yield, relative to hydrogenation, and the coal-derived liquids are high in heteroatoms and in fine-particulate matter, both organic and inorganic. Pyrolysis, sometimes called destructive distillation, essentially involves heating the coal in an inert atmosphere, followed by recovery of coal-derived tars and distillates in the off-gas stream (46). Pyrolysis carried

Table 7. **Activity and Selectivity Using as Catalysts Iron Oxides and Hydroxide Made via Flow-Through Techniques**[a,b]

Catalyst sample no.	Identified phase(s)	Model compound consumed (%) ($\pm$ 4%)	Selectivity[c] (%) ($\pm$ 3%)
no catalyst		<5	40–60
sulfur		<20	50–70
RTDS[d] *products*			
62-79-4	hematite	23	83
54-53-5	2-line ferrihydrite	20	84
54-50-3	hematite/6-line ferrihydrite	81	96
48-20-1	6-line ferrihydrite	90	96
48-19-7	6-line ferrihydrite	<90	96
54-54-2	magnetite	<90	98
54-56-1	ferric oxyhydroxysulfate	<90	94
54-56-4	ferric oxyhydroxysulfate	<90	96
MRM[d] *products*			
MRM-7-2	2-line ferrihydrite	51	91
MRM-39-39	2-line ferrihydrite	41	90
MRM-7-2A	magnetite/maghemite	77	89
MRM-39-39A	maghemite	32	92
MRM-39-49	goethite/lepidocrocite	44	93

[a]Reprinted with permission from Ref. 43. Copyright © 1994 American Chemical Society.
[b]Naphthyl bibenzylmethane (NBBM), 9,10-dihydrophenanthrene (DHN), sulfur, 400°C; sealed tube, 25-mg NBBM, 10 mg S, 100 mg DHN, 10-mg catalyst precursor.
[c]Cleavage of "primary" bonds (arrows in structure above) as a percentage of total bonds cleaved.
[d]See text for acronyms.

out in a hydrogen atmosphere is termed hydropyrolysis; pyrolysis at extremely rapid heating rates is termed flash pyrolysis. These processes are not in use for producing transportation fuels because of the yield and purity disadvantages noted above.

Table 8. **Comparison of Three Iron-Based Catalysts with Pyrite as Catalyst and No Catalyst (Thermal)**[a,b]

Catalyst	THF Conv. (%)[c]	DHP (%)[d]
1wt% WVU impregnated catalyst	93.0	13.4
thermal	51.7	1.73
1wt% PNL cat. precursor + 1wt% sulfur	89.4	8.41
thermal + 1wt% sulfur	63.6	2.35
1wt% U. of Pitt. cat. precursor + 2wt% sulfur	82.3	5.35
thermal + 2wt% sulfur	63.0	2.43
1wt% pyrite	73.4	3.88
thermal	54.9	1.08

[a]From Ref. 45.
[b]DECS-17 coal, 400°C, 800 psig hydrogen (cold) pressure; 43-mL microautoclave reactor, 60 min, 1.67 g coal, 3.34 g phenanthrene, 1 wt% catalyst.
[c]THF soluble.
[d]A measure of hydrogenating ability of the catalyst.

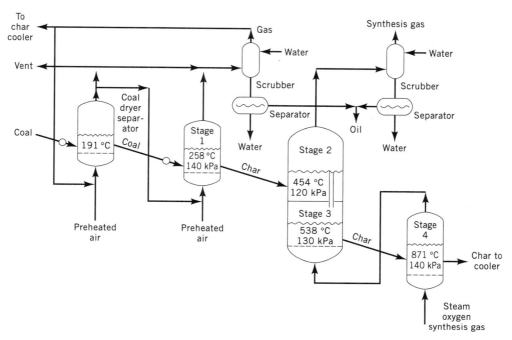

Fig. 13. COED process. To convert kPa to psi, multiply by 0.145.

Pyrolysis. Large-scale research and development on coal pyrolysis was carried out on the Char Oil Energy Development (COED) process (47). This scheme involved temperature-staged pyrolysis in a dryer–separator and three interacting fluidized beds, as shown in Figure 13, and was tested in a 36-t/day process demonstration unit during the early 1970s. Pyrolysis temperatures ranged from 191 to 871°C in the COED process, and the long residence times associated with the fluid beds mandated low yields of liquid products. Typical product yields for four different U.S. coals are shown in Table 9. The yield structure is heavily weighted toward production of char and gas. Production of coal-derived liquids ranged from 0.04 to 0.21 m^3/t of coal as compared to 0.61–0.79 m^3/t for direct hydrogenation. Further, the liquids produced were high in heteroatoms (especially nitrogen) and required extensive hydrotreating before use as a synthetic crude oil.

Table 9. **FMC/COED Process Product Distribution for Four U.S. Coals**

Coal composition, wt%	ND lignite	Utah	Illinois	West Kentucky
Yields, dry coal basis				
char	55.8	54.5	59.5	63.0
tar	5.3	21.5	19.3	17.3
gas	37.6	18.3	15.1	13.0
liquor (aq)[a]	1.3	5.7	6.1	6.7

[a]Water containing water-soluble organics produced during pyrolysis.

Process development on fluidized-bed pyrolysis was also carried out by the Consolidation Coal Co., culminating in operation of a 32-t/day pilot plant (48). The resulting CONSOL pyrolysis process incorporated a novel stirred carbonizer as the pyrolysis reactor, which made operation of the system feasible even when using strongly agglomerating Eastern-U.S. bituminous coals. This allowed the process to bypass the normal preoxidation step that is often used with caking coals, and resulted in a nearly 50% increase in tar yield. Use of a sweep gas to remove volatiles rapidly from the pyrolysis reactor gave overall tar yields of nearly 25% for a coal that had tar yields of only 15% as measured by the Fischer assay, a standardized test to measure the amount of liquids produced by pyrolysis.

Other large-scale coal pyrolysis process developments were carried out by the Tosco Corp., with its TOSCOAL process (49). Essentially a direct copy of Tosco's rotating kiln technology that was developed for pyrolysis of oil shale, this slow-heating scheme achieved tar yields at maximum temperatures of 482–521°C that were essentially identical to those obtained by a Fischer Assay.

Hydropyrolysis. Process development of the use of hydrogen as a radical quenching agent for the primary pyrolysis was conducted (50). This process was carried out in a fluidized-bed reactor at pressures of 3.7–6.9 MPa (540–1000 psi), and a temperature of 566°C. The reactor was designed to minimize vapor residence time in order to prevent cracking of coal volatiles, thus maximizing yield of tars. Average residence times for gas and solids were quoted as 25 s and 5–10 min. A typical yield structure for hydropyrolysis of a subbituminous coal at 6.9 MPa (1000 psi) total pressure was char 38.4, oil 29.0, water 19.2, and gas 16.2, on a wt% maf coal basis. Tar yields of $\sim$0.32 m^3/t were quoted. Because the scheme used hydrogen, the liquids generally exhibited lower heteroatom contents than conventional tars derived from coal pyrolysis in an inert atmosphere. Process development proceeded through a 270-t/day semiworks plant that was operated successfully on noncaking coals. Operability for caking coals was difficult, however.

Flash Pyrolysis. Development of a rapid, ie, flash, pyrolysis process was carried out in the late 1960s and early 1970s (51). The process was designed to heat coal at rates in excess of 5000 °C/s. Process development proceeded through to a 2.7-t/day process development unit (PDU) using a variety of caking and noncaking coals. The reactor section facilitated rapid heating by direct contact with hot char from the char burner. Gas residence times were brief (<2 s) and carefully controlled in order to minimize secondary cracking reactions and to maximize the yield of coal-derived liquids. Typical yield structures for pyrolysis at 580°C for two coals are shown in Table 10. Rapid heating, and hence high tar

Table 10. **Product Distribution for the Occidental Flash Pyrolysis Process for Two U.S. Coals**

Coal	Western Kentucky bituminous	Wyoming bituminous
Yield, wt%		
tar	35	27
char	56	52
gas	7	13

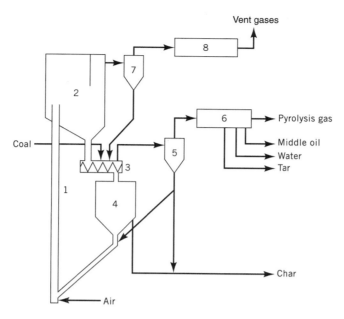

Fig. 14. Lurgi–Ruhrgas flash-pyrolysis system, where 1 is a lift pipe; 2, primary pyrolysis reactor; 3, screw feeder; 4, secondary pyrolysis reactor; 5 and 7, cyclones; and 6 and 8, product recovery and tailgas cleaning.

yields, could be obtained with this system. However, rapid quenching of reaction products proved to be a significant problem, especially as the process was scaled up from the laboratory.

Development of a flash-pyrolysis reaction system was also carried out by Lurgi–Ruhrgas (52). Between 1940 and 1960, units processing 10 t/h were operated, and a small commercial plant was built and has operated in the former Yugoslavia since 1963. As shown in Figure 14, coal is rapidly heated by mixing with hot recycled char in a screw-conveyor-type reactor. Volatiles recovery is completed at 750°C in vessel number 4. A typical product distribution for this system operating on a high volatile West Virginia bituminous coal gave a tar yield of 28 wt%, char of 58 wt%, and gas + liquor of 14 wt%, all on a basis of maf coal.

A novel high pressure flash hydropyrolysis reaction system was designed and operated by Rockwell Corp. during the mid-1970s (53). The process was operated in a 1-t/h pilot plant, where the technology was successfully demonstrated for a variety of different feed coals. The reactor was designed to mix hot high pressure hydrogen and coal in a highly turbulent zone such that extremely rapid heating rates could be obtained, >10,000 °C/s. A schematic of the reactor is shown in Figure 15(**a**). In this system, the energy required to heat coal to temperatures of 871–1038°C was generated by combustion of a portion of the hydrogen feed to the reactor. Rapid heating then was facilitated by direct contact with hot hydrogen and the combustion gases. The rapid heating, coupled with extremely fast transition through the coal's plastic regime, prevented problems associated with operation using agglomerating coals. Further, the extremely short

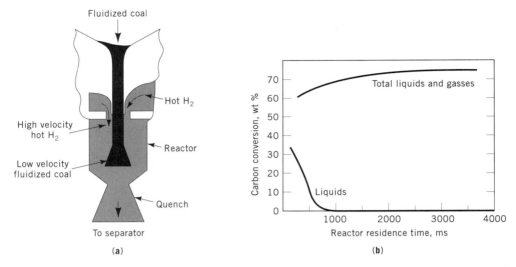

Fig. 15. (**a**) Rockwell flash-hydropyrolysis reactor; (**b**) carbon conversion as a function of reactor residence time. The reaction was run at 1038°C and 10.3 MPa (1500 psig).

residence times for coal-derived volatiles and the activity of hydrogen as a radical scavenger helped minimize secondary cracking reactions, thus permitting yields of coal-derived tars to exceed greatly that predicted by the Fischer assay. Whereas total coal conversion was relatively insensitive to reactor residence time, the yield of liquid or oil was a maximum at ∼0.1s for a U.S. bituminous coal. Longer residence times favored formation of gases. Operating pressure also had an effect on coal conversion and product distribution. Higher pressures favored production of liquids.

2.3. Coprocessing. The main difference between coprocessing and hydroliquefaction is that the solvent is not simply a recycled stream from the process but is a separate feed stream, either a resid fraction (or other fraction, typically heavy) of petroleum or a waste (such as postconsumer plastic material, tire rubber or even municipal waste). The motivations for the additional feed are to reduce the severity of the liquefaction conditions compared to coal-alone hydroliquefaction, to recycle to extinction the heavy fraction, to take advantage of synergies of operation, and to take advantage of the favorable economics and/or politics in eliminating an unneeded stream.

Coal–Oil Coprocessing. Chevron, CANMET, and Ohio-Ontario Clean Fuels are among the organizations that developed strategies and technologies for coal–oil coprocessing. The two-stage coprocessing scheme of HRI (now HTI/Headwaters) illustrated in Figure 16 was used on coal ranks from lignite through high volatile bituminous and with a variety of resids. As an example of the synergistic benefits of coprocessing, resid-based organometallic Ni and V compounds (that would serve as poisons for downstream processing if present in the liquid product) were found to be included in the solid (ash) phase contributed by the coal, and thereby removable before downstream processing. Further, the conversion to the heaviest products (liquids boiling above ∼225°C) is greater in coprocessing than the value expected for individual processing of the feedstocks (54).

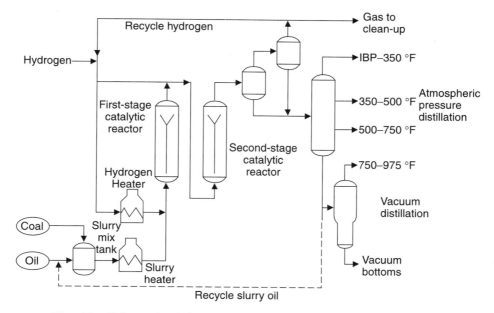

Fig. 16. Schematic of the HRI Two-Stage Coprocessing Scheme (54).

Coal–Waste Coprocessing.　The use of catalysts, iron-based and others, on coal-waste coprocessing has been quantified in tests at the bench scale and larger. Bench-scale tests have been carried out on a standard commingled waste plastic developed by the American Plastics Council, as well as on pure low density polyethylene, high density polyethylene, polypropylene and poly-(vinyl chloride); these were used alone, with coal, and with coal and resid (55,56). Sawdust and farm manure have also been used in bench-scale coprocessing (57). A two-stage process was suggested for coprocessing waste rubber (from tires) with coal (58)—the tire would be liquefied noncatalytically at relatively low severity conditions to obtain a tire oil and (marketable) carbon black, and the tire oil would be combined with coal containing *in situ* ferric sulfide-based catalyst at higher temperatures and pressures.

HTI used a proof-of–concept bench-scale unit to study the effect of adding waste plastics to either coal/resid feedstock, or resid alone (59). The plastics were obtained from curbside recycling in northern New Jersey, the coal was subbitu-minous (Wyoming Black Thunder) and the resid was Hondo-VTB. A proprietary iron catalyst was combined with Molyvan-A and used in a dispersed slurry in a first-stage reactor. An interstage separator operated at high pressure. After the second-stage reactor, the product was flashed and the light ends hydrotreated to yield a naphtha-like fraction. The addition of the waste plastics was found to increase the yield of distillate and to decrease the consumption of hydrogen, regardless of whether coal/resid or resid alone was used as the feedstock.

The Duales System Deutschland (DSD) has supported the recycling of ~300,000 t/year of waste plastic, including mechanical recycling as well as con-version to oil, chemical feedstocks or synthesis gas in Germany. In the United

States, a feasibility study for a demonstration plant for coprocessing of waste plastics, tires, and coal was carried out (60). Base-case amounts were 200 t/day of plastic and 100 t/day of tires. Under these conditions, using typical tipping fees and with oil priced at $20/barrel, the return on investment was found to range between 9 and 20%.

3. Indirect Liquefaction

The second category of coal liquefaction involves those processes that first generate synthesis gas (syngas), a mixture of CO and H_2, by steam gasification of coal:

$$C(s) + H_2O \rightarrow CO + H_2 \tag{1}$$

followed by production of solid, liquid, and gaseous hydrocarbons and oxygenates via catalytic reduction of CO in subsequent stages of the process (61). Whereas coal is usually the preferred feedstock, other carbon-containing materials such as coke, biomass, or natural gas can also be used (see FUELS FROM BIOMASS; GAS, NATURAL).

Processes whereby coal is gasified to syngas are not discussed here; these involve commercial gasifiers such as those of Lurgi, Kellogg, or Koppers-Totzek (see COAL CONVERSION PROCESSES, GASIFICATION). Processes to obtain fuels and/or chemicals from syngas are discussed below, regardless of the feedstock used for syngas. Wender (62) illustrates the principal paths for fuels and chemicals as shown in Figure 17.

In the general process, syngas from the gasifier is first cleaned to remove gasifier tars, hydrogen sulfide and organic sulfur. The composition of the gas is then adjusted in a catalytic shift converter to increase the H_2/CO ratio via the water–gas shift reaction:

$$CO + H_2O \rightleftharpoons CO_2 + H_2 \tag{2}$$

This clean and shifted gas is finally converted to hydrocarbons and/or other products in a series of catalytic reactors. The synthesis reaction is usually carried out using two or three reactors in series because of the highly exothermic nature of the overall reaction.

The first demonstration of catalytic conversion of synthesis gas to hydrocarbons was accomplished in 1902 using a nickel catalyst (63). The fundamental research and process development on the catalytic reduction of carbon monoxide was carried out by Fischer, Tropsch, and Pichler (64). Generalized stoichiometric relationships such as those below are often used to represent the fundamental aspects of the formation of hydrocarbons and oxygenates:

$$n\,CO + 2n\,H_2 \rightarrow (-CH_2-)_n + nH_2O \tag{3}$$

$$2n\,CO + n\,H_2 \rightarrow (-CH_2-)_n + nCO_2 \tag{4}$$

$$n\,CO + 2n\,H_2 \rightarrow H(-CH_2-)n\,OH + (n-1)\,H_2O \tag{5}$$

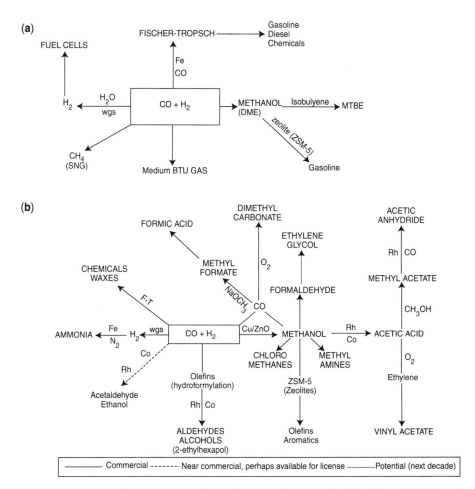

Fig. 17. Schematics for production from Syngas of (**a**) Fuels (**b**) Chemicals. Reprinted from Ref. 62 with permission from Elsevier Science.

However, the chemistry of the synthesis reactions is complex, and the formation of hydrocarbons is fundamentally different in many respects from the formation of oxygenates. For example, the side products vary, depending upon the H_2/CO ratio. Further, the formation of hydrocarbons requires a catalyst upon which CO can adsorb dissociatively, with rupture of the C–O bond, while CO may adsorb on a single site, without rupture of the bond, for oxygenate formation. By proper selection of catalyst and reaction conditions, including the H_2/CO ratio, hydrocarbons and oxygenates ranging from methane and methanol through paraffin waxes of high molecular weight ($>10,000$) can be synthesized, as indicated in Figure 18 (65).

3.1. Production of Hydrocarbon Fuels. By convention, only the production of hydrocarbons is termed Fischer-Tropsch (FT) synthesis. Hydrocarbons are typically used as fuels or fuel enhancers, generally diesel fuel.

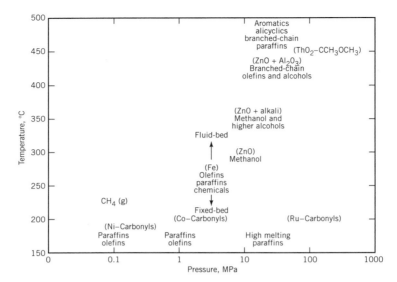

Fig. 18. Optimum pressure–temperature ranges for indirect synthesis processes showing the various catalysts in parentheses. To convert MPa to psi, multiply by 145.

Processes that operated at relatively low pressures, in the range of 100–200 kPa (1–2 atm) dominated commercial applications of FT synthesis in Germany prior to 1939 (66). Catalysts were primarily cobalt based. However, catalyst lifetimes were short and deactivation was difficult to reverse. At the other extreme, high pressure synthesis has been carried out at pressures in the range 5–100 MPa (50–1000 atm) and temperatures of 100–400°C. Supported ruthenium catalysts are used, and the products are typically straight-chain paraffin waxes (67).

The greatest successes, including the processes used by the South African Coal Oil and Gas Corporation Ltd. (SASOL), have occured at medium pressures, typically in the range 0.5–5 MPa (5–50 atm). Cobalt catalysts, similar to those used for the low pressure synthesis, were typically used at temperatures of 170–200°C. Iron catalysts, usually promoted, have also been used in the SASOL process, but at temperatures of 220–340°C. The primary differences between low and medium pressure synthesis are increased catalyst life for the medium pressure process, more diesel fuel, and a slightly higher hydrocarbon yield.

SASOL.   The SASOL plants are worthy of mention as probably the only commercial facilities currently operating. They have supplied between one-third and one-half of South Africa's fuel requirements.

SASOL-I.   This was the first plant, put into operation at Sasolburg, South Africa in 1955 (68). An overall flow schematic for the original setup of SASOL-I is shown in Figure 19. The product slate from this facility comprised materials ranging from FT products (hydrocarbons) to oxygenates, including alcohols and acids.

The plant utilizes iron catalysts. The catalyst is manufactured by precipitation from an iron nitrate solution using sodium carbonate. Copper and potassium

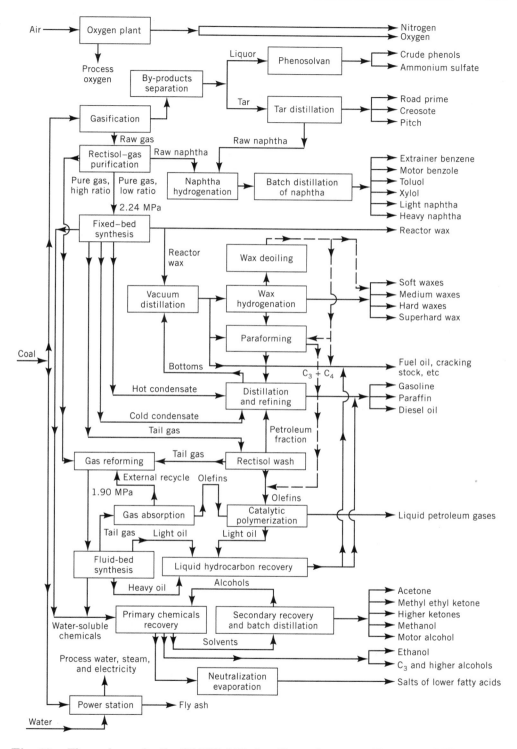

Fig. 19. Flow scheme for the SASOL-I Fischer-Tropsch process. To convert MPa to psig, multiply by 145.

are added as promoters, and the final material is pelletized and reduced with hydrogen prior to use. Catalyst life is variously reported to be 100 days to 6 months.

The overall processing scheme at SASOL-I involves steam-oxygen gasification of coal using high pressure (3 MPa, 30 atm) Lurgi gasifiers producing 22,500 m^3 each of raw gas having a H$_2$/CO ratio of 1.7. The feed to the plant is coal of high ash (35 wt%) and low energy content (23 MJ/kg) from mines near Sasolburg. SASOL-I consumes ~5.5 million tons per year of coal, with 60% going for gasification and synthesis and 40% for generation of onsite power. The raw gas is purified using Rectisol (chilled methanol) technology for removal of gasification tars, H$_2$S, CO$_2$, and some methane. The purified gas is then sent to the reactors.

Originally, both fixed-bed reactors (ARGE) and fluidized-bed reactors (Synthol) were used. The fixed-bed reactors, designed by Lurgi, contain approximately 40 m^3 of catalyst in over 2000 vertical tubes having diameters of 4.5 cm OD. There are six fixed-bed reactor trains in parallel, each reactor processing 30,000 m^3/day of feed at relatively low temperature (220–255°C) and medium pressure (2.5 Mpa, 25 atm), and producing 87.4 m^3 (550 barrels) of product per day (69). A flowsheet showing one fixed-bed reactor train is given in Figure 20.

In 1992, the three fluidized-bed reactors of SASOL-I were shut down and replaced by a single low temperature slurry-bed reactor. The SASOL Slurry-Bed Reactor (SSBR) is 5 m in diameter and 22 m high and has a capacity of 2400 barrels/day (70). It contains the catalyst suspended as a slurry in a FT wax or other liquid. The SSBR is cheaper to build, can be scaled up, permits

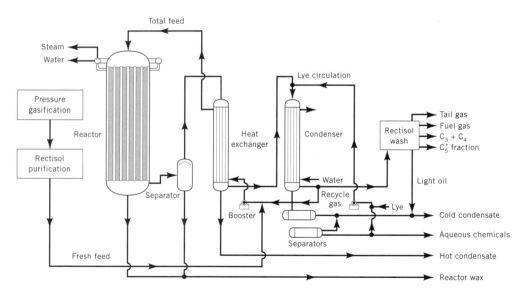

Fig. 20. Flowsheet of medium pressure synthesis, fixed-bed reactor (Lurgi-Ruhrchemie-Sasol) having process conditions for SASOL-I of an alkaline, precipitated-iron catalyst, reduction degree 20–25%; having a catalyst charge of 32–36 t, at 220–255°C and 2.48 MPa (360 psig) at a fresh feed rate of 20,000–22,000 m^3/h in the reactor.

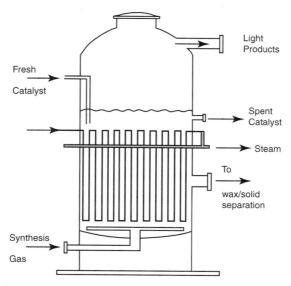

Fresh
Catalyst

Light
Products

Spent
Catalyst

Steam

To

wax/solid
separation

Synthesis

Gas

Fig. 21. Sketch of SSBR, operating at low temperatures (220–270°C) and medium pressures (2–3 MPa, 20–30 atm) for conversion of syngas to FT waxes (62).

near-isothermal behavior and results in improved catalyst economy. A sketch of the SSBR is given as Figure 21.

As a result of the revamp of the facility and the introduction of the SSBR, the primary products of SASOL-I are now waxes and waxy products. The idea is to crack the waxes to obtain diesel fuel. Oxygenates and other products are also formed.

SASOL-II and-III. Two additional plants were built near Secunda, South Africa: SASOL-II in 1980, and SASOL-III, essentially identical to SASOL-II, in 1983. A block flow diagram for the original SASOL-II and -III processes is shown in Figure 22.

The catalysts are made from millscsale from a steelworks, ground, combined with alkali and other promoters, and fused in an open-arc furnace. The consumption of coal for these two plants combined is ~35 million tons/year, and these plants together produce ~$1.6 \times 10^4 m^3$ (100,000 barrels) per day of transportation fuels. As can be seen in Figure 22, the original design of the SASOL-II and SASOL-III plants used only the fluid-bed Synthol reactor, and extensive secondary catalytic processing of intermediates (alkylation, polymerization, etc) maximized the production of transportation fuels. Product selectivities of the fixed-bed reactor and the Synthol reactor are given in Table 11. As shown, the fixed-bed system is more selective for middle distillates as well as heavy oils and waxes, whereas the fluidized bed system is considerably more selective for formation of C_2–C_4 olefins as well as products in the gasoline (C_5–C_{11}) boiling range.

A new design of fluidized-bed reactor has recently been coupled to the existing production facility. The SASOL Advanced Synthol (SAS) reactor is less than one-half of the size and one-half of the cost of the Synthol reactor, and eliminates

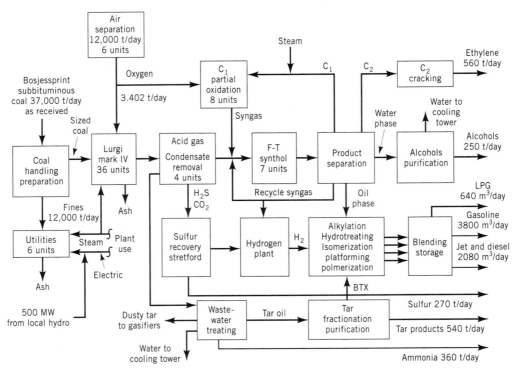

Fig. 22. Block flow diagram, SASOL-II and -III, where BTX is benzene–toluene–xylene. To convert m^3 to barrels, multiply by 6.29.

the recirculation of the catalyst. The SAS reactor results in a higher conversion and higher selectivity to oil; scale-up is also easier than for the conventional Synthol reactor (71). The Synthol reactor and the SAS reactor are shown in Figure 23. Recently the 16 Synthol reactors were replaced by eight SAS reactors, with a ninth SAS reactor under commission. Both the Synthol and SAS reactors are operated at relatively high temperatures, around 340°C. Work on the SAS reactor led to the development of the SSBR now used in SASOL-I.

Table 11. **Product Selectivities of SASOL Commercial Reactors**[a]

Product	Fixed bed	Synthol (Fluidized bed)
CH_4	4	7
C_2–C_4 olefins	4	24
C_2–C_4 paraffins	4	6
gasoline	18	36
middle distillate	19	12
heavy oils and waxes	48	9
water soluble oxygenates	3	6

[a]Reprinted with permission from Ref. 71.

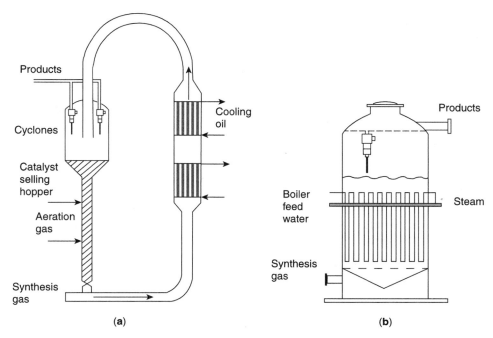

Fig. 23. Synthol reactor (**a**) and SAS reactor used in SASOL-II and-III (62).

Indirect Liquefaction from Natural Gas. These are mentioned here for completeness. The SASOL Slurry-Phase Distillate (SSPD) process has recently been developed. The idea here is to convert natural gas (rather than coal) to syngas and hence to naphtha and diesel fuel. A cobalt-based catalyst is used in the slurry-phase FT reactor. The methanol-to-gasoline (MTG) process developed by Mobil (now ExxonMobil) was used in New Zealand to convert natural gas to methanol and hence to gasoline, using a ZSM-5 zeolite catalyst. Currently, however, the process stops at methanol production. The Shell Middle-Distillate Synthesis (SMDS) process produces fuel from natural gas. Proprietary catalysts are used to convert the syngas to long-chain paraffins, which are then hydrocracked to give the required middle distillates. Conditions can be altered to maximize diesel fuel or kerosene.

3.2. Production of Alcohols and Other Oxygenates. Methanol is used as a fuel in its own right, as an octane extender for gasoline, and as a feed stock for the production of polymers and other chemicals. Methanol has been obtained from syngas since the Bayer patent in 1923. The original process operated at high temperatures and pressures (350–450°C, 25–35MPa) using a zinc oxide/chromium catalyst. Current ICI plants operate at low temperatures and pressures (220–280°C, 5–10 MPa) using a $Cu/ZnO/Al_2O_3$ catalyst in a multi-quench reactor. Lurgi plants operate under similar conditions except in a multi-tubular reactor. In both cases, naphtha or natural gas is preferred to coal as a feed material for the syngas. Space–time yields of 1 kg methanol/Ltr of catalyst per hour are typical.

Table 12. **Composition of Some Fuel Alcohols from Syngas**[a]

Alcohol (%)	C_1	C_2	C_3	C_4	C_5	Catalyst
MAS (SEHT)	69	3	4	13	9	K/Zn/Cr
Substifuel (IFP)	64	25	6	2	2.5	K/Cu/Co/Al
Octamix (Lurgi)[b]	62	7	4	8	19	alkali/Cu/Zn/Cr
HAS (Dow)[c]	26[d]	48	14	3.5	0.5	CoS/MoS$_2$/K

[a]See Ref. 72.
[b]Isobutanol is 70% of C_4 alcohols.
[c]Straight-chain alcohols.
[d]Methanol can be recycled to extinction, increasing the amount of ethanol.

Higher molecular weight alcohols ("higher alcohols," HAs) are preferred as fuel additives because of their lower vapor pressure. The compositions of various mixed alcohols from syngas using various processes are given in Table 12.

3.3. Production of Other Chemicals. Many of the chemicals attributed to the indirect liquefaction of coal are formed from methanol; see Figure 17(**b**). However, it is worth noting the hydroformylation (oxo) reaction. Here aldehydes are produced by reacting olefins with CO using complexes of groups 8–10 (Group VIII) metals such as Co or Rh as a homogeneous catalyst. Hydroformylation is the fourth-largest use of synthesis gas, after the production of hydrogen, methanol synthesis and FT synthesis (62). As an example of hydroformylation, propylene can be converted to n-butyraldehyde:

$$CH_3CH{=}CH_2 + CO + H_2 \rightarrow CH_3CH_2CH_2CHO + CH_3CH(CH_3)CHO \qquad (6)$$

used in the synthesis of 2-ethyl-hexanol, a plasticizer. This is a multimillion t/year operation. More details on the hydroformylation reaction can be found in, eg, Ref. 73.

3.4. Developments in Indirect Liquefaction. Much of the research and process development on indirect liquefaction after the 1990s is aimed at matching the synthesis conditions with modern, efficient coal gasifiers such as those developed by Texaco, Dow, and Shell (see COAL CONVERSION PROCESSES, GASIFICATION). Whereas the newer gasifiers are considerably more efficient, the gas produced has a much lower H_2/CO ratio. The slurry reactor has been shown to be capable of using this type of feedstock, under the right conditions. Optimization of the performance of the slurry-bed reactor requires work on improved catalysts and on the separation of catalyst and wax in the product stream (74).

In the production of oxygenates and chemicals from syngas, the Alternate Fuels Development Unit in LaPorte, Texas has been used by Air Products and Chemicals Inc. to test several strategies for the USDOE. This includes development of the slurry-phase methanol reactor and the formation of dimethyl ether. A slurry bubble-column reactor is also used. In bench-scale tests, molybdenum- or nickel-based catalysts have been used for the production of high molecular weight alcohols. A promising development is the introduction of a high boiling inert solvent, such as tetraglyme, in concurrent flow with the syngas in a fixed-bed reactor (75). The solvent absorbs methanol as it is produced, and shifts the chemical equilibrium to the "right" so that more methanol is produced.

BIBLIOGRAPHY

"Hydrogenation" under "Coal Chemicals and Feedstocks" in *ECT* 3rd ed., Supplement, by J. Langhoff, pp. 216–228. "Coal Conversion Processes, Liquefaction," in *ECT* 4th ed., Vol. 6, pp. 568–594, by Robert M. Baldwin, Colorado School of Mines; "Coal Conversion Processes Liquefaction" in *ECT* online, posting date: December 4, 2000, by Robert M. Baldwin, Colorado School of Mines.

CITED PUBLICATIONS

1. DOE Fossil Energy–Vision 21 Energy Plant of the Future, www.fe.doe.gov/coal_ power/vision21/index.shtml.
2. R. F. Probstein and R. E. Hicks, *Synthetic Fuels*, McGraw-Hill Book Co., New York, 1976.
3. D. D. Whitehurst, T. O. Mitchell, and M. Farcasiu, *Coal Liquefaction*, Academic Press, New York, 1980.
4. M. L. Gorbaty, D. F. McMillen, R. Malhotra, B. H. Davis, F. Burke, H. D. Schindler, R. F. Sullivan, H. Frumkin, D. Gray, G. Tomlinson, and B. Wilson, "Review of Direct Liquefaction," Chap. 4, in *Coal Liquefaction: A Research Needs Assessment Technical Background*, Vol. II, H. D. Schindler, Chair COLIRN, DOE/ER - 0400, 1989.
5. N. Berkowitz, *An Introduction to Coal Technology*, 2nd ed., Academic Press, New York, 1994.
6. A. L. Hammond, W. D. Metz, and T. M. Maugh II, *Energy and the Future*, AAAS, Washington D.C., 1973.
7. S.-C. Shin, R. M. Baldwin, and R. L. Miller, *Energy and Fuels* **3**, 71 (1989).
8. D. F. McMillen, R. Malhotra, S. J. Chang, and E. S. Nigenda, *Energy and Fuels* **1**, 193 (1987).
9. Ger. Pat. 301,231 (Nov. 26, 1919), F. Bergius and J. Billwiller.
10. E. E. Donath, in H. H. Lowry, ed., *Chemistry of Coal Utilization, Supplementary Volume*, p. 1041, John Wiley & Sons, Inc., New York, 1963.
11. D. L. Kloepper, T. F. Rogers, C. H. Wright, and W. C. Bull, *Office of Coal Research R&D Report No. 9*, U.S. Dept. of the Interior, Washington, D.C., 1965.
12. S. B. Alpert and R. H. Wolk, in M. A. Elliott, ed., *Chemistry of Coal Utilization, Second Supplementary Volume*, John Wiley & Sons, Inc., New York, 1981.
13. J. C. Tao, *Proceedings of the 3rd International Coal Utilization Conference*, Houston, Tex., 1980.
14. C. H. Wright and D. E. Severson, *ACS Div. Fuel Chem. Preprints* **16**(2), 68 (1972).
15. B. F. Alexander and R. P. Anderson, *ACS Div. Fuel Chem. Preprints* **27**(2), 18 (1982).
16. W. R. Epperly and J. W. Taunton, *Sixth International Conference of Coal Gasification, Liquefaction, and Conversion to Electricity*, Pittsburgh, Pa., 1979.
17. S. J. Hsia, *Proceedings of the Eleventh Biennial Lignite Symposium*, Grand Forks, N.D., 1981.
18. G. C. Lahn, *Proceedings of the Exxon Engineering Symposium 1981*, Exxon Engineering and Research Company, Florham Park, N.J., 1982
19. C. C. Kang and E. S. Johanson, in R. T. Ellington, ed., *Liquid Fuels from Coal*, Academic Press, New York, 1977.
20. M. Farcasiu, *PETC Rev.*, Issue 3, p. 4, March 1991.

21. E. L. Huffman; *Proceedings of the Third Annual International Conference on Coal Gasification and Liquefaction*, Pittsburgh, Pa., 1976.
22. H. D. Schindler, J. M. Chen, and J. D. Potts, *Final Technical Report on DOE Contract No. DE-AC22-79ET14804*, 1983.
23. H. D. Schindler; "Coal Liquefaction: A Research Needs Assessment," Vol. 2, Technical Background, *Final Report on DOE Contract No. DE - AC01 - 87ER30110*, 1989.
24. S. R. Hart, Jr., and E. L. Huffman, *Electric Power Research Institute Report No. AP-4257-SR* Vol. 2, paper 34, 1985.
25. S. R. Hart, Jr. and E. L. Huffman, *Proceedings of the Joint Conference on Coal Gasification and Synthetic Fuels for Power Generation*, San Francisco,, Calif., 1985.
26. A. G. Comolli, E. S. Johanson, J. B. McLean, and T. O. Smith, *Proceedings of the DOE Direct Liquefaction Contractors'* Review Meeting, Pittsburgh Pa. 1986.
27. K. Hirano, *Fuel Proc. Technol.* **62**, 109 (2000).
28. M. Onozaki, Y. Namiki, H. Ishibashi, M. Kobayashi, H. Itoh, M. Hiraide, and S. Morooka, *Fuel Proc. Technol.* **64**, 253 (2000).
29. Department of Trade and Industry, Cleaner Coal Technology Programme, Technology Status Report 010, *Coal Liquefaction*, October 1999, www.dti.gov.uk/cct/pub/tsr010.pdf.
30. *Gasoline from Coal*, National Coal Board, Coal House, Harrow, Middlesex, UK, 1986.
31. S. Chung, in *Proceedings of the Conference on Materials Problems and Research Opportunities in Coal Conversion*, Ohio State University, Columbus, OH, 1974, p. 263.
32. E. Gorin, H. E. Leibowitz, C. H. Rice, and R. J. Struck, *Proceedings of the 8th World Petroleum Congress*, Moscow, 1971.
33. L. Xu, A. P. Raje, R. A. Keogh, S. Lambert, R. L. Spicer, D. E. Sparks, S.-J. Liaw, and B. H. Davis, *Catalysis Today* **19**, 421 (1994).
34. C. W. Zielke, R. T. Struck, J. M. Evans, C. P. Costanza, and E. Gorin, *Ind. Eng. Chem. Proc. Des. Devel.* **5**(2), 158 (1966).
35. N. B. Moll and G. J. Quarderer, *Chem. Eng. Prog.* **75**, 46 (1979).
36. V. R. Pradhan, J. W. Tierney, I. Wender, and G. P. Huffman, *Energy and Fuels* **5**, 497 (1991).
37. V. R. Pradhan, J. Hu, J. W. Tierney, and I. Wender, *Energy and Fuels* **7**, 446 (1993).
38. P. G. Stansberry, J.-P. Wann, W. R. Stewart, J. Yang, J. W. Zondlo, A. H. Stiller, and D. B. Dadyburjor, *Fuel* **72**, 793 (1993).
39. Z. Liu, J. Yang, J. W. Zondlo, A. H. Stiller, and D. B. Dadyburjor, *Fuel* **75**, 51 (1996).
40. D. B. Dadyburjor, A. H. Stiller, C. D. Stinespring, A. Chadha, D. Tian, S. B. Martin, Jr., and S. Agarwal, "Use of an Aerosol Reactor to Prepare Iron Sulfide Based Catalysts for Direct Coal Liquefaction," in *Advanced Catalysts and Nanostructured Materials*, Academic, New York, 1996.
41. D. B. Dadyburjor, T. E. Fout, and J. W. Zondlo, *Catalysis Today* **63**, 33 (2000).
42. R. K. Sharma, J. S. MacFadden, A. H. Stiller, and D. B. Dadyburjor, *Energy and Fuels* **12**, 312 (1998).
43. D. W. Matson, J. C. Linehan, J. G. Darab, and M. F. Buehler, *Energy and Fuels* **8**, 10 (1994).
44. M. Farcasiu, G. P. Huffman, and I. Wender, eds., "Proceedings ACS Division of Fuel Chemistry Symposium on Iron-Based Catalysts for Coal Liquefaction," *Energy and Fuels* **8**(1) (1994).
45. F. V. Stohl, K. V. Diegert, and D. C. Goodnow, *Proceedings Coal Liquefaction and Gas Conversion Contractors Review Conference*, Pittsburgh, PA (1995), PETC/USDC p. 679.
46. C. Y. Wen and S. Dutta, in C. Y. Wen and E. S. Lee, eds., *Coal Conversion Technology*, Addison Wesley, Inc., Reading, Mass., 1979.
47. *Final ERDA Report*, Report No. FE-1212T9, Vol. 1, Char Oil Energy Development, 1975.

48. L. Seglin and S. A. Bresler, in Ref. 46, Chap. 13.
49. F. B. Carlson, *Proceedings of the 101st Annual AIME Meeting*, San Francisco, Calif., 1972.
50. E. T. Coles, *ERDA Report*, Report No. PER(A)-0, 1975.
51. A. Sass, *Chem. Eng. Prog.* **70**(1), 72 (1974).
52. W. Peters, *Chem. Ing. Tech.* **32**(3), 178 (1960).
53. *Chem. Eng. News*, 27 (Nov. 20, 1978).
54. J. E. Duddy, J. B. McLean, and T. O. Smith, *Proceedings 12th Annual EPRI Contractors Conference on Fuel Science*, Palo Alto, Calif., May 1987; USDOE, *Clean Coal Technology Progress Report*, DOE/FE-0092 (1987).
55. Z. Feng, J. Zhao, J. Rockwell, D. Bailey, and G. P. Huffman, *Fuel Proc. Technol.* **49**, 17 (1996).
56. M. Luo and C. W. Curtis, *Fuel Proc. Technol.* **49**(1–3), 91 (1996); H. K. Joo and C. W. Curtis, *Energy Fuels* **10**(3), 603 (1996).
57. A. H. Stiller, D. B. Dadyburjor, J.-P. Wann, D. Tian, and J. W. Zondlo, *Fuel Proc. Technol.* **49**, 167 (1996).
58. R. K. Sharma, D. Tian, J. W. Zondlo, and D. B. Dadyburjor, *Energy Fuels* **12**(6) 1245 (1998).
59. V. R. Pradhan, A. G. Comolli, L. K. Lee, and G. Popper, in *First Joint Power & Fuel Systems Contractors Conference: Direct Liquefaction*, USDOE-PETC, 1 (1996).
60. G. P. Huffman and N. Shah, *Chemtech* **28**(12), 34 (1998).
61. H. Juntgen, J. Klein, K. Knoblauch, H-J. Schroter, and J. Schulze, in Ref. 46, Chapt. 30.
62. I. Wender, *Fuel Proc. Technol.* **48**(3), 189 (1996).
63. P. Sabatier and J. B. Senderens, *C.R. Acad. Sci. (Paris)* **134**, 514 (1902).
64. F. Fischer and H. Tropsch, *Chem. Ber.* **59**, 830 (1926).
65. J. Schulze, *Chem.-Ing.-Tech.* **46**, 976 (1974).
66. S. S. Penner, ed., *USDOE Working Group on Research Needs for Advanced Coal Gasification Techniques (COGARN)*, DOE Contract No. DE-AC01-85ER30076, Washington, D.C., 1987.
67. H. Pichler and B. Firnhaber, *Brennst. Chem.* **44**, 33 (1963), as cited in Ref. 5, Chap. 12.
68. J. C. Hoogendoorn, in *Clean Fuels from Coal, Institute of Gas Technology Symposium Series*, IGT, Chicago, Ill., 1973, p. 353.
69. P. E. Rosseau, J. W. van der Merwe, and J. D. Louw, *Brennstoff Chem.* **44**, 162 (1963).
70. A. Geertsema, Plenary Lecture, in *Tenth Ann. Mtg., Pittsburgh Coal Conf.* Pittsburgh, PA, Sept 20–24 (1993).
71. B. Jager, M. E. Dry, T. Shingles, and A. P. Steynberg, *Catal. Lett.* **7**, 293 (1990).
72. G. A. Mills, *Final Report: Status and Future Opportunities for Conversion of Synthesis Gas to Liquid Energy Fuels*, NREL, USDOE Subcontract DE-AC-02-83CH10093, (1993).
73. W. Keim, ed., *Catalysis in C_1 Chemistry*, Reidel Publishing Co., Dordrecht 1983; G. W. Parshall and S. D. Ittel, *Homogeneous Catalysis*, John Wiley & Sons, New York, 2nd ed., 1992.
74. G. J. Steigel, *PETC Rev.* **4**, 14 (1991).
75. J. M. Berty, C. Krishman, and J. R. Elliot, Jr., *Chemtech* **20**, 624 (1990).

DADY B. DADYBURJOR
West Virginia University

ZHENYU LIU
Institute of Coal Chemistry